LEAFHOPPERS (Cicadellidae):

A Bibliography, Generic Check-list and Index

to the World Literature

1956 – 1985

LEAFHOPPERS

(Cicadellidae):

A Bibliography, Generic Check-list and Index
to the World Literature
1956 – 1985

P. W. Oman, W. J. Knight and M. W. Nielson

C·A·B International Institute of Entomology

Published August 1990

C·A·B International
Wallingford
Oxon OX10 8DE
UK

Tel: Wallingford (0491) 32111
Telex: 847964 (COMAGG G)
Telecom Gold/Dialcom: 84: CAU001
Fax: (0491) 33508

© C·A·B International 1990. All rights reserved. No part of this publication may be reproduced in any form or by any means, electronically, mechanically, by photocopying, recording or otherwise, without the prior permission of the copyright owners.

British Library Cataloguing in Publication Data

Oman, P. W.
Leafhoppers (cicadellidae): a bibliography, generic check–list and index to the world literature 1956 – 1985.
1. Leafhoppers. Bibliographies
I. Title
II. Knight, W. J. (William James)
III. Nielson, M. W.
016.595752

ISBN 0–85198–690–0

Printed and bound in the UK by BPCC Wheatons, Exeter

CONTENTS

INTRODUCTION 1
    Acknowledgements 1
    A review of the cicadellid literature for the period 1942-1955 2

BIBLIOGRAPHY 5
    List of publications 6

CHECK-LIST OF GENERIC AND FAMILY-GROUP NAMES 173
    Family-group names proposed 1956-1985 174
    Use of family-group names in the Cicadellidae: 1956-1985 175
    Classification 182
    Misassigned generic names 183
    Misidentified type-species 183
    Names placed on the official list of generic names in zoology 183
    Names placed on the official list of family-group names in zoology 184
    Names placed on the official index of rejected and invalid generic
        names in zoology 184
    Names placed on the official index of rejected and invalid
        family-group names in zoology 184
    Check-list 184

INDEX 263
    Subject index 263
    Taxonomic index 287

INTRODUCTION

Catalogues perform an indispensable service to the orderly pursuit of knowledge about species-rich groups such as leafhoppers by providing access to available data. Study of the Auchenorrhyncha in general, and the Cicadellidae in particular, has been greatly accelerated and improved by the organized information contained in the monumental General Catalogue of the Homoptera prepared by Zeno Payne Metcalf and Virginia Wade Burnside. Fascicle VI of the series, dealing with leafhoppers, was published in 17 Parts during the period 1962-1971 (Metcalf 1962a-1968a; Metcalf & Wade 1963a, 1966a; Burnside 1971a) and initiated a surge of taxonomic activity that began in the mid-70's, and continues today. It covered the literature over a 200-year period, through December 1955, and listed the 1392 available generic level names and 13305 names at the species level.

During the ensuing 30 year period, from January 1956 through December 1985, 1084 new genera and nearly 8000 new species have been described. In addition, 55 new family-group names have been proposed, 9 of which are currently accorded subfamily status. Much of the new information deals with faunas of hitherto little known regions, greatly expanding our comprehension of biological diversity in the group, and the geographic ranges of many lineages. This flood of taxonomic actions emphasises anew the great diversity of the group. It also generated a profusion of new combinations, new synonymies, and numerous replacement names at both generic and specific levels, and was accompanied by a comparable increase in non-taxonomic literature as more and more information about the role of leafhoppers as plant pests and vectors of plant pathogens emerged.

This compilation and analysis of literature is designed to provide ready access to published information that deals in any way with the biology, behaviour and classification of leafhoppers. It comprises three major sections: a bibliography of more than 7000 titles for the period 1956-1985; an index to the bibliography to guide users to specified subject-matter fields; and a complete check-list of generic and family-group names used in the Cicadellidae. Other important features are a review of cicadellid literature for the period 1942-1955, a compilation showing usage of family-group names in the taxonomic literature during the years 1956-1985 and an alphabetized list and grouping of subfamily and tribal names within the family.

The present bibliography supplements Metcalf's Bibliography of the Cicadelloidea (Metcalf 1964a). Together, they provide a reference base for the other parts of this work, and contain references to all pertinent literature published before 1986. The index, augmented by the review of the 1942-1955 literature, provides access to a broad spectrum of leafhopper phenomena ranging from behaviour to virus/vector/plant interactions. A comprehensive taxonomic section provides references to genera in which new species have been added, new combinations or new synonymies recorded, or in which other actions with taxonomic implications are noted.

The check-list contains all the generic and family-group names used within the Cicadellidae. Associated with it is a list of family-group names described as new during the 30-year period covered in this report. That list, and the compilation "Use of Family-group Names in the Cicadellidae: 1956-1985", are provided to aid in correlating the various subfamily and tribal concepts that have appeared in the literature. The latter also serves as a convenient additional index to the "Alphabetized List of Subfamily and Tribal Names in the Cicadellidae".

Each of the three main sections is preceded by an account of the relevant methodology adopted for that part of the work.

ACKNOWLEDGEMENTS

This work would not have been possible without the assistance of the many individuals who generously contributed their time and expertise and the support of the authors' respective institutions in providing the necessary facilities and material.

The authors wish to extend their sincere appreciation and gratitude to Mrs. Shirley A. Oman, for keyboarding drafts of the "Check-list of Generic and Family-group Names" and the compilation "Use of Family-group Names in the Cicadellidae: 1956-1985"; Mrs. Leslie Morris, Ms. Nancy Politis and Mrs. JoAnne Abel, for keyboarding various drafts of the Bibliography; Deanna Watkins of Oregon State University Entomology Department, for keyboarding drafts of the taxonomic section of the Index; Miss Pam Gilbert and Miss Julie Harvey of The Natural History Museum Library, London, for deciphering foreign titles and conducting computer searches of the various abstracting journals; Ms. Jody Chandler, Mr. Richard Thomas and Mr. Charles E. Kaluau, for assistance in searching titles in the Brigham Young University Library, Provo, Utah; the librarians of the Brigham Young University Library, for assistance in researching and resolving incomplete citations in titles and publication sources and for access to abstract sources; the William Jasper Kerr Library, Oregon State University, for arranging inter-library loans of certain rare publications; the University of Arizona Science Library, Tucson, for allowing access to various abstract records and assistance in deciphering acronyms and abbreviations of periodicals; Dr. Ralph E. Berry, Chair, and Dr John D. Lattin, Curator of the Systematic Entomology Laboratory, Department of Entomology, Oregon State University, for monetary support in connection with two on-site working sessions by the authors, and for incidental printing costs of some items; the Monte L. Bean Life Science Museum, Brigham Young University, for providing endowment funds, in part, for typing services on the first three drafts of the Bibliography; Dr. Dorald Allred, former director, and Dr. Stanley Welsh, present director, of the Monte L. Bean Life Science Museum, for continuing support. Our thanks are also extended to our several colleagues who have forwarded lists of their publications.

The limitations of the computer system used to compile the initial databases for this work prevented the inclusion of diacritic marks. As a consequence, it was decided to omit them throughout the volume, in the interests of uniformity, Although this does not impede access to published information, the present authors wish, nevertheless, to extend their apologies to the authorities concerned.

The final camera-ready copy was prepared with the assistance of Dr D.B. Williams of The Natural History Museum, London, to whom the authors extend their thanks.

A REVIEW OF THE CICADELLID LITERATURE FOR THE PERIOD 1942-1955

Metcalf's Bibliography of the Cicadelloidea (Metcalf 1964a) lists the 6687 papers published during the period 1758-1955. Four-fifths (5325) of these were published prior to 1942 and are listed also in his earlier bibliography (Metcalf 1944a) and indexed in a companion volume (Metcalf 1944b). The remaining 1362 papers, published between June 1942 and December 1955 are omitted from this index. The present work, which deals only with the period 1956-1985, likewise omits these 1942-1955 publications. The purpose of the present section of the introduction therefore is to review these 1362 papers, to indicate the advances made during the 14 years in question and to draw attention to the key works.

Outside the economic field, the major proportion of the work (495 papers) was directed towards descriptive taxonomy and faunal studies. Most effort concerned the European and Scandinavian fauna (138 papers) as well as that of N. America (119 papers), Mexico (70), Central & South America (24), the Pacific (24), Japan (12), Asia (9), Africa (9) and the Mediterranean (5).

The principle taxonomic workers during this period were R. H. Beamer, B. P. Beirne, D. M. DeLong, D. M. J. Knull, Z. P. Metcalf, P. W. Oman, R. F. Ruppel and D. A. Young in the United States, J. Dlabola, H. Holgersen, P. Kontkanen, V. Lang, H. Lindberg, R. Linnavuori, F. Ossiannilsson, H. Ribaut, W. Wagner and A. A. Zachvatkin in Europe, J. W. Evans in Australia and T. Ishihara in Japan.

Faunal works or check lists, relating to the whole or part of the region concerned, were published for **North America** (Wolfe 1955c, Wray 1951a, Brimley 1942a, Young 1952b, DeLong 1948a, Fattig 1955a, Medler 1943a, Oman 1949a, DeLong et al. 1946a, Procter 1946a, Dutilly 1946a, Foster 1943a, Leech 1945e, Moore 1944a, 1950a,b, Osborn et al. 1947a, Strickland 1953a, Stroud 1950a, Waddell 1952a, Weber 1950a, Wene 1946a, Young 1950a), **Central America** (Wolcott 1950a, Caldwell et al. 1952a, Hartzell 1954a, Metcalf 1954c, Metcalf et al. 1949a, Beatty 1947a, Ramos 1947a), **South America** (Young 1952b, Metcalf 1949b, Soukup 1945a), **Europe** (Ribaut 1952a, Servadei 1952a, Smreczynski 1954a, Zachvatkin 1948a, Boitard 1943a, Kupka 1944a, Dlabola 1954a,c,d, 1943a, 1944a,b, 1945a,b, 1946a, 1952b, Lang 1942a, 1944a, 1945b,c,d, 1946a, Fagel 1949a, Synave 1951b, Franz 1943a, Lindberg 1949a, Caspers 1942a, China 1950a, 1951b, Kloet et al. 1945a, Moosbrugger 1946a, Reclaire 1944a, Seabra 1942a, Servadei 1948a, Wagner 1946a, 1947a, 1951b), **Scandinavia** (Kontkanen 1947a,b, 1949b,c,d, 1952a, 1953b, Lindberg 1943b, 1947a, Linnavuori 1949a, 1950a, 1952e, Ossiannilsson 1942a, 1946c, 1947b, 1951b,c, 1948a, Holgersen 1944a,b, 1945a, 1949a,b, 1954a, Ossiannilsson 1943c), **Iceland** (Fristrup 1945a), **Cyprus** (Ribaut 1948a, Lindberg 1948b), **Palestine** (Linnavuori 1952c, 1953b), **Middle East** (Zachvatkin 1947a,b), **Canary Islands** (Lindberg 1954a), **Azores** (Lindberg 1954b), **Africa** (Villiers 1952a, Evans 1955a,b), **Madagascar** (Evans 1954a, Lallemand 1950b), **India** (Ghani et al. 1946a), **Japan** (Esaki 1950a, Ishihara 1953b, Ishihara et al. 1953a, Kato 1952b), **Palaearctic** (Dlabola 1955a, Jacobi 1943a, Jacobi 1944a, Van Zwaluwenburg 1943a) and **Pacific Islands** (Zimmerman 1948a, Krauss 1944a, Gressitt 1954a, Hawkins 1942a, Metcalf 1946d, Metcalf 1950c, Cole 1951a, Dammerman 1948a).

Only seven papers were published on the phylogeny of the group (Evans 1942b, 1948b; Hennig 1953a; Kramer 1950a; Metcalf 1950a, 1951a; Wagner 1951a), four on classification (Evans 1946a,b, 1947a, 1951a) and six on nomenclature (DeLong 1945b; Knull 1944b; Metcalf 1952a, 1955a; Ossiannilsson 1954a; Wagner 1950a). Only three papers were published on fossils (Evans 1943b, 1950a; Haupt 1943).

The key works on taxonomy during this period are those on the natural classification of leafhoppers by Evans (1946a,b, 1947a) and the faunal works or check lists relating to Czechoslovakia by Dlabola (1954a), Great Britain by China (1950a), Japan by Esaki et al. (1954a) and Ishihara (1953b), Micronesia by Gressitt (1954a), Cyprus by Lindberg (1948b), Canary Islands by Lindberg (1954a), USA by Oman (1949a), Sweden by Ossiannilsson (1946c, 1947b, 1948a), France by Ribaut (1952a), Hawaii by Zimmerman (1948a) and the W Hemisphere by Young (1952b). In phylogeny the key works are those by Kramer (1950a) and Wagner (1951a) whilst the discovery of seasonal variation in the male genitalia by Muller (1954a) is of major significance throughout the entire family. Metcalf's bibliography of the Homoptera (Metcalf 1944a,b) was a major impetus to the study of the group.

The morphology of leafhoppers was dealt with in 25 papers. The most important of these was a study of the comparative morphology in the Auchenorrhynchous Homoptera by Kramer (1950a), the external morphology of leafhoppers by Evans (1946a), the discovery of seasonal forms by Muller (1947a, 1954a,b,c, 1955b) and the evaluation of taxonomic morphological characters by Wagner (1955b). Other papers dealt with the male genitalia (Dulzetto 1942a, Dupuis 1949a, Kontkanen 1950a, Marks 1951a, Qadri 1949b), ovipositor (Trautman 1945a), legs (Muller 1942a, Fennah 1945e), tymbal organs and abdominal apodemes (Ossiannilsson 1946a, 1951d), wing coupling (Ossiannilsson 1950a), head (Evans 1942d), colour mutants (Teitelbaum et al. 1950a,b), external structures (Haupt 1953a) and structural abnormalities (Balazuc 1951a, Kontkanen 1950a).

Very little work was carried out on the anatomy of leafhoppers. Only three papers were published, each dealing with selected species rather than the family as a whole. Fudalewicz (1950a) studied the innervation and sense organs in the wings of *Typhlocyba*, Qadri (1949a) the digestive system and head capsule in two species of *Idiocerus* and Willis (1949a) the head, gut and associated structures of *Typhlocyba ulmi*.

Development studies were limited to only three papers. The effect of temperature on the rate of development in the beet leafhopper was studied by Harries (1943a, 1944a) whilst Qadri (1949b) studied the post-embryonic development of the male genitalia.

Physiological studies were limited to a study of winter vitality of the beet leafhopper in relation to body fat content (Cook 1944a) and functions of the alimentary system (Day *et al.* 1953b).

The genetics of leafhoppers was dealt with in only three papers, one on the mode of bivalent orientation (Helenius 1952a) and two on a colour mutant of *Agalliopsis novella* (Teitelbaum *et al.* 1950a,b).

There were four major publications in the field of symbiosis, three of which were general works (Buchner 1949a, 1953a; Steinhaus 1946a) and one dealing with the systematics and phylogeny of leafhopper symbionts (Muller 1949a). A fifth paper on the symbionts of *Cicadella viridis* in India was published by Mahdihassan (1947c). The relationship between leafhoppers and ants was the subject of papers by Beamer *et al.* (1950a) and Stager (1951a), whilst Wray *et al.* (1943a) published on leafhoppers and pitcher plants.

Of the 11 papers published on leafhopper parasites, the most important was a catalogue of the parasites of Hemiptera by Thompson (1944a). Details on the biology of individual parasite species were published by Buyckx (1948a) on the dryinid *Aphelopus indivisus* and by Lindberg (1946a) on the pipunculid *Pipunculus chlorinae*. A review of the Pipunculidae in North America was published by Hardy (1943a). Kontkanen (1950d) dealt with the leafhopper parasites found in North Karelia and Bohart (1943a) gave a key to the strepsipteren genus *Halictophagus* in North America.

The predators attacking leafhoppers in the United States were documented in 22 separate papers by Knowlton and co-workers. Further papers were published by Aldous (1942a), Bodenheimer (1951b), James (1942a), Schultz (1949a), Wade (1951a), Wray *et al.* (1943a) and York (1944a).

Various aspects of behaviour were studied, with a total of 12 papers. The most important of these were those by Ossiannilsson (1946a, 1948c, 1949a, 1953a) on sound production and one by Muller (1951a) on hatching. The first authentic record of parthenogenesis in the Auchenorrhynchous Homoptera was made by Black *et al.* (1947a). Feeding was dealt with in four papers (Brues 1946a, Day *et al.* 1952a, Houston *et al.* 1947a, Nuorteva 1954a), with only one paper each on secretion (Smith 1950a) and swarming (Steyskal 1945a).

The ecology and bionomics of leafhoppers were the subject of 79 papers, principally in North America (42 papers), Europe (12) and Scandinavia (9). Studies were also carried out in Latin America (2), Africa (2), Japan (3), India (2), Philippines (2) and China (1). The major works were those by Dowdy (1944a, 1947a, 1951a, 1955a), Fautin (1947a), Harries *et al.* (1948a), Hayward (1948a, 1952a), Hutchinson (1955a), McClure (1943a), Romney (1945a, 1946a), Shackleford (1942a) and Whittaker (1952a) in **North America**, Marchand (1953a), Kontkanen (1954a), Morcos (1953a) and Muller (1951a) in **Europe**, Kontkanen (1949a, 1950c), Linnavuori (1952a), Nuorteva (1951d) and Ossiannilsson (1943a) in **Scandinavia**, Cendana *et al.* (1950a) in the **Philippines**, Kato *et al.* (1952a) in **Japan** and Chu *et al.* (1950a) in **China**.

Migration and dispersal were covered in seven papers. Studies on the beet leafhopper were carried out in the United States by Cook (1945a), Douglass *et al.* (1947a), Lawson *et al.* (1951a) and Wallis (1947a), whilst in **Venezuela** Beebe (1951a) investigated insect migration through a mountain pass. The most important papers in this category were those on passive dispersal in the atmosphere by Wellington (1945a) and Wolfenbarger (1946a).

Host plant studies featured prominently in 29 papers, most of which resulted from work in the United States and were related to the host range of individual species or the insect guild on individual plant species. The most comprehensive works were those by Schaffner (1953a) on an index to insects by host plants and an account of the host plants of leafhoppers of the genus *Empoasca* by Poos *et al.* (1943a, 1949a). Nuorteva (1952b) investigated host plant stimuli.

Host plant resistance studies produced 15 papers, the most important and comprehensive of which was by Painter (1951a). Most of the work was on cotton, the major works being those by Afzal *et al.* (1944a) and Afzal *et al.* (1948a) in India, May (1951a,b) in Australia and Parnell *et al.* (1949a,b) in South Africa. Other crops were alfalfa (Davis *et al.* 1953a, Ham *et al.* 1946a, Mumford 1942a) and beans (McFarlane *et al.* 1943a) in the United States and eggplant (Fukuda 1952a).

The greater proportion of the papers (645) are concerned with various aspects of crop production and the role of leafhoppers as pest organisms, either directly or as vectors of plant virus diseases. Most of the work was carried out in the United States (256 papers), followed by Africa (41), Latin America (26), Europe (23), Australia (22), India (22), Japan (9), Pacific (7), Canada (5) and the Middle East (2). The principle crops covered were cotton (72 papers), fruit (68), potatoes (64), vegetables (52), forage crops (36), sugar beet (28), vines (27), sugar cane (11) and cereals (8). The majority of papers refer to the whole complex of insect pests, either on a particular crop or in a particular country or region, and very few exclusively with leafhoppers. Of the latter, the major ones deal with leafhoppers on cotton in the Indian Punjab (Afzal *et al.* 1953a), Australia (May 1950a,b, 1951a,b) and the Sudan (Snow *et al.* 1952a). Other papers deal with the homopterous pests of fruit in Spain (Gomez-Menor 1951a), typhlocybines attacking vine and pepper in Spain (Ruiz Castro 1943a,b), mango-hoppers in Bihar in India (Roy *et al.* 1952a), grape leafhopper in the US (Taschenberg *et al.* 1949a) and beet leafhopper in the US (Romney 1943a).

Among those papers dealing with the whole spectrum of pests, including leafhoppers, on particular crops are those by Box (1953b), Ingram *et al.* (1951a) and Wade (1951a) on **sugar cane**, Annand (1942a) on **sugar beet**, Bodenheimer (1951a) on **citrus**, Breakey *et al.* (1951a), Massee (1946a), Powell *et al.* (1953a), Newcomer (1942a, 1950a), Garman *et al.* (1953a) and Garman *et al.* (1952a) on **fruit**, Ruiz Castro (1944a, 1950a) on **vines**, Brixhe (1949a) on **cotton**, Franklin (1951a), Packard (1952a) and Painter *et al.* (1954a) on **cereals and forage crops**, Holdaway (1944a), Watson *et al.* (1942a), Janes *et al.* (1952a) and Bruner *et al.* (1954a) on **vegetables**, Ishikura (1953a), Kuwayama (1953a), Mitono (1953a), Sakurai *et al.* (1953a), Suenaga (1953a) and Sugawara (1953a) on **soybeans**, Otanes *et al.* (1952a) on **rice**, Jepson (1948a) on **ground nuts**, Hill (1948a) and Dudley *et al.* (1952a) on **potatoes**, Craighead (1950a) on **forest trees** and Schuh *et al.* (1948a) and Weigel *et al.* (1948a) on **ornamentals**.

Broader based regional works cover the agricultural pests of Cuba (Bruner *et al.* 1945a), Alaska (Chamberlin

1949a), Italy (Della Beffa 1949a), Columbia (Figueroa 1952a), Nigeria (Golding 1946a), Argentina (Hayward 1942a), Indonesia (Kalshoven *et al.* 1950a), Fiji (Lever 1946a), Formosa (Miwa 1943a), India (Trehan *et al.* 1946a, Krishnamurti *et al.* 1951a), Peru (Wille 1943a), New Caledonia (Williams 1944a), Tasmania (Evans 1943c), USA (Essig *et al.* 1944a, Fenton 1952a, Sherbakoff *et al.* 1943a) and the British Commonwealth (Evans 1952a).

Textbooks, dealing with the principals of agricultural entomology and pest control, include those by Molinari (1942a), Smith, K.M. (1948a), Balachowsky (1951a), Isely (1942a) and Brown (1951a).

The role of leafhoppers as vectors of plant viruses was dealt with in 178 papers, the majority of the work being carried out in the United States with considerably smaller amounts in Latin America (6 papers), Australia (6), Europe (5), Africa (2), India (1) and Taiwan (1). The principle crops investigated were fruit (31 papers), potatoes (15), vegetables (12), cereals (10), vines (9), fodder (9), sugar beet (6), sugar cane (4), tobacco (3) and flax (1). The major workers in the field of virus transmission were L. M. Black, K. Maramorosch, H. H. P. Severin and H. R. Wolfe, all in the United States, whilst Wade (1951a) published a comprehensive yet selected bibliography of the insects of the world associated with sugar cane. Textbooks covering the field of plant viruses and virus diseases were produced by Bawden 1950a and Smith, K. M. 1948b, 1951a.

Pesticide resistance in leafhoppers, involving DDT, was dealt with in only 3 short papers in the United States (Barnes *et al.* 1954a, Stafford *et al.* 1953a,b).

Thirteen papers on a variety of techniques were published. A comprehensive paper on the collection and preservation of insects in general was published by Oman *et al.* (1946a). Kretzschmar (1948a) also wrote on sampling techniques, whilst major works on light traps and reaction to light were published by Frost (1952a, 1953a) and Milne *et al.* (1944a,b) respectively. The use of carbon dioxide as an anaesthetic in virus vector work was described by Maramorosch (1953b), and Braun *et al.* (1951a) described a method for obtaining saliva. A method for the location of eggs in plant tissue was described by Curtis (1942a). A major study on the estimation of foliage area damage was carried out by Hartzell (1946a) whilst a smaller work dealt with treatment evaluation and infestation analysis (Hartzell *et al.* 1944a). Data analysis was also covered by Kontkanen (1950b). The remaining paper (Scholl *et al.* 1947b) dealt with the use of trap strips for insect control.

Some of the catalogues published during this period have been mentioned already, such as Metcalf's bibliography of the Homoptera (Metcalf 1944a,b) and Wade's bibliography of sugar cane insects (Wade 1951a). Only two further catalogues were published, one by Esaki *et al.* (1954a) on the leafhoppers of Japan and adjacent territories and another by Mathur (1953a) on the entomological collection of the Dehra Dun Forest Research Institute in India.

BIBLIOGRAPHY

The bibliography covers the period 1 January 1956 - 31 December 1985. Pre-1956 references not included in the Metcalf Bibliography 1964a are also listed, except those cited in Metcalf 1967c:2691 and 1968a:1510.

All fields relevant to the taxonomy and biosystematics of the taxa are covered including behaviour, biogeography, biology, fossils, galls, genetics, host plants, host plant resistance, insect collections, natural enemies, nomenclature, pest status, phylogeny, physiology, structure, techniques, variation, vectors of plant pathogens, vector/pathogen interactions.

In addition to surveying current literature, manual and computer searches have been made of the following abstracting journals:

Abstracts of Entomology. Philadelphia: BIOSIS. (1970 (Vol. 1)-1985); **Bibliography of Agriculture.** Washington, D.C.: US Government Printing Office. (1956-1985); **Biological Abstracts.** Philadelphia: BIOSIS. (1956-1985); **Entomology Abstracts.** Bethesda: Cambridge Scientific Abstracts. (1969 (Vol. 1)-1985); **Review of Applied Entomology.** Series A Agriculture. London: C.A.B. International Institute of Entomology. (1956-1985); **Zoological Record.** Philadelphia: BIOSIS and London: Zoological Society of London. (1956-1985).

In the case of bibliographies containing a comprehensive list of references on a specialised topic (e.g. Gyrisco *et al.* 1978a), the bibliography is cited with appropriate annotation to indicate contents. The references contained in these bibliographies have been repeated in the present work. Books containing comprehensive reference lists, e.g. Maramorosch and Harris 1979a, have been extracted and the references, where appropriate, have been repeated in the present work. Review articles providing an introduction and guide to the peripheral literature have been included. Only the more important references cited have been repeated in the present work. Abstracts and dissertations are cited.

All journal titles are given in full, in accordance with the following publications: **Serial Publications in the British Museum (Natural History) Library.** 1980. Vols. 1-3. London: The British Museum (Natural History) [Updated to 1986 on microfiche]; **World List of Scientific Periodicals Published in the years 1900-1960.** Vols. 1-3 (1963-1965). London: Butterworths; **Current Serials Received.** 1985. Boston Spa: The British Library; **CAB Serials Checklist.** 1983. Slough: Commonwealth Agricultural Bureaux; **Chemical Abstracts Service Source Index (CASSI) 1907-1984.** Columbus, Ohio; **Serial Sources for the Biosis Data Base.** Philadelphia: BIOSIS; **Union List of Serials in Libraries in United States and Canada.** Third Edition, 1965. New York: H.W. Wilson & Company.

In the case of those journals which have two titles, often in different languages, the one recommended in the above works is used. Where journals have merged or changed their names, the title given in the bibliography depends on the date of the paper relative to the merger or change. All non-English titles are given in their native language (when possible), followed by an English translation in parentheses (when available). All misspellings and other typographical errors in paper or journal titles are retained, followed by "[sic]".

In the case of those papers in which the date (year) of publication differs from that given on the journal, the date on the journal is given immediately following the author's name with the actual date (year) of publication following immediately in square brackets [ ] without a letter code, e.g. 1964a [1965]. In the rare instances where an article has been inadvertently published twice in a journal, either in the same issue but on different pages or in a subsequent issue, the first occurrence only is listed.

The citation of an author's name follows the format proposed in "Anglo-American Cataloguing Rules", second edition, 1978, London: The Library Association. Forenames are abbreviated to initial letters, except where the full name is necessary to avoid confusion. Different spellings of an author's name, e.g. Ahmed and Ahmad, are indicated in the reference as follows: "Ahmad [sic = Ahmed]". In the case of multiple authorship, only one entry is given; junior authors are not listed separately. References lacking author's initials, e.g. Fachrudin, or with secondary authors referred to as "et al.", and for which the original article could not be traced, are cited as found.

References are annotated, where appropriate, to indicate the number of new genera and new subgenera described, in square brackets [ ] at the end of the entry. The genera and subgenera are named in the case of 5 or less, otherwise a statement only is given. All references indicate the language used, if other than English, and the presence and language of any summary. The entry is annotated with the full citation of the source of any translation.

LIST OF PUBLICATIONS

Abdel-Salam, A. M., Assem, M. A., El-Minshawy, A. M., Yousef, K. H. and Abbasy, A. M.
1971a Studies on potato pests in U.A.R. I. Field evaluation of both granular and sprayable insecticides for the control of certain potato pests. *Zeitschrift fuer Angewandte Entomologie* 69(2):197-204.

Abdel-Salam, A. M., Assem, M. A., Hammad, S. M. and Eid, G.H.
1972a Studies on potato pests in U.A.R. II. Susceptibility of some potato varieties to insect infestation in the field and in the storage. *Zeitschrift fuer Angewandte Entomologie* 70(1):76-82.

Abdel-Salam, F.
1967a Uber die Wirkung von Phosphoraureestern auf einige Arthropoden innerhalb der Apfelbaum-Biozonose in Abhangigkeit von ihrer Dichte. [On the effect of phosphoric-acid esters on some arthropods in the apple-tree biocoenosis in relation to their density.] *Zeitschrift fuer Angewandte Zoologie* 54(2):233-282. [English summary.]

Abdul-Nour, H.
1970a Les Strepsipteres parasites d'Homopteres dans le sud de la France. Description d'une nouvelle espece: *Halictophagus agalliae* n. sp. *Annales de Zoologie—Ecologie Animale* 2:339-344. [In French with English summary.]
1971a Contribution a l'etude des parasites d'Homoptera Auchenorrhynques du sud de la France: Dryinidae (Hymenopteres) et Stresipteres. D. Sc. Thesis, Universite des Sciences et Techniques du Languedoc, Academie Montpellier.
1976a The Dryininae of southern France (Hym., Dryininae). Taxonomic and biological notes: Descriptions of two new genera. *Annales de Zoologie-Ecologie Animale* 8:265-278.
1984a [1983] What do we know about the Auchenorrhyncha fauna of Lebanon? *In*: Drosopoulos, Sakis, ed. *Proceedings of the 1st International Congress Concerning the Rhynchota Fauna of Balkan and Adjacent Regions*, p. 12. Mikrolimni-Prespa, Greece, 29 August - 2 September 1983.
1984b The role of the oak tree in the bio-ecology of some Cicadellidae (Homopt. Auch.). *Mitteilungen der Schweizerischen Entomologischen Gesellschaft* 57(4):403.
1985a Cicadellidae of Lebanon: records and bio-ecological notes. *Marburger Entomologiche Publikationen* 1(10):169-190.

Abdurakhimov, K. A. and Dubovsky, G. K.
1970a Materials on the fauna of cicada (Auchenorrhyncha) of the Issyk Kul hollow. *Nauchnye Doklady Vysshei Shkoly Biologicheskie Nauki* 13(3):30-34. [In Russian.]

Abe, Y. and Okamoto, D.
1975a Effects of non-organochlorine insecticides and analysis of their lethal effect against two major rice pests by the application of insecticides on soil and water. *Bulletin of the Chugoku National Agricultural Experiment Station* 10:1-55.

Abeygunawardena, D. V. W.
1969a [1967] The present status of virus diseases of rice in Ceylon. *In: The Virus Diseases of the Rice Plant. International Rice Research Institute Symposium Proceedings*, pp. 53-57. 354 pp.

Abeygunawardena, D. V. W., Bandaranayaka, C. M. and Karenoawela, C.
1971a Virus diseases of rice and their control. *Review of Plant Pathology* 50 (abstract). 662 pp.

Abney, T. S., Sillings, J. O., Richards, T. L. and Broersman, D.B.
1976a Aphids and other insects as vectors of soybean mosaic virus. *Journal of Economic Entomology* 69(2):254-256.

Abraham, E. V., Natarajan, K. and Jayaraj, S.
1977a Investigations on the insecticidal control of the phyllody disease of sesamum. *Madras Agricultural Journal* 64:379-383.

Abraham, E. V. Padmanabhan, M. E., Mohandoss, A. and Gunasekharan, C. R.
1970a Record of some insects of economic importance on the hill crops in Tamil Nadu. *Madras Agricultural Journal* 57(12):718-722.

Abraham, E. V., Thirumurthi, S., Ali, K. A. and Subramaniam, T. R.
1973a Some new pests of sesamum. *Madras Agricultural Journal* 60:593.

Abu Yaman, I. K.
1967a Population study of the grape leafhopper in Iraq. *Zeitschrift fur Angewandte Entomologie* 60:182-187. [German summary.]
1971a Outbreaks and new records—Saudi Arabia—Tomato stolbur virus: a new record. *FAO Plant Protection Bulletin* 19(6):140-141.

Abul-Nasr, S., El-Nahal, A. K. M. and Sawy, O.
1969a Seasonal population of Hemiptera-Homoptera infesting cotton plants in Egypt. *Bulletin de la Societe Entomologique d'Egypte* 52:371-389.

Adler, P. H.
1982a Nocturnal occurrences of leafhoppers (Homoptera: Cicadellidae) at soil. *Journal of the Kansas Entomological Society* 55(1):73-74.

Adlerz, W. C., and Hopkins, D. L.
1978a Transmission of Pierce's disease of grape by sharpshooters (Homoptera: Cicadellidae) in Florida. *Journal of New York Entomological Society* 85(4):163-164.
1979a Natural infectivity of two sharpshooter vectors of Pierce's disease in Florida. *Journal of Economic Entomology* 72:916-919.

Afzal, M. and Ghani, M. A.
1953a Cotton jassid in the Punjab. *Science Monograph, Pakistan Association for the Advancement of Science* 2:102.

Agarkov, V. A.
1964a Pale-green dwarfing of winter wheat. *In: Zashchita Rastenii ot Vreditelei i Boleznei* 7:17. Moscow.
1972a The specific status of *Psammotettix alienus* Dahlb. [*P. striatus* (L.)] (Cicadellidae), a vector of virus diseases of wheat. *Proceedings of the XIIIth International Congress of Entomology*, Moscow, 2-9 August 1968. [In Russian.]

Agarwal, R. A., Banerjee, S. K. and Katiyar, K. N.
1978a Resistance to insects in cotton. I. To *Amrasca devastans* (Distant). *Coton et Fibres Tropicales* 33:409-414.

Agarwal, R. A., Gupta, G. P. and Katiyar, K. N.
1983a Response of varieties to insecticidal treatment against major insect pests in cotton. *Indian Journal of Entomology* 45:338-341.

Agarwal, R. A. and Krishnananda, N.
1976a Preference to oviposition and antibiosis mechanism to jassids (*Amrasca devastans* Dist.) in cotton (*Gossypium* sp.). *In*: Jermy, T., ed. *The Host-Plant in Relation to Insect Behaviour and Reproduction*. Pp. 13-22. Plenum Publishing Corporation, New York. 322pp.

Agarwal, R. A. and Siddiqi, Z. A.
1964a Sugarcane pests. *In*: Pant, N. C., Prasad, S. K., Vishnoi, H. S., Chatterji, S. N. and Varma, B. K., eds. *Entomology in India*. Pp. 149-186. The Entomological Society of India. 529 pp.

Agarwal, R. A., Wankhede, N. P. and Katiyar, K.
1979a Impact of agronomic practices on the population of jassids *Amrasca devastans* (Distant) in cotton. *Coton et Fibres Tropicales* 34:375-378.

Agarwal, S. C. and Kushwaha, K. S.
1979a Efficacy of insecticidal control schedules against aphids and jassids on tomato. *Indian Journal of Plant Protection* 6:40-43.

Aguda, R. M., Centina, D. B., Heinrichs, E. A. and Dyck, V. A.
1984a Fungicides to control green muscardine fungus, a disease of zigzag leafhopper in rearing cages. *International Rice Research Newsletter* 9(3):14-15.

Aguda, R. M., Litsinger, J. A. and Roberts, D. W.
1984a Pathogenicity of *Beauveria bassiana* on brown planthopper (BPH), white backed planthopper (WPPH) and green leafhopper (GLH). *International Rice Research Newsletter* 9(6):20.

Aguiero, V. M., Daquioag, R. D. and Ling, K. C.
1979a Search for alternate insect vectors of rice ragged stunt disease. *International Rice Research Newsletter* 4(5):15-16.

Aguilar F., P. G.
1964a Especies de artopodos registrados en las lomas de los alrededores de Lima. *Revista Peruana de Entomologia* 7(1):93-95.

Agyen-Sampong, M.
1976a Co-ordinated minimum insecticide trials: Yield performance of insect resistant cowpea cultivars from ITTA compared with Nigerian cultivars. *Tropical Grain Legume Bulletin* No. 5, p. 6.

Ahmad, F. U. and Shafi, M. M.
1966a Concentrates against cotton pests. *World Crops* 18(4):60-62.

Ahmad, I.
1983a Studies on rice insects of Pakistan with reference to their systematics and pheromone glands. *Final Report, Research Project No. FG-PA-310* (PK-ARS-139). Department of Zoology-Entomology, University of Karachi, Pakistan. 388 pp.

Ahmed, Manzoor
1967a A study of genera and species of Typhlocybinae (Cicadellidae: Homoptera) from West Pakistan. Thesis, University of Karachi, Pakistan. 182 pp.
1969a Studies of tribe Dikraneurini (Typhlocybinae, Homoptera) from West Pakistan. *Pakistan Journal of Zoology* 1(1):55-63. [New genera: *Fusiplata, Ayubiana, Karachiota*.]
1969b Some recent studies of *Empoasca* complex in Pakistan, their bearing on tribal classification of sub-family Typhlocybinae (Homoptera: Cicadellidae). *Agriculture Pakistan* 20(4):451-455.
1969c Studies of the genus *Zygina* Fieber (Homoptera, Erythroneurini, Cicadellidae) in West Pakistan. *Pakistan Journal of Zoology* 1(2):157-175.
1969d Studies of the genera of *Eupteryx* complex (Typhlocybini, Homoptera) in West Pakistan. *Pakistan Journal of Forestry* 19(3):311-320. [New genera: *Evansioma, Youngama*.]
1970a Studies of the genus *Erythroneura* Fitch (Erythroneurini: Cicadellidae) in West Pakistan. *Pakistan Journal of Zoology* 2(1):29-42.
1970b Further studies on the leafhoppers of tribe Erythroneurini (Cicadellidae: Typhlocybinae) from West Pakistan. *Pakistan Journal of Zoology* 2(2):167-184. [New genus: *Jalalia*.]
1970c Host and food plants of typhlocybine leafhoppers in Pakistan. *Pakistan Journal of Forestry* 20(2):217-223.
1970d *Andrabia*, new genus and *A. kashmirensis*, new species (Typhlocybinae: Cicadellidae) on the plant Tember (*Zanthoxylum alatum*) in northern areas of West Pakistan. *Pakistan Journal of Scientific and Industrial Research* 13(4):407-409.
1970e Some species of Typhlocyba Germar (Typhlocybinae-Cicadellidae) occurring in West Pakistan. *Pakistan Journal of Science* 22(5&6):269-276.
1971a A preliminary account of the distribution of typhlocybine fauna in forests of West Pakistan. *Pakistan Journal of Forestry* 21(1):53-60.
1971b Studies on the genera and species of tribe Erythroneurini (Typhlocybinae: Typhlocybinae) in East Pakistan. *Pakistan Journal of Zoology* 3(2):175-192. [New genera: *Pakeasta, Mandera*.]
1971c Study of new genus and some species of *Typhlocyba* Germar (Typhlocybini: Cicadellidae) in West Pakistan. *Journal of Science of the Karachi University* 1(1):190-202. [New genus: *Byphlocyta*.]
1971d *Havelia alba*, new genus, new species (Typhlocybinae: Homoptera) on the forest plant *Debregeasia hypoleuca* in parts of West Pakistan. *Pakistan Journal of Forestry* 21(3):277-280.
1972a Some more genera and species of Typhlocybinae (Cicadellidae: Homoptera) from East Pakistan. *Pakistan Journal of Scientific and Industrial Research* 15(1&2):67-71, illust. [New genera: *Hameedia, Sylhetia*.]
1972b Some preliminary observations on the distribution of typhlocybine leafhoppers on woody and fruit trees in West Pakistan. *Journal of Science of the Karachi University* 1:106-111.
1973a Biology, ecology, life history and control of typhlocybine leafhoppers in Pakistan. *Final Technical Report, PL-480 Grant No. FG-PA-158, USDA PL-480 Program*. Department of Zoology, University of Karachi, Pakistan. 228 pp.
1978a Biosystematics of leafhopper pests of vegetable and fruit plants in Pakistan. *Final Technical Report, Project No. PK-ARS-4, Grant FG-PA-220, USDA PL-480 Program*. Department of Zoology, University of Karachi, Pakistan. 86 pp.
1979a [1976] Some genera and species of typhlocybine leafhoppers from Uganda, East Africa (Homoptera, Auchenorrhyncha, Cicadellidae). *Reichenbachia* 17(5):25-41.
1982a Incidence of typhlocybine leafhoppers on vegetable and fruit plants in Baluchistan (Pakistan), 1979. *Pakistan Journal of Scientific and Industrial Research* 25(2&2):31-33.
1982b Evaluation of losses in brinjal (*Solanum melanogena*) by *Amrasca devastans* (Distant). *Pakistan Journal of Agricultural Research* 2:227-280.
1983a Ecology of leafhopper pests of vegetable and fruit plants in Pakistan. *Final Research Progress Report, PL-480 Grant No. FG-PA-319, USDA*. Department of Zoology, University of Karachi, Pakistan. 248 pp.
1983b Biotaxonomy of typhlocybine leafhoppers of Pakistan. In: Knight, W. J., Pant, N. C., Robertson, T. S. and Wilson, M. R., eds. *Proceedings of the 1st International Workshop on Biotaxonomy, Classification and Biology of Leafhoppers and Planthoppers (Auchenorrhyncha) of Economic Importance.* Pp. 179-183. London, 4-7 October 1982. Commonwealth Institute of Entomology, 56 Queen's Gate, London SW7 5JR. 500pp.
1983c Experimental observations on the relationship of leafhopper in infestation and yield of okra (*Abelmoschus esculentus*) in Pakistan. *Journal of Pure and Applied Science* 2:63-68.
1984a Plant association of typhlocybine leafhoppers of Pakistan. *Mitteilungen der Schweizerischen Entomologischen Gesellschaft* 57(4):404.
1984b Interrelationship of leafhopper abundance and yield of potato in Pakistan. *ISI Journal of Science* 8:13-27.
1985a Plant preference of the leafhopper *Amrasca biguttula biguttula* (Ishida) (Typhlocybinae: Cicadellidae) amongst vegetable plants at Karachi-Pakistan. *Tehqique* 3:11-18.
1985b Leafhopper infestations of brinjal and its relationship with factors of growth of the plant at Karachi-Pakistan. *Pakistan Journal of Zoology* 18:209-214.
1985c Typhlocybinae of Pakistan. Fauna of the subfamily Typhlocybinae (Cicadellidae: Homoptera: Insecta). *Pakistan Agricultural Research Council, Islamabad.* 279 pp.

Ahmed, Manzoor, Baluch, Masood Ahmed and Ahmed, Mubarik
1980a Abundance and diversity of typhlocybine leafhoppers on vegetable plants in Pakistan. *Pakistan Journal of Scientific and Industrial Research* 23(1&2):41-47.

Ahmed, Manzoor, Baluch, Masood Ahmed, Ahmed, Mubarik and Akhtar, Pervaiz
    1981a Effects of insecticides used for leafhopper control on the quantity and quality of potato tubers. *Natural Sciences* (Karachi) 3(2):79-82.

Ahmed, Manzoor and Jabbar, Abdul
    1970a [1971] *Typhlocyba karachiensis*, a new species (Typhlocybinae-Homoptera), a pest of *Schinustenem anthifolius* in Karachi, West Pakistan. *Agriculture Pakistan* 22:107-112.
    1972a Some preliminary remarks on *Zyginidia quyumi* (Ahmed), an important pest of wheat and maize in some parts of Pakistan. *Pakistan Journal of Scientific and Industrial Research* 15:382-383.
    1974a Laboratory testing of some organic insecticides against *Zyginidia guyumi* (Ahmed) (Typhlocybinae: Cicadellidae) on maize. *Folia Biologica* 22:353-355.
    1977a Distribution of *Z. quyumi* (Ahmed) (Cicadellidae: Homoptera) a pest of wheat and maize in Punjab (Pakistan) by location, day timings and temperature. *Pakistan Journal of Scientific and Industrial Research* 20(3):188-191.

Ahmed, Manzoor, Jabbar, Abdul and Mohd, Shafiq
    1973a Statistical studies of light trap catches of *Zyginidia quyumi* (Ahmed) (Cicadellidae: Homoptera) on wheat in Punjab. *Pakistan Journal of Scientific and Industrial Research* 16(6):238-240.

Ahmed, Manzoor, Jabbar, Abdul and Samad, K.
    1977a Ecology and behaviour of *Zyginidia quyumi* (Typhlocybinae: Cicadellidae) in Pakistan. *Pakistan Journal of Zoology* 9(1):79-85.

Ahmed, Manzoor and Khokhar, K. F.
    1971a Genus *Helionidia* Zachvatkin (Erythroneurini-Cicadellidae) with four new species from West Pakistan. *Sind University Research Journal (Science Series)* 5(1):63-72.

Ahmed, Manzoor and Lodhi, S.
    1970a [1971] Morphology of the female seventh sternum of Typhlocybinae species studied in Pakistan. *Agriculture Pakistan* 22(1):93-106.

Ahmed, Manzoor and Malik, Kulsoom Fatima
    1972a On the true identification of genus *Chudania* Distant, collected on fig (*Ficus carica*) in forests of Azad Kashmir and Kaptai (Pakistan). *Pakistan Journal of Forestry* 22(2):103-108.

Ahmed, Manzoor and Mahmood, S. H.
    1970a (1969) A new genus and two new species of Nirvaninae (Cicadellidae – Homoptera) from Pakistan. *Pakistan Journal of Scientific and Industrial Research* 12:260-263. [New genus: *Quercinirvana*.]

Ahmed, Manzoor and Naheed, R.
    1981a On some species of the genera *Edwardsiana* and *Farynala* from Pakistan (Homoptera, Cicadellidae, Typhlocybinae). *Reichenbachia* 19(14):75-83.
    1982a Leafhopper infestations of vegetable and fruit plants in Sind-Pakistan during 1979-1980. *Pakistan Journal of Scientific and Industrial Research* 25(4):124-125.

Ahmed, Manzoor, Naheed, R., Ahmed, Mubarik and Baluch, M. A.
    1980a Some observations on losses to host plants caused by leafhopper pests in Pakistan. *Proceedings of the 1st Pakistan Congress of Zoology*, B, pp. 251-255.

Ahmed, Manzoor, Naheed, Rukhsana and Samad, Kurshid
    1981a On some species of the genus *Typhlocyba* Germar (Homoptera, Cicadellidae, Typhlocybinae) from Pakistan. *Reichenbachia* 19(16):89-95.

Ahmed, Manzoor, Naheed, Rukhsana and Shafiq, M.
    1980a Diversity of typhlocybine leafhoppers affecting fruit plants in Pakistan. *Pakistan Journal of Scientific and Industrial Research* 23(1&2):34-40.

Ahmed, Manzoor and Samad, Khurshid
    1972a An important rice pest, *Thaia oryzivora* Ghauri (Typhlocybinae: Homoptera) and some remarks on its population in East Pakistan. *Pakistan Journal of Scientific and Industrial Research* 15(3):177-180.
    1972b Some new additions to the typhlocybine fauna of East Pakistan. *Pakistan Journal of Scientific and Industrial Research* 15(4&5):285-290.
    1980a Erythroneurine leafhoppers of fruit and vegetable plants in Pakistan. Part I. Genus *Zyginidia* Haupt. *Pakistan Journal of Zoology* 12(1):93-98, illust.

Ahmed, Manzoor, Samad, Khurshid and Malik, K. F.
    1970a Some commonly found leafhoppers in paddy fields of East Pakistan. *Pakistan Journal of Science and Industrial Research* 22:191-198.

Ahmed, Manzoor, Samad, Khurshid and Naheed, Rukhsana
    1981a Empoascan leafhoppers infesting fruit and vegetable plants in Pakistan (Homoptera: Cicadellidae). *Entomologica Scandinavica* 12:1-21.

Ahmed, Manzoor and Waheed, Abdul
    1971a Two new genera of Typhlocybinae (Homoptera: Cicadellidae) feeding on rose (*Rosa indica*), in West Pakistan. *Pakistan Journal of Scientific and Industrial Research* 14(1&2):116-117. [New genera: *Mahmoodiana, Domelia.*]
    1984a Relationship of leafhopper infestation with growth of okra (*Abelmoschus esculentus*) at Karachi. I. Foliage destruction and height of plants. *Pakistan Journal of Agricultural Research* 5:241-244.

Ahmed, Mubarik and Ahmed, Manzoor
    1980a Some aspects of the biology and ecology of *Zygina rubronotata*, a pest of falsa (*Grewia asiatica*) in Pakistan. *Pakistan Journal of Scientific and Industrial Research* 23(1&2):30-33.
    1980b Loss of chlorophyll in leaves of some plants caused by leafhopper feeding. *Pakistan Journal of Scientific and Industrial Research* 23(1&2):48-50.

Ahmed, Mubarik, Baluch, Masood A., Ahmed, Manzoor and Shaukat, S. S.
    1980a Population dynamics of idiocerine leafhoppers on mango (*Mangifera indica*) in Karachi (1978-1979). *Proceedings of the 1st Pakistan Congress of Zoology*, B, pp. 257-270.

Ahmed, Mubarik, Ahmed, Manzoor, Baluch, Masood A. and Naheed, Rukhsana
    1981a Sooty mold on mango plants and its relationship with leafhoppers and climatic factors in Karachi, Pakistan during 1978-79. *Pakistan Journal of Scientific and Industrial Research* 24(4):140-144.
    1984a Assessment of qualitative and quantitative damage to potato in Pakistan. *Tehqique* 2:11-20.

Ahmed, Sajed Sultan, Naheed, Rukhsana and Ahmed, Manzoor
    1980a Three new species of idiocerine leafhoppers. *Proceedings of the 1st Pakistan Congress of Zoology*, B, pp. 221-225.

Aibasov, Kh. A.
    1974a Leafhoppers and scale insects (Homoptera, Cicadinea, Coccinea) damaging poplar. *In: Mityaev, I. D., ed. Fauna, Systematics and Biology of the Insects of Kazakhstan.* [Fauna, sistematika i biology nasekomykh Kazakhstana.] Akademiya Nauk Kazakhskoi SSR 35:126-132. [In Russian.]

Aitchison, C. W.
    1978a Notes on low temperature and winter activity of Homoptera in Manitoba. *Manitoba Entomology* 12:58-60.
    1984a Low temperature feeding by winter-active spiders. *Journal of Arachnology* 12:297-305.

Akingbohungbe, A.E.
    1983a Nomenclatural problems, biology, host plant and possible vector status of Auchenorrhynca associated with crop plants in Nigeria. *In: Knight, W. J., Pant, N. C. Robertson, T. S. and Wilson, M. R., eds. Proceedings of the 1st International Workshop on Biotaxonomy, Classification and Biology of Leafhoppers and Planthoppers (Auchenorrhyncha) of Economic Importance.* Pp. 365-370. London, 4-7 October 1982. Commonwealth Institute of Entomology, 56 Queen's Gate, London SW7 5JR. 500 pp.

Akino, K.
    1969a Effects of temperature & wind on the activity of green rice leafhopper adults, *Nephotettix cincticeps* Uhler (Hemiptera: Deltocephalidae). *Japanese Journal of Applied Entomology and Zoology* 13(2):78-83. [In Japanese with English abstract.]

Akram, M. and Yunis, M.
    1972a Chemical control of leafhoppers attacking maize. *Journal of Agricultural Research* 10(4):239-245.

Alam, M. Z.
    1964a Insect pests of mango in East Pakistan and their control. *Fourth Proceedings of the Pan Indian Ocean Scientific Congress, Section D, Agricultural Science 1960*, pp. 399-407.
    1967a Insect pests of rice in East Pakistan. *In: Pathak, M. D., ed. The Major Insect Pests of the Rice Plant.* Pp. 643-655. Johns Hopkins, Baltimore. 729 pp.

Alam, M. Z. and Mandal, M. C.
    1964a [1965] On the biology of mango hopper, *Idiocerus atkinsoni* Leth. (Jassidae, Hemiptera) in East Pakistan. Dacca, E. Pakistan, pp. 139-147.

Alam, S.
    1971a Population dynamics of the common leafhopper and planthopper pests of rice. *Dissertation Abstracts International* 32(4):2207-2208.
    1977a Rice green leafhoppers. *In: Literature Review of Insect Pest and Diseases of Rice in Bangladesh.* Bangladesh Rice Research Institute, Dacca, Bangladesh. 90 pp.

Alam, S. and Alam, A. S.
    1979a Hopper burn by the orange-headed leafhopper in Bangladesh. *International Rice Research Newsletter* 4:(4)17.

Alam, Z. and Islam, A.
    1959a Biology of the rice leafhopper *Nephotettix bipunctatus* Fabr. in East Pakistan. *Pakistan Journal of Scientific and Industrial Research* 11:20-28.

Alayo, D. P. and Novoa, N.
    1985a El genero *Carneocephala* (Homoptera :Auchenorrhyncha) en Cuba. *Poeyana* 290:1-15.

Albouy, J.
    1966a Le probleme des "germes fins" du glaieul. [The problem of 'germes fins' of *Gladiolus*.] *Annales des Epiphyties* 17():81-91. [With English and German summaries.]

Albouy, J., Cousin, M. T. and Grison, C.
    1967a Etude comparee de 3 maladies a virus: souche "Californienne" de l'aster du glaieul, phyllodie du trefle et stolbur de la tomate sur *Vinca rosea*. *Annales des Epiphyties* 18:157-171.

Albrecht, A.
    1977a Interesting finds of bugs in Finland. *Notulae Entomologicae* 57(2):51-52. [In Swedish with English and Finnish summaries.]
    1980a [1981] Parasitoids of leafhoppers (Homoptera, Auchenorrhyncha) in dry grasslands and open heath forests in southern Finland. *Tvarminne Studies* 1:23.

Alghali, A. M. and Domingo, J. S.
    1982a Weed hosts of some rice pests in Northwestern Sierra Leone. *International Rice Research Newsletter* 7(2):10.

Ali, A. M.
    1978a Ecology of *Anagrus* species, parasitoids of the eggs of leafhopper in *Juncus* stems. *Auchenorrhyncha Newsletter* 1:26.

Aliniazee, M. T., Frost, M. H. and Stafford, E. M.
    1971a Chemical control of grape leafhoppers and Pacific spider mites on grapevines. *Journal of Economic Entomology* 64(3):697-700.

Alivizatos, A. S.
    1982a Feeding behaviour of the spiroplasma vectors *Dalbulus maidis* and *Euscelidius variegatus* in *vivo* and in *vitro*. *Annales de l'Institute Phytopathologique Benaki* 13(2):128-144.
    1983a Acquisition in *vitro* of corn stunt spiroplasma by the leafhopper *Dalbulus maidis*. *Annales de l'Institute Phytopathologique Benaki* 14:101-109.

Alkan, Bekir
    1957a Die wichtigsten und einige neue arten von schdlingen der *Pistacia vera* in der Turkei. *Proceedings of the International Congress of Crop Protection* 1:803-804.

Alkhazishvili, T. V.
    1953a Materials toward the arachnoentomofauna of cotton plants in Georgia. *Academy Science Georgiana SSR, Instituta Zashchita Rastenii, Trudy* 9:43-56. [In Georgian with Russian summary.]

All, J. N., Kuhn, C. S., Gallaher, R. N., Jellum, M. D. and Ussey, R. S.
    1977a Influence of no-tillage-cropping, carbofuran, and hybrid resistance on dynamics of maize chlorotic dwarf and maize dwarf mosaic diseases of corn. *Journal of Economic Entomology* 70:221-225.

All, J. N., Kuhn, C. W. and Jellum, M. D.
    1981a Control strategies for vectors of virus and viruslike pathogens of maize and sorghums. *In: Gordon, D. T., Knoke, J. K. and Scott, G. E., eds. Virus and Viruslike Diseases of Maize in the United States.* Pp. 127-131. Ohio Agricultural Research and Development Center, Wooster.

Allen, A. A.
    1962a *Placotettix taeniatifons* Kbm. (Hem., Cicadellidae) in S.E. London (N.W. Kent). *Entomologist's Monthly Magazine* 98:47.
    1963a *Opsius stactogalus* Fieb. (Hem., Cicadellidae) in S.E. London. *Entomologist's Monthly Magazine* 99:3.
    1963b *Opsius stactogalus* Fieb. (Hem., Cicadellidae) in and near London: a second note. *Entomologist's Monthly Magazine* 99:240.
    1963c The foodplant of *Idiocerus varius* Germ. (Hem., Cicadellidae). *Entomologist's Monthly Magazine* 99(1194-5):233.
    1964a The genus *Idiocerus* (Hem., Hom., Cicadellidae) in suburban North-west Kent. *Entomologist's Record and Journal of Variation* 76(2):55-57.
    1965a *Eupteryx tenella* Fall. (Hem., Cicadellidae) in South-east London — a probable addition to the Kent fauna. *Entomologist's Monthly Magazine* 101:194.
    1978a Records of rare and uncommon British species of *Idiocerus* (Hem.-Hom., Cicadellidae), with special reference to an unnoticed habit of *I. poecilus* Kbm. *Entomologist's Record and Journal of Variation* 90(4):113-115.
    1982a *Eupteryx atropunctata* (Goeze) (Hem., Cicadellidae) on hollyhock and *E. origani* Zachv. and *tenella* (Fall.) incorrectly recorded as new to Kent. *Entomologist's Monthly Magazine* 118:256.
    1985a *Idiocerus herrichi* Kbm. (Hem. Cicadellidae) in S.E. London. *Entomologist's Record and Journal of Variation* 97(3-4):64-65.

Allen, R. M. and Donndelinger, C. R.
    1982a Cultivation in vitro of spiroplasmas from six plant hosts and two leafhopper vectors in Arizona. *Plant Disease* 66:669-672.

Allen, R. T.
    1983a Distribution patterns among arthropods of the north temperate deciduous forest biota. *In* Biogeographical relationships between temperate Asia and temperate North America. *Annals of the Missouri Botanical Garden* 70(4):616-628.

Allen, W. A., Bobb, M. L., Grayson, J. M., Roberts, J. E., Robinson, W. H. and Weidhass, J. A.
    1972a 1972 Virginia insect control guide. *Virginia Polytechnic Institute Extension Division Control Series* 141. 118 pp.

Al-Ne'amy, K. T. and Linnavuori, R. E.
    1982a On the taxonomy of the subfamily Adelungiinae (Homoptera, Cicadellidae). *Suomen Hyonteistieteellinen Aikakauskirja* 48(4):109-115.

Altieri, M. A., Francis, C. A., van Schoonhoven, A. and Doll, J. D.
    1978a A review of insect prevalence in maize (*Zea mays* L.) and bean (*Phaseolus vulgaris* L.) polycultural systems. *Field Crops Research* 1:33-49.

Altieri, M. A., van Schoonhoven, A. and Doll, J. D.
1977a The ecological role of weeds in insect pest management systems: A review illustrated by bean (*Phaseolus vulgaris*) cropping systems. *PANS* 23:195-205.

Alvarenga, M.
1962a A Entomofauna do Arquipelago de Fernando de Noronha, Brasil - 1. *Archivos do Museu Nacional* 52:21-26.

Alverson, D. R.
1979a Epidemiology and control of *Graminella nigrifrons*-borne maize chlorotic dwarf and aphid-borne maize dwarf mosaic virus diseases of corn in Georgia. Doctoral Dissertation, University of Georgia. 148 pp.

Alverson, D. R., All, J. N. and Bush, P. B.
1980a Rubidium as a marker and simulated inoculum for the black-faced leafhopper, *Graminella nigrifrons*, the primary vector of maize chlorotic dwarf virus of corn. *Environmental Entomology* 9(1):29-31.

Alverson, D. R., All, J. N. and Kuhn, C. W.
1980a Simulated intrafield dispersal of maize chlorotic dwarf virus by *Graminella nigrifrons* with a rubidium marker. *Phytopathology* 70(8):734-737.

Alverson, D. R., All, J. N. and Mathews, R. W.
1977a Response of leafhoppers and aphids to variously colored sticky traps. *Journal of the Georgia Entomological Society* 12:337-341.

Ambekar, J. S. and Kalbhor, S. E.
1981a Note on the plant characters associated with resistance to jassid, *Amrasca biguttula biguttula* Ishida, in different varieties of cotton. *Indian Journal of Agricultural Sciences* 51:816-817.

Amen, C. R.
1953a Investigations of insect vectors of potato viruses—1953. *Proceedings of the 13th Annual Meeting Oregon Seed Growers League*, 7-9 December 1953, pp. 77-78.

Ameresekere, R. V. W. E.
1970a *Circulifer tenellus* Baker (Homoptera: Cicadellidae). Control of reproduction by sterilization. *Dissertation Abstracts* 31B:2036.

Ameresekere, R. V. W. E. and Georghiou, G. P.
1971a Sterilization of the beet leafhopper: Induction of sterility and evaluation of biotic effects with a model sterilant (OM-53139) and ^{60}CO irradiation. *Journal of Economic Entomology* 64(5):1074-1080.

Ameresekere, R. V. W. E., Georghiou, G. P. and Sevacherian V.
1971a Histopathological studies on irradiated and chemosterilized beet leafhoppers, *Circulifer tenellus*. *Annals of the Entomological Society of America* 64(5):1025-1031.

Amin, P. W.
1977a Effect of western X-disease on total protein, lipid and triglyceride contents of its leafhopper vector, *Colladonus montanus* (Van Duzee). *Indian Journal of Entomology* 38:325-328.

1977b Effect of western X-disease on the activity of salivary phenolase in its leafhopper vector, *Colladonus montanus* Van Duzee). *Indian Journal of Entomology* 38:379-381.

1982a Jassids (Homoptera: Cicadellidae) as pests of groundnuts (*Arachis hypogaea* L.). *International Crops Research Institute for the Semi-Arid Tropics*. Patancheru, Andhra Pradesh, India. 32 pp.

1983a Resistance of wild species of groundnut to insect and mite pests. *Proceedings of an International Workshop on Cytogenetics of Arachis* 1983:57-60. ICRISAT Center, Patacheru, India.

Amin, P. W. and Jensen, D. D.
1971a The effect of western X-disease on the oxygen consumption of its leafhopper vector, *Colladonus montanus*. *Journal of Invertebrate Pathology* 18(1):108-113.

1971b Effects of tetracycline on the transmission and pathogenicity of the western X-disease agent in its insect and plant hosts. *Phytopathology* 61:696-702.

Ammar, El-Desouky
1975a A light trap modified for catching only smaller insects and its efficiency in catching certain groups compared with a sweeping net. *Zeitschrift fur Angewandte Entomologie* 79:104-109.

1975b Biology of the leafhopper *Cicadulina chinai* Ghauri (Homoptera: Cicadellidae) in Giza, Egypt. *Zeitschrift fur Angewandte Entomologie* 79:337-345.

1977a Biology of *Cicadulina bipunctella zeae* China in Giza, Egypt. *Deutsche Entomologische Zeitschrift* 24(4&5):345-352.

1978a Comparative study on the morphology of the immature stages of *Cicadulina chinai* Ghauri and *Cicadulina bipunctella zeae* China. *Deutsche Entomologische Zeitschrift* 25(1-3):119-127.

1985a Internal morphology and ultrastructure of leafhoppers and planthoppers. In: Nault, L. R. and Rodriguez, J. G., eds. *The Leafhoppers and Planthoppers*. Pp. 127-162. Wiley & Sons, New York. 500 pp.

1985b Muscle cells in the salivary gland of a planthopper, *Peregrinus maidis* (Ashmead) and a leafhopper, *Macrosteles fascifrons* (Stal) (Homoptera: Auchenorrhyncha). *International Journal of Insect Morphology and Embryology* 13(5-6) 1984:425-428.

Ammar, El-D., El-Nahal, A. K. M. and El Bolok, M. M.
1981a Fluctuation of population densities of *Empoasca decipiens* Paoli and *Balchutha hortensis* Lindl. at Giza, Egypt (Homoptera: Cicadellidae). *Bulletin de la Societe Entomologique d'Egypte* 61:245-255.

Ammar, El-D. and Farrag, S. M.
1980a Preliminary survey and relative abundance of leafhoppers and planthoppers (Auchenorrhyncha, Homoptera) at Giza, using a modified light trap. *Bulletin de la Societe Entomologique de Egypte* 60:297-303.

Ammar, El-D. and Hosny, M. M.
1969a Host-plants, symptoms of infestation and certain characteristics of *Empoasca* spp. (Jassidae) in the Cairo area, U.A.R. *Zeitschrift fuer Angewandte Entomologie* 63(3):272-280. [With German summary.]

Ammar, El-D., Kira, M. T. and Abul-Ata, A. E.
1982a Natural occurrence of streak and mosaic diseases on sugarcane cultivars in Upper Egypt, and transmission of sugarcane streak by *Cicadulina bipunctella zeae* (China). *Annales de Virologie* 133:183-185.

Ammar, El-D., Lamie, O. and Khodeir, I. A.
1978a A light-trap survey of the abundance and sex ratio of leafhoppers and planthoppers in Egypt. *International Rice Research Newsletter* 3(3):18.

1978b Populations of leafhoppers and planthoppers in Egypt from 1973 to 1975 as indicated by sweep net samples. *International Rice Research Newsletter* 3(2):12-13.

1983a Population studies of leafhoppers and planthoppers on rice plants at Kafr-el-Sheikh, Egypt (Homoptera: Auchenorrhyncha). *Bulletin de la Societe Entomologique de Egypte* 62:63-70.

Ananthanarayanan, K. P. and Abraham, E. V.
1955a A note on a new jassid pest on sugarcane from the Madras State. *Indian Journal of Entomology* 16(4):372

1956a Bionomics and control of rice jassid, *Nephotettix bipunctatus* F. and the rice fulgorid, *Nilaparvata lugens* S. in Madras State. *Proceedings of the 6th Scientific Work Conference*, Coimbatore, pp. 33-44.

Ancalmo, O., and Davis, W. C.
1961a Achaparramiento. [Corn stunt.] *Plant Disease Reporter* 45:281.

Anderson, B. L.
1984a The effect of dietary iron on the biology of the aster leafhopper, *Macrosteles fascifrons* (Stal) (Homoptera: Cicadellidae). M.S. Thesis, University of Minnesota.

Anderson, L. E., Kurzepa, H. and Jenkin, H. M.
1974a Growth of Japanese encephalitis virus (JEV) in *Aedes aegypti* (AE) cells and JEV stability in fresh and spent leafhopper medium. *Abstracts of the Annual Meeting of the American Society for Microbiology* 74:227.

Andison, H.
1954a Insect pests of fruits and ornamentals on Vancouver Island and in the lower Fraser Valley, 1953 [British Columbia]. *Canadian Insect Pest Review* 32(1):4.

Ando, Y. and Kishino, K.
1981a Resistance of rice plant to the green rice leafhopper, *Nephotettix cincticeps* Uhler (Hemiptera: Cicadellidae). 3. Simple testing method using nonpreference. *Japanese Journal of Applied Entomology and Zoology* 25:196-197.

Andow, D. A.
1984a Microsite of the green rice leafhopper, *Nephotettix cincticeps* (Homoptera: Cicadellidae), on rice plant: Plant nitrogen and leafhopper density. *Researches on Population Ecology* 26:313-329.

Andow, D. A. and Kiritani, K.
1984a Fine structure of trivial movement in the green rice leafhopper, *Nephotettix cincticeps* Uhler (Hemiptera: Cicadellidae). *Applied Entomology and Zoology* 19:306-316.

Andrewes, C. H.
1977a *Idiocerus herrichii* Kirschbaum (Hem., Cicadellidae) in Wiltshire. *Entomologist's Monthly Magazine* 113(Sept.-Dec.):241.

Andrzejewska, L.
1960a Inactivity of a stabilized *Tettigella viridis* L. population in resettling of environments. *Bulletin de l'Academie Polonaise des Sciences. Serie des Sciences Biologiques* 8:581-585.
1960b Results of experimental increases in density of Homoptera in meadows. *Ekologia Polska* (A) 9:439-451. [In Polish with English summary.]
1960c The course of reduction in experimental Homoptera concentrations. *Bulletin de l'Academie Polonaise des Sciences. Serie des Sciences Biologiques* 9:173-178.
1962a *Macrosteles laevis* (Rib.) as an unsettlement index of natural meadow associations of Homoptera. *Bulletin de l'Academie Polonaise des Sciences* 10:221-226.
1964a Differentiation of vertical distribution of *Cicadella viridis* L. (Homopt., Cicadellidae) in meadow habitat. *Polskie Pismo Entomologiczne,* Series B, n. 1-2(33-34), No. 9:93-96.
1965a Stratification and its dynamics in meadow communities of Auchenorrhyncha (Homoptera). *Ekologia Polska* (A) 13:685-715.
1966a Estimation of the effects of feeding of the sucking insect *Cicadella viridis* L. (Homoptera: Auchenorrhyncha) on plants. *In: Petrusewicz, K., ed. Secondary Productivity in Terrestrial Ecosystems.* Pp. 791-805.
1966b An attempt at determining the absolute population numbers of *Cicadella viridis* L. in the light of its layer distribution. *Ekologia Polska* (A) 14(3):73-98.
1971a Productivity investigation of two types of meadows in the Vistula Valley. 6. Production and population density of leafhopper (Homoptera-Auchenorrhyncha) communities. *Ekologia Polska* (A) 19(12):151-172.
1978a The effect of treatment and utilization of meadows on Auchenorrhyncha communities. *TYMBAL. Auchenorrhyncha Newsletter* 1:6.
1981a [1982] Auchenorrhyncha communities on peatsoil meadows influenced by draining. *Acta Entomologica Fennica* 38:5-6.
1984a Ecological structure of Auchenorrhyncha communities under increasing cultivation. *Mitteilungen der Schweizerischen Entomologischen Gesellschaft* 57(4):405-406.

Angelini, A. and Vandamme, P.
1965a Eleven years of experiments with insecticides on cotton in the Ivory Coast. *In: Congress on the Protection of Tropical Crops.* Pp. 359-365.

Anjaneyulu, A.
1975a Rice stubbles and self-sown rice seedlings, the reservoir hosts of tungro virus disease during off-seasons. *Science and Culture* 41:298-299.
1975b *Nephotettix virescens* (Distant) nymphs and their role in the spread of rice tungro virus. *Current Science* 44:357-358.
1980a Life span of and tungro transmission by viruliferous *Nephotettix virescens* on 10 rice varieties at CRRI, India. *International Rice Research Newsletter* 5(5):10.

Anjaneyulu, A. and Chakraborti, N. K.
1977a Geographical distribution of rice tungro virus disease and its vectors in India. *International Rice Research Newsletter* 2:15-16.

Anjaneyulu, A., and John, V. T.
1972a Strains of rice tungro virus. *Phytopathology* 62:1116-1119.

Anjaneyulu, A., Shukla, V. D., Rao, G. M. and Singh, S. K.
1981a Perpetuation of rice tungro virus and its vectors. *International Rice Research Newsletter* 6(1):12-13.
1982a Experimental host range of rice tungro virus and its vectors. *Plant Disease* 66:54-56.

Anjaneyulu, A., Singh, S. K. and Shenoi, M. M.
1982a Evaluation of rice varieties for tungro resistance by field screening techniques. *Tropical Pest Management* 28:147-155.

Anjaneyulu, A., Singh, S. K., Shukla, V. D. and Shenoi, M. M.
1980a Chlorotic streak, a new virus disease of rice. *International Rice Research Newsletter* 5(3):12-13.

Annappan, R. C., Venkataraman, N. and Kamalanathan, S.
1965a An observation on the jassid incidence on different types of Karunganni cotton (*G. arboreum* L.). *Madras Agricultural Journal* 52:412-414.

Annappan, R. S.
1960a Breeding for jassid resistance—a few findings. *Indian Cotton Growing Review* 14:501-508.

Annecke, D. P.
1965a A new species of *Centrodora* Foerster (Hymenoptera: Aphelinidae) parasite in eggs of a cicadellid injurious to citrus in South Africa. *South African Journal of Agricultural Science* 8:1133-1138.

Annecke, D. P. and Mynhardt, M. J.
1968a Citrus leafhopper. *The South African Citrus Journal* 411:1-3.

Anonymous.
[Undated] Annual research report, 1969. Institute of Food and Agricultural Sciences, Florida University, Gainesville. 228 pp.
1928a Injurious insects and other pests. *Kansas Agricultural Experiment Station Directors Report* 1926-1928:67-82.
1956a Insect control for profitable dahlias. *Florists Exchange* 127(10):17, 48.
1957a Tomato big bud and related diseases. *Agricultural Gazette of New South Wales* 68(12):646-647.
1957b Insects not known to occur in the United States. *U.S. Department of Agriculture Cooperative Economic Insect Report* 7:1-67.
1957c Insect pests of soybeans and their control. *Soybean Digest* 17(10):16-18, 20.
1958a Entomology. *Maryland Agricultural Experiment Station Bulletin* A-92:44-48.
1958b Control of insect pests. *Colorado Agricultural Experiment Station Annual Report* 71:5-6.
1959a Insects not known to occur in the United States. *U.S. Department of Agriculture Cooperative Economic Insect Report* 9:1-68.
1959b Recommendaciones para los cultivos de verano [Recommendations for summer cultures (in the dry season).] *El Servicio Shell Para el Agricultor,* Serie A Informe 13:5-66.
1960a Annual report. East Malling Research Station, Kent, England. 144 pp.
1960b Methodes permettant d'evaluer les populations d'insectes. *Revue de Zoologie Agricole et du Pathologie Vegetale* 59(4-6):44-69.
1960c There are some pests that damage soybeans. Recommended measures for their control. *Soybean Digest* 20(8):16-19.
1961a Insects not known to occur in the United States. *Cooperative Economic Insect Report U.S.* 7 [suppl.]. 67 pp.

1961b Distribution maps of pests series A (Agricultural), nos. 130-138. London, Commonwealth Institute of Entomology.

1961c Annual report. East Malling Research Station, Kent, England. 144 pp.

1963a Check list of genus *Nephotettix matsumura*. *Plant Protection Bulletin. Taiwan* 5:206-210.

1963b Insects not known to occur in the United States. *U.S. Department of Agriculture Cooperative Economic Insect Report* 13:1-32.

1964a Entomology in India. *Published by the Entomological Society of India and a Silver Jubilee Number of the Indian Journal of Entomology. S. Pradhan, president; N. C. Pant, chief editor*. Kapoor Art Press, Karol Bagh, New Delhi. 529 pp.

1965a Entomology report of the International Rice Institute, 1964. Pp. 157-184. Los Banos, Laguna, Philippines.

1965b Faunistic collections. *Illinois Natural History Survey Division Annual Report* 1963-64:12-13.

1966a Distribution maps of pests series A (Agricultural), nos. 220-228. London, Commonwealth Institute of Entomology.

1966b Entomology report of the International Rice Institute. 1965. Pp. 233-266. Manila, Philippines.

1967a Entomology report of the International Rice Institute, 1966. Pp. 179-216. Los Banos, Laguna, Philippines.

1967b Annual report. International Rice Research Institute, Los Banos, Laguna, Philippines. 308 pp.

1967c Division of mycology and plant pathology. Section of plant viruses. Sesamum (*Sesamum indicum* DC.) phyllody. *Indian Agricultural Research Institute Annual Science Reports 1964*. 138 pp.

1968a Annual report, 1966-1967. Coca Research Institute (Ghana Academy of Sciences), Accra, Ghana. 105 pp.

1968b Distribution maps of pests series A (Agricultural), nos. 246-252 and nos. 2 & 18 (revised). London, Commonwealth Institute of Entomology.

1968c Entomology circular nos. 78-79. Florida Department of Agriculture, Division of Plant Industry.

1968d Annual report. International Rice Research Institute, Los Banos, Laguna, Philippines. 402 pp.

1968e Asociacion Latinoamericana de entomologia catalogo de insectos de importancia economica en Colombia. Publication No. 1.

1968f Outbreaks and new records—United States. *FAO Plant Protection Bulletin* 16(5):93.

1969a Scientists find the cause of the bean wilt disease. *Journal of Agriculture, Victoria Department of Agriculture* 67(2):49.

1969b Annual report, 1967-1968. Cocoa Research Institute (Ghana Academy of Sciences), Accra, Ghana. 126 pp.

1970a Insects not known to occur in the United States. *Cooperative Economic Insect Report* 20(6):65-66.

1970b Recommended methods for the detection and measurement of resistance of agricultural pests to pesticides. 5. Tentative method for adults of the green rice leafhopper (*Nephotettix cincticeps* Uhler). *FAO Plant Protection Bulletin* 18(3):53-55.

1970c Annual report. International Rice Research Institute. Insecticides for rice. Pp. 237-244.

1970d Annual report, 1969. International Rice Research Institute, Los Banos, Laguna, Philippines. 266 pp.

1971a Pests of wheat. *Technical Bulletin, Rhodesia Ministry of Agriculture* No. 12:18-20.

1971b Distribution maps of pests series A (Agricultural), nos. 280-287 and no. 51 (revised). London, Commonwealth Institute of Entomology.

1971c Annual report, 1970. International Rice Research Institute, Los Banos, Laguna, Philippines. 265 pp.

1971d Pests of wheat. *Technical Bulletin, Rhodesia Ministry of Agriculture* No. 12:18-20.

1972a Insect pests of field crops. *Ohio State University Cooperative Extension Service Bulletin* 545. 20 pp.

1972b Eighty-fourth annual report, 1970-71. *Research Report, Michigan Agricultural Experiment Station* No. 153. 84 pp.

1972c Annual report, 1971. International Rice Research Institute. Insecticides. Pp. 126-136.

1972d Maple pest foiled. *American Nurseryman* 136(1):25.

1973a Annual report, 1972-73. Queensland Department of Primary Industries, Brisbane, Australia. 63 pp.

1974a Alfalfa pest management in Indiana. *Indiana Cooperative Extension Service Publication* ID-93. 54 pp.

1974b Distribution maps of pests Ser. A (Agricultural): *Empoasca flavescens* (F.) (Hemipt., Cicadellidae), no. 326. Commonwealth Institute of Entomology, London. 1 page.

1974c Distribution maps of pests Ser. A (Agricultural): *Amrasca devastans* (Dist.) (Hemipt., Cicadellidae), no. 325. Commonwealth Institute of Entomology, London. 1 page.

1974d Annual report, 1973. Kentville Research Station, Nova Scotia, Canada. 116 pp.

1975a A new leafhopper in California. *Cooperative Economic Insect Report* 25(7):66.

1975b Survey of disease and insect damage. Annual report for 1975. Pp. 163-164. International Institute of Tropical Agriculture, Ibadan, Nigeria.

1976a Annual report, 1975. International Rice Research Institute, Los Banos, Laguna, Philippines. 479 pp.

1976b Leafhoppers (*Scaphoideus* spp.)—Kentucky. *Cooperative Plant Pest Report* 1:11.

1976c A leafhopper (*Scaphoideus opalinus*)—Indiana. *Cooperative Plant Pest Report* 1(48/52):863.

1976d Outbreak of pests and diseases. *Quarterly Newsletter, Plant Protection Committee for the Southeast Asia and Pacific Region* 19(4):3-4.

1976e Sugarcane planthoppers and leafhoppers in Perak, Malaysia. *Entomology Newsletter*, South Africa No. 3:4.

1977a Preliminary report on the integrated control of rice insect pests and diseases. *Acta Entomologica Sinica* 20:135-140.

1978a Pest management strategies for leafhoppers, spittlebugs, and aphids on alfalfa. *Cooperative Regional Project Outline*. 28 pp.

1978b Distribution maps of pests Ser. A. (Agricultural): *Cicadella spectra* (Dist.) (Hemipt., Cicadellidae), No. 385. Commonwealth Institute of Entomology, London. 1 page.

1978c Leafhopper taxonomy—The traditional approach in modern times. *Report of the British Museum (Natural History)* 1975-1977:53-55.

1978d Prospects for biological control of rice hoppers—A status paper. *Commonwealth Institute of Biological Control* 12.

1980a A leafhopper (*Eupteryx atropunctata*) (Goeze)—Connecticut. *Cooperative Plant Pest Report* 5:547.

1980b Our apologies to the grape leafhopper. *California Agriculture* 34:3.

1980c Prospects for biological control of rice hoppers. *Commonwealth Institute of Biological Control* 980:11-21.

1981a Annual Report of the Research Branch, Department of Agriculture, for the year 1980. Ministry of Agriculture and Community Development, Sarawak. 276 pp.

1981b Putting the heat on horseradish disease. *Illinois Natural History Survey Report* 208, June.

1982a Distribution maps of pests, Ser. A. (Agricultural): *Edwardsiana crataegi* (Douglas), (*Edwardsiana australis*) (Froggatt), (*Typhlocyba froggatti*) (Baker) (Hem., Cicadellidae) (apple leaf-hoppers), No. 432. Commonwealth Institute of Entomology, London. 1 page.

1982b Preliminary studies on the bionomics of *Thaia subrufa* (Motschulsky). *Insect Knowledge (Kunchong Khishi)* 18:247-250.

1983a Annual report. Bean program. *Centro International de Agricultura Tropical*, Cali, Colombia. 238 pp.

1984a A summary of major achievements during the period 1977-1978. *Centro International de Agricultura Tropical*, Cali, Colombia. 96 pp.

1984b *In: Institute de Reserches Agronomiques Tropicales et des Cultures Vivieres,* Reunion. Pp. 115-130.

1985a Identifying and selecting *Cicadulina* species for effective mass rearing and maize streak resistance screening. *Research Highlights, International Institute of Tropical Agriculture* 1984:66-69.

1985b Rice pests. *Annual Report of the Research Branch, Department of Agriculture for the year 1983.* Sarawak Department of Agriculture. Pp. 37-43.

1985c Resistance to invertebrate pests. Bean program. *Annual Report, Central International de Agricultura Tropical* 7:67-77.

1985d Genetic evaluation and utilization (GEU) program. Insects. *Annual Report of the International Rice Research Institute* 1985:189-216.

1985e Genetic evaluation and utilization (GEU) program. Insect resistance. *Annual Report of the International Rice Research Institute* 1984:65-83.

1985f Leafhopper news. *Newsletter Auchenorrhyncha Record Scheme* No. 3(1983). 2 pp.

1985g Yellowing in rice. *Tamil Nadu Agriculture University Publication* 107:1-51.

Antova, Y. K.

1958a Insect-pests of non-irrigated plants in Tadzhikistan. *Akademiya Nauk Tadzhikiskoi SSR Instututa Zoologicheskii Parazitologicheskii Trudy* 89:79-109. [In Russian.]

Anufriev, G. A.

1967a Notes on the genus *Oncopsis* Burmeister, 1838 (Homoptera, Auchenorrhyncha) with descriptions of new species from the Soviet Far East. *Entomologisk Tidskrift* 88 (3&4):174-184. [New subgenus: *Sispocnis*.]

1968a Study of the genus *Rhytidodus* Fieb. (Homoptera, Auchenorrhyncha) with description of two new species from the Soviet Union. *Entomologisk Tidskrift* 86(1&2):177-187.

1968b New species of genus *Edwardsiana* (Homoptera, Auchenorrhyncha) from the Primor'ya area. *Vestnik Zoologii* 6:38-42, illust. [English summary.]

1968c New and little known species of leafhoppers of the genus *Macrosteles* (Homoptera-Auchenorrhyncha) from the Soviet Far East. *Zoologicheskii Zhurnal* 47(4):555-562, illust. [In Russian with English summary.]

1968d Leafhoppers (Homoptera, Auchenorrhyncha) of the Kurile Islands. *Uchenye Zapiski Gor'kovskogo Universiteta, Seriya Biologiya* 90:68-71.

1969a Description of a new genus of leafhoppers (Homoptera, Auchenorrhyncha) from Japan. *Bulletin de l'Academie Polanaise des Sciences. Serie des Sciences Biologiques* 17(6):403-405. [New genus: *Bambusana*.]

1969b New and little known leafhoppers of the subfamily Typhlocybinae from the Soviet Maritime Territory (Homopt., Auchenorrhyncha). *Acta Faunistica Entomologica Musei Nationalis Pragae* 13(153):163-190. [New genus: *Tautoneura*.]

1969c *Ziczacella*, new subgenus of *Erythroneura* Fitch (Homoptera, Auchenorrhyncha, Cicadellidae). *Bulletin de l'Academie Polanaise des Sciences. Serie des Sciences Biologiques* 17(11&12):697-700.

1970a New leafhoppers of the tribe Opsiini (Homoptera, Cicadellidae) from the East Asia. U.S.S.R. Academy of Science Publishing House. *Entomological Review of the U.S.S.R.* 49:151-153.

1970b Description of a new genus: *Amritodus* for *Idiocerus atkinosoni* Leth. from India (Hemiptera: Cicadellidae). *Journal of Natural History* 4:375-376.

1970c Materials to the knowledge of the fauna of cicads (Homoptera, Auchenorrhyncha) on Kurile Islands. *Trudy Bidogo-Pochvennogo Instituta ANSSSR,* Vladivostok 2:117-148.

1970d New genera of Palaearctic Dikraneurini (Homoptera, Cicadellidae, Typhlocybinae). *Bulletin de l'Academie Polonaise des Sciences. Serie des Sciences Biologiques* 18(5):261-263. [New genera: *Vilbasteana, Micantulina, Wagneriana, Mulsantina, Emelyanoviana*.]

1970e New genus and new species of Cicadinea from the south of the Primor'ya Territory. *In: Entomological Researches in the Far East, 1, Trudy Biologo-Pochvennogo Instituta ANSSSR,* Vladivostok 2:149-158. [In Russian with English title.]

1970f Four new species of *Pagaronia* Ball (Homoptera, Cicadellidae) from the Far East, allied to *Tettigonia guttigera* Uhl. *Bulletin de l'Academie Polonaise des Sciences. Serie des Sciences Biologiques* 18(9):553-557.

1970g Notes on *Empoasca kontkaneni* Oss. (Auchenorrhyncha, Cicadellidae) with description of a new species from the Far East. *Bulletin de l'Academie Polonaise des Sciences. Serie des Sciences Biologiques* 18(10):633-635. [New name: *Wagneriala* for *Wagneriana*.]

1971a New and little known leafhoppers (Homoptera, Auchenorrhyncha) from the Far East of the U.S.S.R. and neighboring countries. *Entomologicheskoe Obozrenie* 50(1):95-116. [In Russian; translation in *Entomological Review,* Washington 50(1):55-67.] [New genera: *Alnella* and *Tautocerus*.]

1971b New species of Cicadellidae (Homoptera) from the Primorsky District. *Zoologicheskii Zhurnal* 50(5):677-685.

1971c New and little known Far Eastern species of leafhoppers (Homoptera, Cicadellidae) of the genus *Pagaronia* Ball, 1902. *Bulletin de l'Academie Polonaise des Sciences. Serie des Sciences Biologiques* 19(5):335-339.

1971d Study of the genus *Matsumurella* Ishihara, 1953 (Homoptera, Auchenorrhyncha, Cicadellidae) with the description of three new species from China and Japan. *Bulletin de l'Academie Polonaise des Sciences. Serie des Sciences Biologiques* 19(7&8):511-516.

1971e Six new Far Eastern species of leafhoppers (Homoptera, Auchenorrhyncha). *Bulletin de l'Academie Polonaise des Sciences. Serie des Sciences Biologiques* 19(7&8):517-522.

1971f Study of leafhoppers from the tribe Signoretiini (Homoptera, Auchenorrhyncha) with descriptions of two new species of *Signoretia* Stal from Western Africa. *Bulletin de l'Academie Polonaise des Sciences. Serie des Sciences Biologiques* 19(11):721-726.

1971g *Straganiassus*—A new genus of leafhoppers of Jassinae subfamily from the Far East. *Vestnik Zoologii* 6:87-88.

1971h Zoogeographical characteristics and associations with Cicadina fauna of Primor'ya area. *Proceedings of XIII International Congress of Entomology,* Moscow 1:105. [In Russian.]

1972a New and little known Palaearctic genera and species of Typhlocybinae (Homoptera, Cicadellidae). *Bulletin de l'Academie Polonaise des Sciences. Serie des Sciences Biologiques* 20(1):35-42. [New genera: *Arbela, Areuna, Schizandrasca, Ussuriasca, Chlorasca*.]

1972b Two new far eastern species of *Aconurella* Rib. previously confused with *Aconurella japonica* (Mats.) (Auchenorrhyncha). *Bulletin de l'Academie Polonaise des Sciences. Serie des Sciences Biologiques* 20(3):203-208.

1972c Notes on the genus *Alnetoidia* Dlabola, 1958 (Homoptera, Cicadellidae, Typhlocybinae) with descriptions of two new species from the Far East. *Bulletin de l'Academie Polonaise des Sciences. Serie des Sciences Biologiques* 20(10):721-726.

1973a Notes on the genus *Typhlocyba* Germ. (Homoptera, Cicadellidae, Typhlocybinae). *Bulletin de l'Academie Polonaise des Sciences. Serie des Sciences Biologique* 21(7&8):505-509, 15 fig.

1973b The genus *Empoasca* Walsh, 1864 (Homoptera, Cicadellidae, Typhlocybinae) in the Soviet Maritime Territory. *Annales Zoologici,* Warszawa 30(18):537-558. [New subgenus: *Matsumurasca*.]

1975a Notes on the genus *Edwardsiana* Zachv. and *Pithyotettix* Rib. (Homoptera, Cicadellidae) with descriptions of two new species. *Bulletin de l'Academie Polonaise des Sciences. Serie des Sciences Biologiques* 23(8):531-536. [New name: *Arbelana* for *Arbela*.]

1976a A review of leafhoppers of the genus *Epiacanthus* Matsumura, 1902 (Homoptera, Auchenorrhyncha, Cicadellidae). *Entomologicheskoe Obozrenie* 55(1):69-75. [In Russian; translated in *Entomological Review* 55(1):48-52.]

1976b Notes on the genus *Psammotettix* HPT with descriptions of two new species from Siberia and the Far East. *Reichenbachia* 16(9):129-134.

1977a Two new species of auchenorrhynchous insects from the temperate Asia. *Reichenbachia* 16(21):211-215.

1977b Leafhoppers (Homoptera, Auchenorrhyncha, Cicadellidae) of the Kurile Islands. *Transactions of the Zoological Institute,* Leningrad 70:10-36. [In Russian.]

1978a Les Cicadellides de le Territoire Maritime. *Horae Societatis Entomologicae Unionis Soveticae* 60. 214 pp. [In Russian.] [New genus: *Paracanthus*; new name: *Arbelana* for *Arbela*; new tribes and new subtribes in the Cicadellinae.]

1979a Notes on some A. Jacobi's species of auchenorrhynchous insects described from North-East China. *Reichenbachia* 17(19):163-170.

1979b Notes on the genus *Notus* Fieb. containing description of a new species from the Kurile Islands (Homoptera, Auchenorrhyncha, Cicadellidae). *Nauchnye Doklady Vysshei Shkoly Biologicheskei Nauki* 12:52-56. [In Russian.]

1979c A new species of cicads of the genus *Kyboasca* Zachv. (Homoptera, Auchenorrhyncha) from the middle zone of the European part of the USSR. *Zoologicheskii Zhurnal* 58(4):499-504.

1979d A new species of leafhopper of the genus *Empoa* Fitch (Homoptera, Auchenorrhyncha) from the Kurile Islands. *Trudy Zoologicheskogo Instituta Akadameii Nauk SSSR* 81:7-9. [In Russian.]

1979e Eurasiatische amphiboreale Areale der Zikaden (Homoptera, Auchenorrhyncha) als Widerspiegelung gemeinsamer Genesisstufen der europaischen fernostlichen Fauna. *Verhandlungen International Symposium Entomofauna Mitteleuropas* 7:127-129.

1981a Homopterological reports I-III. (Homoptera, Auchenorrhyncha). *Reichenbachia* 19(28):159-173.

1981b Key to the Palaearctic species of the sub-genus *Diplocolenus* (Homoptera, Auchenorrhyncha, Cicadellidae). *Trudy Zoologicheskogo Instituta Akademii Nauk SSR,* Leningrad 105:14-20.

1981c [Rare species of Cicadellidae (Homoptera, Auchenorrhyncha) and their protection.] *In:* Bromlej, G. F., Kostenko, V. A., Yudin, V. G. and Nechaev, V. A., eds. *Rare and Endangered Terrestrial Animals of the Far East of the USSR.* Akademiya Nauk SSSR, Vladivostok. Pp. 1-192.

Anufriev, G. A. and Emel'yanov, A. F.
1968a New synonymy of Homoptera, Auchenorrhyncha from the Soviet Far East. *Zoologicheskii Zhurnal* 47:1328-1332, illust.

Anufriev, G. A. and Osychnyuk, A. Z.
1985a Valentina Nikolaevna Logvinenko 2.x.1929-24. viii. 1983. Biography and list of publications. *TYMBAL. Auchenorrhyncha Newsletter* 6:5-12.

Anufriev, G. A. and Zhil'tsova, L. A.
1982a A new east European species of leafhoppers of the genus *Macropsis* (Homoptera, Cicadellidae) from the willow *Salix acutifolia. Entomologicheskoe Obozrenie* 61(2):297-302.

App, B. A.
1960a Insects affecting production of forage seed. *Advances in Agronomy* 12:89-95.

Apple, J. W. and Falter, J. M.
1962a Insect observations in a vegetable garden adjacent to a blacklight insect trap. *Proceedings of the North Central Branch of the Entomological Society of America* 17:41.

Arai, Y.
1977a Mating behaviour of the rhombic-marked leafhopper, *Hishimonus sellatus* Uhler. 1. Mutual communication by abdominal vibration. *Kontyu* 45(1):137-142. [In Japanese and English.]

1978a Discovery of an unrecorded species of the genus *Hishimonus* (Hemiptera: Deltocephalidae) from Japan. *Applied Entomology and Zoology* 22(2):124-126.

Arakelian, A. H. and Actvatsatryan, L. P.
1977a On fauna of vineyard leafhoppers in Armenia. *Biologicheskii Zhurnal Armenii* 30(2):31-36. [Russian with Armenian summary.]

Arakelian, A. H., Terlemezian, H. L. and Manukian, Z. S.
1984a New species of sucking vermins for the fauna of the Armenian SSR. *Biologicheskii Zhurnal Armenii* 37(6):507-508.

Arant, F. S.
1954a Control of thrips and leafhopper on peanut. *Journal of Economic Entomology.* 47(2):257-263.

Archer, D. B., Townsend, R. and Markham, P. G.
1982a Detection of *Spiroplasma citri* in plants and insect hosts by ELISA. *Plant Pathology* 31:299-306.

Arcidiacono, S.
1965a Su una nuova forma di Psen Latr. (Hymenoptera Sphecidae) della Sicilia settentrionale. *Accademi Gioenia di Scienza Naturale in Catania, B. delle Sedute* 8(5):341-346, illust.

Arestegui P., A.
1976a Plagas de la papa en Andahuaylas, Apurimac. *Revista Peruana de Entomologia* 19(1):97-98.

Arizono, T., Miyahara, K. and Abe, K.
1971a Simultaneous control with various insecticide formulations to rice stem borer [*Chilo suppressalis* (Wlk.)] and rice leafhoppers [*Nephotettix cincticeps* (Uhl.) and *Nilaparvata lugens* (Stal)] by pipe-duster. I. Effect of coarse dust formulations. *Proceedings of the Association for Plant Protection of Kyushu* 17:125-128. [In Japanese.]

Arjunan, G., Samiyappan, R., Mariappan, V., Reddy, A. V. R. and Jeyarajan, R.
1985a Rice yellow dwarf in Nadu, India. *International Rice Research Newsletter* 10(4):11.

Armbrust, E. J. and Ruesink, W. G.
1973a Progress report of the (US/IBP-NSF) alfalfa subproject. *Proceedings of the North Central Branch, Entomological Society of America* 28:32-34.

Armitage, H. M.
1955a Current insect notes. *California Department of Agriculture Bulletin* 44(4):164-166.

Arora, G. L. and Singh, Sawai
1962a Morphology and musculature of the head and mouth-parts of *Idiocerus atkinsoni* Leth. (Jassidae, Homoptera). *Journal of Morphology* 110:131-140.

Arzone, A.
1972a Reperti ecologici, etologici ed epidemiologici su *Cicadella viridis* (L.) in piemonte (Hem. Hom. Cicadellidae). *Annali della Facolta di Scienze Agrarie della Universita degli studi di Torino* 8:13-38.

1974a Indagini biologiche sui parassiti oofagi di *Cicadella viridis* (L.) (Hem. Hom. Cicadellidae). I - *Gonatocerus cicadellae* Nik. (Hym. Mymaridae). *Annali della Facolta di Scienze Agrarie della Universita degli studi di Torino* 9:137-160.

1974b Indagini biologiche sui parassiti oofagi di *Cicadella viridis* (L.) (Hem. Hom. Cicadellidae). II - *Oligosita krygeri* Gir. (Hym. Trichogrammatidae). *Annali della Facolta di Scienze Agrarie della Universita degli studi di Torino* 9:193-214.

1974c Indagini biologiche sui parassiti oofagi di *Cicadella viridis* (L.) (Hem. Hom. Cicadellidae). III - *Polynema woodi* Hincks (Hym. Mymaridae). *Annali della Facolta di Scienze Agrarie della Universita degli studi di Torino* 9:297-318.

1975a Descrizione di due nuove specie di *Edwardsiana* dell'Ontano (Hom. Cicad. Typhlocybinae). *Bollettino del Museo di Zoologia della Universita di Torino* 3:85-94.

1976a Revisione del genere *Fagocyba* e descrizione di *F. alnisuga* n. sp. dell'Ontano (Hom. Cicad. Typhlocybinae). *Bollettino del Museo di Zoologie della Universita di Torino* 1:1-12.

Arzone, A. and Vidano, C.
1984a Phytopathological consequences of the migrations of *Zyginidia pulla* from grasses to cereals. *Mitteilungen der Schweizerischen Entomologischen Gesellschaft* 57(4):406-407.

Asahina, S., Ishihara, T. and Yasumatsu, K.
1959a *Iconographia Insectorum Japonicorum colore naturali edita. III.* Hokuzyukan, Tokyo.

Asakawa, M. and Kazano, H.
1976a Resistance to carbamate insecticides in the green rice leafhopper, *Nephotettix cincticeps* Uhler. *Review of Plant Protection Research* 9:101-123.

Asche, M.
1980a *Rhytistylus kaloneri* nov. spec., enie neue Cicadellide aus Nordwest Greichenland (Homoptera, Cicadina Cicadellidae, Deltocephalinae). *Marburger Entomologische Publikationen* 1(3):123-132.
1980b *Lindbergina cretica* nov. spec., eine neue Typhlocybide von Kreta (Homoptera: Auchenorrhyncha: Cicadellidae: Typhlocybinae). *Marburger Entomologische Publikationen* 1(4):37-44.

Asena, N.
1970a Research about the *Erythroneura adanae* Dlabola. *Ministry of Agriculture, Plant Protection Research,* Ankara 4:185.

Asquith, D. and Hull, L. A.
1973a *Stethorus punctum* and pest-population responses to pesticide treatments on apple trees. *Journal of Economic Entomology* 66(5):1197-1203.

Askari, A. and Hussain, M.
1977a *Empoasca* infestation of cotton. *FAO Plant Protection Bulletin* 25:43.

Assa, A. D.
1976a Co-ordinated minimum insecticide trial: Yield performance of insect resistant cowpea cultivars from IITA. *Tropical Grain Legume Bulletin* No. 5:9.

[Atkinson, J. D. et al.]
1956a *Plant Protection in New Zealand*. Wellington. 699 pp.

Attard, G.
1985a Hemipteres du presence dans le sud-ouest de la france de 2 Hemipteres du platane: *Corythuca ciliata* Say 1832 (Heteropteres Tingidae) et *Edwardsiana platanicola* Vidano 1961 (Homopteres, Cicadellidae, Typhlocybinae). *Entomologiste* 41(6):278.

Athwal, D. S. and Pathak, M. D.
1972a Genetics of resistance to rice insects. *In: Rice Breeding.* Pp. 375-386. International Rice Research Institute, Los Banos, Philippines. 738 pp.

Athwal, D. S., Pathak, M. D., Bacalangco, E. H. and Pura, C. D.
1971a Genetics of resistance to brown planthoppers and green leafhoppers in *Oryza sativa* L. *Crop Science* 11:747-750.
1980a Diversity of typhlocybine leafhoppers affecting fruit plants in Pakistan. *Pakistan Journal of Scientific and Industrial Research* 23(1&2):34-40.

Atwal, A. S., Chaudhary, J. P. and Sohi, B. S.
1969a Effect of temperature and humidity on development and population of cotton jassid, *Empoasca devastans* Distant. *Journal of Research, Punjab Agriculture University* 6(1):255-261.

Atwal, A. S. and Singh, K.
1969a Efficacy of various spraying schedules on cotton. *Journal of Research, Punjab Agriculture University* 6(3):661-667.

Auclair, J. L., Baldos, E. and Heinrichs, E. A.
1982a Biochemical evidence for the feeding sites of the leafhopper, *Nephotettix virescens* within susceptible and resistant rice plants. *Insect Science and Its Application* 3:29-34.

Audras, G.
1959a Une poire aspiratrice pratique pour la capture des petits insectes. *Bulletin Mensuel, Societe Linneenne de Lyon* 28:152-153.

Auger, J. G. and Shalla, T. A.
1975a The use of fluorescent antibodies for detection of Pierce's disease bacteria in grapevines and insect vectors. *Phytopathology* 65:493-494.

Auger, J. G., Shalla, T. A. and Kado, C. I.
1974a Pierce's disease of grapevines: Evidence for a bacterial etiology. *Science* 184:1375-1377.
1974b Bacterium discovered to be cause of Pierce's disease of grapevines. *California Agriculture* 28:8-10.

Autrey, L. J. C. and Ricaud, C.
1983a The comparative epidemiology of two diseases of maize caused by leafhopper-borne viruses in Mauritius. *In: Plumb, R. T. and Thresh, J. M., eds. Plant Virus Epidemiology. The Spread and Control of Insect-borne Viruses.* Pp. 277-285. Blackwell Scientific Publications, Oxford, U.K. 337 pp.

Avidov, Z.
1961a *Pests of the Cultivated Plants of Israel*. Magnes Press, Hebrew University, Jerusalem. 546 pp. [In Hebrew.]
1969a *Plant Pests of Israel*. Israel University Press. 549 pp.

Avesi, G. M. and Khush, G. S.
1984a Genetic analysis for resistance to the green leafhopper, *Nephotettix virescens* (Distant), in some cultivars of rice, *Oryza sativa* L. *Crop Protection* 3:41-51.

Avtar Singh and Butani, D. K.
1963a Control of cotton jassids. *Indian Farming* 13(3):9, 12.

Awate, D. G., Naik, L. M. and Pokharkar, R. N.
1978a Efficacies of lower doses of systemic granular insecticides for the control of aphids (*Myzus persicae* Sulzer), jassids (*Amrasca bigutalla* [sic!] *bigutalla* [sic!] Ishida) and thrips (*Hercothrips indicus* Banks) infesting potato in Maharashtra. *Journal of the Maharashtra Agricultural Universities* 3:49-50.

Awate, D. G. and Pokharkar, R. N.
1978a Studies on the residual toxicity of systemic insecticides applied as granules in soil and seed treatment against aphids (*Myzus persicae* Sulzer) and jassids (*Amrasca biguttula biguttula* Ishida) on potato. *Journal of the Maharashtra Agricultural Universities* 3:149-150.

Babu, R. S. H., Rath, S. and Rajput, C. B. S.
1983a Insect pests of cashew in India and their control. *Pesticides* 17:8-16.

Babu, T. H.
1975a Effect of a juvenile hormone analogue on the development of green rice leafhopper, *Nephotettix impicticeps* Ish. *In: VIII International Congress of Plant Protection*, Moscow, 1975. Biological and genetic control. Pp. 24-32.

Babu, T. R. and Azam, K. M.
1982a Relative efficacy of certain new synthetic pyrethroids against pest complex of bhendi. *Indian Journal of Plant Protection* 9(1):13-18.

Backus, E. A.
1984a Host selection and location of feeding tissues by leafhoppers: Behavioral evidence for the importance of precibarial sensilla. *Mitteilungen der Schweizerischen Entomologischen Gesellschaft* 57(4):410.
1985a Anatomical and sensory mechanisms of leafhopper and planthopper feeding behavior. *In: Nault, L. R. and Rodriguez, J. G., eds. The Leafhoppers and Planthoppers.* Pp. 163-194. Wiley & Sons, New York. 500 pp.

Backus, E. A. and McLean, D. L.
1982a Sensory systems and feeding behavior of leafhoppers. I. The aster leafhopper, *Macrosteles fascifrons* (Stal) (Homoptera, Cicadellidae). *Journal of Morphology* 172:361-379.
1983a Sensory systems and feeding behaviour of leafhoppers. II. A comparison of the sensillar morphologies of several species (Homoptera: Cicadellidae). *Journal of Morphology* 176(1):3-14.
1984a Host selection and location of feeding tissues by leafhoppers: Behavioral evidence for the importance of the precibarial sensilla. *XVII International Congress of Entomology* 17:444.

1985a Behavioral evidence that the precibarial sensilla of leafhoppers are chemosensory and function in host discrimination. *Entomologia Experimentalis et Applicata* 37(3):219-228.

Bacon, O. G.
1960a Systemic insecticides applied to cut seed pieces and to soil at planting time to control potato insects. *Journal of Economic Entomology* 53(5):835-839.

Badmin, J. S.
1970a *Pentatoma rufipes* (L.) (Hem., Pentatomidae) and *Limotettix striola* Fall)n (Hem. Cicadellidae) on Rhum Inverness-shire. *Entomologist's Monthly Magazine* 106(1268-70):31.

1979a The rhodendron leafhopper *Graphocephala fennahi* Young 1977, an un-recorded species in Kent. *Bulletin of the Kent Field Club* 24:41.

1981a *Athysanus argentarius* Metcalf (Hem., Cicadellidae) in Kent. *Entomologist's Monthly Magazine* 117(Jan.-Apr.) :14.

1985a Kent leafhopper and psyllid notes. *Bulletin of the Kent Field Club* 30:17-18.

Bae, S. H. and Pathak, M. D.
1969a Common leafhopper populations and incidence of tungro virus in diazinon-treated and untreated rice plots. *Journal of Economic Entomology* 62(4):772-775.

Bae, T. U.
1985a Studies on the population dynamics of green rice leafhopper, *Nephotettix cincticeps* Uhler, in the southern region of Korea rice cultural areas: The population dynamics in the levee. *Korean Journal of Entomology* 15(2):67-76.

Baggiolini, M.
1968a Alterations foliaires proviquees par la cicadelle *Empoasca flavescens* F. (Hom. Typhlocybidae) elevee sur pommiers. *Mitteilungen der Schweizerischen Entomologischen Gesellschaft* 40(3-4):257-262.

Baggiolini, M., Canevascini, V. and Caccia, R.
1972a La cicadelle verte (*Empoasca flavescens* F.), cause d'importants rougissements du feuillage de le vigne. *Bulletin, Organisation Europeenne et Mediterraneenne pour la Protection des Plantes* No. 3:43-49. [In French with English summary.]

Baggiolini, M., Canevascini, V., Caccia, R., Tencalla, Y. and Sobrio, G.
1968a Presence dans le vignoble du Tessin d'une cicadelle nearctique nouvelle pour la Suisse, *Scaphoideus littoralis* Ball. (Hom. Jassidae), vecteur possible de la flavescence doree. *Mitteilungen der Schweizerischen Entomologischen Gesellschaft* 40:270-275.

Baggiolini, M., Canevascini, V., Tencalla, Y. Caccia, R., Sobrio, G. and Cavalli, S.
1968a La cicadelle verte *Empoasca flavescens* F. (Homopt., Typhlocybidae), agent d'alterations foliaires sur vigne. [The green leafhopper *E. flavescens*, the cause of leaf injuries on vines.] *Schweizerische Landwirtschaftliche Forschung* 7(1):43-69. [With German, Italian and English summaries.]

Bailey, J. C.
1982a Influence of plant bug and leafhopper populations on glabrous and nectarless cottons. *Environmental Entomology* 11:1011-1013.

Bailly-maitre, A.
1955a Contribution a l'etude des Typhlocybides. *Laboratoire de Zoologie et Station Agricole Grimaldi de la Faculte des Sciences Travaux, Dijon* 10:1-56, illus.

Bairyamova, V.
1970a Vertreter der Unterordnung Zikaden (Homoptera, Auchenorrhyncha) auf Kulturpflanzen und Obstbumen in der Plane von Sofia. *Bulgarska Akademiya na Naukite Zoologicheski Institut si Muzey. Izvestiya* 32:243-251.

1970b Beitrag zur Untersuchung der Bulgarischen Zikaden-fauna (Homoptera, Auchenorrhyncha). *Bulgarska Akademiya na Naukite Zoologicheski Institut si Muzey. Izvestiya* 32:261-263.

1977a Investigation on the leafhopper fauna (Homoptera, Auchenorrhyncha) of southwestern Bulgaria with a view of its economic importance. *Acta Zoologica Bulgarica* 7:22-26. [In Bulgarian with Russian summary.]

1982a Study on the dynamics of abundance of the cicadellid entomofauna in agricultural crops in the Sofia Plain. *Ekologiya* (Bulgaria) 10:13-21.

Baker, K. K., Perry, S. K., Mowry, T. M., and Russo, J. F.
1983a Association of a mycoplasmalike organism with a disease of annual statice in Michigan. *Plant Disease* 67:699-701.

Bakkendorf, O.
1971a Description of *Oligosita tominici* n. sp. [Hym., Trichogrammatidae] and notes on the hosts of *Anagrus atomus* (L.) and *Anaphes autumnalis* Foerster [Hym. Mymaridae]. *Entomophaga* 16(4):363-366.

Balasubramamian, A.
1972a Morphology and histology of salivary glands of *Idiocerus atkinsoni* L. and *I. clypealis* L. (Homoptera, Jassidae). *Proceedings of the Indian Congress Association* 59:443-444.

Balasubramanian, G.
1979a Prediction of leafhopper population in upland cotton in relation to plant characters. *Indian Journal of Agricultural Sciences* 49:970-973.

Balasubramanian, G. and Chelliah, S.
1985a Chemical control of pests of sunflower. *Pesticides* 19:21-22.

Balasubramanian, G. and Gopalan, M.
1978a Note on the role of phenolics and minerals in cotton varieties in relation to resistance to leafhopper. *Indian Journal of Agricultural Sciences* 48:367-370.

1979a Changes in organic acids content of American cotton varieties in relation to resistance to the leafhopper, *Amrasca biguttula biguttula* Ishida. *Science and Culture* 45:414-416.

1981a Role of carbohydrates and nitrogen in cotton varieties in relation to resistance to leafhopper. *Indian Journal of Agricultural Sciences* 51:795-798.

Balasubramanian, G., Gopalan, M. and Subramaniam, T. R.
1977a Resistance to leafhopper in upland cotton. *Indian Journal of Agricultural Sciences* 47:82-86.

Balasubramanian, M. and Mariappan, V.
1983a Fungal pathogens of *Nephotettix virescens* Dist. and *Nilaparvata lugens* Stal. *International Rice Research Newsletter* 8(3):11.

Balazy, S., Lipa, J. J. and Nowacka, W.
1980a An epizootic caused by the fungus *Zoophthora radicans* (Brefeld) Batko in an insectary rearing of *Aphrodes bicinctus* Schrank (Homoptera, Cicadoidea). *Bulletin de l'Academie Polonaise des Sciences. Serie des Sciences Biologiques* 28(1&2):121-124.

Baldridge, R. S. and Blocker, H. D.
1980a Parasites of leafhoppers (Homoptera: Cicadellidae) from Kansas grasslands. *Journal of the Kansas Entomological Society* 53(2):441-447.

Ball, J. C.
1979a Seasonal patterns of activity of adult leafhopper vectors of phony peach disease in north Florida. *Environmental Entomology* 8(4):686-689.

Ballantyne, B.
1971a Summer death of beans. *Agricultural Gazette of New South Wales* 82(5):295-297.

Baloch, A. A. and Soomro, B. A.
1980a Preliminary studies on plant profile and population dynamics of insect pests of cotton. *Turkiye Bitke Koruma Dergisi* 4:203-217.

Baloch, G. M. and Khan, A. G.
1977a Possibilities for the biological control of Russian thistles, *Salsola* spp. (Chenopodiaceae). *In: Freeman, T. E., ed. Proceedings of the IV International Symposium on Biological Control of Weeds.* Pp. 108-112. University of Florida, Gainesville. 298 pp.

Baltazar, Clare R.
1981a Biological control attempts in the Philippines. *Philippine Entomologist* 4(6):505-523.

Baluch, Masood Ahmed, Ahmed, Mubarik and Ahmed, Manzoor
> 1980a Some ecological studies on *Amrasca devastans*, a pest of brinjal *Solanum melongena* Linn. in the Punjab. *Proceedings of the 1st Pakistan Congress of Zoology*, B. Pp. 275-282.

Banerjee, S. K.
> [Undated] Recent progress in rice insect research in India. Symposium on rice insects. Proceedings of a Symposium on Tropical Agriculture Researches, 19-24 July, 1971. *Tropical Agriculture Research Series* No. 5:83-97.

Banerjee, S. K. and Katiyar, K. N.
> 1984a Insecticidal control of cotton pests. *Indian Journal of Entomology* 46:1-11.

Banerjee, S. K., Katiyar, K. N. and Butani, D. K.
> 1977a Assessment of plant protection requirement of some cotton varieties to control the jassid. *Entomologist's Newsletter* 7:7-8.

Banerjee, S. N.
> 1964a Paddy pests. *In: Pant, N. C., Prasad, S. K., Vishnoi, H. S., Chatterji, S. N. and Varma, B. K., eds. Entomology in India.* pp. 92-98. The Entomological Society of India. 529 pp.

Bang, Y. H. and Kae, B. M.
> 1963a The green rice leafhopper, *Nephotettix bipunctatus cincticeps*, and its control in Korea. *Journal of Economic Entomology* 56(6):773-776.

Bantarri, E. E.
> 1966a Grass hosts of aster yellows virus. *Plant Disease Reporter* 50:17-21.

Bantarri, E. E. and Moore, M. B.
> 1960a Transmission of the aster yellows virus to barley. *Plant Disease Reporter* 44:154.
> 1962a Virus cause of blue dwarf of oats and its transmission to barley and flax. *Phytopathology* 52:897-902.

Bantarri, E. E. and Zeyen, R. J.
> 1968a Natural acquisition compared with mechanical inoculation of six-spotted leafhoppers for transmission of the oat blue dwarf virus. *Phytopathology* 58:1042.
> 1970a Transmission of oat blue dwarf virus by the aster leafhopper following natural acquisition or inoculation. *Phytopathology* 60:399-402.
> 1976a Multiplication of the oat blue dwarf virus in the aster leafhopper. *Phytopathology* 66:896-900.
> 1979a Interactions of mycoplasma-like organisms and viruses in dually infected leafhoppers, planthoppers and plants. *In: Maramorosch, K. and Harris, K. F., eds. Leafhopper Vectors and Plant Disease Agents.* Pp. 327-347. Academic Press, New York. 654 pp.

Barbosa, A. J. da S.
> 1954a *Empoasca fascialis* Jacobi – O jasside do algodoeiro; sua biologia e metodos de combate. *Revista de Faculdade de Ciencias. Universidad de Lisboa. Serie C. Ciencias Naturais* 4:5-28.
> 1954b *Empoasca fascialis* Jacobi – O jasside do algodoeiro; sua biologia e metodos de combate. *Lourenco Marques, Centro de Investigaciones Ciencias Algodoeira, Memoria y Trab.* 15:1-28.

Barnes, D. K., Hanson, C. H., Ratcliffe, R. H., Busbice, T. H., Schillinger, J. A., Buss, G. R., Campbell, W. V., Hemken, R. W. and Blickenstaff, C. C.
> 1970a The development and performance of Team alfalfa. A multiple pest resistant alfalfa with moderate resistance to the alfalfa weevil. *U. S. Department of Agriculture, Agriculture Research Service* 34-115. 41 pp.

Barnes, D. K. and Newton, R. C.
> 1963a Amorphous tumours induced in alfalfa by potato leafhoppers. *Nature* 199(4888):95-96.

Barnes, J. K.
> 1984a The Membracidae and other Homoptera described by Asa Fitch, 1851, and Ebenezer Emmons, 1855: Historical perspective and analysis. *Journal of the New York Entomological Society* 92(1):27-34.

Barnes, P.
> 1954a Biologia, ecologia y distribucion de las chicharritas, *Dalbulus elimatus* (Ball) y *Dalbulus maidis* (Del. & W.). *Folia Technica No. 11.* Secretaria de Agricola y Ganaderia Oficina de Estudiea Especiales, Mexico, D. F. 112 pp.

Barnett, D. E.
> 1975a Distributional records of the genus *Scaphoideus* (Homoptera: Cicadellidae) in Kentucky. *U.S. Department of Agriculture Cooperative Economic Insect Report* 25(44):861-864.
> 1975b A revision of the Nearctic species of the genus *Scaphoideus* (Homoptera: Cicadellidae). *Dissertation Abstracts International,* B 36(3):1054.
> 1976a Some new preparation techniques used in leafhopper identification. *Florida Entomologist* 59(3):321-323.
> 1977a A revision of the Nearctic species of the genus *Scaphoideus* (Homoptera: Cicadellidae). *Transactions of the American Entomological Society* 102:485-593.
> 1977b External morphology of adult leafhoppers of the genus *Scaphoideus. Transactions of the Kentucky Academy of Science* 38(1&2):26-37.
> 1979a Seven new North American species of *Scaphoideus* Uhler (Homoptera: Cidadellidae). *Journal of the Kansas Entomological Society* 52(3):471-482.
> 1980a Seven new African species of *Scaphoidophyes* Kirkaldy (Homoptera: Cicadellidae). *Proceedings of the Washington Entomological Society* 82(3):435-446.

Barnett, D. E. and Freytag, P. H.
> 1976a The genus *Scaphoidophyes* Kirkaldy and description of one new species. *Annals of the Entomological Society America* 69(4):617-619.

Barrett, C. F. and Westdal, P. H.
> 1961a Note on the occurrence of two parasites [*Tomosvaryella sylvatica* (Mg.) and *Pacygonatopus minimus* Fenton] of the six-spotted leafhopper, *Macrosteles fascifrons* (Stal), in Manitoba. *Canadian Journal of Plant Science* 41(2):453.

Barrett, C. F., Westdal, P. H. and Richardson, H. P.
> 1965a Biology of *Pachygonatopus minimus* Fenton (Hymenoptera: Dryinidae) a parasite of the six-spotted leafhopper, *Macrosteles fascifrons* (Stal), in Manitoba. *Canadian Entomologist* 97(2):216-221.

Barrion, A. T. and Litsinger, J. A.
> 1981a *Hippasa holmerae* Thorell (Aranease: Lycosidae): A new predator of rice leafhoppers and planthoppers. *International Rice Research Newsletter* 6(4):15.
> 1981b Nomenclatural changes for some rice insect pests. *International Rice Research Newsletter* 6(4):14.
> 1982a Water striders: New predators of rice leafhoppers and planthoppers in the Philippines. *International Rice Research Newsletter* 7(5):19-20.
> 1983a Parasites of white leafhopper, *Cofana spectra* (Dist.) (Hemiptera: Cicadellidae), in the Philippines. *International Rice Research Newsletter* 8(6):20.

Barteneva, R. D.
> 1963a Rozannaya tsikadka (*Typhlocybas rosae* L.) i borba a neyu na plantatsiyakh rozy efiromas - lichnoi. [The rose leafhopper, *Typhlocyba rosae* L., and its control on rose plantations producing essential oils.] *Sbornik Nauchno-Issledovatel'skikh Rabot Po Maslich-nymi Efiromaslichnym Kul'turam* 2:187-195.

Basden, R.
> 1966a The composition, occurrence and origin of lerp, the sugary secretion of *Eurymela distincta* (Signoret). *Proceedings of the Linnean Society of New South Wales* 91:44-46.

Basilio, R. P. and Heinrichs, E. A.
> 1981a Effect of carbofuran root zone application rate on feeding activity and mortality of the green leafhopper. *International Rice Research Newsletter* 6(6):20.
> 1981b Evaluation of commercial and coded insecticides to control brown planthopper and green leafhopper. *International Rice Research Newsletter* 6(6):21.
> 1982a Activity of insecticides applied in the root zone to control brown planthopper and green leafhopper. *International Rice Research Newsletter* 7(1):12-13.

Baskaran, P., Narayanasamy, P., Balasubramaniam, M. and Ragunathan, V.
  1976a Field evaluation of insecticide-fungicide spray combinations against rice pests. *Rice Entomology Newsletter* 4:37.

Bastos Cruz, B. P., Figueriredo Barreto, M. and Almeida, E.
  1962a Principais doencas e pragas do amedoim no Estado de Sao Paulo [The principal diseases and pests of groundnut in the state of Sao Paulo.] *Biologica* 28(7):189-195.

Basu, A. N., Ghosh, A. Mishra, M. D. Niazi, F. R. and Raychaudhuri, S. P.
  1974a The joint infection of rice tungro and yellow dwarf. *Annals of the Phytopathological Society of Japan* 40(1):67-69.
  1976a The over-wintering of rice tungro virus and *Nephotettix virescens* (Distant) in Delhi. *Indian Journal of Entomology* 38:57-62.

Basu, A. N., Mishra, M. D., and Niazi, F. R.
  1982a Intensive observations on the effect of some insecticides on *Nephotettix virescens* (Distant), a vector of virus and MLO disease, with special reference to its transmission ability. *European Journal of Forest Pathology* 12:391-394.

Batiashvili, I. D. and Dekanoidze, G. I.
  1967a Contribution to the fauna of cicades (Cicadinea) of the subfamily Typhlocybinae, injuring crops in Georgia. In: *Materials on the Georgian Fauna.* Akademiya Nauk Gruzinskoi SSR, Instuta Zoologichiski 2:252-262. [In Russian.]
  1967b A contribution to the fauna of harmful cicades (Cicadinea) of the fruit and vine cultures of Georgia. *Zoologicheski Zhurnal* 46(6):873-882. [In Russian with English summary.]

Batra, H. N. and Jangyani, L. B.
  1960a Record of some new host plants of important crop pests. *Indian Journal of Entomology* 22(4):304.

Batra, S. W. T.
  1979a Insects associated with weeds of the north-eastern United States: Quickweeds, *Galinsoga ciliata* and *G. parviflora* (Compositae). *Environmental Entomology* 8(6):1078-1082.

Bazlul Huw, S.
  1985a Uber *Eudorylas subterminalis* Coll. (Dipt., Pipunculidae) als parasit bei Cicadellidae (Homoptera) in Berlin (West). *Anzeiger fuer Schaedlingskunde, Pflanzenschutz, Umweltschutz* 58(8):147-149.

Beamer, R. H.
  1940a Important methods of mounting. *Rocky Mountain Conference of Entomologists Report* 16:24-25.
  1945a A new species of *Dikraneura* from Arizona (Homoptera-Cicadellidae). *Journal of the Kansas Entomological Society* 18(2):83-84.

Beardsley, J. W.
  1956a Notes and exhibitions. *Proceedings of the Hawaiian Entomological Society* 16(1):9; 16.
  1958a The use of light traps for the early detection of newly established immigrant insect pests in Hawaii. *Hawaii Planters' Record* 55(3):237-242.
  1961a *Circulifer tenellus* (Baker). *Proceedings of the Hawaiian Entomological Society* 17(3):314 & 317.
  1964a Notes and exhibitions: Mar. 11, 1963 – New Molokai Records; Oct. 14, 1963 – New insect records for Molokai and Lanai. *Proceedings of the Hawaiian Entomological Society* 18(3):346.
  1965a Not recorded from Maui before – beet leafhopper *Circulifer tenellus* (Baker). *Proceedings of the Hawaiian Entomological Society* 19(1):22.
  1966a Insects and other terrestrial arthropods from the Leeward Hawaiian Islands. *Proceedings of the Hawaiian Entomological Society* 19(2):157-185.
  1969a Notes and exhibitions – March. *Proceedings of the Hawaiian Entomological Society* 20(3):484.
  1969b Notes and Exhibitions – April. *Proceedings of the Hawaiian Entomological Society* 20(3):487.
  1976a *Carneocephala sagittifera* (Uhler): November notes and exhibitions. *Proceedings of the Hawaiian Entomological Society* 22(2):177.

Beardsley, J. W. and Funasaki, G.
  1976a *Zygina penapacha* (Beamer): March notes and exhibitions. *Proceedings of the Hawaiian Entomological Society* 22(2):162.

Beccari, F.
  1970a Gli insectti nocivi all'erba medica (*Medicago sativa* L.) in Arabia Saudita. Specie segnalate e loro diffusione. *Rivista di Agricoltura Subtropicale e Tropicale* 64(10-12):272-295.

Becker, M.
  1974a The biology and population ecology of *Macrosteles sexnotatus* (Fallen) (Cicadellidae: Hemiptera). Ph. D. Thesis, University of London.
  1979a The nymphal stages and internal reproductive organs of *Macrosteles sexnotatus* (Fall.) (Hem., Cicadellidae). *Entomologist's Monthly Magazine* 115(1376-1379):11-16.
  1981a The wing venation in *Macrosteles sexnotatus* (Fallen, 1806) (Homoptera: Cicadellidae) and the occurrence of a short winged form. *Anais da Sociedade Entomologica do Brasil* 10(2):141-148.

Bednarczyk, J.
  1983a Structure of the male reproductive system in *Alnetoidia alneti* (Dahlb.) and *Populicerus populi* (L.) (Homoptera, Auchenorrhyncha). *Prace Naukowe Uniwersytetu Slaskiego w Katowicach* 627:71-82.

Begley, J. W. and Butler, L.
  1980a Distribution of leafhoppers in subfamilies Cicadellinae and Gyponinae (Homoptera: Cicadellidae) in West Virginia. *Entomological News* 91(5):165-169.

Behncken, G. M.
  1984a *Orosius lotophagorum* subsp. *ryukyuensis* (Hemiptera: Cicadellidae), a new vector of little leaf in Australia. *Australian Plant Pathology* 13:35-36.

Beingolea G. O. D.
  1957a Metasystox en el control de *Empoasca* sp. y su accion sobre la fauna benefica en el algodonero. *Informe Mensual de la Estacion Experimental Agricola de la Molina* 31(361):1-10.
  1958a Rotenona en el control de la "cigarrita verde" del algodonero *Empoasca* spp. *Informe Mensual de la Estacion Experimental Agricola de la Molina* 107:21-37.

Beirne, B. P.
  1955a Collecting, preparing and preserving insects. *Canada Department of Agriculture Publication* 932. 133 pp.
  1955b New species of leafhoppers from Canada (Homoptera: Cicadellidae). *Canadian Entomologist* 87(9):373-375.
  1956a Leafhoppers (Homoptera: Cicadellidae) of Canada and Alaska. *Canadian Entomologist* 88:1-180.
  1972a Pest insects of annual crop plants in Canada. 4. Hemiptera-Homoptera. 5. Orthoptera. 6. Other Groups. *Memoirs of the Entomological Society of Canada* 85:1-73.

Bekker-Migdisova, E. E.
  1962a Homoptera. In: Rodendorf, B. B., ed. *Fundamentals of Paleontology. A Manual for Paleontologists and Geologists.* v. [9]. Pp. 162-194. Moscow. [In Russian.]
  1967a Tertiary Homoptera of Stavropol and a method of reconstruction of Continental palaeobiocoenoses. *Palaeontology* 10:542-553.

Belli, G.
  1974a Virus diseases of plant – how they manifest themselves, how they are transmitted. *Italia Gricola* 111:71-86.

Belli, G., Amici, A., Corbetta, G. and Osler, R.
  1974a The 'giallume' disease of rice (*Oryza sativa* L.). *Acta Biologica Jugoslavica* 11:101-107.

Belli, G., Amici, A. and Osler, R.
1972a The mycoplasmas as a cause of plant disease, with reference to the Italian situation. International Symposium over Fytofarmacie en Fytiatrie, 9 Mei 1972. *Mededelingen Fakulteit Landbouwwetenschappen Gent* 37(2):441-449.

Belli, G., Corbetta, G. and Osler, R.
1975a Investigation and observation on the epidemiology and on the possibility of prevention of 'giallume' of rice. *Rise* 24:359-363.

Belli, G., Fortusini, A., Osler, R. and Amici, A.
1973a Presenza di una malattia del tipo "flavescence doree" in vigneti dell'Oltrepo Pavese. *Rivista di Patologia Vegetale*, IV 9 [Supplement]:50-56. [In Italian with English summary.]

Belli, G. and Osler, R.
1975a Diseases of the grapevine attributed to mycoplasmas and rickettsiae. *Coltivatore e Giornale Vinicolo Italiano* 1-9.

Benavides, M.
1955a Efectividad de varios insecticidas en el control del "Lorito verde" *Empoasca fabae* (Harris) del frijol. *Agricultura Tropical* (Colombia) 0(10):817.

Bennett, C. W.
1953a Evidence of lack of interference between strains of the curly-top virus. *6th Congresso Internationale di Microbiological Riassunti delle Comunita* 2(512):159-160.
1962a Curly top virus content of the beet leafhopper influenced by the virus concentration in diseased plants. *Phytopathology* 52(6):538-541.
1962b Acquisition and transmission of curly top virus by artificially-fed beet leafhoppers. *Journal of American Sugar Beet Technology* 11(8):637-648.
1967a Apparent absence of cross-protection between strains of the curly top virus in the beet (*Beta vulgaris*) and leafhopper, *Circulifer tenellus*. *Phytopathology* 57(2):207-209.
1967b Epidemiology of leafhopper-transmitted viruses. *Annual Review of Phytopathology* 5:87-108.
1971a The curly top disease of sugarbeets and other plants. *American Phytopathological Society Monograph* 7:1-81.

Bennett, C. W. and Costa, A. S.
1949a The Brazilian curly top of tomato and tobacco resembling North American and Argentina curly top of sugarbeet. *Journal of Agricultural Research* 78(12):675-693.

Bennett, C. W. and Tanrisever, A.
1957a Sugar beet curly-top disease in Turkey. *Plant Disease Reporter* 41(9):721-725.
1958a Curly top disease in Turkey and its relationship to curly top in North America. *Journal of the American Society of Sugar Beet Technologists* 10:189-211.

Bennett, S. H.
1959a A preliminary note on the leafhoppers occurring on apples at Long Ashton. *Long Ashton Research Station Report* 1958:139-141.

Bentur, J. S. and Kalode, M. B.
1980a Biocontrol studies on leaf and planthoppers. *Proceedings of the 3rd Workshop of All India Coordinated Research Project on Biological Control of Crop Pests and Weeds, Ludhiana*. Indian Council of Agricultural Research, New Delhi. Pp. 103-108.

Ben-Ze'ev, I. and Kenneth, R. G.
1981a *Zoophthora radicans* and *Zoophthora petchi* sp. nov. (Zygomycetes: Entomophthorales), two species of the "spareosperma group" attacking leafhoppers and froghoppers (Hom.). *Entomophaga* 26:131-142.

Bergman, B. H. H.
1956a On three Jassidae of *Arachis hypogaea* in Java. *Entomologische Berichten* (Amsterdam) 16:64-71.
1956b Het mozaiek en de heksenbezemziekte van de aardnoot (*Arachis hypogaea* L.) in West Java en hun vector, de Jassidae *Orosius argentatus* (Evans). [Mosiac I and witches' broom disease of peanut in West Java and the vector.] *Tijdscrift over Plantenziekten* 62:291-304.
1957a A new dryinid parasite of leafhoppers in Java. *Entomologische Berichten* 17(1):9-12

Bergmann, E. C., Moreti, A. C. de C., D'Andretta, M. A., Pereira, F. S. and Mendonca, N. T. de
1984a Survey of insects in a mixed orchard in Ibiuma, Sao Paulo state. *Biologico* 50:59-64.

Bergmann, E. C., Ramiro, Z. A. and Mendonca, N. T. de
1984a Survey of rice arthropod fauna of pasture in Sao Paulo state. *Biologico* 50:27-31.

Bergonia, H. T.
1978a Control measure to prevent tungro virus outbreak. *Plant Protection News* 7:4-16.

Berlin, L. C.
1962a The morphology and salivary enzymes of the digestive system of the potato leafhopper, *Empoasca fabae* (Harris). M. S. Thesis, Iowa State University. 91 pp.

Berlin, L. C. and Hibbs, E. T.
1963a Digestive system morphology and salivary enzymes of the potato leafhopper, *Empoasca fabae* (Harris). *Proceedings of the Iowa Academy of Science* 70:527-540.

Bernard, R. L. and Singh, B. B.
1968a Inheritance of pubescence type in soybeans: Glabrous, curly, dense, sparse and puberulent. *Crop Science* 9(2):192-197.

Bernhard, R., Marenaud, C., Eymet, J., Sechet, J., Fos, A. and Moutous, G.
1977a A complex disease of certain *Prunus*: 'Molieres decline'. *Comptes Rendus des Seances de l'Academie d'Agriculture de France* 3:178-188.

Bertels, A.
1962a Insetos-hospedes de solanaceas. *Iheringia Serie Zoologia* (Porto Alegre) 25:1-11.

Bertels, A. and Baucke, O.
1966a Segunda relacao das pragas das plantas cultivadas no Rio Grande do Sul. *Pesquisa Agropecuaria Brasileira* 1:17-46.

Betsch, W. D.
1978a A biological study of three hemipterous insects on honey locust in Ohio. M. S. Thesis, Ohio State University, Columbus.

Bhaktavatsalam, G. and Anjaneyulu, A.
1984a Greenhouse evaluation of synthetic pyrethroids for tungro (RTV) control. *International Rice Research Newsletter* 9(5):11.

Bhalani, P. A. and Patel, R. M.
1981a Effect of different types of food on the development of cotton jassid, *Amrasca biguttula biguttula* (Ishidu) [sic!]. *Gujarat Agricultural University Research Journal* 7(1):45-46.

Bhalla, O. P. and Pawar, A. D.
1975a Homopterans injurious to rice in Himachal Pradesh, India. *Rice Entomologist Newsletter* 2:40-41.

Bharadwaj, R. K., Reddy, D. V. R. and Sinha, R. C.
1966a A reinvestigation of the alimentary canal in the leafhopper *Agallia constricta* (Homoptera, Cicadellidae). *Annals of the Entomological Society of America* 59(3):616-617.

Bharadwaj, S. C. and Yadav, S. R.
1980a Record of *Pruthiana sexnotata* (Izzard) (Cicadellidae Homoptera), a hopper pest of sugarcane from Haryana. *Indian Journal of Entomology* 42(2):317.

Bhat, M. G., Joshi, A. B. and Mehta, S. L.
1981a Biochemical basis of resistance to jassid in cotton. *Crop Improvement* 8:1-6.
1984a Relative loss of seed cotton yield by jassids and bollworms in some cotton genotypes (*Gossypium hirsutum* L.). *Indian Journal of Entomology* 46:169-173.

Bhatia, G. N.
1972a Control of sugar-cane leafhopper epidemic. *Pesticides* 6(2):11-12

Bhatnagar, V. S.
1974a New hosts of the mango jassid. *FAO Plant Protection Bulletin* 22(2):50-51.

Bhattacharya, A. K.
1972a Gonial and spermatocyte chromosomes of *Empoasca devastans* (Homoptera, Typhlocybinae). *Current Science* (Bangalore) 41(20):739-740.

1973a Mitotic and meiotic chromosomes of three species of leaf-hoppers in Bythoscopinae (Cicadellidae, Homop.) *Proceedings of the National Academy of Sciences*, India (Section B) 43:180-182.

1973b Cytology and inter-relationship between two species of *Zizyphoides* (Jassinae, Homoptera). *Proceedings of the National Academy of Science,* India (Section B) 43:187-189.

1973c Chromosomes of a jassid, *Penthimia scapularis* (Gyponinae, Cicadellidae, Homoptera). *Proceedings of the National Academy of Sciences,* India (Section B) 43:291-292.

1973d The karyotype of *Zizyphoides punctatus* sp. nov. *Current Science* (Bangalore) 42(11):391-392.

1973e Morphology, behaviour and metrical studies of the germinal chromosomes in three species of jassids (Tettigoniellinae, Homoptera). *Chromosome Information Service* 14:23-25.

1974a Chromosome studies in four species of *Phrynomorphus* (Cicadellidae, Homoptera). *Proceedings of the Indian Science Congress* 63(3):134 [abstract].

1975a Multivalents in *Eutettix apricus* (Melichar) with notes on the chromosomes of five near relatives (Jassidae, Homoptera). *Caryologia* 28(4):437-443.

1975b Interlocked bivalents in the male germ line of *K. strigicollis* (Spin.) (Jassidae, Selenocephalaria). *Proceedings of the Indian Science Congress* 62(3):131.

1980a An analysis of the chromosomes of leafhoppers (Cicadellidae, Homoptera). *Genetica* 52-53:49-55.

Bhirud, K. M. and Pitre, H. N.

1972a Comparative susceptibility of three cicadellid vectors of the corn stunt disease agent to carbofuran and disulfoton in greenhouse test. *Journal of Economic Entomology* 65(5):1236-1238.

1972b Bioactivity of systemic insecticides in corn: Relationship to leafhopper vector control and corn stunt disease incidence. *Journal of Economic Entomology* 65(4):1134-1140.

1972c Influence of soil class and soil moisture on bioactivity of carbofuran and disulfoton in corn in greenhouse test: Relationship to leafhopper vector control and corn stunt disease incidence. *Journal of Economic Entomology* 65(2):324-329.

Bhola, R. K. and Srivastava, K. P.

1979a Studies on the neuroendocrine system of the mangohopper, *Idiocerus atkinsoni* Leth. (Homoptera: Jassidae). *Acta Biologica Academiae Scientiarum Hungaricae* 30(4):363-372.

Bianchi, F. A.

1955a Notes and exhibitions. *Proceedings of the Hawaiian Entomological Society* 15(3):380.

Bigi, F.

1953a Gli ambienti, i parassiti e le malattie del cotone in Africa orientale (Eritrea, Ethiopia, Somalia Italiana). *Rivista di Agricoltura Subtropicale e Tropicale* 47(4-6):162-176.

Bigornia, A. E.

1963a Toxemic symptoms on some weeds caused by the leafhopper *Empoasca formosana* in the Philippines. *FAO Plant Protection Bulletin* 11(5):103-106.

Bindra, O. S.

1973a Cicadellid vectors of plant pathogens. *Final Report of the PL-480 Project, No. A7-ENT-22, Grant No. FG-In-300,* October 31, 1971. Punjab Agricultural University, Ludhiana, India. 56 pp.

Bindra, O. S. and Deol, G. S.

1972a Bionomics of *Circulifer tenellus* (Baker), the cicadellid vector of sugarbeet curly top virus. *Indian Journal of Agricultural Science* 42:513-519.

Bindra, O. S. and Kaur, P.

1969a Influence of temperature on the rate of development of *Circulifer opacipennis* (Lethierry). *Indian Journal of Entomology* 31:174-177.

Bindra, O. S., Khatri, H. L. and Sohi, A. S.

1973a Surveys for the presence of cicadellid borne viruses/mycoplasma-like agents. *In: O. S. Bindra, ed. Cicadellid Vectors of Plant Pathogens.* Pp. 39-41.

Bindra, O. S., Khatri, H. L., Sohi, A. S. and Deol, G. S.

1973a Insecticidal control of sesame phyllody. *In: Bindra, O. S., ed. Cicadellid Vectors of Plant Pathogens.* Pp. 46-49.

Bindra, O. S., Khatri, H. L., Sohi, A. S., Deol, G. S. and Sajjan, S. S.

1973a Studies on the epidemiology and control of brinjal little-leaf disease. *In: Bindra, O. S., ed. Cicadellid Vectors of Plant Pathogens.* Pp. 44-46.

Bindra, O. S. and Mahal, M. S.

1981a Varietal resistance in eggplant (brinjal) (*Solanum melongena*) to the cotton jassid (*Amrasca biguttula biguttula*). *Phytoparasitica* 9:119-131.

Bindra, O. S. and Sagar, P.

1976a Co-ordinated minimum insecticide trial: Yield performance of insect resistant cowpea cultivars from IITA compared with Indian cultivars. *Tropical Grain Legume Bulletin* 5:8-9.

Bindra, O. S., Sawi, S., Sohi, A. S. and Gill, M. I. P. K.

1973a Taxonomic studies on cicadellid species. *In: Bindra, O. S., ed. Cicadellid Vectors of Plant Pathogens.* Pp. 2-10.

Bindra, O. S. and Singh, B.

1969a Biology and bionomics of *Hishimonus phycitis* (Distant), a jassid vector of little leaf disease of brinjal (*Solanum melongena* L.). *Indian Journal of Agricultural Science* 39:912-919.

Bindra, O. S., Singh, B., Chahal, B. S. and Sekhon, S. S.

1971a Aerial application of insecticides for the control of the mango-hopper. *PANS* 17(3):350-353.

Bindra, O. S. and Singh, J.

1970a Biology and bionomics of *Orosius albicinctus* Distant, the jassid vector of sesamum phyllody virus. *Indian Journal of Agricultural Science* 48:340-355.

Bindra, O. S., Singh, S. and Sohi, A. S.

1970a Taxonomy and distribution of Indian species of *Circulifer* (Homoptera: Cicadellidae). *Annals of the Entomological Society of America* 63(3):664-667.

Bindra, O. S. and Sohi, A. S.

1968a Host range of *Hishimonus phycitis* (Distant) (Homoptera: Jassidae), the vector of little leaf of Brinjal. *Journal of Research* (Ludhiana) 5(2):232-236.

1969a Laboratory rearing of vector species of the genus *Circulifer* Zakhvatkin (Cicadellidae: Homoptera). *Indian Journal of Entomology* 31:190-191.

1970a Cross-breeding experiments between *Circulifer opacipennis* (Lethierry) and *C. tenellus* (Baker), cicadellid vectors of sugar-beet curly-top virus. *Indian Journal of Entomology* 32:183-184.

Birat, R. B. S.

1963a Pest for rice in Bihar. *Allahabad Farmer* 37(3):27-39.

Birch, J. B. and Fleischer, S. J.

1984a Diagnostic methods for detecting outliers in regression analysis. *Environmental Entomology* 13:19-25.

Black, L. M.

1953a Occasional transmission of some plant viruses through the eggs of their insect vector. *Phytopathology* 43(1):9-10.

1958a Wound-tumor. *Proceedings of the National Academy of Sciences* 44:364-367.

1958b Transmission of virus by leafhoppers. *Proceedings of the 10th International Congress of Entomology* 3:201-204.

1959a Biological cycles of plant viruses in insect vectors. *In: Burnet, F. M. and Stanley, W. M., eds. The Viruses.* Pp. 157-185. Academic Press, New York.

1962a Some recent advances on leafhopper-borne viruses. *In: Maramorosch, K., ed. Biological Transmission of Disease Agents. A Symposium.* Pp. 1-9. Academic Press, Inc., New York.

1969a Insect tissue cultures as tools in plant virus research. *Annual Review of Phytopathology* 7:73-100.

1979a Vector cell monolayers and plant viruses. *Advances in Virus Research* 25:191-271.

1984a The controversy regarding multiplication of some plant viruses in their insect vectors. In: Harris, K. F., ed. Current Topics in Vector Research II. Pp. 1-32. Praeger, New York. 325 pp.

Black, L. M. and Brakke, M. K.
1952a Multiplication of wound-tumor virus in an insect vector. Phytopathology 42:269-273.

Black, L. M., Wolcyrz, S. and Whitcomb, R. F.
1958a A vectorless strain of wound-tumor virus. VII International Congress of Microbiologists (Abstract 14a).

Blaine, Moore, Stevens, T. R. and McArthur, E. D.
1982a Preliminary study of some insects associated with rangeland shrubs with emphasis on Kochea prostrata. Journal of Range Management 35(1):128-130.

Blair, B. D.
1975a The potato leafhopper on alfalfa. Ohio Cooperative Extension Service, Field and Forage Crop Insect Series 1. 2 pp.

Blair, B. D. and Niemczyk, H. D.
1969a Alfalfa insects in Ohio. Ohio Cooperative Extension Service Leaflet L-130. 8 pp.

Blanchard, E. E.
1966a Nuevos triquiopodinos argentinos, parasitos de hemipteros nocivos (Dipt. Gymnosomatidae). Revista de Investigacions Agropecuarias 3(5):59-95.

Blattny, C., Jr.
1963a Die ubertragung der Hexenbesenkrankheit der Heidelbeeren (Vaccinium myrtillus L.) durch die zikade Idiodonus cruentatus Panz. Phytpathologische Zeitschrift 49:203-205.

Blattny, C., Sr. and Blattny, C., Jr.
1970a Blueberry witches'-broom. In: Frazier, N. W., ed. Virus Diseases of Small Fruits and Grapevines. Pp. 177-179. University of California Division of Agricultural Sciences, Berkeley.

Blickenstaff, C. C. and Huggans, J. L.
1962a Soybean insects and related arthropods in Missouri. Missouri Agricultural Experiment Station Bulletin 803. 51 pp.

Blisard, T. J.
1931a Food plants of the blunt-nosed leafhopper. Convention of the American Cranberry Growers' Association 62:9-16.

Bliven, B. P.
1957a Some Californian mirids and leafhoppers including two new genera and four new species. Occidental Entomologist 1(1):1-7.
1958a Studies on insects of the Redwood Empire II: New Hemiptera and further notes on the Colladonus complex. Occidental Entomologist 1(2):8-24.
1963a New synonymy in the Cicadelloidea and a restoration of Colladonus uhleri (Ball). Occidental Entomologist 1(7):87-89.
1966a Synonymical notes on Idiocerus and Oncopsis. Occidental Entomologist 1(9):108-114.

Blocker, H. D.
1966a A classification of the western hemisphere Balcluthini (Homoptera: Cicadellidae). Dissertations Abstracts 26:4897.
1967a Classification of the western hemisphere Balclutha (Homoptera: Cicadellidae). Proceedings of the United States National Museum 122:1-55.
1967b Athysanella attenuata and a closely related new species from Kansas. Journal of the Kansas Entomological Society 40(4):576-578.
1968a A new Brazilian Balclutha and new records for other species (Homoptera: Cicadellidae). Journal of the Kansas Entomological Society 41(2):207-209.
1970a The genus Stragania in America north of Mexico (Homoptera: Cicadellidae: Iassinae). Annals of the Entomological Society of America 63(5):1424-1433.
1971a Distribution of Balclutha in the Bahama Islands, Newfoundland, and Panama (Homoptera: Cicadellidae). Journal of the New York Entomological Society 79(3):158-160.
1975a The Mexican genus Gargaropsis (Homoptera: Cicadellidae: Iassinae). Annals of the Entomological Society of America 68(3):561-564.
1975b The genus Grunchia (Homoptera: Cicadellidae: Iassinae). Journal of the Kansas Entomological Society 48(3):327-330.
1976a Three new genera of Neotropical Iassinae (Homoptera: Cicadellidae). Annals of the Entomological Society of America 69(3):519-522. [New genera: Garlica, Gehundra, Jivena.]
1979a The Iassinae (Homoptera: Cicadellidae) of the western hemisphere. Journal of the Kansas Entomological Society 52(1):1-70. [Describes 16 new genera and one new subgenus.]
1979b A new subgenus and species of Iassinae (Homoptera: Cicadellidae). Annals of the Entomological Society of America 72(6):856. [New genus: Abdistragania.]
1979c A proposed phylogeny of New World Iassinae (Homoptera: Cicadellidae). Annals of the Entomological Society of America 72(6):857-862.
1981a A new species and new records of Balclutha (Homoptera: Cicadellidae) from Panama. Journal of the Kansas Entomological Society 54(4):678-680.
1982a New species of Iassinae from the Neotropics (Homoptera: Cicadellidae). Journal of the Kansas Entomological Society 55(4):639-650.
1983a Classification and proposed phylogeny of the Neotropical leafhopper genus Icaia Linnavuori (Homoptera: Cicadellidae: Deltocephalinae). Entomography 2:77-87.
1984a Morphological irregularities in the genital structures of Athysanella (Homoptera: Cicadellidae: Deltocephalinae). Mitteilungen der Schweizerischen Entomologischen Gesellschaft 57(4):412.
1985a New Athysanella from Mexico (Homoptera: Cicadellidae: Deltocephalinae). Journal of the Kansas Entomological Society 57(4):732-735.

Blocker, H. D., Harvey, T. L. and Launchbaugh, J. L.
1972a Grassland leafhoppers. 1. Leafhopper populations of upland seeded pastures in Kansas. Annals of the Entomological Society America 65(1):166-172.

Blocker, H. D. and Nixon, P. L.
1978a Balclutha (Homoptera: Cicadellidae) of Paraguay including a new species. Journal of the Kansas Entomological Society 51(3):512-515.
1978b A new species of Balclutha (Homoptera: Cicadellidae) from Paraguay. Journal of the Kansas Entomological Society 51(3):397.

Blocker, H. D. and Reed, R.
1976a Leafhopper populations of a tallgrass prairie (Homoptera: Cicadellidae): Collecting procedures and population estimates. Journal of the Kansas Entomological Society 49(2):145-154.

Blocker, H. D., Reed, R. and Mason C. E.
1971a Leafhopper studies at the Osage site (Homoptera: Cicadellidae). Technical Report 124, Grassland Biome, U.S. International Biological Program. 25 pp.

Blocker, H. D. and Triplehorn, B. W.
1985a External morphology of leafhoppers. In: Nault, L. R. and Rodriguez, J. G., eds. The Leafhoppers and Planthoppers. Pp. 41-60. Wiley & Sons, New York. 500 pp.

Blocker, H. D. and Wesley, C. S.
1985a Distribution of Athysanella (Homoptera: Cicadellidae: Deltocephalinae) in Canada and Alaska, with descriptions of three new species. Journal of the Kansas Entomological Society 58(4):578-585.

Blote, H. C.
1964a On some New Guinean Batrachomorphus species (Hemiptera, Jassidae). Zoologische Mededelingen (Leiden) 39:464-470.

Boccardo, G., Hatte, T., Francki, R. I. B. and Grivell, C. J.
1980a Purification and some properties of reovirus-like particles from leafhoppers and their possible involvement in wallaby ear disease of maize. Virology 100:300-313.

Bode, A.
1953a Die insektenfauna des Ostniederschischen Oberen Lies. *Palaeontographica* 103(1-4):1-375.

Bogavac, M.
1968a A contribution to knowledge of the Auchenorrhyncha on maize. *Zastita Bilja* 19(98):41-45. [In Serbo-Croatian with English summary.]

Bogavac, M. and Antonijevic, A.
1964a The wheat leafhopper, *H. anatolica*. *Zastita Bilja* 15(77):37-50.

Bohm, O.
1963a Die rosenzikade. *Pflanzenarzt* 16(2):21-22.

Boller, E. and Baggiolini, M.
1970a Zum Auftreten der grunen Rebzikade (*Empoasca flavescens* F.) in den ostschweizerischen Rebbergen. *Schweizerische Zeitschrift fur Obst-und Weinbau* 106(13):315-321.

[Boller, E. et al.]
1970a Untersuchungen an der Rebzikade (*Empoasca flavescens* F.) und am einbindigen Traubenwickler (*Clysia ambiguella* Hb.) in der Ostschweiz. *Schweizerische Zeitschrift fur Obst- und Weinbau* 106:651-660. [In German.]

Bolton, M. E., Markham, P. G. and Davies, J. W.
1984a Nucleic acid hybridization techniques for the detection of plant pathogens in insect vectors. *In: British Crop Protection Conference, Pests and Disease.* Brighton Metrople, England. Pp. 181-186. Croyden, U.K. 3 vol., 1207 pp.

Bonfils, J.
1981a Description d'especes nouvelle de Cicadellidae recoltees dans le midi de la France et en Corse (Hom.). *Bulletin de la Societe Entomologique de France* 86(9&10):298-307.

Bonfils, J. and Della Giustina, W.
1978a Inventaire et repartition biogeographique des Homopteres Auchenorhynques de Corse. *Bulletin de la Societe Entomologique de France* 83(1&2):23-29.

1978b Contribution a l'etude des Homopteres Auchenorrhynques (Homoptera Auchenorrhyncha). *Bulletin de la Societe des Sciences historiques et naturelles de la Corse.* Pp. 93-112.

Bonfils, J. and Delplanque, A.
1971a Distribution des principales cicadelles des prairies aux Antilles Francaises (Homoptera). *Annales de Zoologie Ecologie Animale* 3(2):135-150. [In French with English summary.]

Bonfils, J. and Lauriant, F.
1975a Presence en Languedoc de *Synophropsis lauri* (Hom. Cicadellidae). *Entomologiste* 31(2):69-71.

Bonfils, J., Lauriant, F. and Leclant, F.
1974a Sur la presence de la Cicadelle de Flor dans les vergers d'abricotiers et de cerisiers atteints de deperissements. *Progress Agricole et Viticole* 91:1-4.

Bonfils, J. and Schvester D.
1960a Les cicadelles (Homoptera: Auchenorhyncha) dans leurs rapports avec la vigne dans le Sud-Ouest de la France. *Annales des Epiphyties* (Paris) 11(3):325-336.

Bonnefil, L.
1969a A comparative study of Central American *Empoasca* leafhoppers. *Proceedings of the North Central Branch of the Entomological Society of America* 24(1):57.

1972a Some biological interactions of *Empoasca phaseola* Oman (Homoptera: Cicadellidae) with selected leguminous hosts. *Dissertation Abstracts International, B* 32(7):3984.

Borle, M. N., Upalanchiwar, A. R. and Deshmukh, S. D.
1980a Efficacy of insecticidal mixtures in dust formulations for the control of cotton jassid in dryland cultivation. *Indian Journal of Entomology* 42:130-132.

Bortoli, S. A. de and Giacomini, P. C.
1981a Action of some systemic granular insecticides on *Bemisia tabaci* (Gennadius, 1889) (Homoptera: Aleyrodidaee) and *Empoasca kraemeri* Ross & Moore, 1957) (Homoptera: Cicadellidae) and their effects on the productivity of beans (*Phaseolus vulgaris* L.). *Anais da Sociedadae Entomologica do Brasil* 10:97-104.

Bos, L.
1981a Wild plants in the ecology of virus diseases. *In: Maramorosch, K. and Harris, K. F., eds. Plant Diseases and Vectors: Ecology and Epidemiology.* Pp. 1-33. Academic Press, New York. 368 pp.

Bos, L. and Grancini, P.
1965a Some experiments and considerations on the identification of witches' broom viruses, especially in clovers, in the Netherlands and Italy. *Netherlands Journal of Plant Pathology* 71(1):20.

Boulard, M.
1969a Homopteres jassidomorphes nouveaux lies aux colatiers et aux autres plantes stimulantes cultives en Afrique centrale. *Cafe Cacao The* 13(2):151-156.

1969b Hemipteroides nuisibles ou associes aux cacaoyers en Republique Centrafricaine. 2. Homopteres Auchenorhynches. *Cafe Cacao The* 13(4):310-324.

1971a Une tribu nouvelle pour la faune africaine des Homopteres Cicadellidae. *Bulletin de l'Institut Fondamental d'Afrique Noire. Serie A. Sciences Naturelles* 33(3):712-717.

1975a Un nouvel Homoptere jassidomorphe des cacaoyeres camerounaises. *Cafe Cacao The* 19(2):137-138.

1978a Premier cas de "mimetisme ostensible" chez les Homopteres Auchenrhynques (Insecta). *Compte Rendu Hebdomadaire des Seances de l'Academie des Sciences* 287 (Serie D): 1389-1391.

Bournier, A.
1977a Grape insects. *Annual Review of Entomology* 22:355-376.

Bove, J. M., Moutous, G., Saillard, C., Fos, A., Bonfils, J., Vignault, U., Nhami, A., Abassi, M., Kabbage, K., Hafidi, B., Mouches, C. and Viennot-Bourgin, G.
1979a Mise en evidence de *Spiroplasma citri* l'agent causal de la maladie du "stubborn" des argumes dans 7 cicadelles du Maroc. *Comptes Rendu Hebdomadaires des Sciences de l'Academie des Sciences,* D 288:335-338.

Bove, J. M. and Saillard, C.
1979a Cell biology of spiroplasmas. *In: Whitcomb, R. F. and Tully, J. G., eds. The Mycoplasmas. III. Plant and Insect Mycoplasmas.* Pp. 83-153. Academic Press, New York.

Bovey, B.
1957a Une anomalie des fleurs du trefle causee par un virus transmis par des cicadelles. *Revue Romande d'Agriculture de Viticulture et d'Arboriculture* 13:106-108.

1958a Les vecteurs de maladies a virus due Fraisier. *Revue de Pathologie Generale et de Physiologie Clinique* 58:1751-1762.

1958b Le sort des virus de plantes dans les organisms des cicadelles vectrices. *Revue de Pathologie Generale et de Physiologie Clinique* 58:1823-1836.

Bowling, C. C.
1961a Tests with systemic insecticides on rice. *Journal of Economic Entomology* 54(5):937-941.

1970a Lateral movement, uptake, and retention of carbofuran applied to flooded rice plants. *Journal of Economic Entomology* 63(1):239-242.

Bowyer, J. W.
1972a Characterization of some yellows diseases in Australia. *Proceedings of the Fifty-sixth Annual Meeting of the Pacific Division of the American Phytopathological Society* 62:1101.

1974a Tomato big bud, legume little leaf, and lucerne witches' broom: Three diseases associated with different mycoplasma-like organisms in Australia. *Australian Journal of Agricultural Research* 25(3):449-457.

Bowyer, J. W. and Atherton, J. G.
1971a Summer death of French bean: New hosts of the pathogen, vector relationship, and evidence against mycoplasmal etiology. *Phytopathology* 61(12):1451-1455.

1971b Mycoplasma-like bodies in french bean, dodder, and the leafhopper vector of the legume little leaf agent. *Australian Journal of Biological Sciences* 24(4):712-729.

1972a Effects of tetracycline antibiotics on plants affected by legume little leaf disease. *Australian Journal of Biological Sciences* 25(1):43-51.

Boyd, F. J. and Pitre, H. N.
1968a Studies on the field biology of *Graminella nigrifrons*, a vector of corn stunt virus in Mississippi. *Annals of the Entomological Society of America* 61(6):1423-1427.
1969a Greenhouse studies of host plant suitability to *Graminella nigrifrons*, a vector of corn stunt virus. *Journal of Economic Entomology* 62(1):126-129.

Bradfute, O. E., Tsai, J. H. and Folk, B. W.
1985a Maize rayado fino and maize dwarf mosaic in Ecuador. *Plant Disease* 69(12):1078-1080.

Bradfute, O. E., Tsai, J. H. and Gordon, D. T.
1981a Corn stunt spiroplasma and viruses associated with maize disease epidemics in southern Florida. *Plant Disease* 65:837-841.

Bradfute, O. W., Gingery, R. E., Gordon, D. T. and Nault, L. R.
1972a Tissue ultrastructure, sedimentation and leafhopper transmission of a virus associated with a maize dwarfing disease. *Journal of Cell Biology* 55:25.

Brakke, M. K., Maramorosch, K. and Black, L. M.
1953a Properties of the wound-tumor virus. *Phytopathology* 43(7):387-390.

Bram, R. A., Saugstad, E. S. and Nyquist, W. E.
1965a Factors influencing sweep-net sampling of alfalfa insects. *Proceedings of the North Central Branch of the American Association of Economic Entomologists* 20:126.

Brar, J. S.
1974a Studies on the bionomics of maize jassid, *Zyginidia manaliensis* (Singh). M. Sc. Thesis, Punjab Agricultural University, Ludhiana.

Brar, J. S. and Singh, B.
1978a Studies on bionomics of maize jassid *Zyginidia manaliensis* (Singh). *Journal of Research* (Ludhiana) 15(4):463-467.
1981a Morphology and host-range of maize leafhopper, *Zyginidia guyumi* (Ahmed) (Homoptera, Cicadellidae, Typhlocybinae). *Journal of Research* (Ludhiana) 18:25-29.

Brcak, J.
1954a Nova prenasec stoburu/Bezsemenoszti/rajcete a Tabuku krisek *Aphrodes bicinctus* Schrk. *Zoologicke a Entomologicke Listy* 4:231-236.
1979a Leafhopper and planthopper vectors of plant disease agents in central and southern Europe. *In: Maramorosch, K. and Harris, K. F., eds. Leafhopper Vectors and Plant Disease Agents.* Pp. 97-154. Academic Press, New York.

Bremer, K. and Raatikainen, M.
1975a Cereal diseases transmitted or caused by aphids and leafhoppers in Turkey. *Annales Academiae Scientiarum Fennicae A, IV Biologica* 203:1-14.

Brett, C. H. and Brubaker, R. W.
1956a Potato leafhopper control on snap beans. *Journal of Economic Entomology* 49(4):571.

Briansky, R. H., Timmer, L. W., French, W. J. and McCoy, R. E.
1983a Colonization of the sharpshooter vectors, *Oncometopia nigricans* and *Homalodisca coagulata*, by xylem-limited bacteria. *Phytopathology* 73:530-535.

Briansky, R. H., Timmer, L. W. and Lee, R. F.
1982a Detection and transmission of a gram-negative, xylem-limited bacterium in sharpshooters from a citrus grove in Florida. *Plant Disease* 66:590-592.

Bridges, E. T. and Pass, B. C.
1967a The biology of the leafhopper *Draeculacephala mollipes*. *Proceedings of the North Central Branch of the American Association of Economic Entomologists* 22:105.
1970a Biology of *Draeculacephala mollipes* (Homoptera: Cicadellidae). *Annals of the Entomological Society of America* 63(3):789-792.

Broadbent, L.
1957a Insecticidal control of the spread of plant viruses. *Annual Review of Entomology* 2:339-354.
1969a Disease control through vector control. *In: K. Maramorosch, ed. Viruses, Vectors, and Vegetation.* Pp. 593-630. Wiley & Sons, New York. 666 pp.

Broersma, D. B.
1968a The soybean insect situation. *Proceedings of the North Central Branch of the American Association of Economic Entomologists* 23:65-66.

Broersma, D. B., Bernard, R. L. and Luckmann, W. H.
1972a Some effects of soybean pubescence on populations of the potato leafhopper. *Journal of Economic Entomology* 65(1):78-82.

Bronson, T. E. and Rust, R. E.
1951a Mist sprays for control of certain truck crop insects. *Journal of Economic Entomology* 44(2):218-220.

Brooks, J. C.
1979a Natural biological suppression agents of rice pests in the eastern plains of Colombia. *International Rice Research Newsletter* 4(3):19.

Brooks, M. A.
1985a Nutrition, cell culture and symbiosis of leafhoppers and planthoppers. *In: Nault, L. R. and Rodriguez, J. G., eds. The Leafhoppers and Planthoppers.* Pp. 195-216. Wiley & Sons, New York. 500 pp.

Brown, H. E., Hasman, L., Jenkins, L. and Stone, P. C.
1959a Biology and control of field crop insects with special emphasis on the corn ear worm, Hessian fly, grasshoppers, European corn borer, and primary soybean insects. *Missouri Agricultural Experiment Station Bulletin* 728:47.

Brown, H. E., Stone, P. C. and Hasman, L.
1955a The biology and control of field crop insects. *Missouri Agriculture Experiment Station Bulletin* 643:32.

Brown, H. E., Stone, P. C., Jenkins, L. and Hasman, L.
1960a Project 102. Investigations of the biology and control of field crop insects other than corn insects with special emphasis on the Hessian fly, grasshoppers, soil insecticides and primary soybean insects. *Missouri Agricultural Experiment Station Bulletin* 747:53-54.

Brown, H. E., Stone, P. C., Jenkins, L. and Winjo, C. W.
1957a Biology and control of field crop insects. *Missouri Agricultural Experiment Station Bulletin* 695:61.

Brown, L. R. and Eads, C. O.
1965a A technical study of insects affecting the sycamore tree in southern California. *California Agricultural Experiment Station Bulletin* 818:3-38.
1969a Factors affecting the appearance of holly oaks in southern California. *Journal of Economic Entomology* 62(1):114-117.

Bruneau de Mire, P. and Lotode, R.
1974a Behaviour of hybrid cacao families subjected to attacks by Homoptera. *Cafe Cacao The* 18:187-192.

Buchmeier, P. F. and Edwards, C. R.
1979a An evaluation of Lannate and two formulations of Sevin applied by helicopter for activity against the green cloverworm and potato leafhopper on soybeans in northern Indiana. *Proceedings of the North Central Branch of the Entomological Society of America* 33:38.

Budh, D. S.
1975a *Psammotettix striatus* (Linn.), a rare hemipteran ectoparasite of the flying foxes of Poonch. *Current Science* (Bangalore) 42(3):108.

Burgaud, L. and Cessac, M.
1962a Efficacite pratique d'un insecticide acaricide endotherapique nouveau: le vamidothion. *Phytiatrie-Phytopharmacie* 11(3):117-128.

Burnside, V. W.
1971a *General Catalogue of the Homoptera Fasc. VI Cicadelloidea. Index to Genera and Species with Addenda and Corrigenda to Parts 1-17.* U.S. Department of Agriculture, Agricultural Research Service. 269 pp.

Burton, V. E, Grigarick, A. A., Hall, D. H. and Webster, R. K.
1980a Insect and disease control recommendations for rice. *Division of Agricultural Science, University of California Leaflet* 2748.

Bushing, R. W. and Burton, V. E.
    1974a Leafhopper damage to silage corn in California. *Journal of Economic Entomology* 67:656-658.

Bushing, R. W., Burton, V. E., McCutcheon, O. D. and Etchegaray, H.
    1975a Leafhoppers in silage corn. *California Agriculture* 29:11-12.

Butani, D. K.
    1967a Insect pests of cotton. VII. Evaluation studies on insecticides. *Indian Journal of Entomology* 29(4):360-369.
    1975a Crop pests and their control. 1. Cotton. *Pesticides* 9:21-27.
    1976a Insect pests of cotton. XX. Effect of mixtures of insecticides on pest complex and yield of raw cotton. *Coton et Fibres Tropicales* 31:285-287.

Butani, D. K., Das, B. B. and Basu, A. K.
    1977a Comparative efficacy of some new insecticides on insect pests and their effect on the yield of cotton. *Indian Journal of Agricultural Sciences* 45:348-351.

Butani, D. K. and Jotwani, M.
    1976a Crop pests and their control. 2. Rice. *Pesticides* 10:29-35.
    1983a Insects as a limiting factor in vegetable production. *Pesticides* 17:6-13.

Butani, D. K. and Sahni, V. M.
    1966a Insect pests of cotton. IV. Evaluation studies on various spraying schedules. *Indian Cotton Journal* 20(2):107-115.
    1966b Insect pests of cotton. VIII. Evaluation studies on various dusting schedules. *Labdev Journal of Science and Technology* 4(4):271-274.

Butani, D. K. and Surjit Singh.
    1965a A note on the comparative efficacy of newer insecticides against pests of cotton. *Labdev Journal of Science and Technology* 3(1):67-69.

Butani, D. K. and Verma, S.
    1976a Insect pests of vegetables and their control. 3. Lady's finger. *Pesticides* 10:31-37.

Butt, S. and Ahmed, Manzoor
    1981a A note on the biology and ecology of *Empoasca kerri* Pruthi (Typhlocybinae: Cicadellidae), a pest of cluster bean, *Cyamopsis tetragonoloba* (L.), in Pakistan. *Natural Science* (Karachi) 3(1):29-31.

Butt, S., Jabbar, A. and Ahmed, M.
    1981a Biology and ecology of *Empoasca kerri* Pruthi (Typhlocybinae: Cicadellidae), a pest of guar, *Cyamopsis tetragonolobus*, in Pakistan. *Biologia* 27:257-263.

Buzacott, J. H.
    1953a Insects associated with sugar cane in New Guinea. *Technical Communications Bureau, Sugar Experiment Station, Queensland* 1959 No. 2[1+]:23-30.

Byers, R. A. and Bierlein, D. L.
    1984a Continuous alfalfa: Invertebrate pests during establishment. *Journal of Economic Entomology* 77:1500-1503.

Byers, R. A. and Hower, A. A.
    1976a The potato leafhopper and alfalfa quality. *Forage Insects Research Conference* 18:17-18.
    1976b The potato leafhopper and alfalfa quality. *Report of the Alfalfa Improvement Conference* 15:19.

Byers, R. A. and Jung, G. A.
    1979a Insect populations on forage grasses: Effect of nitrogen fertilizers and insecticides. *Environmental Entomology* 8(1):11-18.

Byers, R. A., Neal, J. W., Elgin, J. H., Hill, K. R., McMurtrey, J. E. and Feldmesser, J.
    1977a Systemic insecticides with spring-seeded alfalfa for control of potato leafhopper. *Journal of Economic Entomology* 70(3):337-340.

Cabauatan, P. Q. and Hibino, H.
    1984a Detection of spherical and bacilliform virus particles in tungro-infected rice plants by leafhopper transmission. *International Rice Research Newsletter* 9(1):18-19.

Cabauatan, P. Q. and Ling, K. C.
    1979a Acquisition of rice tungro virus through parafilm membrane by *Nephotettix virescens* (Distant). *International Rice Research Newsletter* 4(4):12.

Cabunagan, R. C., Flores, Z. M. and Hibino, M.
    1985a Reaction of IR varieties to tungro (RTV) under various disease pressure. *International Rice Research Newsletter* 10(6):11.

Cabunagan, R. C., Tiongco, E. R. and Hibino, H.
    1984a Reaction to rice tungro (RTV) complex as influenced by insect pressure. *International Rice Research Newsletter* 9(6):13.

Cadman, C. H.
    1961a Raspberry viruses and virus diseases in Britain. *Horticultural Research* 1:47-61.

Cadou, J.
    1970a Note sur les cicadelles du cotonnier *Empoasca* spp. (Homoptera, Typhlocibidae) en Republique Centrafricaine. *Coton et Fibres Tropicales* 25(3):401-404. [In French with English and Spanish summaries; separate English version available.]

Calavan, E. C., Kaloostian, G. H., Oldfield, G. N., Nauer, E. M. and Gumpf, D. J.
    1979a Natural spread of *Spiroplasma citri* by insect vectors and its implications for control of stubborn disease of citrus. *Proceedings of the International Society of Citriculture*. Pp. 900-1102.

Calavan, E. C., Lee, I. M., Cartia, G., Kaloostian, G. H., Oldfield, G. H. and Pierce, H. D.
    1974a Stubborn disease breakthrough made. *Citrus and Vegetable Magazine* 37(8):17.

Calavan, E. C. and Oldfield, G. N.
    1979a Symptomatology of spiroplasmal plant disease. *In:* Whitcomb, R. F. and Tully, J. G., eds. *The Mycoplasmas. III. Plant and Insect Mycoplasmas.* Pp. 37-64. Academic Press, New York.

Cameron, G. N.
    1972a Analysis of insect trophic diversity in two salt marsh communities. *Ecology* 53(1):58-73.

Campbell, M. B. S. C. and Davies, D. M.
    1985a Redescription of the South African leafhopper, *Distantia frontalis* Signoret (Homoptera: Cicadellidae). *Journal of the Entomological Society of Southern Africa* 48(2):337-340.

Campbell, W. V. and Emery, D. A.
    1971a Resistance of peanut accessions to the potato leafhopper, *Empoasca fabae*. *Journal of the American Peanut Research Education Association* 3(1):219.
    1972a Anatomical and chemical characteristics of peanuts associated with resistance to an insect complex. *Journal Elisha Mitchell Scientific Society* 88(4):195-196.

Campbell, W. V., Emery, D. A. and Wynne, J. C.
    1975a Registration of four germplasm lines of peanuts. *Crop Science* 15:738-739.
    1976a Resistance of peanuts to the potato leafhopper. *Peanut Science* 3(11):40-43.

Cancelado, R. E. and Radcliffe, E. B.
    1979a Action thresholds for potato leafhopper on potatoes in Minnesota. *Journal of Economic Entomology* 72:566-569.
    1979b Manipulations of insect populations for potato pest management studies. *Proceedings of the North Central Branch of the Entomological Society of America* 33:54.

Cancelado, R. E., Radcliffe, E. B. and Stucker, R. E.
    1976a Relationships between potato leafhopper numbers and potato yields in Minnesota. *Proceedings of the North Central Branch of the Entomological Society of America* 31:45.

Cancelado, R. and Yonke, T. R.
    1970a Effect of prairie burning on insect populations. *Journal of the Kansas Entomological Society* 43(3):274-281.

Cantello, W. W. and Sanford, L. L.
    1984a Insect population responses to mixed and uniform plantings of resistant and susceptible plant material. *Environmental Entomology* 13:1443-1445.

Cantoreanu, M.
1959a Especes de Cicadides nouvelles pour la faune de la Republique Populaire Roumaine. III. *Comunicarile Academici Republicii Populare Romane* 9:1159-1164. [In Rumanian with Russian and French summaries.]
1961a Especes de Cicadides nouvelles pour la faune de la Republique Populaire Roumaine. III. *Comunicarile Academici Republicii Populare Romane* 11:195-199. [In Rumanian with Russian and French summaries.]
1963a Specii de Cicadine noi pentru fauna RPR. V. *Comunicarile Academici Republicii Populare Romane* 13(6):567-570.
1963b Specii de Cicadine noi pentru fauna R.P.R. (Homo.-Auchenorrhyncha) VI. *Comunicarile Academici Republicii Populare Romane* 13(12):1063-1067.
1964a Contributii la cunoasterea biologiei speciei *Erythroneura alneti* Dahlb. (Homoptera-Auch.). *Studii si Cercetari de Biologie, Seria Zoologie* 16(6):573-576.
1965a Specii de Cicadine (Homoptera: Auchenorrhyncha) noi pentru fauna R.P.R. VII. *Studii si Cercetari de Biologie, Seria Zoologie* 17(4):325-327.
1965b Specii de Cicadine (Homo.-Auch.) noi pentru fauna Republicii Socialiste Romania VIII. *Analele Universitatii Bucuresti, Seria Stiintele Naturii* 14:163-165.
1965c Observatii asupra dezvoltarii speciei *Oncopsis alni* (Schrk.) (Hom.-Auchen.) *Analele Universitatii Bucuresti, Seria Stiintele Naturii* 14:167-171.
1965d Contributii la cunoasterea faunei de Cicadine din mlastina de turba de la Harman. *Comunicari de Zoologie Bucuresti* 3:137-147.
1968a Eine Neue Cycaden-art: *Aphrodes dobrogicus* n. sp. aus Rumanien. *Revue Roumaine de Biologie Serie de Zoologie* 13(1):53-55.
1968b Ord. Homoptera (Auchenorrhyncah). In: L'Entomogaune de l'ile de Letea (Delta du Danube). *Extrait de Travaux du Museum d'Histoire Naturelle Grigore Antipa* 9:127-131.
1969a Cercetari privind fauna de Cicadine (Hom. Auchen.) in doua biotopuri de munte. *Comunicari de Zoologie*, Part 1-a, pp. 113-118.
1969b Homoptera-Auchenorrhyncha-Arten aus der Dobrogea. *Lucrarile Statiunii de Cercetari Marine "Prof. Ioan Borcea." Agigea III* 1969:259-265.
1971a Cercetari asupra parazitizmului la Cicadine (Homoptera Auchenorrhyncha). *Studii si Comunicari* 1971:49-55.
1971b Cercetari asupra unor specii de Cicadine (Hom. Auchen.) din republica socialista Romania. *Studii si Comunicari* 1971:87-92.
1971c Contributii la cunoasterea faunei de Cicadine din Bazinul Siretului (Homoptera Auchenorrhyncha). *Studii si Comunicari* 1971:43-47.
1973a Studii privind distibutia altitudinala a Cicadinelor (Hom.-Auch.) in Muntii Bucegi. *Cumunicari si Referate, Museul St. Naturii-Ploiest* 1973:195-206.
1975a Homoptera-Auchenorrhyncha. In: Ionescu, M., ed. *Fauna Ser. Monografica.* Pp. 90-95. Academia Republicii Socialiste Romania, Portile de Fier. 316 pp.

Capco, S. R.
1959a A study of Philippine *Bothrogonia* (Homoptera: Cicadellidae) with reference to the female seventh sterna and internal male genitalia. *Philippine Journal of Science* 87(2):159-167.
1959b The Philippine species of *Parabolocratus*. *Philippine Journal of Science* 88(3):325-334.
1960a Philippine species of *Xestocephalus* Van Duzee (Cicadellidae, Homoptera) in the Baker Collection, United States National Museum. *Philippine Journal of Science* 89(1):41-46.

Capel-Williams, G.
1978a The post-embryonic development and functional morphology of the female reproductive system in *Graphocephala fennahi* (Homoptera: Cicadellidae). Doctoral Dissertation, University of London. 337 pp.

Capinera, J. L. and Walmsley, M. R.
1978a Visual responses of some sugarbeet insects to sticky traps and water pan traps of various colors. *Journal of Economic Entomology* 71:926-927.

Cardona, C., van Schoonhoven, A., Gomez, L., Garcia, J. and Garzon, F.
1981a Effect of artificial mulches on *Empoasca kraemeri* Ross and Moore populations and dry bean yields. *Environmental Entomology* 10:705-707.

Cardoso, A. M.
1974a Reconhecimento das cogarrinhas (Homoptera, Cicadelloidea) de Portugal continental. I. *Agronomia Lusitana* 35(2):145-167.

Carle, P.
1964a Adaptation de la pulversation a faible volume/hectare a la lutte chimique dirigee contre *Scaphoideus littoralis* (Homopt. Jassidae). *Phytiatrie-Phytopharmacie* 14(1):39-44.
1964b Essais de pesticides en plein champ contre *Scaphoideus littoralis* (Homopt., Jassidae). *Phytiatrie-Phytopharmacie* 14(1):29-38.
1965a Relations alimentaires entre *Malacocoris chlorizans* Pz. (Hemipt. Heterop. Miridae) et *Scaphoideus littoralis* Ball (Hemipt. Homopt. Jassidae) sur les *Vitis* du sud-ouest de la France. [Alimentary relations between *M. chlorizans* and *S. littoralis* on *Vitis* species in the south-west of France.] *Revue de Zoologie Agricole Applications* 64(7-9):72-78.
1965b Quelques precisions sur les modalites de lutte chimique contre pagent vecteur de la flavescence doree de la vigne. [Some particulars of methods of chemical control of the vector of golden flavescence of vines.] *Revue de Zoologie Agricole Applications* 64(10-12):100-105.
1967a A mite (Erythraeid) parasite of cicadellids. *Revue de Zoologie Agricole Applications* 66(1-3):16-19.

Carle, P. and Amargier, A.
1965a Etude anatomique et histologique des organes internes de *Scaphoideus littoralis* Ball. (Homopt. Jassidae), vecteur du virus de la flavescence doree de la vigne. *Annales des Epiphyties* (Paris) 16(4):355-382.

Carle, P. and Moutous, G.
1966a Observations sur le mode de nutrition sur vigne de quartre especes de cicadelles. [Observations on the method of feeding on grape vine of four species of leafhoppers.] *Annales des Epiphyties* (Paris) 16(4):333-354. [English summary.]
1967a Recherches sur d'eventuels vecteurs de la flavescence doree. *Annales des Epiphyties* (Paris) 18:151-156. [With English and German summaries.]

Carle, P. and Schvester, D.
1964a Nouvelle mise au point sur la lutte contre *Scaphoideus littoralis* Ball, cicadelle vectrice de la flavescence doree de la vigne. *Revue de Zoologie Agricale et de Pathologie Vegetale* 63:107-114.

Carlson, O. V.
1967a Mating and oviposition of *Empoasca fabae* (Harris) (Cicadellidae: Homoptera). Ph. D. Dissertation, Iowa State University. 64 pp. *Dissertation Abstracts International* B 28(9):3737.

Carlson, O. V. and Hibbs, E. T.
1962a Direct counts of potato leafhopper, *Empoasca fabae*, eggs in *Solanum* leaves. *Annals of the Entomological Society of America* 55(5):512-515.
1970a Oviposition by *Empoasca fabae* (Homoptera: Cicadellidae). *Annals of the Entomological Society of America* 63(2):516-519.

Carnegie, A. J. M.
1976a Sugarcane planthoppers and leafhoppers in Perak, Malaysia. *Entomology Newsletter, International Society of Sugarcane Technologists* 3:4.

Carrillo, S. J. L., A. Ortega C., and Gibson, W. W.
1966a Lista de insectos en la coleccion entomologica del Instituto Nacional de Investigaciones Agricolas. [List of insects in the collection of the National Institute of Agricultural Research.] *Mexico, Institute Nacional Investigacione Agricola Folia Miscellanea* 14. 133 pp.

Carter, W.
1961a Ecological aspects of plant virus transmission. *Annual Review of Entomology* 6:347-370.
1962a *Insects in relation to plant diseases.* Interscience Publishers, New York and London. 704 pp.
1973a *Insects in relation to plant diseases.* 2nd ed. Wiley & Sons, New York. 759 pp.

Casale, A.
1981a Cataloghi. II - Colleqione Emitterologica di Massimiliano Spinola. *Museo Regionale di Scienze Naturali.* 120 pp.

Castro, G. D.
1971a Combate quimico de la chicharrita del frijol, *Empoasca fabae* Harris (Homoptera: Cicadellidae), en El Bajio. [Chemical control of the leafhopper *Empoasca fabae* Harris (Homoptera: Cicadellidae).] *Agricultura Tecnico en Mexico* 3(3):93-94.

Catara, A.
1984a Viruses, virus-like organisms and mycoplasmas of citrus in Italy. *Informatore Fitopatologico* 34:15-35.

Cathey, H. M., Smith, F. F., Campbell, J. G., Hartsock, J. G. and McGuire, J. U.
1975a Response of *Acer rubrum* L. to supplemental lighting, reflective aluminum soil mulch, and systemic soil insecticide. *Journal of the American Society of Horticultural Science* 100(3):234-237.

Cattaneo, E. and Arzone, A.
1983a Ciclo biologico di cicadellidi deltocephalini Vettori di MLO. *Atti XIII Congresso Nazionale Italiano di Entomologia* 399-406.

Caudwell, A.
1977a Statistical aspects of infectivity tests for yellows diseases of plants and for viruses transmitted in a persistent manner. Value of broad bean (*Vicia faba*) as a test plant for yellows. *Annales de Phytpathologie* 9:141-159.
1981a The golden flavescence disease of the vine in France. *Phytoma* No. 325, pp. 16-19.

Caudwell, A., Bachelier, J. C., Kuszala, C. and Larrue, J.
1969a Etude de la survie de la cicadelle *Scaphoideus littoralis* Ball sur les plantes herbacees, et utilisation de ces donnees pour transmettre la flavescence doree de la vigne a d'austres especes vegetales. *Compte Rendu Hebdomadaire des Seances de l'Academie des Sciences* 269D:101-103.
1977a Un appareil permettant d'immobiliser les insectes a injecter dans les epreuves d'infectivite des jaunisses. *Annales de Phytopathologie* 9:521-523.

Caudwell, A., Giannotti, J., Kuszala, C. and Larrue, J.
1971a Etude du role de particules de type "mycoplasme" dans l'etiologie de la flavescence doree de la vigne. Examen cytologique des plants malades et de cicadelles infectieuses. *Annales de Phytopathologie* 3:107-123.

Caudwell, A., Kuszala, C., Bachleir, J. C. and Larrue, J.
1970a Transmission de la flavescence doree de la vigne aux plantes herbacees par l'allongement du temps d'utilisation de la cicadelle *Scaphoideus littoralis* Ball et l'etude de sa survie sur un grand nombre d'especes vegetales. *Annales de Phytopathologie* 2(2):415-428.

Caudwell, A, Kuszala, C. and Larrue, J.
1974a Sur la culture in vitro des agents infectieux responsables des jaunisses des plantes (MLO). *Annales de Phytopathologie* 6(2):173-190. [In French with English summary.]

Caudwell, A., Kuszala, C., Larrue, J. and Bacheilier, J. C.
1972a Transmission de la flavescence doree a la feve par des cicadelles des genres *Euscelis* et *Euscelidius* intervention possible de ces insectes dans l'epidemiologies du bois noir en Bourgogne. *Annales de Phytopathologie* 4:181-189.

Caudwell, A. and Larrue, J.
1977a The production of healthy and infective cicadellids for tests of the infectivity of plant yellows caused by mollicutes. The rearing of *Euscelidius variegatus* Kbm. and oviposition on polyurethane foam. *Annales de Zoologie-Ecologie Animale* 9:443-456.

Caudwell, A., Larrue, J. and Kuszala, C.
1978a The loss of infectivity of the leafhopper vectors infected from a long time by yellow type diseases, as a symptom for plants. *Zentralblatt fuer Bakteriologie, Parasitenkunde, Infektionskrankheiten und Hygiene. Este Abteilung Orginale. Reihe A. Medizinische Mikrobiologie und Parasitologie* 241:228.

Caudwell, A., Larrue, J., Moutous, G., Fos, A. and Brun, P.
1978a Transmission by cicadellids of Corsican vine yellowing. Identification of this disease with golden flavescence L.—tests carried out outside Corsica. *Annales de Zoologie-Ecologie Animale* 10(4):613-625.

Caudwell, A., Schvester, D. and Moutous, G.
1972a Variete des degats des cicadelles nuisibles a la vigne—les methodes de lutte. *Progres Agricole et Viticole* 172(24):583-590. [In French.]
1973a Variete des degats des cicadelles nuisibles a la vigne—les methodes de lutte. *Progres Agricole et Viticole* 173(1):8-16. [In French.]

Cavalcante, M. L. S., Cavalcante, R. D. and Castro, Z. B. de
1975a The green leafhopper (*Empoasca* sp.) a pest of cowpea (*Vigna sinensis,* Endl.) in Ceara. *Fitossandidade* 1:83-84.

Cavichioli, R. R.
1984a Nova especie de *Microgoniella* Melichar (Homoptera, Cicadellidae). *Dusenia* 14(1):37-39.

Cavichioli, R. R. and Sakakibara, A. M.
1984a As especies do genero *Sonesimia* Young com descricas duas novas (Homoptera-Cicadellidae). *Revista Brasileira de Entomologia* 28(1):29-38. [In Portuguese with English summary.]

Ceballos, B. I.
1961a Sobre la investigacion de los insectos que atacan al cacao en el Cuzco. *Revista Peruana de Entomologia* 4(1):77-78.

Cendana, S. M. and Calora, F. B.
1967a Insect pests of rice in the Philippines. *In: Pathak, M. D., ed. The Major Insect Pests of the Rice Plant.* Pp. 591-616. Johns Hopkins, Baltimore.

Chaboussou, F.
1971a Le conditionnement physiologique de la vigne et la multiplication des Cicadelles. *Revue de Zoologie Agricole et de Pathologie Vegetale* 70(3):1-66. [In French with English summary.]

Chaduneli, M. D. and Chkheidze, N. Z.
1974a Time necessary for inoculating mulberry with the causal agent of curly dwarf-disease in different methods of inoculation. *Trudy Nauchno-Issledovatal'skogo Zashchity Rastenii Grez SSR* 26:129-130.

Chakravarthy, A. K. and Rao, P. K. A.
1985a Dispersion patterns, sample unit-sizes and techniques for sampling cotton jassid *Amrasca biguttula biguttula* (Ishida) and whitefly *Bemisia tabaci* (Genn.) *Insect Science and Its Application* 6:661-665.

Chakravarthy, A. K., Sidhu, A. S. and Singh, Joginder
1985a Effect of plant phenology and related factors on insect pest infestations in *arboreum* and *hirsutum* cotton varieties. *Insect Science and Its Application* 6:521-532.

Chakravarti, S., Ghosh, A. B. and Mukhopadhyay, S.
1979a Biology of the green leafhopper, *Nephotettix virescens. International Rice Research Newsletter* 4(1):16-17.

Chalfant, R. B.
1965a Resistance of bunch bean varieties to the potato leafhopper and relationship between resistance and chemical control. *Journal of Economic Entomology* 58(4):681-682.

Chand, P.
1984a Management of insect pests in rainfed rice production system. *Oryza* 21:91-94.

Chandler, F. B., Wilcox, R. B., Bain, H. F., Bergman, H. F., and Dermen, H.
1947a Cranberry breeding investigations of the U.S. Department of Agriculture. *Cranberries* 12(2):6-10.

Chandra, G.
1978a A new cage for rearing hopper parasites. *International Rice Research Newsletter* 3(1):12.

1980a Dryinid parasitoids of rice leafhoppers and planthoppers in the Philippines. I. Taxonomy and bionomics. *Acta Oecologica, Series Oecologia Applicata* 1(2):161-172.

1980b Dryinid parasitoids of rice leafhoppers and planthoppers in the Philippines. II. Rearing techniques. *Entomophaga* 25:211-222.

1980c Taxonomy and bionomics of insect parasites of rice leafhoppers and planthoppers in the Philippines and their importance in natural biological control. *Philippine Entomologist* 4:119-139.

Chandramohan, N. and Chelliah, S.

1984a Reaction of yellow stem borer (YSB) resistant accessions to other rice pests. *International Rice Research Newsletter* 9(6):8.

Chandramohan, N. and Kumaraswami, T.

1978a Efficacy of certain dust formulations against the blue jassid, *Zygina maculifrons* (Motsch) on rice. *Pesticides* 12:39.

Chandy, K. C.

1957a Some observations on the incidence and habits of *Pruthiana sexnotata*, Izzard [sic!]. *The Madras Agricultural Journal* 44(8):343-344.

Chang, K. P. and Musgrave, A. J.

1972a Multiple symbiosis in a leafhopper, *Helochara communis* Fitch (Cicadellidae: Homoptera): Envelopes, nucleoids and inclusions of the symbiotes. *Journal of Cell Science* 11(1):275-293.

1975a Endosymbiosis in a leafhopper, *Helochara communis* Fitch (Cicadellidae: Homoptera): Symbiote translocation and auxillary cells in the mycetome. *Canadian Journal of Microbiology* 21(2):186-195.

1975b Conversion of spheroblast symbiotes in a leafhopper, *Helochara communis* Fitch (Cicadellidae: Homoptera). *Canadian Journal of Microbiology* 21(2):196-204.

Chang, S. J. and Oka, H. E.

1984a Attributes of a hopper-predator community in a rice field. *Agriculture, Ecosystems and Environment* 12:73-78.

Chang, T. T., Ou, S. H., Pathak, M. D. and Ling, K. C.

1975a The search for disease and insect resistance in rice germplasm. In: Frankel, O. H. and Hawkes, J. G., eds. *Crop Genetic Resources for Today and Tomorrow.* Pp. 183-200. Cambridge, U.K.

Chang, V.

1974a Leafhopper feeding behavior. *Hawaiian Sugar Planters Association Experiment Station Annual Report* 1974:46.

1975a Leafhopper feeding behavior. *Hawaiian Sugar Planters Association Experiment Station Annual Report* 1975:48-49.

Chang, Y. D.

1980a Egg parasitism of green rice leafhopper, *Nephotettix cincticeps* Uhler by *Gonatocerus* sp. and *Paracentrobia andoi* in southern rice cultural areas. *Korean Journal of Plant Protection* 19:109-112.

Chang, Y. D. and Choe, K. R.

1982a Studies on the insect fauna of Mt. Gyeryong (I). *Research Report, Agricultural Science and Technology, Chungnam National University, Korea* 9:519-539. [In Korean with English summary.]

[Chanoki, N., et al.]

1978a Investigation on the control of mulberry dwarf disease—effect of eradication of leafhopper vector on the disease occurrence in mulberry fields. *Mizazaki Agriculture and Experiment Station*, p. 41. [In Japanese.]

Chapman, J. A. and Kingborn, J. M.

1955a Window flight traps for insects. *Canadian Entomologist* 87(1):46-47.

Chapman, R. F.

1973a Integrated control of aster yellows. *Proceedings of the North Central Branch of the Entomological Society of America* 28:71-92.

Charbonneau, J., Hawthorne, D., Ghiorse, W. C. and Vandemark, P. J.

1979a Isolation of a spiroplasma-like organism from aster yellow infective leafhoppers. *Proceedings of the American Society of Microbiologists* 79:86.

Chatelain, L.

1956a Un nouveau ravageur des orges et des avoines de printemps en Champagne. *Phytoma* 82:31.

Chattopadhyay, K. and Mukhopadhyay, S.

1975a Relative feeding of *Nephotettix* spp. (*N. virescens* and *N. nigropictus*) in artificial media containing leaf sap from different varieties of rice. *Science and Culture* 41:315-316.

1977a Effect of temperature upon the survival of *Nephotettix* spp. *Indian Journal of Entomology* 38:295-296.

Chaudhary, J. P., Yadav, L. S., Poonia, R. S. and Rastogi, K. B.

1980a Some observations on field population of *Empoasca kerri* Pruthi, a jassid pest on mung bean crop in Haryana. *Haryana Agricultural University Journal of Research* 10:250-252.

Chaudhuri, R. P.

1955a Some aspects of insect transmission of plant viruses. *Indian Journal of Entomology* 17(1-3):40-48.

Chelliah, S., Hanifa, A. M., Heinrichs, E. A. and Khush, G. S.

1981a Resistance of rice varieties to the green leafhopper in southern India. *International Rice Research Newsletter* 6(6):8-9.

Chen, C. C.

1969a Studies on the condition of oviposition of rice green leafhoppers. *Plant Protection Bulletin* (Taiwan) 11(2):83-89. [In Chinese with English summary.]

1970a Ecological studies of *Nephotettix impicticeps* Ishihara in Taiwan. *Plant Protection Bulletin* (Taiwan) 12(2):79-90. [In Chinese with English summary.]

1970b Comparative transmission of rice yellow dwarf by three *Nephotettix* leafhoppers in Taiwan. *Plant Protection Bulletin* (Taiwan) 12(4):160-165. [In Chinese with English summary.]

1972a The distribution of *Nephotettix* leafhoppers in Taiwan. *Plant Protection Bulletin* (Taiwan) 14(1):41-45, illust. [Chinese with English summary.]

1975a Studies on the varietal resistance of rice plants to yellow dwarf. I. Factors affecting the varietal resistance test method. *Plant Protection Bulletin* (Taiwan) 17:263-271.

1978a The occurrence and control of homopterous insect vectors of virus diseases on rice in Taiwan. In: Su, C. C., Yen, D. G. and Lin, F. C., eds. *Insect Ecology and Control.* Pp. 113-122. Academia Sinica, Taiwan. 279 pp.

1979a Varietal resistance to yellow dwarf in rice. *Plant Protection Bulletin* (Taiwan) 21:153-160.

Chen, C. C. and Chiu, R. J.

1980a Factors affecting transmission of rice transitory yellowing virus by green leafhoppers. *Plant Protection Bulletin* (Taiwan) 22:297-306.

Chen, C. C. and Ko, W. H.

1978a Survey on the age distribution of rice green leafhoppers and remaining RYD-diseased plants during overwintering period in Tai-chung area. *Plant Protection Bulletin* (Taiwan) 20:1-7.

Chen, C. C., Ko, W. H., Wang, E. S., Yu, S. M. and Hu, D-q

1980a Epidemiological studies on the transitory yellowing with special reference to its transmission by the rice green leafhoppers. *Bulletin Taichung District of Agricultural Improvement Station*, New Series, No. 4.

Chen, C. T.

1973a Insect transmission of sugarcane white leaf disease by single leafhoppers, *Matsumuratettix hiroglyphicus* (Matsumura). *Report of the Taiwan Sugar Research Institute* No. 60:25-33. [In Chinese with English summary.]

1979a Vector-pathogen relationships of sugarcane white leaf disease. *Plant Protection Bulletin* (Taiwan) 21:105-110.

Chen, C. T., Lee, C. S. and Lee, S. M.
1975a Beneficial effects of white leaf infected plants on the leafhopper, *Matsumuratettix hiroglyphicus* Matsumura. In: *Proceedings of the International Society of Sugar Cane Technologists,* Durban, South Africa. Pp. 434-438.

Chen, C. Y., Chiang, C. L., Lin, H., Tsou, B. S. and Tang, C. H.
1978a Studies on the insecticide resistance and synergism in organophosphorus-resistant green leafhopper *Nephotettix cincticeps. Acta Entomologica Sinica* 21:360-368.

Chen, H. T., Lieu, T. L., Ming-Chun-Kao, and Hu, C. C.
1978a Ecological studies on the tea leafhopper, *Empoasca formosana* Paoli and its control. *Plant Protection Bulletin* (Taiwan) 20:93-105.

Chen, J.
1964a The morphological feature of the valvulae of ten common leafhoppers. *Acta Entomologica Sinica* 13:632-636. [In Chinese.]

Chen, L. C.
1975a Studies on the inheritance of resistance to brown planthopper and green leafhopper in rice. *Bulletin of the Taiwan Agricultural Research Institute* No. 32, pp. 17-21.

Chen, M. H. and Sogawa, K.
1969a Three different species of green rice leafhoppers in Taiwan. *Plant Protection Bulletin* (Taiwan) 11(3):109-114. [In Chinese with English summary.]

Chen, M. J. and Shikata, E.
1972a Electron microscopy and recovery of rice transitory yellowing virus from its leafhopper vector, *Nephotettix cincticeps. Virology* 47:483-486.

Chen, S. X.
1985a Verification of an equation for predicting rice transitory yellowing and rice dwarf diseases. *Acta Phytopathologica Sinica* 15:19-24.

Chen, T. A. and Liao, C. H.
1975a Corn stunt spiroplasma: Isolation, cultivation, and proof of pathogenicity. *Science* 188:1015-1017.

Cheng, C. H.
1976a Assessment of rice losses caused by the brown planthopper and the rice green leafhopper. *Plant Protection Bulletin* (Taiwan) 18:147-160.
1979a The occurrence of insect pests and its damage to the yield of rice in the first and second crop seasons in Taiwan. In: Hsieh, S. C. and Liu, D. J., eds. *The Causes of Low Yield of the Second Crop Rice in Taiwan and the Measures for Improvement.* Pp. 191-205. Proceedings of a symposium held at Taiwan Agricultural Research Institute, Tapei, Taiwan, 1978.

Cheng, C. H. and Pathak, M. D.
1971a Bionomics of the rice green leafhopper *Nephotettix impicticeps* Ishihara. *Philippine Entomologist* 2(1):67-74.
1972a Resistance to *Nephotettix virescens* in rice varieties. *Journal of Economic Entomology* 65:1148-1153.

Cheng, L. and Birch, M. C.
1978a Insect flotsam: An unstudied marine resource. *Ecological Entomology* 3:87-97.

Cheng, Y.-J.
1980a New leafhopper taxa (Homoptera: Cicadellidae: Deltocephalinae) from Paraguay. *Journal of the Kansas Entomological Society* 53(1):61-111. [New genera: *Limpica* and *Aplanatus.*]

Chenon, R. D. de
1979a Demonstration of the role of *Recilia mica* Kramer (Homoptera, Cicadellidae, Deltocephalinae) in blast disease of oil palm nurseries in the Ivory Coast. *Oleagineux* 34:107-115.

Cherry, R. H., Wood, K. A. and Ruesink, W. G.
1977a Emergence trap and sweep net sampling for adults of the potato leafhopper from alfalfa. *Journal of Economic Entomology* 70(3):279-282.

Chettanachit, D. and Disthaporn, S.
1982a Life span and tungro virus transmission of viruliferous *Nephotettix virescens. International Rice Research Newsletter* 7(2):10.

Chhabra, K. S. and Kooner, B. S.
1981a Field resistance in black gram, *Vigna mungo* L., against insect-pests complex and yellow mosaic virus. *Indian Journal of Entomology* 43:288-293.

Chhabra, K. S., Kooner, B. S. and Saxena, A. R.
1984a Influence of biochemical components on the incidence of insect pests and yellow mosaic virus in black gram. *Indian Journal of Entomology* 46:148-156.

Chhabra, K. S., Kooner, B. S., Saxena, A. K. and Sharma, A. K.
1981a Effect of biochemical components on the incidence of insect pest complex and yellow mosaic virus in mungbean. *Crop Improvement* 8:56-59.

Chhabra, K. S., Sajjan, S. S. and Singh, Jaswant
1976a Light trap catches at the rice station at Kapurthala, Punjab, India. *Rice Entomology Newsletter* 4:38.

Chiang, H. C.
1977a Pest management in the People's Republic of China—monitoring and forecasting insect populations in rice, wheat, cotton and maize. *FAO Plant Protection Bulletin* 25:1-8.

Chiba, M.
1970a DDT residues in fruit, foliage, and soil of a vineyard following a standard insect control program. *Canadian Journal of Plant Science* 50(3):219-227.

Chiko, A. W.
1973a Failure to transmit barley stripe mosaic virus by aphids, leafhoppers, and grasshoppers. *Plant Disease Reporter* 57(8):639-641.

Childers, W. R. and Dickson, W. D.
1980a Bytown red clover. *Canadian Journal of Plant Science* 60:1041-1043.

Childress, S. A.
1980a The fate of maize chlorotic dwarf virus (MCDV) in the black-faced leafhopper, *Graminella nigrifrons* (Forbes) (Homoptera: Cicadellidae). Doctoral Dissertation, Texas A & M University, College Station. 84 pp.
1981a The fate of maize chlorotic dwarf virus (MCDV) in the black-faced leafhopper, *Graminella nigrifrons* (Forbes) (Homoptera: Cicadellidae). *Disseration Abstracts International* (B) 41(10):3680.

China, W. E.
1955b Hemiptera of the Island of Tromelin. *Naturaliste Malgache* 7(1):13-18.
1957a Corrigendum. Comunicaciones. *Revista Chilena de Entomologia* 5:465. [*Evansiola,* new name for *Evansiella* China 1955, not *Evansiella* Hayward 1948.]

Ching-Chung Chen
1970a Comparative transmission of rice yellow dwarf by three *Nephotettix* leafhoppers in Taiwan. *Plant Protection Bulletin* (Taiwan) 12(4):160-165. [In Chinese with English summary.]

Chiswell, J. R.
1964a Observations on the life history of some leafhoppers (Homoptera: Cicadellidae) occurring on apple trees and their control with insecticides. *Journal of Horticultural Science* 39(1):9-23.

Chiu, R. J.
1982a Virus and viruslike disease of rice in Taiwan with special reference to rice transitory yellowing. *Plant Protection Bulletin* (Taiwan) 24:207-224.

Chiu, R. J. and Black, L. M.
1967a Monolayer cultures of insect cells lines and their inoculation with a plant virus. *Nature* (London) 215:1076-1078.

Chiu, R. J. and Jean, J. H.
1967a Leafhopper transmission of transitory yellowing of rice. In: *Proceedings of the Symposium at the International Rice Research Institute on Virus Diseases of Rice Plants.* Pp. 131-137. John Hopkins, Maryland.

Chiu, R. J., Jean, J. H., Chen, M. H. and Lo, T. C.
1968a Transmission of transitory yellowing virus of rice (*Oryza sativa*) by two leafhoppers. *Phytopathology* 58(6):740-745.

Chiu, R. J., Liu, H. Y., Macleod, R. and Black, L. M.
1970a Potato yellow dwarf virus in leafhopper cell culture. *Virology* 40(2):387-396.

Chiu, R. J., Lo, T. C., Pi, C. L. and Chen, M. H.
1965a Transitory yellowing of rice and its transmission by the leafhopper *Nephotettix apicalis apicalis* (Motsch.). *Botanical Bulletin of the Academia Sinica* (Taipei) 6(1):1-18.

Chiu, R. J., Reddy, D. V. R. and Black, L. M.
1966a Inoculation and infection of leafhopper (*Agallia constricta*) tissue cultures with a plant virus (wound tumor). *Virology* 30(3):562-566.

Chiu, S. C.
1958a Bibliography of Entomology in Taiwan (1684-1957). *Special Publication No. 1. Taiwan Agricultural Research Institute, Taipei, Taiwan, China.* 246 pp. [In Chinese.]

Chiu, S. C. and Cheng, C. H.
1976a Toxicity of some insecticides commonly used for rice insects control to the predators of rice-hoppers. *Plant Protection Bulletin* (Taiwan) 18:254-260.

Chiu, S. C., Chu, Y. I. and Lung, Y. H.
1974a The life history and some bionomic notes on a spider, *Oedothorax insecticeps* Boes et St. (Micryphantidae: Araneae). *Plant Protection Bulletin* (Taiwan) 16:153-161.

Chiu, S. M., Lin, M. H. and Huang, C. S.
1968a A screening test for rice varieties resistant to yellow dwarf disease. *Journal of Taiwan Agricultural Research* 17(4):19-23.

Chiykowski, L. N.
1958a Studies on migration and control of the six-spotted leafhopper, *Macrosteles fascifrons* (Stal), in relation to transmission of aster yellows virus. Ph. D. Thesis, University of Wisconsin, Madison. 134 pp.

1961a Transmission of clover phyllody virus by *Aphrodes bicinctus* (Schrank) in North America. *Nature* 192:581.

1962a Clover phyllody virus in Canada and its transmission. *Canadian Journal of Botany* 40:397-404.

1962b *Scaphytopius acutus* (Say), a newly discovered vector of celery infecting aster-yellows virus. *Canadian Journal of Botany* 40:799-801.

1962c Clover phyllody and strawberry green petal diseases, caused by the same virus in eastern Canada. *Canadian Journal of Botany* 40:1615-1617.

1963a *Endria inimica* (Say), a new leafhopper vector of a celery-infecting strain of aster-yellows virus in barley and wheat. *Canadian Journal of Botany* 41(5):669-673.

1964a Current Canadian research on leafhoppers in relation to virus diseases. *Extrait de Phytoprotection* 45:108-116.

1965a Transmission of clover phyllody virus by the leafhopper, *Paraphlepsius irroratus* (Say). *Canadian Entomologist* 97(11):1171-1173.

1965b The reaction of barley varieties to aster-yellows virus. *Canadian Journal of Botany* 43:373-378.

1965c A yellows-type virus of alsike clover in Alberta. *Canadian Journal of Botany* 43:527-536.

1967a Some factors affecting the acquisition of clover phyllody virus by the aster leafhopper. *Journal of Economic Entomology* 60(3):849-853.

1967b Some host plants of a Canadian isolate of the phyllody virus. *Canadian Journal of Plant Science* 47(2):141-148.

1967c Reaction of some wheat varieties to aster yellows. *Canadian Journal of Plant Science* 47(2):149-151.

1969a A leafhopper transmitted clover disease in the Ottawa area. *Canadian Plant Disease Survey* 49:16-19.

1970a Notes on the biology of the leafhopper *Aphrodes bicincta* (Homoptera: Cicadellidae) in the Ottawa area. *Canadian Entomologist* 102(6):750-758.

1973a The aster yellows complex in North America. *Proceedings of the North Central Branch of the Entomological Society of America* 28:60-66.

1973b Factors affecting the infection of plants with clover phyllody agent transmitted by *Macrosteles fascifrons*. *Annals of the Entomological Society of America* 66(5):987-990.

1974a Yellows diseases and vectors. *Colloque Inserm* (Institut National de al Sante et de la Recherche Medicale) 33:291-298.

1974b Additional host plants of clover phyllody in Canada. *Canadian Journal of Plant Science* 54(4):755-763. [In English with French summary.]

1975a *Aphrodes bicincta* as a vector of the clover phyllody agent. *Annals of the Entomological Society of America* 68:645-648.

1976a Transmission characteristics and host range of the clover yellow edge agent. *Canadian Journal of Botany* 54:1171-1179.

1977a Reduction in the transmissibility of a greenhouse-maintained isolate of aster yellows agent. *Canadian Journal of Botany* 55:1783-1786.

1977b Transmission of a celery-infecting strain of aster-yellows by the leafhopper, *Aphrodes bicinctus*. *Phytopathology* 67:522-524.

1979a *Athysanus argentarius*, an introduced European leafhopper, as a vector of aster yellows in North America. *Canadian Journal of Plant Pathology* 1:37-41.

1981a Epidemiology of disease caused by leafhopper-borne pathogens. In: Maramorosch, K. and Harris, K. F., eds. *Plant Diseases and Vectors: Ecology and Epidemiology.* Pp. 105-159. Academic Press, New York, 368 pp.

1983a Frozen leafhoppers as a vehicle for long-term storage of different isolates of the aster yellows agent. *Canadian Journal of Plant Pathology* 5:101-106.

1985a Biology and rearing of *Paraphlepsius irroratus* (Homoptera: cicadellidae), a vector of peach X-disease. *Canadian Entomologist* 117(6):717-726.

Chiykowski, L. N. and Chapman, R. K.
1965a Part 2. Migration of the six-spotted leafhopper in Central North America. *University of Wisconsin Research Bulletin* 261:21-45.

Chiykowski, L. N. and Craig, D. L.
1978a Plant and insect age as factors in the transmission of clover phyllody (green petal) agent to strawberry by *Aphrodes bicinctus*. *Canadian Journal of Plant Science* 58:467-470.

Chiykowski, L. N. and Hamilton, K. G. A.
1985a *Elymana sulphurella* (Zetterstedt): Biology, taxonomy, and relatives in North America (Rhynchota: Homoptera: Cicadellidae). *Canadian Entomologist* 117:1545-1558.

Chiykowski, L. N. and Sinha, R. C.
1969a Comparative efficiency of transmission of aster yellows by *Elymana virescens* and *Macrosteles fascifrons* and the relative concentration of the causal agent in the vectors. *Journal of Economic Entomology* 62(4):883-886.

1970a Sex and age of *Macrosteles fascifrons* in relation to the transmission of the clover proliferation causal agent. *Annals of the Entomological Society of America* 63(6):1614-1617.

Chiykowski, L. N. and Wolynetz, M. S.
1981a Susceptibility of oat cultivars to aster yellows isolates from eastern Canada. *Canadian Journal of Plant Pathology* 3:53-57.

Choe, K. R.
1980a Three new species of *Pagaronia* Ball (Homoptera, Cicadellidae) from Korea. *Korean Journal of Plant Protection* 19:149-152.

1981a Three new species of leafhoppers (Cicadellidae, Homoptera) from Korea. *Korean Journal of Plant Protection* 20(3):151-154.

1985a Insect fauna of Taesan-myon, Sosan-gun, Chungnam-Province, Korea. *Research Report, Environmental Science and Technology, Chungnam University, Korea* 3:102-126. [In Korean with English summary.]

Choi, S. Y.
1973a Effects of seed treatment with several systemic insecticides to rice, barley and soybean. *Korean Journal of Plant Protection* 12(3):115-120. [In Korean with English summary.]

1975a Varietal resistance of rice to the green rice leafhopper, *Nephotettix cincticeps* Uhler. *Korean Journal of Plant Protection* 14:13-21.

1976a Application of insecticides in the root zone of rice plants. *Korean Journal of Plant Protection* 15:60.

Choi, S. Y., Heu, M. H., Chung, K. Y., Kang, Y. S. and Kim, H. K.
1975a Root-zone application of insecticides in gelatin capsules for the control of rice insect pests. *Korean Journal of Plant Protection* 14:147-153.

Choi, S. Y. and Lee, H. R.
1976a Selective toxicity of insecticides to plant- and leaf-hoppers. *Korean Journal of Plant Protection* 15:1-6.
1976b Host preference by the small brown planthopper and green rice leafhopper on barley and water foxtail (I). *Korean Journal of Plant Protection* 15:178-184.

Choi, S. Y., Lee, J. O., Lee, H. R. and Park, J. S.
1976a Resistance of the new varieties Milyang 21 and 23 to plant- and leaf-hoppers. *Korean Journal of Plant Protection* 15:147-151.

Choi, S. Y., Song, Y. H. and Park, J. S.
1973a Studies on the varietal resistance of rice to the zigzag-striped leafhopper, *Recilia (Inazuma) dorsalis* Motschulsky (II). *Korean Journal of Plant Protection* 12(2):83-87. [In Korean with English summary.]

Choi, S. Y., Song, Y. H. Park, J. S. and Son, B. I.
1973a Studies on the varietal resistance of rice to the green rice leafhopper, *Nephotettix cincticeps* Uhler. *Korean Journal of Plant Protection* 12(1):47-53 [In Korean with English summary.]

Chong, M.
1965a *Circulifer tenellus* (Baker). Species found on the island of Kawaihae, Hawaii: New island record. *Proceedings of the Hawaiian Entomological Society* 19(1):15.

Chopra, N. P.
1973a Cotton jassid: A nomenclatural correction. *Entomologist's Record and Journal of Variation* 85:88-89.

Chou, I. and Ma, N.
1981a On some new species and new records of Typhlocybinae from China (Homoptera: Cicadellidae). *Entomotaxonomia* 3(3):191-210.

Chou, I. and Zhang, Y.
1985a On the tribe Zyginellini from China (Homoptera, Cicadellidae, Typhlocybinae). *Entomotaxonomia* 7(4):287-300. [In Chinese and English.] [New genera: *Dworakowskaia* and *Parazyginella*.]

Chou, M., Chung, C. and Wei, H.
1953a *Compendium of Agricultural Pests in North China.* Chung-hua Book Co., Shanghai. Ed. 1. 274 pp. [In Chinese.]

Chou, T.-g, Yang, S.-j, Huang, P.-y and Chung, S.-j
1975a Studies on loofah (*Luffa cylindrica* Roem.) witches' broom in Taiwan (3) the leafhopper *Hishimonus concavus* Knight as a vector. *Plant Protection Bulletin* (Taiwan) 17(4):384-389.

Chou, W. D. and Cheng, C. H.
1971a Field reactions of rice varieties screened for their resistance to *Nilaparvata lugens* and *Nephotettix cincticeps* in insectary. *Journal of Taiwan Agricultural Research* 20(2):68-75. [In Chinese with English summary.]

Choudhury, M. M. and Rosenkranz, E.
1973a Differential transmission of Mississippi and Ohio corn stunt agent by *Graminella nigrifrons*. *Phytopathology* 63(1):127-133.
1983a Vector relationship of *Graminella nigrifrons* to maize chlorotic dwarf virus. *Phytopathology* 73(5):685-690.

Choudhury, R., Manjit, Singh and Anand, S. K.
1983a Variation in leafhopper burn incidence in *Solanum tuberosum* and *audigena* collections. *Indian Journal of Entomology* 45:493-497.

Chowdhury, M. M. A. and Alam, S.
1979a Effects of diazinon spray on rice pests and their natural enemies. *Bangladesh Journal of Zoology* 7:15-20.

Christensen, C.
1982a Potato leafhopper on new stands of alfalfa. *Pest News Alert* No. 245. University of Kentucky Cooperative Extension Service.

Christian, P. J.
1956a North American species of the genus *Eupteryx* (Homoptera: Cicadellidae). *Transactions of the Kentucky Academy of Science* 17:42-56.
1960a A new North American leafhopper previously confused with *Typhlocyba andromache* McAtee (Homoptera: Cicadellidae). *Transactions of the Kentucky Academy of Science* 21(3&4):73-76.

Chrzanowski, J.
1955a The occurrence of a new beet pest (*Empoasca pteridis*) in Poland. *Proceedings of the Conference on Sugar and Beet in Prague*, 14-19 October 1955. Pp. 515-520.

Chu, H. F., Han, Y. F. and Wang, L. Y.
1961a The populations of wheat insects under different cultural conditions of wheat. *Acta Entomologica Sinica* 10(4-6):411-424. [In Chinese with English summary.)

Chu, H. F. and Teng, K. F.
1950a Life history of the leafhopper, *Cicadella viridis* (L.). (Homoptera: Cicadellidae). *Acta Entomologica Sinica* 1(1):16-40.

Chu, Y. I. and Hirashima Y.
1981a Survey of Taiwanese literature on the natural enemies of rice leafhoppers and planthoppers. *Esakia* 16:33-37.

Chu, Y. I., Ho, C. C. and Chen, B. J.
1975a Relative toxicity of some insecticides to green rice leafhopper, brown planthopper and their predator *Lycosa pseudoannulata*. *Plant Protection Bulletin* (Taiwan) 17:424-430.
1976a The effect of BMPC and Unden on the predation of Lycosa spider (*Lycosa pseudoannulata*). *Plant Protection Bulletin* (Taiwan) 18:42-57.

Chu, Y. I. and Liou, R. F.
1981a The comparison of feeding marks and honeydew and excretion of green rice leafhopper (*Nephotettix cincticeps* Uhler) on various graminaceous plants (Deltocephalidae: Homoptera). *Plant Protection Bulletin* (Taiwan) 23:243-253.

Chu, Y. I., Liou, R. F. and Mu, T.
1981a Evaluation of graminaceous plants as overwintering host plants of green rice leafhopper *Nephotettix cincticeps* Uhler (Deltocephalidae: Homoptera). *Plant Protection Bulletin* (Taiwan) 23:235-242.

Chu, Y. I., Lin, D. S. and Mu, T.
1976a The effect of Padan, Ofunack and Sumithion of the feeding amount of *Lycosa pseudoannulata* (Boes. et Str.) and *Oedthorax insecticeps* Boes. et Str. (Lycosidae and Microphantidae, Arachnida). *Plant Protection Bulletin* (Taiwan) 18:377-390.
1976b Relative toxicity of 9 insecticides against rice insect pests and their predators. *Plant Protection Bulletin* (Taiwan) 18:369-376.
1977a Relative toxicity of 5 insecticides against insect pests of rice and their predators, with the effect of Bidrin on the extent of feeding by *Lycosa pseudoannulata* and *Oedothorax insecticeps*. *Plant Protection Bulletin#* (Taiwan) 19:1-12.

Chu, Y. Y. and Reid, D. C.
1982a The biology of *Microvelia douglasi* Scott, with emphasis on its feeding behaviour. *NTU Phytopathologist & Entomologist* 9:110-141.

Chudzika, E.
1980a Morphological variability in *Streptanus aemulans* (Kbm.) (Homoptera, Cicadellidae). *Annals of Zoology* (Agra) 35(13):205-213.
1981a Leafhoppers (Auchenorrhyndea, Homoptera). *Fragmenta Faunistica* (Warsaw) 26(11):175-191. [In Polish with English and Russian summaries.]
1982a Auchenorrhyncha (Homoptera) of Warsaw and Mazovia. *Memorabilia Zoologica* 36:143-164. [In English.]

Cisneros V. and Fausto H.
1959a Experimento comparativo de insecticidas en el control de la "cigarrita verde" (*Empoasca* sp.) en frijol. *Agronomia* (Lima) 26(3):253-256.

Clancy, D. W. and Mc Alister, H. J.
1956a Selective pesticides as aids to biological control of apple pests. *Journal of Economic Entomology* 49(2):196-202.

Claridge, D. W.
1982a Factors affecting the distribution of tree canopy leafhoppers of the Typhlocybinae (Homoptera: Cicadellidae). Doctoral Dissertation, University of Lancaster. 203 pp.

Claridge, D. W., Derry, N. J. and Whittaker, J. B.
1983a The distribution and feeding of some Typhlocybinae in response to sun and shade. *Acta Entomologica Fennica* 38:8-12.

Claridge, M. F.
1980a Biotaxonomic studies on leafhopper and planthopper pests of rice, with special reference to *Nilaparvata lugens* (Stal). *ODA Research Scheme R32882, Final Report and COPR Research Scheme No. 7.* 54 pp.
1983a Acoustic signals and species problems in the Auchenorrhyncha. In: Knight, W. J., Pant, N. C., Robertson, T. S. and Wilson, M. R., eds. *Proceedings of the 1st International Workshop on Biotaxonomy, Classification and Biology of Leafhoppers and Planthoppers (Auchenorrhyncha) of Economic Importance.* Pp. 111-120. London, 4-7 October 1982. Commonwealth Institute of Entomology, 56 Queen's Gate, London SW7 5JR. 500 pp.
1985a Acoustic signals in the Homoptera: Behavior, taxonomy and evolution. *Annual Review of Entomology* 30:297-317.
1985b Acoustic behavior of leafhoppers and planthoppers: species problems and speciation. In: Nault, L. R. and Rodriguez, J. G., eds. *The Leafhoppers and Planthoppers.* Pp. 103-125. Wiley & Sons, New York. 500 pp.

Claridge, M. F. and Howse, P. E.
1968a Songs of some British *Oncopsis* species (Hemiptera: Cicadellidae). *Proceedings of the Royal Entomological Society of London* 43(4-6):57-61.

Claridge, M. F. and Nixon, G. A.
1981a *Oncopsis* leafhoppers on British trees: polymorphism in adult *O. flavicollis* (L.) *Acta Entomologica Fennica* 38:15-19.

Claridge, M. F. and Reynolds, W. J.
1972a Host plant specificity, oviposition behaviour and egg parasitism in some woodland leafhoppers of the genus *Oncopsis. Transactions of the Royal Entomological Society of London* 124(2):149-166.
1973a Male courtship songs and sibling species in the *Oncopsis flavicollis* species group (Hemiptera: Cicadellidae). *Journal of Entomology, Series B* 42(1):29-39.

Claridge, M. F., Reynolds, W. J. and Wilson, M. R.
1977a Oviposition behavior and food plant discrimination in leafhoppers of the genus *Oncopsis. Ecological Entomology* 2(1):19-25.

Claridge, M. F. and Wilson, M. R.
1976a Diversity and distribution patterns of some mesophyll-feeding leafhoppers of temperate woodland canopy. *Ecological Entomology* 1(4):231-250.
1978a Observations on new and little known species of typhlocybine leafhoppers (Hemiptera: Cicadellidae) in Britain. *Entomologist's Gazette* 29:247-251.
1978b Behaviour and ecology of woodland canopy Typhlocybinae. *Auchenorrhyncha Newsletter* 1:29-30.
1978c Seasonal changes and alternation of food plant preferences in some mesophyll-feeding leafhoppers. *Oecologia* 37(2):247-255.
1978d Ovipositional behaviour as an ecological factor in woodland canopy leafhoppers. *Entomologia Experimentalis et Applicata* 24(3):301-309.
1978e British insects and trees: A study in island biogeography or insect/plant coevolution? *American Naturalist* 112(984):451-456.
1981a Host plant associations, diversity and species-area relationships of mesophyll-feeding leafhoppers of trees and shrubs in Britain. *Ecological Entomology* 6(3):217-238.
1981b Species richness of mesophyll-feeding leafhoppers and leafminers of trees in Britain. *Acta Entomologica Fennica* 38:19-20.
1981c The leafhopper and planthopper fauna of rice fields in South East Asia. *Acta Entomologica Fennica* 38:21-22.
1982a Insect herbivore guilds and species-area relationships: leafminers on British trees. *Ecological Entomology* 7:19-30.
1982b Species-area effects for leafhoppers on British trees: Comments on the paper by Rey et al. *American Naturalist* 119(4):573-575.

Clark, R. L.
1968a Epidemiology of curly top in the Yakima Valley. *Phytopathology* 58:811-813.

Clark, T. B.
1982a Spiroplasmas: Diversity of arthropod reservoirs and host-parasite relationships. *Science* 217:57-59.

Clark, T. B. and Whitcomb, R. F.
1983a Special procedures for demonstration of mycoplasmal pathogenicity in insects. In: Tully, J. G. and Razin, S., eds. *Methods in Mycoplasmology.* Vol. 2. Pp. 369-379. Academic Press, New York.

Clausen, C. P.
1978a Introduced parasites and predators of arthropod pests and weeds: A world review. Cicadellidae. *Agriculture Handbook, U.S. Department of Agriculture* 480:55-57.

Cobben, R. H.
1956a Voorlopige mededeling over enkele cicadenparasieten (Strepsipt.; Hymenopt.; Dipt.) *Entomologische Berichten* 16:160-165.
1956b Bionomie der Jasside *Fieberiella florii* Stal (Hom. Auchenorhyncha). *Publicates van het Natuurhistorisch Genootschap in Limburg* 9:57-82.
1965a Das aero-mikropylare System der Homoptereneier und Revolutionstrends bei Zikadeneiern (Hom. Auchenorhyncha). *Zoologisch Beitrage* (N. S.) 11:13-69
1978a Opening address. *Auchenorrhyncha Newsletter* 1:1-3.
1978b Miscellaneous. *Auchenorrhyncha Newsletter* 1:20-21.
1979a A new *Adarrus* species from Austria (Cicadellidae, Homoptera, Auchenorrhyncha). *Entomologische Berichten* (Amsterdam) 39(11):173-174.

Cobben, R. H. and Gravestein, W. H.
1958a Cicaden, Nieuw voor de Nederlandse fauna (Hom. Auchenorhyncha). *Entomologische Berichten* (Amsterdam) 18:122-124.

Cobben, R. H. and Rozeboom, G. J.
1978a Notes on Auchenorrhyncha (Homoptera) from pitfall traps in the Gerendal Reserve (southern part of Limburg Province). *Publicaties van het Natuurhistorisch Genootschap in Limburg* 28:1-15.

Cohen, G.
1961a Un virus bienfaisant. Du moins pour les insectes. *Nature* (Paris) 3310:61.

Coineau, Y.
1962a Nouvelles methodes de prospection de la faune entomologique des plantes herbacees et ligneuses. *Bulletin Societe Entomologique de France* 67:115-119.

Collins, H. L. and Pitre, H. N.
1969a Corn stunt vector leafhopper attractancy to and oviposition preference for hybrid dent corn. I. *Dalbulus maidis* (Homoptera: Cicadellidae). *Annals of the Entomological Society of America* 62(4):770-773.
1969b Corn stunt vector leafhopper attractancy to and oviposition preference for hybrid dent corn. II. *Graminella nigrifrons* (Homoptera: Cicadellidae). *Annals of the Entomological Society of America* 62(4):773-775.

Colomes, M.
1965a Essai de lutte contre *Scaphoideus littoralis* Ball au moyen de divers produits. [Control tests against *S. littoralis* by means of various products.] *Revue de Zoologie Agricole Applications* 64(10-12):125-128.

Combs, R. L. Jr.
1967a Biological studies of *Dalbulus maidis* (DeLong and Wolcott), a leafhopper vector of corn stunt virus in Mississippi. *Dissertation Abstracts* 28B:1265-1266.

Conner, R. L. and Banttari, E. E.
1979a Storage of the oat blue dwarf virus in whole leafhoppers. *Canadian Journal of Plant Pathology* 1:111-112.

Conti, M.
1981a Wild plants in the ecology of hopper-borne viruses of grasses and cereals. *In: Thresh, J. M., ed. Pests, Pathogens and Vegetation.* Pp. 109-119. Pitman, Boston.

1983a Plant quarantine problems relating to insect vectors of viruses and other plant pathogens. *Informatore Fitopatologico* 33:21-25.

1985a Transmission of plant viruses by leafhoppers and planthoppers. *In: Nault, L. R. and Rodriguez, J. G., eds. The Leafhoppers and Planthoppers.* Pp. 289-307. Wiley & Sons, New York. 500 pp.

Converse, R. H., Clark, R. G., Oman, P. W., Sr. and Milbrath, G.M.
1982a Witches' broom disease of black raspberry in Oregon. *Plant Disease* 66(10):949-951.

Cook, W. C.
1967a Life history, host plants, and migrations of the beet leafhopper in the western U.S. *U.S. Department of Agriculture Technical Bulletin* 1365:1-122.

Corella, H. A., Paz, M. H. and Alvarez, R. J. A.
1969a Insecticidas sistemicos en el control del *Cicadulina pastuasae* Ruppel and Delong en cebada. [Sytemic insecticides in the control of *C. pastusae* on barley.] *Revista de Ciencias Agriconomicas* 1(1):35-49. [With English summary.]

Costa, A. S.
1965a Outbreaks and new records. *FOA Plant Protection Bulletin* 13(6):138-139.

Costa Lima, A. da
1963a Dos cicadelideos brasileiros, um deles especie nova (Homoptera: Cicadellidae, Tettigellinae). *Archuivos do Instituto Biologico Sao Paulo* 30:119-123. [English summary.]

Costilla, M. A., Basco, H. J., Levi, C. and Osores, V. M.
1971a El "bicho llovedor," *Tapajosa rubromarginata* (Signoret) (Homoptera, Cicadellidae) en cultivos de cana de azucar. *Revista Industrial y Agricola de Tucuman* 48(2):49-52. [In Spanish.]

Coudriet, D. L. and Tuttle, D. M.
1963a Seasonal flights of insect vectors of several plant viruses in southern Arizona. *Journal of Economic Entomology* 56(6): 865-868.

Couillaud, R. and Aubertin, F.
1970a Influence de la protection insecticide sur les caracteristiques technologiques de la fibre de coton en Iran. *Coton et Fibres Tropicales* 25(4):489-494. [In French with English and Spanish summaries.]

Couillaud, R. and Daeschner, M.
1971a Premier bilan de l'experimentation cotonniere au Khuzistan (Iran). La date de semis en fonction des facteurs agro-climatiques et entomologiques. *Coton et Fibres Tropicales* 26(4):451-461. [In French with English and Spanish summaries; separate English edition available.]

Coupe, T. R. and Shulz, J. T.
1968a The influence of controlled environments and grass hosts on the life cycle of *Endria inimica* (Homoptera: Cicadellidae). *Annals of the Entomological Society of America* 61(1):74-77.

1968b Biology of *Endria inimica* in North Dakota (Homoptera: Cicadellidae). *Annals of the Entomological Society of America* 61(4):802-806.

Cousin, M. T.
1968a "Phyllodie du trefle," maladie a virus transmise par cicadelles en France. *In: First International Congress of Plant Pathology,* London. 36 pp.

Cousin, M. T. and Moreau, J. P.
1966a Role d'*Euscelis plebejus* Fall. dans la transmission des maladies a virus du trefle blanc: etude de la conservation du virus au cours de l'hiver. [Role of *E. plebejus* in the transmission of virus diseases of white clover: study of the overwintering of the virus.] *Annales des Epiphyties* 17():75-79. [With English and German summaries.]

Cousin, M. T., Moreau, J. P. and Grison, C.
1968a Mise en evidence de discontinuites dans la transmission de la phyllodie du trefle par *Euscelis plebejus* Fall. (Demonstrations of discontinuous transmission of clover phyllody by *E. plebejus*.) *Annales des Epiphyties* 19:115-120. [With English and German summaries.]

Cousin, M. T., Moreau, J. P. and Van Loon, L. C.
1965a Les maladies a virus de trefle transmises par cicadelles en France. [Virus diseases of clover transmitted by cicadellids in France.] *Annales des Epiphyties* 16(1):137-148. [With English and German summaries.]

Crall, J. M. and Stover, L. H.
1957a The significance of Pierce's disease in the decline of bunch grapes in Florida. *Phytopathology* 47:518.

Crane, P. S.
1970a The feeding behaviour of the blue-green sharpshooter *Hordnia circellata* (Baker). Ph. D. Thesis, University of California, Davis. 132 pp.

Cress, D. and Wells, A.
1976a Potato insect pests. *Michigan Cooperative Extension Service Bulletin* E-965. 2 pp.

1976b Snap bean insect pests. *Michigan Cooperative Extension Service Bulletin* E-966. 2 pp.

Cropley, R.
1960a The identification of strawberry viruses transmitted by aphids, leafhoppers and nematodes. *Gembloux Institute Agronomie Buletin de l'Institute Agronomie et des Statione de Recherche de Gembloux,* hors ser. 2:1038-1042.

Cruz, C.
1975a Chemical control of the leafhopper (*Empoasca fabae* (Harris) on snap beans. *Journal of Agriculture University of Puerto Rico* 59(1):82-84.

1976a Resistencia de frijol, *Phaseolus vulgaris*, a *Empoasca* spp. en Puerto Rice. *In: XII Reunion Anual del Programa Cooperativa Centroamericano-para el Mejoramiento de Cultivos Alimenticios, San Jose, Costa Rica.* Pp. L-27-1 to L-27-20.

1981a Effect of soil mulches on leafhopper (*Empoasca* spp.) population and on dry bean yield. *Journal of Agriculture of the University of Puerto Rico* 65:79-80.

Cruz, Y. P.
1974a [1975] Reproductive isolation between *Nephotettix virescens* (Distant) and *N. nigropictus* (Stal) (Euscelidae, Hemiptera-Homoptera). *Philippine Entomologist* 3(1):1-21.

Csiki, E.
1940a Homopteren: Explorationes zoologicae ab E. Csiki in Albania peractae. *Balkan-Kutatasainak Tudomanyos Tudomayos Ered-menyei.* Pp. 289-315.

Cunningham, H. B.
1962a A phylogenetic study of the leafhopper genus *Empoasca* (Homoptera, Cicadellidae). Dissertation DA23, 2631, University of Illinois. 51 pp.

1964a An adventive *Empoasca* (the European *E. flavescens* (F.)] found in North America (Hemiptera: Cicadellidae). *Annals of the Entomological Society of America* 57:263-264.

1967a Annual migration of the potato leafhopper into the Midwest. *Journal of the Alabama Academy Science* 38(4):320.

Cunningham, H. B. and Ross, H. H.
1965a Characters for specific identification of females in the leafhopper genus *Empoasca* (Hemiptera: Cicadellidae). *Annals of the Entomological Society of America* 58(5):620-623.

1965b Twelve new tropical empoascans (Hemiptera: Cicadellidae). *Annals of the Entomological Society of America* 58(6):836-843.

Cunningham, H. B., Decker, G. C. and Ross, H. H.
1965a Adaptation and differentiation of temperate phylogenetic lines from tropical ancestors in *Empoasca. Evolution* 18(4):639-651.

Cuperus, G. W.
1982a Establishment of economic thresholds for pea aphid and potato leafhoppers on alfalfa. Ph. D. Dissertation, University of Minnesota, St. Paul. 105 pp.

Cuperus, G. W. and Radcliffe, E. B.
1983a Thresholds for pea aphid and potato leafhopper on alfalfa. *In: 10th International Congress of Plant Protection.* Volume 1. Proceedings of a conference held at Brighton, England. Coryden, U.K. 106 pp.

Cuperus, G. W., Radcliffe, E. B., Barnes, D. K. and Marten, G. C.
1983a Economic injury levels and economic thresholds for potato leafhopper (Homoptera: Cicadellidae) on alfalfa in Minnesota. *Journal of Economic Entomology* 76(6):1341-1349.

Currado, I.
1983a Observations on *Anteon flavicorne* (Dalman) (Hymenoptera Dryinidae). *Atti XIII Congresso Nazionale Italiano di Entomologia.* Pp. 127-130.

Currado, I. and Olmi, M.
1979a On the identity of some Indian dryinid parasites of rice leafhoppers (Hymenoptera Dryinidae). *Il Riso* 28(2):179-181.

Cwikla, P. S.
1980a A new species of *Paraphlepsius* (Homoptera: Cicadellidae). *Journal of the Kansas Entomological Society* 53(3):639-640.

1984a Description of last nymphal instar of *Xestocephalus ancorifer* (Homoptera: Cicadellidae). *Entomological News* 95(2):40-42.

1985a Classification of the genus *Xestocephalus* (Homoptera: Cicadellidae) for North and Central America including the West Indies. *Brenesia* 24:175-272.

Cwikla, P. S. and Blocker, H. D.
1981a Neotropical genera of Deltocephalinae not included in Linnavuori's 1959 key. *Bulletin of the Entomological Society of America* 27(3):170-178.

1981b An annotated list of leafhoppers (Homoptera: Cicadellidae) from tallgrass prairie of Kansas and Oklahoma. *Transactions of the Academy of Science* 84:89-97.

Cwikla, P. S. and DeLong, D. M.
1985a New species and a new record of Agalliinae from South and Central America (Homoptera: Cicadellidae). *Journal of the Kansas Entomological Society* 58(1):156-162.

Cwikla, P. S. and Freytag, P. H.
1982a Three new leafhoppers (Homoptera: cicadellidae) from Cocos Island. *Proceedings of the Entomological Society of Washington* 84(3):632-635.

1983a External morphology of *Xestocephalus subtessellatus* (Homoptera: Cicadellidae: Xestocephalinae). *Annals of the Entomological Society of America* 76(4):641-649.

Dabek, A. J.
1979a *Cyperus rotundus,* the natural host of the leafhopper *Sanctanus fasciatus* in Jamaica. *FAO Plant Protection Bulletin* 27(4):123-124.

1982a Notes on the biology of *Ollarianus balli* (Van Duzee) (Hemiptera: Cicadellidae), a newly discovered vector of plant mycoplasmas in Jamaica. *Bulletin of Entomological Research* 72(2):207-214.

1982b Transmission experiments on coconut lethal yellowing disease with *Deltocephalus flavicosta* Stal, a leafhopper vector of periwinkle phyllody in Jamaica. *Phytopathologische Zeitschrift* 103(2):109-119. [In English with German summary.]

1982c Natural occurrence and insect transmission of phyllody disease(s) of periwinkle associated with mycoplasma-like organisms in Jamaica. *FAO Plant Protection Bulletin* 30(2):17-22.

1983a Leafhopper transmission of *Rhynchosia* little leaf, a disease associated with mycoplasma-like organisms in Jamaica. *Annals of Applied Biology* 103(3):431-438.

Dabrowski, Z. T.
1985a The biology and behaviour of *Cicadulina triangula* in relation to maize streak virus resistance screening. *Insect Science and Its Application* 6:417-424.

Dabrowski, Z. T. and Okoth, V. A. O.
1985a *Cicadulina*. Species composition in West Africa. IITA Annual Report for 1984, pp. 43-44. Ibadan, Nigeria.

Da Cruz, L. C. and Kitajima, E. W.
1972a The ultrastructure of mature spermatozoa of corn leafhopper, *Dalbulus maidis* Del. and W. (Homoptera: Cicadellidae). *Journal of Submicroscopic Cytology* 4:75-82.

D'Aguilar, J. and Della Giustina, W.
1974a Sur la presence en France de *Graphocephala coccinea* (Hom. Cicadellidae). *Annales de la Societe Entomologique de France* 10(3):747-740. [In French with English summary.]

Dahal, G. and Hibino, H.
1985a Varieties with different resistance to tungro (RTV) and green leafhopper (GLH). *International Rice Research Newsletter* 10(1):5-6.

1985b Relative amounts of tungro (RTV)-associated viruses in selected rices and their relation to RTV symptoms. *International Rice Research Newsletter* 10(6):10-11.

Dahiphale, M. V., Bhirud, K. M. and Chahun
1979a Efficacy of different fertilizer-pesticidal mixtures as soil and foliar application on paddy for the control of stem borer and leafhopper. *Pesticides* 13:20-25.

Dahiya, A. S. and Singh, R.
1982a Bio-efficacy of some systemic insecticides against jassid, thrips and white fly attacking cotton. *Pesticides* 16(12):13-14.

Dahlman, D. L.
1963a Survival of *Empoasca fabae* (Harris) (Cicadellidae) on synthetic media. *Proceedings of the Iowa Academy of Science* 70:498-504.

1965a Responses of *Empoasca fabae* (Harris) (Cicadellidae, Homoptera) to selected alkaloids and alkaloidal glycosides of the *Solanum* species. Ph.D. Dissertation, Iowa State University. 124 pp. [Diss. Abst. 26(10):6245.]

Dahlman, D. L. and Hibbs, E. T.
1967a Responses of *Empoasca fabae* (Cicadellidae: Homoptera) to tomatine, solanine, liptine I., tomatidine, solanidine, and demissidine. *Annals of the Entomological Society of America* 60(4):732-740.

Dahlman, D. L., Schroeder, L. A., Tomhave, R. H. and Hibbs, E. T.
1981a Inhibition and survival response of the potato leafhopper, *Empoasca fabae,* to selected sugars in agar media. *Entomologia Experimentalis et Applicata* 29:228-233.

Dakshinamurthy, A.
1984a The mango leafhopper and its control in Tripura. *Pesticides* 18:13-14.

Damsteegt, V. D.
1980a Investigations of the vulnerability of U.S. maize to maize streak virus. *Protection Ecology* 2:231-238.

1983a Maize streak virus: I. Host range and vulnerability of maize germ plasm. *Plant Disease* 67:734-737.

1984a Maize streak virus: Effect of temperature on vector and virus. *Phytopathology* 74:1317-1319.

Daniels, M. J.
1979a Mechanisms of spiroplasma pathogenicity. *In: Whitcomb, R. F. and Tulley, J. C., eds. The Mycoplasmas. III. Plant and Insect Mycoplasmas.* Pp. 209-227. Academic Press, New York.

1979b The pathogenicity of mycoplasmas for plants. *Zentralblatt fuer Bakteriologie, Parasitenkunde, Infektionskrankheiten und Hygiene. Erste Abteilung Originale. Reihe A. Medizinische Mikrobiologie und Parasitologie* 245:184-199.

Daniels, M. J., Markham, P. G., Meddins, B. M., Plaskitt, A. K., Townsend, R. and Bar-Joseph, M.
1973a Axenic culture of a plant pathogenic spiroplasma. *Nature* 244(5417):523-524.

Danilevskii, A. S.
1965a *Photoperiodism and Seasonal Development of Insects.* Oliver & Boyd, Edinburgh. 283 pp.

Danka, L.
1959a Uber die Fauna der Zikaden in der Lettischen SSR. Fauna of Latvia and adjacent territories II. *Trudy Instituta Biologii. Akademiya Nauk Latviiskoi SSR* 12:95-106. [In Russian with German summary.]
1961a Zikaden der Kuste des Rigaer Meerbusens bei Garciems. *Latvijas Entomologs* 3:59-61. [In Russian with Latvian and German summaries.]
1961b Zikadenfauna der Ackerkulturen in der Lettischen SSR. In: *Fauna Latviiskoi SSR i Copredel. Terr. 3* Akademia Nauk Latviiskoi SSR Instuta Biologicheski Trudy 20:177-185. [In Russian with German summary.]
1964a Zikaden in der Heide bei Vangazi. *Latvijas Entomologs* 8:49-55. [In Russian with Latvian and German summaries.]

Dantsig, E. M., Emeljanov, A. F., Loginova, M. M. and Shaposhnikov, F. K.
1964a Order Homoptera. *In: Bei-Bienko, Keys to Insects of European Part of USSR* 1:335-336.

D'Arcy, C. J. and Nault, L. R.
1982a Insect transmission of plant viruses and mycoplasmalike and rickettsialike organisms. *Plant Disease* 66:99-104.

Dargan, K. S., Butani, D. K. y Sahni, V. M.
1968a Insect pests of cotton: VII. Effect of combined spray of pesticides, fertilizer and hormone on pest population, growth, yield and economic characters of cotton. *Labdev Journal of Science and Technology* 6-B(4):175-179.

Darshan, S., Ramzan, M. and Bindra, O. S.
1982a Determination of economic threshold of cotton jassid, *Amrasca biguttula biguttula* (Ishida) on okra. *Indian Journal of Ecology* 9:113-117.

Das, M. M., Remamony, K. S. and Nair, M. R. G. K.
1969a Biology of a new jassid pest of mango, *Amrasca splendens* Ghauri. *Indian Journal of Entomology* 31:288-290.

Datta, B.
1969a On Indian Cicadellidae (Homoptera). I. *Zoologischer Anzeiger* 182(5&6):391-392.
1972a On Indian Cicadellidae (Homoptera). III. *Zoologischer Anzeiger* 188(1&2):61-67.
1972b On Indian Cicadellidae (Homoptera). IV. *Zoologischer Anzeiger* 188(1&2):67-70.
1972c On Indian Cicadellidae (Homoptera). II. *Zoologischer Anzeiger* 188(3&4):184-189.
1972d On Indian Cicadellidae (Homoptera). V. *Zoologischer Anzeiger* 188(3&4):189-196.
1972e On Indian Cicadellidae (Homoptera). VI. *Zoologischer Anzeiger* 189(1&2):102-108.
1972f On Indian Cicadellidae (Homoptera). VII. *Zoologischer Anzeiger* 189(1&2):109-114.
1972g On Indian Cicadellidae (Homoptera). VIII. *Zoologischer Anzeiger* 189(1&2):114-120.
1972h On Indian Cicadellidae (Homoptera). IX. *Zoologischer Anzeiger* 189(5&6):412-419.
1972i On Indian Cicadellidae (Homoptera). X. *Zoologischer Anzeiger* 189(5&6):419-426.
1972j On Indian Cicadellidae (Homoptera). XI. *Zoologischer Anzeiger* 189(5&6):427-434.
1973a On Indian Cicadellidae (Typhlocybinae) (Insecta: Homoptera). XII. *Zoologischer Anzeiger* 190(3&4):207-213.
1973b On Indian Cicadellidae (Insecta: Homoptera). XIII. *Zoologischer Anzeiger* 190(3&4):214-218.
1973c On Indian Cicadellidae (Insecta: Homoptera). XIV. *Zoologischer Anzeiger* 190(3&4):218-224.
1973d On Indian Cicadellidae (Insecta: Homoptera). XV. *Zoologischer Anzeiger* 190(3&4):225-230.
1973e On Indian Cicadellidae (Insecta: Homoptera). A. *Zoologischer Anzeiger* 190(5&6):381-385.
1973f On Indian Cicadellidae (Insecta: Homoptera). XVI. *Zoologischer Anzeiger* 191(5&6):420-428.
1973g On Indian Cicadellidae (Insecta: Homoptera). XVIII. *Zoologischer Anzeiger* 191(1&2):98-103.
1973h On Indian Cicadellidae (Insecta: Homoptera). XIX. *Zoologischer Anzeiger* 191(1&2):103-108.
1973i On Indian Cicadellidae (Insecta: Homoptera). XX. *Zoologischer Anzeiger* 191(5&6):428-436.
1973j On Indian Cicadellidae (Insecta: Homoptera). XXI. *Zoologischer Anzeiger* 191(5&6):437-443.
1973k On Indian Cicadellidae (Insecta: Homoptera).. XXII. *Zoologischer Anzeiger* 191(5&6):443-447.
1973l On Indian Cicadellidae (Insecta: Homoptera). XXIII. *Zoologischer Anzeiger* 191(5&6):448-454.
1973m On Indian Cicadellidae (Insecta: Homoptera). XXIV. *Zoologischer Anzeiger* 191(5&6):455-462.
1973n On Indian Cicadellidae (Insecta: Homoptera). XXV. *Zoologischer Anzeiger* 191(5&6):462-468.
1980a On Oriental and Nearctic Cicadelloidea (Homoptera). *Revista Espanola de Entomologia* 54(1-4):95-106.

Datta, B. and Dhar, M.
1984a On some collections of cicadellids (Homoptera: Cicadellidae). *Bulletin of the Zoological Survey of India* 6(1-3):181-207.

Datta, B. and Ghosh, L. K.
1973a Two new species of Typhlocybinae (Homoptera: Cicadellidae) from India. *Zoologischer Anzeiger* 191(5&6):410-414.
1973b Three new species of Typhlocybinae (Homoptera: Cicadellidae) from the Himalayan region. *Zoologischer Anzeiger* 191(5&6):415-419.
1974a On a new and two known cicadellids (Homoptera: Cicadellidae). *Indian Journal of Entomology* 36(2):118-124.

Datta, B. and Pramanik, N. K.
1977a New records of leafhoppers (Cicadelloidea: Homoptera) from Subansiri Division, Arunachal Pradesh, India. *Newsletter, Zoological Survey of India* 3(3):127-128.

Davey, K. G. and Manson, G. F.
1958a Chemical control of insects attacking alfalfa in southwestern Ontario. *Canadian Journal of Plant Science* 38:34-38.

David, B. V. and Janagarajan, A.
1969a Notes on pests of safflower in Tamil Nadu. *Madras Agriculture Journal* 56(8):534-538.

David, H. and Alexander, K. C.
1984a Insect vectors of virus diseases of sugarcane. *Proceedings of the Indian Academy of Science* 93:339-347.

David, W. A. L.
1958a Organic phosphorus insecticides for control of field crop insects. *Annual Review of Entomology* 3:377-395.

Davidson, R. H. and Landis, B. J.
1938a *Crabro davidsoni* Sandh., a wasp predacious on adult leafhoppers. *Annals of the Entomological Society of America* 31:5-8.

Davis, B. N. K.
1983a *Empoasca* species (Homoptera, Auchenorrhyncha) in the Chelsea Physic Garden, London. *Entomologist's Gazette* 34(2):97-100.
1984a *Empoasca pteridis* and other Hemiptera from the garden of Buckingham Palace. *Proceedings and Transactions, British Entomological and Natural History Society* 17:37-39.

Davis, M. J., Purcell, A. H. and Thomson, S. V.
1978a Pierce's disease of grape vines: Isolation of the causal organism. *Science* 199:75-77.

Davis, M. J., Whitcomb, R. F. and Gillaspie, Jr., A. G.
1981a Fastidious bacteria of plant vascular tissue and invertebrates (including so-called rickettsia-like bacteria). In: Starr, M. P., Stolp, H., Truper, H. G., Balows, A., and Schlegel, H. G., eds. *The Prokaryotes.* Vol. 2. Pp. 2172-2188. Springer-Verlag, New York.

Davis, R.
1966a Biology of the leafhopper *Dalbulus maidis* (Del. & Wol.) at selected temperatures. *Journal of Economic Entomology* 59(3):766.

Davis, R. B.
1970a Contribution to a classification of the higher categories (family, subfamily and tribe) of the auchenorrhynchous Homoptera (Cicadellidae and Aetalionidae). Doctoral Dissertation, North Carolina State University, Raleigh. 144 pp.

1975a Classification of selected higher categories of auchenorrhynchous Homoptera (Cicadellidae and Aetalionidae). *U.S. Department of Agriculture Technical Bulletin* 1494. 52 pp.

Davis, R. E.
1974a Occurrence of a spiroplasma in corn stunt-infected plants in Mexico. *Plant Disease Reporter* 57:333-337.

1974b Spiroplasma in corn stunt-infected individuals of the vector leafhopper *Dalbulus maidis*. *Plant Disease Reporter* 58(12):1109-1112.

1977a Spiroplasma: Role in the diagnosis of corn stunt disease. *In:* Williams, L. E., Gordon, D. T. and Nault, L. R., eds. *Proceedings, Maize Virus Disease Colloquium and Workshop.* Pp. 92-98. Ohio Agricultural Research Development Center, Wooster.

1979a Spiroplasmas: Newly recognized anthropod-borne pathogens. *In:* Maramorosch, K. and Harris, K. F., eds. *Leafhopper Vectors and Plant Disease Agents.* Pp. 451-484. Academic Press, New York. 654 pp.

Davis, R. E. and Whitcomb, R. F.
1971a Mycoplasmas, rickettsiae, and chlamydiae: Possible relation to yellows diseases and other disorders of plants and insects. *Annual Revue of Phytopathology* 9:11-154.

Davis, R. E., Whitcomb, R. F., Chen, T. A. and Granados, R. R.
1972a Current status of the aetiology of corn stunt disease. *In: Pathogenic Mycoplasmas.* Pp. 205-214. Elsevier-Exerpta Medica-North-Holland.

Davis, R. E., Whitcomb, R. F. and Purcell, R.
1970a Viability of the aster yellows agent in cell-free media. *Phytopathology* 60:573-574.

Davis, R. E., Whitcomb, R. F. and Steere, R. L.
1968a Remission of aster yellows disease by antibiotics. *Science* 161:793-795.

Davis, R. E. and Worley, J. F.
1973a Spiroplasma: Motile, helical microorganism associated with corn stunt disease. *Phytopathology* 63:403-408.

Davletshina, A. G., Avanesova, G. A. and Saparbekov, B.
1981a Characteristics of formation of the entomofauna of the agrobioceoenosis of newly cultivated lands of the Dzhizak region. *Vsesoyuznoe Entomologichskoe Obshchestvo,* pp. 63-67.

Davletishina, A. G. and Radzivilovskaya, M. A.
1965a Entomofauna of ferule. *Akademia Nauk Uzbekskoi SSR Uzbekskoi Biologeski Zhurnal* 9(1):57-62. [In Russian.]

Davoodi, Z.
1980a Some morphological and bio-ecological studies on *Edwardsiana rosae* L. Entomologie et Phytopathologie et Appliq'uees 48(1):53-65. [In Farsi (Persian); English summary, separately paginated, pp. 7-8.]

Day, M. F. and Bennetts, M. J.
1954a A review of problems of specificity in arthropod vectors of plant and animal viruses. *Canberra, Division of Entomology, Commonwealth Science and Industrial Research Organization Australia.* 172 pp.

De, R. K. and Dutta, D. K.
1955a Effect of commercial octamethyl pyrophosphoramides (Schradan), a systemic insecticide, on the mango hopper, *Idiocerus atkinsoni,* Lethierry (Homoptera: Jassidae). *Indian Journal of Horticulture* 12:165-172.

Deal, A. S.
1956a Control of the southern garden leafhopper, a new pest of cotton in Southern California. *Journal of Economic Entomology* 49(3):356-358.

Dean, G. J. W.
1976a Rice pests in Laos. *International Rice Research Newsletter* 1(2):15.

Deay, H. O.
1961a The use of electric light traps as an insect control. *U.S. Department of Agriculture, Agricultural Research Service* 20(10):50-53.

Deay, H. O., Taylor, J. G. and Johnson, E. A.
1959a Preliminary results on the use of electric light traps to control insects in the home vegetable garden. *Proceedings of the North Central Branch of the Entomological Society of America* 14:21-22.

Decker, G. C.
1959a Migration mechanisms of leafhoppers. *Proceedings of the North Central Branch of the Entomological Society of America* 14:11-12.

Decker, G. C. and Cunningham, H. B.
1967a The mortality rate of the potato leafhopper and some related species when subjected to prolonged exposure at various temperatures. *Journal of Economic Entomology* 60(2):373-379.

1968a Winter survival and overwintering area of the potato leafhopper. *Journal of Economic Entomology* 61(1):154-161.

Decker, G. C., Kouskolekas, C. A. and Dysart, R. J.
1971a Some observations on fecundity and sex ratios of the potato leafhopper. *Journal of Economic Entomology* 64(5):1127-1129.

Decker, G. C. and Maddox, J. V.
1967a Cold-hardiness of *Empoasca fabae* and some related species. *Journal of Economic Entomology* 60(6):1641-1645.

Deeming, J. C.
1981a The hemipterous fauna of a northern Nigeria cotton plot. *Samaru Journal of Agricultural Research* 1:211-222.

Deeming, J. C. and Webb, M. D.
1982a A new genus of eusceline leafhopper from West Africa (Homoptera: Cicadellidae: Deltocephalinae). *Archuivos do Museu Bocage,* Series A 1(22):485-494. [New genus: *Moskgha.*]

Deitz, L. L.
1979a Selected references for identifying New Zealand Hemiptera (Homoptera and Heteroptera), with notes on nomenclature. *New Zealand Entomologist* 7(1):21-29.

Deitz, L. L., Van Duyn, F. W., Bradley, J. R., Rabb, R. L., Brooks, W. M. and Stinner, R. E.
1976a A guide to the identification and biology of soybean arthropods in North Carolina. *North Carolina Agricultural Experiment Station Technical Bulletin* 238. 264 pp.

Dekanoidze, G. I.
1962a On the study of the biology of the vine leafhopper (*Erythroneura imeretina* sp.n.) and the development of measures for its control. *Trudy Instituta Sadovostva, Vinogradarstvai Vinodelieisa G. SSR* 14:149-156. [In Georgian with Russian summary.]

Delgadillo-Sanchez, F.
1984a Supervivencia del virus del rayado fino del maiz in Mexico. Tesis, M. C., Colegio de Posgraduados, Chapingo, Mexico.

Delius, H., Darai, G. and Flugel, R. M.
1984a DNA analysis of insect iridescent virus 6: Evidence for circular permutation and terminal redundancy. *Journal of Virology* 49:609-614.

Della Giustina, W.
1977a Presence de *Mongolojassus andorranus* Lindberg dans les Alpes francaises (Homoptera: Auchenorrhyncha). *Bulletin de la Societe Entomologique de France* 82:50-53.

1981a Sur deux cicadelles de Corse: *Opsius spinulosus* n. sp. et description du male d'*Eupteryx corsica* Lethierry (Homoptera: Cicadellidae). *Bulletin de la Societe Entomologique de France* 86(3&4):106-109.

1983a New species of leafhoppers in France from 1952 to 1982. *In: Knight, W. J., Pant, N. C., Robertson, T. S. and Wilson, M. R., eds. Proceedings of the 1st International Workshop on Biotaxonomy, Classification and Biology of Leafhoppers and Planthoppers (Auchenorrhyncha) of Economic Importance.* Pp. 257-262. London, 4-7 October 1982. Commonwealth Institute of Entomology, London, 500 pp.

Della Giustina, W. and Meusnier, S.

1982a Notes de chasse concernant les cicadelles (Hom. Cicadellidae). Distribution geographiques et especies nouvelles pour la France. *Bulletin de la Societe Entomologique de France* 87(7-8):332-334.

DeLong, D. M.

1959a A new genus, *Renonus*, and two new species of Mexican leafhoppers (Homoptera, Cicadellidae). *Ohio Journal of Science* 59:325-326.

1964a A monographic study of the North American species of the genus *Ballana* (Homoptera, Cicadellidae). *Ohio Journal of Science* 64(5):305-370.

1965a Ecological aspects of North American leafhoppers and their role in agriculture. *Bulletin of the Entomological Society of America* 11(1):1-26.

1966a Part 5. Insects. *In: Mirsky, A., ed. Soil Development and Ecological Succession in a Deglaciated Area of Muir Inlet, Southeast Alaska.* Pp. 97-120. Ohio State Research Foundation Institute of Polar Studies Report No. 20. 167 pp.

1967a *Spinulana*, new genus of Mexican Deltocephalinae and two new species of *Spinulana* (Homoptera, Cicadellidae). *Ohio Journal of Science* 67(1):20-22.

1967b *Devolana*, new genus of Mexican Deltocephalinae and a new species of *Devolana* (Homoptera-Cicadellidae). *Ohio Journal of Science* 67(1):22-23.

1967c A new species of *Draeculacephala* from Chile. *Ohio Journal of Science* 67(3):184-185.

1967d *Pseutettix*, a new genus and two new species of Mexican Deltocephalinae. *Pan-Pacific Entomologist* 3:210-212.

1967e Studies of the Mexican Deltocephalinae: A new genus, *Conversana* and three new species. *Proceedings of the Entomological Society of Washington* 69(3):266-269. [New genus: *Conversana.*]

1969a New South American genera related to *Tinobregmus* and *Sandersellus* (Homoptera: Cicadellidae). *Journal of the Kansas Entomological Society* 42(4):462-466. [New genera: *Chilelana, Boliviela.*]

1970a A new genus and species of deltocephaline leafhopper from southern Chile. *Ohio Journal of Science* 70(2):118-119. [New genus: *Nullamia.*]

1970b An Alaskan leafhopper that lives normally beneath icy tidal submergence. *Ohio Journal of Science* 70(2):111-114.

1971a The genus *Prescottia*: A new species and a new related Mexican genus. *Journal of the Kansas Entomological Society* 44:51-54. [New genus: *Soleatus.*]

1971b The bionomics of leafhoppers. *Annual Review of Entomology* 16:179-210.

1973a A new species of *Psammotettix* (Homoptera: Cicadellidae) from Mexico. *Ohio Journal of Science* 73(4):237-238.

1975a A new genus, *Freytagana*, and species of Mexican Gyponinae (Homoptera: Cicadellidae). *Journal of the Kansas Entomological Society* 48(3):409-410.

1975b The genus *Rhogosana* (Homoptera: Cicadellidae) with descriptions of three new species. *Ohio Journal of Science.* 75(3):126-129.

1976a New species of *Osbornellus* from Bolivia and Brazil. *Journal of the Kansas Entomological Society* 49(2):262-265.

1976b Three new species of *Hecalopona* (Homoptera: Cicadellidae) from Panama. *Journal of the Kansas Entomological Society* 49(3):364-366.

1976c A new genus, *Sordana*, for the "*Gypona sordida*" complex (Homoptera: Cicadellidae). *Ohio Journal of Science* 76(2):92-95.

1976d Two new species of *Ponanella* (Homoptera: Cicadellidae) from Panama. *Brenesia* 8:35-39.

1976e New species of *Portanus* (Homoptera: Cicadellidae) from Bolivia. *Brenesia* 9:37-40.

1976f A new genus, *Nullana*, and four new species of Peruvian Gyponinae (Homoptera: Cicadellidae). *Revista Peruana de Entomologia* 19(1):24-25.

1976g *Polana concinna* (Stal) and a new closely related species of Gyponinae (Homoptera: Cicadellidae). *Revista Peruana de Entomologia* 19(1):28-29.

1977a Four new species of British Guiana and Brazil *Curtara* (Homoptera: Cicadellidae). *Journal of the Kansas Entomological Society* 50(1):23-26.

1977b Species of Gyponinae (Homoptera: Cicadellidae) described by Herbert Osborn. *Journal of the Kansas Entomological Society* 50(3):389-393. [New genus: *Bahapona.*]

1977c A new genus, *Tenuacia*, new subgenus, *Rubacea*, and two new species of Gyponinae (Homoptera: Cicadellidae). *Ohio Journal of Science* 77(2):88-90.

1977d Some new species of *Hecalapona* (Homoptera: Cicadellidae) from South America. *Brenesia* 10&11:65-68.

1977e Five new species and two new subgenera of *Ponana* (Homoptera: Cicadellidae) from Peru. *Brenesia* 10&11:109-114. [New subgenera: *Lataponana, Proxaponana.*]

1977f A new subgenus *Elevanosa*, and new species of *Gypona* (Homoptera: Cicadellidae). *Entomological News* 88(1&2):37-38.

1978a The genus *Atanus* (Homoptera: Cicadellidae) in North America. *Journal of the Kansas Entomological Society* 51(3):484-491.

1978b Four new species of *Hecalapona* (Homoptera: Cicadellidae) from Brazil and Peru. *Entomological News* 89(2&3):109-111.

1979a A new subgenus, *Sinchora*, and ten new species of *Curtara* (Homoptera: Cicadellidae). *Journal of the Kansas Entomological Society* 52(2):229-237.

1979b A new genus, *Arapona*, and two new species of Gyponinae (Homoptera: Cicadellidae) related to *Clinonella*. *Ohio Journal of Science* 79(1):45-46.

1979c Four new species of *Culmana* (Gyponinae: Homoptera: Cicadellidae) from Peru and Bolivia. *Entomological News* 90(2):105-109.

1979d Studies of the Gyponinae with six new species of *Polana* (Homoptera: Cicadellidae). *Proceedings of the Entomological Society of Washington* 81(2):298-303.

1979e New species of *Gypona* and *Polana* (Homoptera: Cicadellidae - Gyponinae) from Central and South America. *Brenesia* 16:151-158.

1979f Species of Gyponinae (Homoptera: Cicadellidae) described by Stal and Spangberg. *Brenesia* 16:159-168.

1979g Two new subgenera and three new species of *Polana* (Homoptera: Cicadellidae) from Peru and Colombia. *Entomological News* 90(4):187-190. [New subgenera: *Striapona, Validapona.*]

1980a New species of *Chloroana* and *Reticana* (Homoptera: Cicadellidae: Gyponinae) from Central and South America. *Journal of the Kansas Entomological Society* 53(1):183-188.

1980b New South American Xestocephaline leafhoppers (Homoptera: Cicadellidae). *Entomological News* 91(3):79-84.

1980c A new genus, *Alapona*, a new *Ponana* subgenus, *Peranoa*, and two new species of Gyponinae (Homoptera: Cicadellidae). *Ohio Journal of Science* 80(5):217-219.

1980d New species of *Polana* (Homoptera: Cicadellidae) from Bolivia, Peru, Panama, and Florida. *Entomological News* 91(4):125-129.

1980e New species of *Curtara* (Homoptera: Cicadellidae) from Central and South America. *Brenesia* 17:179-214.

1980f New species of Central and South American *Gypona* (Homoptera: Cicadellidae). *Brenesia* 17:215-250.

1980g New species of *Gypona* (Homoptera, Cicadellidae: Gyponinae) from Central and South America. *Revista Peruana de Entomologia* 23(1):59-62.

1980h New genera and species of Mexican and South American deltocephaline leafhoppers (Homoptera, Cicadellidae, Deltocephalinae). *Revista Peruana de Entomologia* 23(1):63-71. [New genera: *Bardanarana, Bardana, Guadlera, Sanuca, Protranus.*]

1981a New species of *Gypona*, Gyponinae (Homoptera: Cicadellidae) with description of a new subgenus. *Proceedings of the Entomological Society of Washington* 83(3):505-511. [New subgenus: *Carnoseta.*]

1981b New species of *Hecalapona* (Homoptera, Cicadellidae) from Panama, Peru, Brazil and Venezuela. *Annales Entomologica Fennica* 47(1):29-31.

1981c New species of *Rhogosana* and *Ponana* (Homoptera: Cicadellidae) from Central and South America. *Entomological News* 92(1):17-22.

1981d A new genus and species of gyponine leafhopper related to *Gypona* (Homoptera: Cicadellidae). *Entomological News* 92(5):207-208. [New genus: *Woldana.*]

1982a New Central and South American leafhoppers of the "Bahita" group (Homoptera: Cicadellidae: Deltocephalinae). *Proceedings of the Entomological Society of Washington* 84(1):184-190. [New genus: *Angubahita.*]

1982b New species of Xestocephalinae (Homoptera: Cicadellidae) from Mexico, Panama, Peru and Brazil. *Proceedings of the Entomological Society of Washington* 84(2):391-396.

1982c Some new Neotropical leafhoppers of the subfamilies Iassinae and Deltocephalinae (Homoptera: Cicadellidae). *Proceedings of the Entomological Society of Washington* 84(3):610-616.

1982d New species of Central and South American *Gypona* (Homoptera: Cicadellidae). *Entomotaxonomia* 4(4):279-286.

1982e A new genus, *Platypona*, and new species of Gyponinae (Homoptera: Cicadellidae) related to *Gypona* and *Hecalapona*. *Ohio Journal of Science* 82(3):140.

1982f New and little known South American Deltocephalinae (Homoptera: Cicadellidae) with description of a new genus, *Sincholata*. *Brenesia* 19&20:477-485.

1982g New species of *Empoasca* and *Xestocephalus* (Homoptera: Cicadellidae) from Peru. *Brenesia* 19&20:25-29.

1982h New species of Bolivian *Osbornellus* and *Chlorotettix* (Homoptera: Cicadellidae: Deltocephalinae). *Proceedings of the Entomological Society of Washington* 85(2):331-334.

1983a New species of *Curtara* (Homoptera: Cicadellidae) from Central and South America. *Proceedings of the Entomological Society of Washington* 85(3):601-606.

1983b New species of *Loreta* and *Icaia* (Homoptera: Cicadellidae: Deltocephalinae) from Bolivia and Peru. *Entomological News* 94:127-130.

1983c New species of *Gypona* (Homoptera: Cicadellidae: Gyponinae) from Mexico, Jamaica, Colombia and Chile. *Ohio Journal of Science* 83(3):142-144.

1983d New species and distribution notes of Mexican and Bolivian *Idiodonus* (Homoptera: Cicadellidae). *Entomological News* 94(3):89-92.

1983e The genus *Dumorpha* DeLong and Freytag (Homoptera: Cicadellidae: Gyponinae) with descriptions of two new species. *Ohio Journal of Science* 83(4):213-214.

1983f New species of *Gyponana* (Gyponinae, Homoptera: Cicadellidae) from Texas, Mexico, and Ecuador. *Uttar Pradesh Journal of Zoology* 3(2):101-106.

1984a A revised key to the species of *Culumana* (Homoptera: Cicadellidae: Gyponinae) and descriptions of two new species. *Journal of the Kansas Entomological Society* 57(1):127-129.

1984b The use of male genital structures and biological studies in the validation of species of leafhoppers (Homoptera: Cicadellidae). *Journal of the Kansas Entomological Society* 57(3):543-544.

1984c Revised key to *Idiodonus* of Mexico and Bolivia (Homoptera: Cicadellidae). *Entomological News* 95(1):9-15.

1984d Some new species of Bolivian *Amplicephalus* (Homoptera: Cicadellidae: Deltocephalinae). *Uttar Pradesh Journal of Zoology* 4(2):168-174.

1984e New species of *Polana* (Homoptera: Cicadellidae: Gyponinae) from Brazil and Peru. *Brenesia* 22:45-49.

1984f New species of Central American *Polyamia* and *Deltocephalus* (Homoptera: Cicadellidae: Deltocephalinae). *Brenesia* 22:107-114.

1984g New leafhoppers of the "Bahita" group (Homoptera: Cicadellidae, Deltocephalinae) from Central and South America. *Brenesia* 22:115-122. [New genus: *Mairana.*]

1984h New *Gyponana* and *Curtara* Gyponinae (Homoptera: Cicadellidae) from Paraguay, Brazil and Costa Rica. *Brenesia* 22:315-318.

DeLong, D. M. and Bush, M. M.

1971a Studies of the Gyponinae: New species of *Ponanella* and *Acuponana*. *Ohio Journal of Science* 71(6):376-378.

DeLong, D. M. and Currie, N. L.

1959a The genus *Neokolla* (Homoptera, Cicadellidae). *Bulletin of the Brooklyn Entomological Society* (N.S.) 54:60-67.

1960a The genus *Keonolla* (Homoptera, Cicadellidae). *Bulletin of the Brooklyn Entomological Society* (N.S.) 55:4-15.

DeLong, D. M. and Cwikla, P. S.

1984a A new genus and species of deltocephaline leafhopper from Panama (Homoptera: Cicadellidae). *Proceedings of the Entomological Society of Washington* 86(2):432-434. [New genus: *Cumbrenanus.*]

1985a New neotropical Deltocephalinae (Homoptera: Cicadellidae). *Journal of the Kansas Entomological Society* 57(4):725-728.

DeLong, D. M. and Davidson, R. H.

1936a Methods of collecting and preserving insects. Ohio State University Press. 20 pp.

DeLong, D. M. and Foster, D. R.

1981a Six new species of Bolivian *Gypona* (Homoptera: Cicadellidae). *Entomological News* 92(4):141-146.

1982a New species of Bolivian *Polana*, Gyponinae (Homoptera: Cicadellidae). *Journal of the Kansas Entomological Society* 55(2):323-328.

1982b New species of Bolivian Gyponinae (Homoptera: Cicadellidae). *Entomological News* 93(4):114-118.

DeLong, D. M. and Freytag, P. H.

1962a Studies of the Gyponini: The *Gypona glauca* (Fabricius) complex. *Bulletin of the Brooklyn Entomological Society* (N.S.) 57(4):109-132.

1963a Studies of the Gyponinae 1. The genus *Marganana* DeLong (Homoptera: Cicadellidae). *Ohio Journal of Science* 63(6):258-262. [New subgenus: *Declivana.*]

1963b Studies of the Gyponinae 2. A new genus—*Zonana* (Homoptera: Cicadellidae). *Ohio Journal of Science* 63(6):262-265.

1964a Four genera of the world Gyponiae: A synopsis of the genera *Gypona, Gyponana, Rugosana* and *Reticana*. *Bulletin of the Ohio Biological Survey* 2(3):1-227. [New genus: *Reticana*; new subgenera: *Clovana, Obtusana, Paragypona, Sternana, Zerana.*]

1964b Studies of the Gyponinae: A new genus—*Chloronana* (Homoptera: Cicadellidae). *Annals of the Entomological Society of America* 57:503-511.

1964c Studies of the Gyponinae, the genus *Dragonana* Ball and Reeves (Homoptera: Cicadellidae). *Proceedings of the Entomological Society of Washington* 66(2):113-118.

1966a Studies of the Gyponinae: A synopsis of the genus *Acusana* DeLong (Homoptera: Cicadellidae). *Ohio Journal of Science* 66(1):42-63.

1966b Studies of the Gyponinae: Two new primitive genera *Coelogypona* and *Sulcana* (Homoptera, Cicadellidae). *Proceedings of the Entomological Society of Washington* 68(4):309-313.

1966c Studies of the Gyponinae: A synopsis of the genus *Hamana* DeLong (Homoptera, Cicadellidae). *Ohio Journal of Science* 66(6):554-569.

1967a Studies of the world Gyponinae (Homoptera, Cicadellidae): A synopsis of the genus *Ponana*. *Contributions of the American Entomological Institute* (Ann Arbor) 1(7):1-86. [New subgenus: *Neopanana*.]

1967b Studies of the Gyponinae: Two new genera, *Chilenana* and *Chilella* (Homoptera, Cicadellidae). *Ohio Journal of Science* 67(2):105-112.

1969a Studies of the world Gyponinae (Homoptera: Cicadellidae): A synopsis of the genus *Clinonaria*. *Ohio Journal of Science* 69(3):129-182.

1969b Studies of the Gyponinae: A new genus, *Ponanella*, and seven new species (Homoptera: Cicadellidae). *Journal of the Kansas Entomological Society* 42(3):303-312.

1970a Studies of the Gyponinae: A new genus, *Acuponana*, and ten new species. *Journal of the Kansas Entomological Society* 43(3):281-292.

1971a Studies of the Gyponinae: *Rhogosana* and four new genera, *Clinonella, Tuberana, Flexana,* and *Declivara*. *Journal of the Kansas Entomological Society* 44(4):540-543.

1972a Studies of the Gyponinae: a key to the known genera and descriptions of five new genera. *Journal of the Kansas Entomological Society* 45(2):218-235. [New genera: *Folicana, Costanana, Acuera, Curtara, Culumana*.]

1972b Studies of the Gyponinae: The genus *Folicana* and nine new species. *Journal of the Kansas Entomological Society* 45(3):282-295.

1972c Two new subgenera and two new species of *Gyponana* (Homoptera: Cicadellidae). *Ohio Journal of Science* 72(3):158-160. [New subgenera: *Spinanella, Pandara*.]

1972d Studies of the Gyponinae: The genus *Culumana* and seven new species (Homoptera: Cicadellidae). *Journal of the Kansas Entomological Society* 45(4):405-413.

1972e Studies of the Gyponinae: The genus *Costanana* and five new species (Homoptera: Cicadellidae). *Journal of the Kansas Entomological Society* 45(4):491-500.

1972f Studies of the world Gyponinae (Homoptera, Cicadellidae): The genus *Polana*. *Archuivors de Zoologia do Estado de Sao Paulo* 22(5):239-324. [Describes 9 new subgenera.]

1973a A new species of *Rugosana* (Homoptera: Cicadellidae) from Mexico. *Ohio Journal of Science* 73(3):190-191.

1974a Studies of the Gyponinae: The genus *Acuera* (Homoptera: Cicadellidae). *Ohio Journal of Science* 74(3):185-200. [New subgenera: *Tortusana, Parcana*.]

1975a Two new genera, *Proxima* and *Angucephala*, and two new species of Gyponinae (Homoptera: Cicadellidae). *Journal of the Kansas Entomological Society* 48(1):110-113.

1975b A new genus *Regalana* and species of Gyponinae (Homoptera: Cicadellidae). *Journal of the Kansas Entomological Society* 48(1):121-123.

1975c Studies of the Gyponinae: A new genus, *Hecalapona*, and thirty-eight new species. *Journal of the Kansas Entomological Society* 48(4):547-579. [New subgenera: *Carapona, Nulapona*.]

1975d Studies of the Gyponinae (Homoptera: Cicadellidae): Fourteen new species of Central and South American *Gypona*. *Journal of the Kansas Entomological Society* 48(3):308-318.

1975e A new genus *Dumorpha* and new species of Gyponinae (Homoptera: Cicadellidae). *Florida Entomologist* 58(1):33-34.

1976a Studies of the world Gyponinae (Homoptera: Cicadellidae). *Brenesia* 7:1-97. [New subgenera: *Ardasoma, Curtarana, Mysticana, Remarana, Retusana*.]

DeLong, D. M. and C. Guevara, J.

1954a Studies of the genus *Empoasca* (Homoptera: Cicadellidae). XIV. Some new species of Mexican *Empoasca*. *Bulletin of the Brooklyn Entomological Society* 49(3):81-86.

DeLong, D. M. and Hamilton, K. G. A.

1974a The genus *Amblysellus* (Homoptera: Cicadellidae): A key to known species, with descriptions of eight new species. *Canadian Entomologist* 106:841-849.

DeLong, D. M. and Harlan, H. J.

1968a Studies of the Mexican Deltocephalinae: New species of *Eutettix* and two allied new genera (Homoptera: Cicadellidae). *Ohio Journal of Science* 68(3):139-152. [New genera: *Alladanus, Cozadanus*.]

DeLong, D. M. and Knull, D. J.

1971a New species of *Scaphoideus*. *Journal of the Kansas Entomological Society* 44(1):54-59.

DeLong, D. M. and Kolbe, A. B.

1974a Studies of the Gyponinae (Homoptera: Cicadellidae). Four new species of *Gypona* from Panama. *Journal of the Kansas Entomological Society* 47(4):523-526.

1974b A colorful new species of *Ponana* (Homoptera: Cicadellidae) from Mexico. *Journal of the Kansas Entomological Society* 47(3):377-379.

1975a A new genus *Lorellana* (Homoptera: Cicadellidae) and two new species of deltocephaline leafhoppers. *Journal of the Kansas Entomological Society* 48(1):9-11.

1975b Three new species of Neocoelidiinae (Homoptera: Cicadellidae) from Panama and Peru. *Journal of the Kansas Entomological Society* 48(1):124-126.

1975c Studies of the Gyponinae: Six new species of South American *Gypona* (Homoptera: Cicadellidae). *Journal of the Kansas Entomological Society* 48(2):201-205.

DeLong, D. M. and Liles, M. P.

1956a Studies of the genus *Empoasca* (Homoptera: Cicadellidae) XV. Five new species of Mexican *Empoasca*. *Bulletin of the Brooklyn Entomological Society* 51:37-41.

DeLong, D. M. and Linnavuori, R.

1977a Studies of Gyponinae (Homoptera: Cicadellidae): Seven new species of *Gypona* from Central and South America. *Journal of the Kansas Entomological Society* 50(3):335-341.

1978a New tropical *Xestocephalus* (Homoptera: Cicadellidae) and illustrations of little known species. *Journal of the Kansas Entomological Society* 51(1):35-41.

1978b Studies of Neotropical leafhoppers (Homoptera: Cicadellidae). *Entomologica Scandinavica* 9:111-123.

1979a Six new species of Deltocephalini leafhoppers (Homoptera: Cicadellidae) from Mexico and Brazil. *Brenesia* 16:169-174.

DeLong, D. M. and Martinson, C.

1972a Studies of the Gyponinae (Homoptera: Cicadellidae): Fourteen new species of *Gypona* from Central and South America. *Ohio Journal of Science* 72(3):161-170.

1973a The Genus *Knullana* (Homoptera: Cicadellidae) with the description of a new Mexican species. *Journal of the Kansas Entomological Society* 46(2):151-154.

1973b A new genus, *Desertana*, and two new species of leafhoppers belonging to the Deltocephalinae (Homoptera: Cicadellidae). *Ohio Journal of Science* 73(2):125-127.

1973c Two new species of *Polyamia* (Homoptera: Cicadellidae) from Honduras and Chile. *Ohio Journal of Science* 73(3):149-151.

1973d Studies of the Gyponinae (Homoptera: Cicadellidae): Six new species of *Ponana* from Central and South America. *Ohio Journal of Science* 73(3):176-180.

1973e A new species of Mexican *Texananus* (Homoptera: Cicadellidae). *Entomological News* 84:202-204.

1973f A new genus, *Metacephalus*, and new species of Bolivian leafhopper (Homoptera: Cicadellidae). *Entomological News* 84:225-226.

1973g The genus *Maricaona* (Homoptera: Cicadellidae). *Entomological News* 84:251-252.

1974a A new species of *Chlorotettix* (Homoptera: Cicadellidae) from Argentina, Bolivia, Brazil and Mexico. *Journal of the Kansas Entomological Society* 47(2):261-267.

1974b A new genus, *Ancudana*, and a new species of Chilean leafhopper belonging to the Deltocephalinae (Homoptera: Cicadellidae). *Ohio Journal of Science* 74(4):261-263.

1976a Five new species of Bolivian *Osbornellus* (Homoptera: Cicadellidae). *Journal of the Kansas Entomological Society* 49(3):429-432.

1976b New species of Mexican *Osbornellus* (Homoptera: Cicadellidae). *Journal of the Kansas Entomological Society* 49(4):583-588.

1980a New species of *Ponana* and *Nullana* (Homoptera: Cicadellidae) from Central and South America. *Florida Entomologist* 63(4):493-501.

DeLong, D. M. and Thambimuttu, C. C.

1973a A list of the species of *Polyamia* (Homoptera: Cicadellidae) known to occur in Mexico, with descriptions of new species. *Ohio Journal of Science* 73(2):115-125.

1973b Three closely related new genera and five new species of short-winged Chilean leafhoppers (Homoptera: Cicadellidae). *Florida Entomologist* 56(3):165-172. [New genera: *Aequcephalus, Kramerana, Virganana.*]

DeLong, D. M. and Triplehorn, B. W.

1978a Four new species of Gyponinae (Homoptera: Cicadellidae) from Paraguay. *Entomological News* 89(7&8):179-182. [New subgenus: *Lataba.*]

1979a New species of Gyponinae (Homoptera: Cicadellidae) from Peru. *Brenesia* 16:175-188.

DeLong, D. M. and Wolda, H.

1978a New species of *Polana* and *Curtara* (Gyponinae) (Homoptera: Cicadellidae) from Panama. *Entomological News* 89(9&10):227-230.

1982a New species of *Gyponana* (Homoptera: Cicadellidae) from Panama and Mexico. *Entomological News* 93(1):12-14.

1982b New species of *Curtara, Polana* and *Acuera* (Homoptera: Cicadellidae, Gyponinae) from Panama, Peru, Bolivia and Venezuela. *Entomologica Scandinavica* 13:301-311.

1983a New species of *Costanana* and *Acuponana*, Gyponinae (Homoptera: Cicadellidae) from Central and South America. *Journal of the Kansas Entomological Society* 56(4):466-468.

1984a New Panamanian gyponine leafhoppers (Homoptera: Cicadellidae) belonging to the genera *Polana* and *Curtara*. *Uttar Pradesh Journal of Zoology* 4(1):22-30.

1984b New Panamanian gyponine leafhoppers (Homoptera: Cicadellidae) belonging to the genera *Hecalapona, Gypona* and *Gyponana*. *Uttar Pradesh Journal of Zoology* 4(2):135-144.

DeLong, D. M., Wolda, H. and Estribi, M.

1980a The Xestocephaline leafhoppers (Homoptera: Cicadellidae) known to occur in Panama. *Brenesia* 17:251-280.

1983a The *Ponana* (Homoptera, Cicadellidae) of Panama. *Proceedings, Koninklijke Nederlandse Akademie van Wetenschappen* 86(4):455-474.

1983b Some new species of Panama *Xestocephalus* (Homoptera: Xestocephalinae). *Uttar Pradesh Journal of Zoology* 3(2):73-80.

Den Hollander, J.

1982a The chromosomes of *Nilaparavata lugens* Stal, and some other Auchenorrhyncha. *Cytologia* 47(1):227-236.

1984a Morphological and acoustic studies of *Nephotettix virescens, N. nigropictus* and their hybrids. *Mitteilungen der Schweizerischen Entomologischen Gesellschaft* 57(4):415.

Denno, R. F.

1977a Comparison of the assemblages of sap-feeding insects (Homoptera-Hemiptera) inhabiting two structurally different salt marsh grasses in the genus *Spartina*. *Environmental Entomology* 6:359-372.

1980a Ecotype differentiation in a guild of sap-feeding insects on the salt marsh grass, *Spartina patens. Ecology* 61(3):702-714.

Deol, G. S. and Bindra, O. S.

1978a Influence of temperature on rate of development of the beet leafhopper *Circulifer tenellus* (Baker) (Homoptera, Cicadellidae.) *Indian Journal of Ecology* 5(2):261-262.

Deol, G. S., Sohi, A. S. and Bindra, O. S.

1978a Effect of different insecticidal treatments on the control of little-leaf disease and the fruit—and the shoot borer of brinjal. *Entomon* 3:133-134.

Derrick, K. S. and Newsom, L. D.

1984a Occurrence of a leafhopper-transmitted disease of soybeans in Louisiana. *Plant Disease* 68(4):343-344.

Descamps, M.

1956a Insects nuisibles au riz dans le Nord Cameroun. *Agricultura Tropical* 11(6):732-755.

Deseo, V. K.

1959a Beobachtungen uber Luzerneschadlinge mit besonderer Berucksichtigung des *Aphrodes bicinctus* Schrk. (Homopt.), *Apion tenus* Kirby and *Subcoccinella vigintiquatuorpunctata* L. (Coleopt.). *Anzeiger fuer Schadlingskunde* 32(7):97-99.

Deshmukh, S. D. and Akhare, M. D.

1979a Response of early and dwarf varieties of sunflower to jassids. *Indian Journal of Plant Protection* 7:64-65.

Deshmukh, S. D., Akhare, M. D. and Gorle, M. N.

1979a Insecticidal control of the insect pests of castor. *Indian Journal of Plant Protection* 7:66-68.

1980a Testing of some new insecticides against jassids on sunflower. *Pesticides* 14:35.

Deshpande, R. R., Rathore, V. S., Thakur, R. C., Sood, N. K., Raghuwanshi, R. K. and Kaushik, U. K.

1974a Chemical control of jassids on brinjal. *Mysore Journal of Agricultural Sciences* 8:400-406.

De Silva, M. D.

1961a *Empoasca punjabensis* Pruthi, a newly recorded pest of teak in Ceylon. *Tropical Agriculture (Peradeniya)* 117(3):203.

Desmidts, M., Laboucheix, J. and Offeren, A. L. van

1973a Importance economique et epidemiologie de la phyllodie du cotonnier. *In*: Laboucheix, J., Offeren, A. L. van, Desmidts, M. Etude de la transmission par *Orosius cellulosus* (Lindberg) (Hompt., Cicadellidae) de la virescence florale du cotonnier et de Sida sp. *Coton et Fibres Tropicales* 28(4):473-482. [In French.]

Devanesan, S., Jacob, A., Kuruvilla, S. and Mathai, S.

1979a Infection of *Nephotettix virescens* (Stal) (Cicadellidae: Hemiptera) by *Fusarium equiseti* (Corda) Sacc. *Entomon* 4(3):304-305.

Dewan, R. S., Misra, S. S., Handa, S. K. and Beri, Y. P.

1967a Persistence of DDT, gamma BHC and carbaryl residues on bhindi, *Abelmoschus esculentus* fruits. *Indian Journal of Entomology* 29(3):225-228.

Dhamdhere, S. V., Bahadur, J. and Misra, U. S.

1985a Studies on occurrence and succession of pests of okra at Gwalior. *Indian Journal of Plant Protection* 12:9-12.

Dhamdhere, S. V., Deole, J. Y. and Odak, S. C.

1981a Insecticidal control of *Amrasca biguttula biguttula* on okra. *Indian Journal of Plant Protection* 8:147-150.

Dhamdhere, S. V., Singh, O. P., Rajput, S. P. S. and Misra, U. S.

1981a Note on persistence and toxicity of synthetic pyrethroids against *Amrasca biguttula biguttula* Ishida infesting okra. *Indian Journal of Agricultural Sciences* 51:828-830.

Dhawan, A. K. and Sajjan, S. S.

1976a Biology of the rice green leafhopper, *Nephotettix nigropictus* (Stal) (Cicadellidae: Hemiptera). *Journal of Research, Punjab Agricultural University* 13:379-383.

1977a Laboratory studies on the host-range of rice green leafhopper *Nephotettix nigropictus* (Stal). *Journal of Research, Punjab Agricultural University* 14:188-192.

Dhuri, A. V. and Singh, K. M.

1983a Pest complex and succession of insect pests in blackgram *Vigna mungo* (L.) Hepper. *Indian Journal of Entomology* 45:396-401.

Diaz, C. G.
1971a Chemical control of bean leafhopper, *Empoasca fabae* Harris, [Homoptera, Cicadellidae] in El Bajio. *Agricultura Tecnica en Mexico* 3(3):93-94. [In Spanish with English, German, and French summaries.]

1973a Combate quimico de la chicharrita del frijo, *Empoasca fabae* (Harris) (Homoptera: Cicadellidae), en el region del Bajio. [Chemical control of the bean cicada, *Empoasca fabae* (Harris) (Homoptera: Cicadellidae), in the region of the Bajio.] *Folia Entomologica Mexicana* 25-26:64.

Diaz-Chavez, A. J.
1969a Estudio de la poblacion de *Dalbulus* sp., vector del virus causante del achaparramiento del maiz. *XV Reunion Anual Programa Cooperativo Centroamericano Cultivos Alimenticos*, San Salvador, El Salvador.

Dilbagh, Singh and Singh, Harcharan
1978a Effect of wind speeds on the swath width in ULV spraying for the control of jassid on cotton. *Indian Journal of Plant Protection* 6:63-66.

Dillon, L. S.
1954a Some notes on preparing whole insects for sectioning. *Entomological News* 65(3):67-70.

Di Martino, E.
1956a Contributo alla conoscenza biologica dell'*Empoasca decedens* Paoli e la fetola da essa prodotta nei frutti di agrumi in Sicilia. Esperienze di lotta. *Annali della Sperimentazione Agraria* 10(5):1511-1552.

Diniz, M. de Assuncao
1964a Captura, preparacao e conservacao de insectos. *Memorias e Estudos. Museu Zoologica de Universidade de Coimbra* 290. 62 pp.

Dinther, J. B. M. van
1963a Rice-sucking insects in Surinam. *Medelingen Fakulteit Landbouwwetenschappen Ghent* 28(3):815-823.

Dirimanov, M. and Karisanov, A.
1964a Uber die fauna der zikaden (Homoptera, Auchenorhyncha) in Bulgarien. *Plovdiv. Vissh. Selsk. Inst. "Vassil Kolarov." Nauchnye Trudove* 13(2):203-206. [In Bulgarian with German summary.]

1964b The green leafhopper (*Cicadella viridis* L.) – a study of its morphology and bionomics and of means for its control. *Gradinarska i Lozarska Nauka (Izvestiya na Akademiyata ne Selsko-stopanskite Nauki)*, Sofia 1(5):29-38. [In Bulgarian with Russian and English summaries.]

1964c Studies on the morphology and bionomics of the rose leafhopper and means for its control. *Rastenievadni Nauki*, Sofia 1(5):147-144. [In Bulgarian with Russian and English summaries.)

1965a Zweiter Beitrag uber die Fauna der Zikaaden (Homoptera Auchenorhyncha) in Bulgarien. *Visnik Selskostopanski Instituta "Vasil Kolarov." Nauchnye Trudove* 14(2):211-213. [In Bulgarian with German summary.]

Ditman, L. P., Cox, C. E., Burkhardt, G. J. and Todd, S. H.
1961a Results of seven years of spraying potatoes: Equipment, insecticides, fungicides. *Maryland Agricultural Experiment Station Bulletin* A-115. 14 pp.

Ditman, L. P., Owens, H. B. and Harrison, F. P.
1957a Experiments with sprays, dust and aerosols for the home garden. *Journal of Economic Entomology* 50(3):324-328.

Ditman, L. P., Rosenberger, C. R. and Harrison, F. P.
1954a Spraying for control of bean insects. *Journal of Economic Entomology* 47(4):600-603.

Ditman, L. P., and Wiley, R. C.
1958a The effectiveness of several insecticides for control of insects on snap beans. *Journal of Economic Entomology* 51(2):258-259.

Diwakara, M. C.
1975a Current entomological problems of paddy in new agricultural strategy. *Science and Culture* 41:19-22.

Dixit, A. K., Awasthi, M. D., Handa, S. K., Verma, S. and Dewan, R. S.
1977a Residues and residual toxicity of malathion and endosulfan on okra *Abelmoschus esculantus* L. fruits. *Indian Journal of Entomology* 37:251-254.

Dlabola, J.
1956a *Dudanus* gen. nov. and faunistical additions to the fauna of leafhoppers in Czechoslovakia (Homopt. Auchenorrh). *Acta Faunistica Entomologica Musei Nationalis Pragae* 1:31-38. [In Czech with English summary.]

1956b Additions to the knowledge of the European leafhopper fauna (Homo., Auch). *Folia Entomologica Hungarica* 9(17):395-401.

1956c Changes in the genital block of *Rhopalopyx* Rib. (Homoptera, Jassidae). *Acta Entomologicae Bohemoslavca* 53:111-114. [In Czech with English summary.]

1957a Results of the Zoological Expedition of the National Museum in Prague to Turkey. 20. Homoptera, Auchenorrhyncha. *Acta Entomologica Musei Nationalis* Pragae 31:19-68.

1957b Die Zikaden Afghanistans (Homopt.-Auchenorrhyncha) nach der Ergebnisse der von Herrn J. Klapperich in den Jahren 1952-1953 nach Afghanistan untergenommenen expedition. *Mitteilungen der Muenchen Entomologischen Gesellschaft* 47:265-303. [New genus: *Scaphytoceps*.]

1957c The problem of the genus *Delphacodes* and *Calligypona*, three new species and other Czechoslovakian faunistics (Homoptera, Auchenorrhyncha). *Sbornik Entomologickeho Oddeleni Narodniho Musea V Praze* 31(476):113-119.

1957d Das absterben der zikaden-art *Paradorydium lanceolatum* Burm. in der Tschechoslowakei. *Ochrana Prirody* 12(2):50-52. [In Czech with German summary.]

1958a Records of leafhoppers from Czechoslovakia and south European countries (Homoptera Auchenorrhyncha). *Acta Faunistica Entomologica Musei Nationalis Pragae* 3:7-15.

1958b Zikaden-Ausbeute vom Kaukasus (Homoptera Auchenorrhyncha). *Acta Entomologica Musei Nationalis Pragae* 32:317-352. [New genus: *Tbilisica*.]

1958c A reclassification of Palaearctic Typhlocybinae (Homopt., Auchenorrh). *Casopsis Ceskoslovenske Spolecnosti Entomologicke* 55:44-57. [Czech summary.] [Describes 7 new genera and 3 new subgenera.]

1959a Neue palaarkitische Zikaden der Fam. Meenoplidae und der Gattung *Handianus* Rib. (Homopt., Auchenorrhyncha). *Acta Entomologica Musei Nationalis Pragae* 33:445-452.

1959b Zwei neue *Chlorita*-Arten aus Sudeuropa und zoogeographische Bemerkungen (Homopt., Auchenorrhyncha). *Casopsis Ceskoslovenske Spolecnosti Entomologicke* 56:192-196. [New name: *Youngiada* for *Youngia*.]

1959c Funf neue Zikaden-Arten aus dem Gebiet des Mittelmeers. *Bollettino della Societa Entomologica Italiana Genoa* 89:150-155.

1960a Unika und Typen in der Zikaden-sammlung G. Horvaths (Homoptera, Auchenorrhyncha) II. *Acta Zoologica Hungaricae Academiae Scientiarum* 6:237-256. [New genus: *Coexitianus*.]

1960b Einige neue Zikaden aus Dagestan und Zentralasien (Homoptera). *Stuttgarter Beitraege zur Naturkunde aus dem staatliche Museum fur Naturkunde in Stuttgart* 40:1-5.

1960c Iranische Zikaden (Homoptera, Auchenorrhyncha): Ergebnisse der Entomologischen Reisen Willi Richter, Stuttgart, im Iran 1954 and 1956 – Nr. 31. *Stuttgarter Beitraege zur Naturkunde aus dem staatliche Museum fur Naturkunde in Stuttgart* 41:1-24. [New genus: *Concavifer*.]

1960d Pozor na novou chorobu Psenice. [Warning as regards a new disease of wheat.] *Za Vysokou Urodu* 17:403-405.

1961a Eine neue *Platmymetopius*-Art aus Mitteleuropa und weitere tschechoslowakische faunistik. *Acta Faunistica Entomologica Musei Nationalis Pragae* 7:5-9.

1961b Die Zikaden von Zentralasien, Dagestan und Transkaukasien (Homopt. Auchenorrhyncha). *Acta Entomologica Musei Nationalis Pragae* 34:241-358. [New genera: *Kazachstanicus, Tigriculus, Stalinabada, Sagittifer.*]

1961c Neue und bisher unbeschriebene Zikaden-Arten aus Rumanien und Italien. *Casopis Ceskoslovenske Spolecnosti Entomologicke* 58(4):310-323. [In German with Czech summary.]

1963a Zwei neue *Erythroneura*-Arten an der Weinrebe (Homoptera, Typhlocybinae.) *Reichenbachia* 1:309-313.

1963b A revision of the leafhopper genus *Macropsidius* Rib. (Hom., Auchenorrhyncha). *Casopis Ceskoslovenske Spoleonosti Entomologicke* 60:114-124.

1963c Typen und wenig bekannte Arten aus der Sammlung H. Haupt mit Beschreibugen einiger Zikadenarten aus Siberien (Homoptera). *Acta Entomologica Musei Nationalis Pragae* 35:313-331.

1963d Weitere neue Arten der Familie Cicadellidae aus Zentralasien und zoogeographische Bemerkungen zu einzelnen palaarktischen Zikadenarten (Homoptera). *Acta Entomologica Musei Nationalis Prague* 13:381-390.

1964a Die Zikaden Afghanistans (Homoptera, Auchenorrhyncha), Teil II. Ergebnisse der Sammelreisen von Dr. H. G. Amse., G. Ebert, Dr. Erichson, J. Klapperich und Dr. Lindberg. *Mitteilungen der Munchner Entomologischen Gesellschaft* 54:237-255.

1964b Ergebnisse der Zoologischen Nubien-Expedition 1962, Teil XXVI. Homoptera, Auchenorrhyncha. *Annalen des Naturhistorischen Museums* 67:615-626.

1964c Ergebnisse der Albanien-Expedition 1961 des Deutschen Entomologischen Institutes. 22. Beitrag Homoptera: Auchenorrhyncha. *Beitraege zur Entomologie* 14:269-318.

1964d Neue Ergebnisse zur zoogeographischen verbreitung einiger Europaischen zikaden (Homopt., Auchenorrhyncha). *Casopis Ceskoslovenske Spolecnosti Entomologicke* 61(2):173-176.

1965a Zoogeographische arten-gliederung der gattung *Fieberiella* Sign (Homopt., Auchenorrhyncha). *Acta Entomologica Bohemoslovaca* 62:428-442.

1965b Zur Kenntnis der Zikaden gattung *Rhytidodus* Fieber (Auchenorrhyncha: Cicadellidae). *Zoologische Beitrage* 11(1&2):71-75.

1965c Ergebnisse der Zoologischen Forschungen von Dr. Z. Kaszab in der Mongolei. 54. Homoptera – Auchenorrhyncha. *Sbornik Faunistickych Praci Entomologickeho Odelini Narodniho Musea V Prace* 11(100):79-136. [New genera: *Urganus, Kaszabinus;* new subgenus: *Emeljanovianus.*]

1965d Jordanische Zikaden (Homoptera, Auchenorrhyncha): Bearbeitung der von J. Klapperich im Jahre 1956-9 in Jordanien, Libanon und Syrien gesammelten Ausbeute. *Sbornik Entomologickeho Odelini Narodniho Musea V Praze* 36:419-450. [New genus: *Zercanus.*]

1965e Neue Zikadenarten aus Sudeuropa (Homoptera – Auchenorrhyncha). *Sbornik Entomologickeho Odelini Narodniho Musea V Praze* 36:657-699. [New genus: *Limassolla.*]

1966a Ergebnisse der Mongolisch-Deutschen Expeditionen 1962 und 1964. Nr. 16: Homoptera, Auchenorrhyncha. *Acta Entomologica Bohemoslovaca* 63(6):440-452.

1967a Ergebnisse der 1. Mongolisch-tschechoslowakischen entomologisch-botanischen Expedition in der Mongolei. Nr. 1: Reisebericht, Lokalitatenubersicht und Beschreibunen neuer Zikadenarten (Homopt., Auchenorrhyncha). *Sbornik Faunistickych Praci Entomologickeho Odelini Narodniho Musea V Prace* 12(115):1-34.

1967b Ergebnisse der mongolisch-tschechoslowakischen entomologisch-botanischen Expedition in der Mongolei. Nr. 3: Homoptera, Auchenorrhyncha (Erganzung). *Sbornik Faunistickych Praci Entomologickeho Odelini Narodniho Musea V Prace* 12(118):51-102.

1967c Ergebnisse der zoologischen Forschungen von Dr. Z. Kaszab in der Mongolei. Nr. 122: Homoptera – Auchenorrhyncha. *Sbornik Faunistickych Praci Entomologickeho Odelini Narodniho Musea V Prace* 12(123):137-152.

1967d Ergebnisse der 2. Mongolisch-tschechoslowakischen entomologisch-botanischen Expedition in der Mongolei. Nr. 12: Reisebericht, Lokaliteniibersicht und Bearbeitung der gesammelten Zikaden (Homopt., Auchenorrh.) *Sbornik Faunistickych Praci Entomologickeho Odelini Narodniho Musea V Prace* 12(131):207-230. [New genus: *Gobicuellus.*]

1967e Beschreibungen von neuen sudpalaarktischen Zikadenarten (Homoptera, Auchenorrhyncha.) *Acta Entomologica Musei Nationalis Pragae* 37:31-50. [New subgenus: *Maromoustaca.*]

1967f Eine neue *Edwardsiana*-Art von Bohmen und Bulgarien (Homoptera, Cicadellidae.) *Acta Entomologica Musei Nationalis Pragae* 37:251-253.

1967g Drei neue Neotropische Arten der Gattungen *Amplicephalus* und *Osbornellus* (Homo.-Cicadellidae.) *Acta Entomologica Musei Nationalis Pragae* 37:347-350.

1968a Ergebnisse der zoologischen Forschungen von Dr. Z. Kaszab in der Mongolei. Nr. 163: Homoptera, Auchenorrhyncha. *Acta Entomologica Bohemoslovaca* 65(5):364-374. [English summary.]

1968b Ergebnisse der zoologischen Forschungen von Dr. Z. Kaszab in der Mongolei. Nr. 169: Homoptera – Auchenorrhyncha. *Sbornik Faunistickych Praci Entomologickeho Odelini Narodniho Musea V Prace* 13(137):23-36.

1970a Beitrag zur Taxonomie und Chorologie einiger palaearktischer Zikadenarten (Homoptera, Auchenorrhyncha). *Mitteilungen Muenchener Entomologischen Gesellschaft* 59:90-107.

1970b Ergebnisse der Zoologischen Forschungen von Dr. Z. Kaszab in der Mongolei 220. Homoptera: Auchenorrhyncha. *Acta Zoologica Academiae Scientiarum Hungaricae* 16:1-25.

1971a Taxonomische und Chorologische Erganzungen der Zikadenfauna von Anatolien, Iran, Afghanistan und Pakistan (Homoptera, Auchenorrhyncha.) *Acta Entomologica Bohemoslovaca* 68:377-396.

1971b Taxonomische und chorologische Erganzungen zur turkischen und iraischen Zikadenfauna. *Sbornik Faunistickych Praci Entomologickeho Odelini Narodniho Musea V Prace* 14(163):115-138.

1972a Beitrage zur kenntnis der Fauna Afghanistans: Homoptera, Auchenorrhyncha. *Casopsis Maravskaho Musea V Brno* 56-57:189-248.

1974a Ergebnisse der Tschechoslowakisch-Iranischen Entomologischen Expedition Nach dem Iran 1970. Nr. 3: Homoptera, Auchenorrhyncha (1 Teil). *Acta Entomologica Musei Nationalis Pragae* (Supplementum) 6:29-73. [New genera: *Elbirzia, Parafieberialla;* new subgenus: *Quernus.*]

1974b Generische Gliederung der Unterfamilie Idiocerinae in der Palaarktik (Homoptera, Auchenorrhyncha.) *Acta Faunistica Entomologica Musei Nationalis Pragae* 15(174):59-68. [Describes 6 new genera and elevates *Liocratus* to genera rank.]

1974c Ubersicht der Gattungen *Anoplotettix, Goldeus* und *Thamnotettix* mit Beschreibugen von 7 neuen mediterranen Arten (Homoptera, Auchenorrhyncha). *Acta Faunistica Entomologica Musei Nationalis Pragae* 15(177):103-130. [New genera: *Nanosius, Quartausius.*]

1974d Zur Taxonomie und Chlorologie einiger mediterraner Zikaden (Homoptera: Auchenorrhyncha.) *Acta Zoologica Academiae Scientiarum Hungaricae* 20(3&4):289-308.

1974e Faunistische problematik im Spiegel der homopterologischen Taxonomie. *Folia Entomologica Hungarica* 27 (Supplement):291-295.

1975a Neue mediterrane Zikadenarten der Gattung *Hysteropterum* Amyot & Serville, 1843, *Macropsidius* Ribaut, 1952, und *Chlorita* Fieber, 1872 (Homoptera, Auchenorrhyncha). *Beaufortia* 23(299):75-83. [English summary.]

1976a La faune terrestre de l'Ile de Sainte-Helene, troisieme partie, 2. Insectes (suite et fin). 19. Homoptera. 4. Fam. Cicadellidae, subfam. Euscelinae. *Annales du Musee Royale de l'Africaine Centrale. Serie in Quarto Zoologie* 215:278-281. [In German.]

1977a Neue Zikaden-taxone von *Mycterodus, Erythria, Selenocephalus* and *Goldeus* (Homoptera: Auchenorrhyncha). *Acta Zoologica Academiae Scientiarum Hungaricae* 23(3&4):279-292.

1977b Neue iranische Cicadelliden-Gattungen und Arten mit faunistichen Erstfunden (Homoptera: Auchenorrhyncha). *Acta Entomologica Bohemoslovaca* 74(4):242-262. [New genera: *Bampurius, Evinus, Savanicus, Shirazia;* new subgenus: *Sarbazius.*]

1977c Chorologische Erganzungen zur Zikadenfauna das Mittelmeergebietes (Homoptera, Auchenorrhyncha). *Sbornik Narodniho Musea V Praze* 33:21-40.

1977d Homoptera Auchenorrhyncha. In: *Tschechoslowakische Insektenfauna.* Pp. 83-96. *Acta Faunistica Entomologica Musei Nationalis Pragae* (Supplementum) 15(4):1-158.

1979a Neue Zikaden aus Anatolien, Iran und aus Sudeuropaischen Landern (Homoptera: Auchenorrhyncha). *Acta Zoologica Academiae Scientiarum Hungaricae* 25(3&4):235-257. [New genera: *Mirzayansus, Khamiria.*]

1979b Insects of Saudi Arabia. Homoptera. *Fauna of Saudi Arabia* 1:115-139. [In German with Arabic and English summaries.]

1980a Neue griechische Zikadenarten der Fam. Cixiidae, Issidae und Cicadellidae (Homoptera: Auchenorrhyncha). *Acta Faunistica Entomologica Musei Nationalis Pragae* 16(179):5-13.

1980b Neue Zikadenarten der Gattungen *Siculus* gen. n., *Mycterodus* und *Adarrus* aus Sudeuropa und 6 neue *Mycterodus* aus Iran (Homoptera, Auchenorrhyncha). *Acta Faunistica Entomologica Musei Nationalis Pragae* 16(184):55-71.

1980c Drei neue *Diplocolenus*-Arten und taxonomisch-zoogeographische Ubersicht der Gattung in der Palaarktik. *Acta Faunistica Entomologica Musei Nationalis Pragae* 16(185):73-82. [In German.]

1980d Funf neue Issiden- und Cicadelliden-taxa aus Spanien (Homoptera, Auchenorrhyncha). *Annotationes Zoologicae et Botanicae* 136:1-13. [New genus: *Biluscelis.*]

1980e Insects of Saudi Arabia Homoptera: Auchenorrhyncha (Part 2). *Fauna Saudi Arabia* 2 1980:74-94. [In German with English summary.]

1981a Ergebnisse der Tschechoslowakisch-Iranischen entomologischen expeditionen nach dem Iran (1970 und 1973). *Acta Entomologica Musei Nationalis Pragae* 40:127-311.

1982a Fortsetzung der Erganzungen zur Issiden-Taxonomie von Anatolien, Iran und Griechenland (Homoptera, Auchenorrhyncha). *Sbornik Narodniho Muzea V Praze* (B) 38(3):113-169. [German with Czech summary.] [Proposed *Atlantocella* for *Atlantisia* Dlabola 1976a, nec *Atlantisia* Lowe 1923.]

1984a Neue Zikadenarten aus Mediterraneum und dem Iran mit Weiter Beitragen zur Iranischen fauna (Homoptera-Auchenorrhyncha). *Sbornik Narohino Musea V Praze* (B) 40(1):21-64. [German with Czech summary.]

1984b 50 years of taxonomical leafhopper studies and some proposals for the future in the Palaearctic zone. *Mitteilungen der Schweizerischen Entomologischen Gesellschaft* 57(4):416.

1985a Zwei neue *Fieberiella*-Arten aus der Turkie und Spanien (Homoptera, Cicadellidae). *Turkiye Bitki Doruma Dergisi* 9:75-78.

Dlabola, J. and Heller, F.

1962a Iranische Zikaden II (Ergebnisse der Entomologischen Reisen Willi Richter Stuttgart, im Iran 1954 und 1956 — Nr. 42.) *Stuttgarter Beitraege zur Naturkunde aus dem Staatlichen* 90:1-8.

Dlabola, J. and Jankovic, L.

1981a Drei neue *Erythria*-Arten und einige Erganzungen der Jugoslavischen Zikadenfauna. *Bulletin de l'Academie Serve des Sciences et des Arts. Classe des Sciences Naturelles et Mathematiques. Sciences Naturelles* 21:67-79.

Dlabola, J. and Novoa, N.

1976a Doa nuevas especies del genero *Hadria* Metcalf y Bruner, 1936 (Homoptera: Auchenorrhyncha) y revision de otras especies Cubanas. *Poeyana* 157:1-17.

1976b Dos nuevas especies del genero *Arezzia* Metcalf y Bruner, 1936 (Homoptera: Auchenorrhyncha) y revision de otras especies Cubanas. *Poeyana* 158:1-27.

Dobreanu, E. and Manolache, C.

1969a Homoptera, Partea generala. *Fauna Republicii Socialiste Romania (Insecta)* 8(4):1-100.

Dobretsova, V. N.

1935a A survey of insects injurious to utility and forage crops in Transcaucasia. *Vsesoiuznoe Selsskkhoz Akademia im V. I. Lenina. Zakavkaz. Nauchne Issled Khlopkovodstvo Institut Trudy Zaknikhi* 45:1-236.

Dogger, J. R.

1958a Biennial report of the state entomologist. *North Dakota Department of Agriculture and Labor Biennial Report* 1956-1958:77-80.

Doi, Y., Teranaka, M., Yora, K. and Asuyama, H.

1967a Mycoplasma or PLT group-like microorganisms found in the phloem elements of plants infected with mulberry dwarf, potato witches'-broom, aster yellows, or Paulownia witches broom. *Annals of the Phytopathological Society of Japan* 32:259-266.

Dollet, M.

1980a Report on research on the etiology of blast in oil and coconut palms. *Oleagineaux* 35:304.

Dolling, W. R.

1980a Otford: A leafhopper new to Kent. *Transactions Kent Field Club* 8(2):124.

Dookia, B. R. and Poonia, F. S.

1981a Some observations on a field population of *Empoasca flavascens* [sic!] Fabricius, a jassid pest on castor bean crop in Jodhpur (Rajasthan). *Entomon* 6(2):117-119.

Dorairaj, M. S., Savithri, V. and Aiyadurai, S. G.

1963a Population density as a criterion for evaluating varietal resistance of castor (*Ricinus communis* L.) to jassid infestation. *Madras Agricultural Journal* 50:100.

Dorst, H. E.

1960a Experimental control of the beet leafhopper on sugar beets grown for seed. *Journal of the American Society Sugar Beet Technologists* 11(1):12-14.

Douglas, W. A., Whitcomb, W. H., Hepner, L. W., Kirk, V. W. and Davis, R.

1966a Some Cicadellidae (Homoptera) collected from corn in the southeastern U.S. *Annals of the Entomological Society of America* 59(2):393-396.

Douglass, J. R.

1954a Outbreaks of beet leafhoppers north and east of the permanent breeding areas. *Proceedings of the American Society of Sugar Beet Technologists* 8(1):185-193.

Douglass, J. R. and Hallock, H. C.

1956a Studies of four summer hosts of the beet leafhopper. *Journal of Economic Entomology* 49(3):388-391.

1957a Relative importance of various host plants of the beet leafhopper in southern Idaho. *U.S. Department of Agriculture Technical Bulletin* 1155(1+):11.

1958a Russian-thistle distribution in southern Idaho and eastern Oregon in relation to beet leafhopper populations. *U.S. Department of Agriculture Production Research Report* 18, 20:8.

Douglass, J. R., Peay, W. E. and Cowger, J. I.
1956a Beet leafhopper and curly top conditions in the southern great plains and adjacent areas. *Journal of Economic Entomology* 49(1):95-99.

Dourojeanni, R. and Marc, J.
1965a Denominaciones vernaculares de insectos y algunos otros invertebrados en la Selva del Peru. *Revista Peruana de Entomologia* 8(1):131-137.

Doutt, R. L.
1961a The hymenopterous egg parasites of some Japanese leafhoppers. *Acta Hymenopterologica* 1:305-314.

Doutt, R. L. and Nakata, J.
1965a Parasites for control of grape leafhopper (*Erythroneura elegantula*, Homoptera). *California Agriculture* 19(4):3.
1973a The Rubus leafhopper and its egg parasitoid: an endemic biotic system useful in grape-pest management. *Environmental Entomology* 2(3):381-386.

Doutt, R. L., Nakata, J. and Skinner, F. E.
1966a Dispersal of grape leafhopper parasites from a blackberry refuge. *California Agriculture* 20(10):14-15.

Downes, W.
1957a Notes on some Hemiptera which have been introduced into British Columbia. *Proceedings of the British Columbia Entomological Society* 54:11-13.

Downing, R. S.
1962a Polybutenes in orchard pest control. *Canadian Entomologist* 94(11):1222-1227.

Drake, D. C. and Chapman, R. K.
1965a Migration of the six-spotted leafhopper, *Macrosteles fascifrons* (Stal). I. Evidence for long distance migration of the six-spotted leafhopper into Wisconsin. *University of Wisconsin-Madison, Research College of Agricultural and Life Science Research Division Bulletin* 261:1-20.

Drinkwater, T. W., Walters, M. C. and van Rensburg, J. B. J.
1979a The application of systemic insecticides to the soil for the control of the maize stalk borer, *Busseola fusca* (Fuller) (Lep.: Notuidae), and of *Cicadulina mbila* (Naude) (Hem.: Cicadellidae), the vector of maize streak virus. *Phytophylactica* 11:5-11.

Drosopoulos, S.
1980a Hemipterological studies in Greece. Part II. Homoptera-Auchenorrhyncha. A catalogue of the reported species. *Biologia Gallo-Hellenica* 9(1):187-194.
1984a Biosystematic studies of the *Alebra albostriella* complex (Homoptera, Typhlocybinae) in Greece. *Mitteilungen Schweizerischen Entomologischen Gesellschaft* 57(4):417.

Dubovskij, G. K.
1960a [Cicadellids *Kyboasca bipunctatus bipunctatus* (OSH.) and *Empoasca decipiens meridiana* Zachw.—pests of lucerne.] *Akademiya Nauk Uzbekskoi SSR, Doklady* 4:58-61. [In Russian.]
1960b [Materials on the fauna of "Cicads" found on lucerne.] *Akademiya Nauk Uzbekskoi SSR, Doklady* 12:48-49. [In Russian.]
1962a [*Anaceratagallia laevis* Rib.—pest of leguminous plants.] *Akademiya Nauk Uzbekskoi SSR, Biologicheskie Zhurnal* 5:26-28
1962b [*Fieberiella flori*—pest of pomogranate.] *Sadovodstovo* 100(3):42. [In Russian.]
1962c [The cicads injuring maize in East Ferghana.] *Zoologiceskij Zhurnal* 41(6):870-874. [In Russian with English summary.]
1962d [Cicadellidae harmful to rice in eastern Ferghana.] *Akademiya Nauk Uzbekskoi SSR, Biologicheskie Zhurnal* 2:47-52. [In Russian.]
1963a [Cicads occurring in lucerne fields of East Ferghana.] *Zoologiceskij Zhurnal* 42(6):835-841. [In Russian with Engish summary.]
1964a [Cicads injuring grain and forage grasses in East Ferghana.] *Zoologiceskij Zhurnal* 43(10):1560. [In Russian with English summary.]
1964b [Supplement to the work of V. N. Kuznetsova "Note on the Cicadellids (Homoptera), collected by Prof. D. N. Kashkarovy in the region of Arslan-Bob."] *Akademiya Nauk Uzbekskoi SSR, Uzbekskoi Biologicheskie Zhurnal* 8(4):67-70. [In Russian.]
1964c [Cicadellidae which damage cotton.] *Uzbekskii Biologicheskii Zhurnal* 8(2):38-41. [In Russian.]
1965a [Materials on the fauna and ecology of cicads of fruit forests of Arslan-Bob.] *Akademiya Nauk Kirgizii SSR. Vsesoiuznoe Entomologicheskoe Obshchestva Kirgizii Otdel. Entomologicheskoe Issled v Kirgizii.* Pp. 61-66. [In Russian.]
1965b [On the fauna and ecology of cicads (Auchenorrhyncha) inhabiting fruit forests of eastern Fergana.] *Entomologicheskoe Obozrenie* 44(2):324-334. [In Russian; translation in *Entomological Review*, Washington 44(2):182-187.]
1966a Cikadovye (Auchenorrhyncha) Ferganskoy Doliny. [Cicadina (Auchenorrhyncha) of the Fergana Valley.] *Izdatel'stvo "FAN" Uzbekistan SSR, Tashkent.* 256 pp. [In Russian.]
1967a [New species of cicadellids (Auchenorrhyncha) from Uzbekistan.] *In: Diseases and Pests of Invertebrates of Uzbekistan. Akademiya Nauk Uzbekskoi SSR. Institut Zoologiceskij i Parazitologicheskoe.* Pp. 56-59. [In Russian.]
1968a [Data on the fauna of cicadellids (Cicadinea) of western Turkmenia.] *Nauchnye Doklady Vysshei Shkoly. Biologicheskii Nauki* 2(194):50-53.
1968b [New species of leafhoppers (Cicadellidae, Melicharellinae) from Uzbekhistan.] *Zoologicheskii Zhurnal* 47(2):298-301. [In Russian.]
1970a [Materials on the fauna of Cicadinea of southern Uzbekistan.] *Nauchnye Doklady Vysshei Shkoly. Biologicheskii Nauki* 13(11):17-20. [In Russian.]
1970b Cikadoyye Tiktogulskoy vpadiny. *In: Vrednye i Polezyne Zhivotnye.* I. Tashkent. Pp. 32-38. [In Russian.]
1970c [New species of leafhoppers in Middle Asia.] *Uzbekskii Biologicheskii Zhurnal* 14(1):49-53. [In Russian.]
1978a [Cicads (Cicadinea) of the Zarafshan Valley. 2.] *Uzbekskii Biologicheskii Zhurnal* 1:42-46. [In Russian with Uzbek summary.]
1979a [New species of the genus *Platymetopius* Burmeister, 1838 (Homoptera: Cicadellidae) from Central Asia.] *Trudy Vsessoyuznogo Entomologicheskogo Obshchestva* 61:28-29.
1980a [Materials on the Cicadinea of the west Turkmenia.] *Nauchnye Doklady Vysshei Shkoly. Biologicheskii Nauki* 2:50-53.
1984a [New species of the genus *Anaceratagallia* Zachv., 1946 (Cicadellidae) from Central Asia.] *Uzbekskii Biologichekii Zhurnal* (1):65. [In Russian.]

Dubovskij, G. K. and Karimov, N.
1970a Cicadas of the Alaiski valley. *In: Vrednye i Polezne Zhivotnye.* I. Tashkent, pp. 21-31. [In Russian.]

Dubovskij, G. K. and Turgunov, M. T.
1971a [Material on the fauna of Auchenorrhyncha (Homoptera) from inner Tien-Shan.] *Entomologicheskoe Obozrenie* 50:341-346. [In Russian; translation in *Entomological Review*, Washington 50:193-195.]
1978a [Materials on the ecology of cicads (Cicadinea) of the lower reaches of the Amudarya.] *Uzbekskii Biologicheskii Zhurnal* (6):39-42. [In Russian with Uzbek summary.]

Dubovskij, G. K. and Tursunkhodzhaev, T.
1970a On pests of vegetable crops. *In: Vrednye i Polezne Zhivotnye.* I. Tashkent, pp. 19-20. [In Russian.]

Dudley, J. W., Hill, R. R. and Hanson, C. H.
1963a Effects of seven cycles of recurrent phenotypic selection on means and genetic variances of several characters in two pools of alfalfa germ plasm. *Crop Science* 3(6):543-546.

Duffield, C. A. W.
    1963a The genus *Aphrodes* (Homoptera: Auchenorrhyncha). *Transactions of the Kent Field Club* 1(4):155-160.

Duffus, J. E.
    1983a Epidemiology and control of curly top diseases of sugarbeet and other crops. *In: Plumb, R. T. and Thresh, J. M., eds. Plant Virus Epidemiology. The Spread and Control of Insect-borne Viruses.* Blackwell Scientific Publications, Oxford, U.K. 337 pp.

Duffus, J. E. and Gold, A. H.
    1973a Infectivity neutralization used in serological tests with partially purified beet curly top virus. *Phytopathology* 63(9):1107-1110.

Duffus, J. E., Milbrath, G. M. and Perry, R.
    1982a Unique type of curly top virus and its relationship with horseradish bitter root. *Plant Disease* 66:650-652.

Dumbleton, L. J.
    1964a New records of Hemiptera-Homoptera and a key to the leafhoppers (Cicadellidae-Typhlocybinae) in New Zealand. *New Zealand Journal of Science* 7(4):571-578.
    1967a *Cicadella melissae* (Curtis) and *Idiocerus decimusquartus* Schrank established in New Zealand. *New Zealand Entomologist* 3(5):41-42.

Duncan, J. and Genereux, H.
    1960a La transmission par les insectes de *Corynebacterium sepedonicum* (Spieck. & Kott.) Skaptason et Burkholder. *Canadian Journal Plant Science* 40(1):110-116.

Durant, J. A.
    1968a Occurrence of leafhoppers (Homoptera: Cicadellidae) on corn in South Carolina. *Journal of the Georgia Entomological Society* 3(3):77-82.
    1968b Leafhopper populations on ten corn inbred lines at Florence, South Carolina (Homoptera: Cicadellidae). *Annals of the Entomological Society of America* 61(6):1433-1436.
    1973a Notes on factors influencing observed leafhopper (Homoptera: Cicadellidae) population densities on corn. *Journal of the Georgia Entomological Society* 8:1-5.

Durant, J. A. and Hepner, L. W.
    1969a Occurrence of leafhoppers (Homoptera: Cicadellidae) on corn in South Carolina. *Journal of the Georgia Entomological Society* 3(3):77-82.

D'Urso, V.
    1978a Due nuove sottospecie di Deltocephalinae dell'Etna (Sicilia) (Insecta, Homoptera, Cicadellidae). *Animalia (Catania, Italia)* 51(1-3):197-208.
    1980a [1982] *Jassargus dentatus*, nuova specie di Deltocephalinae della brughiera piemontese (Insecta, Homoptera, Cicadellidae). *Animalia (Catania, Italia)* 7(1-3):123-133. [Italian with English summary.]
    1980b Gli omotteri auchenorrinchi della braghiera di Rovasenda (Piemonte). *Quaderni Struttura Zoocenosi Terrestri* AQ1 113:57-70. [Italian with English summary.]
    1981a [1982] Sulla presenza de specie del genere *Oncopsis* in Sicilia (Homoptera, Auchenorrhyncha, Cicadellidae). *Memorie della Societa Entomologica Italiana* 60(2):177-178.
    1981b [1982] Biogeographical structure of the Sicilian Auchenorrhyncha. *Acta Entomologica Fennica* 38:46-47.
    1982a A new species of *Jassargus* (Homoptera, Auchenorrhyncha, Cicadellidae) from Sicily (Italy) and redescriptions of *Jassargus latinus* (Wagner). *Animalia* 9:73-86. [In Italian.]
    1983a Una nuova specie de *Doratura* (Homoptera, Auchenorrhyncha, Cicadellidae) dei Monti Iblei (Sicilia). *Animalia* 10:31-40.
    1984a *Adarrus aeolianus*, nuova species di Cicadellidae (Homoptera, Auchenorrhyncha) delle isole eolie. *Animalia* 11(1-3):31-40.

D'Urso, V. and Ippolito, S.
    1984a The wing coupling-apparatus of the Auchenorrhyncha at the SEM. *Mitteilungen der Schweizerischen Entomologischen Gesellschaft* 57(4):418-419.

D'Urso, V., Ippolito, S. and Lombardo, F.
    1984a Observations on the Heteroptera and Homoptera Auchenorrhyncha of Sicily. *In: Drosopoulos, S., ed. Proceedings of the 1st International Congress Concerning the Rhynchota Fauna of Balkan and Adjacent Regions.* Pp. 13-14. Mikrolimini—Prespa, Greece, 29 August - 2 September 1983.
    1984b Studio faunistico-ecologico sugli eterotteri terrestri ed omotteri Auchenorrinchi di Monte Manfre (Etna, Sicilia). *Animalia* 11(1-3):155-194.

Dutra, J. A. P.
    1966a Homopteros Cicadelideos do Brasil, I: Descricao de duas especies novas do genero *Agallia* Curtis. *Boletim do Museu de Biologia "Prof. Mello-Leitao"* 26:1-3.
    1967a Homopteros Cicadelideos do Brasil, II: Descricao de uma nova especies do genero *Agallia* Curtis. *Boletim do Museu de Biologia "Prof. Mello-Leitao"* 32:1-4.
    1969a Homopteros Cicadelideos do Brasil, III: Descricao de uma especie nova do genero *Agallia* Curtis; ocorrencias novas de Cicadellidae. *Atas da Sociedade de Biologia do Rio de Janeiro* 12(5&6):293-294.
    1970a Homoptera Cicadellidae do Brasil IV: Descricao de uma especie nova do genero *Agallia* Curtis. *Atas da Sociedade de Biologia do Rio de Janeiro* 14(3&4):61-62.
    1971a Homoptera Cicadellidae do Brasil V. Descriacao de uma especie nova do genero *Agallia* Curtis: ocorrencias nova de Cicadellidae. *Atas da Sociedade de Biologia do Rio de Janeiro* 14(5&6):157-158.
    1972a Homopteros Cicadelideos do Brasil, VI: Descricao de uma especie nova do genero *Agallia* Curtis. *Atas da Sociedade de Biologia do Rio de Janeiro* 15(3):159-160.
    1974a Homopteros Cicadelideos do Brasil, VII: Descricao de uma especie nova do genero *Agallia* Curtis. *Avulso 20 Rio de Janeiro. Universidade Federal do Rio de Janeiro. Centro de Ciencas Matematicas e da Netureza. Instituto de Biologia. Departmento de Zoologia.* Pp. 1-3.
    1976a Homopteros Cicadelideos do Brasil, VIII: Descricao de uma especie nova do genero *Agallia* Curtis. *Avulso 25 Rio de Janeiro. Universidade Federal do Rio de Janeiro. Centro de Ciencas Matematicas e da Natureza. Instituto de Biologia. Departmento de Zoologia.* Pp. 1-3.
    1977a Homopteros Cicadelideos do Brasil, IX: Descricao de uma especie nova do genero *Agallia* Curtis. *Avulso 28 Rio de Janeiro. Universidade Federal do Rio de Janeiro. Centro de Ciencas Matematicas e da Natureza. Instituto de Biologia. Departmento de Zoologia.* Pp. 1-4.

Dutra, J. A. P. and Egler, I.
    1982a Descricao de uma especie nova do genero *Agalliota* Oman (Homoptera, Cicadellidae). *Revista Brasileira de Biologia* 42(1):225-228.

Dutt, N. and Biswas, A. K.
    1979a Contribution of antibiosis in locating tolerance of paddy varieties to *Nephotettix virescens* (Dist.) and *N. nigropictus* (Stal). *Journal of Entomological Research* 3:196-211.

Duviard, D.
    1973a Etude, par les pieges a eau, de la faune entomologique d'un champ de coton en Cote d'Ivoire Centrale (Foro-Foro). *Annales de la Societe Entomologique de France* 9(1):147-172. [In French with English summary.]

Dworakowska, I.
    1967a A new species of the genus *Doratura* Shlb. (Homop., Cicadell.) from Mongolia. *Bulletin de l'Academie Polonaise des Sciences. Serie des Sciences Biologiques* 15(3):159-160.
    1967b Some *Typhlocybinae* (Homoptera, Cicadellidae) from Mongolia with descriptions of two new species. *Bulletin de l'Academie Polonaise des Sciences. Serie des Sciences Biologiques* 15(10):633-637.
    1968a Notes on the genus *Elymana* DeLong (Homoptera, Cicadellidae). *Bulletin de l'Academie Polonaise des Sciences. Serie des Sciences Biologiques* 16(4):233-238.

1968b Some Typhlocybinae (Homoptera, Cicadellidae) from Mongolia with description of a new species and a new subspecies. *Bulletin de l'Academie Polonaise des Sciences. Serie des Sciences Biologiques* 16(6):365-371.

1968c Contributions to the knowledge of Polish species of the genus *Doratura* Shlb. (Homoptera, Cicadellidae). *Annales Zoologici* (Warsaw) 25(7):381-401.

1968d Some Typhlocybinae (Homoptera, Cicadellidae) from Korea, with descriptions of six new species from Korea and one from Vietnam. *Bulletin de l'Academie Polonaise des Sciences. Serie des Sciences Biologiques* 16(9):565-572.R

1969a Some Cicadellidae (Homoptera, Auchenorrhyncha) from Mongolia, with redescription of one species. *Bulletin de l'Academie Polonaise des Sciences. Serie des Sciences Biologiques* 17(1):51-55.

1969b A new species of the genus *Dikraneura* Hardy and redescription of *Motschulskyia inspirata* (Motsch.) from South India (Homoptera, Typhlocybinae). *Bulletin de l'Academie Polonaise des Sciences. Serie des Sciences Biologiques* 17(4):247-249.

1969c Contribution to the taxonomy of genera related to *Eupteryx* Curt. with description of two new genera (Homoptera, Typhlocybinae). *Bulletin de l'Academie Polonaise des Sciences. Serie des Sciences Biologiques* 17(6):381-385. [New genera: *Wagneriunia, Almunisna*.]

1969d On the genera *Zyginella* Low and *Limassolla* Dlab. (Cicadellidae, Typhlocybinae). *Bulletin de l'Academie Polonaise des Sciences. Serie des Sciences Biologiques* 17(7):433-438.

1969e Contribution to the taxonomy of the *Eupteryx* complex with descriptions of one new subgenus, one new genus and four new species (Homoptera, Typhlocybinae). *Bulletin de l'Academie Polonaise des Sciences. Serie des Sciences Biologiques* 17(7):439-445. [New subgenus: *Stacla;* new genus: *Knightipsis.*]

1969f Two new typhlocybine genera from the Oriental region with a remark on synonymy (Homoptera, Cicadellidae, Typhlocybinae). *Bulletin de l'Academie Polonaise des Sciences. Serie des Sciences Biologiques* 17(8):487-490. [New genera: *Britimnathista, Ledeira.*]

1969g Revision of the Palaearctic and Oriental species of the genus *Eurhadina* Hpt. (Homoptera, Cicadellidae, Typhlocybinae). *Annales Zoologici* (Warsaw) 27(5):67-68.

1970a Remarks on the tribe Bakerini Mahm. with description of one new genus of Typhlocybini (CIcadellidae, Typhlocybinae). *Bulletin de l'Academie Polonaise des Sciences. Serie des Sciences Biologiques* 17(11&12):691-696. [New genus: *Mahmoodia.*]

1970b On the species of *Eupteryx artemisiae* (Kbm.) group (Homoptera, Cicadellidae, Typhlocybinae). *Annales Zoologici* (Warsaw) 27(16):361-372.

1970c On the genus *Thaia* Ghauri (Homoptera, Cicadellidae, Typhlocybinae). *Bulletin de l'Academie Polonaise des Sciences. Serie des Sciences Biologiques* 18(2):87-92.

1970d A new subgenus of *Erythroneura* Fitch (Auchenorrhyncha, Cicadellidae, Typhlocybinae). *Bulletin de l'Academie Polonaise des Sciences. Serie des Sciences Biologiques* 18(6):347-354. [New subgenus: *Balila.*]

1970e On some species of the genus *Helionidia* Zachv. with descriptions of three new species (Auchenorrhyncha, Cicadellidae, Typhlocybinae). *Bulletin de l'Academie Polonaise des Sciences. Serie des Sciences Biologiques* 18(2):147-152.

1970f On some East Palaearctic and Oriental Typhlocybini (Homoptera, Cicadellidae, Typhlocybinae). *Bulletin de l'Academie Polonaise des Sciences. Serie des Sciences Biologiques* 18(4):211-217. [New genera: *Agnesiella, Farynala, Warodia, Zorka.*]

1970g On the genus *Arboridia* Zachv. (Auchenorrhyncha, Cicadellidae, Typhlocybinae). *Bulletin de l'Academie Polonaise des Sciences. Serie des Sciences Biologiques* 18(10):607-615.

1970h On some genera of Typhlocybini and Empoascini (Auchenorrhyncha, Cicadellidae, Typhlocybinae). *Bulletin de l'Academie Polonaise des Sciences. Serie des Sciences Biologiques* 18(11):707-716. [New genera: *Sundapteryx, Ishiharella.*]

1970i On some genera of Empoascini (Cicadellidae, Typhlocybinae). *Bulletin de l'Academie Polonaise des Sciences. Serie des Sciences Biologiques* 18(5):269-275.

1970j On the genus *Zygina* Fieb. and *Hypericiella* sgen. n. (Auchenorrhyncha, Cicadellidae, Typhlocybinae). *Bulletin de l'Academie Polonaise des Sciences. Serie des Sciences Biologiques* 18:559-567.

1970k Three new genera of Erythroneurini (Auchenorrhyncha, Cicadellidae, Typhlocybinae). *Bulletin de l'Academie Polonaise des Sciences. Serie des Sciences Biologiques* 18:617-624. [New genera: *Ciudadrea, Hauptidia, Kropka;* new subgenus: *Melicharidia.*]

1970l On the genera *Zyginidia* Hpt. and *Lublinia* gen. n. (Auchenorrhyncha, Cicadellidae, Typhlocybinae). *Bulletin de l'Academie Polonaise des Sciences. Serie des Sciences Biologiques* 18:625-632. [New genus: *Lublinia.*]

1970m On some genera of Erythroneurini (Auchenorrhyncha, Cicadellidae, Typhlocybinae). *Bulletin de l'Academie Polonaise des Sciences. Serie des Sciences Biologiques* 18:697-705. [New genera: *Imbecilla, Sempia.*]

1970n On the genera *Asianidia* Zachv. and *Singapora* Mahm. with the description of two new genera (Auchenorrhyncha, Cicadellidae, Typhlocybinae). *Bulletin de l'Academie Polonaise des Sciences. Serie des Sciences Biologiques* 18:759-765. [New genera: *Anufrievia, Mitjaevia.*]

1971a On the North African species of the genus *Eupteryx* Curt. (Hom., Cicadellidae, Typhlocybinae). *Suomen Hyonteistieteellinen Aikakauskirja* 37(1):14-20.

1971b On the genera related to *Tamaricella* Zachv. and some other Erythroneurini (Hom., Cicadellidae, Typhlocybinae). *Suomen Hyonteistieteellinen Aikakauskirja* 37(2):99-121. [New genera: *Accacidia, Amicula.*]

1971c On Eastern Hemisphere Alebrini (Auchenorrhyncha, Cicadellidae, Typhlocybinae). *Bulletin de l'Academie Polonaise des Sciences. Serie des Sciences Biologiques* 19(7&8):493-500. [New genera: *Asialebra, Orientalebra.*]

1971d On some genera of the tribe Dikraneurini (Auchenorrhyncha, Cicadellidae, Typhlocybinae). *Bulletin de l'Academie Polonaise des Sciences. Serie des Sciences Biologiques* 19(9):579-586. [New genera: *Javadikra, Kirkaldykra, Uzeldikra.*]

1971e On some genera of Erythroneurini (Cicadellidae, Typhlocybinae) from the Oriental Region. *Bulletin de l'Academie Polonaise des Sciences. Serie des Sciences Biologiques* 19(5):341-350. [New genera: *Seriana, Coloana.*]

1971f *Dayus takagii* sp. n. and some Empoascini (Auchenorrhyncha, Cicadellidae, Typhlocybinae). *Bulletin de l'Academie Polonaise des Sciences. Serie des Sciences Biologiques* 19(7&8):501-509.

1971g *Opamata* gen. n. from Viet-Nam and some Typhlocybini (Auchenorrhyncha, Cicadellidae, Typhlocybinae). *Bulletin de l'Academie Polonaise des Sciences. Serie des Sciences Biologiques* 19(10):647-657. [New genera: *Draberiella, Opamata.*]

1972a Revision of the genus *Aguriahana* Dist. (Auchenorryncha, Cicadellidae, Typhlocybinae). *Polskie Pismo Entomologiczne* 42(2):273-312.

1972b On some Oriental Erythroneurini. *Bulletin de l'Academie Polonaise des Sciences. Serie des Sciences Biologiques* 20(6):395-405. [Descriptions of 8 new genera.]

1972c On several Central African Empoascini. *Bulletin de l'Academie Polonaise des Sciences. Serie des Sciences Biologiques* 20(7):477-486. [New genera: *Atucla, Habenia, Jacobiella, Theasca.*]

1972d On some species of the genus *Eupteryx* Curt. (Auchenorrhyncha, Cicadellidae, Typhlocybinae). *Bulletin de l'Academie Polonaise des Sciences. Serie des Sciences Biologiques* 20(10):727-734.

1972e *Aaka* gen. n. and some other Erythroneurini (Auchenorrhyncha, Cicadellidae, Typhlocybinae). *Bulletin de l'Academie Polonaise des Sciences. Serie des Sciences Biologiques* 20(11):769-778. [New genera: *Aaka, Salka, Sapporoa*.]

1972f *Stehliksia* gen. n. and some other African Typhlocybinae (Auchenorrhyncha, Cicadellidae). *Bulletin de l'Academie Polonaise des Sciences. Serie des Sciences Biologiques* 20(12):845-855. [Descriptions of 6 new genera.]

1972g *Zyginoides* Mats. and some other Typhlocybinae (Auchenorrhyncha, Cicadellidae). *Bulletin de l'Academie Polonaise des Sciences. Serie des Sciences Biologiques* 20(12):857-866. [New genera: *Bunyipia, Dilobonota*.]

1972h On some East Asiatic species of the genus *Empoasca* Walsh (Auchenorrhyncha, Cicadellidae, Typhlocybinae). *Bulletin de l'Academie Polonaise des Sciences. Serie des Sciences Biologiques* 20(1):17-24.

1972i On some Oriental and Ethiopian genera of Empoascini (Auchenorrhyncha, Cicadellidae, Typhlocybinae). *Bulletin de l'Academie Polonaise des Sciences. Serie des Sciences Biologiques* 20(1):25-34. [New genera: *Distantasca, Epignoma, Jacobiasca, Laokayana, Quartasca*.]

1972j Five new Oriental genera of Erythroneurini (Auchenorrhyncha, Cicadellidae, Typhlocybinae). *Bulletin de l'Academie Polonaise des Sciences. Serie des Sciences Biologiques* 20(2):107-115. [New genera: *Gredzinskiya, Hepneriana, Kaukania, Raabeina, Ramania*.]

1972k On some Oriental genera of Typhlocybinae (Auchenorrhyncha, Cicadellidae). *Bulletin de l'Academie Polonaise des Sciences. Serie des Sciences Biologiques* 20(2):117-125. [New genera: *Kanguza, Koperta, Lectotypella, Michalowskiya*.]

1972l Australian Dikraneurini (Auchenorrhyncha, Cicadellidae, Typhlocybinae). *Bulletin de l'Academie Polonaise des Sciences. Serie des Sciences Biologiques* 20(3):193-201. [New genus: *Dziwneono*.]

1973a On the Genus *Molopopterus* Jac. (Auchenorrhyncha, Cicadellidae, Typhlocybinae). *Bulletin de l'Academie Polonaise des Sciences. Serie des Sciences Biologiques* 21(1):39-47.

1973b *Baguoidea rufa* (Mel.) and some other Empoascini (Auchenorrhyncha, Cicadellidae, Typhlocybinae). *Bulletin de l'Academie Polonaise des Sciences. Serie des Sciences Biologiques* 21(1):49-58.

1973c On some Palaearctic species of the genus *Kybos* Fieb. (Auchenorrhyncha, Cicadellidae, Typhlocybinae). *Bulletin de l'Academie Polonaise des Sciences. Serie des Sciences Biologiques* 21(3):235-244.

1973d New species and some interesting records of leafhoppers from Mongolia and Korea (Auchenorrhyncha, Cicadellidae). *Bulletin de l'Academie Polonaise des Sciences. Serie des Sciences Biologiques* 21(6):419-424.

1974a Contribution a la faune du Congo (Brazzaville). Mission A. Villiers et A. Descarpentries. 109. Hemipteres Typhlocybinae. *Bulletin de l'Institut Fondamental d'Afrique Noire*, Serie A 36(1):132-243. [Descriptions of 32 new genera.]

1976a *Kybos* Fieber, subgenus of *Empoasca* Walsh (Auchenorrhyncha, Cicadellidae, Typhlocybinae) in Palearctic. *Acta Zoologica Cracoviensia* 21(13):387-463.

1976b On some Oriental and Ethiopian Typhlocybinae. *Reichenbachia* 16(1):1-51. [Descriptions of 7 new genera.]

1976c Two new species of *Empoasca* Walsh (Auchenorrhyncha, Cicadellidae, Typhlocybinae). *Bulletin de l'Academie Polonaise des Sciences. Serie des Sciences Biologiques* 24(3):151-159.

1977a On the genus *Erythria* Fieb. (Auchenorrhyncha, Cicadellidae, Typhlocybinae). *Bulletin de l'Academie Polonaise des Sciences. Serie des Sciences Biologiques* 24(10):597-605.

1977b On several of Dlabola's Typhlocybinae species (Auchenorrhyncha: Cicadellidae). *Acta Zoologica Academiae Scientiarum Hungaricae* 23(1&2):29-36.

1977c On the genus *Epignoma* Dwor. (Homoptera, Cicadellidae, Typhlocybinae). *Revue de Zoologique Africaine* 91(1):212-236.

1977d On some species of the subgenus *Kybos* Fieb., with remarks on some other Empoascini (Auchenorrhyncha, Cicadellidae, Typhlocybinae). *Bulletin de l'Academie Polonaise des Science. Serie des Sciences Biologiques* 25(9):609-617.

1977e *Krameriata* gen. n. and some other Empoascini. *Revue de Zoologie Africaine* 91(4):845-874. [New subgenera: *Marolda, Ociepa*.]

1977f On some north Indian Typhlocybinae. *Reichenbachia* 16(29):283-306. [New genera: *Bakshia, Kamaza, Watara*.]

1977g On some African species of *Empoasca* Walsh and *Jacobiasca* Dwor. *Reichenbachia* 16(35):327-334.

1977h *Jimara* gen. n. of Dikraneurini from Africa. *Reichenbachia* 16(37):337-346.

1977i On some Typhlocybinae from Vietnam (Homoptera: Cicadellidae). *Folia Entomologica Hungarica* 30(2):9-47. [New genera: *Amurta, Goifa, Lowata, Sobrala, Usharia*.]

1978a On the genera *Empoascanara* Dist. and *Seriana* Dwor. (Auchenorrhyncha, Cicadellidae, Typhlocybinae). *Bulletin de l'Academie Polonaise des Sciences. Serie des Sciences Biologiques* 26(3):151-160.

1978b On some Typhlocybini (Auchenorrhyncha, Cicadellidae, Typhlocybinae). *Bulletin de l'Academie Polonaise des Sciences. Serie des Sciences Biologiques* 26(10):703-713.

1979a On some Typhlocybinae from India and adjoining areas (Homoptera, Auchenorrhyncha, Cicadellidae). *Reichenbachia* 17(18):143-162. [New genera: *Cubnara, Dattasca, Kabakra*.]

1979b On some Erythroneurini from Vietnam (Typhlocybinae, Cicadellidae). *Annotationes Zoologicae et Botanicae* 131:1-50. [Descriptions of 9 new genera.]

1979c The leafhopper tribe Zyginellini (Homoptera, Auchenorrhyncha, Cicadellidae, Typhlocybinae). *Revue de Zoologie Africaine* 93(2):299-331. [New genera: *Ahimia, Mordania, Muluana, Narta, Tafalka*.]

1979d Five new species of the genus *Typhlocyba* (Auchenorrhyncha, Cicadellidae, Typhlocybinae). *Bulletin de l'Academie Polonaise des Sciences. Serie des Sciences Biologiques* 27(3):195-203.

1979e Eight new species of eastern Hemisphere Dikraneurini (Auchenorrhyncha, Cicadellidae, Typhlocybinae). *Bulletin de l'Academie Polonaise des Sciences. Serie des Sciences Biologiques* 27(4):263-272. [New genus: *Afrakra*.]

1979f Nine new species of Asiatic Erythroneurini (Auchenorrhyncha, Cicadellidae, Typhlocybinae). *Bulletin de l'Academie Polonaise des Sciences. Serie des Sciences Biologiques* 27(5):371-380.

1980a Review of the genus *Naratettix* Mats. (Auchenorrhyncha, Cicadellidae, Typhlocybinae). *Bulletin de l'Academie Polonaise des Sciences. Serie des Sciences Biologique* 27(8):645-652.

1980b On some Typhlocybinae from India (Homoptera, Auchenorrhyncha, Cicadellidae). *Entomologische Abhandlungen und Berichte aus dem Staatlichen Museum fur Tierkunde in Dresden* 43(8):151-201. [New genera: *Burara, Faiga, Gindara, Shamala, Vermara*.]

1980c On some species of the genus *Empoascanara* Dist. (Homoptera, Auchenorrhyncha, Cicadellidae, Typhlocybinae). *Reichenbachia* 18(26):173-188.

1980d Contribution to the taxonomy of the genus *Empoascanara* Dist. (Homoptera, Auchenorrhyncha, Cicadellidae, Typhlocybinae). *Reichenbachia* 18(27):189-197.

1981a *Kusala* gen. n. and some other Erythroneurini (Auchenorrhyncha, Cicadellidae, Typhlocybinae). *Bulletin de l'Academie Polonaise des Sciences. Serie des Sciences Biologiques* 28(5):317-325.

1981b On the genera *Acia* McAtee and *Omiya* gen. n. (Typhlocybinae, Cicadellidae). *Annotationes Zoologicae et Botanicae* 141:1-47. [New subgenera: *Africia, Icaiana, Naracia, Paoliella.*]

1981c On the genus *Diomma* Motschulsky (Auchenorrhyncha, Cicadellidae, Typhlocybinae). *Bulletin de l'Academie Polonaise des Sciences. Serie des Sciences Biologiques* 28(6):363-370.

1981d On some Palearctic Erythroneurini (Auchenorrhyncha, Cicadellidae, Typhlocybinae). *Bulletin de l'Academie Polonaise des Sciences. Serie des Sciences Biologiques* 28(6):371-379.

1981e On some Typhlocybinae from India, Sri Lanka and Nepal (Homoptera, Auchenorrhyncha, Cicadellidae). *Entomologische Abhandlungen und Berichte aus dem staatlichen Museum fur Tierkunde in Dresden* 44(8):153-202. [New genus: *Pasara.*]

1981f *Badylessa* gen. n. and some other African Empoascini (Auchenorrhyncha, Cicadellidae, Typhlocybinae). *Bulletin de l'Academie Polonaise des Sciences. Serie des Sciences Biologiques* 28(10&11):583-592. [New name: *Paolicia* for *Paoliella.*]

1981g On some Typhlocybini from India and Nepal (Auchenorrhyncha, Cicadellidae, Typhlocybinae). *Bulletin de l'Academie Polonaise des Sciences. Serie des Sciences Biologiques* 28(10&11):593-602. [New genus: *Baaora.*]

1981h *Rhusia* Theron and some other Erythroneurini (Auchenorrhyncha, Cicadellidae, Typhlocybinae). *Revue de Zoologie Africaine* 95(2):315-341. [New genera: *Iseza, Tamaga.*]

1981i The genus *Wiata* Dwor. (Auchenorrhyncha, Cicadellidae, Typhlocybinae) in the Ethiopian region. *Acta Zoologica Cracoviensis* 25(20):51-544. [New subgenus: *Czecza.*]

1981j *Proskura* gen. n. and some Erythroneurini from southern India (Homoptera, Auchenorrhyncha, Cicadellidae, Typhlocybinae). *Reichenbachia* 19(37):225-246. [Descriptions of 8 new genera.]

1982a Typhlocybini of Asia (Homoptera, Auchenorrhyncha, Cicadellidae). *Entomologische Abhandlungen und Berichte aus dem Staatlichen Museum fur Tierkunde in Dresden* 45(6):99-181. [New genera: *Gratba, Mahmoba, Meketia, Sannella, Sharmana.*]

1982b Empoascini of Japan, Korea and north-east part of China (Homoptera, Auchenorrhyncha, Cicadellidae, Typhlocybinae). *Reichenbachia* 20(1):33-57. [New genus: *Okubasca.*]

1983a The leafhopper *Ranbara* gen. n. (Cicadellidae, Typhlocybinae) and remarks on *Singapora* Mahm. with descriptions of one new species from Thailand. *Thai Journal of Agricultural Science* 16:115-124.

1984a Studies on Typhlocybinae of Malaysia and Singapore. *Reichenbachia* 22(1):1-21. [New genera: *Gratba, Meketia, Sharmana, Sannella, Mahmoba.*]

Dworakowska, I. and Lauterer, P.

1975a Two new genera and five new species of Typhlocybinae (Auchenorrhyncha, Cicadellidae). *Bulletin de l'Academie Polonaise des Sciences. Serie des Sciences Biologiques* 23(1):33-40. [New genera: *Afrasca, Raunoia.*]

Dworakowska, I., Nagaich, B. B. and Singh, S.

1978a *Kapsa simlensis* sp. n. from India and some other Typhlocybinae (Auchenorrhyncha, Cicadellidae). *Bulletin de l'Academie Polonaise des Sciences. Serie des Sciences Biologiques* 26(4):243-249.

Dworakowska, I. and Pawar, A. D.

1974a Six new Oriental species of Typhlocybinae (Auchenorrhyncha, Cicadellidae). *Bulletin de l'Academie Polonaise des Sciences. Serie des Sciences Biologiques* 22(9):583-590. [New genus: *Ifugoa.*]

Dworakowska, I., Singh, S. and Nagaich, B. B.

1979a Two new species of *Dikraneura* Hardy (Auchenorrhyncha, Cicadellidae, Typhlocybinae) from India with remarks on the genus. *Entomon* 4(3):289-293.

Dworakowska, I. and Sohi, A. S.

1978a On some Typhlocybinae with remarks on synonymy of Indian species (Auchenorrhyncha, Cicadellidae). *Bulletin de l'Academie Polonaise des Sciences. Serie des Sciences Biologiques* 26(1):35-44.

1978b *Kadrabia* gen. n. and some Typhlocybinae (Auchenorrhyncha, Cicadellidae) from India. *Bulletin de l'Academie Polonaise des Sciences. Serie des Sciences Biologiques* 26(7):463-471.

Dworakowska, I., Sohi, A. S. and Viraktamath, C. A.

1980a One new genus and two new species of Zyginellini (Cicadellidae: Typhlocybinae) from India. *Oriental Insects* 14(3):271-277. [New genus: *Borulla.*]

Dworakowska, I. and Trolle, L.

1976a On some Typhlocybinae (Auchenorrhyncha, Cicadellidae) from Africa. *Bulletin de l'Academie Polonaise des Sciences. Serie des Sciences Biologiques* 24(6):365-372.

Dworakowska, I. and Viraktamath, C. A.

1975a On some Typhlocybinae from India (Auchenorrhyncha, Cicadellidae). *Bulletin de l'Academie Polonaise des Sciences. Serie des Sciences Biologiques* 23(8):521-530. [New genera: *Anaka, Mandola, Takama.*]

1978a On some Indian Typhlocybinae (Auchenorrhyncha, Cicadellidae). *Bulletin de l'Academie Polonaise des Sciences. Serie des Sciences Biologiques* 26(8):529-548. [New genus: *Livasca.*]

1979a On some Indian Erythroneurini (Auchenorrhyncha, Cicadellidae, Typhlocybinae). *Bulletin de l'Academie Polonaise des Sciences. Serie des Sciences Biologiques* 27(1):49-59. [New genera: *Ahmedra, Meremra, Tuzinka.*]

Dyck, V. A., Htun, T., Dulay, A. C., Salinas, G. D. and Orlido, G.C.

1981a Economic injury levels for rice insect pests. *Agricultural Research Journal of Kerala* 19(2):75-85.

Dysart, R. J.

1962a Local movement of the potato leafhopper in alfalfa. *Proceedings of the North Central Branch of the Entomological Society of America* 17:100-101.

Eastman, C. E., Schultz, G. A., Fletcher, J., Hemmati, K. and Oldfield, G. N.

1984a Virescence of horseradish in Illinois. *Plant Disease* 68(11):968-971

Easwaramoorthy, S. Chelliah, S. and Uthamasamy, S.

1976a Efficacy of certain insecticides against the sucking pests of bhendi, *Abelmoschus esculentus* (L.) Moench. *Madras Agricultural Journal* 63:254-256.

Eckenrode, C. J.

1973a Foliar sprays for control of the aster leafhopper on carrots. *Journal of Economic Entomology* 66(1):265-266.

1981a Influence of potato leafhopper control on kidney beans in New York. *Journal of Economic Entomology* 74:510-513.

Eckenrode, C. J. and Ditman, L. P.

1963a An evaluation of potato leafhopper damage to lima beans. *Journal of Economic Entomology* 56(5):551-553.

Eden-Green, S. J.

1979a Attempts to transmit lethal yellowing disease of coconuts in Jamaica by leafhoppers (Homoptera: Cicadelloidea). *Tropical Agriculture* 56:185-192.

Eden-Green, S. J., Markham, P. G., Townsend, R., Archer, D. B., Clark, M. F. and Tully, T. G.

1985a Plant pathogenicity and transmission tests with *Acholeplasma* spp. isolated from coconut palms in Jamaica. *Annals of Applied Biology* 106:439-449.

Edwards, C. R.
1974a Potato leafhopper control of alfalfa. *Indiana Cooperative Extension Service Publication* E-36. 3 pp.

Einyu, P. and Ahmed, M.
1977a Plant associations and abundance of typhlocybine leafhoppers in Rwenzori National Park, Uganda. *Proceedings of the Entomological Society of Karachi* 7-8:37-43.

1979a Typhlocybine leafhoppers of Rwenzori National Park, Uganda. 1. Tribe Dikraneurini (Homoptera, Auchenorrhyncha, Typhlocybinae). *Reichenbachia* 17(35):303-307. [New genera: *Ramsisa* and *Afrakeura*.]

1980a Typhlocybine leafhoppers of Rwenzori National Park, Uganda. Part II. Tribe Empoascini (Homoptera, Auchenorrhyncha, Cicadellidae). *Reichenbachia* 18(2):13-21.

1982a Erythroneurine leafhoppers from Rwenzori National Park, Uganda (Homoptera: Cicadellidae, Typhlocybinae). I. *Entomologica Scandinavica* 13(2):237-244.

1983a Erythroneurine leafhoppers from Rwenzori National Park, Uganda (Homoptera, Auchenorrhyncha, Typhlocybinae). II. *Entomologica Scandinavica* 14(3):330-336.

El-Bolok, M. M.
1981a Specific and nonspecific transmission of spiroplasmas and mycoplasma-like organisms by leafhoppers (Cicadellidae: Homoptera) with implications for etiology of aster yellows disease. Ph.D. Thesis, Cairo University, Giza. 125 pp.

El-Dine, A. M. N. and Rizkallah, R.
1971a Unrecorded insects and pests injurious to some ornamental plants in Egypt. *Bulletin de la Societe Entomologique de Egypte 1970* 54:123-127.

Elerdashvili, N. L. and Dekanoidze, G. I.
1961a Homoptera, pests on maize in Georgia. *Zaschchita Rastenii ot Vreditelei i Boleznei* 6(12):46.

Elias, S. A.
1982a Paleoenvironmental interpretation of Holocene insect fossils from northeastern Labrador, Canada. *Arctic and Alpine Research* 14(4):311-319.

El-Kady, E., Badawy, A. and Herakly, F.
1974a Differentiation between the nymphal instars of certain species of leafhoppers. (Hemiptera: Cicadellidae). *Bulletin de la Societe Entomologique d'Egypte* 58:243-247.

1974b Population studies on different species of leafhoppers infesting certain truck crops in Egypt (Hemiptera: Cicadellidae). *Bulletin de la Societe Entomologique d'Egypte* 57:387-400.

El-Kifel, A., El-Dessouki, S. A. and El-Khouly, A.
1976a The susceptibility of certain ornamental plant varieties to jassid infestation (Cicadellidae, Homoptera-Hemiptera) in Egypt. *Zeitschrift fur Angewandte Zoologie* 63:1-18.

Elliott, D. R.
1981a Species diversity of leafhoppers and vascular plants observed in old field successions. Doctoral Dissertation, University of Missouri, Columbia. 156 pp.

Ellis, C. R.
1984a Injury by *Empoasca fabae* (Homoptera: Cicadellidae) to peanuts in southwestern Ontario. *Canadian Entomologist* 116(12):1671-1673.

Ellis, C. R. and Roy, R. C.
1980a The potato leafhopper, *Empoasca fabae* (Homoptera: Cicadellidae) and other pests of peanuts in Ontario. *Proceedings of the Entomological Society of Ontario* 110:41-45.

Ellsbury, M. M. and Nielson, M. W.
1978a A revision of the genus *Spathanus* DeLong (Homoptera: Cicadellidae). *Journal of the Kansas Entomological Society* 51(2):276-282.

El-Nahal, A. K. M., Ammar, E. D. and El-Bolok, M. M.
1979a Population studies on nine leafhopper species on various host plants at Giza, Egypt, using light trap and sweeping net (Hom., Cicadellidae). *Deutsche Entomologische Zeitschrift* 26:321-327.

1981a Survey and population density of leafhoppers, planthoppers and froghoppers (Homoptera: Auchenorrhyncah) on field and vegetable crops at Giza. *Bulletin de la Societe Entomologique d'Egypte* 61:99-108.

El-Saadany, G. and Abdel-Fattah, M. I.
1980a Fluctuations of population densities of three homopterous pests, *Myzus persicae* (Sulzer), *Aphis gossypii* Glover and *Empoasca decipiens* Paoli, attacking potato plants in Egypt. *Bulletin of the Entomological Society of Egypt* 60:389-394.

El-Saadany, G., Abdel-Fattah, M. I. and Marzouk, I.
1976a On the effect of fertilizers on aphids and leafhoppers on potatoes in Egypt. *Anzeiger fur Schadlingskunde, Pflanzenschutz, Umweltschutz* 49:167-169.

Emeljanov, A. F.
1959a [New genera and species of leafhoppers (Auchenorrhyncha) from Kazakhstan.] *Entomologicheskoe Obozrenie* (Moscow) 38:833-839. [In Russian; translation in *Entomological Review* 38(4):751-757.] [New genera: *Achaetica, Chloothea, Calamotettix, Cleptochiton,* and *Condolytes*.]

1960a [Order Homoptera. *In: Corn Damaging Insects in USSR*. Reference book.] *Akademiya Nauk SSR. Zoologicheskii Institut*, pp. 7-228. [In Russian.]

1961a [New genera and species of leafhoppers (Auchenorrhyncha, Jassidae) from the USSR.] *Entomologicheskoe Obozrenie* (Moscow) 40:120-130. [In Russian; translation in *Entomological Review* 40(1):57-62.] [Descriptions of 9 new genera.]

1962a [Materials on taxonomy of Palaearctic leafhoppers (Auchenorrhyncha, Euscelinae).] *Trudy Zoologicheskogo Instituta Akademii Nauk SSR* 30:156-184. [In Russian.] [Descriptions of 12 new genera.]

1962b [New tribes of leafhoppers of the subfamily Euscelinae (Auchenorrhyncha, Cicadellidae).] *Entomologicheskoe Obozrenie* (Moscow) 41:388-397. [In Russian with English summary; translation in Entomological Review (Washington) 41(2):236-240.]

1962c [A new leafhopper genus—*Selachina* (Cicadellidae, Auchenorrhyncha).] *Zoologicheskii Zhurnal* 41(5):769-770. [In Russian with English summary.]

1963a [A new leafhopper genus from the subfamily Ulopinae (Auchenorrhyncha, Cicadellidae).] *Zoologicheskii Zhurnal* 42:1581-1582. [In Russian with English summary.] [New genus: *Bufonaria*.]

1964a [A new species of *Macropsis* Lewis (Homoptera: Cicadellidae) found on eleaster (*Eleagnus*).] *Doklady Akademii Nauk Tadzhikskoi SSR* 7(1):47-48. [In Russian.]

1964b [New genera and species of Cicadellidae from Central Asia (Homoptera).] *Doklady Akademii Nauk Tadzhikskoi SSR*, Dushanbe 7(2):52-56. [New genera: *Allotapes, Asiotuxum, Ophionotum, Tapetia*.]

1964c Suborder Cicadinea (Auchenorrhyncha). *In: Bei-Bienko, G. Y., ed. Keys to the Insects of the European USSR: Apterygota, Palaeoptera, Hemimetabola* 1:337-437. [New subgenera of *Paralimnus*: *Paragyrus* and *Anthocallis*.]

1964d [New species of steppe leafhoppers (Homoptera, Cicadellidae) from Transbaikalia and other eastern regions of the USSR.] *Entomologicheskoe Obozrenie* (Moscow) 43:626-632. [In Russian; translation in *Entomological Review* (Washington) 43(3):321-324.]

1964e [Food specialization of cicadellids based on material about the fauna of central Kazakhstan.] *Zoologicheskii Zhurnal* 43(7):1000-1010. [In Russian.]

1964f [New Auchenorrhyncha from Kazakhstan (Homoptera).] *Trudy Zoologicheskogo Instituta Akademii Nauk SSR* 34:3-51. [In Russian.] [New genera: *Cyanidius, Durgula*; seven new subgenera.]

1966a [New Palaearctic and certain Nearctic cicads (Homoptera, Auchenorrhyncha).] *Entomologicheskoe Obozrenie* (Moscow) 45:95-133. [In Russian; translation in *Entomological Review* (Washington) 45:53-74.] [New genera: *Acharis, Dorycara, Matuta, Sigista, Zelenius.*]

1966b [On the tribe Stirellini trib. n. and its taxonomic position (Homoptera, Cicadellidae).] *Zoologicheskii Zhurnal* 45:609-610. [In Russian with English summary.]

1966c [The genus *Coulinus* Beirne in the Palearctic (Homoptera, Cicadellidae).] *Zoologicheskii Zhurnal* 45(2):299-300. [In Russian with English summary.]

1967a [Some features of the distribution of oligophagous insects on food plants.] Reports on the 19th annual lecture in memory of N. A. Kholodovskii :28-65. [In Russian.]

1967b Suborder cicadinea (Auchenorrhyncha). *In: Bei-Bienko, G. Y., ed. Keys to the insects of the European U.S.S.R. I. Apterygota, Palaeoptera, Hemimetabola.* Pp. 421-551. Israel Program for Scientific Translations, Jerusalem.

1968a [Two new genera of leafhoppers (Homoptera, Cicadellidae) from Mongolia.] *Entomologicheskoe Obozrenie* (Moscow) 47:147-150. [In Russian; translation in *Entomological Review* (Washington) 47:83-84.] [New genera: *Acharista, Zapycna.*]

1968b [Notes on the tribe Stirellini (Homoptera, Cicadellidae).] *Zoologicheskii Zhurnal* 47:249-253.

1969a [New Palaearctic leafhoppers of the tribe Opsiini (Homoptera, Cicadellidae, Deltocephalinae).] *Zoologicheskii Zhurnal* 48:1100-1104. [In Russian.] [New genus: *Norva.*]

1969b [Cicadina (Homoptera, Auchenorrhyncha). *Phytocoenoses and Fauna of the Steppes and Deserts of Central Kazakhstan. Pt. 3. Composition and Analysis of the Fauna of the Western Part of Central Kazakhstan.*] Pp. 358-381.

1970a [Palaearctic leafhoppers of the genus *Athysanella* Baker (Homoptera, Cicadellidae).] *Entomologicheskoe Obozrenie* (Moscow) 49(1):161-164.

1972a [Auchenorrhyncha. *In:* Narchuk, E., et al., eds. *Insects and Arachnids. Pests of Agricultural Crops. I. Insects with Incomplete Metamorphosis.*] Pp. 117-138. *Nasekomye i kleshchi: vrediteli sel-skokhozyaistvennykh kul'tur.* Leningrad. [In Russian.]

1972b [New leafhoppers from the Mongolian People's Republic (Homoptera, Auchenorrhyncha). *Insects of Mongolia Leningrad*] 1(2):199-260. [In Russian with English title.]

1972c [Review of viewpoints on origin of desert biota of Central Asia.] *Mongol Orny Shavizh. I-r Tsuvral. Otdelnyi Ottisk. Nasekomye Mongolii. Vypusk* 1:11-49. [In Russian.]

1972d [New Palaearctic leafhoppers of the subfamily Deltocephalinae (Homoptera, Cicadellidae).] *Entomologicheskoe Obozrenie* (Moscow) 51(1):102-111. [Descriptions of 8 new genera and 1 new name.]

1972e [Short note.] Insects and Acari: pests of agricultural crops 1:283. [In Russian.]

1975a [Revision of the tribe Adelungiini.] *Entomologicheskoe Obozrenie* (Moscow) 54(2):383-390. [New genera: *Homogramma, Dalus, Pleopardus;* new name: *Parapotes* for *Paranestus.*]

1976a [New genera and species of leafhoppers (Homoptera, Auchenorrhyncha) from north-east of the USSR.] *Entomologicheskoe Obozrenie* (Moscow) 50(2):357-363.

1977a [Leafhoppers (Homoptera, Auchenorrhyncha) from the Mongolian People's Republic based mainly on materials of the Soviet-Mongolian zoological expeditions (1967-1969).] *Nosekomye Mongolii* 5:96-195. [In Russian.]

1977b [Homology of wing structures in Cicadina and primitive Polyneuroptera. Terminology and homology of venation in insects.] *Trudy Vsesoyuznogho Entomologicheskogo Obshchestva* 58:3-48. [In Russian.]

1979a [New species of the leafhoppers (Homoptera, Auchenorrhyncha) from the asiatic part of the USSR.] *Entomologicheskoe Obozrenie* (Moscow) 58(2):322-332. [In Russian.]

1981a [Morphology and origin of endopleural structures of the metathorax of the Auchenorrhyncha (Homoptera).] *Entomologicheskoe Obozrenie* (Moscow) 60(4):732-744.

Emeljanov, A. F. and Kuznetsova, V. G.

1983a [The number of seminal follicles as a phylogenetic and taxonomic feature in the Dictyopharidae (Homoptera) and other leafhoppers.] *Zoologicheskii Zhurnal* (3):79-85. [In Russian; translation in *Paleontological Journal* 17(3):1984.]

Emeljanov, A. F., Zaytsev, V. F. and Korzhner, I. M.

1968a [Entomological expedition of the Zoological Institute, USSR Academy of Sciences, to the Mongolian People's Republic in 1967.] *Entomologicheskoe Obozrenie* (Moscow) 47:942-946. [In Russian; translation in *Entomological Review* (Washington) 47(4):575-579.] [In Russian.]

Emmrich, R.

1966a Beobachtungen uber die Parasitierung von Zikadenpopulationen verschiedener Grunlandflachen der Greifswalder Umgebung (Homoptera, Auchenorhyncha). *Deutsche Entomologische Zeitschrift* 13:173-181.

1966b Faunistic-ecological investigations on the Cicada fauna of pasture land and agricultural areas in the region of Germany. *Mitteilungen aus dem Zoologischen Museum in Berlin* 42(1):61-126. [In German.]

1973a Das Typen-Material der Zikaden (Homoptera, Auchenorrhyncha) des Staatlichen Museums fur Tierkunde Dresden. I und II Teil: Cicadelloidea und Membracoidea. *Entomologische Abhandlungen Staatlichen Museum fur Tierkunde in Dresden* 39(5):247-267.

1974a Ein Beitrag zur Synonymie in der Gattung *Bathysmatophorus* J. Sahlberg (Homoptera, Auchenorrhyncha, Cicadellidae). *Reichenbachia* 15(4):33-35.

1975a Zur Kenntnis der Gattung *Oncometopia* Stal, 1869 (Homoptera, Cicadellidae, Cicadellinae). *Entomologische Abhandlungen Museum Staatliches fur Tierkunde in Dresden* 40(9):277-303.

1975b Faunistische Daten von Zikaden aus dem Gebeit der DDR. *Faunistische Abhandlungen Staatlidnen Museum Tierkunde in Dresden* 5(10):241-254.

1980a Zur taxonomischen Gliederung sowie Verbreitung von *Aphrodes bicinctus* (Schrk.) *sensu* Rib. (Homoptera, Auchenorrhyncha, Cicadellidae). *Faunistische Abhandlungen Staatlichen Museum fur Tierkunde in Dresden* 7(31):279-284.

1984a Weiteres zur Kenntnis der Gattung *Oncometopia* Stal (*s. str.*) (Homoptera, Auchenorrhyncha, Cicadellidae, Cicadellinae). *Reichenbachia* 22(15):113-124. [In German.]

Emmrich, R. and Lauterer, P.

1975a Ein Beitrag zur Kenntnis der Gattung *Hyogonia* China, 1927 (Homoptera, Cicadellidae, Cicadellinae). *Reichenbachia* 15(41):309-314.

Endo, S. and Masuda, T.

1979a Uptake and translocation of disulfoton and propaphos [4-(Methylothio) phenyl dipropl phosphate] in rice plant by seedling-box application and the insecticidal effect on the green rice leafhopper. *Proceedings of the Association for Plant Protection of Kyushu* 25:77-78.

Enkerlin, S. D. and Medina, M. B.
1979a Susceptibility of 99 lines and varieties of beans *Phaseolus vulgaris* L. to attack by *Empoasca* spp. and their adaptation to the environmental conditions of Apadaca, N. L. *Informe de Investigacion* No. 16, pp. 93-95.

Esau, K., Magyarosy, A. C. and Breazeale, V.
1976a Studies of the mycoplasma-like organism (MLO) in spinach leaves affected by the aster yellows disease. *Protoplasma* 90:189-203.

Escalante G., J. A., Del Castillo E., M. and Ochoa M., O.
1981a Catalogo preliminar de las plazas insectiles de papa, maiz, y frutales en el Departamento del Cuzco, Peru. *Revista Peruana Entomologia* 24(1):87-90.

Escobedo, B. L., Vera, G. J., Rodrigues, M. R. and Bravo, M. H.
1985a Tablas de vida y fertilidad de poblaciones de *Dalbulus maidis* de Long and Wolcott y *Dalbulus elimatus* Ball (Homoptera: Cicadellidae) transmisoras y no transmisoras del virus del rayado fino del maiz. *Agrociencia* 57:195-205.

Esipenko, P. A.
1973a The biology of *Leptogaster cylindrica* Deg. (Diptera, Asilidae) in the Khabarovsk territory. *Trudy Biologo-pochvennogo Instituta, Dal'nevostochnyi Nauchnyi Tsentr, Akademiya Nauk SSR* No. 5, pp. 130-134. [Predaceous on leafhoppers.]

Eskafi, F. M.
1982a Leafhoppers and planthoppers feeding on coconut palm in Jamaica. *Tropical Agriculture* 59:289-292.

Eskafi, F. M. and Van Schoonhoven, A.
1978a Comparison of greenhouse and field resistance of bean varieties to *Empoasca kraemeri* (Homoptera: Cicadellidae). *Canadian Entomologist* 110:853-858.
1981a Interactions of leafhopper population, varietal resistance, insecticide treatment, and plant growth on dry bean yields in a tropical environment. *Journal of Economic Entomology* 74(1):7-12.

Espinoza, A. M. and Gamez, R.
1980a La ultrastructura de la superficie foliar de cultivares de maiz infectados con el virus del rayado fino. *Turrialba* 30:413-420.

Estrada, R. F. A.
1960a Lista preliminar de insectos asociados al maiz en Nicaragua. *Turrialba* 10(2):68-73.

Eto, M., Oshima, Y., Kitakata, S., Tanaka, F. and Kojima, K.
1966a Studies on saligenin cyclic phosphorus esters with insecticidal activity. Part X. Synergism of malathion against susceptible and resistant insects. *Botyu-Kagaku* 31(1):33-38. [With Japanese summary.]

Evans, D. E.
1965a Jassid populations on three hairy varieties of Sakel cotton. *Empire Cotton Growing Review* 42(3):211-217.
1966a The distribution of *Empoasca lybica* (de Berg) (Hemiptera, Cicadellidae) on cotton in the Sudan. *Bulletin of Entomological Research* 56(4):635-647.

Evans, F.
1981a The Tartessinae of Australia, New Guinea and some adjacent islands (Homoptera: Cicadellidae). *Pacific Insects* 23(1&2):112-188. [Descriptions of 32 new genera.]

Evans, H. E.
1968a Notes on some digger wasps that prey upon leafhoppers. *Annals of the Entomological Society of America* 61(5):1343-1344.

Evans, J. W.
1956a Palaeozoic and Mesozoic Hemiptera. *Australian Journal of Zoology* 4:165-258.
1956b Some ancient bugs. *Australian Museum Magazine* 12:76-80.
1956c Australian insects. *Australian Museum Magazine* 12:129-132.
1957a Los insectos de las Islas Juan Fernandez (Cicadellidae Homoptera). *Revista Chilena de Entomologia* 5:365-374. [New genera: *Agalita*, *Kuscheliola* and *Stenagallia*.]
1957b Some aspects of the morphology and inter-relationships of extinct and recent Homoptera. *Transactions of the Royal Entomological Society of London* 109:275-294.
1957c The distribution of animals and plants in the southern hemisphere. *Australian Museum Magazine* 12:270-272.
1958a Character selection in systematics with special reference to the classification of leafhoppers (Insecta, Homoptera, Cicadelloidea). *Systematic Zoology* 7(3):126-131.
1958b Insect distribution and continental drift. *Continental Drift, a Symposium*. Geology Department, University of Tasmania 1956:134-161.
1958c Some Upper Permian Homoptera from the Belmont beds. *Record of the Australian Museum* 24:109-114.
1959a Quelques nouveaux Cicadellides (Homopteres) de Madagascar: Observations sur la signification evolutive de la speciation dans le genre *Coloborrhis* Germar. *Memoires de l'Institut Scientifique de Madagascar - Tananarive* (E) 11:481-507. [In French.] [New genus: *Ledracorrhis*.]
1959b The zoogeography of some Australian insects. In: *Biogeography and Ecology in Australia*. Pp. 150-163. W. Junk, The Hague.
1960a Factors associated with the possible origin of some of the higher categories of leafhoppers (Homoptera - Cicadellidae). In: *Proceedings of the XI International Congress of Entomology* I:43-45.
1960b Fauna Australiensis. *Australian Journal of Science* 22:433-435.
1961a [1962] Leafhoppers from Chile collected by the Royal Society expedition to southern Chile, 1958-1959 (Homoptera, Cicadelloidea). *Annals and Magazine of Natural History* 13(4):513-517. [New genus: *Holdgatiella*.]
1961b Some Upper Triassic Homoptera from Queensland. *Memoirs of the Queensland Museum* 14:13-23.
1961c Fossil insects. *Australian Museum Magazine* 13:294-296.
1962a Evolution in the Homoptera. In: *The Evolution of Living Organisms*. Pp. 250-259. Melbourne University Press, Melbourne, Australia.
1963a The phylogeny of the Homoptera. *Annual Review of Entomology* 8:77-94.
1963b The zoogeography of New Zealand leafhoppers and froghoppers (Insecta, Homoptera, Cicadelloidea and Cercopoidea). *Transactions of the Royal Society of New Zealand Zoology* 3(9):85-91.
1963c The systematic position of the Ipsviciidae (Upper Permian Hemiptera) and some new Upper Permian and Middle Triassic Hemiptera from Australia (Insecta). *Journal of the Entomological Society of Queensland* 2:17-23.
1964a The periods of origin and diversification of the superfamilies of the Homoptera-Auchenorhyncha (Insecta) as determined by a study of the wings of Palaeozoic and Mesozoic fossils. *Proceedings of the Linnean Society of London* 175(2):171-181.
1965a The distribution of the Ulopinae (Homoptera, Cicadellidae). *Proceedings of the XII International Congress of Entomology,* London (1964):469.
1965b A new genus and species of Eurymelidae (Homoptera, Cicadelloidea). *Proceedings of the Linnean Society of New South Wales* 90:85-86. [New genus: *Aloeurymela*.]
1966a The leafhoppers and froghoppers of Australia and New Zealand (Homoptera: Cicadelloidea and Cercopoidea). *Memoirs of the Australian Museum* 12:1-347. [Descriptions of 30 new genera.]
1968a Some relict New Guinea leafhoppers and their significance in relation to the comparative morphology of the head and prothorax of the Homoptera-Auchenorrhyncha (Homoptera-Cicadellidae: Ulopinae). *Pacific Insects* 10(2):215-229. [New genus: *Monteithia*.]

1969a Characteristics and components of Ledrinae and some new genera and new species from Australia and New Guinea. *Pacific Insects* 11(3&4):735-754. [New genera: *Cololedra, Microledrella, Putoniessiella,* and *Thymbrella.*]

1969b Notes on the Eurymelidae (Homoptera: Cicadelloidea) and some new species from eastern and western Australia. *Proceedings of the Royal Society of Queensland* 81(4):51-55.

1971a Two new genera and species of Oriental Cicadellidae and remarks on the significance of male genitalia in leafhopper classification (Homoptera: Cicadelloidea). *Journal of Entomology,* Series B 40(1):43-48. [New genera: *Placidellus* and *Kasinella.*]

1971b Leafhoppers from New Guinea and Australia belonging to the sub-families Macropsinae and Agalliinae with notes on the position of *Nionia* Ball and *Magnentius* Pruthi. *Pacific Insects* 13(2):343-360.

1971c Some new African Ulopini (Homoptera, Cicadellidae, Ulopinae). *Journal of Natural History* 5:441-445. [New genera: *Conlopa* and *Delopa.*]

1971d Some Upper Triassic insects from Mt. Crosby, Queensland. *Memoirs of the Queensland Museum* 16(1):145-162.

1972a Characteristics and relationships of Penthimiinae and some new genera and new species from New Guinea and Australia; also new species of Drabescinae from New Guinea and Australia. (Homoptera: Cicadellidae). *Pacific Insects* 14(1):169-200. [Descriptions of 16 new genera.]

1972b Some leafhoppers from New Guinea, Australia and Thailand belonging to the subfamily Jassinae and a new genus from New Guinea referred to a new subfamily the Acostemminae (Homoptera: Cicadellidae). *Pacific Insects* 14(4):647-662. [Descriptions of 6 new genera.]

1972c Some remarks on the Family Jascopidae (Homoptera, Auchenorrhyncha). *Psyche* 79(1&2):120-121.

1973a Some new genera and species of Cicadelloidea from Australia and New Guinea (Homoptera). *Pacific Insects* 15(2):185-197. [New genera: *Alospangbergia, Aloxestocephalus,* and *Micrelloides.*]

1973b The maxillary plate of Homoptera-Auchenorrhyncha. *Journal of Entomology Series (A)* 48(1):43-47.

1974a New Caledonian leafhoppers and the systematic position of *Kosmiopelix* Kirkaldy and *Eucanthella* Evans (Homoptera: Cicadelloidea). *Pacific Insects* 16(2&3):165-175. [New genus: *Selenomorphus.*]

1975a The structure, functions and possible origin of the subgenital plate of leafhoppers (Homoptera: Cicadelloidea). *Journal of the Australian Entomological Society* 14:77-80.

1975b The external features of the heads of leafhoppers (Homoptera, Cicadelloidea). *Records of the Australian Museum* 29(14):407-440.

1977a The leafhoppers and froghoppers of Australia and New Zealand. (Homoptera: Cicadelloidea and Cercopoidea). Part 2. *Records of the Australian Museum* 31(3):83-129. [Descriptions of 6 new genera.]

1982a Biogeography of New Guinea leafhoppers (Homoptera: Cicadelloidea). *Monographiae Biologicae* 42:639-644.

Evans, J. W., Woodward, T. E. and Eastop, V. F.

1970a Hemiptera. In: *The Insects of Australia.* Chapter 26. Melbourne University Press, Melbourne, Australia.

Evenhuis, H. H.

1955a Over de Cicadellidenfauna van de kers. [Observations on leafhoppers of cherry.] *Tijdschrift over Plantenziekten* 61(2):56-59. [With English summary.]

1955b Second conference on potato virus diseases, proceedings. 12. Observations on some leafhopper-borne virus diseases in the Netherlands. *Lisse Laboratory for Flowerbulb Research* 104:84-88.

1958a Investigations on leafhopper-borne clover virus. *Proceedings of the Conference on Potato Virus Diseases,* Wageningen-Lisse, pp. 251-254.

1958b The vectors of the virus causing phyllody (virescence) in clover flowers. *Tijdschrift over Plantenziekten* 64(4):335:336.

Eyles, A. C. and Linnavuori, R.

1974a Cicadellidae and Issidae (Homoptera) of Niue Island, and material from the Cook Islands. *New Zealand Journal of Zoology* 1(1):29-44.

Ezueh, M. I.

1976a An evaluation of ULV sprays for the control of cowpea insect pests in southern Nigeria. *Tropical Grain Legume Bulletin* No. 4, pp. 15-18.

Faan, H. C., Lai, W. J., Chan, D. S. and Tom, W. F.

1964a A preliminary study on the incidence and transmission of mulberry dwarf disease in Kwangtung province. *Acta Phytopathologica Sinica* 7:151-156.

Faan, H. C., Zhang, S. G., He, X. Z., Zie, S. D., Liu, C. Z., Zhou, L. G., Liu, X. R. and Zhu, D.

1983a Rice gall dwarf—a new virus disease epidemic in the west of Guangdong Province of South China. *Acta Phytopathologica Sinica* 13:1-4.

Fabellar, L., Heinrichs, E. A. and Rapusas, H.

1981a Response to insecticides of green leafhoppers from four sites in the Philippines. *International Rice Research Newsletter* 6(6):8.

Fabellar, L. T. and Mochida, O.

1985a Sensitivity levels of green leafhopper (GLH) populations to insecticide at IRRI in 1983-1984. *International Rice Research Newsletter* 10(3):30-31.

[Fachrudin]

1980a Bionomi *Nephotettix virescens* (Distant) (Homoptera: Cicadellidae, Euscelidae). Disertasi, Sekolah Pasco Sarjana, Institut Pertanian, Bogor. 181 pp.

Fajardo, T. G., Bergonia, H. T., Capule, N. and Novero, E.

1964a Studies on rice diseases in the Philippines I. Progress report on "tungro" disease of rice. In: *10th Meeting of FAO-IRC Working Party on Rice Production and Protection,* Manila, Philippines.

Fajemisin, J. M., Cook, G. E., Okusanya, F. and Shoyinka, S. A.

1976a Maize streak epiphytotic in Nigeria. *Plant Disease Reporter* 60:443-447.

Fajemisin, J. M. and Shoyinka, S. A.

1977a Maize streak and other maize virus diseases in West Africa. In: Williams, L. E., Gordon, D. T. and Nault, L. R., eds. *Proceedings of the International Maize Virus Disease Colloquium and Workshop,* Ohio Agricultural Research and Development Center, Wooster, Ohio, pp. 52-60.

Faleiro, J. R., Rai, S. and Vasisht, A. K.

1982a Sequential sampling plan for the management of leafhopper population on okra. *Annals of Agricultural Research* 3(1-2):164-170.

Falk, B. W., Weathers, L. G. and Greer, F. C.

1981a Identification of potato yellow dwarf virus occurring naturally in California. *Plant Disease* 65:81-83.

Falk, J. H.

1982a Response of two turf insects, *Endria inimica* and *Oscinella frit,* to mowing. *Environmental Entomology* 11:29-31.

Faris, M. A., Baenziger, H. and Terhune, R. P.

1981a Studies on potato leafhopper (*Empoasca fabae*) damage in alfalfa. *Canadian Journal of Plant Science* 61:625-632.

Farrar, M. D. and Woddworth, C. M.

1939a New strains of alfalfa studies for leafhopper resistance. *Illinois Agriculture Experiment Station Annual Report* 50:167.

Feng, H. T., Kao, L. R., Chou, S. M., Tseng, Y. J., Chung, T. C., Chen, C. F., Kao, H. L. and Sun, C. N.

1979a Rice green leafhopper susceptibility to vamidothion in central Taiwan. *International Rice Research Newsletter* 4(2):19-20.

Feng, Y. X.

1981a Rice resistance to the rice leafhopper *Nephotettix cincticeps* Uhler. *Acta Entomologica Sinica* 24:471-474.

Fennah, R. G.
> 1959a A new species of *Cicadulina* (Homoptera: Cicadellidae) from East Africa. *Annals and Magazine of Natural History* 13(2):757-758.
> 1983a Prospects for the use of biotaxonomy in characterisation of homopterous pests in relation to cost and other factors affecting taxonomic output. *In: Knight, W. J., Pant, N. C., Robertson, T. S. and Wilson, M. R., eds. Proceedings of the 1st International Workshop on Biotaxonomy, Classification and Biology of Leafhoppers and Planthoppers (Auchenorrhyncha) of Economic Importance.* Pp. 493-496. London, 4-7 October, 1982. Commonwealth Institute of Entomology, 56 Queen's Gate, London SW 7 5JR. 500 pp.

Fenton, F. A.
> 1959a The effect of several insecticides on the total arthropod population in alfalfa. *Journal of Economic Entomology* 52:428-432.

Fenton, F. A. and Howell, D. E.
> 1957a A comparison of five methods of sampling alfalfa fields for arthropod populations. *Annals of the Entomological Society of America* 50(6):606-611.

Fernando, H. E.
> 1959a Studies of *Empoasca devastans* Dist. (fam. Jassidae, ord. Hemiptera), a new pest of cacao causing defoliation, and its control. *Tropical Agriculturist* (Colombo) 115(2):121-144.

Fernando, H. E. and Manickavasagar, P.
> 1958a Investigations on potato insects and their control with special references to *Dorylus orientalis*. *Tropical Agriculturalist* (Colombo) 114(2):127-139.

Fernando, H. E., Weerawardebs, G. A. and Manickavasgar, P.
> 1954a Paddy pests control in Ceylon. *Tropical Agriculturist* (Colombo) 110(3):159-174.

Finkner, R. E. and Scott, P. R.
> 1972a Sugarbeet cultivar and systemic insecticide interrelationships in the control of curly top virus. *Journal of the American Society of Sugar Beet Technologists* 17(2):97-104.

Fiori, A.
> 1983a New or little known hosts of dryinid Hymenoptera (Hymenoptera: Dryinidae). *Frustula Entomologica* 6:141-145.

Fiori, B. J. and Dolan, D. D.
> 1981a Field tests for *Medicago* resistance against the potato leafhopper (Homoptera: Cicadellidae). *Canadian Entomologist* 113:1049-1053.

Fisher, J. B. et al.
> 1973a Report of the lethal yellowing symposium at Fairchild Tropical Garden, Miami. *Principes* 17(4):151-159.

Fisher, R. W., Menzies, D. R., Sutton, J. C. and Stevenson, A. B.
> 1980a Comparative efficacy of hydraulic boom and compressed-air boom sprayer for spray coverage and control of leafhoppers and blight in carrots. *Proceedings of the Entomological Society of Ontario* 111:3-6.

Flaherty, D. L., Peacock, W. L. and Jensen, F. L.
> 1978a Grape pest management in the San Joaquin Valley. *California Agriculture* 32:17-18.

Fleischer, S. J. and Allen, W. A.
> 1982a Field counting efficiency of sweep-net samples of adult potato leafhoppers (Homoptera: Cicadellidae) in alfalfa. *Journal of Economic Entomology* 75(5):837-840.

Fleischer, S. J., Allen, W. A., Luna, J. M. and Pienkowski, R. L.
> 1982a Absolute-density estimation from sweep sampling techniques for adult potato leafhoppers in alfalfa. *Journal of Economic Entomology* 75:425-430.

Fleischer, S. J., Allen, W. A. and Pienkowski, R. L.
> 1983a Relationship between absolute density and sticky trap catches of adult potato leafhoppers in alfalfa. *Journal of the Georgia Entomological Society* 18:213-218.

Fletcher, M. J.
> 1982a Australian leafhoppers. *Entomology Society of New South Wales Circular* No. 327:34-37.
> 1984a The identification of leafhoppers of economic importance in Australia. *In: Bailey, P. and Swincer, D., eds. Proceedings of the Fifth Australian Applied Entomological Research Conference. Pest Control: Recent Advances and Future Prospects.* Pp. 418-421. Adelaide, Australia.

Fletcher, J., Schultz, G. A., Davis, R. E., Eastman, C. E. and Goodman, R. M.
> 1981a Brittle root disease of horseradish: Evidence for an etiology role of *Spiroplasma citri*. *Phytopathology* 71:1073-1079.

Flinn, P. W.
> 1981a Effects of density and age class of the potato leafhopper on seedling alfalfa development. M. S. Thesis, Pennsylvania State University, State College.

Flinn, P. W. and Hower, A. A.
> 1984a Effects of density, stage and sex of the potato leafhopper, *Empoasca fabae* (Homoptera: Cicadellidae), on seedlings and alfalfa growth. *Canadian Entomologist* 116(11):1543-1548.

Flinn, P. W., Hower, A. A. and Taylor, R. A. J.
> 1985a Preference of *Reduviolus americoferus* (Hemiptera: Nabidae) for potato leafhopper nymphs and pea aphids. *Canadian Entomologist* 117:1503-1508.

Flint, J. H.
> 1963a The insects of the Malhan Tarn area. Hemiptera. *Proceedings, Leeds Philosophical and Literary Society, Scientific Section* 9(2):25-29.
> 1976a *Eupteryx origani* Zach, (Hem., Cicadellidae) in Yorkshire. *Entomologist's Monthly Magazine* 111:204.

Flock, R. A.
> 1977a Citrus stubborn disease in relation to the beet leafhopper control program in Imperial County, California. *Proceedings of the American Phytopathological Society* 4:205.

Flock, R. A. and Deal, A. S.
> 1959a A survey of beet leafhopper populations on sugar beets in the Imperial Valley, California, 1953-1958. *Journal of Economic Entomology* 52(3):470-473.

Flock, R. A., Doutt, R. L., Dickson, R. C. and Laird, E. F., Jr.
> 1962a A survey of beet leafhopper egg parasites in the Imperial Valley, Calif. *Journal of Economic Entomology* 55(3):277-281.

Flores, R. F. and Aguirre, F. R. V.
> 1972a Estudio del ciclo biologico de la chicharrita de le vid *Erythroneura* spp. y su control por medio de insecticidas sistemicos aplicados en riego por goteo. Halffter, G., ed. VII Congreso Nacional de Entomologia, 13-16 octubre 1970, Mexico, D. F. Entomologia agricola: cultivos diversos. *Folia Entomologica Mexicana* No. 23/24:47-48. [In Spanish.]

Fluiter, H. J., de
> 1958a Leafhoppers as vectors of plant viruses. *Archives Neerlandaises de Zoologie* 12:557-562.

Fluiter, H. J., de, Evenhuis, H. H. and van der Meer, F. A.
> 1955a Observations on some leafhopper-borne virus diseases in the Netherlands. *Proceedings of the 2nd Conference of Potato Virus Diseases* 1954:84-88.

Fluiter, H. J., de and van der Meer, F. A.
> 1956a Bestrijdings mogelijkheden van de vector van de dwergziekte van de framboos. *Tijdschrift over Plantenziekten* 62:26.
> 1958a The biology and control of *Macropsis fuscula* Zett., the vector of the *Rubus* stunt virus. *Proceedings of the 10th International Congress of Entomology* 3:341-345.

Folliot, R. and Maillet, P. L.
> 1967a Problemes souleves par l'infection virale du testicule chez un Insecte Homoptere. *Compte Rendu des Seances de la Societe de Biologie* 161:913-914.

Forbes, A. R. and MacCarthy, H. R.
> 1969a Morphology of the Homoptera with emphasis on virus vectors. *In: Maramorosch, K., ed. Viruses, Vectors, and Vegetation.* Pp. 211-234. Wiley & Sons, New York. 666 pp.

Forbes, A. R. and Raine, J.
1973a The stylets of the six-spotted leafhopper, *Macrosteles fascifrons* (Homoptera: Cicadellidae). *Canadian Entomologist* 105:559-567.

Forsythe, H. Y. and Gyrisco, G. G.
1961a Determining the appropriate transformation of data from insect control experiments for use in the analysis of variance. *Journal of Economic Entomology* 54(5):859-861.

Forsythe, H. Y., Hardee, D. D. and Gyrisco, G. G.
1962a Field tests for the control of certain alfalfa insect pests in New York. *Journal of Economic Entomology* 55(6):828-830.

Fos, A.
1977a Preliminary entomological observations on cherry decline in Tarn-et-Garonne. *Revue de Zoologie Agricole et de Pathologie Vegetale* 73:134-140.

Foster, J. A.
1982a Plant quarantine problems in preventing the entry into the United States of vector-borne plant pathogens. *In: Harris, K. F. and Maramorosch, K., eds. Pathogens, Vectors, and Plant Diseases: Approaches to Control.* Pp. 151-185. Academic Press, New York. 310 pp.

Fox, J. W.
1963a Reports on the Margaret M. Cary and Carnegie Museum expedition to Baja California, Mexico, 1961. 3. A portable ultraviolet insect trap. *Annals Carnegie Museum* 36:205-212.

Frankel, O. H. and Hawkes, J. G., eds.
1975a *Crop Genetic Resources for Today and Tomorrow.* Cambridge University. 492 pp.

Fraval, A.
1968a Procedes et materiels mis au point pour l'elevage et la recolte d'Homopteres (Aphididae et Jassidae) a l'occasion d'etudes sur la transmission de virus. *Revue de Zoologie Agricole et Appliquee* 67(7-9):87-94.

Frazier, N. W.
1965a Xylem viruses and their insect vectors. *Proceedings of the International Conference of Viruses and Vectors on Perennial Hosts, with Special Reference to Vitis.* Pp. 91-99. University of California.

1975a Possible transmission of strawberry pallidosis by the leafhopper *Coelidia olitoria*. *Plant Disease Reporter* 59(1):40-46.

Frazier, N. W. and Jensen, D. D.
1970a Strawberry, an experimental host of peach Western X-disease. *Phytopathology* 60:1527-1528.

Frazier, N. W. and Posnette, A. F.
1956a Leafhopper transmission of a clover virus causing green petal disease in strawberry. *Nature* 177:1040-1041.

1957a Transmission and host-range studies of strawberry green-petal virus. *Annals of Applied Biology* 45:580-588.

Frederiksen, R. A.
1962a Evidence for the simultaneous transmission of two viruses by the aster leafhopper (*Macrosteles fascifrons*). *Phytopathology* 52:732.

1964a Simultaneous infection and transmission of two viruses in flax by *M. fascifrons*. *Phytopathology* 54(8):1028-1030, illust.

1964b Aster yellow of flax. *Phytopathology* 54(1):44-48.

Frederiksen, R. A. and Christensen, J. J.
1957a Aster yellows destructive to flax in the upper Mississippi Valley in 1957. *Plant Disease Reporter* 41(12):994.

Frederiksen, R. A. and Goth, R. W.
1959a Crinkle, a new virus disease of flax. (Abstract.) *Phytopathology* 49(9):538.

Frediani, D.
1955a Appunti sulla variabilita di un carattere chetotattico dei femori posteriori in una popolazione di *Cicadella viridis* L. *Memorie della Societa Entomologica Italiana* 33:141-146.

1956a Appunti sulla variabilita di un carattere chetotattico dei femori posteriori in una populazione di *Cicadella viridis* L. (II nota). *Memorie della Societa Entomologica Italiana* 35:35-42.

1956b Note morfo-biologiche sulla *Cicadella viridis* L. (Homoptera Jassidae) nell'Italia Centrale. *Bollettino del Laboratorio di Entomologia Agraria Portici* 14:1-47.

1958a Reperti sulla costituzione dell'apparato boccale della *Cicadella viridis* L. (Homoptera, Jassidae). *Bollettino del Laboratorio di Entomologia Agraria "Filippo Silvestri"* 16:148-159. [English summary.]

Freitag, J. H.
1956a Western aster yellows virus infection of squash, pumpkin, and cucumber. *Phytopathology* 46(6):323-326.

1958a Cross-protection tests with 3 strains of the aster yellow virus in host plants and in the aster leafhopper. *Phytopathology* 48(8):393.

1962a Leafhopper transmission of Western aster-yellows virus to legumes and solanaceous plants. *Phytopathology* 52(2):128-133.

1962b Leafhopper transmission of three strains of aster yellows virus to barley. *Proceedings of the Czechoslakian Conference of Plant Virologists*, Prague 5:136-142.

1963a Cross protection of three strains of the aster-yellows virus in the leafhopper and in the plant. *Netherlands Journal of Plant Pathology* 69:215.

1964a Interaction and mutual suppression among three strains of aster-yellows virus. *Virology* 24(3):401-413.

1967a Interaction between strains of aster yellows virus in the six-spotted leafhopper, *Macrosteles fascifrons*. *Phytopathology* 57:1016-1024.

1969a Interactions of plant viruses and virus strains in their insect vectors. *In: Maramorosch, K., ed. Viruses, Vectors, and Vegetation.* Pp. 303-325. Wiley & Sons, New York. 666 pp.

Freitag, J. H., Aldrich, T. M. and Drake, R. M.
1962a The control of the spread of aster yellows virus to celery. *Mededelingen van de Landbouwhogeschool ende Opzoekingsstations van de Staatte Gent* 27:1047-1052.

Freitag, J. H., Jensen, D. D. and Lateer, R. W.
1954a Aster yellows of gladiolus. *University of California Agricultural Experiment Station Service. Alemeda County Circular* 37. 5 pp.

Freitag, J. H. and Smith, S. H.
1969a Effects of tetracyclines on symptom expression and leafhopper transmission of aster yellows. *Phytopathology* 59(12):1820-1823.

French, W. J. and Feliciano, A.
1982a Distribution and severity of plum leaf scald disease in Brazil. *Plant Disease* 66:515-517.

Freytag, P. H.
1962a A new species of *Idiocerus* from the Southwest and a review of the related species. *Ohio Journal of Science* 62:244-252.

1964a A revision of the Nearctic species of the genus *Idiocerus* (Homoptera: Cicadellidae: Idiocerinae). Thesis, Ohio State University.

1965a A revision of the Nearctic species of the genus *Idiocerus* (Homoptera, Cicadellidae, Idiocerinae). *Transactions of the American Entomological Society* 91(4):391-430.

1965b The Nearctic species of the genus *Idiocerus* (Homoptera) in relation to their host plants. *Proceedings of the XII International Congress of Entomology*, London (1964):104-105.

1967a A new species of *Idiocerus* from Honduras. *Ceiba* 13(2):81-82.

1969a *Jamacerus* and *Optocerus*, two new genera of Central American and Jamaican Idiocerinae. (Homoptera: Cicadellidae). *Annals of the Entomological Society of America* 62(2):348-353.

1970a A new species of *Jamacerus* from Florida. *Ohio Journal of Science* 70(5):304.

1971a The type specimens of the genus *Chunroides*. *Entomological News* 82:39-41.

1971b Distribution of leafhoppers in Kentucky. *University of Kentucky Agricultural Experiment Station Annual Report* [1970], p. 87.

1974a A new genus of leafhoppers—*Nanopsis* (Homoptera: Cicadellidae). *Annals of the Entomological Society of America* 67(4):605-606.

1975a A new species of *Idiocerus* from Panama (Homoptera: Cicadellidae). *Ohio Journal of Science* 75(1):25-26.

1976a The leafhoppers of Kentucky. Part 1: Agalliinae, Idiocerinae and Macrospinae. *University of Kentucky, College of Agriculture, Agricultural Experiment Station Progress Report* 223. 63 pp.

1976b A new species of *Scaphoideus* from Formosa (Homoptera: Cicadellidae). *Entomological News* 87(5&6):171-175.

1976c A review of the genus *Neogonatopus* for North America (Hymenoptera: Dryinidae). *Annals Entomological Society of America* 70:569-576.

1979a Additions to the genus *Folicana* (Homoptera-Cicadellidae- Gyponinae). *Journal of the Kansas Entomological Society* 52(4): 810-819.

1983a A new species of *Unerus* from Honduras (Homoptera: Cicadellidae). *Entomological News* 94(5):187-190.

1985a The insect parasites of leafhoppers and related groups. In: Nault, L. R. and Rodriguez, J. G., eds. *The Leafhoppers and Planthoppers*. Pp. 423-467. Wiley & Sons, New York. 500 pp.

Freytag, P. H. and Cwikla, P. S.

1982a New species and records of Gyponinae from Dominica (Homoptera: Cicadellidae). *Journal of the Kansas Entomological Society* 55(4):658-664.

1984a Two new species of Idiocerine leafhoppers from Malagasy Republic (Homoptera: Cicadellidae). *Pan-Pacific Entomologist* 60(4):341-344.

Freytag, P. H. and Back, D. W.

1977a Redescription, synonymy and host of *Gonatopus bicolor* (Hymenoptera: Dryinidae). *Entomology News* 88:221-227.

Freytag, P. H. and DeLong, D. M.

1968a Corrective note on *Gypona decorata* Fowler. *Ohio Journal of Science* 68(5):333.

1971a Studies of the Gyponinae: Additions to the genus *Chilenana*. *Journal of the Kansas Entomological Society* 44(4):540-543.

1975a *Gyponana* (*Zerana*) *secunda*, a new species of leafhopper from Panama and Grand Cayman Island (Homoptera: Cicadellidae). *Entomological News* 86(7&8):141-143.

1982a Additional records and new species of the leafhopper genus *Scaris* (Homoptera, Cicadellidae). *Ohio Journal of Science* 82(1):2-13.

Freytag, P. H. and Knight, W. J.

1966a The Idiocerinae of Madagascar (Hom., Cicadellidae). *Annales de la Societe Entomologique de France* 11(1):75-103. [New genus: *Nesocerus.*]

Freytag, P. H. and Morrison, W. P.

1969a A preliminary study of the Idiocerinae of Chile (Homoptera: Cicadellidae). *Entomological News* 80(11):285-292. [New genus: *Chileanoscopus.*]

1972a A new genus of leafhoppers from New Guinea. *Entomological News* 83:41-44. [New genus: *Balocerus.*]

1973a Two new species of *Balocerus* from Thailand and China (Homoptera: Cicadellidae). *Ohio Journal of Science* 73(2):111.

Frick, K. E. and Hawkes, R. B.

1970a Additional insects that feed upon tansy ragwort, *Senecio jacobaea*, an introducted weedy plant, in western United States. *Annals of the Entomological Society of America* 63(4):1085-1090.

Fritzche, R., Karl, E., Lehmann, W. and Proesler, G.

1972a *Tierische Vektoren Pflanzenpathogener Viren*. G. Fischer, Stuttgart. 531 pp.

Frost, S. W.

1956a Traps and lights to catch night-flying insects. *Proceedings of the 10th International Congress of Entomology* 2:583:587.

1957a The Pennsylvania insect light trap. *Journal of Economic Entomology* 50:287-292.

Fujimura, T. and Somasundaram, P. H.

1984a Comparative observations of rice insect populations on indica and japonic rices. *International Rice Research Newsletter* 9(3):16.

Fukamachi, S.

1976a Field resistance to the brown planthopper and green rice leafhopper in the rice lines bred at TARC. *Proceedings of the Association for Plant Protection of Kyushu* 22:102-103.

Fukuda, H.

1966a Rice insect control by granular insecticide. *JARQ* 1(1):17-20.

Fukushi, T.

1969a Relationships between propagative rice viruses and their vectors. In: Maramorosch, K., ed. *Viruses, Vectors, and Vegetation*. Pp. 279-301. Wiley & Sons, New York. 666 pp.

Fukushi, T. and Shikata, E.

1963a Localization of rice dwarf virus in its insect vector. *Virology* 21:503-505.

Fukushi, T., Shikata, E., Shioda, H., Sekiyam, A. E., Takaka, L., Oshima, N. and Nishia, Y.

1955a Insect transmission of potato witches' broom in Japan. *Proceedings of the Japan Academy* 31:234-236.

Fullaway, D. T.

1956a Notes and exhibitions. *Proceedings of the Hawaiian Entomological Society for 1955* 16(1):3.

Fullerton, D.

1964a Beet leafhopper and curly-top disease survey in Washakie County (Wyoming). *Journal of the American Society of Sugar Beet Technology* 13(1):62-67.

Furniss, R. L. and Carolin, V. M.

1977a Western forest insects. *Miscellaneous Publication* No. 1339. U.S. Department of Agriculture Forest Service. 654 pp.

Furuta, T.

1977a Rice waika, a new virus disease found in Kyushu, Japan. *Review of Plant Protection Research* 10:70-82.

Gaborjanyi, R. and Nagy, F.

1972a Termesztett gyogynovenyeink virus-es mikoplazma betegsegei. [Virus and mycoplasma disease of cultivated medicinal plants in Hungary.] *Herba Hungarica* 11(2):39-51. [In Hungarian with English and Russian summaries.]

Gaborjanyi, R. and Saringer, G.

1967a A sztolbur virust ferjeszio kabocak biologiaja es a vedekezes lehetosegei Magyarorszagon. [The biology of leafhopper vectors of stolbur virus and possibilities for their control in Hungary]. *Kiserletugyi Kozlemenyek* (A) 60(1-3):3-12. [In Hungarian with Russian and English summaries.]

Gaddoura, W. M. and Venkatraman, T. V.

1967a A simple method for counting leafhopper eggs inserted in plant tissue. *Current Science* 36(22):619.

Gaedike, H.

1971a Katalog der in den Sammlungen des ehemaligen Deutschen Entomologischen Institutes aufbewahrten Typen. VI. (Homoptera [exklusive Aphidina]). *Beitraege zur Entomologische* 21(3-6):315-339.

Gajewskaja, N. S.

1965a On food relations of Homopteran insects (Insecta, Homoptera) with higher aquatic plants. *Biulleten' Moskovskogo Obshchestva Ispytatelei Prirody, Otdel Biologicheski* 70(5):30-35. [In Russian with English summary.]

Gajewski, A.

1961a Polish species of the genus *Macrosteles* Fieb. (Homoptera, Jassidae). *Fragmenta Faunistica* (Warsaw) 9:87-106. [In Polish with English summary.]

Gallun, R. L.
 1972a Genetic interrelationships between host plants and insects. *Journal of Environmental Quality* 1(3):259-265.

Galvez, G. E. and Miah, M. S. A.
 1969a Virus and mycoplasma-like diseases of rice in East Pakistan. *International Rice Commission Newsletter* 18(4):18-26.

Galvez, G. E., Thurston, H. D. and Bravo, G.
 1963a Leafhopper transmission of enanismo of small grains. *Phytopathology* 53:106-108.

Galvez, G. E., Shikata, E. and Miah, M. S. A.
 1971a Transmission and electron microscopy of a rice tungro virus strain. *Phytopathologische Zeitschrift* 70:53-61.

Galwey, N. W.
 1983a Characteristics of the common bean, *Phaseolus vulgaris*, associated with resistance to the leafhopper *Empoasca kraemeri*. *Annals of Applied Biology* 102:161-175.

Galwey, N. W. and Evans, A. M.
 1982a The inheritance of resistance to *Empoasca kraemeri* Ross and Moore in the common bean, *Phaseolus vulgaris* L. *Euphytica* 31:933-952.

Galwey, N. W., Temple, S. R. and Schoonhoven, A. van
 1985a The resistance of genotypes of two species of *Phaseolus* beans to the leafhopper *Empoasca kraemeri*. *Annals of Applied Biology* 107:147-150.

Gambrell, F. L. and Gilmer, R. M.
 1956a Insects and diseases of fruit nursery stocks and their control. *New York Agricultural Experiment Station Bulletin* (Geneva) 776. 50 pp.
 1960a The influence of insecticide-fungicide spray programs on the growth of apple nursery trees. *Journal of Economic Entomology* 53(5):717-719.

Gameel, O. I.
 1974a Field evaluation of insecticides for jassid, *Empoasca lybica* De Berg, and whitefly, *Bemisia tabaci* (Gennadius), control on cotton (Hemiptera: Jassidae and Aleyrodidae). *Bulletin of the Entomological Society of Egypt, Economic Series 1973*, 7:113-122.

Gamez, R.
 1969a A new leafhopper-borne virus of corn in Central America. *Plant Disease Reporter* 53:929-932.
 1975a Transmission of rayado fino virus of maize (*Zea mays*) by *Dalbulus maidis*. *Annals of Applied Biology* 73:285-292.
 1977a Leafhopper-transmitted maize rayado fino virus in Central America. *Proceedings of Maize Virus Disease Colloquium Workshop, 16-19 August 1976*. Pp. 15-19. Ohio Agricultural Research and Development Center, Wooster, Ohio.
 1980a Maize rayado fino virus. *In: Descriptions of Plant Viruses. Commonwealth Mycological Institute-Association of Applied Biologists*. Kew, Surrey, England.
 1980b Rayado fino virus disease of maize in the American tropics. *Tropical Pest Management* 26:26-33.
 1983a The ecology of rayado fino virus in the American tropics. *In: Plumb, R. T. and Thresh, J. M., eds. Plant Virus Epidemiology*. Pp. 267-275. Blackwell Scientific, Oxford, England. 337 pp.
 1983b Maize rayado fino disease: The virus-host-vector interaction in Neotropical environments. *In: Gordon, D. T., Knoke, J. K., Nault, L. R. and Ritter, R. M., eds. Proceedings International Maize Virus Disease Colloquium and Workshop 2-6 August 1982*. Pp. 62-68. Ohio Agricultural Research and Development Center, Wooster, Ohio. 261 pp.

Gamez, R. and Black, L. M.
 1967a Application of particle-counting to a leafhopper-borne virus. *Nature* 215:173-174.
 1968a Particle counts of wound tumor virus during its peak concentration in leafhoppers. *Virology* 34:444-451.

Gamez, R., Kitajima, E. W. and Lin, M. T.
 1979a The geographical distribution of maize rayado fino virus. *Plant Disease Reporter* 63:830-833.

Gamez, R. and Leon, P.
 1983a Maize rayado fino virus: Evolution with plant host and insect vector. *In: Harris, K. F., ed. Current Topics in Vector Research*. Pp. 149-168. Praeger, New York.
 1985a Ecology and evolution of a neotropical leafhopper-virus-maize association. *In: Nault, L. R. and Rodriguez, J. G., eds. The Leafhoppers and Planthoppers*. Pp. 331-350. Wiley & Sons, New York. 500 pp.

Gamez, R., Rivera, C. and Kitajima, E. W.
 1981a The biological cycle of maize rayado fino virus in its insect vector *Dalbulus maidis*. *In: 5th International Congress of Virology, Strasbourg*. P. 213.

Gamez, R. and Saavedra, F.
 1985a A model of a leafhopper-borne virus disease in the neotropics. *In: Garret, R., MacLean, G. and Ruesink, W., eds. Plant Virus Epidemiology*. Pp. 315-326. Academic, Melbourne, Australia.

Gandhale, D. N., Naik, L. M. and Darekar, K. S.
 1975a Chemical control of mango hoppers (*Idioscopus clypealis* Leth.). *Journal of Plantation Crops* 3:9010.

Garcia, J., Cardona, C. and Raigosa, J.
 1979a Evaluation of populations of insect pests in mixed sugar-cane and beans and their relation to yield. *Revista Colombiana de Entomologia* 5:17-24.

Garcia, J. E., Cardon, C. and Schoonhoven, A. van
 1981a Resistencia del frijol comun, *Phaseolus vulgaris* L. al *Empoasca kraemeri* Ross and Moore. *Revista Colombiana de Entomologia* 7:15-21.

Gardner, D. E. and Cannon, D. S.
 1972a Curly top viruliferous and nonviruliferous leafhopper feeding effects upon tomato seedlings. *Phytopathology* 62(1):183-186.

Gardner, M. E., Schmidt, R. and Stevenson, F. J.
 1945a The Sequoia potato: A recently introduced insect resistant variety. *American Potato Journal* 22:97-103.

Garg, A. K. and Sethi, G. R.
 1980a Succession of insect pests in Kharif paddy. *Indian Journal of Entomology* 42:482-487.
 1982a Distribution of phorate, disulfoton, dimethoate and chlorpyrifos in paddy. *Indian Journal of Entomology* 44:194-197.
 1983a First record of predatory beetle, *Brumoides suturalis* (F.), feeding on rice pests. *Bulletin of Entomology* 24:138-140.

Gao, D. M.
 1983a Detection of viruliferous green leafhopper, *Nephotettix cincticeps*, with RTYV by enzyme-linked immunosorbent assay (ELISA). *Zhejiang Agricultural Science* 4:194-198.
 1985a Studies on the joint detection of viruliferous insects of RDV and RTYY with ELISA. *Zhejiang Agriculture Science* 3:133-137.

Garman, P.
 1959a Experiments in apple insect control. *Journal of Economic Entomology* 52(5):826-282.

Garneau, A.
 1984a Liste de Cicadaires recoltes en 1983 sur de la rhubarbe (*Rheum rhaponticum* L.), a granby, comte de shefford. *Fabreries* 10(5-6):77-79.

Gartel, W.
 1965a Investigations on the occurrence and course of flavescence doree in the Moselle and Rhine vineyards. *Weinberg und Keller* 12:347-376.

Gates D.
 1945a Varietal resistance of beans to the potato leafhopper. *Nebraska Agricultural Experiment Station Annual Report* 58:61.

Gatoria, G. S. and Harcharan, S.
 1984a Effect of insecticidal and fertilizer applications on the jassid complex in green gram, *Vigna radiata* (L.) Wilczek. *Journal of Entomological Research* 8:154-158.

Gayen, A. K.
 1975a Studies on persistence of some insecticides against jassid, *Amrasca biguttula biguttula* (Ishida), an imprint pest of 'bhindi'. *Entomologist's Newsletter* 5:16-17.

Gebicki, C.
1979a The association of leafhoppers (Homoptera: Auchenorrhyncha) in selected environments of Huta 'Katowice' region. *Prace Naukowe Uniwersystetu Slaskiego w Katowicach* No. 297:29-44.
1983a The associations of leafhoppers (Homoptera, Auchenorrhyncha) of pine and deciduous forests in Pinczow neighborhood. *Prace Naukowe Uniwersystetu Slaskiego w Katowicach* No. 627:83-98.

Gebicki, C., Bartnicka, I., Boklak, E. and Malkowski, E.
1982a Leafhoppers (Homoptera, Auchenorrhyncha) of Biebrza Valley. *Prace Naukowe Uniwersystetu Slaskiego w Katowicach* No. 505:13-21.

Gentsch, B. J.
1982a The influence of weeds on the incidence of potato leafhopper in alfalfa. M. S. Thesis, Southern Illinois University, Carbondale.

Genung, W. G.
1956a Insects reduce quantity and quality of pasture grasses. *Florida Sunshine State Agricultural Research Report* 1(2):8-9.
1957a Some possible cases of insect resistance to insecticides in Florida. *Proceedings of the Florida State Horticulture Society* 70:148-151.

Genung, W. G. and Mead, F. W.
1969a Leafhopper populations (Homoptera: Cicadellidae) on five pasture grasses in the Florida Everglades. *Florida Entomologist* 52(3):165-170.

Genyte, L. and Staniulis, J.
1976a Rearing of leafhopper *Aphrodes bicinctus* Schrank species under laboratory conditions. *Trudy Akademii Nauk Litovskoi SSR* (Ser. C) 76:65-69. [In Russian with English and Lithuanian summaries.]

George, J. A.
1959a Notes on two species of leafhoppers (Homoptera: Cicadellidae) new to Canada. *Canadian Entomologist* 91(4):253.
1959b Note on *Epigonatopus plesius* (Fenton) (Homoptera, Dryinidae), a parasite of the six-spotted leafhopper, *Macrosteles fascifrons* (Stal), in Ontario. *Canadian Entomologist* 91(4):256.

George, J. A. and Davidson, T. R.
1959a Notes on life-history and rearing of *Colladonus clitellarius* (Say) (Homoptera: Cicadellidae). *Canadian Entomologist* 91(6):376-379.

George, J. A. and Richardson, J. K.
1957a Aster yellows on celery in Ontario. *Canadian Journal of Plant Science* 37(2):132-135.

Georghiou, G. P.
1957a A catalogue of Cyprus insects. *Cyprus Department of Agriculture Technical Bulletin* 7:1-65.

Gerard, B. M.
1972a Two new species of Typhlocybinae (Hemiptera: Cicadellidae) from South Africa, and a redescription of two of Cogan's species. *Journal of the Entomological Society of Southern Africa* 35(1):139-147.

Gerhardt, P. D. and Turley, D. L.
1961a Control of certain potato insects in Arizona with soil applications of granulated phorate. *Journal of Economic Entomology* 54(6):1217-1221.

Ghai, S. and Ahmed, R.
1975a Larvae of *Bochartia* sp. (Acarina: Erythraeidae) parasitizing jassids. *Entomologist's Newsletter* 5:29-30.

Ghandhi, J. R.
1978a Utilization of food constituents in a leafhopper, *Amrasca devastans* (Distant). *Indian Journal of Entomology* 39:297-299.
1980a Duration of feeding by *Amrasca devastans* Distant and *A. kerri motti* on different plant leaves. *Indian Journal of Entomology* 43(1):49-58.
1980b Duration of feeding by *Amrasca devastans* Distant and *A. kerri motti* on different plant leaves. *Indian Journal of Entomology* 42:296-297.
1982a Preference for acceptance of different plants by *Amrasca devastans* and *Amrasca kerri motti*. *Indian Journal of Entomology* 44:190-191.

Ghauri, M. S. K.
1961a A new East African *Cicadulina* resembling *C. zeae* China infesting maize in southern Rhodesia. *Annals and Magazine of Natural History* (Series 13) 4:369-370.
1962a A new typhlocybid genus and species (Cicadelloidea: Homoptera) feeding on rice in Thailand. *Annals and Magazine of Natural History* (Series 13) 5:253-256. [New genus: *Thaia*.]
1963a A new grape-vine leafhopper (Homoptera: Cicadelloidea) from Iraq. *Annals and Magazine of Natural History* (Series 13) 6:381-383.
1963b A new species of *Zygina* Fieber (1866) (Homoptera: Cicadelloidea) from New Zealand. *Annals and Magazine of Natural History* (Series 13) 39-42.
1963c A new citrus leafhopper from Tchad. *Annals and Magazine of Natural History* (Series 13) 6:391-393.
1963d Distinctive features and geographical distribution of two closely similar pests of cotton (*Empoasca devastans* Dist. and *E. terraereginae* Paoli). *Bulletin of Entomological Research* 53(4):653-656.
1963e New fig leafhoppers (Homoptera: Cicadelloidea) from India with redescription of allied species under new genera. *Annals and Magazine of Natural History* (Series 13) 6:465-475. [New genera: *Rawania, Velu, Lankasca* and *Ficiana*.]
1963f A new genus based on *Orosius maculatus* Pruthi and generic re-assignment of some of Pruthi's species of Cicadelloidea (Auchenorrhyncha: Homoptera). *Annals and Magazine of Natural History* (Series 13) 6:559-563. [New genus: *Pruthiorosius*.]
1964a A new species of *Empoasca* Walsh (Homoptera: Cicadelloidea) attacking tea in Argentina. *Annals and Magazine of Natural History* (Series 13) 7:189-192.
1964b A new species of *Cicadulina* China (Homoptera: Cicadelloidea) from Kenya. *Annals and Magazine of Natural History* (Series 13) 7:205-208.
1964c A new species of *Zygina* Fieber (Homoptera: Cicadelloidea) attacking cotton in Sudan. *Annals and Magazine of Natural History* (Series 13) 7:397.
1964d Two new species of Cicadelloidea (Homoptera) attacking coffee in New Guinea. *Annals and Magazine of Natural History* (Series 13) 7:635-639.
1965a Notes on the Hemiptera from Pakistan and adjoining areas. *Annals and Magazine of Natural History* (Series 13) 7:673-688.
1966a Revision of the genus *Orosius* Distant (Homoptera: Cicadelloidea). *Bulletin of the British Museum (Natural History). Entomology* 18(7):231-252.
1967a New mango leafhoppers from the Oriental and Austro-oriental regions (Homoptera: Cicadelloidea). *Proceedings of the Royal Entomological Society of London* (B) 36:159-166. [New genera: *Amrasca* and *Mangganeura*.]
1968a The African and Malagasian species of *Nephotettix* Matsumura (Homoptera: Cicadelloidea.) *Bulletin of Entomological Research* 57(4):643-650.
1969a Two new sympatric species of a new leafhopper genus (Homoptera, Cicadelloidea) attacking cocoa in Ghana. *Journal of Natural History* 3(1):35-39. [New genus: *Afroccidens*.]
1971a Revision of the genus *Nephotettix* Matsumura (Homoptera: Cicadelloidea: Euscelidae) based on the type material. *Bulletin of Entomological Research* 60:481-512.
1971b A remarkable new species of *Cicadulina* China (Hom., Cicadelloidea) from East Africa. *Bulletin of Entomological Research* 60:631-633.
1971c A new genus of Euscelinae from the Lower Himalayas, and a new species of *Balclutha* Kirkaldy (Homoptera, Cicadelloidea). *Bulletin of Entomological Research* 61:113-118. [New genera: *Subhimalus*.]
1972a A new genus and species of Coelidiinae (Homoptera, Cicadelloidea) from Iraq. *Bulletin of Entomological Research* 62(2):207-209. [New genus: *Iraquerus*.]
1972b New species of *Exitianus* Ball (Homoptera, Cicadelloidea, Euscelidae) from Africa. *Bulletin of Entomological Research* 61(4):689-692.

1972c Notes on Hemiptera from Pakistan and adjoining areas. *Journal of Natural History* 6:279-288.

1974a The *solana*-group of *Empoasca* Walsh (Homoptera, Cicadelloidea): Its generic status and a new species from pawpaw. *Bulletin of Entomological Research* 63:425-429. [New genus: *Solanasca*.]

1974b Two new species of *Exitianus* Ball (Homoptera, Cicadelloidea, Euscelidae) related to *E. turneri* Ross, one infesting rice in Ghana. *Bulletin of Entomological Research* 63:531-534.

1974c New genera and species of Cicadelloidea (Homoptera, Auchenorrhyncha) from economic plants in India. *Bulletin of Entomological Research* 63(11):551-559. [New genera: *Jilinga, Pseudosubhimalus*, and *Sarejuia*.]

1974d A new genus and species of Erythroneurini (Homoptera, Cicadelloidea) and a key to the African genera. *Bulletin of Entomological Research* 64:637-641. [New genus: *Zanjoneura*.]

1975a A new species of *Singapora* Mahmood (Homoptera, Cicadelloidea) attacking *Pterocarpus macrocarpus* in Thailand. *Bulletin of Entomological Research* 65:373-375.

1975b Taxonomic notes on a collection of Cicadellidae from maize and light traps in the vicinity of crop fields in Nigeria. *Journal of Natural History* 9(5):481-493. [New genera: *Moorada, Erinwa, Olokemeja, Ifeneura*, and *Orucyba*.]

1975c A new genus and species of idiocerine leafhopper (Cicadelloidea, Auchenorrhyncha, Homoptera) on pear tree from Simla. *Nouvelle Revue d'Entomologie* 5(3):287-290. [New genus: *Tasnimocerus*.]

1978a A new genus and species of Erythroneurini (Hemiptera: Cicadelloidea) causing severe damage to flamboyant trees (*Delonix regia*) in Mauritius and Reunion. *Bulletin of Entomological Research* 68:203-206. [New genus: *Cerneura*.]

1979a The identity of *Empoasca dolichi* Paoli (Hemiptera: Cicadellidae) and descriptions of new species of *Empoasca* from light-traps near crops in Malawi and Nigeria. *Bulletin of Entomological Research* 69:343-355.

1980a A new species of *Zygina* from Papua New Guinea, and records of some leafhopper and planthopper species (Homoptera: Cicadelloidea and Delphacidae). *New Zealand Journal of Zoology* 7:207-209.

1980b Illustrated redescription of two of Pruthi's species of Cicadelloidea from India. *Reichenbachia* 18(25):165-171.

1980c The identity of a suspected vector of coconut lethal yellowing disease in Jamaica and notes on *Caribovia intensa* (Walker) (Homoptera: Cicadelloidea). *Bulletin of Entomological Research* 70:411-415.

1981a A new species of *Aconurella* Ribaut on maize in Mozambique. *Reichenbachia* 19(8):47-50.

1983a Scientific name of the Indian cotton jassid. *In: Knight, W. J., Pant, N. C., Robertson, T. S. and Wilson, M. R. eds. Proceedings of the 1st International Workshop on Biotaxonomy, Classification and Biology of Leafhoppers and Planthoppers (Auchenorrhyncha) of Economic Importance.* Pp. 97-103. London, 4-7 October 1982. Commonwealth Institute of Entomology, 56 Queen's Gate, London SW7 5JR. 500 pp.

1983b A case of long-distance dispersal of a leafhopper. *In: Knight, W. J., Pant, N. C., Robertson, T. S. and Wilson, M. R., eds. Proceedings of the 1st International Workshop on Biotaxonomy, Classification and Biology of Leafhoppers and Planthoppers (Auchenorrhyncha) of Economic Importance.* Pp. 249-253. London, 4-7 October 1982. Commonwealth Institute of Entomology, 56 Queen's Gate, London SW7 5JR. 500 pp.

1985a *Namiocerus*, a new genus for *Bythoscopus cephalotes* Walker, 1857 (Homoptera, Cicadellidae, Idioceridae). *Reichenbachia* 23(10):41-46.

1985b A new species of *Singapora* Mahmood attacking *Derris robusta* in Assam, with a key to species (Homoptera: Cicadellidae). *Turkiye Bitki Koruma Dergisi* 9:3-11.

1985c *Muinocerus qadirii* gen. et sp. nov. (Idiocerinae: Cicadelloidea: Homoptera) from Malaysia. *Turkiye Bitki Koruma Dergisi* 9:67-73.

Ghauri, M. S. K. and Onder, F.

1980a Heteroptera and Homoptera with keys to economic families and pests. *Ziraat Fakultesi Yaymlari, Ege Universitesi* No. 371. 101 pp.

Ghorpade, K. D.

1979a *Ballia eucharis* (Coleoptera: Coccinellidae) breeding on Cicadellidae (Homoptera) at Shillong. *Current Research* 8:113.

Ghosh, A.

1980a Observations on a 'blue form' of green leafhopper. *International Rice Research Newsletter* 5(6):15.

Ghosh, A. and John, V. T.

1979a The role of *Nephotettix* spp., harbouring on grasses, and the perpetuation of rice tungro virus in fields. *Indian Phytopathology* 32(3):458-460.

Ghosh, A. B. and Mukhopadhyay, S.

1978a Varietal preferences of the green leafhopper. *International Rice Research Newsletter* 3(6):13.

Ghosh, A. B, Tarafder, P. and Muhopadhyay, S.

1979a Effect of transplanting time on the spread of tungro in Boro. *International Rice Research Newsletter* 4(1):13.

Ghosh, A. K., Medler, J. T. and Hildebrandt, A. C.

1968a Maintenance of *Empoasca fabae* (Homoptera: Cicadellidae) on plants and tissue cultures of three bean varieties. *Journal of the Kansas Entomological Society* 41(3):331-334.

Ghosh, M.

1974a On a small collection of jassids from Darjeeling District, West Bengal, with description of 2 new species of the genera *Iassus* Fabricius, 1803, and *Scaphoideus* Uhler, 1888, and redescription of *Kolla maculifrons* Schmidt (Insecta: Homoptera: Cicadelloidea). *Journal of the Zoological Society of India* 26(1&2):77-81.

Ghosh, S. K.

1981a Association of mycoplasmas and allied pathogens with tree diseases in India. *In: Maramorosch, K. and Raychandhuri, S. P., eds. Mycoplasma Diseases of Trees and Shrubs.* Pp. 231-243. Academic Press, New York. 362 pp.

1985a Dual infection with anisometric viruses. *International Journal of Tropical Plant Diseases* 3:15-18.

Giannotti, J.

1969a Transmission of clover phyllody by a new leafhopper vector, *Euscelidius variegatus. Plant Disease Reporter* 53:173.

1969b Lesions cellulaires chez deux cicadelles vectrices de la phyllodie du trefle. *Annals Societe Entomologique de France* (N.S.) 5(1):155-160. [With English summary.]

1972a La culture 'in vitro' des mycoplasmes de plantes et leur transmission par des vecteurs. Vierentwintigste International Symposium over Fytofarmacie en Fytiatrie, 9 Mei 1972. *Mededelingen Fakulteit Landbouwwetenschappen Gent* 37(2):429-440. [In French with English summary.]

Giannotti, J. and Devauchelle, G.

1969a Lesions cellulaires dues a des mycoplasmes lies a la phyllodie du trefle chez la cicadelle vectrice *Euscelis plebejus* Fall. [Cellular lesions caused by the mycoplasma associated with clover phyllody in the cicadellid vector *E. plebeja*.] *Annales de Zoologie, Ecologie Animale* 1(1):31-38. [With English summary.]

Giannotti, J., Devauchelle, G. and Vago, C.

1968a Microorganismes de type mycoplasmes chez une cicadelle et une plante infectees par la phyllodie. *Compte Rendu de l'Acadamie d'Agriculture de France* 266:2168-2170.

Giannotti, J., Kuhl, G. and Czarnecky, D.

1969a Transmission of yellows by cicadellids injected with extracts of tissues of diseased plants. *Revue de Zoologie Agricole et de Pathologie Vegetale* 68(7-9):111-115.

Gianotti, J., Quiot, J.-M., Schwemmler, W. and Vago, C.

1970a Multiplication des mycoplasmes de la phyllodie en culture organotypique de l'insecte vectour. *Compte Rendu de l'Academie d'Agriculture de France* 56:59-62.

Giannotti, J., Vago, C., Devauchelle, G. and Marchoux, G.
1968a Recherches sur les microorganismes de type mycoplasme dans les cicadelles vectrices et dans les vegetaux attents de jaunisses. *Entomologica Experimentalis et Applicata* 11:470-474.

Giannotti, J., Vago, C. and Duthoit, J. L.
1968a Isolement et purification de microorganismes a structure de mycoplasmes a partir de cicadelles et de plantes infectees de jaunisses. [Isolation and purification of mycoplasma-like microorganisms from cicadellids (*Euscelis plabeja* [Fall.]) and plants (clover and tomato) infected with yellows (clover phyllody and stolbur, respectively).] *Revue de Zoologie Agricole Applique* 67(4-6):69-72. [With English summary.]

Gibson-Hill, G. A.
1950a Hemiptera collected on the Cocos-Keeling Islands, January-October 1941. *Bulletin Raffles Museum (Singapore)* 23:206-211.

Gibson, K. E. and Fallini, J. T.
1963a Beet leafhopper control in southern Idaho by seeding breeding areas to range grass. *U.S. Department of Agriculture, Agricultural Research Service* 33-83:1-5.

Gibson, K. E. and Oliver, W. N.
1970a Comparison of nymphal and adult beet leafhoppers as vectors of the virus of curly top. *Journal of Economic Entomology* 63(4):1321.

Gibson, L. P.
1973a An annotated list of the Cicadellidae and Fulgoridae of elm. *USDA Forest Service Research Paper* NE-279. Pp. 1-5.

Gibson, W. W. and Carrillo, S., J. L.
1959a Lista de Insectos en la Coleccion Entomologica de la Oficina de Estudios Especiales, S. A. G. *Folleto Miscelaneo* 9:1-254.

Giddings, N. J.
1954a Two recently isolated strains of curly top virus. *Phytopathology* 44(3):123-125.

Gifford, J. R. and Trahan, G. B.
1966a Flight activity of rice water weevil and green rice leafhopper, 1966. *Louisiana State University Agricultural Experiment Station Rice Experiment Station Annual Progress Report* 58:161-165.

Gil-Fernandez, C. and Black, L. M.
1965a Some aspects of the internal anatomy of the leafhopper *Agallia constricta* (Homoptera-Cicadellidae). *Annals of the Entomological Society of America* 58(3):275-284.

Gill, C. C., Westdal, P. H. and Richardson, H. P.
1969a Comparison of aster yellows and barley yellow dwarf of wheat and oats. *Phytopathology* 59(5):527-531.

Gill, R. J. and Oman, P. W.
1982a A new species and new distributional records for megophthalmine leafhoppers, genus *Tiaja* (Homoptera: Cicadellidae). *Entomography* 1:281-288.

Gilmer, R. M. and McEwen, F. L.
1958a Insect transmission of X-disease virus. *Phytopathology* 48:262.

Gilmer, R. M., Palmiter, D. H., Schaefers, G. S. and McEwen, F. L.
1966a Insect transmission of X-disease virus of stone fruits in New York. *New York State Agricultural Experiment Station Bulletin* 813. 22 pp.

Giorgadze, D. C.
1968a Previous data on the study of the biology of virus vectors of mulberry disease in Georgei—cicadas, *Hishimonus sellatus* Uhler. *Soobschcheniya Akademii Nauk Gruzinskoi SSR* 51:441-444. [In Russian with Georgian summary]

Giorgadze, D. C. and Tulashvili, N. D.
1973a Results of study on the interrelationships between the pathogen of mulberry 'curly leaf disease' and its leafhopper vector *Hishimonus sellatus* Uhler under conditions in West Georgia. *Trudy Nauchno-Issledovatel'skogo Instituta Zashchity Rastenii Gruz SSR* 24:209-213.

Gist, G. R., Lyon, W. F., Blair, R. D., Staubus, J. R., Garter, W. G., Goleman, L. and Niemczyk, H. D.
1968a Alfalfa and alfalfa insects. *Ohio Cooperative Extension Service Leaflet* L-130. 7 pp.

Giunchi, P.
1952-53a Contributi alla conoscenza dell'entomofauna dell'erba medica. *Bollettino della Instituto Entomologia Recerche Universita Bologna* 19:1-30.

Glass, E. H., Reyes, S. L. and Colona, F. B.
1966a Notes on the biology of three leafhoppers on mango. *Philippine Agriculturalist* 50(7):739-743.

Glen, D. M.
1975a Searching behavior and prey-density requirements of *Blepharidopterus angulatus* (Fall.) (Heteroptera: Miridae) as a predator of the lime aphid, *Eucallipterus tiliae* (L.), and leafhopper, *Alnetoidea alneti* (Dahlbom). *Journal of Animal Ecology* 44(1):115-134.

Glick, P. A.
1957a Collecting insects by airplane in southern Texas. *U.S. Department of Agriculture Technical Bulletin* 1158:28.

1960a Collecting insects by airplane, with special reference to dispersal of the potato leafhopper. *U.S. Department of Agriculture Technical Bulletin* 1222.

Goble, H. W.
1967a Insects of the season 1967 related to fruits, vegetables and ornamentals. *Proceedings of the Entomological Society of Ontario* 98:5-6.

Godoy, C.
1985a Efecto de la infeccion del virus del rayado fino del maize en la biologia reproductiva de su insecto vector *Dalbulus maidis* DeLong and Wolcott. Tesis, Lic. Biol., Universidad de Costa Rica.

Goeden, R. D. and Ricker, D. W.
1968a The phytophagous insect fauna of Russian thistle (*Salsola kali* var. *tenuifolia*) in southern California. *Annals of the Entomological Society of America* 61:67-72.

Goel, S. C.
1978a Bioecological studies on two years capture of Hemiptera in western Uttar Pradesh. *Oriental Insects* 12(3):369-376.

Goheen, A. C., Nyland, G. and Lowe, S. K.
1973a Association of a rickettsialike organism with Pierce's disease of grapevines and alfalfa dwarf and heat therapy of the disease in grapevines. *Phytopathology* 63(3):341-345.

Gold, R. E.
1974a Pathogen strains and leafhopper species as factors in the transmission of western X-mycoplasma under varying temperature and light conditions. Ph. D. Thesis, University of California, Berkeley.

1979a Leafhopper vectors and western X-disease. In: Maramorosch, K. and Harris, K. F., eds. *Leafhopper Vectors and Plant Disease Agents.* Pp. 587-602. Academic Press, New York. 654 pp.

Gold, R. E. and Sylvester, E. S.
1982a Pathogen strains and leafhopper species as factors in the transmission of western X-disease agent under varying light and temperature conditions. *Hilgardia* 50(3):1-43.

Gomez-Menor, J.
1956a Las tribus de Hemipteros de Espana. *Trabajos del Instituto Espanol de Entomologia.* 147 pp.

Gomez, L. A. and Schoonhoven, A. van
1977a Oviposicion del *Empoasca kraemeri* en frijol y evaluacion del parasitismo por *Anagrus* sp. *Revista Colombiana de Entomologia* 3:29-38.

Gondran, J.
1967a Importance d'*Euscelis plebejus* Fall. et variation des symptomes viraux transmis par cet insecte sur le trefle blanc. [Importance of *E. plebeja* and variations in the viral symptoms transmitted by this insect to white clover.] *Annales des Epiphyties* 18(1):75-83. [With English summary.]

Gonot, K. and Purcell, A. H.
1981a Seasonal acquisition of X-disease agent (cherry buckskin) from cherry by the leafhopper vector, *Colladonus montanus*. *Phytopathology* 71:105.

Gonzalez, A.
1955a Siete nuevas especies del genero *Empoasca* de Mexico. *Annales del Instituto de Biologia Universidad Nacional Autonoma de Mexico* 16:211-221.

Gonzalez B., J. E.
1959a Datos preliminares sobre la distribucion y control de los insectos del frijol en el Peru. *Revista Peruana de Entomologia Agricola* 2(1):84-86.
1960a Control quimico de *Empoasca kraemeri* Ross and Moore (Homoptera: Jassidae) en el frijol. *Revista Peruana de Entomologia Agricola* 3(1):59-62.

Gonzalez, V. and Gamez, R.
1974a Algunos factores que afectau la transmision del virus del rayado fino del maiz por *Dalbulus maidis* DeLong & Walcott. *Turrialba* 24(1):51-57. [English summary.]

Goodman, A. and Toms, A. M.
1956a Effect of raubstirns upon the cotton jassid in the Sudan Gezira. *Nature* 178(4530):436.

Gopalkrishnan, P., Gopalan, N., George, K. M., Thomas, B. and Shanmughan, S. N.
1974a Occurrence of "tungro" virus disease in Kerala. *Rice Pathology Newsletter* 1:4.

Gordon, D. T. and Nault, L. R.
1977a Involvement of maize chlorotic dwarf virus and other agents in stunting diseases of *Zea mays* in the United States. *Phytopathology* 67:27-36.

Gould, G. E.
1960a Problems in the control of mint insects. *Journal of Economic Entomology* 53(4):526-531.

Gouranton, J.
1967a Presence d'une phosphomonoesterase alcaline liee brochosomes dans les tubes de malpighi de la cicadelle verte. *Compte Rendu des Seances de la Societe de Biologie* 161:907-909.
1968a Observations histochimiques et histoenzymologiques sur le tube digestif de quelques Homopteres Cercopides et Jassides. *Journal of Insect Physiology* 14:569-579. [English summary.]
1968b Ultrastructures en rapport avec un transit d'eau. Etude de la "Chambre filtrante" de *Cicadella viridis* L. (Homoptera, Jassidae). *Journal de Microscopie* (Paris) 7:559-574. [English summary.]
1968c Etude ultrastructurale du system cryptonephridien de *Cicadella viridis* L. (Homoptera Jassidae). *Compte Rendu Hebdomadaire des Sceances de l'Academie des Sciences* 226D:1403-1406.

Gouranton, J. and Maillet, P. L.
1966a Sur la production de corpuscules lipoproteins par les es de Malpighi de certains Insectes. *Compte Rendu des Seances de la Societe de Biologie* 160:1724-1726.
1973a High resolution autoradiography of mycoplasmalike organisms multiplying in some tissues of an insect vector for clover phyllody. *Journal of Invertebrate Pathology* 21:158-163.

Gourlay, E. S.
1964a Notes on New Zealand insects and records of introduced species. *New Zealand Entomologist* 3(3):45-51.

Gourret, J. P., Maillet, P. L. and Gouranton, J.
1973a Virus-like particles associated with the mycoplasmas of clover phyllody in the plant and in the insect vector. *Journal of General Microbiology* 74:241-249.

Govindan, R., Kumar, C. T. A., Rangaswamy H. R. and Thimmaiah, G.
1979a Efficacy of certain systemic insecticides on the population density of castor jassid, *Sundapteryx biguttula biguttula* Ishida (Hemiptera: Cicadellidae). *Current Research* 8:178-179.

Govindu, H. C., Harris, H. M., and Yaraguntaiah, R. C.
1968a Possible occurrence of tungro and yellow dwarf viruses on rice in Mysore. *Mysore Journal of Agricultural Sciences* 2:125-127.

Gowda, C., Jayaram, K. R. and Nagaraju
1983a *Thaia subrufa* (Homoptera: Cicadellidae) occurrence in Karnataka, India. *International Rice Research Newsletter* 8(5):17.

Gracia, O. and Feldman, J. M.
1972a Studies on weed plants as sources of viruses. 1. *Datura ferox* as a source of beet curly top virus. *Phytopathologische Zeitschrift* 73(1):69-74.

Graham, C. L.
1979a Inability of certain vectors in North America to transmit maize streak virus. *Environmental Entomology* 8(2):228-230.

Graham, K. M., Hodgson, W. A., Munro, J. and Pond, D. D.
1967a Control of diseases and pests of potatoes. *Canada Department of Agriculture Publication* 1215. 63 pp.

Granada, G. A.
1979a Machismo disease of sorghum: I. Symptomatology and transmission. *Plant Disease Reporter* 63:47-50.

Granados, R. R.
1965a Strains of aster-yellows virus and their transmission by the six-spotted leafhopper, *Macrosteles fascifrons* (Stal). Ph.D. Dissertation, University of Wisconsin, Madison. 150 pp.
1969a Electron microscopy of plant and insect vectors infected with the corn stunt disease agent. *Contributions of the Boyce Thompson Institute for Plant Research* 24:173-187.
1969b Maize viruses and vectors. In: Maramorosch, K., ed. *Viruses, Vectors, and Vegetation*. John Wiley & Sons, New York. Pp. 327-359.
1972a Viruses in plant and insect cells. In: Tinsley, T. W. and Harrap, K. A., eds. *Moving Frontiers in Invertebrate Virology. Monographs in Virology* 6:42-49. Karger, Basel and New York.

Granados, R. R. and Chapman, R. K.
1968a Identification of some new aster yellows virus strains and their transmission by the aster leafhopper, *Macrosteles fascifrons*. *Phytopathology* 58:1685-1692.
1968b Heat inactivation and interactions of four aster yellows virus strains in their vector, *Macrosteles fascifrons* (Stal). *Virology* 36:333-342.

Granados, R. R., Granados, J. S., Maramorosch, K. and Reinitz, J.
1968a Corn stunt virus: Transmission by three cicadellid vectors. *Journal of Economic Entomology* 61:1282-1287.

Granados, R. R., Gustin, R. D., Maramorosch, K. and Stoner, W. N.
1968a Transmission of corn stunt virus by the leafhopper *Deltocephalus sonorus* Ball. *Contributions of the Boyce Thompson Institute for Plant Research* 24:57-60.

Granados, R. R., Hirumi, R. R. and Maramorosch, K.
1967a Electron microscopic evidence for wound-tumor virus accumulation in various organs in an inefficient leafhopper vector, *Agalliopsis novella*. *Journal of Invertebrate Pathology* 9:147-159.

Granados, R. R. and Maramorosch, K.
1967a Transmission of corn stunt virus by three leafhopper vectors. *Phytopathology* 57:340.

Granados, R. R., Maramorosch, K., Everett, T. and Pirone, P. T.
1966a Leafhopper transmission of a corn stunt virus from Louisiana. *Phytopathology* 56:584.
1966b Transmission of corn stunt virus by a new leafhopper vector, *Graminella nigrifrons* (Forbes). *Contributions of the Boyce Thompson Plant Institute for Plant Research* 23:275-280.

Granados, R. R., Maramorosch, K. and Shikata, E.
1968a Mycoplasma: Suspected etiologic agent of corn stunt. *Proceedings of the National Academy of Sciences* (U.S.A.) 60(3):841-844.

Granados, R. R. and Meehan, D. J.
1973a Morphology and differential counts of hemocytes of healthy and wound tumor virus-infected *Agallia constricta*. *Journal of Invertebrate Pathology* 22(1):60-69.

1975a Pathenogenicity of the corn stunt agent to its insect vector *Dalbulus elimatus*. *Journal of Invertebrate Pathology* 26(3):313-320.

Granados, R. R., Ward, L. and Maramorosch, K.

1967a Electron microscopy of leafhopper blood cells infected with wound tumor virus. *Bulletin of the Entomological Society of America* 13:199.

1968a Insect viremia caused by a plant-pathogenic virus: Electron microscopy of vector hemocytes. *Virology* 34:790-796.

Granados, R. R. and Whitcomb, R. F.

1971a Transmission of corn stunt mycoplasma by the leafhopper, *Baldulus tripsaci*. *Phytopathology* 61:240-241.

Granett, P. and Reed, J. P.

1960a Field evaluation of Sevin as an insecticide for pests of vegetables in New Jersey. *Journal of Economic Entomology* 53(3):388-395.

Gravestein, W. H.

1953a Faunistische mededelingen over Cicaden. 1. Hemiptera-Homoptera (Jassidae). *Entomologische Berichten (Amsterdam)* 14(336):280-281.

1965a New faunistic records on Homoptera–Auchenorhyncha from the Netherlands North Sea island Terschelling. *Zoologische Beitraege* (N.S.) 11:103-111.

Greathead, D. J.

1983a Natural enemies of *Nilaparvata lugens* and other leaf- and planthoppers in tropical agroecosystems and their impact on pest populations. In: Knight, W. J., Pant, N. C., Robertson, T. S. and Wilson, M. R., eds. *Proceedings of the 1st International Workshop on Biotaxonomy, Classification and Biology of Leafhoppers and Planthoppers (Auchenorrhyncha) of Economic Importance*. Pp. 371-383. London, 4-7 October 1982. Commonwealth Institute of Entomology, 56 Queen's Gate, London SW7 5JR. 500 pp.

Greber, R. S.

1979a Cereal chlorotic mottle virus–a rhabdovirus of Gramineae in Australia transmitted by *Nesoclutha pallida* (Evans). *Australian Journal of Agriculture Research* 30:433-443.

1983a Leafhopper-borne viruses of grasses and cereals in northern Australia. *Australian Microbiologist* 4:100.

1984a Leafhopper and aphid-borne viruses affecting subtropical cereal and grass crops. In: Harris, K. F., ed. *Current Topics in Vector Research*. II. Pp. 141-184. Praeger, New York.

1984b Relationship of rhabdoviruses reported on maize in Australia. *Maize Virus Diseases Newsletter* 1:46-47.

Greber, R. S. and Gowanlock, D. H.

1979a Rickettsia-like and mycoplasma-like organisms associated with yellows-type diseases of strawberries in Queensland. *Australian Journal of Agricultural Research* 30:1101-1109.

1979b Cereal chlorotic mottle virus–purification, serology and electron microscopy in plant and insect tissues. *Australian Journal of Biological Sciences* 32:399-408.

Greene, G. L.

1967a Biology and control of insects affecting winter vegetable crops. *Florida Agricultural Experiment Station Annual Report* 1967:213-214.

Greene, J. F.

1971a A revision of the Nearctic species of the genus *Psammotettix* (Homoptera: Cicadellidae). *Smithsonian Contributions to Zoology* 74:1-40.

Gressitt, J. and Gressitt, M. K.

1962a An improved Malaise trap. *Pacific Insects* 4:87-90.

Gressitt, J. L. and Nakata, S.

1958a [1957] Trapping of air-borne insects on ships in the Pacific. *Proceedings of the Hawaiian Entomological Society* 16(3):363-365.

Gressitt, J. L., Sedlacek, J., Wise, K. A. J. and Yoshimoto, C. M.

1961a A high speed airplane trap for airborne organisms. *Pacific Insects* 3:549-555.

Grigarick, A. A.

1984a General problems with rice invertebrate pests and the control in the United States. *Protection Ecology* 7:105-114.

Grigorov, S.

1967a The effects of chemicals used for the control of cereal bugs and other cereal pests. *Rastitelna Zashtita* 15(6):23-27. [In Bulgarian.]

Groen, N. P. A., and Slogteren, D. H. M. van

1974a Symptomen en bestrijding van vergelings-heksenbezemziekte bij gladiolen. [Symptoms and control of aster yellows in gladioli.] *Praktijkmedeling, Laboratorium voor Bloembollenonderzoek, Lisse* No. 41. 17 pp. [In Dutch.]

Gromadzka, J.

1970a Observations on the biology and occurrence of leafhoppers, *Eupteryx atropunctata* (Goeze) and *Empoasca pteridis* (Dhlb.) (Homoptera, Typhlocybidae), on potatoes. *Polskie Pismo Entomologiczne* 40:829-840. [In Polish with English summary.]

1970b The occurrence of leafhoppers (Homoptera, Auchenorrhyncha) on rye grown near shelterbelts. *Ekologia Polska* 18(13):291-306.

1971a Energy assimilation in *Eupteryx atropunctata* (Goeze) and *Empoasca pteridis* (dhlb) (Homoptera, Typhlocybidae). *Ekologia Polska* 19(24):325-332. [Polish summary.]

Grossmann, R. E.

1957a Preliminary studies on the effectiveness of insecticides in controlling insect pests of pasture and meadow crops. M. S. Thesis, University of Illinois. 116 pp.

Grylls, N. E.

1954a Rugose leaf curl–a new virus disease transovarially transmitted by the leafhopper *Austroagallia torrida*. *Australian Journal of Biological Sciences* 7(1):47-58.

1955a An outbreak of rugose leaf curl disease on clover in Queensland. *Journal of the Australian Institute of Agricultural Science* 21(3):187-188.

1961a A leafhopper-borne virus of cereals and grasses new to Australia. *Plant Pathology Conference, Adelaide, S. A.* 1(30):1-2.

1963a A striate mosaic virus disease of grasses and cereals in Australia, transmitted by the cicadellid *Nesoclutha obscura*. *Australian Journal of Agricultural Research* 14:143-153.

1963b Insects and viruses. Plant viruses: Wheat striate mosaic virus. *Australia Commonwealth Science and Industrial Research Organization, Division of Entomology Annual Report*. 20 pp.

1975a Leafhopper transmission of a virus causing maize wallaby ear disease. *Annals of Applied Biology* 79:282-296.

1979a Leafhopper vectors and the plant disease agents they transmit in Australia. In: Maramorosch, K. and Harris, K. F., ed. *Leafhopper Vectors and Plant Disease Agents*. Pp. 179-214. Academic Press, New York, San Francisco and London. Vols. I-XVI. 654 pp.

Grylls, N. E. and Waterford, C. J.

1976a Transmission of the causal agent of *Chloris* striate mosaic disease by insect injection and membrane feeding. *Australian Plant Pathological Society Newsletter* (Suppl.) 5(1):89.

Grylls, N. E., Waterford, C. J., Filshie, B. K. and Beaton, C. D.

1974a Electron microscopy of rugose leaf curl virus in red clover, *Trifolium pratense*, and in the leafhopper vector *Austroagallia torrida*. *Journal of General Virology* 23(2):179-183.

Gu, D.-X. and Ito, Y.

1981a Distribution patterns of the green rice leafhopper, *Nephottetix cincticeps* Uhler (Hemiptera: Deltocephalidae), in Hokuriku and Tokai-Nisinippon districts of Japan. *Japanese Journal of Applied Entomology and Zoology* 25:276-279. [In Japanese with English summary.]

1982a Longevity of female adults of green rice leafhopper, *Nephotettix cincticeps* Uhler (Hemiptera: Deltocephalidae), in field cages. *Japanese Journal of Applied Entomology and Zoology* 26:228-231. [In Japanese with English summary.]

1982b Diagram for the determination of sample size for direct counting of the green rice leafhopper *Nephotettix cincticeps* (Hemiptera: Deltocephalidae), on rice hills. *Japanese Journal of Applied Entomology and Zoology* 26:305-306.

Guagliumi, P.

1965a Contributo alla conoscenza dell'entomofauna nociva del Venezuela. [Contribution to knowledge of the injurious insect fauna of Venezuela.] *Revista di Agricolture, Subtropical et Tropical* 59(7-8):376-408.

1965b Contributo alla conoscenza dell'entomofauna nociva del Venezuela (continuazione fine). [Contribution to knowledge of the injurious insect fauna of Venezuela.] *Revista di Agricolture, Subtropical et Tropical* 59(10-12):447-472.

Guenthart, H., see Gunthart, H.

Guevara, C. J.

1949a New chlorinated nitroparaffin insecticides against the potato leafhopper (*Empoasca fabae*). M. S. Thesis, Ohio State University.

1957a The development and use of varieties of beans resistant to certain insect pests of legumes. Ph. D. Dissertation, Ohio State University. 124 pp. [Diss. Abst. 18(6):1931.]

1966a Experiencias mexicanas sobre la obtencion de variedades de cultivos resistentes a plagas. *Revista Peruana de Entomologia* 9(1):71-74.

Gui, H. L.

1945a Susceptibility of bean varieties to insect infestation. *Ohio Agriculture Experiment Station Bulletin* 659:112-113.

Gulab, Singh, Balan, J. S. and Naresh, J. S.

1983a Comparative susceptibility of cotton varieties to cotton jassid *Amrasca biguttula biguttula* Ishada [sic!]. *Indian Journal of Plant Protection* 11:45-48.

Gulab, Singh and Chopra, N. P.

1979a Chemical control of major insect pests of okra. *Haryana Journal of Horticultural Science* 8:64-68.

Gullyev, A.

1965a Insect pests of agricultural crops of the Tedzhen Oasis. *Akademiya Nauk Turkmenskoi SSR Izvestiya Seriya Biologicheskii* 5:71-75. [In Russian.]

Gumpf, D. J. and Calavan, E. C.

1981a Stubborn diseases of citrus. In: Maramorosch, K. and Raychandhuri, S. P., eds. Mycoplasma Diseases of Trees and Shrubs. Pp. 97-134. Academic Press, New York. 362 pp.

Gunathilagaraj, K. and Jayaraj, S.

1977a Ovicidal effects of some chemicals on the white rice leafhopper, *Tettigella spectra* (Dist.), and red cotton bug, *Dysdercus cingulatus* F. *Madras Agricultural Journal* 64:369-374.

1979a Influence of age of eggs on the ovicidal action of phorate and carbofuran. *Science and Culture* 45:330-331.

Gunthart, H.

1971a Beitrag zur Kenntnis der Kleinzikaden (Typhlocybinae, Hom., Auch.) der Schweiz. *Mitteilungen der Schweizerischen Entomologischen Gesellschaft* 43(3&4):218-224.

1971b Kleinzikaden (Typhlocybinae) an Obstbaumen in der Schweiz. *Schweizerische Zeitschrift fuer Obst-und Weinbau* 107(80):285-306.

1971c La cicadelle de rhododendron (*Graphocephala coccinea*) pour la premiere fois Suisse. *Revue Horticole Suisse* 44(12):358-359.

1974a Beitrag zur Kenntnis der Kleinzikaden (Typhlocybinae, Hom., Auch.) der Schweiz. 1. Ergaenzung. *Mitteilungen der Schweizerischen Entomologischen Gesellschaft* 47(1&2):15-27.

1977a Einfluss des Insektenalters auf Bestimmungsmerkmale. Biotaxonomische und rasterelektronenmikroskopsiche Untersuchungen bei Kleinzikaden (Hom., Auchenorrhyncha, Cicadellidae). *Mitteilungen der Schweizerischen Entomologischen Gesellschaft* 50:189-201.

1978a Problems in the identification of *Zygina* (*Flammigeroidia*) *pruni* Edwards 1924 (Cicadellidae). *Auchenorrhyncha Newsletter* 1:18.

1979a Biotaxonomic experiments proving *Zygina pruni* Edwards 1924 is a synonym of *Zygina* (*Flammigeroidia*) *flammigera* (Fourcroy 1785) (Hom., Auch., Cicadellidae, Typhlocybinae). *Mitteilungen der Schweizerischen Entomologischen Gesellschaft* 52:13-17.

1984a Zikaden (Hom. Auchenorrhyncha) aus der alpinen Hohenstufe der Schweizer Zentralalpen. *Mitteilungen der Schweizerischen Entomologischen Gesellschaft* 57:129-130.

1984b Zoogeographical and ecological investigations of Auchenorrhyncha in the Lower Engadine valley. *Mitteilungen der Schweizerischen Entomologischen Gesellschaft* 57(4):420-421.

1985a *Adarrus ernesti* n. sp., a new leafhopper species from the Lower Engadine, Switzerland (Hom., Auchenorrhyncha, Cicadellidae). *Mitteilungen der Schweizerischen Entomologischen Gesellschaft* 58:401-404.

Gunthart, H. and Gunthart, E.

1967a Schaden von Kleinizikaden, besonders von *Empoasca flavescens* F. an Reben in der Schweiz. *Schweizerische Zeitschrift fuer Obst-und Weinbau* 103(76):602-610.

1967b Determinazione de Alterazioni causate da Cicaline su viti nel Ticino. *Agricolture Ticinese* 20 [not paginated].

Gunthart, H. and Gunthart, M. S.

1981a Biology and feeding behaviour of *Aguriahana germari* (Zett.) (Homoptera, Auchenorrhyncha, Typhlocybinae). *Acta Entomologica Fennica* 38:24.

1983a *Aguriahana germari* (Zett.) (Hom. Auch. Cicadellidae, Typhlocybinae): Breeding and specific feeding behaviour on pine needles. *Mitteilungen der Schweizerischen Entomologischen Gesellshaft* 56:33-44.

Gunthart, H. and Thaler, K.

1981a Fallenfange von Zikaden (Hom., Auchenorrhyncha) in zwei Grundlandparzellen des Innsbrucker Mittelgebirges (Nordtirol, Osterreich). *Mitteilungen der Schweizerischen Entomologischen Gesellschaft* 54:15-31.

Gunthart, M.

1975a Die kleinzikaden *Empoasca decipiens* Paoli und *Eupteryx atropunctata* Foetze (Homoptera, Auchenorrhyncha) auf Ackerbohnen (*Vicia faba* L.). Anatomische und physiologische Untersuchungen. Ph. D. Thesis, Zurich. 125 pp. [In German with English summary.]

Gunthart, M. S. and Wanner, H.

1981a The feeding behavior of two leafhoppers on *Vicia faba*. *Ecological Entomology* 6(1):17-22.

Guppy, J. C.

1958a Insect surveys of clovers, alfalfa, and birdsfoot trefoil in eastern Ontario. *Canadian Entomologist* 90(9):523-531.

Gupta, H. C. C. and Kavadia, V. S.

1984a Bioefficacy and resistance of permethrin and acephate in cotton. *Pesticides* 18:39-42.

Gupta, R. N. and Dhari, K.

1978a Evaluation of some insecticides for the control of shoot and fruit borer (*Earias* species) and jassid (*Amrasca devastans* Distant) on rainfed okra crop. *Pesticides* 12:21-24.

Gurdip Singh and Sandhu, G. S.

1976a New records of spiders as predators of maize borer and maize jassid. *Current Science* 45:642.

Gurdip Singh, Simwat, G. S. and Sidhu, A. S.

1974a Comparative efficacy of some new insecticides for the control of cotton pests. *Indian Journal of Agricultural Sciences* 43:653-658.

Gurney, A. B., Kramer, J. P. and Steyskal, G. C.
1964a Some techniques for the preparation, study and storage in microvials of insect genitalia. *Annals of the Entomological Society of America* 57:240-242.

Gustin, R. D.
1974a Termination of diapause in the painted leafhopper, *Endria inimica. Annals of the Entomological Society of America* 67(4):607-609.

Gustin, R. D. and Stoner, W. N.
1968a Biology of *Deltocephalus sonorus* (Homoptera: Cicadellidae). *Annals of the Entomological Society of America* 61(1):77-82.

1973a Life history of *Exitianus exitiosus* (Homoptera: Cicadellidae) in the laboratory. *Annals of the Entomological Society of America* 66(2):388-389.

Guthrie, E. J.
1978a Measurement of yield losses caused by maize streak disease. *Plant Disease Reporter* 62:839-941.

Guyer, G., Wells, A., Anderson, A. L. and Dezeeuw, D. J.
1960a An evaluation of systemic insecticides for control of insects on snap and field beans. *Michigan Agricultural Experiment Station Quarterly Bulletin* 42(4):827-835.

Gyllensvard, N.
1963a Nagra for Sverige nya eller sallsynta Hemiptera II. *Opuscula Entomologica* 28:198-200.

1964a *Typhlocyba sundholmi* n. sp. (Hem. Hom.) *Opuscula Entomologica* 29(1&2):170-173.

1965a Nagra for Sverige nya eller sallsynta Hemiptera III. *Opuscula Entomologica* 30(3):227-230.

1969a Nagra for Sverige nya eller sallsynta Hemiptera. 4. *Opuscula Entomologica* 34(1-2):162-166.

Gyorffy, G. and Pollak, T.
1983a Habitat specialization of leafhopper community living in a sandy soil grassland. *Acta Biologica* (Szeged) 29(1-4):153-158.

Gyrisco, G. G.
1958a Forage insects and their control. *Annual Review of Entomology* 3:421-448.

Gyrisco, G. G. and Armbrust, E. J.
1966a Forage and cereal insect recommendations. *New York College of Agriculture Miscellaneous Bulletin* (Cornell) 71. 8 pp.

Gyrisco, G. G., Landman, D., York, A. C., Irwin, B. J. and Armbrust, E. J.
1978a The literature of arthropods associated with alfalfa. IV. A bibliography of the potato leafhopper, *Empoasca fabae* (Harris) (Homoptera: Cicadellidae). *Illinois Agricultural Experiment Station Special Publication* 51. 75 pp.

Habeck, D. H.
1960a Insects associated with sweet potato foliage and their relation to the dissemination of internal cork virus. Ph. D. Dissertation, North Carolina State University. 63 pp. [Diss. Abst. 20(7):2976.]

Habib, A., Badawi, A. and Herakly, F.
1972a Biological studies on certain species of leafhoppers (Hemiptera-Cicadellidae) in Egypt. *Zeitschrift fur Angewandte Entomologie* 71(2):172-178.

Habib, A., El-Kady, E. and Herakly, F. A.
1975a Taxonomy of jassids infesting truck crops in Egypt (Hemiptera: Jassidae). 1. Taxonomy of subfamily Typhlocybinae. *Bulletin, Societe Entomologique d'Egypte* 59:331-343.

1975b Taxonomy of jassids infesting truck crops in Egypt (Hemiptera: Jassidae). 2. Taxonomy of subfamily Euscelinae, tribe Euscelini. *Bulletin, Societe Entomologique d'Egypte* 59:363-380.

1976a Taxonomy of jassids infesting truck crops in Egypt (Homoptera: Jassidae). 3. Taxonomy of subfamily Euscelinae, tribe Macrostelini. *Bulletin, Societe Entomologique d'Egypte* 60:197-213.

Hafidi, B., Bencheqroun, N., Fischer, H. U. and Vanderveken, J.
1979a Application of the immuno-enzymatic 'ELISA' technique for the detection of *Spiroplasma citri*, the causal agent of citrus stubborn disease, in Moroccan orange orchards. *Parasitica* 35:141-151.

Hagel, G. T.
1969a Aphids and leafhoppers found on red clover in eastern Washington. *Journal of Economic Entomology* 62(1):273.

Hagel, G. T. and Hampton, R. O.
1970a Dispersal of aphids and leafhoppers from red clover to red Mexican beans, and the spread of bean yellow mosaic by aphids. *Journal of Economic Entomology* 63(4):1057-1060.

Hagel, G. T. and Landis, B. J.
1967a Biology of the aster leafhopper *Macrosteles fascifrons* (Homoptera: Cicadellidae) in eastern Wash., and some overwintering sources of aster yellows. *Annals of the Entomological Society of America* 60(3):591-595.

1974a Leafhopper transmission of western aster yellows agent to potato and carrot in eastern Washington. *Western Region, Agricultural Research Service, United States Department of Agriculture* ARS W-13. 7 pp.

Hagel, G. T., Landis, B. J. and Abrens, M. C.
1973a Aster leafhopper: Source of infestation, host plant preference, and dispersal. *Journal of Economic Entomology* 66:877-881.

Hainzelin, E.
1982a Virus diseases of maize: A bibliographical account. *Agronomie Tropicale* 37:393-404.

Halffter, G.
1957a Plagas que afectan a las distintas especies de *Agave* cultivadas en Mexico. Mexico, D.F., Direccion General Departmiento de Agricola, Secretariat de Agricola y Ganaderia 135:5.

Halkka, O.
1959a Chromosome studies on the Hemiptera Homoptera Auchenorrhyncha. *Suomalaisen Tiedeakatemian Toimituksia*, Series A 4(43):1-72.

1960a The structure of bivalents in the Homoptera Auchenorrhyncha. *Chromosoma* 11(2):245-262.

1960b Chromosomal evolution in the Cicadellidae. *Hereditas* 46:581-591.

1962a Meiotische Mosaikzysten in der Zikadengattung *Idiocerus*. *Naturwissenschaften* (Gottingen) 49:310-311.

1965a X-ray induced changed in the chromosomes of *Limotettix* (Homoptera). *Chromosoma* 16:185-191.

Halkka, O. and Heinonen, L.
1964a The chromosomes of 17 Nearctic Homoptera. *Hereditas* 52:75-80.

1966a Chromosome numbers in 20 American or Sundanese leafhoppers, with notes on the role of cytology in homopteran taxonomy. *Suomalaisen Tiedeakatemian Toimituksia* A 32(1):11-19.

Hall, W. J.
1956a Insect pests in British Colonial Dependencies: A half-yearly report. *FAO Plant Protection Bulletin* 5(1):4.

Hallock, H. C. and Deen, O. T.
1956a Greenhouse tests on control of the beet leafhopper. *Journal of Economic Entomology* 49(1):123-126.

Hallock, H. C. and Douglass, J. R.
1956a Studies of four summer hosts of the beet leafhopper. *Journal of Economic Entomology* 49(3):388-391.

Halteren, P. van
1970a Some observations on the biology of *Draeculacephala clypeata* Osb. (Cicadellidae, Homoptera) in Surinam. *Surinaamse Landbouw* 18(2):84-86.

Halteren, P. van, Shagir, Sama and Saleh, K. M.
1974a The tungro virus disease in Sulawesi. *In: Agricultural Cooperation Indonesia—The Netherlands. Research Reports 1968-1974. Section II. Technical Contributions*. Pp. 175-184.

Hama, H.
1975a Resistance to insecticides in the green rice leafhopper. *Japan Pesticide Information* No. 23, pp. 9-12.

1975b Toxicity and antiacetylcholinesterase activity of propaphos, O,O-di-(n)-propyl-O-4-methylthiophenyl phosphate, against the resistant green rice leafhopper, *Nephotettix cinticeps* Uhler. *Botyu-Kagaku* 40:14-19. [In Japanese with English summary.]

1976a Modified and normal cholinesterases in the respective strains of carbamate-resistant and susceptible green rice leafhoppers, *Nephotettix cincticeps* Uhler (Hemiptera: Cicadellidae). *Applied Entomology and Zoology* 11:239-247.

1977a Cholinesterase activity and its sensitivity to inhibitors in resistant and suceptible strains of the green rice leafhopper, *Nephotettix cincticeps* Uhler. *Botyu-Kagaku* 42:82-88. [In Japanese with English summary.]

1978a Preliminary report on the existence of butyrylcholinsterase-like enzyme in the green rice leafhopper *Nephotettix cincticeps* Uhler (Hemiptera: Cicadellidae). *Applied Entomology and Zoology* 13:324-325.

1980a Mechanism of insecticide resistance in green rice leafhopper and brown planthopper. *Review of Plant Protection Research* 13:54-73.

1980b Studies on mechanism of resistance to insecticides in the green rice leafhopper, *Nephotettix cincticeps* Uhler, with particular reference to reduced sensitivity of acetylcholinesterase. *Bulletin of the National Institute of Agricultural Sciences* No. 34, pp. 75-138.

1983a Changed acetylcholinesterase and resistance in leaf- and planthoppers. *In:* Miyamoto, J., Kearney, P. C., Matsunaka, D. H. and Murphy, S. D., eds. *Pesticide Chemistry: Human Welfare and the Environment. Proceedings of the 5th International Congress of Pesticide Chemistry, Kyoto, Japan.* Pp. 203-208. Pergamon Press, Oxford, U.K.

1984a Mechanism of fenitrothion- and diazinon-resistance in the green rice leafhopper, *Nephotettix cincticeps* Uhler (Hemiptera: Deltocephalidae)—the role of aliesterase. *Japan Journal of Applied Entomology and Zoology* 28:143-149.

Hama, H. and Iwata, T.

1971a Insensitive cholinesterase in the Nakagawara strain of the green rice leafhopper, *Nephotettix cincticeps* Uhler (Hemiptera: Cicadellidae), as a cause of resistance to carbamate insecticides. *Applied Entomology and Zoology* 6:183-191.

1972a Sensitive aliesterase to a carbamate-insecticide, propoxur, in the resistant strains of the green rice leafhopper. *Applied Entomology and Zoology* 7:177-179.

1973a Resistance to carbamate insecticides and its mechanism in the green rice leafhopper *Nephotettix cincticeps* Uhler. *Japanese Journal of Applied Entomology and Zoology* 17:154-161.

1973b Synergism of carbamate and organophosphorus insecticides against insecticide-resistant green rice leafhoppers, *Nephotettix cincticeps* Uhler. *Japanese Journal of Applied Entomology and Zoology* 17:181-186.

1978a Studies on the inheritance of carbamate-resistance in the green rice leafhopper, *Nephotettix cincticeps* Uhler (Hemiptera: Cicadellidae). Relationships between insensitivity of acetylcholinesterase and cross-resistance to carbamate and organophosphate insecticides. *Applied Entomology and Zoology* 13:196-202.

Hama, H., Iwata, T., Miyata, T. and Saito, R.

1980a Some properties of acetylcholinesterase partially purified from susceptible and resistant green rice leafhoppers, *Nephotettix cincticeps* Uhler (Hemiptera: Deltocephalidae). *Applied Entomology and Zoology* 15:249-261.

Hama, H., Iwata, T., Tomizawa, C. and Murai, T.

1977a Mechanism of resistance to malathion in the green rice leafhopper, *Nephotettix cincticeps* Uhler. *Botyu-Kagaku* 42:188-197.

Hama, H., Iwata, T. and Tomizawa, C.

1979a Absorption and degradation of propoxur in susceptible and resistant green rice leafhoppers, *Nephotettix cincticeps* Uhler (Homoptera: Cicadellidae). *Applied Entomology and Zoology* 14:333-339.

Hama, H. and Yamasaki, Y.

1981a Individual variation of acetylcholinesterase sensitivity to propoxur in the green rice leafhopper, *Nephotettix cincticeps* Uhler (Hemiptera: Deltocephalidae). *Applied Entomology and Zoology* 16:52-54.

Hamad, N. E. F., Hanna, H. M. and Herakly, F. A.

1981a Studies of catches of leaf-hoppers in a light trap at Assiut (Hemiptera: Cicadellidae). *Bulletin de la Societe Entomologique d'Egypte* 61:21-29.

Hamilton, C. C.

1959a The use of systemic insecticides to control nursery insects. *Florists Exchange* 133(11):14, 20-25.

Hamilton, C. D.

1957a Systematic insecticides: Their nature and use in arboriculture. *Proceedings of the National Shade Tree Conference* 33:128-145.

Hamilton, D. W. and Cleveland, M. L.

1958a Control of the codling moth and other apple pests with Ryania. *Journal of Economic Entomology* 50(6):756-759.

Hamilton, D. W. and Fahey, J. E.

1954a DDT fails to control *Erythroneura lawsoniana*. *Journal of Economic Entomology.* 47(2):361-362.

Hamilton, K. G. A.

1967a A new species of *Draeculacephala* (Homoptera: Cicadellidae) from Manitoba. *Canadian Entomologist* 99:767-769.

1970a The genus *Cuerna* (Homoptera: Cicadellidae) in Canada. *Canadian Entomologist* 102(4):425-441.

1970b Morphology and phylogeny of the insect suborder Homoptera (Hemiptera.) Thesis, University of Georgia, Athens. 127 pp.

1971a A remarkable fossil homopteran from Canadian cretaceous amber representing a new family. *Canadian Entomologist* 103:943-946.

1972a The leafhopper genus *Empoasca* subgenus *Kybos* in the southern interior of British Columbia. *Journal of the Entomological Society of British Columbia* 69:58-67.

1972b The Manitoban fauna of leafhoppers I. Descriptions of new species and colour forms. *Canadian Entomologist* 104:825-831.

1972c The classification, morphology and phylogeny of the family Cicadellidae (Homoptera). Dissertation, University of Georgia, Athens. 220 pp.

1972d The Manitoban fauna of leafhoppers II. The fauna of macro-leafhoppers. *Canadian Entomologist* 104:1137-1148.

1975a Revision of the genera *Paraphlepsius* Baker and *Pendarus* Ball (Rhynchota: Homoptera: Cicadellidae). *Memoirs of the Entomological Society of Canada* 96:1-129. [New subgenera: *Strephonius, Sabix, Gamarex,* and *Paraphysius.*]

1975b Review of the tribal classification of the leafhopper subfamily Aphrodinae (Deltocephalinae of authors) of the Holarctic region (Rhynchota: Homoptera: Cicadellidae). *Canadian Entomologist* 107:477-498.

1975c Additional characters for specific determinations in Nearctic *Xerophloea* (Rhynchota: Homoptera: Cicadellidae). *Canadian Entomologist* 107:943-946.

1975d A review of the Northern Hemisphere Aphrodina (Rhynchota: Homoptera: Cicadellidae), with special reference to the Nearctic fauna. *Canadian Entomologist* 107:1009-1027. [New genus: *Planaphrodes.*]

1976a Cicadellidae (Rhynchota: Homoptera) described by Provancher, with notes on his publications. *Le Naturaliste Canadien* 103:29-45.

1980a Review of the Nearctic Idiocerini, excepting those from the Sonoran subregion (Rhynchota: Homoptera: Cicadellidae). *Canadian Entomologist* 112:811-848.

1980b Contributions to the study of the World Macropsini (Rhynchota: Homoptera: Cicadellidae). *Canadian Entomologist* 112:875-932. [Descriptions of 5 new genera and 4 new subgenera.]

1982a A review of the Nearctic species of the nominate subgenus of *Gyponana* Ball (Rhyncota: Homoptera: Cicadellidae). *Journal of the Kansas Entomological Society* 55(3):547-562.

1982b Two new species of arboricolous Typhlocybinae (Rhynchota: Homoptera: Cicadellidae). *Canadian Entomologist* 114:1129-1132.

1983a Synopsis of *Typhlocyba* subgenus *Empoa* Fitch, with a key to the species of the *gillettei* complex (Rhynchota: Homoptera: Cicadellidae). *Canadian Entomologist* 115:67-73.

1983b Introduced and native leafhoppers common to the Old and New worlds (Rhynchota: Homoptera: Cicadellidae). *Canadian Entomologist* 115:473-511.

1983c Classification, morphology and phylogeny of the family Cicadellidae (Rhynchota: Homoptera). *In: Knight, W. J., Pant, N. C., Robertson, T. S. and Wilson, M. R., eds. Proceedings of the 1st International Workshop on Biotaxonomy, Classification and Biology of Leafhoppers and Planthoppers (Auchenorrhyncha) of Economic Importance.* Pp. 15-37. London, 4-7 October 1982. Commonwealth Institute of Entomology, 56 Queen's Gate, London SW7 5JR. 500 pp.

1983d Revision of the Macropsini and Neopsini of the New-World (Rhynchota: Homoptera: Cicadellidae), with notes on intersex morphology. *Memoirs of the Entomological Society of Canada* 123:1-223. [New genus: *Nollia.*]

1985a Review of *Draeculacephala* Ball (Homoptera, Auchenorrhyncha, Cicadellidae). *Entomologische Abhandlungen und Berichte aus dem Staatlichen Museum fur Tierkunde in Dresden* 49(5):83-103. [New genera: *Xphon.*]

1985b The *Graphocephala coccinea* complex in North America (Homoptera, Auchenorrhyncha, Cicadellidae). *Entomologische Abhandlungen und Berichte aus dem Staatliche Museum fur Tierkunde in Dresden* 49(6):105-111.

1985c Taxa of *Idiocerus* Lewis new to Canada (Rhynchota: Homoptera: Cicadellidae). *Journal of the Entomological Society of British Columbia* 82:59-65.

1985d Leafhoppers of ornamental and fruit trees in Canada. *Agriculture Canada.* Publication 1779/E. 71 pp.

Hamilton, K. G. A. and Ross, H. H.

1972a New synonymy and descriptions of Nearctic species of the genera *Auridius* and *Hebecephalus*. *Journal of the Georgia Entomological Society* 7(2):133-139.

1975a New species of grass-feeding deltocephaline leafhoppers with keys to the Nearctic species of *Palus* and *Rosenus* (Rhynchota: Homoptera: Cicadellidae). *Canadian Entomologist* 107:601-611.

Hamilton, R. E.

1964a Wheat striate mosaic observed in Montana. *Plant Disease Reporter* 48:68.

Hammad, S. M.

1978a Pests of grain legumes and their control in Egypt. *In: Singh, S. R., Van Emden, H. F. and Ajibola Taylor, T., eds. Pests of Grain Legumes: Ecology and Control.* Pp. 135-137. Academic Press, New York. 454 pp.

Hamon, C.

1969a Etude de l'ovaire acrotrophique d'un homoptere auchenorhynche *Ulopa reticulata* Fab. *Bulletin de la Societe Scientifique de Bretagne* 44:49-63.

1969b Le filament terminal de l'ovaire de divers homopteres auchenorhynches. *Compte Rendu Hebdomadaire des Seances de l'Academie des Sciences* 269D:1860-1862.

Hamon, C. and Folliot, R.,

1969a Ultrastructure des cordons trophiques de l'ovaire de divers homopteres auchenorhynches. *Compte Rendu Hebdomadaire des Seances de l'Academie des Sciences* 268D:577-580.

Hanna, A. D.

1969a A review of the use of insecticides on the main agricultural crops of the world—Section 3: Cotton. *World Review of Pest Control* 8(1):23-44.

Hanna, M. A.

1970a Relative toxicity of phorate and its metabolites to *Tetranychus urticae* Koch (Tetranychidae, Acarina) and *Empoasca fabae* Harris (Cicadellidae: Homoptera). Ph. D. Dissertation, Iowa State Univiversity. 68 pp. [Diss. Abst. Int. B 31(6):3460.]

Hansen, H. P.

1961a *Laerbogi Systematisk Plantevirologi med saerligt henblok pa Viroseri Danske Landbrugsagroder.* [Textbook of systematic plant virology, with special reference to viruses in Danish agricultural crops.] Kobenhavn. 161 pp.

Hanson, C. H.

1969a Registration of alfalfa germplasm (Reg. Nos. Gr 1 and 7). *Crop Science* 9(4):526-527.

Hanson, C. H., Busbice, T. H., Hill, R. R., Hunt, O. J. and Oakes, A. J.

1972a Directed mass selection for developing multiple pest resistance and conserving germplasm in alfalfa. *Journal of Environmental Quality* 1(1):106-111.

Hanson, C. H., Loper, G. M., Kohler, G. O., Bickoff, E. M., Taylor, K. W., Kehr, W. R., Standord, E. H., Dudley, J. W., Pedersem, M. W., Sorensen, E. L., Carnahan, E. L. and Wilsie, C. P.

1965a Variation in coumestrol content of alfalfa as related to location, variety, cutting, stage of growth, and disease. *U.S. Department of Agriculture Technical Bulletin* 1333. 72 pp.

Hanson, C. H., Norwood, B., Blickenstaff, C. C. and Van Denburgh, R. S.

1963a Recurrent phenotypic selection for resistance to potato leafhopper yellowing in alfalfa. *Agronomy Abstracts* 1963:80.

Hanson, E. W., Hanson, C. H., Frosheiser, F. I., Sorensen, E. L., Sherwood, R. T., Graham, J. H., Elling, L. J., Smith, D. and Davis, R. L.

1964a Reactions of varieties, crosses and mixtures of alfalfa to six pathogens and the potato leafhopper. *Crop Science* 4:273-276.

Hao, T. F. G. and Pitre, H. N.

1970a Relationship of vector numbers and age of corn plants at inoculation to severity of corn stunt disease. *Journal of Economic Entomology* 63(3):924-927.

Haque, S. Q. and Parasram, S.

1973a *Empoasca stevensi*, a new vector of bunchy top disease of papaya. *Plant Disease Reporter* 57(5):412-413.

Harakly, F. A.

1974a Preliminary survey of pests infesting solanaceous truck crops in Egypt. *Bulletin de la Societe Entomologique d'Egypte* 58:133-140.

Harakly, F. A. and Assem, M. A. H.

1978a Ecological studies on the truely pests of leguminous plants in Egypt. II. Piercing and sucking pests. *In: Proceedings of the Fourth Conference of Pest Control*, September 30 - October 3, 1978 (Part 1). Academy of Scientific Research and Technology. Pp. 237-242. 1356 pp.

Harakly, F. A. and Shalaby, F. M.

1981a Utilization of attractant traps for ecological studies. *Agriculture Research Review* 59:65-77.

Harcharan, S. and Mann, V. S.

1979a Comparative bio-efficacy of nozzles for the control of jassid *Sundapteryx biguttula biguttula* Ishida on cotton. *Journal of Research, Punjab Agricultural University* 16:295-299.

Hardee, D. D., Forsythe, H. Y., Jr. and Gyrisco, G. G.

1962a *Allygidius atomarius* (Fab.) (Homoptera: Cicadellidae), a new unreported species for North America. *Proceedings of the Entomological Society of Washington* 64(3):202.

1963a A survey of the Hemiptera and Homoptera infesting grasses (Graminae) in New York. *Journal of Economic Entomology* 56:555-559.

Harder, D. E. and Westdal, P. H.

1971a A cereal enation disease in Kenya. *Plant Disease Reporter* 55(9):802-803.

Harding, R. M. and Teakle, D. S.
　1985a Mycoplasma-like organisms as causual agent of potato purple top wilt in Queensland. *Australian Journal of Agricultural Research* 36:443-449.

Hardy, D. E.
　1971a Pipunculidae (Diptera) parasitic on rice leafhoppers in the Oriental region. *Proceedings of the Hawaii Entomological Society* 21:79-91.

Harizanov, A.
　1968a Cicadas as pests and carriers of viruses among plants. *Priroda* [Sofia] 17(4):49-55. [In Bulgarian.]
　1969a A new enemy of grape vine in Bulgaria. *Rastitelna Zashchita* 17(11):21-23. [In Bulgarian.]
　1970a Harmful cicada on wheat in Bulgaria. *Rastitelna Zashchita* 18(2):18-21. [In Bulgarian.]

Harrell, J. C. and Hoizapfel, E.
　1966a Trapping air-borne insects on ships in the Pacific: 6. *Pacific Insects* 8(1):33-42.

Harries, F. H. and Valcarce, A. C.
　1957a Laboratory toxicity tests against insects affecting sugar beets grown for seed. *Journal of Economic Entomology* 50(2):120-122.

Harris, K. F.
　1979a Leafhoppers and aphids as biological vectors: Vector-virus relationships. In: Maramorosch, K. and Harris, K. F., eds. *Leafhoppers, Vectors and Plant Disease Agents.* Pp. 217-308. Academic Press, New York. 654 pp.
　1980a Aphids, leafhoppers and planthoppers. In: Harris, K. F. and Maramorosch, K., eds. *Vectors of Plant Pathogens.* Pp. 1-13. Academic Press, New York. 467 pp.
　1981a Sucrose stimulation of leafhopper probing and feeding: The sensory transduction mechanism. *Phytopathology* 71:879.
　1981b Arthropod and nematode vectors of plant viruses. *Annual Review of Phytopathology* 19:391-426.
　1981c Horizontal transmission of plant viruses. In: McKelvey, J. J., Eldridge, B. F. and Maramorosch, K., eds. *Vectors of Disease Agents: Interaction with Plants, Animals and Man.* Pp. 92-108. Praeger, New York.
　1983a Auchenorrhynchous vectors of plant viruses: Virus-vector interactions and transmission mechanisms. In: Knight, W. J., Pant, N. C., Robertson, T. S. and Wilson, M. R., eds. *Proceedings of the 1st International Workshop on Biotaxonomy, Classification and Biology of Leafhoppers and Planthoppers (Auchenorrhyncha) of Economic Importance.* Pp. 405-413. London, 4-7 October 1982. Commonwealth Institute of Entomology, 56 Queen's Gate, London SW7 5JR. 500 pp.

Harris, K. F., ed.
　1984a *Current Topics in Vector Research.* Praeger, New York. 325 pp.

Harris, K. F. and Childress, S. A.
　1980a Fate of maize chlorotic dwarf virus (MCDV) in its black-faced leafhopper vector, *Graminella nigrifrons.* In: *16th International Congress of Entomology*, Kyoto, Japan 1980:47.

Harris, K. F. and Maramorosch, K., eds.
　1980a *Vectors of Plant Pathogens.* Academic Press, New York. 467 pp.
　1982a *Pathogens, Vectors, and Plant Diseases: Approaches to Control.* Academic Press, New York. 310 pp.

Harris, K. F., Truer, B., Tsai, J. and Toler, R.
　1981a Observations on leafhopper ingestion-egestion behavior: Its likely role in the transmission of noncirculative viruses and other plant pathogens. *Journal of Economic Entomology* 74(4):446-453.

Harrison, B. D.
　1981a Plant virus ecology: Ingredients, interactions and environmental influences. *Annals of Applied Biology* 99:195-209.

Harrison, R. G.
　1980a Dispersal polymorphisms in insects. *Annual Review of Ecology and Systematics* 11:95-118.

Hartwig, E. E. and Edwards, A. J.
　1970a Effects of morphological characteristics upon seed yield in soybeans. *Agronomy Journal* 62(1):64-65.

Hasabe, B. M. and Moholkar, P. R.
　1981a Studies on the efficacy of systemic insecticides as seed dressers against the pests of bhendi. *Pestology* 5:9-11.

Hasanuddin, A. and Ling, K. C.
　1980a Effect of different tungro-infected varieties as virus sources on the infectivity of *Nephotettix virescens. International Rice Research Newsletter* 5(1):5-6.
　1980b Effect of age of tungro-disease plants on GLH [green leafhopper] infectivity. *International Rice Research Newsletter* 5(1):9.

Hashioka, Y.
　1952a Varietal resistance of rice to the brown spot and yellow dwarf. Studies on pathological breeding of rice. VI. *Japanese Journal of Breeding* 2:14-16.

Hashizume, B.
　1964a Studies on forecasting and control of the green rice leafhopper *Nephotettix cincticeps* Uhler with special reference to eradication of the rice dwarf disease. *Memoirs of the Association for Plant Protection of Kyushu* 2:1-77 [sic].

Hashmi, A. A., Hussain, M. M. and Ulfat, M.
　1983a Insect pest complex of wheat crop. *Pakistan Journal of Zoology* 15:169-176.

Hashmi, A. A. and Nizam-ud-Din, M.
　1978a Lethal exposure time as substitute for economic threshold. *Pakistan Journal of Zoology* 10:229-234.

Hassan, A. A., Soliman, A. A. and Hosny, M. M.
　1960a Chemical control of some cotton insects under field conditions, in Egypt. *Bulletin de la Societe Entomologique d'Egypte* 44:393-404.

Hassan, S. M., Saad, A. S. and Mansour, M. H.
　1975a Evaluation of certain insecticides against aphids, jassids, whiteflies and red spider mites attacking cotton. *Bulletin of the Entomological Society of Egypt, Economic Series* 8:41-45.

Hassanein, M. H., Khalil, F. M. and Eisa, M. A.
　1970a The effectiveness of certain insecticides on the population density of jassids in the cotton fields of Upper Egypt (Hemiptera-Homoptera: Jassidae). *Bulletin of the Entomological Society of Egypt, Economic Series* 4:197-202.
　1971a Preliminary studies on jassids infesting cotton plants in Assuit region (Hemiptera-Homoptera). *Bulletin de la Societe Entomologique d'Egypte* 54:159-164.

Hatta, T. and Francki, R. I. B.
　1982a Similarity in the structure of cytoplasmic polyhedrosis virus, leafhopper A virus and Fiji disease virus particles. *Intervirology* 18:203-208.

Haupt, H., Schmutterer, H. and Muller F. P.
　1969a Homoptera-Gleichflugler In: Stresemann, E., ed. *Exkursionfauna von Deutschland. Insekten-zweiter Halband.* Pp. 28-141. Wirbellose 11/2. Volk und Wissen volkseigener Verlag, Berlin.

Hawkins, B. A. and Cross, E. A.
　1982a Patterns of refaunation of reclaimed strip mine spoils by nonterricolous arthropods. *Environmental Entomology* 11(3):762-775.

Hawkins, J. A.
　1979a Leafhoppers and planthoppers infesting coastal bermudagrass: Their effect on yield and quality; their control by varying frequency of harvest. *Dissertation Abstracts International* (B) 39(8):3676.

Hawkins, J. A., Wilson, B. H., Mondart, C. L., Nelson, B. D., Farlow, R. A. and Schilling, P. E.
　1979a Leafhoppers and planthoppers in coastal bermudagrass: Effects on yield and quality and control by harvest frequency. *Journal of Economic Entomology* 72:101-104.

Hayashi, M. and Hayakawa, M.
　1962a Malathion tolerance in *Nephotettix cincticeps* Uhler. *Japan Journal of Applied Entomology and Zoology* 6(3):250-252.

Hay, C. J. and Meyer, W. C.
　1961a Use of P32 as an aid in biological studies of the leafhopper, *Scaphoideus luteolus*. *Journal of Economic Entomology* 54(6):1260-1261.

Hayflick, L. and Arai, S.
　1973a Failure to isolate mycoplasmas from aster-yellows diseased plants and leafhoppers. *Annals of the New York Academy of Science* 225:494-502.

Haynes, H. L., Lambrech, J. A. and Moorefield, H. H.
　1957a Insecticidal properties and characteristics of 1-Naphthyl N-Methylcarbamate. *Contributions from Boyce Thompson Institute for Plant Research* 18(11):507-513.

Heady, S. E., Madden, L. V. and Nault, L. R.
　1985a Oviposition behavior of *Dalbulus* leafhoppers (Homoptera: Cicadellidae). *Annals of the Entomological Society of America* 78(6):723-727.

Heady, S. E. and Nault, L. R.
　1984a Leafhopper egg microfilaments (Homoptera: Cicadellidae). *Annals of the Entomological Society of America* 77(5):610-615.
　1985a Escape behavior of *Dalbulus* and *Baldulus* leafhoppers (Homoptera: Cicadellidae). *Environmental Entomology* 14(2):154-158.
　1985b Discovery of microfilaments on eggs of leafhoppers. *Ohio Report* 70:41-43.

Hegab, A. M.
　1985a Biological studies on leafhopper *Streptanus* species (Homoptera: Cicadellidae). *Folia Entomologica Hungarica* 46(2):57-62.

Hegab, A. M., Orosz, A. and Jenser, G.
　1980a Observations on the larvae and imagoes of some *Allygus* species (Homoptera). *Folia Entomologica Hungarica* 33(1):61-66.

Hegab, A. M., Shahein, A. and Rabeh, M.
　1984a Biological studies on the leafhopper *Streptanus aemulans* (Kirschb.) (Hemiptera: Cicadellidae). *Proceedings of the International Congress of Entomology* 17:326. [In English.]

Heggestad, H. and Moore, E. L.
　1959a Occurrence of curly top on tobacco in Maryland and North Carolina. *Plant Disease Reporter* 43:682-684.

Heinrichs, E. A.
　1979a Control of leafhopper and planthopper vectors of rice viruses. In: Maramorosch, K. and Harris, K. F., eds. *Leafhopper Vectors and Plant Disease Agents*. Pp. 529-560. Academic Press, New York.
　1984a Resistance to leafhoppers and planthoppers in the world collection of rices. *Mitteilungen der Schweizerischen Entomologischen Gesellschaft* 57(4):421-422.

Heinrichs, E. A. and Arceo, M.
　1978a Residual activity of acephate sprays on rice as influenced by spreader-stickers. *International Rice Research Newsletter* 3(4):18.

Heinrichs, E. A., Basilio, R. P. and Valencia, S. L.
　1984a Buprefezin, a selective insecticide for the management of rice planthoppers (Homoptera: Delphacidae) and leafhoppers (Homoptera: Cicadellidae). *Environmental Entomology* 13:515-521.

Heinrichs, E. A., Medrano, F. G. and Rapusas, H. R.
　1985a *Genetic evaluation for insect resistance in rice*. International Rice Research Institute, Los Banos, Lagunos, Philippines. 356 pp.

Heinrichs, E. A., Medrano, F. G., Rapusas, H. R., Vega, C., Medina, E., Romena, A., Viajante, V., Sunio, L., Domingo, I. and Camanag, E.
　1985a Insect pest resistance of IR5-IR62. *International Rice Research Newsletter* 10(6):12-13.

Heinrichs, E. A., Medrano, F., Sunio, L., Rapusas, H., Romena, A., Vega, C., Viajante, V., Centina, D. and Domingo, I.
　1982a Resistance of IR varieties to insect pests. *International Rice Research Newsletter* 7(3):9-10.

Heinrichs, E. A. and Rapusas, H. R.
　1983a Correlation of resistance to the green leafhopper, *Nephotettix virescens* (Homoptera: Cicadellidae), with tungro virus infection in rice varieties having different genes for resistance. *Environmental Entomology* 12(1):201-205.
　1984a Feeding, development, and tungro virus transmission by the green leafhopper, *Nephotettix virescens* (Distant) (Homoptera: Cicadellidae), after selection on resistant rice cultivars. *Environmental Entomology* 13(4):1074-1078.
　1984b Feeding activity of the green leafhopper (GLH) and tungro (RTV) infection. *International Rice Research Newsletter* 9(5):15.
　1985a Cross-virulence of *Nephotettix virescens* (Homoptera: Cicadellidae) biotypes among some rice cultivars with the same major-resistance gene. *Environmental Entomology* 14(6):696-700.
　1985b Virulence of green leafhopper (GLH) *Nephotettix virescens* colonies on rice cultivars with Glh 2 or Glh 5 gene for resistance. *International Rice Research Newsletter* 10(4):4-5.

Heinze, K. G.
　1959a *Phytopathogene Viren und ihre Ubertrager*. Berlin. 291 pp.
　1972a Report to the government of Jamaica on lethal yellowing disease of coconut. *United Nations Development Programme Report* No. TA 3152. 28 pp.

Heinze, K. and Kunze, L.
　1955a Die europaische asterngelbsucht und ihre ubertragung durch zwergzikaden. *Nachrichtenblatt des Deutscher Pflanzenschutzdienstes* 7(8):161-164.

Helaly, M. M., Ibrahim, A. E. and Saleh, R. A.
　1985a Fluctuations of population densities of *Empoasca* sp., *Aphis cracciyora* Koch, and *Tetranychus arabicus* Attiah attacking cowpea at Zagazig, Egypt. *Bulletin de la Societe Entomologique d'Egypte* 64:35-43.

Heller, F. R.
　1961a Zur Synonymie von *Deltocephalus lindneri* Fahr. (Homoptera - Deltocephalidae). *Entomologische Zeitschrift* 71:82-84.
　1969a Eine neue Idioceridae aus Chile: *Idiocerus hichinsi* n. sp. (Homopt.). *Entomologische Zeitschrift* 79:155-157.
　1972a Zwei neue Nirvaniden aus Kamerun (Homopt., Cicad.). *Stuttgarter Beitrage zur Naturkunde aus dem Staatlichen Museum fur Naturkunde in Stuttgart*, Series A 246:1-7. [New genus: *Afrokana*.]
　1975a *Adarrus ocellaris* (Fall.) ssp. *tatraensis* nova (Homoptera, Cicadellidae). *Stuttgarter Beitrage zur Naturkunde aus dem Stattlichen Museum fur Naturkunde in Stuttgart*, Series A 288:1-3.

Heller, R. and Linnavuori, R.
　1968a Cicadelliden aus Athiopien. *Stuttgarter Beitrage zur Naturkunde aus dem Stattlichen Museum fur Naturkunde in Stuttgart*, Series A 186:1-42. [New genera: *Alemaia* and *Awasha*.]

Helm, C. G., Kogan, M. and Hill, R. G.
　1980a Sampling leafhoppers on soybean. *Springer Series Experimental Entomology* 260-282.

Helms, T. J.
　1967a Post embryonic development of reproductive systems in *Empoasca fabae* (Harris) (Homoptera, Cicadellidae). Ph. D. Dissertation, Iowa State University. 97 pp. [Diss. Abst. Int. B 28(4):1562.]
　1968a Postembryonic reproductive-systems development in *Empoasca fabae* (Homoptera: Cicadellidae). *Annals of the Entomological Society of America* 61(2):316-332.
　1968b Anatomy of the alimentary canal of *Empoasca fabae* (Homoptera: Cicadellidae). *Annals of the Entomological Society of America* 61(6):1604-1606.

Helson, G. A. H.
　1950a Yellow dwarf of tobacco in Australia. V. Transmission by *Orosius argentatus* (Evans) to some alternative host plants. *Australian Journal of Agricultural Research* 1(2):144-147.

Hemmati, K.
1979a Localization and pathogenicity of dwarf aster yellows mycoplasma-like agent in the midgut of the leafhopper vector *Macrosteles fascifrons*. *Iranian Journal of Plant Pathology* 15:45-51.

Hemmati, K. and McLean, D. L.
1980a Ultrastructure and morphological characterisitcs of mycoplasma-like organisms associated with Tulelake aster yellows. *Phytopathologische Zeitschrift* 99:146-154.

Henderson, C. F.
1955a Parasitization of the beet leafhopper in relation to its dissemination in southern Idaho. *U.S. Department of Agricultural Circular* 968:1-16.

Henne, R. C.
1970a Effect of five insecticides on populations of the six-spotted leafhopper and the incidence of aster yellows in carrots. *Canadian Journal of Plant Science* 50(2):169-174.

Henriquez M. R.
1971a Variacion poblacional de homoptera en estrato herbaceo de clima mediterraneo en sabana chilena. *Revista Peruana de Entomologia* 14(2):334-339.

Hepner, L. W.
1946a A new subgenus and several new species of *Scaphytopius* (Homoptera-Cicadellidae). *Journal of the Kansas Entomological Society* 19(3):87-109. [New subgenus: *Vertanus*.]
1966a Twenty new species of *Erythroneura* related to *Erythroneura bigemina* (Homoptera: Cicadellidae). *Journal of the Kansas Entomological Society* 39(1):78-89.
1966b New species of *Erythroneura* related to *lenta* (Homoptera: Cicadellidae). *Florida Entomologist* 49(2):95-100.
1966c New species of *Erythroneura* (Cicadellidae) related to *inepta*. *Journal of the Georgia Entomological Society* 1(4):1-5.
1966d New species of *Erythroneura* related to *campora* (Homoptera: Cicadellidae). *Florida Entomologist* 49(2):101-106.
1967a New species of *Erythroneura* related to *E. dira* (Homoptera: Cicadellidae). *Journal of the Kansas Entomological Society* 40(1):17-24.
1967b New species of *Erythroneura* (Homoptera: Cicadellidae). *Entomological News* 78(3):59-73.
1969a New species of *Erythroneura* from oaks and hickories (Homoptera: Cicadellidae). *Journal of the Kansas Entomological Society* 42(2):126-133.
1972a Five new species of *Erythroneura* (Homoptera: Cicadellidae). *Journal of the Kansas Entomological Society* 45(4):430-433.
1972b A new species of *Erythroneura* (Homoptera: Cicadellidae) and some characteristics of the nymph. *Journal of the Georgia Entomological Society* 7(3):216-218.
1972c New species of *Erythroneura* (Homoptera: Cicadellidae). *Florida Entomologist* 55(4):267-272.
1973a New species of *Erythroneura* (Homoptera: Cicadellidae). *Journal of the Kansas Entomological Society* 46(2):184-186.
1975a New species of *Erythroneura* (Homoptera: Cicadellidae). *Journal of the Kansas Entomological Society* 48(1):4-7.
1976a Thirteen new species of *Erythroneura* (*Erythridula*) (Homoptera: Cicadellidae). *Journal of the Kansas Entomological Society* 49(2):204-211.
1976b Sixteen new species of *Erythroneura* (*Erythridula*) from eastern North America (Homoptera, Cicadellidae). III. *Journal of the Georgia Entomological Society* 11(2):119-126.
1976c Fifteen new species of *Erythroneura* (*Erythridula*) (Homoptera, Cicadellidae). II. *Florida Entomologist* 59(3):293-300.
1976d Seventeen new species of *Erythroneura* (*Erythridula*) (Homoptera, Cicadellidae). VI. *Journal of the Georgia Entomological Society* 11(4):309-316.
1977a Fourteen new species of *Erythroneura* (*Erythridula*) (Homoptera: Cicadellidae). IV. *Journal of the Kansas Entomological Society* 50(2):247-255.
1977b Fourteen new species of *Erythroneura* (*Erythridula*) (Homoptera, Cicadellidae). V. *Florida Entomologist* 60(1):49-55.
1977c Fifteen new species of *Erythroneura* (*Erythridula*) (Homoptera: Cicadellidae). VIII. *Journal of the Georgia Entomological Society* 12(4):359-365.
1978a Sixteen new species of *Erythroneura* (*Erythridula*) (Homoptera: Cicadellidae). VII. *Journal of the Kansas Entomological Society* 51(1):131-139.

Hepner, L. W., Patrick, C. R., III and Norton, W. N., Jr.
1970a Scattering of male leafhoppers of the genus *Erythroneura*. *Journal of the Georgia Entomological Society* 5(3):147.

Herman, K. W. and Maramorosch, K.
1977a Feeding of the leafhopper *Macrosteles fascifrons* on an artificial diet containing the aster yellows spiroplasm. *Proceedings of the American Phytopathological Society* 4:197.

Hermoso de Mendoza, A. and Medina, V.
1979a Preliminary study of cicadellids (Homoptera, Cicadellidae) in the citrus groves of the Valencia region. *Anales del instituto Nacional de Investigaciones Agrarias, Proteccion Vegetal* 10:43-68.

Herms, D. A.
1984a Biology, population dynamics, and host impact of selected insects inhabiting honeylocust, *Gleditsia triacanthos* L. and London planetree, *Platanus x acerifolia* (Ait.) Wild. M. S. Thesis, Ohio State University, Columbus.

Hernandez, J. C.
1982a Efecto de la asociacion maiz-frijol sobre poblaciones de insectos plagas, con enfasis en *Empoasca kraemeri* Ross and Moore. Thesis of Colegio de Posgraduados, Entomology Department, Chapingo, Mexico.

Herne, D. H. C., Simpson, C. M. and Townshend, J. L.
1973a Aldicarb: A systemic soil pesticide useful for establishing apple plantings. *Proceedings of the Entomological Society of Ontario* 103:27-31.

Herrera A., J. M.
1963a Problemas insectiles del cultivo de la papa en el valle de Canete. *Revista Peruana de Entomologia* 6(1):1-9.

Heskova, D., Jermoljec, E. and Chad, J.
1961a Studium virove zakyslosti u obilivon a spenatu. [Studies of virus dwarfing in cereals and spinach.] *Ceskoslovenska Akademie Zemedelska Vedeckotechnickych Sbornik Rostlinna Vyroba* 34(10):1343-1350.

Hespenheide, H. A. and Rubke, M. A.
1977a Prey, predatory behaviour and the daily cycle of *Holopogon wilcoxi* Martin. *Pan-Pacific Entomologist* 53:277-285.

Hewitt, G. B. and Burleson, W. H.
1976a A preliminary survey of the arthropod fauna of sainfoin in central Montana. *Montana Agricultural Experiment Station Bulletin* No. 693. 11 pp.

Hewitt, W. B., Loomis, N. H., Overcash, J. P. and Parris, G. K.
1958a Pierce's disease virus in Mississippi and other southern states. *Plant Disease Reporter* 42(2):207-210.

Heyde, J. von der, Saxena, R. C. and Schmutterer, H.
1984a Effect of neem derivatives on growth and fecundity of the rice pest *Nephotettix virescens* (Distant). *Mitteilungen der Schweizerischen Gesellschaft* 57(4):423.
1985a Effects of neem derivatives on growth and fecundity of the rice pest *Nephotettix virescens* (Homoptera: Cicadellidae). *Zeitschrift fur Pflanzenkrankheiten und Pflanzenschutz* 92:346-354.

Heyer, W., Chiang Lok, M. C. and Cruz, B.
1984a Some aspects of the development of *Empoasca fabae* (Harris) in bean crops under the influence of temperature. *Centro Agricola* 11:114.
1984b Dynamics of *Empoasca fabae* (Harris) in plantations of beans (*Phaseolus vulgaris* L.). *Centro Agricola* 11:115.

Hibben, C. R., Hagar, S. S. and Karpel, M. A.
  1973a Infection of declining ash trees by virus and mycoplasmalike bodies: Identification and implications. *Abstracts of Papers, 2nd International Congress of Plant Pathology,* Minneapolis.

Hibbs, E. T., Dahlman, D. L. and Rice, R. L.
  1964a Potato foliage sugar concentration in relation to infestation by the potato leafhopper, *Empoasca fabae. Annals of the Entomological Society of America* 57(5):517-521.
  1964b Potato leafhopper, *Empoasca fabae* (Harris), infestation and potato foliage sugars. *American Potato Journal* 41(9):297.

Hibino, H.
  1980a Transmission of two kinds of rice tungro associated virus particles by insect vectors. *Annals of the Phytopathological Society of Japan* 46:412. [In Japanese.]
  1983a Relations of rice tungro bacilliform and rice tungro spherical viruses with their vector *Nephotettix virescens. Annals of the Phytopathological Society of Japan* 49:545-553.
  1983b Transmission of two rice tungro-associated viruses and rice waika virus from doubly or singly infected source plants by leafhopper vectors. *Plant Disease* 67:774-777.

Hibino, H. and Cabuatan, P. Q.
  1985a Purification and serology of rice tungro spherical virus (RTSV). *International Rice Research Newsletter* 10(4):10-11.

Hibino, H., Roechan, M., and Sudarisman, S.
  1978a Association of two types of virus particles with penyakit habang (tungro disease) of rice in Indonesia. *Phytopathology* 68:1412-1416.

Hibino, H., Saleh, N., Jumanto, H., Sudarisman, S. and Roechan, M.
  1978a Two kinds of virus particle associated with tungro disease of rice. *Annals of the Phytopathological Society of Japan* 44:394.

Hibino, H., Saleh, N. and Roechan, M.
  1979a Transmission of two kinds of rice tungro-associated viruses by insect vectors. *Phytopathology* 69:1266-1288.

Hibino, H., Tiongco, E. R. and Cabunagan, R. C.
  1983a Rice tungro virus complex in tungro-resistant IR varieties. *International Rice Research Newsletter* 8(4):12-13.

Hill, B. G.
  1969a Comparative morphological study of selected higher categories of leafhoppers (Homoptera: Cicadellidae.) Ph.D. Dissertation, North Carolina State University, Raleigh. 187 pp.
  1973a A new genus of Cicadellidae from Brazil. *Proceedings of the Entomological Society of Washington* 75(1):78-79. [New genus: *Evanirvana.*]

Hill, M. G.
  1976a The population and feeding ecology of 5 species of leafhoppers (Homoptera) on *Holcus mollis* L. Doctoral Dissertation, University of London. 266 pp.
  1982a Feeding strategies of grassland leafhoppers (Homoptera). *Zoologische Jahrbucher Abteillung fuer Systematik Oekologie und Geographie der Tiere* 109(1):24-32.

Hill, R. R., Hanson, C. H. and Busbice, T. H.
  1969a Effect of four recurrent selection programs on two alfalfa populations. *Crop Science* 9(3):363-365.

Hill, S. A.
  1980a Virus and mycoplasm disease of strawberry. *Ministry of Agriculture, Fisheries and Food Leaflet* No. 530. 8 pp.

Hills, O. A.
  1958a Effect of curly top-infective beet leafhopper on watermelon plants in different stages of development. *Journal of Economic Entomology* 51:434-436.

Hills, O. A. and Brubaker, R. W.
  1968a Comparison of the effect on beet seed production of spring and fall infestations of beet leafhoppers carrying curly top virus. *Journal of American Society of Sugar Beet Technologists* 15:214-220.

Hills, O. A., Coudriet, D. L., Bennett, C. W., Jewell, H. K., and Brubaker, R. W.
  1963a Effect of three insect-borne virus diseases on sugar beet seed production. *Journal of Economic Entomology* 56(5):690-693.

Hills, O. A., Coudriet, D. L. and Brubaker, R. W.
  1964a Phorate treatments against the beet leafhopper on cantaloupes for prevention of curly top. *Journal of Economic Entomology* 57(1):85-89.

Hills, O. A., Coudriet, D. L., Brubaker, R. W. and Keener, P. D.
  1961a Effects of curly top and cucumber mosaic viruses on cantaloupes. *Journal of Economic Entomology* 54(5):970-972.

Hills, O. A. and Taylor, E. A.
  1954a Effect of curly top-infective beet leafhoppers on cantaloupe plants in varying stages of development. *Journal of Economic Entomology* 47:44-46.

Hills, O. A., Valcarce, A. C., Jewell, H. K. and Coudriet, D. C.
  1960a Beet leafhopper control in sugar beets by seed or soil treatment. *Journal of the Society of Sugar Beet Technologists* 11(1):15-24.

Hinckley, A. D.
  1963a Insect pests of rice in Fiji. *FAO Plant Protection Bulletin* 11:31-33.

Hincks, W. K.
  1963a Mymaridae. *Proceedings of the Leeds Philosophical and Literary Society* (Scientific Section) 9:75.

Hino, T., Wathanakul, L., Nabheerong, N., Surin, P., Chaimongkoi, U., Disthaporn, S., Putta, M., Kerdochokchai, D. and Surin, A.
  1974a Studies on rice yellow orange leaf disease in Thailand. *Tropical Agricultural Research Centre Technical Bulletin TARC* No. 7 (Tokyo). 67 pp.

Hirai, T., Saito, T., Onda, H., Kitani, K. and Kiso, A.
  1968a Inhibition by Blasticidin S of the ability of leafhoppers to transmit rice stripe virus. *Phytopathology* 58:602-604.

Hirao, J. and Inoue, H.
  1978a Transmission efficiency of rice waika virus by the green rice leafhoppers, *Nephotettix* spp. (Hemiptera: Cicadellidae). *Applied Entomology and Zoology* 13:264-273.
  1978b Bionomics of the green rice leafhopper, *Nephotettix cincticeps,* in relation to the incidence of rice yellow dwarf disease in Japan. *In: Plant Disease Due to Mycoplasma-Like Organisms.* Pp. 143-157. Food and Fertilizer Technology Center, Taipei, Taiwan. 178 pp.
  1979c Transmission characteristics of rice waika virus by the green rice leafhopper, *Nephotettix cincticeps* Uhler (Hemiptera, Cicadellidae). *Applied Entomology and Zoology* 14:44-50.
  1980a Infective capacity of *Nephotettix* spp. (Hemiptera: Cicadellidae) in transmission of rice waika virus. *Applied Entomology and Zoology* 15(4):378-384.

Hirao, J., Satomi, H. and Okada, T.
  1974a Transmission of the "Waisei" disease of rice plant by the green rice leafhopper, *Nephotettix cincticeps* Uhler. *Proceedings of the Association of Plant Protection,* Kyushu 20:128-133.

Hirashima, Y.
  1981a Field studies on the biological control of leafhoppers and planthoppers (Hemiptera: Homoptera) injurious to rice plants in South-East Asia. An account for the year 1979. *Esakia* 16:1-4.

Hirashima, Y., Aizawa, K., Miura, T. and Wongsiri, T.
  1979a Field studies on the biological control of leafhoppers and planthoppers (Hemiptera: Homoptera) injurious to rice plants in South-East Asia: Progress report for the year 1977. *Esakia* 13:1-20.

Hirashima, Y. and Kifune, T.
  1978a Strepsipterous parasites of Homoptera injurious to the rice plant in Sarawak, Borneo, with description of a new species (Notulae-Strepsipterologicae-III). *Esakia* 11:53-58.
  1985a On the identity of *Halictophagus munroei* Hirashima et Kifune with *H. bipunctatus* Yang (Notulae Strepsipterologicae-IV.). *Esakia* 23:58-59.

Hirata, M. and Sogawa, K.
  1976a Antifeeding activity of chlordimeform for plant-sucking insects. *Applied Entomology and Zoology* 11:94-99.

Hiremath, S. C.
  1979a Studies on the bionomics of the mango hoppers (Cicadellidae; Hemiptera) and their control by different methods of application of insecticides. Thesis Abstr., Haryana Agricultural University 5(1):42-43.

Hiruki, C. and Chen, M. H.
  1978a Fababean yellows incited by the aster yellows agent. *Zentralblatt fuer Bakteriologie, Parasitenkunde, Infektionskrankheiten und Hygiene. Ester Abteillung Originale. Reihe A. Medizinische Mikrobiologie und Parasitologie* 241:188.

Hirumi, H.
  1965a Study on insect tissue culture. Cultivation of embryonic leafhopper tissues *in vitro*. *Wakayama Igaku* 15:325-334. [In Japanese with English summary.]

Hirumi, H., Granados, R. R. and Maramorosch, K.
  1967a Electron microscopy of a plant-pathogenic virus in the nervous system of its insect vector. *Journal of Virology* 1:430-444.

Hirumi, H. and Maramorosch, K.
  1963a Cultivation of leafhopper (Cicadellidae) tissues and organs *in vitro*. *Annals des Epiphyties* 14:77-79.
  1963b Recovery of aster yellows virus from various organs of the insect vector, *Macrosteles fascifrons*. *Contributions from Boyce Thompson Institute for Plant Research* 22:141-151.
  1964a Insect tissue culture: Use of blastokinetic stage of leafhopper embryo. *Science* 144:1465-1467.
  1964b Insect tissue culture: Further studies on the cultivation of embryonic leafhopper tissues in vitro. *Contributions from Boyce Thompson Institute for Plant Research* 22:343-352.
  1964c The *in vitro* cultivation of embryonic leafhopper [*Macrosteles fascifrons* (Stal)] tissue. *Experimental Cell Research* 36:625-631.
  1968a Electron microscopy of wound tumor virus in cultured embryonic cells of the leafhopper *Macrosteles fascifrons*. 2nd International Colloquium of Invertebrate Tissue Culture. Pp. 203-217.
  1969a Mycoplasma-like bodies in the salivary glands of insect vectors carrying aster yellows agent. *Journal of Virology* 3:82-84.
  1969b Further evidence for a mycoplasma etiology of aster yellows. *Phytopathology* 59:1030-1031 (Abstr.).

Ho, H. S. and Chen, C. C.
  1968a Ecological studies on rice green leaf hoppers in Taiwan (I). *Plant Protection Bulletin*, Taiwan 10(1):15-36. [In Chinese with English summary.]

Hodjat, S. H.
  1970a Experiments on waterless spraying with systemic insecticides in controlling the bean pests. *Plant Pests Disease Research Institute Tehran CENTO Research Project* 2(3):25-29.
  1972a Experiments on waterless spraying with systemic insecticides in controlling bean pests. *Entomologie et Phytopathologie Appliquees* No. 32:En 15-18; Pe 46-50. [In English and Persian.]

Hoebeke, E. R.
  1980a A leafhopper [*Eupteryx atropunctata* (Goeze)]. *Cooperative Plant Pest Report* 5(29):547.

Hoebeke, E. R. and Wheeler, A. G., Jr.
  1983a *Eupteryx atropunctata*: North American distribution, seasonal history, host plants, and description of the fifth-instar nymph (Homoptera: Cicadellidae). *Proceedings of the Entomological Society of Washington* 85(3):528-536.

Hoelscher, C. E.
  1966a A leafhopper, *Ollarianus strictus* (Cicadellidae) and a hymenopterous parasite, *Lymaenon marylandicus* (Mymaridae) reared from Texas citrus. *Annals of the Entomological Society of America* 59(4):820-823.

Hoffman, J. R.
  1952a Leafhopper control to prevent the spread of the virus disease of aster yellows in commercial lettuce production. *Michigan Agricultural Experiment Station Quarterly Bulletin* 34:262-265.

Hoffman, J. R. and Taboada, O.
  1960a A three-year inventory of the leafhoppers of the East Lansing area (1954-1956). *Michigan Agricultural Experiment Station Quarterly Bulletin* 42(3):607-614.
  1961a 1957-1959 survey, leafhoppers of the East Lansing, Michigan area. *Michigan Agricultural Experiment Station Quarterly Bulletin* 43(4):714-721.
  1962a Leafhopper vectors of virus diseases in Michigan. *Plant Disease Reporter* 46:114.
  1965a Distribution of leafhopper vectors of plant diseases in Michigan. *Proceedings, North Central Branch, American Association of Economic Entomologists* 20:151-152.

Hofmaster, R. N.
  1958a The potato leafhopper and its control. *Vegetable Growers News* 12(12):3.
  1959a Effectiveness of new insecticides on the potato leafhopper and the influence of leafhopper control and potato variety on tubeworm infestations. *Journal of Economic Entomology* 52(5):908-910.
  1959b Potato insect investigations. *Agricultural Chemicals* 14(6):79, 81.
  1959c Some new aspects of potato insect investigations in southeastern Virginia. *Virginia Journal of Science* 10(4):237.
  1963a Insect pest control. *Virginia Joint Agricultural Publication* 5:39-49.

Hofmaster, R. N. and Dunton, E. M.
  1961a Soil application of insecticides for the control of foliage pests of the Irish potato. *American Potato Journal* 38(10):341-345.

Hofmaster, R. N., Landis, B. J., Schulz, J. T., Sleesman, J. P., Shands, W. A. and Simpson, G. W.
  1968a The use of insecticides for insect control in potato production. *American Potato Journal* 45(8):300-311.

Hofmaster, R. N. and Waterfield, R. L.
  1965a The Colorado potato beetle problem in Virginia and present status of control measures. *Transactions Peninsula Horticultural Society* 55(5):20-26.

Hofmaster, R. N., Waterfield, R. L. and Boyd, J. C.
  1967a Insecticides applied in the soil for control of eight species of insects on Irish potatoes in Virginia. *Journal of Economic Entomology* 60(5):1311-1318.

Hogg, D. B.
  1985a Potato leafhopper (Homoptera: Cicadellidae) immature development, life tables, and population dynamics under fluctuating temperature regimes. *Environmental Entomology* 14(3):349-355.

Hohmann, C. L., Schoonhoven, A. van, and Cardona, C.
  1980a Management of pests of beans (*Phaseolus vulgaris* Linnaeus, 1753) through the use of diversification of the crops with weeds in conjunction with varietal resistance. *Anais da Sociedade Entomologica do Brasil* 9:143-153.

Hokkanan, H. and Raatikainen, M.
  1977a Faunal communities of the field stratum and their succession in reserved fields. *Maataloustieteelinen Aikakauskirja* 49:390-405.

Hokyo, N.
  1971a Applicability of the temperature-sum rule in estimating the emergence time of the overwintering population of the green rice leafhopper, *Nephotettix cincticeps* Uhler (Hemiptera: Deltocephalidae). *Applied Entomology and Zoology* 6(1):1-10.
  1972a Studies on the life history and the population dynamics of the green rice leafhopper, *Nephotettix cincticeps* Uhler. *Bulletin of the Kyushu Agricultural Experiment Station* 16:282-382.
  1976a Population dynamics, forecasting and control of the green rice leafhopper, *Nephotettix cincticeps* Uhler. *Review of Plant Protection Research* 8:1-13.

Hokyo, N. and Kuno, E.
1977a Life table studies on the paddy field populations of the green rice leafhopper *Nephotettix cincticeps* Uhler (Hemiptera: Cicadellidae), with special reference to the mechanism of population regulation. *Researches on Population Ecology* 19:107-124.

Hokyo, N., Lee, M. H., Park, J. S.
1976a Some aspects of population dynamics of rice leafhoppers in Korea. *Korean Journal of Plant Protection* 15:111-126.

Hokyo, N., Otake, A. and Okada, T.
1977a Species composition of *Nephotettix* (Homoptera: Cicadellidae) in west Malaysia and Indonesia. *Applied Entomology and Zoology*, Tokyo 12(1):83-85.

Holdeman, Q. L. and McCartney, W. O.
1965a Virus disease of corn, *Zea mays* L. (a plant disease detection aid). *California Department of Agriculture, Bureau of Plant Pathology.* 28 pp.

Hollebone, J. W., Medler, J. T. and Hildebrandt, A. C.
1966a The feeding of *Empoasca fabae* (Harris) on broad bean callus in tissue culture. *Canadian Entomologist* 98(12):1259-1264.

Holthuis, L. B.
1960a Proposed addition of the generic name *Jasus* Parker, 1883, to the official list of generic names in Zoology (Class Crustacea, Order Decapoda) Z.M. (S.) 620. *Bulletin of Zoological Nomenclature* 17:193-196.

Hong, You-Chong
1980a Granulidae, a new family of Homoptera from the Middle Triassic of Tongchuan, Shanxi Province. *Acta Zootaxonomica Sinica* 5(1):63-70. [In Chinese and English.] [New genus: *Granulus.*]

Hongsaprug, W.
1983a Studies on the genus *Empoascanara* Distant (Homoptera: Cicadellidae, Typhlocybinae) in Thailand. *Thai Journal of Agricultural Science* 16:35-45.
1983b Taxonomy of *Nephotettix* in Thailand. In: Knight, W. J., Pant, N. C., Robertson, T. S. and Wilson, M. R., eds. *Proceedings of the 1st International Workshop on Biotaxonomy, Classification and Biology of Leafhoppers and Planthoppers (Auchenorrhyncha) of Economic Importance.* Pp. 87-93. London, 4-7 October 1982. Commonwealth Institute of Entomology, 56 Queen's Gate, London SW7 5JR. 500 pp.
1984a Taxonomic study of mango leafhoppers in Thailand. *Mitteilungen der Schweizerischen Entomologischen Gesellschaft* 57(4):423-424.

Hongsaprug, W. and Wilson, M. R.
1985a Species of the genera *Bakera* and *Mangganeura* (Homoptera: Auchenorrhyncha: Typhlocybinae) occurring on mango and other crops plants in the Oriental region. *Journal of Natural History* 19(1):173-183.

Hopkins, D. L.
1977a Diseases caused by leafhopper-borne rickettsia-like bacteria. *Annual Review of Phytopathology* 17:277-294.

Hopkins, D. L. and Adlerz, W. C.
1980a Pierce's disease bacterium causes a disease of rough lemon citrus. *Phytopathology* 70:568.

Hopkins, D. L., Adlerz, W. C. and Bistline, F. W.
1978a Pierce's disease bacterium occurs in citrus trees affected with blight (young tree decline). *Plant Disease Reporter* 62:442-445.

Hopkins, D. L. and Mollenhauer, H. H.
1973a Rickettsia-like bacterium associated with Pierce's disease of grapes. *Science* 179:298-300.

Hopkins, D. L., Mollenhauer, H. H. and French, W. J.
1973a Occurrence of a rickettsia-like bacterium in the xylem of peach trees with phony disease. *Phytopathology* 63:1422.

Hopkins, L. C. and Johnson, D. T.
1984a A survey of the arthropods associated with blueberries with emphasis on the abundance and dispersal of *Scaphytopius* species (Homoptera: Cicadellidae) in northwestern Arkansas. *Journal of the Georgia Entomological Society* 19(2):248-264.

Hoppe, W.
1972a Dalsze obserwacje nad wystepowaniem wirozy pszenicy o objawach paskowanej mozaiki i wyniki niektorych badan nad jej przenoszeniem. Komunikat. [Further observations on the appearance of virosis of wheat with symptoms of streak mosaic and results of some investigations on its transmission. Report.] *Zeszyty Problemove Postepow Nauk Rolniczych* No. 133:143-144. [In Polish with Russian and English summaries.]
1974a Researches on wheat virosis caused by winter wheat mosaic virus. *Prace Naukowe Instyltutu Ochrony Roslin* 16:105-180.
1976a Investigations on virus disease of cereals. *Biuletyn Instytutu Ochrony Roslin* 59:269-276.

Horber, E.
1972a Plant resistance in insects. *Agricultural Science Review* 10(2):1-10.
1974a Resistenz gegen Insekten an Futterleguminosen. [Insect resistance in forage legumes.] *Schweizerische Landwirtscheftliche Forschung* 13(1-2):221-234.
1974b Techniques, accomplishments and potential of insect resistance in forage legumes. In: Maxwell, F. G. and Harris, F. A., eds. *Proceedings of the Summer Institute on Biological Control of Plant Insects and Diseases.* Pp. 312-343. University Press of Mississippi, Jackson.
1976a Selecting alfalfa resistant to the potato leafhopper. *Forage Insects Research Conference* 18:18-19.
1976b Selecting alfalfa resistant to potato leafhopper. *Report Alfalfa Improvement Conference* 25:20.

Horber, E., Leath, K. T., Berrange, B., Marcarian, V. and Hanson, C. H.
1974a Biological activities of saponin components from DuPuits and Lahontan alfalfa. *Entomologia Experimentalis et Applicata* 17:410-424.

Hori, Y.
1967a Seasonal prevalence of some leafhoppers in a sugar beet field and their feeding behaviour on the plant. *Research Bulletin, Obihiro Zootechnical University* (Ser. 1) 5(2):43-199. [Japanese summary.]
1969a Two species of Homoptera, unrecorded from Shikoku, Japan. *Transactions of the Shikoku Entomological Society* 10(2):64.
1982a The Homoptera-Auchenorrhyncha of the Ozegahara Moor. In: Hara, H., ed. *Ozegahara: Scientific Researches of the Highmoor in Central Japan.* Pp. 335-346. Japan Society for the Promotion of Science, Tokyo. 456 pp. [In English.]

Horimoto, M., Imamura, M., Oshikawa, M., Ikeda, K., Higo, S., Sakae, S., Kamino, K., Kami, T., Fukamachi, S. and Kamiwada, H.
1976a Bell-jar dusting method for evaluation of insecticide susceptibility in the green rice leafhopper. *Proceedings of the Association for Plant Protection of Kyushu* 22:113-115.

Horning, D. S., Jr. and Barr, W. F.
1970a Insects of Craters of the Moon National Monument. *University of Idaho, College of Agriculture, Miscellaneous Series* No. 8. 117 pp.

Horvath, J.
1969a Green petal. A new disease of rape in Hungary. *Acta Phytopatholica* (Budapest) 4(4):363-367.
1974a New data on the properties and occurrence of mycoplasmas and mycoplasma diseases. *Novenyvedelem Korszerusitese* 8:103-148.

Hosny, M. M. and El-Dessouki, S. A.
1967a Ecological and biological studies on *Empoasca* spp. (Jassidae) in the Cairo area. *Zeitschrift fuer Angewandte Entomologie* 60:397-411.
1968a The susceptibility of certain cotton varieties to *Empoasca* spp. (Jassidae)-infestation under some different agricultural practices in U.A.R. *Zeitschrift fuer Angewandte Entomologie* 62(3):252-286.
1969a Vertical and horizontal distribution of *Empoasca* spp. (Jassidae) on cotton plants in Cairo area, U.A.R. *Zeitschrift fuer Angewandte Entomologie* 63(1):53-61.

Hosoda, A. and Fujiwara, A.
1977a Studies on the control countermeasure of the insecticide-resistant insect pest. 2. Changes of susceptibility in the carbamate insecticide-resistant green rice leafhopper, *Nephotettix cincticeps* Uhler, during the continuous selection with diazinon, propaphos and propaphos NAC mixture. *Bulletin of the Hiroshima Prefectural Agricultural Experiment Station* 39:21-26. [In Japanese with English summary.]

Hou, R. F.
1976a Artificial rearing and nutrition of the aster leafhopper, *Macrosteles fascifrons* Homoptera: Cicadellidae). Ph. D. Thesis, University of Minnesota, Minneapolis.

Hou, R. F. and Brooks, M. A.
1975a Continuous rearing of the aster leafhopper, *Macrosteles fascifrons*, on a chemically defined diet. *Journal of Insect Physiology* 21(8):1481-1483.
1977a Effects of cholesterol on growth and development of the aster leafhopper, *Macrosteles fascifrons* (Stal) (Hemiptera: Deltocephalidae). *Applied Entomology and Zoology* 12:248-254.
1978a Trace metals especially Fe^{3+} as dietary factors for leafhoppers. *Experimentia* 34:465-466.

Hou, R. F. and Lin, L.-C.
1979a Artificial rearing of the rice green leafhopper, *Nephotettix cincticeps*, on a holidic diet. *Entomologia Experimentalis et Applicata* 25:158-164.

Houk, E. J. and Griffiths, G. W.
1980a Intracellular symbiotes of the Homoptera. *Annual Review of Entomology* 25:161-187.

Howe, W. L. and Manglitz, G. R.
1959a Insect-resistant crops. *Nebraska Agricultural Experiment Station Quarterly* 6(2):3-5.

Howe, W. L. and Miller, L. L.
1954a Effects of demeton soil drenches on peanut pests. *Journal of Economic Entomology* 47(4):711-712.

Howe, W. L. and Rhodes, A. M.
1976a Phytophagous insect associations with *Cucurbita* in Illinois. *Environmental Entomology* 5:747-751.

Howell, J. O. and Pienkowski, R. L.
1971a Spider populations in alfalfa, with notes on spider prey and effect of harvest. *Journal of Economic Entomology* 64(1):163-168.

Hower, A. A., Jr.
1979a Relationship of potato leafhopper density to alfalfa quality. *Proceedings of the 14th Northeastern Regional Alfalfa Insects Conference.*
1979b Potato leafhopper density and its relationship to alfalfa quality. *Journal of the New York Entomological Society* 86:298-299.

Hower, A. A. and Byers, R. A.
1977a Potato leafhoppers reduce alfalfa quality. *Science in Agriculture* 24:10-11.

Hower, A. A. and Davis, G. A.
1984a Selectivity of insecticides that kill the potato leafhopper (Homoptera: Cicadellidae) and alfalfa weevil (Coleoptera: Curculinoidae) and protect the parasite *Microctonus aethiopoides* Loan (Hymenoptera: Braconidae). *Journal of Economic Entomology* 77:1601-1607.

Hower, A. A. and Muka, A. A.
1975a The role of insects as factors affecting alfalfa stand persistence. *Proceedings of the Annual Alfalfa Symposium* 5:88-95.

Howse, P. E. and Claridge, M. F.
1970a The fine structure of Johnston's organ of the leafhopper, *Oncopsis flavicollis. Journal of Insect Physiology* 16:1665-1675.

Hsieh, C. Y.
1975a Seasonal population fluctuations of the green rice leafhopper and its natural enemies. *Rice Entomology Newsletter* No. 3, pp. 29-30.
1976a Field re-evaluation of commercial insecticides for control of the green rice leafhopper and the brown planthopper. *Rice Entomology Newsletter* No. 4, p. 23.

Hsieh, C. Y. and Dyck, V. A.
1975a Influence of predators on the population density of the rice green leafhopper. *Plant Protection Bulletin* (Taiwan) 17:346-352.

Hsieh, S. P. Y.
1969a Multiplication of the rice transitory yellowing virus in its vector, *Nephotettix apicalis* Motch. *Plant Protection Bulletin* (Taiwan) 11(4):159-170.
1969b An analysis of field collections of the rice green leafhoppers for infection with transitory yellowing virus. *Plant Protection Bulletin* (Taiwan) 11(4):171-174. [In Chinese.]

Hsieh, S. P., Chiu, R. R. and Chen, C. C.
1970a Transmission of rice transitory yellowing virus by *Nephotettix impicticeps. Phytopathology* 60:1534.

Hsieh, S. P. Y. and Liao, C. H.
1974a Effect of transitory yellowing virus on agronomic characters of rice plant infected at different ages. *Plant Protection Bulletin* (Taiwan) 16(1/2):35-41. [In Chinese with English summary.]

Hsieh, S. P. Y. and Roan, S. C.
1967a Mechanical transmission of rice transitory yellowing virus to its leafhopper vector, *Nephotettix cincticeps* Uhler. *Plant Protection Bulletin* (Taiwan) 9(1/2):23-30. [In English with Chinese summary.]

Hsu, H. T. and Black, L. M.
1974a Multiplication of potato yellow dwarf virus on vector cell monolayers. *Virology* 59:331-334.

Hsu, H. T., McBeath, J. H. and Black, L. M.
1977a The comparative susceptibilities of cultured vector and nonvector leafhopper cells to three plant viruses. *Virology* 81:257-262.

Hsu, H. T., Nuss, D. and Adam, G.
1983a Utilization of insect tissue culture in the study of the molecular biology of plant viruses. In: Harris, K. F., ed. *Current Topics in Vector Research. I.* Pp. 189-214. Praeger, New York.

Hsu, T. P. and Banttari, E. E.
1979a Dual transmission of the aster yellows mycoplasmalike organism and the oat blue dwarf virus and its effect on longevity and fecundity of the aster leafhopper vector. *Phytopathology* 69:843-845.

Huang, Pang-Kan and Lo, Shao-Nan
1964a Studies on the leafhopper, *Empoasca subrufa* Melichar. *Acta Entomologica Sinica* 13(1):101-117. [In Chinese with summary in English.]

Huber, R. T. and Giese, R. L.
1973a The Indiana alfalfa pest management program. *Proceedings of the North Central Branch Entomological Society of America* 28:139-143.

Huber, R. T. and Osmun, J. T.
1966a Insects and other arthropods of economic importance in Indiana during 1966. *Proceedings of the Indiana Academy Science* 76:291-307.

Huddleston, E. W. and Gyrisco, G. G.
1961a Residues of phosdrin on alfalfa and its effectiveness on the insect complex. *Journal of Economic Entomology* 54(1):209-210.

Hudon, M., Martel, P. and Ritchot, C.
1980a Status of the insect pests of certain crops in southwestern Quebec in 1978. *Annals of the Entomological Society of Quebec* 25:68-71.

Huff, F. A.
1963a Relation between leafhopper influxes and synoptic weather conditions. *Journal of Applied Meteorology* 2(1):39-43.

Huffaker, C. B., Holloway, J. K., Doutt, R. L. and Finney, G. L.
1954a Introduction of egg parasites of the beet leafhopper. *Journal of Economic Entomology* 47(5):785-789.

Hulbert, D. von and Schaller, G.
1972a Der Einfluss kunstlicher Ernahrung auf Entwicklung und Formenbildung bei *Euscelis plebejus* Fall. (Homoptera Auchenorrhyncha) unter verschiedenen photoperiodischen Bedingungen. *Zoologische Jahrbuecher Abteilung fuer Systematik Oekologie und Geographie der Tiere* 99(4):545-560.

Hulden, L.
  1982a Records of Heteroptera and Auchenorrhyncha (Hemiptera) from northern Norway. *Notulae Entomologicae* 62:66-68.
  1983a *Streptanus okaensis* Zachv. (Cicadellidae) new for Finland. *Notulae Entomologicae* 63(4):212. [In Swedish.]
  1984a Observations on an egg parasite of *Cicadella viridis* (Homoptera, Auchenorrhyncha). *Notulae Entomologicae* 64(2):84-85.
  1984b Two species of cicadellids (Homoptera, Cicadellidae) new for Finland. *Notulae Entomologicae* 64(4):197. [In Swedish.]

Hulden, L. and Albrecht, A.
  1984a Auchenorrhyncha (Homoptera) of Inari Lapland. *Kevo Notes* 7:47-52. [In English.]

Hulden, L. and Heikinheimo, O.
  1984a Checklist of Finnish insects. Hemipteroidea. *Notulae Entomologicae* 64:97-124.

Hull, R.
  1972a Mycoplasma and plant diseases. *PANS [Pest Articles & News Summaries]* 18(2):154-164.

Hummelen, P. J. and Soenarjo, E.
  1977a Light trap studies on the rice gall midge, *Orseolia oryzae* (Woo-Mason), and some other insects. *Contributions, Central Research Institute for Agriculture,* Bogor, Indonesia, No. 29. 16 pp.

Hunang, B. and Shichen, Q.
  1981a A list of rice pests from Fujian Province. *Wuji Science Journal* 1, Supplement 19.

Huq, S.
  1984a Breeding methods for Pipunculidae (Diptera), endoparasites of leafhoppers. *International Rice Research Newsletter* 9(4):14-15.
  1985a On *Eudorylas subterminalis* Coll. (Diptera: Pipunculidae) as a parasite of Cicadellidae in West Berlin. *Anzeiger fur Schadlingskunde Pflanzenshutz Umweltschutz* 58:147-149.

Hurd, L. E. and Eisenberg, R. M.
  1984a Experimental density manipulation of the predator *Tenodora sinensis* (Orthoptera: Mantidae) in an old-field community. II. The influence of mantids on arthropod community structure. *Journal of Animal Ecology* 53:955-967.

Ichikawa, T.
  1979a Studies on the mating behavior of the four species of Auchenorrhynchous Homoptera which attack the rice plant. *Memoirs of Faculty of Agriculture, Kagawa University* 34:1-60.

Ichikawa, T. and Kiritani, K.
  1973a Influence of parental age upon the offspring in the green rice leafhopper, *Nephotettix cincticeps* Uhler (Hemiptera: Deltocephalidae), with special reference to environmental conditions. *Kontyu* 41(1):1-9.

Igarashi, A., Lin, W. J. and Hayashi, K.
  1982a An enveloped virus isolated from leafhoppers captured on East China Sea. *Tropical Medicine* 24:179-186.

Iida, T.
  1965a Geographical distribution of leafhopper-borne viruses of cereals. *In: Conference on Relationships Between Arthropods and Plant-Pathogenic Viruses,* Tokyo, Japan.

Iida, T. and Shinkai, A.
  1950a Transmission of rice yellow dwarf by *Nephotettix cincticeps* (Uhl.). *Annals of the Phytopathological Society of Japan* 14:113-114. [In Japanese.]

Imbe, T.
  1981a Field evaluation of rice varietal resistance to the green rice leafhopper with the sticky board method. *Proceedings of the Association for Plant Protection of Kyushu* 27:78-80.

Ingram, B. F.
  1969a Insect pests of sunflowers. *Queensland Agricultural Journal* 95(10):703-705.

Inoue, H.
  1966a Larval development of the green rice leafhopper, *Nephotettix cincticeps* Uhler, and the white back planthopper, *Sogatella furcifera* Horvath, on Japonica and Indica rice varieties. *Odokon-Chugoku* 8:17-19. [In Japanese.]
  1977a A new leafhopper vector of rice waika virus, *Nephotettix malayensis* Ishihara and Kawase (Homoptera: Cicadellidae). *Applied Entomology and Zoology* 12:197-199.
  1978a Strain S, a new strain of leafhopper-borne rice waika virus. *Plant Disease Reporter* 62:867-871.
  1979a Transmission efficiency of rice transitory yellowing virus by the green rice leafhoppers, *Nephotettix* spp. (Hemiptera: Cicadellidae). *Applied Entomology and Zoology* 14:123-126.
  1982a Species-specific calling sounds as a reproductive isolating mechanism in *Nephotettix* spp. (Hemiptera: Cicadellidae). *Applied Entomology and Zoology* 17(2):253-262.
  1983a Reproductive isolation mechanisms and hybridization of *Nephotettix* spp. *In: Knight, W. J., Pant, N. C., Robertson, T. S. and Wilson, M. R. eds. Proceedings of the 1st International Workshop on Biotaxonomy, Classification and Biology of Leafhoppers and Planthoppers (Auchenorrhyncha) of Economic Importance.* Pp. 339-349. London, 4-7 October 1982. Commonwealth Institute of Entomology, 56 Queen's Gate, London SW7 5JR. 500 pp.
  1985a On the control of insecticide resistant rice leafhoppers and planthoppers, Kumihop, a mixture of IBP and malathion. *Japan Pesticide Information* No. 47:14-16.

Inoue, H. and Hirao, J.
  1980a Effects of temperature on the transmission of rice waika virus by *Nephotettix cincticeps* Uhler (Hemiptera: Cicadellidae). *Applied Entomology and Zoology* 15:433-438.
  1981a Transmission of rice waika virus by green rice leafhoppers, *Nephotettix* spp. (Hemiptera: Cicadellidae). *Bulletin of the Kyushu National Agricultural Experiment Station* 21(4):509-552. [In English with Japanese summary.]

Inoue, H., Hirao, J. and Kawai, A.
  1979a Interspecific hybridization between *Nephotettix virescens* Distant and *N. cincticeps* Uhler (Hemiptera: Cicadellidae). *Applied Entomology and Zoology* 14:293-302.

Inoue, H. and Omura, T.
  1982a Transmission of rice gall dwarf virus by the green rice leafhopper. *Plant Disease* 66(1):57-59.

Inoue, H., Omura, T., Morinaka, T., Saito, Y., Putta, M., Chettanachit, D., Parejarearn, A., Disthaporn, A. and Kadkoao, S.
  1980a A new record of rice transitory yellowing virus in northern Thailand. *International Rice Research Newsletter* 5(2):11-12.

Ionica, M. and Grigorescu, R.
  1978a Biological activity of several insecticides on the cicadellids *Psammotettix alienus* and *Macrosteles laevis*. *Analele Institutului de Cercetari pentru Protectia Plantelor* 14:225-235.

Iqbal, Singh
  1983a Studies on the biology of *Cassianeura cassiae* (Ahmed) (Homoptera: Cicadellidae) on Indian Laburnum. *Thesis Abstr., Haryana Agricultural University* 9(1):52-53.

Iren, Z.
  1976a Investigations to determine the most important pests of vineyards in central Anatolia. *Bitki Koruma Bulteni* 16:211-222.

Isenhour, D. J.
  1985a Efficiency of insecticides against *Spissistilus festinus* (Say), *Empoasca fabae* (Harris) and *Lygus lineolaris* (Palisot de Beauvois) in alfalfa in Georgia. *Journal of Entomological Science* 20:121-128.

Ishiguro, T. and Saito, T.
1970a Fundamental research on the application of systemic insecticides. (I). The absorption, translocation and penetration of P-vamidothion in the rice plant. *Botyu-Kagaku* 35(1)1-6. [In Japanese with English summary.]

Ishiguro, T., Saito, T. and Toyoda, I.
1971a The fundamental research to the application of systemic insecticides. (III). Fate of vamidothion in rice plant and citrus plant. *Botyu-Kagaku* 36(4):159-168. [In Japanese with English summary.]

Ishihara, T.
1956a The genus *Idiocerus* Lewis of Japan (Hemiptera: Idioceridae). *Memoirs of the Ehime University*, Section 6 (Agriculture) 1955:15-21.

1957a *Bathysmatophorus*, a genus of family Evacanthidae, found in Japan (Insecta: Hemiptera). *Zoological Magazine* (Tokyo) 66:337-340.

1958a The superfamily Cicadelloidea of Niigata Prefecture, North Honshu, Japan. *Kontyu* 26:225-232.

1958b Description of a peculiarly differentiated species of the family Tettigellidae found in Niigata Prefecture, N. Honshu, Japan (Insecta: Hemiptera). *Zoological Magazine* (Tokyo) 67:155-157. [In Japanese with English summary.] [New genus: *Babacephala*.]

1959a The genus *Parabolocratus* of Japan (Hemiptera). *Kontyu* 27:77-80.

1959b A valid species *Thamnotettix sellatus* Uhler (Hemiptera, Deltocephalidae). *Transactions of the Shikoku Entomological Society* 6(3):48.

1961a Homoptera of Southeast Asia collected by the Osaka City University Biological Expedition to Southeast Asia 1957-1958. *Nature & Life in Southeast Asia* 1:225-257. [New genera: *Umesaona*, *Ikomella*, *Phlogothamnus*.]

1961b The family Xestocephalidae of Japan (Hemiptera). *Transactions of the Shikoku Entomological Society* 7:19-25.

1962a The black-tipped leafhopper *Bothrogonia ferruginea* Auct., of Japan and Formosa. *Japanese Journal of Applied Entomology and Zoology* 6:289-292.

1963a Some genera, especially "*Eutettix*" of Japan and Formosa (Hemiptera: Deltocephalidae.) *Transactions of the Shikoku Entomological Society* 7:119-124. [New genera: *Onukiades* and *Paraonukia*.]

1963b Genus *Onukia* and new Formosan allied genera (Hemiptera: Evacanthidae). *Transactions of the Shikoku Entomological Society* 8:1-5.

1964a Revision of the genus *Nephotettix* (Hemiptera: Deltocephalidae). *Transactions of the Shikoku Entomological Society* 8:39-44.

1964b A noteworthy species of Japanese Xestocephalidae (Hemiptera). *Transactions of the Shikoku Entomological Society* 8(2):44.

1965a Taxonomic position of some leafhoppers known as virus vectors. In: *Seminar of U.S.-Japan Cooperative Science. Relationships Between Arthropods and Plant-Pathogenic Viruses*, Tokyo. Pp. 1-16.

1965b Two new cicadellid-species of agricultural importance (Insecta: Hemiptera). *Japanese Journal of Applied Entomology and Zoology* 9(1):19-22. [New genus: *Hishimonoides*.]

1965c Some species of Formosan Homoptera. *Special Bulletin of Lepidopterological Society of Japan* No. 1:201-221. [New genus: *Formotettigella*.]

1965d Homoptera Auchenorrhyncha. In: *Iconographia Insectorum. Japonicorum Colore Naturali Edita* 3:117-118, 120-122. Hokurgukan, Tokyo, Japan. 358 pp. [In Japanese with English index.]

1966a Homoptera of the Kurile Islands. *Transactions of the Shikoku Entomological Society* 9(2):31-40.

1966b *An Introduction to the Study of Agricultural Insects of Japan.* Yokendo, Co., Tokyo. 310 pp. [In Japanese.]

1967a Three cicadellid-species injurious to agriculture (Hemiptera). *Transactions of the Shikoku Entomological Society* 9(3):63-68.

1967b Japanese species having been recorded as *Tettigonia albomarginata* (Hemiptera: cicadellidae). *Transactions of the Shikoku Entomological Society* 9(3):63-68.

1968a Hemipterous fauna of the Japan archipelago. *Transactions of the Shikoku Entomological Society* 10(1):18-28.

1969a Families and genera of leafhopper vectors. In: Maramorosch, K., ed. *Viruses, Vectors, and Vegetation.* Pp. 235-254. John Wiley & Sons, New York. 666 pp.

1971a Several species of the genus *Kolla* Distant (Hemiptera: Cicadellidae). *Transactions of the Shikoku Entomological Society* 11(1):14-20. [New genus: *Yasumatsuus*.]

1972a A valid genus, *Hishimonus* Ishihara, 1953 (Hemiptera, Euscelidae). *Transactions of the Shikoku Entomological Society* 11(3):84.

1976a Some notes on two leafhoppers injurious to the sugarcane (Homoptera). *Transactions of the Shikoku Entomological Society* 13:25-27.

1978a A new species of leafhopper injurious to sweet poteto [sic!] (Hemiptera: Tettigellidae). *Transactions of the Shikoku Entomological Society* 14(1&2):11-12.

1979a Some notes on four Indian species of Cicadelloidea (Hemiptera). *Transactions of the Shikoku Entomological Society* 14(3&4):99-103.

1982a Some notes on a leafhopper of economic importance, *Orosius orientalis* (Matsumura, 1914) (Hemiptera: Cicadellidae). *Applied Entomology and Zoology* 17(3):364-367.

1983a Biotaxonomy of Cicadelloidea in Japan. In: Knight, W. J., Pant, N. C., Robertson, T. S. and Wilson, M. R., eds. *Proceedings of the 1st International Workshop on Biotaxonomy, Classification and Biology of Leafhoppers and Planthoppers of Economic Importance.* Pp. 457-467. London, 4-7 October 1982. Commonwealth Institute of Entomology, 56 Queen's Gate, London SW7 5JR. 500 pp.

Ishihara, T. and Edwards, C. A.
1973a A control project for insect pests and diseases injurious to the rice-plants in Japan and India. Association of Rice Research Workers: Symposium on Rice Production under Environmental Stress. *Oryza* 8:399-402.

Ishihara, T. and Kawase, E.
1968a Two new Malayan species of the genus *Nephotettix* (Hemiptera: Cicadellidae). *Applied Entomology and Zoology* (Tokyo) 3(3):119-123.

Ishihara, T. and Nasu, S.
1966a Leafhoppers transmitting plant-viruses in Japan and adjacent countries. In: Ishikura, Hidetsugu and Iwata, Yoshito. *11th Pacific Science Congress, Division Meeting on Plant Protection.* Pp. 159-170. Tokyo. 318 pp.

Ishii, M.
1973a Control of rice virus diseases. *Japan Pesticide Information* No. 17, pp. 11-16.

Ishii, M. and Ono, K.
1966a Decrease of viruliferous individuals in *Nephotettix cincticeps* congenitally infected with rice dwarf virus. *Annals of the Phytopathological Society of Japan* 32:317. [In Japanese.]

Ishiie, T., Doi, Y., Yora, K. and Asuyama, H.
1967a Suppressive effects of antibiotics of tetracycline group on symptom development of mulberry dwarf disease. *Annals of the Phytopathological Society of Japan* 33:267-275. [In Japanese.]

Ishiie, T. and Matsuno, M.
1971a Studies on the transmission of mulberry dwarf disease by a vector leafhopper, *Hishimonus sellatus* Uhler. I. On the acquisition of the causal agent in non-viruliferous vector. *Annals of the Phytopathological Society of Japan* 37(2):136-140. [In Japanese with English summary.]

1971b Studies on the transmission of mulberry dwarf disease by a vector leafhopper, *Hishimonus sellatus* Uhler. 2. On the inoculation of the causal agent by viruliferous vector. *Annals of the Phytopathological Society of Japan* 37(2):141-146. [In Japanese with English summary.]

Ishijima, T.

1969a Studies on the transmission of the mulberry dwarf disease by insects and its control. (2) Some epidemical observations on the features of the prevalence of the disease. *Sericulture Experiment Station Bulletin* 23(4):411-440. [In Japanese with English summary.]

1971a Transmission of the possible pathogen of mulberry dwarf disease by a new leafhopper vector, *Hishimonoides sellatiformis* Ishihara. *Journal of Sericultural Science* (Japan) 40(2):136-140. [In Japanese with English summary.]

Ishijima, T. and Ishiie, T.

1981a Mulberry dwarf: First tree mycoplasm disease. *In: Maramorosch, K. and Rayschandhuri, S. P., eds. Mycoplasma Diseases of Trees and Shrubs.* Pp. 147-184. Academic Press, New York. 362 pp.

Ishizaki, T.

1980a Histopathological observations on the vector leafhopper, *Hishimonus sellatus* Uhler, sucking the sap of the dwarf-diseased mulberry tree. *Journal of Sericultural Science* (Japan) 49(4):342-346.

1980b On the relationship between the electrophoretic patterns of body proteins and the multiplication of mycoplasma-like organisms of mulberry dwarf disease in the leafhopper, *Hishimonus sellatus* Uhler, living in mulberry plantations. *Journal of Sericultural Science* (Japan) 49(4):367-368.

Israel, P., Kalode, M. B. and Misra, B. C.

1968a Toxicity and duration of effectiveness of some insecticides against *Sogatella furcifera* (Horv.), *Nilaparvata lugens* (Stal) and *Nephotettix impicticeps* Ish. on rice. *Indian Journal of Agricultural Science* 38(3):427-431.

Israel, P. and Misra, B. C.

1968a Occurrence of blue leafhopper *Typhlocyba maculifrons* Motsch. on rice. *Current Science* (Bangalore) 37(4):113-114.

Ito, Y.

1982a Difference in population dynamics of the green rice leafhopper, *Nephotettix cincticeps* Uhler, in two districts of Japan. *Applied Entomology and Zoology* 17:337-349.

Ito, Y. and Gu, D.-X. and Johraku, T.

1983a Population processes of *Nephotettix cincticeps* Uhler (Hemiptera: Deltocephalidae) in Hokuriku and Tokai-Nisinippon districts of Japan. *Applied Entomology and Zoology* 18(2):200-210. [In English.]

Ito, Y. and Johraku, T.

1982a Differences in the population dynamics of the green rice leafhopper, *Nephotettix cincticeps* Uhler (Hemiptera, Deltocephalinae), in two districts of Japan. *Applied Entomology and Zoology* 17(3):337-349.

Ito, Y. and Miyashita, K.

1961a Studies on the dispersal of leaf- and plant-hoppers. I. Dispersal of *Nephotettix cincticeps* Uhler on paddy fields at the flowering stage. *Japanese Journal of Ecology* 11:181-186.

1965a Studies on the dispersal of leaf- and plant-hoppers. III. An examination of the distance-dispersal rate curves. *Japanese Journal of Ecology* 15:85-89.

Ivancheva G, Todora, S., Valdivieso, A., Becquer, A. and Saenz, Braulio.

1967a Las enfermedades virosas de la fruta bomba (*Carica papaya* L.) en Cuba. *Revista de Agricola* (Cuba) 1(2):1-21.

Ivliev, L. A. and Kononov, D. G.

1966a Insect-pests of forest stands of the Magadan oblast. *In: Harmful Forest Insects in the Soviet Far East.* Akademiya Nauk SSR Sibirskoo Otdelenie Dal'nevost. Filial im V. L. Komarova, Biol. -Pochv. Inst. Pp. 65-96. [In Russian.]

Iwaki, M.

1979a Virus and mycoplasma disease of leguminous crops in Indonesia. *Review of Plant Protection Research* 12:88-97.

Iwasaki, M., Maejima, I. and Shinkai, A.

1978a Reaction of tungro differential varieties to rice waika disease. *Proceedings of the Association of Plant Protection* (Kyushu) 24:7-8. [In Japanese.]

1979a Overwintering of rice waika virus in rice stubbles. *Annals of the Phytopathological Society of Japan* 45:91. [In Japanese.]

Iwata, T.

1972a Black type in the green rice leafhopper, *Nephotettix cincticeps* Uhler. *Japanese Journal of Applied Entomology and Zoology* 16(3):162.

1981a Effect of pesticide combinations on the development of resistance in green rice leafhopper, *Nephotettix cincticeps* Uhler. *Pesticide Information* 39:307.

Iwata, T. and Hama H.

1971a Green rice leafhopper, *Nephotettix cincticeps* Uhler, resistant to carbamate insecticides. *Botyu-Kagaku* 36:174-179. [In Japanese with English summary.]

1971b Insensitive cholinesterase in the Nakagawara strain of the green rice leafhopper, *Nephotettix cincticeps* Uhler (Homoptera: Cicadellidae), as a cause of resistance to carbamate insecticides. *Applied Entomology and Zoology* 6:183-191.

1972a Insensitivity of cholinesterase in *Nephotettix cincticeps* resistant to carbamate and organophosphorus insecticides. *Journal of Economic Entomology* 65:643-644.

1977a Comparison of susceptibility to various chemicals between malathion-selected and methyl parathion-selected strains of the green rice leafhopper, *Nephotettix cincticeps* Uhler. *Botyu-Kagaku* 42:181-188.

1981a Selection of multiple resistant green rice leafhopper, *Nephotettix cincticeps* Uhler (Hemiptera: Deltocephalidae), with Propaphos. Reversion of resistance to Propoxur. *Applied Entomology and Zoology* 16:37-44.

Iwaya, K. and Kollmer, G.

1975a Effectiveness of Curaterr granular against rice pests. *Pflanzenschutz-Nachrichten* 28:137-143.

Izadpanah, K. and Shepherd, R. J.

1973a Extraction of the curly top agent from segmented petioles of beet. *Phytopathology* 63(9):1209-1210.

Izzard, H. J.

1955a A new genus and species of Nirvaniinae [sic!] (Homoptera, Cicadellidae) from South India. *Indian Journal of Entomology* (Pruthiana) 17:186-188.

Jabbar, A.

1984a Studies of typhlocybine leafhopper species on sugarbeet in Pakistan. *Mitteilungen der Schweizerischen Entomologische Gesellschaft* 57(4):424.

Jabbar, A. and Ahmed, M.

1974a Light trap studies of *Zyginidia quyumi* (Ahmed), a leafhopper pest of wheat and maize in Pakistan. *Mushi* 48:79-85.

1975a Life history and morphology of *Zyginidia quyumi* (Ahmed) an important leafhopper pest of wheat and maize in Pakistan. *Pakistan Journal of Scientific Research* 27(1-4):127-135.

1976a Population studies of *Zyginidia quyumi* (Ahmed), (Typhlocybinae: Cicadellidae) on maize in NWFP. *Proceedings of the Entomological Society of Karachi* 6:69-93.

1980a Feeding effects of *Zyginidia quyumi* (Ahmed) on foliage of wheat and maize in Pakistan. *Pakistan Journal of Scientific and Industrial Research* 32(1&2):21-24.

Jabbar, A., Ahmed, Manzoor and Ahmed, Munir

1977a Distribution of *Zyginidia quyumi* on wheat in Punjab (Pakistan) by location day timings and temperatures. *Pakistan Journal of Scientific and Industrial Research* 20:188-191.

1982a Sex ratio in field populations of *Zyginidia quyumi* (Ahmed) (Homoptera: Cicadellidae). *Pakistan Journal of Agricultural Research* 3(3):178-181.

Jabbar, A., Ahmed, Manzoor and Samad, K.
1977a Population fluctuation of *Zyginidia quyumi* (Ahmed) on wheat in Punjab-Pakistan. *Pakistan Journal of Science and Industrial Research* 20:275-276.
1980a Population studies of *Zyginidia quyumi* (Ahmed) (Typhlocybinae: Cicadellidae) in the maize field at Abbottabad. *Pakistan Journal of Science* 32(1&2):117-120.

Jackle, H. and Schmidt, O.
1978a Characterization of ribosomal RNA from insect eggs (*Euscelis plebejus*, Cicadina: *Smittia* spec., Chironomidae, Diptera). *Experientia* (Basel) 34(10):1260-1261.

Jackson, G. V. H. and Zettler, F. W.
1983a Sweet potato witches' broom and legume little-leaf disease in the Solomon Islands. *Plant Disease* 67:1141-1144.

Jackson, J. E., Razoux Schultz, L. and Faulkner, R. C.
1965a Effects of jassid attack on cotton yield and quality in the Sudan Gezira. *Empire Cotton Growing Review* 42(4):295-299.

James, F. R. and Granovsky, A. A.
1927a Yellowing of alfalfa caused by leafhoppers. *Phytopathology* 17:39 (Abstract).

Janjua, N. A.
1957a Insect pests of paddy in Pakistan. *Agriculture Pakistan* 8(1):5-21.

Jankovik, L.
1966a Fauna Homoptera: Auchenorrhyncha in Serbia. *Glasnik Prironjackog Muzeja u Beogradu* 21B:137-166. [In Serbian with Russian summary].
1971a Fauna Homoptera: Auchenorrhyncha Srbije. 3. *Glasnik Prioronjackog Muzeja u Beogradu* (Ser. B) 26:125-134. [German summary.]
1977a [1978] Beschreibung einer neuen Cicadellidae (Homoptera, Auchenorrhyncha) aus Mazedonien (Jugo-slawien). *Bulletin, Academie Serbe des Sciences et des Arts* 60(16):111-114.
1978a Description of a new species from the family Cicadellidae (Homoptera, Auchenorrhyncha) from Macedonia (Yugoslavia). *Glasnik Srpska Akademiije Naukaiumetnosti odeljenje Prirodno-Matematichikh Nauka* 306(43):27-30.

Jansky, V.
1983a Homoptera, Delphacidae. Homoptera, Cicadellidae. *Biologia* (Bratislavia) 38(6):613.
1984a Faunistical notes, Homoptera, Cixiidae. Homoptera, Cicadellidae. *Biologia* (Bratislavia) 39(6):659.
1985a Faunistical notes. Homptera, Cicadellidae. *Biologia* (Bratislavia) 40(10):1097.
1985b Auchenorrhyncha of Male Karpaty Mountains. *Zbornik Slovenskcho Narodneho Muzea* 31:105-122. [In Slovak with English summary.]

Janson, B. F. and Ellet, C. W.
1963a A new corn disease in Ohio. *Plant Disease Reporter* 47:1107-1108.

Jansson, A. M.
1973a Svepelektronmikroskopiska studier av nagra arter tillhorande sl. *Macrosteles* Fieber (Homoptera, Cicadellidae). *Entomologisk Tidskrift* 94(3-4):199-207. [In Swedish with English abstract.]

Janvier, H.
1956a Observations sur deux predateurs chasseurs d'Homoptera (Hym. Sphegidae). *Annales de la Societe Entomologique de France* 124:195-208.

Jarvis, J. L. and Kehr, W. R.
1964a Evaluating alfalfa for resistance to the potato leafhopper. *Proceedings of the North Central Branch of the Entomological Society of America* 19:64-65.
1966a Population counts vs. nymphs per gram of plant material in determining degree of alfalfa resistance to the potato leafhopper. *Journal of Economic Entomology* 59(2):427-430.

Jasinska, J.
1980a Leafhoppers (Homoptera, Auchenorrhyncha) of the Bledowska wilderness. *Acta Biologica Uniwersytet Slaski w Katowicach* 8:40-49.

Jat, N. R.
1981a Insecticidal efficacy of malathion against insect pests of okra. *Indian Journal of Entomology* 43:137-139.

Javaid, I.
1979a Efficacy of some insecticides against cotton insects. *East African Agricultural and Forestry Journal* 42:125-126.

Jayaraj, S.
1964a Investigation on the mechanism of resistance in Castor (*Ricinus communis* L.) to the leafhopper, *Empoasca flavescens* (F.). Ph. D. Thesis, University of Madras.
1965a Investigations on the mechanism of resistance in castor (*Ricinus communis* L.) to the leafhopper, *Empoasca flavescens* (F.) (Jassidae: Homoptera). *Madras Agricultural Journal* 52(8):369-371.
1966a Resistenz von kulturpflanzen gegenuber Jassiden in India. *Zeitschrift fuer Angewandte Entomologie* 58:95-102.
1966b Influence of sowing times of castor varieties on their resistance to the leafhopper, *Empoasca flavescens* (F.) (Homoptera). *Entomologia Experimentalis et Applicata* 9(3):359-369.
1966c The influence of feeding by the leafhopper, *Empoasca flavescens* (F.), on transpiration and moisture content of certain varieties of *Ricinus communis*. *Phytopathologische Zeitschrift* 57:121-126.
1966d The effect of leafhopper infestation on the respiration of castor bean varieties in relation to their resistance to *Empoasca flavescens* (F.) (Homoptera, Jassidae). *Experimentia* 22:445.
1966e Organic acid contents in castor varieties in relation to their preference by the leafhopper *Empoasca flavescens* (F.). *Naturwissenschaften* 53:511.
1967a Studies on the resistance of castor plants (*Ricinus communis* L.) to the leafhopper, *Empoasca flavescens* (F.) (Homoptera, Jassidae). *Zeitschrift fuer Angewandte Entomologie* 59:117-126.
1967b Antibiosis mechanism of resistance in castor varieties to the leafhopper, *Empoasca flavescens* (F.) (Homoptera: Jassidae). *Indian Journal of Entomology* 29(1):73-78.
1967c Influence of leafhopper feeding on the shoot root ratio and mortality of certain castor varieties. *Indian Journal of Entomology* 29(3):301-303.
1967d Hopperburn disease of castor bean varieties caused by *Empoasca flavescens* (F.) in relation to the histology of leaves. *Phytopathologische Zeitschrift* 58:397-406.
1967e Effect of leafhopper infestation on the metabolism of carbohydrate and nitrogen in castor varieties in relation to their resistance to *Empoasca flavescens* (F.) *Indian Journal of Experimental Biology* 5:156-162.
1968a Preference of castor varieties for feeding and oviposition by the leafhopper, *Empoasca flavascens* (F.), with particular reference to its honeydew excretion. *Journal of the Bombay Natural History Society* 65:64-74.
1968b Studies on the plant characters of castor associated with resistance to *Empoasca flavescens* (Fabr.) (Homoptera: Jassidae) with reference to selection and breeding of varieties. *Indian Journal of Agricultural Science* 38(1):1-16.
1969a Influence of a phytotoxemia on the activities of catalaze enzyme and free auxins of castor bean varieties in relation to their resistance to *Empoasca flavescens* (F.) (Homoptera). *Zeitschrift fuer Angewandte Entomologie* 61:32-39.
1976a *Research on Resistance in Plants to Insects and Mites in Tamil Nadu*. Tamil Nadu Agricultural University, Coimbatore. 65 pp.

Jayaraj, S. and Basheer, M.
1964a An analysis of the effects of weather factors upon the population of *Empoasca devastans* Dist. on bhendi. (*Abelmoschus esculentus* Moench.). *Madras Agricultural Journal* 51(2):90 (Abstract).

1964b Biological observations on the castor leafhopper, *Empoasca flavescens* Fab. (Jassidae: Homoptera). *Madras Agricultural Journal* 51(2):89 (Abstract).

Jayaraj, S. and Seshadri, A. R.

1966a Influence of leafhopper infestation on the pigment contents of castor varieties in relation to their resistance to *Empoasca flavescens* (F.) (Homoptera, Jassidae). *Current Science* 35:572-573.

1967a Preference of the leaf-hopper *Empoasca kerri* Pruthi (Homoptera, Jassidae) to pigeon pea (*Cajanus cajan* [L.] Millsp.) plants infected with sterility mosaic virus. *Current Science* 36(13):353-355.

Jayaraj, S. and Venugopal, M. S.

1964a Observations of the effect of manuring and irrigation on the incidence of the cotton leafhopper, *Empoasca devastans* Dist., and the cotton aphid, *Aphis gossypii* G., at different periods of crops growth. *Madras Agricultural Journal* 51:189-196.

Jayaswal, A. P. and Saini, R. K.

1982a Cotton pest management in Haryana. *Indian Journal of Plant Protection* 9(1):29-33.

Jenkins, J. W. and Smith, J. C.

1977a Control of the potato leafhopper, *Empoasca fabae* (Harris), and the effect on peanut yields. *Virginia Journal of Science* 28:51.

Jensen, D. D.

1953a Leafhopper-virus relationships of peach yellow leaf roll. *Phytopathology* 43(10):561-564.

1955a Evidence that celery is a host of peach yellow leaf roll virus. *Phytopathology* 45(2):694.

1956a Insect transmission of virus between tree and herbaceous plants. *Virology* 2:249-260.

1956b Leafhopper transmission of peach yellow leaf roll virus and cherry buckskin virus to celery. *Pest Control Review*, p. 2.

1957a Transmission of peach yellow leaf roll virus by *Fieberiella florii* (Stal) and a new vector, *Osbornellus borealis* De L. & M. *Journal of Economic Entomology* 50:668-672.

1957b Differential transmission of peach yellow leaf roll virus to peach and celery by the leafhopper, *Colladonus motanus*. *Phytopathology* 47:575-578.

1958a Reduction in longevity of leafhoppers carrying peach yellow leaf roll virus. *Phytopathology* 48:394.

1959a A plant virus lethal to its insect vector. *Virology* 8:164-175.

1959b Insects, both hosts and vectors of plant viruses. *Pan-Pacific Entomologist* 35:65-82.

1962a Pathogenicity of western-X-disease virus of stone fruits to its leafhopper vector, *Colladonus montanus* (Van Duzee). *Proceedings of the 11th International Congress of Entomology*, Vienna 1960:790-791.

1963a Effects of plant-viruses on insects. *Annals of the New York Academy of Science* 105:685-712.

1968a Influence of high temperature on the pathogenicity and survival of western X-disease virus in leafhoppers. *Virology* 36:662-667.

1969a Comparative transmission of western X-disease virus by *Colladonus montanus*, *C. geminatus*, and a new leafhopper vector *Euscelidius variegatus*. *Journal of Economic Entomology* 62:1147-1150.

1969b Insect disease induced by plant-pathogenic viruses. *In: Maramorosch, K., ed. Viruses, Vectors and Vegetation.* Pp. 505-525. John Wiley & Sons, New York.

1971a Herbaceous host plants of western X-disease agent. *Phytopathology* 61:1465-1470.

1971b Vector fecundity reduced by western X-disease. *Journal of Invertebrate Pathology* 17:389-394.

1972a Temperature and transmission of western X-disease agent by *Colladonus montanus*. *Phytopathology* 62:452-456.

Jensen, D. D., Frazier, N. W. and Thomas, H. E.

1952a Insect transmission of yellow leaf roll virus of peach. *Journal of Economic Entomology* 45:335-337.

Jensen, D. D., Hukuhara, T. and Lanada, Y.

1972a Lethality of *Chilo* iridescent virus to *Colladonus montanus* leafhoppers. *Journal of Invertebrate Pathology* 19:276-278.

Jensen, D. D. and Nasu, S.

1970a Effects of plant mycoplasma on their insect vectors. *In: Perez-Miravete, A. Xth International Congress for Microbiology*, Mexico, D.F., p. 223. Plenum Press, New York. 301 pp.

Jensen, D. D. and Richardson, J.

1972a Effects of bacteria on leafhopper transmission of the western X-disease agent. *Journal of Invertebrate Pathology* 19:131-138.

Jensen, D. D. and Thomas, H. E.

1954a Leafhopper transmission of the Napa strain of cherry buckskin virus from cherry to peach. *Phytopathology* 44(9):494.

1955a Transmission of the green valley strain of cherry buckskin virus by means of leafhoppers. *Phytopathology* 45:694.

Jensen, D. D., Whitcomb, R. F. and Richardson, J.

1967a Lethality of injected peach western X-disease virus to its leafhopper vector. *Virology* 31:532-538.

Jensen, F. L.

1969a Microbial insecticides for control of grape leaf folder. *California Agriculture* 23(4):5-6.

Jensen, F. L., Flaherty, D. D. and Chiarappa, L.

1969a Population densities and economic injury levels of grape leafhopper. *California Agriculture* 23(4):9-10.

Jensen, F. L., Lynn, C. D., Stafford, E. M. and Kido, H.

1965a Insecticides for control of grape leafhopper. *California Agriculture* 19(4):10-11.

Jensen, F. L., Stafford, E. M., Kido, H. and Flaherty, D.

1961a Field tests for control of grape leafhoppers resistant to insecticides. *California Agriculture* 15(7):13-14.

1965a Surveying leafhopper populations. *California Agriculture* 19(4):7.

Jensen, J. O.

1983a Management of the aster leafhopper and aster yellows in Wisconsin. *Dissertation Abstracts International* (B) 43(7):2102.

Jensen, J. O. and Chapman, R. K.

1979a A curvilinear method for the determination of the base temperature and heat units associated with aster leafhopper development. *Proceedings of the North Central Branch of the Entomological Society of America* 33:53.

Jenser, G., Hegab, A. and Kollanyi, L.

1981a Vectors of rubus stunt in Hungary. *In: Polak, J., Brcak, J., Chod, J., Ulrychova, M., Petru, E. and Vacke, J., eds. Plant Pathology, Ninth Conference of the Czechoslovak Plant Virologists.* Pp. 73-75. Praha, Czechoslovakia.

Jeppson, L. R. and Carman, G. E.

1960a Citrus insects and mites. *Annual Review of Entomology* 5:353-378.

Jervis, M. A.

1978a Studies on the parasite complex associated with typhlocybine leafhoppers. *Auchenorrhyncha Newsletter* 1:25.

1980a Life history studies on *Aphelopus* species (Hymenoptera, Dryinidae) and *Chalarus* species (Diptera, Pipunculidae), primary parasites and typhlocybine leafhoppers (Homoptera, Cicadellidae). *Journal of Natural History* 14(6):769-780.

1980b Ecological studies on the parasite complex associated with typhlocybine leafhoppers (Homoptera, Cicadellidae). *Ecological Entomology* 5(2):123-136.

1980c Studies on the oviposition behavior and larval development in species of *Chalarus* (Diptera, Pipunculidae), parasites of typhlocybine leafhoppers (Homoptera, Cicadellidae). *Journal of Natural History* 14(6):759-768.

Jhoraku, T.

1966a Seasonal prevalence and control of the green rice leafhopper, *Nephotettix cincticeps* Uhler. *Agriculture and Horticulture* 41:1214-1218. [In Japanese.]

Jhoraku, T. and Kato, S.
1974a Effect of the snow on the survival of green rice leafhopper, *Nephotettix cincticeps* Uhler. *Proceedings of the Association of Plant Protection Hokuriku* 22:30-31. [In Japanese.]

Jhoraku, T., Sekiguchi, W., Kato, S., Naruse, H., Imai, F. and Wakamatsu, T.
1983a Fluctuation of population of the green rice leafhopper, *Nephotettix cincticeps* Uhler (Hemiptera: Deltocephalidae). *Japanese Journal of Applied Entomology and Zoology* 27:146-151.

Jin, K. X., Liang, C. J. and Deng, D. L.
1981a A study of the insect vectors of witches' broom in Paulownia trees. *Linye Keji Tongxun* 12:23-24.

Joginder Singh, Gatoria, G. S., Sindhu, A. S. and Bindra, O. S.
1982a Studies on the supervised control of cotton jassid (*Amrasca biguttula biguttula* Ishida) in the Punjab. *Entomon* 7(3):275-279.

Johansson, S.
1962a Insects associated with *Hypericum* L. 2. Lepidoptera, Diptera, Hymenoptera, Homoptera, and general remarks. *Opuscula Entomologica* 27(3):175-192.

John, B. and Claridge, M. F.
1974a Chromosome variation in British populations of *Oncopsis* (Hemiptera: Cicadellidae). *Chromosoma* 46:77-89.

John, V. T.
1965a On the antigenicity of virus causing "tungro" disease of rice. *Plant Diseases Reporter* 49:305-306.
1966a Insecticidal control of *Nephotettix* spp., the vector of tungro and yellow dwarf diseases of rice in the Philippines. *Indian Phytopathology* 19(2):150-154.

John, V. T., Freeman, W. H. and Shahi, B. B.
1979a Occurrence of tungro disease in Nepal. *International Rice Research Newsletter* 4(5):16.

John, V. T. and Ghosh, A.
1981a Life span of and tungro transmission by viruliferous *Nephotettix virescens* on 10 rice varieties at AICRIP, India. *International Rice Research Newsletter* 5(6):6-7.

Johnson, C. G.
1969a Migration and Dispersal of Insects by Flight. Methuen & Co., Ltd. xxii + 763 pp.

Johnson, C. G., Southwood, T. R. E. and Entwistle, H.
1957a A new method of extracting arthropods and molluscs from grassland and herbage with a suction apparatus. *Bulletin of Entomological Research* 48(1):211-218.

Johnson, D. H.
1970a Classification of the subgenus *Amphipyga*, genus *Athysanella* (Homoptera: Cicadellidae). Thesis, Kansas State University. 93 pp. (microfilm).

Johnson, D. H. and Blocker, H. D.
1979a Eight new species of the subgenus *Amphipyga*, genus *Athysanella* (Homoptera: Cicadellidae). *Journal of the Kansas Entomological Society* 52(2):377-385.

Jones, A. L., Whitcomb, R. F., Williamson, D. L and Coan, M. E.
1977a Comparative growth and primary isolation of spiroplasmas in media based on insect tissue culture formulations. *Phytopathology* 67:738-746.

Joomaye, A.
1976a Studies on the movements of *Empoasca devastans* Distant (Homoptera: Jassidae) within cotton and 'bhindi' crops. *Entomologist Newsletter* 6:25.

Joplin, C. E.
1974a Pulse crops of the world and their important insect pests. M. S. Thesis, Simon Fraser University, Barnaby, B.C., Canada. 134 pp.

Joshi, A. B. and Rao, S. B.
1959a The problem of breeding jassid resistant varieties of cotton in India. *Indian Cotton Growers Revue* 13:270-279.

Joshi, G.
1983a Integrated pest management strategy for rice for an area—X. In: *10th International Congress of Plant Protection. Vol. 3. Proceedings of a Conference Held at Brighton, England*, p. 935. Croydon, U.K. 1228 pp.

Jotwani, M. G., Sarup, Prakash, Sircar, P. and Singh, D. S.
1967a Control of jassid (*Empoasca devastans* Dist.) on okra (*Abelmoschus esculentus*) seedlings by treatment of seeds with different insecticides. *Indian Journal of Entomology* 28(4):477-481.

Joyce, C. R.
1955a Notes and exhibitions. *Proceedings of the Hawaiian Entomological Society for 1954* 15(3):389.
1962a Notes and exhibitions—Feb. 13, 1961. *Proceedings of the Hawaiian Entomological Society for 1961* 18(1):4.
1963a Notes and exhibitions. *Proceedings of the Hawaiian Entomological Society for 1962* 18(2):195.

Joyce, R. J. V.
1955a Cotton spraying on the Sudan Gezira. I. Yield increase from spraying and spraying methods. *FAO Plant Protection Bulletin* 3(6):86-92.
1961a Some factors affecting numbers of *Empoasca lybica* (de Berg.) (Homoptera: Cicadellidae) infesting cotton in the Sudan Gezira. *Bulletin of Entomological Research* 52(1):191-232.

Jubb, G. L., Jr., Danko, L. and Haeseler, C. W.
1983a Impact of *Erythroneura comes* Say (Homoptera: Cicadellidae) on caged "concord" grapevines. *Environmental Entomology* 12(5):1576-1580.

Judge, F. D., McEwen, F. L. and Rinick, H. B.
1970a Field testing candidate insecticides on beans and alfalfa for control of Mexican bean beetle, potato leafhopper and plant bugs in New York state. *Journal of Economic Entomology* 63(1):58-62.

Julia, J. F.
1979a Demonstration and identification of the insects responsible for the diseases of young coconut and oil palms in the Ivory Coast. *Oleagineaux* 34:385-393.

Jurisoo, V.
1964a Agro-ecological studies on leafhoppers (Auchenorrhyncha, Homoptera) and bugs (Heteroptera) at Ekensgard farm in the province of Halsingland, Sweden. *Contributions of the National Institute of Plant Protection* 13(101):1-147.

Kabir, S. M. H. and Choudhury, J. C. S.
1975a The rice leafhoppers and planthoppers in the Mymensingh area with notes on their seasonal abundance. *Bangladesh Journal of Zoology* 3:29-42.

Kaiser, B.
1980a Licht-und electronemikroskopische untersuchung der symbionten von *Graphocephala coccinea* Forstier (Homoptera: Jassidae). *International Journal of Insect Morphology and Embryology* 9:79-88.

Kajak, A., Breymeyer, A., Petal, J. and Olechowicz, E.
1972a The influence of ants on the meadow invertebrates. In: Brian, M. V. and Petal, J., eds. *Proceedings of the IBP (PT Section) Meeting on Methods of Investigating the Productivity of Social Insects and Their Role in the Ecosystems*. Institute of Ecology, Warsaw, June 22-28, 1970. *Ekologia Polska* 20(17):163-171. [In English with Polish summary.]

Kakiya, N. and Kiritani, K.
1972a The influence of maternal age, and rearing density upon flight activity of the green rice leafhopper, *Nephotettix cincticeps* Uhler (Hemiptera: Deltocephalidae). *Japanese Journal of Applied Entomology and Zoology* 16(2):79-85.

Kalandadze, L. P., Tulashvili, N. D. and Shavkatsipvili, L. D.
1954a Results of studies of insect pests of cereals in western Georgia. *Akademiya Nauk Gruzinskoi SSR, Instituta Zashchita Rastenii, Trudy* 10:3-31. [Russian summary.]

Kalkandelen, A.
1972a Four new species and a subspecies of Cicadellidae: Euscelinae from Central Anatolia. *Istanbul Universitesi Fen Fakultesi Mecmuasi* (Serie B), *Sciences Naturelles* 37(3&4):153-163.
1974a *Orta Anadolu'da Homoptera: Cicadellidae. Familyasi turlerinin Taksonomilei; Uzerinde Arastirmalar*. Ankara University, Ankara. [Taxonomic study on the species of Homoptera: Cicadellidae from central Anatolia.] 221 pp. [English summary.]

1980a Contributions to the families Delphacidae and Cicadellidae (Homoptera) from Turkey. *Turkiye Bitki Koruma Dergisi* 4(3):147-154. [Turkish summary.]

1985a Four new species of genus *Zyginidia* (Zyginidia) Haupt (Homptera: Cicadellidae) and with notes on the taxonomy and distributions of the species of this genus in Turkey. *Turkiye Bitki Koruma Dergisi* 9(1):13-25.

Kalkandelen, A. and Fox, R. C.

1968a Distribution of phony peach vectors in South Carolina. *Journal of Economic Entomology* 61(1):65-67.

Kalode, M. B.

1983a Leafhopper and planthopper pests of rice in India. *In: Knight, W. J., Pant, N. C., Robertson, T. S. and Wilson, M. R., eds. Proceedings of the 1st International Workshop on Biotaxonomy, Classification and Biology of Leafhoppers and Planthoppers (Auchenorrhyncha) of Economic Importance.* Pp. 225-245. London, 4-7 October 1982. Commonwealth Institute of Entomology, 56 Queen's Gate, London SW7 5JR. 500 pp.

Kaloostian, G. H.

1956a Overwintering habits of the geminate leafhopper in Utah. *Journal of Economic Entomology* 49:272.

1956b *Ufens niger* (Ashm.), an egg parasite of the geminate leafhopper. *Journal of Economic Entomology* 49:140.

Kaloostian, G. H., Oldfield, G. N., Calavan, E. C. and Blue, R. L.

1976a Leafhopper transmits disease to weed host. *Citrograph* 61:389-390.

Kaloostian, G. H., Oldfield, G. N., Gumph, D. and Calavan, E. C.

1979a Control of citrus stubborn vectors in the laboratory. *Citrograph* 65:17-18.

Kaloostian, G. H., Oldfield, G. N., Pierce, H. D. and Calavan, E. C.

1979a *Spiroplasma citri* and its transmission to citrus and other plants by leafhoppers. *In: Maramorosch, K. and Harris, K. F., eds. Leafhopper Vectors and Plant Disease Agents.* Pp. 447-450. Academic Press, New York. 654 pp.

Kaloostian, G. H., Oldfield, G. N., Pierce, H. D., Calavan, E. C., Granett, A. L., Rana, G. L. and Gumpf, D. J.

1975a Leafhopper—natural vector of citrus stubborn disease? *California Agriculture* 29(2):14-15.

Kaloostian, G., Oldfield, G. Pierce, H., Calavan, G., Granett, A. L., Rana, G. L., Gumpf, D. J. Blue, R. L. and Harjung, M. K.

1975a Suspect in stubborn disease found guilty. *Citrograph* 60(4):99, 112, 113.

Kaloostian, G. H. and Pierce, H. D.

1972a Notes on *Scaphytopius nitridus* on citrus in California. *Journal of Economic Entomology* 65(3):880.

Kaloostian, G. H. and Pollard, H. N.

1962a Experimental control of phony peach virus vectors with di-syston. *Journal of Economic Entomology* 55:566-567.

Kaloostian, G. H., Pollard, H. N. and Turner, W. F.

1962a Leafhopper vectors of Pierce's disease virus in Georgia. *Plant Disease Reporter* 46:292.

Kamal, El Din M. H.

1964a List of localities, dates and hosts of order Hemiptera-Homoptera as recorded in the entomological collection, Ministry of Agriculture. *Agricultural Research Review* 42(3):116-128.

Kamel, S. A. and Farouk, Y. E.

1965a Relative resistance of cotton varieties in Egypt to spider mites, leafhoppers, and aphids. *Journal of Economic Entomologist* 58:209-212.

Kameswara Rao, P.

1976a Studies on Indian Cicadelloidea (Homoptera). *Entomologist's Newsletter* 6(2):11-12.

Kameswara Rao, P. and Ramakrishnan, U.

1978a Studies on Indian Cicadelloidea (Homoptera): The genus *Moonia*. *Oriental Insects* 12(2):221-233.

1978b Studies on Indian Cicadelloidea (Homoptera): Three new species of *Agallia*. *Oriental Insects* 12(2):235-242.

1978c A new species of the genus *Durgades* of the family Agallidae [sic!] (Homoptera: Cicadelloidea). *Proceedings of the Indian Academy of Sciences* (Section B) 87(12):357-360.

1979a Two new species of *Balocha* (Idioceridae, Homoptera). *Journal of the Bombay Natural History Society* 76(2):342-345.

1979b A new species of *Moonia* (Homoptera: Ulopidae). *Journal of the Bombay Natural History Society* 76(2):346-347.

1979c The validity of *Idiocerus nigroclypeatus* Melichar and its transfer to *Idioscopus* Baker. *Indian Journal of Entomology* 41:296-298.

1979d [1980] Two new species of *Macropsis* (Homoptera: Cicadellidae: Macropsinae) from northern India. *Oriental Insects* 13(3&4):307-309.

1980a Two new species of the genus *Batracomorphus* (Homoptera: Cicadelloidea: Iassidae). *Entomon* 5(1):43-46.

1983a Description of *Balocha tricolor* Distant (Idiocerinae: Homoptera). *Bulletin of Entomology. (Entomological Society of India)* New Delhi 24(1):21-23.

1983b Description of a new species of the genus *Symphypyga* Haupt (Homoptera: Agallidae) [sic!]. *Bulletin of Entomology, Entomological Society of India* (New Delhi) 24(1):21-23.

Kameswara Rao, P., Ramakrishnan, U. and Ghai, S.

1979a A new species of *Balocha* (Homoptera: Cicadellidae) from Delhi. *Entomon* 4(3):295-298.

1979b Description of *Austroagallia afganistanensis* sp. nov. (Agallidae [sic!]: Homoptera). *Current Science* (Bangalore) 48:655-656.

Kamiwada, H., Hara, K. and Fukamachi, S.

1976a Timing of insecticide application to control rice waika disease in Japan. *Proceedings of the Association for Plant Protection of Kyushu* 22:107-109.

Kamm, J. A. and Ritcher, P. O.

1972a Rapid dissection of insects to determine ovarial development. *Annals of the Entomological Society of America* 65(1):271-274.

Kamm, J. A. and Swenson, K. G.

1972a Termination of diapause in *Draeculacephala crassicornis* with synthetic juvenile hormone. *Journal of Economic Entomology* 65(2):364-367.

Kandoria, J. L. and Haracharan, Singh

1980a Influence of ultra-low-volume sprayers on the insecticidal control of *Amrasca biguttula biguttula* (Ishida) on cotton. *Indian Journal of Agricultural Sciences* 50:961-964.

1982a Effect of ULV sprayers on the efficacy of spray formulations of malathion against *Amrasca biguttula biguttula* (Ishida) infesting cotton. *Journal of Entomological Research* 6:76-89.

1982b Effect of ULV sprayers on the efficacy and persistence of low dosages of malathion LVC against *Amrasca biguttula biguttula* (Ishida) on cotton. *Indian Journal of Entomology* 44:230-251.

1984a Effect of ULV sprayers on the size and density of droplets of malathion LVC. *Indian Journal of Entomology* 46:92-100.

Kang, J. and Kiritani, K.

1978a Winter mortality of the green rice leafhopper (*Nephotettix cincticeps* Uhler) caused by predation. *Japanese Journal of Applied Entomology and Zoology* 22:243-249.

Kannaiyan, S., Jayaraman, V., Jagannathan, R. R. and Palaniyandi, V. C.

1978a Occurrence of rice tungro virus in Tamil Nadu. *International Rice Research Newsletter* 3:8.

Kantack, B. H. and Berndt, W. L.

1972a 1972 South Dakota insecticide recommendations. *South Dakota Cooperative Extension Service Handbook* EC-683. 119 pp.

Kantack, E. J., Martin, W. J. and Newsom, L. D.

1960a Relation of insects to internal cork of sweet potato in Louisiana. *Phytopathology* 50(6):447-449.

Kao, Hue-Lien, Liu, Ming-Yie and Sun, Chih-Ning
  1981a Green rice leafhopper resistance to malathion, methyl parathion, carbaryl, permethrin and fevalerate in Taiwan. *International Rice Research Newsletter* 6(5):19.
  1982a *Nephotettix cincticeps* (Homoptera: Cicadellidae) resistance to several insecticides in Taiwan. *Journal of Economic Entomology* 75(3):495-496.

Kapoor, V. C. and Grewal, J. S.
  1984a Prospects of *pipunculids* (Diptera: Pipunculidae) as biocontrol agents of leafhoppers (Homoptera) pests. *In: XVII International Congress of Entomology* 17:34. 968 pp.

Kapoor, V. C. and Sohi, A. S.
  1972a The nomenclature status of the cotton jassid *Empoasca devastans*. *Entomologist's Record and Journal of Variation* 84:51.

Karaman, S.
  1966a Nouvel homoptere pour la Macedoine *Doratura heterophyla* Horv. (Homopt. Auchenorhynch-Jassidae). *Bulletin de la Societe Entomologique de Mulhouse* 30-31.

Kareem, A. A., Thangavel, P., Jayaraj, S. and Subramaniam, T. R.
  1977a Efficacy of toxaphene, DDT and the combination product in the control of cotton pests. *Madras Agricultural Journal* 64:340-342.

Karel, A. K. and Malinga, Y.
  1980a Leafhopper and aphid resistance in cowpea varieties. *Tropical Grain Legume Bulletin* 20:10-11.

Kariappa, B. K.
  1978a Studies on control of the jassid *Empoasca flavescens* (F.) (Homptera-Jassidae) on mulberry. *Indian Journal of Sericulture* 17(1):1-6.

Karim, A. N. M. R.
  1978a Varietal resistance to the green rice leafhopper *Nephotettix virescens* (Distant): Sources, mechanisms, and genetics of resistance. Ph. D. Thesis, University of Philippines, Los Banos. 169 pp.

Karim, A. N. M. R. and Pathak, M. D.
  1979a An alternate biotype of green leafhopper in Bangladesh. *International Rice Research Newsletter* 4(6):7-8.
  1982a New genes for resistance to green leafhopper *Nephotettix virescens* (Distant) in rice, *Oryza sativa* L. *Crop Protection* 1:483-490.

Kartal, V.
  1982a Neue Homopteren aus der Turkei:-1. *Priamus* 1(1):24-30.
  1983a Neue Homopteran aus der Turkei:-2. (Homoptera: Auchenorrhyncha). *Marburger Entomologische Publikationen* 1(8):235-248.

Katayama, E.
  1975a The relation between ovarial development and population in a leafhopper and planthoppers of rice plant (Homoptera: Auchenorrhyncha). *Japanese Journal of Applied Entomology and Zoology* 19:176-181.

Kathirithamby, J.
  1974a Key for the separation of larval instars of some British Cicadellidae (Hem., Homoptera). *Entomologist's Monthly Magazine* 109:214-216.
  1974b Genital abnormalities in adult Cicadellidae (Homoptera). *Entomologist's Monthly Magazine* 110:193-201.
  1974c Development of the external male and female genitalia in the immature stages of Cicadellidae (Homoptera). *Journal of Entomology* 48:193-197.
  1976a Further abnormalities found in the external genitalia of *Eupteryx urtica* (F.) (Homoptera, Cicadellidae). *Entomologist's Monthly Magazine* 112:77-82.
  1977a Stylopisation in *Ulopa reticulata* (F.) (Homoptera: Cicadellidae). *Entomologist's Monthly Magazine* 113(Jan.-Apr.) 1977(1978):89-91.
  1978a Studies of the *Nephotettix* spp. (Cicadellidae: Homoptera) in the Krian District, peninsular Malaysia. *Malaysian Agricultural Journal* 51(3):273-279.
  1979a The occurrence of sex-mosaics in parasitized *Eupteryx urtica* (F.) (Homoptera: Cicadellidae). *Entomologist's Monthly Magazine* 114:147-148.
  1981a Common Auchenorrhyncha (Homoptera) in rice fields in southeast Asia. *Bulletin, Ministry of Agriculture, Federation of Malaya* 155:1-36.

Kato, S. and Wakamatsu, T.
  1978a Occurrence of the green rice leafhopper in Toyama Prefecture. *Proceedings of the Association of Plant Protection* (Kokuriku) 26:12-17. [In Japanese.]

Kato, Y., Kawauchi, N., Asaka, S. and Koyama, T.
  1979a Relation of metabolism and translocation of propaphos in rice plant to insecticidal activity against two strains of green rice leafhoppers. *Journal of Pesticide Science* 4:439-446.

Kaur, P., Bindra, O. S. and Singh, S.
  1971a Biology of *Circulifer opacipennis* (Lethierry), a leafhopper vector of sugar-beet curly-top virus. *Indian Journal of Agricultural Science* 41(1):1-10.

Kaushik, C. D., Tripathi, N.N. and Satyavir
  1978a Effect of date of sowing of toria on phyllody incidence and estimation of losses. *Haryana Agricultural University Journal of Research* 8:28-30.

Kawabe, S.
  1978a The electronic measurement of feeding behavior of the green rice leafhopper and the mechanism of rice resistance to the leafhopper. *In: 22nd Conference of the Japanese Society of Applied Entomology and Zoology*, p. 124. [In Japanese.]
  1979a Feeding behavior of the green rice leafhopper and resistance of rice. *Plant Protection* (Tokyo) 33:193-199.
  1985a Mechanism of varietal resistance to the rice green leafhopper (*Nephotettix cincticeps* Uhler). *Japan Agricultural Research Quarterly* 19:115-124.

Kawabe, S., Fukumorita, T. and Chino, M.
  1980a Stylectomy of plant-sucking insects using a YAG laser to collect rice phloem sap. *International Rice Research Newsletter* 5(5):19-20.

Kawabe, S. and McLean, D. L.
  1978a Electronically recorded waveforms associated with salivation and ingestion behavior of the aster leafhopper, *Macrosteles fascifrons* (Stal) (Homoptera: Cicadellidae). *Applied Entomology and Zoology* 13(3):143-148.
  1980a Electronic measurement of probing activities of the green leafhopper of rice. *Entomologia Experimentalis et Applicata* 27:77-82.

Kawabe, S., McLean, D. L., Tatsuki, S. and Ouchi, T.
  1981a An improved electronic measurement system for studying ingestion and salivation of leafhoppers. *Annals of the Entomological Society of America* 74:222-225.

Kawahara, S. and Kiritani, K.
  1975a Survival and egg-sac formation rates of adult females of *Lycosa pseudoannulata* (Boes. et Str.) (Araneae: Lycosidae) in the paddy field. *Applied Entomology and Zoology* 10:232-234.

Kawahara, S., Kiritani, K. and Sasaba, T.
  1971a The selective activity of rice-pest insecticides against the green rice leafhopper and spiders. *Botyu-Kagaku* 36(3):121-128.

Kawai, A.
  1977a Inheritance of blue-type in the green rice leafhopper, *Nephotettix cincticeps* Uhler. *Proceedings of the Association for Plant Protection of Kyushu* 23:83-84. [In Japanese with English summary.]

Kawase, E.
  1971a The genus *Nephotettix* in Thailand (Hemiptera: Cicadellidae). *Japanese Journal of Applied Entomology and Zoology* 15(2):70-75.

Kazano, H.
  1983a Studies on insecticidal characteristics of carbamate compounds and their metabolism in insects, soils and model ecosystem. *Bulletin of the National Institute of Agricultural Sciences* 37:31-89.

Kazano, H., Asakawa, M. and Fukunaga, K.
1975a Evaluation methods of insect-sterilizing compounds and search for new chemicals. *Bulletin of the National Institute of Agricultural Sciences* 29:1-43.

Kazano, H., Asakawa, M. and Tomizawa, C.
1978a Metabolism of some carbamate insecticides in the green rice leafhopper and the smaller brown planthopper. *Journal of Pesticide Science* 3(4):419-425.

Kazano, H., Koyama, S. and Sutrison
1983a Mechanism of insecticidal selectivity of propaphos between the green rice leafhopper, *Nephotettix cincticeps* Uhler and the common cutworm, *Spodoptera litura* Fabricius. *Journal of Pesticide Science* 8:561-565.

Kazano, H., Kurosu, Y., Asakawa, M., and Fukunaga, K.
1969a Studies on carbamate insecticides. II. Insecticidal activities of substituted phenyl N. methylcarbamates against the smaller brown planthopper, *Laodelphax striatellus* Fallen, and the green rice leafhopper, *Nephotettix cincticeps* Uhler. *Japanese Journal of Applied Entomology and Zoology* 13(3):117-123. [In Japanese with English summary.]

1969b Studies on carbamate insecticides. III. Insecticidal characteristics of carbamates to planthoppers and leafhoppers. *Japanese Journal of Applied Entomology and Zoology* 13(4):191-199. [In Japanese with English summary.]

1970a Studies on carbamate insecticides. IV. Insecticidal activities of butyl- and amyl-substituted phenyl carbamates on planthoppers and leafhoppers, and their insecticidal characteristics. *Japanese Journal of Applied Entomology and Zoology* 14(4):173-181.

Keener, P. D.
1956a Virus diseases of plants in Arizona. 2. Field and experimental observations on curly-top affecting vegetable crops. *University of Arizona Agricultural Experiment Station Bulletin* 271:1-28.

Kehr, W. R.
1970a Registration of N.S. 16 alfalfa germplasm (Reg. No. GP15). *Crop Science* 10(6):731.

1970b Registration of N.S. 30 alfalfa germplasm (Reg. No. GP16). *Crop Science* 10(6):731.

Kehr, W. R., Kindler, S. D., Schalk, J. M. and Ogden, R. L.
1967a Breeding alfalfa with resistance to insects and diseases. *Nebraska Agricultural Experiment Station Farm Ranch Home Quarterly* 1967(Spring):22-23.

Kehr, W. R. and Manglitz, G. R.
1984a Registration of N.S. 76 PA1 and N.S. 86 alfalfa germplasms resistant to potato leafhopper yellowing. *Crop Science* 24:1003-1004.

Kehr, W. R., Manglitz, G. R. and Ogden, R. L.
1968a Dawson alfalfa—a new variety resistant to aphids and bacterial wilt. *Nebraska Agricultural Experiment Station Bulletin* SB-497. 23 pp.

Kehr, W. R., Ogden, R. L. and Kindler, S. D.
1970a Diallel analyses of leafhopper injury to alfalfa. *Crop Science* 10(5):584-586.

1974a Varieties and cutting practices in integrated alfalfa pest management. *Proceedings of the Forage Insect Research Conference* 17:3-4.

1975a Management of four alfalfa varieties to control damage from potato leafhoppers. *Nebraska Agricultural Experiment Station Research Bulletin* 275. 42 pp.

Keldysh, M. A. and Pomazkov, Y. I.
1967a Vector of raspberry dwarf virus. *Zashchita Rastenii* 12(7):55. [In Russian]

Kelly, R. E. and Klostermeyer, L. E.
1984a Relative abundance and insecticidal control of leafhoppers (Homoptera: Cicadellidae) in a mixed tall fescue pasture in Tennessee. *Journal of Kansas Entomological Society* 57:657-661.

Kennedy, G. G.
1971a Reproduction of *Scaphytopius delongi* (Homoptera: Cicadellidae) in relation to age of males and females. *Annals of the Entomological Society of America* 64(5):1180-1182.

Kennedy, G. G., Kishaba, A. N. and Bohn, G. W.
1975a Response of several pest species to *Cucumis melo* L., lines resistant to *Aphis gossypii* Glover. *Environmental Entomology* 4:653-657.

Kenneth, R. and Olmert, Y.
1975a Endopathogenic fungi and their hosts in Israel: Additions. *Israel Journal of Entomology* 10:105-112.

Kenneth, R., Wallis, G., Olmert, I. and Halperin, J.
1971a A list of entomogenous fungi of Israel. *Israel Journal of of Agricultural Research* 21:63-66.

Kerr, T. W.
1957a Leafhoppers associated with forage crops in Rhode Island. *Journal of Economic Entomology* 50(3):271-273.

1957b Leafhoppers infesting lawns in Rhode Island. *Journal of Economic Entomology* 50(3):372.

Kerr, T. W. and Stuckey, I. H.
1956a Insects attacking red clover in Rhode Island and their control. *Journal of Economic Entomology* 49(3):371-375.

Keshavamurthy, K. V. and Yaraguntaiah, R. C.
1969a Further studies on the transmission of the virus component of the ragi disease-complex in Mysore. *Mysore Journal of Agricultural Science* 3(4):480.

Khaire, V. M. and Bhapkar, D. G.
1971a Some new records of rice hoppers in Maharashtra. *Current Science* 40(4):91-92.

Khan, A. M.
1975a Observations on the mycetome of *Idiocerus clypealis* Leth. (Hemiptera: Jassidae). *Indian Journal of Entomology* 37(2):194-196.

1976a Histochemical observation of the mycetomes of *Idiocerus clypealis* Leth. *Acta Histochemica* 55(2):241-244.

1977a Isolation of bacterial symbiont from mycetomes of *Idiocerus clypealis* Leth. and effect of antibiotics and sulpha drugs on its growth. *Journal of Entomological Research* 1:103-105.

Khan, A. M., Jabber, A. and Qadri, M. A. H.
1964a Some new records of the insect pests and their relative abundance in Rajshabi (East Pakistan). *Proceedings of Pakistan Science Conference* 16(Pt. 3):B13-B14.

Khan, I. A., Zuberi, R. I., Khan, Z. I., Ali, A. and Naqvi, S. N. H.
1971a Effectiveness of Petkolin, Endrin and their combination against jassids and thrips on cotton crop at Multan. *Science and Industry* 8(2):147-151.

Khan, Z. R. and Agarwal, R. A.
1981a Relationship between gossypol glands and incidence of some important pests on different genotypes of cotton. *Journal of Entomological Research* 5:169-172.

1984a Oviposition preference of jassid, *Amrasca biguttula biguttula* Ishida, on cotton. *Journal of Entomological Research* 8:78-80.

Khan, Z. R. and Saxena, R. C.
1984a Technique for demonstrating phloem or xylem feeding by leafhoppers (Homoptera: Cicadellidae) and planthoppers (Homoptera: Delphacidae) in rice plant. *Journal of Economic Entomology* 77:550-552.

1984b Electronically recorded wave forms associated with feeding behaviour of green leafhopper (GLH). *International Rice Research Newsletter* 9(4):8-9.

1984c A simple technique for locating feeding sites of green leafhopper in rice plants. *International Rice Research Newsletter* 9(2):16-17.

1985a Effect of steam distillate extract of a resistant rice variety on feeding behavior of *Nephotettix virescens* (Homoptera: Cicadellidae). *Journal of Economic Entomology* 78(3):562-566.

1985b Mode of feeding and growth of *Nephotettix virescens* (Homoptera: Cicadellidae) on selected resistant and susceptible rice varieties. *Journal of Economic Entomology* 78(3):583-587.

1985c Behavior and biology of *Nephotettix virescens* (Homoptera: Cicadellidae) on tungro virus-infected rice plants: Epidemiology implications. *Environmental Entomology* 14(3):297-304.

Khanna, V. M. (editor)
 1973a Investigations on Hairy Sprout Disease of Potato. Simla, Indian Council of Agricultural Research. Central Potato Research Institute. 35 pp.

Kharizanov, A.
 1969a A new pest of grape vine in Bulgaria. *Rastitelna Zashchita* 17(11):21-23.
 1970a Injurious Auchenorrhyncha on cereal crops in Bulgaria. *Rastitelna Zaschita* 18(2):18-21.

Kheyri, M.
 1969a The leafhoppers of sugarbeet in Iran and their role in curly-top virus disease. Sugarbeet Seed Institute, Karaj. Entomology Research Division, Tehran. 50 pp. [In Farse with English summary]

Khokhlova, S. Y. and Anufriev, G. A.
 1981a Variation of colour and pattern in two remote populations of *Macropsis prasina* (Boh., 1852) (Homoptera, Auchenorrhyncha). *Vestnik Zoologii* 1:47-51. [In Russian with English summary.]

Kholmuminov, A. and Dubovskii, G. K.
 1979a On the fauna of Cicadina of the Golodnostep plain. *Uzbekskii Biologicheskii Zhurnal* 4:61-63.

Khristova, E. and Loginova, E.
 1975a New pests of greenhouse crops and possible methods of control. *Rastitelna Zaschita* 23:19-21.

Khurana, S. M. P., Deshmukh, U. and Ramakrishnan, U.
 1974a Serious infestation of bajra by the jassid, *Balclutha* sp. *Labdev Journal of Science and Technology* 12:34.

Khush, G. S.
 1973a Rice breeding for disease and insect resistance at IRRI. Association of Rice Research Workers: Symposium on rice production under environmental stress. *Oryza* 1971 [publ. 1973] (8):111-119.
 1977a Breeding for resistance in rice. *Annals of the New York Academy of Sciences* 287:296-308.
 1980a Breeding rice for multiple disease and insect resistance. *In:* Hargrove, T. R., ed. *Rice Improvement in China and Other Asian Countries.* Pp. 219-238. International Rice Research Institute, Los Banos, Philippines.

Khush, G. S. and Beachell, H. M.
 1972a *Breeding for Disease and Insect Resistance at IRRI. Rice Breeding.* Pp. 309-322. International Rice Research Institute, Los Banos, Philippines.

Kiauta, B.
 1962a Contribution a la connaissance de la faune des Cicadides (Homoptera-Cicadinea) dans les environs de Skofja Loka. *Loski Razgledi, Skofja Loka* 9:59-66. [In Serbo-Croatian with French summary.]

Kidokoro, T.
 1979a Geographic trend in the annual population fluctuation of the green rice leafhopper, *Nephotettix cincticeps* Uhler (Hemiptera: Deltocephalidae). *Applied Entomology and Zoology* 14:127-129.
 1980a Geographic trend in population fluctuation of green rice leafhopper (*Nephotettix cincticeps*). *In: XVI International Congress of Entomology* (Kyoto) 16:130 (Abstract).

Kido, H.
 1980a A succession of insect pests. *California Agriculture* 34:25-27.

Kido, H., Flaherty, D. L., Bosch, D. F. and Valero, K. A.
 1983a Biological control of grape leafhopper. *California Agriculture* 37:4-6.
 1984a The variegated grape leafhopper in the San Joaquin valley. *California Agriculture* 38(1-2):31-32.

Kido, H. and Stafford, E. M.
 1965a Feeding studies on the grape leafhopper. *California Agriculture* 19(4):6-7.

Kieckhefer, R. W.
 1962a Some factors affecting populations of the potato leafhopper *Empoasca fabae* (Harris) Ph. D. Dissertation, University of Wisconsin. 99 pp. [Diss. Abst. 23(6):1884.]

Kieckhefer, R. W. and Medler, J. R.
 1960a Toxicity of cellulose acetate sheeting to luguminous plants. *Journal of Economic Entomology* 53(3):484.
 1964a Some environmental factors influencing oviposition by the potato leafhopper, *Empoasca fabae. Journal of Economic Entomology* 57(4):482-484.
 1966a Aggregations of the potato leafhopper in alfalfa fields in Wisconsin. *Annals of the Entomological Society of America* 59(1):180-182.

Kifune, T.
 1983a A new species of the genus *Halictophagus* from Thailand with a proposition of a new subgenus *Allohalictophagus* (Strepsiptera, Halictophagidae) (Notulae Strepsiterogicae-IX). *Kontyu* 51:165-168.

Kim, C. J.
 1966a Witches' broom of jujube tree (*Zizyphus jujube* Mill. var. *intermis* Rehd.) *In: Papers presented at the divisional meeting on plant protection, the Eleventh Pacific Science Congress,* Tokyo, 1966. Pp. 188-194.

Kim, H. S., Heinrichs, E. A. and Rapusas, H. R.
 1985a Mass rearing of *Nephotettix malayanus. International Rice Research Newsletter* 10(4):20.

Kim, J.
 1963a Notes and exhibitions. *Proceedings of the Hawaiian Entomological Society for 1962* 18(2):209.

Kim, J. B.
 1984a Studies on the egg parasite, *Paracentrobia andoi* Ishii (Hymenoptera: Trichogrammidae), of green rice leafhopper, *Nephotettix cincticeps* Uhler. *Korean Journal of Plant Protection* 23:237-241.

Kim, J. B. and Kim, C. H.
 1984a Studies on bionomics of *Gonatocerus* sp. y (Hymenoptera: Mymaridae), an egg parasite of the green rice leafhopper, *Nephotettix cincticeps* Uhler. *Korean Journal of Plant Protection* 23:158-165.

Kim, K. C.
 1978a Studies on the resistance of leading rice varieties to leaf- and planthoppers. *Korean Journal of Plant Protection* 17:53-63.

Kim, S. S. and Hyun, J. S.
 1979a Some considerations on the population regulation of the green rice leafhopper, *Nephotettix cincticeps* Uhler. *Korean Journal of Plant Protection* 18:15-21.

Kim, Y. S., and Lamey, H. A.
 1973a *Literature Review of Korean Rice Pests.* Institute of Agricultural Science, Office of Rural Development, Suweon. 46 pp.

Kimura, I.
 1962a Further studies on the rice dwarf virus. *Annals of the Phytopathological Society of Japan* 27:197-263. [In Japanese.]
 1976a Loss of vector-transmissibility in an isolate of rice dwarf virus. *Annals of the Phytopathological Society of Japan* 42:322-324.
 1980a Cultivation of cells of the leafhopper (*Nephotettix cincticeps* Uhler), a carrier of rice dwarf virus (RDV), and inoculation of RDV on its vector cell monolayers. *Annals of Phytopathological Society of Japan* 46:413. [In Japanese.]
 1982a Cultivation of cells of the leafhopper (*Nephotettix nigropictus* Stal), a carrier of rice dwarf virus and infectivity assays of the virus on its vector cell monolayers. *Annals of the Phytopathological Society of Japan* 48:389. [In Japanese.]
 1985a Quick detection of rice dwarf virus infection employing its vector cell monolayers. *Japan Agricultural Research Quarterly* 19:109-114.

Kimura, I. and Black, L. M.
 1971a Some factors affecting assays of wound tumor virus on cell monolayers from an insect vector. *Virology* 46:266-276.
 1972a Growth of wound tumor virus in vector cell monolayers. *Virology* 48:852-854.
 1972b The cell-infecting unit of wound tumor virus. *Virology* 49:549-561.

Kimura, I. and Fukushi, T.
 1960a Studies on the rice-dwarf virus. *Annals of the Phytopathological Society of Japan* 25:131-135.

Kimura, T.
1975a Percentage of the leaf hopper *Nephotettix cincticeps* (Uhler) viruliferous for rice dwarf in Kyushu. *Proceedings of the Association for Plant Protection of Kyushu* 21:143-146.

Kimura, T., Maejima, I. and Nishi, Y.
1975a Transmission of the rice waika virus by *Nephotettix virescens* Distant. *Annals of the Phytopathological Society of Japan* 41:115.

Kincade, R. T., Hepner, L. W. and Laster, M. L.
1970a A survey of leafhoppers in soybean fields in Mississippi. *Journal of Economic Entomology* 63(6):1991-1993.

Kindler, S. D. and Kehr, W. R.
1970a Field tests of alfalfa selected for resistance to potato leafhopper in the greenhouse. *Journal of Economic Entomology* 63(5):1463-1467.

1970b Selection for potato leafhopper resistance. *Report on the Alfalfa Improvement Conference* 22:7-8.

1974a Evaluating potato leafhopper yellowing resistance. U.S. Department of Agriculture. *Agricultural Research Service* NC-19:19.

Kindler, S. D., Kehr, W. R., Ogden, R. L. and Schalk, J. M.
1973a Effect of potato leafhopper injury on yield and quality of resistant and susceptible alfalfa clones. *Journal of Economic Entomology* 66(6):1298-1302.

Kindler, S. D., Manglitz, G. R. and Schalk, J. M.
1968a Insecticides for control of insects attacking alfalfa seed in eastern Nebraska. *Journal of Economic Entomology* 61(6):1636-1639.

Kindler, S. D., Ogden, R. L., Kehr, W. R. and Schalk, J. M.
1968a Effects of the potato leafhopper on the yield and chemical composition of alfalfa clones. *Proceedings of the North Central Branch of the Entomological Society of America* 23(1):42-43.

King, T. H.
1968a Occurrence and distribution of diseases and pests of rice and their control in Thailand. *FAO Plant Protection Bulletin* 16:41-44.

Kira, M. T., Ammar, E. D. and Abul-Ata, A. E.
1978a Survey and seasonal fluctuation of leafhoppers and planthoppers on sugarcane and other gramineous crops in upper Egypt. *African Journal of Agricultural Science* 5(2):67-73.

Kirejtshuk, A. G.
1975a Four new species of leafhoppers of the subfamily Typhlocybinae (Auchenorrhyncha: Cicadellidae) from the Ukraine and the Caucasus. *Doklady Akademiyi Nauk Ukrayins'Koyi SSR* (Seriya B) 11:1033-1037. [In Ukranian with English summary.]

1977a Note on the subfamily Typhlocybinae (Homoptera, Cicadellidae) of Kharkov region. In: Skarlato, O. A., Kryzhanovskii, O. L. and Galkin, A. K., eds. *Systematics and Faunistics of Insects.* Pp. 3-26. Akademiya Nauk SSR, Leningrad. 114 pp. [In Russian.]

1979a A new species of Cicadina of the genus *Austroasca* Lower (Homoptera, Cicadellidae, Typhlocybinae) from south Ukraine. *Trudy Vsesoyuznogo Entomologicheskogo Obshchestva* 61:24-25.

Kiritani, K.
1976a Systems approach for management of rice pests. In: Packer, J. S. and White, D., eds. *Proceedings of XV International Congress of Entomology,* Washington, D.C. Pp. 591-598. 824 pp.

1977a Recent progress in the pest management for rice in Japan. *Japan Agricultural Research Quarterly* 11:40-49.

1979a Pest management in rice. *Annual Review of Entomology* 24:279-312.

1981a Spatio-temporal aspects of epidemiology in insect borne rice virus disease. *JARQ* 15:92-99.

1983a Changes in cropping practices and the incidence of hopper-borne diseases of rice in Japan. In: Plumb, R. T. and Thresh, J. M., eds. *Plant Virus Epidemiology. The Spread and Control of Insect-borne Viruses.* Pp. 239-247. Blackwell Scientific Publications, Oxford, U.K. 337 pp.

Kiritani, K., Hokyo, N., Sasaba, T. and Nakasuji, F.
1970a Studies on population dynamics of the green rice leafhopper, *Nephotettix cincticeps* Uhler. Regulatory mechanism of the population density. *Researches on Population Ecology* 12:137-153.

Kiritani, K., Inoue, T., Nakasuji, F., Kawahara, S. and Sasaba, T.
1972a An approach to the integrated control of rice pests: Control with selective, low dosage insecticides by reduced number of applications. *Japanese Journal of Applied Entomology and Zoology* 16(2):94-106. [In Japanese with English summary.]

Kiritani, K. and Kakiya, N.
1975a An analysis of the predator-prey system in the paddy field. *Researches on Population Ecology* 17:29-38.

Kiritani, K. and Kawahara, S.
1973a Food-chain toxicity of granular formulations of insecticides to a predator, *Lycosa pseudoannulata*, of *Nephotettix cincticeps*. *Botyu-Kagaku* 38(2):69-75. [In English with Japanese summary.]

Kiritani, K., Kawahara, S., Sasaba, T. and Nakasuji, F.
1971a An attempt of rice pest control by integration of pesticides and natural enemies. *Gensei* 22:19-23.

1972a Quantitative evaluation of predation by spiders on the green rice leafhopper, *Nephotettix cincticeps* Uhler, by a sight-count method. *Researches on Population Ecology* 13:187-200.

Kiritani, K. and Kono, T.
1975a Strategy of integrated control of rice insect pests in Japan. In: *VIII International Plant Protection Congress. Integrated Plant Protection.* Pp. 302-309.

Kiritani, K. and Nakasuji, F.
1977a A systems model for the prediction of rice dwarf virus infection of the middle-season rice. *Applied Entomology and Zoology* 118-123.

Kiritani, K. and Sasaba, T.
1978a An experimental validation of the systems model for the prediction of rice dwarf virus infection. *Applied Entomology and Zoology* 13:209-214.

Kirkpatrick, B., Stenger, D. C., Morris, T. J. and Purcell, A. H.
1985a Detection of X-disease mycoplasmalike organisms in plant and insect hosts using cloned, disease-specific DNA. *Phytopathology* 75:1351 (Abstract).

Kirollos, J. Y. and Hibbs, E. T.
1971a Viability of *Empoasca fabae* (Homoptera: Cicadellidae) eggs in sterile media with added monosaccharides, amino acids, or phorate. *Annals of the Entomological Society of America* 64(1):32-36.

Kisha, J. S. A.
1978a The relative efficiency of foliar sprays and granular insecticides in control of the jassid *Empoasca lybica* on eggplant. *Annals of Applied Biology* 89:451-457.

1981a Effect of jassid *Empoasca lybica* on the growth and yield of two eggplant cultivars. *Insect Science and Its Application* 2:223-225.

Kishino, K. and Ando, Y.
1978a Insect resistance of the rice plant to the green rice leafhopper *Nephotettix cincticeps* Uhler. 1. Laboratory technique for testing the antibiosis. *Japanese Journal of Applied Entomology and Zoology* 22:169-177.

1979a Resistance of rice plant to the green rice leafhopper, *Nephotettix cincticeps* Uhler. 2. Fluctuation of antibiosis with the growing stages of the resistant rice varieties. *Japanese Journal of Applied Entomology and Zoology* 23:129-133.

Kishore, N., Kashyap, R. K. and Khankhar, B. S.
1983a Field resistance of okra lines against jassid and fruit borer. *Annals of Applied Biology* 102:130-131.

Kisimoto, R.
1959a Differences in several morphological and physiological characters between two species of the green rice leafhoppers, *Nephotettix cincticeps* Uhler and *N. apicalis* Motschulsky (Homoptera, Jassidae). *Japanese Journal of Applied Entomology and Zoology* 3(2):128-135.

1959b Studies on the diapause in the planthoppers and leafhoppers (Homoptera). II. Arrest of development in the fourth and fifth larval stage induced by short photoperiod in the green rice leafhopper, *Nephotettix bipunctatus cincticeps* Uhler [*N. apicalis* var. *cincticeps*]. *Japanese Journal of Applied Entomology and Zoology* 3(1):49-55.

1959c Studies on the diapause in the planthoppers and leafhoppers. III. Sensitivity of various larval stages to photoperiod and the forms of ensuing adults in the green rice leafhopper, *Nephotettix cincticeps* Uhler. *Japanese Journal of Applied Entomology and Zoology* 3(2):200-207.

1973a Leafhoppers and planthoppers. *In: Gibbs, A. J., ed. Viruses and Invertebrates.* Vol. 8:137-156. American Elsevier, New York. 673 pp.

1984a Insect pests of the rice plant in Asia. *Protection Ecology* 7:83-104.

Kitagawa, Y. and Shikata, E.

1969a On some properties of rice black-streaked dwarf virus. *Memoirs of the Faculty of Agriculture Hokkaido University* 6:439-445. [In Japanese with English summary.]

1973a Effect of corn extract and organic solvents on the infectivity of rice black-streaked dwarf virus. *Reports of the Tottori Mycological Institute* 10:787-795.

1974a Multiplication of rice black-streaked dwarf virus in its plant and insect hosts. *Annals of the Phytopathological Society of Japan* 40(4):329-336. [In Japanese with English summary.]

Kitajima, E. W. and Costa, A. S.

1972a Microscopia electronica de microrganismos do tipo micoplasma nos tecidos de milho afetado pelo enfezamento e nos ogaes da cigarrinha vectora portadora. *Bragantia* 31(6):75-82. [In Portuguese with English summary.]

Kitajima, E. W. and Gamez, R.

1977a Histological observations on maize leaf tissue infected with maize rayado fino virus. *Turrialba* 27:71-74.

1983a Electron microscopy of maize rayado fino virus in the internal organs of its leafhopper vector. *Intervirology* 19:129-134.

Kitajima, E. W. and Landim, C. C.

1972a An electron microscopic study of the process of differentiation during spermiogenesis in the corn leafhopper *Dalbulus maidis* Del. and W. (Homoptera: Cicadellidae). *Revista de Biologia* (Lisbon) 8:5-19.

Kitbamroong, N. A. and Freytag, P. H.

1978a The species of the genus *Scaphoideus* (Homoptera: Cicadellidae) found in Thailand, with descriptions of new species. *Pacific Insects* 18(1&2):9-31.

Kitching, R. L., Grylls, N. E. and Waterford, C.

1973a The identity of the Australian species of *Cicadulina* China (Homoptera: Cicadellidae). *Journal of the Australian Entomological Society* 12:139-143.

Kittur, S. U., Kaushik, U. K. and Pophady, D. J.

1985a The role of phototropism in controlling the major pests of rice. *In: Regupathy, A. and Juyaraj, S., eds. Behavioural and Physiological Approaches in Pest Management.* Pp. 100-103. Coimbature, Tamil Nadu, India. 218 pp.

Klashorst, G., van de and Tingey, W. M.

1979a Effect of seedling age, environmental temperature, and foliar total glycoalkaloids on resistance of five *Solanum* genotypes to the potato leafhopper. *Environmental Entomology* 8:690-693.

Klein, M.

1970a Safflower phyllody—a mycoplasma disease of *Carthamus tinctorius* in Israel. *Plant Disease Reporter* 54:735-738.

1984a Are the green leafhoppers which breed on peaches, grapevines and cotton related to each other? *In: Drosopoulos, Sakis, ed. Proceeding of the 1st International Congress Concerning the Rhynchota Fauna of Balkan and Adjacent Regions.* Pp. 19-20. Mikrolimni-Prespa, Greece, 29 August - 2 September 1983. 24 pp.

Klein, M. and Raccah, B.

1980a The effect of temperature and hosts on the population dynamics of *Neoaliturus fenestratus* (Herrich-Schaffer) (Hemiptera: Euscelidae). *Bulletin of Entomological Research* 70(3):471-473.

1984a [1983] The complexity of the *Circulifer* genus in Israel. *In: Drosopoulos, Sakis, ed. Proceeding, 1st International Congress Concerning the Rhynchota Fauna of Balkan and Adjacent Regions.* P. 19. Mikrolimni-Prespa, Greece, 29 August - 2 September 1983. 24 pp.

Klein, M., Raccah, B. and Oman, P. W.

1982a The occurrence of a member of the *Circulifer tenellus* species complex (Homoptera: Cicadellidae: Euscelini) in Israel. *Phytoparasitica* 10(4):237-240.

Klein, M., Zelcer, A., Fleischer, Z. and Loebenstein, G.

1976a Mycoplasma-like organisms associated with witches-broom disease of *Rubus sanguineus*. *Hassadeh* 56:1103-1104.

Kleinhempel, H., Karl, E., Lehmann, W., Proeseler, G. and Spaar, D.

1974a The experimental transmission of mycoplasmas by insects (short communication). *Archiv fur Phytopathologie und Pflazenschutz* 10:351-352.

Klimaszewski, S. M., Wojciechowski, W., Czylok, A., Gebicki, C., Herczek, A. and Jasisnka, J.

1980a The association of selected groups of homopterans (Homoptera) and true bugs (Heteroptera) in the forests of the region of the "Katowice" ironworks. *Acta Biologica, Uniwersytet Slaski w Katowicah* 8:22-39.

Klimaszewski, S. M., Wojciechowski, W., Gebicki, C., Czylok, A., Jasinska, J. and Glowacka, E.

1980a The associations of sucking insects (Homoptera and Heteroptera) in the grassland and herb communities in the region of the "Katowice" ironworks. *Acta Biologica, Uniwersytet Slaski w Katowicach* 8:9-21.

Knight, K. G.

1961a Spread of phony disease into Georgia peach orchards. *Phytopathology* 51(6):345-349.

Knight, R. L.

1952a The genetics of jassid resistance in cotton. I. The gene H_1 and H_2. *Journal of Genetics* 51:47-66.

Knight, W. J.

1965a Techniques for use in the identification of leafhoppers (Homoptera: Cicadellidae). *Entomologist's Gazette* 16(4):129-136.

1965b A re-description of *Dikraneura micantula* (Zett.) (Homoptera, Cicadellidae) and a closely related new species from southern Finland. *Annals and Magazine of Natural History* 8(89&90):345-350.

1966a A description of the previously unknown male of *Wolfella evansi* Kramer (Homoptera, Cicadellidae). *Annals and Magazine of Natural History* 9(103-105):437-438.

1966b A preliminary list of leaf-hopper species (Homoptera: Cicadellidae) occurring on plants of economic importance in Great Britain. *Entomologist's Monthly Magazine* 101(1211-1213):94-109.

1967a A taxonomic revision of the holarctic genus *Dikraneura* (Homop.: Cicadell.). 275 pp. Diss. Abst. 28(3):930B.

1968a A revision of the Holarctic genus *Dikraneura* (Homoptera: Cicadellidae). *Bulletin of the British Museum (Natural History) Entomology* 21(3):99-201.

1970a A revision of the genus *Hishimonus* Ishihara (Hom., Cicadellidae). *Suomen Hyonteisteelinen Aikakauskirja* 36(3):125-139.

1970b Two new genera of leafhoppers related to *Hishimonus* Ishihara (Hom., Cicadellidae). *Suomen Hyonteisteelinen Aikakauskirja* 36(4):174-182. [New genera: *Litura* and *Naevus*.]

1973a Hecalinae of New Zealand (Homoptera: Cicadellidae). *New Zealand Journal of Science* 16:957-969.

1973b Ulopinae of New Zealand (Homoptera: Cicadellidae). *New Zealand Journal of Science* 16:971-1007.

1973c A new species of *Hishimonus* Ishihara (Hom., Cicadellidae) attacking *Terminalia* spp. in India, with comments on the relationship of the genus to *Cestius* Distant. *Suomen Hyonteisteelinen Aikakauskirja* 39:153-156.

1974a Revision of the New Zealand genus *Novothymbris* (Homoptera: Cicadellidae). *New Zealand Journal of Zoology* 1(4):453-573.

1974b Leafhoppers of New Zealand: Subfamilies Aphrodinae, Jassinae, Xestocephalinae and Macropsinae (Homoptera: Cicadellidae). *New Zealand Journal of Zoology* 1(4):475-493.

1974c Evolution of the Holarctic leafhopper genus *Diplocolenus* Ribaut, with descriptions and keys to subgenera and species (Homoptera: Cicadellidae). *Bulletin of the British Museum (Natural History) Entomology* 29(7):357-413.

1975a Deltocephalinae of New Zealand (Homoptera: Cicadellidae). *New Zealand Journal of Zoology* 2(2):169-208. [New genera: *Arawa*, *Arahura*, and *Horouta*.]

1976a Typhlocybinae of New Zealand (Homoptera: Cicadellidae). *New Zealand Journal of Zoology* 3(2):71-87. [New genus: *Matatua*.]

1976b The leafhoppers of Lord Howe, Norfolk, Kermadec and Chatham Islands and their relationship to the fauna of New Zealand (Homoptera: Cicadellidae). *New Zealand Journal of Zoology* 3(2):89-98.

1981a The Cicadellidae of S.E. Asia and S.W. Pacific: An account of taxonomic and faunistic work in progress. *Acta Entomologica Fennica* 38:25-26.

1982a Cicadellidae (Homoptera, Auchenorrhyncha) from Rennell Island. *The Natural History of Rennell Island, British Solomon Islands* 8:73-78.

1983a The Cicadellidae of S.E. Asia—Present knowledge and obstacles to identification. *In: Knight, W. J., Pant, N. C., Robertson, T. S. and Wilson, M. R., eds. Proceedings of the 1st International Workshop on Biotaxonomy, Classification and Biology of Leafhoppers and Planthoppers (Auchenorrhyncha) of Economic Importance.* Pp. 197-224. London, 4-7 October 1982. Commonwealth Institute of Entomology, 56 Queen's Gate, London SW7 5JR. 500 pp.

1983b The leafhopper genus *Batracomorphus* (Cicadellidae, Iassinae) in the eastern Oriental and Australian regions. *Bulletin of the British Museum (Natural History) Entomology* 47(2):27-210.

Knight, W. J., Pant, N. C., Robertson, T. S. and Wilson, M. R., eds.

1983a *Proceedings of the 1st International Workshop on Biotaxonomy, Classification and Biology of Leafhoppers and Planthoppers (Auchenorrhyncha) of Economic Importance,* London, 4-7 October 1982. Commonwealth Institute of Entomology, 56 Queen's Gate, London SW7 5JR. 500 pp.

Knoke, J. D., Anderson, R. F. and Louie, R.

1977a Virus disease epiphytology: Developing field tests for disease resistance. *In: Williams, L. E., Gordon, D. T. and Nault, L. R., eds. Proceedings of the International Maize Virus Disease Colloquium and Workshop.* Pp. 116-121. Ohio Agricultural and Research Development Center, Wooster. 145 pp.

Knoke, J. D., Louie, R., Madden, L. V. and Gordon, D. T.

1983a Spread of maize dwarf mosaic virus from johnsongrass to corn. *Plant Disease* 67:367-370.

Knoke, J. K.

1962a Systemic insecticides for the control of insects attacking potatoes. Ph. D. Dissertation, University of Wisconsin. 196 pp. [Diss. Abst. 23(6):1844.]

Knowlton, G. F.

1955a Some Utah insects of 1954. Pt. 4, Supplement. *Utah State Agricultural College Extension Service, Mimeo Series* 138:1-5.

1955b Some Hemiptera and Homoptera of Utah—1955. *Utah State Agricultural College Extension Service, Mimeo Series* 145:1-9.

1955c Uinta Mountain leafhoppers. *Bulletin of the Brooklyn Entomological Society,* n.s. 50(3):69.

1958a Insects—a big problem in Utah, 1958. *Utah Agricultural Extension Service Mimeograph* 170. 2 pp.

1962a Observations on Utah insects—1962. *Utah Agricultural Extension Service Mimeograph* 200. 3 pp.

Knull, J. N. and Knull, D. J.

1960a Observation on behavior of some male Gyponini (Homoptera: Cicadellidae). *Entomologist's Newsletter* 71(3):78.

Knutson, L.

1976a Preparation of specimens submitted for identification of the Systematic Entomology Laboratory, USDA. *Bulletin of the Entomological Society of America* 22(2):130.

Knutson, L. and Murphy, W. L.

1984a Directory of systematic entomologists and acarologists in the People's Republic of China. *Journal of the New York Entomological Society* 92(1):2-18.

Kobayashi, A., Supaad, M. A. and Othman, B. O.

1983a Inheritance of resistance of rice to tungro and biotype selection of green leafhopper in Malaysia. *JARQ* 16:306-311.

Kobayashi, T.

1961a Effect of insecticidal applications to the rice stem borer on leafhopper populations. *Special Report on Predication of Pests, Ministry of Agriculture and Forestry* 13:1-126.

Kobayashi, T., Noguchi, Y., Hiwada, T., Kanayama, K. and Maruoka, N.

1973a Studies on the arthropod associations in paddy fields, with particular reference to insecticidal effect on them. I. General composition of the arthropod fauna in paddy fields revealed by sweep-netting in Tokushima Prefecture. *Kontyu* 41:359-373.

1974a Studies on the arthropod associations in paddy fields, with particular reference to insecticidal effect on them. II. Seasonal fluctuations of species diversity and abundance in arthropod associations in paddy fields by net-sweeping in Tokushima Prefecture. *Kontyu* 42:87-106.

Koblet-Gunthardt, M.

1975a Die Kleinzikaden *Empoasca decipiens* Poli und *Eupteryx atropunctata* Goetze (Homoptera, Auchenorrhyncha) auf Ackerbohnen (*Vicia faba* L.). Anatomische und physiologische Untersuchungen. Inaugural Dissertation, Universitat Zurich. 125 pp.

Kocak, A. O.

1981a List of genera of Turkish Auchenorrhyncha (Homoptera), with some replacement names for genera existing in other countries. *Priamus* 1(1):30-40.

1981b Nomenclatural note on Homoptera. *Priamus* 1(1):41.

1981c Two replacement names for the genera of Homoptera. *Priamus* 1(3):124-125.

1983a List of the genera of Turkish Auchenorrhyncha (Homoptera), with some replacement names for the genera existing in other countries. *Priamus* 3(2):43-85.

Kochman, J. and Ksiazek, D.

1967a Investigations on the transmission of viruses: Aster yellows—*Callistephus* virus 1A. Smith-onion yellow dwarf—*Allium* virus 1. Melhus-leafhoppers—*Macrosteles laevis* Rib. In: Virus Diseases of Plants and Terminology of Virus Diseases of Plants. Polskiej Akademia Nauk Zescyty. *Postepy Nauk Rolniczych* 70:165-167.

Koesnang, S. and Rao, P. S.

1980a Life span of and tungro transmission by viruliferous *Nephotettix virescens* on 10 rice varieties at LPPM, Indonesia. *International Rice Research Newsletter* 5(5):9-10.

Koga, K., Kiso, A. and Nomura, Y.

1982a Occurrence of onion-biwadama like symptoms and lettuce yellows caused by a leafhopper, *Macrosteles orientalis*, in developed areas of lettuce yellows, mycoplasmalike organism. *Proceedings of the Association for Plant Protection of Kyushu* 28:48-49.

Kohno, M. and Hashimoto, S.
1978a Studies on the smaller green leafhopper, *Empoasca* sp. in the citrus orchard. II. Application of yellow cylindrical sticky traps to record the seasonal prevalence. *Proceedings of the Association for Plant Protection of Kyushu* 24:154-155.

Kohno, M. and Nagahama, M.
1970a Studies on injuries caused by the smaller green leafhopper *Chlorita flavescens* Fabricius, on Ponkan orange fruits. *Proceedings of the Association for Plant Protection of Kyushu* 16:75-77.

Koinzan, S. D. and Pruess, K. P.
1975a Effects of a wide-area application of ULV malathion on leafhoppers on alfalfa. *Journal of Economic Entomology* 68:267-268.

Koizumi, K.
1959a On four dorilaid parasites of the green rice leafhopper, *Nephotettix cincticeps* Uhler (Diptera). *Science Reports of the Faculty of Agriculture, Okayama University* 13:37-45.
1960a A new dorilaid parasite of the zigzag-striped leafhopper *Inazuma dorsalis* (Motschulsky) and notes on other paddy-field inhabiting Dorilaidae (Diptera). *Science Reports of the Faculty of Agriculture, Okayama University* 16:33-42.

Kojima, K. and Ishizuka, T.
1960a On the potentiation of the effectiveness of malathion by DDVP against the adults of the green rice leafhopper, *Nephotettix bipunctatus cincticeps* Uhler. *Botyu-Kagaku* 25:16-22.

Kojima, K., Ishizuka, T. and Kitakata, S.
1963a Mechanism of resistance to malathion in the green rice leafhopper, *Nephotettix cincticeps*. *Botyu-Kagaku* 28:17-25. [In Japanese with English summary.]

Kojima, K. S., Kitakata, A., Shiino and Yoshida, T.
1963a On the development and decline of resistance to malathion of the green rice leafhopper, *Nephotettix cincticeps*. *Botyu-Kagaku* 28:13-17.

Kono, Y., Kawabe, S., Sakai, M., Sato, Y. and Suzuki, T.
1982a Effect of cartap on sucking activity of the green rice leafhopper (Homoptera: Deltocephalidae). *Japanese Journal of Applied Entomology and Zoology* 26:41-47.

Kono, Y., Nagaarashi, D. and Sakai, M.
1975a Effects of cartap, chlordimefrom and diazinon on the probing frequency of the green rice leafhopper (Hemiptera: Deltocephalidae). *Applied Entomology and Zoology* 10:58-60.

Kono, Y., Sakai, M. and Moriya, S.
1976a Effect of cartap on the transmission of the rice dwarf disease virus by the vector *Nephotettix cincticeps*. *Japanese Journal of Applied Entomology and Zoology* 20:191-197.

Kontkanen, P.
1960a Suomelle uusi kaskaslaji, *Dikraneura variata* Hardy (Hom., Auchenorrhyncha). *Suomen Hyonteistieteellinen Aikakauskirja* 26:56-58.

Kooner, B. S. and Deol, G. S.
1982a Comparative rate of development of *Orosius albicinctus* (Cicadellidae: Homoptera) on different host plants. *Journal of Agricultural Science Camb.* 98(3):613-614.

Kooner, B. S., Sidhu, H. S. and Bindra, O. S.
1978a Effect of sesamum phyllody disease on the longevity and fecundity-cum-fertility of its leafhopper vector, *Orosius albicinctus* Distant (Cicadellidae: Homoptera). *Journal of Agricultural Science Camb.* 91(2):509-510.
1980a Rate of development of *Orosius albicinctus* Distant at different constant temperatures. *Indian Journal of Entomology* 42:294-296.

Koponen, S.
1978a Notes on herbivorous insects of the birch in southern Greenland. *Report from the Kevo Subarctic Research Station* 14:13-17.

Koppany, T.
1969a Fragen der notigen Grosse der zonologische Aufnahme fur Untersuchung der Heteroptera und Cicadinea Population. *Folia Entomologica Hungaricae* 22(2):279-309. [In Hungarian with German summary.]

Koppany, T. and Wolcsanszky, E. S.
1956a Diozonologische Untersuchungen der Weiden-und Wiesentypen des Hortobagy. *Acta Zoologica Academiae Scientiarum Hungaricae* 2:357-378.

Korner, H. K.
1969a Die embryonale Entwicklung der symbiontenfuhrenden Organe von *Euscelis plebejus* Fall. (Homoptera: Cicadina). *Oecologia Berlin* 2:319-346.
1969b Entwicklung der Symbiontenorgane einer Kleinzikade (*Euscelis plebejus* Fall.) nach Ausschaltung der Symbionten. *Experientia* 25:767-768. [English summary.]
1972a Elektronenmikroskopische Untersuchungen am embryonalen Mycetom der Kleinzikade *Euscelis plebejus* Fall. (Homoptera, Cicadina). 1. Die Feinstruktur der a-Symbionten. *Zeitschrift fur Parasitenkunde* 40(3):203-226. [In German with English summary.]
1974a Elektronenmikroscopische Untersuchungen am embryonalen Mycetom der Kleinzikade *Euscelis plebejus* Fall. (Homoptera, Cicadina). II. Die Feinstruktur der t-Symbionten. *Zeitschrift fur Parasitenkunde* 44(2):149-164. [In German with English summary.]
1976a On the host-symbiont-cycle of a leafhopper (*Euscelis plebejus*) endosymbiosis. *Experientia* 32(4):463-464.
1978a Intraovarially transmitted symbionts of leafhoppers. *Zoologische Beitraege* 24(1):59-68. [German summary.]

Korolevskaya, L. L.
1963a New species from the genus *Macropsis* (Auchenorrhyncha, (Cicadellidae) from Tadzhihistan. *Doklady Akademii Nauk Tadzhikskoi SSR* 6(7):40-43.
1964a Two new cicadellid species of genus *Rhytidodus* Fieb. (Auchenorrhyncha, Cicadellidae) from Tadzhikistan. *Doklady Akademii Nauk Tadzhikskoi SSR* 11:39-41. [In Russian.]
1968a A new species of *Psammotettix* Hpt. (Auchenorrhyncha, Cicadellidae). *Doklady Akademii Nauk Tadzhikskoi SSR* 11(12):57-58. [In Russian.]
1968b Data on the study of dendrophilous Cicadina fauna on the southern slopes of the Gussar Mountains. *In: Narzikulov, M. N., ed. The Gorge of Kondara. Akademii Nauk Tadzhikskoi SSR* 2:100-117. [In Russian.]
1971a Toward a biology of 2 species of leafhoppers injurious to vegetable crops in Tadzhikistan. *News of the Academy of Science of Tadzhikistan SSR, Division of Biological Sciences* 1(42):1-3.
1971b On the biology of two cicada species, pests of vegetable cultivations in Tadzhikistan. *Izvestiya Akademii Nauk Tadshikskoi SSR* (1):88-90. [In Russian with Tadzhik summary.]
1973a Leafhoppers of vegetables and curcubits in central Tadzhikistan. *News of the Academy of Science of Tadzhikistan SSR, Division of Biological Sciences* 2(51):44-51.
1974a On the fauna of leafhoppers of the subfamily Typhlocybinae, Auchenorrhyncha of central Tadzhikistan. *News of the Academy of Science of Tadzhikistan SSR, Division of Biological Sciences* 1(54):40-44.
1975a New species of Cicadas (Auchenorrhyncha, Cicadellidae) in the fauna of Tadzhikistan and adjacent republics. *Izvestiya Akademii Nauk Tadzhikskoi SSR, Otdelenie Biologicheskikh Nauk* 3:17-22. [In Russian.]
1976a Leafhoppers of the genus *Mityaevia* (Homoptera, Typhlocybinae) from Tadzhik SSR, USSR. *Izvestiya Akademii Nauk Tadzhikskoi SSR, Otdelenie Biologicheskikh Nauk* 3(64):41-44. [In Russian.]

1976b Cikadki roda *Asianidia* Zachvatkin, 1946 iz Tadzhikistana. *Izvestiya Akademii Nauk Tadzhikskoi SSR, Otdelenie Biologicheskikh Nauk* 2(63):33-36. [In Russian.]

1977a New cicadids (Auchenorrhyncha, Typhlocybinae) for the fauna of central Asia. *Izvestiya Akademii Nauk Tadzhikskoi SSR, Otdelenie Biologicheskikh Nauk* 3:39-42. [In Russian with Tadzhik summary.]

1978a Novyj i maloizvestnye vidy cikadok podsemejstav Typhlocybinae iz Srednej Azii. *Izvestiya Akademii Nauk Tadzhikskoi SSR, Otdelenie Biologicheskikh Nauk* 4(73):48-52.

1978b [New species of cicadids (Auchenorrhyncha, Deltocephalinae) of Central Asia.] *Izvestiya Akademii Nauk Tadzhikskoi SSR, Otdelenie Biologichekikh Nauk* 3(72):45-51. [In Russian.]

1979a A new species of the Cicadina genus *Arboridia* Zachvatkin, 1946 (Homoptera, Cicadellidae, Typhlocybinae) from Tadzhikstan. *Trudy Vsesoyuznogo Entomologicheskogo Obshchestva* 61:26-28.

1979b New cicadids (Auchenorrhyncha, Deltocephalinae) from the fauna of Tadzhikistan. *Izvestiya Akademii Nauk Tadzhikskoi SSR, Otdelenie Biologicheskikh Nauk* 2(75):46-53. [In Russian.]

1980a New species of leafhoppers (Auchenorrhyncha, Cicadellidae) from Tajikistan. *Entomologicheskoe Oborzrenie* 59(4):800-810. [In Russian with English summary.] [New subgenera: *Anaemotettix* and *Asthenotettix*.]

1981a New species of the genus *Phlepsidius* Emeljanov 1961 (Auchenorrhyncha, Cicadellidae) from Kirgizia. *Izvestiya Akademii Nauk Tadzhikskoi SSR, Otdelenie Biologicheskikh Nauk* 2(83):85-86.

1984a A new species of *Oncopsis* (Auchenorrhyncha, Cicadellidae) from Tadzhkistan. *Zoologicheskii Zhurnal* 63(4):622-623. [In Russian.]

Korytkowski G. and Torres B., M.

1966a Insectos que atacan al cultivo del frijol de palo (*Cajanus cajan*) en el Peru. *Revista Peruana de Entomologia Agricole* 9(1):3-9.

Koshihara, T.

1971a Resistance of rice varieties against the green rice leafhopper, *Nephotettix cincticeps* Uhler. *Tropical Agriculture Research Series* 5:221-225.

1972a Characteristics in occurrences of the green rice leafhopper, *Nephotettix cincticeps* Uhler, in the Tohoku district. *Annual Report of the Society of Plant Protection* (North Japan) 23:17-77. [In Japanese with English summary.]

Kouskolekas, C. A.

1964a Biological studies on the potato leafhopper, *Empoasca fabae* (Harris), as an alfalfa pest. University of Illinois Dissertation Abstract 25:41. 123 pp.

Kouskolekas, C. A. and Decker, G. C.

1966a The effect of temperature on the rate of development of the potato leafhopper, *Empoasca fabae* (Homoptera: Cicadellidae). *Annals of the Entomological Society of America* 59(2):292-298.

1968a A quantitative evaluation of factors affecting alfalfa yield reduction caused by the potato leafhopper attack. *Journal of Economic Entomology* 61(4):921-927.

Koyama, K.

1971a Choice experiments of *Inazuma dorsalis* (Hemiptera: Deltocephalida) on some sugars. *Japanese Journal of Applied Entomology and Zoology* 15:269-271.

1972a Experiments on the preservation of eggs of *Inazuma dorsalis* (Hemiptera: Deltocephalidae) at low temperatures. *Japanese Journal of Applied Entomology and Zoology* 16(1):50-51. [In Japanese.]

1973a Preference for colour of diets in *Inazuma dorsalis* (Homoptera: Deltocephalidae). *Japanese Journal of Applied Entomology and Zoology* 17(2):49-53.

1973b Rearing of *Inazuma* and *Nephotettix cincticeps* on a synthetic diet. *Japanese Journal of Applied Entomology and Zoology* 17(3):163-166.

Koyama, T.

1971a Lethal mechanisms of granulated insecticides. *PANS* 17(2):198-201.

Kozhevnikova, A. G.

1975a A contribution to the biology of a leafhopper, *Kyboasca bipunctata*, a cotton pest. *Zoologicheskii Zhurnal* 54(6):1257-1259.

Kramer, J. P.

1958a Six new species of *Chinaia* from Central America (Homoptera: Cicadellidae). *Proceedings of the Biological Society of Washington* 71:69-74.

1959a An elucidation of the Neotropical genus *Chinaia* with key to males and a new allied genus (Homoptera: Cicadellidae: Neocoelidiinae). *Proceedings of the Biological Society of Washington* 72:23-32. [New genus: *Xenocoelidia*.]

1960a A remarkable new species of Neotropical *Agalliopsis* and the previously unknown male of *Agalliopsis inscripta* Oman. *Proceedings of the Biological Society of Washington* 73:63-65.

1961a New Venezuelan leafhoppers of the subfamilies Xestocephalinae and Neocoelidiinae. *Proceedings of the Biological Society of Washington* 74:235-239. [New genus: *Deltocoelidia*.]

1961b The phylogeny of *Deltocephalus* and allied genera (Homoptera: Cicadellidae: Deltocephalinae). *Dissertation Abstracts* 23(5):1755.

1962a A synopsis of *Biza* and a new allied genus (Neocoelidiinae). *Proceedings of the Biological Society of Washington* 75:101-106. [New genus: *Tichocoelidia*.]

1962b New Liberian leafhoppers of the genus *Recilia* (Homoptera: Cicadellidae: Detocephalinae). *Proceedings of the Biological Society of Washington* 75:259-268.

1963a A key to the New World genera of Iassinae with reviews of *Scaroidana* and *Pachyopsis* (Homoptera: Cicadellidae). *Bulletin of the Brooklyn Entomological Society* 58:37-50. [New genus: *Grunchia*.]

1963b New Neotropical Neobalinae with key to the genera and to the species of *Conala*. *Proceedings of the Entomological Society of Washington* 65:201-210. [New genera: *Rhobala* and *Psibala*.]

1963c New and little known Mexican and Neotropical Deltocephalinae (Homoptera: Cicadellidae). *Proceedings of the Biological Society of Washington* 76:37-46. [New genus: *Bolotheta*.]

1964a New World leafhoppers of the subfamily Agalliinae: A key to genera with records and descriptions of species (Homoptera: Cicadellidae). *Transactions of the American Entomological Society* 89:141-163.

1964b A review of the Neotropical Nirvaninae (Homoptera: Cicadellidae). *Entomological News* 75:113-128. [New genera: *Krocodona, Krocozzota, Krocobella, Pentoffia*, and *Jassosqualus*.]

1964c A review of the Oriental leafhopper genus *Sudra* Distant (Homoptera: Cicadellidae: Hylicinae). *Proceedings of the Biological Society of Washington* 77:47-52.

1964d A generic revision of the leafhopper subfamily Neocoelidiinae (Homoptera: Cicadellidae). *Proceedings of the United States National Museum* 115(3484):259-288. [New genera: *Tozzita* and *Xiqilliba*.]

1964e A key for *Portanus* with new records and descriptions of new species (Homoptera: Cicadellidae: Xestocephalinae). *Proceedings of the Entomological Society of Washington* 66:5-11.

1964f The leafhopper vectors of corn stunt and some related species in southern United States (Homoptera: Cicadellidae). *Cooperative Economic Insect Report* 14(8):105-106.

1965a Studies of Neotropical leafhoppers. I. (Homoptera: Cicadellidae). *Proceedings of the Entomological Society of Washington* 67(2):65-74. [New genera: *Goblinaja, Tungurahuala, Perugrampta*.]

1965b New species of Deltocephalinae from the Americas (Homoptera: Cicadellidae). *Proceedings of the Biological Society of Washington* 78:17-31. [New genus: *Quaziptus.*]

1965c A review of the African leafhopper genus *Wolfella* Spinola (Homoptera: Cicadellidae). *Transactions of the American Entomological Society* 91(2):167-180.

1966a A revision of the New World leafhoppers of the subfamily Ledrinae (Homoptera: Cicadellidae). *Transactions of the American Entomological Society* 92:469-502. [New genera: *Hespenedra* and *Xedreota.*]

1967a A taxonomic study of *Graminella nigrifrons*, a vector of corn stunt disease, and its congeners in the United States (Homoptera: Cicadellidae: Deltocephalinae). *Annals of the Entomological Society of America* 60:604-616.

1967b New Neotropical Neocoelidiinae with keys to the species of *Coelidiana, Xenocoelidia,* and *Nelidina* (Homoptera, Cicadellidae). *Proceedings of the Entomological Society of Washington* 69(1):31-46. [New genera: *Chinchinota* and *Coelindroma.*]

1967c A key to the species of *Lystridea* Baker with description of a new species from California (Homoptera: Cicadellidae: Errhomenellini). *Proceedings of the Entomological Society of Washington* 69(3):292-294.

1967d A taxonomic study of the brachypterous North American leafhoppers of the genus *Lonatura* (Homoptera: Cicadellidae: Deltocephalinae). *Transactions of the American Entomological Society* 93(4):433-462.

1971a North American deltocephaline leafhoppers of the genus *Amblysellus* Sleesman. *Proceedings of the Entomological Society of Washington* 73(1):83-98.

1971b North American deltocephaline leafhoppers of the genus *Amplicephalus* DeLong with a new genus and new generic combinations. Proceedings of the Entomological Society of Washington 73(2):198-210. [New genus: *Kansendria.*]

1971c A taxonomic study of the North American leafhoppers of the genus *Deltocephalus* (Homoptera: Cicadellidae: Deltocephalinae). *Transactions of the American Entomological Society* 97:413-439.

1971d North American deltocephaline leafhoppers of the genus *Planicephalus* with new generic segregates from *Deltocephalus. Proceedings of the Entomological Society of Washington* 73(3):255-268. [New genera: *Tideltellus* and *Deltazotus.*]

1976a A revision of the new Neotropical leafhopper subfamily Phereurhininae (Homoptera: Cicadellidae). *Proceedings of the Entomological Society of Washington* 78(2):117-131. [New genus: *Dayoungia.*]

1976b North American deltocephaline leafhoppers of the genus *Destria* and a new species of *Lonatura* from Arizona (Homoptera: Cicadellidae). *Proceedings of the Entomological Society of Washington* 78(1):51-57.

1976c Studies of Neotropical leafhoppers. II. (Homoptera: Cicadellidae). *Proceedings of the Entomological Society of Washington* 78(1):38-50.

Kramer, J. P. and DeLong, D. M.
1968a Studies of the Mexican Deltocephalinae: *Aligia* and some new allied genera and species (Homoptera: Cicadellidae). *Ohio Journal of Science* 68(3):169-175. [New genera: *Pseudaligia* and *Ilagia.*]
1969a Studies of the Mexican Deltocephalinae. II. Seven new species of *Norvellina* (Homoptera: Cicadellidae). *Ohio Journal of Science* 69(2):115.

Kramer, J. P. and Linnavuori, R.
1959a A new genus and two new species of leafhoppers from South America (Homoptera: Cicadellidae: Neocoelidiinae). *Proceedings of the Biological Society of Washington* 72:55-58. [New genus: *Megacoelidia.*]

Kramer, J. P. and Whitcomb, R. F.
1968a A new species of *Baldulus* from gamagrass in Eastern United States with its possible implications in the corn stunt virus problem. *Proceedings of the Entomological Society of Washington* 70(1):88-92.

Krczal, H.
1960a Eine vom Weissklee auf *Fragaria vesca* (L.) ubertragbare Virose. [A virus transmissible from white clover to F. vesca.] *Zeitschrift fuer Pflanzenkrankheiten* [English summary.]

Krishen, K., Jai, Kishan, Rai, B. K. and Rattan, L.
1966a Duration of the effectiveness of insecticides on cotton in the field for controlling *Empoasca devastans* Distant. *Indian Journal of Agricultural Science* 36(5):273-277.

Krishnaiah, K.
1975a Resistance to green leaf hoppers in rice. *Entomologist's Newsletter* 5:30-31.

Krishnaiah, K. and Bari, M. A.
1977a Note on the influence of nitrogen levels on the susceptibility of paddy variety Taichung (Native) 1 to green leaf hopper, *Nephotettix nigropictus* Stal. *Indian Journal of Entomology* 37:311-313.

Krishnaiah, K. and Ramachander, P. R.
1979a Sampling technique for estimation of jassid population on okra. *Indian Journal of Entomology* 41:200-202.

Krishnaiah, K., Tandon, P. L., Mathur, A. C. and Mohan, N. J.
1976a Evaluation of insecticides for the control of major insects pests of okra. *Indian Journal of Agricultural Sciences* 46:178-186.

Krishnananda, N. and Agarwal, R. A.
1979a Sources of resistance to jassid, *Amrasca devastans* Distant in cotton. *Indian Journal of Entomology* 41:177-180.
1980a Preference to different leaves by *Amrasca devastans* (Distant) in cotton. *Coton et Fibres Tropicales* 34:333-335.

Krishnasamy, N., Chauhan, O. P. and Das, R. K.
1984a Some common predators of rice insect pests in Assam, India. *International Rice Research Newsletter* 9(2):15-16.

Kristensen, N. P.
1965a Cikader (Homoptera, Auchenorrhyncha) fra Hansted Reservatet. *Entomologiske Meddelelser* 30(3):269-287.
1965b Cikaden *Eupteroidea stelulata* (Burmeister 1841) i Danmark (Hemiptera, Cicadellidae). *Flora og Fauna* 71:81-82.

Ku, T. Y. and Wang, S. C.
1978a Further studies on the resistance of green rice leafhoppers to insecticides (II). *Plant Protection Bulletin* (Taiwan) 20:21-32.
1978b The efficacy of pesticides for the control of major insect pests on rice. *In:* Su, C. C., Yen, D. F. and Lin, F. C., eds. *Insect Ecology and Control.* Pp. 143-162. Academic Sinica, Taiwan. 279 pp.
1981a Insecticidal resistance of major insect rice pests, and the effect of insecticides on natural enemies on non-target animals. *NTU Phytopathologist and Entomologist* No. 8, pp. 1-18.

Kuc, J.
1966a Resistance of plant to infectious agents. *Annual Review of Microbiology* 20:337-370.

Kuhn, C. W., Jellum, M. D. and All, J. N.
1975a Effect of carbofuran treatment on corn yield, maize chlorotic dwarf and maize mosaic virus diseases, and leafhopper populations. *Phytopathology* 65:1017-1020.

Kukalova-Peck, J.
1978a Origin and evolution of insect wings and their relation to metamorphosis, as documented by the fossil record. *Journal of Morphology* 156:53-126.

Kuklarni, S., Hegde, R. K. and Basavarajaiah, A. B.
1980a Occurrence of wheat streak in Karnataka. *Current Research* 9(8):135.

Kuklarni, Y. S. and Saoji, N. T.
1961a An unrecorded jassid pest of groundnut. *Poona Agricultural College* 51(3&4):39.

Kullenberg, B. and Wallin, L.
1963a Bioakustic, vetenskapen om djurens laten. *Svensk Naturvetenskap* 16:330-342.

Kumar, A. R. V.
1983a A revision of Indian *Batracomorphus* Lewis (Homoptera: Cicadellidae: Iassinae). Thesis Abstract, Harayana Agricultural University 9(4):331.

Kumar, D., Roy, C. S., Khan, Z. R., Yazdani, S. S., Hameed, S. F. and Mahmood, M.
1983a An entomogenous fungi *Isaria tax* parasitizing mango hopper *Idioscopus clypealis*. *Science and Culture* 49:253-254.

Kumar, H. and Sazena, K. N.
1978a Mating behaviour of the leafhopper, *Empoasca devastans*, in relation to its age, ovarian development, diurnal cycle and CO_2 treatment. *Annals of the Entomological Society of America* 71:108-110.
1985a Certain factors influencing the acoustic communication in the sexual behavior of the leafhopper *Amrasca devastans* (Distant) (Homoptera: Cicadellidae). *Applied Entomology and Zoology* 20:199-209.

Kunkel, L. O.
1954a Maintenance of yellow-type viruses in plant and insect reservoirs. In: Hartman, F. W., Horsfall, F. A., Jr., and Kidd, J. D., eds. *The Dynamics of Virus and Rickettsial Infections* (international symposium). Pp. 150-163. 461 pp.
1957a Acquired immunity from infection by strains of aster yellows virus in the aster leafhopper. *Science* 126:1233.

Kuno, E.
1968a Studies on the population dynamics of rice leafhopper in a paddy field. *Bulletin of the Kyushu Agricultural Experiment Station* 14:131-246.
1973a Population ecology of rice leafhoppers in Japan. *Review of Plant Protection Research* 6:1-16.
1980a Population dynamics of rice leafhoppers: A comparative study. In: *XVI International Congress of Entomology* 16:118. Kyoto. 265 pp.
1984a Pest status, dynamics and control of rice planthopper and leafhopper populations in Japan. *Protection Ecology* 7:129-145.

Kuno, E. and Hokyo, N.
1970a Comparative analysis of the population dynamics of rice leafhoppers, *Nephotettix cincticeps* Uhler and *Nilaparvata lugens* Stal, with special reference to natural regulation of their numbers. *Researches on Population Ecology* 12:154-184.
1976a Population regulation and dispersal of adults in the green rice leafhopper, *Nephotettix cincticeps* Uhler (Hemiptera: Deltocephalidae). *Physiology and Ecology Japan* 17:117-123. [In Japanese with English summary.]

Kunze, L.
1959a Die funktionsanatomische Grundlagen der Kopulation der Zwergzikaden, untersucht an *Euscelis plebejus* (Fall.) und einigen Typhlocybinen (Homoptera, Auchenorrhyncha). *Deutsche Entomologische Zeitschrift* (N. F.) 6:322-387.

Kuoh, C. L.
1964a Studies on the species and distribution of rice leafhoppers in China. *Journal of Anhui Agricultural College* 8:98-110.
1966a Economic insect fauna of China. Fasc. 10, Cicadellidae. P. 170. Science Press, Beijing, China.
1973a Two new species of *Pseudonirvana* (Hom.: Cicadellidae) *Acta Entomologica Sinica* 16(2):180-187. [In Chinese with English summary.]
1976a Some new species of Chinese *Hishimonus* and *Hishimonoides* (Homoptera: Cicadellidae). *Acta Entomologica Sinica* 19(4):431-437. [In Chinese with English summary.]
1981a Six new species of Cicadellidae from Qinghai, China. *Entomotaxonomia* 3(2):111-117. [In Chinese with English.]
1981b Homoptera: Cicadelloidea. *Insects of Xizang* 1:195-219. [New genera: *Duanjina* and *Yisiona*.]
1981c Distinctions between four new species of *Nephotettix* from China. *Kunchong Zhishi* 18(5):221-223.
1982a Two new genera and five new species of the *Thaia* group (Homoptera: Cicadelloidae). *Acta Zootaxonomica Sinica* 7(4):396-404. [New genera: *Parathaia* and *Pseudothaia*.]
1983a The leafhoppers and planthoppers of rice in China. *In:* Knight, W. J., Pant, N. C., Robertson, T. S. and Wilson, M. R., eds. *Proceedings of the 1st International Workshop on Biotaxonomy, Classification and Biology of Leafhoppers and Planthoppers (Auchenorrhyncha) of Economic Importance.* Pp. 277-278. London, 4-7 October 1982. Commonwealth Institute of Entomology, 56 Queen's Gate, London SW7 5JR. 500 pp.
1984a Six new species of the genus *Petalocephala* (Homoptera: Ledridae). *Entomotaxonomia* 6(4):271-278. [In Chinese and English.]
1985a New species of *Drabescus* and a new allied genus (Homoptera: Iassidae). *Acta Zoologica Sinica* 31(4):377-383. [In Chinese with Chinese and English summaries.] [New genus: *Paradrabescus*.]
1985b Organisms of Mt. Tuomuer area in Tianshan (Homoptera: Cicadelloidea). Pp. 82-89. Xinjiang People Press, Urumqi.

Kuoh, C. L. and Fang, Q. Q.
1984a The Idioceridae leafhoppers in China (Homoptera: Cicadellidae: Idioceridae). In: *Proceedings of the XVII International Congress of Entomology*, p. 7. Hamburg. 968 pp. [In English.]
1985a A new genus and new species of the Idioceridae (Homoptera: Cicadelloidea). *Acta Entomologica Sinica* 28(1):91-93. [New genus: *Flexocerus*.]
1985b Two new species of *Idioscopus* from China (Homoptera: Cicadelloidea: Idioceridae). *Acta Zootaxonomica Sinica* 10(2):189-192.

Kuoh, C. L. and Kuoh, J. L.
1983a A new species of the genus *Mukaria* (Homoptera: Cicadelloidea: Nirvanidae). *Acta Entomologica Sinica* 26(1):78-79. [In Chinese with English summary.]
1983b New species of *Pseudonirvana* (Homoptera: Nirvanidae). *Acta Entomologica Sinica* 26(3):316-325.

Kurata, S. and Sogawa, K.
1976a Sucking inhibitory action of aromatic amines for the rice plant- and leafhoppers (Homoptera: Delphacidae, Deltocephalidae). *Applied Entomology and Zoology* 11:89-93.

Kurosu, Y.
1972a Insecticidal activity of carbamate insecticides against leafhoppers and planthoppers. *Japan Pesticide Information* No. 10, pp. 60-65.

Kuruvilla, S., Jacob, A. and Mathal, S.
1980a A new host for the entomogenous fungus *Penicillium oxalicum* Currie and Thom. *Entomon* 5:357-358.

Kuwahara, M.
1974a Studies on the diel activity of the rhombic-marked leafhopper, *Hishimonus sellatus* Uhler. *Japanese Journal of Applied Entomology and Zoology* 18:89-93.

Kwon, Y. J.
1980a *Changwhania* gen. n., new Palearctic genus of leafhoppers from the subtribe Deltocephalina (Homoptera: Cicadellidae). *Commemoration Papers for Professor C.-W. Kim's 60th Birthday Anniversary*, pp. 95-102.
1980b A new genus of leafhoppers belonging to the subtribe Platymetopiina (Homoptera: Cicadellidae). *Korean Journal of Entomology* 10(2):3-7. [New genus: *Tongdotettix*.]
1981a Description of a new species of the genus *Xestocephalus* Van Duzee (Homoptera: Cicadellidae). *Korean Journal of Entomology* 9(1):1-4.
1981b Contributions to the knowledge of the genus *Pagaronia* Ball from Korea (Homoptera: Cicadellidae). *Korean Journal of Entomology* 11(2):1-5.
1983a Classification of leafhoppers of the subfamily Cicadellinae from Korea (Homoptera: Auchenorrhyncha). *Korean Journal of Entomology* 13(1):15-25. [New genera: *Bannalgaechungia* and *Malmaemichungia*.]

1985a Classification of the leafhopper pests of the subfamily Idiocerinae from Korea. *Korean Journal of Entomology* 15(1):61-74. [New subgenera: *Nabicerus, Bicenarus, Podulmorinus, Pugnostilus, Koreocerus.*]

Kwon, Y. J. and Lee, C. E.

1978a Revision of the genus *Pagaronia* of the Palaearctic region with descriptions of one new subgenus and three new species (Homoptera: Cicadellidae). *Korean Journal of Entomology* 8(1):7-15. [New subgenus: *Parapagaronia.*]

1978b A review of Ledrinae of Korea (Homoptera: Cicadellidae). *Nature and Life in Southeast Asia (Kyungpook Journal of Biological Sciences)* 8(2):65-72.

1978c Penthimiinae of Korea with descriptions of two new species (Homoptera: Cicadellidae). *Nature and Life in Southeast Asia (Kyungpook Journal of Biological Sciences)* 8(2):73-77.

1978d A new species of *Scaphoideus* Uhler from Is. Hongdo, Korea (Homoptera: Auchenorrhyncha). *Korean Journal of Entomology* 8(2):21-24.

1979a Revision of the tribe Hecalini Distant from Korea. *Nature and Life in Southeast Asia (Kyungpook Journal of Biological Sciences)* 9(1):41-48.

1979b On some new and little known Palearctic species of leafhoppers (Homoptera: Auchenorrhyncha: Cicadellidae). *Nature and Life in Southeast Asia (Kyungpook Journal of Biological Sciences)* 9(2):69-97.

1979c Some new genera and species of Cicadellidae of Korea (Homoptera: Auchenorrhyncha.) *Nature and Life in Southeast Asia (Kyungpook Journal of Biological Sciences)* 9(1):49-61. [New genera: *Drabescoides, Futasujinoidella, Koreanopsis,* and *Paralimnoidella.*]

1979d Some new genera and species of Cicadellidae from Korea (Homoptera: Auchenorrhyncha). *Korean Journal of Entomology* 9(1):48 (Abstract).

1979e The genus *Doratulina* Melichar from Korea (Homoptera: Cicadellidae). *Nature and Life in Southeast Asia (Kyungpook Journal of Biological Sciences)* 9(2):99-105.

1979f New genus and new species of Homoptera: Cicadellidae from the Palaearctic region. *Korean Journal of Entomology* 9(2):86 (Abstract).

1979g Revision of the tribe Hecalini Distant from Korea. *Korean Journal of Entomology* 9(1):48 (Abstract).

1979h The genus *Doratulina* Melichar from Korea (Homoptera: Cicadellidae). *Korean Journal of Entomology* 9(1):48 (Abstract).

1980a Two new species of the genus *Pagaronia* Ball from far east Asia (Homoptera: Cicadellidae). *Korean Journal of Entomology* 10(1):1-5.

1980b Two new species of leafhoppers of the genus *Pagaronia* Ball from Korea (Homoptera: Cicadellidae). *Nature and Life in Southeast Asia (Kyungpook Journal of Biological Sciences)* 10(1):43-45.

1980c On the synonymy of *Xestocephalus sjaolinus* Dlabola (Cicadellidae). *Nature and Life in Southeast Asia (Kyungpook Journal of Biological Sciences)* 10(1):45.

Kyomura, N. and Takahashi, Y.

1979a Joint insecticidal effect of N-propyl and N-methylcarbamates on the green rice leafhopper, resistant to N-methylcarbamates. *Journal of Pesticide Research* 4(3):401-409.

Labouchiex, J., Offeren, A. L. van, and Desmidts, M.

1972a Mise en evidence du role vecteur d'*Orosius cellulosus* (Lindberg) (Homoptera, Cicadelloidea) dans la virescence florale du cotonnier en Haute Volta. *Coton et Fibres Tropicales* 27:393-394.

1973a Etude de la transmission par *Orosius cellulosus* (Lindberg) (Hompt., Cicadellidae) de la virescence florale du cotonnier et de *Sida* sp. *Coton et Fibres Tropicales* 28(4):461-471. [In French with English and Spanish summaries.]

Lacy, G. H.

1982a Occurrence and seasonal distribution of leafhopper vectors of the X-disease causal agent in methoxychlor-sprayed and unsprayed peach orchards. *Crop Protection* 1:333-340.

Lacy, G. H., McClure, M. S. and Andreadis, T. G.

1979a Reducing populations of vector species is a new approach to X-disease control. *Frontiers of Plant Science* 32:2-4.

Ladd, T. L.

1963a The effects of the feeding of the potato leafhopper, *Empoasca fabae* (Harris), and the potato flea beetle, *Epitrix cucumeris* (Harris), and transpiration in the potato plant. Ph. D. Dissertation, Cornell University. 171 pp. [Diss. Abst. 23(12):4787-4788.]

Ladd, T. L. and Rawlins, W. A.

1965a The effects of the feeding of the potato leafhopper on photosynthesis and respiration in the potato plant. *Journal of Economic Entomology* 58(4):623-628.

Laffoon, J. L.

1960a Common names of insects. *Bulletin of the Entomological Society of America* 6:175-211.

Lal, R.

1946a Biological observations on *Empoasca kerri* var. *motti* Pruthi on potato plant. *Indian Journal of Entomology* 8:195-201.

Lal, S. S.

1985a A review of pests of mungbean and their control in India. *Tropical Pest Management* 31:105-114.

Lall, B. S.

1964a Vegetable pests. In: Pant, N. C., Prasad, S. K., Vishnoi, H. S., Chatterji, S. N. and Varma, B. K., eds. *Entomology in India.* Pp. 187-212. The Entomology Society of India. 529 pp.

Lamborn, W. A.

1914a On the relationship between certain West African insects, especially ants, Lycaenidae and Homoptera. With an appendix containing descriptions of new species. *Transactions of the Royal Entomological Society,* London, pp. 436-498.

Lamey, H. A., Surin, P. and Leewangh, H.

1967a Transmission experiments on the tungro virus in Thailand. *International Rice Commission Newsletter* 16(4):15-19.

Lamey, H. A., Surin, P., Disthaporn, S. and Wathanakul, L.

1967a The epiphytotic of yellow orange leaf disease of rice in 1966 in Thailand. *FAO Plant Protection Bulletin* 15(4):67-69.

Lamp, W. O., Barney, R. J., Armbrust, E. J. and Kapusta, G.

1984a Selective weed control in spring-planted alfalfa: Effect on leafhoppers and planthoppers (Homoptera: Auchenorrhyncha), with emphasis on potato leafhopper. *Environmental Entomology* 13(1):207-213.

Lamp, W. O., Morris, M. J. and Armbrust, E. J.

1984a Suitability of common weed species as host plants for the potato leafhopper. *Entomologia Experimentalis et Applicata* 36(2):125-131.

Lamp, W. O., Roberts, S. J., Steffy, K. L. and Armbrust, E. J.

1985a Impact of insecticide applications at various alfalfa growth stages on potato leafhopper (Homoptera: Cicadellidae) abundance and crop damage. *Journal of Economic Entomology* 78(6):1393-1398.

Landis, B. J. and Hagel, G. T.

1969a Aphids and leafhoppers found on red clover in Eastern Washington. *Journal of Economic Entomology* 62(1):273.

Landis, B. J., Powell, D. M. and Hagel, G. T.

1970a Attempt to suppress curly top and beet western yellows by the green peach aphid with insecticide-treated sugarbeet seed. *Journal of Economic Entomology* 63:493-496.

Lange, W. H. and Grigarick, A. A.

1970a Insects and other animal pests of rice. *California Agricultural Experiment Station Extension Service Circular* 555:6-8.

Langlitz, H. O.
1964a The economic species of *Empoasca* in the coastal and Sierra regions of Peru (Homoptera: Cicadellidae). *Revista Peruana de Entomologia* 7(1):54-70.

Lapierre, H.
1980a The viruses of small-grain cereals. *Phytoma* 321:34-38.

Laporte, M.
1966a Comparaison de la structure des bacteries symbiotiques chez divers insectes suceurs de seve. [Comparison of the structure of symbiotic bacteria in various sap-sucking insects.] *In: Insect Pathology and Microbial Control. Proceedings of the International Colloquium on Insect Pathology and Microbial Control*, Wageningen, The Netherlands, September 5-10, 1966. Pp. 69. 360 pp.

Larson, R. H.
1945a Resistance in potato varieties to yellow dwarf. *Journal of Agricultural Research* 71:441-451.

Lastra, R. and Trujillo, G. E.
1976a Diseases of maize in Venezuela caused by viruses and mycoplasmas. *Agronomia Tropical* 26:441-445.

Latif, A. and Qayyum, A.
1950a The use of synthetic insecticide Guesarol 550 against the mango hopper. *Punjab Fruit Journal* 14:6-7.

Lattin, J. D. and Oman, P.
1983a Where are the exotic insect threats? *In: Wilson, C. L. and Graham, C. L., eds. Exotic Plant Pests and North American Agriculture.* Pp. 137. Academic Press, New York. 522 pp.

Lauer, F. and Radcliffe, E.
1967a Insect resistance in the potato. *Minnesota Science* 24(1):22-23.

Lauterer, P.
1957a Contribution a la connaissance des Cicadines de la Tchecoslovaquie (Hom., Auchenorrhyncha). *Casopis (Acta Societatis Entomologicae Cechosloveniae)* 54(1):81-83.
1958a A contribution to the knowledge of the leafhoppers of Czechoslovakia (Hom., Auchenorrhyncha). II. *Casopis Moravskeho Musea Brno* 43:125-136. [In Czech with English summary.]
1972a A homopterological note. *Sbornik Slovensko Narodneho Musea* 18(1):157.
1973a A new species of Typhlocybinae (Homoptera, Cicadellidae) from the collections of the Moravian museum. *Casopis Moravskeho Musea v Brno* 58:109-112.
1978a New records of leafhoppers from Czechoslovakia (Homoptera, Auchenorrhyncha). *Casopis Moravskeho Musea v Brno* 63:111-116.
1980a New and interesting records of leafhoppers from Czechoslovakia (Homoptera, Auchenorrhyncha). *Casopis Moravskeho Musea v Brno* 65:117-140.
1980b Die Zirpe *Empoasca vitis* Goethe (Homoptera, Cicadellidae) als schadling der *Paulownia tomentosa*. *Entomologicke Problemy* (Bratislava) 16:31-33.
1981a Leafhoppers and psyllids in the food of young house martins (Delichon urbica) in the Krkonose Mountains (Homoptera: Auchenorrhyncha & Psyllidea). *Acta Musei Reginaehradensis, Series A: Scientiae Naturales* 16:183-193.
1981b Contribution to the knowledge of the family Pipunculidae of Czechoslovakia (Diptera). *Casopis Moravskeho Musea v Brno* 66:123-150.
1983a Contribution to the knowledge of distribution and bionomics of some representatives of the family Pipunculidae in central and southern Europe (Diptera). *Casopsis Moravskeho Musea v Brno* 68:131-138. [Leafhopper and other Auchenorrhyncha hosts named.]
1983b *Fagocyba cerricola* sp. n. and new and interesting records of leafhoppers from Czechoslovakia (Homoptera, Auchenorrhyncha). *Casopis Moravskeho Musea v Brno* 68:139-152.
1984a New and interesting records of leafhoppers from Czechoslovakia (Homoptera, Auchenorrhyncha). II. *Casopis Moravskeho Musea v Brno* 69:143-162.

Lauterer, P. and Anufriev, G.
1969a Contribution to the knowledge of the genus *Oncopsis* Burm. (Homoptera, Cicadellidae) from China and Far East. *Casopis Moravskeho Musea v Brno* 54:161-168.

Lauterer, P. and Bures, S.
1984a Notes on the bionomics of *Errhomenus brachypterus* (Homoptera, Cicadellidae) and its occurrence in the food of *Ficedula albiocollis albiocollis* (Aves: Muscicapidae). *Zpravy Krajskebo Vlastivedneho Musea v Olomouci* 227:18-20. [In Czech with English summary.]

Lauterer, P. and Okali, I.
1974a Auchenorrhyncha. Fauna Tatraskeho narodneho parku. [Fauna of Tatra National Park.] *Zbornik TANAP* 16:115-131.

Lauterer, P. and Schroder, H.
1970a Types of Cicadellinae (Homoptera, Cicadellidae) in the Moravian Museum. *Casopis Moravskeho Musea v Brno* 55:127-132.

Lavallee, A. G. and Shaw, f. R.
1969a Preferences of golden-eye lacewing larvae for pea aphids, leafhopper and plant bug nymphs, and alfalfa weevil larvae. *Journal of Economic Entomology* 62:1228-1229.

Lavigne, R.
1966a *Parabolocratus viridis* (Homoptera: Cicadellidae) and ants associated with it. *Journal of the Kansas Entomological Society* 39(1):65-67.

Lazarevic, B. M.
1970a Effects of the number of sprayings and different chemicals in controlling cotton pests in the Sudan. *Journal of Economic Entomology* 63(2):629-633.

Leath, K. T. and Byers, R. A.
1977a Interaction of Fusarium root rot with pea aphid and potato leafhopper feeding on forage legumes. *Phytopathology* 67(2):226-229.

Lee, C. E.
1971a Leafhoppers of Korea. *Nature and Life in Southeast Asia* 2:9-20.
1979a *Illustrated Flora and Fauna of Korea. 23. Insecta 7.* Samhwa Publishing Co., Ltd. Seoul. 1070 pp. [In Korean.]

Lee, C. E. and Kwon, Y. J.
1976a Faunistic investigation on a collection of Auchenorrhyncha from Mt. Chuwangsan (Homoptera). *Korean Journal of Entomology* 6(2):1-9.
1977a A new species of *Xestocephalus* from Korea (Homoptera: Cicadellidae). *Nature and Life in Southeast Asia* 7(1):1-4.
1977b Studies on spittlebugs, leafhoppers and planthoppers (Auchenorrhyncha, Homoptera, Hemiptera). *Nature and Life in Southeast Asia* 7(2):55-111.
1979a Comparative morphology and phylogeny of the Korean Auchenorrhyncha (Homoptera). *Korean Journal of Entomology* 9(1):7-27.
1979b A check list of Auchenorrhyncha from Korea (Homoptera). *In: Lee, C. E. Illustrated Flora and Fauna of Korea. 23. Insecta 7.* Samhwa Publishing Co., Ltd., Seoul. Pp. 779-1018.

Lee, C. E., Lee, S. M. and Kwon, Y. J.
1976a An insect list of the Auchenorrhyncha preserved in the National Science Museum of Korea (Homoptera). *Nature and Life in Southeast Asia* 6:51-64.

Lee, C. S. and Chen, C. T.
1972a Preliminary studies on transmission characteristics of the sugar cane white leaf disease by *Matsumuratettix hiroglyphicus*. *Report of the Taiwan Sugar Experiment Station* No. 56:57-62. [In Chinese with English summary.]

Lee, I. M., Calavan, E. C., Cartia, G. and Kaloostian, G. H.
1973a Stubborn disease organism cultured from leafhopper. *Citrograph* 59(2):39.
1973b Citrus stubborn disease organism cultured from beet leafhopper. *California Agriculture* 27(11):14-15.

Lee, P. E.
1962a Acquisition time and inoculation time in transmission of an aster-yellows virus by single leafhoppers. *Virology* 17:394-396.
1963a Mechanical transmission of wheat striate mosaic virus [by injection] to its leafhopper vector *Endria inimica* Say. *Virology* 19(1):88-91.

Lee, P. E. and Chiykowski, L. N.
1963a Mechanical transmission of clover phyllody virus to its leafhopper vectors. *Canadian Journal of Botany* 41:311-312.
1963b Infectivity of aster-yellow virus preparation after differential centrifugations of extracts from viruliferous leafhoppers [*Macrosteles fascifrons* (Stal)]. *Virology* 21(4):667-669.

Lee, P. E. and Jensen, D. D.
1963a Crystalline inclusions in *Colladonus montanus* (Van Duzee), a vector of western X-disease virus. *Virology* 20(2):328-332.

Lee, P. E. and Robinson, A. G.
1958a Studies on the 6-spotted leafhopper, *Macrosteles fascifrons* (Stal) and aster yellows in Manitoba. *Canadian Journal of Plant Science* 38:320-327.

Lee, P. W.
1963a The extraction of wheat striate mosaic virus from diseased wheat plants. *Canadian Journal of Botany* 41:1617-1621.

Lee, P. W. and Bell, W.
1963a Some properties of wheat striate mosaic virus. *Canadian Journal of Botany* 41:767-771.

Lee, R. F., Timmer, L. W. and Tucker, D. P. H.
1981a Populations of sharpshooters in healthy and blight affected citrus groves in Florida. *Phytopathology* 71:235 (Abstract).

Lee, S. C. and Yoo, J. K.
1975a Chemical resistance of striped rice borer, *Chilo suppressalis*, and green rice leafhopper, *Nephotettix cincticeps*. *Korean Journal of Plant Protection* 14:65-70.

Lee, Y. I.
1983a The potato leafhopper, *Empoasca fabae*, soybean pubescence, and hopperburn resistance. Ph. D. Thesis, University of Illinois.

Leech, H. B.
1966a The cicadellid *Parabolocratus viridis* attended by the ant *Formica* (F.) *altipetens* (Homoptera: Cicadellidae and Hymenoptera: Formicidae). *Wasmann Journal of Biology* 24(2):279-280.

Leeuwangh, J.
1968a Leafhoppers and planthoppers on rice in Thailand. Final report of work in Thailand 1965-1968, as FAO Entomologist. 141 pp.

Leeuwangh, J. and Leuamsang, P.
1967a Observations on the ecology of *Thaia oryzivora*, a leafhopper found on rice in Thailand. *Plant Protection Bulletin FAO* 15(2):30-31.

Legal, J.
1961a Phytosanitary note on the cultivation of cotton in Morocco. Present state of pest infestation and the possibilities of control. *Phytiatrie- Phytopharmacie* 10(1):27-37. [In French.]

Lehker, G. E.
1957a The ten most important plant feeding pests in Indiana. *Proceedings of the Indiana Academy of Science* 67:173-174.

Lehmann, W.
1969a Beobachtungen uber Blutenvergrunungsviren bei einigen Zierpflanzen. [Investigations on phyllody viruses of some ornamental plants.] *Archiv vur Pflanzenschutz* 5(4):233-244.
1971a Identifizierung von Vergilbungsviren und ihre Reaktion auf Antibiotika. Probleme der pflanzlichen Virusforschung und der Erforschung pflanzenpathogener Mikroorganismen. *Tagungsbericht Deutsche Akademie der Landwirtschaftswissenschaften zu Berlin* No. 115:99-114. [In German with Russian and English summaries.]
1973a *Euscelis plebejus* (Fallen) und *Macrosteles cristatus* Ribaut als Ubertrager von Pflanzenkrankheiten von vermutter Mycoplasma-Atiologie. *Archiv fuer Phytopathologische und Pflanzenschutz* 9:363-370.
1973b Untersuchung der Zikadenfauna von Obstanlagen mit Hilfe von Lichtfallen. *Biologisches Zentralblatt* 92(5):625-635. [In German with English summary.]
1973c Untersuchungen der Zikadenfauna von Obstgeholzen. [Studies on the leaf-hopper fauna of fruit trees.] *Biologisches Zentralblatt* 92(1):75-94. [In German with English summary.]
1982a The culture of insect cells and tissues and possibilities for its application in plant virus research. A review. *Archiv fur Phytopathologie und Pflanzenschutz* 18:369-380.

Lehmann, W. and Schmidt, H. E.
1969a Zum Vorkommen von Zikaden am Hopfen (*Humulus lupulus* L.). *Biologisches Zentralblatt* 88:179-189.

Lehmann, W. and Skadow, K.
1970a Die Eignung von Pflanzenarten fur die Ernahrung von *Euscelis plebejus* Fall. in Beziehung zu ihrer Bedeutung als Viruswirte. *Biologisches Zentralblatt* 89:9-21.
1971a Untersuchungen zur Verbreitung, Atiologie und Vektorubertragbarkeit der Blutenvergrunung des Rapses. *Archiv fuer Pflanzenschutz* 7(5):323-336, illus.

Leite Filho, A. S. and Ramalho, F. S.
1979a Biologia da cigarrinha verde, *Empoasca kraemeri* Ross & Moore 1957 em feijao e em feijao-de-corda. *Anais de Sociedade Entomologica do Brasil* 8(1):93-101.

Leon, M. J. R., de
1978a Behaviour of 350 lines and varieties of beans in the face of attack by cicadellids and beetles in Veracruz. *Informe Tecnico de la Coordinacion Nacional del Apoyo Entomologico* 3:34-38.

Le Quesne, W. J.
1959a [1960] Some additions to the British Deltocephalinae (Hemiptera, Cicadellidae) with notes on synonymy. *Entomologist's Monthly Magazine* 95:281-287.
1961a An examination of the British species of *Empoasca* Walsh sensu lato (Hem., Cicadellidae), including some additions to the British list. *Entomologist's Monthly Magazine* 96:233-239.
1961b An examination of the British species of *Macropsis* Lewis and *Hephathus* Ribaut including some additions to the British list. *Entomologist's Monthly Magazine* 96:247-255, 41 figures.
1961c Studies in the British species of *Oncopsis* Burmeister including a discussion whether *O. subangulata* Sahlberg is a valid species. *Entomologist's Monthly Magazine* 97:164-170.
1962a Taxonomic studies in the British and some European species of *Scleroracus* Van Duzee (Hem., Cicadellidae). *Entomologist's Monthly Magazine* 97:260-264.
1964a Some taxonomic changes and additions in the British Cicadellidae (Hemiptera) including a new species and subspecies. *Proceedings of the Royal Entomological Society of London* (B) 33:73-82.
1964b Some observations on *Macropsis marginata* (Herrich-Schaeffer) and *M. albar* Wagner (Hem., Cicadellidae). *Entomologist's Monthly Magazine* 99:128.
1964c Auchenorrhyncha. In: Kloet and Hincks, A Check List of British Insects. Pp. 53-65. 2nd ed., Part 1. Royal Entomological Society of London.
1964d The macropterous form of *Ulopa trivia* Germar (Hem., Cicadellidae.) *Entomologist's Monthly Magazine* 99:201.
1964e On the association of the Rosaceae with catkin-bearing trees as insect host-plants. *Entomologist's Monthly Magazine* 100(1206-7):252.
1965a A preliminary list of the Auchenorrhyncha of Woodwalton Fen, Hutingdonshire. *Entomologist's Monthly Magazine* 100:252.
1965b *Athysanus argentarius* Metcalf (Hem., Cicadellidae) in the Isle of Wight. *Entomologist's Monthly Magazine* 101:288.

1965c Handbooks for the identification of British insects, Hemiptera, Cicadomorpha (excluding Deltocephalinae and Typhlocybinae). *Royal Entomological Society of London* 2(A):1064.

1965d The establishment of the relative status of sympatric forms, with special reference to cases among the Hemiptera. *Zoologische Beitraege* 11:117-128.

1968a *Macrosteles ossiannilssonei* (Hem., Cicadellidae), a new species previously confused with *M. sexnotatus* (Fallen). *Entomologist's Monthly Magazine* 103:190-192.

1968b *Macrosteles ossiannillssoni* Le Quesne (Hem., Cicadellidae) — a correction. *Entomologist's Monthly Magazine* 104:3.

1969a Leafhoppers (Auchenorrhyncha). *Amateur Entomologists Society*. Leaflet 32, 10 pp.

1969b Leafhoppers (Auchenorrhyncha). *Bulletin of the Amateur Entomologists Society* 28(284):83-92.

1969c Handbook for the identification of British insects. Hemiptera, Cicadomorpha, Deltocephalinae. *Royal Entomological Society of London* 2(2B):65-148.

1972a Studies on the coexistence of three species of *Eupteryx* (Hemiptera: Cicadellidae) on nettle. *Journal of Entomology* (Series A) 47(1):37-44.

1974a *Eupteryx origani* Zakhvatkin (Hem., Cicadellidae) new to Britain, and related species. *Entomologist's Monthly Magazine* 109:203-206.

1976a [1977] A new species of *Alebra* Fieber (Hem., Cicadellidae). *Entomologist's Monthly Magazine* 112:49-52.

1977a A new species of *Lindbergina* (Hemiptera: Cicadellidae) from Jersey. *Annual Bulletin, Societe Jersiaise* 22(1):87-90.

1978a Genital structures — accident or design? *Auchenorrhyncha Newsletter* 1:12-15.

1983a Problems in identification of species of leafhoppers and planthoppers. In: Knight, W. J., Pant, N. C., Robertson, T. S. and Wilson, M. R., eds. *Proceedings of the 1st International Workshop on Biotaxonomy. Classification and Biology of Leafhoppers and Planthoppers (Auchenorrhyncha) of Economic Importance.* Pp. 39-47. London, 4-7 October 1982. Commonwealth Institute of Entomology, 56 Queen's Gate, London SW7 5JR. 500 pp.

1983b Use of the International Code in naming of species. In: Knight, W. J., Pant, N. C., Robertson, T. S. and Wilson, M. R., eds. *Proceedings of the 1st International Workshop on Biotaxonomy., Classification and Biology of Leafhoppers and Planthoppers (Auchenorrhyncha) of Economic Importance.* Pp. 165-171. London, 4-7 October 1982. Commonwealth Institute of Entomology, 56 Queen's Gate, London SW7 5JR. 500 pp.

1983c *Cicadula flori* (Sahlberg), new to Britain (Hem., Cicadellidae). *Entomologist's Monthly Magazine* 119:177.

1984a The biological records scheme for British Auchenorrhyncha. *Mitteilungen der Schweizerischen Entomologischen Gesellschaft* 57(4):429.

Le Quesne, W. J. and Morris, M. G.

1971a Auchenorrhyncha from pitfall traps at Weeting heath National Nature Reserve Norfolk. *Entomologist's Monthly Magazine* 107:39-44.

Le Quesne, W. J. and Payne, K. R.

1981a Cicadellidae (Typhlocybinae) with a checklist of the British Auchenorrhyncha (Hemiptera, Homoptera). *Handbook for the Identification of British Insects* II(2C):1-95.

Le Quesne, W. J. and Woodroffe, G. E.

1976a Geographic variation in the genitalia of three species of Cicadellidae (Hemiptera). *Systematic Entomology* 1:169-172.

Leston, D.

1955a The stethoscopic cage: A device for hearing the sounds produced by small arthropods. *Entomologist* 88(1107):188-189.

Lezhava, V. V.

1953a List of species found on yew *(Taxus baccata* L.) *Tbilisskogo botanicheslogo sada, Vestnik* 61:201-204. [In Georgian with Russian summary.]

Lherault, P.

1968a From clover to strawberry: Transmission of a virus disease by cicadellids. *Phytoma* 20(195):41.

Li, H. K.

1982a Experiments on insecticidal effects on *Serratia marcescens*. *Weishengwuxue Tongbao* 9(2):55-58.

1985a Entomopathogenic microorganisms of rice planthoppers and leafhoppers in China. *International Rice Research Newsletter* 10(2):13-14.

Li, Z. Z.

1983a Three new species of the genus *Bothrogonia* from Guizhou province. *Entomotaxonomia* 5(2):145-148.

1985a A new species of the genus *Evacanthus* (Homoptera: Evacanthidae). *Acta Entomologica Sinica* 28(4):435-436.

1985b A new species of the genus *Bothrogonia* (*Obothrogonia*) from Guizhou province (Homoptera: Tettigellinae). *Zoologicical Research* 6(4):11-13. [In Chinese with English summary.]

Liang, W. F.

1981a The distinctions between the eggs of some leafhoppers and planthoppers on rice. *Kunchong Zhishi* 18(3):119-121.

Libby, J. L. and Hartberg, T. J.

1972a. Effect of soil applied systemic insecticides on yield of Norland potato. *American Potato Journal* 49(11):438-443.

Licha-Baquero, M.

1979a The witches' broom disease of pigeon pea (*Cajanus cajan* [L.] Millsp.) in Puerto Rico. *Journal of Agriculture of the University of Puerto Rico* 63:424-441.

Lim, G. S.

1969a The bionomics and control of *Nephotettix impicticeps* Ishihara and transmission studies on its associated viruses in West Malaysia. *Malaysia Ministry of Agriculture Cooperative Bulletin* 121:62.

1972a Studies of Penyakit merah disease of rice. III. Factors contributing to an epidemic on North Krian, Malaysia. *Malaysia Agricultural Journal* 48:278-294.

1972b Chemical control of rice insects and diseases in Malaysia. *Japan Pesticide Information* No. 10, pp. 27-36.

1973a Control of rice insects using ULV concentrate and high spreading oil insecticides in Malaysia. *Malaysian Agriculture Journal* 49:122-130.

1976a Evaluation for alternative host plants of *Nephotettix virescens*. *Rice Entomology Newsletter* 4:21-22.

Lim, G. S. and Jusho, M.

1979a Relative attraction of rice hoppers and their common predators to different light sources. *International Rice Research Newsletter* 4(6):14-15.

Lin, K. J.

1979a *Erythroneura sudra* (Homoptera: Cicadellidae): A new record from Taiwan. *Journal of Agricultural Research of China* 28(4):295-297.

Lin, K. S.

1974a Notes on some natural enemies of *Nephotettix cincticeps* (Uhler) and *Nilaparvata lugens* (Stal) in Taiwan. *Journal of Taiwan Agricultura Research* 23:91-115.

Lin, L. C. and Hou, R. F.

1980a Some dietary requirements of the rice green leafhopper, *Nephotettix cincticeps*. In: *XVI International Congress of Entomology* 16:82 (Abstract). 265 pp.

1981a Dietary requirements of *Nephotettix cincticeps* for amino acids and cholesterol. *Chinese Journal of Entomology* 1(1):41-53.

Lin, Q. Y., Zie, L. H. and Wang, H.

1984a Study on the resistance of rice cultivars to tungro disease and its vectors. *Fujian Nonye Keji* 4:34-35.

Lin, T. L. and Liu, T. H.
1984a Autoregressive integrated moving average (ARIMA) models for population data of four insects in a rice field. *Plant Protection Bulletin* 26:121-133.

Lindberg, H.
1950a Notes on the biology of dryinids. *Commentationes Biologicae Societas Scientiarum Fennica* 10(15):1-19.
1953a Bemerkungen uber Arten der Jassiden Gattung *Selenocephalus* Germ sowie Beschreibung einer Neuen nahestenden Gattung *Levantotettix*. *Notulae Entomologicae* 33:109-114. [New genus: *Levantotettix*.]
1956a Contribution a l'etude de la zone d'inondation du Niger (Mission G. Remaudiere). XII. Homoptera-Zikaden. *Bulletin de l'Institut Francais d'Afrique Noire* (Series A) 18:1200-1211.
1956b Uber einige Zikaden aus Marokko und Rio De Oro. *Notulae Entomologicae* 36:11-17.
1958a Hemiptera Insularum Caboverdensium. Systematik, Okologie und Kapverdischen Inseln. Ergebnisse der Zoologischen Expedition von Professor Hakan Lindberg nach den Kapverde-Inseln im Winter 1953-1954, Nr. 22. *Commentationes Biologicae Societas Scientiarum Fennica* 19(1):1-246. [New genus: *Nicolaus.*]
1958b On the Arctic and subarctic Hemiptera of the Palearctic Region. *In: Proceedings of the 10th International Congress of Entomology* 1:715-717. Montreal.
1960a Eine Zikadenausbeute aus Portugal 1959. *Notulae Entomologicae* 40:45-55. [New genera: *Miraldus* and *Ericotettix*.]
1960b Ueber Zikaden von Sowjetarmenien. *Notulae Entomologicae* 40(2):56-72.
1960c Hemiptera from the Azores and Madeira. *Boletim do Museu Municipal do Funchal* 13(33):85-94.
1960d Supplemetum Hemipterorum Insularum Canariensium. *Commentationes Biologicae Societas Scienterium Fennica* 22(6):1-20.
1961a Hemiptera Insularum Madeirensium. *Commentationes Biologicae Societas Scienterium Fennica* 24(1):1-82.
1962a Weiterer Beitrag zur Kenntnis der Zikadenfauna Portugals. *Notulae Entomologicae* 42:25-26.
1963a Zur Kenntnis der Zikadenfauna von Marokko. I. *Notulae Entomologicae* 43(4):21-37.
1963b *Jassargus (Sayetus) andorranus* n. sp. (Cicadina, Jassidae). *Notulae Entomologicae* 43(4):152-154.
1964a Zur Kenntnis der Zikadenfauna von Marokko. II. *Notulae Entomologicae* 44:53-70.
1965a A small collection of leafhoppers (Homptera) from Aaiun in Spanish Sahara. *Notulae Entomologicae* 45(1):13-16. [In German.]

Lindberg, H. and Wagner, E.
1965a Supplementum secundum ad cognitionem Hemipterorum Insularum Canariensium. *Commentationes Biologicae* 28:1-14.

Lindley, C. D.
1972a Control of pests of cotton, rice and maize with EI 47470. *In: Proceedings of the Sixth British Insecticide and Fungicide Conference*, 15-18 November 1971, Hotel Metropole, Brighton, England. Vols. 1, 2, and 3. Pp. 492-501.

Lindsay, K. L. and Marshall, A. T.
1981a The osmoregulatory role of the filter-chamber in relation to phloem-feeding in *Eurymela distincta* (Cicadelloidea, Homptera). *Physiological Entomology* 6(4):413-419.

Lindsten, K.
1979a Leaf- and planthoppers as virus vectors in Fennoscandia and measures to prevent their spread of virus. *Entomologisk Tidskrift* 100(3&4):159-161.
1985a Occurrences and development of hopper-borne cereal diseases in Sweden. *Mitteilungen Biologische Bundesanstalt fur Land- und Fortswirtschaft* 228:81-85.

Lindsten, K., Vacke, J. and Gerhardson, B.
1970a A preliminary report on three cereal virus diseases new to Sweden spread by *Macrosteles* and *Psammotettix* leafhopppers. *Statens Vaxtskyddanst Meddelanden* 14:285-297.

Ling, K. C.
1966a Nonpersistence of the tungro virus of rice in its leafhopper vector, *Nephotettix impicticeps*. *Phytopathology* 56:1252-1256.
1968a *Virus Diseases of the Rice Plant*. International Rice Research Institute, Los Banos, Philippines. 52 pp.
1968b Hybrids of *Nephotettix impicticeps* Ish. and *N. apicalis* (Motsch.) and their ability to transmit the tungro virus of rice. *Bulletin of Entomological Research* 58:393-398.
1968c Further studies on the non-persistence of the rice tungro virus in its vector. *Philippine Phytopathology* 4:6-6.
1969a Nonpropagative leafhopper-borne viruses. *In: Maramorosch, K., ed. Viruses, Vectors and Vegetation.* Pp. 255-277. Academic Press, New York.
1969b Transmission of rice viruses in southeast Asia. *In: The Virus Diseases of the Rice Plant.* International Rice Research Institute Symposium Proceedings. Pp. 139-153.
1970a Ability of *Nephotettix apicalis* to transmit the rice tungro virus. *Journal of Economic Entomology* 63:582-586.
1972a *Rice Virus Diseases*. International Rice Research Institute, Los Banos, Philippines. 142 pp.
1973a *Synonymies of Insect Vectors of Rice Viruses*. International Rice Research Institute, Los Banos, Philippines. 29 pp.
1975a Experimental epidemiology of rice tungro disease. I. Effect of some factors of vector incidence (*Nephotettix virescens*) on disease incidence. *Philippine Phytopathology* 11:11-20.
1979a *Rice Virus and Virus-like Diseases*. International Rice Research Institute, Los Banos, Philippines. 175 pp.

Ling, K. C. and Carbonell, M. P.
1975a Movement of individual viruliferous *Nephotettix virescens* in cages and tungro infection of rice seedlings. *Philippine Phytopathology* 11-32-45.

Ling, K. C., John, V. T., Rao, P. S., Anajaneyulu, A., Miah, M. S., Ghosh, A., Korsnang, S. and Flores, Z. M.
1981a Report of 1978 rice tungro virus collaborative project. *International Rice Research Newsletter* 6(1):12.

Ling, K. C. and Miah, M. S. A.
1980a Life span of and tungro transmission by viruliferous *Nephotettix virescens* on 10 rice varieties at IRRI, Philippines. *International Rice Research Newsletter* 5(6):7.

Ling, K. C. and Palomar, M. K.
1966a Studies on rice plants infected with the tungro virus at different ages. *Philippine Agriculture* 50(2):165-177.

Ling, K. C. and Tiongco, E. R.
1975a Effects of temperature on the transmission of rice tungro virus by *Nephotettix virescens*. *Philippine Phytopathology* 11:46-57.
1977a Transmission of rice tungro virus at various temperatures: A transitory virus-vector interaction. *IRRI Research Paper Series* No. 4. 26 pp.
1979a Transmission of rice tungro virus at various temperatures: A transitory virus-vector interaction. *In: Maramorosch, K. and Harris, K. F., eds. Leafhopper Vectors and Plant Diseases Agents*. Pp. 349-366. Academic Press, New York.

Ling, K. C., Tiongco, E. R. and Cabunagan, R. C.
1983a Insect vectors of rice virus and MLO-associated diseases. *In: Knight, W. J., Pant, N. C., Robertson, T. S. and Wilson, M. R., eds. Proceedings of the 1st International Workshop on Biotaxonomy, Classification and Biology of Leafhoppers and Planthoppers (Auchenorrhyncha) of Economic Importance*. Pp. 415-437. London, 4-7 October 1982. Commonwealth Institute of Entomology, 56 Queen's Gate, London SW7 5JR. 500 pp.

Ling, K. C., Tiongco, E. R., and Daquioag, R. D.
1979a Tungro propagation. *International Rice Research Newsletter* 4(5):8-9.

Ling, K. C., Tiongco, E. R. and Fores, Z. M.
1983a Epidemiological studies of rice tungro. *In:* Plumb, R. T. and Thresh, J. M., eds. *Plant Virus Epidemiology. The Spread and Control of Insect-borne Viruses.* Pp. 249-257. Blackwell Scientific Publications, Oxford, U.K. 337 pp.

Linn, M. B., Apple, J. W. and Arnold, C.
1948a Effect of leafhopper control with DDT dust on length of growing season, quality and yield of seventeen potato varieties. *American Potato Journal* 25(2):55.

Linnavuori, R.
1955c On some Palearctic Hemiptera. *Suomen Hyonteistieteellinen Aikakaukirja* 21:24-26.

1956a Neotropical Homoptera of the Hungarian National Museum and some other European museums. *Suomen Hyonteistieteellinen Aikakaukirja* 22:5-35. [New genera: *Cixidocoelidia* and *Idionannus*.]

1956b On some Palearctic Homoptera. *Suomen Hyonteistieteellinen Aikakaukirja* 22:136-138.

1956c Cicadellidae from the Phoenix and Danger Islands. *Suomen Hyonteistieteellinen Aikakaukirja* 22:140.

1956d Leafhopper matearial [sic!] from South Spain and Spanish Morocco. *Suomen Hyonteistieteellinen Aikakaukirja* 22:156-165. [New genus: *Muleyrechia*.]

1956e A revision of some of Stal's and Spangberg's cicadellid types. *Suomen Hyonteistieteellinen Aikakaukirja* 22(4):170-181.

1957a The Neotropical Hecalinae (Hom., Cicadellidae). *Suomen Hyonteistieteellinen Aikakaukirja* 23:133-143.

1957b Remarks on the Iassinae (Hom., Cicadellidae). *Suomen Hyonteistieteellinen Aikakaukirja* 23:144-149.

1957c A new *Xestocephalus* species from the Solomon Islands (Hom., Cicadellidae). *Suomen Hyonteistieteellinen Aikakaukirja* 23(2):93.

1958a On some new or little known Mediterranean Homoptera. *Bolletitino della Societe Entomologica Italiana* 88:34-38. [New genus: *Streptopyx*.]

1959a Homoptera Auchenorrhyncha from Mt. Sibillini. *Memorie del Museo Civico di Storia Naturale di Verona* 6:299-304.

1959b Revision of the Neotropical Deltocephalinae and some related subfamilies (Homoptera). *Suomalaisen Elain-ja Kasvitieteelisen Seuran Vanamon Elaintieteellisia Julkaisuja* 20:1-370. [Descriptions of 39 new genera and 15 new subgenera.]

1959c Homoptera Auchenorrhyncha From Mt. Picentini. *Memorie Museo Civico di Storia Naturale di Verona* 6:305-311.

1960a Insects of Micronesia. Homoptera: Cicadellidae. Honolulu, Bishop Museum 6(5):231-344. [New genera: *Oceanopona, Opsianus, Pactana;* new subgenus: *Insulanus.*]

1960b Cicadellidae (Homoptera, Auchenorrhyncha) of Fiji. *Acta Entomologica Fennica* 15:1-71. [New genera: *Lamia, Navaia, Wakaya;* new subgenus: *Idyia.*]

1961a Hemiptera (Homoptera): Cicadellidae. *In: Hanstrom, Brinck and Rudebeck, South African Animal Life* 8:452-486. [Descriptions of 13 new genera.]

1962a Hemiptera of Israel. III. *Suomalaisen Elain-ja Kasviteiteellisen Seuran vanamon Elaintieteellsia Julkaisuja* 24(3):1-108. [New genus: *Jubrinia.*]

1964a Hemiptera of Egypt, with remarks on some species of the adjacent Eremian region. *Annales Zoologici Fennici* 1:306-356.

1965a Studies on the south and east Mediterranean hemipterous fauna. *Acta Entomologica Fennica* 21:5-70. [New genera: *Aindrahamia* and *Wadkufia.*]

1965b On some new or interesting Neotropical Homoptera of the family Cicadellidae. *Zoologische Beitraege* 11:137-150.

1968a On some new or interesting Hemiptera. *Suomen Hyonteistieteellinen Aikakaukirja* 34(4):197-200.

1968b Contribucion al concimiento de la fauna Colombiana de cicadelidos. *Agricultura Tropical* 24(3):147-156.

1969a Contribution a la faune du Congo (Brazzaville). Mission A. Villiers et A. Descarpentries XCIII. Hemipteres Hylicidae et Cicadellidae. *Bulletin de l'Institut Fondamental d'Afrique Noire* (Series A) 31(4):1129-1185. [Descriptions of 9 new genera.]

1969b Contributions to the Hemipterous fauna of Egypt. *Suomen Hyonteistieteellinen Aikakaukirja* 35(4):204-215. [New genera: *Assiuta* and *Pseudocephalelus.*]

1969c Nivelkarsaiset III-IV, Hemiptera III-IV, Kaskaat 1-2. *Suomen Elaimet* 12-13:1-244, 1-312.

1970a On the taxonomic position of two of Spinola's African cicadellid genera. *Suomen Hyonteistieteellinen Aikakaukirja* 36(4):183-185.

1971a A leafhopper material from Tunisia, with remarks on some species of the adjacent countries. *Annales de la Societe Entomologique de France* 7(1):57-73. [New genera: *Bousaada* and *Melillaia.*]

1972a Revision of the Ethiopian Cicadellidae (Hom.), Ulopinae and Megophthalminae. *Suomen Hyonteistieteellinen Aikakaukirja* 38(3):126-149. [New genera: *Dananea* and *Kivulopa.*]

1972b Revisional studies on African leafhoppers (Homoptera Cicadelloidea). *Revue de Zoologie et de Botanique Africaines* 86(3&4):196-252. [New genera: *Afrorubria, Beniledra,* and *Hangklipia.*]

1973a Two new species of the genus *Odomas* Jac. (Homoptera Cicadellidae, Megophthalminae) from Cameroon. *Journal of Natural History* 7:121-123.

1973b Additional notes on the Cicadellidae fauna of Peru. *Revista Peruana de Entomologia* 16:14-16. [New genus: *Icaia.*]

1973c A collection of leafhoppers (Hom., Cicadellidae) from Cuba. *Suomen Hyonteistieteellinen Aikakaukirja* 39(2):94-97.

1973d Two new leafhopper species of the family Cicadellidae from Peru. *Revista Peruana de Entomologia* 16(1):17.

1975a Revision of the Cicadellidae (Homoptera) of the Ethiopian Region. III. Deltocephalinae, Hecalini. *Acta Zoologica Fennica* 143:1-37. [New genera: *Neohecalus* and *Lualabanus.*]

1975b Studies on Neotropical Deltocephalinae (Homoptera, Cicadellidae). *Notulae Entomologicae* 55(2):49-52. [New genus: *Clorindaia.*]

1975c Homoptera: Cicadellidae (Supplement). *Insects of Micronesia* 6(9):611-632.

1975d Studies on the Cicadellidae fauna of Peru. *Revista Peruana de Entomologia* 18(1):10-11.

1977a Revision of the Ethiopian Cicadellidae (Hemiptera-Homoptera): Penthiminae. *Etudes du Continent Africain,* Fascicle 4:1-76. [New genera: *Foroa, Irenaella, Mulungaella, Musosa,* and *Nielsoniella.*]

1978a Revision of the African Cicadellidae: Subfamilies Nioniinae, Signoretiinae and Drabescinae (Homoptera, Auchenorrhyncha). *Suomen Hyonteistieteellinen Aikakaukirja* 44(2):35-48. [New genera: *Abimwa* and *Ndua.*]

1978b Studies on the family Cicadellidae (Homoptera, Auchenorrhyncha). 1. A revision of the Macropsinae of the Ethiopian region. *Acta Entomologica Fennica* 33:1-17. [New genera: *Asmaropsis, Kiamoncopsis, Ruandopsis,* and *Tsavopsis.*]

1978c Studies on the family Cicadellidae (Homoptera, Auchenorryncha). 2. A new cicadellid subfamily from the Neotropical region, Neopsinae, subfam. n. *Acta Entomologica Fennica* 33:18-19.

1978d Revision of the Ethiopian Cicadellidae (Homoptera). Paraboloponinae and Deltocephalinae. Scaphytopiini and Goniagnathini. *Revue de Zoologie et de Botanique Africaines* 92(2):457-500. [New genus: *Odniella.*]

1979a Revision of the African Cicadellidae (Homoptera Auchenorrhyncha). Part I. *Revue de Zoologie et de Botanique Africaines* 93(3):647-747. [New genera: *Lycisca, Paraphrodes, Penedorydium,* and *Phlogis;* Descriptions of new subfamilies: Drakensberginae and Phlogisinae.]

1979b Revision of the African Cicadellidae (Homoptera Auchenorrhyncha). Part II. *Revue de Zoologie et de Botanique Africaines* 93(4):929-1010. [New genera: *Bassareus, Kasunga,* and *Yaoundea.*]

Linnavuori, R. E. and Al-Ne'amy, K. T.

1983a Revision of the African Cicadellidae (subfamily Selenocephalinae) (Homoptera, Auchenorrhyncha). *Acta Zoologica Fennica* 168:1-105. [Descriptions of 11 new genera and 3 new subgenera.]

Linnavuori, R. E. and Delong, D. M.

1976a New Neotropical leafhoppers from Peru and Bolivia (Homoptera: Cicadellidae). *Revista Peruana de Entomologica* 19(1):29-38. [New genera: *Caphodellus, Napo, Parandanus,* and *Picchusteles.*]

1977a Studies of the Neotropical Mileewaninae (Homoptera: Cicadellidae). *Journal of the Kansas Entomological Society* 50(3):410-421.

1977b New deltocephaline leafhoppers from Central America (Homoptera: Cicadellidae) and illustrations of some Osborn Bolivian species. *Journal of the Kansas Entomological Society* 50(4):558-568. [New genus: *Dariena.*]

1977c The leafhoppers (Homoptera: Cicadellidae) known from Chile. *Brenesia* 12&13:163-267. [Descriptions of 5 new genera and 1 new subgenus.]

1977d The genus *Acinopterus* (Homoptera: Cicadellidae) in Mexico and the Neotropical region. *Entomological News* 88(9&10):249-254.

1978a The genus *Arrugada* in Bolivia and Peru. *Journal of the Kansas Entomological Society* 51(2):174-184.

1978b Neotropical leafhoppers of the *Bahita* group (Homoptera: Cicadellidae: Deltocephalinae). A contribution to the taxonomy. *Brenesia* 14&15:109-169. [Descriptions of 8 new genera.]

1978c Seventeen new species and three new genera of Central and South American Deltocephalini (Homoptera: Cicadellidae). *Brenesia* 14&15:195-226. [New genera: *Comayagua, Neohegira,* and *Vicosa.*]

1978d Some new or little known Neotropical Deltocephalinae (Homoptera: Cicadellidae). *Brenesia* 14&15:227-247. [New genera: *Caruya, Mocoa, Tingopyx, Tingolix,* and *Yuraca.*]

1978e Genera and species of Hecalini (Homoptera, Cicadellidae, Deltocephalinae) known to occur in Mexico. *Suomen Hyonteistieteellinen Aikakaukirja* 44(2):48-53. [New genus: *Jiutepeca.*]

1978f Two new species of South American leafhoppers (Homoptera, Cicadellidae). *Suomen Hyonteistieteellinen Aikakaukirja* 44(2):54-55. [New subgenus: *Fundarus.*]

1979a New genera and species of the tribe Deltocephalini from South America. *Entomologica Scandinavica* 10:43-53. [New genera: *Bolivaia, Cruziella,* and *Picchuia.*]

1979b New species of leafhoppers from Central and South America (Homoptera: Cicadellidae, Deltocephalinae, Neobalinae, Xestocephalinae). *Entomologica Scandinavica* 10:123-138. [New genus: *Lascumbresa.*]

1979c New or little known Agallinae [sic!] from Central America (Homoptera: Cicadellidae). *Journal of the Kansas Entomological Society* 52(2):405-411.

1979d New species of South American Agalliinae leafhoppers (Homoptera: Cicadellidae). *Entomologica Scandinavica* 10:244-256.

1979e Additional notes on the Biturritidae and Cicadellidae (Homoptera) of Chile. *Brenesia* 16:189-196.

Linnavuori, R. E. and Heller, F.

1961a Beitrag zur Cicadelliden-Fauna von Peru. Entomologische Ergebnisse der Stuttgarter Anden-Expedition 1957—Nr. 1. *Stuttgarter Beitraege zur Naturkunde aus dem Staatlichen Museum für Naturkunde in Stuttgart* 67:1-14. [New subgenus: *Mascoitanus.*]

Linnavuori, R. E. and Quartau, J. A.

1975a Revision of the Ethiopian Cicadellidae (Hemiptera-Homoptera): Iassinae and Acroponinae. Foundation pour favoriser les recherches scientifiques en Afrique. *Etudes du Continent Africain.* Fascicle 3:1-170. [New genera: *Acacioiassus, Afroiassus;* new subgenus: *Sudanoiassus.*]

Linnavuori, R. E. and Viraktamath, C. A.

1973a A new cicadellid genus from Africa. *Revue de Zoologie et de Botanique Africaines* 87(3):485-492. [New genus: *Humpatagallia.*]

Linskii, V. G.

1980a *Cicadella viridis. Zashchita Rastenii* 7:62.

Linsley, E. G. and Usinger, R. L.

1966a Insects of the Galapagos Islands. *Proceedings of the California Academy of Sciences* 33:113-196.

Lippold P., Galvez G. E., Miah, M. S. A., and Alam, M. S.

1970a Rice tungro virus in native populations in *Nephotettix impicticeps* in East Pakistan. *International Rice Commission Newsletter* 19(1):18-23.

Lippold, P. C., Hongtrakula, T., Thongdeetaa, S., Banziger, H., Hillerup, P. E., Kelderman, W., Supharngkasen, P. and Deema, P.

1977a Use of colored sticky board traps in insect surveys. *Plant Protection Service Technical Bulletin* No. 29. 60 pp.

Litsinger, J. A. and Ruhendi

1984a Rice stubble and straw mulch suppression of preflowering insect pests of cowpeas sown after puddled rice. *Environmental Entomology* 13:509-514.

Littau, V. C.

1960a Cytological differences between the fat body tissues of adult male and female leafhoppers (Homoptera, Cicadellidae). *Journal of Morphology* 106(2):187-195.

Littau, V. C. and Maramorosch, K.

1956a Cytological effects of aster yellows virus on its insect vector. *Virology* 2:128-130.

1958a Cytopathogenic effect of the aster yellows virus on its insect vector. *Phytopathology* 48(5):263 (Abstract).

1960a A study of the cytological effects of aster yellows virus on its vector. *Virology* 10:483-500.

Liu, H. C., Chen, S. E., Wang, G. C., Liu, C., Chen, Y. and Wang, G. C.

1983a A study of the biological activity of some purified plant oils on some rice insects. *Agricultural Science and Technology* 6:24-27.

Liu, H. Y. and Black, L. M.

1978a Neutralization of infectivity of potato yellow dwarf virus and wound tumor virus assayed on vector-cell monolayers. *Phytopathology* 68:1243-1248.

Liu, H.Y., Gumpf, D. J., Oldfield, G. N. and Calavan, E. C.

1983a Transmission of *Spiroplasma citri* by *Circulifer tenellus. Phytopathology* 73(4):582-585.

1983b The relationship of *Spiroplasma citri* and *Circulifer tenellus. Phytopathology* 73(4):585-590.

Liu, H. Y., Kimura, I. and Black, L. M.

1973a Specific infectivity of different wound tumor virus isolates. *Virology* 51:320-326.

Liu, T. S. and Chang, D. C.

1979a Control of the brown planthopper and the rice green leafhopper with lower dosage of certain chemicals. *Plant Protection Bulletin* (Taiwan) 21:383-390.

1981a Timing of chemical control of the plant- and leaf-hoppers on the first crop of rice at Taichung district. *Plant Protection Bulletin* (Taiwan) 23:169-177.

Liu, T. S. and Cheng, C. H.

1980a Re-evaluation of chemicals for rice green leafhopper control in Taiwan. *Journal of Agricultural Research of China* 29:1-11.

Lo, Shao-Nan and Huang, Pang-Kan

1963a [Tests on the control of *Empoasca subrufa* Melichar with DDT and BHC.] *Acta Phytophylactica Sinica* 2(1):69-74. [In Chinese with a summary in English.]

Lo, T. C.

1966a Transmission of soybean rosette by leafhopper *Nesophrosyne orientalis* Matsumura. *In: Proceedings of the 11th Pacific Science Congress,* Tokyo. Pp. 33-48.

Lockhart, B. E. L., Khaless, N., Lennon, A. M. and El Maatauoi, M.
    1985a Properties of bermuda grass etched-line virus, a new leafhopper-transmitted virus related to maize rayado fino and oat blue dwarf viruses. *Phytopathology* 75(11):1258-1262.

Lodos, N.
    1969a Injury by *Empoasca* spp. (Hemiptera: Cicadellidae) to cocoa in Ghana. *Memoria Conference International Pesquisa Cacau* 2:273-277.
    1969b Minor pests and other insects associated with *Theobroma cacao* L. in Ghana. *Ghana Journal of Agricultural Science* 2(2):61-72.
    1981a Maize pests and their importance in Turkey. *EPPO [European and Mediterranean Plant Protection Organization] Bulletin* 11(2):87-89. [In English with French and Russian summaries.]
    1982a Cicadinea. In: *Turkiye Entomolojisi II. Ege Univeritesi Zirat Fakultesi Yayinlari* 429:56-140. [In Turkish.]

Lodos, N. and Kalkandelen, A.
    1981a Preliminary list of Auchenorrhyncha with notes on distribution and importance of species in Turkey. VII. Family Cicadellidae: Ulopinae, Megophthalminae, Ledrinae, Macropsinae and Agallinae [sic!]. *Turkiye Bitki Koruma Dergisi* 5(4):215-230.
    1982a Preliminary list of Auchenorrhyncha with notes on distribution and importance of species in Turkey. VIII. Family Cicadellidae: Idiocerinae. *Turkiye Bitki Koruma Dergisi* 6(1):15-28.
    1982b Preliminary list of Auchenorrhyncha with notes on distribution and importance of species in Turkey. IX. Family Cicadellidae: Iassinae, Pentheminae [sic!], Dorycephalinae, Hecalinae and Aphrodunae [sic!]. *Turkiye Bitki Koruma Dergisi* 6:147-159.
    1983a Preliminary list of Auchenorrhyncha with notes on distribution and importance of species in Turkey. X. Family Cicadellidae: Xestocephalinae, Stegelytrinae, and Cicadellinae. *Turkiye Bitki Koruma Dergisi* 7:23-28.
    1983b Preliminary list of Auchenorrhyncha with notes on distribution and importance of species in Turkey. XI. Family Cicadellidae, Typhlocybinae: Alebrini and Dikraneurini. *Turkiye Bitki Koruma Dergisi* 7:107-115.
    1983c Preliminary list of Auchenorrhyncha with notes on distribution and importance of species in Turkey. XII. Family Cicadellidae, Typhlocybinae: Empoascini. *Turkiye Bitki Koruma Dergisi* 7:153-165.
    1984a Preliminary list of Auchenorrhyncha with notes on distribution and importance of species in Turkey. XIII. Family Cicadellidae: Typhlocybinae: Typhlocybini (Part 1). *Turkiye Bitki Koruma Dergisi* 8:33-44.
    1984b Preliminary list of Auchenorrhyncha with notes on distribution and importance of species in Turkey. XIV. Family Cicadellidae: Typhlocybinae: Typhlocybini (Part 2). *Turkiye Bitki Koruma Dergisi* 8:87-97.
    1984c Preliminary list of Auchenorrhyncha with notes on distribution and importance of species in Turkey. XV. Family Cicadellidae: Typhlocybinae: Erythroneurini (Part 1). *Turkiye Bitki Koruma Dergisi* 8:159-168.
    1984d Preliminary list of Auchenorrhyncha with notes on distribution and importance of species in Turkey. XVI. Family Cicadellidae: Typhlocybinae: Erythroneurini (Part 2). *Turkiye Bitki Koruma Dergisi* 8:201-210.
    1985a Preliminary list of Auchenorrhyncha with notes on distribution and importance of species in Turkey. XVII. Family Cicadellidae: Deltocephalinae: Grypotini, Goniagnathini and Opsiini (Part 1). *Turkiye Bitki Koruma Dergisi* 9:79-90.
    1985b Preliminary list of Auchenorrhyncha with notes on distribution and importance of species in Turkey. XVIII. Family Cicadellinae [sic!]: Deltocephalinae: Macrostelini (Part 2). *Turkiye Bitki Koruma Dergisi* 9:147-161.
    1985c Preliminary list of Auchenorrhyncha with notes on distribution and importance of species in Turkey. XIX. Family Cicadellidae: Deltocephalinae: Deltocephalini, Scaphytopiini, Doraturini. *Turkiye Bitki Koruma Dergisi* 9:207-215.

Lodos, N. and Onder, F.
    1984a Review of the Rhynchota in Turkey. In: Drosopoulos, S., ed. *Proceedings of the 1st International Congress Concerning the Rhynchota Fauna of Balkan and Adjacent Regions.* Pp. 8-12. Mikrolimni-Prespa, Greece, 29 August - 2 September 1983. 24 pp.

Logvinenko, V. N.
    1956a Study of leafhopper fauna (Fam. Jassidae) in the Dnieper left bank area in the Ukraine. In: *IV aspirantskaya nauchnaya konferentsiya otdeleniya biologicheskikh nauk. Tezisy dokladov, Kiev.* Pp. 17-18. [IV Scientific Conference of Postgraduate Students, Biological Department, Kiev. Abstracts of the addresses.] [In Russian.]
    1957a Tsikadki (Jassidae) stepi i lesostepi Levoberezhnoy Ukrainy. Avtoreferat kand. diss., Kiev. Pp. 1-16. [Leafhoppers (Jassidae) of steppe and forest-steppe in the Dnieper left-bank area of the Ukraine. Synopsis of thesis for a candidate's degree, Kiev.] [In Russian.]
    1957b Jassid fauna of the Dnieper left-bank area of the Ukraine. *Trudy Instytutu Zoolohiyi Kyyiiv* 14:57-73. [In Russian.]
    1957c New data on the fauna of leaf-hoppers (Jassidae) of the Dnieper left-bank area of the Ukraine. *Dopovidi Akademii Nauk Ukrayin'skoi RSR* 2:200-203. [In Ukrainian with Russian and English summaries.]
    1957d On the leafhopper ecology in the Dnieper left-bank area of the Ukraine. In: *V aspirants'ka naukova konferentsiya viddilu biologichnikh nauk. Tezi dopovidey, Kiiv.* Pp. 56-57. [V Scientific Conference of Postgraduate Students, Biological Department, Kiev. Abstracts of the addresses.] [In Ukrainian.]
    1957e Habitats of leafhoppers (Fam. Jassidae) in the Dnieper left-bank area of the Ukraine. In: *Tret'e soveschanie Vsesojuznogo entomologicheskogo obschestva. Tbilisi, 4-9 oktyabrya 1957 g. Tezisy dokladov* 1:90-91. [The 3rd Conference of the All-Union Entomological Society in Tbilisi, 4-9 Oct. 1957. Abstracts of the addresses.] [In Russian.]
    1959a New data on the Transcarpathian Cicadinae fauna. *Dopovidi Akademii Nauk Ukrayin'skoi RSR* 6:686-689.
    1959b Data on the study of the dendrophilic cicadina fauna of the Transcarpathian region. *Dopovidi Akademii Nauk Ukrayin'skoi RSR* 7:800-805. [In Ukrainian with Russian and English summaries.]
    1959c On the study of dendrophilic Cicadinea-fauna of the Transcarpathian Region. In: *Problemi Entomologii na Ukraini, Kiiv.* Pp. 57-58 [Entomological Problems in the Ukraine, Kiev.] [In Ukrainian.]
    1960a Cicadinae of the genus *Mocuellus* Rib. in the Ukraine (Homoptera-Cicadina.) *Dopovidi Akademiyi Nauk Ukrayins'koyi RSR* (Seriya B) 5:663-666. [In Russian with English summary.]
    1960b Carpathian leafhoppers (Auchenorrhyncha, Jassidae) new for the Ukrainian fauna. In: *Konferentsiya po vivchennju flori i fauni Karpat ta prileglikh teritoriy. Tezi dopovidey, Kiiv.* Pp. 286-290. [The Conference on Study of Flora and Fauna of the Carpathians and Adjoining Areas, Kiev. Abstract of the addresses.] [In Russian.]
    1961a A new and little-known species of the genus *Doratura* (Auchenorrhyncha, Jassidae). *Dopovidi Akademiyi Nauk Ukrayins'koyi RSR* (Seriya B) 2:234-241. [In Ukrainian with English and Russian summaries.]
    1961b New species of the genus *Jassargus* (Auchenorrhyncha, Jassidae) from the left-bank steppe of the Ukrainian SSR. *Dopovidi Akademiyi Nauk Ukrayins'koyi RSR* (Seriya B) 3:375-378. [In Ukrainian with English and Russian summaries.]
    1961c Ecological-faunistic survey and distribution of *Cicadinea* (Homoptera-Auchenorrhyncha) in the Carpathians of Ukraine. *Pratsi Instytutu Zoolohiyi* 17:30-50. [In Ukranian with Russian summary.]
    1961d A new species of *Psammotettix* from salt marshes of southern Ukraine. *Pratsi Nauchno-issledovatel'skogo Zoologobiologicheskogo Instituta* 17:51-53. In Ukrainian with Russian summary.]

1962a A new species of leafhoppers of the genus *Stenometopiellus* Hpt. (Homoptera-Auchenorrhyncha) from southern Ukraine. *Dopovidi Akademiyi Nauk Ukrayins'koyi RSR* (Seriya B) 1:119-121. [In Ukrainian with Russian and English summaries.]

1962b On cicadinea ecology of the Black Sea-Sivash steppe. *In: Voprosy ekologii nazemnykh bespozvonochnykh. Po materialam tschetvertoy ekologicheskoy konferentsii* 7:97-98. [*Ecology of Terrestrial Invertebrates. Materials of the 4th Ecological Conference.*] [In Russian.]

1962c Cicadinea as pests of agricultural crops in the Ukrainian Polessye. *In: II zoologitscheskaya konferentsiy Litovskoy SSR. Tezisy dokladov.* Pp. 65-69. [*The 2nd Zoological Conference of the Lithuanian SSR. Abstracts of the addresses.*] [In Russian.]

1962d Materials on Cicadinea-fauna of the Ukrainian Poles'ye. *In: Vtoraya Zoologitseskaya konferentsiya Belorusskoy SSR, Minsk. Tezisy dokladov. Pp. 155-156. [The 2nd Zoological Conference of the Belorussian SSR, Minsk. Abstracts of the addresses.*] [In Russian.]

1962e Rare and little-known Cicadinea-species (Homoptera, Auchenorrhyncha). Of the Ukrainian fauna. *Zbirnik Prats' Zoologicheskii Muzea Akademiya Nauk URSR* 31:82-88 [Trav. Mus. Zool. Acad. Sci. Ukr.]. [In Ukrainian with Russian and English summaries.]

1963a New data on the cicada (Homoptera, Auchenorrhyncha) of the Ukraine. *Dopovidi Akademiya Nauk URSR* 1:120-126 [Rep. Acad. Sci. Ukr.]. [In Ukrainian with Russian and English summaries.]

1963b New data on the cicadellid fauna (Homoptera, Jassidae) of the Ukrainian Carpathians. *Flora i Fauna Karpat Akademii Nauk SSR: Moscow* 2:175-181. [In Russian.]

1964a On trophical connections of cicadinea in the Ukrainian Carpathians. *In: Ekologiya nasekomykh i drugikh nazemnykh bespozvonochnykh Sovetskikh Karpat, Uzhgorod.* Pp. 60-62. [*Ecology of Insects and Other Terrestrial Invertebrates of the Soviet Carpathians, Uzhgorod.*] [In Russian.]

1964b Ecological and faunistic review of leafhoppers of the Ukraine-Polesye. *Pratsi Instytutu Zoologhiyi* 30:79-90. [In Ukranian with Russian summary.]

1965a New forms of leafhoppers from the Crimea. *Dopovidi Akademiyi Nauk Ukrayins'koyi RSR* (Seriya B) 11:1526-1530. [In Ukrainian with English and Russian summaries.]

1966a On Cicadinea-fauna (Homoptera, Auchenorrhyncha) of the Ukrainian Carpathians. *In: Komakhi Ukrains'kikh Karpat i Zakarpattya, Kiiv.* Pp. 14-19. [*Insects of the Ukrainian Carpathians and Transcarpathians, Kiev.*] [*In Ukrainian with Russian summary.*]

1966b New species of leafhoppers (Homoptera: Auch.) from the Caucasus and Moldavia. *Entomologicheskoe Obozrenie* 45(2):401-410. [Translation in *Entomological Review* (Washington) 45(2):218-223.] [In Russian.]

1967a New species of leafhoppers (Homoptera, Auchenorrhyncha) from the south of the European part of the USSR. *Zoologicheskii Zhurnal* 46(5):773-777. [In Russian with English summary.]

1967b Novyi vid tsikadki roda afrodes —*Aphrodes* (Auchenorrhyncha, Cicadellidae) —s Kavkoza. [A new species of genus *Aphrodes* Curt. of the Caucasus.] *Vestnik Zoologii* 1(4):69-71. [English summary.]

1967c Dva novykk vida tsikadok roda *Handiannus* (Homoptera, Cicadellidae). [Two new species of leafhoppers of the genus *Handianus* (Homoptera, Cicadellidae). *Zoologicheskii Zhurnal* 46(11):1720-1721. [English summary.]

1967d On Cicadinea-fauna (Homoptera, Auchenorrhyncha) of the Black Sea Reserve. *In: Tezisy dokladov nauchnoy konferentsii, posvyaschennoy 40-letiju Tschernomorskogo goszapovednika AN SSSR, Kiev.* Pp. 64-66. [*Abstracts of the Addresses of Scientific Conference Dedicated to 40th Anniversary of the Black Sea Reserve,* Academy of Science, USSR, Kiev]. [In Russian.]

1969a New species of Auchenorrhyncha from the south of the USSR. *Zbirnyk Prats Zoologhichnoho Museyu* 33:72-76. [In Ukrainian with English summary.]

1971a New species of cicades (Auchenorrhyncha, Cicadellidae) from Caucasus. *Zoologicheskii Zhurnal* 50:589-592.

1973a Order Homoptera. Suborder Auchenorrhyncha. Suborder Aleyrodoidea. *In: Vrediteli sel'skokhozyaystvennykh kul'tur i lesnykh nasazhdeniy, Kiev* 1:190-207, 212-214. [*Pests of Agricultural Crops and Forest Plantations, Kiev.*] [In Russian.]

1974a *Bilusius valiko* sp. n.—novyy vid tsikodok (Auchenorrhyncha, Cicadellidae) s Kavkaza. [*Bilusius valiko* sp. n., a new species of Cicadellidae (Auchenorrhyncha) from Kavkaz (USSR).] *Zoologicheskii Zhurnal* 53(8):1261-1263.

1975a New species of leafhoppers from the subfamily Euscelinae (Auchenorrhyncha, Cycadellidae [sic!]) from the Transcaucasian Region. *Doklady Akademiyi Nauk Ukrayins'koyi SSR* (Seriya B) 5:462-467.

1975b On studying leafhoppers of the genus *Mesagallia* Zachv. 1946 (Auchenorrhyncha, Cicadellidae). *Doklady Akademiyi Nauk Ukrayins'koyi SSR* (Seriya B) 9:846-848.

1977a A new species of leafhoppers from the *Euscelidius* Ribaut 1942 genus (Homoptera, Auchenorrhyncha, Cicadellidae) from the Ukraine. *Doklady Akademiyi Nauk Ukrayins'koyi SSR* (Seriya B) 8:746-750. [In Ukrainian with English summary.]

1977b New species from the Transcaucasian leafhoppers (Homoptera, Auchenorrhyncha). *Vestnik Zoologii* 5:61-68. [In Russian with English summary.]

1978a New species of leafhoppers (Homoptera, Auchenorrhyncha) from the Caucasus. *Entomologicheskoe Obozrenie* 57(4):797-807. [Translation in *Entomological Review* (Washington) 57(4):547-554.]

1979a *Allygidius wagneri* sp. n.—a new species of leafhopper (Homoptera, Cicadellidae) from eastern Ukraine. *Trudy Vsesoyuznogo Entomologicheskogo Obshchestva* 61:31-32.

1980a *Empoasca* (*Kybos*) *ivanovi* n. sp.—leafhopper (Homoptera: Cicadellidae) from the Ukrainian SSR, USSR. *Vestnik Zoologii* 4:86-89.

1980b New species of Typhlocybinae (Homoptera, Cicadellidae) from the Caucasus. *Entomologicheskoe Obozrenie* 59(3):586-593.

1981a New leafhopper of the genus *Xestocephalus* Van Duzee (Homoptera, Cicadellidae) from the Transcaucasian territory. *Vestnik Zoologii* 5:83-86. [In Russian.]

1981b Leafhoppers of the genus *Macropsidius* (Homoptera, Auchenorrhyncha, Cicadellidae) from the Caucasus. *Vestnik Zoologii* 6:37-43. [In Russian.]

1981c Six new species of leafhoppers of subfamily Typhlocybinae (Homoptera, Auchenorrhyncha, Cicadellidae) from the Caucasus. *Trudy Zoologicheskogo Instituta Leningrad* 105:6-14.

1983a New leafhoppers of the family Cicadellidae (Auchenorrhyncha) from Transcaucasia. *Entomologicheskoe Obozrenie* 62(1):83-90. [Translation in Entomological Review (Washington) 62(1):73-79.]

1984a New data on the fauna of cicadidae in the Ukraine. *In: Taksonomiya i zoogeografia nasekomykh, Kiev.* Pp. 19-27. [*Taxonomy and Zoogeography of Insects, Kiev.*] [In Russian with English summary.]

Lomakina, L. G.
1963a Suctorial insect pests of city decorative plantings of the foothills lowland of Trans-Ili Ala-Tau. *Akademiya Nauk Kazakhskoi SSR Instituta Zoologicheskii Trudy* 21:31-44. [In Russian.]
1967a Insects, pests of city decorative plantings of southeastern Kazakhstan. *Akademiya Nauk Kazakhskoi SSR Instituta Zoologicheskii Trudy*. Pp. 3-141. [In Russian.]

Lomakina, L. Y., Razvyaskina, G. M. and Shubnikova, E. A.
1963a Cytological and histochemical changes in fat body of cicada *Psammotettix striatus* Fall. infected with winter wheat mosaic virus. *Virprosy Virusologii* 2:168-172.

Long, D. L. and Timian, R. G.
1971a Acquisition through artificial membranes and transmission of oat blue dwarf virus by *Macrosteles fascifrons*. *Phytopathology* 61(10):1230-1232.

Loos, C. A.
1965a Leafhopper found cause of shoot dieback in holly oak. Periods of severe dieback coincide with heavy invasion of *Empoasca elongata*. *California Department of Agriculture Bulletin* 54(3):148-149.

Louis, C. and Laporte, M.
1969a Caracteres ultrastructuraux et differenciation de formes migratrices des symbiotes chez *Euscelis plebejus* (Hom. Jassidae). *Annales de la Societe Entomologique de France* 5:799-809. [English summary.]

Louis, C. and Nicolas, G.
1976a Ultrastructure of the endocellular procaryotes of arthropods as revealed by freeze-etching. A study of 'a' type endosymbionts of the leafhopper *Euscelis plebejus* Fall. (Homoptera, Jassidae). *Journal de Microscopie et de Biologie Cellulaire* 26(2-3):121-126.

Louis, C., Nicolas, G. and Pouphile, M.
1976a Ultrastructure of the endocellular procaryotes of arthropods as revealed by freeze-etching. 2. A study of 't' type endosymbionts of the leafhopper *Euscelis plebejus* Fall. (Homoptera, Jassidae). *Journal de Microscopie et de Biologie Cellulaire* 27(1):53-58.

Luckmann, W. H.
1971a The insect pests of soybeans. *World Farming* 13(5):18-19, 22.

Luginbill, P.
1969a Developing resistant plants—the ideal method of controlling insects. *U.S. Department of Agriculture Production Research Report* 111. 14 pp.

Luna, J. M., Fleischer, S. J. and Allen, W. A.
1983a Development and validation of sequential sampling plans for potato leafhopper (Homoptera: Cicadellidae) in alfalfa. *Environmental Entomology* 12(6):1690-1694.

Lyman, J. M. and Cardona, C.
1982a Resistance in lima beans to a leafhopper, *Empoasca kraemeri*. *Journal of Economic Entomology* 75(2):281-286.

Lyman, J. M., Cardona, C. and Garcia, J.
1982a Studies on resistance in lima beans to *Empoasca krameri* Ross & Moore. *Revista Colombiana de Entomologia* 7:27-32.

Lynn, C. D., Jensen, F. L. and Flaherty, D. L.
1965a Leafhopper treatment for Thompson seedless grapes used for raisins or wine. *California Agriculture* 19(4):4-5.

Ma, N.
1982a *Aguriahana triangularis* (Matsumura)—a new record from China (Typhlocybinae). *Entomotaxonomia* 4(1-2):48.
1983a A new species of *Tautoneura* and a new record from China (Typhlocybinae). *Entomotaxonomia* 5(2):149-150.

Mabbett, T.
1980a Management of four major cotton pests in Thailand. *World Crops* 32:101-104.

Mabbett, T. H., Nachapong, M., Monglakul, K. and Mekdaeng, J.
1984a Distribution on cotton of *Amrasca devastans* and *Ayyaria chaetophora* in relation to pest scouting techniques in Thailand. *Tropical Pest Management* 30:133-141.

Maclean, D. B.
1984a An annotated list of leafhoppers (Homoptera: Cicadellidae) from Watercress Marsh, Columbiana County, Ohio. *Ohio Journal of Science* 84(5):252-254.

MacNay, C. G.
1950a A summary of the more important insect infestations and occurrences in Canada in 1949. *Ontario Entomological Society 80th Annual Report for 1949*, pp. 57-77.
1955a Fruit insects. A leafhopper on brambles (*Macropsis fuscula* Zett.) *Canadian Insect Pest Revue* 33(3):192.
1957a Fruit insects—*Rubus* spp.—a leafhopper. *Canadian Insect Pest Revue* 35(6):262.
1961a Summary of important insect infestations of Canada in 1961. Fruit pests—cranberry pests. *Canadian Insect Pest Revue* 39(9):284.
1961b Manitoba—field crop insects. *Canadian Insect Pest Revue* 39(10):331-332.
1961c Some new records in Canada, from the Canadian insect pest record, 1955-1959, of arthropods of real or potential economic importance: A review. *Canadian Insect Pest Revue* 39(1):1-38

Macphee, A. W.
1979a Observations on the white apple leafhopper, *Typhlocyba pomaria* (Hemiptera: Cicadellidae), and on the mirid predator *Blepharidopterus angulatus* and measurements of their cold-hardiness. *Canadian Entomologist* 111(4):487-490.

MacQuillan, M. J.
1975a Pests of rice in the Solomon Islands. *Fiji Agricultural Journal* 37:29-32.

MacRae, T. C. and Yonke, T. R.
1984a Life history of *Forcipita* [sic!] *loca* (Homoptera: Cicadellidae) on three graminaceous hosts, with descriptions of the immature stages. *Journal of the Kansas Entomological Society* 57:69-78.

Madden, L. V.
1985a Modeling the population dynamics of leafhoppers. *In: Nault, L. R. and Rodriguez, J. G., eds. The Leafhoppers and Planthoppers*. Pp. 235-258. Wiley & Sons, New York. 500 pp.

Madden, L. V. and Nault, L. R.
1983a Differential pathogenicity of corn stunting mollicutes to leafhopper vectors in *Dalbulus* and *Baldulus* species. *Phytopathology* 73:1608-1614.

Madden, L. V., Nault, L. R., Heady, S. E. and Styer, W. E.
1984a Effect of maize stunting mollicutes on survival and fecundity of *Dalbulus* leafhopper vectors. *Annals of Applied Biology* 105:431-441.

Maeda, Y. and Moriya, S.
1972a Insecticide susceptibility of the green rice leafhopper, *Nephotettix cincticeps* Uhler. *Proceedings of the Association for Plant Protection of Kyushu* 18:49-51. [In Japanese with English summary.]

Magyarosy, A. C.
1978a A new look at curly top disease. *California Agriculture* 32:13-14.
1980a Beet curly top virus transmission by artificially fed and injected beet leafhoppers (*Circulifer tenellus*). *Annals of Applied Biology* 96:301-305.

Magyarosy, A. C. and Duffus, J. E.
1977a Beet curly top virulence increased. *California Agriculture* 31:12-13.

Magyarosy, A. C. and Sylvester, E. S.
1979a The latent period of beet curly top virus in the beet leafhopper, *Circulifer tenellus*, mechanically injected with infectious phloem exudate. *Phytopathology* 69(7):736-738, Illustr.

Mahadevan, N. R. and Rangarajan, A. V.
  1975a Occurrence of the cicadellid *Zygina* sp. (Homoptera: Cicadellidae) on *Tamarix plumosa* Hort. *Indian Journal of Entomology* 37(2):206.

Mahal, M. S. and Balraj, Singh
  1982a Inheritance of resistance in okra to the cotton jassid *Amrasca biguttula biguttula* (Ishida). I. Field studies. *Indian Journal of Entomology* 44:1-12.
  1982b Inheritance of resistance in okra to the cotton jassid, *Amrasca biguttula biguttula* (Ishida). II. Screenhouse studies. *Indian Journal of Entomology* 44:301-309.
  1982c Genetics of tolerance in okra seedlings to damage by the cotton jassid, *Amrasca biguttula biguttula* (Ishida). *Indian Journal of Agricultural Sciences* 52:225-228.

Mahdihassan, S.
  1957a The symbionts of a jassid and of a coccid. *Zeitschrift fur Angewandte Entomologie* 41:267-271.
  1970a Bacterial origin of some insect pigments and the origin of species through symbiosis. *Pakistan Journal of Science and Industrial Research* 13(4):410-413.
  1976a Symbiosis in a jassid insect with associated problems. *Zoologischer Anzeiger* 197(3&4):212-218.
  1978a Symbiosis in the mangohopper. A study in comparative cytopathology. Karachi, Pakistan. 40 pp.

Maheswariah, B. M.
  1957a Mango-hoppers and their control. *Mysore Agriculture Calendar and Year Book 1956-57*, pp. 125-126.

Mahmood, S. H.
  1962a A study of the typhlocybine leafhopper genera of the Oriental Region (Homoptera: Cicadellidae). Dissertation DA23, 1459, North Carolina State College. 159 pp.
  1967a A study of the typhlocybine genera of the Oriental region (Thailand, the Philippines and adjoining areas). *Pacific Insects Monograph* 12:1-52. [Descriptions of 27 new genera and 1 new subgenus.]
  1968a Taxonomic studies of the Oriental leafhoppers (Typhlocybinae). *Final Technical Report of the Department of Zoology, University of Karachi, Pakistan*. Mimeo. 23 pp.
  1973a "Why systematics." *Proceedings of the Entomological Society of Karachi* 3:11-16.
  1973b Taxonomic studies of Deltocephalinae (Cicadellidae: Homoptera) of Pakistan and adjoining countries of the Oriental region. *Final Technical Report, Department of Zoology, University of Karachi, Pakistan.* Mimeo. 41 pp.
  1975a A revision of the leafhoppers (Cicadellidae: Homoptera) of Pakistan and adjoining countries of the Oriental region. *First Annual Report, Department of Zoology, University of Karachi, Pakistan.* Mimeo. 24 pp.
  1975b A new species of *Hishimonoides* Ishihara (Homoptera Cicadellidae) from Chittagong hill tracts (Bangladesh). *Proceedings of the Pakistan Academy of Science* 12(1):15-18.
  1980a Faunistic studies of Cicadellidae (Homoptera) of the Indo-Pakistan region. *Proceedings of the Pakistan Congress of Zoology* 1(A):157-172.

Mahmood, S. H. and Ahmed, Manzoor
  1968a Problems of higher classification of Typhlocybinae (Cicadellidae-Homoptera). *University Studies, University of Karachi* 5(3):72-79.
  1969a Studies of tribe *Alebrini* (Typhlocybinae: Cicadellidae) in East Pakistan. *Sind University Research Journal Science Series* 4(1&2):85-91. [New genera: *Benglebra* and *Hussainiana*.]

Mahmood, S. H., Ahmed, Manzoor and Aslam, M.
  1969a *Empoasca albizziae*, new species (Typhlocybinae, Homoptera), a pest of *Albizza lebbek*, in Pakistan. *Pakistan Journal of Zoology* 1(1):49-54.

Mahmood, S. H., Ahmed, M. and Khan, I. A.
  1964a Preliminary studies on the population and host plants of leafhoppers (Cicadellidae-Typhlocybinae) – in some districts of East Pakistan. *Pakistan Science Conference Proceedings* 16(3):B-28.

Mahmood, S. H. and Aziz, S.
  1979a Taxonomic studies of the genus *Nephotettix* (Homoptera-Cicadellidae) from Pakistan and Bangladesh. *Proceedings of the Pakistan Academy of Sciences* 16(2):53-69.

Mahmood, S. H. and Meher, K.
  1973a A new species of *Paramesodes* Ishihara from Pakistan (Hemiptera: Cicadellidae). *Transactions of the Shikoku Entomological Society* 11(4):135-137.

Mahmood, S. H., Sultana, S. and Waheed, A.
  1972a Two new species of *Stirellus* Osborn & Ball (Homoptera, Cicadellidae, Deltocephalinae) from West Pakistan. *Pakistan Journal of Zoology* 4(1):79-84.

Maillet, P. L.
  1956a Contribution a l'etude des Homopteres Auchenorrhynques. I. Jassides recoltes en 1956 dans le Perigord noir. *Cahiers des Naturalistes* 12:97-100.
  1959a Essai sur l'ecologie des Jassides praticoles du perigord noir. Contribution a l'etude des Homopteres Auchenorrhynques II(1). *Vie et Milieu* 10(2):117-134.
  1959b Sur la reproduction des Homopteres Auchenorrhynques. *Compte Rendu Hebdomadaire des Seances de l'Academie des Sciences* 249:1945-1947.
  1960a Sur le parasitisme d'oeufs de la cicadelle verte (*Cicadella viridis* L.) par un hymenoptere Mymaridae: *Anagrus atomus* (L.) forme *incarnatus* Hal. *Revue de Pathologie Vegetale et d'Entomologie Agricole de France* 39:197-203.
  1962a Influence of non-biotic factors on the development cycle of the vine phylloxera (*Dactylosphaera vitifolii* Schin) and on a population of Jassidae Homoptera, Auchenorrhyncha, Jassidae). *Casopis Ceskoslovenske Spolecnosti Entomologicke* 59(2):101-110. [In French.]
  1970a Infection simultanee par des particules de type PLT (Rickettsiales) et de type PPLO (Mycloplasmatales) chez un insecte vector de la phyllodie due trefle *Euscelis lineolatus* Brulle (Homoptera: Jassidae). *Journal de Microscopie* (Paris) 9:827-832.
  1970b Presence de particules de type Rickettsien dans la salive d'un homptere vecteur de la phyllodie du trefle *Euscelis lineolatus* Brulle (Homoptera Jassidae). *Revue Canadienne de Biologie* 29(4):391-393.

Maillet, P. L. and Folliot, R.
  1967a Nouvelles observations sur le transport de micro-organismes intranucleaires, appeles particules Phi, par les spermatozoides chez des insectes Homopteres. *Compte Rendu Hebdomadaire des Seances de l'Academie des Sciences* 264D:965-968.
  1967b Sur la presence d'un virus dans le testicule chez un insecte Homoptere. *Compte Rendu Hebdomadaire des Seances de l'Academie des Sciences* 264D:2828-2831.
  1968a Sur le transport de micro-organismes (particules Phi) par les cellules germinales femelles chex l'Homoptere *Typhlocyba douglasi* Edw. *Compte Rendu Hebdomadaire des Seances de l'Academie des Sciences* 266D:923-925.

Maillet, P. L. and Gouranton, J.
  1971a Etude de cycle biologique du mycoplasme de la phyllodie du trefle dans l'insecte vecteur *Euscelis lineolatus* Brulle (Homoptera: Jassidae). *Journal de Microscopie* (Paris) 11(1):143-161.

Maillet, P. L., Gourret, J. P. and Hamon, C.
  1968a Sur la presence de particules de type mycoplasme dans le liber de plantes atteintes de maladies du type "jaunisse" (aster yellows, phyllodie due trefle, stolbur de la tomate) et sur la parente ultrastructurale de ces particles avec celles trouvees chez divers insectes homopteres. *Compte Rendu de l'Academie d'Agriculture de France* 266:2309-2311.

Maiti, R. K., Ghosh, G. C. and Chakraborty, D. P.
  1980a Studies on the evaluation and economics of some important insecticides in rice (var. Jaya). *Pesticides* 14:11-14.

Malabuyoc, L. and Heinrichs, E. A.
1981a Honeydew excretion, feeding activity, and insect weight gain as criteria in determining levels of varietal resistance to the green leafhopper. *International Rice Research Newsletter* 6(1):9.

Malaguti, B. G and Ordosgoitty, F. A.
1969a El achaparramiento del maiz en Venezuela. [Maize stunt in Venezuela.] *Agronomia Tropical* 19(2):85-87. [With English summary.]

Malbrunot, P. and Francois, B.
1965a Essai du formothion en 1965 contre *Scaphoideus littoralis*. [Test of formothion in 1965 against *S. littoralis*.] In: *Report of Tests of Insecticides Carried Out in 1965 Against S. littoralis, the Vector of Golden Flavescence of Vines*. Pp. 106-110.

Maldonado-Capriles, J.
1957a A study of some Neotropical leafhoppers (Homoptera: Cicadellidae: Idiocerinae). *Dissertation Abstract* 17(5):1160-1161.

1961a Studies on Idiocerinae leafhoppers: I. *Idiocerinus* Baker, 1915, synonym of *Balocha* Distant, 1908, and notes on the species of *Balocha*. *Proceedings of the Entomological Society of Washington* 63(4):300-308. [New genus: *Paraidioscopus*.]

1964a Studies of Idiocerinae leafhoppers: II. The Indian and Philippine species of *Idiocerus* and the genus *Idioscopus* (Homoptera: Cicadellidae). *Proceedings of the Entomological Society of Washington* 66:89-100.

1965a Studies on Idiocerinae leafhoppers: III. On Singh-Pruthi's Indian species of *Idiocerus* (Homoptera, Cicadellidae). *Proceedings of the Entomological Society of Washington* 67(4):244-246.

1968a A new genus and species of Deltocephaline from Puerto Rico. *Proceedings of the Entomological Society of Washington* 70(1):35-37. [New genus: *Krameraxus*.]

1968b Studies on Idiocerinae leafhoppers: IV. A new species of *Balocha* and one of *Pedioscopus*, mimics. *Proceedings of the Entomological Society of Washington* 70(2):97-100.

1970a Studies on idiocerine Leafhoppers: VI. New species of *Balocha* from the Papuan Subregion. *Pacific Insects* 12(2):297-302.

1971a About Idiocerinae leafhoppers: V. *Balcanocerus*, a new genus for *Chunrocerus balcanicus* Zachvatkin, 1946. *Proceedings of the Entomological Society of Washington* 73(2):184-187.

1971b Studies on idiocerine leafhoppers: VII. Concerning the Ethiopian genus *Rotifunkia* China 1926. *Suomen Hyonteistieteellinen Aikakouskirja* 37(4):202-204.

1972a Studies on idiocerine leafhoppers: VIII. The Papuan genus *Pedioscopus* and two allied new genera from the Philippine Islands. *Pacific Insects* 14(3):529-551. [New genera: *Philipposcopus* and *Angusticella*; new subgenus: *Upsicella*.]

1972b Studies on idiocerine leafhoppers: IX. Three new genera from the eastern Oriental region. *Pacific Insects* 14(3):627-633. [New genera: *Brachylorus*, *Meroleucocerus* and *Philippocerus*.]

1973a Studies on idiocerine leafhoppers: X. *Idioscopus nitidulus* (Walker), new combination (Homoptera: Cicadellidae). *Proceedings of the Entomological Society of Washington* 75(2):179-181.

1974a Studies on idiocerine leafhoppers: XII. *Idioscopus clavosignatus* spec. nov. (Homoptera, Cicadellidae). *Zoologische Mededelingen* (Leiden) 48:163-167.

1975a Studies on Idiocerinae leafhoppers: XVI. *Pachymetopius* Matsumura transferred to Coelidiinae (Cicadellidae: Homoptera). *Proceedings of the Entomological Society of Washington* 77(3):306-307.

1975b Studies on Idiocerinae leafhoppers: XI. The Neotropical genus *Chunroides* Evans. *Entomologist's Monthly Magazine* 110:233-236.

1976a Studies on Idiocerinae Leafhoppers: XIII. *Idioceroides* Matsummura and *Anidiocerus*, a new genus from Taiwan (Agallinae [sic!]: Idiocerinae). *Pacific Insects* 17(1):139-143. [New genus: *Anidiocerus*.]

1977a Studies on Idiocerinae leafhoppers: XVII. Three new Neotropical genera (Homoptera: Cicadellidae). *Proceedings of the Entomological Society of Washington* 79(3):317-325. [New genera: *Adchrunroides*, *Parachunroides* and *Rotundicerus*.]

1977b Studies on Idiocerinae leafhoppers: XVIII. Four new genera and species from Guyana, South America (Homoptera: Cicadellidae). *Proceedings of the Entomological Society of Washington* 79(3):358-366. [New genera: *Corymbonotus*, *Hyalocerus*, *Luteobalmus*, and *Pseudoidioscopus*.]

1977c Studies on Idiocerinae leafhoppers: XIX. A new genus from Surinam, northern South America. *Proceedings of the Entomological Society of Washington* 79(4):605-608. [New genus: *Nannicepus*.]

1977d Studies on Idiocerinae leafhoppers: XV. *Busonia* Distant and an allied new genus from the Oriental region (Cicadellidae: Idiocerinae). *Pacific Insects* 17(4):491-501. [New genus: *Busoniomimus*.]

1984a Studies of idiocerine leafhoppers: XVI. *Tomopennis*, a new genus for Guyana, and a key to the Neotropical genera of the subfamily. *Caribbean Journal of Science* 20(3&4):97-100. [New genus: *Tomopennis*.]

1985a Studies on idiocerine leafhoppers: XX. *Gressitocerus* and *Dolichopscerus*, new genera from New Guinea (Homoptera: Cicadellidae). *International Journal of Entomology* 27(3):270-276. [New genera: *Gressitocerus* and *Dolichopserus*.]

1985b Studies on idiocerine leafhoppers: XXI. Color variations of *Idioscopus clypealis* (Homoptera: Cicadellidae). *International Journal of Entomology* 27(3):277-279.

Malhotra, Y. R. and Sharma, B.
1974a *Ushamenona* subgen n., a new subgenus of *Apheliona* Kirkaldy, 1907, from India (Homoptera: Cicadellidae: Typhlocybinae). *Bulletin de l'Academie Polonaise des Sciences, Serie des Sciences Biologiques* 22(4):245-248.

1977a A new species of *Litura* from Jammu, India (Homoptera: Cicadellidae). *Oriental Insects* 11(1):27-29.

1977b Key to identification of common leafhoppers (Homoptera, Cicadellidae) of Jammu division of J & K State. *Journal of the University of Jammu* 6:1-19.

1977c On some immature stages of *Apheliona* (*Ushamenona*) *aryavartha* (Ramakrishnan and Menon) and ecological notes. *Journal Assam Science Society* 20(2):141-144.

Malik, N., Baluch, M. A. and Ahmed, Manzoor
1979a Biology and ecology of *Zygina binotata* (Distant) (Typhlocybinae: Cicadellidae) a pest of kachnar, *Bauhinia variegata* in Pakistan. *Pakistan Journal of Scientific and Industrial Research* 22(3):135-137.

Mal'kooskiy, M. P.
1956a *Cicadella viridis* L. (Cicadoidea, Cicadellidae) and its control in young orchards. *Trudy Respublikanskoi Stantsii Zashchity Rastenii*, Alma-Ata 3:3-34. [In Russian.]

Mallamaire, A.
1954a Les Homopteres nuisibles aux plantes utiles en Afrique Occidentale. *Bulletin, Protection des Vegetale* 2:91-153.

Mamet, J.
1957a A revised and annotated list of the Hemiptera (Heteroptera and Homoptera, excluding Sternorhyncha) of Mauritius. *Bulletin, Mauritius Institute* 5(1&2):31-81.

Mancia, J. E., Bruno, G. O., Cortez, M. R. and Amaya, G. M.
1973a Combate de la cigarrita del frijol. *XII Reunion Anual del Programa Cooperativo Centroamericano para el Majoramiento de Cultivos Alimenticios*, Managua, Nicaragua.

Manglitz, G. R. and Gorz, H. J.
1972a A review of insect resistance in the clovers (*Trifolium* spp.). *Bulletin of the Entomological Society of America* 18(4):176-178.

Manglitz, G. R., Gorz, H. J., Haskins, F. A., Akeson, W. R. and Beland, G. L.
1976a Interactions between insects and chemical components of sweet clover. *Journal of Environmental Quality* 5(4):347-352.

Manglitz, G. R. and Jarvis, J. L.
1966a Damage to sweetclover varieties by potato leafhopper. *Journal of Economic Entomology* 59(3):750-751.

Manglitz, G. R. and Kreitlow, K. W.
1960a Vectors of alfalfa and bean yellow mosaic viruses in Ladino white clover. *Journal of Economic Entomology* 53(1):113-115.

Mani, M. S.
1974a *Modern Classification of Insects.* Satish Book Enterprise, Moti Katra, Agra, India. 331 pp.

Mani, M. and Jayaraj, S.
1976a Biochemical investigation on the resurgence of rice blue leafhopper *Zygina maculifrons* (Motch.) [sic!]. *Indian Journal of Experimental Biology* 14(5):636-637.

1976b Control of rice leafhopper, *Nephotettix virescens* (Dist.) with insecticides in three methods of application. *Madras Agricultural Journal* 63:312-316.

1976c Laboratory studies on the control of rice green leafhopper *Nephotettix virescens* (Dist.) by seed and seedling dip treatments with systemic insecticides. *Labdev Journal of Science and Technology* 13:14-18.

Manjunath, T. M., Rai, P. S. and Gavi, Gowda
1978a Natural enemies of brown planthopper and green leafhopper in India. *International Rice Research Newsletter* 3(2):11.

Manjunath, T. M. and Urs, K. C. D.
1979a *Empoascanara indica* in paddy nurseries. *FAO Plant Protection Bulletin* 27(3):93-94.

Manley, G. V.
1976a Immature stages and biology of *Orius tantillus* (Motschulsky), (Hemiptera: Anthocoridae), inhabiting rice fields in West Malaysia. *Entomological News* 87:103-110.

1977a *Paederus fuscipes* (Col.: Staphylinidae): A predator of rice fields in West Malaysia. *Entomophaga* 22:47-59.

Manna, G. K. and Bhattacharya, A. K.
1973a Meiosis in two species of *Parabolocratus* with polymorphic sex chromosomes in males of *P. albomaculatus* (Homoptera). *Caryologia* 26(1):1-11.

Mannheims, B.
1965a Die Cicaden-Typen im Zoologischen Forschungsinstitut und Museum A. Koenig (Insecta, Homopterra). *Bonner Zoologische Beitrage* 16:352-356.

Maramorosch, K.
1954a A leafhopper-borne disease from western Europe. *Phytopathology* 44(2):111 (Abstract).

1955a Mechanical transmission of clover club-leaf virus to its insect vector. *Bulletin, Torrey Botanical Club* 82(5):339-342.

1955b Mechanical transmission of California celery-yellows virus to aster leafhoppers. *Phytopathology* 45(3):185 (Abstract).

1956a Semiautomatic equipment for injecting insects with measured amount of liquids containing viruses or toxic substances. *Phytopathology* 46:188-190.

1956b Two distinct types of corn disease in Mexico. *Phytopathology* 46(4):241 (Abstract).

1956c Multiplication of aster yellows virus in vitro preparations of insect tissues. *Virology* 2(3):369-376.

1957a Cross-protection studies of two types of corn stunt virus. *Phytopathology* 47(1):23 (Abstract).

1958a Studies of aster yellows virus transmission by the leafhopper species *Macrosteles fascifrons* Stal and *M. laevis* Ribaut. *In: Proceedings of the 10th International Congress of Entomology,* Montreal 3:221-227. 4 vols.

1958b Beneficial effect of virus diseased plants on non-vector insects. *Tijdschrift over Plantenziekten* 63:383-391.

1958c Cross protection between two strains of corn stunt virus in an insect vector. *Virology* 6:448-459.

1959a Reversal of virus-caused stunting by gibberellic acid in plants infected with corn stunt, aster yellows, and wound tumor. *In: Proceedings of the 6th International Congress of Crop Protection* 1:271-272.

1959b An ephemeral disease of maize transmitted by *Dalbulus elimatus. Entomologia Experimentalia et Applicata* 2:169-170.

1959c Leafhoppers (Cicadellidae) as vectors and reservoirs of phytopathogenic viruses. *In: Bucharest Academia Republicii Populare Romine.* Pp. 421-442. Omagiu lui Traian Savulescu cu Prilejul Implinirii a 70 de ani, Bucharest.

1960a Leafhopper-transmitted plant viruses. *Protoplasma* 52:457-466.

1960b Experimental transmission of corn stunt virus to *Vinca rosea. Bulletin of the Entomological Society of America* 8:159.

1962a Differences in incubation periods of aster-yellows virus strains. *Phytopathology* 52:925.

1962b Acquisition and transmission of aster yellows virus. *Phytopathology* 52(11):1219 (Abstract).

1963a The occurrence in Arizona of corn stunt diseases and of the leafhopper vector *Dalbulus maidis. Plant Disease Reporter* 47:858.

1963b Arthropod transmission of plant viruses. *Annual Review of Entomology* 8:396-414.

1963c Corn stunt virus transmissions (*Dalbulus maidis,* vector) and attenuation in new host. *Phytopathology* 53:350.

1964a Virus-vector relationships: Vectors of circulative and propagative viruses. Pp. 175-193. *In: Corbett and Sisler, Plant Virology.* University of Florida Press, Gainesville. 525 pp.

1964b Interrelationships between plant pathogenic viruses and insects. *Annals of the New York Academy of Science* 118:363-370.

1965a New techniques for the study of insect-transmitted plant viruses. Pp. 712-713. *In: Proceedings of the 12th International Congress of Entomology.* 842 pp.

1965b New applications of tissue culture in the study of leafhopper-borne viruses. Pp. 541-546. *In: White, P. R. and Grove, A. R., eds. Proceedings of the International Conference on Plant Tissue Culture,* 1963. 553 pp.

1967a Viruses of two insect and plant hosts and their meaning for the study of the origin of viruses. *Biologicheskii Zhurnal Armenii* 42:58-67.

1968a Plant pathogenic viruses in insects. *Current Topics in Microbiology* 42:94-107.

1969a *Viruses, Vectors and Vegetation.* Wiley & Sons, New York. 666 pp.

1969b Effects of rice-pathogenic viruses on their insect vectors. *In: Proceedings of a Symposium of the International Rice Research Institute, Los Banos. The Virus Disease of Rice.* Pp. 179-203. Laguna, Philippines. Johns Hopkins, Baltimore.

1970a Insect infection caused by a plant tumor virus. *World Revue of Pest Control* 9(1):29-41.

1973a (Editor) Mycoplasma and mycoplasma-like agents of human, animal and plant diseases. *Annals of the New York Academy of Sciences* 225. 532 pp.

1974a Ovipositing of *Circulifer tenellus* (Baker) (Homoptera, Cicadellidae). *Journal of the New York Entomological Society* 82(1):42-44.

1979a Leafhopper tissue culture. *In: Maramorosch, K. and Harris, K. F., eds. Leafhopper Vectors and Plant Disease Agents.* Pp. 485-511. Academic Press, New York. 654 pp.

1980a Insects and plant pathogens. *In: Maxwell, F. G. and Jennings, P. R., eds. Breeding Plants Resistant to Insects.* Pp. 137-155. Wiley & Sons, New York. 683 pp.

1981a Spiroplasmas: Agents of animal and plant diseases. *BioScience* 31:374-380.

1982a Control of vector-borne mycoplasms. *In: Harris, K. F. and Maramorosch, K., eds. Pathogens, Vectors and Plant Diseases: Approaches to Control.* Pp. 265-295. Academic Press, New York. 310 pp.

Maramorosch, K., Calica, C. A., Agati, J. A. and Pableo, G.
1961a Further studies on the maize and rice leaf galls induced by *Cicadulina bipunctella*. *Entomologia Experimentalis et Applicata* 4:86-89.

Maramorosch, K., Govindu, H. C. and Kondo, F.
1977a Rhabdovirus particles associated with a mosaic disease of naturally infected *Eleusine coracana* (finger millet) in Karnatake State (Mysore), South India. *Plant Disease Reporter* 61:1029-1031.

Maramorosch, K., Granados, R. R. and Hirumi, H.
1970a Mycoplasma diseases of plants and insects. *Advances in Virus Research* 16:135-193.

Maramorosch, K. and Harris, K. F., eds.
1979a *Leafhopper Vectors and Plant Disease Agents*. Academic Press, Inc., New York. 654 pp.
1981a *Plant Diseases and Vectors: Ecology and Epidemiology*. Academic Press, New York. 368 pp.

Maramorosch, K., Hirumi, H., Kimura, M. and Bird, J.
1975a Mollicutes and rickettsia-like plant disease agents (Zoophytomicrobes) in insects. *Annals of the New York Academy of Science* 266:276-292.

Maramorosch, K., Hirumi, H. Kimura, M. Bird, J. and Vakili, N. G.
1974a Disease of pigeon pea in the Caribbean area: An electron microscopy study. *FAO Plant Protection Bulletin* 22(2):32-36.

Maramorosch, K. and Jensen, D. D.
1963a Harmful and beneficial effects of plant viruses in insects. *Annual Review of Microbiology* 17:495-530.

Maramorosch, K. and Jernberg, N.
1964a A device for rapid inoculation of arthropods. *Bulletin of the Entomological Society of America* 10:171.

Maramorosch, K. and Kondo, F.
1978a Aster yellows spiroplasma: Infectivity and association with a rod-shaped virus. *Zentralblatt fuer Bakeriologie, Parasitenkunde, Infektionskrankheiten und Hygiene. Erste Abteilung Originale. Reihe A. Medizinische Mikrobiologie und Parasitologie* 241:196.

Maramorosch, K., Martinez, A. L. and Maisey, S.
1962a Translocation of aster yellows virus in aster plants. *Phytopathology* 52:22.

Maramorosch, K., Mitsuhashi, J. Streissle, G. and Hirumi, H.
1965a Animal and plant viruses in insect tissues *in vitro*. *Bacteriological Proceedings*, p. 120.

Maramorosch, K. and Oman, P.
1966a U.S.-Japan joint conference on arthropod-borne plant viruses. *Bioscience* 16(9):608-610.

Maramorosch, K., Orenski, S. W. and Suykerbuyk, J.
1965a New corn stunt isolates from field-infected corn in Florida, Louisiana and Mississippi. *Phytopathology* 55:129 (Abstract).

Maramorosch, K. and Raychandhuri, S. P., eds.
1981a *Mycoplasma Diseases of Trees and Shrubs*. Academic Press, New York. 362 pp.

Maramorosch, K. and Shikata, E.
1965a The fate of wound tumor virus in vectors and non-vectors: An electron microscopy study. In: *Conference on the Relationships of Arthropod Plant-Pathogenic Viruses*. Pp. 70-72. Tokyo, Japan.

Maramorosch, K., Shikata, E. and Granados, R. R.
1968a Structures resembling mycoplasma in diseased plants and in insect vectors. *Transactions of the New York Academy of Science* 30(6):841-855.
1969a The fate of plant-pathogenic viruses in insect vectors: Electron microscopy observations. In: Maramorosch, K., ed. *Viruses, Vectors and Vegetation*. Pp. 417-431. Wiley & Sons, New York. 666 pp.

Maramorosch, K., Shikata, E., Hirumi, H. and Granados, R. R.
1969a Multiplication and cytopathology of a plant tumor virus in insects. *National Cancer Institute Monograph* 31:493-507.

Marchoux, G., Leclant, F. and Mathai, P. J.
1970a Maladies de type jaunisse et maladies voisines affectant principalement les Solanacees et transmises par des insectes. *Annales de Phytopathologie* 2(4):735-773.

Marfo, K. O.
1985a Evolving insect pest resistant cowpea varieties in Ghana. *Insect Science and its Application* 6:385-388.

Mariappan, V., Hibino, H., Mew, T. and Pathak, M. D.
1983a Effect of azolla on rice tungro virus disease. *International Rice Research Newsletter* 8(3):8-9.

Mariappan, V., Hibino, H. and Shanmugan, N.
1984a A new rice virus disease in India. *International Rice Research Newsletter* 9(6):9-10.

Mariappan, V. and Saxena, R. C.
1983a Effects of custard-apple oil and neem oil on survival of *Nephotettix virescens* (Homoptera: Cicadellidae) and rice tungro virus transmission. *Journal of Economic Entomology* 76(3):573-576.
1983b Effect of blends of custard-apple oil and neem oil on survival of *Nephotettix virescens* and tungro virus transmission. *International Rice Research Newsletter* 8(4):15-16.
1984a Effects of custard-apple oil and neem oil on survival of *Nephotettix virescens* (Homoptera: Cicadellidae) and on rice tungro virus transmission. *Journal of Economic Entomology* 77(2):519-522.

Mariappan, V., Saxena, R. C. and Ling, K. C.
1982a Effect of custard-apple oil and neem oil on the life span of and rice tungro transmission by *Nephotettix virescens*. *International Rice Research Newsletter* 7(3):13-14.

Marino de Remes Lenicov, A. M.
1982a Aportes al concimiento de los Agallinae [sic!] Argentinos (Homoptera-Cicadellidae.) *Neotropica* (La Plata) 28(80):125-138. [In Spanish with English summary.]

Marino de Remes Lenicov, A. M. and Teson, A.
1975a Notas sobre estrepsipteros argentinos parasitos de homopteros I (Insecta). *Neotropica* (L Plata) 21:65-71.
1983a Designacion del lectotipoly observaciones sobre *Ciminius platensis* (Berg) (Homoptera-Cicadellidae.) *Neotropica* (La Plata) 29(81):49-50.

Markelova, E. M.
1961a *Edwardsiana* — pest of the apple orchards. *Zashchita Rastenii ot Vreditelei i Bolezni* 6(3):47. [In Russian.]

Markelova, Y. M.
1962a Leafhoppers of the genus *Edwardsiana* (Homoptera, Eupterygidae) which are apple pests in the USSR. *Entomologicheskoe Obozrenie* 41(4):737-740. [In Russian; English translation in *Entomological Review* (Washington) 41(4):461-464.]
1968a Leafhoppers of the genus *Edwardsiana* Zachv. (Homoptera, Cicadellidae) and their control in apple orchards of the Tambov Province. [In Russian; English translation in *Entomological Review* (Washington) 48(3):301-306.

Markham, P. G.
1983a Spiroplasmas in leafhoppers: A review. *Yale Journal of Biology and Medicine* 56(5&6):745-751.

Markham, P. G., Clark, T. B. and Whitcomb, R. F.
1983a Culture techniques for spiroplasmas from arthropods. In: *J. G. Tully and S. Razin, ed. Diagnostic Mycoplasmology*. Vol. 2. Pp. 217-223. Academic Press, New York. 440 pp.

Markham, P. G. and Pinner, M. S.
1984a Spiroplasmas in leafhopper vectors: Multiplication and pathogenicity studies. *Mitteilungen der Schweizerischen Entomologischen Gesellschaft* 57(4):429-430.

Markham, P. G., Pinner, M. S. and Boulton, M. I.
1984a The transmission of maize streak virus by leafhoppers, a new look at host adaptation. *Mitteilungen der Schweizerischen Entomologischen Gesellschaft* 57(4):431-432.
1984b *Dalbulus maidis* and *Cicadulina* species as vectors of disease in maize. *Maize Virus Disease Newsletter* 1:33-34.

Markham, P. G. and Townsend, R.
1974a Transmission of *Spiroplasma citri* to plants. *INSERM* (Paris) 33:201-206.

1977a Transmission of prokaryotic plant pathogens by insect vectors. *John Innes Institute Annual Report* 68:17-19.

1979a Experimental vectors of spiroplasmas. *In: Maramorosch, K. and Harris, K. F., eds. Leafhopper Vectors and Plant Disease Agents.* Pp. 413-445. Academic, New York. 654 pp.

Markham, P. G., Townsend, R., Bar-Joseph, M., Daniels M. J., Plaskitt, A. and Meddins, B. M.

1974a Spiroplasmas are the causal agents of citrus little-leaf disease. *Annals of Applied Biology* 78(1):49-57.

Markham, P. G., Townsend, R., Plaskitt, K. and Saglio, P.

1977a Transmission of corn stunt to dicotyledonous plants. *Plant Disease Reporter* 61:342-345.

Marletto, O. O. and Maggiora, S.

1983a Dueteromycetes entomoparasites of *Zyginidia pullala* Boh. *Atti XIII Congresso Nazionale Italiano di Entomologia.* Pp. 539-544.

Marshall, J. and Morgan, C. V. G.

1956a Notes on limitations of natural control of phytophagous insects and mites in a British Columbia orchard. *Canadian Entomologist* 88(1):1-5.

Martel, C. G.

1958a Influencia de alguno insecticidas en el desarrollo y rendimiento del frijol en Chapingo, Mexico. [The influence of some insecticides in the development and yield of beans in Chapingo, Mexico.] *Congreso Nacional de Entomologia y Fitopatologia Memoria* 1:185-188.

Martin, A. D., Frederiksen, R. A. and Westdahl, P. H.

1961a Aster yellows resistance in flax. *Canadian Journal of Plant Science* 41:316-319.

Martinez, D. G. and Pienkowski, R. L.

1982a Laboratory studies on insect predators of potato leafhopper eggs, nymphs and adults. *Environmental Entomology* 11(2):361-366.

1983a Comparative toxicities of several insecticides to an insect predator, a nonpest species and a pest prey species. *Journal of Economic Entomology* 76:933-935.

Martinez, M. R.

1975a Estudio de la distribucion y de la densidad de las cigarritas (Homoptera, Cicadellidae), y su posible relacion con el dano de la punta morada en el cultivo de papa (12 Variedades), del Ciclo Agricola 1971 en Chapingo, Mexico. Professional Thesis, Universidad Nacional Autonoma de Mexico. 269 pp.

Martinez, M. R. and Ramos-Elorduy de Concini, J.

1978a Populations of leafhoppers (Homoptera-Cicadellidae) on 12 varieties of potato in Chapingo, Mexico, and their possible relation to 'purple-top wilt' disease. *Agrociencia* 34:79-90.

Martinez-Lopez, G. and Black, L. M.

1974a New medium and conditions for the culture of monolayers of agallian leafhopper (*Agallia constricta* Van D.) cells used in studies of certain plant viruses. *Phytopathology* 64:1040-1041.

Martinez Lopez, G., Cujia, L. M. R. de, Luque, C. S. de

1974a Una nueva enfermedad del maiz en Colombia transmitida por el saltahojas *Dalbulus maidis* (De Long and Wolcott). *Fitopatologia* 9(2):93-99. [In Spanish with English summary.]

Martinson, C.

1978a Revision of the genus *Osbornellus* Ball (Homoptera: Cicadellidae). *Dissertation Abstracts International* (B) 38(8):3542.

Marwitz, R., Petzold, H. and Kuhne, H.

1984a Mycoplasmas in primulas. Report on new types of disease symptoms. *Biologische Bundesanstalt fur Land- und Forstwirtschaft* 84:608-612.

Mason, C. E. and Yonke, T. R.

1970a A preliminary report of the leafhoppers of central Missouri with emphasis on the subfamily Cicadellinae (Homoptera: Cicadellidae). *Transactions of the Missouri Academy of Science* 4:157.

1971a Key to species of Cicadellidae (Homoptera: Cicadellidae) of Missouri with notes on their taxonomy, distribution and biology. *Transactions of the Missouri Academy of Science* 5:93-120.

1971b Life history of four *Draeculacephala* species and *Paraulacizes irrorata* (Homoptera: Cicadellidae). *Annals of the Entomological Society of America* 64(6):1393-1399.

Massee, A. M.

1953a Notes on some interesting insects observed in 1952. Fruit tree leafhopper. *East Malling Research Station Annual Report* 40:148.

Matesova, G. I.

1960a Insects and mites—pests of apple in central and northern Kazakhstan. *Akademiya Nauk Kazakhskoi SSR Instituta Zoologichiskii Trudy* 11:24-31. [In Russian.]

Mathai, S., Kuruvilla, S. and Jacob, A.

1979a *Syncephalastrum racemosum* Cohn ex Schroeter, an entomogenus fungus of rice leafhopper *Cicadella spectra* (Dist.). *Entomon* 4:215-216.

Mathew, J. and Abraham, J.

1977a Occurrence of rice yellow dwarf in Kerala. *Agricultural Research Journal of Kerala* 15:172-173.

Mathur, B. P. and Bhandari, D. R.

1978a Studies on relative resistance of cotton varieties (*G. hirsutum*) against jassids (*Empoasca devastans* Dist.). *Pesticides* 12:51-52.

Mathur, K. C. and Chaturvedi, D. P.

1980a Biology of leaf and planthoppers, the vectors of rice virus disease in India. *Proceedings Indian National Science Academy* (B) 46(6):797-812.

Mathur, R. S. and Singh, B. R.

1975a A viruslike disorder of yellow mustard in India. *Plant Disease Reporter* 59(2):174-175.

Matolcsy, G., Saringer, G., Gaborjanyi, R. and Jermy, T.

1968a Antifeeding effect of some substituted phenoxy compounds on chewing and sucking phytophagous insects. *Acta Phytopathologica, Academiae Scientiarum Hungaricae* 3(2):275-277.

Matsuda, R.

1963a Some evolutionary aspects of the insect thorax. *Annual Review of Entomology* 8:59-76.

1965a Morphology and evolution of the insect head. *Memoirs American Entomological Institute* (Ann Arbor) 4:1-334.

Matsumoto, T., Lee, C. H. and Teng, W. S.

1968a Studies on white leaf disease of sugarcane—tranmission by *Epitettix hiroglyphicus* Matsumura. *Plant Protection Bulletin* (Taiwan) 10(4):3-9. [In Chinese with English summary.]

1968b Studies on sugar cane white leaf disease of Taiwan, with special reference to transmission by a leafhopper *Epitettix hieroglyphicus. Proceedings of the International Society of Sugar Cane Technologists* 13:1090-1099.

Matsuzaki, M.

1975a Ultrastructural changes in developing oocytes, nurse cells, and follicular cells during oogenesis in the telotrophic ovarioles of *Bothrogonia japonica* Ishihara (Homoptera, Tettigellidae). *Kontyu* 43:75-90.

Matthews, G. A. and Tunstall, J. P.

1967a Insect attack and crop loss on cotton in Rhodesia. *Cotton Growers Review* 44(4):269-283.

Maughan, F. B.

1937a Varietal differences in insect populations and injuries to potatoes. *American Potato Journal* 14:157-161.

Maung, M. M.

1985a Impact of levels of varietal resistance on populations of the green leafhopper *Nephotettix virescens* (Distant) and its natural enemies. M. S. Thesis, University of the Philippines, Los Banos, Philippines. 123 pp.

Mavi, G. S. and Harcharan, Singh

1975a Comparative efficacy of some insecticides for the control of jassid on potato. *Plant Protection Bulletin* (India) 23:40-43.

Mavi, G. S. and Sidhu, A. S.

1982a Studies on the use of nymphs for rapid screening of cotton germ-plasm for resistance to jassid. *Entomon* 7(1):85-89.

Mayhew, D. E. and Flock, R. A.

1981a Sorghum stunt mosaic. *Plant Disease* 65:84-86.

Mayse, M. A.
1978a Effects of spacing between rows on soybean arthropod populations. *Journal of Applied Ecology* 15:439-450.
1981a Observations on the occurrence of chalky deposits on forewings of *Oncometopia orbona* (F.) (Homoptera: Cicadellidae). *Proceedings Arkansas Academy of Science* 35:84-86.

Mayse, M. A., Kogan, M. and Price, P. W.
1978a Sampling abundance of soybean arthropods: Comparison of methods. *Journal of Economic Entomology* 71:135-141.

Mazzone, P.
1976a Notizie sugli stadi giovanili di *Alnetoidia alneti* (Dahl.) [sic!] e di *Alebra albostriella* (Fall.) (Homoptera, Typhlocybidae). *Bollettino del Laboratorio di Entomologia Agraria Filippo Silvestri* 33:236-240. [In Italian with English summary.]

Mbondji, P. M.
1982a The morphology of *Coloborrhis corticina camerunensis* Mbondji (Hom., Cicadellidae). *Revue Science et Technique* 2:17-24.
1982b Description and biology of *Coloborrhis corticina camerunensis* n. sp. (Homoptera Cicadellidae). *Revue Science et Technique* 2(4):25-34.
1983a Description and biology of *Coloborrhis corticina camarunensis* n. subsp. (Hom., Cicadellidae). *Bulletin de la Societe Entomologique de France* 88(3&4):347-356.
1984a *Prionomastix wonjeae* new species, parasites of *Coloborrhis corticina camarunensis* (Hymenoptera, Encyrtidae: Homoptera, Cicadellidae). *Revue Francaise d'Entomologie (Nouvelle Serie)* 6(1):17-20. [In French with English summary.]

McBride, D. K.
1972a *North Dakota Insect Control Guide for Safe Use of Pesticides.* North Dakota Cooperative Extension Service. 68 pp.

McCabe, T. L. and Johnson, L. M.
1980a Catalogue of the types in the New York State Museum insect collection. *Bulletin of the New York State Museum and Science Service* 434. 38 pp.

McCarthy, H. R.
1956a Insect populations in Cariboo Potato Field. *Proceedings of the Entomological Society of British Columbia* 52:8-11.

McClanahan, R. J.
1963a Food preferences of the six-spotted leafhopper, *Macrosteles fascifrons* (Stal). *Proceedings of the Entomological Society of Ontario* 93:90-92.

McClure, M. S.
1974a Biology of *Erythroneura lawsoni* (Homoptera: Cicadellidae) and coexistence in the sycamore leaf-feeding guild. *Environmental Entomology* 3:59-72.
1975a Key to the eight species of coexisting *Erythroneura* (Homoptera: Cicadellidae) on American sycamore. *Annals of the Entomological Society of America* 68(6):1039-1043.
1980a Role of the wild host plants in the feeding, oviposition, and dispersal of *Scaphytopius acutus* (Homoptera: Cicadellidae), a vector of peach X-disease. *Environmental Entomology* 9(2):265-274.
1980b Spatial and seasonal distribution of leafhopper vectors of peach X-disease in Connecticut. *Environmental Entomology* 9(5):668-672.
1982a Factors affecting colonization of an orchard by leafhopper (Homoptera: Cicadellidae) vectors of peach X-disease. *Environmental Entomology* 11(3):695-700.

McClure, M. S., Andreadis, T. G. and Lacy, G. H.
1982a Manipulating orchard ground cover to reduce invasion by leafhopper vectors of peach X-disease. *Journal of Economic Entomology* 75(1):64-68.

McClure, M. S. and Price, P. W.
1975a Competition among sympatric *Erythroneura* leafhoppers (Homoptera: Cicadellidae) on American sycamore. *Ecology* 56(6):1388-1397.
1976a Ecotype characteristics of coexisting *Erythroneura* leafhoppers (Homoptera: Cicadellidae) on sycamore. *Ecology* 57:928-940.

McCoy, R. E.
1979a Mycoplasmas and yellows diseases. In: Whitcomb, R. F. and Tully, J. G., eds. *The Mycoplasmas. III. Plant and Insect Mycoplasmas.* Pp. 229-264. Academic Press, New York.

McCoy, R. E., Thomas, D. W., Tsai, J. H. and French, W. J.
1978a Periwinkle wilt, a new disease associated with xylem delimited rickettsialike bacteria transmitted by a sharpshooter. *Plant Disease Reporter* 62:1022-1026.

McCoy, R. E., Tsai, H. H., Norris, R. C. and Gwin, G. H.
1983a Pigeon pea witches' broom in Florida. *Plant Disease* 67:443-445.

McCrae, A. W. R.
1975a An instance of man-biting by *Empoasca* sp. (Hem., Cicadellidae) in Britain. *Entomologist's Monthly Magazine* 109:238-239.

McEwen, F. L. and Kawanishi, C. Y.
1967a Insect transmission of corn mosaic: Laboratory studies in Hawaii. *Journal of Economic Entomology* 60(5):1413-1417.

McIntosh, A. H., Maramorosch, K. and Rechtoris, C.
1973a Adaptations of an insect cell line (*Agallia constricta*) in a mammalian cell culture medium. *In Vitro* 8:375-378.

McKenzie, L. M. and Beirne, B. P.
1972a A grape leafhopper, *Erythroneura ziczac* (Homoptera: Cicadellidae), and its mymarid (Hymenoptera) egg-parasite in the Okanagan Valley, British Columbia. *Canadian Entomologist* 104(8):1229-1233.

Mead, F. W.
1957a Additions to the United States list of Cicadellidae. *Florida Entomologist* 40(2):62.
1965a The leafhopper genus *Carneocephala* in Florida (Homoptera: Cicadellidae). *Florida Department of Agriculture. Entomology Circular* No. 42. 2 pp.
1966a "Sharpshooter" leafhoppers in Florida. *Biennial Report, Division of Plant Industry* (Florida) 26(1964-1966):49-50.
1981a Tabebuia leafhopper, *Rabela tabebuiae* (Dozier). *Florida Department of Agriculture Consumer Services, Entomology Circular* 228. 3 pp. [unnumbered].

Meade, A. B.
1962a The origin and development of populations of the six-spotted leafhopper, *Macrosteles fascifrons* (Stal), on an area basis. University of Minnesota, DA 23, 3570 Dissertation. 145 pp.

Meade, A. B. and Peterson, A. G.
1964a Origin of populations of the six-spotted leafhopper, *Macrosteles fascifrons*, in Anoka County, Minnesota. *Journal of Economic Entomology* 57(6):885-888.
1967a Some factors influencing population growth of the aster leafhopper in Anoka County, Minnesota. *Journal of Economic Entomology* 60(4):936-941.

Mebes, Hans-Detlef
1974a The biophysics of sound production of leafhoppers. *Forma et Functio* 7:95-118. [In German with English summary]

Medina, R. M.
1974a Reaccion de catorce variedades de frijol al ataque de chicharrita, picudo del ejote y conchuela del frijol en Calera, Zac. [Reaction of 14 varieties of beans to attack by potato leafhoppers, weevils and Mexican bean beetles in Calera, Zacatecas, Mexico.] *Folia Entomologica Mexicana* 29:36.

Medina, V., Archelos, D., Llacer, G., Coanova, R., Sanchez-Capuchino, J. A., Martinez, A. and Garrido, A.
1981a Contribution to the study of leafhoppers (Homoptera, Cicadellidae) in the provinces of Valencia and Murcia. *Annales del Instituto Nacional de Investigaciones Agronomicas* (Madrid) 15:157-179.

Medina, V., Garrido, A. and Al Faro, A.
1982a Nota preliminar sobre cicadellidos (Homoptera: Cicadellidae) de los Arrozals Valencinnos. *Anales del Instituto Nacional de Investigaciones Agrarias, Agricola* 19:125-137.

Medler, J. T.
1956a Control of common alfalfa insects in Wisconsin. *Journal of Economic Entomology* 48(6):718-723.

1957a Migration of the potato leafhopper — a report on a cooperative study. *Journal of Economic Entomology* 50(4):493-497.

1958a Four new *Scleroracus* from the western United States (Homoptera: Cicadellidae). *Pan-Pacific Entomologist* 34:13-16.

1958b A review of the genus *Scleroracus* in North America (Homoptera: Cicadellidae). *Annals of the Entomological Society of America* 51:230-241.

1958c Seed production and certain growth characteristics of insect-free alfalfa. *Journal of Economic Entomology* 51(5):729-733.

1960a *Agrosoma*, a new genus for *Tettigonia pulchella* Guerin and related species (Homoptera: Cicadellidae). *Annals of the Entomological Society of America* 53:18-26.

1960b Long-range displacement of Homoptera in the Central United States. *Proceedings of the XI International Congress of Entomology* 3:30-35.

1963a A review of the genus *Erythrogonia* Melichar (Homoptera, Cicadellidae). *Miscellaneous Publications of the Entomological Society of America* 4:3-30.

1966a Leafhoppers and membracids in yellow pan water traps (Homoptera). *Journal of the Kansas Entomological Society* 39(3):492-494.

Medler, J. T. and Brooks, G. N.
1957a Insect control in relation to alfalfa seed production in central Wisconsin. *Journal of Economic Entomology* 50(3):336-337.

Medler, J. T., Pinkowski, R. L. and Kieckhefer, R. W.
1966a Biological notes on *Empoasca fabae* in Wisconsin. *Annals of the Entomological Society of America* 59(1):178-180.

Medrano, F., Heinrichs, E. A., Khush, G. S. and Bacalangco, E.
1984a Hot water treatment to remove insects from rice plants used in the planthopper and leafhopper rearing program. *International Rice Research Newsletter* 9(3):20-21.

Megenasa, T.
1971a Responses of aster yellows to antibiotics and systemic insecticides. *Dissertation Abstracts International* B 32(3):1636.

Mehtre, S. S., Thombre, M. V. and Patil, A. K.
1981a Evaluation of different mutants in M3 and M4 generation for jassid (*Amrasca biguttula biguttula* Ishida) resistance in cotton. *Journal of Haharashtra Agricultural Universities* 6(1):28-29.

Meinhardt, H.
1981a Pattern formation and the activation of particular genes. *In: Progress in Developmental Biology. Fortscritte der Zoologie* 26:163-174.

Meleshko, R.
1965a Notes of plant protection. *Zashchita Rastenii ot Vrediteli i Boleznei* 10(6):42-43. [In Russian.]

Mendes, M. A.
1959a A entomofauna do castanheiro (*Castanea sativa* Miller) no concelho de Moimenta da Beira. *Portugal Direccion General dos Servico Florestais e Aquicolas, Pubs.* 26(1&2):119-278.

Menezes, M. de
1972a Contribuciao ao estudo da subfamilia Deltocephalinae (Homoptera: Cicadellidae) no Estado Sao Paulo. *Tese de doutoramento, Escola Superior de Agricultura "Luiz de Queiroz,"* Paricicaba. 146 pp.

1973a *Haldorellus*, subg. n., e tres especies novas de *Haldorus* Oman do Brasil (Homoptera, Cicadellidae, Deltocephalinae). *Revista Brasileira de Entomologia* 17(17):115-120.

1973b Notas sobre Deltocephalinae da regiao Neotropical (Homoptera, Cicadellidae). *Revista Brasileira de Entomologia* 17(20):131-136.

1974a Notas sobre Deltocephalinae da regiao Neotropical (Homoptera, Cicadellidae). II. *Revista Brasileira de Entomologia* 18(3):107-116.

1975a Notas sobre o genero *Neomesus* Linnavuori (Homoptera, Cicadellidae, Deltocephalinae). *Revista Brasileira de Entomologia* 18(3):133-137. [English summary.]

1978a Notes on the oviposition habits and the host plants of *Apogonalia grossa* (Signoret, 1854) (Homoptera, Cicadellidae, Cicadellinae). *Revista Brasileria de Entomologia* 22(2):61-64.

Menezes, M. de and Coelho, M. I. P.
1982a Chromosome studies on leafhoppers (Homoptera: Cicadellidae). *Revista Brasileira Genetica* 5(1):69-93. [In English with Portuguese summary.]

Mercer, P. C.
1977a Pests and diseases of groundnuts in Malawi. I. Virus and foliar diseases. *Oleagineaux* 32:483-488.

Merr, F. A.
1981a Mozaiekvirus, heksenbezem en knobbelziekte bij populier, en een virusachtige groeiremmeng bij wilg. *Populier* 18(3):51-59.

Merzheevskaia, O. I.
1955a Material on the investigation of insect pests of cereal crops of Polesie. *Izvestiya Akademii Nauk Belorusskoi SSR* (Minsk) 3:109-117. [In Russian.]

Metcalf, Z. P.
1956a The cataloguing of insects. *Bollettino del Laboratorio di Zoologia Generale e Agraria della Facolta Agraria in Portici* 33:673-682.

1962a *General Catalogue of the Homoptera. Fascicle VI. Cicadelloidae. Part 2. Hylicidae.* U.S. Department of Agriculture, Agriculture Research Service. 18 pp.

1962b *General Catalogue of the Homoptera. Fascicle VI. Cicadelloidea. Part 3. Gyponidae.* U.S. Department of Agriculture, Agriculture Research Service. 229 pp.

1962c *General Catalogue of the Homoptera. Fascicle VI. Cicadelloidea. Part 4. Ledridae.* U.S. Department of Agriculture, Agriculture Research Service. 147 pp.

1962d *General Catalogue of the Homoptera. Fascicle VI. Cicadelloidea. Part 5. Ulopidae.* U.S. Department of Agriculture, Agriculture Research Service. 95 pp. [New name: *Evansiola* for *Evansiella*.]

1963a *General Catalogue of the Homoptera. Fascicle VI. Cicadelloidea. Part 6. Evacanthidae.* U.S. Department of Agriculture, Agriculture Research Service. 63 pp.

1963b *General Catalogue of the Homoptera. Fascicle VI. Cicadelloidea. Part 7. Nirvanidae.* U.S. Department of Agriculture, Agriculture Research Service. 35 pp.

1963c *General Catalogue of the Homoptera. Fascicle VI. Cicadelloidea. Part 8. Aphrodidae.* U.S. Department of Agriculture, Agriculture Research Service. 268 pp.

1963d *General Catalogue of the Homoptera. Fascicle VI. Cicadelloidea. Part 9. Hecalidae.* U.S. Department of Agriculture, Agriculture Research Service. 123 pp.

1964a *General Catalogue of the Homoptera. Fascicle VI. Cicadelloidea. Bibliography of the Cicadelloidea (Homoptera: Auchenorrhyncha).* U.S. Department of Agriculture, Agriculture Research Service. 349 pp.

1964b *General Catalogue of the Homoptera. Fascicle VI. Cicadelloidea. Part 11. Coelidiidae.* U.S. Department of Agriculture, Agriculture Research Service. 182 pp.

1965a *General Catalogue of the Homoptera. Fascicle VI. Cicadelloidea. Part 1. Tettigellidae.* U.S. Department of Agriculture, Agriculture Research Service. 730 pp.

1965b *General Catalogue of the Homoptera. Fascicle VI. Cicadelloidea. Part 12. Eurymelidae.* U.S. Department of Agriculture, Agriculture Research Service. 43 pp.

1966a *General Catalogue of the Homoptera. Fascicle VI. Cicadelloidea. Part 14. Agalliidae.* U.S. Department of Agriculture, Agriculture Research Service. 173 pp.

1966b *General Catalogue of the Homoptera. Fascicle VI. Cicadelloidea. Part 13. Macropsidae.* U.S. Department of Agriculture, Agriculture Research Service. 261 pp.

1966c *General Catalogue of the Homoptera. Fascicle VI. Cic adelloidea. Part 15. Iassidae.* U.S. Department of Agriculture, Agriculture Research Service. 229 pp.

1966d *General Catalogue of the Homoptera. Fascicle VI. Cicadelloidea. Part 16. Idioceridae.* U.S. Department of Agriculture, Agriculture Research Service. 237 pp.

1967a *General Catalogue of the Homoptera. Fascicle VI. Cicadelloidea. Part 10. Section I. Euscelidae.* U.S. Department of Agriculture, Agriculture Research Service. 1077 pp.

1967b *General Catalogue of the Homoptera. Fascicle VI. Cicadelloidea. Part 10. Section II. Euscelidae.* U.S. Department of Agriculture, Agriculture Research Service. Pp. 1078-2074.

1967c *General Catalogue of the Homoptera. Fascicle VI. Cicadelloidea. Part 10. Section III. Euscelidae.* U.S. Department of Agriculture, Agriculture Research Service. Pp. 2075-2695. [New name: *Zepama* for *Metcalfiella*.]

1968a *General Catalogue of the Homoptera. Fascicle VI. Cicadelloidea. Part 17. Cicadellidae.* U.S. Department of Agriculture, Agriculture Research Service. 1513 pp.

Metcalf, Z. P. and Wade, V.
1963a *A Bibliography of the Membracoidea and Fossil Homoptera (Homoptera: Auchenorrhyncha).* Waverly Press, Inc. Baltimore. 200 pp.
1966a *General Catalogue of the Homoptera. A Supplement to Fascicle 1—Membracidae, of the General Catalogue of the Hemiptera. A Catalogue of the Fossil Homoptera (Homoptera: Auchenorrhyncha).* Waverly Press, Inc. Baltimore. 245 pp.

Meusnier, S.
1982a Deux Typhlocybinae nouveaux de France: *Zyginidia cornicula*, n. sp. et *Eupteryx genesteri*, n. sp. (Homoptera, Cicadellidae). *Revue Francaise d'Entomologie* 4(3):142-144.

Meyer, J. R.
1984a Life history of the sharpnosed leafhoppers (*Scaphytopius magdalensis* (Provancher) and four related species in southeastern North Carolina. *Journal of the Georgia Entomological Society* 19(1):72-87.

Meyer, J. R. and Colvin, S. A.
1985a Diel periodicity and trap bias in sticky trap sampling of sharpnosed leafhopper populations. *Journal of Environmental Entomology* 20(2):237-243.

Meyer, M. de and Bruyn, L. de
1984a On the phenology of some Pipunculidae (Diptera) in Belgium. *Bulletin et Annales de la Societe Royale Belge d'Entomologie* 120:123-131.

Meyer, R. W. and Osmun, J. V.
1970a Insects and other arthropods of economic importance in Indiana during 1970. *Proceedings of the Indiana Academy of Science* 80:286-298.

Meyerdirk, D. E. and Hessein, N. A.
1985a Population dynamics of the beet leafhopper, *Circulifer tenellus* (Baker), and associated *Empoasca* spp. (Homoptera: Cicadellidae) and their egg parasitoids on sugar beets in southern California. *Journal of Economic Entomology* 78(2):346-353.

Meyerdirk, D. E. and Oldfield, G. N.
1985a Evaluation of trap color and height placement for monitoring *Circulifer tenellus* (Baker) (Homoptera: Cicadellidae). *Canadian Entomologist* 117(4):505-511.

Meyerdirk, D. E., Oldfield, G. N. and Hessein, N. A.
1983a Bibliography of the beet leafhopper, *Circular [sic!] tenellus* (Baker), and two of its transmitted plant pathogens, curly top virus and *Spiroplasma citri* Saglio et al. *Bibliographies of the Entomological Society of America* 2:17-55.

Miah, M. A. H. and Husain, M.
1981a Assessment of soybean (*Glycine max* [L.] Merril) cultivars for resistance to insect pests. *Science and Culture* 47:72-79.
1983a Tagging green leafhopper of rice *Nephotettix virescens* (Distant). *Journal of Nuclear Agriculture and Biology* 12:81.

Miah, M. S. A.
1971a A short report on virus diseases of rice in East Pakistan. *Agriculture Pakistan* 22(1):117-129.

Miah, S. A. and Ling, K. C.
1978a Effect of temperature on the movement of *Nephotettix virescens*. *International Rice Research Newsletter* 3(5):13-14.

Michelsen, A., Fink, F., Gogala, M. and Traue, D.
1982a Plants as transmission channels for insect vibrational songs. *Behavioral Ecology and Sociobiology* 11(4):269-281.

Miles, P. W.
1972a The saliva of hemiptera. *Advances in Insect Physiology* 9:183-255.

Miller, L. A.
1956a Changes in insect populations in southwestern Ontario. *Entomological Society of Ontario Annual Report* 87:15-19.
1956b Field crop and garden insects. *Entomological Society of Ontario Annual Report* 87:64-69.
1960a Control of the six-spotted leafhopper in southern Ontario. *Canada Department of Agriculture Publication* 1076:3-7.

Miller, L. A. and De Lyzer, A. J.
1960a A progress report on studies of biology and ecology of the six-spotted leafhopper, *Macrosteles fascifrons* (Stal), in southwestern Ontario. *Proceedings of the Entomological Society of Ontario* 90:7-13.

Miller, P. R.
1954a A new leafhopper-borne virus disease from the Netherlands. *FAO Plant Protection Bulletin* 2(5):66-67.
1966a International usefulness of an isolation laboratory for plant pathogens, especially viruses and their vectors. *Plant Disease Reporter* 50:803-805.

Miller, R. L.
1962a Seasonal distribution of the potato leafhopper, *Empoasca fabae* (Harris) among *Solanum* clones. Iowa State University of Science and Technology DA 23, 757 Dissertation. 222 pp.

Miller, R. L. and Hibbs, E. T.
1963a Distribution of eggs of the potato leafhopper, *Empoasca fabae*, on *Solanum* plants. *Annals of the Entomological Society of America* 56(6):737-740.

Milliron, H. E.
1958a Economic insect and allied pests of Delaware. *Delaware Agricultural Experiment Station Bulletin* 321. 87 pp.

Milne, R. G. and Lovisolo, O.
1977a Maize rough dwarf and related viruses. *Advances in Virus Research* 21:267-341.

Minoranskiy V. A. and Logvinenko V. N.
1968a Cicadinea on beet in Rostov area. *Science Report Higher School, Biological Science* 3:25-278. [In Russian.]

Miranda Colin, S.
1971a Efecto de las malezas, plagas y fertilizantes en la produccion de frijol. [The effect of weeds, pests and fertilizers on the yield of beans.] *Agricultura Tecnico en Mexico* 3(2):61-66.

Mircetich, S. M., Lowe, S. K., Moller, W. J. and Nyland, G.
1976a Etiology of almond leaf scorch disease and transmission of the causual agent. *Phytopathology* 66:17-24.

Mirzayans, H.
1976a List of Auchenorrhyncha (Homoptera) from Province of Fars. (1). *Journal of the Entomological Society of Iran* 3(1&2):110-112. [In English with Farsi summary.]

Mishra, M. D., Ghosh, A., Niazi, F. R., Basu, A. M. and Raychaudhuri, S. P.
1973a The role of gramineceous weeds in the perpetuation of rice tungro virus. *Journal of the Indian Botanical Society* 52:176-183.

Mishra, M. D., Niazi, F. R., Basu, A. M., Ghosh, A. and Raychaudhuri, S. P.
1976a Detection and characterization of a new strain of rice tungro virus in India. *Plant Disease Reporter* 60:23-25.

Mishra, P. M. and Saxens, H. P.
    1983a Phytotonic effects of insecticides in pigeon pea *Cajanus cajan* (L.) Millsp. *Madras Agriculture Journal* 70:309-311.

Mishra, R. K.
    1979a Male reproductive organs of five species of Auchenorrhyncha (Homoptera). *Acta Entomologica Bohemoslovaca* 76(3):162-168.

Mishra, R. K. and Shankar, A.
    1980a The 'knockdown speed' of insecticides against brown planthopper and green leafhopper. *International Rice Research Newsletter* 5(6):15-16.

Mishra, U. S.
    1977a Insect pest problems of rice in Chhattisgarh. *Pesticides* 11:24-26.

Mishra, U. S. and Kaushik, U. K.
    1976a Phosphamidon as an aerial spray against *Nephotettix virescens* in Madhya Pradesh, India. *Rice Entomology Newsletter* 4:22-23.

Misiga, S., Musil, M. and Valenta, V.
    1960a Niektore Hostitelske Rastliny Virusu Zelenokvetosti Dateliny. [Some host plants of strawberry green petal virus.] *Biologia* 15:538-542.

Misra, B. C.
    1980a *The leaf and plant hoppers of rice*. Central Rice Research Institute, Cuttack, India. 182 pp.

Misra, B. C. and Israel, P.
    1968a Leaf and planthoppers of rice. *International Rice Commission Newsletter* 17:7-12.
    1968b Anatomical studies of the ovipositional site of plant hoppers and leaf hoppers on rice. *Indian Journal of Entomology* 30(2):178.
    1970a The leaf and plant hopper problems in high yielding varieties of rice. *Journal of the Association of Rice Research Workers* 7(2):127-130.

Misra, B. C., Rajamiani, S., Prasad, K., Hansda, N. N. and Chatterji, S. M.
    1983a Isolation of donors resistant to major insect pests of rice. *Journal of Entomological Research* 7:84-87.

Misra, D. S. and Reddy, K. D.
    1985a Seasonal population of rice leafhoppers and planthoppers at Varansi, India. *International Rice Research Newsletter* 10(4):21.

Misra, M. P. and Krishna, S. S.
    1982a Ovipositional behavior of *Epipyrops melanoleuca* Fletcher (Lepidoptera: Epipyropidae) parasitizing sugarcane leafhoppers affected by ambient temperature. *Mitteilungen aus dem Zoologischen Museum in Berlin* 58(2):229-231.

Misra, M. P., Pawar, A. D., Gupta, B. N., Samujh, R. and Sahu, A. K.
    1984a New records of insect pests and their natural enemies on paddy from Gorakhpur, Utter Pradesh, India. *Journal of Advanced Zoology* 5(2):126-127.

Misra, S. S. and Lal, L.
    1981a Evaluation of foliar systemic insecticides against leaf hopper, *Amrasca devastans* (Dist.), on potato crop. *Entomon* 6(3):201-205.
    1985a Evaluation of granular systemic insecticides against *Amrasca devastans* (Dist.) on potato crop. *Indian Journal of Plant Protection* 12:53-54.

Mitjaev, I. D.
    1960a Concerning the insect fauna—pests of *Eleagnus* in Kazakhstan. *Akademii Nauk Kazakhskoi SSR, Institut Zoologicheskii Trudy* 11:108-128. [In Russian.]
    1960b Pests of strawberries and raspberries in central and northern Kazakhstan. *Akademii Nauk Kazakhskoi SSR, Institut Zoologicheskii Trudy* 11:32-35. [In Russian.]
    1962a Typhlocybidae (Auchenorrhyncha) as pests of fruit trees. *Sbornik Entomologicheskogo Rabot Akadamii Nauk Kirgizsk SSR* 1:45-54. [In Russian; translation in *Collected Entomological Papers of the Kirgiz Academy of Sciences* 1:45-54. Frunze, Kirgiz Academy of Science Press.]
    1962b The leafhopper fauna of crops in northeastern Kazakhstan. *Trudy Instituta Zoologii Akademii Nauk Kazakhskoi SSR* 18:142-149. [In Russian.]
    1963a On the mass outbreak and fungus disease of *Cicadella viridis* L. in eastern Kazakhstan. *Akademii Nauk Kazakhskoi SSR, Institut Zoologicheskii Trudy* 21:19-24. [In Russian.]
    1963b Data on the fauna and biology of Typhlocybinae (Homoptera; Auchenorrhyncha) Kazakhstan. *Trudy Instituta Zoologii Akademii Nauk Kazakhskoi SSR* 21:49-73. [In Russian.]
    1963c New and little-known species of leafhoppers (Auchenorrhyncha, Typhlocybinae) from Kazakhstan. *Entomologicheskoe Obozrenie* 42(2):399-409. [In Russian; translated in *Entomological Review* (Washington) 42(2):221-225.]
    1963d Materials for the study of Kazakhstanian insects: Mass production of fungoid disease in the reproductive organs of *Cicadella virdis* L. found in eastern Kazakhstan. *Trudy Instituta Zoologii Akademii Nauk Kazakhskoi SSR*, Alma Ata 21:19-24.
    1965a A new subgenus and species of *Pithyotettix* (Auch.-Cicadellidae) from the southern Altai. *Zoologicheskii Zhurnal* 44(8):1260-1262. [In Russian with English summary.] [New subgenus: *Abietotettix*.]
    1967a New and little-known species of planthoppers and leafhoppers (Homoptera, Auchenorrhyncha) from eastern Kazakhstan. *Entomologicheskoe Obozrenie* 46(3):712-723. [Translation in *Entomological Review* (Washington) 46(3):422-428.]
    1967b On systematics and ecology of cicades [sic!] of the genus *Linnavuoriana* Dlab., 1958 (Homoptera, Typhlocybinae) of south-eastern Kazakhstan. *Zoologicheskii Zhurnal* 46:710-714. [In Russian.]
    1967c New species of leafhoppers of the subfamily Euscelinae (Homoptera, Cicadinea) from south eastern Kazakhstan. *Zoologicheskii Zhurnal* 46(8):1203-1208. [In Russian.]
    1968a Several new species of leafhoppers (Homoptera, Cicadellidae) from south-western Altai and Zaisan hollow. *Zoologicheskii Zhurnal* 47:635-637. [In Russian with English summary.]
    1968b Cicadas of eastern Kazakhstan. *Trudy Instituta Zoologii Akademii Nauk Kazakhskoi SSR.* Alma Ata 30:5-57.
    1968c A gall-forming leafhopper. *Transactions of the Insect of Zoology of the Academy of Sciences of the Kazakh SSR* 30:205-206.
    1968d The hop cicadellid. *Zashchita Rastenii* 13(5):40. [In Russian.]
    1969a New leafhoppers (Homoptera, Cicadinea) from southern and western Kazakhstan. *Zoologicheskii Zhurnal* 48(3):363-370.
    1969b New species of leafhoppers (Homoptera, Cicadinea) from Tien Shan and Karatau. *Zoologicheskii Zhurnal* 48(7):1041-1048.
    1969c New species of cicades [sic!] (Homoptera, Cicadellidae) from Zailiisky Alatau. *Zoologicheskii Zhurnal* 48(11):1635-1640.
    1970a Cikady (Homoptera, Idiocerinae), povrezhdayushchie derevya; kustarniki v Alma-Atinskom zapovednike. [Leafhoppers (Homoptera, Idiocerinae) which injure trees and shrubs in Alma-Ata National Park.] *Trudy Altaiskogo Gosudarstvennogo Zapovednika* 9:217-227. [In Russian.]
    1970b Characteristics of the vertical distribution of leafhoppers in the montane-forest and alpine zones of the Zaiylskiy Alatau. *Transactions of the Alma-Ata National Park* 9.
    1971a *Leafhoppers of Kazakhstan (Homoptera-Cicadinea): The determinant.* Academy of Sciences of the Kazakh SSR. 209 pp. [In Russian.]
    1971b Novye vidy cikadovykh iz aridnykh landshaftov yuzhnogo Kazakhstana. [New species of leafhoppers from arid terrains in southern Kasakhstan.] *Trudy Instituta Zoologii Akademii Nauk Kazakhskoi SSR*, Alma Ata 32:56-66. [In Russian.]

1971c Zonal and spatial distribution of cicadina in south and east Kazakhstan. *Trudy Instituta Zoologii Akademii Nauk Kazakhskoi SSR,* Alma Ata 32:44-55. [In Russian.]

1971d Some peculiarities of feeding specializations of cicadas from southern Kazakhstan. *Proceedings of the Academy of Sciences of the Kazakh SSR, Biological Series* 2:42-47.

1973a A description of some species of the genus *Macropsidius* Rib., 1952 (Homoptera, Cicadellidae) from Kazakhstan. *Entomologicheskoe Obozrenie* 50(2):347-351. [English translation in *Entomological Review* (Washington) 50:239-242.]

1973b Pathological variability in leafhoppers. *Izvestiya Akademii Nauk Kazakhskoi SSR* (Seriya Biologicheskikh Nauk) 6:18-23. [In Russian.]

1974a Zoogeographical review of the Cicadina fauna of S. Kazakhstan. *Trudy Instituta Zoologii Akademii Nauk Kazakhskoi SSR,* Alma Ata 35:5-19.

1974b The fauna and ecological geographical characteristics of leafhoppers in the west of Kazakhstan. *Deposited in the All-Union Institute of Scientific and Technical Information* No. 1565.

1975a New species of Cicadinea (Homoptera) from Kazakhstan. *Entomologicheskoe Obozrenie* 54(3):577-586.

1975b Fauna and biology of the leafhoppers of Kazakhstan. *Deposited in the All-Union Institute of Scientific and Technical Information* No. 1577.

1979a New species of *Austroasca* Lower, 1952 (H,C,T) from North Kazakhstan. *Trudy Vsesoyuznogo Entomologicheskogo Obshchestva* 61:29-31.

1979b A new species and subspecies of leafhoppers (Homoptera, Cicadellidae) from north Kazakhstan. *Zoologicheskii Zhurnal* 58(11):1738-1741.

1979c Zoogeographical review of the leafhopper fauna of southern Kazakhstan. *Transactions of the Institute of Zoology of the Academy of Sciences of Kazakh SSR* 35.

1979d Leafhoppers of Northern Kazakhstan. *Deposited in the All-Union Institute of Scientific and Technical Information* No. 1190.

1979e Characteristics of the zonal distribution of leafhoppers in Northern Kazakhstan. Sbornik "Novosti entomologii Kazakhstana." *Deposited in the All-Union Institute of Scientific and Technical Information* No. 1949.

1979f Leafhoppers of the floodlands of the Aley River. *Deposited in the All-Union Institute of Scientific and Technical Information* No. 1939.

1980a Novye vidy cikadovykh (Homoptera, Auchenorrhyncha) iz severnogo Kazakhstana. [New species of cicadid (Homoptera, Auchenorrhyncha) from northern Kazakhstana.] *Izvestiya Akademii Nauk Kazakhskoi SSR* (Seriya Biologicheskikh Nauk), Alma Ata 1:36-39.

1980b Obzor vidov *Eremochlorita* Zachvatkin, 1946 (Auchenorrhyncha, Typhlocybidae). [Review on the species of *Eremochlorita* Zachvatkin, 1946 (Auchenorrhyncha, Typhlocybidae) of Kazakhstan.] *Trudy Instituta Zoologii Akademii Nauk Kazakhskoi SSR,* Alma Ata 39:32-43

1980c Dva novykh vida tiflocibin (Auchenorrhuncha [sic!], Typhlocybidae) iz severnogo Kazakhstan. [Two new species of leafhoppers (Auchenorrhyncha, Typhlocybinae) of Kazakhstan.] *Trudy Instituta Zoologii Akademii Nauk Kazakhskoi SSR,* Alma Ata 39:107-109.

1980d Two new species of the tribe Paralimnini Distant, 1908 (Homoptera, Cicadellidae) from northern Kazakhstan. *Entomologicheskoe Obozrenie* 59(1):128-131.

Mitra, D. K. and John, V. T.
1973a Response of rice seedlings towards feeding by the rice leafhopper, *Nephotettix* spp., vector of rice tungro virus and yellow dwarf. *Indian Phytopathology* 26(2):368-370.

Mitra, D. K., Raychaudhuri, S. P., Everett, T. R., Ghosh, A. and Niazi, F. R.
1970a Control of the rice green leafhopper with insecticidal seed treatment and pretransplant seedling soak. *Journal of Economic Entomology* 63:1958-1961.

Mitsuhashi, J.
1964a Axenic rearing of insect vectors of plant viruses. *Annals of the New York Academy of Science* 118:384-386.

1964b Insect tissue culture up to present. *Plant Protection* 18:359-366. [In Japanese.]

1965a Aseptic rearing and tissue culture of leafhoppers. *Japanese Journal of Applied Entomology and Zoology* 9:69.

1965b Preliminary report on plant virus multiplication in the leafhopper vector (*Nephotettix cincticeps* (Uhl.), *Macrosteles fascifrons* (Stal) and *Agallia constricta* Van D.) cells grown in vitro. *Japanese Journal of Applied Entomology and Zoology* 9:137-141.

1965c *In vitro* cultivation of the embryonic tissues of the green rice leafhopper, *Nephotettix cincticeps* Uhler (Homoptera: Cicadellidae). *Japanese Journal of Applied Entomology and Zoology* 9:107-114.

1965d Insect transmission. 3. Multiplication of rice dwarf virus in the green rice leafhopper cells grown in vitro. *Annals of the Phytopathological Society of Japan* 31:385. [In Japanese.]

1965e Aseptic rearing of leafhoppers and planthoppers (Homoptera: Cicadellidae and Delphacidae). *Kontyu* 33:271-274.

1965f Multiplication of plant virus in the leafhopper vector cells grown *in vitro*. *U.S.-Japan Joint Seminar,* Tokyo, Japan, pp. 60-69.

1965g Insect tissue culture: A review. *Biological Science* (Tokyo) 17:105-117.

1965h Multiplication of rice dwarf virus in the green rice leafhopper cells grown *in vitro*. *Annals of the Phytopathological Society of Japan* 31:385. [In Japanese.]

1966a Chromosome numbers of the green rice leafhopper, *Nephotettix cincticeps* Uhler. (Homoptera: Cicadellidae). *Applied Entomology and Zoology* 1(2):103-104.

1966b Tissue culture in insects. *Kagaku-to-Seibut-su* 4:608-613. [In Japanese.]

1967a Infection of leafhopper and its tissues cultivated *in vitro* with Chilo iridescent virus. *Journal of Invertebrate Pathology* 9:432-434.

1967b An observation on the formation of a binucleate cell in the primary cultures of leafhopper embryonic cells. *Experimental Cell Research* 48(1):93-96.

1969a Plant pathogenic viruses in insect vector tissue culture. In: Maramorosch, K., ed. *Viruses, Vectors and Vegetation.* Pp. 475-503. Wiley & Sons, New York. 666 pp.

1970a A device for collecting planthopper and leafhopper eggs (Hemiptera: Delphacidae and Deltocephalidae). *Applied Entomology and Zoology* 5(1):47-49.

1974a Methods for rearing leafhoppers and planthoppers on artificial diets. *Review of Plant Protection Research* 7:57-67.

1975a Primary culture of leafhopper embryonic tissues. *Tissue Culture Association and Management* 1(3):151-154.

1979a Artificial rearing and aseptic rearing of leafhopper vectors: Applications in virus and MLO research. *In: Maramorosch, K. and Harris, K. F., eds. Leafhopper Vectors and Plant Disease Agents.* Pp. 369-412. Academic Press, New York. 654 pp.

Mitsuhashi, J. and Kono, Y.
1975a Intracellular microorganisms in the green rice leafhopper, *Nephotettix cincticeps* Leher [sic!] (Hem., Deltocephalidae). *Applied Entomology and Zoology* 10(1):1-9.

Mitsuhashi, J. and Maramorosch, K.
   1963a Aseptic cultivation of four virus transmitting species of leafhoppers (Cicadellidae). *Contributions from Boyce Thompson Institute for Plant Research* 22(4):165-173.
   1963b A successful method for rearing leafhopper vectors of plant viruses under aseptic conditions. *Proceedings of the XVI International Congress of Zoology* 16(1):3.
   1964a Inoculation of [carrot] plant tissue cultures with aster yellows virus [by *M. fascifrons*]. *Virology* 23(2):277-279.
   1964b Leafhopper tissue culture: Embryonic, nymphal, and imaginal tissues from aseptic insects. *Contributions from Boyce Thompson Institute for Plant Research* 22(8):435-460.

Mitsuhashi, J. and Nasu, S.
   1967a An evidence for the multiplication of rice dwarf virus in the vector cell cultures inoculated *in vitro*. *Applied Entomology and Zoology* 2(2):113-114.

Mitsuhashi, J., Ringle, S. M., Martin, J. F. and Maramorosch, K.
   1965a A presumptive bacterial symbiont from the six-spotted leafhopper, *Macrosteles fascifrons* Stal. *Contributions from Boyce Thompson Institute for Plant Research* 23:123-126.

Miura, T.
   1976a Parasitism of *Gonatocerus* sp. (Hymenoptera: Mymaridae), an egg parasite of the green rice leafhopper, *Nephotettix cincticeps* Uhler, in the paddy field. *Bulletin of the Faculty of Agriculture, Shimane University* 10:43-48. [In Japanese with English summary.]
   1976b Parasitic activity of *Paracentrobia andoi* (Ishii) and *Gonatocerus* sp. (Hymenoptera: Mymaridae), two egg parasites of the green rice leafhopper, *Nephotettix cincticeps* Uhler in the paddy field. *Bulletin of the Faculty of Agriculture, Shimane University* 10:49-55. [In Japanese with English summary.]
   1978a On the parasitic activity and the age in days of adult *Paracentrobia andoi* (Ishii) (Hymenoptera: Trichogrammatidae), an egg parasite of the green rice leafhopper *Nephotettix cincticeps* Uhler (Homoptera: Deltocephalidae). *Bulletin of the Faculty of Agriculture, Shimane University* 12:36-40. [In Japanese with English summary.]
   1979a On the longevity and parasitic activity of adult *Gonatocerus* sp. (Hymenoptera: Mymaridae). *Bulletin of the Faculty of Agriculture, Shimane University* 13:156-162.

Miura, T., Hirashima, Y., Chujo, M. T. and Chu, Yau-i
   1981a Egg and nymphal parasites of rice leafhoppers and planthoppers: A result of field studies in Taiwan in 1979. Part 1. *Esakia* 16:39-50.

Miura, T., Hirashima, Y. and Wongsiri, T.
   1979a Egg and nymphal parasites of rice leafhoppers and planthoppers: A result of field studies in Thailand in 1977. *Esakia* 13:21-44.

Miyahara, K., Matsuzaki, M., Tanaka, K. and Sako, N.
   1982a A new disease on onion caused by mycoplasma-like organisms in Japan. *Annals of the Phytopathological Society of Japan* 48:551-554.

Miyahara, K., Nakamura, H. and Hashimoto, E.
   1976a Influence of the second crop after rice cropping on occurrence of rice dwarf virus disease transmitted by the green rice leafhopper (*Nephotettix cincticeps* Uhler). *Proceedings of the Association for Plant Protection of Kyushu* 22:109-111.

Miyahara, K., Wakibe, H., Matsuzaki, M. and Tanaka, K.
   1983a Studies on the influencing factors for occurrence of onion yellows caused by mycoplasma-like organisms and its control method. (VI.) Seasonal fluctuation of the population of *Macroseteles orientalis* and its transmission rates on onion yellows in Japan. *Proceedings of the Association for Plant Protection of Kyushu* 29:100-102.

Miyai, S., Kiritani, K. and Sasaba, T.
   1978a An empirical model of *Lycosa*-hoppers interaction system in the paddy field. *Protection Ecology* 1:9-21.

Miyamoto, S. and Miyatake, Y.
   1963a Homoptera taken by the Kyushu University Expedition to the Yaeyama Group, 1962. *Report of the Committee on Foreign Scientific Research, Kyushu University* 1:83-90.

Miyashita, K., Ito, Y., Yasuo, S., Yamaguchi, A. and Ishi, M.
   1964a Studies on the dispersal of leafhoppers and planthoppers. III. Dispersal of *Delphacodes striatella* Fallen, *Nephotettix cincticeps* Uhler, and *Deltocephalus dorsalis* Motschulsky in nursery and paddy fields. *Japanese Journal of Ecology* 14:233-241.

Miyata, T. and Saito, T.
   1976a Mechanism of malathion resistance in the green rice leafhopper, *Nephotettix cincticeps* Uhler (Hemiptera: Deltocephalidae). *Journal of Pesticide Science* 1:23-29.
   1978a Adaptation and resistance to insecticides in the planthopper and leafhopper. *Journal of Pesticide Science* 3(2):179-184.
   1982a Mechanism of selective toxicity of malathion and pyridafenthion against insect pests of rice and their natural enemies. In: Heong, K. L., Lee, B. S., Lim, T. M., Tech, C. H. and Ibrahim, Y., eds. *Proceedings of the International Conference on Plant Protection in the Tropics.* Pp. 391-397. Kuala Lumpur, Malaysia.

Miyata, T., Saito, T., Fukamachi, S., Kiritani, K., Kawahara, S., Yoshioka, K., Ozaki, K., Sasaki, Y., Tsuboi, A., Hama, H. and Iwata, T.
   1981a On the relation between insecticide resistance and aliesterase activity in the green rice leafhopper, *Nephotettix cincticeps* Uhler (Hemiptera: Deltocephalidae). *Japanese Journal of Applied Entomology and Zoology* 25:150-155.

Miyata, T., Saito, T., Hama, H., Iwata, T. and Ozaki, K.
   1980a A new and simple detection method for carbamate resistance in the green rice leafhopper, *Nephotettix cincticeps* Uhler (Hemiptera: Deltocephalidae). *Applied Entomology and Zoology* 15:351-352.

Miyata, T., Sakai, H., Saito, T., Yoshioka, K., Ozaki, K., Sakai, Y. and Tsuboi, A.
   1981a Mechanism of joint action of Katajin P with Malathion in the Malathion resistant green rice leafhopper, *Nephotettix cincticeps* Uhler (Hemiptera: Deltocephalidae). *Applied Entomology and Zoology* 16:258-263.

Mochida, O.
   1970a Discrimination of stadium and sex of nymphs in *Nephotettix cincticeps* (Uhler) (Hemiptera: Cicadellidae). *Applied Entomology and Zoology* 5(1):44-47.
   1973a Auchenorrhynchous insects (Homoptera) collected on the island of Ishigaki (Ryukyu, Japan) late June, 1972. *Kontyu* 41(4):475-476.

Mochida, O. and Kuno, E.
   1962a Relations of seasonal prevalences of four species of planthoppers and leafhoppers determined by light trap to their seasonal abundances in a rice field. *Proceedings of the Kyushu Association for Plant Protection* 8:6-9.

Mochida, O., Wahyu, A., Tatang, S. and Rahayu, A.
   1979a The occurrence of the brown planthopper (*Nilaparvata lugens*), the white-backed planthopper (*Sogatella furcifera*), and the green leafhopper (*Nephotettix* spp.) in Sumatra, Java, Bali, Lombok, South Kalimantan and South Sulawesi during February to April 1978. *Kongres Entomologi* 2:1-15 [articles individually paginated].

Mochida, O. and Valencia, S. L.
   1984a Evaluation of eight synthetic pyrethroids for delphacid and cicadellid pest control on rice. *Mitteilungen der Schweizerishchen Entomologischen Gesellschaft* 57(4):435.

Moczar, L.
   1979a Some dryinids from Malaysia (Hymenoptera). *Acta Biologica Szeged* 25(1-2):77-83.

Moffitt, H. R.
   1968a The bionomics of *Empoasca solana* Delong on cotton in Southern California. *Dissertation Abstracts* 28B:3956-3957.

Moffitt, H. R. and Reynolds, H. T.
1972a Bionomics of *Empoasca solana* DeLong on cotton in southern California. *Hilgardia* 41(11):247-298.

Mohamed Hamed, K. el Din
1964a List of localities, dates and hosts of order Hemiptera-Homoptera as recorded in the entomological collection, Ministry of Agriculture. *Agricultural Research Review* (Cairo) 42:116-128.

Mohammad, S., Hussain, M. and Sagar, G. C.
1980a Preferential infestation of the leaf-hopper of sunflower in relation to nitrogen nutrition and plant density. *Indian Journal of Agricultural Sciences* 50:955-957.

Mohammad Zaky, S. H. F.
1982a Damage potential of potato leafhopper nymphs, *Empoasca fabae* Harris (Homoptera: Cicadellidae) on established stand alfalfa. *Dissertation Abstracts International* (B) 42(9):3516.

Mohan, N. J., Krishnaiah, K. and Prasad, V. G.
1983a Chemical control of insect pests of okra, *Abelmoschus esculentus* Moench. *Indian Journal of Entomology* 45:152-158.

Mohan, N. J., Kumar, N. K. L. and Prasad, V. G.
1980a Control of leaf hopper (*Amrasca biguttula biguttula* Ishida) and fruit borer (*Leucinodes orbonalis* Guen.) on brinjal. *Pesticides* 14:19-21.

Mohan, S. and Janartharan, R.
1985a On certain behavioural responses of major rice pests to different light sources. *In: Regupathy, A. and Juyaraj, S., eds. Behavioural and Physiological Approaches in Pest Management.* Pp. 94-99. Coimbature, Tamil Nadu, India.
1985b Light trap attraction of major pests of rice and a mirid predator during rice growing season. *In: Regupathy, A. and Juyaraj, S., eds. Behavioural and Physiological Approaches in Pest Management.* Pp. 107-109. Coimbature, Tamil Nadu, India.

Moiseeva, N. V.
1963a Insect pests of *Artemisia cina* Berg. "ex Poljak." *Akademii Nauk Kazakhskoi SSR Institut Zoologicheskii Trudy* 21:26-28. [In Russian.]

Moiz, S. A. and Naqvi, K. M.
1971a Insecticidal trial against cotton jassid, *Empoasca devastans* Dist. *Agriculture Pakistan* 22(1):71-76.

Mokrotovarov, S.
1965a Pests and diseases of rice in Indonesia. *Zashchita Rastenii ot Vrediteli Boleznei* 11:49-50. [In Russian.]

Molina Valero, L. A. and Viana, G. B.
1970a Mancha clorotica de la curuba (*Passiflora mollissima* [H. B. K.] Bailey) causada por *Empoasca* sp., en Narino. *Revista de Ciencias Agricolas* 2(1/2):5-16, 12 ref. [In Spanish.]

Monge Casillas, J.
1981a Contribution to the study of the cicadellid *Erythroneura* sp., the flower thrips, *Frankliniella* sp. and the leaf skeletonizer *Harrisona* sp., the main pests of grapevine in Caborca, Sonora. *Agricultura Tecnica en Mexico* 7(1):37-50.

Monsef, A.
1981a Life-cycle and toxicogenic role of *Austroasca* (s.g. *Jacobiasca*) *lybica* Berg & San. in cotton fields in Fars province. *Entomologie et Phytopathologie Appliquees* 49:11-17.

Monty, J.
1977a Entomological news from the agricultural services of the ministry of agriculture, natural resources and the environment. *Revue Agricole et Sucriere de l'Ile Maurice* 56:107-109.

Moore, D. H.
1959a Field evaluation of Thiodan as an insecticide for potatoes. *Journal of Economic Entomology* 52(4):564-567.

Moore, G. D.
1968a Evaluation of feeding injury to alfalfa by the potato leafhopper, *Empoasca fabae* (Harris). Ph. D. Dissertation, University of Minnesota. 123 pp. [Diss. Abst. Int. B 29(8):2931.]
1971a A resume of the effects of feeding by the potato leafhopper, *Empoasca fabae* (Harris), on alfalfa. *Proceedings of the North Central Branch of the Entomological Society of America* 26(1-2):82-83.
1971b Selecting for resistance to the potato leafhopper in alfalfa. *Proceedings of the North Central Branch of the Entomological Society of America* 26(1-2):83-84.

Moore, K. M.
1965a Observations on some Australian forest insects. 20. Insects attacking *Hakea* spp. in New South Wales. *Proceedings of the Linnean Society of New South Wales* 89(406):295-306.

Moore, L.
1973a Biological relationships of selected species of *Carneocephala* (Homoptera, Cicadellidae). *Dissertation Abstracts International* (B) 33(5):2135.

Moore, T. E.
1961a Audiospectrographic analysis of sounds of Hemiptera and Homoptera. *Annals of the Entomological Society of America* 54(2):273-291.

Moore, T. E. and Ross, H. H.
1957a The Illinois species of *Macrosteles*, with an evolutionary outline of the genus (Homoptera, Cicadellidae). *Annals of the Entomological Society of America* 50(2):109-118.

Moraes, G. J. and Oliveira, C. A. V. de
1981a Behaviour of varieties of *Vigna unguiculata* Walp in relation to attack by *Empoasca kraemeri* Ross & Moore, 1957. *Anais da Sociedada Entomologica do Brasil* 10:255-259.

Moraes, G. J. de, Oliveira, C. A. V. de, Albuquerque, M. M., de Salviano, L. M. C. and Possidio, P. L. de
1980a The appropriate time to control the green leafhopper on cowpea crops. *Comunicado Tecnico* No. 1. 2 pp.
1980b Effect of the time of infestation by *Empoasca kraemeri* Ross & Moore, 1957 (green bean leafhopper) (Homoptera: Typhlocibidae) on crops of *Vigna unguiculata* Walp (Macassar bean). *Anais da Sociedade Entomologica do Brasil* 9:67-74.

Moraes, G. J. and Ramalho, F. de S.
1980a Some insects associated with *Vigna unguiculata* Walp in the northeast. *Boletim de Pesquisa* No. 1. 10 pp.

Morallo-Rejesus, B. and Eroles, L. C.
1980a Two insecticidal principles from marigold (*Tagetes* spp.) roots. *Philippine Entomologist* 4:87-97.

Moravskaja, A. S.
1948a To the knowledge of the genus *Zyginidia* (Homoptera-Cicadina). *Nauchno-metodicheskye Zapiski* 11:198-207. [In Russian.]
1956a On the taxonomy and ecology of the leafhoppers of the genus *Psammotettix* (Homoptera, Cicadina). *Zoologicheskii Zhurnal* (Moscow) 35:709-718. [In Russian with English summary.]

Moreau, J. P.
1963a A propos de l'elevage permanent d'un Homoptere Jasside du genre Macrosteles. *Revue de Zoologie Agricole et Appliquee* 62(10-12):114-117.
1969a Presence d'*Orosius* sp. (Homopt. Auchenorrhynques) dans une cotonnerie atteinte de virescence en Haute-Volta. Gourret, J. P. and Maillet, P. L., Ultrastructure des mycoplasmes dans le phloeme du cotonnier atteint de virescence. *Coton et Fibre Tropicales* 24(4):471-472. [In French with English and Spanish summaries.]

Moreau, J. P. and Boulay, C.
1967a Mode de piqure de trois cicadelles vectrices de virus, *Euscelis plebejus* Fall., *Macrosteles sexnotatus* Fall. et *Aphrodes bicinctus* Schrk.: Etude histologique. [Mechanisms of the feeding of some virus-vector leafhoppers. Histological study of the tracks.] *Annales des Epiphyties* 18(horo-serie):133-141. [In French with English and German summaries.]

Moreau, J. P., Cousin, M. T. and Lacote, J. P.
1968a Role de jasside *Aphrodes bicinctus* Schrk. (Homoperes -Auchenorrhynques) dans la transmission de la jaunisse due trefle blanc. *Annals des Epiphyties* 19:103-110.

Moreau, J. P. and Leclant, F.
1973a Contribution a l'etude de deux insectes du Lavandin, *Hyalesthes obsoletus* Sign. et *Cechenotettix martini* Leth. (Hom. Auchenorrh.). *Annales de Zoologie Ecologie Animale* 5(3):361-364.

Moretti, M. F. D.
1956a Studies on the potato leafhopper *Empoasca fabae* as a nursery pest. Ph. D. Dissertation, Rutgers University. 227 pp. [Diss. Abst. 16(12):2263.]

Morinaka, T., Inoue, H., Omura, T., Saito, Y., Putta, M., Chettanachit, D., Panejarean, A. and Disthaporn, S.
1980a Leafhopper transmission of rice gall dwarf disease in Thailand. *Annals of the Phytopathological Society of Japan* 46:412. [In Japanese.]

Morinaka, T., Putta, M., Chettanachit, D., Parejarearn, A., Disthaporn, S., Omura, T. and Inoue, H.
1982a Transmission of rice gall dwarf virus by cicadellid leafhoppers *Recilia dorsalis* and *Nephotettix nigropictus* in Thailand. *Plant Disease* 66(8):351-358.

Morinaka, T. and Sakurai, Y.
1970a Varietal resistance to yellow dwarf of the rice plant and the method of testing resistance. *Bulletin of the Chugoku National Agricultural Experiment Station* (Series A) 6:57-79.

Moriya, S.
1978a Control of green rice leafhopper, virus vector by application of insecticides to the soil in nursery box. *Review of Plant Protection Research* 11:53-61.

Moriya, S. and Maeda, Y.
1975a Bell-jar dusting method for the evaluation of insecticide susceptibility of the green rice leafhopper. *Proceedings of the Association for Plant Protection of Kyushu* 21:65-67.

1976a Penetration and metabolism of substituted phenyl methylocarbamates in the green rice leafhopper, *Nephotettix cincticeps* Uhler. *Japanese Journal of Applied Entomology and Zoology* 20:198-202.

Morris, M. G.
1966a Leafhoppers of the genus *Idiocerus* (Homoptera: Cicadellidae) in Huntingdonshire. *Report, Huntingdonshire Fauna and Flora Society* 19:8-10.

1971a Differences between the invertebrate faunas of grazed and ungrazed chalk grassland. IV. Abundance and diversity of Homoptera: Auchenorrhyncha. *Journal of Applied Ecology* 8:37-52.

1972a Distributional and ecological notes on *Ulopa trivia* Germar (Hem., Cicadellidae). *Entomologist's Monthly Magazine* 107:174-181.

1974a Associations of grassland Auchenorrhyncha. *Proceedings of the Royal Entomological Society of London* 39(1):1-2.

1974b Auchenorrhyncha (Hemiptera) of the Burren with special reference to species associations of the grasslands. *Proceedings of the Royal Irish Academy* 74(2):7-30.

1978a Studies of Auchenorrhyncha in the Spanish Pyrenees (Jacetania). *Auchenorrhyncha Newsletter* 1:30-31.

1978b The effects of cutting on grassland Hemiptera: A preliminary report. *Scientific Proceedings of the Royal Dublin Society* 6:285-295.

1982a Some responses of *Arthaldeus pascuellus* (Hem., Cicadellidae) to changes in an *Arrhenatherum* grassland. *Zeitschrift fuer Angewandte Entomologie* 94(4):351-358.

1984a Leafhopper populations in the Park Grass Experiment, Rothamsted. *Annual Report of the Institute of Terrestrial Ecology* 1983:38-40.

1984b Colonization of sown calcareous grasslands by leafhoppers (Homoptera, Auchenorrhyncha). *Annual Report of the Institute of Terrestrial Ecology* 1983:40-41.

Morrison, W. P.
1973a A revision of the Hecalinae (Homoptera: Cicadellidae) of the Oriental region. *Pacific Insects* 15(3&4):379-438.

Mote, U. N.
1978a Chemical control of brinjal jassids (*Amrasca devastans* Dist.) and shoot and fruit borer (*Leucinodes orbonalis* Guen.). *Pesticides* 12:20-23.

1980a Varietal resistance of okra to jassid. I. Screening under field conditions. *Bulletin of Entomology* 21:126-131.

1981a Control of brinjal pests. *Indian Journal of Entomology* 43:229-232.

1982a Varietal susceptibility of brinjal (*Solanum melongena* L.) to jassid (*Amrasca biguttula biguttula* Ishida). *Journal of the Maharashtra Agricultural Universities* 7:59-60.

Motoyama, N., Kao, L. R., Lin, P. T. and Dauterman, W. C.
1984a Dual role of esterases in insecticide resistance in the green rice leafhopper. *Pesticide Biochemistry and Physiology* 21:139-147.

Mound, L. A.
1983a For a taxonomist you seem to know a lot about biology. In: Knight, W. J., Pant, N. C., Robertson, T. S. and Wilson, M. R., eds. *Proceedings of the 1st International Workshop on Biotaxonomy, Classification and Biology of Leafhoppers and Planthoppers (Auchenorrhyncha) of Economic Importance.* Pp. 9-11. London, 4-7 October 1982. Commonwealth Institute of Entomology, 56 Queen's Gate, London SW7 5JR. 500 pp.

Moustafa, M. A., Salem, M. M., Badr, M. A., Radwan, H. S. and Assal, D. M.
1985a Field evaluation of various synthetic attractants in luring homopterous insects. *Bulletin de la Societe Entomologique d'Egypte* 64:173-181.

Moutous, G.
1977a Definitions of the symptoms of 'golden flavescence' on vinestock varieties. Study of their susceptibility. *Revue de Zoologie Agricole et de Pathologie Vegetale* 76:90-98.

Moutous, G. and Carle, P.
1965a Determination of the residual effectiveness in 48 hours of pesticides effective against *Scaphoideus littoralis*. In: *Report of Tests of Insecticides Carried Out in 1965 Against S. littoralis, the Vector of Golden Flavescence of Vines*. Revue Zoologie. Agricole et Applicata 64:120-123.

Moutous, G. and Fos, A.
1971a Essais de lutte chimique contre la cicadelle de la vigne *Empoasca flavescens* Fabr. Resultats 1970. *Revue de Zoologie Agricole et de Pathologie Vegetale* 70(2)):48-56. [In French with English summary.]

1973a Influence des niveaux de populations de cicadelles de la vigne (*Empoasca flavescens* Fab.) sur le symptome de la "grillure" des feuilles. *Annales de Zoologie—Ecologie Animale* 5(2):173-185. [In French with English summary.]

1975a Damage by *Empoasca vitis* Goethe (Homoptera Typhlocybidae) on *Actinidia chinesis* in south-western France. *Revue de Zoologie Agricole et de Pathologie Vegetale* 74:43-44.

1976a Specification of a permanent rearing method for *Empoasca vitis* Goethe, Homoptera Typhlocybidae. Influence of photoperiod. *Revue de Zoologie Agricole et de Pathologie Vegetale* 75:149-152.

Moutous, G., Fos, A., Besson, J., Joly, E. and Biland, P.
1977a Results of ovicide tests against *Scaphoideus littoralis* Ball, a leafhopper vector of golden flavescence. *Revue de Zoologie Agricole et de Pathologie Vegetale* 76:37-49.

Mowry, T. M. and Whalon, M. E.
1984a Comparison of leafhopper species complexes in the ground cover of sprayed and unsprayed peach orchards in Michigan (Homoptera: Cicadellidae). *Great Lakes Entomologist* 17:205-209.

Muir, R. C.
1966a The effect of sprays on the fauna of apple trees. IV. The colonization of orchard plots by the predatory Mirid *Blepharidopterus angulatus* and its effect on populations of *Panonychus ulmi*. *Journal of Applied Ecology* 3(2):269-276.

Muka, A. A.
  1975a Little insect (potato leafhopper, *Empoasca fabae*) makes big trouble on alfalfa. *Hoards' Dairyman* 120(11):685.

Mukharji, S. P.
  1959a The structure and histology of the alimentary canal of Homoptera (Hemiptera). Ph. D. Thesis, Banaras Hindu University, Varanasi, India.
  1962a The structure and histology of the alimentary canal of *Idiocerus clypealis* Leth. (Homoptera: Jassidae). *Science and Culture* 28:29-31.

Mukhopadhyay, S. and Chowdhury, A. K.
  1970a Incidence of tungro virus of rice in West Bengal. *International Rice Commission Newsletter* 19(2):9-12.
  1973a Some epidemiological aspects of tungro virus disease of rice in West Bengal. *International Rice Research Commission Newsletter* 22:44-57.

Mukhopadhyay, S., Ghosh, A. B. and Chakravarti, S.
  1979a Population dynamics of green leafhopper with respect to time and space. *International Rice Research Newsletter* 4(1):16.

Mukhopadhyay, S., Ghosh, A. B., Tarafder, P. and Chakravarti, S.
  1978a Studies on rice tungro virus diseases and its vector *Nephotettix* spp. in West Bengal, India, in 1976-77. *International Rice Research Newsletter* 3(4):14.

Mukhopadhyay, S. and Saha, P. K.
  1981a Differential response of *Nephotettix virescens* (Distant) to few rice varieties. *Indian Agriculturist* 25:125-129.

Mulla, M. S.
  1956a Two mymarid egg parasites attacking *Typhlocyba* species in California. *Journal of Economic Entomology* 49:438-441.
  1957a The biology of *Typhlocyba prunicola* Edwards and *T. quercus* (Fabricius) (Cicadellidae-Homoptera). *Annals of the Entomological Society of America* 50(1):76-87.

Muller, H. J.
  1955a Uber das Zahlenverhaltnis der Geschlechter bei der Jasside *Streptanus marginatus* Kbm. *Wissenschaftliche Zeitschrift Karl-Marx-Universitat Leipzig, Mathematisch-Naturwissenschaftliche Reihe* 4:27-30.
  1955b Die bedeutung der tageslange fur die saisonformenbildung der insekten, insbesondere bei den zikaden. *Deutches Entomologische* (Berlin) 8-10:102-120.
  1955c Uber den Einfluss der Tageslange auf die Saisonformenpragung von *Euscelis plebejus* Fall. *Verhandlungen der Deutschen Zoologischen Gesellschaft*(Tubingen) 1954:307-316.
  1956a Homoptera. *In: Sorauer, ed. Handbuch der Pflanzenkrankheiten* 5(3):150-359. Paul Parey, Berlin u. Hamburg.
  1956b The taxonomic value of the male genitalia in leafhoppers in the light of new studies on the seasonal forms of *Euscelis*. *Proceedings of the 10th International Congress of Entomology,* Montreal 1:357-362.
  1956c Uber die Wirkung von Umwelfaktoren auf die Variabilitat saisondimorpher Insekten, insbesondere der Gattung *Euscelis*. *Verhandlunger der Deutschen Zoologischen Gesellschaft* 450-462.
  1957a Die Wirkung exogener Faktoren auf die zyklische Formenbildung der Insekten, insebesondere der Gattung *Euscelis* (Hom. Auchenorrhyncha). *Zoologische Jabrbuecher. Abteilung fuer Systematik Oekologie und Geographie der Tiere* 85:317-430.
  1959a Tageslange als Regulator des Gestaltwandels bei Insekten. *Umschau* 2:36-39.
  1960a Uber morphologische Folgen der Parasitierung von *Euscelis*-Mannchen (Homoptera Auchenorrhyncha) mit Dryiniden-Larven. *Zeitschrift fuer Morphologie und Oekologie der Tiere* 49:32-46.
  1960b Uber photoperiodisch bedingte Okomorphosen bei Insekten. [The ontogeny of insects.] *Acta Symposii de Evolutione Insectorum* (Praha) 1959:297-304.
  1960c Der Honigtau als Nahrung der hugelbauenden Waldameisen. *Entomaphaga* 5:55-75.
  1961a Erster Nachweis einer Eidiapause bei den Jassiden *Euscelis plebjus* Fall. und *lineolatus* Brulle (Homoptera Auchenorrhyncha). *Zeitschrift Angewandte Entomologie* 48:233-241.
  1962a Neuere Vorstellungen uber Verbreitung und Phylogenie der Endosymbiosen der Zikaden. *Zeitschrift fur Morphologie und Okologie der Tiere* 51:190-210.
  1962b Die variabilitat der genitalstrukturen bei zikaden und ihre biologische und taxonomische bedeutung. *Mitteilungsblatt fuer Insektenkunde* 6(2):36-42.
  1964a Uber die Wirkung verschiedener Spektralbereiche bei der photoperiodischen Induktion der Saisonformen von *Euscelis plebejus* Fall. (Homoptera: Jassidae). *Zoologische Jahrbuecher. Abteilung fuer Allgemeine Zoologie und Physiologie der Tiere* 70:411-426.
  1965a Zur weiteren Analyse der Okomorphosen von *Euscelis plebejus* Fall. (Homoptera Auchenorrhyncha). *Zoologische Beitraege* (n. s.) 11(1&2):151-182.
  1973a Die postembryonale Entwicklung der Saisonformen von *Euscelis incisus* (Kmb. 1858) (Homoptera Jassidae) unter dem Einfluss von Temperatur und Tajeslange. 3. Beitrag zur weiteren Analyse der Okomorphosen von *Euscelis incisus* (Kmb.) (Homoptera Auchenorrhyncha). *Zoologische Jahrbucher (Systematics)* 100(3):321-350.
  1974a Farb-Polymorphismus bei Larven der Jasside *Myocydia crocea* H. S. (Homoptera, Auchenorrhyncha). *Zoologischer Anzeiger* 192(5&6):303-315.
  1976a Uber die Parapause also Dormanzform am Beispiel der Imaginal Diapause von *Mocydia crocea* H. S. (Homoptera Auchenorrhyncha). *Zoologische Jahrbuecher, Abteilung fuer Allgemeine Zoologie und Physiologie der Tiere* 80(3):231-258. [In German with English summary.]
  1979a Effects of photoperiod and temperature on leafhopper vectors. *In: Maramorosch, K. and Harris, K. F., eds. Leafhopper Vectors and Plant Disease Agents.* Pp. 29-94. Academic Press, New York. 654 pp.
  1979b Zur weiteren Analyse des larvalen Polymorphismus der Jasside *Mocydia crocea* H.S. (Homoptera, Auchenorrhyncha). *Zoologische Jahrbucher (Systematics)* 106(3):311-343.
  1980a Die Bedeutung abiotischer Faktoren fur die Einnischung der Organismen in Raum und Zeit. Autokologie als Auftrag der Okosystemforschung. [Importance of abiotic factors for fitting in organisms in space and time. Autecology on duty of ecosystem research.] *Biologiche Rundschau* 18(6):373-388.
  1981a On the larval polymorphism in *Mocydia crocea* H.-S. *Acta Entomologica Fennica* 38:28-29.
  1981b Die bedeutung der dormanzform fur die populationsdynamik der zwergzikade *Euscelis incisus* (Kbm.) (Homoptera, Cicadellidae). *Zoologische Jahrbuecher, Abteilung fuer Systematik Oekologie und Geographie der Tiere* 108(3):314-334.
  1983a Effects of extradian daylengths on the allometric growth of *Euscelis incisus* (Kbm.) (Homoptera: Auchenorrhyncha). *Zoologische Jahrbuecher, Abteilung fuer Systematik Oekologie und Geographie der Tiere* 110(3):301-322.
  1984a On the dormancy forms of Auchenorrhyncha. *Mitteilungen der Schweizerischen Entomologischen Gesellschaft* 57(4):435-436.

Muller, J.
  1969a Investigations on the intracellular symbiosis of some Aetalionidae, Eurymelidae and Cicadellidae (Homoptera-Auchenorrhyncha). *Zoologische Jahrbuecher, Abteilung fuer Systematik Oekologie und Geographie der Tiere* 96(4):558-608.

Mumford, D. L.
  1982a Using enzyme-linked immunisorbent assay to identify beet leafhopper populations carrying curly top virus. *Plant Disease* 66:940-941.

Mumford, D. L. and Peay, W. E.
1970a Curly top epidemic in western Idaho. *Journal of the American Society of Sugar Beet Technologists* 16:185-187.

Muniyappa, V. and Raju, B. C.
1981a Response of cultivars and wild species of rice to yellow dwarf disease. *Plant Disease* 65:679-680.

Muniyappa, V. and Ramakrishnan, K.
1976a Epidemiology of yellow dwarf disease of rice. *Mysore Journal of Agricultural Science* 10:259-270.
1976b Reaction of cultivars and varieties of rice and species of Oryza to yellow dwarf disease of rice. *Mysore Journal of Agricultural Science* 10:270-279.
1980a Transmission studies on yellows dwarf diseases of rice. *Mysore Journal of Agricultural Sciences* 14:55-59.

Muniyappa, V. and Veeresh, G. K.
1980a Leafhoppers as vectors of plant mycoplasma diseases. *Proceedings Indian National Science Academy* (B) 46(6):827-831.

Muniyappa, V. and Viraktamath, C. A.
1981a Transmission of rice yellow dwarf with the blue race of *Nephotettix virescens*. *International Rice Research Newsletter* 6(2):15.

Munk, R.
1967a Zur Morphologie und Histologie des Verdauungstraktes ziveier Jassiden (Homoptera, Auchenorrhyncha) unter besonderer Berucksichtigung der sogenannten Filterkammer. [On morphology and histology of the alimentary tracts of two Jassidae with special consideration of the so-called filter chambers.] *Zeitschrift fuer Wissenschafrliche Zoologie* 175(3&4):405-423.
1967b Licht-und elektronenmikroskopische Befunde an der Filterkammer der kleinzikade *Euscelidius variegatus* Kbm. (Jassidae). *Verhandlungen der Deutschen Zoologischen Gesellschaft* 31:519-527.
1968a Die Richtungen des Nahrungsflusses in Darmtrakt der Kleinzikade: *Euscelidius variegatus* Kbm. (Jassidae) [The directions of the flow of food in the alimentary tract of the leafhopper *Euscelidius variegatus* Kbm. *Zeitschrift fuer Vergleichende Physiologie* 58(4):423-428.
1968b Autoradiographische Untersuchungen des Transportes einiger Nahrungsbestandteile im Darmstrakt zweier Klunzikaden: *Euscelidius variegatus* Kbm. (Jassidae) und *Triecphora vulnerata* Germ. (Cercopidae.) [Autoradiographic experiments on the transport of some nutritional elements in the digestive tract of two leafhoppers.] *Zeitschrift fuer Vergleichende Physiologie* 61(1):129-136.
1968c Uber den Feinbau der FilterKammer der Kleinzikade *Euscelidius variegatus* Kbm. (Jassidae). *Zeitschrift fur Zellforschung und Mikroskopische Anatomie* 85:210-244. [English summary.]

Munro, J. A.
1954a Entomology problems in Bolivia. *FAO Plant Protection Bulletin* 2(7):97-121.

Munteanu, I., Muresan, T. and Radulescu, E.
1983a Some aspects concerning the aetiology and ecology of wheat yellow dwarf in Romania. *In: 10th International Congress of Plant Protection*. Vol. 3. Proceedings of a conference held at Brighton, England. Croydon, U.K.

Murai, M. and Kiritani, K.
1970a Influence of parental age upon the offspring in the green rice leafhopper, *Nephotettix cincticeps* Uhler (Hemiptera: Deltocephalidae). *Applied Entomology and Zoology* 5(4):189-201.

Muramatsu, Y., Sugino, T. and Nakanura K.
1970a Sampling efficiency of methods to estimate the number of *Nephotettix cincticeps* Uhler in the resting paddy field. *Japanese Journal of Applied Entomology and Zoology* 14(1):19-24. [In Japanese with English abstract.]

Murata, T., Noda, M., Takasaki, T. and Tateisi, T.
1965a Overwintering habitats of the green rice leafhopper, *Nephotettix cincticeps* Uhler in northern Ksushu. *Kyushu Agricultural Research* 27:136-137. [In Japanese]

Muratomaa, A.
1967a Aster yellows-type virus infecting grasses in Finland. *Annales Agriculturae Fennicae* 5:324-333.
1969a Aster yellows pa gramineer. *Nordisk Jordbruksforskning* 51:290-292.

Muratomaa, A. and Valenta, V.
1968a Transmission of a leafhopper-borne virus from naturally infected *Pao annua* in Czechoslovakia. *Biologia* (Bratislava) 23:389-392.

Murayama, D.
1966a On the witches' broom diseases of sweet potato and leguminous plants in the Ryukyu Islands. *Memoirs of the Faculty of Agriculture, Hokkaido University* 6(1):81-103. [In Japanese with English summary.]

Murguido, C. A.
1982a Effectiveness of new insecticides for the control of leafhopper (*Empoasca fabae*) order Homoptera, family Cicadellidae, applied up to 30 and 50 days after germination of beans. *Ciencia y Tecnica en la Agricultura* 5:31-42.
1983a Comparative effect of some organophosphorus insecticides for the control of the whitefly (*Bemisia tabaci*) and the leafhopper (*Empoasca* sp.) in bean crops. *Ciencia y Tecnica en la Agricultura* 6:59-65.
1983b Effectiveness of various insecticides on the bean leafhopper (*Empoasca* sp., Homoptera: Cicadellidae). *Ciencia y Tecnica en la Agricultura* 6:67-77.

Murguido, C. and Beltran, C.
1983a Incidence of and damage by the leafhopper (*Empoasca* sp.) (Homoptera: Cicadellidae) and other pests in six bean varieties. *Ciencia y Tecnica en la Agricultura* 6:31-58.

Murguido, C. and Izquierdo, D.
1982a Determination of indices of larval attack for indicating the application of insecticides to control the bean leafhopper (*Empoasca fabae*). *Ciencia y Tecnica en la Agricultura* 5:97-104.

Murguido, C. and Ruiz, I.
1982a Technical and economic assessment of a new method of issuing warnings of leafhoppers (*Empoasca fabae*, Homoptera: Cicadellidae) in bean crops. *Ciencia y Tecnica en la Agricultura* 5:39-48.

Murisier, F. and Jelmini, G.
1985a Perilimbar necrosis: An aspect of precocious russetting of vine foliage in Tessia. *Revue Suisse de Viticulture d'Arboriculture et d'Horticulture* 17:305-306.

Murthy, K. V. K., Yaraguntaiah, R. C. and Govindu, H. C.
1975a Studies on insect transmission of ragi mosaic from Karnataka. *University of Agricultural Sciences Technical Series* 10:1-12.

Murugesan, S., Ramakrishnan, D., Kandaswamy, T. K. and Murugesan, M.
1973a Forecasting phyllody disease of sesamum. *Madras Agricultural Journal* 60:492-495.

Musgrave, C. A.
1974a Biological and morphological studies of the *Scaphytopius acutus* complex (Homoptera: Cicadellidae). Ph.D. Dissertation, Oregon State University. 146 pp.
1975a Taxonomy of the *Scaphytopius* (*Cloanthanus*) *acutus* complex (Homoptera: Cicadellidae). *Annals of the Entomological Society of America* 68(3):434-438.
1979a *Scaphytopius* in Florida (Homoptera: Cicadellidae). *Entomology Circular, Florida Department of Agriculture* 204:1-3.

Musil, M.
1956a Uber das vorkommen der zikade *Typhlocyba horvathiana* Dlabola 1955 in der Slowakei. *Biologia* (Bratislava) 11(10):621-624. [In Czech with German summary.]
1958a Zvirena krisu okoli Bratislavy. I. [The leafhopper fauna (Homoptera Auchenorrhyncha) of the region of Bratislava.] *Zoologicke Listy* 2(2):122-134.
1958b Transmission of the stolbur and related viruses by some leafhoppers (preliminary report). *Biologia* (Bratislava) 13(2):133-136.

1958c Beitrag zur Kenntnis der Zikadenfauna der Slowakei. I. Zikadenfauna der Steppenbiotope. *Biologia* (Bratislava) 13(6):419-427.

1958d Cikadofauna Pastvin Potiske Niziny (Prispevek k Poznani Cikadofauny Slovenska II). *Biologia* 13:502-508.

1959a Zvirena Krisu Nekterych Biotopu V Okoli Sladkovicova (Prispevek K Poznani Cikadofauny Slovenska III). *Biologia* 14:737-748.

1959b Ubertragung des Stolburvirus durch die Zikade *Eiscelis plebejus* (Fallen). [Transmission of the stolbur virus by the cicadellid, *E. plebeja*.] *Biologia* 14(6):410-417.

1960a Vyskyt Rozsireni A Skodlvost Krisku *Aphrodes bicinctus* (Schrk.), *Macrosteles laevis* (Rib.), A *Euscelis plebjus* (Fall.) Na Slovenska. [Occurrence, distribution and injuriousness of the leafhoppers *Aphrodes bicinctus* (Schrk.); *Macrosteles laevis* (Rib.), and *Euscelis plebjus* (Fall.) in Slovakia.] *Zoologicke Listy* 23:39-46.

1960b Ubertragung des Stolbur, Kleeverzergungs-, und Kleeverlaubungsvirus durch die Zikade *Aphrodes bicinctus* (Schrank). *Biologia* 15:721-728.

1961a Transmission of the clover phyllody virus by means of the leafhopper *Euscelis plebejus* (Fallen). *Biologia Plantarum* (Prague) 3(1):29-33.

1961b Zikadenfauna einiger Wiesen-und Weidenbiotope in der Slowakei (Beitrag zur Kenntnis der Zikadenfauna der Slowakei IV). *Biologicke Prace* 7(10):58-77. [In Czech with German summary.]

1962a An attempt to pass the clover dwarf virus by serial transfers in its vector. *Acta Virologica* 6:93.

1962b Prenos viru Parastoburu Krisken *Euscelis plebejus* (Fallen). *Biologia* 17:332-339.

1962c Investigations on the multiplication of yellows type viruses in the leafhopper, *Euscelis plebejus* (Fallen). *In: Plant Virology, 5th Conference of Czechoslovakia Plant Virologists*, pp. 152-154.

1963a Uber das Vorkommen einiger Zikadenarten (Homopt. Auchenorrhyncha) in der Slowakei. *Biologia* (Bratislava) 18(9):693-697. [In Czech with German and Russian summaries.]

1964a Persistence of infectivity yellows-type in frozen viruliferous leafhoppers. *Acta Virologica* 8:92.

1964b Multiplication of yellows-type plant viruses in *Euscelis plebejus* (Fallen) leafhoppers. *Acta Virologica* 8:230-238.

1964c Persistence of infectivity of yellows-type plant viruses in extracts from viruliferous *Euscelis plebejus* (Fallen) leafhoppers. *Acta Virologica* 8:239-242.

1965a Ubertragung der Gelbsuchtiviren durch die Zwerzikade *Euscelis plebejus* (Fallen). [Transmission of the jaundice viruses by the dwarf cicada *Euscelis plebejus* (Fallen).] *Biologicke Prace* 11:1-86.

1966a Thermal inactivation of parastolbur virus in extracts from viruliferous *Eusceliy plebejus* leafhoppers. *Acta Virologica* (Prague) 10(3):273-275.

1966b Uber die fahigkeit der zikade *Macrosteles laevis* (Ribaut), das virus der zwergkrankheit des klees zu ubertragen. *Biologia* (Bratislava) 21(1):45-49. [In Czech with German summary.]

Musil, M. and Valenta, V.

1958a Prenos Stolburu A Bribuznych Virusov Promocou Niektorych Cikad. [Transmission of the stolbur and related viruses by some leafhoppers.] *Biologia* 13:133-136.

Mzira, C. N.

1984a Maize streak. *Zimbabwe Agricultural Journal* 81:187.

1984b Cultural control of maize streak virus in wheat by spacing and time of planting. *Zimbabawe Agricultural Journal* 81:189-191.

Naba, K.

1981a Regional difference in the feeding damage caused by the green rice leafhopper to paddy rice in Japan. *In: International Symposium on Problems of Insect Pest Management in Developing Countries.* Pp. 73-81. Symposium on Tropical Agricultural Research, 1980, Kyoto, Japan.

1982a Studies on the direct feeding damage due to the green rice leafhopper, *Nephotettix cincticeps* Uhler, on rice plants. 1. Methods of population estimation of *N. cincticeps* on rice plants at the reproductive age. *Bulletin Hokkaido Prefecture Agricultural Experiment Station* 45:35-42.

1983a Studies on the direct feeding damage due to the green rice leafhopper, *Nephotettix cincticeps* Uhler, on rice plants. 2. Changes in the vertical distribution of *N. cincticeps* and its infesting parts on rice plants at the reproductive age. *Bulletin Hokkaido Prefecture Agricultural Experiment Station* 46:13-20.

Nachiappan, R. M. and Baskaran, P.

1984a Quantum of feeding and survival of mango leafhopper adults on the inflorescence of certain varieties of mango. *Indian Journal of Agricultural Sciences* 54:312-314.

Nachev, P.

1960a The finding of new pests of vines in Bulgaria. *Rasteni Zashchita* 8(4):63-64. [In Bulgarian.]

Nagai, K., Tabaru, M. and Nonaka, K.

1979a Application of cartap granular insecticide to the seedling box before seeding for control of the green rice leafhopper *Nephotettix cincticeps* Uhler. *Proceedings of the Association for Plant Protection of Kyushu* 25:77-78.

Nagaich, B. B.

1960a Simultaneous transmission of wound-tumor and potato yellow dwarf viruses by *Agallia constricta* Van Duzee. *Science and Culture* 25:591-592.

1975a The problem of purple top-roll and marginal flavescence of potatoes in the north Indian hills. Progress in the study of biology of pest organisms and the development of forecast methods. *In: VIII International Plant Protection Congress*, 1975. Pp. 243-251.

Nagaich, B. B. and Giri, B. K.

1971a Marginal flavescence-a leafhopper transmitted disease of potatoes. *Indian Phytopathology* 24:824-826.

1973a Purple top roll disease of potato. *American Potato Journal* 50(3):79-85, 4 fig., 1 tab.

Nagaich, B. B., Puri, B. K., Sinha, R. C., Dhingra, M. K and Bhardwaj, V. P.

1974a Mycoplasma-like organisms in plants affected with purple top-roll, marginal flavescence and witches' broom diseases of potatoes. *Phytopathologische Zeitschrift* 81(3):273-279. [In English with German summary.]

Nagaraj, A. N. and Black, L. M.

1961a Localization of wound-tumor antigen in plant tumors by the use of fluorescent antibodies. *Virology* 15:289-294.

1962a Hereditary variation in the ability of a leafhopper to transmit two unrelated plant viruses. *Virology* 16:152-162.

Nagaraj, A. N., Sinha, R. C. and Black, L. M.

1961a A smear technique for detecting virus antigen in individual vectors by the use of fluorescent antibodies. *Virology* 15:205-208.

Nagaraju, S. V.

1981a Studies on the relationship of ragi streak virus and its vector *Cicadulina chinai*. *Indian Phytopathology* 34:458-460.

Nagaraju, S. V., Reddy, H. R. and Channamma, K. A. L.

1984a Raji streak – a leafhopper transmitted virus disease in Karnataka. *Mysore Journal of Agricultural Science* 16(3):301-305.

Nagata, T.
1983a Insecticide resistance in rice pests, with special emphasis on the brown planthopper (*Nilaparvata lugens* Stal). *In: 10th International Congress of Plant Protection.* Vol. 2. Proceedings of a conference held at Brighton, England. Pp. 599-607. Croydon, U.K. 3 vols. 1228 pp.

Nagata, T. and Mochida, O.
1984a Development of insecticide resistance and tactics for prevention. *In: Smith, W. H., ed. Proceedings of the FAO/IRRI Workshop on Judicious and Effecient Use of Insecticides on Rice.* Pp. 93-105. International Rice Research Institute, Manila, Philippines.

Nagel, H. G.
1973a Effects of spring prairie burning on herbivorous and non-herbivorous arthropod populations. *Journal of the Kansas Entomological Society* 46(4):485-496.

Naheed, R. and Ahmed, Manzoor
1974a Bioecological studies of *Amrasca devastans* (Distant) on okra (*Abelmoschus esculentus*) in Karachi, Pakistan. *Pakistan Journal of Science and Industrial Research* 17:127-129.

1980a Some new species of leafhopper genus *Empoasca* (Typhlocybinae: Cicadellidae) from Pakistan. *Pakistan Journal of Zoology* 12(1):77-84.

1980b Biology and life history of *Empoasca signata* (Typhlocybinae: Cicadellidae) on castor and potato in Karachi, Pakistan. *Proceedings of the 1st Pakistan Congress of Zoology* (B) 1:271-273.

Naheed, R., Ahmed, Manzoor and Ahmed, Mubarik
1981a 'Kachnar' leafhopper, *Zygina binotata* (Distant) (Typhlocybinae: Cicadellidae), a study of its population dynamics in Karachi, Pakistan. *Pakistan Journal of Agricultural Research* 2(3):192-195.

Nair, K. K.
1983a Natural enemies of *Nephotettix* species, the paddy pests. *Indian Journal of Entomology* 45:484-487.

Nair, M. R. G. K., Christudas, S. P. and Mathal, S.
1977a On the control of the jassid *Amrasca biguttula biguttula* (Ishida) on bhindi using some new insecticides. *Agricultural Journal of Kerala* 14:171-172.

Naito, A.
1964a Methods of detecting feeding marks of planthoppers and leafhoppers. *Plant Protection Tokyo* 18:482-484. [In Japanese.]

1965a Collecting method of salivary sheath material of leafhopper (*Nephotettix cincticeps* (Uhl.) and planthoppers (*Laodelphax striatellus* (Fall.)). *Japanese Journal of Applied Entomology and Zoology* 9:142-144.

1976a Studies on the feeding habits of some leafhoppers attacking the forage crops. I. Comparison of the feeding habits of the adults. *Journal of Applied Entomology and Zoology* 20:1-8.

1976b Studies on the feeding habits of some leafhoppers attacking the forage crops. II. A comparison of the feeding habits of the green rice leafhoppers in different development stages. *Japanese Journal of Applied Entomology and Zoology* 20:51-54.

1977a Feeding habits of leafhoppers. *Japanese Agriculture Research Quarterly* 11:115-119.

1977b Studies on the feeding habits of some leafhoppers attacking forage crops. III. Relations between the infestation by leafhopper and its feeding habits. Interruption of assimilates in the plant body by the feeding of *Empoasca sakii* Dworakowska (Homoptera: Cicadellidae). *Japanese Journal of Applied Entomology and Zoology* 21:105.

Naito, A. and Masaki, J.
1967a Studies on the feeding behavior of the green rice leafhopper, *Nephotettix cincticeps* Uhler. I. Insertion of the stylets into host plant. *Japanese Journal of Applied Entomology and Zoology* 11(2):50-56. [In Japanese with English summary.]

1967b Studies on the feeding behavior of green rice leafhopper, *Nephotettix cincticeps* Uhler. II. Probing frequency of the adult leafhopper. *Japanese Journal of Applied Entomology and Zoology* 11:150-156. [In Japanese with English summary.]

Nakamura, H., Mizumati, M., Yamaguchi, H., Kitazima, T., Isakari, Y., Zintake, M., Arizono, T., Mizuta, H., Seki, M., Miyahara, K. Onizuka, S., Abe, K. and Eto, R.
1971a Control of rice leafhoppers and drift of the chemicals by aerial application of Sumi-ace fine granules. *Proceedings of the Association for Plant Protection of Kyushu* 17:131-133. [In Japanese.]

Nakamura, K.
1977a A model for the functional response of a predator to varying prey densities, based on the feeding ecology of wolf spiders. *Bulletin of the National Institute of Agricultural Science* 31:28-29.

Nakamura, K., Ito, Y., Miyashita, K. and Takai, A.
1967a The estimation of population density of the green rice leafhopper, *Nephotettix cincticeps* Uhler, in spring field by the capture-recapture method. *Researches on Population Ecology* (Kyoto University) 9:113-129. [Japanese summary.]

Nakasuga, T. and Higuchi, T.
1972a Insecticide susceptibility of the green rice leafhopper *Nephotettix cincticeps* Uhler in Nagasaki Prefecture. *Proceedings of the Association for Plant Protection of Kyushu* 18:46-48. [In Japanese.]

Nakasuji, F.
1974a Epidemiological study on rice dwarf virus transmitted by the green rice leafhopper *Nephotettix cincticeps*. *Japanese Agricultural Research Quarterly* 8:84-91.

Nakasuji, F. and Kiritani, K.
1970a Ill-effects of rice dwarf virus upon its vector, *Nephotettix cincticeps* Uhler (Hemiptera: Deltocephalidae), and its significance for changes in relative abundance of infected individuals among vector populations. *Applied Entomology and Zoology* 5:1-12.

1971a Inter-generational changes in relative abundance of insects infected with rice dwarf virus in populations of *Nephotettix cincticeps* Uhler (Hemiptera: Deltocephalidae). *Applied Entomology and Zoology* 6:75-83.

1972a Descriptive models for the system of the natural spread of infection of rice dwarf virus (RDV) by the green rice leafhopper, *Nephotettix cincticeps* Uhler (Hemiptera: Deltocephalidae). *Researches on Population Ecology* 14:18-35.

1977a Epidemiology of rice dwarf virus in Japan. *Tropical Agriculture Research* Series 10:93-98.

Nakasuji, F., Kiritani, K. and Tomida, E.
1975a A computer simulation of the epidemiology of the rice dwarf virus. *Researches on Population Ecology* 16:245-251.

Nakasuji, F., Miyai, S., Kawamota, H. and Kiritani, K.
1985a Mathematical epidemiology of rice dwarf virus transmitted by green rice leafhoppers: A differential equation. *Journal of Applied Ecology* 11:839-847.

Nakasuji, F. and Nomura, S.
1968a A study on the injury by the green rice leafhopper, *Nephotettix cincticeps* Uhler. *Proceedings of the Association of Plant Protection* (Shikoku) 3:21-26. [In Japanese with English summary.]

Namba, R.
1956a A revision of the *Balclutha* species found in Hawaii, with descriptions of five new species (Homoptera: Cicadellidae). *Proceedings of the Hawaiian Entomological Society* 16(1):101-112.

1956b Notes and exhibitions. *Proceedings of the Hawaiian Entomological Society* 16(1):4

Nanda, U. K., Shi, N., Naik, R. and Das, K. C.
1976a Field screening of rice varieties against gall midge and rice tungro virus. *International Rice Research Newsletter* 1(2):6.

Nangju, D., Flinn, J. C. and Singh, S. R.
1979a Control of cowpea pests by utilization of insect-resistant cultivars and minimum insecticide application. *Field Crops Research* 2:373-385.

Napompeth, B. and Nishida, T.
1971a The number of *Draeculacephala* species in Hawaii (Homoptera, Cicadellidae). *Proceedings of the Hawaiian Entomological Society* 21(2):239-246.

Napompeth, B. and Tanagsnakod, C.
1976a An investigation on the beneficial organisms for biological control of leafhoppers of economic importance in Thailand. *Research Report, Kasetsart University*, pp. 37-38.

Narayanasamy, P.
1972a Influence of age of rice plants at the time of inoculation on the recovery of tungro virus by *Nephotettix impicticeps* (Ishihara). *Phytopathologische Zeitschrift* 74:109-114.
1972b Distribution of tungro virus in infected rice plant. *Zeitschrift fur Pflanzenkrankheiten und Pflanzenschutz* 79(8/9):540-543.

Narayanasamy, P., Balasubramaniam, M. and Baskaran, P.
1979a Biological studies of the population dynamics of rice brown planthopper and green leafhopper. *International Rice Research Newsletter* 4(3):21.

Narayanasamy, P., Baskaran, P., Balasubramaniam, M. and Adaicklalam, V.
1976a Different levels of N, P and K on the incidence of rice insect pests. *Rice Entomology Newsletter* No.4, p. 37.

Narayanasamy, P., Raghunathan, V. and Bhaskaran, P.
1979a Incidence of brinjal mosaic virus and brinjal little leaf diseases in Dipelorganic insecticide combination treatments. *Indian Journal of Plant Protection* 6:31-34.

Nast, J.
1955a New and little known Polish species of Homoptera. III. *Fragmenta Faunistica* 7(6):213-231.
1972a *Palearctic Auchenorrhyncha (Homoptera): An annotated Check List*. Institute of Zoology, Polish Academy of Sciences, Polish Scientific Publisher, Warsaw. 550 pp.
1973a Some additions and modifications in the Polish list of Auchenorrhyncha (Homoptera). *Fragmenta Faunistica* 19(4):39-53.
1976a Auchenorrhyncha (Homoptera) of the Pieniny Mts. *Fragmenta Faunistica* 21(6):145-183. [In Polish with English summary.]
1976b Katalog Fauny Polski Piewiki Auchenorrhyncha (Cicadodea). *Poliska Akademia Nauk Institut Zoologii* 25:1-255.
1977a Homopterological notes XIII-XX. *Annales Zoologici* (Warsaw) 34(2):27-37.
1979a Palearctic Auchenorrhyncha (Homoptera). Part 2. Bibliography, addenda and corrigenda. *Annales Zoologici* (Warsaw) 34(18):481-499.
1981a Homopterological Notes XXI-XXV. *Annales Zoologici* (Warsaw) 36(14):255-263. [New genus: *Anufrievella*.]
1982a Palaearctic Auchenorrhyncha (Homoptera). Part 3. New taxa and replacement names introduced till 1980. *Annales Zoologici* (Warsaw) 36(17):290-362.
1984a Notes on some Auchenorrhyncha Homoptera. 1-5. *Annales Zoologici* (Warsaw) 37(13-18):391-398.
1985a Remarks on the genus *Tetartostylus* (Homoptera, Cicadellidae) with descriptions of new species from Africa. *Annales Zoologici* (Warsaw) 39(2&7):147-152. [In English with English, Polish, and Russian summaries.]

Nasu, S.
1963a Studies on some leafhoppers and planthoppers which transmit virus diseases of rice plant in Japan. *Bulletin of the Kyushu Agricultural Experiment Station* 8:153-349. [In Japanese.]
1965a Electron microscopic studies of transovarial passage of rice dwarf virus. *Japanese Journal of Applied Entomology and Zoology* 9:225-237.
1967a Rice leafhoppers. In: Pathak, M. D., ed. *The Major Insect Pests of the Rice Plant*. Pp. 443-523. Johns Hopkins, Baltimore.
1969a Electron microscopy of the transovarial passage of rice dwarf virus. In: Maramorosch, K., ed. *Viruses, Vectors and Vegetation*. Pp. 433-448. Wiley & Sons, New York. 666 pp.
1969b Vectors of rice viruses in Asia. In: *The Virus Diseases of the Rice Plant. Proceedings of the International Rice Research Institute Symposium*. Pp. 93-109.

Nasu, S., Jensen, D. D. and Richardson, J.
1970a Electron microscopy of mycoplasma-like bodies associated with insect and plant host of peach western X-disease. *Virology* 41:583-595.
1974a Isolation of western X mycoplasmalike organism from infections extracts of leafhoppers and celery. *Applied Entomology and Zoology* 9:199-203.

Nasu, S., Jensen, D. D., Richardson, J., Chiu, R. J. and Black, L M.
1974a Extraction of western X mycoplasmalike organism from leafhoppers and celery infected with peach western X-disease. *Applied Entomology and Zoology* 9:53-75.
1974b Primary culturing of the western X mycoplasmalike organism from *Colladonus montanus* leafhopper vectors. *Applied Entomology and Zoology* 9:115-126.

Nasu, S., Kono, Y. and Jensen, D. D.
1974a The multiplication of western X mycoplasmalike organism in the brain of a leafhopper vector, *Colladonus montanus* (Homoptera: Cicadellidae). *Applied Entomology and Zoology* 9:277-279.

Nasu, S. and Nakasuka, M.
1973a Some ecological aspects of *Macrosteles orientalis* Vilbaste which transmits witches' broom disease of *Cryptotaenia japonica* Hassk. *Japanese Journal of Applied Entomology and Zoology* 17:221-223.

Nasu, S., Suguira, M., Wakimoto, S. and Iida, J. T.
1967a Pathogen of rice yellow dwarf disease. *Annals of Phytopathological Society of Japan* 33:343-344.

Natchev, P.
1965a New enemies of the plum culture and the walnut-tree in Bulgaria. *Gradinarska i Lozarska Akademiia Naukite* 2(5):581-587. [In Bulgarian with English summary.]

Nath, D. K.
1975a Note on the insect pests of sesame (*Sesamum indicum* L.) of West Bengal. *Indian Journal of Agricultural Research* 9:151-152.

Nault, L. R.
1980a Maize bushy stunt and corn stunt: A comparison of disease symptoms, pathogen host ranges, and vectors. *Phytopathology* 70(7):659-665.
1983a Origins in Mesoamerica of maize viruses and mycoplasmas and their leafhopper vectors. In: Plumb, R. T. and Thresh, J. M., eds. *Plant Virus Epidemiology. The Spread and Control of Insect-borne Viruses*. Pp. 259-266. Blackwell Sci. Publ., Oxford, London. 377 pp.
1983b Origins of leafhopper vectors of maize pathogens in Mesoamerica. In: Gordon, D. T., Knoke, J. K., Nault, L. R. and Ritter, R. M., eds. *Proceedings of International Maize Virus Disease Colloquium and Workshop, 2-6 August 1982*. Pp. 75-82. Ohio State University, Ohio Agricultural Research and Development Center, Wooster. 261 pp.
1984a *Dalbulus* leafhopper vectors of maize pathogens. *Maize Virus Disease Newsletter* 1:61-63.
1984b Corn leafhopper: The making of an insect pest. *Ohio Report* 69:77-78.
1985a Evolutionary relationships between maize leafhoppers and their host plants. In: Nault, L. R. and Rodriguez, J. G., eds. *The Leafhoppers and Planthoppers*. Pp. 309-330. Wiley & Sons, New York. 500 pp.

Nault, L. R. and Bradfute, O. E.
1977a Reevaluation of leafhopper vectors of corn stunting pathogens. *Proceedings of the American Phytopathological Society* 4:172.

1979a Corn stunt: Involvement of a complex of leafhopper-borne pathogens. *In: Maramorosch, K. and Harris, K. F., eds. Leafhopper Vectors and Plant Disease Agents.* Pp. 561-586. Academic Press, New York. 654 pp.

Nault, L. R. and DeLong, D. M.
1980a Evidence for co-evolution of leafhoppers in the genus *Dalbulus* (Cicadellidae: Homoptera) with maize and its ancestors. *Annals of the Entomological Society of America* 73(4):349-353.

Nault, L. R., Delong, D. M., Triplehorn, B. W., Styer, W. E. and Doebley, J. F.
1983a More on the association of *Dalbulus* (Homoptera: Cicadellidae) with Mexican *Tripsacum* (Poaceae), including the description of two new species of leafhoppers. *Annals of the Entomological Society of America* 76(2):305-309.

Nault, L. R., Gingery, R. E. and Gordon, D. T.
1980a Leafhopper transmission and host range of maize rayado fino virus. *Phytopathology* 70(8):709-712.

Nault, L. R., Gordon, D. T., Damsteegt, V. D. and Iltis, H. H.
1982a Response of annual and perennial teosintes (*Zea*) to six maize viruses. *Plant Disease* 66:661-662.

Nault, L. R., Gordon, D. T., Gingery, R. E., Bradfute, O. E. and Loayza Castillo, J.
1979a Identification of maize viruses and mollicutes and their potential insect vectors in Peru. *Phytopathology* 69:824-828.

Nault, L. R. and Knoke, J. K.
1981a Maize vectors. *In: Gordon, D. T., Knoke, J. K. and Scott, G. E., eds. Virus and Viruslike Diseases of Maize in the United States.* Pp. 77-84.

Nault, L. R. and Madden, L. V.
1985a Ecological strategies of *Dalbulus* leafhoppers. *Ecological Entomology* 10(1):57-63.

Nault, L. R., Madden, L. V., Styer, W. E., Triplehorn, B. W., Shambaugh, G. F. and Heady, S. C.
1984a Pathogenicity of corn stunt spiroplasma and maize bushy stunt mycoplasma to its vector *Dalbulus longulus*. *Phytopathology* 74:977-979.

Nault, L. R. and Rodriguez, J. G., eds.
1985a *The Leafhoppers and Planthoppers.* Wiley & Sons, New York. 500 pp.

Nault, L. R., Styer, W. E., Knoke, J. K. and Pitre, H. N.
1973a Semipersistent transmission of leafhopper-borne maize dwarf virus. *Journal of Economic Entomology* 66:1271-1273.

Nayak, P. and Srivastava, R. P.
1979a Occurrence of *Beauveria bassiana* (Bals.) Vuilli on certain rice pests. *Indian Journal of Entomology* 41:99-100.

Nelson, J. M.
1971a The invertebrates of an area of Pennine Moorland within the Moor House Nature Preserve in northern England. *Transactions of the Society for British Entomology* 19(2):173-235.

Neparidze, N. N.
1969a Materialy k izuceniju cikadovyh (Cicadinae) rasprostranennyh v okresnostjah Chiatura. *Soobsheniya Academii Nauk Gruzinskoi SSR* 55:201-203.

Neparidze, N. N. and Dekanoidze, G. I.
1971a On the vertical-zonal distribution of cicadas (Homoptera: Cicadinea) in Bakhazia. *Soobsheniya Akademii Nauk Gruzinskoi SSR* 61(2):461-464. [In Georgian with English summary.]

Newcomer, E. J.
1966a Insect pests of deciduous fruits in the west. *Agriculture Handbook No. 306.* Agricultural Research Service, United States Department of Agriculture. 57 pp.

Newton, R. C. and Barnes, D. K.
1965a Factors affecting resistance of selected alfalfa clones to the potato leafhopper. *Journal of Economic Entomology* 58:435-439.

Newton, R. C., Hill, R. R. and Elgin, J. H.
1970a Differential injury to alfalfa by male and female potato leafhoppers. *Journal of Economic Entomology* 63(4):1077-1079.

Ngoan, N. D.
1971a Recent progress in rice insect research in Vietnam. Symposium on rice insects. Proceedings of a Symposium on Tropical Agriculture Researches, 19-24 July 1971. *Tropical Agriculture Research Series* 5:133-141.

Nguyen, Cong Thuat
1982a Rice insect pests in Vietnam. *International Rice Research Newsletter* 7(2):10-11.

Niazi, F. R., Lakshman, D. K., Mishra, M. D., Basu, A. N. and Raychaudhuri, S.
1985a Interaction of rice tungro virus strains and its vector *Nephotettix virescens* and in cultivars Tn-1 and Pacita. *International Journal of Tropical Plant Diseases* 3:73-77.

Nichiporich, W.
1965a The aerial migration of the six-spotted leafhopper and the spread of the virus disease aster yellows. *International Journal of Biometerology* 9:219-227.

Nickel, J. L.
1962a Notes on the feeding of two leafhoppers on bananas in Honduras. *Tropical Agriculture* 39(4):321-325.

Nicolaus, M.
1957a Zikaden und Blattflohe aus Ost-Thuringen. *Entomologische Mitteilungen aus dem Zoologischen Staatsinstitut und Zoologischen Museum Hamburg* 1(11):307-341.

Nielson, M. W.
1949a Leafhoppers (Cicadellidae) associated with the stone fruit orchards of northern Utah. M. S. Thesis, Utah State Agricultural College, Logan. 63 pp.

1955a A revision of the genus *Colladonus* (Homoptera, Cicadellidae). Ph. D. Dissertation, Oregon State College. 251 pp.

1957a A revision of the genus *Colladonus* (Homoptera, Cicadellidae). *U.S. Department of Agriculture Technical Bulletin* 1156:1-52.

1962a New species of leafhoppers in the genus *Colladonus*. *Annals of the Entomological Society of America* 55:143-147.

1962b A revision of the genus *Xerophloea* (Homoptera, Cicadellidae). *Annals of the Entomological Society of America* 55(2):234-244.

1962c A synonymical list of leafhopper vectors of plant viruses (Homoptera, Cicadellidae). *U.S. Department of Agriculture, Agriculture Research Service, Mimeo Series* 33-74. 11 pp.

1965a A revision of the genus *Cuerna* (Homoptera, Cicadellidae). *U.S. Department of Agriculture Technical Bulletin* 1318:1-48.

1966a A synopsis of the genus *Colladonus* (Homoptera, Cicadellidae). *Journal of the Kansas Entomological Society* 39(2):333-336.

1968a Biology of the geminate leafhopper, *Colladonus geminatus*, in Oregon (Homoptera: Cicadellidae). *Annals of the Entomological Society of America* 61(3):598-610.

1968b The leafhopper vectors of phytopathogenic viruses (Homoptera, Cicadellidae). Taxonomy, biology and virus transmission. *U.S. Department of Agriculture Technical Bulletin* No. 1382. 386 pp.

1969a A revision of the genus *Pasadenus* (Homoptera: Cicadellidae). *Journal of the Kansas Entomological Society* 42(2):141-154.

1975a A revision of the subfamily Coelidiinae (Homoptera: Cicadellidae). Tribes Tinobregmini, Sandersellini and Tharrini. *Bulletin of the British Museum (Natural History) Entomology* 24:1-197. [New genera: *Haranthus, Neotharra*.]

1977a A revision of the subfamily Coelidiinae (Homoptera: Cicadellidae). II. Tribe Thagriini. *Pacific Insects Monograph* 34. 218 pp. [New genus: *Tahara*.]

1979a A new genus, *Tantulidia*, in the tribe Tinobregmini with a review of the species and the generic limitations of the tribe (Homoptera: Cicadellidae). *Journal of the Kansas Entomological Society* 52(4):653-661.

1979b A revision of the subfamily Coelidiinae (Homoptera: Cicadellidae). III. Tribe Teruliini. *Pacific Insects Monograph* 35. 329 pp. [Describes 38 new genera.]

1979c Taxonomic relationships of leafhopper vectors of plant pathogens. *In: Maramorosch, K. and Harris, K., eds. Leafhopper Vectors and Plant Disease Agents.* Pp. 3-27. Academic Press, New York. 654 pp.

1979d Distribution, importance and control of the leafhopper vectors of X-disease organism. *In: Proceedings of the 38th Annual Convention, National Peach Council,* pp. 56-64.

1980a New Oriental species of leafhoppers of the genus *Thagria* (Homoptera: Cicadellidae: Thagriini). *Journal of the Kansas Entomological Society* 53(1):123-131.

1980b Seven new species of thagriine leafhoppers from Southeast Asia (Homoptera: Cicadellidae: Thagriini). *Journal of the Kansas Entomological Society* 53(2):305-319.

1980c New leafhopper species of *Thagria* from Malaysia (Homoptera: Cicadellidae: Thagriini). *Journal of the Kansas Entomological Society* 53(2):343-349.

1980d Four new leafhopper species of *Thagria* from the Australian Region with notes on *Thagria sumbawensis* (Jacobi) (Homoptera: Cicadellidae: Thagriini). *Journal of the Kansas Entomological Society* 53(3):607-616.

1981a Taxonomy of *Jassus varicolor* Spangberg with description of a new species in the Genus *Conbalia* Nielson (Homoptera: Cicadellidae: Coelidiinae). *Journal of the Kansas Entomological Society* 54(1):27-31.

1981b A new leafhopper species of *Sandersellus* from Panama (Cicadellidae: Coelidiinae: Sandersellini). *Journal of the Kansas Entomological Society* 54(3):658-660.

1982a A new genus *Corilidia* and a new species of the tribe Tinobregmini with a revised key to the genera (Homoptera: Cicadellidae: Coelidiinae). *Journal of the Kansas Entomological Society* 55(3):423-426.

1982b New species of leafhoppers in the genus *Tharra* from Oriental and Australian regions (Cicadellidae: Coelidiinae: Tharrini). *Journal of the Kansas Entomological Society* 55(3):447-460.

1982c Some additional new species of Thagriine leafhoppers from Malaysia and Indonesia (Cicadellidae: Coelidiinae: Thagriini). *Journal of the Kansas Entomological Society* 55(3):461-473.

1982d New leafhopper species of *Terulia* from Central America with a revised key to the species (Cicadellidae: Coelidiinae: Teruliini). *Journal of the Kansas Entomological Society* 55(3):489-493.

1982e A revision of the subfamily Coelidiinae (Homoptera: Cicadellidae). IV. Tribe Coelidiini. *Pacific Insects Monograph* 38. 318 pp. [Descriptions of 35 new genera.]

1982f New species of Brazilian leafhoppers in the genus *Docalidia* (Cicadellidae: Coelidiinae: Teruliini). *Entomography* 1:237-256.

1982g New species of leafhoppers of *Docalidia* from Peru (Cicadellidae: Coelidiinae: Teruliini). *Entomography* 1:289-302.

1982h Some additional new leafhopper species of *Docalidia* from South America (Cicadellidae: Coelidiinae: Teruliini). *Entomography* 1:439-445.

1983a Descriptions of three new species of *Labocurtidia* with a revised key to the species (Cicadellidae: Coelidiinae: Teruliini). *Journal of the Kansas Entomological Society* 56(3):315-319.

1983b New Neotropical species of teruliine leafhoppers (Cicadellidae: Coelidiinae: Teruliini). *Journal of the Kansas Entomological Society* 56(3):365-370.

1983c Descriptions of two new species of *Perulidia* with a revised key to the species (Cicadellidae: Coelidiinae: Teruliini). *Journal of the Kansas Entomological Society* 56(3):371-374.

1983d A new species of *Crepluvia* with a revised key to the species (Cicadellidae: Coelidiinae: Teruliini). *Journal of the Kansas Entomological Society* 56(3):375-376.

1983e A revision of the subfamily Coelidiinae (Homoptera: Cicadellidae). V. New tribes Hikangiini, Youngolidiini, and Gabritini. *Pacific Insects Monograph* 40. 78 pp. [Describes 7 new genera.]

1983f Descriptions of two new species of *Peayanus* with a revised key to the species (Cicadellidae: Coelidiinae: Teruliini). *Journal of the Kansas Entomological Society* 56(4):469-472.

1983g Two new species of *Noritonus* from Brazil with a revised key to the species (Cicadellidae: Coelidiinae: Teruliini). *Journal of the Kansas Entomological Society* 56(4):473-476.

1983h New genera in the tribe Teruliini with descriptions of new species (Homoptera: Cicadellidae: Coelidiinae). *Journal of the Kansas Entomological Society* 56(4):560-570. [Describes 6 new genera.]

1983i New leafhopper species of *Coelidia* with a revised key and notes on homonymy and distribution (Homoptera: Cicadellidae, Coelidiinae). *Great Basin Naturalist* 43(4):669-674.

1983j Biosystematics and breeding experiments for resolution of species problems. *In: Knight, W. J., Pant, N. C., Robertson, T. S. and Wilson, M. R., eds. Proceedings of the 1st International Workshop on Biotaxonomy, Classification and Biology of Leafhoppers and Planthoppers (Auchenorrhyncha) of Economic Importance.* Pp. 105-109. London, 4-7 October 1982. Commonwealth Institute of Entomology, 56 Queen's Gate, London SW7 5JR. 500 pp.

1985a Leafhopper Systematics. *In: Nault, L. R. and Rodriguez, J. G., eds. The Leafhoppers and Planthoppers.* Pp. 11-39. Wiley & Sons, New York. 500 pp.

Nielson, M. W. and Bleak, E. E.

1963a Relationship of sex and population densities of the leafhopper *Aceratagallia curvata* to damage of seedling alfalfa. *Journal of Economic Entomology* 56:93-95.

Nielson, M. W. and Currie, W. E.

1962a Leafhoppers attacking alfalfa in the Salt River Valley of Arizona. *Journal of Economic Entomology* 55:803-804.

Nielson, M. W. and Freytag, P.

1976a Biological evidence for the taxonomic suprression of *Gyponana hasta* (Homoptera: Cicadellidae). *Journal of the Kansas Entomological Society* 49(3):401-404.

Nielson, M. W. and Gill, R. J.

1984a *Amphigonalia bispinosa*, a new leafhopper species from California and the replacement vector species for *Amphigonalia severini* (DeLong) (Homoptera: Cicadellidae: Cicadellinae). *Journal of the Kansas Entomological Society* 57(3):400-404.

Nielson, M. W. and Jones, L. S.

1954a Insect transmission of western X - little-cherry virus. *Phytopathology* 44(4):218-219.

Nielson, M. W. and Kaloostian, G. H.

1956a Leafhoppers collected in and near stone fruit orchards in northern Utah. *Utah State Agricultural College, Mimeo Series* 427. 14 pp.

Nielson, M. W. and Lehman, W. F.

1980a Breeding approaches in alfalfa. *In: Maxwell, F. G. and Jennings, P. R., eds. Breeding Plants Resistant to Insects.* Pp. 277-311. Wiley & Sons, New York. 683 pp.

Nielson, M. W. and May, C. J.

1975a Comparative developmental biology of *Cuerna arida* and *C. balli* in Arizona. *Annals of the Entomological Society of America* 68(2):346-348.

Nielson, M. W., May, C. J. and Tingey, W. M.

1975a Developmental biology of *Oncometopia alpha*. *Annals of the Entomological Society of America* 68(3):401-403.

Nielson, M. W. and Morgan, L. A.
1982a Developmental biology of the leafhopper, *Scaphytopius nitridus* (Homoptera: Cicadellidae), with notes on distribution, hosts and interspecific breeding. *Annals of the Entomological Society of America* 75(3):350-352.

Nielson, M. W. and Schonhorst, M. W.
1965a Screening alfalfa for resistance to some common insect pests in Arizona. *Journal of Economic Entomology* 58(1):147-150.

Nielson, M. W. and Toles, S. L.
1968a Observations on the biology of *Acinopterus angulatus* and *Aceratagallia curvata* in Arizona (Homoptera: Cicadellidae). *Annals of the Entomological Society of America* 61(1):54-56.
1970a Interspecific hybridization in *Carnecephala* (Homoptera, Cicadellidae). *Journal of the Kansas Entomological Society* 43(1):1-10.

Niemczyk, H. D. and Guyer, G. E.
1963a The distribution, abundance and economic importance of insects affecting red and mammoth clover in Michigan. *Michigan Agricultural Experiment Station Technical Bulletin* 293. 38 pp.

Nikolova, V.
1969a Entomocenological and biological studies in *Rosa damascena* Mill. plantations. 3. Homoptera. *Izvestiya na Zoologicheskiya Institut* (Sofia) 29:191-208. [In Bulgarian with Russian and English summaries]

Nishi, Y., Kimura, T. and Maejima, I.
1975a Causal agent of 'waika' disease of rice plants in Japan. *Annals of the Phytopathological Society of Japan* 41:228-231.

Nishida, T., Wongsiri, T. and Wongsiri, N.
1976a Species composition, population trends and egg parasitism of planthopper and leafhopper rice pests of Thailand. FAO *Plant Protection Bulletin* 24:22-26.

Nixon, G. A.
1984a Biotaxonomy of the birch feeding *Oncopsis flavicollis* (L.) species complex. *Mitteilungen der Schweizerischen Entomologischen Gesellschaft* 57(4):436.

Noda, H.
1980a Sterol biosynthesis by symbiotes of leafhoppers. *Proceedings of the XVI International Congress of Entomology* 16:69 (Abstract).

Noda, H. and Mittler, T. E.
1983a Sterol biosynthesis by symbiotes of aphids and leafhoppers. 41-55, Illustr. In: *Mittler, T. E. and Dadd, R. H., eds. Metabolic Aspects of Lipid Nutrition.* Westview Press, Boulder, Colorado. 255 pp.

Noda, H., Sogawa, K. and Saito, T.
1973a Amino acids in honeydews of the rice planthoppers and leafhoppers (Homoptera: Delphacidae, Deltocephalidae). *Applied Entomology and Zoology* 8:191-197.

Noda, M., Takasaki, T. and Murata, T.
1968a On the forecast of numerical populations of the second generation larvae of the small brown planthopper and the green rice leafhopper in the paddy field. *Proceedings of the Association of Plant Protection of Kyushu* 14:12-15. [In Japanese.]

Noguchi, H., Tamaki, Y. and Sugimoto, A.
1968a Fatty acid composition of the green rice leafhopper, *Nephotettix cincticeps* Uhler. *Japanese Journal of Applied Entomology and Zoology* 12:100-102. [In Japanese.]

Noon, Z. B.
1962a Environmental influences on changes in populations of potato leafhopper, *Empoasca fabae*, in relation to alfalfa. Ph. D. Dissertation, University of Illinois. 164 pp. [Diss. Abst. 23(2):373.]

Nordlander, G.
1977a Observations on the insect fauna in apple trees in connection with tests on insecticides for integrated control. *Vaxtskyddsnotiser* 41:39-48.

Norment, B. R., Haskins, J. R. and Hepner, L. W.
1972a A comparative electrophoretic study of Cicadellidae and Membracidae. *Annals of the Entomological Society of America* 65(5):1149-1153.

Norris, D. O.
1954a Purple-top wilt, a disease of potato caused by tomato big-bud virus. *Australian Journal of Agricultural Research* 5(1):1-8.

Nour, M. A.
1962a Witches' broom and phyllody in some plants in Khartoum Province, Sudan. *FAO Plant Protection Bulletin* 10(3):49-56.

Novak, J. B.
1961a Onemocnemi Cibuloue, Korenove, Kostalove Zeleniny A Hlavkoveho Salatu Zloutenkous Ze Skupiny Virovych Bazsemenosti. Predbezna Zprava. [Infection of bulbous root, and stem vegetables, and head lettuce with yellows from the group of aspermy viruses.] *Ceskoslovenska Akademie Zemedelska Vedeckotechnickych Sbornik Rostlinna Vyroba* 34:855-870.

Novak, P. and Wagner, W.
1962a Beitrag zur Kenntnis der Homopteren-Fauna Dalmatiens. *Godisnjak Bioloskog Instituta u Sarajevu* 15:31-53. [In Serbo-Croatian with German summary.]

Novoa, N. and Alayo D. P.
1985a El genero *Tylozygus* Fieber 1866 (Homoptera: Auchenorrhyncha) en Cuba. *Peoyana* 289:1-14.

Nowacka, W.
1964a From present studies on the occurrence and bionomics of the leafhopper *Aphrodes bicinctus* in the vicinity of Pozan. *Biuletyn, Instytutu Ochrony Roslin* 28:27-41. [With Russian and English summaries.]
1966a The leafhopper fauna occuring on alfalfa in the district of Poznan. *Roczniki Wyzszej Szkoly Rolniczej w Poznaniu* 29:189-196. [In Polish with English summary.]
1978a Leafhoppers of the genus *Macrosteles* Fieb. (Homoptera, Cicadoidea) occurring on some crops in Poland. *Roczniki Nauk Rolniczych* 7:143-160.
1978b Observations on the morphology and biology of the rose leafhopper (*Edwardsiana rosae* L., Homoptera, Cicadellidae, Typhlocybinae). *Roczniki Akademii Rolniczej w Poznaniu* 98:155-165.
1982a Leafhoppers (Homoptera, Auchenorrhyncha) occurring on cereal and grass crops in Poland. *Roczniki Adademii Rolniczej w Poznaniu* No. 122. 82 pp.

Nowacka, W. and Adamska-Wilczek, J.
1972a Leafhoppers (Homoptera, Cicadodea), pests of the medical plants. *Polskie Pismo Entomologiczne* 44:393-404. [In Polish with English summary.]

Nowacka, W. and Bielejewski, J.
1978a Leafhoppers (Homoptera, Cicadoidea), pests of the sunflower crops. *Roczniki Nauk Rolniczych Serie E Ochrona Roslin* 8:203-214. [In Polish with English summary.]

Nowacka, W. and Hoppe, W.
1969a Investigations of vectors of cereal virus diseases. In: *IX Scientific Session of the Institute of Plant Protection. Biuletyn, Instytutu Ochrony Roslin.* Pp. 109-121.

Nowacka, W. and Zoltanska, E.
1974a Leafhoppers (Cicadodea [sic!], Homoptera) on umbelliferous and alliaceous vegetables. *Roczniki Nauk Rolniczych* 4:33-46.

Nuorteva, P.
1956a Notes on the anatomy of the salivary glands and the occurrence of proteases in these organs in some leafhoppers (Hom., Auchenorrhyncha). *Annales Entomologici Fennici* 22:103-108.

Nuque, F. L. and Miah, S. A.
1969a A rice virus disease resembling tungro in East Pakistan. *Plant Disease Reporter* 53(11):888-890.

Nyland, G., Goheen, A. C., Lowe, S. K. and Kirkpatrick, H. C.
1973a The ultrastructure of a rickettsialike organism from a peach tree affected with phony disease. *Phytopathology* 63(10):1275-1278.

Nyland, G., Raju, B. C. and Purcell, A. H.
1981a An epidemic of peach yellow leaf roll (X-disease) in northern California: Association with pear orchards. *Phytopathology* 71:107.

O'Brien, M. F. and Kurczewski, F. E.
1982a Further observations on the ethology of *Alysson conicus* Provancher (Hymenoptera: Sphecidae). *Proceedings of the Entomological Society of Washington* 84(2):225-231.

Obrtel, R.
1969a The insect fauna of the herbage stratum of lucerne fields in southern Moravia (Czechoslovakia). *Prirodovedne Prace ustavu Ceskoslovenske Akaemie Ved v Brne* 3:1-49.

Ofori, F. A. and Francki, R. I. B.
1983a Evidence that maize wallaby ear disease is caused by an insect toxin. *Annals of Applied Biology* 103:185-189.
1984a The etiology of maize wallaby ear disease (MWED). *Biennial Report of the Waite Agricultural Research Institute, 1982-1983.* Pp. 153-154.
1985a Transmission of leafhopper A virus, vertically through eggs and horizontally through maize in which it does not multiply. *Virology* 144(1):152-157.

Ogane, Z., Takita, Y. and Naito, A.
1979a Location of parts of rice plants sucked by green rice leafhopper, *Nephotettix cincticeps* Uhler (Hemiptera: Cicadellidae). *Japanese Journal of Applied Entomology and Zoology* 23:11-16.

Ogunlana, M. O. and Pedigo, L. P.
1974a Economic injury levels of the potato leafhopper on soybeans in Iowa. *Journal of Economic Entomology* 67(1):29-32.
1974b Pest status of the potato leafhopper on soybeans in central Iowa. *Journal of Economic Entomology* 67(2):201-202.
1984a Newly discovered plant hosts of *Spiroplasma citri. Plant Disease* 68:336-338.

O'Hayer, K. W., Schultz, G. A., Eastman, C. E., Fletcher, J. and Goodman, R. M.
1983a Transmission of *Spiroplasma citri* by the aster leafhopper, *Macrosteles fascifrons* (Homoptera: Cicadellidae). *Annales Applied Biology* 102:311-318.

Ohmart, C. P., Stewart, L. G. and Thomas, J. R.
1983a Phytophagous insect communities in canopies of three forest types in southeastern Australia. *Australian Journal of Ecology* 8(4):395-404.

O'Keeffe, L. W.
1965a The influence of environmental factors upon oviposition of *Empoasca fabae* (Harris) (Cicadellidae, Homoptera). Ph. D. Dissertation, Iowa State University. 162 pp. [Diss. Abst. 26(10):6251-6252.]

Okada, T.
1971a Leafhopper genus *Empoasca* in Kyushu with some ecological surveys (Homoptera, Cicadellidae). *Bulletin of the Kyushu Agricultural Experiment Station* 15:693-735.
1976a Three new species of *Pagaronia* Ball (Homoptera: Cicadellidae) from Japan. *Kontyu* 44:138-141.
1977a Redescription of four Japanese *Scaphoideus* species (Homoptera: Cicadellidae). *Kontyu* 45(2):192-198.
1978a Twelve new species of *Pagaronia* Ball (Homoptera: Cicadellidae) from Japan. *Kontyu* 46(3):371-384.
1978b A new species, *Hishimonus araii*, from Japan and Korea (Homoptera: Cicadellidae: Deltocephalinae: Opsiini). *Applied Entomology and Zoology* 13(4):308-311.

Okali, I.
1959a Contribution to the knowledge of the cicadas of Slovakia. *Casopis Ceskoslovenske Spolecnosti Entomologicke* 56(1):109. [In Russian.]
1960a Homoptera Auchenorrhyncha einige Biotope in der Umgegend von Bratislava. *Acta Facultatis Rerum Naturalium Universitatis Comenianae Zoologia* 4(6-8):353-363.
1963a Bericht uber das *Macrosteles viridigriseus-Vorkommen* (Edwards, 1924) in der Slowakei. *Biologia* [Bratislava] 18(4):313.
1964a Historische Ubersicht der Erforschung von Zikaden in der Slowakei. *Entomologicke Symposium, Sleske Museum, Opava* 179-197. [In Slovakian with German summary.]
1968a Die Verbreitung der Arten der Gattungen *Macrosteles, Sonronius* und *Sagatus* in Mitteleuropa. *Folia Entomologica Hungarica* 21:325-334.

Okamoto, D.
1970a Granular insecticide application in paddy field. *Japan Pesticide Information* 2:15-18.

Okoth, V. A. O., Dabrowski, Z. T. and Efron Y.
1985a *Cicadulina* species, comparative biology and virus transmission. *IITA Annual Report for 1984.* Pp. 44-46. Ibdan, Nigeria.

Okuma, C., Lee, M. H. and Hokyo, N.
1978a Fauna of spiders in a paddy field in Suweon, Korea. *Esakia* 11:81-88.

Oldfield, G. N.
1980a A virescence agent transmitted by *Circulifer tenellus* (Baker): Aspects of its host range and association with *Spiroplasma citri. In: 3rd Conference of International Organization of Mycoplasmologists,* p. 46.

Oldfield, G. N. and Kaloostian, G. H.
1979a Vectors and host range of the citrus stubborn disease pathogen, *Spiroplasma citri. Plant Protection Bulletin* (Taiwan) 21:119-125.

Oldfield, G. N., Kaloostian, G. H., Pierce, H. D., Calavan, E. C., Granett, A. L. and Blue, R. L.
1976a Beet leafhopper transmits citrus stubborn disease. *California Agriculture* 30:15.

Oldfield, G. N., Kaloostian, G. H., Pierce, H. D., Calavan, E. C., Granett, A. L., Blue, R. L., Rana, G. L. and Gumpf, D. J.
1977a Transmission of *Spiroplasma citri* from citrus to citrus by *Scaphytopius nitridus. Phytopathology* 67:763-765.

Oldfield, G. N., Kaloostian, G. H., Pierce, H. D., Granett, A. L. and Calavan, E. C.
1977a Beet leafhopper transmits virescence of periwinkle. *California Agriculture* 31:14-15.

Oldfield, G. N., Kaloostian, G. H., Sullivan, D. A., Calavan, E. C. and Blue, R. L.
1978a Transmission of the citrus stubborn disease pathogen, *Spiroplasma citri,* to a monocotyledonous plant. *Plant Disease Reporter* 62:758-760.

Olivares, F. M. and San Juan, M. O.
1966a The transmission, virus-vector relationship and host range of tobacco leaf curl virus. *In: Papers Presented at the Divisional Meeting on Plant Protection, the Eleventh Pacific Science Congress,* Tokyo, 1966. Pp. 283-299. 318 pp.

Oliveira, J. V. de and Arauja, A. D. de
1979a The use of coloured sticky traps to catch *Empoasca kraemeri* Ross & Moore, 1957, on cowpea *Vigna unguiculata* (L.) Walp. *Fitossanidade* 3:10-11.

Oliveira, J. V. de, Silva, I. P. da and Fernandez, M. B. D.
1981a Dinamica populacional de "cigarrinha verde" *Empoasca kraemeri* Ross and Moore, 1957, em cultivares de feijao. *Anais da Sociedade Entomologica do Brasil* 10:21-26.

Olmi, M.
1968a Cicaline della risaia da vicenda vercellese (Homoptera: Auchenorhyncha). Studi del gruppo di lavoro del C.N.R. per le virosi: CXXI. [Rice-field leaf- and planthoppers in the neighbourhood of Vercelli (Homoptera: Auchenorrhyncha). Studies of the C.N.R. working party for virus diseases: CXXI.] *Annali della Facolta di Scienze Agrarie della Universita degli Studi di Torino* 4:247-260.
1975a Scetta della vittima in insetti emenotteri e connessioni con il sistema degli ospitiomotteri. *Atti della Accademia delle Scienze di Torino* 109(1-2):69-77.
1976a Variabilita morfologica di un cicadellide dannoso alla Graminacea foraggere negli alti pascoli Piemontesi *Diplocoelenus (Verdanus) abdominalis* (F.) Hemiptera Cicadellidae. *Fragmenta Entomologica* 12:103-112. [In Italian with English summary.]

Oman, P. W.
1967a Problems in the taxonomy and zoogeography of leafhoppers. *Proceedings of the Washington State Entomological Society* 25:220-221.
1969a Criteria of specificity in vector-virus relationships. In: Maramorosch, K., ed. *Viruses, Vectors and Vegetation.* Pp. 1-22. Wiley & Sons, New York. 666 pp.
1970a Taxonomy and nomenclature of the beet leafhopper *Circulifer tenellus* (Homoptera: Cicadellidae). *Annals of the Entomological Society of America* 63(2):507-511.
1970b Leafhoppers of the *Agalliopsis novella* complex (Homoptera: Cicadellidae.) *Proceedings of the Entomological Society of Washington* 72(1):1-29.
1970c Leafhoppers of the *Agalliopsis variabilis* group with description of a new species (Homoptera: Cicadellidae). *Proceedings of the Entomological Society of Washington* 72(1):30-32.
1971a The female of *Thatuna gilletti* Oman, with biological notes. *Proceedings of the Entomological Society of Washington* 73(4):368-372.
1971b The leafhopper subfamily Koebelinae. In: Asahima, S., Gressitt, J. L., Hikada, Z., Nishida, T. and Namura, K., eds. *Entomological Essays to Commemorate the Retirement of Professor K. Yasumatsu.* Pp. 129-139. Hokuryukan, Tokyo. 389 pp.
1971c A new *Agallia* from the western United States. *Journal of the Kansas Entomological Society* 44(3):325-328.
1972a A new megophthalmine leafhopper from Oregon, with notes on its behaviour (Homoptera: Cicadellidae). *Journal of Entomology* (B) 41(1):60-76.
1976a World catalogues of the Homoptera: Auchenorrhyncha. *Bulletin of the Entomological Society of America* 22(2):161-164.
1985a A synopsis of the Nearctic Dorycephalinae (Homoptera: Cicadellidae). *Journal of the Kansas Entomological Society* 58(2):314-336.

Oman, P. W. and Krombein, K. V.
1968a Systematic entomology: Distribution of insects in the Pacific. *Science* 161:78-79.

Oman, P. W. and Musgrave, C. A.
1975a The Nearctic genera of Errhomenini (Homoptera: Cicadellidae). *Melanderia* 21:1-14. [New subgenera: *Ankosus* and *Hylaius.*]

Omura, T., Hibino, H., Usugi, T., Inoue, H., Morinaka, T., Tsurumachi, S., Ong, A., Putta, M., Tsuchizaki, T. and Saito, Y.
1984a Detection of rice viruses in plants and individual insect vectors by latex flocculation tests. *Plant Disease Reporter* 68:374-378.

Omura, T., Inoue, H., Morinaka, T., Saito, Y., Chettanacit, D., Putta, M., Parajarean, A. and Disthaporn, S.
1980a Rice gall dwarf, a new virus disease. *Plant Disease Reporter* 64:795-797.
1980b Observations on rice gall dwarf, a new virus disease. *International Rice Research Newsletter* 5(3):11-12.

Omura, T., Inoue, H., Pradhan, R. B., Thapa, B. J. and Saito, Y.
1982a Identification of rice dwarf virus in Nepal. *Japanese Agricultural Research Quarterly* 15(3):218-220.

Omura, T., Kimura, I., Tsuchizaki, T. and Sato, Y.
1982a Infectivity assays of rice gall dwarf virus on its vector cell monolayers. *Annals of the Phytopathological Society of Japan* 48:389-390. [In Japanese.]

Omura, T., Saito, Y., Usugi, T. and Hibino, H.
1982a Purification and serology of rice tungro spherical virus and rice tungro bacilliform virus. *Annals of the Phytopathological Society of Japan* 49:74-77.

Onishchenko, A. N.
1984a The effect of different factors on the incubation period of phytopathogenic mycoplasmas in plants and leafhopper vectors. *Mikrobiologicheskii Zhurnal* 46:52-56.

Onstad, D. W., Shoemaker, C. A. and Hansen, B. C.
1984a Management of potato leafhopper, *Empoasca fabae* (Homoptera: Cicadellidae), on alfalfa with the aid of systems analysis. *Environmental Entomology* 13(4):1046-1058.

Orenski, S. W.
1964a Effects of a plant virus on survival, food acceptability, and digestive enzymes of corn leafhoppers. *Annals of the New York Academy of Science* 118:374-383.

Orenski, S. W., Mitsuhashi, J., Ringel, S. M., Martin, J. F. and Maramorosch, K.
1965a A presumptive bacterial symbiont from the eggs of the six-spotted leafhopper, *Macrosteles fascifrons* Stal. *Contributions from the Boyce Thompson Institute for Plant Research* 23:123-126.

Orenski, S. W., Murray, J. R. and Maramorosch, K.
1965a Further studies on the feeding habits of aster yellows virus-carrying corn leafhoppers. *Contributions from the Boyce Thompson Institute for Plant Research* 23(3):47-50.

Orenski, S. W., Staples, R. C. and Maramorosch, K.
1962a The uptake of C_{14} and P_{32} from labeled leaves by two species of leafhopper vectors. *Phytopathology* 52(11):1220 (Abstract).

Orgell, W. H., Hamilton, E. W., Hibbs, E. T. and Carlson, O. V.
1959a Cholinesterase-inhibitory compounds occurring naturally in potatoes, *Solanum tuberosum* (L.) and their relationship to leafhopper, *Empoasca fabae* (Harris), resistance. *Proceedings of the North Central Branch of the Entomological Society of America* 14:5.

Orita, S.
1969a *Gonatocerus* sp. (Hymenoptera: Mymaridae), a natural enemy of green rice leafhopper, *Nephotettix cincticeps* Uhler. *Proceedings of the Association of Plant Protection* (Hokuriku) 17:67-69. [In Japanese.]
1971a Effect of depth and duration of snow cover to survival of green rice leafhopper, *Nephotettix cincticeps* Uhler. *Proceedings of the Association of Plant Protection* (Hokuriku) 19:42-44. [In Japanese.]
1972a Some notes on *Lymaenon* sp. (Hymenoptera: Mymaridae), an egg parasite of the green rice leafhopper, *Nephotettix cincticeps* Uhler (Homoptera: Cicadellidae), and its distribution in Hokuriku district. *Bulletin of the Hokuriku National Agricultural Experiment Station* 14:91-124. [In Japanese.]

Orosz, A.
1977a Beitrag zur Kenntnis der Gattung *Ulopa* Fallen, 1814 (Homoptera: Ulopidae). [Contributions to the knowledge of the genus *Ulopa* Fallen 1814 (Homptera: Ulopidae.] *Folia Entomologica Hungarica* 30(2):95-103. [English and Hungarian summaries.]
1979a *Iassus mirabilis* sp. n. und einige fur die Fauna Ungarns neue Zikaden (Homoptera). *Folia Entomologica Hungarica* 32(1):65-69.
1981a Cicadellidae of the Hortobagy National Park. *Natural History of the National Parks* (Hungary) 1:65-76.

Osler, R., Amici, A., Belli, G. and Corbetta, G.
1973a Research on the epidemiology of rice yellows. I. Demonstration of the presence of winged vectors. *Riso* 22:105-110.

Osler, R., Fortusini, A. and Belli, G.
1975a Presence of *Scaphoideus littoralis* in vineyards in the Pavese district beyond the Po affected by a disease of the golden-flavescence type. *Informatore Fitopatologico* 25:13-15.

Osmelak, J. A.
1984a Aspects of the control of tomato big bud disease and the vector *Orosius argentatus* (Evans) in Victoria. Masters Thesis. Melbourne University.

Osmun, J. V.
1957a Insects and other arthropods of economic importance in Indiana in 1957. *Proceedings of the Indiana Academy of Science* 67:150-154.
1958a Insects and other arthropods of economic importance in Indiana in 1958. *Proceedings of the Indiana Academy of Science* 68:190-195.
1959a Insects and other arthropods of economic importance in Indiana in 1959. *Proceedings of the Indiana Academy of Science* 69:167-174.

Ossiannilsson, F.

1955a A few leafhoppers (Hom. Auchenorrhyncha) new to Sweden with a synonymic note. *Entomologick Tidskrift* 76(2):131-133.

1957a Preservation in "Celodal." *Entomologisk Tidskrift* 78:178-179.

1958a Is tobacco mosaic virus not imbibed by aphids and leafhoppers? *Kungliga Lantbrukshogskolans Annaler* 24:369-374.

1958b "Celochraoal" — a new mounting medium for insects. *Entomologisk Tidskrift* 79:2-5.

1961a *Balclutha calamagrostis*, n. sp., a new Swedish leafhopper (Hem., Hom., Auchenorrh.) *Opuscula Entomologica* 26(1&2):59-60.

1961b Anmarknigar och tillagg till Sveriges hemipterfauna (Hemipterologiska notiser VIII). *Opuscula Entomologica* 26:228-234.

1962a Hemipterfyndi Norge 1960. *Norsk Entomologist Tidskrift* 12:56-62.

1966a Insects in the epidemiology of plant viruses. *Annual Review of Entomology* 11:213-232.

1974a Hemiptera (Heteroptera, Auchenorrhyncha and Psylloidea). *Fauna of the Hardangervidda* 5:13-35.

1976a Two new species of leafhoppers from Fennoscandia (Homoptera: Cicadellidae). *Entomologica Scandinavica* 7:31-34.

1978a *The Auchenorrhyncha (Homoptera) of Fennoscandia and Denmark. Part 1: Introduction, Infraorder Fulgoromorpha. Fauna Entomologica Scandinavica* 7. Scandinavian Science Press Ltd., Klampenborgm Denmark. 222 pp.

1981a *The Auchenorrhyncha (Homoptera) of Fennoscandia and Denmark. Part 2: The Families Cicadidae, Cercopidae, Membracidae, and Cicadellidae (exc. Deltocphalinae). Fauna Entomologica Scandinavica* 7. Scandinavian Science Press Ltd., Klampenborg, Denmark. Pp.223-593.

1982a Designation of lectotypes of *Allygus* Fieber (Homoptera, Auchenorrhyncha). *Entomologica Scandinavica* 13(2):140.

1983a *The Auchenorrhyncha (Homoptera) of Fennoscandia and Denmark. Part 3: The Family Cicadellidae: Deltocphalinae, Catalogue, Literature and Index. Fauna Entomologica Scandinavica* 7. Scandinavian Science Press Ltd., Klampenborg, Denmark. Pp. 594-979.

1983b *Allygus* Fieber, 1872 (Insecta, Homoptera): Proposed designation of type species, Z. N. (S.) 2431. *Bulletin Zoological Nomenclature* 40(2):119-121.

Ossowski, L. L. J.

1957a Forstentomologische probleme im verbreitungsgebiet der schwarzakazie, *Acacia mollissima* Willd., in der Sud-afrikanischen Union. *Anzeiger fuer Schadlingskunde* 30(9):133-137.

Otake, A.

1966a Analytical studies of light trap records in the Kokuriku district. II. The green rice leafhopper, *Nephotettix cincticeps*. *Researches on Population Ecology* 8:62-68.

1983a Leafhoppers and planthoppers as 'virus' vectors in Japan. *In: Knight, W. J., Pant. N. C., Roberston, T. S. and Wilson, M. R., eds. Proceedings of the 1st International Workshop on Biotaxonomy, Classification and Biology of Leafhoppers and Planthoppers (Auchenorrhyncha) of Economic Importance.* Pp. 439-455. London, 4-7 October 1982. Commonwealth Institute of Entomology, 56 Queen's Gate, London SW7 5JR. 500 pp.

Otake, A. and Hokyo, N.

1976a Rice plant - leafhopper incidence in Malaysia and Indonesia. *Report of a Research Tour January to March 1976. Report, Tropical Agricultural Research Center, Ministry of Agricultural and Forestry, Japan. Shiryo* No. 33. 64 pp.

Ou, S. H.

1972a *Rice diseases.* IRRI. Commonwealth Mycological Institute, Los Banos, Philippines.

Ou, S. H. and Ling, K. C.

1966a Virus diseases of rice in the South Pacific. *FAO Plant Protection Bulletin* 14(5):113-121.

Ou, S. H. and Rivera, C. T.

1969a Virus diseases of rice in southeast Asia. *In: The Virus Diseases of the Rice Plant. Proceedings of the International Rice Research Institute Symposium.* Pp. 23-34.

Ou, S. H., Rivera, C. T., Vavaratnam, S. J. and Goh, K. G.

1965a Virus nature of "penyakit merah" diseases of rice in Malaysia. *Plant Disease Reporter* 49:778-782.

Ouchi, Y. and Suenaga, H.

1963a On the transmissibility by the leafhopper *Nephotettix apicalis* of rice yellow dwarf virus. *Proceedings of the Association of Plant Protection* (Kyushu) 9:60-61.

1963b Ability of *Nephotettix apicalis* (Motschulsky) to transmit rice yellow dwarf virus. *Proceedings of the Association for Plant Protection of Kyushu* 9(1963):60-61; 10(1964):10-12.

1968a Ability of *Nephotettix apicalis* (Motschulsky) to transmit rice yellow dwarf virus. *Revue of Plant Protection Research* (Japan) 1:76-78.

Ouyang, J. J., Li, K. and Lu, H. C.

1984a Trials on the control of *Empoasca flavescens* (Fabricius). *Guangdong Nongye Kexue* 4:30-32.

Oya, S.

1978a Effects of photoperiod on the induction of diapause in the green rice leafhopper, *Nephotettix cincticeps* Uhler (Hemiptera: Cicadellidae). *Japanese Journal of Applied Entomology and Zoology* 22:108-114.

1979a Survival tests on submerged green rice leafhoppers, *Nephotettix cincticeps* Uhler, in relation to winter mortality under snow cover in the Hokuriku district. *Applied Entomology and Zoology* 14:319-325.

1980a Feeding habits and honeydew components of the green rice leafhopper, *Nephotettix cincticeps* Uhler (Hemiptera: Deltocephalidae). *Applied Entomology and Zoology* 15(4):393-399.

Oya, S. and Sato, A.

1973a Influence of low temperature, host-plant and submerged conditions to the green rice leahopper, *Nephotettix cincticeps* Uhler. *Proceedings of the Association of Plant Protection* (Hokuriku) 21:65-68. [In Japanese.]

1980a Antibiosis and non-preference in resistant rice varieties to the green rice leafhopper, *Nephotettix cincticeps* Uhler. *Proceedings of the Association of Plant Protection* (Hokuriku) 28:23-29. [In Japanese with English summary.]

1981a Differences in feeding habits of the green rice leafhopper, *Nephotettix cincticeps* Uhler (Hemiptera: Deltocephalidae), on resistant and susceptible rice varieties. *Applied Entomology and Zoology* 16:451-457.

Ozaki, K.

1966a Some notes on the resistance to malathion and methyl parathion of the green rice leafhopper, *Nephotettix cincticeps* Uhler (Homoptera: Cicadelliae). *Applied Entomology and Zoology* 1(4):189-196.

1983a Suppression of resistance through synergistic combinations with emphasis on planthoppers and leafhoppers infecting rice in Japan. *In: Georghiou, G. P. and Saito, T., eds. Pest Resistance to Pesticides.* Pp. 595-613. Plenum Press, New York. 889 pp.

Ozaki, K. and Kassai, T.

1984a The insecticidal activity of pyrethroids against insecticide-resistant strains of planthoppers, leafhoppers and the housefly. *Journal of Pesticide Science* 9:61-66.

Ozaki, K., Kassai, T. and Sasaki, Y.

1984a Insecticide activity of pyrethroids against the green rice leafhopper, *Nephotettix cincticeps* Uhler. *Journal of Pesticide Science* 9:155-157.

Ozaki, K. and Kurosu, Y.

1967a Resistance pattern in four strains of insecticide-resistant green rice leafhopper, *Nephotettix cincticeps* Uhler, collected in field. *Japanese Journal of Applied Entomology and Zoology* 11:145-149.

Ozaki, K., Kurosu, Y. and Koike, H.
1966a The relation between malathion resistance and esterase activity in the green rice leafhopper, *Nephotettix cincticeps* Uhler. *SABCO Journal* 2:98-106.

Ozaki, K., Saski, Y. and Kassai, T.
1984a The insecticidal activity of mixtures of pyrethroids and organophosphates or carbamates *against* the insecticide-resistant green rice leafhopper, *Nephotettix cincticeps* Uhler. *Journal of Pesticide Science* 9:67-72.

Ozer, M.
1958a Investigations on some pests of wild pistachio in the regions of Balikesir and Utahya. *Ankara Universitesi Ziraat Fakultesi Yilligi* 8(2):111-120. [In Turkish with French summary.]
1964a Preliminary studies on mites and insects harmful to mint in the Samsun and Istanbul areas and to sesame and vegetables in Antalya. *Ankara Universitesi Ziraat Fakultesi Yilligi* 14(3-4):205-222.

Paddick, R. G. and French, F. L.
1964a Control of yellow dwarf of tobacco by treatment with systemic insecticides. *Proceedings Australian Tobacco Research Conference*, 1964. Pp. 304-310.
1968a Some observations on tobacco yellow dwarf control. *Australian Tobacco Grower's Bulletin* 13:9-10.
1972a Suppression of tobacco yellow dwarf with systemic organophosphorous insecticides. *Australian Journal of Experimental Agriculture and Animal Husbandry* 12:331-334.

Page, F. D.
1979a The immature stages of *Austroasca viridigrisea* (Paoli) (Homoptera: Cicadellidae: Typhlocybinae). *Journal of the Australian Entomological Society* 18:111-114.
1983a Biology of *Austroasca viridigrisea* (Paoli) (Hemiptera: Cicadellidae). *Journal of the Australian Entomological Society* 22(2):149-153.

Paik, W. H.
1967a Insect pests of rice in Korea. In: Pathak, E. D., ed. *The Major Insect Pests of the Rice Plant*. Pp. 657-674. Johns Hopkins, Baltimore.

Painter, R. H.
1955a Insects on corn and teosinte in Guatemala. *Journal of Economic Entomology* 48(1):36-42.
1958a Resistance of plants to insects. *Annual Review of Entomology* 3:267-290.
1968a Crops that resist insects provide a way to increase world food supply. *Kansas Agricultural Experiment Station Bulletin* 520. 22 pp.

Palis, F. G., Jackman, J. A., Benigno, E. A., Gayabyab, B. F., Ebuenga, M. D. and Bayot, R. G.
1984a A computer simulation model of the green rice leafhopper population. *Philippine Entomologist* 6:161-179.

Paliwal, Y. C.
1968a Changes in relative virus concentration in *Endria inimica* in relation to its ability to transmit wheat striate mosaic virus. *Phytopathology* 58:386-387.

Palmer, L. T. and Rao, P. S.
1981a Grassy stunt, ragged stunt and tungro diseases of rice in Indonesia. *Tropical Pest Management* 27:212-217.

Palmiter, D. H. and Adams, J. A.
1957a Seasonal occurrence of leafhopper vectors of X-disease virus in sprayed and unsprayed peach blocks. *Phytopathology* 47:531.

Palmiter, D. H., Coxeter, W. J. and Adams, J. A.
1960a Seasonal history and rearing of *Scaphytopius acutus*. (Homoptera: Cicadellidae). *Annals of the Entomological Society of America* 53(6):843-846.

Palo, M. A. and Garcia, C. E.
1935a Further studies on the control of leafhoppers and tip borers on mango inflorescence. *Philippine Journal of Agriculture* 6(4):425-464.

Palomar, M. K. and Ling, K. D.
1966a Growth and yield of rice plants infested with tungro virus. *Philippine Phytopathology* 2:17.

Pan, Y. S.
1981a Management of sugar cane insect pests in Taiwan. *Chinese Journal of Entomology* 1:115-116.

Panda, N., Heinrichs, E. A. and Hibino, H.
1984a Resistance of the rice variety Utri Rajapan to ragged stunt and tungro viruses. *Crop Protection* 3:491-500.

Pandya, P. S. and Patel, C. T.
1964a Possibilities of imparting resistance to pests in cotton by use of wild species of *Gossypium*. *Indian Cotton Growing Review* 18:175-176.

Paniagua, R. and Gamez, R.
1976a El virus del rayado fino del maiz: Estudios adicionales sobre la relacion del virus y su insecto vector. *Turrialba* 26:39-43.

Pant, J. C. and Gupta, M.
1984a Effect of trace elements on the growth and survival of cotton jassid, *Amrasca devastans* (Distant), through chemically defined diets. *Journal of Entomological Research* 8:171-173.

Papovic, R. M.
1977a New records of species of cicadas (Homoptera, Auchenorrhyncha) in the fauna of Yugoslavia. *Acta Biologiae Experimentalis* (Warsaw) 2:47-51.

Pareek, B. L. and Noor, A.
1980a Evaluation of some insecticides against *Amrasca biguttula biguttula* Ishida infesting ridge gourd. *Entomon* 5:55-57.

Parencia, C. R.
1968a Control of cotton insects with an insect-collecting machine. *Journal of Economic Entomology* 61(1):274-279.

Parh, I. A.
1982a The immature stages of *Empoasca dolichi* Paoli (Homoptera: Cicadellidae). *Revue de Zoologie Africaine* 96:61-74.
1983a Species of *Empoasca* associated with cowpea, *Vigna unguiculata* (L.) Walp., in Ibadan and 3 ecological zones in south-western Liberia. *Revue de Zoologie Africaine* 97(1):202-210.
1983b Greenhouse studies on the feeding damage of two species of *Empoasca* (Cicadellidae) on six cultivars of cowpea. *Zimbabwe Agricultural Journal* 80:111-113.
1983c The effects of *Empoasca dolichi* Paoli (Hemiptera: Cicadellidae) on the performance and yield of two cowpea cultivars. *Bulletin of Entomological Research* 73:25-32.

Parh, I. A. and Taylor, T. A.
1981a Studies on the life-cycle of the cicadellid bug *Empoasca dolichi* Paoli, in southern Nigeria. *Journal of Natural History* 15:829-835.

Parida, B. B. and Dalua, B. K.
1981a Preliminary studies on the chromosome constitution in 72 species of Auchenorrhynchan Homoptera from India. *Chromosome Information Service* 31:13-16.
1981b Meiosis in four species of Indian leafhoppers with a review of chromosome numbers in jassids (Homoptera: Insecta). *Prakruti Utkal University Journal of Science* 13(1-2):111-130.

Park, J. S. and Lee, J. O.
1976a Studies on varietal resistance of rice to insect pests. *Rice Entomology Newsletter* 4:9-10.

Parker, H. L.
1967a Notes on the biology of *Tomosvaryella frontata* (Diptera: Pipunculidae), a parasite of the leafhopper *Opsius stactogalus* on *Tamarix*. *Annals of the Entomological Society of America* 60(2):292-295.

Parsons, F. S.
1956a Cotton in Yugoslavia. *Empire Cotton Growing Review* 33(2):105-113.

Passlow, T.
1969a Insect pests of peanuts in southern Queensland. *Queensland Agricultural Journal* 95(7):449-451.

Passlow, T. and Waite, G. K.
1969a Control of lucerne pests on dairy farms. *Queensland Agricultural Journal* 95(12):843-845.

Patange, D. S., Patel, R. K., Rawat, R. R. and Verma, R.
1981a Note on the assessment of efficiency of common light sources and their operating time in the predilection of rice green leafhoppers, *Nephotettix* spp. *Indian Journal of Agricultural Sciences* 51:817-818.

Patel, B. K., Rote, N. B. and Mehta, N. W.
1985a Comparative efficacy of some insecticides against sucking pests of hybrid-4 cotton. *Indian Journal of Plant Protection* 12:139-141.

Patel, B. R. and Vora, V. J.
1981a Efficacy of different insecticides for the control of groundnut jassid, *Empoasca kerri* Pruthi. *Pesticides* 15:33-34.

Patel, G. A. and Hadli, S. N.
1953a Experiments with some new insecticides for the control of mango-hoppers (*Idiocerus atkinsoni*, Leth.). *Indian Journal of Entomology* 15(2):107-114.

Patel, G. A., Katarki, H. V. and Patel, N. G.
1957a Field experiments on insecticidal control of the cotton jassid, *Empoasca devastans* Distant. *Indian Journal of Entomology* 19(1):23-30.

Patel, H.K. and Patel, V. C.
1967a Control of castor jassids by Telodrin. *Indian Oilseeds Journal* 9(2):96-97.

Patel, J. R., Patel, R. C. and Amin, P. R.
1980a Field evaluation of some insecticides for the control of *Amrasca biguttula biguttula* Ishida and *Aphis gossypii* Glover on okra. *Indian Journal of Entomology* 42:776-779.

Patel, R. C., Patel, H. K., and Patel, J. C.
1961a Field testing of insecticides for control of cotton aphids (*Aphis gossypii* Glov.) and cotton jassids (*Empoasca sp.*). *Indian Cotton Grower's Review* 15(6):370-373.

Patel, R. K.
1976a Occurrence of the blue leafhopper, *Typhlocyba maculifrons* Motschulsky, on wheat in Madhya Pradesh. *Indian Journal of Entomology* 36(4):354-355.

Patel, R. K., Patel, S. R. and Shah, A. H.
1975a Biology of mango-hopper *Amritodes atkinsoni* (Leth.) (Jassidae: Hemiptera) in south Gugarat. *Indian Journal of Entomology* 37(2):150-153.
1975b Studies on sex-ratio, longevity and seasonal incidence of the mango-hopper, *Amritodus atkinsoni* Leth. (Jassidae, Homoptera) in South Gujarat. *Indian Journal of Entomology* 35:255-257.

Pathak, M. D.
[Undated] Resistance to leafhoppers in rice varieties. Symposium on Rice Insects. Proceedings of a Symposium on Tropical Agriculture Researches, 19-24 July 1971. *Tropical Agriculture Research Series* 5:179-193.
1966a Application of insecticides to the paddy water for rice pest control. *Papers presented at the Divisional Meeting on Plant Protection, the Eleventh Pacific Science Congress, Tokyo, Japan Plant Protection Association.* Pp. 108-122. 318 pp.
1967a *The Major Insect Pests of the Rice Plant.* Proceedings. Johns Hopkins, Baltimore. 729 pp.
1968a Ecology of common insect pests of rice. *Annual Review of Entomology* 13:257-294.
1969a Stem borer and leafhopper-planthopper resistance in rice varieties. *Entomologia Experimentalis et Applicata* 12:789-800.
1971a Resistance to insect pests in rice varieties. In: *Rice Breeding.* Pp. 325-341. International Rice Research Institute, Manila, Philippines. 738 pp.
1977a Defense of rice crop against insect pests. *Annals of the New York Academy of Sciences* 287:287-295.

Pathak, M. D., Cheng, C. H. and Fortuno, M. E.
1969a Resistance to *Nephotettix impicticeps* and *Nilaparvata lugens* in varieties of rice. *Nature* 223:502-504.

Pathak, M. D., Encarnacion, D. and Dupo, H.
1974a Application of insecticides in the root zone of rice plants. *Indian Journal of Plant Protection* 1:1-16.

Pathak, M. D. and Saxena, R. C.
1980a Breeding approaches in rice. In: Maxwell, F. G. and Jennings, P. R., eds. *Breeding Plants Resistant to Insects.* Pp. 421-455. Wiley, New York. 683 pp.

Pathak, M.D., Vea, E. and John, V. T.
1967a Control of insect vectors to prevent virus infection of rice plants. *Journal of Economic Entomology* 60:218-225.

Pathak, P. K.
1983a The prevalence and degree of host specificity in leafhoppers and planthoppers of rice and their importance to taxonomy and pest control. In: Knight, W. J., Pant, N. C., Robertson, T. S. and Wilson, M. R., eds. *Proceedings of the 1st International Workshop on Biotaxonomy, Classification and Biology of Leafhoppers and Planthoppers (Auchenorrhyncha) of Economic Importance.* P. 335 (abstract only). London, 4-7 October 1982. Commonwealth Institute of Entomology, 56 Queen's Gate, London SW7 5JR. 500 pp.

Patrick, C. R.
1970a Morphological and behavioural studies of the *gemina* complex of *Erythroneura* (Homoptera: Cicadellidae). Thesis, Mississippi State University (microfilm).

Patterson, R. S.
1962a Use of phorate and di-syston for potato insect control and a study of factors which influence phorate absorption by plants and loss in the soil. Ph. D. Dissertation, Cornell University. 134 pp. [Diss. Abst. 23(5):1471-1472.]

Patterson, R. S. and Rawlins, W. A.
1964a Evaluation of phorate and di-syston for potato insect control in New York. *American Potato Journal* 41(7):196-200.

Pawar, A. D.
1975a *Cyrtorhinus lividipennis* Reuter (Miridae: Hemiptera) as a predator of the eggs and nymphs of the brown planthopper and green leafhoppers in Himachal Pradesh, India. *Rice Entomology Newsletter* 3:30-31.

Pawar, A. D. and Bhalla, O. P.
1967a Pest complex of rice in Himachal Pradesh. *Indian Journal of Entomology* 36:358-359.

Payne, K.
1981a The life history and host plant relationships of *Eupteryx notata* Curtis (Homoptera: Cicadellidae). *Entomologist's Monthly Magazine* 117(May-Aug.):167-173.
1981b A comparison of catches of Auchenorrhyncha (Homoptera) obtained from sweep netting and pitfall trapping. *Entomologist's Monthly Magazine* 117(1408-1411):215-223.
1981c Notes on leafhoppers (Homoptera: Auchenorhyncha) found in Northwest Yorkshire during 1979. *Entomologist's Monthly Magazine* 116(May-Aug.):108.

Pearson, E. O.
1958a *The Insect Pest of Cotton in Tropical Africa.* London, Empire Cotton Growers Corporation and Commonwealth. 355 pp.

Peay, W. E.
1950a Laboratory tests for control of the beet leafhopper on snap beans grown for seed. *Journal of Economic Entomology* 52(4):700-703.

Peay, W. E. and Oliver, W. N.
1964a Curly top prevention by vector control on snap beans grown for seed. *Journal of Economic Entomolgy* 57(1):3-5.

Pedigo, L. P.
1972a Economic levels of insect pests. *Iowa Cooperative Extension Service* EC-713E. 4 pp.
1974a Bio-economics of Iowa soybean insects. *Proceedings of the North Central Branch of the Entomological Society of America* 29:56-61.

Pelet, F., Bovey, R. and Baggiolini, M.
1968a The control of the sharka disease of plum. *Agriculture Romande* 7(4):51-52.

Pelov, V.
1968a Prinos k'm prouchbaneto na b'lgarskata cikadna fauna (Homoptera: Auchenorrhyncha). [Beitrag zur Erforschung Bulgarischen Zikadenfauna (Homoptera: Auchenorrhyncha).] *Izvestiya na Zoologicheskiya Institut* (Sofia) 26:157-171.

Pena, M. B. and Shepard, B. M.
1985a Parasitism by nematodes on three species of hopper pests of rice in Laguna, Philippines. *International Rice Research Newsletter* 10(1):19-20.

Penny, N. D. and Arias, J. R.
1982a *Insects of an Amazon Forest.* Columbia University, New York. 269 pp.

Perfect, T. J. and Cook, A. G.
1982a Diurnal periodicity of flight in some Delphacidae and Cicadellidae associated with rice. *Ecological Entomology* 7(3):317-326.

Perfect, T. J., Cook, A. G. and Ferrer, E. R.
1983a Population sampling for planthoppers, leafhoppers (Hemiptera: Delphacidae and Cicadellidae) and their predators in flooded rice. *Bulletin of Entomological Research* 73:345-355.

Perkes, R. R.
1970a *Circulifer tenellus* (Baker) (Homoptera: Cicadellidae): mating behavior, ecology of reproduction and attraction to monochromatic electromagnetic radiation. Doctoral Dissertation, University of California, Riverside. 227 pp.

Perrin, R. M. and Gibson, R. W.
1985a Control of some insect-borne plant viruses with the Pyrethroid PP. 321 (Karate). *International Pest Control* 27:142-143.

Perron, J. P. and Crete, R.
1968a Comparison of two methods of sampling populations of adults of *M. fascifrons* (Stal) (Hemiptera: Cicadellidae) on lettuce on organic soils. *Phytoprotection* 49(2):55-60. [With English abstract.]

Peswani, K. M., Jain, H. K., Agnihotri, N. P., Bose, B. N., Saxena, A. N. and Pandey, S. Y.
1979a Persistence of disulfoton and phorate against the cotton jassid, *Amrasca devastans* Distant. *Journal of Entomological Research* 3:84-86.

Peter Ooi, A. C.
1973a Some insect pests of green gram, *Phaseolus aureus*. *Malaysian Agricultural Journal* 49:131-142.

Peters, D.
1971a Relationships between plant viruses and their aphid and leafhopper vectors. *Annales de Parasitologie Humaine et Comparee* 46(3 bis):233-242. [In English with French summary.]

Peterson, A. G.
1950a A study of the resistance of some varieties of potato to the potato leafhopper, *Empoasca fabae* (Harris). Ph. D. Dissertation, University of Minnesota. 165 pp. [Amer. Doc. Diss. 1950:100.]
1958a Potato insect control. *Proceedings of the North Central Branch of the Entomological Society of America* 13:33.
1971a Observations on soybean insects in Minnesota. *Proceedings of the North Central Branch of the Entomological Society of America* 26(1-2):82.
1973a Host plant and aster yellow leafhopper relationships. *Proceedings of the North Central Branch of the Entomological Society of America* 28:66-71.

Peterson, A. G., Bates, J. D. and Saini, R. S.
1969a Spring dispersal of some leafhoppers and aphids. *Journal of the Minnesota Academy of Science* 35(2/3):98-102.

Peterson, A. G. and Saini, R. S.
1964a Relations of aster yellows infection in host plants to increase of six-spotted leafhopper. *Proceedings of the North Central Branch of the Entomology Society of America* 19:99.

Peterson, G. D., Jr.
1957a Recent additions to the list of insects which attack crops and other important plants in Guam. *Proceedings of the Hawaiian Entomological Society* 16(2):203-207.

Peterson, G. W.
1984a Spread and damage of western X-disease of chokecherry in eastern Nebraska plantings. *Plant Disease Reporter* 68:103-104.

Petty, H. B. and Bigger, J. H.
1966a Control of certain legume insects with low-volume dimethoate applied by airplane. *Journal of Economic Entomology* 59(5):1309-1310.

Phillips, J. H. H.
1951a An annotated list of Hemiptera inhabiting sour cherry orchards in the Niagra peninsula, Ontario. *Canadian Entomologist* 83(8):194-205.

Pholboon, P.
1965a *A Host List of the Insects of Thailand*. Department of Agriculture, Royal Thai Government, and the United States Operations Mission to Thailand. 149 pp. [In Thai and English.]

Pienkowski, R. L.
1962a Factors affecting populations of the potato leafhopper, *Empoasca fabae* (Harris). Ph. D. Dissertation, University of Wisconsin. 170 pp. [Diss. Abst. 29(9):3307-3308.]
1970a Potato leafhopper dispersal, buildup, damage, and control. *Report on the Alfalfa Improvement Conference* 22:6.

Pienkowski, R. L. and Medler, J. T.
1962a Effects of alfalfa cuttings on the potato leafhopper *Empoasca fabae*. *Journal of Economic Entomology* 55(6):973-978.
1963a Air masses associated with long range movement of the potato leafhopper into Wisconsin. *Proceedings of the XVI International Congress of Zoology* 16(2):12.
1964a Synoptic weather conditions associated with long-range movement of the potato leafhopper, *Empoasca fabae*, into Wisconsin. *Annals of the Entomological Society of America* 57(5):588-591.
1966a Potato leafhopper trapping studies to determine local flight activity. *Journal of Economic Entomology* 59(4):837-843.

Pierce, H. D.
1972a *Scaphytopis nitridus:* Parasitization by *Tomasvaryella appendipes*. *Environmental Entomology* 1(6):796-797.
1976a A new, lightweight, transparent insect cage. *U.S. Department of Agriculture, Agricultural Research Service* W-34:1-4.

Pierce, W. D.
1963a Fossil arthropods of California: 25 silicified leafhoppers from California mountains nodules. *Bulletin of the Southern California Academy of Sciences* 62(2):69-82.

Pieters, A. J.
1929a Red clover's hairiness in American types is due to the leafhopper. *U.S. Department Agriculture Yearbook*, 1928. Pp. 521-524.

Pillai, K. S., Saradamma, K. and Das, N. M.
1983a Field evaluation of two newer insecticides against some rice pests in Kerala. *Pesticides* 17:15-18.

Pillemer, E. A. and Tingey, W. M.
1976a Hooked trichomes: A physical plant barrier to a major agricultural pest. *Science* 193:482-484.
1978a Hooked trichomes and resistance of *Phaseolus vulgaris* L. to *Empoasca fabae* (Harris). *Entomologia Experimentalis et Applicata* 24:83-94.

Pimentel, D. and Wheeler, A. G.
1973a Influence of alfalfa resistance on a pea aphid population and its associated parasites, predators and competitors. *Environmental Entomology* 2(1):1-11.
1973b Species and diversity of arthropods in the alfalfa. *Environmental Entomology* 2:659-668.

Pinto, J. D. and Frommer, S. I.
1980a A survey of the arthropods on Jojoba (*Simmondsia chinensis*). *Environmental Entomology* 9(1):137-143.

Pirone, T. P., Bradfute, O. E., Freytag, P. H., Lung, M. C. Y. and Poneleit, C. G.
1972a Virus-like particles associated with a leafhopper-transmitted disease of corn in Kentucky. *Plant Disease Reporter* 56:652-656.

Pitre, H. N.
1966a Corn virus diseases of recent occurrence in southeastern United States with emphasis on the corn stunt virus-vector-plant interrelationships in Mississippi. *Proceedings, North Central Branch, American Association of Economic Entomologists* 21:43-47.

1967a Greenhouse studies of the host range of *Dalbulus maidis*, a vector of the corn stunt virus. *Journal of Economic Entomology* 60(2):417-421.

1968a Systemic insecticides for control of the black-faced leafhopper, *Graminella nigrifrons*, and effect on corn stunt diseases. *Journal of Economic Entomology* 61:765-768.

1968b A preliminary study of corn stunt vector populations in relation to corn planting dates in Mississippi. Notes on disease incidence and severity. *Journal of Economic Entomology* 61(3):847-849.

1968c Effect of insecticidal sprays on stunt and mosaic virus diseases of corn in small field plots in Mississippi. *Journal of Economic Entomology* 61(2):585-587.

1970a Observations on the life cycle of *Dalbulus maidis* on three plant species. *Florida Entomologist* 53(1):33-37.

1970b Notes on the life history of *Dalbulus maidis* on gama grass and plant susceptibility to the corn stunt agent. *Journal of Economic Entomology* 63:1661-1662.

Pitre, H. N. and Boyd, F. J.
1970a A study of the role of weeds in corn fields in the epidemiology of corn stunt disease. *Journal of Economic Entomology* 63(1):195-197.

Pitre, H. N., Coombs, R. L. and Douglas, W. A.
1966a Gama grass, *Tripsacum dactyloides*: A new host of *Dalbulus maidis*, vector of corn stunt virus. *Plant Disease Reporter* 50:570-571.

Pitre, H. N., Douglas, W. A., Combs, R. L. and Hepner, L. W.
1967a Annual movement of *Dalbulus maidis* into the southeastern United States and its role as a vector of corn stunt virus. *Journal of Economic Entomology* 60:616-617.

Pitre, H. N., Jr. and Hepner, L. W.
1967a Seasonal incidence of indigenous leafhoppers (Homoptera: Cicadellidae) on corn and several winter crops in Mississippi. *Annals of the Entomological Society of America* 60(5):1044-1055.

Piza, S. de Toledo, Jr.
1968a *Insectos de Piracicaba.* Illus. Escola Superior de Agricultura "Luiz de Queiroz," Universidad de Sao Paulo.

Pizzamiglio, M. A.
1979a Aspectos da biologia de *Empoasca kraemeri* Ross and Moore 1957 (Homoptera: Cicadellidae) em *Phaseolus vulgaris* L. 1753 e ocorrencia de parasitismo em ovos. *Anais da Sociedade Entomologicado Brasil* 8:367-368.

Ploaie, P. G.
1967a Electron microscopic evidence of virus-like particles in both plant and vector infected with clover phyllody virus. *In: Blattny, C., ed. Plant Virology.* Proceedings of the 6th Conference of Czechoslovakian Plant Virologists, Olomoue. Pp. 134-137. Academia, Prague. 346 pp.

1981a Mycoplasmalike organisms and plant diseases in Europe. *In: Maramorosch, K. and Harris, K. F., eds. Plant Diseases and Vectors: Ecology and Epidemiology.* Pp. 61-104. Academic Press, New York. 368 pp.

Pollard, D. G.
1968a Stylet penetration and feeding damage of *Eupteryx melissae* Curtis (Hemiptera, Cicadellidae) on sage. *Bulletin of Entomological Research* 58:55-71.

1969a Directional control of the stylets in phytophagous Hemiptera. *Proceedings of the Royal Entomological Society of London* 44:173-185.

1971a The use of polyporus for the investigation of stylet behaviour in the Hemiptera. *Entomologia Experimentalis et Applicata* 14(3):283-296.

1972a The stylet structure of a leafhopper (*Eupteryx melissae* Curtis: Homoptera: Cicadellidae). *Journal of Natural History* 6(3):261-271.

Pollard, D. G. and Saunders, J. H.
1956a Relations of some cotton pests to jassid resistant Sakel. *Empire Cotton Growing Revue* 33(3):202.

Pollard, H. N.
1962a Sex determination of fifth-instar nymphs of leafhoppers (Cicadellidae: Proconiini). *Annals of the Entomological Society of America* 55:141.

1965a Fecundity of *Homalodisca insolita* (Cicadellidae-Hemiptera), a leafhopper vector of phony peach virus disease. *Annals of the Entomological Society of America* 58(6):935-936.

1965b Description of stages of *Homalodisca insolita*, a leafhopper vector of phony peach virus disease. *Annals of the Entomological Society of America* 58:699-702.

Pollard, H. N. and Kaloostian, G. H.
1961a Overwintering habits of *Homalodisca coagulata*, the principal natural vector of phony peach disease virus. *Journal of Economic Entomology* 54(4):810-811.

Pollard, H. N., Turner, W. F. and Kaloostian, G. H.
1959a Invasion of the Southeast by a western leafhopper, *Homalodisca insolita*. *Journal of Economic Entomology* 52(2):359-360.

Pollard, H. N. and Yonce, C. E.
1965a Significance of length of tibial spines relative to oviposition processes by some leafhoppers (Hemiptera-Cicadellidae). *Annals of the Entomological Society of America* 58(4):594-595.

Pollini, C., Poggi, and Giunchedi, L.
1984a Focus on: Mycoplasmas and diseases of plants. *Difesa delle Piante* 7:177-204.

Polyakova, G. P.
1972a Electron microscopic study of the winter wheat mosaic virus transmitted by leafhopper *Psammotettix striatus* L. *In: Proceedings of XIII International Congress of Entomology,* Moscow, 2-9 August 1968. Vol. 3. Rafes, P. M., ed. *Symposium F. Insects as Vectors of Virus-Borne Plant Diseases.* Pp. 445-446.

Pomazkov, Y. I.
1966a The bionomics and control of the vectors of the black currant reversion and raspberry stunt viruses. *Entomologist* 11:452-456.

Pophaly, D. J., Rao, T. B. and Kalode, M. B.
1979a Biology and predation of the mirid bug, *Cyrtorhinus lividipennis* Reuter, on plant and leafhoppers in rice. *Indian Journal of Plant Protection* 6:7-14.

Popov, Y. A.
1963a Ecological and faunistic survey of current Hemiptera from the southern regions of Tyan'-Shan'. *Akademya Nauk SSSR* 42-43. [In Russian.]

Port, G. R.
1981a Auchenorrhyncha on roadside verges. A preliminary survey. *Acta Entomologica Fennica* 38:29-30.

Posnette, A. J. and Ellenberger, C. E.
1963a Further studies of green petal and other leafhopper-transmitted viruses infecting strawberry and clover. *Annals of Applied Biology* 51:69-83.

Poston, F. L. and Ogunlana, M. O.
1973a Preliminary evaluation of selected mirids and the potato leafhopper in soybean alfalfa complex. *Proceedings of the North Central Branch of the Entomological Society of America* 28:179.

Poston, F. L. and Pedigo, L. P.
1975a Migration of plant bugs and the potato leafhopper in a soybean-alfalfa complex. *Environmental Entomology* 4(1):8-10.

Prabhakar, B., Rao, P. K. and Rao, B. H. K. M.
1981a Note on hemipterous species complex on sorghum at Hyderabad. *Indian Journal of Agricultural Sciences* 51:818-819.

Pradhan, S.
1964a Assessment of losses caused by insect pests of crops and estimation of insect population. *In: Pant, N. C., Prasad, S. K., Vishnoi, H. S., Chatterji, S. N. and Varma, B. K., eds. Entomology in India.* Pp. 17-58. The Entomological Society of India. 529 pp.

Prasad, S. K.
1957a The nature and extent of damage to crucifers and potatoes in relation to density of insect populations. Ph. D. Dissertation, University of Minnesota. 214 pp. [Diss. Abst. 17(1):2717.]
1960a Effect of insecticides on potato leafhopper, *Empoasca fabae* (Harris), abundance and yield of potatoes. *Indian Journal of Entomology* 22(4):283-286.
1961a Quantitative estimation of damage to potatoes caused by potato leafhopper, *Empoasca fabae* (Harris). *Indian Potato Journal* 3(2):105-107.

[Premusekar]
1985a Induced mutant for jassid resistance in cotton. *Mutation Breeding Newsletter* 26:7.

Prestidge, R. A.
1982a Instar duration, adult consumption, oviposition and nitrogen utilization efficiency of leafhoppers feeding on different quality food (Auchenorrhyncha: Homoptera). *Ecological Entomology* 7(1):91-101.
1982b The influence of nitrogenous fertilizer on the grassland Auchenorrhyncha (Homoptera.) *Journal of Applied Ecology* 19(3):735-749.

Prestidge, R. A. and McNeill, S.
1982a Plant food quality and leafhopper ecology. In: Lee, K. E., ed. *Proceedings of the 3rd Australasian Conference on Grassland Invertebrate Ecology*. Pp. 273-281. Adelaide, Australia. 402 pp.
1983a Auchenorrhyncha-host plant interactions: Leafhoppers and grasses. *Ecological Entomology* 8(3):331-339.
1983b The role of nitrogen in the ecology of grassland Auchenorrhyncha. In: Lee, J. A., McNeill, S. and Rorison, I. H., eds. *Nitrogen as an Ecological Factor Symposium*. Pp. 257-281. British Ecological Society No. 22, Blackwell Scientific, Oxford, England. 470 pp.

Pridantseva, E. A.
1972a Leafhoppers *Psammotettix striatus* L. (Cicadellidae) — vectors of cereal virus diseases. In: *Proceedings of XIII International Congress of Entomology, Moscow, 2-9 August 1968. Vol. 3.* Rafes, P. M., ed. *Symposium F. Insects as Vectors of Virus-Borne Plant Diseases*. Pp. 444-445.

Prigent, J. P.
1961a Contribution a l'etude du systeme nerveux central des Homopteres: les ganglion cerebriodes de *Cicadella viridis* L. *Verhandlungen XI Internationaler Kongress fur Entomologie*, Vienna 1:388-390.
1962a Sur les "corps pedoncules" des ganglions cerebroides des Insectes Homopteres Auchenorhynches. *Compte Rendu Hebdomadaire des Seances de l'Academie des Sciences* 254:1146-1147.

Prior, R. N. B.
1964a Two new techniques used in leafhopper taxonomy which may be applicable to other orders of small insects requiring maceration and partial dissection. *Entomologist's Monthly Magazine* 100:246-249.
1966a Occurrence of *Macrosteles cristatus* (Ribaut) (Homoptera, Auchenorrhyncha) in Britain with notes on related species. *Entomologist's Monthly Magazine* 102:286-288.
1972a A note on the capturing and feeding behavior of *Platypalpus notata* Meigen (Dipt., Empididae) on the leafhopper prey *Macrosteles sexnotatus* (Fall.) (Hem., Cicadellidae). *Entomologist's Monthly Magazine* 107:183-184.

Probst, A. H. and Everly, R. T.
1957a Effect of insecticides on growth, yield, and chemical composition of soybeans. *Agronomy Journal* 49(11):577-581.

Proctor, J. H.
1958a A note on cotton insects in British Guiana. *Empire Cotton Growing Revue* 35(2):107-110.
1961a Minor pests of cotton in the Aden Protectorate. *Empire Cotton Growing Revue* 38(3):172-181.

Proeseler, G. and Eisbein, K.
1974a The amputation of the stylets of vectors (Homoptera and Heteroptera) by thermic and mechanical methods and with the aid of laser beams (short communication). *Archiv fur Phytopathologie und Pflazenschutz* 10:409-411.

Pruess, K. P., Saxena, K. M. L. and Koinzan, S.
1977a Quantitative estimation of alfalfa insect populations by removal sweeping. *Environmental Entomology* 6:705-708.

Pruess, K. P. and Whitmore, R. W.
1976a A D-Vac calibration technique. *Journal of Economic Entomology* 69:51-52.

Purcell, A. H.
1974a Spatial patterns of Pierce's disease in the Napa Valley. *American Journal of Enology and Viticulture* 25(3):162-167.
1975a Role of the blue-green sharpshooter, *Hordnia circellata*, in the epidemiology of Pierce's disease of grapevines. *Environmental Entomology* 4:745-752.
1976a Seasonal changes in host plant preference of the blue-green sharpshooter *Hordnia circellata* (Homoptera: Cicadellidae). *Pan-Pacific Entomologist* 52(1):33-37.
1978a Lack of transtadial passage by Pierce's disease vector. *Phytopathology News* 12:217-218.
1979a Transmission of X-disease agent by the leafhoppers, *Scaphytopius nitridus* and *Acinopterus angulatus*. *Plant Disease Reporter* 63:549-552.
1979b Leafhopper vectors of xylem-borne plant pathogens. In: Maramorosch, K. and Harris, K. F., eds. *Leafhopper Vectors and Plant Disease Agents*. Pp. 603-625. Academic Press, New York. 654 pp.
1979c Control of the blue-green sharpshooter and effects on the spread of Pierce's disease of grapevines. *Journal of Economic Entomology* 72:887-892.
1980a Almond leaf scorch: Leafhopper and spittlebug vectors. *Journal of Economic Entomology* 73(6):834-838.
1981a Vector preference and inoculation efficiency as components of resistance to Pierce's disease in European grape cultivars. *Phytopathology* 71(4):429-435.
1982a Evolution of the insect vector relationship. In: Mount, M. S. and Lacy, G. H., eds. *Phytopathogenic Procaryotes*. Pp. 121-156. Academic Press, New York.
1982b Insect vector relationships with procaryotic plant pathogens. *Annual Review of Entomology* 20:397-417.
1985a The ecology of bacterial and mycoplasma plant diseases spread by leafhoppers and planthoppers. In: Nault, L. R. and Rodriguez, J. G., eds. *The Leafhoppers and Planthoppers*. Pp. 351-380. Wiley & Sons, New York. 500 pp.

Purcell, A. H. and Elkinton, J. S.
1980a A comparison of sampling methods for leafhopper vectors of X-disease in California cherry orchards. *Journal of Economic Entomology* 73(6):854-860.

Purcell, A. H. and Finlay, A.
1979a Evidence for noncirculative transmission of Pierce's disease bacterium by sharpshooter leafhoppers. *Phytopathology* 69:393-395.
1979b Acquisition and transmission of bacteria through artificial membranes by leafhopper vectors of Pierce's disease. *Entomologia Experimentalis et Applicata* 25:188-195.

Purcell, A. H., Finlay, A. H. and McLean, D. L.
1979a Pierce's disease bacterium: Mechanism of transmission by leafhopper vectors. *Science* 206:839-841.

Purcell, A. H. and Frazier, N. W.
1985a Habitats and dispersal of the principal leafhopper vectors of Pierce's disease bacterium in the San Joaquin Valley. *Hilgardia* 53(4):1-32.

Purcell, A. H. and Loher, W.
1976a Acoustical and mating behavior of two taxa in the *Macrosteles fascifrons* species complex. *Annals of the Entomological Society of America* 69(3):513-518.

Purcell, A. H., Raju, B. C. and Nyland, G.
1980a Transmission by injected leafhoppers of spiroplasma isolated from plants with X-disease. *Proceedings of the 61st Annual Meeting of the Pacific Division of AAAS*, p. 13.
1981a Transmission by injected leafhoppers of spiroplasma isolated from plants with X-disease. *Phytopathology* 71:108.

Purcell, A. H., Richardson, J. R. and Finlay, A. H.
1981a Multiplication of X-disease agent in a nonvector leafhopper, *Macrosteles fascifrons*. *Annals of Applied Biology* 99:283-289.

Purcell, A. H. and Suslow, K. G.
1982a Dispersal behavior of *Colladonus montanus* (Homoptera: Cicadellidae) in cherry orchards. *Environmental Entomology* 11(6):1178-1182.
1984a Surveys of leafhoppers (Homoptera: Cicadellidae) and pear psylla (Homoptera: Psyllidae) in pear and peach orchards and the spread of peach yellow leaf roll disease. *Journal of Economic Entomology* 77(6):1489-1494.

Putta, M., Chettanachit, D., Morinaka, T., Parejarearn, A. and Disthaporn, S.
1980a Gall dwarf—a new rice virus disease in Thailand. *International Rice Research Newsletter* 5(3):10-11.

Putta, M., Chettanachit, D., Omura, T., Inoue, H., Morinaka, T., Honda, Y., Saito, Y. and Disthaporn, S.
1982a Host range of rice gall dwarf virus. *International Rice Research Newsletter* 7(6):13.

Qadri, M. A. H.
1967a Phylogenetic study of Auchenorrhyncha. *University Studies* (Karachi) 4(3):1-16.

Quartau, J. A.
1968a Two new species of the genus *Batracomorphus* Lewis from the Cape Verde Islands (Homoptera, Cicadellidae). *Archivos do Museu Bocage* (Lisbon) 2:1-8.
1970a A new leafhopper genus and species from Portugal, *Lusitanocephalus sacarraoi* gen. and sp. n. (Homoptera: Cicadellidae, Deltocephalini). *Ciencias Biologicas* (Luanda) 1(1):17-23.
1971a *Agallia linnavuorii* n. sp., a new leafhopper from Portugal (Homoptera: Cicadellidae, Agalliinae). *Ciencias Biologicas* 1(2):73-79.
1972a *Goldeus dlabolai* n. sp., a new leafhopper from Portugal (Homoptera: Cicadellidae, Deltocephalini). *Arquivos do Museu Bocage* (Lisbon) 3(16):435-446.
1978a Character associations in *Batracomorphus* Lewis (Cicadellidae). *In: Cobben, R. H., ed. Auchenorrhyncha Newsletter* 1:19.
1978b A taxonomic study of the Ethiopian species of *Batracomorphus* Lewis (Homoptera: Cicadellidae) with an application of some numerical methods. Ph.D. Thesis, University of London. 492 pp.
1979a An annotated checklist of the species of leafhoppers known to occur in the Azores (Homoptera: Cicadellidae). *Arquivos do Museu Bocage* (Lisbon) 7(45):1-6.
1980a Notes on *Acacioiassus serenus* (Melichar) in Southern Africa (Homoptera: Cicadellidae: Iassinae.) *Arquivos do Museu Bocage* 7(47):1-6.
1980b A contribution to the knowledge of the Cicadellidae (Homoptera: Auchenorrhyncha) of the Azores. *Bocagiana* (Funchal) 49:1-3.
1981a New and redescribed species of African *Batracomorphus* Lewis (Homoptera: Cicadellidae), with a key to all known Ethiopian species. *Memoirs of the Entomological Society of Southern Africa* 14. 134 pp.
1981b A new species of *Iassomorphus* Theron from Africa (Homoptera, Cicadellidae, Iassinae). *Boletim da Sociedade Portuguesa de Ciencias Naturais* 20:63-66.
1981c On some numerical methods applied to the taxonomy of some leafhoppers (Homoptera, Cicadellidae): An empirical investigation and critical appraisal. *Acta Entomologica Fennica* 38:30-41.
1981d On a new species of the genus *Brachypterona* Lindberg (Homoptera: Cicadellidae) from the Salvage Islands. *Arquivos do Museu Bocage* (Lisbon) 1(10):125-131.
1981e *Hauptidia maroccana* (Melichar, 1907) (Homoptera: Cicadellidae, Typhlocybinae) new to Portugal. *Arquivos do Museu Bocage* (Lisbon) 1(14):169-171.
1981f Missao zoologica aos Arquipelagos da Madeira e das Selvagens. *Arquivos do Museu Bocage* (Lisbon) 1(1):1-29.
1981g Ecological notes on *Batracomorphus* Lewis (Insecta, Homoptera, Cicadellidae) in Africa. *Boletim da Sociedade Portuguesa de Entomologia* 14:1-8.
1982a A study of some Palaearctic leafhoppers (Insecta, Homoptera, Cicadellidae, Aphrodinae) using the methods of numerical taxonomy. *Boletim da Sociedade Portuguesa de Entomologia* 24:1-14.
1982b A numerical analysis of character correlations in *Batracomorphus* Lewis (Insecta: Cicadellidae: Iassinae). *Boletim da Sociedade Portuguesa de Ciencias Naturais* 21:51-58.
1982c A preliminary faunistic analysis of the Cicadellidae (Homoptera, Auchenorrhyncha) of the Azores. *Boletim da Sociedade Portuguesa de Entomologia* 7(Supl. A):145-149.
1982d Reflexoes sobre a Zoogeografia dos Arquipelagos da Madeira e das Selvagens. *Boletim do Museu Municipal do Funchal* 34(149):124-141.
1982e The effect of using ratios instead of raw measurements on the classification of a group of leafhoppers (Insecta, Homoptera, Cicadellidae). *Boletim de Sociedade Portuguesa de Entomologia* 30:1-12.
1983a An evaluation of several methods of principal component and coordinate analysis applied to the taxonomy of *Batracomorphus* (Homoptera, Cicadellidae). *In: Knight, W. J., Pant, N. C., Robertson, T. S. and Wilson, M. R., eds. Proceedings of the 1st International Workshop on Biotaxonomy, Classification and Biology of Leafhoppers and Planthoppers (Auchenorrhyncha) of Economic Importance*. Pp. 135-163. London, 4-7 October 1982. Commonwealth Institute of Entomology, 56 Queen's Gate, London SW7 5JR. 500 pp.
1983b A comparison of different numerical methods applied to the taxonomy of *Batracomorphus* Lewis (Insecta, Homoptera, Cicadellidae, Iassinae): Ordination techniques. *Journal of the Entomological Society of Southern Africa* 46(1):9-35.
1983c A numerical classification of the afrotropical species of genus *Batracomorphus* Lewis (Homoptera, Cicadellidae). *Arquivos do Museu Bocage* (Lisbon) 2(4):41-55.
1983d Sobre a aplicacao de tecnicas de computacao na classificacao de cigarrinhas (Insecta, Auchenorrhyncha, Cicadellidae). *Actas Congresso Iberico Entomologia* 1(2):623-626.
1984a Two new records of leafhoppers (Homoptera, Auchenorrhyncha, Cicadellidae) from the Small Salvage Island. *Bocagiana* (Funchal) 72:1-7.
1984b Preparacao e Preservacao de Insectos: Sinopse dos metodos a seguir. *Archivos do Museu Bocage* (Ser. D) 11(2):25-39.
1984c Classificao e Sinopse do Hexapodes Actuais (Hexapoda ou Insecta *sensu lato*). *Colecrao "Natura,"* Nova Serie 12:1-42.
1985a A new subspecies of *Iassomorphus drakensteini* (Naude, 1926) from Angola (Homoptera, Cicadellidae, Iassinae). *Journal of the Entomological Society of Southern Africa* 48(1):103-106.
1985b On some objections to the paranotal theory on the origin of the insect wings. *Boletim da Sociedade Portuguesa de Entomologia Supplement* 1:359-371.
1985c On the origin of the insect wing. *Rostrum* 10(2):[1-2].

Quartau, J. A. and Davies, R. G.
1983a A comparison of different numerical methods applied to the taxonomy of *Batracomorphus* Lewis (Insecta, Homoptera, Cicadellidae, Iassinae): Cluster analysis. *Revista de Biologia* (Lisbon) 12(3&4):550-596.

1984a On the selection of taxonomic characters by numerical methods. *Mitteilungen der Schweizerschen Entomologischen Gesellshaft* 57(4):437.

1985a Character selection by information content in the numerical taxonomy of some male *Batracomorphus* (Homoptera: Cicadellidae). *Zeitschrift fuer Zoologische Systematik und Evolutions Forschung* 23:100-115.

Quartau, J. A. and Rodrigues, D. P.
1969a Contribution a la connaissance des Cicadellidae du Portugal. *Boletim da Sociedade Portuguesa de Ciencias Naturais* 12:187-207.

Quezada, J. R.
1979a The finding of *Agonatopus* sp. (Hymenoptera: Dryinidae), a parasite of *Dalbulus maidis* (Homoptera: Cicadellidae) in El Salvador. *Ceiba* 23:1-2.

Quisenberry, S., MacRae, T. C. and Yonke, T. R.
1983a Preference of *Forcipata loca* (Homoptera: Cicadellidae) adults to forage plants. *Environmental Entomology* 12(4):1149-1153.

Quisenberry, S. and Yonke, T. R.
1981a Responses of 'Kentucky-31' tall fescue to varying *Forcipata loca* DeLong & Caldwell infestation levels: Growth chamber study (Homoptera: Cicadellidae, Typhlocybinae). *Environmental Entomology* 10(4):550-553.

1981b Responses of Kentucky-31 tall fescue to varying *Forcipata loca* DeLong and Caldwell infestation levels: Field Study. *Environmental Entomology* 10(5):650-653.

1981c Effects of *Forcipata loca* feeding on tissue of Kentucky-31 tall fescue. *Annals of the Entomological Society of America* 74(6):521-524.

Quisenberry, S. S., Yonke, T. R. and Huggans, J. L.
1979a Leafhoppers associated with mixed tall fescue pastures in Missouri (Homoptera: Cicadelidae). *Journal of the Kansas Entomological Society* 52:421-437.

Raatikainen, M.
1968a *Macrosteles ossiannilssoni* Le Quesne (Hom., Cicadellidae), a species new to Finland. *Suomen Hyonteistieteellinen Aikakauskirga* 34(4):243-244.

1971a Seasonal aspects of leafhopper (Homoptera, Auchenorrhyncha) fauna in oats. *Annales Agriculturae Fenniae* 10(49):1-8.

1972a Dispersal of leafhoppers and their enemies to oatfields. *Annales Agriculturae Fenniae* 11:146-153.

Raatikainen, M., Halkka, O., Halkka, L., Hovinen, R. and Vasarainen, A.
1976a Aberrant spermatogenesis in the leafhopper *Macrosteles laevis* (Rib.), infected by the aster yellows agent. *Annales Agriculturae Fenniae* 15:97-100.

Raatikainen, M. and Heikinheimo, O.
1974a The flying times of Strepsiptera males at different latitudes in Finland. *Annales Entomologici Fennici* 40(1):22-25.

Raatikainen, M. and Vasarainen, A.
1971a Comparison of leafhopper faunae in cereals. *Annales Agriculturae Fenniae* 10:119-124.

1973a Early- and high-summer flight periods of leafhoppers. *Annales Agriculturae Fenniae* 12:77-94.

1976a Composition, zonation and origin of the leafhopper fauna of oatfields in Finland. *Annales Zoologici Fennici* 13:1-24.

Raccah, B. and Klein, M.
1982a Transmission of the safflower phyllody mollicute by *Neoaliturus fenestratus*. *Phytopathology* 72:230-232.

Rachie, K. O., Singh, S. R., Williams, R. J., Watt, E., Nangju, D., Wien, H. C. and Luse, R. A.
1976a Vita-4 cowpea. *Tropical Grain Legume Bulletin* 5:40-41.

Rack, K.
1984a Relationship between the green cicadellid and the fungus *Myxosporium devastans* on the bark of young birches. *Nachrichtenblatt des Deutschen Pflanzenschutzdienstes* 36:33-36.

Radcliffe, E. B.
1982a Insect pests of potato. *Annual Revue of Entomology* 27:173-204.

Radcliffe, E. B. and Lauer, F. I.
1967a Insect resistance in the wild *Solanum* species. *Proceedings of the North Central Branch of the Entomological Society of America* 22:165-167.

1968a Resistance to *Myzus persicae* (Sulzer), *Macrosiphum euphorbiae* (Thomas) and *Empoasca fabae* (Harris) in the wild tuber-bearing *Solanum* (Tourn.) L. species. *Minnesota Agricultural Experiment Station Technical Bulletin* 259. 27 pp.

Radulescu, E. and Munteanu, L.
1970a Recherches sur la symptomatologie, l'etiologie et la lutte contre le rabougrissement jaune du ble en Roumanie. [Investigations on the symptomatology, aetiology and control of yellow stunt of wheat in Rumania.] *Annales de Phytopathologie* 2(2):403-414. [With English summary.]

Raheja, A. K.
1976a Co-ordinated minimum insecticide trials: Yield performance of insect resistant cowpea cultivars from IITA compared with Ghanian cultivars. *Tropical Grain Legume Bulletin* 5:5.

Rahman, M. M., Hahar, M. A. and Miah, S. A.
1985a Preventing tungro (RTV) by applying insecticides to control green leafhopper (GLH) in Bangladesh. *International Rice Research Newsletter* 10(4):19-20.

Rahman, M. M. and Hibino, H.
1985a Recovery of virus from tungro (RTV)-infected leaves with or without leafhopper infestations. *International Rice Research Newsletter* 10(5):18.

1985b Reactions of eight rices to tungro (RTV). *International Rice Research Newsletter* 10(6):12.

Rai, S., Janin, H. K. and Agnihotri, N. P.
1980a Bioefficacy and dissipation of synthetic pyrethroids in okra. *Indian Journal of Entomology* 42:657-660.

Raine, J.
1960a Life history and behavior of the bramble leafhopper *Ribautiana tenerrima* (H.-S.) (Homoptera: Cicadellidae). *Canadian Entomologist* 92(1):10-20.

1967a Leafhopper transmission of witches' broom and clover phyllody viruses from British Columbia to clover, alfalfa and potato. *Canadian Journal of Botany* 45:441-445.

Raine, J. and Forbes, A. R.
1969a Mycoplasma-like bodies in the saliva of the leafhopper *Macrosteles fascifrons* (Stal) (Homoptera: Cicadellidae). *Canadian Journal of Microbiology* 15(9):1105-1107.

1971a The salivary syringe of the leafhopper *Macrosteles fascifrons* (Homoptera: Cicadellidae) and the occurrence of mycoplasma-like organisms in its ducts. *Canadian Entomologist* 103(1):110-116.

Raine, J., Forbes, A. R. and Skelton, F. E.
1976a Mycoplasma-like bodies, rickettsia-like bodies and salivary bodies in the salivary glands and saliva of the leafhopper *Macrosteles fascifrons* (Homoptera: Cicadellidae). *Canadian Entomologist* 108:1009-1019.

Raine, J. and Tonks, N. V.
1960a Control of three species of leafhoppers, on *Rubus* in British Columbia. *Proceedings of the Entomological Society of British Columbia* 57:22-26.

Rains, B. D. and Christensen, C. M.
1983a Effect of soil-applied carbofuran on transmission of maize chlorotic dwarf virus and maize dwarf mosaic virus to susceptible field corn hybrid. *Journal of Economic Entomology* 76:290-293.

Rajagopal, D. and Channabasavanna, G. P.
1975a Insect pests of maize in Karnataka. *Mysore Journal of Agricultural Sciences* 9:110-121.

Rajamohan, N., Ramakrishnan, C., Krishnaraj, J. and Subramanaiam, T. R.
1974a The incidence of jassid *Amrasca biguttula biguttula* (Ishida) on sunflower in relation to manuring and spacing. *Madras Agricultural Journal* 61:486-489.

Rajendra, Singh
- 1978a Field observations on copulation, pre-oviposition and oviposition periods of *Amrasca biguttula biguttula* (Ishida). *Indian Journal of Entomology* 40:100.
- 1982a Now [sic!] record of an alternate host of cotton jassid, *Amrasca biguttula biguttula* (Ishida). *Indian Journal of Entomology* 44:407.

Rajendra, Singh and Teotia, T. P. S.
- 1978a Relative toxicity of some insecticides to the adults of cotton jassid, *Amrasca biguttula biguttula* (Ishida). *Indian Journal of Entomology* 40:82-85.

Raju, A. K., Veni, G. P. and Rao, P. A.
- 1983a Combination of some pesticides and urea on the control of jassid, *Amrasca biguttula biguttula* Ishida, and yield of mesta. *Indian Journal of Plant Protection* 10:87-88.

Raju, B. C., Goheen, A. C. and Frazier, N. W.
- 1983a Occurrence of Pierce's disease bacteria on plants and vectors in California. *Phytopathology* 73:1309-1313.

Raju, B. C. and Nyland, G.
- 1981a Enzyme-linked immunosorbent assay for the detection of corn stunt spiroplasma in plant and insect tissues. *Current Microbiology* 5:101-104.

Raju, B. C., Purcell, A. H. and Nyland, G.
- 1979a Isolation of spiroplasmas from leafhoppers exposed to aster yellows and X-disease. *Florida Agricultural Research Publication* FL-80-1.
- 1984a Spiroplasmas from plants with aster yellows disease and X-disease: Isolation and transmission by leafhoppers. *Phytopathology* 74:925-931.

Ram, S. and Gupta, M. P.
- 1976a Assessment of losses in the fodder yield of lucerne due to leaf-hopper and weevil. *Indian Journal of Agricultural Sciences* 46:278-280.

Ram, S., Patil, B. D. and Purohit, M. L.
- 1984a Cowpea varieties resistant to major insect pests. *Indian Journal of Agricultural Sciences* 54:307-311.

Ramachandra Rao, K.
- 1967a On a new species of *Zizyphoides* Distant (Homoptera: Jassidae) from India. *Oriental Insects* 1(3&4):239-241.
- 1967b A note on *Arya rubrolineata* Dist. (Jassidae, Homoptera) from Kalyani. *Journal, Bombay Natural History Society* 64(1):130.
- 1969a On the female of *Paralimnus confuscus* Pruthi (Homoptera: Cicadellidae.) *Oriental Insects* 3(2):187-188.
- 1973a Studies on a small collection of jassids from Poona (India) (Homoptera: Cicadellidae). *Zoologischer Anzeiger* 191(1&2):93-98.
- 1973b A note on *Idioscopus clypealis* (Leth.) (Hemiptera: Cicadellidae). *Journal, Bombay Natural History Society* 70(1):217-218.
- 1973c A note on the venation of a freak of *Phrynomorphus indicus* (Distant) (Hemiptera: Cicadellidae). *Indian Journal of Entomology* 35(4):350.

Ramakrishnan, U.
- 1982a New genus and new species of leafhopper (Empoascini: Typhlocybinae: Cicadelloidea: Homoptera) from India. *Journal of Entomological Research* (New Delhi) 6(1):10-12. [New genus: *Daluana*.]
- 1983a A new species of the genus *Subhimalus* Ghauri (Euscelidae: Cicadelloidea: Homoptera) from India. *Journal of Entomological Research* 7(2):190-191.
- 1983c Morphometrics of various populations of *Nephotettix* species in India. *In: Knight, W. J., Pant, N. C., Robertson, T. S. and Wilson, M. R., eds. Proceedings of the 1st International Workshop on Biotaxonomy, Classification and Biology of Leafhoppers and Planthoppers (Auchenorrhyncha) of Economic Importance.* Pp. 319-333. London, 4-7 October 1982. Commonwealth Institute of Entomology, 56 Queen's Gate, London SW7 5JR. 500 pp.

Ramakrishnan, U. and Ghauri, M. S. K.
- 1979a Probable natural hybrids of *Nephotettix virescens* (Distant) and *N. nigropictus* (Stal) (Hemiptera: Cicadellidae) from Sabah, Malaysia. *Bulletin of Entomological Research* 69:357-361.
- 1979b New genera of the *Empoascanara* complex (Homoptera, Cicadellidae, Typhlocybinae). *Reichenbachia* 17(24):193-213. [Fifteen new generic level taxa proposed.]

Ramakrishnan, U. and Ramdas Menon, M. G.
- 1971a Studies on Indian Typhlocybinae (Homoptera: Cicadellidae). 1. Five new genera and a new record of Dikraneurini. *Oriental Insects* 5(4):455-469. [New genera: *Murreeana, Pusaneura, Duttaella, Pusatettix, Basuaneura.*]
- 1972a Studies on Indian Typhlocybinae (Homoptera: Cicadellidae). 2. Ten new species of Typhlocybini. *Oriental Insects* 6(1):111-129.
- 1972b Studies on Indian Typhlocybinae (Homoptera: Cicadellidae). 3. Three new genera with four new species of Typhlocybini. *Oriental Insects* 6(2):183-192. [New genera: *Sabourasca, Viridasca, Sundara.*]
- 1973a Studies on Indian Typhlocybinae (Homoptera: Cicadellidae). 4. Seven new genera with fourteen new species of Erythroneurini. *Oriental Insects* 7(1):15-48. [Describes 7 new genera.]
- 1974a Studies on Indian Typhlocybinae (Homoptera: Cicadellidae). 5. The tribe Erythroneurini. *Oriental Insects* 8(4):433-452.

Ramalho, F. S.
- 1978a Water traps to determine the height and periodicity of flight in *Empoasca kraemeri* Ross & Moore, 1957 (Homoptera, Typhlocibidae [sic!]). *Turrialba* 28:283-285.

Ramalho, F. S. and Albuquerque, M. M.
- 1979a Influence of shades of yellow used in water traps to catch the green leafhopper, *Empoasca kraemeri* Ross & Moore, 1957. *Ciencia e Cultura* 31:305-306.

Ramalho, F. S. and Ramos, J. K.
- 1979a Distribuicao de ovos de *Empoasca kraemeri* Ross and Moore, 1957, na planta de feijao. *Anais da Sociedade Entomologica do Brasileira* 8:85-91.

Raman, K. V.
- 1977a Leafhopper resistance in cowpea varieties. *Tropical Grain Legume Bulletin* 8:14.
- 1978a Resistance of 11 tropical grain legumes to *Empoasca dolichi* (Paoli). *Tropical Grain Legume Bulletin* 11/12:3-7.

Raman, K. V., Singh, S. R. and Emden, H. F., van
- 1978a Yield losses in cowpea following leafhopper damage. *Journal of Economic Entomology* 71:936-938.
- 1980a Mechanisms of resistance to leafhopper damage in cowpea. *Journal of Economic Entomology* 73:484-488.

Raman, K. V., Tingey, W. M. and Gregory, P.
- 1979a Potato glycoalkaloids: Effect on survival and feeding behaviour of the potato leafhopper. *Journal of Economic Entomology* 72:337-341.

Ramirez, J. L., DeLeon, C., Garcia, C. and Granados, G.
- 1975a *Dalbulus guevarai* (DeL.) nuevo vector del achaparramiento del maiz in Mexico: Incendencia de la enfermedad y su relation con el vector *Dalbulus maidis* (Del. & Wol.) en Muna, Yucatan. *Agrociencia* 22:39-49.

Ramos-Elorduy de Conconi, J.
- 1970a Biologia de *Marathonia nigrifascia* (Walker) (Homoptera: Cicadellidae). Universidad Nacional Autonoma de Mexico, Tesis Professional, Graciela Serrano Limon. 62 pp.
- 1972a Tres especies nuevas del genero *Gyponana* Ball (Homoptera-Cicadellidae) para Mexico. *Anales Instituto de Biologia Universidad Nacional Autonoma de Mexico, Serie Zoologia* 43(1):31-46.
- 1972b Nota sobre el ciclo biologico de *Dikrella scinda* Ruppel y DeLong (Homoptera: Cicadellidae). *Anales Instituto de Biologia Universidad Nacional Autonoma de Mexico, Serie Zoologia* 43(1):47-50.

1972c Lista de algunos Homopteros colectados en el Cerro "El Vigia," Vera Cruz, Mexico. *Anales Instituto de Biologia Universidad Nacional Autonoma de Mexico, Serie Zoologia* 43(1):131-138.

1973a Variacion altitudinal y extacional de poblaciones de algunos Homopteros de la region del Valle de Bravo, EDO de Mexico. *Folia Entomologica Mexicana* 34:37-60.

Rana, G. L., Kaloostian, G. H., Oldfield, G. N., Granett, A. L., Calavan, E. C., Pierce, H. D., Lee, I.-M. and Gumpf, D. J.
1975a Acquisition of *Spiroplasma citri* through membranes by homopterous insects. *Phytopathology* 65:1143-1145.

Raney, H. G. and Yeargan, K. V.
1977a Seasonal abundance of common phytophagous and predaceous insects in Kentucky soybeans. *Transactions of the Kentucky Academy of Science* 38(1-2):83-87.

Rangaiah, P. F. and Sehgal, V. K.
1984a Insects of T21 pigeonpea and losses caused by them at Pantnagar, northern India. *International Pigeonpea Newsletter* 3:40-43.

Rao, B. R. M., Raju, A. K., Rao, R. V. A. and Azam, K.
1983a The effect of sowing dates on the yield of mesta and the incidence of jassids *Amrasca biguttula biguttula* Ishida. *Madras Agricultural Journal* 70:328-330.

Rao, D. V. and John V. T.
1971a An easy and efficient method of transporting rice leafhoppers. *Indian Phytopathology* 24(2):408-410.

Rao, G. M. and Anjaneyulu, A.
1977a Alternate hosts of rice tungro virus an its vectors. *International Rice Research Newsletter* 2(5):15.

1978a Host range of rice tungro virus. *Plant Disease Reporter* 62:955-957.

1979a Rice tungro virus acquisition by the vector, *Nephotettix virescens*, from rice cultivars. *Plant Disease Reporter* 63:855-858.

1979b Carbofuran prevents rice tungro virus infection. *Current Science* 48:116-117.

1980a Feeding damage by tungro virus vector, *Nephotettix virescens*, in rice. *Oryza* 17:205-209.

1980b Estimation of yield losses due to tungro virus infection in rice cultivars. *Oryza* 17:210-214.

1982a Effect of meteorological factors on symptomatology and acquisition of rice tungro virus by *Nephotettix virescens*. *International Rice Research Newsletter* 7(6):9.

1983a Effect of meteorological factors on oviposition and nymphal hatching of rice tungro virus vector *Nephotettix virescens*. *International Rice Research Newsletter* 8(1):12-13.

Rao, K. R.
1980a Studies on cicadellid collections from Manipur (Hemiptera: Cicadellidae). *Records of the Zoological Survey of India* 76:189-194.

1981a Studies on the leafhopper fauna of Meghalaya and adjacent areas. II. Cicadellid collections from Manas and Dibrugarh (Assam) surveys. *Records of the Zoological Survey of India* 78:1-6.

Rao, N. H. P. and Agarwal, R. A.
1981a Effect of different placements of granular insecticides on the economic control of cotton jassid. *Pesticides* 15:18-20.

Rao, N. V. and Reddy, P. S.
1982a Impact of natural enemies on rice pest populations. *Indian Farming* 32:20-21.

Rao, P. Nageswara and Rao, B. V. Ramana
1984a Effect of jassid incidence on some qualitative and quantitative characteristics of cotton. *Pesticides* 18:38-39.

Rao, P. R. M., Raju, A. K., Rao, R. V. A. and Rao, B. H. K. M.
1981a Note on a new record of spider predators of *Amrasca biguttula biguttula* Ishida, a serious pest on nesta from Andhra Pradesh. *Indian Journal of Agricultural Sciences* 51:203-204.

Rao, P. S.
1975a Widespread occurrence of *Beauveria bassiana* on rice pests. *Current Science* 44:441-442.

Rao, P. S. and Halteren, P. van
1976a Green leafhopper infestation in South Sulawesi, Indonesia. *Rice Entomology Newsletter* 4:5.

Rao, R. D. and John, V. T.
1974a Alternate hosts of rice tungro virus and its vector. *Plant Disease Reporter* 58:856-860.

Rao, V. P.
1965a Occurrence of *Pruthiana sexnotata* on sugarcane in Assam. *FAO Plant Protection Bulletin* 13(6):138.

Rapusas, H. R., Chen, J. M. and Heinrichs, E. A.
1985a Behaviour of two green leafhopper (GLH) colonies on three rice varieties. *International Rice Research Newsletter* 10(1):7-8.

Rapusas, H. and Heinrichs, E. A.
1981a Green leafhopper resistance and tungro infection in IR varieties. *International Rice Research Newsletter* 6(6):7-8.

1982a Plant age and levels of resistance to green leafhopper, *Nephotettix virescens* (Distant), and tungro virus in rice varieties. *Crop Protection* 1:91-98.

1982b Comparative virulence of 15 Philippine cultures of green leafhopper, *Nephotettix virescens* (Distant), on selected varieties. *Philippine Entomologist* 5:213-226.

1985a Virulence of *Nephotettix virescens* colonies on resistance rice. *International Rice Research Newsletter* 10(5):10-11.

Rasmy, A. H. and Hassib, M.
1974a Influence of plant nitrogen supply on the populations of some cotton pests. *Applied Entomology and Zoology* 9:48-49.

Rassel, A. and Desmidts, M.
1976a Graft transmission of a naturally occurring "virescens" of *Vinca rosea* in Upper Volta and its relationship to cotton phyllody. *FAO Plant Protection Bulletin* 24:90-93.

Ratcliffe, R. H., Ditman, L. P. and Young, J. R.
1960a Field experiments on the insecticidal control of insects attacking peas, snap and lima beans. *Journal of Economic Entomology* 53(5):818-820.

Raupp, M. J. and Denno, R. F.
1979a The influence of patch size on a guild of sap-feeding insects that inhabit the marsh grass *Spartina patens*. *Environmental Entomology* 8(3):412-417.

Raven, J. A.
1983a Phytophages of xylem and phloem: A comparison of animal and plant sap-feeders. *Advances in Ecological Research* 136-234.

Rawat, R. R. and Jakhmola, S. S.
1977a Relative efficacy of some granular insecticides for the control of *Amrasca devastans* Dist. on okra. *Indian Journal of Entomology* 38:293-295.

Rawat, R. R. and Sahu, H. R.
1969a A new record of *Phyllotetra chotanica* Duviver (Chrysomelidae: Coleoptera) and *Typhlocyba* [*Zygina*] *maculifrons* Motsch. (Jassidae: Homoptera) as pests of wheat in Madhya Pradesh [India]. *JNKVV* [*Jawarharlal Nehru Krishi Vishwa Vidyalaya*] *Research Journal* 3(1):68.

1975a Estimation of losses in growth and yield of okra due to *Empoasca devastans* Distant and *Earias* species. *Indian Journal of Entomology* 35:252-254.

Rawat, R. R., Vaishampayan, S. M. and Veda, O. P.
1980a Evaluation of effective and economic insecticides for the control of wheat jassid *Empoasca nagpurensis* (Dist.) infesting wheat. *Pesticides* 14:17-18.

Rawlins, W. A. and Glidden, W. C.
1969a Spray less with systemic insecticides. *New York Food and Life Science* 2(2):10-11.

Rawlins W. A. and Gonzales, D.
1966a Incidence of aster yellows in lettuce as affected by placement of systemic insecticides. *Journal of Economic Entomology* 59(1):226-227.

Raychaudhuri, S. P., Mishra, M. D. and Basu, A. N.
1973a Investigations on tungro and yellow dwarf diseases of rice. Association of Rice Research Workers: Symposium on Rice Production Under Environmental Stress. *Oryza* 8:361-363.

Raychaudhuri, S. P., Mishra, M. D. and Ghosh, A.
 1967a Preliminary note of the occurrence and transmission of rice yellow dwarf virus in India. *Plant Disease Reporter* 51:1040-1041.
 1967b Virus disease that resembles 'tungro'. *Indian Farming* 17(3):29, 33.
 1969a Occurrence of paddy virus and viruslike symptoms in India. In: *The Virus Diseases of the Rice Plant. Proceedings of the International Rice Research Institute Symposium.* Pp. 59-65.

Raymer, W. B.
 1956a Identity and host relations of the potato late-breaking virus. *Phytopathology* 46(11):639 (Abstract).

Razvyazkina, G. M.
 1955a In: Pavlovskii, E. N. and Shtakel'berg, A. A., eds. *Pests of the Forest* [reference book]. Akademiya Nauk Zoologichiskogo Institute. 2 vols. (vol. 1, pp. I-XV, 1-421; vol. 2, pp. 422-1097). [In Russian.]
 1957a New and little known species of the genus *Macrosteles* (Homoptera-Cicadoidea). *Zoologicheskii Zhurnal* 36:521-528. [In Russian with English summary.]
 1959a Cikada *Aphrodes bicinctus* Schrank Perenoscik Novogo Virus Nogo Zabolevanija Kleverapozelenenija Cvetkov. *Zoologicheskii Zhurnal* 38:494-495.
 1959b The cicadellid *Aphrodes bicinctus* Schrank—a vector of a new virus disease of clover—greening of the flowers. *Zoologicheskii Zhurnal* 38(3):494-495. [In Russian.]
 1960a Biology of cicads belonging to the genus *Macrosteles* and their epiphytological importance. *Zoologicheskii Zhurnal* 39:1855-1865. [In Russian with English summary.]
 1962a Insect carriers of phytopathogenic viruses. *Zoologicheskii Zhurnal* 41:481-492.

Razvyazkina, G. M. and Polyakova, G. P.
 1967a Electron-microscope studies of wheat mosaic virus transmitted by the leafhopper *Psammotettix striatus* L. *Doklady Akademiya Nauk SSSR (Biol.)* 174(6):1435-1436. [In Russian.]
 1967b Electron microscopic study of wheat mosaic virus transmitted by the cicada *Psammotettix striatus* L. *Doklady (Proceedings) of the Academy of Sciences of the USSR (Biological Sciences Sections)* 174:397-399.

Razvyazkina, G. M. and Pridantseva, E. A.
 1968a Leafhoppers of the group *Psammotettix striatus* (Homoptera, Cicadellidae) vectors of virus diseases of cereals, their systematics and distribution. *Zoologicheskii Zhurnal* 47(5):690-696. [In Russian with English summary.]

Razzaque, Q. M. A. and Heinrichs, E. A.
 1985a Screening for resistance of IR varieties to green leafhopper (GLH). *International Rice Research Newsletter* 10(3):10-11.
 1985b Evaluation of germplasm accessions for green leafhopper (GLH) resistance. *International Rice Research Newsletter* 10(3):9-10.
 1985c Screening wild rices for resistance to green leafhopper (GLH). *International Rice Research Newsletter* 10(3):11-12.
 1985d Screening for green leafhopper (GLH) resistance. *International Rice Research Newsletter* 10(3):12.

Razzaque, Q. M. A., Heinrichs, E. A. and Rapusas, H. R.
 1985a Mass rearing technique for the rice green leafhopper *Nephotettix nigropictus* (Stal). *Philippine Entomologist* 6:398-404.

Reddy, D. B.
 1968a The symposium on virus diseases of rice April 1967. *Information Letter, FAO Plant Protection Commission S.E. Asia* 55:7.

Reddy, D. K. and Garg, G. D.
 1983a Light trap catches of green leafhoppers by time of day. *International Rice Research Newsletter* 8(4):19.

Reddy, D. V. R.
 1966a New methods and applications for studying injections of a plant virus into its insect vector. *Dissertation Abstracts* 27:668B-669B.

Reddy, D. V. R. and Black, L. M.
 1966a Production of wound-tumor virus and wound-tumor soluble antigen in the insect vector. *Virology* 30(3):551-561.
 1972a Increase of wound tumor virus in leafhoppers as assayed on vector cell monolayers. *Virology* 50:412-421.

Reddy, H. R.
 1975a Miniplant-tubes for studies on virus transmission with leafhopper vectors. *Current Science* 44:593.

Reddy, K. D. and Mishra, D. S.
 1983a Light trap catches of green leafhoppers by time of day. *International Rice Research Newsletter* 8(4):19.

Reddy, M. S., Rao, P. K., Rao, B. H. K. and Rao, G. N.
 1983a Preliminary studies on the seasonal prevalence of certain Homoptera occurring on rice at Hyberbad. *Indian Journal of Entomology* 45:20-28.

Reed, W.
 1974a Selection of cotton varieties for resistance to insect pests in Uganda. *Cotton Growing Review* 51:106-123.

Rees, C. J. C.
 1983a Microclimate and the flying Hemiptera fauna of a primary lowland rain forest in Sulawesi. In: Sutton, S. L., Whitmore, T. C. and Chadwick, A. C., eds. *Tropical Rain Forest: Ecology and Management.* Pp. 121-136. Blackwell Scientific Publications, Oxford, U.K. 498 pp.

Regupathy, A. and Jayaraj, S.
 1972a Physiology of yellow-vein mosaic virus disease in okra, *Abelmoschus esculentus* (L.), in relation to its preference by *Aphis gossypii* G. & *Amrasca devastans* (Dist.) (Homoptera). *Indian Journal of Experimental Biology* 10:436-438.
 1973a Protection of bhendi plants from the attack of aphid *Aphis gossypii* G. and leafhopper *Amrasca devastans* (Dist.) with systemic insecticides. *Madras Agricultural Journal* 60:519-524.
 1973b Physiology of sesamum phyllody disease and its influence on the infestation of the leafhopper vector *Orosius albicinctus* Dist. *Phytopathologische Zeitschrift* 78(1):86-88. [In English with German summary.]
 1974a Physiology of systemic insecticide-induced susceptibility of bhendi plants to *Aphis gossypii* G. and *Amrasca devastans* (D.). *Madras Agricultural Journal* 61:76-80.

Regupathy, A., Rathnasamy, R., Venkatnarayanan, D. and Subramanaim, T. R.
 1975a Physiology of yellow mosaic virus in green gram, *Phaseolus aureus* Roxb., with reference to its preference by *Empoasca kerri* Pruthi. *Current Science* 44:577-578.

Regupathy, A. and Subramaniam, T. R.
 1980a Efficacy of aldicarb in relation to the time of its application to control aphids and leafhoppers on 'MCU' cotton. *Indian Journal of Agricultural Sciences* 50:82-83.

Reissig, W. H. and Kamm, J. A.
 1975a Reproductive development of adult male *Draeculacephala crassicornis*. *Annals of the Entomological Society of America* 68(1):58-60.

Reitzel, J.
 1964a Iagttagelser over forekomster af cikaden *Macropsis fuscula* (Zett.) i hindbaerkulturer. *Statens Plantepatologiske Forsog Manedsoversigt over Plantesygdomme* 413:112-114.
 1971a Cikaden *Macropsis fuscula* (Zett.) i hindbaerkulturer. [The leafhopper *Macropsis fuscula* (Zett.) in raspberry plantings.] *Tidsskrift foer Planteavl* 75:577-580. [In Danish with English summary.]

Reling, D.
 1983a Aerial sampling to determine some factors affecting flight of the potato leafhopper (Homoptera: Cicadellidae). M. S. Thesis, Pennsylvania State University, University Park.

Reling, D. and Taylor, R. A. J.
 1984a A collapsible tow net used for sampling arthropods by airplane. *Journal of Economic Entomology* 77:1615-1617.

Remane, R.
1958a Die Besiedelung von Grunlandflachen verschiednener Herkunft durch Qanzen und Zikaden im Weser-Ems-Gebiet. *Zeitschrift fur Angewandte Entomologie* 42:353-400.

1959a *Lebradea calmagrostidis* gen. et spec. nov., eine neue Zikade aus Norddeutchland (Hom. Cicadina Cicadellidae). *Zoologischer Anzeiger* 163:386-387.

1960a Zur Kenntnis der Gattung *Arthaldeus* Ribaut (Hom. Cicadina, Cicadellidae). *Mitteilungen der Munchener Entomologischen Gesellschaft Munich* 50:72-82.

1961a *Endria nebulosa* (Ball), comb. nov. eine nearktische Zikade in Deutschland. *Nachrichtenblatt der Bayerischen Entomologen* (Munich) 10:73-76, 90-98.

1961b Revision der Gattung *Mocydiopsis* Ribaut. *Abhandlungen der Akademie der Mathematisch-Natur Wissenschaftlichen Klasse. Akademie der Wissenschaften und der Literatur*, pp. 101-149. [New genus: *Dlabolaracus*.]

1961c Die systematische Position vor *Deltocephalus aurantiacus* Forel. *Nachrichtenblatt der Bayerischen Entomologen* (Munich) 10(1):1-6.

1961d Zur Kenntnis der Verbreitung einiger Zikadenarten (Homoptera, Cicadina). *Nachrichtenblatt der Bayerischen Entomologen* (Munich) 10(12):111-114.

1962a Einige bemerkenswerte Zikaden-Funde in Nordwestdeutschland. *Faunistische Mitteilungen aus Norddeutschland* 2(2):23-26.

1965a Beitrage zur Kenntnis der Gattung *Psammotettix* Hpt. *Zoologische Beitrage* 11:221-245.

1966a Zur kenntnis der Gattung *Anoplotettix* Ribaut. (Homoptera). *Reichenbachia* 6(23):181-190.

1967a Zur Kenntnis der Gattung *Euscelis* Brulle (Homoptera, Cicadina, Jassidae). *Entomologische Adhandlungen und Berichte aus dem Staatliche Museum fur Tierkende in Dresden* 36(1):1-36.

1968a Erganzungen und Kritische anmerkungen zu der Heteropteren- und cicadinen-fauna der Makaronesischen Inseln. *Bocagiana* (Funchal) 16:1-14.

1983a The biogeography of Auchenorrhyncha of the Mediterranean region. In: Knight, W. J., Pant, N. C., Robertson, T. S. and Wilson, M. R., eds. *Proceedings of the 1st International Workshop on Biotaxonomy, Classification and Biology of Leafhoppers and Planthoppers (Auchenorrhyncha) of Economic Importance*. Pp. 291-295. London, 4-7 October 1982. Commonwealth Institute of Entomology, 56 Queen's Gate, London SW7 5JR. 500 pp.

1984a Adaptive radiation in leafhoppers: Erythroneurini on Madeira and the Canary Islands (Homoptera, Auchenorrhyncha, Cicadellidae, Typhlocybinae). *Mitteilungen der Schweizerischen Entomologische Gesellschaft* 57(4):438-439.

Remane, R. and Asche, M.
1980a Neue Zikaden-Taxa aus dem tribus Paralimnini Distant, 1908, aus dem Mittelmeergebiet (Homoptera Cicadina Cicadellidae). *Marburger Entomologische Publikationen* 1(4):67-166.

Remane, R. and Koch, J.
1977a Merkmalsverschiebungen im Bau der genitalarmatur der mannlichen zentraliberischer Populationen des *Euscelis—incisus* Kbm.—*alsius* Rib. Formenkreises—ein Indiz fur Introgressions-phanomene? *Zoologische Beitrage* 23(1):133-167. [In German.]

Remane, R. and Schulz, K.
1973a Storungen in der Ausbildung der ektoderalen weiblicher Genitalarmatur im Zusammenhang mit parasitarer Kastration bei Zikaden der Gattung *Jassargus* Zachv. (Homoptera, Cicadelloidea, Cicadellidae). *Zeitschrift fuer Wissenschaftliche Zoologie* 186(1&2):108-117.

1977a Uber bisher wenig beachtete merkmale von taxonomischer und phylogenetischer bedeutung im bereich der weiblichen ektodermalen genitalarmaturen bei cicadelliden (Homoptera, Auchenorrhyncha). *Sitzungberichte der Gesellschaft Naturforschender Freundezu Berlin* (N. F.) 17:116-123.

Renard, J. L., Quillec, G. and Ollagnier, M.
1982a Possible incidence of infections by *Recilia mica*, vector of oil-palm blast, on the metabolism of amino acids in the plant. *Oleagineaux* 37:43-48.

Rensner, P. E., Lamp, W. O., Barney, R. J. and Armbrust, E. J.
1983a Feeding tests of *Nabis roseipennis* (Hemiptera: Nabidae) on potato leafhopper, *Empoasca fabae* (Homoptera: Cicadellidae), and their movement into spring alfalfa. *Journal of the Kansas Entomological Society* 56(3):446-450.

Reyes, G. M.
1957a Rice dwarf disease in the Philippines. *FAO Plant Protection Bulletin* 6(2):17-19.

Reyes, T. M. and Gariel, B. P.
1975a The life-history and consumption habits of *Cyrtorhinus lividipennis* Reuter (Hemiptera: Miridae). *Philippine Entomologist* 3:79-88.

Reynolds, H. T. and Deal, A. S.
1956a Control of the southern garden leafhopper, a new pest of cotton in southern California. *Journal of Economic Entomology* 49(3):356-358.

Reynolds, H. T., Dickson, E. C., Hannibal, R. M. and Laird, E. F., Jr.
1967a Effects of the green peach aphid, southern garden leafhoppper, and carmine spider mite population upon yield of sugar beets in the Imperial Valley, California. *Journal of Economic Entomology* 60(1):1-7.

Reynolds, H. T., Fukuto, T. R., Metcalf, R. L. and March, R. B.
1957a Seed treatment of field crops with systemic insecticides. *Journal of Economic Entomology* 50(5):527-539.

Reynolds, H. T., Fukuto, T. R. and Peterson, G. D., Jr.
1960a Effect of topical applications of granulated systemic insecticides and of conventional applications of other insecticides on control of insects and spider mites on sugar beet plants. *Journal of Economic Entomology* 53(5):725-729.

Reynolds, H. T., Stern, V. M., Fukuto, T. R. and Peterson, G. D., Jr.
1960a Potential use of Dylox and other insecticides in a control program for field crop pests in California. *Journal of Economic Entomology* 53(1):72-78.

Ribaut, H.
1959a Homopteres nouveaux pour la France. *Bulletin de la Societe d'Histoire Naturelle de Toulouse* 94(3&4):393-399.

1959b Nouvelles especes francaises d'Homopteres. *Bulletin of the Societe d'Histoire Naturelle de Toulouse* 94:400-405.

Ricaud, C. and Felix, S.
1976a Identification and relative importance of virus diseases of maize in Mauritius. *Revue Agricole et Sucriere de l'Ile Maurice* 55:163-169.

Rice, R. E. and Jones, R. A.
1972a Leafhopper vectors of the western X-disease pathogen: Collections in central California. *Environmental Entomology* 1(6):726-730.

1974a Biology of *Colladonus mendicus* (Ball) in central California. *Environmental Entomology* 3(3):511-514.

Richards, O. W.
1961a An introduction to the study of polymorphism in insects. *Royal Entomological Society of London Symposium* 1:2-10.

1964a The entomological fauna of southern England with special reference to the country round London. *Transactions of the Society for British Entomology* 16(1):1-48.

1972a Two new species of Hymenoptera Dryinidae, with notes on some other species. *Bulletin of Entomological Research* 61(3):539-546.

Richards, W. R. and Varty, I. W.
1964a A new species of *Erythroneura* Fitch (Homoptera: Cicadellidae). *Canadian Entomologist* 96(3):515-516.

Richardson, B. H. and Raabe, R. D.
1956a Transmission of a spinach virus by the beet leafhopper in the winter garden area. *Journal of the Rio Grande Valley Horticultural Society* 10:84-86.

Richardson, H. P. and Westdal, P. H.
1963a Control of the six-spotted leafhopper, *Macrosteles fascifrons*, and aster yellows on head lettuce in Manitoba. *Canadian Journal of Plant Science* 48:12-17.
1964a Experiments on control of the six-spotted leafhopper, *Macrosteles fascifrons*, and aster yellows on head lettuce with contact and systemic insecticides in Manitoba. *Canadian Journal of Plant Science* 44:393-396.
1967a Disposable cage and pot for virus transmission studies with leafhoppers. *Canadian Entomologist* 99(7):769-770.

Richardson, J. and Jensen, D. D.
1971a Tissue culture of monolayer cell lines of *Colladonus montanus* (Homoptera: Cicadellidae), a vector of the casual agent of Western X-disease of peach. *Annals of the Entomological Society of America* 64:722-729.

Richter, F. T.
1970a The biology and control of *Erythroneura aclys* McAtee (Cicadellidae). *Proceedings of the North Central Branch of the Entomological Society of America* 25:63.

Rico de Cujia, L. A. and Martinez-Lopez, G.
1977a Efecto de la temperatura en la incubacion del virus del rayado colombiano del maiz. *Revista del Instituto Colombiano Agropecuario* 12:13-25.

Ricou, G.
1960a Methodes d'elevage au laboratoire de quelques Cicadelles vivant sur graminees. *Annales des Epiphyties* (Paris) 11(3):419-422.

Ricou, G. and Duval, E.
1969a Influence of leafhoppers upon some meadow grasses. *Zeitschrift fuer Angewandte Entomologie* 63:163-173.

Riddell, J. A.
1982a Carbosulfan, a versatile new insecticide. *In: Heong, K. L., Lee, B. S., Lim, T. M., Teoh, C. H. and Ibrahim, Y., eds. Proceedings of the International Conference on Plant Protection in the Tropics.* Pp. 712. Kuala, Lumpur, Malaysia. 743 pp.

Rikhter, V. A.
1968a A list of new taxa and new synonyms published in the Entomological Review in the period 1945-1967. *Entomological Review* (Washington) 47(2):222-260.

Rioux, G.
1963a Essais d'insecticides pour la repression des insects de la pomme de terre en sol organique. [Insecticide trials for control of potato insect pests in organic soil.] *Recherches Agronomiques* 9:23-24.

Ripper, W. E.
1956a Effect of pesticides on balance of arthropod populations. *Annual Review of Entomology* 1:403-438.

Ripper, W. E. and George, L.
1965a *Cotton Pests of the Sudan, Their Habits and Control.* Blackwell Science, Oxford. 345 pp.

Ritenour, G., Hills, F. J. and Lange, W. H.
1970a Effect of plant date and vector control on the suppression of curly top and yellows in sugarbeet. *Journal of the American Society of Sugar Beet Technologists* 16:78-84.

Rivera, C.
1981a Multiplication del virus del rayado fino del maiz en el insecto vector *Dalbulus maidis* (Homoptera: Cicadellidae). M. S. Thesis, Universidad de Costa Rica, San Jose, Costa Rica.

Rivera, C., Kozuka, Y. and Gamez, R.
1981a Rayado fino virus: Detection in salivary glands and evidence of increase in virus titre in the leafhopper vector *Dalbulus maidis*. *Turrialba* 31:78-80.

Rivera, C. T., Ling, K. C. and Ou, S. H.
1969a Susceptible range of rice tungro virus. *Phytopathology* 5:16-17 (Abstract).

Rivera, C. T., Ling, K. C., Ou, S. H. and Aguiero, V. M.
1972a Transmission of two strains of rice tungro virus by *Recilia dorsalis*. *Philippine Phytopathology* 5:17.

Rivera, C. T. and Ou, S. H.
1965a Leafhopper transmission of "Tungro" disease of rice. *Plant Disease Reporter* 49:127-131.
1967a Transmission studies on the two strains of rice tungro virus. *Plant Disease Reporter* 51:877-881.

Rivera, C. T., Ou, S. H. and Pathak, M. D.
1963a Transmission studies of the orange-leaf disease of rice. *Plant Disease Reporter* 47:1045-1048.

Revilla M., V. A.
1965a Una "virosis" destructiva del pasto pangola en el Peru. *Vida Agricola* 42(504-505):521-530.

Rivnay, E.
1962a *Field Crop Pests in the Near East. Monographiae Biologicae, Volumen X.* Uitgeverij Dr. W. Junk, Den Haag. 450 pp.

Rizk, G. N. and Ahmed, K. G.
1981a Population dynamics of some insect pests attacking squash plants, *Cucurbite pepo* L., in Iraq. *Research Bulletin, Faculty of Agriculture, Ain Shams University* No. 1653. 8 pp.

Robbins, J. C., Daugherty, D. M. and Hatchett, J.
1979a Ovipositional and feeding preference of leafhoppers (Homoptera: Cicadellidae) on Clark soybeans in relation to plant pubescence. *Journal of the Kansas Entomological Society* 52:603-608.

Rocha-Pena, M.
1981a Algunos aspectos relacionados con el virus del maiz en Mexico. M. S. Thesis, Colegio de Postgraduados, Chapingo, Mexico.

Rodriguez, J. G.
1985a Dwight Moore DeLong—A tribute. *In: Nault, L. R. and Rodriguez, J. G., eds. The Leafhoppers and Planthoppers.* Pp. 1-9. Wiley & Sons, New York. 500 pp.

Rodrigues, P. D.
1968a Notes Homopterologiques I. Remarques sur quelques Cicadellidae du Portugal avec la description de quatre especes nouvelles. *Archivos do Museu Bocage* (Lisbon) (Serie 2) 2(2):9-23. [New genus: *Heliotettix*.]

Rogers, C. E.
1981a Resistance of sunflower species to the western potato leafhopper. *Environmental Entomology* 10(5):697-700.

Romankow, W.
1963a Insect pests on lucerne in Poland. *Prace Naukowe Instytutu Ochrony Roslin* 5(2):89-207.

Roof, M. E.
1974a Mechanisms of potato leafhopper resistance in alfalfa. Ph. D. Dissertation, Kansas State University. 82 pp. [Diss. Abst. Int. B 35(8):3960-3961.]

Roof, M. E., Horber, E. and Sorenson, E. L.
1972a Bioassay technique for potato leafhopper, *Empoasca fabae* (Harris). *Proceedings of the North Central Branch of the Entomological Society of America* 27:140-143.
1974a Effect of saponin on the potato leafhopper *Empoasca fabae* (Homoptera, Cicadellidae). *Journal of the Kansas Entomological Society* 47(4):538-539.
1976a Evaluating alfalfa cuttings for resistance to the potato leafhopper. *Environmental Entomology* 5(2):295-301.

Room, P. M. and Wardhaugh, K. G.
1977a Seasonal occurrence of insects other than *Heliothis* spp. feeding on cotton in the Namoi Valley of New South Wales. *Journal of the Australian Entomological Society* 16:165-174.

Rose, D. J. W.
1962a Insect vectors of maize streak. *Zoological Society of the South African News Bulletin* 3(3):11.
1962b Pests of groundnuts. *Rhodesia Agriculture Journal* 59(4):197-198.
1972a Times and sizes of dispersal flights of *Cicadulina* species (Homoptera: Cicadellidae), vectors of maize streak disease. *Journal of Animal Ecology* 41(2):495-506.
1972b Dispersal and quality in population of *Cicadulina* spp. (Cicadellidae). *Journal of Animal Ecology* 41(3):589-609.
1973a Laboratory observations on the biology of *Cicadulina* spp. (Homoptera: Cicadellidae) with particular reference to the effects of temperature. *Bulletin of Entomological Research* 62:471-476.

1973b Field studies in Rhodesia on *Cicadulina* spp. (Hom. Cicadellidae), vectors of maize streak disease. *Bulletin of Entomological Research* 62(3):477-495.

1973c Distances flown by *Cicadulina* spp. (Hom. Cicadellidae) in relation to distribution of maize streak disease, in Rhodesia. *Bulletin of Entomological Research* 62(3):497-505.

1973d Management of *Cicadulina* leaf hopper populations to reduce streak disease in maize crops in the highveld in Rhodesia. *Rhodesia Agricultural Journal* 70:63-64.

1974a The epidemiology of maize streak disease in relation to population densities of *Cicadulina* spp. *Annals of Applied Biology* 76:199-207.

1978a Epidemiology of maize streak disease. *Annual Review of Entomology* 23:259-282.

1983a The distribution of various species of *Cicadulina* in different African countries, frequency of their attack and impact on crop production. *In: Knight, W. J., Pant, N. C., Robertson, T. S. and Wilson, M. R., eds. Proceedings of the 1st International Workshop on Biotaxonomy, Classification and Biology of Leafhoppers and Planthoppers (Auchenorrhyncha) of Economic Importance.* London, 4-7 October 1982. Pp. 297-304. Commonwealth Institute of Entomology, 56 Queen's Gate, London SW7 5JR. 500 pp.

Rosenberger, D. A. and Jones, A. L.

1978a Leafhopper vectors of peach X-disease pathogen and its seasonal transmission from chokecherry. *Phytopathology* 68(5):782-790.

Rosenkranz, E.

1969a A new leafhopper-transmissible corn stunt disease agent in Ohio. *Phytopathology* 59:1344-1346.

Ross, H. H.

1953b Polyphyletic origin of the leafhopper fauna of *Ilex decidua*. *Transactions of the Illinois State Academy of Science* 46:186-192.

1956a New Nearctic species of *Erythroneura* (Homoptera Cicadellidae). *Entomological News* 67(4):85-90.

1956b How to collect and preserve insects. *Illinois State Natural History Survey Circular* 39:1-159.

1957a New oak-inhabiting species of *Erythroneura* from Illinois (Hemiptera, Cicadellidae). *Entomological News* 68:183-190.

1957b Evolutionary developments in the leafhoppers, the insect family Cicadellidae. *Systematic Zoology* 6(2):88-98.

1957c Principles of natural coexistence indicated by leafhopper populations. *Evolution* 7(2):113-129.

1958a Evidence suggesting a hybrid origin for certain leafhopper species. *Evolution* 12(3):337-446.

1958b The relationship of systematics and the principles of evolution. *Proceedings of the 10th International Congress of Entomology* 1:423-429.

1958c Further comments on niches and natural coexistence. *Evolution* 12:112-113.

1959a A survey of the *Empoasca fabae* complex. *Annals of the Entomological Society of America* 52:304-316.

1959b Records and new species of leafhoppers belonging to the "*Empoasca fabae*" complex from Argentina. *Acta Zoologica Lilloana* 17:429-435.

1959c Migration mechanisms in leafhoppers, delineation of the study area. *Proceedings of the North Central Branch of the Entomological Society of America* 14:11.

1962a How to collect and preserve insects. *Illinois State Natural History Survey Circular* 39:1-79.

1963a An evolutionary outline of the leafhopper genus *Empoasca* subgenus *Kybos*, with a key to the Nearctic fauna. *Annals of the Entomological Society of America* 56:202-223.

1965a The phylogeny of the leafhopper genus *Erythroneura* (Hemiptera, Cicadellidae.) *Zoologische Beitrage* 11:247-270.

1965b Pleistocene events and insects. *In: Wright, H. D., Jr. and Frey, D. G, eds. The Quaternary of the United States.* Pp. 583-596. Princeton University Press. 922 pp.

1968a The evolution and dispersal of the grassland leafhopper genus *Exitianus*, with keys to the Old World species (Cicadellidae-Hemiptera). *Bulletin of the British Museum (Natural History) Entomology* 22(1):1-30.

1970a The ecological history of the Great Plains: Evidence from grassland insects. *In: Dort, W. and Jones, J. K., eds. Pleistocene and Recent Environments of the Great Plains.* Pp. 225-240. University of Kansas Press, Lawrence. 433 pp.

Ross, H. H. and Cooley, T. A.

1969a A new Nearctic leafhopper of the genus *Flexamia* (Hemiptera: Cicadellidae). *Entomological News* 80:246-248.

Ross, H. H. and Cunningham, H. B.

1960a A key to the *Empoasca solana* complex with descriptions of new species. *Ohio Journal of Science* 60:309-317.

Ross, H. H., Decker, G. C. and Cunningham, H. B.

1965a Adaptation and differentiation of temperate phylogenetic lines from tropical ancestors in *Empoasca*. *Evolution* 18(4):639-651.

Ross, H. H. and Hamilton, K. G. A.

1970a Phylogeny and dispersal of the grassland leafhopper genus *Diplocolenus* (Homoptera; Cicadellidae). *Annals of the Entomological Society of America* 63(1):328-331.

1970b New Nearctic species of the genus *Mocuellus* with a key to Nearctic species (Hemiptera: Cicadellidae). *Journal of the Kansas Entomological Society* 43(2):172-177.

1972a New species of North American Deltocephaline leafhoppers. *Proceedings of the Biological Society of Washington* 84(51):439-444.

1972b A review of the North American leafhopper genus *Laevicephalus* (Hemiptera: Cicadellidae). *Annals of the Entomological Society of America* 65(4):929-942.

Ross, H. H. and Moore, T. E.

1957a New species in the *Empoasca fabae* complex (Hemiptera, Cicadellidae). *Annals of the Entomological Society of America* 50:118-122.

Rossel, H. W.

1984a On geographical distribution and control of maize mottle chlorotic stunt in Africa. *Maize Virus Diseases Newsletter* 1:17-19.

Rossel, H. W., Buddenhagen, I. W. and Thottappilly, G.

1980a Storey's maize mottle virus rediscovered? *IITA Research Briefs* 1(2):2-4.

Rossel, H. W., Nabukenya, R., Thottapilly, G. and Zagre, M'B. B.

1984a Maize. Virology. *Annual Report, International Institute of Tropical Agriculture*, 1983. Pp. 42-43.

Rossiter, P. D.

1975a Pests of potatoes. *Queensland Agricultural Journal* 101:291-298.

1977a Control of jassids (f. Cicadellidae) on potatoes in southern Queensland. *Queensland Journal of Agricultural and Animal Sciences* 34:141-145.

Roy, R. S. and Ram, K. G.

1952a Control of mango hoppers in Bihar. *Indian Journal of Horticulture* 9:32-40.

Ruan, Y. L.

1985a On the population dynamic of the rice virus vector green leafhopper *Nephotettix cincticeps* Uhler and its chemical control. *Kunchong Zhishi* 22(2):54-57.

Ruhendi, and Litsinger, J. A.

1982a Effect of rice stubble and tillage methods on the preflowering insects pests of grain legumes. *In: Zandstra, H. G., Morris, R. A., Price, E. C., Litsinger, J. A., Moody, K., Jayasuriya, S., Rockwood, W. G. and Argosino, G. Report of a Workshop on Cropping Systems Research in Asia.* Pp. 85-98. Bogor, Indonesia.

Ruppel, R. F.

1958a The Typhlocybinae of Mexico (Homoptera: Cicadellidae.) *Dissertation Abstracts* 18(3):1164-1165.

1959a Especies nuevas de *Alebrini columbiana* (Homoptera, Cicadellidae). *Revista de la Academia Colombiana de Ciencias Exacta Fisicas y Naturales* 10(41):367-370.

1965a A review of the genus *Cicadulina* (Hemiptera, Cicadellidae). *Publications of the Museum of Michigan State University, Biological Series* 2(8):385-428.

1966a New Dikraneurini from Colombia (Hemiptera: Cicadellidae). *Journal of the Kansas Entomological Society* 39(1):102-104.

1969a *Cicadulina bipunctella* and *C. chinai* from India (Hemiptera, Cicadellidae). *Journal of the Kansas Entomological Society* 42(3):257-260.

1974a Diurnal sampling of the insect complex of alfalfa. *Great Lakes Entomologist* 7(4):113-116.

1975a Two new species of *Beamerana* (Hemiptera, Cicadellidae) from Panama (Hom.). *Michigan Academician* 7(4):437-440.

1975b Insect control in hay, forage and pasture crops. *Michigan Cooperative Extension Service Extension Bulletin* E-827:1-5.

Ruppel, R. F. and DeLong, D. M.

1956a A new species of *Cicadulina* (Homoptera: Cicadellidae) from Colombia. *Bulletin of the Brooklyn Entomological Society* 51:82-84.

1956b *Empoasca* (Homoptera: Cicadellidae) from highland crops of Colombia. *Bulletin of the Brooklyn Entomological Society* 51:85-92.

Ruppel, R. F. and Idrobo, E.

1962a Lista preliminar de insectos y otros animales que danan frijoles en America. [Preliminary list of insects and other animals that damage beans in America.] *Agricultura Tropical* (Colombia) 18(11):650-678.

Ruppel, R. F. and Janes, R. L.

1970a Insect control in forages, field corn and small grains. *Michigan Cooperative Extension Service Extension Bulletin* 672. 7 pp.

Ruppel, R. F., Romero, S. and J. I.

1972a A new sexually dimorphic *Empoasca*. *Journal of the Kansas Entomological Society* 45(1):125-127.

Russell, L. M.

1973a The correct citations for the reports on Homoptera collected during the Harriman Alaska expedition. *Journal of the Washington Academy of Science* 63(4):154-155.

Russo, M., Rana, G. L., Granett, A. L. and Calavan, E. C.

1976a Visualization of *Spiroplasma citri* in the leafhopper *Scaphytopius nitridus* (DeLong). *Proceedings of the 7th Conference of the International Organization of Citrus Virologist* 1975:1-6.

Ryan, J. G., O'Connor, J. P. and Beirne, B. P.

1984a *A Bibliography of Irish Entomology*. Flyleaf Press, Glenageary, Dublin. 363 pp.

Rygg, T.

1981a Occurrence, damage and control of the potato leafhopper, *Empoasca vitis* (Gothe), in potatoes. *Forskning og Forsok in Landbruket* 32:75-84.

Ryu, J. K., Choi, S. Y., Lee, H. R. and Song, Y. H.

1977a Root-zone placement of carbofuran for control of rice insect pests. *Korean Journal of Plant Protection* 16:217-220.

Saavedra, F.

1982a Epidemiologia del virus del rayado fino en plataciones de maiz en Alajuela, Costa Rica. M. S. Thesis, Universidad de Costa Rica.

Saboo, K. C. and Puri, S. N.

1978a Effect of insecticides on incidence of sucking pests and yield of groundnut, *Arachis hypogaea* Linn. *Indian Journal of Entomology* 40:311-315.

Sabrosky, C. W.

1966a Mounting insects from alcohol. *Bulletin of the Entomological Society of America* 12(4):349.

1971a Packing and shipping pinned insects. *Bulletin of the Entomological Society of America* 17(1):6-8.

Sachan, J. N.

1980a Evaluation of incidence of insect pests on hybrid napier strains under different dosages of fertilizers. *Annals of Arid Zone* 19:82-91.

Sackston, W. E.

1959a Aster yellows — a challenging problem in plant pathology. *Proceedings of the Entomological Society of Manitoba* 15:23-30.

Sagar, P. and Mehta, S. K.

1982a Field screening of exotic cowpea cultivars against jassid *Amrasca bigutulla* [sic!] *bigutulla* [sic!] (Ishida) (Cicadellidae: Homoptera) in Punjab. *Journal of Research, Punjab Agricultural University* 19:222-223.

Saglio, P. H. M. and Whitcomb, R. F.

1979a Diversity of wall-less prokaryotes in plant vascular tissue, fungi, and invertebrate animals. In: Whitcomb, R. F. and Tully, J. G., eds. *The Mycoplasmas. III. Plant and Insect Mycoplasmas.* Pp. 1-36. Academic Press, New York.

Sahad, K. A.

1982a Biology and morphology of *Gonatocerus* sp. (Hymenoptera, Mymaridae), an egg parasitoid of the green rice leafhopper, *Nephotettix cincticeps* Uhler (Homoptera, Deltocephalidae). I. Biology. *Memoirs of the Faculty of Agriculture* (Kyushu University) 50:246-260.

1982b Biology and morphology of *Gonatocerus* sp. (Hymenoptera, Mymaridae), an egg parasitoid of the green rice leafhopper, *Nephotettix cincticeps* Uhler (Homoptera, Deltocephalidae). II. Morphology. *Memoirs of the Faculty of Agriculture* (Kyushu University) 50:467-476.

1982c Descriptions of new species of *Gonatocerus* Nees and *Anagrus* Haliday from Japan (Hymenoptera, Mymaridae). *Esakia* 19:191-204.

Sahad, K. A. and Hirashima, Y.

1984a Taxonomic studies on the genera *Gonatocerus* Nees and *Anagrus* Haliday of Japan and adjacent regions, with notes on their biology (Hymenoptera, Mymaridae). *Bulletin of the Institute of Tropical Agriculture* 7:1-78.

Sahambi, H. S.

1970a Studies on sesamum phyllody virus — virus-vector relationship and host range. In: Raychaudhuri, S. P., ed. *Plant Disease Problems. Proceedings of the 1st International Symposium on Plant Pathology.* Pp. 340-351. New Delhi. 915 pp.

Sahtiyanci, S.

1966a The potato stolbur virus and its first occurrence in Turkey. *Bitki Koruma Bulteni* 6(1):24-30.

Saillard, C., Garcia-Jurado, O., Bove, J. M., Vignault, J. C., Moutous, G., Fos, A., Bonfils, J., Nhami, A., Vogel, R. and Viennot-Bourgin, G.

1980a Application of ILISA to the detection of *Spiroplasma citri* in plants and insects. *Proceedings of the 8th Conference of International Organization of Citrus Virologists* 1979:145-152.

Saillard, C., Vignault, J. C., Gadeau, A., Carle, P., Garnier, M., Fos, A., Bov, J. M., Tully, J. G. and Whitcomb, R. F.

1984a Discovery of a new plant-pathogenic spiroplasma. *Israel Journal of Medical Science* 20:1013-1015.

Saini, M. L. and Chhabra, K. S.

1968a A technique for evaluating varietal resistance of castor to castor jassid (*Empoasca flavescens* Fabricius). *Journal of Research, Punjab Agriculture University* 5(1):72-74.

Saini, R. S.

1967a Low temperature tolerance of some insects. *Indian Journal of Entomology* 29(2):135-138.

1967b Low temperature tolerance of aster leafhopper. *Journal of Economic Entomology* 60(2):619-620.

Saini, R. S. and Cutkomp, L. K.

1967a Effects of phorate on eggs of six-spotted leafhopper. *Journal of Economic Entomology* 60(1):191-195.

Saint-Aubin, P., Lamothe, M. and Garrabos, J. P.

1965a Essai de lutte contre *Scaphoideus littoralis* avec une specialite huileuse de D.D.T. *Revue de Zoologie Agricole Applique* 64:111-117.

1966a Nouvel essai de lutte contre *Scaphoideus littoralis* Ball. au moyen de divers produits. [A new test on the control of *S. littoralis* by means of various products.] *Revue de Zoologie Agricole Appliquee* 65(10-12):161-164.

Saito, Y.
1977a Interrelationship among waika disease, tungro and other similar diseases of rice in Asia. *Tropical Agriculture Research Series* 10:129-135.
1977b Rice viruses, with special reference to particle morphology and relationship with cells and tissues. *Review of Plant Protection Research* 10:83-90.

Saito, Y., Chaimongkol, U., Singh, K. G. and Hino, T.
1976a Mycoplasmalike bodies associated with rice orange leaf disease. *Plant Disease Reporter* 60:649-651.

Saito, Y., Hibino, H., Omura, T. and Usugi, T.
1981a Transmission of rice tungro baciliform virus and rice tungro spherical virus by leafhopper vectors. *Proceedings of the 5th International Congress of Virology*, p. 213.

Saito, Y., Inoue, H. and Satomi, H.
1978a Occurrence of rice transitory yellowing virus in Okinawa, Japan. *Annals of the Phytopathological Society of Japan* 44:666-669. [In Japanese with English summary.]

Saito, Y. and Miyata, T.
1982a Studies on insecticide resistance in *Nephotettix cincticeps*. In: Heong, K. L., Lee, B. S., Lim, T. M., Ttoh, C. H. and Ibrahim, Y., eds. *Proceedings of the International Conference on Plant Protection in the Tropics*. Pp. 377-382. Kuala Lumpur, Malysia.

Saito, Y., Roechan, M., Tantera, D. M. and Iwaki, M.
1975a Small baciliform particles associated with penyakit habang (tungro-like) disease of rice in Indonesia. *Phytopathology* 75:793-796.

Saitoh, K., Kudoh, K. and Mochida, O.
1970a A study of the male meiotic chromosomes in two Homopteran species of economic importance. *Science Reports of the Hirosaki University* 17(1/2):31-37.

Sajian, S. S. and Dhawan, A. K.
1977a Description of various stages of the rice green leafhopper *Nephotettix nigropictus* (Stal.) (Hemiptera; Cicadellidae). *Journal of Research, Punjab Agriculture University* 14(4):439-444.

Sajian, S. S., Sekhon, S. S. and Kanta, U.
1982a Control of jassid, *Zygnidia* [sic!] *manaliensis* (Singh), and thrips, *Anaphotrips sudanensis* Trybom, infesting maize. *Journal of Entomological Research* 6:18-21.

Sakai, M., Sato, Y. and Kato, M.
1967a Insecticidal activity of 1,3-bis(carbamoylthio) -2-(N,N-dimethylamino) propane hydrochloride, cartap, with special references to the effectiveness for controlling the rice stem borer. *Japanese Journal of Applied Entomology and Zoology* 11(3):125-134. [In Japanese with English summary.]

Sakakibara, A. M. and Cavichioli, R. R.
1982a Duas especies novas de *Diedrocephala* Spinola (Homoptera, Cicadellidae). *Revista Brasileira Entomologia* 26(3&4):241-246.

Sakurai, Y. and Morinaka, T.
1970a Effects of tetracyclines on symptom development and leafhopper transmission of rice yellow dwarf disease. *Bulletin of the Chugoku Agricultural Experiment Station* (E) 5:1-14, 27. [In Japanese with English summary.]

Salas, A. and Hansen, A. J.
1963a Una cigarrita (Homoptera: Cicadellidae) toxicogenica observada en cacao (*Theobroma cacao* L.). *Cacao* 8(1):6-12.

Sales, F. M.
1979a Insects and related forms in the soybean agroecosystem in the State of Ceara. *Fitossanidade* 3:57.

Salmon, M. A.
1954a A new species of *Cicadula* (Homoptera, Cicadellidae) from Hertfordshire. *Entomologist's Gazette* 5:15.
1954b A catalogue of the Hertfordshire leafhoppers. *Entomologist's Gazette* 5(1):11-14.

1959a On the rediscovery of *Athysanus argentatus* (Fab.) (Hem. Cicadellidae) in Britain. *Entomologist's Gazette* 10(1):51-53.

Sam, M. D. and Chelliah, S.
1984a Biology of the white leafhopper on rice. *International Rice Research Newsletter* 9(1):22.
1984b Influence of weather on populations of rice white leafhopper in light traps. *International Rice Research Newsletter* 9(1):26-27.
1984c Chemical control of the rice white leafhopper *Cofana spectra* (Distant). *International Rice Research Newsletter* 9(2):18.

Samad, K. and Ahmed, Manzoor
1979a Some genera and species of dikraneurine and typhlocybine leafhoppers studied from fruit and vegetable plants in Pakistan. *Reichenbachia* 17(30):249-260. [New genus: *Hazaraneura*.]
1979b Erythroneurine leafhoppers of fruit and vegetable plants in Pakistan. Part 2. Genera *Zygina* Fieber, *Erythroneura* Fitch and *Jalalia* Ahmed. *Islamabad Journal of Science* 6(1):1-4.

Samal, P. and Misra, B. C.
1983a The water measure bug, *Hydrometra* sp. (Hydrometridae: Hemiptera), a new predator of rice and plant hoppers. *International Rice Research Newsletter* 4:3.

Sameshima, T. and Nagai, K.
1962a Relation between the life-history of green rice leafhopper, *Nephotettix cincticeps* Uhler, and rice yellow dwarf propagation [in Tokyo]. *Japanese Journal of Applied Entomology and Zoology* 6(4):267-273.

Samsinak, K. and Dlabola, J.
1980a *Suidasia pontifica* Oudemans, 1905, a pest of insect collections in tropical regions. *Acta Faunistica Entomologica Musei Nationalis Pragae* 16:15-18.

Samyn, G., Stalle, J. van and Welvaert, W.
1982a The presence of different vectors of mycoplasma like organisms (MLO) in the Gent region. *Mededelingen van de Faculteit Landbouwwetenschappen, Rijksuniversiteit Gent* 47:1071-1977.

Sanchez, F. F.
1977a The current status of rice pest management in the Philippines. *Bulletin of the Entomological Society of America* 23:29-31.

Sander, F. W.
1984a Faunistic-ecological investigations of the neuer Muellberg Leipzig-Moeckern, East Germany. *Entomologische Nachrichten EER* 28(1):1-4.

Sander, K.
1959a Analyse des ooplasmatischen Reaktionssystems von *Euscelis plebejus* Fall. (Cicadine) durch isolieren und kombinieren von Keimteilen. *Wilhelm Roux Archiv fur Entwicklungsmechanik der Organismen* 151:430-497.
1960a Analyse der ooplasmatichen Reaktionssystems von *Euscelis plebejus* Fall. (Cicadina) durch Isolieren und Kombinieren von keimteilen. II. Mitteilugn. Die Differenzierungsleistungen nach Verlagern von Hinterpolmaterial. *Wilhelm Roux Archiv fur Entwicklungsmechanik der Organismen* 151:660-707.
1962a Uber den Einflus von verlagertem Hinterpolmaterial auf das metamere Organisationsmuster im Zikaden-Di. *Verhandlungen der Deutschen Zoologischen Gesellschaft Zoologischer Anzeiger Suppl.* 25:315-322.
1967a Mechanismen der Keimeseinrollung (Anatrepsis) im Insekten-Ei. *Verhandlungen der Deutschen Zoologischen Gesellschaft* 31:81-89. [English summary.]
1968a Entwicklungsphysiologische Untersuchugen am embryonalen Mycetom von *Euscelis plebejus* F. (Homoptera, Cicadina). 1. Ausschaltung und abnorme Kombination einzelner Komponenten des symbiontischen Systems. *Developmental Biology* 17:16-38. [English summary.]

Sandhu, G. S., Chahal, B. S. and Kanta, U.
1977a Chemical control of 'cereal jassid' on maize in Punjab. *Indian Journal of Entomology* 37:243-246.

Sandhu, G. S. and Harcharan, Singh
1975a Comparative persistent toxicity of LVC and EC/WP formulations of some insecticides to important pests of cotton. *Journal of Research, Punjab Agricultural University* 12:387-393.
1979a Influence of simulated rain and dew on the persistent toxicity of LVC and EC/WP formulations of some insecticides to cotton jassid, *Amrasca biguttula biguttula* (Ishida) on cotton. *Indian Journal of Plant Protection* 5:131-143.
1980a Influence of seasons on the persistent toxicity of LVC and EC/WP formulations on some insecticides to cotton jassid *Amrasca biguttula biguttula* on cotton. *Indian Journal of Plant Protection* 8:3-10.

Sandhu, M. S., Gatoria, G. S. and Sukhdev, Singh
1978a Control of cotton jassid by aerial application of some systemic insecticides. *Pesticides* 12:33.

Sanford, L. L.
1973a Selection for resistance to potato leafhopper infestation I. Selection methods. *American Potato Journal* 50(10):382 (Abstract).
1982a Effects of plant age on leafhopper infestations of resistant and susceptible potato clones. *American Potato Journal* 59:9-16.

Sanford, L. L., Carlson, O. V. and Hibbs, E. T.
1972a Genetic variation in a population of teraplid potatoes: Foliar resistance to oviposition of the potato leafhopper. *American Potato Journal* 49(3):98-108.

Sanford, L. L. and Ladd, T. L.
1979a Selection for resistance to potato leafhopper in potatoes. II. Progress after two cycles of phenotypic selection. *American Potato Journal* 56:541-547.

Sanford, L. L., Ladd, T. L., Sinden, S. L. and Cantelo, W. W.
1984a Early generation selection of insect resistance in potato. *American Potato Journal* 61:405-418.

Sanford, L. L. and Sleesman, J. P.
1970a Genetic variation in a population of tetraploid potatoes: Response to the potato leafhopper and the potato flea beetle. *American Potato Journal* 47(1):19-34.
1974a Genetic variation in a population of tetraploid potatoes: Response to the potato leafhopper and the potato flea beetle. *Ohio Agricultural and Development Research Center Research Summary* 72:33-34.
1974b Selection for resistance to potato leafhopper in potatoes. I. Selection methods. *American Potato Journal* 51(2):44-50.
1975a Selection for resistance to potato leafhopper in potatoes: Selection methods. *Ohio Agricultural and Development Center Research Summary* 81:45-46.

Sankaran, T. and Rao, V. P.
1966a Insects attacking witchweed (Strige) in India. *Commonwealth Institute of Biological Control Technical Bulletin* 7:63-73.

Santa Cruz, S.
1965a Dos parasitos en *Deltocephalus glaucus* Blanchard (Homoptera, Cicadellidae). *Agricultura Tecnica* (Santiago) 25(2):89-90.

Santok Singh, Gupta, V. K., Mathew, K. and Krishna, S. S.
1955a Entomological survey of the Himalayas. 12. Second annotated check-list of insects from the northwest (Punjab) Himalayas. *Agra University Journal of Research (Science)* 4:657-716.

Sardar Singh, Sidhu, A. S. and Sidhu, H. S.
1958a Field control of cotton jassid and whitefly in the Punjab. *Indian Cotton Growing Review* 12(6):391-405.

Saringer, G.
1958a Revision und Erganzungen zum Homopteren-Teil des Werkes Fauna Regni Hungariae (Gattung *Aphrodes*). *Folia Entomologica Hungarica* 11:479-489.
1958b A Fauna Regni Hungariae Homoptera fejezetenek revizioja. (*Aphrodes* genus). *Folia Entomologica Hungarica* 11:490-492.
1959a Zwei neue *Aphrodes*-Arten aus dem Karpatenbecken (Homoptera, Auchenorrhyncha). *Annales Historica-Naturales Musei Nationalis Hungarici* 1:425-428.
1961a Adatok az *Aphrodes bicinctus* Schr. es a *Hyalesthes obsoletus* Sign. Virusterjeszto Kabocak Elterjedesenek es Eletmodjanak Ismeretehez. [Data to the knowledge of the distribution and life history of two virus-transmitting leafhoppers, *Aphrodes bicinctus* Schr. and *Hyalesthes obsoletus* Sign.] *Kutato Intezet Evkon* 8:249-252. [In Hungarian.]

Sarkar, K. R. and Kulshreshtha, K.
1978a Nymphs of *Eutettix physitis* [sic!] as vector of brinjal little leaf disease. *Indian Journal of Agricultural Research* 12:99-100.

Sarma, P. V. and Rao, P.V. R.
1979a Studies on the effect of certain granular insecticides in the control of bhendi jassid *Amrasca biguttula biguttula* (Ishida). *Indian Journal of Plant Protection* 6:35-39.

Sarma, P. V., Sharma, V. K., Bahadur, U. and Chakravarti, K. B.
1981a Evaluation of phenyl carbamate insecticides for the control of mango hopper, *Idioscopus* spp. *Pestology* 5:13-14.

Sarmiento M., Jorge and F. Cisneroa, V.
1966a Control quimico de la "cigarrita verde" (*Empoasca* spp.) y tripidos [thrips] en plantas de vainitas. *Anales Cientificos de la Universidad Nacional Agraria la Molina* 4(1-2):31-34.

Saroja, R.
1981a Use of colored lights to attract rice insects. *International Rice Research Newsletter* 6(5):17-18.

Sasaba, R.
1974a Computer simulation studies on the life system of the green rice leafhopper *Nephotettix cincticeps* Uhler. *Review of Plant Protection Research* 7:81-98.

Sasaba, T. and Kiritani, K.
1971a Factors responsible for the variation of egg-mass size in the green rice leafhopper, *Nephotettix cincticeps* Uhler. *Kontyu* 39(1):54-60. [In Japanese with English summary.]
1972a Evaluation of mortality factors with special reference to parasitism of the green rice leafhopper, *Nephotettix cincticeps* Uhler (Hemiptera: Deltocephalidae). *Applied Entomology and Zoology* 7(2):83-93.
1974a Simulations of the population changes of *Lycosa* in the paddy field (Lycosidae: Lycosa). *Applied Entomology and Zoology* 9:273-275.
1975a A system model and computer simulation of the green rice leafhopper populations in control programmes. *Researches in Population Ecology* 16:231-244.

Sasaba, T., Kiritani, K. and Urabe, T.
1973a A preliminary model to simulate the effect of insecticides on a spider-leafhopper system in the paddy field. *Researches on Population Ecology* 15(1):9-22. [In English with Japanese summary.]

Sasamoto, K., Kobayoshi, M. and Shiraishi, H.
1968a Insect control by light trap. I. Attracting effectiveness of various lamps of different wave lengths against the green rice leafhopper (*Nephotettix cincticeps* Uhler). *Japanese Journal of Applied Entomology and Zoology* 12(3):164-170. [In Japanese with English summary.]

Sastry, K. S. S.
1957a Some potential insect pests of sugarcane in Mysore. *Mysore Agriculture Calendar and Year Book, 1956-57.* Pp. 129-135.
1957b Occurrence of *Tettigella spectra* (Distant) (Cicadellidae, Homoptera) on sugarcane in Mysore. *Current Science* (India) 26(4):122-123.

Sastry, M. V. S. and Rao, P. S. P.
1976a Promising multiple insect resistant rice varieties. *Current Science* 45:424-425.

Satapathy, M. K. and Anjaneyulu, A.
1982a Insecticide control of rice tungro virus disease. *International Rice Research Newsletter* 7(6):9-10.
1982b Greenhouse evaluation of granular, wettable powder, and flowable insecticide formulations for tungro prevention. *International Rice Research Newsletter* 7(6):11.

1982c Greenhouse evaluation of emulsifiable concentrate insecticides against tungro virus infection. *International Rice Research Newsletter* 7(6):11-12.

1983a Evaluation of emulsifiable concentrate and wettable powder insecticides for control of rice tungro virus diseases and its vector. *Zeitschrift fur Pflanzenkrankheiten und Pflanzenschutz* 90:269-277.

1985a Optimum dose of cypermethrin for tungro protection and its control. *Zeitschrift fur Pflanzenkrankheiten und Pflanzenschutz* 92:370-375.

Satari, G.
1983a Integrated control program on major rice pests in Indonesia. Status, problems and prospects. *In: 10th International Congress of Plant Protection. Vol. 3. Proceedings of a conference held at Brighton, England.* Pp. 895-903. Croyden, U.K.

Sathiyandandam, V. K. and Subramanian, A.
1982a The attraction of brown planthoppers and green leafhoppers to colored lights. *International Rice Research Newsletter* 7(3):13.

Sato, A. and Sogawa, K.
1981a Biotypic variations in the green rice leafhopper, *Nephotettix cincticeps* Uhler (Homoptera: Deltocephalidae), in relation to rice varieties. *Applied Entomology and Zoology* 16(1):55-57.

Satomi, H., Hirao, J. and Kimura, T.
1975a Transmission of the rice waika virus by a new leafhopper vector, *Nephotettix nigropictus* (Stal) (Homoptera: Cicadellidae). *Proceedings of the Association of Plant Protection* (Kyushu) 21:60-63.

Satpathy, J. M. and Mishra, B.
1969a Field tests with insecticides to control jassids and fruit borer of okra (bhendi). *Pesticides* 3(5):27.

Sattar, A., Ullah, K., Ahad, and Yousaf, M.
1984a Insect pests of sunflower in N.W.F.P., Pakistan. *Pakistan Journal of Agricultural Research* 5:239-240.

[Satyavir]
1983a Efficacy of some important insecticides for the control of leafhopper *Empoasca kerri* Pruthi on mothbean crop. *Indian Journal of Plant Protection* 11:130-133.

Saugstad, E. W., Bram, R. A. and Nyquish, W. E.
1967a Factors influencing sweep-net sampling of alfalfa. *Journal of Economic Entomology* 60(2):421-426.

Savage, J. M.
1958a The concept of ecologic niche, with reference to the theory of natural coexistence. *Evolution* 12:111-112.

Savinov, A. B.
1983a Morphology of the thoracic skeleton in the nymph and adult of *Cicadella viridis* L. (Homoptera, Cicadellidae). *Entomologicheskoe Obozrenie* 62(4):673-689. [In Russian; translation in *Entomological Review* (Washington) 62(4):9-27.]

1984a Thoracic musculature in the nymph and adult green leafhopper *Cicadella viridis* L. (Homoptera, Cicadellidae). *Entomologicheskoe Obozrenie* 63(4):672-684. [In Russian.]

Savio, C. and Conti, M.
1983a Epidemiology and transmission of some plant mycoplasmas. *Atti XIII Congresso Nazionale Italiano di Entomologia.* Pp. 407-414.

Savulescu, A. and Ploaie, P.
1962a Studies on the incidence of the 'stolbur' virus disease in relation to the vectors, host plants and ecological conditions. *In: Plant Virology. Proceedings of the 5th Conference of Czechoslovakian Plant Virologists, Prague.* Pp. 195-202.

Sawai Singh, G.
1969a Fifteen new species of jassids (Cicadellidae) from Himachal Pradesh and Chandigarh. *Research Bulletin of the Punjab University* 20:339-361.

1971a Morphology of the head of Homoptera. *Research Bulletin of the Punjab University,* N. S. (Science) 22(3&4):261-316.

1971b *Hishimonus* Ishihara, 1953 and *Hishimonoides* Ishihara, 1965 as synonyms of *Cestius* Distant, 1908 (Cicadellidae: Hemiptera). *Journal of Natural History* 5:569-571.

Sawai Singh, G. and Gill, M. I. P. K.
1973a Taxonomy of Indian species of *Austroagallia* Evans. *In: Bindra, O. S., ed. Cicadellid Vectors of Plant Pathogens.* Pp. 11-15. Final Project Report PL-480, Punjab Agricultural University, Ludhiana. 56 pp.

Sawbridge, J. R.
1975a A revision of the deltocephaline leafhopper genus *Giprus* Oman (Homoptera: Cicadellidae: Deltocephalinae). *Occasional Papers of the California Academy of Science* 114:1-27.

1975b *Tiaja insula,* a new megophthalmine leafhopper from the Santa Barbara Channel Islands. *Pan-Pacific Entomologist* 51(4):268-270.

1977a The leafhopper genus *Tiaja* Oman (Homoptera: Cicadellidae), with a contribution to the systematics of the group. *Dissertation Abstracts International* (B) 37(9):4310-4311.

Saxena, K. N.
1954a Alimentary canal and associated structures of the Jassidae (Homoptera). *Current Science* (India) 23(6):198-199.

1954b Feeding habits and physiology of certain leafhoppers (Homoptera, Jassidae). *Experimentia* 10:383.

1955a The anatomy and histology of the digestive organs and Malpighian tubes of the Jassidae (Homoptera). *Journal of the Zoological Society of India* 7:41-52.

1969a Patterns of insect-plant relationships determining susceptibility or resistance of different plants to an insect. *Entomologia Experimentalis and Applicata* 12:751-766.

1979a Physiology of leafhoppers: Their behaviour and nutrition. *Final Technical Report PL-480, Project No. A7-ENT-109,* Department of Zoology, University of Delhi-110007, India. 95 pp.

1981a Control of behavior of leafhoppers. *Final Technical Report PL-480, Project No. IN-ARS-90, Grant No. FG-IN-551,* Department of Zoology, University of Delhi, Delhi-110007, India.

1985a Certain factors influencing the acoustic communication in sexual behavior of the leafhopper *Amrasca devastans* (Distant) (Homoptera: Cicadellidae). *Applied Entomology and Zoology* 20(2):199-209.

Saxena, K. N. and Basit, A.
1982a Interference with the establishment of the leafhopper *Amrasca devastans* on its host plants by certain non-host plants. *In: Visser, J. H. and Minks, A. K., eds. Proceedings of the 5th International Symposium on Insect-Plant Relationships.* Pp. 153-162. Wageningen, Netherlands.

1982b Inhibition of oviposition by volatiles of certain plants and chemicals in the leafhopper *Amrasca devastans* (Distant). *Journal of Chemical Ecology* 8:329-338.

Saxena, K. N., Gandhi, J. R. and Saxena, R. C.
1974a Patterns of relationships between certain leafhoppers and plants. Part 1. Responses to plants. *Entomologia Experimentalis et Applicata* 17(2):303-317.

Saxena, K. N. and Kumar, H.
1980a Interruption of acoustic communication and mating in a leafhopper and a planthopper by aerial sound vibrations picked up by plants. *Experientia* (Basel) 36(8):933-936.

1984a Acoustic communication in the sexual behaviour of the leafhopper, *Amrasca devastans. Physiological Entomology* 9(1):77-86.

1984b Interference of sonic communications and mating in leafhopper *Amrasca devastans* (Distant) by certain volatiles. *Journal of Chemical Ecology* 10(10):1521-1532.

Saxena, K. N. and Saxena, R. C.
1974a Patterns of relationships between certain leafhoppers and plants. Part 2. Role of sensory stimuli in orientation and feeding. *Entomologia Experimentalis et Applicata* 17(4):493-503.

1975a Patterns of relationships between certain leafhoppers and plants. Part 3. Range and interaction of sensory stimuli. *Entomologia Experimentalis et Applicata* 18(2):194-206.

1975b Patterns of relationships between certain leafhoppers and plants. Part 4. Sequence of stimuli determining arrival on a plant. *Entomologia Experimentalis et Applicata* 18(2):207-212.

Saxena, R. C. and Barrion, A. A.

1985a Meiotic chromosomes of male green leafhopper (GLH) [*Nephotettix virescens*]. *International Rice Research Newsletter* 10(3):29.

Saxena, R. C., Barrion, A. A. and Soriano, M. V.

1985a Comparative morphometrics of male and female genital and abdominal characters in *Nephotettix virescens* (Distant) populations from Bangladesh and the Philippines. *International Rice Research Newsletter* 10(3):27-28.

Saxena, R. C. and Khan, Z. R.

1984a Neem oil disturbs *Nephotettix virescens* feeding. *Neem Newsletter* 1:28-29.

1985a Electronically recorded disturbances in feeding behavior of *Nephotettix virescens* (Homoptera: Cicadellidae) on neem treated rice plants. *Journal of Economic Entomology* 78(1):222-226.

Saxena, R. C. and Saxena, K. N.

1971a Growth, longevity and reproduction of *Empoasca devastans* on okra fruit for laboratory-rearing. *Journal of Economic Entomology* 64:424-425.

Schaefers, G. A. and Brann, J. L.

1974a Suppression of X-disease in peach orchards. *Proceedings of the New York State Horticultural Society* 119:110-112.

Schafer, A. and Taubert, S.

1984a A new portable, battery-powered insect suction trap. Intersect 1A. *Mitteilungen der Schweizerischen Entomologischen Gesellschaft* 57(4):440-441.

Schalk, J. M., Plaiste, R. L. and Sanford, L. L.

1975a Progress report: Resistance to the Colorado potato beetle and potato leafhopper in *Solanum tuberosum* subsp. *andigena*. *American Potato Journal* 52(6):175-177.

Schalk, J. M. and Radcliffe, R. H.

1976a Evaluation of ARS program on alternative methods of insect control: Host plant resistance to insects. *Bulletin of the Entomological Society of America* 22(1):7-10.

Schauff, M. E.

1981a A review of Nearctic species of *Acmopolynema* Oglobin (Hymenoptera: Mymaridae). *Proceedings of the Entomological Society of Washington* 83:444-460.

Schefer-Immel, V.

1957a Uber ein massenauftreten der zikade *Idiocerus laminatus* Flor an *Populus tremula*, mit einigen bemerkungen zur biologie und zum honigtau der art. *Anzeiger fuer Schadlingskunde* 30(10):165-169.

Schengeliya, E. S.

1956a On the cicada (Auchenorrhyncha) fauna of the suburban zone of Tbilisi. *Trudy Instituta Zoologii Akademii Nauk Gruzinskoi SSR* 15:35-49.

1961a Contributions to fauna of leafhoppers, silkworm moths and geometrids from the Trusovskoe canyon. *Trudy Instituta Zoologii Akademii Nauk Gruzinskoi SSR* 18:163-165. [In Russian.]

1964a Auchenorrhyncha of the fauna of the high mountains of Greater Caucasus. In: *Fauna of the High Mountains of Greater Caucasus in Georgia. Tbiliskii Akademiya Nauk Gruzinskoi SSR Instituta Zoologicheskogo.* Pp. 99-115.

1966a Auchenorrhyncha. In: *Invertebrates of the Trialets Range. Tbiliskii Akademiya Nauk Gruzinskoi SSR Instituta Zoologicheskogo.* Pp. 48-64. [In Russian.]

Schengeliya, E. S. and Dlabola, G. A.

1964a *Svanetia* – a new cicadid genus from Svanetia (GSSR). *Soobshcheniya Akademii Nauk Gruzinskoi SSR* 34:659-663. [In Russian.]

1964b Auchenorrhyncha in the alpine areas of the Great Caucasus in Georgia SSR. *Metaniereba: Tillis* 99-115.

Schieber, E. and Costillo, M.

1960a Corn stunt disease in Guatemala. *Plant Disease Reporter* 44:764.

1962a Corn stunt, a severe disease in Guatemala. *Phytopathology* 52:287.

Schiemenz, H.

1964a Beitrag zur kenntnis der zikadenfauna (Homoptera, Auchenorrhyncha) und ihrer okologie in feldhecken, restwaldern und den angrenzenden fluren. *Archiv fuer Naturschutz und Landschaftsforschung* 4(4):163-189.

1965a Zur Zikadenfauna des Geisings und Pohlberges im Erzgebirge (Hom., Auchenorrhyncha): Eine faunistisch-okologische Studie. *Zoologische Beitrage* 11(1&2):271-288.

1969a Die Zikadenfauna (Homoptera Auchenorrhyncha) mitteleuropaischer Trockenrasen - Untersuchungen zu ihrer Phanologie, Okologie, Bionomie und Chorologie. *Abhandlungen und Berichte des Naturkundemuseums Forschungsstelle Goerlitz* (Leipzig) 44:(2):195-205.

1970a Beitraege zur Insekten-fauna der DDR: Verzeichnis der im Gebeit der Deutschen Demokratischen Republik bisher festgestellten Zikaden (Homoptera: Auchenorrhyncha). *Beitrage zur Entomologie* 20(5&6):481-502.

1971a Die zikadenfauna (Homoptera, Auchenorrhychen) der Erzgebirgshochmoore. *Zoologische Jahrbuecher Abteilung Systematik Oekologie und Geographie der Tiere* 98(3):397-417.

Schillinger, J. A., Elliot, F. C. and Ripel, R. F.

1964a A method for screening alfalfa plants for potato leafhopper resistance. *Michigan Agriculture Experiment Station Bulletin* 46:512-517.

Schindler, U.

1960a Die gruene Zikade (*Cicadella viridis* L.) ein Schadling an jungen Laubholzern. *Forschungen und Holzweise* 15(3):48-50.

Schlottfeldt, C. S.

1944a Insectos encontrados em plantas cultivadas e comuns Vicosa, Minas Gerais. [Insects infesting cultivated and common plants around Vicosa, Minas Gerais.] *Revista Ceres* 6(31):52-65.

Schmalscheidt, W.

1985a Bud death in rhododendrons. *Immergrune Blatter* 26:35-37.

Schmutterer, H.

1956a Grune Labheuschrecke *Tettigonia viridissima* L., eine Ubertragerin des Tabakmosaikvirus. *Zeitschrift fuer Pflanzenkrankheiten und Pflanzenschutz* 63:6-9.

1974a Ecological studies on entomophagous syrphid species and their parasites in the Kenya Highlands (East Africa). *Zeitschrift fur Angewandte Entomologie* 75:42-67.

Schneider, C. L.

1959a Occurrence of curly top of sugar beets in Maryland in 1958. *Plant Disease Reporter* 43:681.

Schneider, F.

1962a Dispersal and migration. *Annual Review of Entomology* 7:223-242.

Schoenfeld, P.

1984a Die Zikaden-Typen Jacobis der Sunda-Expedition Rensch 1927 im zoologischen Museum Berlin (Auchenorrhyncha, Homoptera). *Mitteilungen aus dem Zoologischen Museum in Berlin* 60(1):69-76.

Schoonhoven, A. van and Cardona, C.

1980a Insectos y otras plagas del frijol en America Latina. In: Schwartz, H. F. and Galvez, G. E., eds. *Problemas de Produccion del Frijol*. CIAT, Palmira, Colombia.

Schoonhoven, A. van, Cardona, C., Garcia, J. and Garzon, F.

1981a Effect of weed covers on *Empoasca kraemeri* Ross and Moore populations and dry bean yields. *Environmental Entomology* 10:901-905.

Schoonhoven, A. van, Gomez, L. A. and Avalos, F.
1978a The influence of leafhopper (*Empoasca kraemeri*) attack during various bean (*Phaseolus vulgaris*) plant growth stages on seed yield. *Entomologia Experimentalis et Applicata* 23:115-120.

Schoonhoven, A. van, Hallman, G. H. and Temple, S. R.
1985a Breeding for resistance to *Empoasca kraemeri* Ross and Moore in *Phaseolus vulgaris* L. *In: Nault, L. R. and Rodriguez, J. G., eds. The Leafhoppers and Planthoppers.* Pp. 405-422. Wiley & Sons, New York. 500 pp.

Schowalter, T. D.
1985a Adaptations of insects to disturbances. *In: Pickett, S. T. A. and White, P. S., eds. The Ecology of Natural Disturbance and Patch Dynamics.* Pp. 235-253. Academic Press, New York. 472 pp.

Schroder, H.
1957a Neues zur Gattung *Homalodisca* (Insecta, Homoptera, Cicadellidae). *Senckenbergiana Biologica* 38:251-258.
1959a Taxonomische und tiergeographische Studien an Neotropischen Zikaden (Cicadellidae, Tettigellinae). [Taxonomic and zoogeogeogrical studies on Neotropical Cicadellidae.] *Abhandlungen Senckenbergischen Naturforchenden Gesellschaft* 499:1-93.
1960a Drei neue *Oncometopia*-Arten aus Costa Rica (Insecta, Homopt., Cicad.). *Senckenbergiana Biologica* 41:97-101.
1960b Neue und wenig bekannte neotropische Tettigellinae (Ins., Homopt., Cicad.). *Senckenbergiana Biologica* 41:315-324.
1961a Zur Kenntnis der Gattung *Cuerna* (Insecta, Homoptera, Cicadellidae). *Senckenbergiana Biologica* 42:87-91.
1962a Neue und wenig bekannte neotropische Tettigellinae (Ins., Homopt., Cicadellidae). 2. *Senckenbergiana Biologica* 43:153-164. [New subgenus: *Pygometopia*.]
1972a *Egidemia gracilis* n. sp., eine neue Zikaden-Art aus Sudamerika (Homoptera, Cicadellidae, Cicadellinae). *Entomologische Zeitschrift* 82(24):278-280.
1975a *Oncometopia amseli* n. sp., eine neue Zikaden-Art aus Mexico (Homoptera, Cicadellidae). *Entomologische Zeitschrift* 85(7):77-79.

Schruft, G.
1962a On the occurrence of *Empoasca flavescens* (Homoptera, Jassidae) on grapevines in the Bodensee region. *Wein-Wissenschaft* 17(12):304-308.

Schuler, L.
1954a Quelques procedes de chasse et de preparation. *L'Entomologiste* 10(4):75-78.
1956a Notes sur les moeurs des Homopteres. *Bulletin de la Societe Entomologique de France* (Paris) 61:157-160.

Schultz, G. A.
1979a Epidemiology of the aster leafhopper and aster yellows in relation to disease control. Doctoral Dissertation, University of Wisconsin, Madison. 574 pp.

Schultz, G. A. and Chapman, R. K.
1979a Carrot resistant to aster yellows. *Proceedings of the North Central Branch of the Entomological Society of America* 33:52-53.

Schultz, J. T.
1976a Insects infecting potatoes in the Red River Valley. *North Dakota Agricultural Extension Bulletin* 26:50-52.

Schulz, C. A.
1976a Homoptera, Auchenorrhyncha from Norway. *Norwegian Journal of Entomology* 23(2):159-160.

Schulz, C. A. and Meijer, J.
1978a Migration of leafhoppers (Homoptera: Auchenorrhyncha) into a new polder. *Holarctic Entomology* 1:73-78.

Schumann, G. L., Tingey, W. M. and Thurston, H. D.
1980a Evaluation of six insect pests for tranmission of potato spindle tuber viroid. *American Potato Journal* 57:205-211.

Schvester, D.
1962a Perspectives de lutte contre la flavescence doree par destruction de son vecteur: *Scaphoideus littoralis* Ball. *Revue de Zoologie Agricole et de Pathologie Vegetale* 10-12:135-143.
1965a *Cicadellidae and Flavescence Doree of Grapevine.* Institute National de la Recherche Agronomique, Gironde, France.
1966a Phloem feeding insect vectors (with special reference to *Scaphoideus littoralis* Ball vector of the *flavescence doree*). *Proceedings of the International Conference on Virus and Vector on Perennial Hosts with Special Reference to Vitis.* University of California, Division of Agriculture Science. Pp. 107-116.
1970a Insecticide treatments and the recovery of vines affected by golden flavescence. *Annales de Zoologie-Ecologie Animale* 1(4):467-494. [Summaries in English and German.]
1972a Cicadelles de la vigne. *Organization Europeene et Mediterraneene pour la Protection des Plantes Publications Bulletin* 3:37-42.

Schvester, D., Carle, P. and Moutous, G.
1961a Sur la transmission de la flavescence doree des vignes par une cicadelle. [On the transmission of vine golden flavescence by a leafhopper.] *Compte Rendu de l'Academie d'Agriculture de France* 47:1021-1024.
1963a Transmission de la flavescence doree de la vigne par *Scaphoideus littoralis* Ball. (Homopt. Jassidae). *Annales des Epiphyties* 14(3):175-198.
1963b Sur la transmission de la flavescence doree de la vigne par *Scaphoideus littoralis* Ball. *Compte Rendu de l'Academie d'Agriculture de France* 49:130-144.
1967a Testing the susceptibility of vine varieties to golden flavescence by means of inoculation by *S. littoralis. Annales des Epiphyties* 18:143-150.
1970a Nouvelles donnees sur la transmission de la flavescence doree de la vigne par *Scaphoideus littoralis* Ball. *Annales de Zoologie-Ecologie Animale* 1(4):445-465. [English and German summaries.]

Schvester, D., Moutous, G., Bonfils, J. and Carle, P.
1962a A biological study of grape leafhoppers in south-west France. *Annales des Epiphyties* 13(3):205-237. [In French with English summary.]

Schvester, D., Moutous, G. and Carle, P.
1962a *Scaphoideus littoralis* Ball (Homopt., Jassidae) cicadelle vectrice de la flavescence doree de la vigne. *Revue de Zoologie Agricole et de Pathologie Vegetale* 10-12:118-131.
1962b Perspectives de lutte contre la flavescence doree par destruction de son vecteur: *Scaphoideus littoralis* Ball. *Revue de Zoologie Agricole et de Pathologie Vegetale* 10-12:118-131.
1963a Tests insecticides de plein champ contre *Scaphoideus littoralis* Ball (Homopt. Jassidas), cicadelle vectrice de la flavescence doree. *Phytiatrie-Phytopharmacie* 12(1):51-56.

Schwartz, P. H., Osgood, C. E. and Ditman, L. P.
1961a Experiments with granulated systemic insecticides for control of insects on potatoes, lima beans and sweet corn. *Journal of Economic Entomology* 54(4):663-665.

Schwarz, R.
1959a Erhohte Anlockung von *Macrosteles laevis* Rib. (Hom. Cicadina) durch Attraktivflachen. [Increased trapping of *M. laevis* by attractive surfaces.] *Zeitschrift fuer Pflkrankheiten* 66(9):589-590.

Schwemmler, W.
1978a Leafhopper endocytobiosis: Dates and models. *Auchenorrhyncha Newsletter* 1:21-24.
1984a Leafhopper endocytobiosis as an intracellular ecosystem. *Mitteilungen der Schweizerischen Entomologischen Gesellschaft* 57(4):441.

Schwemmler, W. and Kemner, G.
1984a Fine structure analysis of leafhopper egg cell. *Mitteilungen der Schweizerischen Entomologischen Gesellschaft* 57(4):442.

Schwemmler, W. , Quiot, J. M. and Amargier, A.
1971a A study of intracellular symbionts in organotypic and cellular cultures of *Euscelis plebeja* (Jassidae) in media of standard composition. *Annales de la Societe Entomologique de France* (N.S.) 7(2):423-438.

Scott, G. E. and Rosenkranz, G.
1981a Effectiveness of resistance to maize dwarf mosaic and maize chlorotic dwarf viruses in maize. *Phytopathology* 71:937-941.

Scudder, G. G. E.
1961a Additions to the list of Cicadellidae (Homoptera) of British Columbia, with one genus and four species new to Canada. *Proceedings of the Entomological Society of British Columbia* 58:33-35.

Seetharaman, R. and John, J. T.
1981a Studies on rice virus diseases with special emphasis on breeding for resistance. *Final Technical Report PL-480 Project, All-India Coordinated Rice Improvement Project (I.C.A.R.)*, Rajendranager, Hyderabad-500 030, Andra Pradesh.

Sekar, P. and Cheliah, S.
1983a Varietal resistance to white leafhopper. *International Rice Research Newsletter* 8(2):7.

Seki, M. and Onizuka, S.
1964a Detection of rice yellow dwarf virus in leafhopper vectors, by use of hemagglutination tests. *Annals of the Phytopathological Society of Japan* 29:276. [In Japanese.]

Sekido, S. and Sogawa, K.
1976a Effects of salicylic acid on probing and oviposition of the rice plant- and leafhoppers (Homoptera: Delphacids and Deltocephalidae). *Applied Entomology and Zoology* 11:75-81.

Sekiyama, E. and Fukushi, T.
1955a Insect transmission of potato witches' broom in Japan. *Proceedings, Japan Academy* 31(4):234-236.

Sekizawa, K. and Ogawa, T.
1980a Studies on the breeding of rice varieties resistant to the green rice leafhopper. 1. Varietal differences of resistance to the green rice leafhopper. *Bulletin of the Chugoku National Agricultural Experiment Station* (Series A) 27:37-48. [In Japanese with English summary.]

Selim, A. A.
1977a Insect pests of safflower (*Carthamus tinctorius*) in Mosul Northern Iraq. *Mesopotamia Journal of Agriculture* 12:75-78.

Seliskar, C. E. and Wilson, C. L.
1981a Yellows diseases of trees. In: Maramorosch, K. and Raychandhuri, S. P., eds. *Mycoplasma Diseases of Trees and Shrubs*. Pp. 35-96. Academic Press, New York. 362 pp.

Selsky, M. I.
1961a A comparison of wound-tumor and tobacco club-root. *Phytopathology* 51:581-582.

Selsky, M. I. and Black, L. M.
1961a Effect of high and low temperature on the survival of wound-tumor virus in sweet clover. *Virology* 16:190-198.

Semede, C. M. B.
1961a Alguns insectos da biocenose do ulmeiro em Portugal. A Galerucela do ulmeiro (*Galerucella luteola* Mull.). *Broteria* 30(57):99-148.

Senapati, B. and Hohanty, G. B.
1980a A note on the population fluctuation of sucking pests of cotton. *Madras Agricultural Journals* 67:624-630.

Senapati, B. and Khan, S. R.
1978a A note on population fluctuation of *Amrasca biguttula biguttula* (Ishida) at Bhubaneswar. *Indian Journal of Agricultural Research* 12:97-98.

Senboku, T. and Shikata, E.
1980a Embryonic tissue culture of *Nephotettix nigropictus* Stal and *Nilaparvata lugens* Stal. *Memoirs of the Faculty of Agriculture, Hokkaido University* 12:101-108.

Sen-Sarma, P. K.
1982a Insect vectors of sandal spike disease. *European Journal of Forest Pathology* 12:297-299.

1984a Mycoplasma and allied diseases of forest trees in India and vector-host-pathogen interactions. *Proceedings of the Indian Academy of Science, Animal Science* 93(4):323-333.

Serrano, F. B.
1957a Rice "Accep Na Pula" or stunt disease, a serious menace to Philippine rice industry. *Philippine Journal of Science* 86:203-230.

Serrano-Limon, G.
1970a Biologia de *Marathonia nigrifascia* (Walker) (Homoptera Cicadellidae). Tesis Profesional, Facultad de Ciencias, Universidad Nacional Autonoma de Mexico. 72 pp.

Serrano-Limon, G. and Ramos Elorduy de Conconi, J.
1972a Biologia de *Marathonia nigrifascia* (Walk.) (Homoptera, Cicadellidae). *Anales del Instituto de Biologia, Universidad de Mexico* 43:67-88.

Servadei, A.
1957a Gli Omotteri (Hemiptera Homoptera Auchenorrhyncha) del promontorio Garganico. *Memorie di Biogeografia Adriatica, Istituto di Studi Adriatici* 3:197-243.

1958a Considerazioni sugli Omotteri italiani (Heteroptera et Homoptera Auchenorrhyncha). *Atti della Accademia Nazionale Italiana di Entomologia* 6:20-27.

1960a Gli Omotteri (Hemiptera Homoptera Auchenorrhyncha) della Calabria. *Memorie del Museo Civico di Storia Naturale di Verona* 8:301-333. [In Italian with English summary.]

1967a Rhynchota (Heteroptera: Homoptera Auchenorrhyncha). *Catalogo topografico e sininonimico. Fauna d'Italia* 9:1-851. [In Italian.]

1968a Contributo alla Corologia dei Rhynchota Homoptera Auchenorrhyncha d'Italia. *Annali del Museo Civico di Storia Naturale Giacomo Doria Genova* 77:138-183. [In Italian with English summary.]

1969a I Rincoti Endemici d'Italia. *Memorie della Societa Entomologica Italiana* 48:417-439. [In Italian with English summary.]

1971a I Rincoti (Eterotteri ed Omotteri Auchenorrinchi) dell'Appennino Abruzzese. *Lavori della Societa Italiana di Biogeografia* 2:170-217. [In Italian with English summary.]

1972a I Rincoti di valmalenco (Heteroptera et Homoptera Auchenorrhyncha). *Bollettino dell'Instituto di Entomologia della Universita degli Studi di Bologna* 31:13-26. [In Italian with English summary.]

Seshadri, C. R. and Seshu, K. A.
1956a Preliminary observations on jassid injury in castor. *Madras Agricultural Journal* 43:197-199.

Seth, M. L., Raychaudhuri, S. P. and Sing, D. V.
1972a Bajra (pearl millet) streak: A leafhopper-borne cereal virus in India. *Plant Disease Reporter* 56:424-428.

Sethi, G. R., Prasad, H. and Singh, K. M.
1978a Incidence of insect pests on different varieties of sunflower *Helianthus annuus* Linnaeus. *Indian Journal of Entomology* 40:101-103.

Seyedoleslami, H.
1978a Aspects of the temporal and spatial coincidence of the white apple leafhopper (*Typhlocyba pomaria* McAtee. Cicadellidae: Homoptera) and two parasitic Hymenoptera. Ph. D. Dissertation, Michigan State University. 184 pp.

Seyedoleslami, H. and Croft, B. A.
1980a Spatial distribution of overwintering eggs of the white apple leafhopper, *Typhlocyba pomaria*, and parasitism by *Anagrus epos*. *Environmental Entomology* 9(5):624-629.

Shade, R. E., Doscil, M. J. and Maxon, N. P.
1979a Potato leafhopper resistance in glandular-haired alfalfa species. *Crop Science* 19:287-289.

Shagir, Sama and Halteren, P. van
1976a Additional investigations on the application of insecticides to the root-zone of rice plants. *Contributions from the Central Research Institute for Agriculture* (Bogor) No. 18. 15 pp.

Shagir, Sama, Hasanuddin, A. and Suprihatno, B.
1983a Risalah Lokakarya Penelitian Padi, Cibogo, Bogor. 1983. Pp. 111-129.

Shah, A. H., Jhala, R. C. and Patel, G. M.
1984a Studies on the need-based chemical control of the mango borer [sic!] *Amritodus atkinsoni* (Leth.). *International Pest Control* 26:15-17.

Shah, A. H., Jhala, R. C., Patel, G. M. and Patel, S.
1983a Evaluation of effective dose of monocrotophos for the control of mango hopper *Amritodus atkinsoni* Lethierry (Cicadellidae: Homoptera) by injection method and its comparison with foliar application. *Gujarat Agricultural University Research Journal* 9:14-18.

Shah, A. H., Purohit, M. S., Jhala, R. C. and Patel, M. B.
1984a Efficacy of granular insecticides against jassids, aphids, and shoot and fruit borer of brinjal. *Gujarat Agricultural University Research Journal* 9:25-28.

Shaheen, A. H.
1976a Survey and chemical control of insects attacking *Luffa aegyptiaca*. *Agricultural Research Review* 54:143-152.

Shaheen, A. H., Elezz, A. A. and Assem, M. A.
1973a Chemical control of cucurbits pests at Komombo. *Agricultural Research Review* 51(1):103-107.

Shalaby, F., El-Haidari, H. and Derwesh, A. I.
1968a Contribution to the insect fauna of Iraq. *Bulletin, Societe Entomologique d'Egypte* 50:77-89.

Shands, W. A.
1964a Some leafhopper problems in the United States. *Proceedings of the XII International Congress of Entomology* 1:527-528.

Shantha, P. R. and Lakshmanan, M.
1984a Vector-borne MLOs of brinjal little leaf. *Current Science* 53:265-267.

Sharma, B.
1977a A new species of genus *Farynala* Dworakowska 1970, from India (Homoptera: Cicadellidae: Typhlocybinae). *Entomon* (India) 2(2):241-242.
1977b Leafhopper (Homoptera: Cicadellidae) fauna of Jammu Division of J. & K. State. *Entomology Newsletter* 6(11&12):62.
1978a *Elbelus tripunctatus* Mahmood: A new record and description of a new species of *Agnesiella* Dworakowska, from India, along with some generic synonymies (Homoptera, Cicadellidae, Typhlocybinae). *Entomologist's Record and Journal of Variation* 90(1):11-14.
1979a *Indodikra* gen. nov. along with four new Typhlocybinae from Nagaland, India (Auchenorrhyncha: Homoptera: Cicadellidae). *Proceedings of the Indian Science Congress* 66(3):5.
1984a On some Typhlocybinae from Kishtwar Jammu and Kashmir India with description of 5 new species. *Reichenbachia* 22(3):31-40. [In English.]

Sharma, B. and Badan, P.
1985a On genus *Balelutha* [sic!] (Homoptera: Cicadellidae) from Jammu. *Entomotaxonomia* 7(2):149-156.
1985b Light trap catches of macrosteline leafhoppers in relation to climatic factors at Jammu. *Indian Journal of Ecology* 12(1):29-34.

Sharma, B. and Malhotra, Y. R.
1981a Two new species of *Agnesiella* (*Draberiella*) (Homoptera: Cicadellidae: Typhlocybinae) from Jammu and Kashmir infesting *Alnus nitida*. *Entomon* (India) 6(1):41-45.

Sharma, G. K., Intodia, S. C. and Khanvilkar, V.
1981a Developmental studies of *Amrasca devastans* Dist. and *Earias vitella* Febr. on different cotton varieties. *Zeitschrift fur Angewandte Entomologie* 92:529-533.

Sharma, H. C. and Agarwal, P.
1983a Role of some chemical components and leaf hairs in varietal resistance in cotton to jassid, *Amrasca biguttula biguttula* Ishida. *Journal of Entomological Research* 7:145-149.
1983b Relationship between insect population and the proline content in some cotton genotypes. *Indian Journal of Entomology* 45:244-246.

Sharma, O. P. and Gupta, S. C.
1980a On the chromosomes of two species of jassids (Cicadellidae: Homoptera). *Cytobios* 28(113):37-41.
1980b Studies on the chromosomes of four species of the genus *Deltocephalus* (Cicadellidae: Homoptera). *National Academy of Science and Letters* 3(8):255-257.
1980c Karyotypic studies on the two species of jassids (Deltocephalinae: Cicadellidae). *Chromosome Information Service* 29:12-13.

Sharma, S. K., Mathur, Y. K., Verma, J. P. and Batra, R. C.
1968a Outbreaks and new records. India. Jassid pest of pea. *FAO Plant Protection Bulletin* 16(6):115.

Sharonova, M. V.
1969a The species composition of the pests of oil-bearing rose in Moldavia. *Vrednaia i Poleznaia Fauna Bespozvonochnykh Moldavii* 4-5:84-86. [In Russian.]

Shaskolskaya, D. D.
1962a Transovarial'naya Pecedocka Tsikadoi *Psammotettix striatus* L. Virusa Mozaiki Ozimoi Pshenitsy. [The transovarial transmission of winter wheat mosaic virus by the leafhopper *Psammotettix striatus* L.] *Zoologicheskii Zhurnal* 41:717-720.

Shatrughna Singh and Nagaich, B. B.
1979a *Alebroides dravidanus*: A new vector for purple top roll and witches'-broom of the potato. *Journal of the Indian Potato Association* 4:71-73.

Shaw, F. R. and Zeiner, W. H.
1964a The disappearance of dimethoate and SD-7438 from alfalfa. *Journal of Economic Entomology* 57(6):997-998.

Shaw, K. C.
1976a Sounds and associated behavior of *Agallia constricta* and *Agalliopsis novella* (Homoptera: Auchenorrhyncha: Cicadellidae). *Journal of the Kansas Entomological Society* 49(1):1-17.

Shaw, K. C. and Carlson, O. V.
1979a Morphology of the tymbal organ of the potato leafhopper *Empoasca fabae* Harris (Homoptera: Cicadellidae). *Journal of the Kansas Entomological Society* 52(4):701-711.

Shaw, K. C., Vargo, A. and Carlson, O. V.
1973a Sounds and associated behavior of some species of *Empoasca* (Homoptera: Cicadellidae). *Journal of the Kansas Entomological Society* 47(3):284-307.

Shcherbakov, D. Ye.
1981a Diagnostics of the families of the Auchenorrhyncha (Homoptera) on the basis of the wings. 1. Forewings. *Entomologicheskoe Obozrenie* 60(4):828-843. [In Russian]
1982a Diagnostics of the families of the Auchenorrhyncha (Homoptera) on the basis of the wings. 2. Hindwing. *Entomologicheskoe Obozrenie* 61(3):528-536. [In Russian]

Shepard, M.
1984a A splitter device for subsampling small insects. *Journal of Plant Protection in the Tropics* 1:111-115.

Shepard, M., Aquino, G., Ferrer, E. R. and Heinrichs, E. A.
1985a Comparison of vacuum and carbondioxide-cone sampling devices for arthropods in flooded rice. *Journal of Agricultural Entomology* 2:364-369.

Sherman, K. E. and Maramorosch, K.
1977a Feeding of the leafhopper *Macrosteles fascifrons* on an artificial diet containing the aster yellows spiroplasm. *Proceedings of the American Phytopathological Society* 4:197.

Sherman, M. & Tamashiro, M.
1957a Control of insect and mite pests of beans in Hawaii. *Journal of Economic Entomology* 50(3):236-237.

Sherman, M., Tamashiro, M. and Fukunaga, E. T.
1954a Control of greenhouse whitefly and other insects on beans in Hawaii. *Journal of Economic Entomology* 47(3):530-535.

Shibang, Q.
1983a Biological control of insect pests with indigenous natural enemies in the People's Republic of China. *In: 10th International Congress of Plant Protection. Vol. 2. Proceedings of a conference held at Brighton, England.* Pp. 777. Croydon, U.K.

Shibuya, M.
1956a Effect of organic phosphorous insecticides applied for rice stem borer control on the leafhopper-association in paddy fields. *Memoirs, Faculty of Agriculture Kagoshima* 2(2):145-152.

Shikata, E.
1966a Electron microscopic study of multiplication of rice dwarf virus in plant and insect hosts. *In: Papers to be presented at the symposium on plant diseases in the Pacific. XI Pacific Science Congress*, Tokyo, 25-27 August 1966. Pp. 71-79.
1979a Cytopathological changes in leafhopper vectors of plant viruses. *In: Maramorosch, K. and Harris, K. F., eds. Leafhopper Vectors and Plant Disease Agents.* Pp. 309-325. Academic Press, New York. 654 pp.
1979b Rice viruses and MLO's and leafhopper vectors. *In: Maramorosch, K. and Harris, K. F., eds. Leafhopper Vectors and Plant Disease Agents.* Pp. 515-525. Academic Press, New York. 654 pp.

Shikata, E. and Maramorosch, K.
1965a Electron microscopic evidence for the systemic invasion of an insect host by a plant pathogenic virus. *Virology* 27:461-475.
1965b Plant tumor virus in arthropod hosts: Microcrystal formation. *Nature* 208:507-508.
1965c Wound tumor virus in insect hosts inoculated parenterally. *Journal of Applied Physiology* 36:2620.
1965d Electron microscopy of wound-tumor virus in ultrathin sections of plants and insects. *Phytopathology* 55(10):1076.
1967a Electron microscopy of wound tumor virus assembly sites in insect vectors and plants. *Virology* 32(3):363-377.
1967b Electron microscopy of the formation of wound tumor virus in abdominally inoculated insect vectors. *Journal of Virology* 1(5):1052-1073.
1969a Electron microscopy of insect borne viruses *in situ. In: Maramorosch, K., ed. Viruses, Vectors and Vegetation.* Pp. 393-415. Wiley & Sons, New York. 666 pp.

Shikata, E., Maramorosch, K. and Ling, K. C.
1969a Presumptive mycoplasma etiology of yellows diseases. *FAO Plant Protection Bulletin* 17(6):121-128.

Shikata, E., Maramorosch, K., Ling, K. C. and Matsumoto, T.
1968a On the mycoplasma like structures encountered in the phloem cells of American aster yellows, corn stunt, Philippine rice yellow dwarf, and Taiwan sugar cane white leaf diseased plants. *Annals of the Phytopathological Society of Japan* 34:208-209. [In Japanese.]

Shikata, E., Orenski, S. W., Hirumi, H., Mitsuhashi, J. and Maramorosch, K.
1964a Electron micrographs of wound-tumor virus in an animal host and in a plant tumor. *Virology* 23:441-444.

Shikata, E., Orenski, S. W. and Maramorosch, K.
1966a Wound-tumor virus in vector hemolymph. *Phytopathology* 56(6):586.

Shim, J. W.
1978a Inheritance of insecticide resistance in plant- and leaf-hoppers. I. Inherited properties of MEP resistance in small brown plant-hopper (*Laodelphax striatellus* Fallen). *Korean Journal of Plant Protection* 17:75-80.

Shimada, K.
1972a On dorilaid parasites of the green rice leafhopper. *Proceedings of the Association for Plant Protection of Kyushu* 18:41-43. [In Japanese.]
1975a On the insecticide resistance of the green rice leafhopper in Kumamoto Pref. *Proceedings of the Association for Plant Protection of Kyushu* 21:63-64.

Shimada, K. and Araki, M.
1976a Control of the rice dwarf disease by the combination of herbicide and insecticide applications. *Proceedings of the Association for Plant Protection of Kyushu* 22:111-113.

Shinkai, A.
1956a Difference in rice dwarf virus transmitting ability of leafhoppers from various localities. *Annals of the Phytopathological Society of Japan* 21:127. [In Japanese.]
1958a Transovarial passage of rice dwarf virus in *Inazuma dorsalis* (Motsch.). *Annals of the Phytopathological Society of Japan* 23:26.
1959a Transmission of rice yellow dwarf by taiwan-tsumaguro-yokobai. *Annals of the Phytopathological Society of Japan* 25:36. [In Japanese.]
1960a Epidemics of rice dwarf and yellows with special reference to the infectivity of *Nephotettix cincticeps*. *Plant Protection* 14:146. [In Japanese.]
1960b Premature death of *Inazuma dorsalis* Motschulsky which received the rice dwarf virus through eggs. *Annals of the Phytopathological Society of Japan* 25:42. [In Japanese.]
1962a Studies on insect transmission of rice virus diseases in Japan. *Bulletin of the National Institute of Agricultural Sciences* (Series C) 14:1-112.
1964a Studies on insect transmission of sweet potato witches' broom disease in the Ryukyu Islands. *In: Studies on Sweet Potato Witches' Broom Disease in the Ryukyu Islands. Government of Ryukyu Island Economic Department of Agriculture Section Special Research Report.* 44 pp.
1965a Transmission of four rice viruses by leafhoppers. *Annals of the Phytopathological Society of Japan* 31:380-383. [In Japanese.]
1968a Studies on insect transmission of sweet potato witch's broom disease in the Ryukyu Islands. *Review of Plant Protection Research* (Japan) 1:65-72.
1975a Insect vectors and hosts of several witches' brooms and aster yellows. *Annals of the Phytopathological Society of Japan* 40:231-233.
1977a Rice waika, a new virus disease, and problems related to its occurrence and control. *Japan Agricultural Research Quarterly* 11:151-155.

Shiomi, T. and Choi, Y. M.
1983a Occurrence and host range of sickle hare's ear (*Bupleurum falcatum* L.) yellows in Japan. *Annals of the Phytopathological Society of Japan* 49:228-238.

Shiomi, T. and Sugiura, M.
1983a Occurrence and host range of strawberry witches' broom in Japan. *Annals of the Phytopathological Society of Japan* 49:727-730.
1984a Groupings of mycoplasma-like organisms transmitted by the leafhopper vector, *Macrosteles orientalis* Vibaste [sic!], based on host range. *Annals of the Phytopathological Society of Japan* 50:149-157. [In Japanese with English summary.]
1984b Differences among *Macrosteles orientalis*-transmitted MLO, potato purple-top wilt MLO in Japan and aster yellows MLO from USA. *Annals of the Phytopathological Society of Japan* 50:455-460. [In Japanese with English summary.]

Shukla, S. P.
1961a A new type of anatomical association of the jassid Malpighian tubules in the mango-hopper, *Idiocerus atkinsoni* Leth. (Homoptera, Jassidae). *Entomologist's Monthly Magazine* 96(1154-6):169-170.

Shukla, V. D. and Anjaneyulu, A.
1980a Evaluation of systemic insecticides to control rice tungro. *Plant Disease* 64:790-792.
1981a Adjustment of planting date to reduce rice tungro disease. *Plant Disease* 65:409-411.
1981b Plant spacing to reduce rice tungro incidence. *Plant Disease* 65:584-568.
1981c Spread of tungro virus disease in different ages of rice crop. *Zeitschrift fuer Pflanzenkrankheiten und Pflansenschutz* 88:614-620.

1982a Spread of tungro virus disease in irrigated and rainfed rice. *Indian Phytopathology* 35:47-51.

1982b Effects of number of leafhoppers and amount and source of virus inoculum on the spread of rice tungro. *Zeitschrift fuer Pflansenkrankheiten und Pflansenschutz* 89:325-331.

1982c Population and sex ratios of rice green leafhoppers in the paddy field. *Science and Culture* 48:65-66.

Shukla, R. P. and Prasad, V. G.

1984a Evaluation of some insecticides against mango hopper, *Idioscopus clypealis* (Lethierry) (Homoptera: Cicadellidae). *Indian Journal of Agricultural Sciences* 54:677-681.

Shumiya, A., Komura, T., Shaku, I., Takamatsu, M., Kudo, S., Nakajima, Y. and Kato, T.

1984a Breeding for resistance to rice dwarf disease and its mechanism. 8. Breeding of Aichi 421 resistant to rice dwarf disease and green leafhoppers. *Research Bulletin of the Aichi-Ken Agricultural Research Center* No. 16. 14 pp.

Shurovenkov, Y. B.

1964a Sucking pests of cereals. *Zashchita Rastenii ot Vreditelei i Boleznei* 9(4):48-49. [In Russian.]

Sidhu, A. S. and Dhawan, A. K.

1976a Developing economic threshold of spraying against cotton jassid *Amrasca devastans* (Dist.). *Journal of Research, Punjab Agricultural University* 13:186-189.

1980a Incidence of some insect pests on different varieties of cotton. *Journal of Research, Punjab Agricultural University* 17:152-156.

1981a Seasonal abundance of different insect pests on desi cotton (*Gossypium arboreum* L.). *Journal of Research, Punjab Agricultural University* 17:275-281.

Sidhu, A. S. and Simwat, G. S.

1975a Evaluation of some new insecticides for the control of *Amrasca devastans* (Distant) infesting okra, *Abelmoschus esculentus* (L.). *Indian Journal of Entomology* 35:297-299.

Sidhwi, G. S. and Khush, G. S.

1984a Allelic relationship of dominant genes for resistance to green leafhopper in four varieties of rice. *Crop Improvement* 11:142-144.

Sikka, S. M., Sahni, V. M. and Butani, D. K.

1966a Studies on jassid resistance in relation to hairiness of cotton leaves. *Euphytica* 15:383-388.

Silveira-Guido, A. and Carbonell Bruhn, J.

1965a Los insectos enemigos del girasol en el Uruguay. *Boletin, Universidad de la Republica, Universidad Facultad de Agronomia Montevideo* 81:3-64.

Silveira Neto, S., Braz, A. J. B., Zucchi, R. A., Chagas, E. F. and Menezes, M.

1983a Survey of sucking insects on citrus by means of a knapsack suction collector. *Anais da Sociedad Entomologica do Brasil* 12:165-173.

Silvere, A. P. and Giorgadze, D.

1978a Mycoplasma and virion-like structures in the cells of salivary gland of leafhopper *Hishimonus sellatus* Uhler infected with mulberry dwarf disease. *Eesti NSV Teaduste Akadeemia Toimetised (Biologia)* 27(1):9-15.

Simmons, A. M., Godfrey, L. D. and Yeargan, K. V.

1985a Ovipositional sites of the potato leafhopper (Homoptera: Cicadellidae) on vegetative stage of soybean plants. *Environmental Entomology* 14(2):165-169.

Simmons, A. M., Pass, B. C. and Yeargan, K. V.

1984a Influence of selected legumes on egg production, adult survival, and ovipositional preference by the potato leafhopper. *Journal of Agricultural Entomology* 1:311-317.

Simmons, A. M., Yeargan, K. V. and Pass, B. C.

1984a Development of the potato leafhopper on selected legumes. *Transactions of the Kentucky Academy of Science* 45:33-35.

Simonet, D. E.

1978a Population studies of the potato leafhopper, *Empoasca fabae* (Harris), on alfalfa, *Medicago sativa* L. *Dissertation Abstracts International* (B) 39(6):2654-2655.

Simonet, D. E. and Pienkowski, R. L.

1977a Sampling and distribution of potato leafhopper eggs in alfalfa stems. *Annals of the Entomological Society of America* 70:933-936.

1977b Sampling and distribution of potato leafhopper nymphs in alfalfa. *Journal of the New York Entomological Society* 85(4):200 (Abstract).

1979a A sampling program developed for potato leafhopper nymphs, *Empoasca fabae* (Homoptera: Cicadellidae), on alfalfa. *Canadian Entomologist* 111:481-486.

1979b Impact of alfalfa harvest on potato leafhopper populations with emphasis on nymphal survival. *Journal of Economic Entomology* 72:428-431.

1980a Temperature effect on development and morphometrics of the potato leafhopper. *Environmental Entomology* 9(6):798-805.

Simonet, D. E., Pienkowski, R. L., Martinez, D. G. and Blakeslee, R. D.

1978a Laboratory and field evaluation of sampling techniques for nymphal stages of the potato leafhopper. *Journal of Economic Entomology* 71:840-842.

1979a Evaluation of sampling techniques and development of a sampling program for potato leafhopper adults on alfalfa. *Environmental Entomology* 8:398-399.

Simonet, D. E., Pienkowski, R. L. and Wolf, D. D.

1979a Factors associated with oviposition by the potato leafhopper on alfalfa. *Journal of the Georgia Entomological Society* 14:315-317.

Simons, J. N.

1958a Los insecticidas sistemicos en el control de las plagas del algodonero en la costa del Peru. *Boletin, Estacion Experimental Agricola "La Molina"* 71(3-4):41.

1962a The pseudo-curly top diseases in South Florida. *Journal of Economic Entomology* 55(3):358-363.

Singh, B. B., Hadley, H. H. and Bernard, R. L.

1971a Morphology of pubescence in soybeans and its relationship to plant vigor. *Crop Science* 11(1):13-16.

Singh, J., Sinha, M. M. and Prasad, A. R.

1977a Light trap catches of paddy pests at Agricultural Research Institute Farm. Kholi, Muzaffarpur (Bihar). *Entomologists' Newsletter* 7:35.

Singh, K. G.

[Undated] Recent progress in rice insect research in Malaysia. Symposium on rice insects. Proceedings of a Symposium on Tropical Agriculture Research, 19-24 July 1971. *Tropical Agriculture Research Series* 5:109-121.

1971a Studies on transmission of penyakit merah virus disease of rice. *Malayan Agricultural Journal* 48:93-103.

1971b Transmission studies on orange leaf virus disease of rice in Malaysia. *Malaysian Agricultural Journal* 48(2):122-132.

1979a Resistance to rice blast disease (*Pyricularia oryzae* Cav.) in penyakit merah virus infested plants. *Malaysian Agricultural Journal* 52:65-67.

Singh, K. G., Saito, Y. and Nasu, S.

1970a Mycoplasma-like structures in rice plant infected with 'padi jantan' disease. *Malaysian Agricultural Journal* 47(3):333-337.

Singh, K. M. and Singh, R. N.

1978a Succession of insect-pests in green gram and black gram under dryland conditions at Delhi. *Indian Journal of Entomology* 39:365-370.

Singh, K. M., Singh, R. N., Kailashnathan, K. and Rao, G. G. S. N.

1975a Succession of pests in cowpea (*Vigna sinensis* Savi) under dryland conditions at Delhi. *Indian Journal of Entomology* 36:69-72.

Singh, M. P. Satyavir and Lodha, S.

1984a Insect pest and disease management in arid zone. *Indian Farming* 34:71, 73, 79.

Singh, R. N. and Singh, K. M.
1978a Succession of insect-pests in early varieties of red gram, *Cajanus cajan* (L.) Millsp. *Indian Journal of Entomology* 40:1-6.

Singh, R. P. and Parshad, B.
1967a An alternate host for *Anagrus empoascae* Dozier in India. *Indian Journal of Entomology* 29(3):306.

Singh, S. R.
1967a Systemic insecticides solve the problem of leaf hoppers on paddy. *Indian Farming* 17(3):24.
1976a Co-ordinated minimum insecticide trials: Yield performance of insect resistant cowpea cultivars from IITa compared with Nigerian cultivars. *Tropical Grain Legume Bulletin* No. 5, p. 4.
1977a Cowpea cultivars resistant to insect pests in world germplasm collection. *Tropical Grain Legume Bulletin* No. 9, pp. 3-7.

Singh, S. R., Williams, R. J., Rachie, K. O., Watt, E., Nangju, D., Wien, H. C. and Luse, R. A.
1976a Vita-5 cowpea. *Tropical Grain Legume Bulletin* No. 5, pp. 41-42.

Singh, T. H., Gurdip, S., Sharma, K. P. and Gupta, S. P.
1972a Resistance of cotton (*Gossypium hirsutum* L.) to cotton-jassid, *Amrasca devastans* (Distant) (Homoptera: Jassidae). *Indian Journal of Agricultural Sciences* 42(5):421-425.

Sinha, R. C.
1963a Effect of age of vector and of abdomen punctures on virus transmission. *Phytopathology* 53:1170-1173.
1964a Distribution of wound-tumor virus antigens in internal organs of an insect vector [*Agallis constricta* Van D.]. *Phytopathology* 54(4):908.
1965a Sequential infection and distribution of wound-tumor virus in the internal organs of vectors after ingestion of virus. *Virology* 26:673-686.
1965b Recovery of potato yellow dwarf virus from hemolymph and internal organs of an insect vector. *Virology* 27(1):118-119.
1967a Response of wound-tumor virus infection in insects to vector age and temperature. *Virology* 31(4):746-748.
1968a Mechanism of persistence of wound tumor virus in its vector. *1st International Congress of Plant Pathology*, London. Pp. 185.
1968b Recent work on leafhopper-transmitted viruses. *Advances in Virus Research* 13:181-223.
1968c Serological detection of wheat striate mosaic virus in extracts of wheat plants and vector leafhoppers. *Phytopathology* 58(4):452-455.
1969a Localization of viruses in vectors: Serology and infectivity tests. In: Maramorosch, K., ed. *Viruses, Vectors and Vegetation*. Pp. 379-391. Wiley & Sons, New York. 666 pp.
1970a *Elymana virescens*, a newly described vector of wheat striate mosaic virus. *Canadian Plant Disease Survey* 50:118-120.
1973a Virus-like particles in salivary glands of leafhoppers *Endria inimica*. *Virology* 51:244-246.
1973b Viruses and leafhoppers. In: Gibbs, J. A., ed. *Viruses and Invertebrates*. Pp. 493-511. North Holland, New York. 673 pp.
1974a Purification of mycoplasma-like organisms from China aster plants affected with clover phyllody. *Phytopathology* 64:1156-1158.
1979a Chemotherapy of mycoplasmal plant diseases. In: Whitcomb, R. F. and Tully, J. G., eds. *The Mycoplasmas*. Pp. 309-335. Academic Press, New York.
1981a Vertical transmission of plant pathogens. In: McKelvey, J. J., Eldridge, B. F. and Maramorosch, K., eds. *Vectors of Disease Agents: Interaction with Plants, Animals and Man*. Pp. 109-121. Praeger, New York. 229 pp.
1983a Relative concentration of mycoplasma-like organisms in plants at various times after infection with aster yellows. *Canadian Journal of Plant Pathology* 5:7-10.
1984a Transmission mechanisms of mycoplasma-like organisms by leafhopper vectors. *Current Topics in Vector Research* 2:93-110.

Sinha, R. C. and Black, L. M.
1962a Studies on the smear technique for detecting virus antigens in an insect vector by use of fluorescent antibodies. *Virology* 17:582-587.
1963a Wound-tumor virus antigens in the internal organs of an insect vector. *Virology* 21:183-187.
1965a Effects of age on the resistance of the vector's intestinal tract to infection by wound tumor virus. *Phytopathology* 55(10):1077.
1967a Distribution of aster yellows virus in the vector *Macrosteles fascifrons* and the sequences of infection sites. *Phytopathology* 57:831.

Sinha, R. C. and Chiykowski, L. N.
1966a Multiplication of aster yellows virus in a nonvector leafhopper. *Phytopathology* 56(8):902.
1967a Multiplication of aster yellows virus in a nonvector leafhopper. *Virology* 31(3):461-466.
1967b Multiplication of wheat striate mosaic virus in its leafhopper vector *Endria inimica*. *Virology* 32(3):402-405.
1967c Initial and subsequent sites of aster yellows virus infection in a leafhopper vector. *Virology* 33(4):702-708.
1968a Distribution of clover phyllody virus in the leafhopper *Macrosteles fascifrons* (Stal). *Acta Virologica* (Prague) 12:546-550.
1969a Synthesis, distribution and some multiplication sites of wheat striate mosaic virus in a leafhopper vector. *Virology* 38:679-684.
1980a Transmission and morphological features of mycoplasmalike bodies associated with peach X-disease. *Canadian Journal of Plant Pathology* 2:119-124.

Sinha, R. C. and Paliwal, Y. C.
1970a Localization of a mycoplasma-like organism in tissues of a leafhopper vector carrying clover phyllody agent. *Virology* 40(3):665-672.

Sinha, R. C. and Petersen, E. E.
1972a Uptake and persistence of oxytetracycline in aster plants and vector leafhoppers in relation to inhibition of clover phyllody agent. *Phytopathology* 62:377-383.

Sinha, R. C. and Reddy, D. V. R.
1964a Improved fluorescent smear technique and its application in detecting virus antigens in an insect vector. *Virology* 24:626-634.

Sinha, R. C., Reddy, D. V. R. and Black, L. M.
1964a Survival of insect vectors after examination of hemolymph to detect virus antigens with fluorescent antibody. *Virology* 24:666-667.

Sinha, R. C. and Shelley, S.
1965a The transovarial transmission of wound tumor virus. *Phytopathology* 55:324-327.

Siva Shankara Sastry, K. S.
1958a Some observations on *Tettigella spectra* (Distant) on paddy in Mysore. *Mysore Agriculture Journal* 33(1):11-12.

Sivaramakrishnan, V. R. and Sen-Sarma, P. K.
1978a Experimental transmission of spike disease of sandal, *Santalum album*, by the leafhopper, *Nephotettix virescens* (Distant) (Homoptera: Cicadellidae.) *Indian Forester* 104(3):202-205.

Siwi, B. H.
1983a Penelitian padi dalam dasawarsa 1980-an. Badan Penelitian dan Pengembangan Pertanian, Lokakarya, Ciawi, Bogor. Pp. 9-15.

Siwi, B. H. and Khush, G. S.
1977a New genes for resistance to the green leafhopper in rice. *Crop Science* 17:17-20.

Siwi, B. H., Sidhu, G. S. and Khush, G. S.
1978a Genetic analysis of rice cultivar Ptb 8 for resistance to green leafhopper and brown planthopper. *Contributions from the Central Research Institute for Agriculture* 43:7-10.

Siwi, S. S.
1979a Identification of green leafhopper in Indonesia. *Kongres Entomologi* 3:1-10.

1985a Studies on green leafhoppers genus *Nephotettix* Matsumura in Indonesia with special reference to morphological aspects. Thesis Dissertations, Tokyo University of Agriculture. 238 pp. Unpublished.

Siwi, S. S. and Roechan, M.
1983a Species composition and distribution of green leafhoppers, *Nephotettix* spp., and the spread of rice tungro disease in Indonesia. *In: Knight, W. J., Pant, N. C., Robertson, T. S. and Wilson, M. R., eds. Proceedings of the 1st International Workshop on Biotaxonomy, Classification and Biology of Leafhoppers and Planthoppers (Auchenorrhyncha) of Economic Importance.* Pp. 263-276. London, 4-7 October 1982. Commonwealth Institute of Entomology, 56 Queen's Gate, London SW7 5JR. 500 pp.

Sleesman, J. P.
1940a Developing potato resistance to leafhoppers. *Ohio Agriculture Experiment Station Bulletin* 617:26-27.
1945a Search for a leafhopper resistant potato. *Ohio Agriculture Experiment Station Bulletin* 659:114-115.
1956a New insecticides for controlling insects on potatoes. *Proceedings of the Ohio Vegetable and Potato Growers Association* 41:115-118.
1957a Soil treatments for the control of foliage pests. *Proceedings of the North Central Branch of the Entomological Society of America* 12:43-44.
1958a Where are we with systemic insecticides? *Vegetable Growers Association of America Annual Report* 1958:23-26.
1970a Insect resistance in vegetables. *Ohio Report on Research and Development* 55:96.

Sleesman, J. P. and Hedden, O.
1958a Thimet provides built-in protection against potato leafhopper. *Ohio Farm Home Research* 43:39.

Slogteren, D. H. M. van and Muller, P. J.
1972a 'Lissers', a yellows disease in hyacinths, apparently caused by a mycoplasma. Fakulteit van de Landbouwwetenschappen, Gent. Vierentwintigste International Symposium over Fytofarmacie en Fytiatrie, 9 Mei 1972. *Mededelingen Fakulteit Landbouwwetenschappen*, Gent, 1972 37(2):450-457. [In English with German summary.]
1972b Pathogens, probably mycoplasmas, isolated from gladiolus plants with symptoms of aster yellows and from hyacinths with the so-called "Lisser" syndrome and their behavior on the test plant *Vinca rosea. Acta Botanica Neerlandica* 21(1):111 (Abstract).

Slykhuis, J. T.
1951a The relation of leafhoppers to certain streak symptoms on wheat in South Dakota. *Plant Disease Reporter* 35(10):439-440.
1953a Striate mosaic, a new disease of wheat in South Dakota. *Phytopathology* 43(10):537-540.
1961a The causes and distribution of mosaic diseases of wheat in Canada in 1961. *Canadian Plant Disease Survey* 41:330-343.
1963a Vector and host relations of North American wheat striate mosaic virus. *Canadian Journal of Botany* 41:1171-1185.

Slykhuis, J. T. and Sherwood, P. L.
1964a Temperature in relation to the transmission and pathogenicity of wheat striate mosaic virus. *Canadian Journal of Botany* 42:1123-1133.

Smith, A. J., McCoy, R. E. and Tsai, J. H.
1981a Maintenance in vitro of aster yellows mycoplasmalike organism. *Phytopathology* 71:819-822.

Smith, D.
1962a Alfalfa cutting practices. Part 1. *Wisconsin Agricultural Experiment Station Research Report* 11. 11 pp.

Smith, D. and Medler, J. T.
1959a Influence of leafhoppers on the yield and chemical composition of alfalfa hay. *Agronomy Journal* 51(2):118-119.

Smith, D. S. and Littau, V. C.
1960a The chemical nature of brochosomes, an excretory product of leafhoppers (Cicadellidae, Homoptera). *Journal of Histochemistry and Cytochemistry* 8(5):312.
1960b Cellular specialization in the excretory epithelia of an insect *Macrosteles fascifrons* Stal (Homoptera). *Journal of Biophysical and Biochemical Cytology* 8:103-133.

Smith, F. F. and Brierley, P.
1956a Insect transmission of plant viruses. *Annual Review of Entomology* 1:299-322.

Smith, J. W., Jr.
1970a Studies on the acoustics of the beet leafhopper, *Circulifer tenellus* (Baker) (Homoptera: Cicadellidae). Doctoral Dissertation, University of California, Riverside. 153 pp.

Smith, J. W. and Georghiou, G. P.
1972a Morphology of the tymbal organ of the beet leafhopper, *Circulifer tenellus. Annals of the Entomological Society of America* 65(1):221-226.

Smith, J. W., Jr., Sams, R. L., Agnew, C. W. and Simpson, C. E.
1985a Methods of estimating damage and evaluating the reaction of selected peanut cultivars to the potato leafhopper, *Empoasca fabae* (Homoptera: Cicadellidae). *Journal of Economic Entomology* 78(5):1059-1062.

Smith, K. M.
1957a *A Textbook of Plant Virus Diseases.* 2nd ed. London. 652 pp.
1958a Transmission of plant viruses by arthropods. *Annual Review of Entomology* 3:469-482.

Smith, R. C. and Franklin, W. W.
1961a Research notes on certain species of alfalfa insects at Manhattan (1904-1956) and at Fort Hays, Kansas (1948-1953). *Kansas Agricultural Experiment Station Progress Report* 54. 121 pp.

Smith, S. M. and Ellis, C. R.
1982a Sampling the potato leafhopper (Homoptera: Cicadellidae) on alfalfa. *Proceedings of the Entomological Society of Ontario* 113:35-41.
1983a Economic importance of insects on regrowths of established alfalfa fields in Ontario. *Canadian Entomologist* 115:859-868.

Smol'yannikov, V. V.
1980a Homopterous insects as pests of orchards. *Zashchita Rastenii* 4:62-63.

Smreczynski, S.
1955a Supplement aux "Materiaux pour la faune d'Hemipteres de Pologne." *Fragmenta Faunistica* 7(5):209-211. [In Polish with English summary.]

Snodgrass, R. E.
1957a A revised interpretation of the external reproductive organs of male insects. *Smithsonian Miscellaneous Collections* 135(6):1-60.

Soehardjan, M.
1971a Recent progress in rice insect research in Indonesia. Symposium on Rice Insects. Proceedings of a Symposium on Tropical Agriculture Research, 19-24 July 1971. *Tropical Agriculture Research Series* 5:99-108.
1973a Observations of leaf and planthoppers on rice in West Java. *Central Research Institute of Agriculture* (Bogor) 3:1-10.

Soejitno, J., Breden, G., van and Suartini
1974a Biology and mass rearing of rice insects. *In: Agricultural Cooperation Indonesia—The Netherlands. Research Reports 1968-1974.* Section II. Technical contributions. Pp. 139-145.

Sogawa, K.
1965a Studies on the salivary glands of rice plant leafhoppers. I. Morphology and histology. *Japanese Journal of Applied Entomology and Zoology* 9:275-290.
1967a Chemical nature of the sheath materials secreted by leafhoppers (Homoptera). *Applied Entomology and Zoology* 2:13-21.
1967b Studies on the salivary glands of rice plant leafhoppers. 2. Origins of the structural precursors of the sheath material. *Applied Entomology and Zoology* 2:195-202.
1968a Studies on the salivary glands of rice plant leafhoppers. III. Salivary phenolase. *Applied Entomology and Zoology* 13:13-25.

1968b Studies on the salivary glands of rice plant leafhoppers. 4. Carbohydrase activities. *Applied Entomology and Zoology* 3:67-73.

1973a Feeding of the rice plant- and leafhoppers. *Review of Plant Protection Research* 6:31-43.

1976a Rice tungro virus and its vectors in tropical Asia. *Review of Plant Protection Research* 9:21-46.

Sogawa, K. and Saito, A.

1981a The green rice leafhopper, *Nephotettix cincticeps* Uhler (Hemiptera: Deltocephalidae), populations with differential reactions to rice varieties. *Japanese Journal of Applied Entomology and Zoology* 25:280-285. [In Japanese with English summary.]

1983a Differences in morphological and physiological characters between Joetsu and Chikugo green rice leafhopper populations with differential reactions to rice varieties. *Japanese Journal of Applied Entomology and Zoology* 27(1):22-27.

Sohi, A. S.

1972a *Cestius (Cestius) indicus*, new species (Homoptera: Cicadelliae) from Punjab, India. *Oriental Insects* 6(1):131-132.

1975a Taxonomy of the typhlocybines (Homoptera, Cicadellidae, Typhlocybinae) of North Western India. Ph. D. Dissertation, Punjab University. 123 pp.

1976a New combinations of some typhlocybines (Homoptera, Cicadellidae, Typhlocybinae) from India. *Entomon* (India) 1(2):203-205.

1976b Indian cotton jassid—*Amrasca biguttula biguttula* (Ishida). *Entomologist's Record and Journal of Variation* 88:236-237.

1977a New genera and species of Typhlocybinae (Homoptera: Cicadellidae) from north-western India. *Oriental Insects* 11(3):347-362. [New genera: *Apetiocellata* and *Plumosa*.]

1977b New records of some Typhlocybinae (Homoptera: Cicadellidae) from India. *Entomon* (India) 2(1):101-102.

1977c Typhlocybines (Homoptera, Cicadellidae, Typhlocybinae) as pests of plants in north-western India. *2nd Oriental Entomology Symposium*, Loyola College, Madras, 23-27 March 1977. Pp. 99-100.

1978a Keys to the tribes and Indian species of Dikraneurini of subfamily Typhlocybinae (Homoptera, Cicadellidae). *Journal of Research Punjab Agricultural University* 15(2):182-186.

1983a The Oriental Typhlocybinae with special reference to the pests of cotton and rice: A review. *In: Knight, W. J., Pant, N. C., Robertson, T. S. and Wilson, M. R., eds. Proceedings of the 1st International Workshop on Biotaxonomy, Classification and Biology of Leafhoppers and Planthoppers (Auchenorrhyncha) of Economic Importance*. Pp. 49-74. London, 4-7 October 1982. Commonwealth Institute of Entomology, 56 Queen's Gate, London SW7 5JR. 500 pp.

Sohi, A. S., Bindra, O. S. and Deol, G. S.

1974a Studies on the control of the brinjal little-leaf disease and insect pests of brinjal. *Indian Journal of Entomology* 36(4):362-364.

Sohi, A. S. and Dhaliwal, C. S.

1985a Comparative biology of *Hishimonus phycitis* (Distant) (Homoptera: Cicadellidae) on different host plants. *Annals of Biology* (Ludhiana) 1(2):165-178.

Sohi, A. S. & Dworakowska, I.

1979a New species of *Alebroides* Matsumura from India and Tibet (Auchenorrhyncha, Cicadellidae, Typhlocybinae). *Entomon* (India) 4(4):367-372.

1981a The Typhlocybinae fauna of India. *In: Manali, H. P., ed. Workshop on Advances in Insect Taxonomy in India and the Orient*. Pp. 50-51. Association for the Study of Oriental Insects, New Delhi. 106 pp.

1984a A review of the Indian Typhlocybinae (Homoptera: Cicadellidae) from India. *Oriental Insects* 17:159-213.

Sohi, A. S. and Kapoor, V. C.

1973a A note on the cicadellid *Erythroneura cassiae* Ahmed. *Entomologist's Record and Journal of Variation* 85(9):217-218.

1973b New record of *Exitianus taeniaticeps* (Kirschbaum) with description of a new species of *Erythroneura* Fitch (Cicadellidae: Homoptera). *Entomologist's Record and Journal of Variation* 85(12):294-297.

1974a A new species of the genus *Erythroneura* Fitch (Homoptera: Cicadellidae) from India. *Indian Journal of Entomology* 36(1):42-43.

Sohi, A. S. and Sandhu, P. K.

1971a *Arborifera*—a new subgenus of *Arboridia* Zachv. (*Typhlocybinae, Cicadellidae*) from Punjab, India, with description of its immature stages. *Bulletin de la Academie Polonaise des Sciences (Serie des Sciences Biologiques)* 19(6):401-406.

Sohi, A. S., Viraktamath, C. A. and Dworakowska, I.

1980a *Kalkiana bambusa*, gen. et sp. nov. (Homoptera: Cicadellidae), a dikraneurine leafhopper breeding on bamboo in northern India. *Oriental Insects* 14(3):279-281.

Sohi, B. S., Bindra, O. S. and Sohi, A. S.

1975a Effectiveness of insecticidal sprays against grapevine cicadellid, *Arboridia (Arborifera) viniferata* Sohi & Sandhu in the Punjab, India. *Indian Journal of Horticulture* 32(1&2):68-70.

Sohi, G. S.

1964a Pests of cotton. *In: Pant, N. C., Prasad, S. K., Vishnoi, H. S., Chatterji, S. N. and Varma, B. K., eds. Entomology in India*. Pp. 111-148. The Entomological Society of India. 529 pp.

Somadikarta, S., Kadarsan, S. and Djajasasmita, M.

1966a Primary type specimens of the Museum Zoologicum Bogoriense (III). *Treubia* 27:33-44.

Somchoudhury, A. K.

1981a Seasonal fluctuations of *Amblyseius delhiensis* (Acari: Phytoseiidae), a predator on eggs of cotton jassid. *In: Channabasavanna, G. P. Contributions to Acarology in India* (Proceedings of the 1st All India Symposium in Acarology, 23-25 April 1979, Bangalore). Pp. 179-184. Acarological Society of India, Bangalore, India. 256 pp.

Son, B. I.

1973a *Nephotettix cincticeps* (Uhler) rice green leafhopper. *In: Kim, Y. S. and Lamey, H. A., eds. Literature Review of Korean Rice Pests*. Pp. 12-14. Institute of Agricultural Science, Office of Rural Development, Suweon. UNDP/FAO/Korea. 46 pp.

1973b *Recilia dorsalis* (Motschulsky) zigzag-striped leafhopper. *In: Kim, Y. S. and Lamey, H. A., eds. Literature Review of Korean Rice Pests*. Pp. 27-28. Institute of Agricultural Science, Office of Rural Development, Suweon. UNDP/FAO/Korea. 46 pp.

Song, Y. H.

1978a The effect of temperature and daylength conditions on the growth and fecundity of green rice leafhopper, *Nephotettix cincticeps* Uhler. *Korean Journal of Plant Protection* 17:187-191.

Sonku, U.

1973a Studies on the varietal resistance to rice stripe disease, the mechanism of infection and multiplication of the causal virus in plant tissues. *Bulletin of the Chugoku National Agricultural Experiment Station* 8:1-86. [In Japanese with English summary.]

Soon, L. G. and Guan, G. K.

1968a Leafhopper transmission of a virus disease of rice locally known as "Padi Jantan" in Krian, Malaysia. *Malaysian Agriculture Journal* 46(4):435-450.

Soos, A.

1956a Revision und Erganzungen zum Homopteren-Teil des Werkes "Fauna Regni Hungariae." III. *Folia Entomologica Hungarica* (n. s.) 9:418.

Soper, F. J., McIntosh, M. S. and Elden, T. C.

1984a Diallel analyses of potato leafhopper resistance among selected alfalfa clones. *Crop Science* 24:667-670.

Soper, R. S.

1985a Pathogens of leafhoppers and planthoppers. *In: Nault, L. R. and Rodriguez, J. G., eds. The Leafhoppers and Planthoppers*. Pp. 469-488. Wiley & Sons, New York. 500 pp.

Sorenson, E. L. and Horber, E.
1974a Selecting alfalfa seedlings to resist the potato leafhopper. *Crop Science* 14(1):85-86.

Sorenson, J. T., Kinn, D. N., Doutt, R. L. and Cate, J. R.
1976a Biology of the mite, *Anystis agilis* (Acari: Anystidae): A California vineyard predator. *Annals of the Entomological Society of America* 69:905-910.

Soto, P. E.
1978a A new vector of maize streak virus. *East African Agricultural and Forestry Journal* 44:70-71.

Soto, P. E., Perez, A. T. and Buddenhangen, I. W.
1976a Survey of insect pests and disease of rice in different ecological zones in Nigeria. *Rice Entomology Newsletter* No. 4, pp. 35-36.

Soto, P. W. and Siddiqui, Z.
1978a Insect pests and rice production in Africa. In: Buddenhagen, I. W. and Bersley, G. J., eds. *Rice in Africa*. Pp. 175-179. Proceedings of the Conference Held at I.I.T.A., Ibadan, Nigeria, 1977. Academic Press, New York. 356 pp.

Southern, P. S.
1979a A taxonomic study of the leafhopper genus *Empoasca* (Homoptera: Cicadellidae) in eastern Peru. Doctoral Dissertation, North Carolina State University, Raleigh. 326 pp.

1982a A taxonomic study of the leafhopper genus *Empoasca* (Homoptera: Cicadellidae) in eastern Peru. *North Carolina Agricultural Research Service Technical Bulletin* 272. 194 pp.

Southwood, T. R. E. and Pleasance, H. J.
1962a A hand-operated suction apparatus for the extraction of arthropods from grassland and similar habitats, with notes on other models. *Bulletin of Entomological Research* 53:125-128.

Spendlove, R. S., Whitcomb, R. F. and Jensen, D. D.
1967a Enhancement of infectivity of a mammalian virus preparation after injection into insects. *Virology* 34(1):182-184.

Sperka, C. and Freytag, P. H.
1975a Auchenorrhynchous hosts of mermithid nematodes in Kentucky. *Transactions of the Kentucky Academy of Science* 36(3&4):57-62.

Sprague, G. F. and Dahms, R. G.
1972a Development of crop resistance to insects. *Journal of Environmental Quality* 1(1):28-34.

Squire, F. A.
1972a Entomological problems in Bolivia. *Pest Articles and News Summaries* 18(3):249-268.

1972b Insect pests of grain crops. In: Squire, F. A. Entomological problems in Bolivia. *Pest Articles and News Summaries* 18:253-254.

Sridhar, R., Mohanty, S. K. and Anjaneyulu, A.
1972a General insect pests. In: Squire, F. A. Entomological problems in Bolivia. *Pest Articles and News Summaries* 18:251-252.

1978a Physiology of rice tungro virus disease: Increased cytokinin activity in tungro-infected rice cultivars. *Physiologia Plantarum* 43:363-366.

Sriharan, S. and Garg, A. K.
1975a Assessment of populations of white-backed plant hopper and green leaf-hopper by light trap. *Entomologists' Newsletter* 5:2-3.

1976a Effectiveness of granular placement of carbofuran, phorate and sevidol on the absorption by rice plant and the control of green leaf-hopper. *Indian Journal of Agricultural Sciences* 44:540-544.

Srinivasan, K.
1978a Studies on the relationship between the mycoplasma causing little leaf of brinjal and its vector *Cestius phycitis* (Dist.) (Homptera: Cicadellidae). *Thesis Abstracts Haryana Agricultural University* 4(3):200.

Srinivasan, K. and Chelliah, S.
1980a The mechanism of preference of the leafhopper vector, *Hishimonus phycitis* (Distant), for egg plants infected with little leaf disease. *Proceedings Indian National Science Academy* (B) 46(6):786-796.

1981a Preference of the leafhopper vector *Hishimonus phycitis* (Distant) for egg plants infected with little leaf mycoplasma with reference to certain biochemical and histological changes. In: Govindu, H. C., Maramorosch, K., Raychaudhuri, S. P and Miniyappa, V., eds. *Mycoplasma and Allied Pathogens of Plants, Animals and Human Beings*. Pp. 14-21. University of Agricultural Sciences, Bangalore. 117 pp.

Srinivasan, K. and Krishnakumar, N. K.
1983a Studies on the extent of loss and economics of pest management in okra. *Tropical Pest Management* 29:363-370.

Srinivasan, K., Krishnakumar, N. K., Ramachander, P. R. and Krishnamurthy, A.
1981a Spatial distribution pattern of leafhopper (*Amrasca biguttula biguttula* Ishida) on okra (*Abelmoschus esculentus* Moench). *Entomon* 6(3):261-264.

Srinivasan, S.
1980a Ripcord—a synthetic pyrethroid against rice tungro virus. *International Rice Research Newsletter* 5(4):17.

1980b Varietal and insecticidal trial against rice tungro virus. *International Rice Research Newsletter* 5(6):11.

Srisamudh, N. and Rejesus, R. S.
1972a Evaluation of PARC 1,2, C4-63G and IR20 rice varieties for tungro and leafhopper resistance. *Araneta Research Journal* 19(2):115-136.

Srivastava, B. K.
1958a On the external morphology of *Idiocerus clypealis* Leth. (Homoptera: Jassidae). *Beitraege zur Entomologie* 8(5&6):732-744.

Srivastava, O. S.
1973a Effect of insecticidal treatment on the yield of Bragg variety of soy bean. *JNKVV Research Journal* 7(2):96-97.

Srivastava, O. S., Malik, D. S. and Thakur, R. C.
1972a Estimation of losses in yield due to the attack of arthropod pests in soybean. *Indian Journal of Entomology* 33(2):224-225.

Srivastava, R. P. and Bisaria, A.
1981a Life history of *Cicadulina* sp. (Homptera: Cicadellidae) and comparative susceptibility of rice varieties to its attack. *Indian Veterinary and Medical Journal* 5(1):46.

Srivastava, R. P., Mathur, S.B. and Sahani, M. L.
1965a An entomogenous fungus *Entomophthora* sp. (Entomophthoraceas: Entomophthorales) pathogenic on *Empoasca devastans* Distant (Jassidae: Hemiptera), a bhindi jassid [in India]. *Indian Journal of Entomology* 27(3):367-368.

Srivastava, R. P. and Misra, S. M.
1981a Study of life cycle of *Exitianus* sp. (family Cicadellidae) and effect of temperature variations on its life cycle. *Indian Veterinary and Medical Journal* 5(1):46.

Stafford, E. M., Jensen, F. L. and Kido, M.
1960a Control of the grape leaf folder in California. *Journal of Economic Entomology* 53(4):531-534.

Stafford, E. M. and Kido, M.
1969a Newer insecticides for the control of grape insect and spider mite pests. *California Agriculture* 23(4):6-8.

Stafford, E. M. and Summer, F. M.
1963a Insects not known to occur in the United States. *Co-operative Economic Insect Report* 12[suppl.]:24.

Stanarius, A., Kleinhempel, H. and Muller, H. M.
1980a Electron microscope studies on the behaviour of mycoplasmas of different origin after injection into the cicadellid *Euscelis plebejus* Fall. (Homoptera: Jassidae). *Archiv fur Phytopathologie und Pflanzenshutz* 16:233-249.

Stanarius, A., Muller, H. M., Kleinhempel, H. and Spaar, D.
1976a Electron-microscopic evidence of mycoplasma-like organisms in the salivary glands of *Euscelis plebejus* Fall. after injection of crude sap from proliferation-affected *Catharanthus roseus* (L.) G. Don. *Archiv fur Phytopathologie und Pflanzenshutz* 12:245-252.

Staples, R., Jansen, W. P. and Andersen, L. W.
1970a Biology and relationship of the leafhopper *Aceratagallia calcaris* to yellow vein disease of sugarbeets. *Journal of Economic Entomology* 63:460-463.

Stathopoulos, D. G.
1967a Studies on the identification and bio-ecology of *Aphis* spp., *Thrips tabaci* Lind., *Bemisia tabaci* Genn., *Empoasca* sp. and *Tetranychus urticae*. *Thessaloniki* 3:41-49. [In Greek with English summmary.]

Statz, G.
1950a Cicadariae (Zikaden) aus den oberoligocanen ablagerunen von Rott. *Palaeontographica* 98(1-4):1-46.

Steel, W. O. and Woodroffe, G. E.
1969a The entomology of the Isle of Rhum National Nature Reserve. *Transactions of the Society for British Entomology* 18(6):91-167.

Steinhauer, A. L., Blickenstaff, C. C. and Adler, V. E.
1962a Experiments on alfalfa insect control in Maryland. *Journal of Economic Entomology* 55(5):718-722.

Stevens, M. M.
1985a Notes on the status, taxonomy and biology of the genus *Eurymeloides* Ashmead (Homoptera: Cicadellidae: Eurymelinae). *General Applied Entomology* 17:9-16.

Stevens, W. A. and Spurdon, C.
1972a Green petal disease of primula. *Plant Pathology* 21(4):195-196.

Stevenson, A. B. and Pree, D. J.
1985a Toxicity of insecticides to the aster leafhopper, *Macrosteles fascifrons* (Homoptera: Cicadellidae), in the laboratory and field. *Proceedings of the Entomological Society of Ontario* 115:89-92.

Stevenson, F. J.
1956a Breeding varieties of potato resistant to diseases and insect injuries. *American Potato Journal* 33(2):37-46.

Stevenson, N. D., Brown, A. H. D. and Latter, B. D. H.
1972a Quantitative genetics of sugarcane. IV. Genetics of Fiji disease resistance. *Theoretical and Applied Genetics* 42(6):262-266.

Stewart, A. J. A.
1981a Nymphal polymorphism in two species of *Eupteryx* (Curt.) *Acta Entomologica Fennica* 38:43.
1983a Studies on the ecology and genetics of certain species of *Eupteryx* (Curt.) leafhopper. Doctoral Dissertation, University College, Cardiff, Wales. 264 pp.

Stiling, P. D.
1980a Colour polymorphism in nymphs of the genus *Eupteryx* (Hemiptera: Cicadellidae). *Ecological Entomology* 5(2):175-178.
1980b Host plant specificity, oviposition behaviour and egg parasitism in some leafhoppers of the genus *Eupteryx* (Hemiptera: Cicadellidae.) *Ecological Entomology* 5(1):79-85.
1980c Competition and coexistence among *Eupteryx* leafhoppers (Hemiptera: Cicadellidae) occurring on stinging nettles (*Urtica dioica*). *Journal of Animal Ecology* 49(3):793-805.

Stilling, D.
1978a Ecological studies on leafhoppers occurring on stinging nettles (*Urtica dioica* L.). *Auchenorrhyncha Newsletter* 1:8-9.

Stoner, W. N.
1953a Leafhopper transmission of a degeneration of grape in Florida and its relation to Pierce's disease. *Phytopathology* 43(11):611-615.
1958a Field symptoms indicate occurrence of "alfalfa dwarf" or "Pierce's disease virus" of Rhode Island. *Plant Disease Reporter* 42:573-580.
1965a A review of corn stunt disease (Achaparramiento) and its insect vectors, with resumes of other virus diseases of maize. *U.S. Department of Agriculture, Agriculture Research Service* 33-99. 35 pp.

Stoner, W. N. and Gustin, R. D.
1967a Biology of *Graminella nigrifrons* (Homoptera, Cicadellidae) a vector of corn (maize) stunt virus. *Annals of the Entomological Society of America* 60(3):496-505.

1968a Biology of *Deltocephalus sonorus* (Homoptera, Cicadellidae). *Annals of the Entomological Society of America* 61(1):77-82.

Stoner, W. N. and Ullstrup, A. J.
1964a Corn stunt disease. *Mississippi Agriculture Experiment Station Information Sheet* 844. 4 pp.

Storey, H. H., Howland, A. K. and Prosser, S. C.
1966a Insect vectors of the maize streak virus. *East Africa Agricultural and Forest Research Organization Records and Research Annual Report, 1965.* 123 pp.

Story, G. E. and Halliwell, R. S.
1969a Association of a mycoplasma-like organism with the bunchy top disease of papaya. *Phytopathology* 59(9):1336-1337.

Streissle, G., Bystricky, V., Granados, R. R. and Strohmaier, K.
1968a Number of virus particles in insects and plants infected with wound tumor virus. *Journal of Virology* 2(3):214-217.

Strong, R. G. and Rawlins, W. A.
1957a Mass rearing of six-spotted leafhoppers. *Journal of Economic Entomology* 50(1):76-79.
1958a Aster yellows virus infection and transmission by *Macrosteles fascifrons* with reference to lettuce yellows prevention. *Journal of Economic Entomology* 51:512-515.
1959a Field evaluation of four insecticides for the prevention of lettuce-yellows disease with known populations of viruliferous six-spotted leafhoppers. *Journal of Economic Entomology* 52:686-689.
1959b Evaluation of four insecticides for the control of six-spotted leafhoppers on lettuce. *Journal of Economic Entomology* 51(6):903-905.

Strubing, H.
1956a *Neogonatopus ombrodes* Perkins (Hymenoptera, Dryinidae) als Parasit an *Macrosteles laevis* Rib. (Homoptera-Auchenorrhyncha). *Zoologische Beitrage* 2:145-158.
1956b Beitrage zur Okologie einiger Hochmoorzikaden (Homoptera-Auchenorrhyncha). *Osterreichische Zoologische Zeitschrift* 6(3-5):566-596.
1958a Lautausserung der entscheidende Faktor fur das Zusammenfinden der Geschlechter bei Kleinzikaden (Homoptera-Auchenorrhyncha). (Vorlaufige Mitteilung.) *Zoologische Beitrage* (N. S.) 4:15-21.
1963a Lautausserungen von *Euscelis basterden* (Homoptera: Auchenorrhyncha). *Verhandlungen der Deutschen Zoologischen Gesellschaft* 27:268-281.
1965a Das Lautverhalten von *Euscelis plebejus* Fall. und *Euscelis ohausi* Wagn. (Homoptera-Cicadina). *Zoologische Beitraege* (Neue Folge) 11(1&2):289-341.
1966a Ein Vergleich von Lautaeusser ungen verschiedener *Euscelis*-Arten (Homoptera-Cicadina). *Deutsche Entomologische Zeitschrift* 13(4&5):351-378.
1967a Zur Untersuchungsmethodik der Lautausserungen von Kleinzikaden (Homoptera-Cicadina). *Zoologische Beitrage* (Neue Folge) 13(2&3):265-284.
1970a Zur Artberechtigung von *Euscelis alsius* Ribaut gegenuber *Euscelis plebejus* Fall. (Homoptera-Cicadina). Ein Beitrag zur Neuen Systematik. *Zoologische Beitrage* (Neue Folge) 16(2&3):441-478.
1976a *Euscelis ormanderensis* Remane 1968. 1. Saisonformenbildung und akustische Signalgebung. *Zitzungsberichte Gesellschaft Naturforschung Freunde Berlin* 16(2):151-169. [In German with English summary.]
1978a *Euscelis lineolatus* Brulle, 1832 and *Euscelis ononidis* Remane, 1967. 1. An ecologic, morphologic and bioacoustic comparison. *Zoologische Beitraege* (Neue Folge) 24(1):123-154.
1978b *Euscelis ohausi* Wgn. 1939 and *Euscelis singeri* Wgn. 1951: Separate species or not? *Auchenorrhyncha Newsletter* 1:15.
1980a *Euscelis remanei*, eine neue *Euscelis*-Art aus Sudspanien im Vergleich zu anderen *Euscelis*-Arten (Homoptera-Cicadina). *Zoologische Beitrage* 26:383-404.

1981a *Euscelis remanei* Strubing, 1980, from southern Spain compared with other *Euscelis* species. *Acta Entomologica Fennica* 38:44.

1983a Die Bedeutung des Kommunikationssignals fuer die Diagnose von *Euscelis*-Arten (Homoptera: Cicadina). *Zoologische Jahrbuecher* 87(2&3):343-351.

1984a Eine neue *Euscelis*-art aus marokko und ihre verwandtschaftliche stellung zu anderen *Euscelis-arten* (Homoptera, Cicadina). *Verhandlungen International Symposium Entomofauna Mitteleuropa* 10:209-213.

Strubing, H. and Hasse, A.

1974a Zur Artberechtigung von *Euscelis alsius* Ribaut (Homoptera-Cicadina). II. Die Aedeagus-Variabilitat. [On the specific status of *Euscelis alsius* Ribaut (Hom., Cicadina). II. Aedeagus variability.] *Zoologische Beitrage* 20(3):527-542.

Stuben, M.

1973a Differing attractiveness of yellow and blue pans to different insect orders. *Mitteilungen aus der Biologischen Bundesanstalt fur Land- und Forstwirtschaft*, Berlin-Dahlem No. 151. Pp. 313-314.

Sturani, M.

1948a Un nouveau modele d'aspirateur. *L'Entomologiste* 4(5-6):195-197.

Subba Rao, B. R.

1966a Records of known and new species of mymarid parasites of *Empoasca devastans* Distant from India. *Indian Journal of Entomology* 28(2):187-196.

1983a A catalogue of enemies of some important planthoppers and leafhoppers. In: Knight, W. J., Pant, N. C., Robertson, T. S. and Wilson, M. R., eds. *Proceedings of the 1st International Workshop on Biotaxonomy, Classification and Biology of Leafhoppers and Planthoppers (Auchenorrhyncha) of Economic Importance*. Pp. 385-403. London, 4-7 October 1982. Commonwealth Institute of Entomology, 56 Queen's Gate, London SW7 5JR. 500 pp.

1984a The identity of some mymarid and trichogrammatid parasitoids of leaf and plant hoppers. *XVII International Congress of Entomology* 17:44 (Abstract).

Subba Rao, B. R., Parshad, B., Ram, A., Singh, R. P. and Srivastava, M. L.

1965a Studies on the parasites and predators of *Empoasca devastans* Distant (Homoptera: Jassidae). *Indian Journal of Entomology* 27(1):104-106.

1968a Distribution of *Empoasca devastans* and its egg parasites in the Indian Union. *Entomologia Experimentalis et Applicata* 11(2):250-254.

Subbaratnam, G. V. and Butani, D. K.

1982a Chemical control of insect pest complex of brinjal. *Entomon* 7(1):97-100.

1984a Persistent toxicity of certain insecticides to jassid *Amrasca biguttula biguttula* (Ishida) on brinjal. *Entomon* 9:15-17.

Subbaratnam, G. V., Butani, D. K. and Rao, B. H.

1983a Leaf characters of brinjal governing resistance to jassid, *Amrasca biguttula biguttula* (Ishida). *Indian Journal of Entomology* 45:171-173.

Subramanian, R. and Balasubramanian, M.

1976a Effect of potash nutrition on the incidence of certain insect pests of rice. *Madras Agricultural Journal* 63:561-564.

Subramanian, T. R.

1957a Insects affecting Sea-Island cotton on Malabar. *Madras Agricultural Journal* 44(3):97-100.

Suehiro, A.

1961a Notes and exhibitions—Oct. 10, 1960. *Proceedings of the Hawaiian Entomological Society 1960* 17(3):326.

Suenaga, H.

1962a The occurrence of the rice dwarf disease and the ecology of the green rice leafhopper. *Symposium on Vectors of Plant Viruses*. Hokkaido University, Sapporo, Japan.

Suenaga, H. and Nakatsuka, K.

1958a Critical review of forecasting the occurrence of planthoppers and leafhoppers of rice in Japan. *Byogaityu Hasseiyosatu Tokobetu Doloka* 1:1-453.

Sugimoto, A.

1969a Rearing cage for mass rearing of green rice leafhopper, *Nephotettix cincticeps* Uhler (Hemiptera, Deltocephalidae). *Bulletin of the Agricultural Chemicals Inspection Station, Ministry of Agriculture and Forestry, Tokyo* 9:19-24. [In Japanese with English summary.]

1977a A method of mass-rearing rice green leafhoppers. In: *Seminar on The Rice Brown Planthopper*, Tokyo, Japan, October, 1976. Pp. 248-256.

1981a Mass rearing of green rice leafhopper, *Nephotettix cincticeps* Uhler (Hemiptera: Deltocephalidae). *Japanese Journal of Applied Entomology* 25:77-83.

Sugiyama, H., Kanki, T. and Shigematsu, H.

1971a Difference in the metabolic rate of carbaryl between the silkworm, *Bombyx mori* L. (Lepidoptera: Bombycidae), and the black-tipped leafhopper, *Bothrogonia japonica* Ishihara (Hemiptera; Tettigellidae.) *Applied Entomology and Zoology* 6(2):57-62.

Suhardjan, M., J. Leeuwangh, and Houten, A. V.

1973a Notes on the occurrence of rice stemborer, gallmidge, leafhoppers and planthoppers in Java during the 1970 dry season. *Contributions of Central Research Institute of Agriculture*, Bogor,Indonesia, No. 2. 22 pp.

Sulochana, C. B.

1984a Leafhopper and planthopper transmitted viruses of cereal crops. *Proceedings of the Indian Academy of Science* 93:335-338.

Sun, C. N., Feng, H. T., Kao, L. R., Chou, S. M., Tzeng, Y. I., Chung, T. C., Chen, C. F. and Kao, H. L.

1975a The susceptibility of green rice leafhopper, *Nephotettix cincticeps* Uhler, to vamidothion in central Taiwan. *Journal of Pesticide Science* 5:263-266.

Sun, S. C.

1963a A preliminary study of the life history and chemical control of the peach leafhopper, *Erythroneura* sp., in Kiangsu. *Acta Entomologica Sinica* 12(2):209-219. [In Chinese with English summary.]

Sundararaju, D. and Jayaraj, S.

1977a The biology and the host range of *Orosius albicinctus* Dist. (Homoptera: Cicadellidae), the vector of sesame phyllody disease. *Madras Agricultural Journal* 64:442-446.

Sunder, M. S. and Ali, H.

1961a A trial on chemical control of mango leafhoppers (*Idiocerus atkinsonii* L. and *I. clypealis* L.). *Madras Agricultural Journal* 48(5):174-179.

Suslow, K. G. and Purcell, A. H.

1982a Seasonal transmission of X-disease agent from cherry by leafhopper *Colladonus montanus*. *Plant Disease* 66:28-30.

Suzuki, Y. and Kiritani, K.

1974a Reproduction of *Lycosa pseudoannulata* (Boesenberg et Strand) (Araneae: Lycosidae) under different feeding conditions. *Japanese Journal of Applied Entomology and Zoology* 18:166-170.

Sweeney, R. C. H.

1961a Insect pests of cotton in Nyasaland. I. Hemiptera (Bugs). *Bulletin, Department of Agriculture* (Nyasald) 18(27):14.

Swenson, K. G.

1971a Environmental biology of the leafhopper *Scaphytopius delongi*. *Annals of the Entomological Society of America* 64(4):809-812.

1971b Relation of age, sex, and mating of *Macrosteles fascifrons* to transmission of aster yellows. *Phytopathology* 61:657-659.

1974a Host plants and seasonal cycle of a leafhopper *Fieberiella florii* in western Oregon. *Journal of Economic Entomology* 67(2):299-300.

Swenson, K. G. and Kamm, J. A.

1975a Voltinism in the leafhopper *Draeculacephala crassicornis* Van Duzee. *Melanderia* 21:18-30.

Swincer, D. E.
1984a Insect vectors—an interdisciplinary approach. *In:* Bailey, P. and Swincer, D., eds. *Proceedings of the 4th Australian Applied Entomological Research Conference. Pest Control, Recent Advances, and Future Prospects.* Pp. 463-468. Adelaide, Australia. 520 pp.

Synave, H.
1976a La faune terrestre de l'Ile de Sainte-Helene, Troisieme partie. 2. Insectes (suite et fin). 19. Homoptera. 4. Fam. Cicadellidae, subfam. Aphrodinae. *Annales du Musee Royal de l'Afrique Centrale* (Serie B, Sciences Zoologiques) 215:271-275.

Synave, H. and Dlabola, J.
1976a La faune terrestre de l'Ile Sainte-Helene, Troisieme partie. 2. Insectes (suite et fin). 19. Homoptera. 4. Fam. Cicadellidae, subfam. Agalliinae. *Annales du Musee Royal de l'Afrique Centrale* (Serie B, Sciences Zoologiques) 215:275-277.

Szekessy, V.
1969a Facherflugler—Zikaden—Pflanzenvirosen. *Acta Zoologica* (Budapest) 15(1-2):203-212.

Szent-Ivany, J. J. and Barrett, J. H.
1960a Major insect pests in the highlands of New Guinea. *Highlands Quarterly Bulletin* (New Guinea), pp. 10-11.

Szymkowski, W. and Fernandez Yepex, F.
1963a Insecta y arachnida relacionados con *Gossypium* en Venezuela. Lista preliminar. [Insects and arachnida associated with cotton in Venezuela. Preliminary list.] *L'Agronomie Tropicale* 13(2):83-88.

Taboada, O.
1959a The leafhoppers of Michigan excluding the subfamilies Athysaninae and Cicadellinae (Homoptera, Cicadellidae). *Dissertation Abstracts* 20(4):1495.

1964a An annotated list of the Cicadellidae of Michigan. *Michigan Agricultural Experiment Station Quarterly Bulletin* 47(1):113-122.

1979a New records of leafhoppers (Homoptera: Cicadellidae) for Michigan, including a vector of X-disease. *Great Lakes Entomologist* 12(3):99-100.

1982a First report of the leafhopper *Graminella fitchii* for Michigan (Homoptera: Cicadellidae). *Great Lakes Entomologist* 15(3):223.

Taboada, O. and Burger, T. L.
1967a A new record of *Scaphytopius magdelensis*: Another plant disease vector in Michigan (Homoptera: Cicadellidae). *Michigan Entomologist* 1(7):249-252.

Taboada, O. and Hoffman, J. R.
1965a Distribution of leafhopper vectors of plant diseases in Michigan. *Transactions of the American Microscopical Society* 84(2):201-210.

Taboada, O., Rosenberger, D. A. and Jones, A. L.
1975a Leafhopper fauna of X-diseased peach and cherry orchards in southwest Michigan. *Journal of Economic Entomology* 68:255-257.

Taguchi, H.
1975a Two new *Chaetomymar* species from Japan and Taiwan (Hymenoptera: Mymaridae). *Transactions of the Shikoku Entomological Society* 12:111-114.

Tahama, Y.
1963a Studies on the dwarf disease of mulberry tree. 4. Transmission by leafhopper, *Hishimonus disciguttus* (Walker). *Annals of the Phytopathological Society of Japan* 28(1):49-52. [In Japanese with English summary.]

1964a Studies on the dwarf disease of mulberry tree. X. Further experiments on the transmission by leafhopper, *Hishimonus disciguttus* Walker. *Annals of the Phytopathological Society of Japan* 29:185-188. [In Japanese.]

1968a Studies on the dwarf disease of the mulberry tree. *Bulletin of the Kumamoto Sericulture Experiment Station* 40:184. [In Japanese.]

Taimr, L. and Dlabola, J.
1965a Application of radioactive phosphorus for the marking of leafhoppers in the study of their migration. *Sbornik Ustav Vedeckotechnickych Informaci Ochrana Rostlin* 38(3):75-83.

Takagi, K.
1981a The injuries to citrus fruit caused by green leafhoppers, *Empoasca* spp., tea yellow thrips, *Scirtothrips dorsalis* Hood, citrus red mite, *Panonychus citri* (McGregor) and citrus rust mite, *Aculops pelekassi* (Keifer). *Bulletin of the Fruit Tree Research Station*, Japan 3:101-108.

Takahashi, Y.
1963a Detecting method of the viruliferous green rice leafhopper, *Nephotettix cincticeps* Uhler, a vector of the yellow dwarf disease of the rice plant. *Japanese Journal of Applied Entomology and Zoology* 7(3):200-206. [In Japanese.]

1963b Globules in the adipose cells of the green rice leafhopper (*Nephotettix cincticeps*) infected with the virus of the yellow dwarf disease of the rice plant. *Japanese Journal of Applied Entomology and Zoology* 7(4):350-351. [In Japanese.]

1979a Present status of insecticides for controlling the resistant green rice leafhopper. *Japan Pesticide Information* 36:22-27.

Takahashi, Y. and Kiritani, K.
1973a The selective toxicity of insecticides against insect pests of rice and their natural enemies. *Applied Entomology and Zoology* 8:220-226.

Takahashi, Y., Kyomura, N. and Yamamoto, I.
1977a Synergistic insecticidal action of N-methyl and N-propylcarbamates to the green rice leafhopper, *Nephotettix cincticeps*, resistant to aryl N-methyl-carbamates. *Journal of Pesticide Science* 2:467-470.

Takahashi, Y. and Sekiya, I.
1962a Adipose tissue of the green rice leafhopper, *Nephotettix cincticeps* Uhler, infected with the virus of the yellow dwarf disease of the rice plant. *Annals of Applied Entomology and Zoology* 6:90-94.

Takai, A., Hara, K. and Ino, M.
1972a Study on methods for estimating the number of the overwintering green rice leafhopper, *Nephotettix cincticeps* Uhler (Hemiptera: Deltocephalidae) and their population dynamics. *Japanese Journal of Applied Entomology and Zoology* 16(2):67-74. [In Japanese with English summary.]

Takai, A., Ito, Y., Nakamura, K. and Miyashita, K.
1965a Estimation of population density of the green rice leafhopper (*Nephotettix cincticeps* Uhler) by the mark-recapture and the sweeping methods. *Japanese Journal of Applied Entomology and Zoology* 9(1):5-12. [In Japanese.]

Takita, T. and Habibuddin, H.
1985a Relationship between laboratory-developed biotypes of green leafhopper and resistant varieties of rice in Malaysia. *Japan Agricultural Research Quarterly* 19:219-223.

Taksdal, G.
1977a Auchenorrhyncha and Psylloidea collected in strawberry fields. *Norwegian Journal of Entomology* 24:107-110.

Talitskii, V. I. and Logvinenko, V. N.
1965a Cicadinea-fauna of the Moldavian SSR. *In: Materialy zoologicheskogo soveschaniya po probleme "Biologicheskie osnovy rekonstruktsii, ratsional'nogo ispolzovaniya i okhrany fauny juzhnoy zony evropeyskoy chasti SSSR."* Kishinev. Pp. 451-452. [Materials of zoological conference "Biological Foundations of Reconstruction, Rational Use and Protection of Fauna in the Southern Areas of the USSR European Part." Kishinev.] [In Russian.]

1966a Survey of cicadid fauna (Homoptera, Cicadinea) of Moldavia. *Trudy Moldavskogo nauchno-Issledovatel'skogo Innstituta Sadovodstva, Vinogradarstava i Vinodeliya (Ent.)* 13:231-269. [In Russian.]

Tallamy, D. W. and Denno, R. F.
1979a Responses of sap-feeding insects (Homoptera-Hemiptera) to simplification of host plant structure. *Environmental Entomology* 8(6):1021-1028.

Tanaka, K. and Ito, Y.
1982a Decrease in respiratory rate in a wolf spider, *Pardosa astrigera* (L. Koch), under starvation. *Researches on Population Ecology* 24:360-374.

Tanasijevic, N.
1963a Prilog poznavanju faune cikada (Homoptera, Auchenorrhyncha) na crvenoj detelini u Jugoslaviji. [Leafhopper fauna in red clover.] *Zastita Bilja* [Plant Protection] 14(69&70):93-106. [English summary.]
1964a Stalni i povremeni clanovi faune cikada (Homoptera, Auchenorrhyncha) lucerke u Jugoslaviji. [Constant and temporary members of leafhopper fauna in alfalfa in Yugoslavia (Hom., Auch.).] *Zastita Bilja* [Plant Protection] 15(80):378-388.
1965a Prilog poznavanju faune cikada (Homoptera, Auchenorrhyncha) strnih zita i nekih drugih biljaka. [A contribution to knowledge of the Auchenorrhyncha on cereals and some other plants.] *Zbornik Radova Poljoprivrednog Fakulteta, Universitet u Beogradu* 13(387):14. [English summary.]
1966a A new contribution to the study of leafhopper fauna (Homoptera: Auchenorrhyncha). *Zastita Bilja* 17(91-92):205-212. [In Serbo-Croatian with English summary.]

Tandon, R. L. and Lal, B.
1979a Studies on the chemical control of the mango hopper, *Idioscopus clypealis* Lethierry (Cicadellidae: Homoptera). *International Pest Control* 21:6-7.
1983a Predatory spiders associated with insect pests of mango in India. *Bulletin of Entomology* 24:144-147.

Tandon, R. L., Lal, B. and Rao, G. S. P.
1983a Prediction of the mango hopper, *Idioscopus clypealis* Leth., population in relation to physical environmental factors. *Entomon* 8:257-261.

Tanner, V. M. and Harris, D. R.
1969a List of insect type specimens in the entomological collection of Brigham Young University, Provo, Utah, No. IV. *Great Basin Naturalist* 29(4):183-205.

Tantera, D. M. and Roechan, R.
1978a Virus-vector relationship on penyakit habang of rice. *Ringkasan Publikasi dan Laporan Penelitian Pertanian* 8:28-29.

Tao, C. H.
1958a New insecticides for control of the cotton insects in 1957. *Agricultural Research* 8(1):40-43.
1960a Control of cotton pests with some new insecticides after typhoon. *Agricultural Research* 9(1):40-42.

Tapia, H. and Saenz, L.
1971a Information basica para el control del achaparramiento del maiz en Nicaragua. *XVIII Reunion Anual del Programa Cooperativo Cultivos Alimenticios*, Panana, Panama.

Tarafder, P., Ghosh, A. B. and Mukopadhyay, S.
1980a Effect of transplanting date and seedling age at transplanting on spread of tungro. *International Rice Research Newsletter* 5(2):12-13.

Tasch, P. and Riek, E. F.
1969a Permian insect wing from Antarctic Sentinel Mountains. *Science* 164:1529-1530.

Taschenberg, E. F.
1957a Evaluation of organic phosphorus insecticides for grape leafhopper control. *Journal of Economic Entomology* 50(4):411-414.
1973a Economic status and control of the grape leafhopper in western New York. *Search Agriculture (Entomology)* 3(4&5):1-9.

Tashev, D. G.
1968a Investigaciones sobre los hospedantes de las especies del genero *Empoasca* Walsh (Cicadellidae, Homoptera). *Revista de Agricola* (Cuba) 2(1):24-30.

Tashiro, H.
1973a Evaluation of soil applied systemic insecticides on insects of white birch in nurseries. *Search Agriculture (Entomology)* 3(9):1-11.

Tay, E. B.
1972a Population ecology of *Cicadella viridis* (L.) and bionomics of *Graphocephala coccinea* (Forster) (Homoptera: Cicadellidae). Ph. D. Thesis, University of London.

Taylor, J. H.
1971a The development of malathion ultra-low-volume concentrate in Asia. *Pest Articles and News Summaries* 17(1):12-17.

Taylor, N. L.
1955a Screening for leafhopper resistance in alfalfa. *Association of Southern Agricultural Workers Proceedings* 52:190.
1956a Pubescence inheritance and leafhopper resistance relationships in alfalfa. *Agronomy Journal* 48(2):78-81.

Taylor, R. A. J.
1985a Migratory behavior in the Auchenorrhyncha. In: Nault, L. R. and Rodriguez, J. G., eds. *The Leafhoppers and Planthoppers*. Pp. 259-288. Wiley & Sons, New York. 500 pp.

Taylor, W. E.
1976a Co-ordinated minimum insecticide trial: Yield performance of insect resistant cowpea cultivars from IITA compared with Sierra Leone cultivars. *Tropical Grain Legume Bulletin* 5:7.

Taylor, W. E. and Kamara, S. B.
1974a Insect succession on wet and dry season rice in Sierra Leone. *Ghana Journal of Agricultural Science* 7:109-115.

Teli, V. S. and Dalaya, V. P.
1981a Varietal resistance in okra to *Amrasca biguttula biguttula* (Ishida). *Indian Journal of Agricultural Sciences* 51:729-731.

Teraguchi, S., Stenzel, J., Sedlacek, J. and Deininger, R.
1981a Arthropod-grass communities: Comparison of communities in Ohio and Alaska. *Journal of Biogeography* 8:53-66.

Tereshko, L. I.
1965a Some data on the cicadellids of Moldavia. *Nauchnye Doklady Vysshei Shkoly, Biologicheskie Nauki* 2:25-26. [In Russian.]

Teson, A.
1971a Notas sobre Giponinos neotropicales: I. (Homoptera, Cicadellidae). *Revista de la Sociedad Entomologica Argentina* 33(1-4):105-108.
1972a [1973] Notas sobre Giponinos neotropicales. II. (Homoptera, Cicadellidae). *Revista de la Sociedad Entomologica Argentina* 34(1&2):57-60.
1972b [1973] Notas sobre Giponinos neotropicales. III. (Homoptera, Cicadellidae). *Revista de la Sociedad Entomologica Argentina* 34(1&2):131-135.

Teulon, D. A. and Penman, D. R.
1984a Spray timing for control of Froggatt's apple leafhopper in Canterbury. In: Hartley, M. J., Popay, A. J. and Popay, A. I., eds. *Proceedings of the 37th New Zealand Weed and Pest Control Conference*, Christchurch, New Zealand. Pp. 245-247. Canterbury, New Zealand.

Tewari, G. C. and Moorthy, P. N. K.
1983a Effectiveness of synthetic pyrethroids against the pest complex of brinjal. *Entomon* 8:365-368.

Thapa, V. K.

1983a Descriptions of two new genera and a few new records of leafhoppers (Typhlocybinae, Cicadellidae, Homoptera) from Nepal. *In: Knight, W. J., Pant, N. C., Robertson, T. S. and Wilson, M. R., eds. Proceedings of the 1st International Workshop on Biotaxonomy, Classification and Biology of Leafhoppers and Planthoppers (Auchenorrhyncha) of Economic Importance.* Pp. 173-177. London, 4-7 October 1982. Commonwealth Institute of Entomology, 56 Queen's Gate, London SW7 5JR. 500 pp. [New genera: *Omanesia, Guheswaria.*]

1984a A new species and new records of genus *Agnesiella* (*Draberiella*) Dwor., 1971 (Homoptera: Cicadellidae) from Kathmandu Valley, Nepal. *Zoologica Orientalis* 1(1):15-16.

1984b Some erythroneurine leafhoppers (Homoptera: Cicadellidae, Typhlocybinae) from the Kathmandu Valley, Nepal. *Journal of Entomological Research* 8(1):46-52. [New genus: *Lokia.*]

1985a Some empoascan leafhoppers (Homoptera, Cicadellidae, Typhlocybinae) from the Kathmandu Valley, Nepal. *Journal of Entomological Research* (New Delhi) 9(1):65-74. [New genus: *Ghauriana;* and new subgenus: *Greceasca.*]

Thapa, V. K. and Sohi, A. S.

1982a A new species of *Thaia* Ghauri, 1962 (Erythroneurini, Typhlocybinae, Cicadellidae) from the Kathmandu Valley, Nepal. *Journal of Natural History Museum, Tribhuvan University* 6(4):97-100.

1984a Description of the male of *Bolanusoides heros* Distant (Cicadellidae: Typhlocybinae: Typhlocybini). *Zoologica Orientalis* 1(2):47-48.

Theron, J. G.

1970a A new species of *Narecho* (Hemiptera: Cicadelloidea: Nirvanidae) from South Africa. *Novos Taxa Entomologicos* 83:1-7.

1970b Redescription of ten of Cogan's species of South African Cicadelloidea (Hemiptera). *Journal of the Entomological Society of Southern Africa* 33(2):303-323.

1971a Two new species of Cicadellidae (Hemiptera) from the Western Cape. *Novos Taxa Entomologicos* 94:1-6.

1972a The Naude species of South African Cicadellidae (Hemiptera). I. Species assigned to the genera *Bythoscopus* Germar, 1833, *Parabolitus* Naude, 1926, and *Gcaleka* Naude, 1926. *Journal of the Entomological Society of Southern Africa* 35(2):201-210. [New genus: *Iassomorphus.*]

1973a The Naude species of South African Cicadellidae (Hemiptera). II. Species assigned to the genera *Eugnathodus* Baker and *Cicadula* Zetterstedt. *Journal of the Entomological Society of Southern Africa* 36(1):25-35. [New genera: *Ragia, Stellana.*]

1973b Southern African species of the genus *Tetartostylus* Wagner (Hemiptera: Cicadellidae). *Journal of the Entomological Society of Southern Africa* 36(2):229-234.

1974a A new species of *Empoasca* (Hemiptera: Cicadellidae) injurious to citrus in South Africa. *Journal of the Entomological Society of Southern Africa* 37(1):1-3.

1974b The Naude species of South African Cicadellidae (Hemiptera). III. Species assigned to the genus *Chlorotettix* Van Duzee, *Thamnotettix* Zetterstedt, *Euscelis* Brulle, *Scaphoideus* Uhler and *Selenocephalus* Germar. *Journal of the Entomological Society of Southern Africa* 37(1):147-166. [Six new genera.]

1974c Class Insecta. Order Hemiptera. Suborder Homoptera. Cicadelloidea. *In: Coaton, W. G. H. Status of the Taxonomy of the Hexapoda of Southern Africa. Entomology Memoirs.* Department of Agricultural Technical Services, Republic of South Africa 38:54-55.

1975a The Naude species of South African Cicadellidae (Hemiptera). IV. Species assigned to the genera *Aconura* Lethierry and *Deltocephalus* Burmeister. *Journal of the Entomological Society of Southern Africa* 38(2):189-206. [Eight new genera.]

1975b *Afrosteles,* a new genus of South African Macrostelini (Hemiptera: Cicadellidae). *Journal of the Entomological Society of Southern Africa* 38(2):207-209.

1976a The Naude species of South African Cicadellidae (Hemiptera). V. Species assigned to the genera *Dorydium* Burmeister, *Penthimia* Germar, *Equeefa* Distant, *Petalocephala* Stal and *Idiocerus* Lewis. *Journal of the Entomological Society of Southern Africa* 39(2):247-260. [New genera: *Rhusia, Rhinocerotis.*]

1977a The Naude species of South African Cicadellidae (Hemiptera). VI. Some species assigned to the genera *Erythroneura* Fitch and *Empoasca* Walsh. *Journal of the Entomological Society of Southern Africa* 40(1):109-118.

1978a Southern African species of the genus *Molopopterus* Jacobi (Hemiptera: Cicadellidae), with a note on the species attacking rooibostee. *Journal of the Entomological Society of Southern Africa* 41(1):31-44.

1978b *Coganus* gen. nov. (Hemiptera: Cicadellidae), with a note on alary dimorphism. *Journal of the Entomological Society of Southern Africa* 41(2):257-258.

1979a Cicadellidae (Hemiptera) associated with the ganna bush, *Salsola esterhuyseniae* Botsch. *Journal of the Entomological Society of Southern Africa* 42(1):77-88. [New genera: *Gannachrus, Gannia, Salsolibia, Salsolicola.*]

1980a Notes on some southern African Cicadellidae described by Stal in "Hemiptera Africana." *Journal of the Entomological Society of Southern Africa* 43(2):275-292. [New genera: *Nataretus, Platentomus.*]

1980b South African species of *Dagama* Distant (Hemiptera Cicadellidae). *Journal of the Entomological Society of Southern Africa* 43(2):293-298.

1980c Economically important Cicadellidae in South Africa. *Proceedings of the Congress of Entomological Society of South Africa* 3:16-17 (Abstract).

1981a Ekonomies belangrike blaarspringers (Cicadellidae) van Suid-Afrika. *Proceedings of the 3rd Entomological Congress,* Pretoria, 1980. Pp. 16-17.

1982a Leafhopper (Hemiptera: Cicadellidae) on grapevines in South Africa. *Journal of the Entomological Society of Southern Africa* 45(1):23-26.

1982b Grassland leafhoppers (Hemiptera: Cicadellidae) from Natal, South Africa, with descriptions of new genera and species. *Phytophylactica* 14:17-30. [Describes six new genera.]

1983a Cicadellidae (Hemiptera) collected by Darwin at the Cape of Good Hope, South Africa, with description of a related species. *Journal of the Entomological Society of Southern Africa* 46(1):147-151. [New genus: *Kaapia.*]

1984a Leafhoppers (Hemiptera: Cicadellidae) associated with the renosterbos, *Elytropappus rhinocerotis* Less. I. The genus *Renosteria* Theron. *Journal of the Entomological Society of Southern Africa* 47(1):83-97.

1984b Leafhoppers (Hemiptera: Cicadellidae) associated with the renosterbos, *Elytropappus rhinocerotis* Less. II. The genera *Cerus* Theron and *Refrolix* gen. nov. *Journal of the Entomological Society of Southern Africa* 47(2):217-230.

1984c Coelidiinae (Hemiptera: Cicadellidae) described from South Africa in the genera *Palicus* Stal, *Equeefa* Distant and *Aletta* Metcalf, with descriptions of new species. *Journal of the Entomological Society of Southern Africa* 47(2):313-327. [New genera: *Keia, Mgenia, Modderena.*]

1985a Order Hemiptera (bugs, leafhoppers, cicadas, aphids, scale-insects, etc.). Suborder Homoptera. *In: Scholtz, C. H. and Holm, E., eds. Insects of Southern Africa.* Pp. 152-164. Butterworths, Durban. 502 pp.

1985b An improved pooter (aspirator). *Antenna* 9:163.

1985c Staining genitalia of leafhoppers. *Tymbal* 6:21.

Thimmaiah, G.
1977a Chemical control of leafhoppers and bollworms on Varalaxmi hybrid cotton by soil and foliar applications. *Mysore Journal of Agricultural Sciences* 11:386-391.

Thimmaiah, G., Thippeswamy, C., Govindan, R. and Hanumanna, M.
1980a Comparative efficacy of some insecticidal sprays in controlling cotton leafhopper *Sundapteryx biguttula biguttula* Ishida (Hemiptera: Cicadellidae). *Current Research* 9:156-158.

Thomas, J. and John, V. T.
1980a Suppression of symptoms of rice tungro virus disease by carbendazim. *Plant Disease* 64:402-403.

Thomas, P. E.
1972a Mode of expression of host preference by *Circulifer tenellus*, the vector of curly top virus. *Journal of Economic Entomology* 65:119-123.

Thomas, P. E. and Boll, R. K.
1977a Effect of host preference on transmission of curly top virus to tomato by the beet leafhopper. *Phytopathology* 67:903-905.

Thomas, P. E. and Martin, M. W.
1971a Vector preference, a factor of resistance to curly top virus in certain tomato cultivars. *Phytopathology* 61:1257-1260.

Thompson, L. S.
1967a Reduction of lettuce yellows with systemic insecticides. *Journal of Economic Entomology* 60(3):716-718.
1968a Strawberry green petal virus transmission in the field. *Plant Disease Reporter* 52:393-394.

Thompson, L. S., Cutcliffe, J. A., Gourley, C. O. and Murray, R. A.
1973a Evaluation of several insecticides for control of strawberry green petal disease. *Canadian Plant Disease Survey* 53(1):16-18.

Thompson, L. S. and Rawlins, W. A.
1961a Systemic insecticides for six-spotted leafhopper control and reduction of incidence of lettuce yellows. *Journal of Economic Entomology* 54:1137-1176.

Thompson, P.
1978a The oviposition sites of five leafhopper species (Homoptera, Auchenorrhyncha) on *Holcis mollis* and *H. lanatus*. *Ecological Entomology* 3(3):231-240.

Thontadarya, T. S. and Devaiah, M. C.
1975a Effect of aerial spraying of fenthion and fenitrothion on some insect pests of paddy. *Current Research* 4:107.

Thontadarya, T. S., Viswanath, B. N. and Viremath, S. C.
1978a Preliminary studies on the control of mango hoppers by stem injection with systemic insecticides. *Current Science* 47:378.

Thresh, J. M.
1974a Vector relationships and the development of epidemics: The epidemiology of plant viruses. *Phytopathology* 64(8):1050-1056.
1980a An ecological approach to the epidemiology of plant viruses. *In*: Palti, J. and Kranz, J., eds. *Comparative Epidemiology*. Pp. 57-70. Pudoc, Wageningen. 122 pp.
1980b The origins and epidemiology of some important plant virus diseases. *In*: Coaker, T. H., ed. *Applied Biology*. Pp. 2-65. Academic Press, London. 407 pp.

Thung, T. H. and Hadiwidjaja, T.
1957a De Heksenbezemzieke bij Leguminosen. [Witches' broom disease of Leguminosea.] *Tijdschrift over Plantenziekten* 63:58-63.

Tidke, P. M. and Sane, P. V.
1962a Jassid resistance and morphology of cotton leaf. *Indian Cotton Growing Review* 16:324-327.

Tiivel, T.
1984a Ultrastructural aspects of endocytobiosis in leafhopper (Insecta: Cicadinea) cells. *Eesti NSV Teaduste Akadeemia Toimetised (Bioloogia)* 33(4):244-255.

Timian, R. G.
1960a A virus of durum wheat in North Dakota transmitted by leafhoppers. *Plant Disease Reporter* 44:771-773.
1978a Recovery of wheat striate mosaic virus by leafhoppers from liquid nitrogen stored plant tissue. *Phytopathology News* 12:194.
1985a Oat blue dwarf virus in its plant host and insect vectors. *Plant Disease* 69(8):706-708.

Timian, R. G. and Alm, K.
1973a Selective breeding of *Macrosteles fascifrons* for increased efficiency in virus transmission. *Phytopathology* 63(1):109-112.

Timmer, L. W. and Lee, R. E.
1985a Survey of blight-affected citrus groves for xylem-limited bacteria carried by sharpshooters. *Plant Disease* 69(6):497-498.

Timmer, L. W., Lee, R. E., Allen, J. C. and Tucker, D. P. H.
1982a Distribution of sharpshooters in Florida citrus groves. *Environmental Entomology* 11(2):456-460.

Ting, W. P.
1971a Studies on Penyakit merah disease of rice. II. Host range of the virus. *Malaysian Agricultural Journal* 48:10-12.

Ting, W. P. and Ong, C. A.
1974a Studies on Penyakit merah disease of rice. IV. Additional hosts of the virus and its vector. *Malaysian Agricultural Journal* 49:269-274.

Ting, W. P. and Paramsothy, S.
1970a Studies on Penyakit merah disease of rice. I. Virus-vector interaction. *Malaysian Agricultural Journal* 47:290-298.

Tingey, W. M.
1981a Potential for plant resistance in management of arthropod pests. *In*: Laschomb, J. and Casagrande, R. A., eds. *Advances in Potato Pest Management*. Pp. 201-245. Academic Press, New York.
1985a Plant defensive mechanisms against leafhoppers. *In*: Nault, L. R. and Rodriguez, J. G., eds. *The Leafhoppers and Planthoppers*. Pp. 217-234. Wiley & Sons, New York. 500 pp.
1985b Techniques for evaluating plant resistance to insects. *In*: Miller, T. A. and Miller, J. A., eds. *Methods for Studying Mechanistic Interactions Between Insects and Plants*. Springer, Berlin.

Tingey, W. M. and Gibson, R. W.
1978a Feeding and mobility of the potato leafhopper impaired by glandular trichomes of *Solanum berthaultii* and *S. polyadenium*. *Journal of Economic Entomology* 71:856-858.

Tingey, W. M. and Laubengayer, J. E.
1981a Defense against the green peach aphid and potato leafhopper by glandular trichomes of *Solanum berthaultii*. *Journal of Economic Entomology* 74:721-725.

Tingey, W. M., Mackenzie, J. D. and Gregory, P.
1978a Total foliar glycoalkaloids and resistance of wild potato species to *Empoasca fabae* (Harris). *American Potato Journal* 55:577-585.

Tingey, W. M. and Plaisted, R. L.
1976a Tetraploid sources of potato resistance to *Myzus persicae*, *Macrosiphum euphorbiae*, and *Empoasca fabae*. *Journal of Economic Entomology* 69(5):673-676.

Tingey, W. M. and Sinden, S. L.
1982a Glandular pubescence, glycoalkaloid composition, and resistance to the green peach aphid, potato leafhopper, and potato fleabeetle in *Solanum berthaultii*. *American Potato Journal* 59:95-106.

Tissot, A. N.
1932a Resistance of beans to *Empoasca*. *Florida Agriculture Experiment Station Annual Report*. Pp. 690-691.

Tokmakoglu, C. and Celik, Y.
1972a An insecticidal experiment on pistachio leafhopper (*Idiocerus stali* Fieb.). *Plant Protection Research Annual*. Antepfistiklarinda zararli sirali zenk (*Idiocerus stali* Fieb.) e karsi ilac denemeleri. Tarim Bakanligi Zirai Mucadele ve Zirai Karantina Genel Mudurlugu Arastirma Subesi. *Zirai Mucadele Arastirma Yilligi* 57:186. [In English and Turkish.]

Tokumitsu, T. and Maramorosch, K.
1967a Cytoplasmic protrusions in insect cells during mitosis in vitro. *Journal of Cell Biology* 34:677-683.
1968a Bubbling of dividing leafhopper cells *in vitro*. In: *Second International Colloquium on Invertebrate Tissue Culture.* Pp. 10-13. Antonio Baselli, Villa Carlotta, 9-10 September 1967.

Tomeu, A. and Moseley, E.
1972a Insect resistance in hybrid sorghum. 2. Rainy season. *Revista Cubana de Ciencia Agricola* 6(3):365-370.

Toms, A. M. and Goodman, A.
1957a The importance of drift in insecticide spraying experiments: some observations from insect behavior in cotton. *Empire Cotton Growing Review* 34(3):177-188.

Tonks, N. V.
1960a Life history and behavior of the leafhopper *Macropsis fuscula* (Zett.) (Homoptera: Cicadellidae) in British Columbia. *Canadian Entomologist* 92(9):707-713.
1963a Leafhopper investigations at Victoria, 1947-1960. *Canada Department of Agriculture Research Branch, Entomology Newsletter* 41(4):2-3.

Tormala, T.
1977a Effects of mowing and ploughing on the primary production and flora and fauna of a reserved field in central Finland. *Acta Agriculturae Scandinavica* 27:253-264.
1982a Evaluation of 5 methods of sampling field layer arthropods, particularly the leafhopper community in grasslands. *Suomen Hyonteistieteellinen Aikakauskirja* 48(1):1-16.
1983a Evaluation of five methods of sampling leafhoppers in a meadow. *Acta Entomologica Fennica* 38:46.

Tormala, T. and Vanninen, I.
1983a Leafhopper (Homoptera, Auchenorrhyncha) communities in two urban lawns in central Finland. *Suomen Hyonteistieteellinen Aikakauskirja* 49(4):111-114.

Torres, B. A.
1960a Los tipos de Typhlocybinae de Carlos Berg pertenecientes actualmente al genero *Empoasca* Walsh 1862. Actas y Trabajos. *I Congresso Sudamerica Zoologia* (La Plata) 3:213-221.
1960b Nueva especie de Typhlocybinae de la Puna Jujena, (Homoptera-Cicadellidae). *Neotropica* (La Plata) 5(A):65-69.

Touzeau, J.
1968a La cicadelle *Empoasca flavescens* et le 'grillage' de la vigne dans le sud-ouest de la France. [The cicadellid *E. flavescens* and the 'scorching' of vines in the south-west of France.] *Phytoma* 20(200):31-35.

Townes, H.
1962a Design for a Malaise trap. *Proceedings of the Entomological Society of Washington* 64:253-262.

Townsend, R., Markham, P. G. and Plaskitt, K. A.
1977a Multiplication and morphology of *Spiroplasma citri* in the leafhopper *Euscelis plebejus. Annals of Applied Biology* 87:307-313.

Toyoda, S., Kimura, I. and Suzuki, N.
1964a Rice viruses with special reference to rice dwarf virus. *Protein Nucleic Acid Enzyme* 9:861-867. [In Japanese.]
1965a Purification of rice dwarf virus. *Annals of the Phytopathological Society of Japan* 30:225-230.

Tracy, R. K.
1977a Pierce's disease: Feeding preference of *Hordnia circellata* (Baker) for selected grapevine cultivars and treatment of disease virus with solar heating. M. S. Thesis, University of California, Davis. 73 pp.

Trammel, K.
1974a The white apple leafhopper in New York — insecticide resistance and current control status. *Search Agriculture* (Geneva) 4(8):1-10.

Traue, D.
1978a Zur Biophysik der Schallabstrahlung bei Kleinzikaden am Beispiel von *Euscelis incisus* Kb. (Homoptera-Cicadina: Jassidae). *Zoologische Beitraege* 24:155-164.

1981a New results of substrate-borne sounds of small cicadinea (Homoptera, Cicadellidae). *Acta Entomologica Fennica* 38:44-45.

Tribout, P.
1956a Contribution a l'etude des Jassides. *Travaux du Laboratoire de Zoologie et de la Station Aquicole Grimaldi de la Faculte des Science de Dijon* 16:1-32.

Trinh, T. T.
1980a New rice diseases and insects in the Senegal River Basin in 1978/79. *International Rice Commission Newsletter* 29(2):37.

Tripathi, G. M., Sandhu, J. S. and Tewari, G. C.
1978a Occurrence of *Tettigella spectra* Dist. as a new pest of sugarcane in Punjab. *Indian Journal of Entomology* 40:447.

Triplehorn, B. W. and Nault, L. R.
1985a Phylogenetic classification of the genus *Dalbulus* (Homoptera: Cicadellidae), and notes on the phylogeny of the Macrostelini. *Annals of the Entomological Society of America* 78(3):291-315.

Triplehorn, B. W., Nault, L. R. and Horn, D. J.
1984a Feeding behavior of *Graminella nigrifrons* (Forbes). *Annals of the Entomological Society of America* 77(1):102-107.

Trojanowski, H.
1963a Nowy Kierunek w zwalczaniu szkodnikow ziemniaka. [New method of control of potato pests.] *Biuletyn Instytutu Ochrony Roslin* 24:261-278.

Trolle, L.
1966a Nye danske cikader (Hemiptera, Cicadellidae). *Flora og Fauna* 72:93-100.
1968a Nye danske cikader 2. (Hemiptera, Cicadellidae). *Flora og Fauna* 74:113-121.
1974a Danske Typhlocybiner (Auchenorrhyncha, Cicadellidae, Typhlocybinae). *Entomologiske Meddelelser* 42:53-62.

Troxclair, N. N. and Boethel, D. J.
1984a Influence of tillage practices and row spacing on soybean insect populations in Louisiana. *Journal of Economic Entomology* 77:1571-1579.

Trujillo, G. E., Acosta, J. M. and Pinero, A.
1974a A new corn virus found in Venezuela. *Plant Disease Reporter* 58(2):122-126.

Tsai, J. H.
1979a Vector transmission of mycoplasmal agents of plant diseases. In: *Whitcomb, R. F. and Tully, J. G., eds. The Mycoplasmas. Vol. 3. Plant and Insect Mycoplasmas.* Pp. 265-307. Academic Press, New York.

Tsai, J. H. and Anwar, M.
1977a Mating and longevity of *Oncometopia nigricans* (Homoptera: Cicadellidae), a suspected vector of lethal yellowing of coconut palms, on various host plants. *Florida Entomologist* 60(2):105-108.

Tsai, J. H. and Mead, F. W.
1982a Rotary trap survey of homopterans in palm plantings in South Florida. *Journal of Economic Entomology* 75(5):809-812.

Tsai, P., Shinkai, A., Mukoo, H. and Nakamura., S.
1972a Distribution of mycoplasma-like organism and ultrastructural changes in host cells in witches' broom diseased sweet potato and morning glory. *Annals of the Phytopathological Society of Japan* 38(1):81-85. [In Japanese with English summary.]

Tseng, Y. H., Yang, S. L. and Pan, Y. S.
1976a Two new erythraeid mites from Taiwan (Acarina: Prostigmata). *Report on the Taiwan Sugar Research Institute* 74:63-74.

Tshmir, P. G.
1977a A new species of the family Cicadellidae from Caucasus. *Zoologicheskii Zhurnal* 56(4):944-945.

Tsugawa, C., Yamada, M., Shirasaki, S. and Oyama, N.
1966a Biology and control of the leafhopper *Typhlocyba ishidai* Matsumura in apple orchards. *Bulletin, Aomori Agricultural Experiment Station* No. 10. 45 pp. [In Japanese with English summary.]

Tsuji, H. and Fujita, K.
1978a Synergistic action of formothion and MTMC against a resistant strain of the green rice leafhopper, *Nephotettix cincticeps* Uhler. *Japanese Journal of Applied Entomology and Zoology* 22:33-37.

Tugwell, P., Rouse, E. P. and Thompson, R. G.
1973a Insects in soybeans and a weed host (*Desmodium* sp.). *Arkansas Agricultural Experiment Station Report* Series 214. 18 pp.

Tunstall, J. P. and Matthews, G. A.
1966a Large scale spraying trials for the control of cotton insect pests in central Africa. *Empire Cotton Growing Review* 43(2):121-139.

Turian, G.
1960a Mycose a *Empusa apiculta* Thaxt. chez les Cicadelles du genre *Deltocephalus*. *Mitteilungen der Schweizerischen Entomologischen Gesellschaft* 33:88-90.

Turner, W. F. and Pollard, H. N.
1956a Additional leafhopper vectors of phony peach. *Journal of Economic Entomology* 48(6):771-772.

1959a Life histories and behavior of five insect vectors of phony peach disease. *U.S. Department of Agriculture Technical Bulletin* 1188. 28 pp.

1959b Insect transmission of phony peach disease. *U.S. Department of Agriculture Technical Bulletin* 1193. 27 pp.

Turnipseed, S. G.
1972a Management of insect pests of soybeans. *Proceedings of the Tall Timbers Conference on Ecological Animal Control by Habitat Management* 4:189-202.

1977a Influence of trichome variations on populations of small phytophagous insects in soybean. *Environmental Entomology* 6:815-817.

Turnipseed, S. G. and Kogan, M.
1976a Soybean entomology. *Annual Review of Entomology* 21:247-282.

Turnipseed, S. G. and Sullivan, M. J.
1976a Plant resistance in soybean insect management. In: Hill, L. D., ed. *World Soybean Research. Proceedings of the World Soybean Research Conference*. Pp. 549-560. Interstate Press, Danville, Illinois.

Twine, P. H.
1971a Insect pests of maize. *Queensland Agricultural Journal* 97(10)523-531.

Twinn, D. C.
1964a Suction bulb aspirator for collecting small insects. *Entomologist* 97:8-10.

Uchida, M. and Ushiyama, K.
1969a Studies on injuries caused by some species of green leafhopper on Unshiu orange fruit. 2. On the period of infestation, appearance of injuries on fruit and the control of leafhopper. *Kanagawa Horticulture Experiment Station Bulletin* 17:39-44. [In Japanese with English summary.]

Ulenberg, S. A., Burger, H. C., de Goffau, L. J. W. and van Rossem, G.
1983a Bijzondere aantastigen door insekten in 1982. *Entomologische Berichten* 43(11):164-168.

Upadhyay, R. K.
1984a Ecological studies on some of the beneficial insects and pests of paddy crop in Chhatishgarh, Madhya Pradesh. *Plant Protection Bulletin* 35:1013.

Upadhyay, V. R., Shah, A. H. and Desai, N. D.
1981a Light trap, a tool for forecasting the incidence of some insect pests of rice. *Gujarat Agricultural University Research Journal* 6:111-114.

Uppal, B. N. and Wagle, P. V.
1944a Control of mango hoppers in Bombay province. *Indian Farming* 5:401-403.

Urbauer, D. and Pruess, K. P
1973a Drift of terrestrial arthropods in an irrigation canal following a wide-area application of ULV malathion. *Journal of Economic Entomology* 66(6):1267-1268.

Uribe, M. and L. Valencia
1970a *Dalbulus maidis* (DeLong and Wolcott) (Homoptera: Cicadellidae), una nueva "cigarrita" para el valle de Ica y su importancia como portadora del virus del "acha parrarriento" del maiz. *Promocion* (Ica) 1(1):48-50.

Usman, S.
1953a *Empoasca devastans* Distant – a little known potential insect pest in Mysore. *Mysore Agricultural Journal* 29(3-4):104-107.

Uthamasamy, S.
1969a Studies on the varietal resistance of bhendi, *Abelmoschus esculentus* (L.), to the leafhopper, *Amrasca* (*Empoasca*) *devastans* (Dist.) (Homoptera: Jassidae). M. Sc. Dissertation, University of Madras.

1980a Studies on host-resistance in certain okra (*Abelmoschus esculentus* [L.] Moench) varieties of the leafhopper, *Amrasca devastans* (Dist.) (Cicadellidae: Homoptera). *Pesticides* 14:39.

1985a Influence of leaf hairiness on the resistance of bhendi or lady's finger, *Abelmoschus esculentus* (L.) Moench, to the leafhopper, *Amrasca devastans* (Dist.). *Tropical Pest Management* 31(4): 294-295, 341, 345.

Uthamasamy, S. and Balasubramanian, M.
1978a Efficacy of some insecticides in controlling the pests of bhendi, *Abelmoschus esculentus* (L.) Moench. *Pesticides* 12:39-41.

Uthamasamy, S., Jayaraj, S. and Subramaniam, T. R.
1971a Biochemical mechanisms of resistance in bhendi (*Abelmoschus esculentus* [L.]) to the leafhopper, *Amrasca* (*Empoasca*) *devastans* (Dist.) *South Indian Horticulture* 19:53-59.

1972a Studies on the varietal resistance of bhendi (*Abelmoschus esculentus* [L.] Moench) to the leafhopper *Amrasca* (*Empoasca*) *devastans* (Dist.) (Homoptera: Jassidae). IV. Antibiosis mechanism in bhendi varieties under insectary conditions. *South Indian Horticulture* 20(1/4):71-75.

Uthamasamy, S. and Subramaniam, T. R.
1985a Inheritance of leafhopper resistance in Okra. *Indian Journal of Agricultural Sciences* 66:159-160.

Uthamasamy, S., Subramaniam, T. R. and Jayaraj, S.
1972a Studies on the plant characters of bhendi (*Abelmoschus esculentus* [L.] Moench) associated with the resistance to the leafhopper, *Amrasca* (*Empoasca*) *devastans* (Dist.) (Homoptera: Jassidae). *Progressive Horticulture* 3(4):25-31.

1973a Studies on the varietal resistance of bhendi (*Abelmoschus esculentus* [L.] Moench) to the leafhopper *Amrasca* (*Empoasca*) *devastans* (Homoptera: Jassidae). *Madras Agricultural Journal* 60:27-31.

1976a Preference of the leafhopper, *Amrasca biguttula biguttula* for okra (*Abelmoschus esculentus* [L.] Moench) varieties. *Indian Journal of Entomology* 37:100, 102.

Vacke, J.
1966a Die durch die Zikade *Macrosteles laevis* Rib. ubertragbaren Krankheiten. *Sbornik UVTI, Ochrana Rostlin* 2(39):79-80.

Vacke, J. and Hoppe, W.
1975a Leafhopper-borne mosaic disease of wheat new to Czechoslovakia. *Sbornik UVTI, Ochrana Rostlin* 11:335-336.

Vago, C. and Flandre, O.
1963a Culture prolongee de tissus d'insectes et de vecteurs de maladies en coagulum plasmatique. *Annals des Epiphyties* 14:127-139.

Vago, C. and Giannotti, J.
1972a Les mycoplasme chez les vegetaux et chez les vecteurs. *Physiolgoie Vegetale* 10:87-101.

Vaithilingam, C., Balasubramanian, M. and Bakaran, P.
1979a K-induced crop resistance against certain insect pests of paddy. I. Sap sucking insects. *Indian Journal of Plant Protection* 7:82-88.

Valencia, S. L., Mochida, O. and Basilio, R. P.
1983a Efficacy of burofezin (NNI-750) for brown planthopper (*N. lugens*), green leafhopper (*Nephotettix* sp.) and white-backed planthopper (*S. furcifera*) control. *International Rice Research Newsletter* 8(4):18-19.

Valenta, V.
1958a The problem of stolbur and related yellows type virus diseases in Czechoslovakia. *Agronomski Glasnik* 8:87-102.

Valenta, V., Musil, M. and Misiga, S.
1961a Investigations on European yellows-type viruses. *Phytopathologische Zeitschrift* 42:1-38.

Valentine, E. W.
1963a New records of hymenopterous parasites of Homoptera in New Zealand. *New Zealand Journal of Science* 6:6-13.

Valle, R. R.
1985a A comparative study of the bionomic and demographic parameters of 4 green rice leafhoppers, *Nephotettix* species (Homoptera: Cicadellidae). M. S. Thesis, Kyoto University, Kyoto, Japan.
1985b Biology of the rice green leafhopper (FLH). *International Rice Research Newsletter* 10(1):23.

Valle, R. R. and Kuno, E.
1984a A comparative study of some bionomic parameters of three species of green leafhoppers. *International Rice Research Newsletter* 9(4):17.

Valley, K. R. and Wheeler, A. G., Jr.
1985a Leafhoppers (Homoptera: Cicadellidae) associated with ornamental honey locust: Seasonal history, habits and descriptions of eggs and fifth instars. *Annals of the Entomological Society of America* 78(6):709-716.

Vanderveken, J.
1964a Studies on a virus of white clover transmitted by *Euscelis plebeja*. Sixteenth *International Symposium on Phytopharmacy and Phytiatry*, Ghent. Pp. 917-925.
1965a Studies on the hibernation of the clover phyllody virus. *Seventeenth International Symposium on Phyopharmacy and Phytiatry*, Ghent. Pp. 1796-1800.

Van Der Laan, P. A.
1961a Toxaphene and Delnav as insecticides on cotton in the Sudan Gezira. *Empire Cotton Growing Review* 38(2):111-118.

Van Der Meer, F. A. and Fluiter, H. J. de
1962a Control of Rubus stunt in raspberry by means of chemical control of the vector. *Fourteenth International Symposium on Phytopharmacy and Phytiatry, Ghent* 27(3)1053-1059.
1970a Rubus stunt. *In: Frazier, N. W., ed. Virus Diseases of Small Fruits and Grapevines.* Pp. 128-132. University of California, Division of Agricultural Science. 290 pp.

Van Dinther, J. B. M.
1963a Rice-sucking insects in Surinam. *Mededelingen van de Landbouwhogeschool en de Opzoekingsstations van de Staat te Gent* 28:815-823.

Van Halteren, P.
1979a The insect pest complex and related problems of lowland rice cultivation in South Sulawesi, Indonesia. *Mededelingen Landbouwrhogeschool Wageningen* 79:1-112.

Van Halteren, P. and Sama, S.
1973a *The Tungro Disease in South Sulawesi.* Ministry of Agriculture, Central Research institute for Agriculture. Makassar Research Station for Agriculture, Makassar, Bogor, Indonesia.

Van Rensburg, G. D. J.
1979a Maize streak disease: A new infection technique for use under field conditions. *In: Du Plessis, J. G., ed. Proceedings of the 3rd South African Maize Breeding Symposium.* Department of Agricultural Technical Services Technical Communication No. 152. Pp. 80-89.
1981a Effect of plant age at the time of infection with maize streak virus on yield of maize. *Phytophylactica* 13:197-198.
1982a Laboratory observations on the biology of *Cicadulina mbila* Naude (Homoptera: Cicadellidae). A vector of maize streak virus. I. The effect of temperature. *Phytophylactica* 14:99-107.
1982b Laboratory observations on the biology of *Cicadulina mbila* (Naude) (Homoptera: Cicadellidae), a vector of maize streak disease. 2. The effect of selected host plants. *Phytophylactica* 14(3):109-111.
1983a Southern African species of the genus *Cicadulina* China (Homoptera: Cicadellidae) with descriptions of new species. *Entomology Memoirs, Department of Agriculture Technical Services Republic South Africa* 57:1-22.

Van Rensburg, G. D. J. and Kuhn, H. C.
1977a Maize streak disease. South Africa Department of Agriculture Technical Service. *Maize Series Leaflet* No. 152.

Van Rensburg, G. D. J. and Walters, M. C.
1977a A method for long-distance transport of *Cicadulina mbila* (Naude) (Homoptera: Cicadellidae), a vector of maize streak virus. *Phytophylactica* 9:115-116.
1978a The efficacy of systemic insecticides applied to the soil for the control of *Cicadulina mbila* (Naude) (Hem: Cicadellidae), the vector of maize streak disease, and the maize stalk borer *Busseola fusca* (Fuller) (Lep: Noctuidae). *Phytophylactica* 10:49-52.

Van Stalle, J.
1982a Scientific results of the Belgian Mount Cameroon expedition (February-April 1981). V. Ulopinae and Megophthalminae (Homoptera, Cicadellidae). *Bulletin et Annales de la Societe Royale Entomologique de Belgique* 118(7-12):291-294.
1983a On some new and interesting West African Ulopinae and Megophthalminae (Homoptera, Cicadellidae). *Suomen Hyonteistieteellinen Aikakauskirja* 49(3):75-78.
1983b On some new and interesting East African Megophthalminae. *Revue de Zoologie Africaine* 97(2):304-312. [New genus: *Coronophtus*.]
1983c New and interesting African Ulopinae. *Revue de Zoologie Africaine* 97(4):816-820.

Van Velsen, R. J.
1967a *Axonopus* chlorotic streak, a leafhopper transmitted virus of *Axonopus affinis* in New Guinea. *Papua New Guinea Agricultural Journal* 18:139-141.

Vargo, A.
1970a The sounds and associated behavior of *Empoasca fabae* (Harris) (Homoptera Cicadellidae). M.S. Thesis, Iowa State University.

Varisai Muhammad, S. and Stephen Doriraj, M.
1967a Inheritance studies in *Ricinus communis* L. 1. Resistance to jassid infestation. *Madras Agricultural Journal* 54(6):323-325.

Varmah, J. C.
1981a Sandal (*Santalum album* L.) spike disease. *In: Maramorosch, K. and Raychaudhuri, S. P., eds. Mycoplasma Disease of Trees and Shrubs.* Pp. 253-258. Academic Press, New York. 362 pp.

Varney, E. H.
1977a Viruses and mycoplasmalike disease of blueberry. *Horticultural Science* (Stuttgart) 12:476-478.
1981a Blueberry stunt. *In: Maramorsch, K. and Raychandhuri, S. P., eds. Mycoplasma Diseases of Trees and Shrubs.* Pp. 245-252. Academic Press, New York. 362 pp.

Varty, I. W.
1963a A survey of the sucking insects of the birches in the maritime provinces. *Canadian Entomologist* 95(10):1097-1106.
1964a *Erythroneura* leafhoppers from birches in New Brunswick. I. Subgenus *Erythridula* (Homoptera, Cicadellidae). *Canadian Entomologist* 96:1244-1255.
1967a Leafhoppers of the subfamily Typhlocybinae from birches. *Canadian Entomologist* 99(2):170-180.
1967b *Erythroneura* leafhoppers from birches in New Brunswick: II. Subgenus *Eratoneura*. *Canadian Entomologist* 99(6):570-573.

Varzimska, R.
1983a Leafhoppers of the Maritime zone of the Gulf of Riga. *Latvijas Entomologs* (Riga) 26:68-82.

Vasudeva, R. S. and Sahambi, H. S.
1959a Phyllody diseases: Transmitted by a species of *Deltocephalus* Burmeister. *Proceedings of the 6th International Congress of Crop Protection* 1:359-360.

Vaszai-Virag, E.
1984a Occurrence of cicadellids (Homoptera: Auchenorrhyncha) in insect material collected in light traps in maize fields. *Novenyvedelem* 20:213.

Veeravel, R. and Baskaran, P.
1976a Efficacy of quinalphos and other insecticide sprays against insects pests (of brinjal). *Madras Agricultural Journal* 63:338-340.

Velikan, V. S., Gegechkori, A. M., Golub, V. B., Danzig, E. M., Dorokhova, G. I., Emelyanov, A. F., Eromolenko, V. M., Zerova, M. D. and Kuslitskij, V. S.
1984a *Key to Harmful and Useful Insects and Mites of Fruit and Berry Cultures in the USSR*. Kolos, Leningrad. 288 pp.

Velikan, V. S., Golub, V. B., Gureva, E. L., Dzhanokmen, K. A., Dorokhova, G. I., Emelyanov, A. F., Ermolenko, V. M., Zerova, M. K., and Kasparyan, D. R.
1980a *Key to the Identification of Harmful and Useful Insects and Acari of Industrial Crops in the USSR*. Kolos, Leningrad. 335 pp.

Velimirovic, V.
1966a *Erythroneura Arboridia adanae vitisuga-diabloa*, novi stetni insekt na lozi u Crnoj Gori. *Poljoprivredna i Sumarstvo* 3:53-80.
1980a The leafhopper *Empoasca decedens* Paoli—a pest of citrus cultures in the Montenegran Littoral Zone. *Zastita Bilja* 31:273-276.

Velusamy, R., Janaki, I. P., Swaminathoan, R. and Subramanian, T. R.
1975a Varietal resistance in rice to the white leafhopper (*Cicadella spectra* Distant). *Madras Agricultural Journal* 62:305-307.

Venkatanarayanan, D.
1971a Studies on resistance of paddy to three species of leafhoppers (Jassidae: Homoptera). M. Sc. Dissertation, Madurai University.

Venkataraman, A., Abraham, E. V. and Samuel, J. C.
1973a Control of rice pests with granular insecticides in Thanjavur District. *Madras Agricultural Journal* 60:428-430.

Venkataraman, A. and Abraham, E. V. and Srinivasan, D.
1971a Ultra low volume application of phosphamidan 100 for control of paddy pests in Thanjavur. *Madras Agriculture Journal* 58(4):267-274.

Venkatarao, A. and Shanmugam, N.
1983a Diseases of sesamum and their control. *Pesticides* 17:34-35.

Vereshchagina, V. V.
1962a Rose leafhopper (*Typhlocyba rosae* L.) in the gardens and nurseries of Moldavia (Homoptera, Typhlocybinae). *Zoologicheskii Zhurnal* 41(11):1637-1645.

Vereshchagina, V. V. and Vereshchagina, B. V.
1969a On insects that damage berry-fruit bushes in Moldavia. Vrednaig i Poleznaia Fauna Bespozvonochnykyh. *Moldavii* 4-5:167-183. [In Russian.]

Verma, K. L.
1976a Some leafhopper records from Simla Hills. *Entomologist's Newsletter, Division of Entomology, Indian Agricultural Research Institute* 6(6&7):43.

Verma, S. and Pant, N. C.
1979a Movement of phorate in soil and its residues and residual toxicity of phorate against the pests of green gram (moong) and red gram (arhar) under laboratory conditions. *Pesticides* 13:27-30.

Veronica, B. K. and Kalode, M. B.
1983a Rice varieties having multiple resistance to leafhoppers. *Indian Journal of Agricultural Sciences* 53:378-380.

Vidano, C.
1958a Le Cicaline Italiane della vite (Hemiptera, Typhlocybidae). *Bollettino di Zoologia Agraria e Bachicoltural* 2(1):61-115.
1959a Sulla identificazione specifica di alcuni Erythroneurini europei (Hemiptera, Typhlocybidae). *Annali del Museo Civico di Storia Naturale Genoa* 71:328-348.
1959b Revisione delle *Erythria* ed *Erythridea* alpine con descrizione di specie nuove (Hemiptera, Typhlocybidae). Contributi scientifici alla conoscenza del Parco Nazionale del Gran Paradiso: N. 11. *Bollettino dell' Instituto di Entomologia della Universita degli Studi di Bologna* 23:293-343.
1959c Possibili rapporti tra Emitteri Tiblocibidi e degenerazione infettiva della Vite. *Organo della Societa Italiana de Fitoiatria* 47-48 (N.S. 26-27):219-226.
1960a Dioecia obbligata in *Typhlocyba* (*Ficocyba* n. subg.) *ficaria* Horvath. *Bollettino dell' Instituto di Entomologia della Universita degli studi di Bologna* 24:121-145.
1960b Indagubu siora yb deoerunebti dekka *Vigna sinensis* Endlicher in coltura italiana. *Bollettino di Zoologia Agraria e di Bachicoltura* (2)3:3-97
1961a Descrizione di una nuova specie di *Typhlocyba* (subg. Edwardsiana) del *Platanus*. *Memorie della Societa Entomologica Italiana* 40:44-50.
1961b Variazioni infraspecifiche in *Erythroneura* provocate da fattori ambientali abiotici (Hemiptera, Homoptera, Typhlocybidae). *Atti della Accademia delle Scienze di Torino* 95:527-533.
1961c L'influenza microclimatica sui caratieri tassonomici in tiflocibidi sperimentalmente saggiati. *Memorie della Societa Entomologica Italiana* 40:144-167.
1961d Una Cicalina nuova nemica dei platani in Italia la *Typhlocyba* (*Edwardsiana*) *platanicola* Vidano. *L'Italia Agricola* 98(12):1157-1165.
1961e Scoperta della dioecia obbligata in *Typhlocyba ficaria* Horvath (Hem. Hom. Typhlocybidae) (Nota riassuntiva). *Verhandlungen XI Internationaler Kongres fur Entomologie*, Vienna 1:46-48.
1962a Ampelopotie da Tifloabidi. La *Empoasca libyca* Bergevin nuovo nemico della Vita in Italia. Studi del Gruppo di Lavoro del C.N.R. per le virosi: XXXIV. *L'Italia Agricola* 4:329-346.
1962b Sulla alterata morfogenesi in auchenorinchi parassitizzati e sulle sue interferenze speciografiche (Hemiptera, Homoptera, Jassidae). *Atti della Accademia delle Scienze di Torino* 96:557-574.
1962c Massiccio attacco alle viti insulari dell'*Empoasca libyca*. [G. Agric.] 72(45):427.
1963a Appunti comparativi sui danni da Cicaline alla Vite. *Informatore Fitopatologico* (Bologna) 3:173-177.
1963b Esiste in Europa la Virosi della Vite detta "Pierce's disease?" *Atti della Accademia delle Scienze di Torino* 97:289-324.
1964a Scoperta in Italia dellos *Scaphoideus littoralis* Ball, Cicalina americana collegata alla "Flavecence doree" della Vite. *L'Italia Agricola* 101(10):1031-1049.
1964b Contributo alla conoscenza dei Typhlocybidae di Sardegna. *Archivo Botanico e Biogeografico Italiano* 40:120-130.
1965a A contribution to the chorological and oceological knowledge of the European Dikraneurini. *Zoologische Beitrage* (Neue Folge) 2:343-367.
1965b Sulle *Forcipata* transalpine e cisalpine con descrizione di specie nuove (Homoptera Typhlocybidae). *Bollettino di Zoologia Agraria e di Bachicoltura* 6:37-60.
1965c Responses of vitis to insect vector feeding. *Proceedings International Conference on Virus and Vector on Perennial Hosts, with Special Reference to Vitis.* Pp. 73-80.

1966a Alterazioni provocate da insetti in *Vitis* osservate, sperimentate e comparate. Studi del gruppo di lavoro del C.M.R. perle virosi: LI. [Observations, experiments and comparisons of the injuries caused by insects on *Vitis*. Studies of the working group of the National Council of Research for virus diseases: LI.] *Annali della Facolta di Science Agrarie della Universita degli Torino* 1:513-642. [English summary.]

1966b Scoperta della ecologia ampelofila del cicadellide *Scaphoideus littoralis* Ball nella regione neartica originaria. [Discovery of the ecology of the cicadellid *S. littoralis* Ball on vines in its Nearctic region of origin.] *Annali della Facolta di Science Agrarie della Universita degli Torino* 3:297-302. [English summary.]

1967a Microecologia e miltiplicazione di virus fitopatogeni dioce nell'insetto vettore. Studi del Gruppo di lavoro del C.N.R. per le Virosi: CX. Publicazione No. 125 del Centro di Entomologia alpina e forestale del Consiglio Nazionale delle Ricerche (Diretto dal Prof. Athos Goidanich). *Annali della Facolta di Scienze Agrarie della Universita degli Studi Torino* 5:1-16. [English summary.]

1968a Recensione. *Bollettino della Societa Entomologica Italiana Colume XVCIII*, N. 7-8. [*Announcement of the Publication of a Catalog of the Rhynchota by A. Servadei; Item Servadei, A. 1967a.*]

1972a Electron microscopy of leafhopper-transmitted viruses. p. 450. *Proceedings of the XIII International Congress of Entomology,* Moscow, 2-9 August 1968. Vol. 3. 494 pp. Rafes, P. M., ed. Symposium F. Insects as Vectors of Virus-Borne Plant Disease.

1981a Contributo alla conoscenza delle *Zyginidia* d'Italia. *Memorie della Societa Entomologica Italiana* 60:343-355.

Vidano, C. and Arzone, A.

1976a Tiflocibini infestanti piante officianali coltivate in piemonte. *Annali dell'Accademia di Agricoltura di Torino* 118:1-14.

1976b Sulla collezione Spinola conservata nel Castello di Tassarolo. *Atti XI Congresso Nazionale Italiano di Entomologia.* Pp. 253-260.

1978a Typhlocybinae on officinal plants. *Auchenorrhyncha Newsletter* 1:27-28.

1981a Typhlocybinae of broad-leaf trees in Italy. 1. *Alnus. Acta Entomologica Fennica* 38:47-49.

1983a Biotaxonomy and epidemiology of Typhlocybinae on vine. *In: Knight, W. J., Pant, N. C., Robertson, T. S. and Wilson, M. R., eds. Proceedings of the 1st International Workshop on Biotaxonomy, Classification and Biology of Leafhoppers and Planthoppers (Auchenorrhyncha) of Economic Importance.* Pp. 75-85. London, 4-7 October 1982. Commonwealth Institute of Entomology, 56 Queen's Gate, London SW7 5JR. 500 pp.

1984a "Wax area" in cicadellids and its connection with brochosomes from Malpighian tubules. *Mitteilungen der Schweizerischen Entomologischen Gesellschaft* 57(4):444-445.

1984b *Zyginidia* infesting maize in Italy. *Proceedings of the XVII International Congress of Entomology* Hamburg:541.

1984c Phytopathological consequences of the migrations of *Zyginidia pullula* from grasses to cereals. *Mitteilungen der Schweizerischen Entomologischen Gesellschaft* 57(4):406-407.

1985a *Zyginidia pullula*: Distribuzione nel territorio e ciclo biologico. *Redia* 68:135-150.

1985b Cicaline paleartiche cerealicole del genere *Zyginidia*. Indagini speciografiche, corologiche ed ecologiche. *Apicoltore moderno* 76(3):71-82.

Viennot-Bourgin, G.

1981a Simultaneous observation in France of rhododendron bud blast and a cicadellid playing the role of vector. *Agronomie* 1:87-92.

Viggiani, G.

1971a Ricerche sulla entomofauna del Nocciolo. 3. Le cicaline (Homoptera, Typhlocybidae). 1. Identificazione delle specie e reperti biologici preliminari. *Bolletino del Laboratorio di Entomologia Agraria Filippo Silvestri* 29:149-173.

1973a Studies on the insect fauna of hazel. IX. Notes on some Homoptera (*Ceresa bubalus* Fbrc., *Ledra aurita* O. L., and *Cicadella viridis* L.). *Annali della Facolta di Scienze Agrarie della Universita degli Studi di Napoli* 7:153-160.

1981a Notes on some species of *Oligosita* Walker (Hym. Trichogrammitidae) and descriptions of four new species. *Bollettino del Laboratorio di Entomologia Agraria Filippo Silvestri* 38:125-132.

1985a Parassiti oofagi delie cicaline delle querce: *Epoligosita vera* Viggiani. *Proceedings XIV Italian National Congress of Entomology, Accademia Nazionale di Entomologia.* Pp. 867-872.

[Vignault, J. C. et al.]

1980a Culture of spiroplasmas from plant material and insects from Mediterrean and Near-eastern countries. *Comptes Rendus Hebdomadaries des Seances de l'Academie des Sciences* 290:775-778.

Vilbaste, J.

1955a Eesti NSV soode rohurinde nokaliste faunast. *Loodusuurijate Seltsi Aastaraamat* 48:104-121.

1958a Markmeid eesti NSV madalsoode Tsikaadide Faunast. *Eesti NSV Teaduste Akadeemia Toimetised* (Bioloogiline Seeria No. 1) 7:48-51.

1958b Bemerkungen zur Zikadenfauna der Niederungsmoore in der Estnischen SSR. *Akademiya Nauk Estonskoi SSR, Izvestiya Seriya Biologicheskogo* 7(1):47-52. [In Estonian with German summary.]

1959a Eine neue Art der Gattung *Anaceratagallia* Zachw. aus Estland. *Eesti NSV Teaduste Akadeemia Toimetised* (Bioloogiline Seeria No. 3) 8:199-203. [In Russian with Estonian and German summaries.]

1959b Notes on the fauna of the Cicadinea of the Soviet Carpathians. *In: Fauna and Animal Life of the Soviet Carpathians.* Science Reports, Uzhgorod State University 40:173-179.

1960a A new species of the genus *Psammotettix* Haupt (H: Jassidae) from Uzbekistan. *Tashkent Agricultural Institute and Institute of Zoology and Parasitology, Academy of Science, Uzbek SSR.* Pp. 119-121.

1961a New species of cicadellids (Homoptera: Jassidae). *Uzbekskii Biologicheskii Zhurnal* 1:42-50. [In Russian.] [New genus: *Platytettix.*]

1961b Neue Zikaden (Homoptera: Cicadina) aus der Umgebung von Astrachan. *Eesti NSV Teaduste Akadeemia Toimetised* (Bioloogiline Seeria No. 4) 10:315-331. [In Russian with German summary.]

1962a Uber die Arten *Rhopalopyx preyssleri* (H.-S.) und *Rh. adumbrata* (C. R. Sahlb.) (Homoptera, Jassidae). *Notulae Entomologicae* 42:62-66.

1962b Uber die Zikadenfauna des ostlichen Teiles des kaspischen Tieflandes. *Eesti NSV Teaduste Akademia Juures Asuva Loodusuurijate Seltsi Aastaraamat* 55:129-151. [In Russian with Estonian and German summaries.]

1964a Uber die Zikadenfauna der Auwiesen Estlands. *Eesti NSV Teaduste Akadeemia Toimetised* (Bioloogiline Seeria No. 13):302-318. [In Estonian with German and Russian summaries.]

1965a On the Genus *Aconura* Leth. (Homoptera, Jassidae). *Notulae Entomologicae* 45:3-12. [New genera: *Anareia, Altaiotettix.*]

1965b Uber die zikadenfauna Altais. *Tartu.* 144 pp. [In Russian with German summary.]

1966a Neue Zikadenarten aus dem primorje-gebiet. 1. *Eesti NSV Teaduste Akadeemia Toimetised XV Koide* (Bioloogiline Seeria No. 15):61-71. [In Russian with German summary.]

1967a On some East-Asiatic leafhopper genera (Homoptera: Cicadina: Jassidae). *Insecta Matsumurana* 30(1):44-51.

1968a Systematic treatise of Cicadas found on the edge of the coastal regions. Uber die Zikadenfauna des Primorje Gebietes. *Izdatel'stvo "Valgus,"* Tallin. 195 pp. [Descriptions of 6 new genera and 1 new subgenus.] [In Russian with German summary.]

1969a On some East-Asiatic leafhoppers described by Professor S. Matsumura (Homoptera: Cicadinea: Iassidae). *Insecta Matsumurana* — Supplement 6. 12 pp. [New genera: *Takagiella, Alishania, Peitouellus, Hengchunia, Watanabella* and *Wanutettix.*]

1969b On the fauna of Homoptera Cicadina of Taimyr. *Eesti NSV Teaduste Akadeemia Toimetised* 18, Biologia No. 3:258-268. [English summary.]

1973a Revision of the collection of G. Flor. II. Homoptera: Cicadinea: Cicadelloidea. *Eesti NSV Teaduste Akadeemia Toimetised* 22, Bioloogia No. 1:15-28.

1973b On North-European species of the genus *Limotettix* J. Sb., with notes on North-American species (Homoptera, Cicadellidae). *Eesti NSV Teaduste Akadeemia Toimetised* 22, Biologia No. 3:199-209. [In English with Estonian summary.]

1974a Preliminary list of Homoptera-Cicadinea of Latvia and Lithuanian. *Eesti NSV Teaduste Akadeemia Toimetised* 23, Bioloogia No. 2:131-163. [In English with Latvian and Russian summaries.]

1975a On some species of Homoptera-Cicadinea described by V. Motschulsky. *Eesti NSV Teaduste Akadeemia Toimetised* 24, Biologia No. 3:228-236. [In English with Estonian summary.] [New genus: *Oneratulus.*]

1975b Zur Larvalsystematik der Zikaden. *Verhandlungen des Sechsten Internationalen Symposiums uber Entomofaunistik in Mitteleuropa.* Pp. 343-346.

1976a A revision of Homoptera-Cicadinea described by S. Matsumura from Europe and the Mediterranean area. *Eesti NSV Teaduste Akadeemia Toimetised* 25, Bioloogia No. 1:25-36. [In English with Estonian summary.]

1979a The Hemipteroidea of the Vooremaa hardwood-spruce forest. *Progress Report, Estonian Contributions to the International Biological Programme* 12:70-84. [In English.]

1980a On the Homoptera-Cicadinea of Kamchatka. *Annales Zoologici* (Warsaw) 35(24):367-418.

1980b *Homoptera Cicadinea of Tuva.* "Valgus," Tallinn. 217 pp. [New genus: *Parocerus*] [In Russian with English summary.]

1982a Preliminary key for the identification of nymphs of North European Cicadinea. II. Cicadelloidea. *Annales Zoologici Fennici* 19:1-20.

1983a Zoogeography of the Auchenorrhyncha of the USSR and adjoining territories. *In: Knight, W. J., Pant, N. C., Robertson, T. S. and Wilson, M. R., eds. Proceedings of the 1st International Workshop on Biotaxonomy, Classification and Biology of Leafhoppers and Planthoppers (Auchenorrhyncha) of Economic Importance.* Pp. 279-289. London, 4-7 October 1982. Commonwealth Institute of Entomology, 56 Queen's Gate, London SW7 5JR. 500 pp.

Villiers, A.

1956a Contribution a l'etude du peuplement de la Mauritanie: Description de nouveaux Hemipteres. *Bulletin de l'Institut Francais d'Afrique Noire* (Serie A) 18:834-842.

1971a Homopteres recoltes dans la haute vallee du Giffre. *Entomologiste* (Paris) 27(3):62-64.

Vincent, J. J.

1975a Study of the entomological fauna of an apple orchard in the Paris basin by means of coloured pan traps. *Annales de la Societe Entomologique de France* 11:303-334.

Viraktamath, C. A.

[Undated] Oriental Agalliinae, Macropsinae, Iassinae and Idiocerinae — a review (Homoptera: Cicadellidae). *In: Workshop on Advances in Insect Taxonomy in India and the Orient,* Manali (H. P.), 9-12 October 1979. Pp. 78-79 (Abstract). Association for the Study of Oriental Insects, Delhi. 106 pp.

1972a A new species of *Austroagallia* Evans from the Galapagos Islands (Homoptera: Cicadellidae). *Occasional Papers of the California Academy of Science* 101:1-4.

1973a A new species of Idiocerinae (Cicadellidae: Homoptera) on *Semecarpus anacardium* L. *Oriental Insects* 7(1):133-135.

1973b Some species of Agalliinae (Cicadellidae, Homoptera) described by Dr. S. Matsumura. *Kontyu* 41(3):307-311.

1976a New species of *Doratulina* and *Bumizana* (Homoptera: Cicadellidae) from Karnataka. *Oriental Insects* 10(1):79-86. [New subgenus: *Cymbopogonella.*]

1976b Four new species of Idiocerine leafhoppers from India with a note on male *Balocha astuta* (Melichar) (Homoptera: Cicadellidae: Idiocerinae). *Mysore Journal of Agricultural Science* 10:234-244.

1978a A new species of *Cestius* Distant (Homoptera: Cicadellidae) from southern India. *Journal of Natural History* 12:241-244.

1979a Studies on the Iassinae (Homoptera: Cicadellidae) described by Dr. S. Matsumura. *Oriental Insects* 13(1&2):93-108.

1979b *Jogocerus* gen. nov. and new species of idiocerine leafhoppers from southern India (Homoptera: Cicadellidae). *Entomon* (India) 4(1):17-26.

1979c Four new species of *Idioscopus* (Homoptera: Cicadellidae) from southern India. *Entomon* (India) 4(2):173-181.

1980a Indian Macropsinae (Homoptera: Cicadellidae). I. New species of *Macropsis* from South India. *Journal of Natural History* 14(3):319-329.

1980b Four new species of Agalliinae (Homoptera: Cicadellidae) from Juan Fernandez. *Journal of Natural History* 14(5):621-628.

1980c Notes on *Idioscopus* species (Homoptera: Cicadellidae) described by Dr. H. S. Pruthi, with description of a new species from Meghalaya, India. *Entomon* (India) 5(3):227-231.

1981a Indian species of *Grammacephalus* (Homoptera: Cicadellidae). *Colemania* 1(1):7-12.

1981b Indian Macropsinae (Homoptera: Cicadellidae). II. Species described by W. L. Distant and descriptions of new species from the Indian subcontinent. *Entomologica Scandinavica* 12:295-310.

1983a Genera to be revised on a priority basis. The need for keys and illustrations of economic species of leafhoppers and preservation of voucher specimens in recognized institutions. *In: Knight, W. J., Pant, N. C., Robertson, T. S. and Wilson, M. R., eds. Proceedings of the 1st International Workshop on Biotaxonomy, Classification and Biology of Leafhoppers and Planthoppers (Auchenorrhyncha) of Economic Importance.* Pp. 471-492. London, 4-7 October 1982. Commonwealth Institute of Entomology, 56 Queen's Gate, London SW7 5JR. 500 pp.

Viraktamath, C. A. and Dworakowska, I.

1979a Indian species of *Singapora* (Homoptera: Cicadellidae: Typhlocybinae). *Oriental Insects* 13(1&2):87-91.

Viraktamath, C. A. and Murphy, D. H.

1980a Description of two new species with notes on some Oriental Idiocerinae (Homoptera: Cicadellidae). *Journal of Entomological Research* 4(1):83-90.

Viraktamath, C. A. and Sohi, A. S.

1980a Notes on the Indian species of *Austroagallia* (Homoptera: Cicadellidae). *Oriental Insects* 14(3):283-289.

Viraktamath, S. and Viraktamath, C. A.

1980a Redescriptions of *Allectus, Divitiacus* and *Lampridius* (Homoptera: Cicadellidae) described by W. L. Distant. *Entomon* (India) 5(2):135-140.

1981a Biology of two species of *Austroagallia* (Homoptera: Cicadellidae) from India with description of one new species. *Colemania* 1(2):79-87.

1982a Biology of *Agallia campbelli* (Homoptera: Cicadellidae) in south India. *Colemania* 1:155-162.

1985a New species of *Busoniomimus* and *Idioscopus* (Homoptera: Cicadellidae: Idiocerinae) breeding on mango in South India. *Entomon* 10(4):305-311.

Viswanathan, P. R. K. and Kalode, M. B.

1981a Studies on varietal resistance and host specificity of rice green leafhoppers. *International Rice Research Newsletter* 6(3):7-8.

1984a Comparative study of varietal resistance to rice green leafhoppers, *Nephotettix virescens* (Distant) and *N. nigropictus* (Stal). *Proceedings of the Indian Academy of Science* 93:55-63.

Visvanathan, T. and Abdul-Kareem, A.

1983a Field evaluation of some insecticides against jassids, on cotton. *Pesticides* 17:33-34.

Vlasov, Y. I. and Rodina, K. I.

1969a On the classification of some cereal virus diseases. *Byulleten, Vsesoyuznogo Nauchno-issledovatelels skogo Instituta Zashchityi Rastenii* 1(13):61-65. [In Russian with English summary.]

Vlasov, Y. I., Rodina, K. I. and Larina, E. I.

1971a Vectors of winter wheat mosaic virus. Perenoschiki virusa mozaiki ozimoi Pshenitsy. *Byulleten' Vsesoyuznogo Nauchno-issledovatelel' skogo Instituta Zashchityi Rastenii* 20:35-38. [In Russian with English summary.]

Vogel, O.

1978a Pattern formation in the leafhopper *Euscelis plebejus* Fall. (Homoptera): Developmental capacities of fragments isolated from the polar egg regions. *Developmental Biology* 67(2):357-370.

1982a Development of complete embryos in drastically deformed leafhopper eggs. Wilhelm Roux's Archives of *Developmental Biology* 191(2):134-136.

1982b Experimental test fails to confirm gradient interpretation of embryonic patterning in leafhopper eggs. *Developmental Biology* 90(1):160-164.

1983a Test for intercalary regeneration of the metameric pattern of the leafhopper *Euscelis pelbejus* Fall. (Homoptera). Wilhelm Roux's Archives of *Developmental Biology* 192(5):295-298.

1983b Pattern formation by interaction of three cytoplasmic factors in the egg of the leafhopper *Euscelis plebejus*. *Developmental Biology* 99(1):166-171.

Volk, J.

1958a Ubertragung durch Insekten und das Virus-Insekt Verhaltnis. *Pflanzliche Virologie* 1:1-279. Berlin.

1967a Ubertragung durch tierische Vektoren und das Virus-Vektor-Verhaltnis. *In: Klinkowski, M., ed. Pflanzliche Virologie*. Pp. 94-140. Akademie-Verlag, Berlin.

Vora, V. J., Bharodie, R. K. and Kapadia, M. N.

1985a Pest oilseed crops and their control. *Pesticides* 19:19-22.

Voss, G.

1983a The biochemistry of insecticide resistant hopper races. *In: Knight, W. J., Pant, N. C., Robertson, T. S. and Wilson, M. R., eds. Proceedings of the 1st International Workshop on Biotaxonomy, Classification and Biology of Leafhoppers and Planthoppers (Auchenorrhyncha) of Economic Importance*. Pp. 351-356. London, 4-7 October 1982. Commonwealth Institute of Entomology, 56 Queen's Gate, London SW7 5JR. 500 pp.

Vungsilabutr, P.

1978a Biological and morphological studies on *Paracentrobia andoi* (Ishii) (Hymenoptera: Trichogrammatidae), a parasite of the green leafhopper, *Nephotettix cincticeps* Uhler (Homoptera Delotcephalidae). *Esakia* 11:29-51.

Vyas, H. N. and Saxena, H. P.

1981a Some observations on the effectiveness of disulfoton and phorate seed treatment against galerucid beetle, jassid, plant height and yield of green gram. *Pesticides* 15:44.

1983a Bioefficacy of some insecticides against jassid *Empoasca kerri* Pruthi in green gram, *Vigna radiata* (L.) Wilczek. *Pesticides* 17:15-17.

Wade, V.

1966a *General Catalogue of the Homoptera. A Species Index of the Membracoidea and Fossil Homoptera (Homoptera: Auchenorrhyncha). A Supplement to Fascicle 1—Membracidae, of the General Catalogue of the Hemiptera*. Waverly Press, Inc., Baltimore. 40 pp.

Wadhi, S. R. and Batra, H. N.

1964a Pests of tropical and sub-tropical fruit trees. *In: Pant, N. C., Prasad, S. K., Vishnoi, H. S., Chatterji, S. N. and Varma, B. K., eds. Entomology in India*. Pp. 227-260. The Entomological Society of India. 529 pp.

Wagle, P. V.

1934a The mango hoppers and their control in the Konkan, Bombay presidency. *Agriculture Livestock in India* 4:176-188.

Wagner, W.

1958a Uber eine Zikaden-Ausbeute vom Grossen Belchen im Schwarzwald (Homoptera, Auchenorrhyncha). *Entomologische Mitteilungen aus dem Zoologischen Staatsinstitut und Zoologischen Museum Hamburg* 14:3-11 (435-443).

1958b Uber die Variabilitat der Penisform bei der Deltocephalide *Psammotettix helvolus* Kirschb. (Homopt). *Mitteilungen der Deutschen Entomologischen Gesellschaft* 17:90-92.

1958c Eine neue *Idiocerus*-Art (Homoptera, Jassidae) als Schadling auf *Pistacia vera*. *Entomologische Mitteilungen aus dem Zoologischen Staatsinstitut und Zoologischen Museum Hamburg* 16:1-4.

1959a Ueber neue und schon bekannte Zikadenarten aus Italien (Hemiptera-Homoptera). *Fragmenta Entomologica* 3 fasc. 4:67-86.

1959b *Chlorita tamaninii*, eine neue Zikade aus Nord-Italien (Homoptera-Auchenorrhyncha). *Bollettino de la Societa Entomologica Italiana* (Genoa) 89:51-54.

1959c Zoologische Studien in West-Griechenland von Max Beier, Wien. IX Teil: Homoptera. *Sitzungsberichte der Osterreichen Akademie der Wissenschaften* (Vienna) 168:583-605.

1960a Proposed use of the plenary powers to designate a type-species for the nominal genus *Macropsis* Lewis, 1834 in accordance with accustomed use (Class Insecta, Order Hem.) *Bulletin of Zoological Nomenclature* 17:185-188.

1960b Proposed use of the plenary powers to suppress the generic name *Promecopsis* Dumeril, 1806 (Class Insecta, Order Hemiptera) Z. M. (S.) 483. *Bulletin of Zoological Nomenclature* 17:191-192.

1963a Eine neue Art der Gattung *Fieberiella* Signoret 1880 (Insecta, Homoptera) aus Griechenland. *Beufortia* (Amsterdam) 10:48-50.

1963b Revision der europaischen Arten dreier Gattungen der Homoptera-Cicadina *Dryodurgades* Zakhvatkin, *Fieberiella* Signoret und *Phlepsius* Fieber. *Entomologische Mitteilungen aus dem Zoologischen Staatsinstitut und Zoologischen Museum Hamburg* 2(45):1-14.

1964a Die auf Rosaceen Lebenden *Macropsis*-Arten der Niederlande. *Entomologische Berichten* (Berlin) 24(7):123-136.

1967a Taxonomie der Gattung *Paluda* DeLong 1937 (Homoptera Euscelidae). *Zoologische Beitrage* 13(2&3):479-499.

1968a Die Grundlagen der dynamischen Taxonomie, zugleich ein Beitrag zur Struktur der Phylogenese. *Abhandlungen und Verhandlungen des Naturwissenschaftlichen Vereins in Hamburg* 12:27-66.

1968b 20. Cicadina (Auchenorrhynchi) (2. Unterordnung der Homoptera). *In: Weidner, H. and Wagner, W. Die Entomologischen Sammlungen des Zoologischen Staatsinstitutes und Zoologischen Museums Hamburg*. Pp. 134-156. *Mitteilungen aus dem Hamburgischen Zoologischen Museum und Institut* 65:123-179.

Wagner, W. and Duzgunes, Z.
1960a An *Idiocerus* sp. (Homoptera, Jassidae) as pest on *Pistacia vera*. *Bitki Koruma Bulteni* (Ankara) 1(4&5):58-61. [In Turkish.]

Wagner, W. and Franz, H.
1961a Unterfamilie *Auchenorrhyncha* (Zikaden). pp. 74-158. In: Franz, H. Die Nordost-Alpen im Spiegel ihrer Landtierwelt. II. Innsbruck. 792 pp.

Waite, G. K.
1976a Effect of various temperatures on development of the lucerne jassid *Austroasca alfalfae* (Evans) (Homoptera: Cicadellidae) with reference to population levels in lucerne. *Queensland Journal of Agricultural and Animal Sciences* 33:39-42.

1976b The economic status of lucerne jassids (*Austroasca* spp.) in south-eastern Queensland. *Queensland Journal of Agricultural and Animal Sciences* 33:67-72.

1977a Seasonal history of lucerne jassids in south-east Queensland. *Queensland Journal of Agricultural and Animal Sciences* 34:163-168.

1978a Insecticidal control of lucerne jassids in south-east Queensland. *Queensland Journal of Agricultural and Animal Sciences* 35:133-138.

Wakibe, H. and Miyahara, K.
1984a Host range of MLOs collected from various diseased plants in Saga Prefecture. *Proceedings of the Association for Plant Protection of Kyushu* 30:48-51.

Wakibe, H., Shiomi, T., Matzuzaki, M., Miyahara, K. and Sugiura, M.
1983a Survival of *Macrosteles orientalis* (Hemiptera, Deltocephalidae) on host plants of onion yellows. *Proceedings of the Association for Plant Protection of Kyushu* 29:24-26.

Walgenbach, J. F.
1982a Management of the potato leafhopper on potatoes in Wisconsin. M. S. Thesis, University of Wisconsin, Madison.

Walgenbach, J. F. and Wyman, J. A.
1984a Dynamic action threshold levels for potato leafhopper (Homoptera: Cicadellidae) on potatoes in Wisconsin. *Journal of Economic Entomology* 77(5):1335-1340.

1985a Potato leafhopper (Homoptera: Cicadellidae) feeding damage on various potato growth stages. *Journal of Economic Entomology* 78(3):671-675.

Walgenbach, J. F., Wyman, J. A. and Hogg, D. B.
1985a Evaluation of sampling methods and development of sequential sampling plan for potato leafhopper (Homoptera: Cicadellidae) on potatoes. *Environmental Entomology* 14(3):231-236.

Walker, I.
1979a Some British species of *Anagrus* (Hymenoptera: Mymaridae). *Zoological Journal of the Linnaean Society* 67:181-202.

Wallace, C. R.
1967a The identity of a cicadellid infesting lucerne crops in New South Wales—a correction. *Journal of the Entomological Society of Australia* (N.S.W.) 3:29.

Wallis, R. L.
1960a Host plants of the six-spotted leafhopper and the aster yellows virus and other vectors of the virus. *United States Department of Agriculture, Agriculture Research Service*, ARS-33-55. 15 pp.

1962a Spring migration of the six-spotted leafhopper in the western Great Plains. *Journal of Economic Entomology* 55:871-874.

1962b Host plant preference of the six-spotted leafhopper. *Journal of Economic Entomology* 55(6):998-999.

Waloff, N.
1973a Dispersal by flight of leafhoppers (Auchenorrhyncha: Homoptera). *Journal of Applied Ecology* 10(3):705-730.

1974a Biology and behaviour of some species of Dryinidae (Hymenoptera). *Journal of Economic Entomology* 49:97-109.

1975a The parasitoids of the nymphal and adult stages of leafhoppers (Auchenorrhyncha: Homoptera) of acid grasslands. *Transactions of the Royal Entomological Society of London* 126(4):637-686.

1978a Some characteristics of populations of grassland Auchenorrhyncha. *Auchenorrhyncha Newsletter* 1:7-8 (Abstract).

1979a Partitioning of resources by grassland leafhoppers (Auchenorrhyncha, Homoptera). *Ecological Entomology* 4(4):379-385.

1980a Studies on grassland leafhoppers (Auchenorrhyncha: Homoptera) and their natural enemies. *Advances Ecological Research* 11:81-215.

1981a The life history and descriptions of *Halictophagus silwoodensis* sp. n. (Strepsiptera) and its host *Ulopa reticulata* (Cicadellidae) in Britain. *Systematic Entomology* 6(1):103-113.

1982a Halictophagus silwoodensis Waloff (Strepsiptera, Halictophagidae) and its host *Ulopa reticulata* (F.) (Auchenorrhyncha, Cicadellidae). *Acta Entomologica Fennica* 38:52-53.

1983a Absence of wing polymorphism in the arboreal phytophagous species of some taxa of temperate Hemiptera: An hypothesis. *Ecological Entomology* 8:229-232.

1984a Flight in grassland and arboreal Auchenorrhyncha. *Mitteilungen der Schweizerischen Entomologischen Gesellschaft* 57(4):446-447.

Waloff, N. and Solomon, M. G.
1973a Leafhoppers (Auchenorrhyncha: Homoptera) of acidic grassland. *Journal of Applied Ecology* 10:189-212.

Waloff, N. and Thompson, P.
1980a Census data of populations of some leafhoppers (Auchenorrhyncha, Homoptera) of acidic grassland. *Journal of Animal Ecology* 49(2):395-416.

Walstrom, R. J.
1961a Insecticides increase alfalfa seed production in South Dakota. *Bulletin of the South Dakota Agricultural Experiment Station* 499(13):7.

Walter, S.
1978a Larvenformen mitteleuropaischen Euscelinen (Homoptera, Auchenorrhyncha). Teil 2. *Zoologische Jahrbuecher Abteilung fuer Systematik Oekologie und Geographie der Tiere* 105(1):102-130.

Wan, M. T. K.
1969a Entomology, 2. Crop pest studies. *Sarawak Department of Agriculture, Research Branch Annual Report*. Pp. 162-169.

1972a Observations on rice leaf and plant hoppers in Sarawak (Malaysian Borneo). *Malaysian Agriculture Journal* 48(4):308-335.

Ward, C. R., O'Brien, C. W., O'Brien, L. B., Foster, D. E. and Huddleston, E. W.
1977a Annotated checklist of New World insects associated with *Prosopis* (mesquite). *U.S. Department of Agriculture Technical Bulletin* 1577:1-115.

Wathanakul, L.
1964a A study on the host range of tungro and orange leaf viruses of rice. M. S. Thesis, College of Agriculture, University of the Philippines. 35 pp.

Wathanakul, L. and Weerapat, P.
1969a Virus diseases of rice in Thailand. In: The Virus Diseases of the Rice Plant. Proceedings of the International Rice Research Institute Symposium. Pp. 79-85.

Watts, J. G.
1963a Insects associated with black grama grass, *Bouteloua eriopoda*. *Annals of the Entomological Society of America* 56:374-379.

Way, M. O.
1983a The aster leafhopper, *Macrosteles fascifrons* Stal: conditions affecting abundance, tactics for control and effects on yield of rice in California. *Dissertation Abstr. Int.* (B) 43(10):3130

Way, M. O., Grigarick, A. A. and Mahr, S. E.
1983a Effects of rice plant density, rice water weevil (Coleoptera: Curculionidae), damage to rice, and the aquatic weeds on aster leafhopper (Homoptera: Cicadellidae) density. *Environmental Entomology* 12(3):949-952.

1984a The aster leafhopper (Homoptera: Cicadellidae) in California rice: Herbicide treatment affects population density and induced infestations reduced grain yield. *Journal of Economic Entomology* 77:936-942.

Way, M. O., Grigarick, A. A. and Rice, S. E.
1982a The questionable role of aster leafhopper feeding on rice plants. *Annual Meeting of the Entomological Society of America* (Pacific Branch) 65:149 (Abstract).

Weaver, C. R.
1954a Yellowing in alfalfa is caused by attacks of potato leafhopper. *Ohio Farm Home Research* 39:54, 62.

1959a What is an adequate sample for determining populations of forage insects in the field? *Proceedings of the North Central Branch of the Entomological Society of America* 14:74-75.

Weaver, J. E.
1984a Field evaluation of insecticides for control of alfalfa weevil and potato leafhopper on alfalfa in West Virginia, 1981-1983. *West Virginia Agricultural and Forestry Experiment Station Current Report* 76:1-10.

Webb, D. W.
1980a Primary insect types in the Illinois Natural History Survey Collection, exclusive of Collembola and Thysanura. *Illinois Natural History Survey Bulletin* 32:56-191.

Webb, M. D.
1975a A review of the genus *Idiocerus* Lewis (Homoptera: Cicadellidae) in the Ethiopian region, with description of eight new species. *Journal of the Entomological Society of Southern Africa* 35(2):165-184.

1976a A review of the genus *Idioscopus* Baker (Homoptera: Cicadellidae) in the Ethiopian region, with descriptions of twenty-seven new species and a comparison with the genus *Idiocerus* Lewis, *sensu* Ribaut (1952). *Journal of the Entomological Society of Southern Africa* 39(2):291-331.

1979a Revision of Rambur's Homoptera species from types in the British Museum. *Annales de la Societe Entomologique de France* 15(1):227-240.

1980a The Cicadellidae from Aldabra, Astove and Cosmoledo atolls collected by the Royal Society Expedition 1967-1968 (Hemiptera, Homoptera). *Journal of Natural, History* 14:829-863. [New genera: *Caloduferna, Gullifera, Jakarellus, Tuakamara.*]

1981a A new species of *Dryadomorpha* from the Central African Republic with a key to the known African species of the genus (Homoptera, Cicadellidae). *Revue Francaise d'Entomologie* (Nouvelle Serie) 3(2):57-58.

1981b The Asian, Australasian and Pacific Paraboloponinae (Homoptera: Cicadellidae). *Bulletin of the British Museum (Natural History) (Entomology)* 43(2):39-76. [New genera: *Favintiga, Karoseefa, Parohinka, Rhutelorbus.*]

1983a Revision of the Australian Idiocerinae (Hemiptera: Homoptera: Cicadellidae). *Australian Journal of Zoology* Supplementary Series 92. 147 pp. [Describes 17 new genera.]

1983b The Afrotropical idiocerine leafhoppers (Homoptera: Cicadellidae). *Bulletin of the British Museum (Natural History) (Entomology)* 47(3):211-257. [Describes 12 new genera.]

Webb, R. E. and Schultz, E. S.
1955a Ladino clover—a possible source of the virus causing purple-top wilt in potatoes. *Plant Disease Reporter* 39(4):300-301.

Webster, J. A.
1975a Association of plant hairs and insect resistance. An annotated bibliography. *U.S. Department of Agriculture Miscellaneous Publication* 1297. 17 pp.

Webster, J. A., Sorensen, E. L. and Painter, R. H.
1968a Resistance of alfalfa varieties to the potato leafhopper: Seedling survival and field damage after infestation. *Crop Science* 8(1):15-17.

1968b Temperature, plant growth stage, and insect-population effects on seedling survival of resistant and susceptible alfalfa infested with potato leafhoppers. *Journal of Economic Entomology* 61(1):142-145.

Wegorek, W.
1967a Insect vectors of virus diseases of various forage legumes. *Institute of Plant Protection* (Poznan), FG-PO-135-62, E-21-ENT-9. 178 pp.

Wei, C. S., Peng, Z. J., Yang, G. Q., Cao, Y., Huang, B. Z. and Chen, X.
1984a On the flower bug, *Orius similis* Zheng. *Natural Enemies of Insects* 6:32-40.

Wei, L.-Y.
1978a Symbiotes and nutrition of the aster leafhopper, *Macrosteles fascifrons* (Stal). M. S. Thesis, University of Minnesota, Minneapolis.

Wei, L.-Y. and Brooks, M. A.
1978a Intracellular symbiotes of the aster leafhopper, *Macrosteles fascifrons* (Stal). *Bulletin of the Institute of Zoology Academia Sinica* 17(1):61-66. [Chinese summary.]

1979a Aseptic rearing of the aster leafhopper, *Macrosteles fascifrons* (Stal), on a chemically defined diet. *Experimentia* 35:476-477.

Weidner, H.
1977a Dr. H. C. Wilhelm Wagner zum Gedachtnis. *Mitteilungen aus den Hamburgischen Zoologischen Museum und Institute* 74:7-10.

Weidner, H. and Wagner, W.
1968a Die Entomologischen Sammlungen des Zoologischen Staatsinstitutes und Zoologischen Museums Hamburg. *Mitteilungen aus dem Hamburgischen Zoologischen Museum und Institut* 65:123-179.

Welbourn, W. C.
1983a Potential use of Trombidioid and Erythraeoid mites as biological control agents of insect pests. pp. 103-140. In: Hoy, M. A., Cunningham, G. L. & Knutson, L. (eds). *Biological Control of Pests by Mites.* Proceedings of Conference. University of California, Berkeley. 185 pp.

Wells, P. W., Dively, G. P. and Schalk, J. M.
1984a Resistance and reflective foil mulch as control measures for the potato leafhopper (Homoptera: Cicadellidae) on *Phaseolus* species. *Journal of Economic Entomology* 77(4):1046-1051.

Wen, H. C. and Lee, H. S.
1978a Bionomics, observation and control of mango brown leafhopper (*Idiocerus niveosparsus* Leth.). *Journal of Agricultural Research of China* 27(1):47-52.

1980a Investigations on the oviposition of mango brown leafhopper (*Chunrocerus niveosparsus*) and its egg parasite (*Gonatocerus* sp.). *Journal of Agricultural Research of China* 29(3):245-250.

1985a Seasonal occurrence of the shoot insects and their control on guava. *Journal of Agricultural Research of China* 34:105-109.

Wenzil, H.
1959a Zikaden Ubertragbare Virus Krankheiten von Kulturpflanzen. *Pflanzenarzt* 12:76-77.

Wesley, C. S.
1983a Taxonomic studies on Nirvaninae (Homoptera: Cicadellidae) of the Indian subcontinent. *Thesis Abstract, Haryana Agricultural University* 9(4):332.

Wesley, C. S. and Blocker, H. D.
1985a *Athysanella* of Mexico including descriptions of a new subgenus and six new species (Homoptera: Cicadellidae). *Entomography* 3:163-180.

Westdal, P. H., Barrett, C. F. and Richardson, H. P.
1959a The six-spotted leafhopper, *Macrosteles fascifrons* (Stal), and aster yellows, *Chlorogenus callistephi* H. in Manitoba. *Proceedings of the Entomological Society of Manitoba* 15:32.

1961a The six-spotted leafhopper, *Macrosteles fascifrons*, and the aster yellows in Manitoba. *Canadian Journal of Plant Science* 41:320-331.

Westdal, P. H. and Richardson, H. P.

1966a The painted leafhopper, *Endria inimica* (Say), a vector of wheat striate mosaic virus in Manitoba. *Canadian Entomologist* 98(9):922-931.

1972a Control of the aster leafhopper in relation to incidence of aster yellows and effects on seed yield of barley. *Canadian Journal Plant Science* 52:177-182.

Whalley, P. E. S.

1955a Notes on some Homoptera, Auchenorhyncha found in Caernarvonshire and Anglesey. *Entomologist's Monthly Magazine* 91(1097):243-245.

1956a On the identity of species of *Anagrus* (Hym. Mymaridae) from leafhopper eggs. *Entomologist's Monthly Magazine* 92:147-149.

1970a The mymarid (Hym.) egg-parasites of *Tettigella viridis* L. (Hem., Cicadellidae) and embryo-parasitism. *Entomologist's Monthly Magazine* (1969) 105(1265-1267):239-243.

1971a The effect of temperature on the development and diapause of the eggs of leafhoppers (Hem., Auchenorrhyncha), with notes on its ecological effects. *Entomologist's Monthly Magazine* 107:249-253.

Wheeler, A. G.

1971a A study of the arthropod fauna of alfalfa. Ph. D. Dissertation, Cornell University. 342 p. [*Diss. Abst. Int.* (B) 32(4):2216-2217.]

1974a Phytophagous arthropod fauna of crown vetch in Pennsylvania. *Canadian Entomologist* 106(9):897-908.

1974b Studies in the arthropod fauna of alfalfa. VI. Plant bugs (Miridae). *Canadian Entomologist* 106(12):1267-1275.

1977a Studies on the arthropod fauna of alfalfa. VII. Predaceous insects. *Canadian Entomologist* 109:423-427.

Wheeler, A. G., Jr. and Walley, K.

1980a *Graphocephala coccinea*: Seasonal history and habits on ericaceous shrubs, with notes on *G. fennahi* (Homoptera: Cicadellidae). *Melsheimer Entomological Series* 29:23-27.

1980b Seasonal history of the leafhopper complex (Homoptera: Cicadellidae) on ornamental honey locust. *Journal of the New York Entomological Society* 88(1):81-82 (Abstract).

Whellan, J. A.

1964a Outbreaks and new records. Rhodesia and Nyasaland. Vectors of viruses. *FAO Plant Protection Bulletin* 12(1):22.

Whitaker, J. O., Jr.

1972a Food habits of bats from Indiana. *Canadian Journal of Zoology* 50(6):877-883. [In English with French summary.]

Whitcomb, R. F.

1958a Host relationships of some grass-inhabiting leafhoppers. M. S. Thesis, University of Illinois. 36 pp.

1964a A comparison of serological scoring with test plant scoring of leafhoppers infected by wound tumor virus. *Virology* 24:488-492.

1968a Proliferative symptoms in leafhoppers infected with western X-disease virus. *Virology* 35:174-177.

1969a Bioassay of plant viruses transmitted persistently by their vectors. In: Maramorosch, K., ed. *Viruses, Vectors, and Vegetation.* Pp. 449-462. Wiley & Sons, New York. 666 pp.

1972a Transmission of viruses and mycoplasma by Auchenorrhychous Homoptera. In: Kado, C. I. and Agrawal, H. O., eds. *Principles and Techniques in Plant Virology.* Pp. 168-169. Van Nostrand Reinhold, New York. 688 pp.

1972b Bioassay of clover wound tumor virus and the mycoplasmalike organisms of peach western X and aster yellows. *USDA Technical Bulletin* 1438:1-32.

1973a Ecology of leafhoppers. *Principes [Journal of Palm Society]* 17:155 (Abstract).

1981a The biology of spiroplasmas. *Annual Review of Entomology* 26:397-425.

1983a Special techniques for isolation and identification of mycoplasmas from plants and insects. Introductory remarks. In: J. G. Tully and S. Razin, eds. *Diagnostic Mycoplasmology. (Methods in Mycoplasmology.* Vol. 2.) Pp. 211-216. Academic Press, New York. 440 pp.

Whitcomb, R. F. and Black, L. M.

1959a Serological measurement of wound-tumor soluble antigen in *Agallia constricta*. *Phytopathology* 49:334 (Abstract).

1961a Synthesis and assay of wound-tumor soluble antigen in an insect vector. *Virology* 15:136-145.

1969a Demonstration of exvectorial wound-tumor virus. *Annual Revue Phytopathology* 7:86-87.

1982a Plant and arthropod mycoplasmas: A historical perspective. In: Daniels, M. J. and Markham, P. G., eds. *Plant and Insect Mycoplasma Techniques.* Pp. 40-81. Halsted, England. 369 pp.

Whitcomb, R. F. and Bov, J. M.

1983a Mycoplasma-plant-insect interrelationships. In: S. Razin and J. G. Tully, eds. *Methods in Mycoplasmology.* Vol. 1. Pp. 21-25. Academic Press, New York.

Whitcomb, R. F., Clark, T. B. and Vaughn, J. L.

1983a Pathogenicity of mycoplasmas for arthropods and its possible significance in biological control. In: J. G. Tully and S. Razin, eds. *Diagnostic Mycoplasmology. (Methods in Mycoplasmology.* Vol. 2.) Pp. 361-367.

Whitcomb, R. F. and Coan, M. E.

1982a Blind passage: A potentially useful technique for vector searches. *Journal of Economic Entomology* 75(5):913-915.

Whitcomb, R. F., Coan, M. E., Kramer, S. and Ross, H. H.

1973a Host relationships of cicadellids in the grassland biome. *First International Congress of Systematics and Evolutionary Biology*, Boulder, Colorado.

Whitcomb, R. F. and Davis, R. E.

1970a Mycoplasma and phytoboviruses as plant pathogens persistently transmitted by insects. *Annual Review of Entomology* 15:405-464.

1970b Evidence on possible mycoplasma etiology of aster yellows disease. II. Suppression of aster yellows in insect vectors. *Infection and Immunity* 2:209-215.

Whitcomb, R. F. and Jensen, D. D.

1968a Proliferative symptoms in leafhoppers infected with western X-disease virus. *Virology* 35:174-177.

Whitcomb, R. F., Jensen, D. D. and Richardson, J.

1966a The infection of leafhoppers by the western X-disease virus. I. Frequency of transmission after injection or acquisition feeding. *Virology* 28(3):448-453.

1966b The infection of leafhoppers by the western X-disease virus. II. Fluctuation of virus concentration in the hemolymph after injection. *Virology* 28(3):454-458.

1967a The infection of leafhoppers by western X-disease virus. III. Salivary, neural and adipose histopathology. *Virology* 31:539-549.

1968a The infection of leafhoppers by western X-disease virus. IV. Pathology in the alimentary tract. *Virology* 34(1):69-78.

1968b The infection of leafhoppers by western X-disease virus. V. Properties of the infectious agent. *Journal of Invertebrate Pathology* 12(2):192-201.

1968c The infection of leafhoppers by western X-disease virus. VI. Cytopathological interrelationships. *Journal of Invertebrate Pathology* 12(2):202-221.

Whitcomb, R. F., Kramer, J. P. and Coan, M. E.

1972a *Stirellus bicolor* and *S. obtutus* (Homoptera: Cicadellidae): Winter and summer forms of a single species. *Annals of the Entomological Society of America* 65(4):797-798.

Whitcomb, R. F., Shapiro, M. and Richardson, J.

1966a An *Erwinia*-like bacterium pathogenic to leafhoppers. *Journal of Invertebrate Pathology* 8(3):299-307.

Whitcomb, R. F. and Tully, J. G.

1973a Multiplication of mycoplasmas and mycoplasmalike organisms in leafhoppers. *Annual Meeting, ASM* (Abstract M45).

1973b Effect of antibiotics on multiplication of wall-free procaryotes in insects. *Proceedings 2nd International Congress Plant Pathologists* (Abstract 647).

1979a *The Mycoplasmas: Plant and Insect Mycoplasmas*. Vol. 3. Academic Press, New York.

Whitcomb, R. F., Tully, J. G., Bode, J. M. and Saglio, P.

1973a Spiroplasmas and acholeplasmas: Multiplication in insects. *Science* 182(4118):1251-1253.

Whitcomb, R. F. and Williamson, D. L.

1975a Helical wall-free prokaryotes in insects: multiplication and pathogenicity. *Annals New York Academy of Science* 266:260-275.

1979a Pathogenicity of mycoplasms in arthropods. *Zentralblatt fuer Bakteriologie, Parasitenkunde, Infektionskrankheiten und Hygiene. Erste Abteilung Originale. Reihe A. Medizinische Mikrobiologie und Parasitologie* 245:200-221.

Whitcomb, R. F., Williamson, D. L., Rosen, J. and Coan, M.

1974a Relationship of infection and pathogenicity in the infection of insects by wall-free prokaryotes. *Colloquium Institute National Sante Recherche Medicine* 33:275-282.

Whitfield, G. H. and Ellis, C. R.

1977a The pest status of foliar insects on soybeans and white beans in Ontario. *Proceedings of the Entomological Society of Ontario* 107:47-55.

Whitmore, R. W., Pruess, K. P. and Nichols, J. T.

1981a Leafhopper and planthopper populations on eight irrigated grasses grown for livestock forage. *Environmental Entomology* 10(1):114-118.

Whittaker, J. B.

1964a Auchenorrhyncha (Homoptera) of the Moor House Nature Reserve, Westmorland, with notes on *Macrosteles alpinus* (Zett.), a species new to Britain. *Entomologist's Monthly Magazine* 100:168-171.

1965a Some Homoptera Auchenorrhyncha from the Ainsdale Sand Dunes National Reserve, Lancashire. *Entomologist's Monthly Magazine* 101:124.

1967a Estimation of production in grassland froghoppers and leafhoppers (Homoptera: Insecta). *In: Petrusewicz, K., ed. Secondary Productivity in Terrestrial Ecosystems.* Pp. 779-789. Institute of Ecology, Polish Academy of Sciences, Warsaw.

1969a Quantitative and habit studies of the froghoppers and leafhoppers (Homoptera, Auchenorrhyncha) of Wytham Woods, Berkshire. *Entomologist's Monthly Magazine* 105:27-37.

1984a Responses of sycamore (*Acer pseudoplatanus*) leaves to damage by a typhlocybine leafhopper, *Ossiannilssonola callosa*. *Journal of Ecology* 72:455-462.

Whittaker, J. B. and Warrington, S.

1984a Typhlocybine damage to sycamore trees. *Mitteilungen der Schweizerischen Entomologischen Gesellshaft* 57(4):447-448.

1985a An experimental field study of different levels of insect herbivory induced by *Formica rufa* predation on sycamore (*Acer pseudoplatanus*). 3. Effects on tree growth. *Journal Applied Ecology* 22(3):797-811.

Whitten, M. J.

1965a Chromosome numbers in some Australian leafhoppers (Homoptera, Auchenorrhyncha). *Proceedings of the Linnaean Society of New South Wales* 90(1):78-86.

1968a An unusual chromosome system in a leafhopper (Homoptera Auchenorrhyncha). *Chromosoma* 24(1):37-41.

Whitten, M. J. and Taylor, W. C.

1969a Chromosomal polymorphism in an Australian leafhopper (Homoptera, Cicadellidae). *Chromosoma* 26(1):1-6.

Wilbur, D. A.

1954a Host plants of the leafhopper *Endria inimica* (Say) (Homoptera, Cicadellidae). *Transactions of the Kansas Academy of Science* 57(2):139-146.

Wilcox, R. B.

1951a Tests of cranberry varieties and seedlings for resistance to the leafhopper vector of false-blossom disease. *Phytopathology* 41:722-735.

Wilde, G. and Apostol, R.

1983a Resistance to *Nilaparvata lugens* (Homoptera) and *Nephotettix virescens* (Homoptera: Cicadellidae) in ratoon rice. *Environmental Entomology* 12:451-454.

Wilde, G. and van Schoonhoven, A.

1976a Mechanisms of resistance to *Empoasca kraemeri* in *Phaseolus vulgaris*. *Environmental Entomology* 5:251-255.

Wilde, G., van Schoonhoven, A. and Gomez-Laverde, L.

1976a The biology of *Empoasca kraemeri* on *Phaseolus vulgaris*. *Annals of the Entomological Society of America* 69(3):442-444.

Wilde, W. H. A.

1960a Insect transmission of the virus causing little cherry disease. *Canadian Journal of Plant Science* 40:707-712.

1962a A note on color preference of some Homoptera and Thysanoptera in British Columbia. *Canadian Entomologist* 92(1):107.

1962b Incidence of leafhoppers inhabiting sweet cherry orchards in the Kootenay Valley of British Columbia. *Proceedings of the Entomological Society of British Columbia* 59:15-17.

Wilkins, R. M., Batterby, S., Heinrichs, E. A., Aquino, G. B. and Valencia, S. L.

1984a Management of the rice tungro virus vector *Nephotettix virescens* (Homoptera: Cicadellidae) with controlled-release formulations of carbofuran. *Journal of Economic Entomology* 77(2):495-499.

Wilks, J. M. and Welsh, M. F.

1964a Apparent reduction of little cherry disease spread in British Columbia. *Canadian Plant Disease Survey* 44(2):126-130.

Williams, D. W.

1982a Ecology of the blackberry leafhopper, *Dikrella californica* (Lawson), and its role in California grape agro-ecosystems. *Dissertation Abstracts International* (B) 43(6):1729-1730.

1984a Ecology of the blackberry-leafhopper-parasite system and its relevance to California grape ecosystems. *Hilgardia* 52:1-32.

Williamson, D. L. and Whitcomb, R. F.

1974a Helical, wall-free prokaryotes in *Drosophila*, leafhoppers, and plants. *Colloquium Institute National Sante Recherche Medicine* 33:283-290.

1975a Plant mycoplasmas: A cultivable spiroplasma causes corn stunt disease. *Science* 188:1018-1020.

Wilson, B. H., Phillips, S. and Harris, H. M.

1973a Species and seasonal occurrence of leafhoppers and planthoppers in a coastal bermudagrass pasture in the Macon Ridge area of Louisiana. *Journal of Economic Entomology* 66(6):1346-1347.

Wilson, H. L.

1955a Effects of residual DDT on beet leafhoppers. *Bulletin of the California Department of Agriculture* 44(4):147-150.

1961a The nonresistance of beet leafhoppers to DDT. *Bulletin of the California Department of Agriculture* 49(4):242-245.

Wilson, J. W. and Genung, W. G.

1956a Insect problems in the production of southern peas (cowpeas). *Proceedings of the Florida State Horticultural Society* 69:217-223.

Wilson, K. J.

1964a Outbreaks and new records. *FAO Plant Protection Bulletin* 12(1):22

Wilson, M. C.

1958a Problems in selecting alfalfa for resistance to the meadow spittlebug and potato leafhopper. *Report of the Alfalfa Improvement Conference* 16:103-106.

1984a Management strategies for leafhoppers, spittlebugs, and aphids on alfalfa. *1983 Reports of the NC-149 Technical Committee*. USDA, Washington, D.C.

Wilson, M. C., Armburst, E. J., Fenwick, J. R. and Lien, R. M.
1968a Progress report on the effect of stubble flaming on alfalfa growth and yield under conditions of intensive insect control. *Proceedings of the Annual Symposium on Thermal Agriculture* 5:31-34.

Wilson, M. C. and Cleveland, M. L.
1955a Residual versus non-residual insecticides to control leafhoppers on alfalfa. *Proceedings of the Indiana Academy of Science* 65:145-146.

Wilson, M. C. and Davis, R. L.
1958a Development of an alfalfa having resistance to the meadow spittlebug. *Journal of Economic Entomology* 51(2):219-222.
1962a Screening methods for meadow spittlebug and potato leafhopper resistance in alfalfa. *Report of the Alfalfa Improvement Conference* 18:4-12.

Wilson, M. C., Davis, R. L. and Williams, G. G.
1955a Multiple effects of leafhopper infestation on irrigated and non-irrigated alfalfa. *Journal of Economic Entomology* 48(3):323-326.

Wilson, M. C., Stewart, J. K. and Vail, H. D.
1979a Full season impact of the alfalfa weevil, meadow spittlebug, and potato leafhopper. *Journal of Economic Entomology* 72:830-834.

Wilson, M. R.
1978a Descriptions and key to the genera of the nymphs of British woodland Typhlocybinae (Homoptera). *Systematic Entomology* 3(1):75-90.
1981a Identification of European *Iassus* species (Homoptera: Cicadellidae) with one species new to Britain. *Systematic Entomology* 6(1):115-118.
1981b Directory of research workers on Homoptera-Auchenorrhyncha. University College, Cardiff, Wales. 31 pp.
1983a The nymphal stages of some Auchenorrhyncha associated with rice in south east Asia. In: Knight, W. J., Pant, N. C., Robertson, T. S. and Wilson, M. R., eds. *Proceedings of the 1st International Workshop on Biotaxonomy, Classification and Biology of Leafhoppers and Planthoppers (Auchenorrhyncha) of Economic Importance*. Pp. 121-134. London, 4-7 October 1982. Commonwealth Institute of Entomology, 56 Queen's Gate, London SW7 5JR. 500 pp.
1983b A revision of the genus *Paramesodes* Ishihara (Homoptera, Auchenorrhyncha: Cicadellidae) with descriptions of eight new species. *Entomologica Scandinavica* 14:17-32.
1985a Sulawesi collecting notes. *Tymbal* 6:13-16.

Wilson, M. R. and Claridge, M. F.
1985a The leafhopper and planthopper fauna of rice fields. In: Nault, L. R. and Rodriguez, J. G., eds. *The Leafhoppers and Planthoppers*. Pp. 381-404. Wiley & Sons, New York. 500 pp.

Wilson, N. L. and Oliver, A. D.
1969a Food habits of the imported fire ant in pasture and pine forest areas in southeastern Louisianna. *Journal of Economic Entomology* 62(6):1268-1271.

Windsor, I. M. and Black, L. M.
1973a Evidence that clover club leaf is caused by a rickettsia-like organism. *Phytopathology* 63:1139-1148.

Witsack, W.
1985a Dormanzformen bei zikaden (Homoptera Auchenorrhyncha) und ihre okologische bedeutung. *Zoologische Jahrbuecher (Systematics)* 112(1):71-139.

Wolanski, B. S. and Maramorosch, K.
1979a Rayado fino virus and corn stunt spiroplasma: Phloem restriction and transmission by *Dalbulus elimatus* and *D. maidis*. *Fitopatologia Brasiliera* 4:47-54.

Wolda, H.
1977a Ecologia de insectos en la provincia de Chirique. *Con Ciencia* 4:3-5.
1979a Fluctuaciones en abundancia de insectos en el bosque tropical. pp. 519-539. *In: Actas del IV Simposium Internacional de Ecologica Tropical*, Panama, 1977.
1979b Abundance and diversity of Homoptera in the canopy of a tropical forest. *Ecological Entomology* 4:181-190.
1980a Seasonality of tropical insects. I. Leafhoppers (Homoptera) in Las Cumbres, Panama. *Journal of Animal Ecology* 49(1):277-290.
1980b Fluctuaciones estacionales de insectos en Hornitos ("Fortuna"), Provincia de Chiriqui. *Con Ciencia* 7:8-10.
1982a Seasonality of Homoptera on Barro Colorado Island. In: Leigh, E. G., Rand, E. S. and Windsor, D. M., eds. *Ecology of a Tropical Forest: Seasonal Rhythms and Long-term Changes*. pp. 319-330. Smithsonian Institution Press, Washington, D. C.
1983a "Long-term" stability of tropical insect populations. *Researches on Population Ecology*. Suppl. 3:112-126.

Wolfe, H. R.
1955a Transmission of the western X-disease virus by the leafhopper, *Colladonus montanus* (Van D.). *Plant Disease Reporter* 39(4):298-299.
1955b Relation of leafhopper nymphs to the western X-disease virus. *Journal of Economic Entomology* 48:588-590.
1955c Leafhoppers of the state of Washington. *Washington Agricultural Experiment Station Circular* 277:1-37.
1956a Acaricides in insect vector virus research. *Journal of Economic Entomology* 48(6):749-750.
1958a Effect of Endrin sprays for mouse control on insects found on orchard cover crops. *Journal of Economic Entomology* 50(6):837-838.
1958b *Dorilas (Eudorylas) subopacus* (Loew), a parasite of the leafhopper, *Scaphytopius acutus* (Say). *Journal of Economic Entomology* 1957 50(6):831.

Wolfenbarger, D. A.
1961a Resistance of beans (*Phaseolus, Glycine max, Vigna sinesis, Vicia fabae* and *Dolechos lablab*) to the Mexican bean beetle and the potato leafhopper. Ph. D. Dissertation, Ohio State University, Columbus. 145 pp. [*Diss. Abst. Int.* 22(2):686.]

Wolfenbarger, D. A. and Sleesman, J. P.
1959a Resistance in beans to attack by the potato leafhopper and the Mexican bean beetle. *Proceedings of the North Central Branch of the Entomological Society of America* 14:7-8.
1961a Plant characteristics of *Phaseolus vulgaris* associated with potato leafhopper nymphal infestation. *Journal of Economic Entomology* 54(4):705-707.
1961b Resistance in common bean lines to potato leafhopper. *Journal of Economic Entomology* 54(5):846-849.
1961c Resistance of the potato leafhopper in lima bean lines, interspecific crosses, *Phaseolus* crosses, *Phaseolus* spp., the cowpea and Bonavisk bean. *Journal of Economic Entomology* 54(6):1077-1079.
1963a Variation in susceptibility of soybean pubescent types, broad bean and runner bean varieties and plant introduction to the potato leafhopper. *Journal of Economic Entomology* 56(6):895-897.

Wolfenbarger, D. O.
1963a Control measures for the leafhopper, *Empoasca kraemeri*, on beans. *Journal of Economic Entomology* 56(3):417-419.

Womack, C. L.
1984a Reduction in photosynthetic and transpiration rates of alfalfa caused by potato leafhopper (Homoptera: Cicadellidae) infestations. *Journal of Economic Entomology* 77(2):508-513.

Wonn, L.
1956a Okologische studien uber die Zikadenfauna der Mainzer sande. *Jahrbucher des Nassauischen Vereins fuer Naturkunde* 92:80-122.

Woodroffe, G. E.
1967a Further notes on some Hemiptera and Coleoptera from the Isles of Scilly. *Entomologist's Monthly Magazine* 102(1229-1231):285-286.
1968a *Euscelis venosus* (Kirschbaum) (Hem., Cicadellidae) new to Britain. *Entomologist's Monthly Magazine* 104(1247-1249):131.
1971a Records of two uncommon species of *Macrosteles* (Hem., Cicadellidae) in Britain. *Entomologist's Monthly Magazine* 106(1277-1279):223.

1971b The first British record of *Eurhadina kirschbaumi* (Wagner) (Hem., Typhlocybidae). *Entomologist's Monthly Magazine* 107(1280-1292):44.

1971c *Youngiada pandellei* (Leth.) and *Ribautiana cruciata* (Rib.) (Hem. Cicadellidae) associated with cultivated blackberry and raspberry (*Rubus*). *Entomologist's Monthly Magazine* 107(1280-1282):64.

1972a Hemiptera from the Braemar area (Aberdeenshire), including the first British record of *Dikraneura contraria* Ribaut (Hem., Cicadellidae). *Entomologist's Monthly Magazine* 107:172-173.

1972b Further notes on the Hemiptera of the Braemar area (Aberdeenshire), including the first British record of *Orthops basalis* (Cost.) (Miridae). *Entomologist's Monthly Magazine* 107(1289-91):255-256.

Woodward, T. E., Eastop, V. F. and Evans, J. W.

1974a Hemiptera (bugs, leafhoppers, etc.). In: CSIRO, ed. *The Insects of Australia.* Supplement 1974. Pp. 52-57. 2nd ed. Melbourne University Press, Melbourne, Australia. 146 pp.

Wootton, R. J.

1981a Palaeozoic insects. *Annual Review of Entomology* 26:319-344.

Wouters, L. J. A.

1963a The control of pests and diseases at the Wageningen rice-project in Surinam. *Bulletin, LandbProefstatlon in Suriname* 82:369-407. [French summary.]

Wressell, H. B.

1960a Resistance of alfalfa varieties to potato leafhopper. *Forest Notes* 6(1):13.

1971a A survey of insects infecting vegetable crops in south-western Ontario, 1969. *Proceedings of the Entomological Society of Ontario* 101:13-23.

Wressell, H. B. and Driscoll, G. R.

1964a The use of systemic insecticides for control of the potato leafhopper, *Empoasca fabae*, and effect on potato yield. *Journal of Economic Entomology* 57(6):992-993.

Wressell, H. B. and Miller, L. A.

1960a Control of the potato leafhopper in field beans. *Canada Department of Agriculture Publication* 1073. 4 pp.

Wright, W. E.

1970a Cotton pests and their control. *Agricultural Gazette of New South Wales* 81(5):268-278.

Wu, G. R. and Ruan, Y. L.

1982a Bionomics and the appropriate time for chemical control of the white leafhopper *Thaia subrufa* (Motschulsky). *Acta Entomologica Sinica* 25:178-184.

Wysoki, M. and Izhar, Y.

1978a A list of arthropod pests of avocado and pecan trees in Israel. *Phytoparasitica* 6:89-93.

Xie, L. H. and Lin, J. Y.

1983a On the bunchy stunt disease of rice. III. Stability in vitro and distribution in host plant of rice bunchy stunt virus (RBSV). *Acta Phytopathologica Sinica* 13:15-19.

Xie, L. H., Lin, J. Y. and Guo, J. R.

1981a A new insect vector of rice dwarf virus. *International Rice Research Newsletter* 6(5):14.

1982a *Nephotettix virescens* – vector of rice dwarf virus. *Fujian Nongye Jeji* 3:24, 50.

Xie, L. H., Lin, J. Y., Xie, L. M. and Nai, G. B.

1984a On the bunchy stunt disease of rice. IV. The experiments of occurrence, development and control of rice bunchy stunt. *Acta Phytopathologica Sinica* 14:33-38.

Yadav, L. S. and Yadav, P. R.

1983a Pest complex of cowpea (*Vigna sinensis* Savi) in Haryana. *Bulletin of Entomology* 24:57-58.

Yadava, C. P. and Shaw, F. R.

1968a The preferences of certain coccinellids for pea aphids, leafhoppers and alfalfa weevil larvae. *Journal of Economic Entomology* 61(4):1104-1105.

Yadava, H. N., Mittal, R. K. and Singh, H. G.

1967a Correlation studies between leaf midrib structure and resistance to jassids (*Empoasca devastans* Dist.) in cotton. *Indian Journal of Agriculture Science* 37:495-497.

Yalovitasyn, M. V.

1962a An *Entomophthora* causing an epizootic disease of leafhoppers. *Doklady Nauchne Uchrezhdenii Min Sel'skokhoz Kazakhok SSR* 2:27-30.

Yamamoto, I., Kyomura, N. and Takahashi, Y.

1977a Aryl N-propylcarbamates, a potent inhibitor of acetylcholinesterase from the resistant green rice leafhopper, *Nephotettix cincticeps*. *Journal of Pesticide Science* 2:463-466.

1978a Joint insecticidal effect of N-propyl and N-methylcarbamates on the green rice leafhopper, resistant to aryl N-methylcarbamates. *Review of Plant Protection Research* 11:79-92.

Yamamoto, I., Takahashi, Y. and Kyomura, N.

1983a Suppression of altered acetylcholinesterase of the green rice leafhopper by N-propyl and N-methyl carbamate combinations. In: Georghiou, G. P. and Saito, T., eds. *Pest Resistance to Pesticides.* Pp. 579-594. Plenum Press, New York.

Yang, C.

1965a Two new leafhoppers of the genus *Pyramidotettix* (Homoptera: Jassidae) injuring apple trees. *Acta Entomologica Sinica* (Peking) 14:196-202. [In Chinese with English summary.]

1984a A new species of *Namsangia* from west Yunnan injuring bamboo (Homoptera: Cicadellidae). *Entomotaxonomia* 6(2&3):197-199. [In Chinese and English.]

Yang, C. and Li, F.

1980a Notes on *Bothrogonia* (Homoptera, Cicadellidae) with descriptions of 22 new species from China. *Entomotaxonomia* 2(3):191-213. [In Chinese with English summary.] [New subgenera: *Abothrogonia, Obothrogonia.*]

Yang, I. L. and Chou, L. Y.

1982a Transmission of sweet potato witches' broom by *Orosius orientalis* in Taiwan. *Journal of Agricultural Research of China* 31:169.

Yang, O.

1955a Description of a new species of Strepsiptera parasitic on the rice leaf-hoppers. *Acta Entomologica Sinica* 5:327-333. [In Chinese with English summary.]

Yang, S., Huang, J. and Jin, M.

1982a Studies on Dryinidae (Hymenoptera) – the natural enemies of leafhoppers in Guangxi Province. *Natural Enemies of Insects* 4(2):1-12.

Yang, S. L. and Pan, Y. S.

1979a Ecology of *Matsumuratettix hiroglyphicus* (Matsumura), an insect vector of sugarcane white leaf disease. *Plant Protection Bulletin* (Taiwan) 21:111-117. [In English with Chinese summary.]

Yano, K.

1979a Faunal and biological studies on the insects of paddy fields in Asia. Part II. Illustrated key to the Thai species of Pipunculidae (Diptera). *Esakia* 13:45-54.

Yano, K., Ishitani, M., Asai, I. and Satoh, M.

1984a Faunal and biological studies on the insects of paddy fields in Asia. XIII. Pipunculidae from Japan (Diptera). *Transactions of the Shikoku Entomological Society* 16:53-74.

Yano, K., Miyamoto, S. and Gabriel, B. P.

1981a Faunal and biological studies on the insects of paddy fields in Asia. Part IV. Aquatic and semiaquatic Heteroptera from the Philippines. *Esakai* 16:5-32.

Yano, K., Morakote, R., Satoh, M. and Asai, I.

1985a An evidence for behavioral change in *Nephotettix cincticeps* Uhler (Hemiptera: Deltocephalidae) parasitized by pipunculid flies (Diptera: Pipunculidae). *Applied Entomology and Zoology* 20:94-96.

Yashiro, N.
1979a Studies on the Japanese species of *Oligosita* (Hymenoptera: Trichogrammatidae). *Transactions of the Shikoku Entomological Society* 14:195-203.

Yasumatsu, K., Wongsiri, T., Nananichit, S. and Tirawat, C.
1975a Approaches towards an integrated control of rice pests. Part I: Survey of natural enemies of important rice pests in Thailand. *Plant Protection Service Technical Bulletin Department of Agriculture* (Thailand) 24:1-21.

Yazdani, S. S. and Mehto, D. N.
1980a A note on the control of mango leafhopper, *Amritodus atkinsoni* Leth. at Dholi. *Indian Journal of Entomology* 42:275-276.

Yein, B. R.
1981a Relative efficacy of certain insecticides against cotton pests. *Journal of Research, Assam Agricultural University* 2(2):196-201.

1982a Effect of carbofuran and fertilizers on the incidence of insect pests and on growth and yield of green gram. *Journal of Research, Assam Agricultural University* 3:197-203.

Yein, B. R., Borthakur, D. and Dwivedi, J. L.
1980a Relative susceptibility of some ahu rice cultivars to stem borer, whorl maggot, leaf roller and green leafhopper. *Journal of Research, Assam Agricultural University* 1:103-106.

Yen, D. F.
1973a *A Natural Enemy List of Insects of Taiwan*. College of Agriculture, National Taiwan University. 106 pp.

Yen, D. F. and Tsai, Y. T.
1970a An Entomophthora infection in the rice leaf hopper, y *Nephotettix cincticeps* (Uhler). *Plant Protection Bulletin* (Taiwan) 12(1):15-20. [In Chinese with English summary.]

Ying, S. H.
1982a The ovicidal activity of some new insecticides. *Acta Entomologica Sinica* 25:289-293.

Ylonen, H. and Raatikainen, M.
1984a Uber die Deformierung mannlicher Kopulationsorgane zweier *Diplocolenus*-Arten (Homoptera, Auchenorrhyncha) beeinflust durch Parasitierung. [On the deformation of male genitalia of two *Diplocolenus* species (Homoptera, Auchenorrhyncha) induced by parasitism.] *Suomen Hyenteistieteellinen Aikakauskirja* 50:13-16. [In German.]

Yokayama, S. and Sakai, H.
1975a Transmission by green leafhopper, *Nephotettix cincticeps* Uhler, of a new dwarf disease of rice plants. *Annals of the Phytopathological Society of Japan* 41:219-222. [In Japanese.]

Yokayama, S., Sakai, H., Hidari, M. and Inoue, T.
1974a Studies on the "Waisei" disease of rice plant. 3. Transmission of the causal agent by *Nephotettix cincticeps* Uhler. *Proceedings of the Association of Plant Protection* (Kyushu) 20:140.

Yonce, C. E.
1983a Geographical and seasonal occurrence, abundance, and distribution of phony peach disease vectors and vector response to age and condition of peach orchards and a disease host survey of Johnsongrass for rickettsia-like bacteria in the southeastern United States. *Journal of the Georgia Entomological Society* 18(3):410-418.

Yoshii, H.
1959a Recent studies on the virus diseases in Japan. *Proceedings of the 6th International Congress of Crop Protection* 1:383-387.

1959b Studies on the nature of insect-transmission in plant viruses. V. On the abnormal metabolism of the virus-transmitting green rice leafhopper, *Nephotettix bipunctatus cincticeps* Uhler, as affected with the rice stunt virus. *Virus* 9:415-422. [In Japanese.]

Yoshii, H., Kiso, A., Yamaguchi, T. and Miyauchi, S.
1961a Plant virus transmission by insects. VI. The incorporation of P-32 into stunt virus-infected rice plant and virus-transmitting green rice leafhopper (*Nephotettix apicalis*). *Virus* 9:453-462.

Yoshimeki, M.
1966a Ecological and physiological studies on the dynamics of the migratory population of rice planthoppers and leafhoppers. *Bulletin of the Kyushu Agricultural Experiment Station* 12(1&2):1-78. [In Japanese with English summary.]

Yoshimoto, C. M. and Gressitt, J. L.
1960a Trapping of air-borne insects on ships on the Pacific. *Pacific Insects* 2:239-243.

1963a Trapping of air-borne insects in the Pacific-Antarctic area, 2. *Pacific Insects* 5:873-883.

Yoshimoto, C. M., Gressitt, J. L. and Mitchell, C. J.
1962a Trapping of air-borne insects in the Pacific-Antarctic area, 1. *Pacific Insects* 4:847-858.

Yoshioka, E.
1965a Notes and exhibitions—May 1964. *Proceedings of the Hawaiian Entomological Society* 19(1):15.

Yoshioka, K., Matsumoto, M., Bekku, I. and Kanamori, S.
1975a Synergism of IBP and insecticides against insecticide resistant green rice leafhopper, *Nephotettix cincticeps* Uhler. *Proceedings of the Association of Plant Protection* (Shikoku) 10:49-58.

Young, D. A., Jr.
1956a Three new Neotropical Typhlocybine leafhoppers from economic plants. *Bulletin of the Brooklyn Entomological Society* 51:72-75.

1957a A new genus of economic leafhoppers and notes on the genus *Atanus* Oman (Homoptera Cicadellidae). *Revista Chilena Entomologia* 5:13-17. (With summary in Spanish.) [New genus: *Paratanus*.]

1957b Description of the Osborn types of Typhlocybinae (Homoptera, Cicadellidae) in the Carnegie Museum collection. *Revista Brasileira de Entomologia* 7:183-226.

1957c The leafhopper tribe Alebrini (Homoptera: Cicadellidae). *Proceedings of the United States National Museum* 107:127-277. [Describes 10 new genera.]

1958a A synopsis of the species of *Homalodisca* in the United States. *Bulletin of the Brooklyn Entomological Society* 53(1):7-13.

1958b On lectotype proposals. *Systematic Zoology* 7(3):120-122.

1963a Types of Cicadellinae (Homoptera, Cicadellidae) in the Naturhistoriska Riksmuseet in Stockholm. *Entomologisk Tidskrift* 84:224-227.

1964a Some cicadelline types of species described by Signoret from Berlin collections (Homoptera Cicadellidae). *Mitteilungen aus dem Zoologischen Museum in Berlin* 40:9-13.

1965a Western Hemisphere Mileewanini (Homoptera, Cicadellidae.) *Zoologische Beitrage* 2:369-382.

1965b Some cicadelline lectotypes of Breddin species. *Beitraege zur Entomologie* 15(1/2):11-14. [German and Russian summaries.]

1965c Cicadelline types in the British Museum (Natural History) (Homoptera: Cicadellidae). *Bulletin of the British Museum (Natural History) Entomology* 17(4):163-199.

1965d Notes on Fabrician species of Cicadellinae (Homoptera, Cicadellidae) in Copenhagen and Kiel with lectotype designations. *Entomologiske Meddelelser* 34(1):10-18.

1967a *Homoptera*. McGraw-Hill Encyclopedia of Science and Technology. Pp. 469-476.

1968a Taxonomic study of the Cicadellinae (Homoptera, Cicadellidae). Part 1. Proconiini. *U.S. National Museum Technical Bulletin* 261. 287 pp. [Describes 15 new genera.]

1974a Types of Cicadellinae (Homoptera, Cicadellidae) in the Museum National d'Histoire Naturelle, Paris, France. *Bulletin Museum National d'Histoire Naturelle*, Serie III (Paris) 263:1697-1699.

1977a Taxonomic study of the Cicadellinae (Homoptera: Cicadellidae). Part 2. New World Cicadellini and the genus *Cicadella*. *North Carolina Agricultural Experiment Station Technical Bulletin* 239. 1135 pp. [Describes 91 new genera.]

1979a A review of the leafhopper genus *Cofana* (Homoptera: Cicadellidae). *Proceedings of the Entomological Society of Washington* 81(1):1-21.

Young, D. A., Jr. and Beier, M.

1963a Types of Cicadellinae (Homoptera: Cicadellidae) in the Natural History Museum in Vienna. *Annalen des Naturhistorischen Museums in Wien* 67:565-575.

Young, D. A., Jr. and Beirne, B. P.

1958a A taxonomic revision of the leafhopper genus *Flexamia* and a new related genus. *U.S. Department of Agriculture Technical Bulletin* 1173:1-53.

Young, D. A., Jr. and Davidson, R. H.

1959a A review of leafhoppers of the genus *Draeculacephala*. *U.S. Department of Agriculture Technical Bulletin* 1198:1-32.

Young, D. A., Jr. and Lauterer, P.

1964a Cicadelline lectotypes from the A. Jacobi collection (Homoptera, Cicadellidae). *Reichenbachia* 2:293-296.

1966a Types of Cicadellinae (Homoptera, Cicadellidae) in the Moravian Museum. *Casopis Moravskeho Musea v Brne* 51:261-270.

Young, D. A., Jr. and Nast, J.

1963a Cicadelline types of species described by Edmund Schmidt (Homoptera, Cicadellidae). *Annales Zoologici* (Warsaw) 21:265-271.

Young, D. A., Jr. and Soos, A.

1964a Types of Cicadellinae (Homoptera, Cicadellidae) in the Hungarian Natural History Museum. *Annales Historico-Naturales Musei Nationalis Hungarici* 56:465-467.

Young, S. A. and Cannon, O. S.

1973a Vector preference among varieties of *Lycopersicon peruvianum* and *L. esculentum* as related to curly top virus resistance. *Proceedings of the Utah Academy of Sciences, Arts and Letters* 50(1):53 (Abstract).

Young, V. L.

1961a Effects of Systox seed treatment on the control of early rice pests. *Acta Entomologica Sinica* 10(4-6):425-428. [In Chinese.]

Yunus, A. and Rothschild, G. H. L.

1967a Insect pests of rice in Malaysia. *In: Pathak, M. D., ed. The Major Insect Pests of the Rice Plant*. Pp. 617-641. Johns Hopkins, Baltimore.

Yusof, O. H.

1982a Biological and taxonomic studies on some leafhopper pests of rice in southeast Asia. Ph. D. Thesis, University of Wales, Cardiff.

Zabel, U.

1978a Analysis of the embryonic development of a leafhopper: *Euscelidius variegatus* Kbm. *Auchenorrhyncha Newsletter* 1:21.

1980a Das Experiment: Insekten-embryogenese am beispiel einer kleinzikade. *Biologie in Unserer Zeit* 10(4):120-123.

Zachvatkin, A. A.

1948a Novye cikady (Homoptera: Cicadina) Srednerusskoj Fauny. *Nauchno-Metodicheskie Zapiski Glavnogo Upravlenie po Zapovednikam* 11:177-185.

1948b New and little known leafhoppers from the Oka reserve. *Nauchno-Metodicheskie Zapiski glavnogo Upravlenie po Zapovednikam* 11:186-197. [In Russian.]

1953a Biological and systematic notes on Cicadina fauna of central Russia. *Collection of scientific works Izdat'stvo, Moscow*. Pp. 205-209. [In Russian.]

1953b Cikadiny okskogo gosundarstvennogo zapovednika. *In: Sbornik nauchnykh rabot, Moskava*: 211-223. [Cicadina from the state national forest of Oka district, central Russia. Collection of scientific works Izdatal'stvo, Moscow. Pp. 211-223. [In Russian.]

1953c [Cicadina from Trans-Volga sand areas, Astrakhan.] *In: Smirnov, E. S., ed. Collection of Scientific Works Izdatel'stvo, Moscow*. Pp. 225-236. 418 pp.] [In Russian.] [New genus: *Xerochlorita*.]

1953d [Faunistic notes on Eupterygidae (Homoptera, Cicadina) from Central Asia. Empoascinae. *Collection of Scientific Works Izdatel'stvo, Moscow*. Pp. 237-245.] [In Russian.]

1953e [Cicadina collected by V. S. Elpatrvskij on the shores of Lake Khubsugul (B. Kosogol.) *Collection of Scientific Works Izdatel'stvo, Moscow*. Pp. 247-250.] [In Russian.]

Zack, R. S.

1984a Catalog of types in the James Entomological Collection. *Melanderia* 42:1-41.

Zahradnik, J.

1984a Dr. Jiri Dlabola, CSc-zum sechzigsten Geburtstag. *Acta Entomologica Bohemoslovaca* 81(4):314-320. [In German.]

Zajac, M. A. and Wilson, M. C.

1983a Evaluation of a method of estimating yields of alfalfa under insect stress. *Proceedings of the Indiana Academy of Science* 92:238.

Zaky, S. J.

1981a Damage potential of potato leafhopper nymphs. *Empoasca fabae* Harris (Homoptera: Cicadellidae), on established stand alfalfa. Ph. D. Dissertation, Pennsylvania State University, State College.

Zalom, F. G.

1981a The influence of reflective mulches and lettuce types on the incidence of aster yellows and abundance of its vector, *Macrosteles fascifrons* (Homoptera: Cicadellidae), in Minnesota. *Great Lakes Entomologist* 14:145-150.

Zaman, M. Q.

1980a Mycoplasma—a new causal entity for plant diseases. *Bangladesh Journal of Agricultural Research* 5:47-57.

Zang, Mu and Luo, Hengwen

1976a Preliminary observations on *Entomophthora sphaerosperma* Fres. attacking *Empoasca flavescens* Fab. *Acta Microbiologica Sinica* 16:256-257.

Zanol, K. M. R.

1980a A cigarrinha e sua importancia na agricultura. *Natureza Revista* 7:10-14.

Zanol, K. M. R. and Menezes, M. de

1982a Lista preliminar dos cicadelideos (Homoptera, Cicadellidae) do Brasil. *Iheringia Serie Zoologia* (Porto Alegre) 61:9-65. [In Portugese.]

Zanol, K. M. R. and Sakakibara, A. M.

1984a As Especies do Genero *Copididonus* Linnavuori E Descricao de Tres Novas (Homoptera, Cicadellidae). *Dusenia* 14(1):11-21. [In Portuguese with English summary.]

Zelazny, B. and Pacumbaba, E.

1982a Phytophagous insects associated with cadang cadang infected and healthy coconut palms in South-eastern Luzon, Philippines. *Ecological Entomology* 7(1):113-120.

Zeyen, J. R. and Banttari, E. E.

1972a History and ultrastructure of oat blue dwarf virus infected oats. *Canadian Journal of Botany* 50:2511-2519.

Zhang, C. N., Shu, D., Liu, Z. J. and Deng, L. Z.

1983a Studies on the causal agents of witches' broom disease of *Casuarina equisetifolia*. *Acta Phytopathologica Sinica* 13:37-42.

Zhizhilashvili, T. I.

1954a Materials toward the investigation of the bio-ecology of *Edwardsiana platani* A. Zach. in litt. under the conditions of Tbilisi and its environs. *Soobshcheniia Akademii Nauk Gruzinskoi SSR* 15(6):371-376. [In Russian.]

Zhong, L.

1983a Utilization of the natural enemies for the control of planthoppers and leafhoppers in the paddy fields. *Kunchong Tiandi* 5(3):158-165.

Zhou, B. L., Du, S. L. and Jiang, Z. M.
    1981a Rapid investigation method on the population fluctuation of major pests in cotton fields. *Kunchong Zhishi* 18:126-128.

Zhou, H. H. and Tang, J. Q.
    1982a Serological assessement of arthropod predation on insect pests of rice-fields. *Natural Enemies of Insects* 4(4):40-41.

    1984a Identification of predators of insect pests of rice using the precipitin test. *Acta Scientiarum Naturalium Universitatis Sunyatseni* 2:118-121.

Zhu, R. L. and Zheng, S. X.
    1984a A brief summary of the utilization of spiders for insect control in rice fields in the Taizhou region of Zhejiang. *Natural Enemies of Insects* 6:87-90.

Zile Singh and Vaishampayan, S. M.
    1976a A new technique for sampling populations of jassids. *JNKVV Research Journal* 1(2):176-177.

Zverezomb-Zubovskii, E. V.
    1957a Pests of sugar beet. *Akademiya Nauk Ukrainskoy SSR, Instituta Entomologicheskii i Fitopatologicheskii.* 276 pp. [In Russian.]

CHECK-LIST OF GENERIC AND FAMILY-GROUP NAMES

The check-list includes, in alphabetical sequence, all generic and subgeneric names published up to 31 December 1985, including those of fossil taxa. The list includes invalid names, unavailable names, suppressed names, emendations and mispellings, all with appropriate comment and cross-reference. All generic and subgeneric names in the list are in bold, except for invalid names, unavailable names, suppressed names, emended names and mispellings which are in normal type. Pre-1956 names omitted from the Metcalf catalogue are annotated accordingly. Generic names of taxa other than leafhoppers, yet which are included in the Zoological Record under Cicadellidae, are not included in the check-list but are listed below, prior to the check-list, under the heading "Misassigned generic names". References dealing with such names are not included in the bibliography.

Each entry, invalid names and mispellings excepted, includes, in the following sequence:

    a. genus name
    b. author
    c. coded year of publication and page number
    d. type-species
    e. type locality of type-species, in parentheses ( ). Names of countries are cited as they appear in the original publication.
    f. Metcalf reference, e.g. 1:643, in case of pre-1956 names
    g. references which add considerably to the knowledge of the genus, e.g. revisions, synonymies, status changes, faunal works
    h. subfamily and tribe (if known).
    i. geographical distribution
    j. subgenera and synonyms

Explanatory comments, required in some genera to draw attention to problems or unusual procedures, are given at the end of the entry in square brackets. The anglicized rather than the latinized version is used in all annotations, e.g. "replacement for" instead of "nom. nov. pro."

Family-group names, either adopted in Metcalf's catalogues or proposed since 1955, are also included in the check-list. Such family-group names are given in CAPITALS, each with author if post-1955 (e.g. ACHAETICINA Emeljanov 1962b: 389, subtribe of OPSIINI) or, if pre-1956, with reference to appropriate Metcalf volume that contains the relevant literature sources (e.g. ADELUNGIINAE 14: 169). Only the first use of post-1955 family-group names is included in the check-list. Other uses of family-group names are noted in the following section "Use of family-group names in the Cicadellidae: 1956-1985".

The status of generic and subgeneric names, as of 31 December 1985, are as originally published except where a change is so indicated in the entry. Names published as subgenera are so indicated in the entry. All changes made in the status of generic level names are based on published works, or on nomenclatorial grounds, and not on unpublished opinions. Coded citations at the end of entries indicate the authority for the change, except for pre-1956 names which have usually been resolved in the Metcalf catalogue. Any exceptions to this generalisation are discussed in bracketed comments in the check-list. Where published opinions vary, regarding the status of a name, this is so indicated, with coded citations where necessary. When listing subgenera under the generic names, the nominotypical subgenus name is omitted in all cases.

In the case of those names published in non-English speaking journals which are subsequently translated, e.g. Entomologicheskoe Obozrenie, the pagination following the author's name refers to the original paper and not the English translation, or reprint if paginated differently.

Those genera, either Recent or fossil, for which a type-species has not been designated are annotated accordingly. The present authors have not undertaken to provide such designations. Misidentified type-species and related nomenclatorial problems present in the 1956-1985 literature are listed below, prior to the check-list, together with generic and family-group names placed on the Official List in Zoology.

All synonymies are taken from the literature and are not necessarily the views of the present authors. In addition to inclusion under the name of the senior synonym, all junior synonyms are listed separately in the alphabetical sequence, with appropriate cross reference. In the case of pre-occupied names, the author and date (year) of the senior homonym is included.

References to post-1955 literature dealing with synonymies, homonymy, changes in status, etc., are given under the appropriate generic entry in the check-list. They are not repeated in the taxonomic index unless containing additional information.

Geographical distribution records are based on faunal regions as defined in the following standard works: Essig, E.O., 1947, College Entomology, New York: Macmillan; Darlington, P.J., 1957, Zoogeography: The Geographical Distribution of Animals, New York: Wiley; Brown, J.H. & Gibson, A.C. 1983, Biogeography, London: Mosby; Bartholomew's Physical Atlas, Vol. V, Atlas of Zoogeography, 1911, Edinburgh: Bartholomew & Co.

When deciding upon the terms of reference for distribution, the authors have been guided by the overriding need to communicate effectively. For this reason a combination of terms (zoogeographic, political, geographical, etc.) have been used, together, if necessary, with additional comments to define a particular area. Transition zones between biogeographical regions are indicated as, for example, Palearctic/Oriental or Oriental/Palearctic, the first mentioned region in each case being the one whose faunal elements predominate in the area in question. Detailed information about occurrence records within the range of a genus can be obtained from the literature cited in the taxonomic index. The repetitive use of a country's name is avoided where possible, only the state, province, etc. being given. In such cases, the full geographical location may be obtained from the index and bibliography. The names of fossil genera show the distribution by horizon and geographical location to distinguish them from Recent taxa. Any extension of the distribution of a genus to take account of a fossil record is clearly indicated by an appropriate comment.

When citing distributional records for genera, the following policies have been adopted: the stated distribution reflects the aggregate range of all included species; to distinguish between endemic and adventitious occurrence records, the latter are enclosed in square brackets [ ] and sometimes commented upon more fully; the type locality of the type-species is given in parentheses ( ) immediately after the name of the type-species.

The subfamily and tribal assignment of genera in the check-list is based on work published up to 31 December 1985, the action followed in the most recent publication being adopted in most cases. Exceptions to this rule apply in the case of changes made between 31 December 1955, the cut-off date of the Metcalf Catalogue, and the actual publication date of the relevant part of the Catalogue, e.g. the genus *Alloproctus* was transferred to the AGALLIINAE by Evans 1957a and the genus *Bythonia* referred to a new subfamily, BYTHONIINAE, by Linnavuori 1959b, yet both are still listed in the family IASSINAE by Metcalf 1966c; Linnavuori 1957b suppressed *Eurinoscopus* as a junior synonym of *Batracomorphus*, the currently accepted status, yet is still shown in Metcalf 1966c as a separate genus. In all such cases, the action of the other author takes precedence over Metcalf for present purposes.

The tribal assignment of pre-1956 generic names follows the Metcalf catalogue, subject to post-1955 changes in assignment or tribal synonymy. For post-1955 generic names, the original tribal assignment is followed where indicated by the author, except when re-assigned by a subsequent author. If no placement has been published, a presumptive assignment has been made by relationship with genera with which it has been compared in the literature. In the case of those subfamilies where there are no tribal names, the subfamily name only is given in the generic check-list entry. Subfamily and tribal placements are not indicated for genera known only as fossils.

Subtribal names appear in the check-list only as carry-overs from the Metcalf catalogues, or as names proposed subsequent to 1955. Subtribal placements, when indicated in post-1955 publications, are also shown in the taxonomic index.

The check-list entry for senior synonyms gives the aggregated distribution of the combined taxa (synonyms and subgenera). In the case of junior synonyms and subgenera, only the type locality of the type-species is given where the taxon is listed separately in the alphabetical sequence in the check-list.

FAMILY-GROUP NAMES PROPOSED 1956-1985

Following is an alphabetized list of family-group names proposed during the period 1956-1985. The rank accorded family-group names reflects subjective judgements. Under the Principle of Coordination (ICZN Article 36 (a)), a name established for a taxon at any rank in the family-group is deemed to be simultaneously established with the same author and date for taxa based on the same name-bearing type at other ranks in the family-group. Current placements for names of junior rank in this list, if differing from the initial placement, are indicated in square brackets at the end of the entry. The use of family-group names at various ranks during the period 1956-1985 is indicated in the following compilation, "Use of family-group names in the Cicadellidae: 1956-1985".

[Note: The status of DORATURINI Emeljanov 1962b and STIRELLINI Emeljanov 1966b is questionable because of prior use of the names by other authors. Both names appear in Metcalf's catalogues; the former attributed to Ribaut 1952a:68, the latter to Evans, J.W. 1947a:240. Neither Ribaut nor Evans validated the name in question because their actions do not meet the requirements of the Code (ICZN Article 11(f)(i)(2) and example cited). Since Emeljanov's publications antedated Metcalf's 1963c and 1967a catalogues, the names should be attributed to him. The names are accepted on that basis although the matter should be referred to the ICZN.]

ACHAETICINA Emeljanov 1962b:389; subtribe of OPSIINI.
ACHRINI Davis, R.B. 1975a:12; tribe of ADELUNGIINAE.
ACOSTEMMINAE Evans, J.W. 1972b:660; subfamily.
ACROPONINAE Linnavuori & Quartau 1975a:159; subfamily.
ADAMINI Linnavuori & Al-Ne'amy 1983a:27; tribe of SELENOCEPHALINAE.
ARRUGADIINAE (sic) (ARRUGADINAE) Linnavuori 1965b:141; subfamily.
BAKERINI Mahmood 1967a:21; tribe of TYPHLOCYBINAE.
BATHYSMATOPHORINA Anufriev 1978a:58; subtribe of ERRHOMENINI.
BHATIINI Linnavuori & Al-Ne'amy 1983a:21; tribe of SELENOCEPHALINAE.
BOTHROGONIINA Kwon 1983a:16; subtribe of CICADELLINI.
BYTHONIINAE Linnavuori 1959b:12; subfamily.
CERILLINI (sic) (CERRILLINI) Linnavuori 1975b:49; tribe of DELTOCEPHALINAE.
CIRCULIFERINA Emeljanov 1962b:389; subtribe of OPSIINI.
CORYPHAEINI Emeljanov 1962b:390; tribe of EUSCELINAE [DELTOCEPHALINAE]. Invalid; based on a homonym (ICZN Article 39).
CORYPHAELINI Nast 1972a:432; tribe of DELTOCEPHALINAE.
DORATURINI Emeljanov 1962b:391; tribe of EUSCELINAE [DELTOCEPHALINAE]. See Note above.
DRAKENSBERGENINAE Linnavuori 1979a:673; subfamily.
DWIGHTIINI Linnavuori & Al-Ne'amy 1983a:23; tribe of SELENOCEPHALINAE.
EUACANTHELLINI Evans, J.W. 1966a:142; tribe of APHRODINAE [EUACANTHELLINAE].
EVANSIOLINAE Linnavuori & DeLong 1977c:164; subfamily.
FRISCANINA Anufriev 1978a:59; subtribe of PAGARONIINI.
GABRITINI Nielson 1983e:68; tribe of COELIDIINAE.
GRANULIDAE Hong 1980a:63; family. (Fossil).
HELOCHARINI Metcalf 1965a:356; tribe of TETTIGELLINAE [CICADELLINAE].
HIKANGIINI Nielson 1983e:9; tribe of COELIDIINAE.
HYALOJASSINI Evans, J.W. 1972b:648; tribe of JASSINAE [IASSINAE].
HYPACOSTEMMINI Linnavuori & Al-Ne'amy 1983a:54; tribe of SELENOCEPHALINAE.
IANEIRINI Linnavuori 1978a:43; tribe of DRABESCINAE [SELENOCEPHALINAE].
JASCOPIDAE Hamilton 1971a:943; family. (Fossil).
JASSARGINI Emeljanov 1962b:393; tribe of EUSCELINAE [DELTOCEPHALINAE].
LISTROPHORINI Boulard 1971a:712; tribe of EUPELICINAE.
LUHERIINI Linnavuori 1959b:60; tribe of DELTOCEPHALINAE.
MAGNENTIINI Linnavuori 1978a:34; tribe of NIONIINAE.
MELLIOLINI Metcalf 1962a:16; tribe of HYLICINAE.
MONTEITHIINI Evans, J.W. 1968a:221; tribe of ULOPINAE.
MYERSLOPIINI Evans, J.W. 1957a:367; tribe of ULOPINAE.
NEOBALIINAE (sic) (NEOBALINAE) Linnavuori 1959b:17; subfamily.
NEOKILLINI (sic) (NEOKOLLINI) Metcalf 1965a:265; tribe of TETTIGELLINAE [CICADELLINAE].
NEOPSINAE Linnavuori 1978c:18; subfamily. [Tribe of MACROPSINAE].

OCCINIRVANINI Evans, J.W. 1966a:151; tribe of NIRVANINAE.
OPSIINI Emeljanov 1962b:389; tribe of EUSCELINAE [DELTOCEPHALINAE].
PAGARONIINI Anufriev 1978a:65; tribe of CICADELLINAE.
PARAPHRODINI Linnavuori 1979a:680; tribe of APHRODINAE.
PETALOCEPHALINI Metcalf 1962c:58; tribe of LEDRINAE.
PEYERIMHOFFIOLINI Al-Ne'amy & Linnavuori 1982a:111; tribe of ADELUNGIINAE.
PHEREURHININAE Kramer 1976a:117; subfamily.
PHLOGISINAE Linnavuori 1979a:683; subfamily.
PORTANINI Linnavuori 1959b:45; tribe of XESTOCEPHALINAE.
REUPELEMMELINI Evans, J.W. 1966a:209; tribe of JASSINAE [IASSINAE].
SELENOMORPHINI Evans, J.W. 1974a:166; tribe of JASSINAE [IASSINAE].
STIRELLINI Emeljanov 1966b:609; tribe of EUSCELINAE [DELTOCEPHALINAE]. See Note above.
TERULIINI Nielson 1979b:10; tribe of COELIDIINAE.
THARRINI Nielson 1975a:31; tribe of COELIDIINAE.
YOUNGOLIDIINI Nielson 1983a:18; tribe of COELIDIINAE.
ZYGINELLINI Dworakowska 1977i:24; tribe of TYPHLOCYBINAE.

USE OF FAMILY-GROUP NAMES IN THE CICADELLIDAE: 1956-1985

The validity and status of many subfamily groups are uncertain and one of the main problems facing leafhopper taxonomists today is the lack of a clearly defined and universally accepted classification at the subfamily level. The present section documents departures from the Metcalf classification and shows the use of family-group names at different rank or, in the case of subordinate names, in different combination with names of higher rank. It illustrates the need for a revised classification and why changes from Metcalf's system are adopted in our tentative and provisional list of subfamilies and tribes. When compiling the present section the following policies have been applied: subtribal names are not included except as needed to show the disposition of down-graded names; when two or more nominotypical family-group names appear in a cited publication, references to subordinate names are not repeated; references enclosed in square brackets [ ] deal exclusively with the taxa under which cited, although family-group names are not specifically indicated; bracketed annotations following the family-group name headings indicate the placement in our alphabetized list; non-taxonomic references are omitted; the repetitious use of the name Cicadellidae is omitted, only departures from the name are listed.

ACHRINI (tribe of ADELUNGIINAE).
Al-Ne'amy & Linnavuori 1982a. Hamilton, K. G. A. 1983c (=ADELUNGIINI).
ACINOPTERINI (tribe of DELTOCEPHALINAE).
Emeljanov 1972d. Hamilton, K. G. A. 1975b (=APHRODINAE, DELTOCEPHALINI, PLATYMETOPIINA). Linnavuori 1956a, 1959b. Metcalf 1967a (tribe of EUSCELINAE).
ACOSTEMMINAE
Evans, J. W. 1972b. Hamilton, K. G. A. 1983c (=APHRODINAE, KRISNINI). Knight, W. J. 1983a. Linnavuori & Al-Ne'amy 1983a. Linnavuori & Quartau 1975a.
ACROPONINAE (= ACOSTEMMINAE).
Linnavuori & Al-Ne'amy 1983a (=ACOSTEMMINAE). Linnavuori & Quartau 1975a (= ACOSTEMMINAE).
ADAMINI (tribe of SELENOCEPHALINAE).
Linnavuori & Al-Ne'amy 1983a.

ADELUNGIINAE
Al-Ne'amy & Linnavuori 1982a. Metcalf 1966a (subfamily of AGALLIIDAE). Theron 1979a.
ADELUNGIINI (tribe of ADELUNGIINAE).
Emeljanov 1975a. Hamilton, K. G. A. 1983c (tribe of EURYMELINAE). Nast 1972a (tribe of AGALLIINAE).
AGALLIIDAE
Ishihara 1958a, 1961a. Kristensen 1965a. Kameswara Rao, Ramakrishnan, & Ghai 1979a. Lee, C. E. & Kwon 1976a. Metcalf 1966a.
AGALLIINAE
Anufriev 1970e, 1978a. Dlabola 1964c, 1967b, 1967c, 1967d, 1968b, 1972a, 1974d, 1976a, 1981a. Dubovskij 1966a. Emeljanov 1964c. Evans, J. W. 1966a, 1971b, 1977a. Freytag 1976a. Heller & Linnavuori1968a. Jankovic 1971a. Kalkandelen 1974a. Knight, W. J. 1983a. Korolevskaya 1975a. Kramer 1960a, 1964a, 1965a,1976c. Lee, C. E. & Kwon 1977b. LeQuesne 1965c. Lindberg 1958a. Linnavuori 1956a, 1956d, 1956e, 1959a, 1960a, 1960b, 1961a, 1962a, 1964a, 1965b, 1969a, 1975d. Linnavuori & DeLong 1977a, 1979c, 1979d. Linnavuori & Heller1961a. Linnavuori & Viraktamath 1973a. Lodos & Kalkandelen 1981a. Mitjaev 1971a. Nast 1972a, 1982a. Ossiannilsson 1981a. Quartau 1971a. Quartau & Rodrigues 1969a. Sawai Singh 1969a. Schiemenz 1970a. Synave & Dlabola 1976a. Theron 1980a. Vilbaste 1968a, 1969b, 1982a. Viraktamath, C. A. 1983a. Webb, M. D. 1980a.
AGALLIINI (tribe of AGALLIINAE).
Hamilton, K. G. A. 1983c (=EURYMELINAE, MEGOPHTHALMINI).
ALEBRINAE
Metcalf 1968a (subfamily of CICADELLIDAE, sensu Metcalf).
ALEBRINI (tribe of TYPHLOCYBINAE).
Ahmed, Manzoor 1985c. Anufriev 1978a. Dworakowska 1976b, 1979c. Emeljanov 1964c. Kramer 1965c. Lee, C. E. & Kwon 1977b. LeQuesne & Payne 1981a. Linnavuori 1962a. Lodos & Kalkandelen 1983b. Mahmood 1967a. Mahmood & Ahmed 1969a. Malhotra & Sharma 1974a. Nast 1972a. Ossiannilsson 1981a. Schiemenz 1970a. Sharma, B. 1979a. Vilbaste 1980b. Young 1957c.
AMBLYCEPHALINAE (=CICADELLINAE).
Hamilton, K. G. A. 1983c (=CICADELLINAE, CICADELLINI, CICADELLINA).
ANIPOINI (=IPOINI).
Evans, J. W. 1966a (=IPOINI). Hamilton, K. G. A. 1983c (=EURYMELINAE, EURYMELINI, IPOINA).
ANOTEROSTEMMINI (tribe of CICADELLINAE).
Dlabola 1981a (as ANOTEROSTEMMATINI), (tribe of XESTOCEPHALINAE?). Emeljanov 1964c. Hamilton, K. G. A. 1975b, 1983c (=APHRODINAE, APHRODINI, ANOTEROSTEMMINA). Metcalf 1963c (tribe of APHRODINAE). Nast 1972a.
APHRODIDAE
Ishihara 1958a, 1966a, Kuoh, C. L. 1981b. Lee, C. E. & Kwon 1976a. Metcalf 1963c.
APHRODINAE
Anufriev 1978a. Dlabola 1964c, 1967b, 1967c, 1967d, 1968a, 1968b, 1970b, 1971b, 1972a, 1981a. Dubovskij 1966a. Emeljanov1964c. Evans, J. W. 1966a, 1977a. Hamilton, K. G. A. 1975b (including DELTOCEPHALINAE of Authors); 1975d, 1983c (sensu Hamilton 1975b). Jankovic 1971a. Kalkandelen 1974a. Knight, W. J. 1983a. Korolevskaya 1975a. Kwon & Lee 1979a. LeQuesne 1965c. Linnavuori 1956d, 1959a, 1959c, 1962a, 1979a. Lodos & Kalkandelen 1982b. Logvinenko 1966b. Mitjaev1971a. Nast 1972a, 1982a. Nielson 1962c, 1968b, 1979c, 1985a. Ossiannilsson 1981a. Quartau & Rodrigues 1969a. Schiemenz 1970a. Servadei 1968a (for some genera of CERCOPIDAE; error for APHROPHORIDAE ?). Synave 1976a. Theron 1982b (sensu Hamilton, K. G. A. 1975b). Vilbaste 1968a, 1980b, 1982a. Viraktamath 1983a.
APHRODINI (tribe of APHRODINAE).
Lee, C. E. & Kwon 1977b.

ARRUGADINAE
  Hamilton, K. G. A. 1983c (=EUPELICINAE, EUPELICINI). Linnavuori1965b. Linnavuori & DeLong 1978a.
ATHYSANINI (tribe of DELTOCEPHALINAE).
  Anufriev 1978a. Emeljanov 1972d. Hamilton, K. G. A. 1975b, 1983c (=APHRODINAE, DELTOCEPHALINI, ATHYSANINA). Linnavuori & DeLong 1977c. Metcalf 1967a (tribe of EUSCELINAE). Nast 1972a. Ossiannilsson 1983a.
AUSTROAGALLOIDINAE
  Davis, R. B. 1975a. Evans, J. W. 1966a, 1977a. Hamilton, K. G. A. 1983c (=EURYMELINAE, ADELUNGIINI). Knight, W. J. 1983a. Metcalf 1966a (subfamily of AGALLIIDAE).
BAKERINI (tribe of TYPHLOCYBINAE).
  Dworakowska 1979c (not retained). Hamilton, K. G. A. 1983c(=TYPHLOCYBINI, ERYTHRONEURINA). Mahmood 1967a.
BALBILLINI (tribe of NIRVANINAE).
  Hamilton, K. G. A. 1983c (tribe of TYPHLOCYBINAE). Linnavuori 1979b.
BALCLUTHIDAE
  Ishihara 1958a.
BALCLUTHINI (tribe of DELTOCEPHALINAE).
  Anufriev 1978a. Blocker 1967a. Hamilton, K. G. A. 1975b. (=APHRODINAE, DELTOCEPHALINI, MACROSTELINA). Mahmood 1973b. Metcalf 1967c. Nast 1972a. Ossiannilsson 1983a. Schiemenz 1970a. Vilbaste 1982a.
BHATIINI (tribe of SELENOCEPHALINAE).
  Linnavuori & Al-Ne'amy 1983a.
BOBACELLINI (=DELTOCEPHALINAE, ATHYSANINI).
  Metcalf 1962d (tribe of ULOPINAE).
BYTHONIINAE
  Hamilton, K. G. A. 1983c (=EURYMELINAE; TARTESSINI, tribe of EURYMELINAE). Linnavuori 1959b.
BYTHOSCOPIDAE (=IASSINAE).
  Hamilton, K. G. A. 1983c (=JASSINI LePeletier & Serville). Zachvatkin 1953b.
CEPHALELINAE (see ULOPINAE).
  Linnavuori 1961a. Metcalf 1962d (subfamily of ULOPIDAE).
CEPHALELINI (tribe of ULOPINAE).
  Evans, J. W. 1966a, 1967a, 1977a. Knight, W. J. 1973a, 1973b. Linnavuori 1972a.
CERRILLINI (tribe of DELTOCEPHALINAE).
  Hamilton, K. G. A. 1983c (=APHRODINAE, DELTOCEPHALINI, COCHLORHININA). Linnavuori 1975b.
CHIASMUSINI (tribe of DELTOCEPHALINAE).
  Hamilton, K. G. A. 1975b (as CHIASMUSARIA); 1975d (=EUPELICINI). Metcalf 1963c (=ERRHOMENINI).
CICADELLIDAE (*sensu* Metcalf 1968a = TYPHLOCYBINAE).
  Ishihara 1958a, 1961a, 1965a. Kuoh, C. L. 1981b. Lee, C. E. & Kwon 1976a, 1977b. Lindberg 1958a.
CICADELLIDAE (*sensu* ICZN Opinion 647).
  Emeljanov 1964c. Ghauri 1961a, 1963d, 1965a, 1972e, 1975b, 1979a, 1985b. Lindberg 1965a. Lindberg & Wagner 1965a. Logvinenko 1967b, 1967c, 1974a. Mbondji 1983a. Schroder 1960b, 1961a, 1962a. Zachvatkin 1953b. [NOTE: See comment under IASSIDAE].
CICADELLINAE
  Anufriev 1978a. DeLong & Currie [1959a, 1960a]. Dlabola 1961a, 1964c, 1970b, 1972a, 1981a. Dubovskij 1966a. Emmrich 1975a, 1984a. Evans, J. W. 1966a, 1974a, 1977a. Hamilton, K. G. A. 1983c. Heller & Linnavuori 1968a. Ishihara 1967b, 1971a. Ishihara & Kawase 1968a. Jankovic 1971a. Kalkandelen 1974a. Knight, W. J. 1983a. Korolevskaya 1975a. Kramer 1976c. Kwon 1983a. Lauterer & Schroder 1970a. Lee, C. E. & Kwon 1977b. LeQuesne 1965c. Linnavuori 1959a, 1959b, 1959c, 1960b, 1961a, 1962a, 1979a, 1979b. Linnavuori & DeLong 1978f (error), 1979b. Lodos & Kalkandelen 1983a. Mbondji 1983a. Menezes 1978a. Metcalf 1968a (based upon *Cicadella* Dumeril 1806a, invalid =TYPHLOCYBINAE). Mitjaev 1971a. Nast 1972a, 1982a, 1985a. Nielson 1968b. Nielson & Gill 1984a. Ossiannilsson 1981a. Schiemenz 1970a. Schroder 1972a. Sharma, B. 1977b. Vilbaste 1968a, 1975a, 1980b, 1982a. Young 1963a, 1964a, 1965d, 1968a, 1974a, 1977a. Young & Beier 1963a. Young & Lauterer 1964a, 1966a. Young & Nast 1963a. Zachvatkin 1953b.
CICADELLINI
  Metcalf 1968a (tribe of CICADELLINAE *sensu* Metcalf =TYPHLOCYBINAE). Ossiannilsson 1981a.
CICADELLOIDEA
  Ghauri 1962a, 1963a, 1963b, 1963c, 1963e, 1963f, 1964a, 1964b, 1964c, 1964d, 1966a, 1967a, 1968a, 1969a, 1971a, 1971b, 1971c, 1972a, 1972b, 1974a, 1974b, 1974c, 1975a, 1975c, 1978a, 1980a, 1980b, 1980c, 1985a, 1985c. Kameswara Rao & Ramakrishnan 1980a. Ramakrishnan 1983a.
CICADULINI (tribe of DELTOCEPHALINAE).
  Hamilton, K. G. A. 1975b (=APHRODINAE, DELTOCEPHALINI, CICADULINA). Metcalf 1967b.
CICCIANINI (=PROCONIINI).
  Hamilton, K. G. A. 1983e (=CICADELLINAE, CICADELLINI, PROCONIINA). Metcalf 1965a (tribe of PROCONIINAE).
CICCINI (=PROCONIINI).
  Hamilton, K. G. A. 1983c (=CICADELLINAE, CICADELLINI, PROCONIINA). Metcalf 1965a (tribe of PROCONIINAE).
COCHLORHININI (tribe of DELTOCEPHALINAE).
  Hamilton, K. G. A. 1975b, 1983c (=APHRODINAE, DELTOCEPHALINI, COCHLORHININA). Linnavuori & DeLong 1977c. Metcalf 1967b.
COELIDIIDAE
  Ishihara 1965c. Metcalf 1964b.
COELIDIINAE
  DeLong [1969a]. Dlabola 1972a, 1974a. Evans, J. W. 1966a, 1971a, 1974a. Eyles & Linnavuori 1974a. Ghauri 1972a. Hamilton, K. G. A. 1983c. Heller & Linnavuori 1968a. Knight, W. J. 1982a, 1983a. Linnavuori 1956a, 1960a, 1960b, 1961a, 1975c, 1976c. Linnavuori & DeLong 1977c. Nast 1972a, 1982a. Nielson 1962c, 1968b, 1975a, 1977a, 1979b, 1979c, 1981a, 1981b, 1982a, 1982b, 1982c, 1982d, 1982f, 1982g, 1982h, 1983a, 1983b, 1983c, 1983e, 1983f, 1983h, 1983i, 1985a. Theron 1984c. Webb, M. D. 1980a.
COLLADONINI (=ATHYSANINI).
  Hamilton, K. G. A. 1975b (=APHRODINAE, DELTOCEPHALINI, PLATYMETOPIINA). Metcalf 1967b.
CORNUTIPOINI (=IPOINI).
  Evans, J. W. 1966a (=IPOINI). Hamilton, K. G. A. 1983c (=EURYMELINI subtribe IPOINA).
CORYPHAELINI (tribe of DELTOCEPHALINAE).
  Emeljanov 1964c. Kalkandelen 1974a. Nast 1972a. Ossiannilsson 1983a. Schiemenz 1970a. Vilbaste 1982a.
DELTOCEPHALIDAE
  Ishihara 1958a, 1959b, 1961a, 1963a, 1964a, 1965a, 1966a. Lee, C. E., Lee & Kwon 1976a. Linnavuori & DeLong 1978b. Villiers 1956a. Wagner 1958b.
DELTOCEPHALINAE
  Anufriev 1978a. Blocker & Wesley 1985a. Cwikla & Freytag 1982a. DeLong [1959a, 1964a], 1967a, 1967b, 1967d, 1967e, 1970a, [1971a, 1976a, 1978a], 1980h, [1982a], 1982c, 1982f, [1982h, 1983b, 1983d, 1984c], 1984d, 1984f, 1984g. DeLong & Cwikla 1984a. DeLong & Hamilton [1974a]. DeLong & Harlan 1968a. DeLong & Knull [1971a]. DeLong & Kolbe 1975a. DeLong & Linnavuori 1978b, 1979a. DeLong & Liles [1956a]. DeLong & Martinson [1973a], 1973b, [1973c, 1973g, 1974a, 1976a, 1976b]. Dlabola 1977b. Evans, J. W. 1966a, 1976a. Eyles & Linnavuori 1974a. Haupt, Schmutterer & Muller 1969a. Knight, W. J. 1975a, 1976b, 1982a, 1983a. Kramer 1962b, 1963c, 1965b, 1967a, 1967d, 1971a, 1971b, 1971c, 1976b. Kramer & DeLong 1968a, 1969a. Kramer & Whitcomb 1968a. LeQuesne 1965c, 1969c. Linnavuori 1956a, 1959a, 1959b, 1959c, 1960a, 1960b, 1961a, 1969a, 1973b, 1975a, 1975b, 1975c, 1975d, 1978a, 1978d. Linnavuori & DeLong 1976a, 1977b, 1977c, 1978b, 1978c, 1978e, 1979a, 1979b. Mahmood 1973a, 1973b, 1975a. Mahmood, Sultana & Waheed 1972a. Maldonado-Capriles 1968a.

Menezes 1973a, 1974a, 1975a. Metcalf 1967b (subfamily of EUSCELIDAE). Mitjaev 1971a. Nast 1972a, 1982a. Nielson 1962c, 1968b, 1979c, 1985a. Nielson & Kaloostian 1956a. Okada 1978b. Ossiannilsson 1983a. Sawbridge 1975a. Schiemenz 1970a. Trolle 1966a, 1968a. Vilbaste 1968a, 1975a, 1980b, 1982a. Viraktamath, C. A. 1983a. Webb, M. D. 1980a.

DELTOCEPHALINI (tribe of DELTOCEPHALINAE).
Evans, J. W. 1977a. Hamilton, K. G. A. 1975b, 1983c (tribe of APHRODINAE). Kalkandelen 1974a. Lee, C. E. & Kwon 1977b. Linnavuori & DeLong 1978c. Lodos & Kalkandelen 1985a, 1985b, 1985c. Metcalf 1967a. Quartau 1970a, 1972a.

DIKRANEURINI (tribe of TYPHLOCYBINAE).
Ahmed, Manzoor 1969a, 1985c. Anufriev 1978a. Dlabola 1964b. Dubovskij 1966a. Dworakowska 1977h, 1979e. Einyu & Ahmed 1979a. Emeljanov 1964c. Lee, C. E. & Kwon 1977b. LeQuesne & Payne 1981a. Lodos & Kalkandelen 1983b. Linnavuori 1962a. Mahmood 1967a. Metcalf 1968a (tribe of CICADELLINAE *sensu* Metcalf = TYPHLOCYBINAE). Nast 1972a. Ossiannilsson 1981a. Ramakrishnan & Menon 1971a. Schiemenz 1970a. Sharma, B. 1979a. Sohi 1978a. Vidano 1976a. Webb, M. D. 1980a.

DORATURINI (tribe of DELTOCEPHALINAE).
Anufriev 1978a. Blocker 1983a. Dlabola 1964c, 1967b, 1976c. Hamilton, K. G. A. 1975b, 1983c (= APHRODINAE, APHRODINI, DORATURINA). Kalkandelen 1974a. Metcalf 1963c (tribe of APHRODINAE). Nast 1972a. Ossiannilsson 1983a. Vilbaste 1982a. Webb, M. D. 1980a.

DORYCEPHALINAE (see EUPELICINAE).
Emeljanov 1964c. Dlabola 1961b, 1967b, 1968b, 1970b, 1976b, 1976c, 1981a. Kalkandelen 1974a. Linnavuori 1979a. Mitjaev 1971a. Nast 1972a. Oman 1985a. Ossiannilsson 1981a. Vilbaste 1980b.

DORYCEPHALINI (tribe of EUPELICINAE).
Hamilton, K. G. A. 1983c (tribe of CICADELLINAE). Metcalf 1963d (tribe of HECALIDAE).

DRABESCIDAE (see SELENOCEPHALINAE).
Hamilton, K. G. A. 1975b (= APHRODINAE, SELENOCEPHALINI). Ishihara 1958a. Lee, C. E. & Kwon 1976a. Lee, C. E., Lee & Kwon 1976a. Nast 1972a. Vilbaste 1968a.

DRABESCINAE (see SELENOCEPHALINAE).
Anufriev 1978a. Evans, J. W. 1966a, 1972a, 1977a. Knight, W. J. 1983a. Linnavuori 1960b, 1961a, 1975c. 1978a. Nast 1972a, 1982a.

DRABESCINI (tribe of SELENOCEPHALINAE).
Linnavuori & Al-Ne'amy 1983a.

DRAECULACEPHALINI (= CICADELLINI).
Hamilton, K. G. A. 1983c (= CICADELLINI, CICADELLINA). Metcalf 1965a (tribe of TETTIGELLINAE).

DRAKENSBERGENINAE
Hamilton, K. G. A. 1983c (= TYPHLOCYBINAE, MACROCERATOGONIINI). Linnavuori 1979a.

DWIGHTIINI (tribe of SELENOCEPHALINAE).
Linnavuori & Al-Ne'amy 1983a.

EMPOASCINAE (see TYPHLOCYBINAE). Zachvatkin 1953c, 1973d.

EMPOASCINI (tribe of TYPHLOCYBINAE).
Ahmed, Manzoor 1969b, 1985c. Anufriev 1978a. Dworakowska 1972c, 1973b, 1977e, 1981f, 1982b. Einyu & Ahmed 1980a. LeQuesne & Payne 1981a. Linnavuori 1975c. Lodos & Kalkandelen 1983c. Mahmood 1967a. Metcalf 1968a (tribe of CICADELLINAE *sensu* Metcalf = TYPHLOCYBINAE). Nast 1972a. Ramakrishnan 1982a. Webb, M. D. 1980a.

ERRHOMENELLIDAE (see ERRHOMENINI).
Ishihara 1958a.

ERRHOMENELLINAE (see ERRHOMENINI).
Lee, C. E. & Kwon 1977b. Nast 1972a.

ERRHOMENELLINI (= ERRHOMENINI).
Hamilton, K. G. A. 1983c (= CICADELLINAE, ERRHOMENINI).

ERRHOMENINI (tribe of CICADELLINAE).
Anufriev 1978a. Emeljanov 1964c. Hamilton, K. G. A. 1983c. Lee, C. E. & Kwon 1977b. Metcalf 1963c (tribe of APHRODINAE). Nast 1972a. Oman & Musgrave 1975a. Ossiannilsson 1981a. Schiemenz 1970a.

ERYTHRONEURINI (tribe of TYPHLOCYBINAE).
Ahmed, Manzoor 1971d, 1985c. Anufriev 1978a. Dworakowska 1971e, 1972b, 1979b, 1979f, 1981a, 1981d, 1981e, 1981f, 1981g, 1981h. Emeljanov 1964c. Haupt, Schmutterer & Muller 1969a. Kuoh, C. L. 1981b. Lee, C. E. & Kwon 1977b. LeQuesne & Payne 1981a. Lodos & Kalkandelen 1984c, 1984d. Linnavuori 1960b, 1962a. Mahmood 1967a. Metcalf 1968a (tribe of CICADELLINAE *sensu* Metcalf = TYPHLOCYBINAE). Nast 1972a. Ossiannilsson 1981a. Ramakrishnan & Menon 1973a, 1974a. Schiemenz 1970a. Vilbaste 1980b. Webb, M. D. 1980a.

EUACANTHIDAE (= CICADELLINAE, EVACANTHINI).
Hamilton, K. G. A. 1983c (= CICADELLINAE, EVACANTHINI).

EUACANTHELLINAE
Evans, J. W. 1974a. Knight, W. J. 1983a.

EUACANTHELLINI (tribe of EUACANTHELLINAE).
Evans, J. W. 1966a. Hamilton, K. G. A. 1983c (tribe of CICADELLINAE). Knight, W. J. 1974b (= APHRODINAE).

EUACANTHINI
Emeljanov 1964c.

EUPELICINAE
Dlabola 1964c, 1967b, 1976b. Hamilton, K. G. A. 1983c. Kristensen 1965a. LeQuesne 1965c. Lindberg 1958a. Linnavuori 1956d, 1962a, 1979a. Quartau & Rodrigues 1969a. Schiemenz 1970a. Vilbaste 1982a.

EUPELICINI (tribe of EUPELICINAE).
Dubovskij 1966a. Emeljanov 1964c. Hamilton, K. G. A. 1975b (tribe of APHRODINAE). Linnavuori 1979a (tribe of DORYCEPHALINAE). Metcalf 1963d (tribe of HECALIDAE).

EUPTERYGIDAE (= TYPHLOCYBINAE).
Hamilton, K. G. A. 1983c (= TYPHLOCYBINAE, TYPHLOCYBINI, TYPHLOCYBINA). Zachvatkin 1953b, 1953d, 1955a.

EURYMELIDAE
Evans, J. W. 1959b, 1965b, 1966a, 1969b, 1973a, 1974a, 1977a. Logvinenko 1957c. Metcalf 1965b. Ramakrishnan 1983a.

EURYMELINAE
Hamilton, K. G. A. 1983c. Knight, W. J. 1983a.

EUSCELIDAE (see DELTOCEPHALINAE, ATHYSANINI).
Dlabola 1970a. Ghauri 1971a, 1972b, 1974b. Kristensen 1965b. Kuoh, C. L. 1981b. Logvinenko 1957c. Metcalf 1967a, 1967b, 1967c. Ramakrishnan 1983a.

EUSCELINAE
Anufriev 1970e. Dlabola 1964c, 1966a, 1967b, 1967c, 1967d, 1968a, 1968b, 1970b, 1971b, 1972a, 1974a, 1976a, 1980d. Dubovskij 1966a. Emeljanov 1964c. Ghauri 1971c. Habib 1976a. Habib, El-Kady & Herakly 1975b, 1976a. Haupt, Schmutterer & Muller 1969a. Heller & Linnavuori 1968a. Jankovic 1971a. Kalkandelen 1974a. Lindberg 1958a. Linnavuori 1961a, 1962a, 1964a. Linnavuori & Heller 1961a. Logvinenko 1960b, 1975a. Metcalf 1967a. Mitjaev 1967c. Quartau & Rodrigues 1969a. Sawai Singh 1969a. Theron 1979a, 1980a.

EUSCELINI (= ATHYSANINI).
Dlabola 1974c. Ghauri 1975b. LeQuesne 1969c. Linnavuori 1956d, 1959a, 1959b, 1959c, 1960a, 1961a, 1962a. Linnavuori & DeLong 1977c, 1978c. Metcalf 1967a. Schiemenz 1970a. Vilbaste 1982a.

EVACANTHIDAE (see CICADELLINAE, EVACANTHINI).
Datta & Pramanik 1977a. Ishihara 1963b, 1965c, 1966a. Kuoh, C. L. 1981b. Li, Z. Z. 1985a. Metcalf 1963a.

EVACANTHINAE
Dlabola 1964c, 1967b, 1967c, 1967d, 1968a, 1968b, 1970b, 1972a. Ishihara 1957a. Knight, W. J. 1983a. Lee, C. E. & Kwon 1977b. LeQuesne 1965c. Vilbaste 1971f.

EVACANTHINI (tribe of CICADELLINAE).
Anufriev 1978a. Dubovskij 1966a. Hamilton, K. G. A. 1983c. Hill, B. G. 1973a. Nast 1972a. Ossiannilsson 1981a.

EVANSIOLINAE
Linnavuori & DeLong 1977c.

FIEBERIELLINI (tribe of DELTOCEPHALINAE).
Anufriev 1978a. Dlabola 1964c, 1977b. Emeljanov 1964c, 1972d (= ACINOPTERINI). Hamilton, K. G. A. 1975b (= APHRODINAE, DELTOCEPHALINI,

PLATYMETOPIINA). Kalkandelen 1974a. Linnavuori 1959c, 1962a. Linnavuori & Al-Ne'amy 1983a. Metcalf 1967b. Nast 1972a. Quartau & Rodrigues 1969a. Schiemenz 1970a.

GNATHODINI (=BALCLUTHINI).
Hamilton, K. G. A. 1975b (=APHRODINAE, DELTOCEPHALINI, MACROSTELINA).

GONIAGNATHINI (tribe of DELTOCEPHALINAE).
Anufriev 1978a. Dlabola 1964c, 1967b, 1968b, 1981a (tribe of CICADELLINAE). Dubovskij 1966a. Emeljanov 1964c. Hamilton, K. G. A. 1975b (=APHRODINAE, SELENOCEPHALINI). Kalkandelen 1974a. Linnavuori 1956d, 1959c, 1962a, 1978d. Linnavuori & Al-Ne'amy 1983a. Lodos & Kalkandelen 1985a. Metcalf 1967b. Nast 1972a. Webb, M. D. 1980a.

GRAPHOCEPHALINI (=CICADELLINAE, CICADELLINI).
Metcalf 1965a (tribe of TETTIGELLINAE).

GRYPOTINI (tribe of DELTOCEPHALINAE).
Emeljanov 1964c. Haupt, Schmutterer & Muller 1969a. Kalkandelen1974a. Linnavuori 1956d, 1962a. Lodos & Kalkandelen 1985a. Metcalf 1967b. Nast 1972a. Ossiannilsson 1983a. Schiemenz 1970a. Vilbaste 1982a.

GYPONIDAE
Metcalf 1962b. Zachvatkin 1953b.

GYPONINAE
DeLong 1975a, [1975b, 1976b, 1976c, 1976d], 1976f, 1976g, [1977a], 1977b, 1977c, [1977d, 1977e, 1977f, 1978b, 1979a], 1979b, 1979c, 1979d, 1979e, 1979f, [1979g], 1980a, 1980c, [1980d, 1980e, 1980f], 1980g, 1981a, [1981b, 1981c], 1981d, 1982a, [1982d], 1982e, [1983a], 1983c, 1983e, 1983f, 1984a, 1984e, 1984h. DeLong & Bush 1971a. DeLong &Foster 1982a, 1982b. DeLong & Freytag 1963a, 1963b, 1964a, 1964b, 1964c, 1966a, 1966b, 1966c, 1967a, 1967b, 1969a, 1969b, 1970a, 1971a, 1972a, 1972b, [1972c], 1972d, 1972e, 1972f, [1973a], 1974a, 1975a, 1975b, 1975c, 1975d, 1975e, 1976a. DeLong & Kolbe 1974a, [1974b], 1975c. DeLong & Linnavuori 1977a. DeLong & Martinson 1972a, 1973d, [1980a]. DeLong & Triplehorn 1978a, 1979a. DeLong & Wolda 1978a, [1982a], 1982b, 1983a, 1984b. DeLong, Wolda & Estribi [1983a]. Dlabola 1972a (for *Penthimia*). Freytag & Cwikla 1982a. Freytag& DeLong 1971a. Kramer 1963a. Linnavuori & DeLong 1977c. Nielson 1968b, 1979c, 1985a.

GYPONINI (tribe of GYPONINAE).
Evans, J. W. 1959a, 1961a (tribe of JASSINAE). DeLong & Freytag 1962a. Knull & Knull 1960a. Linnavuori & Quartau 1975a (tribe of IASSINAE).

HECALIDAE (see DELTOCEPHALINAE, HECALINI).
Ishihara 1961a. Metcalf 1963d.

HECALINAE
Anufriev 1978a. Dlabola 1966a, 1967b, 1967c, 1968a, 1970b, 1972a,1974d,1977b,1980d. Dubovskij 1966a. Emeljanov 1964c. Evans, J. W. 1959a, 1966a, 1973a, 1977a. Kalkandelen 1974a. Knight, W. J. 1973a. Linnavuori 1956a, 1956d, 1957a, 1960a. Lodos & Kalkandelen 1982b. Mitjaev 1971a. Morrison 1973a. Nast 1972a, 1982a. Nielson & Kaloostian 1956a. Vilbaste 1975a, 1980b.

HECALINI (tribe of DELTOCEPHALINAE).
Hamilton, K. G. A. 1972d, 1975b, 1983c (=APHRODINAE, HECALINI). Kwon & Lee 1979g. Lee, C. E. & Kwon 1977b. Linnavuori 1975c. Linnavuori & DeLong 1977c, 1978c, 1978e. Linnavuori & Heller 1961a. Sawai Singh 1969a. Webb, M. D. 1980a.

HELIONINAE
Hamilton, K. G. A. 1983c (=TYPHLOCYBINAE, TYPHLOCYBINI, EMPOASCINA).

HELIONINI (tribe of TYPHLOCYBINAE).
Metcalf 1968a (tribe of CICADELLINAE *sensu* Metcalf = TYPHLOCYBINAE).

HELOCHARINI (=CICADELLINI).
Hamilton, K. G. A. 1983c (= CICADELLINAE, CICADELLINI, CICADELLINA). Metcalf 1965a (tribe of TETTIGELLINAE).

HYALOJASSINI (tribe of IASSINAE).
Evans,J. W. 1972a. Hamilton,K. G. A. 1983c (= SCARINAE, REUPLEMMELINI). Linnavuori & Quartau 1975a.

HYLICIDAE
Linnavuori 1969a, 1972b. Metcalf 1962a.

HYLICINAE
Knight, W. J. 1983a. Kramer 1964c, 1965c.

HYLICINI (tribe of HYLICINAE).
Hamilton, K. G. A. 1983c (tribe of CICADELLINAE).

HYPACOSTEMMINI (tribe of SELENOCEPHALINAE).
Linnavuori & Al-Ne'amy 1983a.

IANERINI (tribe of SELENOCEPHALINAE).
Hamilton, K. G. A. 1983c (=SCARINAE, SELENOCEPHALINI). Linnavuori 1978a. Linnavuori & Al-Ne'amy 1983a.

IASSIDAE
Cantoreanu 1959a, 1965d. Dlabola 1957b. Metcalf 1966c. Vilbaste 1959a, 1959b, 1961a, 1961b, 1962a, 1965a, 1965b, 1967a, 1968a,1969a,1969b. [NOTE: In the Cantoreanu and Vilbaste articles, IASSIDAE is used in the sense of CICADELLIDAE].

IASSINAE
Anufriev 1978a. Blocker 1975a, 1975b, 1976a, 1979a, 1979b, 1979c, 1982a. DeLong 1982c. Dlabola 1964c, 1965c, 1967b, 1967d, 1970b, 1971b, 1972a, 1976c, 1981a. Heller & Linnavuori 1968a. ICZN Opinion 612, 1966. Knight, W. J. 1983a. Kramer 1963a, 1965a. Lee, C. E. & Kwon 1977b. Linnavuori 1957b, 1959a, 1960b, 1961a, 1962a, 1964a, 1965b, 1969a, 1975c. Linnavuori & DeLong 1977c. Linnavuori & Quartau 1975a. Lodos & Kalkandelen 1982b. Mahmood 1975a. Nast 1972a, 1982a. Nielson & Kaloostian 1956a. Ossiannilsson 1981a. Quartau 1980a, 1981b, 1982b, 1983b, 1985e. Quartau & Davies 1983a. Schiemenz 1970a. Theron 1980a. Vilbaste 1982a. Viraktamath, C. A. 1979a, 1983a. Webb, M. D. 1980a.

IDIOCERIDAE
Ghauri 1975c, 1985c. Heller 1969a. Ishihara 1956a, 1958a, 1961a, 1966a. Kameswara Rao & Ramakrishnan 1979a. Kristensen 1965a. Kuoh, C. L. & Fang 1984a, 1985a, 1985b.

IDIOCERINAE
Ahmed, S. S., Naheed & Ahmed 1980a (Idiocerine). Anufriev 1971a, 1978a. Dlabola 1964c, 1967b, 1967d, 1968a, 1968b, 1970b, 1971b, 1972a, 1974a, 1974b, 1976a, 1976b, 1977b, 1981a. Dubovskij 1966a. Emeljanov 1964c. Evans, J. W. 1959a, 1966a,1974a,1977a. Freytag 1965a, 1969a, 1975b, 1976a. Freytag & Knight, W. J. 1966a. Freytag & Morrison 1967a, 1972a. Heller 1969a. Heller & Linnavuori 1968a. Jankovic 1971a. Kalkandelen 1974a. Kameswara Rao, Ramakrishnan & Ghai 1979a. Knight, W. J. 1974b, 1983a. Korolevskaya 1975a. Kwon 1985a. Lee, C. E. & Kwon 1977b. LeQuesne 1965c. Linnavuori 1956a, 1959c, 1960a, 1960b, 1961a, 1962a. Linnavuori & DeLong 1977c. Lodos & Kalkandelen 1982a. Mahmood 1975a. Maldonado-Capriles 1961a, 1964a, 1965a, 1968a, 1970a, 1971a, 1971b, 1972a, 1972b, 1973a, 1974a, 1975a, 1976a, 1977a, 1977b, 1977c, 1977d, 1984a, 1985a, 1985b. Mitjaev 1971a. Nast 1972a, 1982a. Nielson & Kaloostian 1956a. Ossiannilsson 1981a. Quartau & Rodrigues 1969a. Schiemenz 1970a. Vilbaste 1968a, 1980b, 1982a. Viraktamath, C. A. 1973a, 1976b, 1979b. Viraktamath, C. A. & Murphy 1980a. Webb, M. D. 1980a, 1983a, 1983b.

IDIOCERINI (tribe of IDIOCERINAE).
Hamilton, K. G. A. 1983c (=EURYMELINAE, IDIOCERINI). Ghauri 1975c, 1985c.

IPOINAE (see EURYMELINAE).
Evans, J. W. 1959b. Metcalf 1965b (subfamily of EURYMELIDAE).

IPOINI (tribe of EURYMELINAE).
Evans, J. W. 1966a. Hamilton, K. G. A. 1983c (=EURYMELINAE, EURYMELINI, IPOINA).

JASCOPIDAE
Evans, J. W. 1972a. Hamilton, K. G. A. 1971a.

JASSARGINI (=PARALIMNINI).
Dlabola 1964c, 1967b, 1967c, 1967d, 1968b, 1976b. Dubovskij 1966a. Emeljanov 1964c, 1968a, 1972d. Kalkandelen 1974a. Schiemenz 1970a. Webb, M. D. 1980a.

JASSIDAE
Bergman 1956a. Gourlay 1964a. Habib, El-Kady & Herakly 1975a, 1975b, 1976a. Ishihara 1958a, 1961a, 1966a. Lee, C. E. & Kwon 1976a. Lindberg 1956a, 1956b, 1958a, 1960a, 1960b, 1960c, 1961a, 1962a, 1963a, 1964a. Logvinenko 1956a, 1956b, 1957b, 1957c, 1961a, 1961b, 1961d, 1963b. Ramachandra Rao 1967b. Servadei 1957a, 1958a, 1960a. Srivastava, B. K. 1958a.

Vidano 1962b. Vilbaste 1960a. Villiers 1956a. Yang, C. K. 1965a (= CICADELLIDAE). Zachvatkin 1953b, 1953e.

JASSINAE
Dubovskij 1966a. Emeljanov 1964c. Evans, J. W. 1959a, 1961a, 1966a, 1972b, 1974a, 1977a. Hamilton, K. G. A. 1983c. Kalkandelen 1974a. Knight, W. J. 1974b, 1976b. LeQuesne 1965c. Linnavuori 1956a, 1956e. Linnavuori & Heller 1961a. Mahmood 1975a. Mitjaev 1971a. Ramachandra Rao 1967b. Remane 1961a, 1967a. Villiers 1956a. Zachvatkin 1953b, 1953e.

JORUMINI (tribe of TYPHLOCYBINAE).
Hamilton, K. G. A. 1983c. (= TYPHLOCYBINAE, TYPHLOCYBINI, EMPOASCINA). Metcalf 1968a (tribe of CICADELLINAE *sensu* Metcalf = TYPHLOCYBINAE).

KAHAVALUINI (= ULOPINI).
Evans, J. W. 1966a (= ULOPINI).

KOEBELIINAE
Evans, J. W. 1969a. Hamilton, K. G. A. 1983c (= tribe of EUPELICINAE). Kramer 1966a. Metcalf 1962c (subfamily of LEDRIDAE). Oman 1971b.

KRISNIDAE
Ishihara 1961a.

KRISNINAE
Evans, J. W. 1972b. Knight, W. J. 1983a.

KRISNINI (tribe of IASSINAE).
Evans, J. W. 1959a (tribe of JASSINAE). Hamilton, K. G. A. 1983c(tribe of SCARINAE). Linnavuori 1959b. Linnavuori & Al-Ne'amy 1983a. Linnavuori & DeLong 1977a. Linnavuori & Quartau 1975a. Metcalf 1966c.

LEDRIDAE ICZN Opinion 647, 1961. Ishihara 1958a. Kuoh, C. L. 1981b, 1984a. Lee, C. E. & Kwon 1976a. Lee, C. E., Lee & Kwon 1976a. Metcalf 1962a. Servadei 1960a.

LEDRINAE
Anufriev 1978a. Datta & Pramanik 1977a. Dlabola 1972a, 1981a. Emeljanov 1964c. Evans, J. W. 1959a, 1959b, 1966a, 1969a. Hamilton, K. G. A. 1983c. Knight, W. J. 1974a, 1976b, 1983a. Kramer 1966a. Kwon & Lee 1978b. Lee, C. E. & Kwon 1977b. LeQuesne 1965c. Linnavuori 1956a, 1959b, 1961a, 1969a, 1972b, 1977a. Linnavuori & DeLong 1977c. Lodos & Kalkandelen 1981c. Nast 1972a, 1982a. Nielson & Kaloostian 1956a. Ossiannilsson 1981a. Schiemenz 1970a. Theron 1980a. Vilbaste 1968a.

LISTROPHORINI (tribe of EUPELICINAE).
Hamilton, K. G. A. 1983c. Linnavuori 1979a (tribe of DORYCEPHALINAE).

LUHERINI (tribe of DELTOCEPHALINAE).
Hamilton, K. G. A. 1975b (= APHRODINAE, DELTOCEPHALINAE, PLATYMETOPIINA). Linnavuori 1959b.

MACROCERATOGONIINI (tribe of NIRVANINAE).
Evans, J. W. 1966a, 1974a. Hamilton, K. G. A. 1983c (tribe of TYPHLOCYBINAE). Linnavuori 1979b. Metcalf 1963b.

MACROPSIDAE ICZN Opinion 603, 1961. Hori 1969a. Ishihara 1961a, 1966a. Kuoh, C. L. 1981b. Metcalf 1966b.

MACROPSINAE
Anufriev 1978a. Dlabola 1964c, 1967b, 1967d, 1968a, 1968b, 1970b, 1974a, 1975a, 1981a. Dubovskij 1966a. Emeljanov 1964c. Evans, J. W. 1966a, 1971b, 1972a, 1977a. Freytag 1974a, 1976a. Jankovic 1971a. Kalkandelen 1974a. Knight, W. J. 1974b, 1983a. Korolevskaya 1975a. Lee, C. E. & Kwon 1977b. LeQuesne 1965c. Lindberg 1958a, 1960b. Linnavuori 1956a, 1959a, 1959c, 1962a, 1965b, 1969a, 1978b. Linnavuori & DeLong 1977c. Lodos & Kalkandelen 1981a. Mitjaev 1971a. Nast 1972a, 1982a. Nielson 1962a, 1968b, 1979c, 1985a. Ossiannilsson 1981a. Quartau & Rodrigues 1969a. Schiemenz 1970a. Theron 1980a. Vilbaste 1968a, 1982a. Viraktamath, C. A. 1980a, 1981b.

MACROPSINI (tribe of MACROPSINAE).
Hamilton, K. G. A. 1980b, 1983c (tribe of EURYMELINAE), 1983d. Metcalf 1966b. Viraktamath, C. A. 1983a.

MACROSTELIDAE
Dlabola 1970a. Ishihara 1958a.

MACROSTELINI (tribe of DELTOCEPHALINAE).
Anufriev 1978a. Dlabola 1964c, 1967b, 1967c, 1967d, 1968b, 1981a. Dubovskij 1966a. Emeljanov 1964c. Evans, J. W. 1966a. Eyles & Linnavuori 1974a. Habib, El-Kady & Herakly 1976a. Hamilton, K. G. A. (= APHRODINAE, DELTOCEPHALINI, MACROSTELINA). Heller & Linnavuori 1968a. Kalkandelen 1974a. LeQuesne 1969c. Linnavuori 1956a, 1956d, 1959b, 1959c, 1960a, 1960b, 1961a, 1962a, 1975c. Linnavuori & DeLong 1977c. Lodos & Kalkandelen 1985b. Mahmood 1973b. Metcalf 1967c. Nast 1972a. Ossiannilsson 1983a. Quartau & Rodrigues 1969a. Schiemenz 1970a. Theron 1975b. Triplehorn & Nault 1985c. Webb, M. D. 1980a.

MAGNENTIINI (tribe of NIONIINAE).
Hamilton, K. G. A. 1983c (= NIONIINI). Linnavuori 1978a.

MAKILINGIINAE
Young 1968a, 1977a.

MAKILINGIINI (tribe of MAKILINGIINAE).
Hamilton, K. G. A. 1983c (= CICADELLINAE, CICADELLINI, CICADELLINA). Linnavuori 1979a. Metcalf 1965a (tribe of PROCONIINAE).

MEGOPHTHALMINAE
Dlabola 1964c, 1981a. Emeljanov 1964c. Gill & Oman 1982a. Hamilton, K. G. A. 1983c (= EURYMELINAE, MEGOPHTHALMINI). Jankovic 1971a. Kalkandelen 1974a. LeQuesne 1965c. Linnavuori 1959a, 1959c, 1962a, 1972a, 1973a. Lodos & Kalkandelen 1981a. Metcalf 1962d (subfamily of ULOPIDAE). Nast 1972a. Ossiannilsson 1981a. Quartau & Rodrigues 1969a. Sawbridge 1975a. Van Stalle 1982a, 1983a, 1983b. Vilbaste 1982a.

MEGOPHTHALMINI (tribe of MEGOPHTHALMINAE).
Evans, J. W. 1968a. Oman 1972a.

MELICHARELLINAE (= ADELUNGIINAE).
Al-Ne'amy & Linnavuori 1982a (= ADELUNGIINAE). Davis, R. B. 1975a. Dlabola 1966a, 1967b, 1968b, 1976a. Dubovskij 1966a, 1968a. Hamilton, K. G. A. 1983c (= ADELUNGIINI). Linnavuori 1962a. Metcalf 1966a (subfamily of AGALLIIDAE). Mitjaev 1971a.

MELICHARELLINI (tribe of MELICHARELLINAE).
Emeljanov 1964c. Linnavuori 1969a (tribe of AGALLIINAE). Nast 1972a (tribe of AGALLIINAE).

MELLIINI (= MELLIOLINI).
Hamilton, K. G. A. 1983c (= CICADELLINAE, HYLICINI). Metcalf 1962a (= MELLIOLINI).

MELLIOINI (tribe of HYLICINAE).
Hamilton, K. G. A. 1983c (= CICADELLINAE, HYLICINI). Metcalf 1962a (tribe of HYLICINAE).

MESAMIINI (= PLATYMETOPIINI).
Hamilton, K. G. A. 1975b (= APHRODINAE, DELTOCEPHALINI, PLATYMETOPIINA).

MILEEWANINAE
Linnavuori & DeLong 1977a. Nast 1972a (as MILEEVANINAE).

MILEEWANINI (tribe of CICADELLINAE).
Anufriev 1978a. Evans, J. W. 1966a. Hamilton, K. G. A. 1983c. Heller & Linnavuori 1968a. Lee, C. E. & Kwon 1977b (as MILEEVANINI). Linnavuori 1961a, 1979a. Metcalf 1965a (tribe of TETTIGELLINAE). Young 1965a (tribe of TYPHLOCYBINAE).

MONTEITHIINI (tribe of ULOPINAE).
Evans, J. W. 1968a. Hamilton, K. G. A. 1983c (= ULOPINI).

MUKARIINAE
Linnavuori 1979b.

MUKARIINI (tribe of MUKARIINAE).
Hamilton, K. G. A. 1983c (= APHRODINAE, MUKARIINI). Metcalf 1963b (tribe of NIRVANIDAE).

MYERSLOPINAE
Linnavuori 1972a. Linnavuori & DeLong 1977c.

MYERSLOPINI (tribe of ULOPINAE).
Evans, J. W. 1961a, 1966a, 1968a, 1977a. Hamilton, K. G. A. 1983c. Knight, W. J. 1973b.

NEHELINI (tribe of AGALLIINAE).
Hamilton, K. G. A. 1983c (= EURYMELINAE, MEGOPHTHALMINI). Metcalf 1966a (= AGALLIINAE).

NEOBALINAE
Hamilton, K. G. A. 1983c (= APHRODINAE, APHRODINI, NEOBALINA). Kramer 1963b. Linnavuori 1956a, 1959b. Linnavuori & DeLong 1976a, 1977c, 1978c, 1979b. Linnavuori & Heller 1961a.

NEOCOELIDIINAE
    DeLong & Kolbe 1975b. Kramer 1958a, 1959a, 1961a, 1962a, 1964d, 1967b. Kramer & Linnavuori 1959a. Linnavuori 1956a, 1959b. Metcalf 1964b (subfamily of COELIDIINAE). Nielson & Kaloostian 1956a.
NEOCOELIDIINI (tribe of NEOCOELIDIINAE).
    Hamilton, K. G. A. 1983c (tribe of TYPHLOCYBINAE). Linnavuori 1965b (tribe of DELTOCEPHALINAE). Linnavuori & Heller 1961a.
NEOKOLLINI (=CICADELLINI).
    Hamilton, K. G. A. 1983c (as NEOKILLINI) (=CICADELLINAE, CICADELLINI, CICADELLINA). Metcalf 1965a (tribe of TETTIGELLINAE).
NEOPSINAE
    Linnavuori 1978c. Linnavuori & DeLong 1976a, 1977c.
NEOPSINI (tribe of MACROPSINAE).
    Hamilton, K. G. A. 1983c, 1983d (tribe of EURYMELINAE).
NIONIINAE
    Linnavuori 1956a, 1959b, 1978a, 1979b.
NIONIINI (tribe of NIONIINAE).
    Evans, J. W. 1971b (tribe of PENTHIMIINAE). Hamilton, K. G. A. 1983c (tribe of EURYMELINAE). Metcalf 1966b (tribe of MACROPSINAE).
NIRVANIDAE
    Heller 1972a. Ishihara 1958a, 1961a. Kuoh, C. L. & Kuoh, J. L. 1983a, 1983b. Metcalf 1963b. Theron 1970a.
NIRVANINAE
    Evans, J. W. 1966a, 1973a, 1974a, 1977a. Heller 1972a. Heller & Linnavuori 1968a. Hill, B. G. 1973a. Izzard 1955a. Knight, W. J. 1983a. Kramer 1964b, 1965a, 1976c. Lee, C. E. & Kwon 1977b. Linnavuori 1959b, 1960a, 1969a, 1975c, 1979b. Linnavuori & DeLong 1978f. Nast 1972a. Theron 1982b. Wesley 1983a.
NIRVANINI (tribe of NIRVANINAE).
    Hamilton, K. G. A. 1983c (tribe of TYPHLOCYBINAE).
OCCINIRVANINI (tribe of NIRVANINAE).
    Evans, J. W. 1966a. Hamilton, K. G. A. 1983c (=TYPHLOCYBINAE, MACROCERATOGONIINI).
OPIOINI (=IPOINI).
    Evans, J. W. 1966a (=IPOINI). Hamilton, K. G. A. 1983c (=EURYMELINAE, EURYMELINI, IPOINA).
OPSIINI (tribe of DELTOCEPHALINAE).
    Anufriev 1978a. Dlabola 1964c, 1967b, 1967d. Emeljanov 1964c, 1968a. Eyles & Linnavuori 1974a. Hamilton, K. G. A. 1975b (=APHRODINAE, DELTOCEPHALINI, PLATYMETOPIINA). Kalkandelen 1974a. Linnavuori 1975c. Linnavuori & DeLong 1977c. Nast 1972a. Okada 1978b. Ossiannilsson 1983a. Lodos & Kalkandelen 1985a. Schiemenz 1970a. Webb, M. D. 1980a.
PAGARONIINI (tribe of CICADELLINAE).
    Anufriev 1978a. Hamilton, K. G. A. 1983c (=CICADELLINAE, EUACANTHELLINI). Kwon 1983a. Lee, C. E. & Kwon 1977b. Nast 1982a. Vilbaste 1982a.
PARABOLOPONINAE
    Eyles & Linnavuori 1974a. Knight, W. J. 1983a. Linnavuori 1978d. Nast 1972a. Webb, M. D. 1980a, 1981b.
PARABOLOPONINI (tribe of PARABOLOPONINAE).
    Hamilton, K. G. A. 1975b (tribe of APHRODINAE); 1983c (tribe of TYPHLOCYBINAE). Lee, C. E. & Kwon 1977b (tribe of APHRODINAE). Linnavuori 1960a (tribe of DELTOCEPHALINAE).
PARADORYDIINAE Heller & Linnavuori 1968a. Lindberg 1958a. Linnavuori 1961a, 1962a. Quartau & Rodrigues 1969a.
PARADORYDIINI (tribe of EUPELICINAE). Dubovskij 1966a (tribe of HECALINAE). Emeljanov 1964c. Evans, J. W. 1966a, 1977a. Hamilton, K. G. A. 1975b (tribe of APHRODINAE), 1983a. Knight, W. J. 1973a. Linnavuori 1979a (tribe of DORYCEPHALINAE). Metcalf 1963d (tribe of HECALIDAE).
PARALIMNINI (tribe of DELTOCEPHALINAE).
    Anufriev 1978a. Dlabola 1972a, 1977b. Hamilton, K. G. A. 1975b (=APHRODINAE, DELTOCEPHALINI). Nast 1972a. Mitjaev 1980d. Ossiannilsson 1983a. Remane & Asche 1980a. Vilbaste 1982a.
PARAPHRODINI (tribe of APHRODINAE).
    Hamilton, K. G. A. 1983c (=SCARINAE, SELENOCEPHALINI). Linnavuori 1979a.

PAROPIDAE (=MEGOPHTHALMINAE).
    Hamilton, K. G. A. 1983c (=EURYMELINAE, MEGOPHTHALMINI). Nast 1972a.
PENTHIMIINAE
    Anufriev 1978a. Dlabola 1981a. Evans, J. W. 1972a, 1973a, 1966a, 1977a. Knight, W. J. 1983a. Kwon & Lee 1978c. Lee, C. E. & Kwon 1977b. Linnavuori 1959c, 1977a. Lodos & Kalkandelen 1982b. Mahmood 1975a. Metcalf 1962b (subfamily of GYPONIDAE). Nast 1972a, 1982a. Schiemenz 1970a. Theron 1980a. Webb, M. D. 1980a.
PENTHIMIINI
    Evans, J. W. 1959a (tribe of JASSINAE). Emeljanov 1964c. Hamilton, K. G. A. 1983c (tribe of SCARINAE). Linnavuori 1961a (tribe of EUSCELINAE); 1959b (tribe of DELTOCEPHALINAE).
PETALOCEPHALINI (tribe of LEDRINAE).
    Hamilton, K. G. A. 1983c (=LEDRINAE, LEDRINI). Metcalf 1962c (tribe of LEDRINAE).
PEYERIMHOFFIOLINI (tribe of ADELUNGIINAE).
    Al-Ne'amy & Linnavuori 1982a.
PHEREURHINAE
    Kramer 1976a.
PHLEPSIINI (=ATHYSANINI).
    Hamilton, K. G. A. 1975b (=SELENOCEPHALINI).
PHLOGISINAE
    Hamilton, K. G. A. 1983c (=EURYMELINAE, NIONIINI). Linnavuori 1979a.
PHRYNOMORPHINI (=ATHYSANINI).
    Hamilton, K. G. A. 1975b (=APHRODINAE, DELTOCEPHALINI, ATHYSANINA).
PLATYJASSINI (tribe of IASSINAE).
    Evans, J. W. 1959a (tribe of JASSINAE), 1972b. Hamilton, K. G. A. 1983c (=SCARINAE, PENTHIMIINI). Linnavuori & Quartau 1975a. Metcalf 1966c.
PLATYMETOPIINI (tribe of DELTOCEPHALINAE).
    Evans, J. W. 1966a, 1977a. Dlabola 1964c, 1977b. Hamilton, K. G. A. 1975b, 1983c (=APHRODINAE, DELTOCEPHALINI, PLATYMETOPIINA). Metcalf 1967c. Nast 1977b. Schiemenz 1970a. Vilbaste 1982a. Webb, M. D. 1980a.
POGONOSCOPINAE (see EURYMELINAE).
    Evans, J. W. 1959b. Metcalf 1965b (subfamily of EURYMELIDAE).
POGONOSCOPINI (tribe of EURYMELINAE).
    Evans, J. W. 1966a. Hamilton, K. G. A. 1983c.
PORTANINI (tribe of XESTOCEPHALINAE).
    Hamilton, K. G. A. 1975b (=SELENOCEPHALINI). Linnavuori 1959b. Linnavuori & Al-Ne'amy 1983a.
PROCONIIDAE
    Kristensen 1965a.
PROCONIINAE
    Metcalf 1965a (subfamily of TETTIGELLIDAE).
PROCONIINI (tribe of CICADELLINAE).
    Kramer 1976c. Linnavuori 1979a. Metcalf 1965a (tribe of PROCONIINAE). Schroder 1959a. Young 1968a, 1977a.
PYTHAMINAE (see CICADELLINAE, EVACANTHINI).
    Metcalf 1963a (subfamily of EVACANTHIDAE).
REUPLEMMELINI (tribe of IASSINAE).
    Evans, J. W. 1966a, 1972b. Hamilton, K. G. A. 1983c (tribe of SCARINAE). Linnavuori & Quartau 1975a.
REUTERIELLINI (=HECALINI).
    Evans, J. W. 1959a. Metcalf 1966c (tribe of IASSINAE).
SANDERSELLINI (tribe of COELIDIINAE).
    Metcalf 1964b. Nielson 1975a, 1981b.
SCAPHOIDEINI (tribe of DELTOCEPHALINAE).
    Hamilton, K. G. A. 1975b (=APHRODINAE, DELTOCEPHALINI, PLATYMETOPINA). Metcalf 1967c.
SCAPHYTOPIINI (tribe of DELTOCEPHALINAE).
    Anufriev 1978a. Dlabola 1977b. Dubovskij 1966a. Emeljanov 1964c. Hamilton, K. G. A. 1975b (=APHRODINAE, DELTOCEPHALINI, PLATYMETOPIINA). Kalkandelen 1974a. Linnavuori 1956a, 1959b, 1961a, 1978d. Linnavuori & DeLong 1978c. Linnavuori & Heller 1961a. Linnavuori & Al-Ne'amy 1983a. Lodos & Kalkandelen 1985c. Metcalf 1967c. Nast 1972a.

SCARINAE (see GYPONINAE).
  Hamilton, K. G. A. 1983c.
SELENOCEPHALINAE
  Linnavuori & Al-Ne'amy 1983a.
SELENOCEPHALINI (tribe of SELENOCEPHALINAE).
  Evans, J. W. 1966a (tribe of DELTOCEPHALINAE). Hamilton, K. G. A. 1975b (tribe of APHRODINAE), 1983c (tribe of SCARINAE). Lee, C. E. &Kwon 1977b. Metcalf 1966c (tribe of IASSINAE). Quartau & Rodrigues 1969a.
SELENOMORPHINI (tribe of IASSINAE).
  Evans, J. W. 1974a (tribe of JASSINAE). Hamilton, K. G. A. 1983c (=SCARINAE, REUPLEMMELLINI).
SIGNORETIINAE
  Linnavuori 1978a. Metcalf 1963a (subfamily of EVACANTHIDAE).
SIGNORETIINI (tribe of SIGNORETIINAE).
  Hamilton, K. G. A. 1983c (tribe of CICADELLINAE).
STEGELYTRINAE
  Dlabola 1974a, 1981a. Hamilton, K. G. A. 1983c (=COELIDIINAE, COELIDIINI, COELIDIINA). Lindberg 1958a. Linnavuori 1962a. Lodos & Kalkandelen 1983a. Nast 1972a, 1982a. Quartau & Rodrigues 1969a.
STEGELYTRINI (tribe of STEGELYTRINAE).
  Hamilton, K. G. A. 1983c (=COELIDIINA). Metcalf 1964b (tribe of COELIDIINAE).
STENOCOTINAE
  Metcalf 1962c (subfamily of LEDRIDAE).
STENOCOTINI (tribe of LEDRINAE).
  Evans, J. W. 1966a. Hamilton, K. G. A. 1983c (tribe of JASSINAE).
STENOMETOPIINI (tribe of DELTOCEPHALINAE).
  Evans, J. W. 1966a. Hamilton, K. G. A. 1983c (tribe of COELIDIINAE). Linnavuori 1979b. Metcalf 1963b (tribe of NIRVANIDAE).
STIRELLINI (=STENOMETOPIINI).
  Eyles & Linnavuori 1974a. Hamilton, K. G. A. 1975b (tribe of APHRODINAE), 1983c (=COELIDIINAE, STENOMETOPIINI). Lee, C. E. & Kwon 1977b. Linnavuori 1975c. Linnavuori & DeLong 1977c, 1978c. Nast 1972a.
SUDRINI (tribe of HYLICINAE).
  Hamilton, K. G. A. 1983c (=CICADELLINAE, HYLICINI). Metcalf 1962a (tribe of HYLICIDAE).
SYNOPHROPSINI (=FIEBERIELLINI).
  Emeljanov 1962b, 1972d (=ACINOPTERINI). Hamilton, K. G. A. 1975b(=APHRODINAE, DELTOCEPHALINI, PLATYMETOPIINA). Linnavuori 1956d, 1962a. Linnavuori & Al-Ne'amy 1983a. Metcalf 1967b (tribe of DELTOCEPHALINAE).
TARTESSIDAE
  Ishihara 1965c.
TARTESSINAE
  Evans, F. 1981a. Evans, J. W. 1966a, 1974a, 1977a. Knight, W. J. 1982a, 1983a. Linnavuori 1956a, 1960a, 1975c. Metcalf 1964b (subfamily of COELIDIINAE). Nast 1972a.
TARTESSINI (tribe of TARTESSINAE).
  Hamilton, K. G. A. 1983c (tribe of EURYMELINAE).
TETARTOSTYLINI (tribe of DELTOCEPHALINAE).
  Dlabola 1964c. Hamilton, K. G. A. 1975b (=APHRODINAE, DELTOCEPHALINI, PLATYMETOPIINA). Heller & Linnavuori 1968a. Kalkandelen 1974a. Linnavuori 1961a. Linnavuori & Al-Ne'amy 1983a. Metcalf 1967b. Nast 1972a. Theron 1973b.
TETTIGELLIDAE (*sensu* Metcalf 1965a).
  Datta & Pramanik 1977a. Ishihara 1958a, 1958b, 1961a, 1965a, 1966a, 1968a. Lee, C. E. & Kwon 1976a. Lee, C. E., Lee & Kwon 1976a.
TETTIGELLINAE
  Dlabola 1967b, 1967c, 1967d, 1968b. Lee, C. E. & Kwon 1977b. Li, Z. Z. 1983a, 1985b. Metcalf 1965a. Nielson 1976c. Nielson & Kaloostian 1956a. Schroder 1959a, 1960b, 1961a, 1962a.
TERULIINI (tribe of COELIDIINAE).
  Hamilton, K. G. A. 1983c (as subtribe of COELIDIINI). Nielson 1979b, 1982d, 1982f, 1982g, 1982h, 1983a, 1983c, 1983d, 1983f, 1983g.
THAGRIINI (tribe of COELIDIINAE).
  Hamilton, K. G. A. 1983c (as subtribe of COELIDIINI). Metcalf 1964b. Nielson 1977a, 1980a, 1980b, 1980c, 1980d, 1982c.

THAMNOTETTIXINI (=ATHYSANINI).
  Hamilton, K. G. A. 1975b (=APHRODINAE, DELTOCEPHALINI, CICADULINA). Metcalf 1967a (tribe of EUSCELINAE).
THARRINI (tribe of COELIDIINAE).
  Hamilton, K. G. A. 1983c (as subtribe of COELIDIINI). Nielson 1975a, 1982b.
THAUMATOSCOPINAE (=PENTHIMIINAE).
  Metcalf 1962b (subfamily of GYPONIDAE).
THYMBRINI (tribe of LEDRINAE).
  Evans, J. W. 1966a, 1969a, 1977a. Metcalf 1962c (tribe of KOEBELIINAE).
TINOBREGMINI (tribe of COELIDIINAE).
  Hamilton, K. G. A. 1983c. Nielson 1975a, 1979a. Linnavuori & DeLong 1977c. Metcalf 1964b.
TROCNADINI (tribe of IASSINAE).
  Evans, J. W. 1966a, 1972b. Hamilton, K. G. A. 1983c (=SCARINAE, PENTHIMIINI). Linnavuori & Quartau 1975a. Metcalf 1966c.
TYPHLOCYBIDAE ICZN Opinion 605, 1961. Kuoh, C. L. 1981b. Mitjaev 1980b, 1980c. Servadei 1957a, 1958a, 1960a, 1968a, 1969a, 1971a, 1972a. Vidano 1958a, 1959a, 1959b, 1960a, 1961a, 1961b, 1961c, 1961d, 1961e, 1962a, 1963a, 1964b, 1965a. Zachvatkin 1953d.
TYPHLOCYBINAE
  Ahmed, Manzoor 1972a, 1983b, 1985c. Anufriev 1970e, 1978a. Cwikla & Freytag 1982a. DeLong & Guevara, C. [1954b]. Dlabola 1958c, 1964c, 1967b, 1967d, 1968a, 1968b, 1970b, 1971b, 1972a, 1974a, 1974d, 1975a, 1977b, 1980d, 1981a. Dubovskij 1966a. Dworakowska 1967b, 1968a, 1968b, 1968d, 1969b, 1969c, 1969d, 1969e, 1969f, 1969g, 1970a, 1970b, 1970c, 1970d, 1970e, 1970f, 1970g, 1970h, 1970i, 1970j, 1970k, 1970l, 1970m, 1970n, 1971a, 1971b, 1971c, 1971d, 1971e, 1971f, 1971g, 1972a, 1972b, 1972c, 1972d, 1972e, 1972f, 1972g, 1972h, 1972i, 1972j, 1972k, 1972l, 1973a, [1973b], 1973c, 1974a, 1976a, 1976b, 1976c, 1977a, 1977b, 1977c, 1977d, 1977f, [1977g, 1977h], 1977i, 1978a, 1978b, 1979a, 1979b, 1979c, 1979d, 1979f, 1980a, 1980b, 1980c, 1980d, 1981a, 1981b, 1981c, 1981d, 1981e, 1981f, 1981g, 1981h, 1981i, 1981j, 1982b, 1983a, 1984a. Dworakowska & Lauterer 1975a. Dworakowska, Nagaich & Singh 1978a. Dworakowska & Pawar 1974a. Dworakowska, Singh & Nagaich 1979a. Dworakowska & Sohi 1978a, 1978b. Dworakowska, Sohi & Viraktamath 1980a. Dworakowska & Trolle 1976a. Dworakowska & Viraktamath 1975a, 1978a, 1979a. Emeljanov 1964c. Evans, J. W. 1966a. Eyles & Linnavuori 1974a. Gunthart, H. 1971a, 1971b, 1974a, 1979a. Gunthart, H. & Gunthart, M. 1981a, 1983a. Habib, El-Kady & Herakly 1976a. Hamilton, K. G. A. 1982b, 1983c. Haupt, Schmutterer & Muller 1969a. Heller & Linnavuori 1968a. Jankovic 1971a. Kalkandelen 1974a. Kirejtshuk 1975a, 1977a, 1979a. Knight, W. J. 1976a, 1976b, 1979a, 1982a, 1983a. Korolevskaya 1979a. Kramer 1965a. Lauterer 1973a. Lee, C. E. & Kwon 1977b. LeQuesne 1965c. LeQuesne & Payne 1981a. Linnavuori 1956a, 1959a, 1960a, 1961a, 1962a, 1964a. Linnavuori & DeLong 1977c. Lodos & Kalkandelen 1983b, 1983e, 1984a, 1984b, 1984c, 1984d. Mahmood 1967a, 1968a, 1975a. Mahmood & Ahmed 1968a, 1969a. Meusnier 1982a. Mitjaev 1963b, 1963c, 1967b, 1971a, 1980b, 1980c. Naheed & Ahmed 1980a. Nast 1972a, 1982a. Nielson 1962c, 1968b, 1979c, 1985a. Ossiannilsson 1981a. Quartau 1981e. Quartau & Rodrigues 1969a. Ramakrishnan 1982a. Ramakrishnan & Ghauri 1979b. Ramakrishnan & Menon 1971a, 1972a, 1972b, 1973a, 1974a. Remane 1984a. Sawai Singh 1969a. Schiemenz 1970a. Sharma, B. 1978a, 1979a, 1984a. Sharma, B. & Malhotra 1981a. Sohi 1975a, 1977a, 1977b, 1977c, 1978a. Sohi & Dworakowska 1979a, 1981a, 1984a. Sohi & Sandhu 1971a. Thapa 1983a, 1984a, 1984b, 1985a. Thapa & Sohi 1982a, 1984a. Theron 1982b. Torres 1960a, 1960b. Trolle 1966a, 1968a, 1974a. Varty 1967a. Vidano & Arzone 1976a, 1978a, 1981a. Viggiani 1971a. Vilbaste 1975a, 1982a. Viraktamath, C. A. & Dworakowska 1979a. Webb, M. D. 1980a. Wilson, M. R. 1978a. Young 1957a, 1957b.
TYPHLOCYBINI
  Metcalf 1968a (tribe of CICADELLINAE *sensu* Metcalf = TYPHLOCYBINAE). Sharma, B. 1979a.
ULOPIDAE
  Kristensen 1965a. Lindberg 1956a. Metcalf 1962d.

ULOPINAE
   Davis, R. B. 1975a. Dlabola 1964c, 1981a. Emeljanov 1963a, 1964c. Evans, J. W. 1959a, 1961a, 1965a, 1966a, 1968a, 1971c, 1977a. Hamilton, K. G. A. 1983c. Kalkandelen 1974a. Kameswara Rao & Ramakrishnan 1979a. Knight, W. J. 1973b, 1983a. LeQuesne 1965c. Linnavuori 1956d, 1961a, 1962a, 1972a. Lodos & Kalkandelen 1981a. Metcalf 1962d. Nast 1972a, 1982a. Ossiannilsson 1968a, 1981a. Schiemenz 1970a. Theron 1980a. Van Stalle 1982a, 1983a, 1983b, 1983c. Vilbaste 1975a.
XEROPHLOEINAE
   Linnavuori 1959b.
XEROPHLOEINI (tribe of LEDRINAE).
   Hamilton, K. G. A. 1983c. Metcalf 1962c.
XESTOCEPHALIDAE
   Ishihara 1961a, 1961b.
XESTOCEPHALINAE
   Anufriev 1978a. Cwikla 1985a. Cwikla & Freytag 1983a. DeLong 1980b, 1982b. DeLong & Linnavuori 1978b. DeLong, Wolda & Estribi 1983a. Dlabola 1981a. Evans, J. W. 1966a, 1973a, 1974a. Eyles & Linnavuori 1974a. Hamilton, K. G. A. 1975b, 1983c (=APHRODINAE, APHRODINI, XESTOCEPHALINA). Heller & Linnavuori 1968a. Knight, W. J. 1974b, 1976b, 1982a, 1983a. Korolevskaya 1974a. Kramer 1961a, 1964e. Linnavuori 1959b, 1960b, 1970b, 1975c. Linnavuori & Al-Ne'amy 1983a. Linnavuori & DeLong 1977c, 1979b. Linnavuori & Heller 1961a. Lodos & Kalkandelen 1983a. Nast 1972a, 1982a. Vilbaste 1968a, 1975a, 1982a.
XESTOCEPHALINI (tribe of XESTOCEPHALINAE).
   Linnavuori 1956a, 1960a (tribe of DELTOCEPHALINAE). Metcalf 1967c (tribe of DELTOCEPHALINAE).
ZYGINELLINI (tribe of TYPHLOCYBINAE).
   Dworakowska 1977i, 1979c.

CLASSIFICATION

In lieu of an acceptable up-to-date classification of the family Cicadellidae to serve as a reference base for the world fauna, we provide a tentative and provisional list and grouping of subfamily and tribal names, together with their junior synonyms, alphabetized for convenient reference. Family-group names used in the check-list are those shown in that list as having priority. The list should not be taken to imply an attempt on the part of the authors to propose a revised scheme. The choice of family-group names is based on a consensus of published opinion up to 31 December 1985. The reader's attention is directed to the following points.

a) Neither the elevation of all subfamilies to the rank of family, referable to the superfamily Jassoidea (Olmi 1975a), nor reducing the number of subfamilies to ten (Hamilton, K. G. A. 1983c), is in accord with general usage. Both schemes are ignored for present purposes.

b) Aphrodinae is used sensu Ossiannilsson 1983c, not sensu Hamilton, K. G. A. 1975a, 1983c.

c) Gyponinae is retained, in view of its long term usage, in preference to the older name Scarinae. An appeal to the International Commission for suspension of the Rules is required.

d) Makilingiinae is treated as a subfamily, in accordance with Young 1968a & 1977a, rather than a tribe of the Cicadellinae as tentatively proposed by Linnavuori 1979a.

Alphabetized list and grouping of subfamily and tribal names in the Cicadellidae

Nominotypical tribal names omitted, except where necessary to show synonymy.

Acostemminae Evans 1972 (= Acroponinae Linnavuori & Quartau 1975)
Adelungiinae Baker 1915 (= Macrocepsinae de Bergevin 1926; Melicharellinae de Bergevin 1927)
   Achrini Davis 1975
   Peyerimhoffiolini Al-Ne'amy & Linnavuori 1982
Agalliinae Kirkaldy 1901
   Nehelini Zachvatkin 1946
Aphrodinae Haupt 1927 (= Acocephalinae Dohrn 1859; Acucephalinae Van Duzee 1916)
   Paraphrodini Linnavuori 1979
Arrugadiinae Linnavuori 1965
Austroagalloidinae Evans 1938
Bythoniinae Linnavuori 1959
Cicadellinae Van Duzee 1916 (= Tettigonides Le Peletier & Serville 1825; Tettigoniellinae Jacobi 1905; Amblycephalinae China 1939; Tettigellinae Evans 1947)
   Anoterostemmini Haupt 1929
   Cicadellini Van Duzee 1916 (= Draeculacephalini Metcalf & Bruner 1936; Graphocephalini Metcalf 1955; Helocharini Metcalf 1965; Neokollini Metcalf 1965)
   Errhomenini Fieber 1872 (= Errhomenellini Baker 1923)
   Evacanthini Metcalf 1939 (= Euacanthidae Crumb 1911; Pythaminae Baker 1915)
   Mileewanini Evans 1947
   Pagaroniini Anufriev 1978
   Proconiini Stal 1869 (= Ciccini Baker 1915; Ciccianini Metcalf 1965)
Coelidiinae Dohrn 1859
   Gabritini Nielson 1983
   Hikangiini Nielson 1983
   Sandersellini DeLong 1945
   Teruliini Nielson 1979
   Thagriini Distant 1908
   Tharrini Nielson 1975
   Tinobregmini Oman 1949
   Youngolidiini Nielson 1983
Deltocephalinae Dallas 1870
   Acinopterini Oman 1943
   Athysanini Van Duzee 1892 (= Bobacellini Kusnezov 1929; Colladonini Bliven 1955; Euscelini Van Duzee 1917; Limotettixini Baker 1915; Phlepsiini Oman 1943; Phrynomorphini Kirkaldy 1907; Thamnotettixini Distant 1908)
   Balcluthini Baker 1915 (= Gnathodini Baker 1915)
   Cerrillini Linnavuori 1975
   Chiasmusini Distant 1908
   Cicadulini Van Duzee 1892
   Cochlorhinini Oman 1943
   Coryphaelini Nast 1972 (= Coryphaeini Emeljanov 1962)
   Doraturini Emeljanov 1962
   Fieberiellini Wagner 1951 (= Synophropsini Ribaut 1952)
   Goniagnathini Wagner 1951
   Grypotini Haupt 1929
   Hecalini Distant 1908 (= Reuteriellini Evans 1947)
   Luheriini Linnavuori 1959
   Macrostelini Kirkaldy 1906
   Opsiini Emeljanov 1962 (= Achaeticini Emeljanov 1962)
   Paralimnini Distant 1908 (= Jassargini Emeljanov 1962)
   Platymetopiini Haupt 1929 (= Mesamiini Oman 1943)
   Scaphoideini Oman 1943
   Scaphytopiini Oman 1943
   Stenometopiini Baker 1923 (= Stirellini Emeljanov 1966)
   Tetartostylini Wagner 1951
Drakensbergeninae Linnavuori 1979
Euacanthellinae Evans 1966
Eupelicinae Sahlberg 1871
   Dorycephalini Oman 1943
   Listrophorini Boulard 1971
   Paradorydiini Evans 1936 (= Doridiini Fieber 1872)
Eurymelinae Amyot & Serville 1843
   Ipoini Evans 1933 (= Anipoini Evans 1934; Cornutipoini Evans 1934; Opioini Evans 1934)
   Pogonoscopini China 1926
Evansiolinae Linnavuori & DeLong 1977
Gyponinae Stal 1870 (= Scarinae Amyot & Serville 1843)
Hylicinae Distant 1908

Melliolini Metcalf 1962 (= Melliini Schmidt 1920)
    Sudrini Schmidt 1920
Iassinae Walker 1870 (= Bythoscopinae Dohrn 1859)
    Hyalojassini Evans 1972
    Krisnini Evans 1947
    Platyjassini Evans 1953
    Reuplemmelini Evans 1966
    Selenomorphini Evans 1974
    Trocnadini Evans 1947
Idiocerinae Baker 1915
Koebeliinae Baker 1897
Ledrinae Fairmaire 1855
    Petalocephalini Metcalf 1962
    Stenocotini Kirkaldy 1906
    Thymbrini Evans 1936 (= Ledrellini Evans 1938)
    Xerophloeini Oman 1943
Macropsinae Evans 1935
    Macropsini Evans 1935 (= Oncopsini Fisher 1952)
    Neopsini Linnavuori 1978
Makilingiinae Evans 1947
Megophthalminae Kirkaldy 1906 (= Paropiinae Dohrn 1859)
Mukariinae Distant 1908
Neobalinae Linnavuori 1959
Neocoelidiinae Oman 1943
Nioniinae Oman 1943
    Magnentiini Linnavuori 1978
Nirvaninae Baker 1923
    Balbillini Baker 1923
    Macroceratogoniini Kirkaldy 1906
    Occinirvanini Evans 1966
Paraboloponinae Ishihara 1953
Penthimiinae Kirschbaum 1868 (= Thaumatoscopinae Baker 1923)
Phereurhininae Kramer 1976
Phlogisinae Linnavuori 1979
Selenocephalinae Fieber 1872
    Adamiini Linnavuori & Al-Ne'amy 1983
    Bhatiini Linnavuori & Al-Ne'amy 1983
    Drabescini Ishihara 1953
    Dwightiini Linnavuori & Al-Ne'amy 1983
    Hypacostemmini Linnavuori & Al-Ne'amy 1983
    Ianeirini Linnavuori 1978
Signoretiinae Baker 1915
Stegelytrinae Baker 1915
Tartessinae Distant 1908
Typhlocybinae Kirschbaum 1868 (= Eupteryginae Kirkaldy 1906; Zyginae Zachvatkin 1946)
    Alebrini McAtee 1926
    Dikraneurini McAtee 1926
    Empoascini Distant 1908
    Erythroneurini Young 1952 (= Bakerini Mahmood 1967)
    Helionini Haupt 1929
    Jorumini McAtee 1926
    Zyginellini Dworakowska 1977
Ulopinae Le Peletier and Serville 1825
    Cephalelini Amyot & Serville 1843
    Monteithiini Evans 1968
    Myerslopiini Evans 1957
    Ulopini Le Peletier and Serville 1825 (= Kahavaluini Kirkaldy 1906)
Xestocephalinae Baker 1915
    Portanini Linnavuori 1959

MISASSIGNED GENERIC NAMES

The following generic names were erroneously listed as Cicadellidae in the Zoological Record for the year indicated. They are not included in the check-list and references dealing with such names are not included in the bibliography.

Ambragaeana Chou & Yao 1985, Vol. 122, Sect. 13f:218. (= Cicadidae)
Babras Jacobi 1960, Vol. 97, Sect. 13:329. (= Cicadidae)
Bubastia Emeljanov 1977, Vol. 114, Sect. 13f:166. (= Issidae)
Bequartina Kato 1985, Vol. 122, Sect. 13f:219. (= Cicadidae)
Cicadatra Kolenati 1958, Vol. 95, Sect. 13:613. (= Cicadidae)
Cicadetta Kolenati 1961, Vol. 98, Sect. 13:368. (= Cicadidae)
Callogaeana Chou & Yao 1985, Vol. 122, Sect. 13f:220. (= Cicadidae)
Cosmopsaltria Stal 1968, Vol. 105, Sect. 13:695. (= Cicadidae)
Gaeana Amyot & Serville 1985, Vol. 122, Sect. 13f:222. (= Cicadidae)
Heterochterus Evans, J.W. 1971, Vol. 108, Sect. 13:747. (= Heteroptera)
Kikihia Dugdale 1973, Vol. 110, Sect. 13f:131 and 1984, Vol. 121, Sect. 13f:199. (= Cicadidae)
Malagasia Distant 1980, Vol. 117, Sect. 13f:186. (= Cicadidae)
Malgachialna Boulard 1980, Vol. 117, Sect. 13f:186. (= Cicadidae)
Malgotilia Boulard 1980, Vol. 117, Sect. 13f:186. (= Cicadidae)
Orientopsaltria Distant 1968, Vol. 105, Sect. 13:698. (= Cicadidae)
Sulphogaeana Chou & Yao 1985, Vol. 122, Sect. 13f:320. (= Cicadidae)

MISIDENTIFIED TYPE-SPECIES

Information on the misidentification of the type-species of the following seven genera is given in the reference indicated in each case.

Alodeltocephalus Evans: Knight 1975a:189-191
Ciccus Latreille: Metcalf 1952a:228; Young 1968a:115, 117
Nanopsis Freytag: Hamilton 1980b: 897
Neophlepsius Linnavuori: Linnavuori 1959b:193
Parunculus Emeljanov: Emeljanov 1964f:44
Sispocnis Anufriev: Hamilton 1980b:897
Yasumatsuus Ishihara: Young 1979a:1

Related problems, concerning the interpretation of type-species, occur in the following three genera. The annotations in each case should be read in conjunction with the check-list entry for the genus.

Coelana DeLong: No type designated.
Neokolla Melichar: Young 1977a:828, 852, 949 placed the name as a junior synonym of *Graphocephala* Van Duzee on the basis of Ball's 1901e:17 interpretation of *Tettigonia hieroglyphica* Say 1830b:313.
Plesiommata Provancher: Hamilton 1976a:29-32, 36 considers Provancher's species to be a cercopid, and proposed *Provancherana* Hamilton 1976a:30, 36 as a replacement name for *Plesiommata* of authors, not Provancher. That interpretation is at variance with Van Duzee's 1912b:327 conclusion, accepted by workers prior to Hamilton 1976a, and by Young 1977a:595, 597. The problem is complicated by the different versions of Provancher's paper(s) known to have been distributed (see Oman 1938c:165 and Hamilton 1976a:31).

NAMES PLACED ON THE OFFICIAL LIST OF GENERIC NAMES IN ZOOLOGY

Allygidius Ribaut 1948; type species *Cicada atomaria* Fabricius 1794. Opinion 1365, Bulletin of Zoological Nomenclature 42:357.
Allygus Fieber 1872; type species *Cicada mixta* Fabricius 1794. Opinion 1365, Bulletin of Zoological Nomenclature 42:357.
Cicadella Latreille 1817; type species *Cicada viridis* Linnaeus 1758. Opinion 647, Bulletin of Zoological Nomenclature 20:35.
Elymana DeLong 1936; type species *Thamnotettix inornatus* Van Duzee 1892. Opinion 603, Bulletin of Zoological Nomenclature 18:249.
Eupteryx Curtis 1831; type species *Cicada picta* Fabricius 1794. Opinion 647, Bulletin of Zoological Nomenclature 20:35.
Iassus Fabricius 1803; type species *Cicada lanio* Linnaeus 1761. Opinion 612, Bulletin of Zoological Nomenclature 18:312.

Macropsis Lewis 1834; type species *Jassus prasina* Boheman 1852. Opinion 603, Bulletin of Zoological Nomenclature 18:249.

Typhlocyba Germar 1833; type species *Cicada quercus* Fabricius 1777. Opinion 605, Bulletin of Zoological Nomenclature 18:254.

NAMES PLACED ON THE OFFICIAL LIST OF FAMILY-GROUP NAMES IN ZOOLOGY

CICADELLIDAE Latreille 1825; type genus *Cicadella* Latreille 1817. Opinion 647, Bulletin of Zoological Nomenclature 20:35.

IASSINAE (correction of JASSIDES) Amyot & Serville 1843; type genus *Iassus* Fabricius 1803. Opinion 612, Bulletin of Zoological Nomenclature 18:312.

LEDRIDAE Kirschbaum [1867]; type genus *Ledra* Fabricius 1803. Opinion 647, Bulletin of Zoological Nomenclature 20:35.

MACROPSIDAE Evans 1938; type genus *Macropsis* Lewis 1834. Opinion 603, Bulletin of Zoological Nomenclature 18:249.

TYPHLOCYBIDAE Kirschbaum 1868; type genus *Typhlocyba* Germar 1833. Opinion 605, Bulletin of Zoological Nomenclature 18:254.

NAMES PLACED ON THE OFFICIAL INDEX OF REJECTED AND INVALID GENERIC NAMES IN ZOOLOGY

Cicadella Dumeril 1806, suppressed under plenary powers. Opinion 647, Bulletin of Zoological Nomenclature 20:35.

Cicadella, all uses of, prior to that of Latreille 1817, suppressed under plenary powers. Opinion 647, Bulletin of Zoological Nomenclature 20:35.

Jasus Megerle 1804 (incorrect spelling for *Iassus* Fabricius 1803). Opinion 612, Bulletin of Zoological Nomenclature 18:312.

Jassus Fallen 1806 (incorrect spelling for *Iassus* Fabricius 1803). Opinion 612, Bulletin of Zoological Nomenclature 18:312.

Macropsis Sars 1876 (a junior homonym of *Macropsis* Lewis 1834). Opinion 603, Bulletin of Zoological Nomenclature 18:249.

Promecopsis Dumeril 1806, suppressed under plenary powers. Opinion 605, Bulletin of Zoological Nomenclature 18:254.

Tettigella China & Fennah 1945 (a junior objective synonym of *Cicadella* Latreille 1817). Opinion 647, Bulletin of Zoological Nomenclature 20:35.

NAMES PLACED ON THE OFFICIAL INDEX OF REJECTED AND INVALID FAMILY-GROUP NAMES IN ZOOLOGY

AMBLYCEPHALINAE China 1939; type genus *Amblycephalus* Curtis, a junior homonym. Opinion 647, Bulletin of Zoological Nomenclature 20:36.

JASSIDES Amyot & Serville 1843; type genus *Iassus* Fabricius 1803. Opinion 612, Bulletin of Zoological Nomenclature 18:312.

TETTIGELLINAE Evans 1947; type genus *Tettigella* China & Fennah 1945. Opinion 647, Bulletin of Zoological Nomenclature 20:36.

TETTIGONIDES Amyot & Serville 1843; type genus *Tetigonia* Geoffroy 1762. Opinion 647, Bulletin of Zoological Nomenclature 20:35.

CHECK-LIST

Aaka Dworakowska 1972e:775. *A. coera* n.sp. (New Guinea: Papua, Kokoda 1200 ft). Typhlocybinae, Erythroneurini. Oriental.

Abana Distant 1908b:72. *Aulacizes dives* Walker 1851b:791 (unknown). 1:643. Young 1968a:148. Cicadellinae, Proconiini. Neotropical; Central America and northern South America.

Synonym:

Mesobana Melichar 1926a:322. *Amblydisca pomposula* Jacobi 1905c:167.

Abdistragania Blocker 1979b:856 as subgenus of *Penestragania* Beamer & Lawson 1945a:50. *P. (A.) noxina* n.sp. (Bolivia: Santa Cruz). Iassinae. Neotropical.

Abelterus Stal 1865b:157. *A. incarnatus* n.sp. (Northern Australia). 15:133. Iassinae, Trocnadini. Northern Australia. Synonym of *Trocnada* Walker 1858a:103.

Abietotettix Mitjaev 1965a:1260 as subgenus of *Pithyotettix* Ribaut 1942a:261. *P.(A.) sibiricus* n.sp. (USSR: Altai Mts). Deltocephalinae, Athysanini. Palaearctic.

Abimwa Linnavuori 1978a:46. *Phlepsius ramulosus* Lv. (sic!) = *A. ramulosa* n.sp. (Zaire: "Tana R."). Drabescinae, Ianeirini. Ethiopian.

Ablycephalus 1:664. Error for *Amblycephalus* Curtis.

Abothrogonia Yang in Yang & Li 1980a:193 as subgenus of *Bothrogonia* Melichar 1926a:341. *Tettigonia impudica* Signoret 1853c:677 (Philippines: Manila). Cicadellinae, Cicadellini. Oriental.

Abrabra Dworakowska 1976b:18. *A. bura* n.sp. (Upper Laos: Nam Hou). Typhlocybinae, Alebrini. Oriental.

Abrela Young 1957c:240. *Alebra robusta* Gillette 1898a:712 (Brazil: Chapada). Typhlocybinae, Alebrini. Neotropical.

Absheta Blocker 1979a:32. *A. coarcta* n.sp. (Peru: Tingo Maria). Iassinae. Neotropical.

Acacimenus Dlabola 1979b:138. *A. makranus* n.sp. (Saudia Arabia). Deltocephalinae, Athysanini. Palaearctic.

Acacioiassus Linnavuori & Quartau 1975a:15. *Macropsis serena* Melichar 1904a:35 (Italian Somaliland). Iassinae, Iassini. Ethiopian; semi-arid savannahs of southern NE and SW Africa.

Acamella Wesley & Blocker 1985a:172 as subgenus of *Athysanella* Baker 1898a:185. *A.(A.) tamara* n.sp. (Mexico: Guerrero, Iguala K-102). Deltocephalinae, Doraturini. Nearctic; Mexico.

Acastroma Linnavuori 1969a:1170. *A. introducta* n.sp. (Cameroon). Deltocephalinae, Athysanini. Ethiopian; Cameroon, Republic of Congo.

Accacidia Dworakowska 1971b:113. *A. dlabolai* n.sp. (Sudan: Ed Damer). Typhlocybinae, Erythroneurini. Afrotropical & Oriental.

Synonym:

Indianella Ramakrishnan & Menon 1973a:22. *I. indraprasthana*.

Accephalus 8:248. Error for *Acocephalus* Burmeister.

Acecephalus 8:248. Error for *Acocephalus* Burmeister.

Acenura 10:2642. Error for *Aconura* Lethierry.

Aceratagallia Kirkaldy 1907d:30. *Bythoscopus sanguinolentus* Provancher 1872c:376 (Canada: Quebec). 14:108, 112. Agalliinae, Agalliini. Nearctic.

Subgenus:

Ionia Ball 1933c:225. *I. triunata* n.sp.

Aceratoagallia 14:170. Error for *Aceratagallia* Kirkaldy.

Aceratogallia 14:170. Error for *Aceratagallia* Kirkaldy.

Acertagallia 14:171. Error for *Aceratagallia* Kirkaldy.

Acertogallia 14:171. Error for *Aceratagallia* Kirkaldy.

Acericerus Dlabola 1974b:61. *Idiocerus rotundifrons* Kirschbaum 1868a:5 (Hesse). Idiocerinae. Palaearctic. Synonym of *Idiocerus* Lewis 1834a:47; Hamilton, KGA 1980a:825.

Achactica Mitjaev 1971b:60. Error for *Achaetica* Emeljanov.

Achaetica Emeljanov 1959a:838. *A. anabasidis* n.sp. (USSR: Kazakhstan). Deltocephalinae, Opsiini. Palaearctic.

ACHAETICINA Emeljanov 1962b:389, subtribe of OPSIINI.

Acharis Emeljanov 1966a:125. *Deltocephalus ussuriensis* Melichar 1902c:144 (USSR: Maritime Territory). Deltocephalinae, Paralimnini. Palaearctic.

Acharista Melichar 1924a:199. *A. variolata* n.sp. (Bolivia). 1:558. Cicadellinae, Cicadellini. Neotropical; Bolivia, Peru. Synonym of *Neiva* Melichar 1925a:368; Young 1977a:30.

Acharista Emeljanov 1968a:149. *A. nudiventris* n.sp. (Mongolia). Deltocephalinae, Paralimnini. Palaearctic.
  [NOTE: Junior homonym of *Acharista* Melichar 1924a:199. Synonym of *Gobicuellus* Dlabola 1967d:223; Nast 1984a:396].

Achetica Mitjaev 1968d:31. Error for *Achaetica* Emeljanov 1959a:838.

Achetica Mitjaev 1971b:59. Error for *Achaetica* Emeljanov.

ACHRINI Davis 1975a:12, tribe of ADELUNGIINAE.

Achrus Lindberg 1925a:108. *A. nigrinervosus* n.sp. (Transkaspien: "Karybent, Tedschen"). 14:165. Adelungiinae, Adelungiini. Palaearctic.

Acia McAtee 1934a:109, 114 as subgenus of *Empoasca* Walsh 1862a:149. *E. (A.) exsaltans* n.sp. (Singapore Island). 17:552. Typhlocybinae, Empoascini. Widely distributed in Oriental and Ethiopian regions.
  Subgenera:
  Icaiana Dworakowska 1981b:18. *A. (I.) tina* n.sp.
  Naracia Dworakowska 1981b:25. *A. (N.) assamensis* n.sp.
  Africia Dworakowska 1981b:27. *Empoasca lineatifrons* Naude 1926a:96.
  Paolicia Dworakowska 1981f:592. *Empoasca brevis* Dworakowska 1972c:483.
  [NOTE: *Acia* raised to generic rank by Dworakowska 1981b].

Acinapterus 10:2642. Error for *Acinopterus* Van Duzee.

ACINOPTERINI 10:1006.

Acinopterus Van Duzee 1892a:307. *A. acuminatus* n.sp. (USA: Maryland). 10:1006. Deltocephalinae, Acinopterini. Nearctic, Neotropical.

Acocepbalus 8:248. Error for *Acocephalus* Burmeister.

Acocephala 8:248. Error for *Acocephalus* Burmeister.

ACOCEPHALIDAE 8.5.

ACOCEPHALINI 8:12, 33.

ACOCEPHINAE 8:11.

Acocephalis 8:249. Error for *Acocephalus* Burmeister.

Acocephalus Burmeister 1835a:111. Invalid emendation of *Acucephalus* Germar. 8:58, 249.

Acocoelidia DeLong 1953a:130. *A. unipuncta* n.sp. (Mexico: Guerrero, Iguala). 11:113. Neocoelidiinae. Neotropical; Mexico to Brazil. Synonym of *Coelidiana* Oman 1938b:397; Kramer 1964d:274.

Acojassus Evans, J.W. 1972b:656. *A. montanus* n.sp. (New Guinea: Irian Barat, Enarotadi). Iassinae, Iassini. Oriental. Synonym of *Batracomorphus* Lewis 1834a:51; Knight 1983b:31.

Aconora 10:2644. Error for *Aconura* Lethierry.

Aconura Lethierry 1876a:9. *A. jakowlefi* n.sp. (USSR: Astrakhan). 10:1586. Deltocephalinae, Doraturini. Palaearctic.
  Subgenera:
  Aconurina Emeljanov 1964f:19. *A. (A.) kamenskii* n.sp.
  Platycina Emeljanov 1964f:19. *A. (A.) depressa* n.sp.
  Synonyms:
  Carinifer Linnavuori 1952b:185. *C. maculiceps* n.sp.
  Stalinabada Dlabola 1961b:344. *S. paraconurae* n.sp.
  Sagittifer Dlabola 1961b:345. *S. optatus* n.sp.
  Jiridlabolina Kocak 1981a:32. *Sagittifer optatus* Dlabola 1961b:346.
  Linnavuorina Kocak 1981a:32. *Carinifer maculiceps* Linnavuori 1952b:185.

Aconurella Ribaut 1948c:57. *Thamnotettix prolixa* Lethierry 1885b:102 (France). 10:1596. Deltocephalinae, Doraturini. Palaearctic.
  [NOTE: Vilbaste 1965a:9-10 redefined the genus as = *Doratulina* of Matsumura 1914 & 1915, not Melichar 1903b:198].

Aconurina Emeljanov 1964f:19 as subgenus of *Aconura* Lethierry 1876a:9. *A. kamenskii* n.sp. (USSR: Kazakhstan). Deltocephalinae, Doraturini. Palaearctic.

Aconuromimus Linnavuori 1954d:83. *Jassus (Deltocephalus) flavidiventris* Stal 1859b:294 (Australia: New South Wales). 10:1598. Deltocephalinae, Deltocephalini. Australian. Synonym of *Euleimonios* Kirkaldy 1906c:342; Evans, J.W. 1966a:222.

Acopis 1:664. Error for *Acopsis* Amyot & Serville.

Acopsis Amyot & Serville 1843a:232. *A. viridicans* n.sp. (Mauritius). 1:427. Cicadellinae, Cicadellini. Malagasian.

Acostemana Evans, J.W. 1954a:117. *A. albida* n.sp. (Madagascar). 15:207. Acostemminae. Malagasian.

Acostemma Signoret 1860a:204. *A. marginalis* n.sp. (Madagascar). 15:193. Acostemminae. Malagasian.
  Synonym:
  Acropona Melichar 1903b:168. *Gypona prasina* Walker 1858b:259.

Acostemmella Evans, J.W. 1954a:120. *A. rubra* n.sp. (Madagascar). 15:207. Acostemminae. Malagasian.

ACOSTEMMINAE Evans, J.W. 1972b:660.

Acouura 10:2642. Error for *Aconura* Lethierry.

Acrinopterus Zoological Record 97, Section 13:328. Error for *Acinopterus* Van Duzee 1892a.

Acrobelus Stal 1869a:60. *Tettigonia reflexa* Signoret 1855b:524 (America) = *Raphirhinus attenuatus* Walker 1851b:806. 1:471. Young 1968a:160. Cicadellinae, Proconini. Neotropical; Costa Rica and Ecuador.

Acrocampoa 1:664. Error for *Acrocampsa* Stal.

Acrocampsa Stal 1869a:66. *Fulgora pallipes* Fabricius 1787a:261 (French Guiana). 1:518. Young 1968a:123. Cicadellinae, Proconini. Neotropical; Atlantic coastal regions of northern South America.

Acrocephalus 8:256. Error for *Acocephalus* Burmeister.

Acrogonia Stal 1869a:61. *Cicada lateralis* Fabricius 1803a:64 (South America) = *Acrogonia fabricii* Metcalf 1965. 1:415. Young 1968a:257. Cicadellinae, Proconiini. Neotropical; Mexico to Brazil.
  Synonyms:
  Pherodes Fowler 1899b:25. *P. flammeicolor* n.sp.
  Orectogonia Melichar 1926a:345. *Tettigonia sparsuta* Signoret 1855c:508.
  Astenogonia Melichar 1926a:345. *Cicada bicolor* Fabricius.
  Sansalvadoria Schroeder 1959a:48. *S. bimaculata* n.sp.

Acropona Melichar 1903b:168. *Gypona prasina* Walker 1858b:258 (Ceylon). 15:193, 209. Acostemminae. Synonym of *Acostemma* Signoret 1860a:204.

ACROPONINAE Linnavuori & Quartau 1975a:159 = Acostemminae Evans; Linnavuori & Al-Ne'amy 1983a:19.
  [NOTE: Evans 1972b:660 previously named the same taxon Acostemminae, a fact noted by Linnavuori and Quartau in connection with their description (p. 159) of Acroponinae, which includes the genus *Acostemma*. Subfamily synonymy noted in Linnavuori & Al-Ne'amy 1983a:19].

Acrosteles 10:2642. Error for *Macrosteles* Fieber.

Acrostigmus Amyot 1847a:422. Invalid, not binominal.

Acrostigmus Thomson 1869a:76 as subgenus of *Jassus*. *Cicada sexnotata* Fallen 1806a:34 (Sweden). 10:2487. Deltocephalinae, Macrostelini. Palaearctic. Synonym of *Macrosteles* Fieber 1866a:504.

Acrulogonia Young 1977a:643. *Poeciloscarta smidti* Metcalf 1965a:82 (South America), replacement for *Cicada tristis* Fabricius 1803a:174, not *Cicada tristis* de Fourcroy 1785. Cicadellinae, Cicadellini. Neotropical; Panama, Costa Rica, and northern South America.

ACUCEPHALIDAE 8:8.

ACUCEPHALINAE 8:12.

ACUCEPHALINI 8:34.

Acucephalus Germar 1833a:181. *Cicada striata* Fabricius 1787a:271 (Bessarabia) = *Cicada bicincta* von Schrank 1776a:75. 8:58, 256. Aphrodinae, Aphrodini. Palaearctic, Nearctic. Synonym of *Aphrodes* Curtis 1833a:195.

Acuera DeLong & Freytag 1972a:229. *Gypona adspersa* Stal 1854b:252 (Brazil). DeLong & Freytag 1974a. Gyponinae, Gyponini. Neotropical; Nearctic; south Texas to Argentina.
  Subgenera:
  Tortusana DeLong & Freytag 1974a:191. *A. (T.) angera* n.sp.
  Parcana DeLong & Freytag 1974a:193. *A. (P.) ultima* n.sp.

Acunasus DeLong 1945f:199. *A. nigriviridis* n.sp. (Mexico: Guerrero, Iguala). 10:2176. Deltocephalinae, Scaphoideini. Neotropical.

Acuponana DeLong & Freytag 1970a:281. *A. consensa* n.sp. (Mexico: Michoacan, Uruapan). Gyponinae, Gyponini. Neotropical; Mexico to Brazil.

Acurhinus Osborn 1920a:158. *Dorydium maculatum* Osborn 1909a:464 (Guatemala). 10:2339. Deltocephalinae, Stenometopiini. [Neotropical?]. Synonym of *Hodoedocus* Jacobi 1910b:126; Linnavuori & DeLong 1978c:208.

Acusana DeLong 1942d:57. *A. veprecula* n.sp. (USA: Arizona, Prescott). 3:120. DeLong & Freytag 1966a, 1972a:225. Gyponinae, Gyponini. Neotropical, Nearctic; Panama to southwest USA.

Actinopterus 10:2642. Error for *Acinopterus* Van Duzee.

Adama Dlabola 1980e:89. *A. buettikeri* n.sp. (Saudi Arabia). Selenocephalinae, Adamini. Linnavuori and Al-Ne'amy 1983a:27-54. Ethiopian; widespread.

Subgenera:

Zinjella Linnavuori & Al-Ne'amy 1983a:29. *Distantia maculithorax* Jacobi 1910b:126.

Paracostemma Linnavuori & Al-Ne'amy 1983a:31. *Selenocephalus variabilis* Evans, J.W. 1955a:34.

Krisnella Linnavuori & Al-Ne'amy 1983a:47. "*A. (K.) elongata* Lv." = *A. (K.) elongata* n.sp.

ADAMINI Linnavuori & Al-Ne'amy 1983a:27, tribe of SELENOCEPHALINAE.

Adarrus Ribaut 1946b:83. *Deltocephalus multinotatus* Boheman 1847b:264 (Gotland). 10:1428. Emeljanov 1966a:119-120. Deltocephalinae, Paralimnini. Palaearctic; European, Mediterranean and Siberian subregions.

Subgenera:

Anargella Emeljanov 1972b:244. *Adarrus praenuntius* n.sp.

Belaunus Ribaut 1952a:268. *Deltocephalus duodecimguttatus* Cerruti 1938a:191.

Batarius Emeljanov 1966a:119. *A. (B.) oshanini* n.sp.

Romanius Emeljanov 1966a:119. *Adarrus servadeinus* Dlabola.

Errastunus Ribaut 1946b:83. *Cicada ocellaris* Fallen 1806a:20.

Adarus Dlabola 1980b:70. Error for *Adarrus* Ribaut.

Adchurnoides Maldonado-Capriles 1977a:320. *A. triangularis* n.sp. (British Guiana: New River). Idiocerinae. Neotropical.

Adelungia Melichar 1902c:134. *A. elegans* n.sp. (Iran). 14:169. Adelungiinae, Adelungiini. Palaearctic.

ADELUNGIINAE 14:169.

ADELLUNGIINI Nast 1972a:214.

Adiaerotoma Spinola 1850a:59; 1850b:130. *A. eupelicoides* Spinola 1850b:131 (Brazil). 16:234. Idiocerinae. Neotropical.

Adiotoma Spinola 9:118. Invalid, *nomen nudum*.

Adoratura Kusnezov 1938a:219-233. *Doratura heros* Melichar 1902c:126 (Persia). 10:955. Deltocephalinae, Doraturini. Palaearctic. Synonym of *Doraturopsis* Lindberg 1935c:423.

Aeocephalus 8:259. Error for *Acocephalus* Germar.

Aequcephalus DeLong & Thambimuttu 1973b:172. *A. gramineus* n.sp. (Chile: P. Santiago, Padre Hurtado). Deltocephalinae, Athysanini. Neotropical.

Aerostigmus 10:2642. Error for *Acrostigmus* Amyot.

Aeternus Distant 1918b:50. *A. hieroglyphicus* n.sp. (Burma: Tenasserim, mountains on Siam border). 11:86. Deltocephalinae; Nielson 1975:11. Oriental.

Aethiopulopa Evans, J.W. 1947c:143. *A. bambiensis* n.sp. (Senegal, Bambey). 5:95. Ulopinae, Bobacellini. Ethiopian; Sudanian.

Aflexia Oman 1949a:167. *Flexamia rubranura* DeLong 1935a:154 (USA: Illinois, Evergreen). 10:1740. Deltocephalinae, Deltocephalini. Nearctic; north central USA.

Afrakeura Einyu & Ahmed 1979a:306. *A. virosae* n.sp. (Uganda). Typhlocybinae, Dikraneurini. Ethiopian.

Afrakra Dworakowska 1979e:263. *A. nigeriana* n.sp. (Nigeria: W. State, Ile-Ife). Typhlocybinae, Dikraneurini. Ethiopian; Ethiopia, Ivory Coast, Nigeria.

Afralebra Paoli 1941a:42 as subgenus of *Alebra* Fieber 1872a:14. *A. (A.) chrysoptera* n.sp. (Italian Somaliland). 17:80. Typhlocybinae, Alebrini. Ethiopian.

[NOTE: Raised to generic rank by Dworakowska 1971c:499].

Afralycisca Kocak 1981a:32, replacement for *Lycisca* Linnavuori 1979a:653, not *Lycisca* Spinola 1840. *Lycisca umbrina* Linnavuori 1979a:654 (South Africa: Cape Province, Drakensbergen, near Rhodes). Eupelicinae, Paradorydiini. Ethiopian; South Africa.

Afrasca Dworakowska & Lauterer 1975a:33. *Empoasca podobna* Dworakowska 1972f:849 (Republic of Congo: Bambesa). Typhlocybinae, Empoascini. Ethiopian; Liberia, Republic of Congo.

Afrascius Linnavuori 1969a:1156. *A. lotis* n.sp. (Congo: Odzala). Deltocephalinae, Scaphytopiini. Ethiopian.

Africia Dworakowska 1981b:27 as subgenus of *Acia* McAtee 1934a:113. *Empoasca lineatifrons* Naude 1926a:96 (South Africa: Natal). Typhlocybinae, Empoascini. Ethiopian.

Africoelidia Nielson 1982e:40. *A. referta* n.sp. (Uganda: Ruwenzari Range). Coelidiinae, Coelidiini. Ethiopian; Afrotropical regions.

Afridonus Nielson 1983e:22. *Jassus piceolus* Melichar 1905a:302 (Tanzania: Amani). Coelidiinae, Youngolidiini. Ethiopian; Central African Republic, Kenya and Tanzania.

Afroccidens Ghauri 1969a:35. *A. lodosi* n.sp. (Ghana: Tafo). Typhlocybinae, Empoascini. Ethiopian.

Afroiassus Linnavuori & Quartau 1975a:20. *Iassus illex* Linnavuori 1969:1131 (Republic of Congo: Brazzaville). Iassinae, Iassini. Ethiopian; Guinean.

Afroideus Linnavuori 1961a:475. *A. cruciatus* n.sp. (South Africa: Cape Province, Tzitzikama Forest, Stormsrivierpiek). Deltocephalinae. Ethiopian.

Afroindica Ramakrishnan & Ghauri 1979b:207. *Empoascanara tagabica* Dworakowska & Trolle 1976a:372 (Kenya: Kericho). Typhlocybinae, Erythroneurini. Ethiopian. Synonym of *Empoascanara* Distant 1918b:94; Dworakowska 1980d:195.

Afrokana Heller 1972a:1. *A. knorri* n.sp. (Cameroon: Kamerunberg, 1200 m). Nirvaninae, Nirvanini. Ethiopian.

Afrolimnus Evans, J.W. 1955a:38. *A. ribauti* n.sp. (Belgian Congo: riv Mubale). 10:1891. Deltocephalinae, Stenometopiini. Ethiopian. Synonym of *Hodoedocus* Jacobi 1910b:126; Linnavuori & DeLong 1978c:208.

Afronirvana Evans, J.W. 1955b:7. *A. abrupta* n.sp. (Belgian Congo: Nyasheke (volc. Nyamuragira), 1800 m). 7:19. Nirvaninae, Nirvanini. Ethiopian.

Afrorubria Linnavuori 1972b:205, 240. *Acocephalus vitticollis* Stal 1855:98 (South Africa: Natal, Caffraria). Ledrinae, Petalocephalini (?). Ethiopian.

Afrosteles Theron 1975b:207. *Dalbulus distans* Linnavuori 1959b:331 (Tucuman, Argentina, error). Deltocephalinae, Macrostelini. Ethiopian.

[NOTE: Linnavuori's specimens of *Dalbulus distans* were recorded as being from Tucuman, Argentina. However, Theron 1975b:207-209 presented evidence that they were probably mislabeled, as *A. distans* appears to be native to Africa].

Afrosus Linnavuori 1959b:93 as subgenus of *Palus* DeLong & Sleesman 1929a:85. *Cicadula unimaculata* Naude 1926a:83 (South Africa: Natal). Deltocephalinae, Deltocephalini. Ethiopian.

[NOTE: Linnavuori 1959b:94 records Naude's *unimaculata* from Tucuman, Argentina, noting that it is a common species in Africa and possibly adventive to Argentina. In view of the similarly questionable record of *Afrosteles distans* Linnavuori from Argentina (q.v.), the occurrence of *Afrosus unimaculatus* Naude in South America is very questionable].

Agalia 14:171. Error for *Agallia* Curtis.

Agallie 14:171. Error for *Agallia* Curtis.

Agalita Evans, J.W. 1957a:370. *A. minuta* n.sp. (Chile: Juan Fernandez Isl., Masatierra). Agalliinae, Agallini. Neotropical; south Chile.

Agallaria Oman 1949a:37 as subgenus of *Agalliopsis* Kirkaldy 1907d:31. *Agallia oculata* Van Duzee 1890d:38 (USA: California). 14:102. Agalliinae, Agalliini. Nearctic.

Agallia Curtis 1833a:193. *A. consobrina* n.sp. (Great Britain). 14:4. Agalliinae, Agalliini. Palaearctic, Nearctic, Neotropical.

Synonyms:

Alloproctus Bergroth 1924a:400. *A. amandatus* n.sp.

Anaceratagallia Zachvatkin 1946b:159. *Cicada venosa* de Fourcroy 1785a:188.
  [NOTE: Opinions of the synonymy of *Anaceratagallia* differ. Nast 1972a:204, and Ossiannilsson 1981a:298 show the name as a junior synonym of *Agallia*. Dubovskij 1966a:105-197, 1984a:65; Dlabola 1984a:34; Vilbaste 1980b:23, 167; and many others use the name in the generic sense].
Agalliana Oman 1933b:70. *Bythoscopus sticticollis* Stal 1859b:291 (Brazil: Rio de Janeiro). 14:151. Agalliinae, Agalliini. Neotropical; Lesser Antilles to Argentina and Chile.
AGALLIIDAE 14:1.
AGALLIINAE 14:3.
Agalliopsis Kirkaldy 1907d:31. *Jassus novellus* Say 1830b:309 (USA: Indiana). 14:87,95. Agalliinae, Agalliini. Nearctic/Neotropical.
  Subgenus:
  Agallaria Oman 1949a:37. *Agallia oculata* Van Duzee 1890d:38.
Agalliota Oman 1938b:351. *Agallia punctata* Oman 1934c:457 (Mexico: Tabasco, Teapa). 14:146. Agalliinae, Agalliini. Neotropical.
Agapellus Dlabola 1967b:70. Error for *Agapelus* Emeljanov.
Agapelus Emeljanov 1961a:127. "*Deltocephalus aurantiacus* Fieb." (Switzerland). Deltocephalinae, Paralimnini. Palaearctic. Synonym of *Cosmotettix* Ribaut 1942a:267.
  [Note: Metcalf 1967b:1459 proposed *fieberi* as a replacement for *Deltocephalus aurantiacus* Fieber 1869a:218 not *Deltocephalus aurantiacus* Forel 1859a:196, and followed Smreczynski 1954a:126 in assignment of the species to the genus *Sorhoanus* Ribaut 1946b:85 while retaining *D. aurantiacus* Forel in the genus *Deltocephalus* Burmeister 1838b:15; Metcalf 1967b:1112. Remane 1961c:2 considers *aurantiacus* Fieber to be a synonym of *aurantiacus* Forel, which was assigned to the genus *Palus* DeLong & Sleesman 1929a:85. Emeljanov 1966a:111 accepts Remane's interpretation and places *Agapelus* as a synonym of *Palus*. However, *Palus* DeLong & Sleesman 1929a is preoccupied by *Palus* Palmer 1928 (Mollusca), so the correct name for the generic concept is *Cosmotettix* Ribaut 1942a:267.]
Agelina Oman 1938b:392. *A. punctata* n.sp. (Argentina: Missiones, Loreto). 10:2462. Deltocephalinae, Macrostelini. Neotropical; Argentina and Paraguay.
Agellus DeLong & Davidson 1933b:210. "*Agellus neglecta* DeL & Dav." = *Eugnathodus neglectus* DeLong & Davidson 1933a:55 (USA: Colorado, Mt. Manitou). 10:2442. Deltocephalinae, Balcluthini. Nearctic, Neotropical. Synonym of *Balclutha* Kirkaldy 1900b:243.
Aggallia 14:171. Error for *Agallia* Curtis.
Aglaenita Spinola 1850a:59. *A. bipunctata* Spinola 1850b:134 (French Guiana). 16:233. Idiocerinae. Neotropical.
  [NOTE: See discussion of generic name spelling under *Aglenita*].
Aglena Amyot & Serville 1843a:575. *Jassus ornata* Herrich-Schaffer 1838b:1, pl.1 (Rumania). 1:91. Deltocephalinae, Athysanini. Palaearctic; Mediterranean region.
Aglenita Spinola 1850b:132. *A. bipunctata* Spinola 1850b:134 (French Guiana).
  [Note: Metcalf 1966d:233 accepted the spelling "*Aglenita*" for this taxon, although the spelling "*Aglaenita*" was used by Spinola in the 1850a publication. Young 1977a:286 used the 1850a spelling without comment, presumably on the assumption that *Aglenita* was an emendation of the 1850a spelling. If the names are of identical publication date, Metcalf 1966d would appear to be the first reviser and have the option of selecting either spelling].
Agnesiella Dworakowska 1970f:211. *Typhlocyba aino* Matsumura 1932a:95 (Japan: Hokkaido). Typhlocybinae, Typhlocybini. Palaearctic and Oriental.
  Subgenus:
  Draberiella Dworakowska 1971g:647. *Chikkaballapura quinquemaculata* Distant 1918b:108.
  Synonym:
  Sarejuia Ghauri 1974c:556. *Chikkaballapura quinquemaculata* Distant 1918b:108.

Agrica Strand 1942a:393, replacement for *Horvathiella* Matsumura 1914a:234, not *Horvathiella* Poppius 1912. *Horvathiella arisana* Matsumura 1914a:235 (Formosa). 8:224. Cicadellinae, Anoterostemmini. Oriental.
Agrosoma Medler 1960a:19. *Tettigonia pulchella* Guerin 1829a:pl.59, fig. 10 (Mexico). Cicadellinae, Cicadellini. Neotropical; Mexico, Central America, and northern South America.
Aguahua Young 1977a:792. *Tettigonia salamandra* Signoret 1855d:787 (Colombia: "Salarnandra" M.). Cicadellinae, Cicadellini. Neotropical; northern South America.
Aguana Melichar 1926a:280. *Tettigonia imbricata* Signoret 1854d:719 (Brazil). 1:603. Young 1977a:37. Cicadellinae, Cicadellini. Neotropical; SE Brazil.
Aguatala Young 1977a:788. *A. compsa* n.sp. (Venezuela?: Alto de Las Cruces). Cicadellinae, Cicadellini. Neotropical.
Agudus Oman 1938b:371. *A. typicus* n.sp. (Argentina: Missiones, Loreto). Deltocephalinae, Deltocephalini. Neotropical; Argentina, Paraguay, and southern Brazil.
Aguriahana Distant 1918b:105. *A. metallica* n.sp. (Eastern Himalayas: Darjeeling). 17:619. Typhlocybinae, Eupterygini. Palaearctic, Oriental. [Nearctic].
  Synonyms:
  Kashitettix Ishihara 1952e:60. *Eupteryx quercus* Matsumura 1916b:394.
  Eupteroidea Young 1952b:92. *Typhlocyba stellulata* Burmeister 1841a:3.
  Asymmetropteryx Dlabola 1958c:52. *Typhlocyba pictilis* Stal 1853a:176.
  Wagneripteryx Dlabola 1958c:52. *Cicadula germari* Zetterstedt 1840a:301.
  Evansioma Ahmed, Manzoor 1969d:313. *E. pini* n.sp.
  Youngama Ahmed 1969d:313. *Y. spinistyla* n.sp.
  [NOTE: Above synonymy indicated by Dworakowska 1972a:278-279].
Ahenobarbus Distant 1918b:28. *A. assamensis* n.sp. (India: Assam). 2:11. Hylicinae, Hylicini. Oriental. Synonym of *Nacolus* Jacobi 1914a:381.
Ahenobarhus 2:17. Error for *Ahenobarbus* Distant.
Ahimia Dworakowska 1979c:304. *A. albonigra* n.sp. (Nigeria: W. State, Ile-Ife). Typhlocybinae, Zyginellini. Ethiopian; Nigeria and Cameroon.
Ahmedra Dworakowska & Viraktamath 1979a:49. *A. distincta* n.sp. (India: Karnataka, Jog Falls). Typhlocybinae, Erythroneurini. Oriental.
Ahngeria Melichar 1902b:76. *A. planifrons* n.sp. (Transcaspia). 14:162. Preoccupied by *Ahngeria* Kokujew 1902, see *Melicharella* Semenov 1902a:353. Adelungiinae, Adelungiini. Palaearctic.
Aidola Melichar 1914b:142. *Typhlocyba orbata* Melichar 1903b:216 (Ceylon: Pattipola). 17:1480. Typhlocybinae, Erythroneurini. Oriental.
  Synonym:
  Basilana Mahmood 1967a:17. *Typhlocyba erota* Distant 1908g:410.
Aindrahamia Linnavuori 1965a:32. *A. silvicola* n.sp. (Tunisia: Ain Drahan). Deltocephalinae, Athysanini. Palaearctic.
Airosius Emeljanov 1966a:111 as subgenus of *Palus* DeLong & Sleesman 1929a:85. *Palus caespitosus* n.sp. (USSR: Chita Prov., Kharanor station). Deltocephalinae, Paralimnini. Palaearctic. Subgenus of *Cosmotettix* Ribaut 1942a:267.
Airosus Ribaut 1952a:126 as subgenus of *Palus* DeLong & Sleesman 1929a:85. *Cicada costalis* Fallen 1826a:32 (Sweden). 10:2048. Deltocephalinae, Paralimnini. Palearctic. Subgenus of *Cosmotettix* Ribaut 1942a:267; Ossiannilsson 1983a:870.
Aisa Dworakowska 1979b:5. *A. terna* n.sp. (Vietnam: Nghe-An Prov., Quy-chau). Typhlocybinae, Erythroneurini. Oriental.
Ajika Dworakowska 1979b:14. *A. nana* n.sp. (Vietnam: Nghe-An Prov., Quy-chau). Typhlocybinae, Erythroneurini. Oriental.
Akotettix Matsumura 1931b:77. *A. akonis* n.sp. (Formosa). 17:267. Typhlocybinae, Empoascini. Oriental. Subgenus of *Dialecticopteryx* Kirkaldy 1907d:71; Dworakowska & Sohi 1978b:463.
  Synonym:
  Banosa Mahmood 1967a:39. *B. sexpunctata* n.sp.; Dworakowska 1970a:691.

Alaca Oman 1938b:386. *A. longicauda* n.sp. (Argentina: Missiones, Loreto). 10:811. Linnavuori 1959b:235. Deltocephalinae. Neotropical.

Aladzoa Linnavuori 1969a:1158. *A. uniformis* n.sp. (Congo: Odzala). Deltocephalinae, Opsiini. Ethiopian.

Alahana Melichar 1926a:342. *Tettigoniella erichsoni* Distant 1908f:137 (Sumatra). 1:260. Cicadellinae, Cicadellini. Oriental.

Alanus DeLong & Hershberger 1947d:231. *A. albidus* n.sp. (Mexico: Sinaloa, Los Mochis). 10:1237. Deltocephalinae, Athysanini. Neotropical. Synonym of *Atanus* Oman 1938b:381; Linnavuori 1959b:296.

Alapona DeLong 1980c:217. *A. elabora* n.sp. (Peru: Tingo Maria). Gyponinae, Gyponini. Neotropical.

Alapus DeLong & Sleesman 1929a:86. *Deltocephalus fraternus* Ball 1911a:201 (USA: Florida). 10:1774. Deltocephalinae, Deltocephalini. Nearctic; southeast USA.

Albelterus 15:209. Error for *Abelterus* Stal.

Albera Young 1957c:143. *Protalebra picea* Osborn 1928a:265 (Brazil: Rio Guapore). Typhlocybinae, Alebrini. Neotropical.

Albicostella Ishihara 1953a:196. *Thamnotettix albicosta* Matsumura 1914a:178 (Japan: Honshu). 10:504. Deltocephalinae, Athysanini. Eastern Palaearctic.
Synonym:
Macednus Emeljanov 1962a:165; Vilbaste 1967b:44.

Albiger Amyot 8:259. Invalid, not binominal.

Alconeura Ball & DeLong 1925a:334. *A. rotundata* n.sp. (USA: Iowa, Ames). 17:208. Typhlocybinae, Dikraneurini. Nearctic, Neotropical.
Subgenus:
Hyloidea McAtee 1926b:162. *Dikraneura (Hyloidea) depressa* n.sp.
Synonym:
Dikraneuroidea Lawson 1929d:307. *O. beameri* n.sp.

Aleba 17:1483. Error for *Alebra* Fieber.

Alebra Fieber 1872a:14, replacement for *Compsus* Fieber 1866a:507, not *Compsus* Schoenherr 1823. *Cicada albostriella* Fallen 1826a:54 (Sweden). 17:25, 43. Typhlocybinae, Alebrini. Palaearctic, Nearctic.
Synonym:
Nesopteryx Matsumura 1931b:60. *N. arisana* Matsumura 1932a:102.

Alebranus Linnavuori 1959b:330 as subgenus of *Dalbulus* DeLong 1950a:105. *Dalbulus cordifer* n.sp. (Colombia: Sierra S. Lorenzo). Deltocephalinae, Macrostelini. Neotropical.
[NOTE: Raised to generic rank; Triplehorn & Nault 1985a:304].

ALEBRINAE 17:18.

Alebroides Matsumura 1931b:68. *A. marginatus* n.sp. (Japan: Honshu). 17:95. Typhlocybinae, Empoascini. Palaearctic, Oriental.

Alemaia Heller & Linnavuori 1968a:6. *A. parviceps* n.sp. (Ethiopia: "Irga Alem"). Deltocephalinae. Ethiopian.

Aletra 17:1483. Error for *Alebra* Fieber.

Aletta Metcalf 1952a:229, replacement for *Palicus* Stal 1866a:120, not *Palicus* Philippi 1838. *Coelidia lineoligera* Stal 1855a:98 (South Africa: Natal). 11:17. Theron 1984c:313. Coelidiinae. Ethiopian; widespread.

Alfralebra Zoological Record 109, Section 13f:139, 1972. Error for *Afralebra* Paoli.

Algothyma Melichar 1926a:343. 1:266. (Of uncertain status; see Young 1977a:1096).

Aligia Ball 1907a:53 as subgenus of *Eutettix* Van Duzee 1892e:307. *Jassus jucundus* Uhler 1877a:469 (USA: Colorado, Manitou). 10:1724. Deltocephalinae, Athysanini. Nearctic.

Alishania Vilbaste 1969a:5. *Thamnotettix formosanus* Matsumura 1914a:181 (Taiwan: Ali-Shan). Deltocephalinae, Opsiini. Oriental.

Alituralis Merino 1934a:77,1936a:388, replacement for *Aliturus* Distant 1908g:398, not *Aliturus* Fairmaire 1902. *Aliturus gardineri* Distant 1908g:398 (Laccadive Islands). 10:1893, 2643. Deltocephalinae, Opsiini. Oriental. Synonym of *Neoaliturus* Distant 1918b:63.

Aliturus Distant 1908g:398. *A. gardineri* n.sp. (Laccadive Islands). 10:1893. Deltocephalinae, Opsiini. Oriental. Preoccupied by *Aliturus* Fairmaire 1902, see *Neoaliturus* Distant 1918b:63.

Alladanus DeLong & Harlan 1968a:147. *A. cephalatus* n.sp. (Mexico: Vera Cruz, Jalapa). Deltocephalinae, Athysanini. Neotropical.

Allebra 17:1483. Error for *Alebra* Fieber.

Allectus Distant 1918b:75. *A. notatus* n.sp. (India: Nilgiri Hills, Ootacamund, Dodabetta Hill). 10:1583. Deltocephalinae, Stenometopiini. Oriental. Synonym of *Doratulina* Melichar 1903b:198; Vilbaste 1965a:10.

Alletta Theron 1984c:314. Error for *Aletta* Metcalf.

Allocha 1:664. Error for *Alocha* Melichar.

Allochus Fieber 1872a:11. 8:259. Invalid.

Allogonia Melichar 1926a:343. *Tettigonia concinnula* Fowler 1900d:287 (Mexico: Guerrero, Omiltama 8000 ft). 1:266. Young 1977a:831. Cicadellinae, Cicadellini. Arizona, USA; central and south Mexico to Panama.

Allonolla Young 1977a:845. *A. smithi* n.sp. (Mexico: Jalisco, Manzanita). Cicadellinae, Cicadellini. Mexico.

Allophleps Bergroth 1920a:27. *A. inspersa* n.sp. (Kenya). 10:1892. Deltocephalinae, Fieberiellini. Ethiopian.

Alloproctus Bergroth 1924a:400. *A. amandatus* n.sp. (Chile: Juan Fernandez Islands). 15:21. Agalliinae, Agalliini. Southern Chile. Synonym of *Agallia* Curtis 1833a:193; Kramer 1964a:153.

Allotapes Emeljanov 1964b:54. *A. nigrifacias* n.sp. (USSR: Tadzhikistan). Deltocephalinae, Athysanini. Palaearctic.
[NOTE: There are two spellings given for the type species; *nigrifacias* for the designated taxon, and *nigrifacies* for the new species described].

Allygianus Ball 1936c:59. *Thamnotettix gutturosa* Ball 1910c:307 (USA: California, Beaumont). 10:933. Deltocephalinae, Athysanini. Nearctic; California.

Allygidius Ribaut 1948c:58. *Cercopis atomaria* Fabricius 1794a:37 (Sweden). 10:957, 985, 2643. Ossiannilsson 1983a:685-688; ICZN Opinion 1365. Deltocephalini, Athysanini. Palaearctic.
Subgenus:
Dicrallygus Ribaut 1952a:214. *Jassus furcatus* Ferrari 1882a:136.

Allygiella Oman 1949a:131. *Allygianus clathratus* Ball 1936g:431 (USA: California, Pine Valley). 10:944. Deltocephalinae, Athysanini. Nearctic; California.

Allygus Fieber 1872a:13. *Cicada mixta* Fabricius 1794a:39 (France). 10:957. ICZN Opinion 1365. Deltocephalinae, Athysanini. Palaearctic; [Nearctic.].
Subgenus:
Syringius Emeljanov 1966a:101. *Allygus syrinx* Dlabola 1961b:330.

Allytus 10:2644. Error for *Allygus* Fieber.

Alminisna Zoological Record 106, Section 13f:551, 1969. Error for *Almunisna* Dworakowska 1969c.

Almunisna Dworakowska 1969c:384. *A. bulbosa* n.sp. (Nepal: Taplejung District, above Sangu). Typhlocybinae, Typhlocybini. Oriental.

Alnella Anufriev 1971b:109. *A. sudzuchenica* n.sp. (USSR: Maritime Territory, Sudzukhe). Typhlocybinae, Erythroneurini. Eastern Palaearctic. Subgenus of *Alnetoidia* Dlabola; Dworakowska 1979b:30.
Synonym:
Sapporoa Dworakowska 1972e:775. *Alnella (Sapporoa) watanabei* n.sp.

Alnetoidia Dlabola 1958c:55. *Cicadula alneti* Dahlbom 1850a:181 Gotland Island). Typhlocybinae, Erthroneurini. Palearctic.
Subgenus:
Alnella Anufriev 1971b:109. *A. sudzuchenica* n.sp.
Synonym:
Sapporoa Dworakowska 1972e:775. *Alnella (Sapporoa) watanabei* n.sp.

Alobaldia Emeljanov 1972d:102. *Thamnotettix tobae* Matsumura 1902a:369 (Japan: Kyushu). Deltocephalinae, Deltocephalini. Palaearctic.

Alocephalus Evans, J.W. 1977a:93. *Dorycephalus ianthe* Kirkaldy 1906c:340 (Australia: Queensland, Bundaberg). Ulopinae, Cephalelini. Australian.

Alocha Melichar 1926a:342. *Tettigonia sordida* Signoret 1855d:781 (Surinam). 1:262. Cicadellinae, Cicadellini. Neotropical. Synonym of *Paromenia* Melichar 1926a:342; Young 1977a:263.

Alocoelidia Evans, J.W. 1954a:108. *A. fulva* n.sp. (Madagascar). 11:82. Unassigned, formerly Coelidiinae; Nielson 1975a:11. Malagasian.

Alodeltocephalus Evans, J.W. 1966a:243. *Phrynomorphus longuinquus* Kirkaldy 1906c:326 (Australia: Queensland, Bundaberg). Deltocephalinae, Deltocephalini. Australian. [Note: Knight 1975a:189 & 191 points out that Evans, J.W. 1966a, misidentified Kirkaldy's species *longuinquus*, which is actually the taxon described as *Deltocephalus obliquus* Evans, J.W. 1938b:16 (Tasmania)].

Aloeurymela Evans, J.W. 1965b:85. *A. gearyi* n.sp. (Australia: Queensland, Cunnamulla). Eurymelinae, Eurymelini. Australian.

Aloipo Evans, J.W. 1966a:35. *Ipoides ooldeae* Evans, J.W. 1934a:156 (South Australia: Ooldea). Eurymelinae, Ipoini. Australian.

Alopenthimia Evans, J.W. 1972a:188. *A. magna* n.sp. (SE New Guinea: Port Moresby, 25 km radius). Penthimiinae. Oriental.

Aloplemmeles Evans, J.W. 1966a:210. *A. gearyi* n.sp. (Australia: Queensland, Cunnamulla). Iassinae, Reuplemmelini. Australian.

Alosarpestus Evans, F. 1981a:179. *A. fakensis* n.sp. (SW New Guinea: Irian, Vogelkop, Fak Fak, S. coast of Bomberi). Tartessinae. Oriental.

Alospangbergia Evans, J.W. 1973a:188. *A. bella* n. sp. (New Guinea: Morobe Dist., Lake Trist 1820 m). Deltocephalinae, Hecalini. Oriental.

Alotartessella Evans, F. 1981a:150. *Tartessella campbelli* Evans, J.W. 1936b:57 (N. Australia: Newcastle Waters). Tartessinae. Australian.

Alotartessus Evans, F. 1981a:124. *Tartessus iambe* Kirkaldy 1907d:46 (Australia: New South Wales, Sydney). Tartessinae. Australian.

Aloxestocephalus Evans, J.W. 1973a:196. *A. nigrobunneus* n.sp. (NE New Guinea: Toricelli Mtns., Mobitei). Xestocephalinae, Xestocephalini. Oriental.

Alseis Kirkaldy 1907d:37. *A. osborni* n.sp. (Australia: Queensland, Brisbane). 4:133. Evans, J.W. 1966a:115. Ledrinae, Thymbrini. Australian.

Altaiotettix Vilbaste 1965b:82. *A. forficula* n.sp. (Mongolia: Altai Mtns.). Deltocephalinae, Paralimnini. Palaearctic.

Alygus 10:2644. Error for *Allygus* Fieber.

Amahuaka Melichar 1926a:344. *Tettigonia angustula* Fowler 1900d:291 (Mexico: Guerrero, Xucumanatlan). 1:420. Young 1968a. Cicadellinae, Mileewanini. Neotropical; Mexico, Costa Rica, Peru.

Amalfia Melichar 1924a:199; 1926a:281. *Scaris latipennis* Walker 1851b:833 (Colombia). 1:605. Young 1977a:1096. Cicadellinae, Cicadellini. Neotropical.

Ambara Dworakowska 1981j:235. *A. acauda* n.sp. (India: Karnataka, Jog Falls, 1600 m). Typhlocybinae, Erythroneurini. Oriental.

Amberbakia Distant 1912d:443. *Petalocephala specularia* Walker 1870b:307 (New Guinea). 3:213. Penthimiinae. Oriental.

Ambigonalia Young 1977a:957. *Tettigonia varicolor* Signoret 1854a:15 (Oahu, Hawaii; error or adventive, see Young 1977a:958). Cicadellinae, Cicadellini. Hawaii and Peru.

Amblyceehalus 1:665. Error for *Amblycephalus* Curtis.

AMBLYCEPHALIDAE 1:21.

AMBLYCEPHALINAE 1:25.

AMBLYCEPHALINI 1:26.

Amblycephalus Curtis 1833a:193. *Cicada viridis* Linnaeus 1758a:438 (Europe). 1:106, 665. Cicadellinae, Cicadellini. Palaearctic. Synonym of *Cicadella* Latreille 1817a.

Amblycephalus Kirschbaum 1858b:357. *A. agrestis* n.sp. (Sweden). 8:35. Aphrodinae, Aphrodini. Palaearctic. Preoccupied by *Amblycephalus* Curtis 1833a:192. Isogenotypic synonym of *Stroggylocephalus* Flor 1861a:210.

Amblydisca Stal 1869a:61. *Tettigonia rubriventris* Signoret 1855a:52 (Mexico). 1:605. Young 1968a:109. Cicadellinae, Proconiini. Neotropical.
Synonym:
Tolua Melichar 1926a:298. *Aulacizes multiguttata* Stal 1864a:80.

Amblydiscea 1:665. Error for *Amblydisca* Stal.

Amblyscarta Stal 1869a:71 as subgenus of *Tettigonia* Olivier 1789a:24. *Cicada modesta* Fabricius 1803a:70 ("Amer. Merid."). 1:210. Young 1977a:176. Cicadellinae, Cicadellini. Neotropical; Mexico to Bolivia; to SE Brazil and northern Argentina.

Amblyscartidia Young 1977a:226. *Tettigonia albofasciata* Walker 1851b:756 ("S. America"). Cicadellinae, Cicadellini. Neotropical.

Amblysellus Sleesman 1929a:131. *Amblycephalus curtisii* Fitch 1851a:61 (USA: New York). 10:117. Kramer 1971a:83-98. DeLong & Hamilton 1974a:841-849. Deltocephalinae, Deltocephalini. Nearctic, Neotropical.

Amblytelinus Lindberg 1954a:213. *A. insularis* n.sp. (Canary Islands). 10:794. Deltocephalinae, Athysanini. Palaearctic.

Amicula Dworakowska 1971b:111 as subgenus of *Frutioidia* Zachvatkin 1946b:152. *Erythroneura (Zygina) amicula* Linnavuori 1965a:16 (Libya: Tobruk). Typhlocybinae, Erythroneurini. Palaearctic. Preoccupied by *Amicula* Gray 1840, see *Dworakowskellina* Kocak 1981a:31.

Amimenus Ishihara 1953b:41. *Scaphoideus mojiensis* Matsumura 1914a:220 (Japan: Honshu). 10:942. Deltocephalinae. Palaearctic; Korea, Japan, Soviet Maritime Territory.

Amphigonalia Young 1977a:825. *Tettigonia gothica* Signoret 1854b:345 ("N. Amer."). Cicadellinae, Cicadellini. Nearctic, Neotropical.

Amphipyga Osborn 1930a:691. *A. balli* n.sp. (USA: Ohio, Marietta). 10:1623. Deltocephalinae, Doraturini. Nearctic. Subgenus of *Athysanella* Baker 1898a:185; Ball & Beamer 1940a:9.

Amphypyga 10:2644. Error for *Amphipyga* Osborn.

Amplicephalus DeLong 1926d:83 as subgenus of *Deltocephalus* Burmeister 1838b:15. *Deltocephalus osborni* Van Duzee 1892d:304 (USA: New York, Lancaster). 10:1573. Kramer 1971b:198-204. Deltocephalinae, Deltocephalini. Nearctic.
Subgenus:
Nanctasus Linnavuori 1959b:99. *A. (N.) bolivicus* n.sp.

Amrasca Ghauri 1967a:p. 159. *A. splendens* n.sp. (India: Vellayani). Typhlocybinae, Empoascini. Oriental, some contiguous Palaearctic areas and Australia, (adventive?).
Subgenus:
Quartasca Dworakowska 1972i:27. *Laokayana (Q.) czerwcowa* n.sp.
Synonyms:
Sundapteryx Dworakowska 1970h:708. *Chlorita biguttula* Shiraki 1912a:96.
Laokayana Dworakowska 1972i:27. *Empoasca bombaxia* Ghauri 1965a:683.

Amritodus Anufriev 1970b:375. *Idiocerus atkinsoni* Lethierry 1889c:252 (India: Bengal). Idiocerinae. Oriental.

Amurta Dworakowska 1977i:39. *A. mirabilis* n.sp. (Vietnam: Laocai, Sa-Pa, 1650 m). Typhlocybinae, Typhlocybini. Oriental; Nepal, Vietnam.

Amylidia Nielson 1983h:561. *A. lynnea* n.sp. (Brazil: Minas Gerais). Coelidiinae, Teruliini. Neotropical.

Anacephaleus Evans, J.W. 1936a:43. *A. minutus* n.sp. (Western Australia: Perth). 5:39. Evans, J.W. 1966a:89, 1977a:91. Ulopinae, Cephalelini. Australian. Synonym of *Notocephalius* Jacobi 1909a:339.

Anaceratagallia Zachvatkin 1946b:159. *Cicada venosa* de Fourcroy 1785a:188 (France). 14:61. Ossiannilsson 1981a:298. Agalliinae, Agalliini. Palaearctic. Synonym of *Agallia* Curtis 1833a:193; Ossiannilsson 1981a:298.

Anacornutipo Evans, J.W. 1934a:163. *Eurymela lignosa* Walker 1858b:166 (Australia: New Holland). 12:28. Evans, J.W. 1966a:48. Eurymelinae, Ipoini. Australian.

Anacotis Evans, J.W. 1937b:161. *A. hackeri* n.sp. (Australia: Queensland, Brisbane). 4:120. Evans, J.W. 1966a:110. Ledrinae, Stenocotini. Australian.

Anacrocampsa Young 1968a:133. *Amblydisca frenata* Melichar 1926a:295 (Brazil: Espirito Santo). Cicadellinae, Proconiini. Neotropical; southeastern Brazil.

Anacuerna Young 1968a:253. *Cuerna centrolinea* Melichar 1925a:365 (Peru). Cicadellinae, Proconiini. Neotropical; Bolivia, Peru.

Anaemotettix Korolevskaya 1980a:800 as subgenus of *Ferganotettix* Dubovskij 1966a:171. *F.(A.) klishinae* n.sp. (USSR: Tajikistan, Karategin Range near Vaydar Kishlak). Nast 1972a:363. Deltocephalinae, Athysanini. Palaearctic.

Anaka Dworakowska & Viraktamath 1975a:521. *A. colorata* n.sp. (India: Dharwar, Karnatak Univ.). Typhlocybinae, Dikraneurini. Oriental.

Anareia Vilbaste 1965b:74. *A. lineiger* n.sp. (Mongolia: Altai Mts). Deltocephalinae, Paralimnini. Palaearctic.
Subgenus:
Calidanus Emeljanov 1966a:122. *Diplocolenus truncatus* Emeljanov 1962a:180.

Anargella Emeljanov 1972b:244 as subgenus of *Adarrus* Ribaut 1946b:83. *Adarrus praenuntius* n.sp. (Mongolia). Deltocephalinae, Paralimnini. Palaearctic.

Anaterostemma Mitjaev 1968b:30. Error for *Anoterostemma* Low.

Anatolidia Zachvatkin 1937a:319, 322 as subgenus of *Zyginidia* Haupt 1929c:268. *Z.(A.) obesa* n.sp. (Turkey). 17:1460. Typhlocybinae, Erythroneurini. Palaearctic.

Anchura Melichar 1926a:344. *Mareba eresia* Distant 1908b:77 (Ecuador: Cachabe). 1:411,666. Preoccupied by *Anchura* Conrad 1860; see
Anchuralia Evans, J.W. 1947a:159. Young 1968b:60. Cicadellinae, Proconiini. Neotropical. Synonym of *Mareba* Distant 1908b:77.

Anchuralia Evans, J.W. 1947a:159, replacement for *Anchura* Melichar 1926a:344, not *Anchura* Conrad 1860. *Mareba eresia* Distant 1908b:77 (Ecuador: Cachabe). 1:411. Young 1968a:60. Cicadellinae, Proconiini. Neotropical. Synonym of *Mareba* Distant 1908b:77.

Ancosus Anufriev 1978a:58, emendation of *Ankosus* Oman & Musgrave 1975a.

Ancudana DeLong & Martinson 1974b:261. *A. cincta* n.sp. (Chile: Chiloe Prov., Ancud). Deltocephalinae, Athysanini. Neotropical.

Andamarca Melichar 1926a:336. *Tettigonia physocephala* Signoret 1854d:720 (Brazil). 1:655. Cicadellinae, Proconiini. Neotropical. Synonym of *Ochrostacta* Stal 1869a:61; Young 1968a:145.

Andamaroa 1:166. Error for *Andamarca* Melichar.

Andanus Linnavuori 1959b:237. *A. bimaculatus* n.sp. (Peru: Madre de Dios). Deltocephalinae, Athysanini. Neotropical.

Andrabia Ahmed, Manzoor 1970d:407. *A. kashmirensis* n.sp. (West Pakistan: Abbotabad). Typhlocybinae, Erythroneurini. Palaearctic.

Anemochrea Kirkaldy 1906c:329. *A. mitis* n.sp. (Australia: New South Wales, Sydney). 10:423. Deltocephalinae, Stenometopiini. Australian.
[Note: Placed as a questionable synonym of *Doratulina* Melichar 1903 by Vilbaste 1965a:10; treated as genus by Evans, J.W. 1966a:231].

Anemolua Kirkaldy 1906c:329. *A. hanuala* n.sp. (Australia: Queensland, Cairns). 10:420. Evans, J.W. 1966a:248. Deltocephalinae, Stenometopiini. Australian.

Anemolus 10:2644. Error for *Anemolua* Kirkaldy.

Aneono Kirkaldy 1906c:358. *A. pulcherrima* n.sp. (Australia: South Wales, Sydney). 17:105. Evans, J.W. 1966a:258. Typhlocybinae, Dikraneurini. Australian.

Angallia Zoological Record 110, Section 13f:126, 1973. Error for *Agallia* Curtis.

Angenus DeLong 1939a:34. 10:2110, 2644. Invalid.

Angenus DeLong & Knull 1971a:54 as subgenus of *Scaphoideus* Uhler 1889a:33. 10:2110, 2644. *"Scaphoideus immistus* (Say)" = *Jassus immistus* Say 1830b:306 (USA: Indiana). Deltocephalinae, Scaphoideini. Nearctic. Synonym of *Scaphoideus* Uhler 1889a:33.

Angheria 14:171. Error for *Ahngeria* Melichar.

Angolaia Linnavuori & Al-Ne'amy 1983a:92. *A. atrata* n.sp. (Angola: Angola). Selenocephalinae, Selenocephalini. Ethiopian.

Angubahita DeLong 1982a:185. *A. arta* n.sp. (Panama: Las Cumbres). Deltocephalinae, Athysanini. Neotropical; Bolivia, Panama.
Subgenus:
Mairana DeLong 1984a:117. *A. (M.) nigrens* n.sp.

Angucephala DeLong & Freytag 1975a:111. *A. mellana* n.sp. (Honduras). Gyponinae, Gyponini. Neotropical; Honduras, Venezuela.

Angulanus DeLong 1946b:30 as subgenus of *Idiodonus* Ball 1936c:57. *I.(A.) incisurus* n.sp. (Mexico, D.F.). 10:1304. Deltocephalinae, Athysanini. Neotropical. Synonym of *Idiodonus* Ball 1936c:57.

Angusana DeLong & Freytag 1972f:254 as subgenus of *Polana* DeLong 1942d:110. *Gypona exornata* Fowler 1903b:315 (Panama: Volcan de Chiriqui). Gyponinae, Gyponini. Neotropical.

Angusticella Maldonado-Capriles 1972a:545. *Pedioscopus coloratus* Baker 1915c:331 (Philippines: Luzon, Los Banos, Malinao). Ghauri 1985c:70. Idiocerinae. Oriental; Philippines, Luzon, Mindanao.
Subgenus:
Upsicella Maldonado-Capriles 1972a:545, 549. *Pedioscopus similis* Baker 1915c:331.

Anidiocerus Maldonado-Capriles 1976a:139. *A. variabilis* n.sp. (Taiwan: Ali-Shan). Idiocerinae. Oriental.

Anipo Evans, J.W. 1934a:159. *Eurymela porriginosa* Signoret 1850c:512 ("Australia"). 12:30. Evans, J.W. 1966a:40. Eurymelinae, Ipoini. Australian. Synonym of *Ipoella* Evans, J.W. 1934a:157; Evans, J.W. 1966a:40.

ANIPOINAE 12:28.

Ankosus Oman & Musgrave 1975a:8. *Errhomus filamentus* Oman 1952a:14-15 (USA: Oregon, Oregon Caves). Cicadellinae, Errhomenini. Nearctic; USA; SW Oregon, NW California.

Anlacizes 1:666. Error for *Aulacizes* Amyot & Serville.

Annidion Kirkaldy 1905a:267. *A. pulcherrimum* n.sp. (Ivory Coast). 9:24. Linnavuori 1975a:35-36. Deltocephalinae, Hecalini. Ethiopian.

Anomia Fieber 1866a:509. *Cicada quercus* Fabricius 1777a:298 (Europe). 17:782. Typhlocybinae, Typhlocybini. Palaearctic. Synonym of *Typhlocyba* Germar 1833a:180.

Anomiana Distant 1918b:109. *A. longula* n.sp. (South India: Chikkaballapura). 10:2442, 2644. Deltocephalinae, Balcluthini. Pan-Tropical. Synonym of *Balclutha* Kirkaldy 1900b:243; Blocker 1967a:4.

Anoplotettix Ribaut 1942a:262. *Thamnotettix fusconervosa* Ferrari 1882a:124 (Italy). 10:215. Deltocephalinae, Athysanini. Palaearctic.

Anoscopsis 8:259. Error for *Anoscopus* Kirschbaum.

Anoscopus Kirschbaum 1858b:357. *Cicada serratulae* Fabricius 1775a:686 (England). 8:58. Hamilton, KGA 1975d:1016. Ossiannilsson 1981a:370. Aphrodinae, Aphrodini. Holarctic. [Nearctic].

Anosterostema 8:260. Error for *Anosterostemma* Puton.

Anosterostemma Puton 8:221, 260. Error for *Anoterostemma* Low 1855a.

Anoterostemma Low 1885a:353. *A. henschii* n.sp. ("Carniola") = *Doratura ivanhofi* Lethierry 1876b:26 (Russia). 8:221. Cicadellinae, Anoterostemmini. Palaearctic.

ANOTEROSTEMMATINI Haupt; Nast 1972a:248.

ANOTEROSTEMMINA Haupt 1929c:243, subtribe of APHRODINAE.

ANOTEROSTEMMINI 8:221.

Antanus Young 1957a:14. Error for *Atanus* Oman.

Anthocallis Emeljanov 1964c:423 as subgenus of *Paralimnus* Matsumura 1902a:386. *Paralimnus minor* Kusnezov 1929a:319 (Turkestan). Deltocephalinae, Paralimnini. Palaearctic. Subgenus of *Paralimnus* Matsumura 1902a:386; Nast 1972a:407.

Antoniellus Linnavuori 1959b:205. *A. irrorellus* n.sp. (Argentina: San Antonio Oesta). Deltocephalinae, Athysanini. Neotropical; Argentina, Paraguay.

Anthysanus 10:2644. Error for *Athysanus* Burmeister.

Anufrievia Dworakowska 1970n:761. *A. rolikae* n.sp. (China: Kiangsu, Hangchow). Typhlocybinae, Erythroneurini. Oriental, Palaearctic.

Anufrieviella Nast 1981a:258. *Macropsis melichari* Oshanin 1906a:69, (China: Szechwan), replacement for *Macropsis scutellaris* Melichar 1902c:120, not *Macropsis scutellaris* Fieber 1868a:451. Iassinae. Palaearctic.

Anufrievola Kocak 1981a:32, replacement for *Mulsantina* Anufriev 1970f:262 not *Mulsantina* Weise 1906. *Typhlocyba stigmatipennis* Mulsant & Rey 1855a:245 (France). Typhlocybinae, Dikraneurini. Palaearctic. Subgenus of *Micantulina* Anufriev 1970d:262.

Apetiocellata Sohi 1977a:347. *A. bifasciata* n.sp. (India: Uttar Pradesh, Dehra Dun, Forest Research Institute). Typhlocybinae, Dikranerini. Oriental.

Aphanalebra McAtee 1926b:147 as subgenus of *Protalebra* Baker 1899b:402. *Protalebra unipuncta* Baker 1899b:404 (Brazil: Minas Gerais). 17:19. Typhlocybinae, Alebrini. Neotropical.

Apheliona Kirkaldy 1907d:67. *Heliona bioculata* Melichar 1903b:216 (Ceylon: Peradeniya). 17:597. Typhlocybinae, Empoascini. Oriental, Palaearctic.
Synonyms:
Chikkaballapura Distant 1918b:107. *C. maculosa* n.sp.
Sujitettix Matsumura 1931b:76. *S. ferrugineus* n.sp.
Ushamenona Malhotra & Sharma 1974a:245. *Sujitettix aryavartha* Ramakrishnan & Menon 1972a:114.

Aphrodes Curtis 1829a:193. *Cicada striata* Fabricius 1787a:271 (Great Britain) = *Cicada bicincta* von Schrank 1776a:75. 8:58. Hamilton, KGA 1975d:1010. Ossiannilsson 1981a:361. Aphrodinae, Aphrodini. Holarctic. [Nearctic].
Synonyms:
Acucephalus Germar 1833a:181. *Cicada striata* Fabricius.
Pholetaera Zetterstedt 1840a:288. *Cercopis rustica* Fallen 1826a:23 = *Cicada bicincta* von Schrank 1776a:75.

Aphroderes 8:260. Error for *Aphrodes* Curtis.

APHRODIDAE 8:5.

APHRODINA Hamilton, KGA 1975b:482, subtribe of APHRODINAE.

APHRODINAE 8:11.

APHRODINI 8:33.

Aplanatus Cheng, Y. J. 1980a:87. *A. pallibandus* n.sp. (Paraguay: "N Cnel Bogado"). Deltocephalinae, Athysanini. Neotropical.

Aplanus Oman 1949a:138. "*Eutettix pauperculus* Ball 1903" = *Phlepsius pauperculus* Ball 1903b:228 (USA: Colorado, Grand Junction). 10:529. Deltocephalinae, Athysanini. Nearctic; southwest USA.

Aplicephalus 10:2644. Error for *Amplicephalus* DeLong.

Apogonalia Evans, J.W. 1947a:159, replacement for *Apogonia* Melichar 1926a:344, not *Apogonia* Kirby 1819. *Tettigonia stalii* Signoret 1855d:787 (Mexico). 1:287. Young 1977a:916. Cicadellinae, Cicadellini. Neotropical and Nearctic; southern Arizona mtns., Mexico and Central America, islands of the Caribbean.
Synonym:
Colimona Oman 1949a:71. Specimens from Santa Rita Mountains, Arizona, identified as *Tettigonia monticola* Fowler 1899.

Apogonia Melichar 1926a:344. *Tettigonia stali* Signoret 1855d:787 (Mexico). 1:287, 666. Young 1977a:916. Preoccupied by *Apogonia* Kirby 1819, see *Apogonalia* Evans, J.W. 1947a:159. Cicadellinae, Cicadellini. Neotropical.

Apphia Distant 1918b:4. *A. burmanica* n.sp. (Upper Burma: Mayenyo, 3500 ft). 6:48. Cicadellinae, Evacanthini. Oriental. Synonym of *Onukia* Matsumura 1912a:44.

Aprodes 8:260. Error for *Aphrodes* Curtis.
Apterodes 8:260. Error for *Aphrodes* Curtis.

Apulia Distant 1908a:525. *Tettigonia quadrimacula* Walker 1851b:741 (Colombia). 1:424. Young 1977a:81. Cicadellinae, Cicadellini. Neotropical; northern South America.
Synonym:
Apulina Melichar 1926a:345. *Tettigonia elongata* Signoret 1854c:495.

Apulina Melichar 1926a:345. *Tettigonia elongata* Signoret 1854:495 (Colombia: Bogota). 1:429. Young 1977a:81. Cicadellinae, Cicadellini. Neotropical; northern South America. Synonym of *Apulia* Distant 1908a:525; Young 1977a:81.

Arahura Knight, W.J. 1975a:185. *A. reticulata* n.sp. (New Zealand: Otago Prov., Hunter Mtns., Mt. Burns 1300 m). Deltocephalinae, Athysanini. Australian; New Zealand only.

Araldus Ribaut 1946b:84. *Deltocephalus propinquus* Fieber 1869a:204 (Spain). 10:1438. Deltocephalinae, Paralimnini. Palaearctic.

Arapona DeLong 1979b:45. *A. vallea* n.sp. (Panama: Cocle Prov., El Valle). Gyponinae, Gyponini. Neotropical; Panama, Peru.

Arawa Knight, W.J. 1975a:176. *A. variegata* n.sp. (New Zealand: Nelson Prov., Richmond Range, Fall Peak 1311 m). Deltocephalinae, Athysanini. Australian; Chatham Island, New Zealand, Tasmania.

Arbela Anufriev 1972a:35. *Alebra ulmi* Anufriev 1969c:163. (Soviet Maritime Territory). Typhlocybinae, Alebrini. Palaearctic. Preoccupied by *Arbela* Stal 1865; see *Arbelana* Anufriev.

Arbelana Anufriev 1978a:70, replacement for *Arbela* Anufriev 1972a:35, not *Arbela* Stal 1865. *Alebra ulmi* Anufriev 1969c:163 (Soviet Maritime Territory). Typhlocybinae, Alebrini. Palaearctic.

Arboridia Zachvatkin 1946b:153 as subgenus of *Zyginidia* Haupt 1929c:268. *Typhlocyba parvula* Boheman 1845b:161 (Germany). Typhlocybinae, Erythroneurini. Palaearctic, Oriental.
Synonym:
Khoduma Dworakowska 1972b:403. *K. jacobii* n.sp.
[NOTE: Raised to generic rank by Dworakowska 1970g, and containing as a subgenus *Erythridula* Young 1952b:81; transferred from *Erythroneura*].

Arborifera Sohi & Sandhu 1971a:401 as subgenus of *Arboridia* Zachvatkin 1946b:153. *A.(A.) viniferata* n.sp. (India: Ludhiana). Typhlocybinae, Erythroneurini. Oriental.

Arceratagallia 14:171. Error for *Aceratagallia* Kirkaldy.

Archijassus Handlirsch 1908a:501 (Unknown). Metcalf & Wade 1966a:215-216. Fossil: Jurassic, Germany, Switzerland, Turkestan.

Arctotettix Linnavuori 1952b:185 as subgenus of *Sorhoanus* Ribaut 1946b:85. *Deltocephalus abiskoensis* Lindberg 1926b:112 (Sweden). 10:1465. Deltocephalinae, Paralimnini. Palaearctic. Synonym of *Rosenus* Oman 1949a:170; Nast 1972a:434.

Ardasoma DeLong & Freytag 1976a:56 as subgenus of *Curtara* DeLong and Freytag 1972a:231. *C.(A.) arda* n.sp. (French Guiana: Gayane, Maroni). Gyponinae, Gyponini. Neotropical.

Areuna Anufriev 1972a:35. Error for *Aruena* Anufriev.

Arezzia Metcalf & Bruner 1936a:957. *A. maestralis* n.sp. (Cuba: Oriente Prov., Sierra Maestro Mts., Turquino Peak). 1:394. Young 1977a:965. Cicadellinae, Cicadellini. Neotropical. Synonym of *Hadria* Metcalf & Bruner 1936a:948; Young 1977a:965.

Argaterma White 1878a:473. *A. alticola* n.sp. (St. Helena). 8:220. Aphrodinae, Aphrodini. Ethiopian.

Argostagum 10:2644. Invalid.

Argyrilla Emeljanov 1972d:107. *Athysanus attenuatus* Emeljanov 1964f:31 (USSR: Kazakhstan). Deltocephalinae, Athysanini. Palaearctic.

Aricanus Linnavuori 1959b:206. *A. filigranus* n.sp. (Chile: Arica). Deltocephalinae, Athysanini. Neotropical.

Aridanus DeLong & Hershberger 1949a:173 as subgenus of *Texananus* Ball 1918b:381. *Phlepsius aerolatus* Baker 1898f:30 (USA: Kansas). 10:705. Deltocephalinae, Athysanini. Nearctic.

Ariellus Ribaut 1952a:275 as subgenus of *Arocephalus* Ribaut 1946b:85. *Jassus (Deltocephalus) punctum* Flor 1861a:247 (Livonia). Deltocephalinae, Paralimnini. Palaearctic.

Arocephalus Ribaut 1946b:85. *Jassus (Deltocephalus) longiceps* Kirshbaum 1868b:135 (Hesse). 10:1439. Ossiannilsson 1983a:806- 811. Deltocephalinae, Paralimnini. Palaearctic; European, Mediterranean, and Siberian subregions.
Subgenus:
Ariellus Ribaut 1952a:275. *Jassus (Deltocephalus) punctum* Flor 1861a:247.

Arrailus Ribaut 1952a:255 as subgenus of *Jassargus* Zachvatkin 1933b:268. 10:1209. *Deltocephalus flori* Fieber 1869a:210 (Livonia). Deltocephalinae, Paralimnini. Palaearctic.

Arrugada Oman 1938b:362. *Huleria rugosa* Osborn 1924c:404 (Bolivia). 8:220. Linnavuori 1965b:141-142. Linnavuori & DeLong 1978a:174-184. Arrugadinae. Neotropical; Bolivia, Peru.

ARRUGADIINAE (sic) Linnavuori 1965b:41 = ARRUGADINAE.

ARRUGADINAE Linnavuori 1965b: 41.

Artemisiella Zachvatkin 1953c:229 as subgenus of *Chlorita* Fieber 1866a:508. *Chlorita nervosa* Fieber 1884a:60 (Spain). Typhlocybinae, Empoascini. Palaearctic. Subgenus of *Chlorita* Fieber; Nast 1972a:270.

Arthaldeus Ribaut 1946b:83. *Cicada pascuella* Fallen 1826a:32 (Sweden). 10:1394. Deltocephalinae, Paralimnini. Palaearctic; adventive in Nearctic.

Arthaldus 10:2645. Error for *Arthaldeus* Ribaut.

Artianus Ribaut 1942a:265. *Jassus interstitialis* Germar 1821a:90 (Prussia). 10:393. Deltocephalinae, Paralimnini. Palaearctic.

Articoelidia Nielson 1979b:97. *Jassus ensiger* Osborn 1924c:435 (Brazil: Hyutanahan, Rio Purus). Coelidiinae, Teruliini. Neotropical; Bolivia, Brazil, Peru.

Artucephalus DeLong 1943e:654. *A. fasciatus* n.sp. (Mexico: Guerrero, Iguala). 10:1740. Deltocephalinae, Athysanini. Neotropical.

Aruena Anufriev 1972a:35. *Dikraneura apicimaculata* Anufriev 1969c:1691 (Soviet Maritime Territory). Typhlocybinae, Dikraneurini. Palaearctic.
 [Note: Two spellings appear in Anufriev's paper - *Aruena* and *Areuna*. Nast 1972a:256 indicates that the former is the intended use].

Arundanus DeLong 1935b:180. *Thamnotettix arundineus* DeLong 1926b:91 (USA: Tennessee). 10:849. Deltocephalinae, Deltocephalini. Nearctic.

Arya Distant 1908g:338. *A. rubrolineata* n.sp. (India: Calcutta, Barrackpore). 10:1895. Deltocephalinae, Stenometopiini. Oriental; India. Ethiopian; Congo. Synonym of *Doratulina* Melichar 1903b:198; Vilbaste 1965a:10.

Ascius DeLong 1943f:250. *A. triangularis* n.sp. (Mexico: Guerrero, Iguala). 10:2248. Deltocephalinae, Scaphytopiini. Neotropical, Nearctic.
 Synonym:
 Vertanus Hepner 1946a:87. *Scaphytopius (Vertanus) ulcus* n.sp.

Asialebra Dworakowska 1971c:494. *Alebra aureovirescens* Dworakowska 1968d:565 (Vietnam: Lao-Kay). Typhlocybinae, Alebrini. Oriental.

Asianidia Zachvatkin 1946b:153 as subgenus of *Zyginidia* Haupt 1929c:268. *Erythroneura asiatica* Kusnezov 1932a:156 (USSR: Uzbekistan). 17:1451. Typhlocybinae, Erythroneurini. Palaearctic.
 [Note: Raised to generic rank by Dworakowska 1970m:697 et seq].

Asiotoxum Emelyanov 1964b:55. *Phlepsius parapulcher* Dlabola 1961b:328 (USSR: Tadzhikistan). Deltocephalinae, Athysanini. Palaearctic.

Asmaropsis Linnavuori 1978b:17. *A. troilos* n.sp. (Eritrea: Asmara-Decamere). Macropsinae, Macropsini. Ethiopian. Synonym of *Hephathus* Ribaut 1952a:437; Hamilton, KGA 1980b:904.

Aspilodora Melichar 1926a:346. *Tettigoniella geminatula* Jacobi 1905c:171 (Peru). 1:27. Young 1977a:1096. Cicadellinae, Cicadellini. Neotropical.

Assina Dworakowska 1979b:35. *A. nulla* n.sp. (Vietnam: Nghe-An, Quy-chau). Typhlocybinae, Erythroneurini. Oriental.

Assiringia Distant 1908g:255. *A. exhibita* n.sp. (Burma: Ruby Mines). 2:9. Hylicinae, Hylicini. Oriental.

Assiuta Linnavuori 1969b:212. *Melicharella salina* Lindberg 1954a:201 (Canary Islands). Adelungiinae, Adelungiini. Palaearctic.

Assymmetrasca Dlabola 1981a:233. Error for *Asymmetrasca* Dlabola 1958c.

Astacotettix Linnavuori 1956a:10. Invalid.

Astenogonia Melichar 1926a:345. *Cicada bicolor* Fabricius 1803a:65 ("Amer. Merid.") = *Capinota virescens* Metcalf 1949b:268 (South America). 1:426. Young 1968a:257. Cicadellinae, Proconiini. Neotropical. Synonym of *Acrogonia* Stal 1869a:67; Young 1968a:257.

Asthenotettix Korolevskaya 1980a:801 as subgenus of *Ferganotettix* Dobovskij 1966a:171. *F.(A.) perarmata* n.sp. (USSR: Tadzhikistan, Hissar Range, Varzob Ravine, near Gazhnekishlak). Deltocephalinae, Athysanini. Palaearctic.

Asthysanus 10:2645. Error for *Athysanus* Burmeister.

Astroasca Thapa 1985a:73. Error for *Austroasca* Lower.

Asythanus 10:2645. Error for *Athysanus* Burmeister.

Asymmetrasca Dlabola 1958c:51. *Empoasca decedens* Paoli 1932a:117 (Italy). Typhlocybinae, Empoascini, Palaearctic.
 [NOTE: Nast's 1972a:264 placement of *Asymmetrasca* Dlabola as a synonym of *Empoasca* Walsh has not been generally accepted by other workers].

Asymmetropteryx Dlabola 1958c:52. *Typhlocyba pictilis* Stal 1853a:176 (Sweden). Typhlocybinae, Typhlocybini. Palaearctic. Synonym of *Aguriahana* Distant 1918b:105; Dworakowska 1972a:278.

Atahysanella 10:2645. Error for *Athysanella* Baker.

Atanus Oman 1938b:381. *Eutettix dentatus* Osborn 1923c:57 (Bolivia: Santa Cruz). 10:1237. Linnavuori 1959b:296. Deltocephalinae, Athysanini. Neotropical, Nearctic.
 Synonyms:
 Alanus DeLong & Hershberger 1947d:231. *A. albidus* n.sp.; Linnavuori 1959b:296.
 Fulvanus Linnavuori 1955a:110. *F. curvilinea* n.sp.; Linnavuori 1959b:296.
 Tubulanus Linnavuori 1955a:109; Linnavuori & DeLong 1977c:209.
 Linnatanus Menezes 1973b:135, replacement for *Tubulanus* Linnavuori 1955a:109, not *Tubulanus* Reinier 1804.
 Nerminia Kocak 1981b:124, replacement for *Tubulanus* Linnavuori 1955a:109.

Athasanus 10:2645. Error for *Athysanus* Burmeister.
Athsanus 10:2645. Error for *Athysanus* Burmeister.
Athyanus 10:2645. Error for *Athysanus* Burmeister.
Athysanas 10:2645. Error for *Athysanus* Burmeister.

Athysanella Baker 1898a:185. *A. magdalena* n.sp. (USA: New Mexico). 10:1609, 1614. Deltocephalinae, Doraturini. Nearctic, Palaearctic, Neotropical.
 Subgenera:
 Amphipyga Osborn 1930a:691. *A. balli* n.sp.
 Gladionura Osborn 1930a:706. *Deltocephalus argenteolus* Uhler 1877a:473.
 Brachydella Ball & Beamer 1940a:9. *A.(B.) abdominalis* n.sp.
 Pedumella Ball & Beamer 1940a:65. *A.(P.) spatulata* n.sp.
 Acamella Wesley & Blocker 1985a:172. *A.(A.) tamara* n.sp.
 Synonym (of *Amphipyga*):
 Pectinapyga Osborn 1930a:689. *P. texana* n.sp.

ATHYSANIDAE 10:3.
ATHYSANINA Hamilton, KGA 1975b:490, subtribe of ATHYSANINI.
ATHYSANINAE 10:10.
ATHYSANINI 10:336

Athysannus 10:2645. Error for *Athysanus* Burmeister.

Athysanopsis Matsumura 1914a:184. *A. salicis* n.sp. (Japan). 10:4209. Linnavuori & Al-Ne'amy 1983a:23. Selenocephalinae, Bhatiini. Palaearctic; Japan, Korea, Manchuria, Soviet Maritime Territory.

Athysansus 10:2645. Error for *Athysanus* Burmeister.

Athysanus Burmeister 1838b:14 as subgenus of *Jassus*. *Cicada argentata* Fabricius 1794a:38 (France) = *Athysanus argentarius* Metcalf 1955a:265. 10:337. Ossiannilsson 1983a:746. Deltocephalinae, Athysanini. Palaearctic. [Nearctic].

Atishania Hamilton 1975b:497. Error for *Alishania* Vilbaste 1969a:5.

Atkinsoniella Distant 1908g:235. *A. decisa* n.sp. (India: Darjeeling). 1:41. Cicadellinae, Cicadellini. Oriental.

Atlantisia Dlabola 1976a:280. *A. leleupi* n.sp. (St. Helena). Deltocephalinae, Athysanini. Preoccupied by *Atlantisia* Lowe 1923, see *Atlantocella* Dlabola 1982.

Atlantocella Dlabola 1982a:162, replacement for *Atlantisia* Dlabola 1976a:280, not *Atlantisia* Lowe 1923. *Atlantisia leleupi* Dlabola 1976a:280 (St. Helena). Deltocephalinae, Euscelini. Ethiopian.

Atractotypus Fieber 1866a:501. *A. bifasciatus* n.sp. (France) = *Chiasmus translucidus* Mulsant & Rey 1855a:216. 8:208. Deltocephalinae, Chiasmusini. Palaearctic, Australian, Oriental. Synonym of *Chiasmus* Mulsant & Rey 1855a:215.

Atritona Melichar 1914g:5. *A. paradoxa* n.sp. (Eastern Africa). 7:23. Nirvaninae, Nirvanini. Ethiopian. Synonym of *Kosasia* Distant 1910e:240; Theron 1982a:17.

Attenuipyga Oman 1949a:30 as subgenus of *Dorycephalus* Kouchakewitch 1866a:102. *Dorycephalus knulli* Sanders & DeLong 1923a:151 (USA: Florida, Cleveland). 9:73. Oman 1985a:319. Eupelicinae, Dorycephalini. Nearctic.
[Note: Raised to generic rank by Emeljanov 1966a:100; page 55 in English translation].
Attractotypus 1:667. Error for *Actrotypus* Fieber.
Attractotypus 8:261. Error for *Atractotypus* Fieber.
Attsanus 10:2646. Error for *Athysanus* Burmeister.
Atucla Dworakowska 1972c:477 as subgenus of *Habenia* Dworakowska 1972c:477. *H.(A.) octave* n.sp. (Ivory Coast: Lamto forest). Typhlocybinae, Empoascini. Ethiopian.
Aucucephalus 8:261. Error for *Acucephalus* Germar.
Augulus Distant 1918b:98. *A. typicus* n.sp. (India: Assam, Margherita). 1:297. Cicadellinae, Mileewanini. Oriental. Synonym of *Mileewa* Distant 1908g:238.
Aulacirus 1:667. Error for *Aulacizes* Amyot & Serville.
Aulaciyes 1:667. Error for *Aulacizes* Amyot & Serville.
Aulaciza 1:667. Error for *Aulacizes* Amyot & Serville.
Aulacizes Amyot & Serville 1843a:571. *Tettigonia quadripunctata* Germar 1821a:59 (Brazil). 1:620. Young 1968a:89. Cicadellinae, Proconiini. Neotropical; Venezuela, southern Brazil, Argentina.
Auridius Oman 1949a:161. *Deltocephalus auratus* Gillette & Baker 1895a:85 (USA: Colorado). 10:1391. Hamilton, KGA & Ross 1972a:137-139. Deltocephalinae, Deltocephalini. Nearctic; intermountain regions of Western USA and Canada.
Aurkius Ribaut 1952a:257 as subgenus of *Jassargus* Zachvatkin 1933b:268. *Deltocephalus repletus* Fieber 1869a:208 (Germany). 10:1216. Deltocephalinae, Paralimnini. Palaearctic.
Aurogonalia Young 1977a:435. *A. dorade* n.sp. (Ecuador). Cicadellinae, Cicadellini. Neotropical.
Auropenthimia Evans, J.W. 1947a:255. *A. aburensis* n.sp. (Gold Coast: Aburi). 3:212. Penthimiinae. Ethiopian; Guinean. Synonym of *Jafar* Kirkaldy 1903f:13; Linnavuori 1977a:53.
Australoscopus China 1962d:296. *A. whitei* n.sp. (Central Australia: Charlotte Waters to Hamilton Bore). 12:42. Evans, J.W. 1966a:78. Eurymelinae, Pogonoscopini. Australian.
Austroagallia Evans, J.W. 1935c:70. *A. torrida* n.sp. (South Australia. Adelaide). 14:143,171. Evans, J.W. 1966a:169, 1977a:111. Agalliinae, Agalliini. Palaearctic; Mediterranean subregion. Also present in central Africa, India and Pakistan .[Australia, New Guinea and several island groups: Aldabra, Canary, Galapagos, Madeira, Cape Verde and Fiji].
Synonym:
Peragallia Ribaut 1948c:59. *Bythoscopus sinuatus* Mulsant & Rey 1855a:222.
Austroagalloides Evans, J.W. 1935c:70. *A. karoondae* n.sp. (South Australia: Murat Bay). 14:158. Evans, J.W. 1966a:170-173, 1977a:111. Austroagalloidinae. Australian.
AUSTROAGALLOIDINAE 14:157.
Austroasca Lower 1952a:202. *Empoasca viridigrisea* Paoli 1936a:12 (Australia: Queensland, Bowen). 17:613. Typhlocybinae, Empoascini. Australian, Oriental, Palaearctic.
Synonym:
Polynia Zachvatkin 1953c:228. *Chlorita vittata* Lethierry 1884c:65.
Austrocerus Evans, J.W. 1941e:37. *A. emarginatus* n.sp. (South Australia: Flinders Chase, Kangaroo Island). 16:215. Idiocerinae. Australian.
Synonym:
Gnatia Evans, J.W. 1941f:150. *G. angustata* n.sp.
Austrolopa Evans, J.W. 1936a:47. *A. brunensis* n.sp. (Tasmania: Bruni Island). 5:85. Evans, J.W. 1966a:86. Ulopinae, Ulopini. Australian.
Austrolpa 5:96. Error for *Austrolopa* Evans.
Austronirvana Evans, J.W. 1941e:41. *A. flavus* n.sp. (Australia: Queensland). 7:17. Nirvaninae, Nirvanini. Australian. Synonym of *Tortor* Kirkaldy 1907d:42; Evans, J.W. 1966a:153.
Austrotartessus Evans, F. 1981a:140. *Tartessus ianassa* Kirkaldy 1907d:47 (Australia: Queensland, Cairns). Tartessinae. Australian. Oriental: New Guinea.
Awasha Heller & Linnavuori 1968a:10. *A. basicornia* n.sp. (Ethiopia: Awash, 900 m). Deltocephalinae, Athysanini. Ethiopian.

Ayubiana Ahmed, Manzoor 1969a:57. *A. lantanae* n.sp. (West Pakistan: Murree). Typhlocybinae, Dikraneurini. Oriental.
Aztrania Blocker 1979a:28. *A. acinaca* n.sp. (British West Indies: Dominica forest, 800 m). Iassinae. Neotropical.
Baaora Dworakowska 1981g:597. *B. borealis* n.sp. (India: Himachal Pradesh, Kulu). Typhlocybinae, Typhlocybini. Oriental; India, northern Pakistan, northern Vietnam.
Babacephala Ishihara 1958b:156. *B. japonica* n.sp. (Japan: Honshu, Niigata Pref., Mt. Futatsumina, 1600 m). Cicadellinae, Errhomenini. Palaearctic.
Backhoffella Schmidt 1928c:83. *B. pulchra* n.sp. (Peru). 1:656. Young 1977a:320. Cicadellinae, Cicadellini. Neotropical.
Baclutha 10:2646 Error for *Balclutha* Kirkaldy.
Badylessa Dworakowska 1981f:583. *B. guenthartae* n.sp. (Liberia: Suakoko). Typhlocybinae, Empoascini. Ethiopian.
Baguoidea Mahmood 1967a:40. *B. rubra* n.sp. (Philippines: Luzon, Baguio). Typhlocybinae, Typhlocybini. Oriental; Philippines and Malaya.
Bahapona DeLong 1977b:392. *Gypona bahia* Osborn 1938a:31 (Brazil: Bahia). Gyponinae, Gyponini. Neotropical.
Bahita Oman 1938b:379. *Eutettix infuscatus* Osborn 1923c:52 (Brazil: Minas Gerias). 10:543. Linnavuori & DeLong 1978b:124- 127. Deltocephalinae, Athysanini. Neotropical; Panama & tropical regions of South America.
Bahitella Linnavuori 1959b:134 as subgenus of *Loreta* Linnavuori 1959b:132. *L. (B.) lineaticollis* n.sp. (Argentina: Loreto, Missiones). Deltocephalinae, Deltocephalini. Neotropical.
Baileyus Singh-Pruthi 1930a:31. *B. brunneus* n.sp. (Eastern Himalayas). 8:34. Aphrodinae, Aphrodini. Oriental.
Bakera Mahmood 1967a:23. *B. vittata* n.sp. (Philippines: Luzon, Mt. Makiling). Hongsaprug & Wilson 1985a. Typhlocybinae, Erythroneurini. Oriental; Philippines.
Bakeriana Evans, J.W. 1954a:129, replacement for *Bakeriola* Evans 1938b:17, not *Bakeriola* Bergroth 1920. *Ipo procurrens* Jacobi 1909a:324 (S.W. Australia: Boyanup). 11:38. Eurymelinae, Ipoini. Australian.
BAKERINI Mahmood 1967a:21, tribe of TYPHLOCYBINAE.
Bakeriola Evans, J.W. 1938b:17, replacement for *Ipocerus* Evans, J.W. 1934a:165, not *Ipocerus* Baker 1915c:319. *Ipo procurrens* Jacobi 1909a:342, (S.W. Australia: Boyanup). 12:38, 42. Eurymelinae, Ipoini. Australian. Preoccupied by *Bakeriola* Bergroth 1920, see *Bakeriana* Evans 1954a:129.
Bakshia Dworakowska 1977f:288. *B. bakshii* n.sp. (India: Asarori Range, Sal forest between Dehra Dun and Mohand). Typhlocybinae, Erythroneurini. Oriental; India.
Balacha Melichar 1926a:343. *Tettigonia melanocephala* Signoret 1854b:27 (Colombia). 1:287. Young 1977a:725. Cicadellinae, Cicadellini. Neotropical; Colombia to northern Argentina.
Balala Distant 1908g:250. *Penthimia fulviventris* Walker 1851b:841 (Unknown). 2:13. Hylicinae, Sudrini. Oriental; Indo- Chinese subregion.
Synonym:
Wania Liu 1939c:297. *W. membracioidea* n.sp.
Balanda Dworakowska 1979b:46. *B. kara* n.sp. (Vietnam: Nghe-An, Quy-Chau). Typhlocybinae, Erythroneurini. Oriental.
Balanus 10:2646 Error for *Bolanus* Distant.
Balbillus Distant 1908g:287. *B. granulosus* n.sp. (Ceylon). 7:4. Nirvaninae, Balbillini. Oriental, and the Guinean subregion of Africa.
Balbilus 7:32. Error for *Balbillus* Distant.
Balcanocerus Maldonado-Capriles 1971a:184. *Idiocerus balcanicus* Horvath 1903a:24 (Serbia). Hamilton, K. G. A. 1980a:824-825. Idiocerinae. Palaearctic.
Balcluta 10:2646. Error for *Balclutha* Kirkaldy.
Balclutha Kirkaldy 1900b:243, replacement for *Gnathodus* Fieber 1866a:505, not *Gnathodus* Pander 1856. *Cicada punctata* Fabricius 1775a:687 (Europe). 10:2382. Deltocephalinae, Balcluthini. Cosmopolitan; all major land masses.
Synonyms:
Gnathodus Fieber 1866a:505. *Cicada punctata* Fabricius 1775a:687.
Eugnathodus Baker 1903a:1. *Gnathodus abdominalis* Duzee 1892a:113.
Nesosteles Kirkaldy 1906c:343. *N. hebe* n.sp.
Agellus DeLong & Davidson 1933b:210. *Eugnathodus neglecta* DeLong & Davidson 1933a:55.

Anomiana Distant 1918b:109. *A. longula* n.sp.
Eusceloscopus Evans, J.W. 1941f:147. *E. yanchepensis* n.sp.
Balcutha 10:2646. Error for *Balclutha* Kirkaldy.
Balcluthina Singh-Pruthi 1930a:46. *B. viridis* n.sp. (India: Orissa). 10:2461. Deltocephalinae, Balcluthini. Oriental.
BALCLUTHINI 10:2380.
Baldriga Blocker 1979a:15. *B. knutsoni* n.sp. (Mexico: Oaxaca, Tehuantepec). Iassinae. Neotropical; Brazil and southern Mexico.
Baldulus Oman 1934a:79. *B. montanus* n.sp. (USA: Arizona, Santa Rita Mtns.) 10:2463. Deltocephalinae, Macrostelini. Nearctic: Mexico, and USA: southern Arizona, Maryland.
Baleja Melichar 1926a:344. *Tettigonia flavoguttata* Latreille 1811a:171 (Mexico: Guerrero). 1:418. Young 1977a:295. Cicadellinae, Cicadellini. Neotropical; Central America & northern South America.
Balelutha Sharma & Badan 1985a:149; error for *Balclutha* Kirkaldy.
Balera Young 1952b:25. *Dikraneura pellucida* Osborn 1928a:271 (Bolivia; Prov. del Sara). 17:21. Typhlocybinae, Alebrini. Neotropical; Bolivia, Brazil, Colombia.
Balila Dworakowska 1970d:347 as subgenus of *Erythroneura*. *Typhlocyba (Zygina) mori* Matsumura 1910e:121 (Japan). Typhlocybinae, Erythroneurini. Palaearctic and Oriental; Ceylon, China, Formosa, Japan, Micronesia, Samoa, Taiwan.
Balilutha 10:2646. Error for *Balclutha* Kirkaldy.
Balkanocerus Lodos 1982a:111-112. Error for *Balcanocerus* Maldonado-Capriles.
Ballana DeLong 1936a:217. *Thamnotettix vetula* Ball 1910c:302 (USA: California, Kelso). 10:820. DeLong 1964a. Deltocephalinae, Athysanini. Nearctic; xeric habitats in western USA, adjacent Canada and northern Mexico.
Synonyms:
Viriosana DeLong 1936a:217. *Thamnotettix viriosa* Ball 1910a:266.
Laterana DeLong 1936a:217. *Thamnotettix dissimilata* Ball 1910c:306.
Ballutha 10:2646. Error for *Balclutha* Kirkaldy.
Balocerus Freytag & Morrison 1972a:41. *B. rozeni* n.sp. (New Guinea: Papua, Sudest Isl., Mt. Rin). Idiocerinae. Oriental.
Balocha Distant 1908g:189. *B. tricolor* n.sp. (Burma: Tenasserim, Myitta). 16:223. Idiocerinae. Oriental; S. India, SE Asia.
Balocharella Webb, M.D. 1983a:60. *B. crosea* n.sp. (Australia: Queensland, Cairns). Idiocerinae. Australian.
Baluba Nielson 1979b:170. *B. retrorsa* n.sp. (Venezuela: El Caura, Suapure). Coelidiinae, Teruliini. Neotropical; Guyana & Venezuela.
Bambusana Anufriev 1969a:403. *Thamnotettix bambusae* Matsumura 1914a:176 (Japan: Honshu). Deltocephalinae, Athysanini. Palaearctic; Japan and Kurile Islands.
Bampurius Dlabola 1977b:252. *Neolimnus eberti* Dlabola 1964a:252 (Afghanistan: Sarobi, 1100 m). Deltocephalinae, Paralimnini. Palaearctic; southern Iran and Afghanistan.
Bandara Ball 1931c:93. *Eutettix johnsoni* Van Duzee 1894c:137 (USA: Pennsylvania, Philadelphia). 10:494. Deltocephalinae, Athysanini. Nearctic; eastern & central No. America, Mexico & regions bordering the Caribbean.
Subgenus:
Bandarana DeLong 1980h:64. *B.(B.) mimica* n.sp.
Bandarana DeLong 1980h:64 as subgenus of *Bandara* Ball 1931c:93. *B.(B.) mimica* n.sp. (Mexico: Orizaba, Vera Cruz). Deltocephalinae, Athysanini. Neotropical.
Bandaromimus Linnavuori 1959b:291. *B. fulvopictus* n.sp. (Peru: Madre de Dios). Deltocephalinae, Athysanini. Neotropical.
Bannalgaechungia Kwon 1983a:22. *B. alticola* n.sp. (Korea: Prov. G. W., Mt.Seolagsan). Cicadellinae, Errhomenini. Palaearctic; Korea.
Banosa Mahmood 1967a:39. *B. sexpunctata* n.sp. (Philippines: Luzon, Los Banos). Typhlocybinae, Empoascini. Oriental. Synonym of *Akotettix* Matsumura 1931b:77, a subgenus of *Dialectocopteryx* Kirkaldy 1907d:71; Dworakowska 1970a:691.
Banus Distant 1908g:353. *B. oblatus* n.sp. (Ceylon: Peradeniya). 10:540. Deltocephalinae, Athysanini. Oriental; Burma, Ceylon & India.

Baramapulana Distant 1910e:235. *B. princeps* n.sp. (South Africa: Johannesburg, 6000 ft). 1:285. Cicadellinae, Cicadellini. Ethiopian.
Synonym:
Tetelloides Evans, J.W. 1955a:19. *T. rubigella* n.sp.; Linnavuori 1979a:688.
Barampulana 1:667. Error for *Baramapulana* Distant.
Barbinolla Young 1977a:906. *Tettigonia costaricensis* Distant 1879b:63 (Costa Rica). Cicadellinae, Cicadellini. Neotropical; Costa Rica and Panama.
Bardana DeLong 1980h:65. *B. depressa* n.sp. (Mexico: D.F.). Deltocephalinae. Neotropical; Mexico.
Subgenus:
Guadlera DeLong 1980h:65. *E. (G.) discapa* n.sp.
Bardera Linnavuori and Al-Ne'amy 1983a:56. *Phlepsius fasciolatus* Melichar 1904a:39 (Somalia: Bardera). Selenocephalinae, Adamini. Ethiopian; East Africa.
Barela Young 1957c:264. *Protalebra decorata* Osborn 1928a:255 (Guatemala: Los Amates). Typhlocybinae, Alebrini. Neotropical; Mexico and Central America.
Barodecus Nielson 1979b:73. *B. panamensis* n.sp. (Panama: Portrerillos). Coelidiinae, Teruliini. Neotropical.
Baroma Oman 1938b:377. *B. reticulata* n.sp. (Argentina: Missiones, Loreto). 10:546. Deltocephalinae, Athysanini. Neotropical.
Bascarhinus 4:139. Error for *Bascarrhinus* Fowler.
Bascarrhinus Fowler 1898a:214. *B. platypoides* n.sp. (Panama: Volcan de Chirique, 4000-6000 ft). 4:65. Kramer 1966a:471-472. Ledrinae, Petalocephalini. Neotropical; Costa Rica, Ecuador & Panama.
Basilana Mahmood 1967a:17. *Typhlocyba erota* Distant 1908g:410 (Ceylon: Pattipola). Typhlocybinae, Erythroneurini. Oriental. Synonym of *Aidola* Melichar 1914b:142; Dworakowska 1972b:395.
Bassareus Linnavuori 1979b:999. *Pseudobalbillus* villiersi Linnavuori 1969a:1147-1148 (Congo: Odzala). Mukariinae. Ethiopian. Preoccupied by *Bassareus* Haldeman 1849, see *Neobassareus* Kocak 1981a:32.
Basuaneura Ramakrishnan & Menon 1971a:465. *B. kalimpongensis* n.sp. (India: West Bengal, Kalimpong). Typhlocybinae, Dikraneurini. Oriental. Synonym of *Erythria* Fieber 1866a:507; Dworakowska 1977a:603.
Basutoia Linnavuori 1961a:471. *B. brachyptera* n.sp. (Basutoland: Makheke Mtns, 10 miles ENE of Teyateyaneng). Deltocephalinae, Athysanini. Ethiopian.
Batarius Emeljanov 1966a:119 as subgenus of *Adarrus* Ribaut 1946b:83. *A.(B.) oshanini* n.sp. (USSR: "Transalay Ridge, Bakmir landmark"). Deltocephalinae, Paralimnini. Palaearctic.
Bathematophorus 8:261. Error for *Bathysmatophorus*.
BATHYSMATOPHORINA Anufriev 1978a:58, subtribe of ERRHOMENINI.
Bathysmatophorus J. Sahlberg 1871a:109. *B. reuteri* n.sp. (Lapland). 8:29. Cicadellinae, Errhomenini. Holarctic; NW North America, North Pacific arc (Japan & Korea), China, & Eurasia.
Subgenus:
Hylaius Oman & Musgrave 1975a:6. *Errhomenus oregonensis* Baker 1898a:262.
Batracomorphus Lewis 1834a:51. *B. irroratus* n.sp. (Great Britain). 15:118. Linnavuori & Quartau 1975a:30-144; Quartau 1981a, 1982b. Iassinae, Iassini. Warm regions of all major land masses except the New World.
Subgenus:
Sudanoiassus Linnavuori & Quartau 1975a:130. *B.(S.) kivuensis* n.sp.
Synonyms:
Eurinoscopus Kirkaldy 1906c:346. *E. lentiginosus* n.sp.
Ossana Distant 1914m:518. *O. bicolor* n.sp.
Acojassus Evans, J.W. 1972c:656. *A. montanus* n.sp.
Edijassus Evans, J.W. 1972c:656. *E. pallidus* n.sp.
Batrachomorphus 15:209. Error for *Batracomorphus* Lewis.
Batrachomorphus 15:209. Error for *Batracomorphus* Lewis.
Baya Dworakowska 1972b:401. *B. dymczata* n.sp. (New Guinea: Papua, Kokoda, 1200 ft). Typhlocybinae, Erythroneurini. Oriental.
Batysmatophorus 8:261. Error for *Bathysmatophorus* Sahlberg.

Beamerana Young 1952b:110. *Erythroneura tropicalis* Osborn 1928a:288 (Bolivia: Prov. del Sara). 17:607. Typhlocybinae, Jorumini. Neotropical; Bolivia, and Panama.

Beamerella 17:1485. Error for *Beamerana* Young.

Beamerulus Young 1957c:242. *B. beameri* n.sp. (Mexico: Morelas, Cuernavaca). Typhlocybinae, Alebrini. Neotropical; Mexico.

Begonalia Young 1977a:171. *Tettigoniella hydra* Distant 1908a:520 (Peru). Cicadellinae, Cicadellini. Neotropical; Bolivia, Brazil, Ecuador, Peru.

Beirneola Young 1977a:547. *Tettigonia anita* Fowler 1900d:281 (Panama: Bugaba). Cicadellinae, Cicadellini. Neotropical; Colombia, Costa Rica, Ecuador, Panama, Venezuela.

Belaunus Ribaut 1952a:268 as subgenus of *Adarrus* Ribaut 1946b:83. *Deltocephalus duodecimguttatus* Cerruti 1938a:191. (Switzerland). 10:1435. Deltocephalinae, Deltocephalini. Palaearctic.

Bella Singh-Pruthi 1930a:44. *B. apicalis* n.sp. (Southern India). 10:2348. Deltocephalinae, Stenometopiini. Oriental. Synonym of *Doratulina* Melichar 1903b:198; Vilbaste 1965a:10.

Benala Oman 1938b:390. *Deltocephalus tumidus* Osborn 1923c:42 (Bolivia: Santa Cruz). 10:1313. Linnavuori 1959b:19, Linnavuori & DeLong 1979b:135-136. Neobalinae. Neotropical; Bolivia.

Benglebra Mahmood & Ahmed 1969a:86. *B. alami* n.sp. (East Pakistan: Cox's Bazar). Deltocephalinae. Oriental; East Pakistan.
[Note: See Dworakowska 1971c:493 for discussion of subfamily placement, not Typhlocybinae].

Bengueta Mahmood 1967a:27. *B. bagua* n.sp. (Philippines: Luzon, Baguio). Typhlocybinae, Erythroneurini. Oriental; Philippines.

Benibahita Linnavuori 1959b:179. *B. furcillata* n.sp. (Bolivia: Rurrenbaque). Deltocephalinae, Athysanini. Neotropical; Bolivia.

Beniledra Linnavuori 1972b:205 & 213. *B. peculiaris* n.sp. (Zaire: Beni forest). Ledrinae, Ledrini. Ethiopian; Zaire.

Bergallia Oman 1938b:350. *Bythoscopus signatus* Stal 1859b:291 (Argentina: Buenos Aires). 14:155. Agalliinae, Agalliini. Neotropical; southern So. America.

Bergevina Evans, J.W. 1947a:189, replacement for *Macroceps* Melichar 1902b:75, not *Macroceps* Signoret 1879a:53. *Macroceps ahngeri* Melichar 1902b:75 ("Transcaspia"). 14:164. Adelungiinae, Achrini. Palaearctic; Turkey & USSR Turkmeniya.
Synonym:
Macroceps Melichar 1902b:74. *M. ahngeri* n.sp.

Bergiella Baker 1897c:157. *Parabolocratus uruguayensis* Berg 1884a:36 (Uruguay). 9:51. Deltocephalinae, Hecalini. Neotropical. Synonym of *Spangbergiella* Signoret 1879b:273.

Bergolix Linnavuori 1959b:288. *B. signatipennis* n.sp. (Brazil:Corumba). Deltocephalinae, Athysanini. Neotropical; Brazil & Argentina.

Bertawolia Blocker 1979a:63. *B. rhebala* n.sp. (Colombia: Valle). Iassinae. Neotropical; Colombia.

Betarmonia Melichar 1926a:342. 1:261. [Of questionable status; see Young 1977a:1096).]

Betawala Blocker 1979a:34. *B. galera* n.sp. (Bolivia: Cochabamba, Chipare Mts., 5000 ft). Iassinae. Neotropical; Bolivia.

Betsileonas Kirkaldy 1903f:13, replacement for *Thaumastus* Stal 1864b:67, not *Thaumastus* Martens 1860. *Ledra marmorata* Blanchard 1840a:194 (Madagascar). 4:109. Ledrinae, Petalocephalini. Malagasian.
Synonym:
Thaumastus Stal 1864b:67. *Ledra marmorata* Blanchard 1840a:194.

Bhandara Distant 1908g:221. *Tettigonia semiclara* Signoret 1853c:666 (Burma: Tenasserim, Myitto). 1:250. Cicadellinae, Cicadellini. Oriental; SE Asia.

Bharata Distant 1918b:12. *B. insignis* n.sp. (Upper Burma). 1:40. Cicadellinae, Cicadellini. Oriental; Burma, Thailand.

Bharinka Webb, M.D. 1983a:70. *B. storeyi* n.sp. (Australia: Queensland, Windsor Tableland via Mt. Carbine). Idiocerinae. Australian.

Bharoopra Webb, M.D. 1983a:83. *B. clavosignata* n.sp. (Australia: New South Wales, Big Hill). Idiocerinae. Australian.

Bhatia Distant 1908g:357. *Eutettix (?) olivaceus* Melichar 1903b:191 (Ceylon: Peradeniya). 10:540. Selenocephalinae, Bhatiini. Oriental; Ceylon, Java, Thailand.

BHATIINI Linnavuori & Al-Ne'amy 1983a:21, tribe of SELENOCEPHALINAE.

Bhooria Distant 1908g:256. *B. modulata* n.sp. (Burma: Ruby Mines). 1:457. Cicadellinae, Cicadellini. Oriental.

Bhythoscopus 15:209. Error for *Bythoscopus* Germar.

Biadorus Nielson 1979b:174. *Cicada nigripes* Fabricius 1794a:34 (Brazil). Coelidiinae, Teruliini. Ethiopian; west coast of Africa: Neotropical; Brazil, Rio Grande de Sul.

Bicenarus Kwon 1985a:66 as subgenus of *Idiocerus* Lewis 1834a:47. *Idiocerus ishiyamae* Matsumura 1905a:67 (Japan). Idiocerinae. Palaearctic; Japan & central South Korea.

Biluscelis Dlabola 1980d:9. *B. hardei* n.sp. (South Spain: Lorca, "an wege nach Velez Rubio"). Deltocephalinae, Athysanini. Palaearctic.

Bilusius Ribaut 1942a:260. *Anosterostemma noualhieri* Melichar 1898a:63 (Pyrenees). 10:221. Deltocephalinae, Athysanini. Palaearctic.

Bithoscopus 15:209. Error for *Bythoscopus* Germar.

Bituitus Distant 1918b:70. *B. projectus* n.sp. (South India: Nandidrug). 10:2349. Vilbaste 1965a:10. Deltocephalinae, Stenometopiini. Oriental. Synonym of *Doratulina* Melichar 1903b:198; Vilbaste 1965a:10.

Biza Walker 1858b:253. *B. crocea* n.sp. (Brazil: Para, Villa Nova). 11:95. Kramer 1962a. Neocoelidiinae. Neotropical; Brazil, Costa Rica.

Blarea Young 1957c:159. *B. brasiliensis* n.sp. (Brazil: Rezende, Estado de Rio). Typhlocybinae, Alebrini. Neotropical.

Bloemia Theron 1974b:159. *Euscelis hieroglyphica* Naude 1926a:63 (South Africa: Bloemfonteim). Deltocephalinae, Athysanini. Ethiopian.

Blythoscopus 15:209. Error for *Bythoscopus* Germar.

Bobacella Kusnezov 1929e:271. *B. teratocera* n.sp. (Southern Russia). 5:94. Deltocephalinae, Athysanini. Palaearctic and USSR: Mongolia, Siberia, Transbaikul, Ukraine.

BOBACELLINI 5:94.

Bohemanella DeLong & Freytag 1972f:242 as subgenus of *Polana*. *Gypona bohemani* Stal 1864:81 (Mexico: Vera Cruz). Gyponinae, Gyponini. Neotropical; Mexico to Brazil.

Bohemania Stal 1855a:97, replacement for *Euryprosopum* (sic) Stal 1853, not *Euryprosopus* White 1853. *B. sobrina* n.sp. (So. Africa: Natal). 5:88, 96. Ulopinae, Ulopini. Ethiopian. Synonym of *Coloborrhis* Germar 1836a:72.

Bohemannia 5:96. Error for *Bohemania* Stal.

Bolanus Distant 1918b:89. *B. baeticus* n.sp. (India: Assam). 10:2106. Deltocephalinae, Scaphoideini. Oriental. Synonym of *Scaphoideus* Uhler 1889a:33.

Bolanusoides Distant 1918b:90. *B. heros* n.sp. (NW India: Kumaon). 17:1044. Typhlocybinae, Typhlocybini. Oriental; northern India, Nepal.
Synonym:
Camulus Distant 1918b:97. *C. ornatus* n.sp.

Bolarga Oman 1938b:366. *Parabolocratus bolivianus* Osborn 1923c:32 (Bolivia: Santa Cruz). 10:1772. Linnavuori 1959b:90. Deltocephalinae, Deltocephalini. Neotropical; Bolivia, Brazil, Paraguay.

Bolidiana Nielson 1979b:311. *Cercopis aurata* Fabricius 1787a:274 (French Guiana). Coelidiinae, Teruliini. Neotropical; Central America to Colombia and Guyana.

Bolinlila Distant 1910e:234. *B. rhodesiana* n.sp. (Rhodesia). 1:298 & 667. Cicadellinae, Mileewanini. Ethiopian. Synonym of *Mileewa* Distant 1908g:238.

Bolivaia Linnavuori & DeLong 1979a:45. *B. lobata* n.sp. (Boliva: Santa Cruz, San Esteban). Deltocephalinae, Deltocephalini. Neotropical.

Boliviela DeLong 1969a:464. *B. inflata* n.sp. (Bolivia: Coroico). Coelidiinae, Coelidiini. Neotropical; Bolivia, Colombia, Ecuador, Peru.

Bolotheta Kramer 1963c:44. *Neocoelidia punctata* Osborn 1923c:77 (Bolivia: Prov. del Sara). Neocoelidiinae. Neotropical; Bolivia, Peru.

Bomolea ("Lv in litt") Heller & Linnavuori 1968a:6, invalid.

Bonamus Oman 1938b:365. *B. lineatus* n.sp. (Argentina: Loreto, Missiones). 9:60. Deltocephalinae, Hecalini. Neotropical; Argentina, Brazil, Peru (?).

Bonaspeia Linnavuori 1961a:468. *B. straminea* n.sp. (South Africa: Cape Prov., Cape Agulhas). Deltocephalinae. Ethiopian; South Africa.

Bonneyana Oman 1949a:127. *Thamnotettix schwarzi* Ball 1911a:197 (USA: Utah, Dewey). 10:1328. Deltocephalinae, Athysanini. Nearctic; Southwestern USA and adjacent Mexico.

Bordesia de Bergevin 1929a:7. *B. mitrata* n.sp. (Algeria). 9:23. Linnavuori 1975a:5 & 19. Deltocephalinae, Hecalini. Ethiopian; Eremian and Sahelian subregions. Palaearctic; Algeria, Egypt, Morocco, Mauritania.

Borditartessus Evans, F. 1981a:145. *Tartessus latus* Evans, J.W. 1941f:156 (West Australia: Dendari, 40 miles west of Coolgardie). Tartessinae. Australian.

Borduria Distant 1908f:147. *Bythoscopus impressus* Walker 1870b:322 (Moluccas: Batjan Island). 11:143. Tartessinae. Oriental.

Boreolettix Zoological Record 103, Section 13F:397, 1966. Error for *Boreotettix* Lindberg 1952a:145.

Boreotettix Lindberg 1952a:145. *Cosmotettix serricauda* Kontkanen 1949c:41 (Finland: Karelia borealis, Hammaslahi) = *Laevicephalus bidentata* DeLong & Davidson 1935b:169. 10:1596. Ossiannilsson 1983a:879-881. Deltocephalinae, Paralimnini. Palaearctic; Finland and Sweden. Nearctic; Colorado.

Borogonalia Young 1977a:1048. *Tettigonia impressifrons* Signoret 1854a:16 (Ecuador). Cicadellinae, Cicadellini. Neotropical; Brazil, Colombia, Ecuador, Peru, Venezuela.

Borulla Dworakowska, Sohi & Viraktamath 1980a:274. *B. gracilis* n.sp. (India: Uttar Pradesh, Dehra Dun). Typhlocybinae, Zyginellini. Oriental; India.

Bothrogonia Melichar 1926a:341. *Cicada ferruginea* Fabricius 1787a:269 ("Cape of Good Hope", error ?). 1:233. Cicadellinae, Cicadellini. Oriental, Palaearctic [and Ethiopian ?].
Subgenus:
Abothrogonia Yang in Yang & Li 1980a: 193,210. *Tettigonia impudica* Signoret 1853c:677.
Synonym:
Megalotettigella Ishihara 1953b:3, 16. "*Cicada ferruginea* Fabricius, 1794" = Fabricius 1787a:269.

Bothrognathus Bergroth 1920a:29, replacement for *Aliturus* Distant 1908g:63, not *Aliturus* Fairmaire 1902, (Laccadive Islands). Deltocephalinae, Opsiini. Palaearctic, Oriental. Synonym of *Neoaliturus* Distant 1918b:63.

BOTHROGONIINA Kwon 1983a:16, subtribe of CICADELLINI.

Boulardus Nielson 1983e:16. *B. concinnus* n.sp. (Central African Republic: Boukoko). Coelidiinae, Hikangiini. Ethiopian; Central African Republic, Ghana.

Bousaada Linnavuori 1971a:62. *B. psapfa* n.sp. (Algeria: near Bou Saada). Deltocephalinae, Hecalini. Palaearctic; Algeria & Spain. Synonym of *Epicephalius* Matsumura 1908a:42; Vilbaste 1976a:35.

Bousadaada Dlabola 1980d:4. Error for *Bousaada* Linnavuori 1971a:62.

Brachydella Ball & Beamer 1940a:64 as subgenus of *Athysanella* Baker 1898a:185. *A. (B.) abdominalis* n.sp. (USA: Arizona, Santa Rita Mtns.). 10:1636. Deltocephalinae, Doraturini. Nearctic.

Brachylope Emeljanov 1962a:162. *Deltocephalus brachypterus* Sidorskii 1938a:126 (Russia). Deltocephalinae. Palaearctic.

Brachylorus Maldonado-Capriles 1972b:630. *B. leucoclavus* n.sp. (British No. Borneo: Tenompok). Idiocerinae. Oriental.

Brachypterona Lindberg 1954a:219. *B. viridissima* n.sp. (Canary Islands). 10:89. Deltocephalinae, Athysanini. Ethiopian; Canary Islands, Salvage Islands.

Brasa Oman 1938b:352. *Macropsis rugicollis* Dozier 1927b:264 (Santa Domingo: Cerro del Cacique). 14:155. Agalliinae, Agalliini. Neotropical; Hispaniola.

Brasilanus Linnavuori 1959b:287. *B. flagellaris* n.sp. (Brazil: Nova Tentoma). Deltocephalinae, Athysanini. Neotropical; Brazil, Peru.
Subgenus:
Mascoitanus Linnavuori & Heller 1961a:12. *B. (M.) lateralis* n.sp.

Brasopsis Linnavuori 1954e:129. *Bythoscopus gilvipes* Stal 1862e:54 (Brazil: Rio de Janeiro). 14:150. Agalliinae, Agalliini. Neotropical.

Brasura Nielson 1982e:27. *B. variabilis* n.sp. (Cameroon: Victoria). Coelidiinae, Coelidiini. Ethiopian; Cameroon, Uganda, Zaire.

Brazoza Oman 1938b:386. *Thamnotettix picturellus* Baker 1923b:532 (Brazil: Minas Gerias). 10:810. Deltocephalinae, Athysanini. Neotropical; Bolivia, Brazil, Peru.

Brazoza 10:2647. Error for *Brazosa* Oman.

Brenda Oman 1941a:205. *Paropulopa arborea* Ball 1909b:184 (USA: California, Colfax). 5:32. Megophthalminae. Nearctic; Central California.

Brevolidia Nielson 1982e:15. *Palicus constricta* Jacobi 1910b:130 (Tanzania: Kilimanjaro). Coelidiinae, Coelidiini. Ethiopian; Central Africa.

Brincadorus Oman 1938b:361. *B. laticeps* n.sp. (Brazil: Chapada). 15:183. Linnavuori 1959b:158. Deltocephalinae, Athysanini. Neotropical; Bolivia and Brazil.

Britimnathista Dworakowska 1969f:487. *Eupteryx sanguinolenta* Distant 1918b:96 (India: Kodiacanal). Typhlocybinae, Dikraneurini. Oriental.

Brunerella Young 1952b:31. *B. magnifica* n.sp. (Mexico: Jalapa). 17:81. Typhlocybinae, Alebrini. Neotropical; Cuba and Mexico.

Brunotartessus Evans, F. 1981a:130. *Bythoscopus fulvus* Walker 1851b:866 (Australia: Van Dieman's Land). Tartessinae. Australian and Oriental; New Guinea.

Brythoscopus 15:209. Error for *Bythoscopus* Germar.

Bubacua Young 1977a:815. *Hortensia filicis* Metcalf & Bruner 1936a:928 (Cuba: Oriente Prov., Sierra Maestra Mts., Palma Mocha Peak). Cicadellinae, Cicadellini. Neotropical.

Bubulcus Dlabola 1961b:320 as subgenus of *Paralimnus* Matsumura 1902a:386. *Paralimnus cingulatus* Dlabola 1960b:2 (USSR: Tadzhikistan: Tigrovaja Balka, south of Stalinabad). Deltocephalinae, Deltocephalini. Palaearctic.

Bucephalogonia Melichar 1926a:341. *Tettigonia xanthophis* Berg 1879d:249 (Argentina). 1:199. Young 1977a:1092. Cicadellinae, Cicadellini. Neotropical; Argentina, Brazil.

Bufonaria Emeljanov 1963a:1581. *B. oshanini* n.sp. (USSR: Tyan Shan). Ulopinae, Ulopini. Palaearctic. Preoccupied by *Bufonaria* Schumacher 1817, see *Neobufonaria* Kocak 1981a:32.

Bugraia Kocak 1981a:32, replacement for *Taeniocerus* Dlabola 1974b:64, not *Taeniocerus* Kaup 1871 (Coleoptera). *Bythoscopus ocularis* Mulsant-Rey 1855a:220 (France: Provence). Idiocerinae. Palaearctic; Mediterranean region.

Buhria Dworakowska 1976b:26 as subgenus of *Empoasca* Walsh 1862a:149. *E. (B.) angusta* n.sp. (Cameroon: Moliwe). Typhlocybinae, Empoascini. Ethiopian.

Bulbana DeLong 1942d:107. *B. pura* n.sp. (USA: Texas, Uvalde). 3:152. Gyponinae, Gyponini. Nearctic. Subgenus of *Ponana* Ball 1920a:93.

Bulbusana DeLong & Freytag 1972f:256 as subgenus of *Polana* DeLong 1942d:110. *P. (B.) plumea* n.sp. (Venezuela: San Esteban). Gyponinae, Gyponini. Neotropical.

Buloria Distant 1908g:271. *B. gyponinoides* n.sp. (India: Calcutta). 7:31. Nirvaninae, Nirvanini. Oriental.

Bulotartessus Evans, F. 1981a:134. *B. cooloolensis* n.sp. (Australia: Queensland, Cooloola). Tartessinae. Australian.

Bumizana Distant 1918b:32. *B. elongata* n.sp. (So. India: Kodiakanal). 9:116. Morrison 1973a:390. Eupelicinae, Paradorydiini. Oriental; southern India, northern Thailand. Synonym of *Paradorydium* Kirkaldy 1901f:339; Theron 1982b:19.

Bundabrilla Webb, M.D. 1983a:84. *B. clovella* n.sp. (Australia: Queensland, Lamington Nat'l. Park). Idiocerinae. Australian.

Bundera Distant 1908g:228. *B. venata* n.sp. (Burma: Tenasserim, Myitta). 6:44. Cicadellinae, Evacanthini. Oriental.

Bunyipia Dworakowska 1972g:860 as subgenus of *Zyginoides* Matsumura 1931b:95. *Z. (B.) ulae* n.sp. (New Guinea: Papua, Kokoda). Dworakowska 1981c. Typhlocybinae, Erythroneurini. Oriental, Australian.

Burakia Kocak 1981a:32, replacement for *Shirazia* Dlabola 1977b:248 not *Shirazia* Amsel 1954. *Shirazia eminens* Dlabola 1977b:249 (Southern Iran: 30 km easterly from Kazerun, 1300 m.). Deltocephalinae, Platymetopiini. Palaearctic.
Synonym:
Shirazia Dlabola 1977b:248. *S. eminens* n.sp.

Burara Dworakowska 1980b:184 as subgenus of *Ratburella* Ramakrishnan & Menon 1973a:20. *R.(B.) maxima* n.sp. (India: Meghalaya, Mawsmai Cave). Typhlocybinae, Erythroneurini. Oriental.

Buritia Young 1952b:67. *Dikraneura lepida* McAtee 1926b:121 (Brazil: Chapada). 17:284. Typhlocybinae, Dikraneurini. Neotropical; Brazil.

Busonia Distant 1908g:198. *B. amentata* n.sp. (India: Assam, Margherita). 16:222. Maldonado-Capriles 1977d:494-500. Idiocerinae. Oriental.

Busoniomimus Maldonado-Capriles 1977d:491. *Idiocerus minor* Bierman 1908a:165 (Indonesia: Java, Semdang). Idiocerinae. Oriental; SE Asia, So. India. Northern Australia.

Byphlocyta Ahmed, Manzoor 1971c:198. *B. nigropunctata* n.sp. (West Pakistan: Swat). Typhlocybinae, Typhlocybini. Oriental.

Bythonia Oman 1938b:358. *Nionia (?) rugosa* Osborn 1923c:32 (Bolivia: Santa Cruz de la Sierra). 15:116. Linnavuori 1959b:13-15. Bythoniinae. Neotropical; Bolivia, Brazil, Peru.

BYTHONIINAE Linnavuori 1959b:12.

Bythoscapus 15:209. Error for *Bythoscopus* Germar.

BYTHOSCOPIDAE 15:5.

BYTHOSCOPINAE 15:19.

BYTHOSCOPINI 15:20.

Bythoscopus Germar 1833a:180. *Cicada lanio* Linnaeus 1761a:242 (Sweden). 15:21,210. Iassinae, Iassini. Palaearctic. Isogenotypic synonym of *Iassus* Fabricius 1803a:85.

Bythosopis 15:221. Error for *Bythoscopus* Germar.
Bythosopus 15:221. Error for *Bythoscopus* Germar.
Bythroscopus 15:221. Error for *Bythoscopus* Germar.
Bytoscopus 15:221. Error for *Bythoscopus* Germar.
Byttioscopus 15:221. Error for *Bythoscopus* Germar.

Bza Dworakowska 1979b:39 as subgenus of *Empoascanara* Distant 1918b:94. *E. (B.) risa* n.sp. (Vietnam: Nghe-An). Typhlocybinae, Erythroneurini. Oriental.

Cabimanus Linnavuori 1959b:204 as subgenus of *Tropicanus* DeLong 1944d:87. "*Tropicanus singularis* DeLong" = *Phlepsius singularis* DeLong 1944d:92 (Mexico: Chiapao, Vergel). Deltocephalinae, Athysanini. Neotropical; Mexico and Panama.

Cabrellus Emeljanov 1964f:50 as subgenus of *Sorhoanus* Ribaut 1946b:85. *S. (C.) minutus* n.sp. (USSR: Kazakhstan). Deltocephalinae, Deltocephalini. Palaearctic.
[Note: Raised to generic status: Emeljanov 1979a:331].

Cabrulus Oman 1949a:172. *Hebecephalus tener* Beamer & Tuthill 1934a:16 (USA: Colorado). 10:1823. Deltocephalinae, Deltocephalini. Nearctic; western USA and western Canada. Subgenus of *Orocastus* Oman 1949a:161; Ross & Hamilton 1972a:443.

Caelidia 11:146. Error for *Coelidia* Germar.
Caelidoides 11:146. Error for *Caelidioides* Signoret.

Caelidioides Signoret 1880b:205. *C. carinatrum* n.sp. = *Scaris tristis* Signoret 1860a:205 (Madagascar). 11:81. Unassigned, formerly Coelidiinae; Nielson 1975a:11. Malagasian.

Caffretus Evans, J.W. 1947a:258. *C. turneri* n.sp. (SW Africa). 10:2209. Deltocephalinae, Scaphoideini. Ethiopian; Sudan, SW Africa and Cape Verde Islands.

Caffrolix Linnavuori 1961a:466. *Jassus (Athysanus) patruelis* Stal 1859b:295 (South Africa: Cape Province). Deltocephalinae, Athysanini. Ethiopian; southern South Africa.

Cafixia Webb, M.D. 1983b:249. *Idiocerus hewitti* Cogan 1916a:180 (South Africa: Cape of Good Hope). Idiocerinae. Ethiopian.

Cahya Linnavuori 1959b:276. *Thamnotettix pulchellus* Osborn 1923a:70 = *Thamnotettix chapadensis* Baker 1923b:532 (Brazil: Minas Gerias) not *Thamnotettix pulchellus* Melichar 1907a:1034. Deltocephalinae, Athysanini. Neotropical; Mexico, Panama, Guatemala, Brazil and Argentina.

Caladonus Oman 1949a:126. *Thamnotettix coquilletti* Van Duzee 1890f:77 (USA: California). 10:1327. Deltocephalinae, Athysanini. Nearctic; western USA.

Calamotettix Emeljanov 1959a:833. *C. viridescens* n.sp. (USSR: Kazakhstan). Deltocephalinae, Deltocephalini. Palaearctic; southern Europe and southwest USSR to Mongolia and Soviet Maritime Territory.

Calana DeLong 1936a:219. *Thamnotettix umbratica* Ball 1910c:309 (USA: southern California). 10:876, 2647. Deltocephalinae, Athysanini. Nearctic. Preoccupied by *Calana* Gray 1847, see *Calanana* DeLong 1945b:272.

Calanana DeLong 1945b:272, replacement for *Calana* DeLong 1936a:218 not *Calana* Gray 1847. *Thamnotettix umbratica* Ball 1910c:309 (USA: southern California). 10:876. Deltocephalinae, Athysanini. Nearctic; western USA and Canada.

Caldwelliola Young 1977a:1039. *Tettigonia reservata* Fowler 1900d:267 (Mexico: Vera Cruz, Atoyac). Cicadellinae, Cicadellini. Neotropical; Mexico, Central America, Panama and northern South America to southwest Brazil.

Calemia Amyot 15:222. Invalid, not binominal.

Calidanus Emeljanov 1966a:122 as subgenus of *Diplocolenus* Ribaut 1946b:82 (USSR: Kazakhstan). *Diplocolenus truncatus* Emeljanov 1962a:180. Deltocephalinae, Paralimnini. Palaearctic. Subgenus of *Anareia* Vilbaste 1965b:74; Knight 1974c:369.

Calidia 11:146. Error for *Coelidia* Germar.

Calistrophia Dlabola 1967c:149, 1970b:18. Error for *Callistrophia* Emeljanov.

Calladonus 10:2647. Error for *Colladonus* Ball.

Calliscarta Stal 1869a:82. *Cicada decora* Fabricius 1803a:69 (South America). 10:946. Linnavuori 1959b:27; Kramer 1963b. Neobalinae, Neotropical; Bolivia, British Guiana, Brazil, Colombia, Ecuador, Peru, Venezuela.
Synonym:
Idiotettix Osborn 1929a:465. *Thamnotettix magnificus* Osborn 1924c:424.

Callistrophia Emeljanov 1962a:165. *Thamnotettix elegans* Melichar 1900a:36 (Northern Mongolia). Deltocephalinae, Athysanini. Palaearctic; Korean peninsula, Mongolia and USSR: Siberia, Altai Mts., Tuva, Soviet Maritime Territory.

Callystrophia Mitjaev 1968b:42. Error for *Callistrophia* Emeljanov.

Calodia Nielson 1982e:140. *C. multipectinata* n.sp. (Mt. Dwlit, Sarawak, Malaysia). Coelidiinae, Coelidiini. Oriental region generally.

Calodicia Nielson 1982e:266. *Jassus maculipennis* Spangberg 1878b:29 (Colombia: Bogota). Coelidiinae, Coelidiini. Neotropical; Colombia and Panama.

Caloduferna Webb, M.D. 1980a:855. *C. minuta* n.sp. (Aldabra, South Island). Deltocephalinae, Paralimnini. Malagasian.

Calonia Beamer 1940a:56. *Uhleriella signata* Ball 1902b:56 (USA: California, Los Angeles Co.). 10:1087. Deltocephalinae, Cochlorhini. Nearctic; southern California.

Calotartessus Evans, F. 1981a:187. *Tartessus stalii* Signoret 1880c:352 (New Caledonia). Tartessinae. Australian.

Calotettix Osborn 1934b:247. *C. metrosideri* n.sp. (Marquesas Islands: Hivaoa, Kopaafaa). 10:2339, 2647. Paraboloponinae. Marquesas and Cook Islands. Synonym of *Dryadomorpha* Kirkaldy 1906a:335; Webb 1981b:49.

Camaija Young 1977a:959. *C. veneris* n.sp. (British West Indies). Cicadellinae, Cicadellini. Neotropical; Jamaica.

Campbellinella Distant 1918b:69. *C. illustrata* n.sp. (India: Chikkaballapura). 10:2347. Deltocephalinae, Stenometopiini. Oriental, Palaearctic. Synonym of *Doratulina* Melichar 903b:198; Vilbaste 1965a:10; Evans 1966a:233.

Campecha Melichar 1925a:363. *Tettigonia semirasa* Fowler 1899e:249 (Mexico: Guerrero, Chilpancingo). 1:541. Young 1977a:840. Cicadellinae, Cicadellini. Neotropical; south-central Mexico.

Camptelasmus Spinola 1850b:110. *C. caffer* n.sp. ("Caffraria"). 4:104. Linnavuori 1972b:246. Ledrinae, Petalocephalini. Ethiopian; southern South Africa.

Camulus Distant 1918b:97. *C. ornatus* n.sp. ("E.Himilayas": Kurseong, Kumaon). 17:1044. Typhlocybinae, Typhlocybini. Oriental. Synonym of *Bolanusoides* Distant 1918b:90.

Canariotettix Lindberg 1954a:214. *C. brachypterus* n.sp. (Canary Islands). 10:1436. Deltocephalinae, Paralimnini. Palaearctic; Canary Islands.

Candulifera Webb, M.D. 1983a:73. *C. rapa* n.sp. (Australia: Northern Territory, 1.5 mi SE of Stuart Point). Idiocerinae. Australian.

Cantura Oman 1949a:121. *Scaphoideus jucundus* Uhler 1889a:34 (North America). 10:2156. Deltocephalinae, Scaphoideini. Nearctic; eastern USA and Canada.

Capeolix Linnavuori 1961a:472. *C. picturatus* n.sp. (South Africa: Hout Bay). Deltocephalinae, Athysanini. Ethiopian; South Africa.

Caphodellus Linnavuori & DeLong 1976a:33. *C. quadrimaculatus* n.sp. (Bolivia: "San Esteban, 49 km N.Sta. Cruz). Deltocephalinae, Deltocephalini. Neotropical.

Caphodus Oman 1938b:375. *C. maculatus* n.sp. (Argentina: Missiones, Loreto). 10:2175. Linnavuori 1959b:225-227. Deltocephalinae. Neotropical; Argentina, Brazil, Paraguay, Panama.

Capinota Melichar 1926a:319. *C. fowleri* n.sp. (Mexico). 1:642. Cicadellinae, Proconiini. Neotropical. Synonym of *Phera* Stal 1864a:77; Young 1968a:184.

Caplopa Evans, J.W. 1947a:252. *C. turneri* n.sp. (South Africa). 5:77. Linnavuori 1972a:136-137. Ulopinae, Ulopini. Ethiopian; Caffrarian.

Capoideus Theron 1974a:162. *Scaphoideus cuprescens* Naude 1926a:43 (South Africa: Stellenbosch). Deltocephalinae. Ethiopian; South Africa.

Caragonalia Young 1977a:244. *Tettigonia carminata* Signoret 1855d:834 (Brazil). Cicadellinae, Cicadellini. Neotropical; SE Brazil.

Caranavia Linnavuori 1959b:278. *C. cruentata* n.sp. (Bolivia: Caranavi). Deltocephalinae, Athysanini. Neotropical; Bolivia and Peru.

Carapona DeLong & Freytag 1975c:565 as subgenus of *Hecalapona* DeLong & Freytag 1975c:547. *H. (C.) vulta* n.sp. (Paraguay: Concepcion). Gyponinae, Gyponini. Neotropical.

Carchariacephalus Montrouzier 1861a:71. *C. forestieri* n.sp. (Loyalty Islands: Lifu). 7:17. Evans, J.W. 1974a:168. Nirvaninae, Nirvanini. Loyalty Islands, New Caledonia.

Cardioscarta Melichar 1932a:285. *Cicada quadrifasciata* Linnaeus 1758a:436 (America). 1:59, 688. Young 1977a:201. Cicadellinae, Cicadellini. Neotropical; Bolivia, Brazil, British Guiana, Colombia, French Guiana, Peru.

Carelmapu Linnavuori 1959b:220. *C. scutellaris* n.sp. (Chile: Prov. Llanquihue). Deltocephalinae, Athysanini. Neotropical; Chile.

Cariancha Oman 1938b:360. *C. cariboba* n.sp. (Brazil: Rio de Janeiro). 10:1916. Linnavuori 1959b:234-235. Deltocephalinae, Athysanini. Neotropical; Brazil.

Caribovia Young 1977a:909. *Tettigonia constans* Walker 1858b:198 (Santa Domingo). Cicadellinae, Cicadellini. Neotropical; Hispaniola, Cuba, Jamaica, Puerto Rico.

Carinifer Linnavuori 1952b:185. *C. maculiceps* n.sp. (Turkestan). 10:1484. Deltocephalinae, Doraturini. Palaearctic. Preoccupied by *Carinifer* Hamm 1881, see *Linnavuorina* Kocak 1981a:32. Synonym of *Aconura* Lethierry 1876a:9; Vilbaste 1965a:9.

Carinolidia Nielson 1979b:15. *Iassus nervosus* Fabricius 1803a:85 ("Amer. Merid."). Coelidiinae, Teruliini. Neotropical; Peru.

Caripuna Melichar 1926a:318. *Tettigonia guerini* Signoret 1855a:51 (French Guiana: Cayenne). 1:641. Cicadellinae, Proconiini. Neotropical. Synonym of *Mareba* Distant 1908b:77; Young 1968a:60.

Carneocephala Ball 1927c:39. *Draeculacephala floridana* Ball 1901b:72 (USA: Florida, Charlotte Harbor). 1:344. Young 1977a:583; Hamilton 1985a:86. Cicadellinae, Cicadellini. Nearctic, Neotropical. Subgenus of *Draeculacephala* Ball 1901b:66; Hamilton, KGA 1985a:86.

Carneophala 1:670. Error for *Carneocephala* Ball.

Carnoseta DeLong 1981a:510 as subgenus of *Gypona*. *G.(C.) colomella* n. sp. (Colombia: "Tunga, Colombia Bay, Atena"). Gyponinae, Gyponini. Neotropical.

Carphosoma Royer 1907a:29, replacement for *Dorydium* Burmeister 1839a:3, not *Dorydium* Burmeister 1835. *Dorydium lanceolatum* Burmeister 1839a:4 (Sicily). 9:106. Eupelicinae, Paradorydiini. Palaearctic. Synonym of *Paradorydium* Kirkaldy 1901f:339.

Carsonus Oman 1938c:173 as subgenus of *Errhomus* Oman 1938c:169. *Acocephalus maculatus* Gillette & Baker 1895a:83 (USA: SW Colorado). 8:26. Oman & Musgrave 1975a:6. Cicadellinae, Errhomenini. Nearctic; USA south of Canada and west of continental divide.

Caruya Linnavuori & DeLong 1978d:230. *Caruya brevicauda* n.sp. (Peru: "Yurac", east of Tingo Maria). Deltocephalinae, Athysanini. Neotropical; Peru.

Carvaka Distant 1918b:40. *C. picturata* n.sp. (India: Nilgiri Hills, Ootacamund). 15:180. Selenocephalinae. Oriental. [Australia].

Caryphaeus 10:2647. Error for *Coryphaeus* Fieber.

Cassassus Amyot 8:261. Invalid, not binominal.

Cassianeura Ramakrishnan & Menon 1973a:27. *C. sexmaculata* n.sp. (India: Delhi). Typhlocybinae, Erythroneurini. Oriental; India & Malaysia.

Castoriella Dworakowska 1974a:164. *C. polleti* n.sp. (Congo: Odzala). Typhlocybinae, Zyginellini. Ethiopian; Congo.

Catagonalia Evans 1947a:160, replacement for *Catagonia* Melichar 1926a:342, not *Catagonia* Brauer & Bergenstamm 1891. *Tettigonia lunata* Signoret 1854b:349 (Mexico). 1:283. Young 1977a:1068. Cicadellinae, Cicadellini. Neotropical.

[Note: Metcalf 1965a:283 shows Mexico as the original source (i.e. type locality) of *lunata* Signoret, with no subsequent records from other countries. Young 1977a:1068, 1069, 1071 lists only South American sources; Ecuador, Peru, Bolivia, Brazil].

Catagonia Melichar 1926a:342. *Tettigonia lunata* Signoret 1854b:349 (Mexico). 1:283, 670. Cicadellinae, Cicadellini. Neotropical. Preoccupied by *Catagonia* Brauer & Bergenstamm 1891, see *Catagonalia* Evans 1947a:160.

Catorthorrhinus Fowler 1898a:213. *C. resimus* n.sp. (Panama: "V. de Chiriqui"). 1:538. Young 1968a:140. Cicadellinae, Proconiini. Neotropical; Costa Rica, Panama.

Caxia Melichar 1924a:201. *C. projecta* Melichar 1925a:359 (New Guinea) 1:539. Ledrinae. Oriental.

Cazenus Oman 1949a:164. *Hebecephalus neomexicanus* Tuthill 1930a:44 (USA: New Mexico, Belen). 10:1650. Deltocephalinae, Deltocephalini. Nearctic; northern latitudes and/or high elevations, primarily in western North America; also western Palaearctic.

Cechenotettix Ribaut 1942a:262. *Thamnotettix martini* Lethierry 1883b:43 (France) = *Athysanus quadrinotatus* Mulsant & Rey 1855a:232. 10:387. Deltocephalinae, Fieberiellini. Palaearctic; Mediterranean basin and North Africa.

Cedarotettix Theron 1975a:197. *Deltocephalus cogani* var. *incisus* Naude 1926a:48 (South Africa: Natal). Deltocephalinae, Deltocephalini. Ethiopian; South Africa.

Celopsis Hamilton 1980b:896 as subgenus of *Pediopsoides* Matsumura 1912b:305. *Macropsis dapitana* Merino 1936a:324 (Philippines: Mindanao). Macropsinae, Macropsini. Oriental.

Celsanus Linnavuori 1954e:137. *Jassus (Thamnotettix) serius* Stal 1862a:52 (Brazil: Rio de Janeiro). 10:710. Linnavuori 1959b:250. Deltocephalinae, Athysanini. Neotropical. Synonym of *Chlorotettix* Van Duzee 1892e:306; Linnavuori 1959b:250.

Cenedaeus Distant 1908g:296. *C. horvathi* n.sp. (India: Bombay). 10:1694. Deltocephalinae, Athysanini. Oriental.

Centrometopea 1:670. Error for *Centrometopia* Melichar.

Centrometopia Melichar 1925a:399 as subgenus of *Oncometopia* Stal 1869a:62. *Tettigonia personata* Signoret 1854b:364 (Brazil: Sao Paulo). 1:591. Cicadellinae, Proconiini. Neotropical. Preoccupied by *Centrometopia* Ragonot 1887, see *Molomea* China 1927d:283.

Centrometopides Strand 1928a:73, replacement for *Centrometopia* Melichar 1925a:399, not *Centrometopia* Ragonot 1887. *Tettigonia personata* Signoret 1854b:364 (Brazil: Sao Paulo). 1:591, 671. Cicadellinae, Proconiini. Neotropical. Synonym of *Molomea* China 1927d:283.

Centrometopoides 1:671. Error for *Centrometopides* Strand.

Cepaleleus 5:96. Error for *Cephalelus* Percheron.

Cephaelus 5:96. Error for *Cephalelus* Percheron.

CEPHALELINAE 5:38.

CEPHALELINI

Cephalelus Percheron 1832b:pl.48. *C. infumatus* n.sp. (Unknown). 5:39. Linnavuori 1972a:141-143; Evans, J.W. 1977a:90, 99. Ulopinae, Cephalelini. Ethiopian; South Africa. Australian; Australia and New Zealand.
Synonym:
Dorydium Burmeister 1835a:106. *D. paradoxum* n.sp.
[Note: Evans, J.W. 1966a:89 lists *Notocephalius* Jacobi 1909a, *Procephaleus* Evans, J.W. 1936a, and *Anacephalus* Evans, J.W. 1936a as synonyms of *Cephalelus* Percheron, but from comments 1977a:99 apparently intends that they apply to a separate generic segregate for which the oldest name is *Notocephalius* Jacobi 1909].

Cephalius Fieber 1875a:402. *Cephalius frontalis* Signoret 1879a:61 (Algeria). 9:115. Deltocephalinae, Hecalini. Palaearctic; Algeria, Mauritania, Spanish Morocco.

Cephalogonalia Evans, J.W. 1947a:160, replacement for *Cephalogonia* Melichar 1926a:342 not *Cephalogonia* Wollaston 1862. *Tettigoniella flabellula* Jacobi 1905c:175 (Peru). 1:265. Young 1977a:80. Cicadellinae, Cicadellini. Neotropical; Bolivia and Peru.

Cephalogonia Melichar 1926a:342. *Tettigoniella flabellula* Jacobi 1905c:175 (Peru). 1:265. Preoccupied by *Cephalogonia* Wollaston 1862, see *Cephalogonalia* Evans 1947a:160. Cicadellinae, Cicadellini. Neotropical.

Ceratagallia Kirkaldy 1907i:61. *Agallia bigloviae* Baker 1896c:26 (USA: New Mexico, Albuquerque). 14:136. Kramer 1964a:143, 152. Agalliinae, Agallini. Nearctic; USA; southwestern states and Baja California.

Ceratogonia Melichar 1926a:350. *Tettigonia recta* Fowler 1900a:264 (Mexico: Guerrero, San Marcos Island). 1:29. Preoccupied by *Ceratogonia* Kolbe 1899, see *Ceratogoniella* Metcalf 1952a:228. Cicadellinae, Cicadellini. Neotropical. Synonym of *Sibovia* China 1927d:283; Young 1977a:696.

Ceratogoniella Metcalf 1952a:228, replacement for *Ceratogonia* Melichar 1926a:350 not *Ceratogonia* Kolbe 1899. *Tettigonia recta* Fowler 1900a:264 (Mexico: Guerrero, San Marcos Island). 1:29. Cicadellinae, Cicadellini. Neotropical. Synonym of *Sibovia* China 1927d:283; Young 1977a:696.

Cerneura Ghauri 1978a:203. *C. delonixia* n.sp. (Mauritius: Soreze). Typhlocybinae, Erythroneurini. Malagasian.

CERILLINI (sic) Linnavuori 1975b:49 = CERRILLINI.

CERRILLINI (sic) Linnavuori 1975b:49, tribe of DELTOCEPHALINAE.

Cerrillus Oman 1938b:362. *Hecalus notatus* Osborn 1923c:27 (Brazil: Para). 9:64. Linnavuori 1975b:49. Deltocephalinae, Cerrillini. Neotropical; Bolivia and Brazil.

Certagallia 14:172. Error for *Ceratagallia* Kirkaldy.

Cerus Theron 1975a:193. *Deltocephalus granarius* Naude 1926a:44 (South Africa, Cape Prov., Ceres). Theron 1984b:217-225. Deltocephalinae. Ethiopian.

Cestius Distant 1908g:309. *C. versicolor* n.sp. (India: Bengal). 15:183. Knight 1973c:153; Viraktamath 1978a. Selenocephalinae. Oriental; India.

Cestocephala Amyot 17:1485. Invalid, not binominal.

Cetexa Oman 1949a:129. *Thamnotettix graecula* Ball 1901a:6 (USA: Colorado). 10:1331. Deltocephalinae, Athysanini. Nearctic; southwestern USA and adjacent Mexico.

Chacotettix Linnavuori 1955a:106. *C. appendiculatus* n.sp. (Paraguay) = *Chlorotettix nigromaculatus* DeLong 1923b:265. 10:1077. Linnavuori 1959b:251, 270-272. Deltocephalinae, Athysanini. Neotropical. Subgenus of *Chlorotettix* Van Duzee 1892e:58; Linnavuori 1959b:251.

Changwhania Kwon 1980a:96. *Aconura terauchii* Matsumura 1915a:163 (Korea). Deltocephalinae, Deltocephalini. Palaearctic; central and southern Korea.

Chaparea Linnavuori 1959b:240. *C. albigena* n.sp. (Bolivia: Yungas, "Chapare".) Deltocephalinae, Athysanini. Neotropical.

Chatura Distant 1908g:176. *C. nigella* n.sp. (Ceylon: Maskeliya). 4:82. Ledrinae, Petalocephalini. Oriental; Ceylon.

Chelidinus Emeljanov 1962a:183. *C. cinerascens* n.sp. (Armenia). Deltocephalinae, Paralimnini. Palaearctic; Mediterranean region, Mongolia and Soviet Maritime Territory.

Chelidnus Vilbaste 1968a:136. Error for *Chelidinus* Emeljanov.

Chelusa Signoret 1879a:51. *Acocephalus madagascariensis* Signoret 1880a:51 (Madagascar). 9:60. Deltocephalinae, Hecalini. Malagasian. Synonym of *Glossocratus* Fieber 1866a:502; Linnavuori 1975a:9.

Chiapasa Schmidt 1928a:40. *Tettigonia rugicollis* Signoret 1855c:525 (Mexico). 1:494. Cicadellinae, Proconiini. Neotropical. Synonym of *Diestostemma* Amyot & Serville 1843a:572; Young 1968a:30.

Chiapasa 1:671. Error for *Chiapasa* Schmidt.

Chiasmes 8:261. Error for *Chiasmus* Mulsant & Rey.

Chiasmus Mulsant & Rey 1855a:215. *C. translucidus* n.sp. (France). 8:208. Evans, J.W. 1974a:172-173. Deltocephalinae, Chiasmusini. Australian, Oriental, southern Palaearctic, southern Ethiopian.
Synonyms:
Atractotypus Fieber 1866a:501. *A. bifasciatus* n.sp.
Kosmiopelix Kirkaldy 1906c:334. *K. varicolor* n.sp.
Kartwa Distant 1908g:394. *K. mustelina* n.sp.
Postumus Distant 1918b:84. *P. fascialis* n.sp.

CHIASMUSINI 8:13.

Chibala Linnavuori & DeLong 1977c:181. *C. modesta* n.sp. (Chile: Bangos). Neobalinae. Neotropical; Chile.

Chichahua Young 1977a:247. *C. stygiana* n.sp. (Ecuador: Mt.Tungurahua). Cicadellinae, Cicadellini. Neotropical; Colombia and Ecuador.

Chickkaballapura 17:1485. Error for *Chikkaballapura* Distant.

Chigallia Linnavuori & DeLong 1977c:167. *C. reticulata* n.sp.(Chile: 51 km E. Osorno). Agalliinae, Agalliini. Neotropical; Chile.

Chikkaballapura Distant 1918b:107. *C. maculosa* n.sp. (India: Chikkaballapura). 17:609. Typhlocybinae, Empoascini. Oriental. Synonym of *Apheliona* Kirkaldy 1907d:67; Dworakowska 1977d:611.

Chileanoscopus Freytag & Morrison 1969a:285. *C. hamulus* n.sp. (Chile: Rio Blanco). Idiocerinae. Neotropical; Chile.

Chilelana DeLong 1969a:462. *C. artigasi* n.sp. (Chile: Concepcion). Nielson 1975b:14. Coelidiinae, Tinobregmini. Neotropical; Bolivia and Chile.

Chilella DeLong & Freytag 1967b:112. *C. rugella* n.sp. (Chile). Gyponinae, Gyponini. Neotropical; Chile.

Chilenana DeLong & Freytag 1967b:106. *C. chilena* n.sp. (Chile: Santiago). Gyponinae, Gyponini. Neotropical; Chile.
Subgenera:
Penaia Freytag & Delong 1971a:541. *Chilenana flexa* DeLong & Freytag 1967b:112.
Rugosella Freytag & DeLong 1971a:543. *C. (R.) obrienorum* n.sp.

Chinaella Evans, J.W. 1935c:76. *C. argentata* n.sp. (South Australia: Everard Range). 15:117. Evans, J.W. 1966a:212-213. Penthimiinae. Australian.

Chinaia Bruner & Metcalf 1934a:120. *C. bella* n.sp. (Costa Rica). 11:112. Kramer 1964d:261, 264. Neocoelidiinae. Neotropical; Central America and northern South America.

Chinchinota Kramer 1967b:41. *C. styx* n.sp. (Colombia). Neocoelidiinae, Neotropical; Colombia.

Chlamydopita Linnavuori 1959b:225 as subgenus of *Garapita* Oman 1938b:369. *G. (C.) aurea* n.sp. (Brazil: Nova Teutonia). Deltocephalinae, Athysanini. Neotropical.

Chlidochrus Emeljanov 1962a:156. *C. ventricosus* n.sp. (USSR: Kazakhstan). Deltocephalinae, Opsiini. Palaearctic.

Chloapala Amyot 17:1486. Invalid, not binominal.

Chlootea Mitjaev 1968b:34. Error for *Chloothea* Emeljanov.

Chloothea Emeljanov 1959a:836. *C. zonata* n.sp. (USSR: Kazakhstan, Akmolinsk region, 6 km NE of Lake Ilektyol). Deltocephalinae, Paralimnini. Palaearctic; Mongolia, Siberia, Altai Mts of Kazakhstan, Algeria.

Chlorate 17:1486. Error for *Chlorita* Fieber.

Chloria Fieber 1866a:508. *Cicada viridula* Fallen 1806a:37 (Sweden). 17:552. Typhlocybinae, Empoascini. Palaearctic. Preoccupied by *Chloria* Schiner 1862, see *Chlorita* Fieber 1872a:14.

Chlorida 17:1486. Error for *Chlorita* Fieber.

Chlorita Fieber 1872a:14, replacement for *Chloria* Fieber 1866a:508 not *Chloria* Schiner 1862. *Cicada viridula* Fallen 1806a:37 (Sweden). 17:552. Dlabola 1958c:46. Typhlocybinae, Empoascini. Palaearctic.
Subgenera:

Artemisiella Zachvatkin 1953c:229. *Chlorita nervosa* Fieber 1884a:60.

Xerochlorita Zachvatkin 1953c:229. *C. prasina* Fieber, Nast 1972a:270.

Chloroasca Anufriev 1972a:39. *C. chloris* n.sp. (USSR: Kuybyshev District). Typhlocybinae, Empoascini. Palaearctic.

Chlorogonalia Young 1977a:573. *Tettigonia coeruleovittata* Signoret 1855d:813, 835 ("Etats Unis" error ?). Young & Beier 1963a:568. Cicadellinae, Cicadellini. Neotropical; southern Mexico to Brazil.

Chloronana DeLong & Freytag 1964b:504. *C. rotunda* n.sp. (Peru: Sinchona). Gyponinae, Gyponini. Neotropical; Panama and northern South America.

Chloroneura Walsh 1862a:149. *C. abnormis* n.sp. (USA: Illinois ?). Knight 1968a:142. 17:129, 1486. Typhlocybinae, Dikraneurini. Nearctic. Synonym of *Dikraneura* Hardy 1850a:423.

Chloropelix Lindberg 1936a:3. *C. canariensis* n.sp. (Canary Islands). 9:114. Deltocephalinae, Hecalini. Palaearctic, Eremian; Canary and Cape Verde Islands, Ethiopia and Israel.

Chlorophlegma 17:1487. Error for *Chloroplegma* Amyot.

Chloroplegma Amyot. Invalid, not binominal.

Chloretettix 10:2648. Error for *Chlorotettix* Van Duzee.

Chloroteltix 10:2648. Error for *Chlorotettix* Van Duzee.

Chlorotettex 10:2648. Error for *Chlorotettix* Van Duzee.

Chlorotettix Van Duzee 1892e:306. *Bythoscopus unicolor* Fitch 1851a:58 (USA: New York). 10:1025. Linnavuori 1959b:251-272. Deltocephalinae, Athysanini. Nearctic, Neotropical.
Subgenus:
Chacotettix Linnavuori 1955a:106. *C. appendiculata* n.sp. = *Chlorotettix nigromaculatus* DeLong 1923b:265.
Synonym:
Celsanus Linnavuori 1954e:137. *Jassus (Thamnotettix) serius* Stal 1862a:52.

Chlorotettrix 10:2648. Error for *Chlorotettix* Van Duzee.

Chonosina Linnavuori & DeLong 1978b:134. *C. flavocostata* n.sp. (Peru: Sinchona). Deltocephalinae, Athysanini. Neotropical; Peru.

Choria 17:1487. Error for *Chloria* Fieber.

Chortophilus Amyot. 15:222. Invalid, not binominal.

Chrolotettix 10:2648. Error for *Chlorotettix* Van Duzee.

Chromagallia Linnavuori 1954e:126. *Bythoscopus flavofasciatus* Stal 1854b:255 (Brazil). 14:106. Kramer 1964a:143, 158-159. Agalliinae, Agalliini. Neotropical; Brazil.

Chromogonia Melichar 1926a:342. 1:263. Of uncertain status; see Young 1977a:1098.

Chroocacus Emeljanov 1962a:171. *C. psittaceus* n.sp. (USSR: Caucasus). Deltocephalinae, Athysanini. Palaearctic.

Chudania Distant 1908g:268. *C. delecta* n.sp. (India: Kurseong). 7:25. Ahmed & Malik 1972a. Nirvaninae, Nirvanini. Oriental and Ethiopian, Guinean subregion.

Chunra Distant 1908g:193. *Iassus puncticosta* Walker 1870b:324 (Morotai). 16:220. Idiocerinae. Tropical Africa, Indonesia, and northeast Australia.

Chunrocerus Zachvatkin 1946b:154. *Idiocerus niveosparsus* Lethierry 1889c:252 (India: Bengal). 16:211. Idiocerinae. Oriental. Synonym of *Idioscopus* Baker 1915c:320; Maldonado-Capriles 1971a:184.

Chunroides Evans, J.W. 1947a:254. *C. wallabensis* n.sp. (British Guiana: Essequibo River). 16:215. Idiocerinae. Neotropical; Brazil and British Guiana.

Ciadula 10:2648. Error for *Cicadula* Zetterstedt.

Cibra Young 1977a:1076. *C. acuta* n.sp. (Dominica, British West Indies). Cicadellinae, Cicadellini. Neotropical; British West Indies.

Cicaclula 10:2648. Error for *Cicadula* Zetterstedt.

Cicadalla 1:671. Error for *Cicadella* Latreille.

Cicadella Dumeril 1806a:266. *Cicada vittata* Linnaeus 1758a:438 (Europe). 1:106.
[Note: Suppressed by the ICZN under suspension of the Rules, Opinion 647, 1963].

Cicadella Latreille 1817a:406. *Cicada viridis* Linnaeus 1758a:438 (Europe). 1:106. Placed in the Official List of Family-Group Names In Zoology, ICZN Opinion 647, 1963. Cicadellinae, Cicadellini. Palaearctic.
Isogenotypic synonyms:

Tetigonia Geoffroy 1762a:429.
Tettigonia Fabricius 1775, a junior homonym of *Tettigonia* Linnaeus.
Tettigonia Olivier 1798a:24.
Amblycephalus Curtis 1833a:193.
Tettigoniella Jacobi 1904a:778.
Tettigella China & Fennah 1945a:711.

CICADELLIDAE 1:7

CICADELLIDAE = Typhlocybinae 17:1.

CICADELLINAE 1:25.

CICADELLINAE = Typhlocybini 17:100.

CICADELLINI 1:26.

CICADELLINI 17:613.

Cicadellites Heer 1853b:119. Metcalf & Wade 1966a:207. Fossil: Tertiary, Croatia.

Cicaduca 10:2648. Error for *Cicadula* Zetterstedt.

Cicadula Zetterstedt 1840a:296. *Cicada quadrinotata* Fabricius 1794a:43 (France). 10:1945. Deltocephalinae, Cicadulini. Holarctic.
Subgenus:
Henriana Emeljanov 1964c:412. *Jassus frontalis* Herrich-Schaeffer 1835a:70.
Synonym:
Cyperana DeLong 1936a:218. *Thamnotettix melanogaster* Provancher 1872c:378.

Cicadulella China 1928b:61, replacement for *Cicadulina* Haupt 1927a:38 not *Cicadulina* China 1926c:43. *Cicadulina pallida* Haupt 1927a:38 (Palestine). 10:2478. Deltocephalinae, Macrostelini. Palaearctic, Eremian.
Synonyms:
Cicadulina Haupt 1927a:38. *C. pallida* n.sp.
Kirrotettix Haupt 1929c:255, *Cicadulina pallida* Haupt 1924c:255.

CICADULINA Hamilton 1975b:491, subtribe of DELTOCEPHALINI.

Cicadulina China 1926c:43. *C. zeae* n.sp. (Kenya). 10:2469. Deltocephalinae, Macrostelini. Pan-tropical.
Subgenus:
Idyia Linnavuori 1960b:58. *C. (I.) fijiensis* n.sp.

Cicadulina Haupt 1927a:38. *C. pallida* n.sp. (Palestine). 10:2478. Deltocephalinae, Macrostelini. Palaearctic. Preoccupied by *Cicadulina* China 1926c:43, see *Cicadulella* China 1928b:61.

CICADULINI 10:1944.

Cicaduloida Osborn 1934b:258. *C. pacifica* n.sp. (Marquesas Islands: Uapou, Hakahetau Valley). 10:1945. Deltocephalinae, Athysanini. Neotropical; Marquesas Islands.

Cicciana Metcalf 1952a:228, replacement for *Ciccus* Stal 1869a:60, not *Ciccus* Latreille 1829a:221. *Ciccus latreillei* Distant 1908b:81 (SE Brazil), replacement for *Ciccus adspersa* of authors, not *Ciccus adspersus* Fabricius 1803a:61. 1:467. Young 1968a:115. Cicadellinae, Proconiini. Neotropical; Guianas and Brazil.

CICCIANINI Metcalf 1965a:467, tribe of CICADELLINAE.

CICCINI 1:611.

Ciccus Latreille 1829a:221. *Cicada adspersa* Fabricius 1803a:61 (South America). 1:611. Young 1968a:80. Cicadellinae, Proconiini. Neotropical.
Synonym:
Coelopola Stal 1869a:65. *Cicada adspersa* Stal 1803a:61.

Cicus 1:677. Error for *Ciccus* Latreille.

Ciminius Metcalf & Bruner 1936a:944. *Tettigonia hartii* Ball 1901b:61 (USA: Ohio). 1:391. Young 1977a:589. Cicadellinae, Cicadellini. Neotropical/Nearctic.

Cinerogonalia Young 1977a:754. *Cicadella lituriceps* Osborn 1926b:180 (Colombia). Cicadellinae, Cicadellini. Neotropical; Colombia and Venezuela.

Circular Meyerdirk, Oldfield & Hessein 1983a:[title]. Error for *Circulifer* Zachvatkin 1935a:111.

Circulifer Zachvatkin 1935a:111. *Jassus haematoceps* Mulsant & Rey 1855a:229 (France). 10:881. Oman 1970a. Deltocephalinae, Opsiini. Palaearctic, Ethiopian, adventive in other regions.
[NOTE: Also treated as a junior synonym or subgenus of *Neoaliturus* Distant 1918b].

CIRCULIFERINA Emeljanov 1962b:389, subtribe of OPSIINI.

Cistocephala 17:1487. Error for *Cestocephala* Amyot.

Citorus Stal 1866a:110. *Selenocephalus decurtatus* Stal 1855a:98 (Caffraria; Theron 1980a:281). 15:171. Linnavuori 1977a:60-72. Penthimiinae. Ethiopian; Guinean subregion, E. Africa and South Africa.
Synonym:
Dlabolaia Linnavuori 1959b:56. *D. rugipennis* n.sp.

Citripo Evans, J.W. 1934a:151. *C. flandersi* n.sp. (Australia: Queensland). 12:34. Evans, J.W. 1966a:40. Eurymelinae, Ipoini. Australian.

Citrous 15:222. Error for *Citorus* Stal.

Ciudadrea Dworakowska 1970k:617. *Zygina punctigera* Horvath 1905b:276 (Spain: "Cuidad Real"). Typhlocybinae, Erythroneurini. Palaearctic; Spain and Canary Islands.

Cixidocoelidia Linnavuori 1956a:34. *C. truncatipennis* n.sp. (Bolivia: Bogota). Coelidiinae, Sandersellini. Neotropical. Synonym of *Sandersellus* DeLong 1945d:414; Nielson 1975a:20.

Clavena Melichar 1902c:127. *C. sulcata* n.sp. (Western China). 9:66. Morrison 1973a:387. Deltocephalinae, Hecalini. Oriental.

Cleptochiton Emeljanov 1959a:836. *C. variegatus* n.sp. (USSR: Kazakhstan, Akmolinsk region, Kokshetau). Deltocephalinae, Deltocephalini. Palaearctic.

Clinonana Osborn 1938a:13. *C. magna* n.sp. (French Guiana). 3:54. Kramer 1966a:471,479. Ledrinae. Neotropical; Guianas, Brazil, Peru.

Clinonaria Metcalf 1949b:277. *C. bicolor* n.sp. (British Guiana). 3:55. DeLong & Freytag 1969a, 1972a:18. Gyponinae, Gyponini. Neotropical. Synonym of *Scaris* LePeletier & Serville 1825a:609; DeLong & Freytag 1972a:18.

Clinonella DeLong & Freytag 1971a:314. *Clinonana declivata* Osborn 1938a:14 (Bolivia: Prov. del Sara). Gyponinae, Gyponini. Neotropical; Bolivia and Peru.

Clinoria Teson 1971a:106. Error for *Clinonaria* Metcalf.

Cloanthanus Ball 1931d:219. *Platymetopius angustatus* Osborn 1905a:518 (USA: New York, Cold Spring Harbor). 10:2263. Linnavuori 1959b:72-83. Deltocephalinae, Scaphytopiini. Nearctic/Neotropical. Subgenus of *Scaphytopius* Ball 1931d:218.

Cloanthauus 10:2649. Error for *Cloanthanus* Ball.
Cloanthus 10:2649. Error for *Cloanthanus* Ball.
Cloothea Vilbaste 1980a:85. Error for *Chloothea* Emeljanov.

Clorindaia Linnavuori 1975b:51. *C. hecaloides* n.sp. (Argentina: Clorinda). Deltocephalinae, Athysanini. Neotropical; Argentina and Chile.

Clorita 17:1487. Error for *Chlorita* Fieber.

Clovana DeLong & Freytag 1964a:118 as subgenus of *Gyponana*. *Gyponana omani* DeLong 1942d:55 (USA: Texas, Brownsville). Gyponinae, Gyponini. Nearctic.

Clydacha Melichar 1926a:345. *Phereurhinus cochlear* Jacobi 1905c:169 (Peru: "N. Rioja"). 1:414. Kramer 1976a:118. Phereurhinae. Neotropical; Peru.

Clypeolidia Nielson 1982e:246. *Jassus brunneus* Osborn 1924c:437 (French Guiana: Manna River). Coelidiinae, Coelidiini. Neotropical; French Guiana, Guyana, Brazil.

Cnlidochrus Mitjaev 1971a:133. Error for *Chlidochrus* Emeljanov.

Cochanga Nielson 1979b:302. *C. spinosa* n.sp. (Peru: Upper Rio Huallaga). Coelidiinae, Teruliini. Neotropical; Peru.

COCHLORHININA Hamilton 1975b:496, subtribe of DELTOCEPHALINAE.

COCHLORHININI 10:1079.

Cochlorhinus Uhler 1876a:358. *C. pluto* n.sp. (USA: California). 10:1079. Deltocephalinae, Cochlorhini. Nearctic; California.
Synonym:
Uhleriella Ball 1902b:54. *Deltocephalus coquilletti* Van Duzee 1890g:95 (homonym) = *Cochlorhinus unispinosus* Beamer 1940a:54.

Cochlorinus 10:2649. Error for *Cochlorhinus* Uhler.
Cochrorinus 10:2649. Error for *Cochlorhinus* Uhler.

Cocoelidia DeLong 1953a:126 as subgenus of *Neocoelidiana* DeLong 1953a:122. *Neocoelidiana antlera* n.sp. (Mexico: Michoacan, Tuxpan). Kramer 1964d:271. Neocoelidiinae. Neotropical.

Codilia Nielson 1982e:220. *C. retrorsa* n.sp. (Peru: "Piches & Perene Vs"). Coelidiinae, Coelidiini. Neotropical; Ecuador and Peru.

Coelana DeLong 1953a:95. (Invalid) 11:112.
[Note: No type designated; the name appears only in the key on p.95 as noted by Metcalf. Kramer 1964d:269 used the name at the generic level, stating that the type was *Neocoelidia modesta* Baker 1898h:290, but did not state that he was using the name and reference to DeLong's description to validate the name].

Coelella DeLong 1953a:125 as subgenus of *Neocoelidiana* DeLong 1953a:122. *Neocoelidia distincta* Oman 1931a:67 (USA: Arizona). 11:109. Kramer 1964d:273. Neocoelidiinae. Nearctic; SW USA.

Coelestinus Emeljanov 1962a:178 as subgenus of *Palus* DeLong & Sleesman 1929a:85. *Deltocephalus hypomelas* Kusnezov 1929b:182 (Mongolia). Deltocephalinae, Paralimnini. Palaearctic; USSR: Kazakhstan, Mongolia, Turinskaya.
[Note: Raised to generic status: Emeljanov 1966a:115].

Coelida 11:147. Error for *Coelidia* Germar.
Coelidea 11:147. Error for *Coelidia* Germar.

Coelidia Germar 1821a:75. *C. venosa* n.sp. (Brazil). 11:31. Nielson 1982e:248-263, 1983i. Coelidiinae, Coelidiini. Neotropical; Central America and south to Brazil and Peru.

Coelidiana Oman 1938b:397. *Neocoelidia rubrolineata* Baker 1898h:290 (Brazil). 11:109,111. Kramer 1964d:274. Neocoelidiinae. Neotropical; southern Mexico to Brazil and Peru.
Synonym:
Acocoelidia DeLong 1953a:130. *A. unipuncta* n.sp.

COELIDIIDAE 11:5.
COELIDIINAE 11:11.
COELIDIINI 11:16.

Coelidiodes 11:147. Error for *Caelidioides* Signoret.
Coelidioides 11:147. Error for *Caelidioides* Signoret.

Coelindroma Kramer 1967b:43. *C. fungosa* n.sp. (Peru: Tingo Maria). Neocoelidiinae. Neotropical; Peru.

Coelogypona DeLong & Freytag 1966b:311. *C. venosella* n.sp. (Peru: Achinamiza). Gyponiinae, Gyponiini. Neotropical; Ecuador, Peru.

Coelopola Stal 1869a:65. *Cicada adspersa* Fabricius 1803a:61 (South America). 1:611. Young 1968a:80. Cicadellinae, Proconiini. Neotropical. Isogenotypic synonym of *Ciccus* Latreille 1829a:221.

Coexitianus Dlabola 1960a:252. *Athysanus albinervosus* Matsumura 1902:372 (Japan: Kyushu). Deltocephalinae, Athysanini. Palaearctic, Oriental. Isogenotypic synonym of *Paramesodes* Ishihara 1953b:14.

Cofana Melichar 1926a:345. *Tettigonia quinquenotata* Stal 1870c:734 ("Ins. Philpp") = *Tettigonia eburnea* Walker 1857b:168. 1:430. Young 1979a:1-21; Linnavuori 1979a:705-715. Cicadellinae, Cicadellini. Oriental, Palaearctic, Ethiopian.
[NOTE: Young 1979a:1 and Linnavuori 1979a:705 placed *Yasumatsuus* Ishihara 1971a:18, type species *Kolla mimica* Distant 1908g:225 (India: Calcutta), as a synonym of *Cofana* Melichar. Inasmuch as that decision involves presumed misidentification of the type species of *Yasumatsuus* by Ishihara, the case should be referred to the Commission under provisions of the Code (Article 70(a) & (b)].

Coganoa Dworakowska 1976b:33. *Typhlocyba purpureatincta* Cogan 1916a:198 (South Africa: Cape of Good Hope). Typhlocybinae, Erythroneurini. Ethiopian and Oriental.

Coganus Theron 1978b:257. *Deltocephalus breviatus* Cogan 1916a:186 (South Africa: Cape Town). Deltocephalinae, Paralimnini. Ethiopian.

Cojana Linnavuori 1979a:687. Error for *Cofana* Melichar.
Coladonus Cantoreanu 1971c:46. Error for *Colladonus* Ball.
Coleopola 1:678. Error for *Coelopola* Stal.

Coleopteropsis Bekker-Migdisova 1949a:29. *C. dolichoptera* n.sp. Metcalf & Wade 1966a:210. Fossil: Mesozoic, Leninabad.

Colimona Oman 1949a:71. *Apogonalia omani* Young 1977a:937 (USA: Arizona, Santa Rita Mtns.). 1:55. Young 1977a:916. Cicadellinae, Cicadellini. Nearctic and Neotropical. Synonym of *Apogonalia* Evans 1947a:159.

COLLADONINI 10:1242.

Colladonus Ball 1936c:57. *Thamnotettix collaris* Ball 1902a:15 (USA: New York, New York City). 10:1242. Nielson 1957a, 1966a. Deltocephalinae, Athysanini. Nearctic, Palaearctic.
Synonyms:

Conodonus Ball 1936c:58. *Thamnotettix flavocapitatus* Van Duzee 1890f:80.
Friscananus Ball 1936c:60. *Thamnotettix intricatus* Ball 1911a:198.
Hypospadianus Ribaut 1942a:264. *Cicada torneela* Zetterstedt 1828a:528.
Coniferadonus Bliven 1955a:4. *Colladonus holmesi* Bliven 1954a:116.
Sequoiatettix Bliven 1955a:3. *Colladonus eurekae* Bliven 1954a:117.
Collandonus 10:2649. Error for *Colladonus* Ball.
Coloana Dworakowska 1971e:346. *C. cinerea* n.sp. (Vietnam: Coloa). Typhlocybinae, Erythroneurini. Oriental.
Coloborrhis Germar 1836a:72. *C. corticinia* n.sp. (South Africa: Cape of Good Hope). 5:88. Linnavuori 1972a:137-140. Ulopinae, Ulopini. Ethiopian, Malagasian and Oriental. Synonyms:
Evryprosopum Stal 1853b:267. *E. sobrina* n.sp.
Bohemania Stal 1855a:97, replacement for *Evryprosopum* Stal 1853b:267 not *Evryprosopus* White 1853.
Oclasma Melichar 1905a:293. *O. degenerata* n.sp.
Gubela Distant 1910e:232. *G. bellicosa* n.sp.
[NOTE: See comments under *Mesargus* Melichar.]
Colobotettix Ribaut 1948c:58. *Thamnotettix morbillosus* Melichar 1896a:293 (Austria). 10:251. Deltocephalinae, Athysanini. Palaearctic; Europe.
Cololedra Evans, J.W. 1969a:744. *C. declivata* n.sp. (New Guinea: Mt. Giluwe, SE). Ledrinae, Ledrini. Oriental.
Columbanus Distant 1916a:224. *C. misranus* n.sp. (India: Bengal). 9:7. Morrison 1973a:410. Deltocephalinae, Hecalini. Oriental. Synonym of *Hecalus* Stal 1864b:65.
Columbonirvana Linnavuori 1959b:34. *C. aurea* n.sp. (Colombia: Sierra S. Lorenzo). Nirvaninae. Neotropical.
Comanopa Blocker 1979a:37. *Stragania puertoricensis* Caldwell 1952a:22 (Puerto Rico: Maricao, Insular forest). Iassinae. Neotropical; Jamaica and Puerto Rico.
Comayagua Linnavuori & DeLong 1978c:206. *C. taeniata* n.sp. (Honduras: Rancho Chiquita, Comayagua). Deltocephalinae, Athysanini. Neotropical; Honduras.
Commellus Osborn & Ball 1902a:245 as subgenus of *Athysanus* Burmeister 1838b:14. *Athysanus comma* Van Duzee 1892a:114 (USA: Iowa). 10:122. Deltocephalinae. Nearctic.
Comellus 10:2650. Error for *Commellus* Osborn & Ball.
Compsus Fieber 1866a:507. *Cicada albostriella* Fallen 1826a:54 (Sweden). 17:26. Preoccupied by *Compsus* Schoenherr 1823. Typhlocybinae, Alebrini. Holarctic. Synonym of *Alebra* Fieber 1872a:14.
Conala Oman 1938b:396. *Spangbergiella fasciata* Osborn 1923c:30 (Brazil: S. Antonio de Guapore'). 10:793. Linnavuori 1959b:18; Kramer 1963b. Neobalinae. Neotropical; Brazil.
Conbalia Nielson 1979b:54. *C. colombiensis* n.sp. (Colombia: 9.7 km W of Cali Val). Coelidiinae, Teruliini. Neotropical; Colombia and Ecuador.
Concavifer Dlabola 1960c:14. *C. marmoratus* n.sp. (Iran: Anbar-Abad). Deltocephalinae, Opsiini. Palearctic.
Concepciona Linnavuori & DeLong 1977c:208. *C. asper* n.sp. (Chile: Concepcion). Deltocephalinae, Athysanini. Neotropical; Chile.
Conchyotettix Oshanin 1891a:45, 67. (*Nomen nudum*). 5:87.
Condylotes Emeljanov 1959a:838. *C. zachvatkini* n.sp. (USSR: Kazakhstan, Akmolinsk region, Kokshetau). Deltocephalinae, Deltocephalini. Palaearctic; Eurasia.
Conemetopius 17:1488. Error for *Conometopus* de Motschulsky.
Confucius Distant 1907g:191. *C. granulatus* n.sp. (Hong Kong Island). 4:84. Linnavuori 1972b:206-211. Ledrinae, Petalocephalini. Oriental, Ethiopian.
Coniferadonus Bliven 1955a:4. *Colladonus holmesi* Bliven 1954a:116 (USA: California, Humboldt Co., Blocksburg). 10:1312. Nielson 1966a:333. Deltocephalinae, Athysanini. Nearctic. Synonym of *Colladonus* Ball 1936c:57.
Coniognathus 10:2650. Error for *Goniognathus* Fieber.
Conlopa Evans, J.W. 1971c:441. *C. bredoni* n.sp. (Congo: Elizabethville). Ulopinae, Ulopini. Ethiopian.
Conodonus Ball 1936c:58. *Thamnotettix flavocapitatus* Van Duzee 1890F:80 (USA: California). 10:1308. Nielson 1957a, 1966a:333. Deltocephalinae, Athysanini. Nearctic. Synonym of *Colladonus* Ball 1936c:57.

Conogonia Breddin 1903c:98. *C. trucidula* n.sp. (Dutch New Guinea). 1:421. Cicadellinae, Cicadellini. Oriental; New Guinea.
Conogonus Osborn 1892a:126. (*Nomen nudum*). 10:163, 2650.
Conogonus Van Duzee 1894c:136. *Conogonus gagates* Van Duzee 1894c:136 (Iowa) = *Athysanus anthracinus* Van Duzee 1894c:136. Deltocephalinae, Athysanini. Nearctic. Synonym of *Scleroracus* Van Duzee 1894c:136.
[Note: Erroneously attributed to Oman 1949a:152; Hamilton 1975b:490].
Conojassus Evans, J.W. 1972b:647 [Abstract]. Error for *Coriojassus* Evans, J.W.
Conomellus 10:2650. Error for *Commellus* Osborn & Ball.
Conometopius 17:1488. Error for *Conometopus* de Motschulsky.
Conometopus de Motschulsky 1863c:103. *C. inspiratus* n.sp. (Ceylon). 17:232. Typhlocybinae, Dikraneurini. Oriental. Preoccupied by *Conometopus* Fieber 1858, see *Motschulskyia* Kirkaldy 1905a:325.
Conosanus Osborn & Ball 1902a:236 as subgenus of *Athysanus* Burmeister 1838b:14. *A. obsoletus* Kirschbaum 1858a:7 (Prussia). 10:12. Ossiannilsson 1983a:771. Deltocephalinae, Athysanini. Palaearctic.
[NOTE: See comment under *Euscelis* Brulle].
Consepusa Linnavuori & Delong 1977c:205. *C. crassistylus* n.sp. (Chile: Maipu). Deltocephalinae, Athysanini. Neotropical.
Convelinus Ball 1931d:220. *Platymetopius nigricollis* Ball 1916b:205 (USA: California, Mojave). 10:2325. Linnavuori 1959b:66-71. Deltocephalinae, Scaphytopiini. Nearctic, Neotropical. Subgenus of *Scaphytopius* Ball 1931d:219.
Conversa Zoological Record 104, Section 13f:670, 1967. Error for *Conversana* DeLong.
Conversana DeLong 1967e:266. *C. reversa* n.sp. (Mexico: Hacienda de Santa Engracia). Deltocephalinae, Athysanini. Neotropical.
Copididonus Linnavuori 1954e:138. *Jassus hyalinipennis* Stal 1854b:255 (Brazil). 10:1332. Linnavuori 1959b:247-250; Zanol & Sakikabara 1985a. Deltocephalinae, Deltocephalini. Neotropical; Brazil, Colombia, Paraguay, Venezuela.
Coratulina 10:2650. Error for *Doratulina* Melichar.
Corilidia Nielson 1982a:424. *C. lenta* n.sp. (Peru: Cajamarca). Coelidiinae, Tinobregmini. Neotropical.
Coriojassus Evans, J.W. 1972b:649. *C. brunneus* n.sp. (Thailand: Ban Yang). Iassinae, Hyalojassini. Oriental.
Cornutipo Evans, J.W. 1934a:150. *C. scalpellum* n.sp. (Australia: Queensland, Duaringa). 12:29. Evans, J.W. 1966a:49-50. Eurymelinae, Ipoini. Australian.
Synonym:
Cornutipoides Evans, J.W. 1934a:164. *C. tricornis* n.sp.
Cornutipoides Evans, J.W. 1934a:164. *C. tricornis* n.sp. (NW Australia). 12:29. Evans, J.W. 1977a:88. Eurymelinae, Ipoini. Australian. Synonym of *Cornutipo* Evans, J.W. 1934a:150.
CORNUTOPOINAE 12:28.
Coronigonalia Young 1977a:1001. *Tettigonia spectanda* Fowler 1900d:285 (Central America). Cicadellinae, Cicadellini. Neotropical; Costa Rica and Panama.
Coronigoniella Young 1977a:1004. *Graphocephala spinosa* Osborn 1926b:232 (Brazil: Rio de Janeiro). Cicadellinae, Cicadellini. Neotropical; Trinidad, Costa Rica and Panama to NE Brazil.
Coronophtus Van Stalle 1983b:304. *C. cervus* n.sp. (Zaire). Megophthalminae. Ethiopian.
Coroticus Distant 1918b:45. *C. tessellatus* n.sp. (India: Kumaon). 10:2350. Deltocephalinae, Scaphytopiini. Oriental; India.
Cortona Oman 1938b:390. *C. minuta* n.sp. (Argentina: Missiones, Loreto). 10:1572. Linnavuori 1959b:328-329. Deltocephalinae, Deltocephalini. Neotropical.
Corupiana Nielson 1979b:115. *C. multidens* n.sp. (Brazil: Santa Catarina, Corupa). Coelidiinae, Teruliini. Neotropical.
Corymbonotus Maldonado-Capriles 1977b:363. *C. bicolor* n.sp. (British Guiana: Amazon-Courantyne Divide, head of Oronoco River). Idiocerinae. Neotropical.
CORYPHAEINI Emeljanov 1962b:390, 397. Invalid, based on homonym.
CORYPHAELINI Nast 1972a:332, tribe of DELTOCEPHINAE.

Coryphaelus Puton 1886b:81, replacement for *Coryphaeus* Fieber 1866a:503, not *Coryphaeus* Gistl 1847. *Jassus gyllenhali* Fallen 1826a:61 (Sweden). 10:1896. Deltocephalinae, Coryphaelini. Palaearctic; Eurasia.

Coryphaeus Fieber 1866a:503. *Jassus gyllenhali* Fallen 1826a:61 (Sweden). 10:1896. Preoccupied by *Coryphaeus* Gistl 1847, see *Coryphaelus* Puton 1886b:81.

Cosmotettix Ribaut 1942a:267. *Jassus (Jassus) caudatus* Flor 1861a:351 (Livonia). 10:2036, 2042. Ossiannilsson 1983a:870. Deltocephalinae, Paralimnini. Holarctic.
Subgenera:
Airosus Ribaut 1952a:126. *Cicada costalis* Fallen 1826a:32.
Agapelus Emeljanov 1961a:127 "*Deltocephalus aurantiacus* Fieb."
Airosius Emeljanov 1966a:111. *Palus caespitosus* Emeljanov 1966a:112.
[Note: See comments under *Agapelus*].

Costamia DeLong 1946a:82. *C. venosa* n.sp. (Mexico: Guerrero, Iguala). 10:2089. Deltocephalinae, Athysanini. Neotropical; Mexico.

Costanana DeLong & Freytag 1972a:227. *C. dunda* n.sp. (Mexico: Guerrero, Iguala). DeLong & Freytag 1972e. Gyponinae, Gyponini. Neotropical; Mexico to Brazil.

Coulinus Beirne 1954e:548. *C. uladus* n.sp. (Alaska: Naknek). 10:303. Deltocephalinae, Athysanini. Holarctic.

Cozadanus DeLong & Harlan 1968a:150. *C. globosus* n.sp. (Mexico: Guerrero, Mazatlan). Deltocephalinae, Athysanini. Neotropical; Mexico.

Crassana DeLong & Hershberger 1947e:76. *Eutettix goniana* Ball 1931a:1 (USA: Arizona, Patagonia). 10:465. Deltocephalinae, Athysanini. Nearctic; Mexico.
Subgenus:
Macrasana DeLong & Hershberger 1947e:78. *C. (M.) marginella* n.sp.

Crassinolanus Nielson 1982e:217. *C. dementius* n.sp. (Bolivia: Corioco). Coelidiinae, Coelidiini. Neotropical; Bolivia.

Crepluvia Nielson 1979b:67. *Coelidia pygmae* Linnavuori 1956a:25 (Brazil: Minas Gerias, Brasilia). Coelidiinae, Teruliini. Neotropical; Argentina, Brazil, Paraguay.

Cribrus Oman 1949a:166. *Laevicephalus shingwauki* Beamer & Tuthill 1934a:19 (USA: Minnesota, Shingwauki). Deltocephalinae, Deltocephalini. Nearctic.

Crinolidia Nielson 1982e:269. *C. simplex* n.sp. (Ecuador: Tina). Coelidiinae, Coelidiini. Neotropical; Ecuador and Peru.

Crinorus Nielson 1982e:283. *C. projectus* n.sp. (Peru: 23 km E of Abancay). Coelidiinae, Coelidiini. Neotropical.

Crispina Distant 1918b:76. *C. nigristigma* n.sp. (S. India: Kodaikanal). 7:25. Nirvaninae, Nirvanini. Oriental.

Crolix 10:2651. Error for *Orolix* Ribaut.

Crossogonalia Young 1977a:996. *Tettigonia hectica* Signoret 1854a:20 (Brazil). Cicadellinae, Cicadellini. Neotropical; SE Brazil.

Cruciatanus DeLong & Hershberger 1946a:208 as subgenus of *Sanctanus* Ball 1932a:10. *Scaphoideus cruciatus* Osborn 1911c:253 (USA: Florida). 10:2096. Deltocephalinae, Deltocephalini. Nearctic.

Crumbana Oman 1949a:172. *Deltocephalus arundineus* Crumb 1915a:191 (USA: Tennessee, Clarksville). 10:1825. Deltocephalinae, Deltocephalini. Nearctic; southeast USA.

Cruziella Linnavuori & DeLong 1979a:49. *C. trispinosa* n.sp. (Bolivia: Sta Cruz, San Esteban). Deltocephalinae, Deltocephalini. Neotropical; Bolivia.

Ctenurella Vilbaste 1968a:140. *C. paludosa* n.sp. (USSR: Maritime Territory). Deltocephalinae, Athysanini. Palaearctic.

Cubnara Dworakowska 1979a:153. *Typhlocyba gemmata* Distant 1918b:99 (India: Kodiakanal, Nilgiri Hills). Typhlocybinae, Erythroneurini. Oriental; south India.

Cubrasa Young 1977a:817. *Poeciloscarta cardini* Metcalf & Bruner 1936a:937 (Cuba: Pinar del Rio Prov., Baracoa, El Yunque Mtn). Cicadellinae, Cicadellini. Neotropical; Cuba.

Cuena 1:678. Error for *Cuerna* Melichar.

Cuerna Melichar 1925a363. *Cercopis lateralis* Fabricius 1798a:524 (USA: "Carolina"). 1:541. Young 1968a:251. Nielson 1965a:1-48. Hamilton 1970a:425-441. Cicadellinae, Proconiini. Nearctic.

Cuitlana Young 1977a:1032. *Tettigonia venusta* Stal 1864a:75 (Mexico). Cicadellinae, Cicadellini. Neotropical; Mexico and Guatemala.

Culumana DeLong & Freytag 1972a:233. *C. torqua* n.sp. (Peru: Sinchona). DeLong & Freytag 1972d. Gyponinae, Gyponini. Neotropical.

Cumbrenanus DeLong & Cwikla 1984a:432. *C. panamus* n.sp. (Panama: Las Cumbres). Deltocephalinae, Deltocephalini. Neotropical.

Cumora Oman 1938b:374. *C. angulata* n.sp. (Argentina: Missiones, Loreto). 10:1313. Linnavuori 1959b:141. Deltocephalinae, Deltocephalini. Neotropical. Synonym of *Haldorus* Oman 1938b:373; Linnavuori 1959b:141.

Cunedda Distant 1918b:46. *C. phaeops* n.sp. (India: Assam). 6:51. Cicadellinae, Evacanthini. Oriental.

Curistuva Blocker 1979a:14 as subgenus of *Gargaropsis* Fowler 1896e:167. *Gargaropsis adibilis* Blocker 1975a:561 (Mexico: Cuernavaca-Acapulco Rd). Iassinae. Neotropical.

Cupelix 9:119. Error for *Eupelix* Germar.

Curta DeLong & Caldwell 1937a:30 as subgenus of *Dikraneura* Hardy 1850a:423. 17:243. *D. (C.) alta* n.sp. (USA: Pennsylvania, Gillette). Typhlocybinae, Dikraneurini. Nearctic. Synonym of *Notus* Fieber 1866a:508.

Curtara DeLong & Freytag 1972a:231. *C. samera* n.sp. (Brazil: Sao Paulo, Piracicaba). DeLong & Freytag 1976a. Gyponinae, Gyponini. Neotropical.
Subgenera:
Ardasoma DeLong & Freytag 1976a:56. *C. (A.) arda* n.sp.
Curtarana DeLong & Freytag 1976:44. *Gypona jansoni* Fowler 1903b:306.
Mysticana DeLong & Freytag 1976a:3. *Gypona mystica* Spangberg 1878a:71.
Remarana DeLong & Freytag 1976a:54. *C. (R.) remara* n.sp.
Retusana DeLong & Freytag 1976a:57. *C. (R.) retusa* n.sp.
Lataba DeLong & Triplehorn 1978a:180. *C. (L.) basala* n.sp.
Sinchora DeLong 1979a:229. *C. (S.) regela* n.sp.

Curtarana DeLong & Freytag 1976a:49 as subgenus of *Curtara* DeLong & Freytag 1972a:231 (Guatemala). *Gypona jansoni* Fowler 1903b:306. Gyponinae, Gyponini. Neotropical.

Cyanidius Emeljanov 1964f:28. *Euscelidius cyanescens* Emeljanov 1962a:168 (USSR: Kazakhstan). Deltocephalinae, Athysanini. Palaearctic.

Cybus Douglas 17:1489. Unnecessary emendation of *Kybos* Fieber 1866.

Cyclogonia Melichar 1926a:348. *Tettigoniella cycopula* Jacobi 1905c:183 (Peru: Rioja). 1:28. Young 1977a:539. Cicadellinae, Cicadellini. Neotropical; Bolivia, Brazil, Ecuador, Peru.

Cyclopherus Emeljanov 1964f:35 as subgenus of *Handianus* 1942a:265. *H. (C.) eurotiae* n.sp. (USSR: Kazakhstan). Deltocephalinae, Athysanini. Palaearctic.

Cymbalopus Kirkaldy 1907d:88. *Tettigonia bigibbosa* Signoret 1855c:510 (French Guiana). 1:649. Young 1968a:118. Cicadellinae, Proconiini. Neotropical. Synonym of *Peltocheirus* Walker 1858b:247.

Cymbopogonella Viraktamath, C.A. 1976a:79 as subgenus of *Doratulina* Melichar 1903b:198. *D. (C.) longivertex* n.sp. (India: Karnataka). Deltocephalinae, Stenometopiini. Oriental.

Cyperana DeLong 1936a:218. *Thamnotettix melanogaster* Provancher 1872c:378 (Canada). 10:1946. Deltocephalinae, Cicadulini. Nearctic. Synonym of *Cicadula* Zetterstedt 1840a:296.
[NOTE: Also used as a subgenus of *Cicadula*, e.g. Ossiannilsson 1983a:719-721].

Cyperacea Beirne 1956a:88.
[NOTE: Cited as synonym of *Cicadula* Zetterstedt, presumably a lapsus for *Cyperana* DeLong 1936a:218].

Cypona 3:221. Error for *Gypona* Germar.

Cyrta Melichar 1902c:136. *C. hirsuta* n.sp. (China: Szechwan). 11:84. Unassigned, formerly Coelidiinae; Nielson 1975a:11. Oriental.

Cyrtodisca Stal 1869a:60. *Tettigonia major* Signoret 1854c:491 (Guatemala). 1:556. Young 1968a:174. Cicadellinae, Proconiini. Neotropical.

Cyrtodisea 1:678. Error for *Cyrtodisca* Stal.

Czecza Dworakowska 1981i:537 as subgenus of *Wiata* Dworakowska 1972:850. *W. (C.) krasna* n.sp. (Zaire: Lumbumbaski "Elizabethville"). Typhlocybinae, Zyginellini. Ethiopian.

Dabrescus 15:222. Error for *Drabescus* Stal.

Dagama Distant 1910e:243. *D. novata* n.sp. (South Africa: Port Natal?) = *Bythoscopus capensis* Walker 1851b:970. 10:422. Theron 1980b. Deltocephalinae, Athysanini. Ethiopian; South Africa.
Synonym:
Paranoplus Linnavuori 1961a:473. *P. maculatus* n.sp.

Daimachus Distant 1916a:225. *D. exemplificatus* n.sp. (southern India). 5:86. Ulopinae, Ulopini. Oriental; India.

Dalagus Amyot 14:172. Invalid, not binominal.

Dalbulus DeLong 1950a:105. *Deltocephalus elimatus* Ball 1900c:345 (Mexico: Santa Fe). 10:2464. Triplehorn & Nault 1985a. Deltocephalinae, Macrostelini. Neotropical; southern USA, Caribbean, and Mexico to Argentina.

Daltocephalus 10:2651. Error for *Deltocephalus* Burmeister.

Daltonia Oman 1949a:176. *Athysanus (Commellus) estacadus* Ball 1911a:200 (USA: Texas). 10:1826. Deltocephalinae, Deltocephalini. Nearctic, Neotropical; southern USA, gulf coastal area of Mexico, Argentina and Bolivia.

Daluana Ramakrishnan 1982a:10. *D. bhubaneswarensis* n.sp. (India: Orissa, Bhubaneswar). Typhlocybinae, Empoascini. Oriental; India.

Dalus Emeljanov 1975a:387. *Symphypyga leopardina* Haupt 1917d:242 (Turkestan). Adelungiinae, Adelungiini. Palaearctic; Kazakhstan, Turkmenia, Uzbekistan.

Dampfiana DeLong & Hershberger 1948c:229. *D. deserta* n.sp. (Mexico: Guerrero, Iguala). 10:503. Deltocephalinae, Athysanini. Neotropical.

Dananea Linnavuori 1972a:144. *D. inflaticeps* n.sp. (Ivory Coast: Copoupleu, near Danane). Megophthalminae. Ethiopian.

Danbara Oman 1949a:114. *Eutettix (Mesamia) aurata* Ball 1909a:81 (USA: Washington, DC). 10:2173. Deltocephalinae, Scaphoideini. Nearctic; southeastern USA.

Dapitana Mahmood 1967a:37. *D. robusta* n.sp. (Philippines: Mindanao, Dapitan). Typhlocybinae, Typhlocybini. Oriental; Philippines, Borneo, Java.

Dardania Stal 1866a:113. *D. granulosa* n.sp. (Tanzania: Zanzibar). 11:89. Acostemminae. Ethiopian; Malagasian.

Daridna Walker 1858c:319. *D. subtangens* n.sp. (Brazil: Rio de Janeiro). 11:31, 147. Nielson 1982e:229-231. Coelidiinae, Coelidiini. Neotropical; Brazil, Bolivia, Peru, Venezuela.

Dariena Linnavuori & DeLong 1977b:561. *D. amabilis* n.sp. (Panama: Darien Prov., Santa Fe). Deltocephalinae, Athysanini. Neotropical.

Darma Walker 1858a:102. *D. bipunctata* n.sp. (Uruguay: Montevideo). 3:69. Gyponinae, Gyponini. Neotropical. Synonym of *Scaris* LePeletier & Serville 1825a:609.

Dasmeusa Melichar 1926a:342. *Cicada pauperata* Fabricius 1803a:71 ("Amer. Merid."). 1:261. Young 1977a:291. Cicadellinae, Cicadellini. Neotropical; Brazil and the Guianas.

Dattasca Dworakowska 1979a:147. *Alebroides orientalis* Datta & Gosh 1973a:410 (India: West Bengal, Darjeeling, Kalimpong, c.1220 m). Typhlocybinae, Empoascini. Oriental.

Davisonia Dorst 1937a:4. *Cicadula major* Dorst 1931a:43 (USA: Minnesota, Ramsey Co.). 10:2611. Deltocephalinae, Macrostelini. Nearctic: central & eastern USA; southern Canada east of the Rocky Mtns.

Davmata Dworakowska 1979b:44. *D. neka* n.sp. (Vietnam: Nghe-An). Typhlocybinae, Erythroneurini. Oriental.

Dayoungia Kramer 1976a:128. *D. magister* n.sp. (Argentina: Missiones, Puerto Bember). Phereurhininae. Neotropical; Argentina, Paraguay, Brazil.

Dayus Mahmood 1967a:39. *D. elongatus* n.sp. (Singapore). Dworakowska 1971f:501. Typhlocybinae, Empoascini. Oriental; Malaya, Fiji, Samoa.

Debrescus 15:222. Error for *Drabescus* Stal.

Dechacona Young 1968a:255. *Tettigonia missionum* Berg 1879d:248 (Argentina: Missiones). Cicadellinae, Proconiini. Neotropical; Argentina, Brazil, Paraguay, Peru.

Declivana DeLong & Freytag 1963a:262 as subgenus of *Marganana* DeLong 1948b:101. *Marganana (Declivana) equata* n.sp. (British Guiana: Shudihar R.). Gyponinae, Gyponini. Neotropical.

Declivara DeLong & Freytag 1971a:321. *D. ornamenta* n.sp. ("St.Paola, coll. Signoret"). Gyponinae, Gyponini. Neotropical; Brazil.

Declivella DeLong & Freytag 1972f:241 as subgenus of *Polana* DeLong 1942d:110. *P. (D.) danesa* n.sp. (Panama: Tocumen). Gyponinae, Gyponini. Neotropical.

Decua Oman 1949a:72. *Cicadella cucurbita* Ball 1936a:20 (USA: Arizona, 12 mi west of Congress Junction). 1:264. Young 1977a:842. Cicadellinae, Cicadellini. Nearctic: USA; Arizona and New Mexico.

Dedra 4:140. Error for *Ledra* Fabricius.

Delongia Young 1952b:48 as subgenus of *Dikraneura* Hardy 1850a:423. *Dikraneura (Notus) luna* DeLong & Caldwell 1937a:24 (USA: California, Mojave). 17:155. Typhlocybinae, Dikraneurini. Nearctic.

Delopa Evans, J.W. 1971c:442. *D. cornuta* n.sp. (Mozambique: "Delagoa Bay" = B. de Lourenco Marques). Ulopinae, Ulopini. Ethiopian, southeast Africa.

Deltanus Oman 1949a:178. *Athysanus texanus* Osborn & Ball 1898f:92 (USA: Texas). 10:1827. Deltocephalinae, Deltocephalini. Nearctic.

Deltazotus Kramer 1971d:265. *Deltocephalus obesus* Osborn & Ball 1898f:81 (USA: Arizona). Deltocephalinae, Deltocephalini. Nearctic; Sonoran.

Deltella Oman 1949a:176. *Deltocephalus decisus* DeLong 1926d:55 (USA: Florida, LaBelle). 10:1828. Deltocephalinae, Deltocephalini. Nearctic; southeast USA.

Deltiocephalus 10:2653. Error for *Deltocephalus* Burmeister.

DELTOCEPHALIDAE 10:1.

DELTOCEPHALINA Hamilton, KGA 1975b:486, subtribe of DELTOCEPHALINI.

DELTOCEPHALINAE 10:1078.

DELTOCEPHALINI 10:11, 1088.

Deltoccphalus 10:2651. Error for *Deltocephalus* Burmeister.
Deltocephalas 10:2651. Error for *Deltocephalus* Burmeister.
Deltocephales 10:2651. Error for *Deltocephalus* Burmeister.
Deltocephalns 10:2651. Error for *Deltocephalus* Burmeister.

Deltocephalus Burmeister 1838b:15 as subgenus of *Jassus*. *Cicada pulicaris* Fallen 1806a:21 (Sweden). 10:1090. Kramer 1971c, Ossiannilsson 1983a:655-658. Deltocephalinae, Deltocephalini. Holarctic.
Subgenus:
Insulanus Linnavuori 1960a:303. *Stirellus subviridis* Metcalf 1964d:125.

Deltocephasus 10:2652. Error for *Deltocephalus* Burmeister.
Deltocephsbus 10:2653. Error for *Deltocephalus* Burmeister.
Deltocephtalus 10:2653. Error for *Deltocephalus* Burmeister.

Deltocoelidia Kramer 1961a:238. *D. maldonadoi* n.sp. (Venezuela: Territory Amazonas, Upper Cunucunuma, Julian). Neocoelidiinae, Neotropical.

Deltodorydium Kirkaldy 1907d:73 as subgenus of *Paradorydium* Kirkaldy 1901f:339. *P. (D.) brighami* n.sp. (Australia: New South Wales, Mittagong). 9:117. Eupelicinae, Paradorydiini. Australian. Synonym of *Paradorydium* Kirkaldy 1901f:339; Evans, J.W. 1966a:136.

Deltolidia Nielson 1982e:263. *Coelidia discolor* Stal 1862b:52 (Brazil). Coelidiinae, Coelidiini. Neotropical; Brazil.

Deltopinus Ball 1931d:218. *Platymetopius nigriviridis* Ball 1909c:163 (USA: California, Tia Juana). 10:2264. Deltocephalinae, Scaphytopiini. Nearctic. Synonym of *Scaphytopius (Cloanthanus)* Ball 1931d:219.

Deltorhynchus DeLong 1943a:79. *D. quadrinotus* n.sp. (Mexico: Guerrero, Iguala). 10:1741. Deltocephalinae, Deltocephalini. Neotropical.

Depanana Young 1968a:85. *Amblydisca bugabensis* Fowler 1898a:210 (Panama: Bugaba, 800-1500 ft). Cicadellinae, Proconiini. Neotropical; Guatemala, Panama.

Depanisca Young 1968a:87. *Tettigonia sulcata* Signoret 1855a:58 (Bolivia). Cicadellinae, Proconiini. Neotropical; Bolivia, Ecuador, Venezuela.

Deptocephalus 10:2653. Error for *Deltocephalus* Burmeister.

Derakandra Blocker 1979a:30. *D. matura* n.sp. (Colombia: Valle, 6 mi. W. Cali). Iassinae. Neotropical.

Deridna 11:147. Error for *Daridna* Walker.
Derogonia Melichar 1926a:353. *Tettigonia pilipennis* Signoret 1854b:342, pl. 11, fig.3. (Bolivia). 1:54. Young 1977a:1098. Cicadellinae, Cicadellini. Neotropical; Bolivia and Colombia.
Derriblocera Nielson 1983h:560. *D. ornata* n.sp. (Peru: Rio Madre de Dios, Atalaya). Coelidiinae, Teruliini. Neotropical.
Desamera Young 1968a:164. *Cicada intersecta* Germar 1821:60 (Brazil: Sao Paulo). Cicadellinae, Proconiini. Neotropical; Brazil, Peru.
Deselvana Young 1968a:151. *Proconia excavata* Le Peletier & Serville 1825a:611 (Brazil). Cicadellinae, Proconiini. Neotropical; Panama to southern Brazil.
Desertana DeLong & Martinson 1973b:125. *D. arida* n.sp. (Chile: P. Valparaiso, 7 km W. Llay-Llay). Deltocephalinae, Athysanini. Neotropical; Chile.
Destinia Nast 1952b:35. *D. maja* n.sp. (Sumatra). 4:65. Ledrinae, Petalocephalini. Oriental; Sumatra.
Destria Oman 1949a:178. *Thamnotettix bisignatus* Sanders & DeLong 1923a:154 (USA: Florida, Cleveland). 10:1828. Deltocephalinae, Deltocephalini. Nearctic; USA and southern Canada.
Dettocephalus 10:2653. Error for *Deltocephalus* Burmeister.
Devolana DeLong 1967b:22. *D. hemicyla* n.sp. (Mexico: Guerrero, Iguala). Deltocephalinae, Athysanini. Neotropical; Mexico.
Dharma 11:147. Error for *Dharmma* Distant.
Dharmma Distant 1908g:323. *D. projecta* n.sp. (Burma: Tenasserim). 11:22. Coelidiinae, Thagriini. Oriental. Synonym of *Thagria* Melichar 1903b:176; Nielson 1977a:9.
Dharrma Nielson 1977a:9. Error for *Dharmma* Distant 1908g.
Dhongariva Webb, M.D. 1983a:90. *D. spinosa* n.sp. (Australia: Queensland, 17 mi. S. Atherton). Idiocerinae. Australian.
Diacra Emeljanov 1961a:120. *D. convexa* n.sp. (USSR: Kazakhstan, 40 km S. Zhan-Ark, Karaganda region). Deltocephalinae, Opsiini. Palaearctic; Eurasia.
Diacraneura 17:1489. Error for *Dikraneura* Hardy.
Diadesmia Amyot 8:261. Invalid, not binominal.
Diadrocephala 1:678. Error for *Diedrocephala* Spinola.
Diakroneura 17:1489. Error for *Dikraneura* Hardy.
Diamachus 5:96. Error for *Daimachus* Distant.
Diastostema 1:678. Error for *Diestostemma* Amyot & Serville.
Dialecticopteryx Kirkaldy 1907d:71. *D. australica* n.sp. (Australia: Queensland, Bundaberg). 17:598. Evans, J.W. 1966a:262; Dworakowska 1970a:691. Typhlocybinae, Empoascini. Australian, Oriental.
Subgenus:
Akotettix Matsumura 1931b:77. *A. akonis* n.sp.
Synonyms (of *Akotettix*):
Banosa Mahmood 1967a:39. *B. sexpunctata* n.sp.
Dzhownica Dworakowska 1974a:157. *Akotettix (Dzhownica) cantoreanuae* n.sp.
Diataemops 8:262. Error for *Diataeniops*.
Diataeniops Amyot 8:262. Invalid, not binominal.
Diateniops Amyot 11:147. Invalid, not binominal.
Diceratalebra Young 1952b:26. *Alebra sanguinolinea* Baker 1903d:5 (Nicaragua: San Marcos). 17:24. Typhlocybinae, Alebrini. Neotropical, Nearctic; south Texas, Mexico, Central America.
Dichometopia Melichar 1925a:406 as subgenus of *Oncometopia* Stal 1869a:62. Preoccupied by *Dichometopia* Schiner 1868. *Tettigonia anceps* Fowler 1899c:234 (Mexico: Las Mercedes, 3000 ft). 1:595. Young 1968a:176. Cicadellinae, Proconiini. Neotropical; Mexico to northern Argentina. Synonym of *Egidemia* China 1927d:283.
Dichrophleps Stal 1869a:62. *Cicada aurea* Fabricius 1803a:63 ("Amer. Merid."). 1:512. Young 1968a:211. Cicadellinae, Proconiini. Neotropical; northern South America.
Dicolecia Nielson 1982e:275. *D. bifurcata* n.sp. (Peru: Cusca, Hacienda Maria). Coelidiinae, Coelidiini. Neotropical; Peru.
Dicodia Nielson 1982e:199. *Coelidia variegata* Germar 1821a:77 (Brazil). Coelidiinae, Coelidiini. Neotropical; Argentina, Bolivia, Brazil, Peru.
Dicrallygus Ribaut 1952a:214 as subgenus of *Allygidius* Ribaut 1948c:58. *Jassus furcatus* Ferrari 1882a:136 (Italy). 10:1000. Deltocephalinae, Athysanini. Palaearctic.
Dicraneura Puton 17:1489. Emendation of *Dikraneura* Hardy.

Dicraneurula Vilbaste 1968a:79. *D. silvicola* n.sp. (USSR: Korean Peninsula). Typhlocybinae, Dikraneurini. Palaearctic; USSR Maritime Territory.
Dicranoneura Douglas 1875b:27. 17:130,1489. Emendation of *Dikraneura* Hardy.
Dicranura 17:1489. Error for *Dicranoneura* Douglas.
Dicroneura 17:1489. Error for *Dikraneura* Hardy.
Dictrocephala 1:678. Error for *Diedrocephala* Spinola.
Dictyodisca Schmidt 1928a:50. *Amblydisca salvini* Fowler 1898e:209 (Panama: Bugaba). 1:619. Young 1968a:50. Cicadellinae, Proconiini. Neotropical; Panama, Costa Rica.
Dictyophorites Heer 1853b:114. *D. tingitinus* n.sp. Metcalf & Wade 1966a:208. Fossil: Tertiary, Croatia.
Dicyphonia Ball 1900a:69. *D. pamentosa* (sic) n.sp. (USA: Colorado, Fort Collins) = *Platymetopius ornatus* Baker 1900c:49. 9:61. Deltocephalinae, Hecalini. Nearctic; central great plains.
Didino 7:32. Error for *Didius* Distant.
Didius Distant 1918b:36. *D. sexualis* n.sp. (India: Calcutta). 7:20. Nirvaninae, Nirvanini. Oriental. Synonym of *Omaranus* Distant 1918b:5.
Diedrocephala Spinola 1850a:57. *Cicada variegata* Fabricius 1775a:684 (Brazil). 1:402. Young 1977a:322. Cicadellinae, Cicadellini. Neotropical; southern Mexico to Argentina.
Diedrocephla 1:679. Error for *Diedrocephala* Spinola.
Diemoides Evans, J.W. 1938b:13. *D. smithtoniensis* n.sp.(Tasmania). 10:2205. Deltocephalinae, Scaphoideini. Australian; Tasmania. Synonym of *Paralimnus* Matsumura 1902a:386; Evans, J.W. 1966a:236.
Diestostemma Amyot & Serville 1843a:572. *Cicada albipennis* Fabricius 1803a:62 ("Amer. Merid."). 1:487. Young 1968a:30. Cicadellinae, Proconiini. Neotropical; Mexico to Bolivia and Paraguay.
Synonyms:
Leucopepla Kirkaldy 1907d:87. *Tettigonia bituberculata* Signoret.
Heterostemma Melichar 1924a:227. *Tettigonia nervosa* Signoret.
Chiapasa Schmidt 1928a:40. *Tettigonia rugicollis* Signoret.
Pibrochoides Haupt 1929c:252. *Tettigonia rugicollis* Signoret.
Dieraneura 17:1490. Error for *Dicraneura* Puton.
Diglenita Spinola 1850b:129. (Name appears in key in 1850a:58). *D. peltastes* n.sp. (South Africa: Cape of Good Hope). 15:142. Iassinae. Ethiopian.
Dikrancura 17:1496. Error for *Dikraneura* Hardy.
Dikraneura Hardy 1850a:423. *D. variata* n.sp. (England). 17:129. Knight 1968a. Typhlocybinae, Dikraneurini. Holarctic.
Subgenus:
Delongia Young 1952b:48. *Dikraneura (Notus) luna* DeLong & Caldwell 1937a:24.
Synonym:
Chloroneura Walsh 1862a:149. *C. abnormis* n.sp.
DIKRANEURINI 17:100.
Dikraneuroidea Lawson 1929d:307. *D. beameri* n.sp. (USA: Texas, Brownsville). 17:224. Typhlocybinae, Dikraneurini. Nearctic. Synonym of *Alconeura (Hyloidea)* McAtee 1926b:162.
Dikrella Oman 1949a:83. *Dicraneura cockerelli* Gillette 1895a:14 (USA: New Mexico, Las Cruces). 17:269, 273. Typhlocybinae, Dikraneurini. Nearctic, Neotropical.
Subgenus:
Readionia Young 1952b:61. *Dikraneura readionis* Lawson 1930e:39.
Dikrellidia Young 1952b:64. *Dikraneura bilineata* Osborn 1928a:270 (Bolivia: Prov. del Sara). 17:282. Typhlocybinae, Dikraneurini. Neotropical.
Dilobonota Dworakowska 1972g:862 as subgenus of *Zyginoides* Matsumura 1931b:59. *Z. (D.) cristata* n.sp. (Sudan: Equatoria, Juba). Typhlocybinae, Erythroneurini. Ethiopian.
Dilobopterus Signoret 1850b:284. *D. decoratus* n.sp. (Brazil). 1:199. Young 1977a:104. Cicadellinae, Cicadellini. Neotropical; Mexico and West Indies, Central and South America to Argentina.
Synonyms:
Pseudoscarta Melichar 1926a:341. *Cicada fastuosa* Fabricius 1803a:70.
Marizella Schmidt 1928c:82. *M. polita* n.sp.

Dio Distant 1918b:56. *D. facialis* n.sp. (Eastern Himalayas). 10:418. Typhlocybinae, Dikraneurini. Palearctic. Synonym of *Erythria* Fieber 1866a:507; Dworakowska 1977a:603.

Diomma de Motschulsky 1863a:102. *D. ochracea* n.sp. (Ceylon). 17:619, 1490. Vilbaste 1975a:232, Dworakowska 1981c. Typhlocybinae, Erythroneurini. Oriental, Palaearctic and Ethiopian; Sudan.
Subgenera:
Bunyipia Dworakowska 1972g:860. *Zyginoides (B.) ulae* n.sp.
Dilobonota Dworakowska 1972g:862. *Zyginoides (D.) cristata* n.sp.
Synonyms:
Zyginoides Matsumura 1931b:59. *Eupteryx taiwanus* Shiraki 1912a:95.
Platytettix Matsumura 1932a:104, preoccupied. *Motschulskia pulchra* Matsumura 1916b:397.
Platytetticis Strand 1942a:393, replacement for *Platytettix* Matsumura 1932a, not *Platytettix* Hancock 1906.
Pakeasta Ahmed, Manzoor 1971b:188. *P. notata* n.sp.

Dioranerva 17:1491. Error for *Dicraneura* Puton.

Diostrostemma 1:679. Error for *Diestostemma* Amyot & Serville.

Diplocenus Kalkandelen 1972a:153. Error for *Diplocolenus* Ribaut.

Diplocolenoidea Linnavuori 1953d:58. *D. turkestanica* n.sp. (Turkestan: Aschabad) = *Thamnotettix macilentus* Horvath 1904b:586. 10:954. Deltocephalinae, Athysanini. Palaearctic. Synonym of *Stenometopiellus* Haupt 1917d:251; Nast 1972a:375.

Diplocolenus Ribaut 1946b:82. *Deltocephalus calceolatus* Boheman 1845a:23 (Sweden) = *Cicada bohemanni* (sic) Zetterstedt 1840a:290. 10:1333. Knight 1974c; Dlabola 1980c. Deltocephalinae, Paralimnini. Holarctic.
Subgenera:
Verdanus Oman 1949a:165. *Deltocephalus evansi* Ashmead 1904a:132.
Erdianus Ribaut 1952a:287. *Acocephalus sudeticus* Kolenati 1860a:390 = *Bythoscopus ? penthopittus* Walker 1851b:964.
Verdanulus Emeljanov 1966a:122. *Jassus nigrifrons* Kirschbaum 1868b:139.
Ribautanus Dlabola 1980c:76. *Diplocolenus convenarum* Ribaut 1946b:85.
Synonyms:
Sabelanus Ribaut 1959a:404. *Diplocolenus nasti* Wagner 1939a:168.
Gelidanus Emeljanov 1966a:122. *Cicada limbatella* Zetterstedt 1828a:522.

Diplocorenus Ishihara 1966a:39. Error for *Diplocolenus* Ribaut 1946b:82.

Discocephalana Metcalf 1952a:229, replacement for *Discocephalus* Kirschbaum 1858b:356 not *Discocephalus* Ehrenberg 1831. *Discocephalus viridis* Kirschbaum 1858b:356 (Germany). 15:141. Selenocephalinae, Selenocephalini. Palearctic.
Synonym:
Discocephalus Kirschbaum 1858b:356, preoccupied.

Discocephalus Kirschbaum 1858b:356. *D. viridis* n.sp. (Germany). 15:141. Preoccupied by *Discocephalus* Ehrenberg 1831, see *Discocephalana* Metcalf 1952a:229. Palaearctic.

Distantasca Dworakowska 1972i:25. *Empoasca terminalis* Distant 1918b:92 (South India: Chikkaballapura). Typhlocybinae, Empoascini. Oriental. Subgenus of *Empoasca* Walsh 1862a:149; Dworakowska & Viraktamath 1975a:529.

Distantessus Evans, F. 1981a:137. *Tartessus iphis* Kirkaldy 1907d:45 (Australia: Queensland, Nelson). Tartessinae. Australian.

Distantia Signoret 1879a:51. *D. frontalis* Signoret 1880a:66 (South Africa: Natal). 15:180. Campbell, M.B.S.C. & Davies 1985a. Selenocephalinae, Selenocephalini. Ethiopian; South Africa.

Distantiella Linnavuori & Quartau 1975a:154. Error for *Distantia* Signoret.

Distomotettix Ribaut 1938a:97. *Jassus fenestratus* Herrich-Schaeffer 1834a:5 (Austria). 10:881. Deltocephalinae, Opsiini. Palaearctic. Synonym of *Neoaliturus* Distant 1918b:63.

Divitiacus Distant 1918b:59. *D. primus* n.sp. (India: Nilgiri Hills, Ootacamund). 10:419. Viraktamath, S. & Viraktamath, C.A. 1980a:137-139. Deltocephalinae, Athysanini. Oriental.

Divus Distant 1908g:365. *Eutettix bipunctatus* Melichar 1903b:192 (Ceylon). 15:137. Selenocephalinae, Selenocephalini. Oriental; Sri Lanka, India.

Dixianus Ball 1918b:387 as subgenus of *Phlepsius* Fieber 1866a:514. *Phlepsius utahnus* Ball 1909a:79 (USA: Utah, "Chads"). 10:572. Deltocephalinae, Athysanini. Nearctic.

Dlabolaia Linnavuori 1959b:56. *D. rugipennis* n. sp (Argentina: Tucuman). Penthimiinae. Neotropical. Synonym of *Citorus* Stal 1866a:110; Theron 1980a:281.

Dlabolaiana Dworakowska 1974a:177. *D. detektata* n.sp. (Congo: Odzala). Typhlocybinae, Zyginellini. Ethiopian.

Dlabolaracus Remane 1961b:138. *Mocydiopsis klapperichi* Dlabola 1957b:284 (Afghanistan: Paghmangebirge). Deltocephalinae, Athysanini. Palaearctic.

Dlabolia Lang 1945a:62. *Deltocephalus ignoscus* Fieber 1869a:208 (Livonia). 10:412. Deltocephalinae, Athysanini. Palaearctic. Subgenus of *Handianus* Ribaut 1942a:265; Emeljanov 1964f:33.

Docalidia Nielson 1979b:179. *Jassus ruficosta* Jacobi 1905c:187 (Peru). Coelidiinae, Teruliini. Neotropical.

Dochmocarus Thomson 1869a:65 as subgenus of *Jassus. Cicada nervosa* Fallen 1826a:39 (Sweden) = *Athysanus obtusifrons* Stal 1853a:175. 10:2062, 2654. Deltocephalinae, Paralimnini. Palaearctic. Synonym of *Paramesus* Fieber 1866a:506.

Docotettix Ribaut 1948a:11. *D. cornutus* n.sp. (Cyprus). 10:390. Deltocephalinae, Fieberiellini. Palaearctic.

Doda Distant 1908f:144. *D. laudata* n.sp. ("Malay States"). 11:31. Unassigned, formerly Coelidiinae, Nielson 1975:11. Oriental.

Doleramus 10:2654. Error for *Doleranus* Ball.

Doleranus Ball 1936c:58. *Thamnotettix longulus* Gillette & Baker 1895a:97 (USA: Colorado, Ft.Collins). 10:1276. Linnavuori 1959b:274-276. Deltocephalinae, Athysanini. Nearctic, Neotropical.

Dolichopscerus Maldonado-Capriles 1985a:273. *D. spectrum* n.sp. (New Guinea NE: Finisterre Range, Saidor Matoko Vill). Idiocerinae. Oriental; New Guinea.

Doliotettix Ribaut 1942a:266. *Cicada pallens* Zetterstedt 1828a:522 (Lapland) = *Thamnotettix lunulata* Zetterstedt 1840a:295. 10:2031. Deltocephalinae, Athysanini. Palaearctic.

Dolyobius Linnavuori 1959b:282. *D. lingulatus* n.sp. (Ecuador: Banos). Deltocephalinae, Athysanini. Neotropical; Ecuador, Paraguay.

Domelia Ahmed, Manzoor & Waheed 1971a:117. *D. nigra* n.sp. (West Pakistan: Azid Kashmir, Domel). Typhlocybinae, Eupterygini. Oriental. Synonym of *Eupteryx* Curtis 1829a:192 or *Eupteryx (Stacla)* Dworakowska 1969e:439; Dworakowska 1978b:707.

Donidea Young 1952b:55. *Typhlocyba verticis* Baker 1903d:8 (Nicaragua: Managua). 17:194. Typhlocybinae, Dikraneurini. Neotropical.

Donleva Blocker 1979a:17. *Batracomorphus sialos* Kramer 1963a:39 (Costa Rica: San Pedro de Montes de Oca). Iassinae. Neotropical.

Donycephalus 9:119. Error for *Dorycephalus* Kouchakewitch.

Dorada Melichar 1902a:276. *D. lativentris* n.sp. (Colombia). 15:140. Preoccupied by *Dorada* Jarocki 1822, see *Doradana* Metcalf 1952a:229. Probably Gyponinae; Linnavuori 1959b:127. Neotropical.

Doradana Metcalf 1952a:229, replacement for *Dorada* Melichar 1902a:276 not *Dorada* Jarocki 1822. *Dorada lativentris* Melichar 1902a:276 (Colombia). 15:140. Probably Gyponinae; Linnavuori 1959b:127. Neotropical.
Synonym:
Dorada Melichar 1902a:276, not *Dorada* Jarocki 1822.

Doratula 10:2654. Error for *Doratulina* Melichar.

Doratulina Melichar 1903b:198. *D. jocosa* n.sp. (Ceylon). 10:1583. Vilbaste 1965a:10. Deltocephalinae, Stenometopiini. Oriental. [Australia].
Subgenus:
Cymbopogonella Viraktamath, C.A. 1976a:79. *D.(C.) longivertex* n.sp.
Synonyms:
Phrynophyes Kirkaldy 1906c:237. *P. phrynophyes* n.sp.
Arya Distant 1908g:338. *A. rubrolineata* n.sp.
Campbellinella Distant 1918b:69. *C. illustrata* n.sp.
Bituitus Distant 1918b:70. *B. projectus* n.sp.

Paternus Distant 1918b:71. *P. pusanus* n.sp.
Volusenus Distant 1918b:72. *V. lahorensis* n.sp.
Nandidrug Distant 1918b:74. *N. speciosum* n.sp.
Allectus Distant 1918b:75. *A. notatus* n.sp.
Galerius Distant 1918b:78. *G. indicatrix* n.sp.
Paivanana Distant 1918b:95. *Typhlocyba indra* Distant 1908g:415.
Bella Singh-Pruthi 1930a:44. *B. apicalis* n.sp.
Sunda Singh-Pruthi 1936a:112. *S. ribeiroi* n.sp.
Pseudaconura Linnavuori 1952b:182. *P. luxorensis* n.sp.
Umesaona Ishihara 1961a:246. *U. asiatica* n.sp.
Trebellius Distant 1918b:52. *T. albifrons* n.sp.; Nast 1972a:354.
[Note: Above indicated synonymy by Vilbaste 1965a:10].
Doratura Sahlberg 1871a:291. *Athysanus stylatus* Boheman 1847a:31 (Sweden). 8:227. Dworakowska 1967a, 1968c. Deltocephalinae, Doraturini. Palaearctic. [Nearctic].
Subgenus:
Doraturina Emeljanov 1964c:403. *Jassus (Athysanus) homophylus* Flor 1861a:276.
Doraturina Emeljanov 1964c:403 as subgenus of *Doratura* Sahlberg 1871a:71 (Livonia). *Jassus (Athysanus) homophylus* Flor 1861a:276. Deltocephalinae, Doraturini. Palaearctic.
DORATURINA Hamilton, KGA 1975b:482, subtribe of DORATURINI.
DORATURINI Emeljanov 1962b:391; tribe of Deltocephalinae. 8:227.
Doraturopsis Lindberg 1935c:423. *D. acutus* n.sp. (Mongolia) = *Doratura heros* Melichar 1902c:126. 10:955. Deltocephalinae, Doraturini. Palaearctic.
Synonym:
Adoratura Kusnezov 1938a:219. *Doratura heros* Melichar 1902a:126.
Dordium 9:119. Error for *Dorydium* Burmeister.
Dortulina Zoological Record 101, Section 13:282-3, 1964. Error for *Doratulina* Melichar.
Dorrotartessus Evans, F. 1981a:123. *D. dorrigensis* n.sp. (Australia: New South Wales, Dorrigo). Tartessinae. Australian.
Dorycara Emeljanov 1966a:100. *Dorycephalus platyrhynchus* Osborn 1894a:216 (USA: Nebraska, West Point). 9:70. Oman 1985:314. Eupelicinae, Dorycephalini. Nearctic.
DORYCEPHALINI 9:67.
Dorycephalus Kouchakewitch 1866a:102. *D. baeri* n.sp. (USSR:"Chkalov"). 9:67,70. Emeljanov 1966a:100. Eupelicinae, Dorycephalini. Palaearctic.
[NOTE: If the subfamily composition shown by Linnavuori 1979a:648-650 is accepted, the subfamily designation becomes Eupelicinae].
Dorycnia Dworakowska 1972b:403. *Typhlocyba funeta* Melichar 1903b:217 (Ceylon: Pattipola). Typhlocybinae, Erythroneurini. Oriental.
Dorydiella Baker 1897c:159. *D. floridana* n.sp. (USA: Florida). 10:2343. Deltocephalinae, Platymetopiini. Nearctic; eastern & central USA.
DORYDIIDAE 9:5.
DORYDIINAE 5:38.
DORYDIINI 9:104.
Dorydium Burmeister 1835a:106. *D. paradoxum* n.sp. (unknown) = *Cephalelus infumatus* Percheron 1832b. 5:39. Ulopinae, Cephalelini. Ethiopian. Synonym of *Cephalelus* Percheron 1832b.
Dorydium Burmeister 1839a:3. *D. lanceolatum* n.sp. (Sicily) = *Jassus paradoxus* Herrich-Schaeffer 1837a:4. 9:106. Preoccupied by *Dorydium* Burmeister 1835, see *Paradorydium* Kirkaldy. Eupelicinae, Paradorydiini. Palaearctic. Synonym of *Paradorydium* Kirkaldy 1901f:339.
Dorydyum 5:97. Error for *Dorydium* Burmeister.
Draberiella Dworakowska 1971g:647 as subgenus of *Agnesiella* Dworakowska 1970f:211. *Chikkaballapura quinquemaculata* Distant 1918b:108 (India: United Prov., Kumaon). Typhlocybinae, Typhlocybini. Oriental.
Synonym:
Sarejuia Ghauri 1974c:556. *Chikkaballapura quinquemaculata* Distant 1918b:108.
Drabescens 15:222. Error for *Drabescus* Stal.
DRABESCIDAE = Selenocephalinae; Linnavuori & Al-Ne'amy 1983a:19.
DRABESCINAE Linnavuori 1978a:41-42.

DRABESCINI 15:137. Linnavuori & Al-Ne'amy 1983a:21.
Drabescoides Kwon & Lee 1979c:53. *Selenocephalus nuchalis* Jacobi 1943a:30 (Manchuria). Selenocephalinae, Drabescini. Palaearctic.
Drabescus Stal 1870b:738 as subgenus of *Selenocephalus* Germar 1833a:180. *Bythoscopus remotus* Walker 1851b:866 (Philippines). 15:171. Linnavuori 1978a:42-43. Selenocephalinae, Drabescini. Oriental, Ethiopian, Japan and Oceania.
Drabrescus 15:222. Error for *Drabescus* Stal.
Draculacephala 1:679. Error for *Draeculacephala* Ball.
Draculaecephala 1:679. Error for *Draeculacephala* Ball.
Draecelacephala 1:679. Error for *Draeculacephala* Ball.
Draecuacephala 1:679. Error for *Draeculacephala* Ball.
Draecucephala 1:679. Error for *Draeculacephala* Ball.
Draeculacephala Ball 1901b:66. *Tettigonia mollipes* Say 1940b:312 (USA). 1:304. Young 1977a:580. Cicadellinae, Cicadellini. Nearctic/Neotropical. [Hawaii, western Africa].
Subgenus:
Carneocephala Ball 1927c:39. *Draeculacephala floridana* Ball 1901b:72.
DRAECULACEPHALINI 1:304.
Draeculocephala 1:679. Error for *Draeculacephala* Ball.
Dragonana Ball & Reeves 1927a:489. *Gypona dracontea* Gibson 1919a:100 (USA: Arizona). 3:154. DeLong & Freytag 1964c, 1972a:219. Gyponinae, Gyponini. Nearctic.
Drakensbergena Linnavuori 1961a:458. *D. fuscovittata* n.sp. (South Africa: Cape Prov., 8 mi ENE of Rhodes). Drakensbergeninae. Ethiopian.
DRAKENSBERGENINAE Linnavuori 1979a:673.
Dremuela Evans, J.W. 1966a:60. *D. hieroglyphica* n.sp. (Australia: Victoria, Stoughton Vale). Eurymelinae, Eurymelini. Australian.
Driatura 10:2654. Error for *Driotura* Osborn & Ball.
Driodurgades Servadei 1960a:236. Error for *Dryodurgades* Zachvatkin.
Drionea 10:2654. Error for *Drionia* Ball.
Drionia Ball 1915a:167. *D. nigra* n.sp. (USA: Oregon, Medford). 10:1086. Deltocephalinae, Cochlorhini. Nearctic; USA, California and southern Oregon.
Driotura Osborn & Ball 1898f:87. *Athysanus gammaroides* Van Duzee 1894f:209 (USA: Kansas, Madison Co.). 10:1661. Deltocephalinae, Doraturini. Nearctic; eastern & central USA and Canada, west to Alberta and Montana.
Drordana Nielson 1983e:20. *Jassus limus* Jacobi 1912a:39 (East Africa, Kutschuru, Ebeue). Coelidiinae, Youngolidiini. Ethiopian.
Druodurgades Lindberg 1960b:60. Error for *Dryodurgades* Zachvatkin.
Dryadomorpha Kirkaldy 1906c:335. *D. pallida* n.sp. (Australia: Queensland, Bundaberg). 10:2174. Webb, M.D. 1981b:49-55. Paraboloponinae. Ethiopian and southern Palaearctic; Asia, Australasia, Pacific Islands.
Synonyms:
Paganalia Distant 1917a:314. *P. virescens* n.sp.
Zizyphoides Distant 1918b:73. *Z. indicus* n.sp.
Rhombopsis Haupt 1927a:22. *R. virens* n.sp.
Calotettix Osborn 1934b:247. *C. metrosideri* n.sp.
Rhombopsana Metcalf 1952a:229. *Rhombopsis virens* Haupt 1927a:22.
Osbornitettix Metcalf 1952a:229. *Calotettix metrosideri* Osborn 1934b:247.
Yakunopona Ishihara 1954f:1. *Y. yakushimensis* n.sp.
Khamiria Dlabola 1979a:252. *K. mangrovecola* n.sp.
Drylix Edwards 1922a:206. *Cicada striola* Fallen 1806a:31 (Sweden). 10:128. Deltocephalinae, Athysanini. Holarctic. Synonym of *Limotettix* Sahlberg 1871a:224.
Dryocyba Vilbaste 1982a:12. *Typhlocyba carri* Edwards 1914a:170 (England). Typhlocybinae, Typhlocybini. Palaearctic. Subgenus of *Fagocyba* Dlabola 1958c:54; Lauterer 1983b:143.
Dryodurgades Zachvatkin 1946b:158 as subgenus of *Durgades* Distant 1912e:608. *Jassus reticulatus* Herrich-Schaeffer 1834d:4 (Germany). 14:5, 172. Agalliinae, Agalliini. Palaearctic.
Dryotura 10:2655. Error for *Driotura* Osborn & Ball.

Duanjina Kuoh, C.L. 1981b:205,217. *D. liangdiana* n.sp. (China: Qinghai-Xizang Plateau, Mangham 3800 m). Typhlocybinae, Erythroneurini. Palaearctic.

Duatartessus Evans, F. 1981a:165. *Bythoscopus badius* Walker 1870b:321 (New Guinea: Irian). Tartessinae. Oriental; New Guinea, Malaysia.

Dudanus Dlabola 1956a:34. *D. pallidus* n.sp. (Slovakia: Cenkov). Deltocephalinae, Athysanini. Palaearctic.

Dumorpha DeLong & Freytag 1975e:33. *D. dedeca* n.sp. (Venezuela: Caracas). Gyponinae, Gyponini. Neotropical.

Duranturopsis 10:2655. Error for *Duraturopsis* Melichar.

Duraturopsis Melichar 1908b:13. *D. katonae* n.sp. (Tanganyika). 10:956. Deltocephalinae, Athysanini. Ethiopian.

Duratopsis 10:2655. Error for *Duraturopsis* Melichar.

Durgades Distant 1912e:608. *D. nigropictus* n.sp. (Eastern Himalayas). 14:107. Agalliinae, Agalliini. Oriental.

Durgula Emeljanov 1964f:16. *D. lycii* n.sp. (USSR: Kazakhstan). Agalliinae, Agalliini. Palaearctic; Iran, Kazakhstan.

Dussana Distant 1908g:322. *D. quaerenda* n.sp. (Ceylon). 6:51. Cicadellinae, Evacanthini. Oriental; south India, Sri Lanka.

Dusuna Distant 1907g:188. *D. mouhoti* n.sp. (Thailand). 4:100. Ledrinae, Petalocephalini. Oriental; southeast Asia.

Duttaella Ramakrishnan & Menon 1971a:461. *D. punjabensis* n.sp. (Pakistan: "Muree Hills"). Typhlocybinae, Dikraneurini. Oriental.

Dwightia Linnavuori and Al-Ne'amy 1983a:24. *D. acutipennis* n.sp. (Central African Republic: "La Maboke"). Selenocephalinae, Dwightiini. Ethiopian; Guinean (Congolese).

DWIGHTIINI Linnavuori & Al-Ne'amy 1983a:23; tribe of SELENOCEPHALINAE.

Dworakowskaia Chou, I. & Zhang 1985a:293, 298. *D. hainanensis* n.sp. (China: Hainan Island, Guangdong Prov.). Typhlocybinae, Zyginellini. Oriental.

Dworakowskellina Kocak 1981a:31, replacement for *Amicula* Dworakowska 1971b:111, not *Amicula* Gray 1840. *Erythroneura (Zygina) amicula* Linnavuori 1965a:16 (Libya: near Tobruk). Typhlocybinae, Erythroneurini. Palaearctic.

Dysmorphoptila Handlirsch 1908a:492. *Belastoma liasina* Giebel 1856. Metcalf & Wade 1966a:219-220. Fossil: Jurassic, England.

Dzhownica Dworakowska 1974a:157 as subgenus of *Akotettix* Matsumura 1931b:77. *A. (D.) cantoreanuae* n.sp. (Congo: Odzala). Typhlocybinae, Empoascini. Ethiopian. Synonym of *Akotettix* Matsumura 1931b:77.

Dziwneono Dworakowska 1972l:197. *D. etcetera* n.sp. (North Australia; Darwin). Typhlocybinae, Dikraneurini. Australian.

Ebarrius Ribaut 1946b:82. *Deltocephalus cognatus* Fieber 1869a:214 (Austria: Carinthia). 16:229. Deltocephalinae, Deltocephalini. Palaearctic; Europe.

Ectipus 1:679. Error for *Ectypus* Signoret.

Ectomops Signoret 1879a:51. *E. chinensis* Signoret 1880a:49 (China). 9:49. Morrison 1973a:397. Deltocephalinae, Hecalini. Oriental. Synonym of *Glossocratus* Fieber 1866a:502.

Ectopiocephalus Kirkaldy 1906c:463. *E. vanduzeei* n.sp. (Australia: New South Wales). 3:160. Evans, J.W. 1966a:211. Penthimiinae. Australian.

Ectupon 1:679. Error for *Ectypus* Signoret.

Ectypus Signoret 1853c:263. *E. coriaceus* n.sp. (Bolivia). 1:656. Young 1977a:1098. Cicadellinae, Proconiini(?). Neotropical.

Ederranus Ribaut 1942a:267. *Athysanus discolor* Sahlberg 1871a:277 (Finland). 10:2029. Deltocephalinae, Athysanini. Palaearctic; Fennoscandia, N. Russia, Siberia, Tuva, Korean Peninsula, Sakhalin, Maritime Territory.
[NOTE: See comment under *Euscelis* Brulle].

Edijassus Evans, J.W. 1972b:656. *E. pallidus* n.sp. (NE New Guinea, 11 km S. of Mount Hagen (town), 2200 m). Iassinae, Iassini. Oriental. Synonym of *Batracomorphus* Lewis 1834a:51; Knight 1983b:31.

Edojassus Evans, J.W. 1982a:643. Error for *Edijassus* Evans.

Edwardsiana Zachvatkin 1929a:262. *Cicada rosae* Linnaeus 1758a:439 (Sweden). 17:966. Ossiannilsson 1981a:460-496. Typhlocybinae, Typhlocybini. Palaearctic. [Nearctic].

Edwardsiastes Kirkaldy 1900b:243. *Jassus (Athysanus) proceps* Kirschbaum 1868b:105 (Prussia). 10:1695. Deltocephalinae, Athysanini. Palaearctic. Synonym of *Rhytistylus* Fieber 1875a:404.

Edwardsina 17:1491. Error for *Edwardsiana* Zachvatkin.

Edwarsiana 17:1491. Error for *Edwardsiana* Zachvatkin.

Egellus 10:2655. Error for *Agellus* DeLong & Davidson.

Egenus Oman 1938b:363. *E. acuminatus* n.sp. (Argentina: Missiones, Loreto). 9:49. Linnavuori 1957a:139. Deltocephalinae, Hecalini. Neotropical; Argentina and Brazil.

Egidemia China 1927d:283, replacement for *Dichometopia* Melichar 1925a:406, not *Dichometopia* Schiner 1868. *Oncometopia anceps* Fowler 1899c:234 (Mexico: Las Mercedes, 3000 ft). 1:595. Young 1968a:176. Cicadellinae, Proconiini. Neotropical; Mexico to Argentina.
Synonyms:
Dichometopia Melichar 1925a:406, preoccupied.
Neometopia Schroeder 1959a:44. *Oncometopia fowleri* Distant 1908b:67.

Ehagua Melichar 1926a:345. *Cicadella dubiosa* Dozier 1931a:5 (Haiti: Port Margot or Fond-des-Negres). 1:429. Young 1977a:1081. Cicadellinae, Cicadellini. Neotropical; Hispaniola, Jamaica.

Ehromenellus 8:262. Error for *Errhomenellus* Puton.

Elabra Young 1952b:35. *Protalebra eburneola* Osborn 1928a:257 (Bolivia: Prov. del Sara). 17:83. Typhlocybinae, Alebrini. Neotropical.

Elbelus Mahmood 1967a:13. *E. tripunctatus* n.sp. (Thailand: Loei, Dansai, Kok Sathon, Phu Lomeo). Typhlocybinae, Erythroneurini. Oriental.

Elbursia Nast 1982a:327. Error for *Elburzia* Dlabola 1974a:7.

Elburzia Dlabola 1974a:71 *E. petrophila* n.sp. (Iran: Elburz Mtns.). Deltocephalinae, Athysanini. Palaearctic.

Eldama Dworakowska 1972b:397. *E. bisetosa* n.sp. (New Guinea: Papua, Kokoda, 1200 ft). Typhlocybinae, Erythroneurini. Oriental.

Eldarbala Young 1977a:261. *E. scrutator* n.sp. (Peru: Pucara). Cicadellinae, Cicadellini. Neotropical.

Eleazara Distant 1908g:182. *E. aedificatura* n.sp. (India: Assam, Dikrang Valley). 4:96. Ledrinae, Petalocephalini. Oriental; India and Sumatra.

Elevanosa DeLong 1977f:37 as subgenus of *Gypona* Germar 1821a:73. *Gypona (E.) vertara* n.sp. (Argentina: La Rioja Prov., Patquia). Gyponinae, Gyponini. Neotropical.

Elginus Theron 1975a:202. *Deltocephalus saltus* Naude 1926a:44 (South Africa: Cape Prov., Viljoen Pass). Deltocephalinae, Deltocephalini. Ethiopian.

Elimana 10:2655. Error for *Elymana* DeLong.

Elmana 10:2655. Error for *Elymana* DeLong.

Elphnesopius Nast 1984a:396, replacement for *Neophlepsius* Dubovskij 1966a:173, not *Neophlepsius* Linnavuori 1955b:118. *Neophlepsius concinnus* Dubovskij 1966a:173 (USSR: Tadzhikistan). Deltocephalinae, Athysanini. Palaearctic.

Elrabonia Linnavuori 1959b:335. *E. cavifrons* n.sp. (Argentina: Santa Fe, El Rabon). Deltocephalinae, Macrostelini. Neotropical.

Elymana DeLong 1936a:218. *Thamnotettix inornata* Van Duzee 1892d:303 (USA: New York, Lancaster). 10:843. Deltocephalinae, Cicadulini. Holarctic.
Synonym:
Solenopyx Ribaut 1939a:273. *Cicada sulphurella* Zetterstedt 1928a:534.
[Note: *Elymana* Placed on Official List of Generic Names in Zoology, Opinion 603, Bulletin of Zoological Nomenclature 18].

Emboasca Young 1957b:209. Error for *Empoasca* Walsh.

Emelyanogramma Kocak 1981a:33, replacement for *Homogramma* Emeljanov 1975a:386, not *Homogramma* Guenee 1854. *Melicharella proxima* Dlabola 1960c:8 (Iran: Iranshar). Adelungiinae, Adelungiini. Palaearctic.
Synonym:
Homogramma Emeljanov 1975a:386. *Melicharella proxima* Dlabola 1960c:8.

Emelyanoviana Anufriev 1970d:263. *Typhlocyba mollicula* Boheman 1845b:160 (Sweden). Typhlocybinae, Dikraneurini. Palaearctic.

Emeljanovianus Dlabola 1965c:126 as subgenus of *Sorhoanus* Ribaut 1946b:85. *S. suncharicus* n.sp. (Mongolia: "Zuun-chara", 1390 m). Deltocephalinae, Deltocephalini. Palaearctic.
[Note: Raised to generic rank by Vilbaste 1980b:50-51].
Empea 17:1491. Error for *Empoa* Fitch.
Empoa Fitch 1851a:63. *E. querci* n.sp. (USA: New York). 17:935. Typhlocybinae, Typhlocybini. Nearctic. Subgenus of *Typhlocyba* Germar 1833a:180; Hamilton, KGA 1983a:67.
Synonym:
Empoides Vilbaste 1968a:93. *E. rubellus* n.sp.
Empoacea 17:1491. Error for *Empoasca* Walsh.
Empoanara Distant 1918b:106. *E. militaris* n.sp. (South India: Nilgiri Hills). 17:609. Typhlocybinae, Erythroneurini. Oriental.
Empoasa 17:1491. Error for *Empoasca* Walsh.
Empoasca Walsh 1862a:149. *E. viridescens* n.sp. = *Tettigonia fabae* Harris 1841a:186 (USA: Illinois). 17:286, 442. Typhlocybinae, Empoascini. All major land masses excepting polar regions.
Subgenera:
Kybos Fieber 1866a:508. *Cicada smaragdula* Fallen 1806a:37.
Hebata DeLong 1931b:32. *Empoasca nigra* Gillette & Baker 1895a:108.
Endeia McAtee 1934a:113. *E. (E.) virgata* n.sp.
Distantasca Dworakowska 1972i:25. *Empoasca terminalis* Distant 1918b:92.
Matsumurasca Anufriev 1973b:438. *Empoasca diversa* Vilbaste 1968a:87.
Ociepa Dworakowska 1977e:850. *E. (O.) medleri* n.sp.
Marolda Dworakowska 1977e:851. *E. (M.) testacea* n.sp.
Livasca Dworakowska & Viraktamath 1978a:540. *E. (L.) malliki* n.sp.
Okubasca Dworakowska 1982b:53. *Empoasca okubella* Matsumura 1931b:81.
Greceasca Thapa 1985a:67. *E. (G.) kapoori* n.sp.
Synonym:
Sabourasca Ramakrishnan & Menon 1972b:183. *S. peculiaris* n.sp.
Empoascanara Distant 1918b:94. *E. prima* n.sp. (South India: Chikkaballapura). 17:1328. Dworakowska 1971e, 1978a, 1979b, 1980c, 1980d. Typhlocybinae, Erythroneurini. Oriental, Ethiopian; Nigeria.
Subgenera:
Kanguza Dworakowska 1972k:121. *K. ibis* n.sp; Dworakowska 1976b:18.
Bza Dworakowska 1979b:39. *E. (B.) risa* n.sp.
Synonyms:
Ratbura Mahmood 1967a:17. *Empoasca nagpurensis* Distant 1918b:93.
Indoformosa Ramakrishnan & Ghauri 1979b:198. *Zygina indica* Datta 1969a:391.
Westindica Ramakrishnan & Ghauri 1979b:199. *Empoascanara stilleri* Dworakowska 1978a:157.
Pantanarendra Ramakrishnan & Ghauri 1979b:199. *Zygina fumigata f. sonani* Matsumura 1931b:66.
Swarajnara Ramakrishnan & Ghauri 1979b:200. *Ratbura unipunctata* Mahmood 1967b:18.
Pradhanasudra Ramakrishnan & Ghauri 1979b:201. *Empoascanara capreola* Dworakowska 1978a:154.
Subbanara Ramakrishnan & Ghauri 1979b:201. *Empoascanara linnavuorii* Dworakowska 1972k:121.
Irenara Ramakrishnan & Ghauri 1979b:204. "*Thamnotettix limbata* Matsumura, 1907b:89" = *Typhlocyba (Zygina) limbata* Matsumura 1910e:120.
Manzoonara Ramakrishnan & Ghauri 1979b:205. *Erythroneura hazarensis* Ahmed 1970b:179.
Sohinara Ramakrishnan & Ghauri 1979b:205. *Empoascanara falcata* Dworakowska 1971e:341.
Afroindica Ramakrishnan & Ghauri 1979b:207. *Empoascanara tagabica* Dworakowska & Trolle 1976a:372.
Sawainara Ramakrishnan & Ghauri 1979b:208. *Empoascanara lutes* Dworakowska 1977f:298.
Ishiharanara Ramakrishnan & Ghauri 1979b:208. *Empoascanara wasata* Dworakowska 1978a:156.
Sayara Ramakrishnan & Ghauri 1979b:209. *Empoascanara mana* Dworakowska & Pawar 1974a:583.
Webbanara Ramakrishnan & Ghauri 1979b:210. *Typhlocyba fumigata* Melichar 1903b:217.
Vietnara Ramakrishnan & Ghauri 1979b:210. *Typhlocyba maculifrons* de Motschulsky 1863a:103.
[Note: Above synonymy by Dworakowska 1980d:195].
EMPOASCINAE 17:100.
EMPOASCINI 17:286.
Empoasco 17:1491. Error for *Empoasca* Walsh.
Empoasea 17:1491. Error for *Empoasca* Walsh.
Empocisco 17:1491. Error for *Empoasca* Walsh.
Empoides Vilbaste 1968a:93. *E. rubellus* n.sp. (USSR: Maritime Provinces). Typhlocybinae, Typhlocybini. Palaearctic. Synonym *Empoa* Fitch 1851a:63; Hamilton, KGA 1983a:67.
Emposca 17:1491. Error for *Empoasca* Walsh.
Enacanthus 6:58. Error for *Evacanthus* Le Peletier & Serville.
Enanthiocephalus 10:2655. Error for *Enantiocephalus* Haupt.
Enantiocephalus Haupt 1926b:308. *Jassus cornutus* Herrich-Schaeffer 1838d:5 (Russia: Moscow). 10:2340. Deltocephalinae, Paralimnini. Palaearctic.
Encanthus 6:59. Error for *Evacanthus* Le Peletier & Serville.
Endeia McAtee 1934a:113 as subgenus of *Empoasca* Walsh 1862a:149. *E. (E.) virgata* n.sp. (Philippines: Mindanao, Zamboanga). 17:551. Typhlocybinae, Empoascini. Oriental.
Endoxoneura Young 1952b:53. *Dikraneura (Hyloidea) splendidula* Osborn 1928a:275 (Bolivia: Prov. del Sara). 17:194. Typhlocybinae, Dikraneurini. Neotropical.
Endria Oman 1949a:175. *Jassus inimicus* Say 1830b:305 (USA: Virginia). 10:1813. Deltocephalinae, Deltocephalini. Nearctic. [Palaearctic].
Enseelis 10:2655. Error for *Euscelis* Brulle.
Entaeniothes Amyot 8:262. Invalid, not binominal.
Enteniothes 1:148. Invalid, not binominal.
Entogonia Melichar 1926a:360. *Tettigonia sagata* Signoret 1854a:27 (Mexico). 1:30. Young 1977a:696. Cicadellinae, Cicadellini. Neotropical. Synonym of *Sibovia* China 1927d:283.
Eogypona Kirkaldy 1901b:38. *E. kirbyi* n.sp. = *Ledropsis kirbyi* Melichar 1903b:143 (Ceylon). 4:81. Ledrinae, Petalocephalini. Oriental.
Eohaldorus Linnavuori 1959b:149 as subgenus of *Haldorus* Oman 1938b:373. *Deltocephalus australis* DeLong 1926d:90 (USA: Florida, Miami). Deltocephalinae, Deltocephalini. Nearctic.
Eohardya Zachvatkin 1946b:171 as subgenus of *Hardya* Edwards 1922a:206. *H. (E.) mira* n.sp. (Persia). 10:265. Deltocephalinae, Athysanini. Palaearctic; Eurasia.
Synonym:
Hardyopsis Ribaut 1948b:218. *Hardya insularis* Lindberg 1948b:155, Linnavuori 1962a:47.
[NOTE: Nast 1972a:375 lists both *Eohardya* and *Hardyopsis* as synonyms of *Stenometopiellus* Haupt 1917d:250, but Dlabola 1981a:290 does not accept that synonymy. There appears to be agreement that *Hardyopsis* is a junior synonym of *Eohardya*].
Eojassus Handlirsch 1939a:145. *E. indistinctus* n.sp. Metcalf & Wade 1966a:216. Fossil: Jurassic, Germany.
Eovulturnops Evans, J.W. 1947a:255. *E. selenocephaloides* n.sp. (British Guiana). 3:212. Penthimiinae. Neotropical.
Ephelodes Emeljanov 1972d:103. *E. prasinus* n.sp. ((USSR: Kazakhstan, Kzyl-Orda Prov., 80 km W of Kzyl-Orda). Deltocephalinae, Acinopterini. Palaearctic.
Ephemerinus Emeljanov 1964f:35 as subgenus of *Handianus* Ribaut 1942a:265. *H. (E.) magnificus* n.sp. (USSR: Kazakhstan). Deltocephalinae, Athysanini. Palaearctic.
Epiacanthus Matsumura 1902a:353. *Deltocephalus stramineus* de Motschulsky 1861a:24 (Japan). 1:397. Anufriev 1976a. Cicadellinae, Pagaroniini. Palaearctic; Japan, Korean Peninsula, Kurile Islands, Sakhalin, Soviet Maritime Territory.
Epicephalius Matsumura 1908a:42. *E. gracilis* n.sp. (Algeria: Oran). 9:114. Deltocephalinae, Hecalini. Palaearctic; Algeria, southern Spain.
Synonym:
Bousaada Linnavuori 1971a:62. *B. psapfa* n.sp.; Vilbaste 1976a:35.
Epicephalus Vilbaste 1976a:35. Error for *Epicephalius* Matsumura.

Epiclinata Metcalf 1952a:228, replacement for *Epiclines* Amyot & Serville 1843a:577, not *Epiclines* Chevrolat 1838. *Membracis planata* Fabricius 1794a:11 (India). 4:105. Ledrinae, Petalocephalini. Oriental.
Synonym:
Epiclines Amyot & Serville 1843a:577, preoccupied.
Epiclines Amyot & Serville 1843a:577. *Membracis planata* Fabricius 1794a:11 (India). 4:105. Ledrinae, Petalocephalini. Oriental. Preoccupied by *Epiclines* Chevrolat 1838, see *Epiclinata* Metcalf 1952a:228.
Epignoma Dworakowska 1972i:32. *Zygina nuchalis* Jacobi 1910b:133 (Tanganyika). Dworakowska 1977c. Typhlocybinae, Empoascini. Ethiopian.
Epimelita 17:1492. Error for *Epimiltia* Amyot.
Epimiltia Amyot 17:1492. Invalid, not binominal.
Epipsychidion Kirkaldy 1906c:345. *E. epipyropis* n.sp. (Australian: New South Wales, Sydney). 4:134. Evans, J.W. 1966a:118. Ledrinae, Thymbrini. Australian.
Epipychidion 4:140. Error for *Epipsychidion* Kirkaldy.
Epirrhaena 15:222. Invalid, not binominal.
Epistictia Amyot 10:2655. Invalid, not binominal.
Epitaenia Amyot 5:97. Invalid, not binominal.
Epitettix Matsumura 1914a:194. *E. hiroglyphicus* n.sp. (Formosa). 10:878. Deltocephalinae, Athysanini. Oriental. Preoccupied by *Epitettix* Hancock 1907, see *Matsumuratettix* Metcalf 1952a:229.
Epoasca 17:1492. Error for *Empoasca* Walsh.
Epteryx 17:1492. Error for *Eupteryx* Curtis.
Equeefa Distant 1910e:242. *E. castelnaui* n.sp. ("Kaffraria"). 11:86. Theron 1984c:315-317. Coelidiinae. Ethiopian.
Erabla Young 1957c:212. *Protalebra lineola* Osborn 1928a:263 (Guatemala: Los Amates). Typhlocybinae, Alebrini. Neotropical.
Erasmoneura Young 1952b:80. *Erythroneura vulnerata* Fitch 1851a:62 (USA: Maryland). 17:1179. Typhlocybinae, Erythroneurini. Nearctic. Subgenus of *Erythroneura* Fitch 1851a:62.
Eratoneura Young 1952b:84. *Erythroneura dira* Beamer 1931d:286 (USA: Kansas, Leavenworth Co.). 17:1257. Typhlocybinae, Erythroneurini. Nearctic. Subgenus of *Erythroneura* Fitch 1851a:62.
Erdianus Ribaut 1952a:287 as subgenus of *Diplocolenus* Ribaut 1946b:82. *Acocephalus sudeticus* Kolenati 1860a:390 (France) = *Bythoscopus ? penthopittus* Walker 1851b:864. 10:1354. Deltocephalinae, Paralimnini. Palaearctic.
Synonyms:
Gelidanus Emeljanov 1966a:122. *Cicada limbatella* Zetterstedt 1828a:522.
Sabelanus Ribaut 1959a:404. *Diplocolenus nasti* Wagner 1939a:168.
Eremitopius Lindberg 1927b:29. *E. albus* n.sp. (Turkestan). 9:67. Deltocephalinae, Platymetopiini. Palaearctic. Synonym of *Platymetopius* Burmeister 1838b:16; Nast 1972a:358. Palaearctic.
Eremochlorita Zachvatkin 1946b:150 as subgenus of *Chlorita* Fieber 1872a:14. *C. (E.) uvaroviana* n.sp. (Turkey). 17:549. Typhlocybinae, Empoascini. Palaearctic.
[NOTE: Metcalf 1968a:549 treats *Eremochlorita* as a subgenus of *Empoasca* Walsh 1862. Nast 1972a and Dworakowska 1977f treat it as a subgenus of *Chlorita* Fieber 1872a. Most other recent workers (e.g. Dlabola 1981a, Emeljanov 1977a, Mitjaev 1971a) use the name in the generic sense].
Eremophlepsius Zachvatkin 1924a:127. *E. rohdendorfi* n.sp. (Turkmenia). 10:543. Deltocephalinae, Opsiini. Palaearctic; USSR: Kazakhstan, Kirghizia, Turkmenia, Uzbekistan.
Erhomenellus 8:262. Error for *Errhomenellus* Puton.
Erhtyroneura 17:1492. Error for *Erythroneura* Fitch.
Ericotettix Lindberg 1960a:54. *E. ericae* n.sp. (Portugal: Penhas da Saude). Deltocephalinae, Fieberiellini. Palaearctic.
Erinwa Ghauri 1975b:483. *E. delicata* n.sp. (Nigeria: Ibadan). Typhlocybinae, Empoascini. Ethiopian. Synonym of *Luvila* Dworakowska 1974a:143; Dworakowska 1977e:848.
Erithria 17:1492. Error for *Erythria* Fieber.

Erotettix Haupt 1929c:255. *Thamnotettix cyane* Boheman 1845b:158 (Sweden). 10:2479. Ossiannilsson 1983a:621. Deltocephalinae, Macrostelini. Palaearctic. Synonym of *Macrosteles* Fieber 1866a:504.
Erottetix 10:2655. Error for *Erotettix* Haupt.
Errastrunus 10:2656. Error for *Errastunus* Ribaut.
Errastunus Ribaut 1946b:83. *Cicada ocellaris* Fallen 1806a:20 (Sweden). 10:1313. Deltocephalinae, Paralimnini. Holarctic. Subgenus of *Adarrus* Ribaut 1946b:83; Emeljanov 1966a:119.
ERRHOMENELLIDAE 8:10.
ERRHOMENELLINI 8:13.
Errhomellus 8:262. Error for *Errhomenellus* Puton.
Errhomenellus Puton 1886b:79, replacement for *Errhomenus* Fieber 1866a:501. *Errhomenus brachypterus* Fieber 1866a:512 (Germany). 8:19. Cicadellinae, Errhomenini. Palaearctic. Synonym of *Errhomenus* Fieber 1866a:501.
Errhomenillus 8:263. Error for *Errhomenellus* Puton.
ERRHOMENINA Anufriev 1978a:58, subtribe of ERRHOMENINI.
ERRHOMENINI 8:12. Oman & Musgrave 1975a; Anufriev 1978a; Kwon 1983a.
Errhomenus Fieber 1866a:501. *E. brachypterus* n.sp.(Germany). 8:19. Anufriev 1978a:59. Cicadellinae, Errhomenini. Palaearctic.
Synonym:
Errhomenellus Puton 1886b:79. *Errhomenus brachypterus* Fieber 1866a:512.
Errhomus Oman 1938c:169. *Errhomenus lineatus* Baker 1898d:261 (USA: Washington, Pullman). 8:23. Oman & Musgrave 1975a:5-6; Anufriev 1978a:58. Cicadellinae, Errhomenini. Nearctic.
Ersaleus Mitjaev 1968b:35. Error for *Erzaleus* Ribaut.
Erthroneura 17:1492. Error for *Erythroneura* Fitch.
Eruthroneura 17:1492. Error for *Erythroneura* Fitch.
Eryapus Evans, J.W. 1954a:116. *E. gibbus* n.sp. (Madagascar). 15:208. Acostemminae. Malagasian.
Erythia 17:1492. Error for *Erythria* Fieber.
Erythoneura 17:1492. Error for *Erythroneura* Fitch.
Erythraneura 17:1492. Error for *Erythroneura* Fitch.
Erythria Fieber 1866a:507. *Cicada aureola* Fallen 1806a:25 (Sweden). 17:181. Dworakowska 1977a. Typhlocybinae, Dikraneurini. Palaearctic, Oriental.
Synonyms:
Dio Distant 1918b:56. *D. facialis* n.sp.
Erythridea Ribaut 1936b:223. *Deltocephalus ferrarii* Puton 1877a.
Basuaneura Ramakrishnan & Menon 1971a:465. *B. kalimpongensis* n.sp.
[NOTE: Ghauri (1974c:558) and Dworakowska (1971d, 1977a) disagree about the concept of *Erythria*, and the status of *Dio*. The above synonymies reflect the opinions of Dworakowska (1977a) and Sohi and Dworakowska (1984a:165)].
Erythridea Ribaut 1936b:223. *Deltocephalus ferrarii* Puton 1877a:XXIII (Italy). 17:1481. Dworakowska 1977a:603. Typhlocybinae, Erythroneurini. Palaearctic. Synonym of *Erythria* Fieber 1866a:507; Dworakowska 1977a:603.
Erythridula Young 1952b:81. *Tettigonia obliqua* Say 1825a:342 (USA). 17:1195. Typhlocybinae, Erythroneurini. Nearctic. Subgenus of *Arboridia* Zachvatkin 1946b:153; Dworakowska 1970g.
Synonym:
Khoduma Dworakowska 1972b:403; Dworakowska & Viraktamath 1975a:529.
Erythrita 17:1493. Error for *Erythria* Fieber.
Erythrogonia Melichar 1926a:373. *Cicada laeta* Fabricius 1787a:271 (French Guiana). 1:42. Medler 1963a; Young 1977a:767. Cicadellinae, Cicadellini. Neotropical; Mexico, Trinidad, Central America, South America to Brazil.
Erythroneaura 17:1493. Error for *Erythroneura* Fitch.
Erythroneura Fitch 1851a:62. *E. tricincta* n.sp. (USA: New York). 17:1050, 1109. Typhlocybinae, Erythroneurini. Cosmopolitan, primarily Nearctic.
Subgenera:
Erasmoneura Young 1952b:80. *Erythroneura vulnerata* Fitch 1851a:62.

Erythridula Young 1952b:81. *Tettigonia obliqua* Say 1825a:342.
Eratoneura Young 1952b:84. *Erythroneura dira* Beamer 1931d:286.
Balila Dworakowska 1970d:347. *Typhlocyba (Zygina) mori* Matsumura 1910e:121.
ERYTHRONEURINI 17:1049.
Erythroneuropsis Ramakrishnan & Menon 1973a:37. *E. indicus* n.sp. (India: Bihar, Sabour). Typhlocybinae, Erythroneurini. Oriental. Synonym of *Singapora* Mahmood 1967a:20; Dworakowski & Sohi 1978a:39.
Erythronura 17:1493. Error for *Erythroneura* Fitch.
Erythroueura 17:1493. Error for *Erythroneura* Fitch.
Erytria 17:1493. Error for *Erythria* Fieber.
Erytroneura 17:1493. Error for *Erythroneura* Fitch.
Eryttiria 17:1493. Error for *Erythria* Fieber.
Eryttoria 17:1493. Error for *Erythria* Fieber.
Erzaleus Ribaut 1952a:296 as subgenus of *Mocuellus* Ribaut 1946b:83. *Jassus (Deltocephalus) metrius* Flor 1861a:264 (Livonia). 10:1383. Deltocephalinae, Paralimnini. Palaearctic.
Esolanus Ribaut 1952a:87 as subgenus of *Laburrus* Ribaut 1942a:268. *Athysanus pellax* Horvath 1903c:472 (Hungary). 10:88. Deltocephalinae, Athysanini. Palaearctic.
Euacanthella Evans, J.W. 1938b:8. *E. palustris* n.sp. (Tasmania: Snug). 8:248. Evans, J.W. 1974a:173. Knight 1974b:476. Euacanthellinae. Australian.
EUACANTHELLINAE Evans, J.W. 1974a:173.
EUACANTHELLINI Evans, J.W. 1966a:142; new tribe for *Euacanthella* Evans, J.W. 1938b; raised to subfamily rank.
EUACANTHIDAE 6:1.
Euacanthus Burmeister 1835a:116. *Cicada acuminata* Fabricius 1794a:36 (Germany). 6:3. Cicadellinae, Evacanthini. Holarctic. Synonym of *Evacanthus* Le Peletier & Serville 1825a:612.
Euacanthvs 6:61. Error for *Euacanthus* Burmeister.
Euaccanthes 6:61. Error for *Euacanthus* Burmeister.
Euacenthus 6:61. Error for *Euacanthus* Burmeister.
Eualebra Baker 1899b:402. *E. smithii* n.sp. (Brazil: Chapada). 17:608. Typhlocybinae, Jorumini. Neotropical.
Eucanthus 6:61. Error for *Euacanthus* Burmeister.
Eucelis 10:2656. Error for *Euscelis* Brulle.
Euetettix 10:2656. Error for *Eutettix* Van Duzee.
Eugnathodes 10:2656. Error for *Eugnathodus* Baker.
Eugnathodus Baker 1903a:1. *Gnathodus abdominalis* Van Duzee 1892a:113 (USA: New Jersey, Jamesburgh). 10:2382. Deltocephalini, Balcluthini. Holarctic. Synonym of *Balclutha* Kirkaldy 1900b:243.
Eugnathoudus Datta 1973d:226. Error for *Eugnathodus* Baker.
Eugonalia Evans, J.W. 1947a:161, replacement for *Eugonia* Melichar 1926a:342, not *Eugonia* Heubner 1819. *Tettigonia diversa* Signoret 1855a:49 (French Guiana). 1:263. Cicadellinae, Cicadellini. Neo-tropical. Synonym of *Iragua* Melichar 1926a:342; Young 1977a:399.
Eugonia Melichar 1926a:342. *Tettigonia diversa* Signoret 1855a:49. 1:263. Preoccupied by *Eugonia* Huebner, see *Eugonalia* Evans, J.W. 1947a:161.
Euleimomos 10:2657. Error for *Euleimonios* Kirkaldy.
Euleimonios Kirkaldy 1906c:342. *E. dimittendus* n.sp. (Australia: New South Wales). 10:2487. Deltocephalinae, Deltocephalini. Australian.
Synonym:
Aconuromimus Linnavuori 1954d:83. *Jassus (Deltocephalus) flavidiventris* Stal 1859b:294.
Eulonus Oman 1949a:97. *Euscelis almus* Van Duzee 1925b:421 (USA: California, Merced Co., Los Banos). 10:1079. Deltocephalinae, Cochlorhini. Nearctic.
Eulopa 5:97. Error for *Ulopa* Fallen.
Eupelex 9:119. Error for *Eupelix* Germar.
Eupelexa 9:120. Error for *Eupelix* Germar.
EUPELICIDAE 9:5.
EUPELICINAE 9:6.
EUPELICINI 9:75.
Eupelis 10:2657. Error for *Euscelis* Brulle.
Eupelis 9:120. Error for *Eupelix* Germar.
Eupelix Germar 1821a:94. *Cicada cuspidata* Fabricius 1775a:687 (England). 9:76. Eupelicinae, Eupelicini. Palaearctic.

Eupenthimia Evans, J.W. 1972a:182. *E. nigra* n.sp. (Northwest New Guinea, Star Mtns., Sibil Valley 1245 m). Penthimiinae. Oriental.
Eupertyx 17:1493. Error for *Eupteryx* Curtis.
Euplex 9:120. Error for *Eupelix* Germar.
Euplix 9:120. Error for *Eupelix* Germar.
Eupoacea 17:1493. Error for *Empoasca* Walsh.
Eupoasca 17:1493. Error for *Empoasca* Walsh.
Euprora Evans, J.W. 1938b:9. *E. mullensis* n.sp. (Western Australia: Mullewa). 11:144. Preoccupied by *Euprora* Busck 1906, see *Stenotartessus* Evans, J.W. 1947a:207. Tartessinae. Australian. Synonym of *Newmaniana* Evans, J.W. 1941f:152.
Eupterella DeLong & Ruppel 1950a:239. *E. mexicana* n.sp. (Mexico: Michoacan, Carapan). 17:1048. Typhlocybinae, Typhlocybini. Neotropical, Oriental?.
Eupterex 17:1494. Error for *Eupteryx* Curtis.
Euptericyba Kirejtchuk 1977a:17. Error for *Eupterycyba* Dlabola.
Eupterix Fieber 1866a:509. 17:621. Emendation of *Eupteryx* Curtis.
Eupteroidea Young 1952b:92. *Typhlocyba stellulata* Burmeister 1841a:3 (Germany). 17:772. Typhlocybinae, Eupterygini. Palaearctic, Nearctic (adventive). Synonym of *Aguriahana* Distant 1918b:105; Dworakowska 1972a:278.
Eupterpx 17:1494. Error for *Eupteryx* Curtis.
Eupterycyba Dlabola 1958c:55 as subgenus of *Typhlocyba* Germar 1833a:180. *Typhlocyba jucunda* Herrich-Schaeffer 1837a:16 (Germany). Typhlocybinae, Typhlocybini. Palaearctic.
EUPTERYGIDAE 17:6.
EUPTERYGINA Anufriev 1978a:107, tribe in TYPHLOCYBINAE.
EUPTERYGINAE 17:100.
EUPTERYGINI 17:613.
Eupteryx Curtis 1829a (1831):192. *Cicada picta* Fabricius 1794a:32 (Germany) = *Cicada atropunctata* Goeze 1778a:161. 17:619. Dworakowska 1970b, 1971a, 1972d, 1978b; Ossiannilsson 1981a:518-541. Typhlocybinae, Eupterygini. Palaearctic. Oriental.
Subgenus:
Stacla Dworakowska 1969e:439. *E. (S.) cristagalli* n.sp.
Synonym:
Domelia Ahmed, Manzoor & Waheed 1971a:117. *D. nigra* n.sp.
[Note: *Eupteryx* Curtis 1831 validated under the plenary powers of the ICZN, Opinion 647, 18 March 1963, Bulletin of Zoological Nomenclature 20:35-38].
Eupteryz 17:1502. Error for *Eupteryx* Curtis.
Euptleryx 17:1502. Error for *Eupteryx* Curtis.
Euragallia Oman 1938b:351. *Agallia furculata* Osborn 1923b:9 (Brazil: Bahia, Bom Fim). 14:85. Agalliinae, Agalliini. Neotropical.
Eurchadina Logvinenko 1957b:69. Error for *Eurhadina* Haupt.
Eurehadina Logvinenko 1959b:804. Error for *Eurhadina* Haupt.
Eurevacanthus Bliven 1955b:10. *E. emilus* n.sp. (USA: California, Eureka). 6:44. Cicadellinae, Evacanthini. Nearctic. Synonym of *Evacanthus* Le Peletier & Serville 1825a:612; Hamilton, KGA 1983b:506.
Eurhadina Haupt 1929b:1075. *Cicada pulchella* Fallen 1806a:36 (Sweden). 17:750. Dworakowska 1969g, 1978b. Typhlocybinae, Eupterygini. Palaearctic.
Subgenus:
Singhardina Mahmood 1967a:32. *S. robusta* n.sp.
Eurhardina 17:1502. Error for *Eurhadina* Haupt.
Eurhodina 17:1502. Error for *Eurhadina* Haupt.
Eurinoscopus Kirkaldy 1906c:346. *E. lentiginosus* n.sp. (Australia: Queensland, Cairns). 15:187. Iassinae, Iassini. Australian. Synonym of *Batracomorphus* Lewis 1834a:51; Linnavuori 1957b:145.
Eurinoscopus 15:223. Error for *Eurinoscopus* Kirkaldy.
Eurmeloides 12:42. Error for *Eurymeloides* Ashmead.
Euronirvanella Evans, J.W. 1966a:155. *E. anomala* n.sp. (Australia: Queensland, Springsure). Nirvaninae, Nirvanini. Australian.
Eurycoelidia DeLong 1953a:112 as subgenus of *Stenocoelidia* DeLong 1953a:94. *Neocoelidia pulchella* Ball 1909c:168 (USA: California, Tia Juana). 11:116. Kramer 1964d:262. Neocoelidiinae. Neotropical, Nearctic. Synonym of *Neocoelidia* Gillette & Baker 1895a:103.
Eurygonia 1:681. Error for *Eurygogonia* Melichar.

Eurymela Le Peletier & Serville 1825a:603. *E. fenestrata* n.sp. (Australia). 12:11. Evans, J.W. 1966a:70-72. Eurymelinae, Eurymelini. Australian.

Eurymelella Evans, J.W. 1939a:47. *E. tonnoiri* n.sp. (Australia: New South Wales, Mt. Kosciusko). 12:27. Evans, J.W. 1966a:47,48. Eurymelinae, Ipoini. Australian.

Eurymelessa Evans, J.W. 1933a:88. *Eurymeloides moruyana* Distant 1917c:188 (Australia: New South Wales, Moruya). 12:27. Evans, J.W. 1966a:60. Eurymelinae, Eurymelini. Australian.

Eurymelias Kirkaldy 1907d:29. *Eurymeloides hyacinthus* Kirkaldy 1906c:351 (Australia: Queensland) = *Eurymela pulchra* Signoret 1850c:508. 12:18,42. Eurymelinae, Eurymelini. Australian. Synonym of *Eurymeloides* Ashmead 1889c:126.

EURYMELIDAE 12:5.

Eurymelidium Tillyard 1919a:884. *E. australe* n.sp. Metcalf & Wade 1966a:220,221. Fossil: Triassic. Australia.

EURYMELINAE 12:7.

EURYMELINI.

Eurymelita Evans, J.W. 1933a:89. *Eurymela terminalis* Walker 1851b:642 (South Australia: Adelaide). 12:26. Evans, J.W. 1966a:61. Eurymelinae, Eurymelini. Australian.

Eurymeloides Ashmead 1889c:126. *Eurymela bicincta* Erichson 1842a:286 (Tasmania). 12:18. Stevens 1985A. Eurymelinae, Eurymelini, Australian.
Synonym:
Eurymelias Kirkaldy 1907d:29. *Eurymeloides hyacinthus* Kirkaldy 1906c:351 = *Eurymela pulchra* Signoret 1850c:508.

Eurymelops Kirkaldy 1906c:354 as subgenus of *Eurymela* Le Peletier & Serville 1825a:603. *Eurymela rubrovittata* Amyot & Serville 1843a:555 (unknown). 12:7. Evans, J.W. 1966a:73-74. Eurymelinae, Eurymelini. Australian.

Eurymosopum 5:97. Error for *Evryprosopum* Stal.

Euryogonia Melichar 1926a:345. (Invalid) 1:423.

Eurypella Evans, J.W. 1966a:58. *Bakeriola tasmaniensis* Evans, J.W. 1947f:227 (Tasmania: Risdon). Eurymelinae, Eurymelini. Australian.

Euryprosopum Stal 1858a:234, lapsus for *Evryprosopum* Stal 1853b:267.

Eurypterix 17:1502. Error for *Eupteryx* Curtis.

Eurypteryx 17:1502. Error for *Eupteryx* Curtis.

Eurythroneura 17:1502. Error for *Erythroneura* Fitch.

Eusallya Evans, J.W. 1972a:189. *E. viridis* n.sp. (SW New Guinea: Vogelkop, Bomberi, 700-900 m). Penthimiinae. Oriental.

Eusama Oman 1949a:128. *Eutettix amanda* Ball 1909a:82 (USA: Arizona). 10:1330. Deltocephalinae, Athysanini. Nearctic.

Eusceles 10:2657. Error for *Euscelis* Brulle.

Euscelia 10:2657. Error for *Euscelis* Brulle'.

EUSCELIDAE 10:1.

Euscelidella Evans, J.W. 1954a:122. *E. malagasiella* n.sp. (Madagascar). 10:78. Deltocephalinae. Malagasian.

Euscelidius Ribaut 1942a:268. *Jassus variegatus* Kirschbaum 1868b:112 (Germany) = *Athysanus variegatus* Kirschbaum 1858a:9. 10:89. Deltocephalinae, Athysanini. Palaearctic. [Nearctic].

EUSCELINAE 10:8.

EUSCELINI 10:10.

Euscelis Brulle 1832a:109. *E. lineolata* n.sp. (Greece). 10:11. Deltocephalinae, Athysanini. Holarctic.
Synonym:
Phrynomorphus Curtis 1833a:194. *P. nitidus* n.sp.
[NOTE: Interpretations of *Euscelis* Brulle and allied generic-level taxa vary. Metcalf 1967a:12 lists *Phrynomorphus* Curtis, *Conosanus* Osborn & Ball 1902a:236, and *Metathysanus* Dahl 1912a:439 as synonyms of *Euscelis*. Nast 1972a:398-399 recognizes *Conosanus* as a separate genus, with the isogenotypic *Metathysanus* Dahl as a junior synonym; and *Phrynomorphus* and *Ederranus* Ribaut 1942a:267 as synonyms of *Euscelis*. Ossiannilsson 1983a:771 and 779, recognizes both *Conosanus* and *Ederranus* as separate genera].

Euscellis 10:2658. Error for *Euscelis* Brulle.

Eusceloidea 10:2658. Error for *Eusceloidia* Osborn.

Eusceloidia Osborn 1923c:46. *E. nitida* n.sp. (Bolivia: Las Juntas). 10:101. Linnavuori 1959b:242-243. Deltocephalinae. Neotropical.

Eusceloscopus Evans, J.W. 1941f:147. *E. yanchepensis* n.sp. (Western Australia: Yanchep). 10:1892. Evans, J.W. 1966a:249. Deltocephalinse, Balcluthini. Australian. Synonym of *Balclutha* Kirkaldy 1900b:243; Evans, J.W. 1966a:249.

Euscelsis 10:2658. Error for *Euscelis* Brulle.

Euscelus 10:2658. Error for *Euscelis* Brulle.

Euscilis 10:2658. Error for *Euscelis* Brulle.

Eusclis 10:2658. Error for *Euscelis* Brulle.

Eusecelis 10:2658. Error for *Euscelis* Brulle.

Eusora Oman 1949a:137. *Eutettix (Mesamia) animana* Ball 1909a:81 (USA: Colorado, Animas, near Durango). 10:527. Deltocephalinae, Athysanini. Nearctic.

Eustollia Goding 1926a:105. *Cicada jubata* Stoll 1781a:30 (Surinam) = *Cicada marmorata* Fabricius 1803a:61. 1:472. Young 1968a:26. Cicadellinae, Proconiini. Neotropical. Synonym of *Proconia* LePeletier & Serville 1825a:610.

Eutambourina Evans, J.W. 1942a:27. *E. punctata* n.sp. (Australia: Queensland, Tambourine Mtns.). 17:1466. Typhlocybinae, Erythroneurini. Australian. Synonym of *Pettya* Kirkaldy 190Evans,J.W. 1966a:272.

Eutandra Webb, M.D. 1983a:71. *E. rhutens* n.sp. (Australia: Queensland, Bundaberg). Idiocerinae. Australian.

Eutartessus Evans, F. 1981a:146. *E. cantrelli* n.sp. (Australia: Queens-land, Crystal Cascades, near Cairns). Tartessinae. Australian.

Eutettix Van Duzee 1892e:307. *Thamnotettix lurida* Van Duzee 1890b:250 (USA: Michigan, Agricultural College, East Lansing). 10:466. Deltocephalinae, Athysanini. Nearctic.

Euttetix 10:2658. Error for *Eutettix* Van Duzee.

Euttettix 10:2659. Error for *Eutettix* Van Duzee.

EVACANTHIDAE 6:1.

EVACANTHINAE 6:2.

EVACANTHINI 6:1.

Evacanthus Le Peletier & Serville 1825a:612. *Cicada interrupta* Linnaeus 1759a:439 (Europe). 6:3. Anufriev 1978a:62-64. Cicadellinae, Evacanthini. Holarctic.
Subgenus:
Paracanthus Anufriev 1978a:63. *Evacanthus ogumae* Matsumura 1911b:21.

Evanirvana Hill, B.G. 1973a:78. *E. aurea* n.sp. (Brazil: "Jussaral (Maranhao) Angra-E. Do Rio"). Neotropical.
[Note: Provisionally placed in the Evacanthini, thought possibly to be annectant between Evacanthini and Nirvaninae].

Evansiella China 1955a:199. *E. kuscheli* n.sp. (Chile: Juan Fernandez Islands, Camote). 5:95. Evansiolinae. Neotropical. Preoccupied by *Evansiella* Hayward 1948, see *Evansiola* Metcalf 1962d:95.

Evansiola China 1957a:465, replacement for *Evansiella* China 1955a:199, not *Evansiella* Hayward 1948. *Evansiella kuscheli* China 1955a:200 (Chile: Juan Fernandez Isls, Camote). Evansiolinae. Neotropical.
Synonyms:
Evansiella China 1955a:199, not *Evansiella* Hayward 1948.
Evansiola Metcalf 1962d:95. *Evansiella kuscheli* China 1955a:200.

Evansiola Metcalf 1962d:95, replacement for *Evansiella* China 1955a:199, not *Evansiella* Hayward 1948. *Evansiella kuscheli* China 1955a:200 (Juan Fernandez Isls, Chile). 5:95. Linnavuori & DeLong 1977a:165-166. Evansiolinae. Preoccupied by *Evansiola* China 1957a:465.

EVANSIOLINAE Linnavuori & DeLong 1977c:164.

Evansioma Ahmed, Manzoor 1969d:313. *E. pini* n.sp. (West Pakistan: Murree). Typhlocybinae, Typhlocybini. Oriental. Synonym of *Aguriahana* Distant 1918b:105; Dworakowska 1972a:278.

Evansolidia Nielson 1982e:279. *E. evansi* n.sp. (Guyana: Kutari Sources). Coelidiinae, Coelidiini. Neotropical.

Evinus Dlabola 1977b:243. *E. graminicolus* n.sp. (SE Iran: 40 km SW Zaboli, 1550 m). Deltocephalinae, Hecalini. Palaearctic.

Evryprosopum Stal 1853b:267. *Bohemania sobrina* Stal 1855a:97 (South Africa: Natal). 5:88. Ulopinae, Ulopini. Ethiopian. Synonym of *Coloborrhis* Germar 1836a:72.

Excavanus DeLong 1946c:446. *E. angusus* n.sp. (Mexico: Guerrero, Iguala). 10:2177. Deltocephalinae, Athysanini. Neotropical.

Excultanus Oman 1949a:142 as subgenus of *Texananus* Ball 1918b:381. *Jassus excultus* Uhler 1877a:467 (USA: Florida). 10:698. Linnavuori 1959b:197-198. Deltocephalinae, Athysanini. Nearctic.

Exitanus 10:2659. Error for *Exitianus* Ball.

Exitianiellus Evans, J.W. 1966a:229. *E. elegantula* n.sp. (Australia: New South Wales, Oxford). Deltocephalinae. Australian.

Exitianus Ball 1929a:5. "*Cicadula exitiosa* Uhl. = *obscurinervis* Stal" = *Cicadula exitiosa* Uhler 1880a:72 (USA). 10:308. Deltocephalinae, Athysanini. All major land masses.
Synonym:
Mimodrylix Zachvatkin 1935a:108. *Athysanus capicola* Stal 1855a:99.
[NOTE: Ross 1968a:4 cites the type species of *Exitianus* Ball as *Cicadula obscurinervis* Stal, but Ball's method of citing the type species indicates his intent that it be Uhler's species, which Osborn (see Ball's footnote 1929a:5) considered a junior synonym of *obscurinervis* Stal 1859b:293, from Argentina and Brazil].

Exobahita Linnavuori 1959b:174 as subgenus of *Bahita* Oman 1938b:379. *Bahita fallaciosa* Linnavuori 1955b:118 (unknown). Deltocephalinae, Athysanini. Neotropical. Synonym of *Frequenamia* DeLong 1947a:63; Linnavuori & DeLong 1978b:115.

Exogonia Melichar 1926a:342. *Tettigonia assimilis* Signoret 1853b:340 (Brazil: Sao Paulo). 1:209. Young 1977a:222. Cicadellinae, Cicadellini. Neotropical; Brazil.

Exolidia Osborn 1923c:75. *E. picta* n.sp. (Brazil: Rio Guapore, near Forte Principe). 16:228. Linnavuori 1959b:25-26. Neobalinae. Neotropical; Brazil, Paraguay.

Extianus 10:2659. Error for *Exitianus* Ball.

Extrusanus Oman 1949a:150. *Athysanus extrusus* Van Duzee 1893a:283 (USA: New York, Portage). 10:266. Deltocephalinae, Athysanini. Nearctic.

Eypteryx 17:1502. Error for *Eupteryx* Curtis.

Eythogonia 1:681. Error for *Erythrogonia* Melichar.

Ezrana Distant 1908g:177. *E. pygmaea* n.sp. (India: Bombay). 4:84. Ledrinae, Petalocephalini. Oriental.

Faenius Distant 1918b:14. *F. lynchi* n.sp. (India: eastern Himalayas, Darjeeling). 1:298. Cicadellinae, Mileewanini. Oriental. Synonym of *Mileewa* Distant 1908g:238.

Fagocyba Dlabola 1958c:54. *Typhlocyba cruenta* Herrich-Schaeffer 1838c:15 (Germany). Typhlocybinae, Typhlocybini. Palaearctic. [Nearctic].

Faiga Dworakowska 1980b:163. *F. dropia* n.sp. (India: Karnataka, 18-25 km E of Mudigere). Typhlocybinae, Empoascini. Oriental.

Falcitettix Linnavuori 1953d:58. *F. sibiricus* n.sp. (Siberia: Minusinsk). 10:2052. Nast 1972a:441. Deltocephalinae, Paralimnini. Palaearctic. Subgenus of *Mocuellus* Ribaut 1946b:83.
[Note: Nast l.c. places *Falcitettix* as a synonym of *Mocuellus*. However, the type species appears to be sufficiently different to justify subgeneric rank].

Faltala Oman 1938b:385. *F. brachyptera* n.sp. (Argentina: Missiones, Loreto). 10:1606. Linnavuori 1959b:159-160. Deltocephalinae, Athysanini. Neotropical.

Farynala Dworakowska 1970f:215. *F. novica* n.sp. (Vietnam: Bao-Ha, south of Fan-si-Pan). Typhlocybinae, Typhlocybini. Oriental.

Favintiga Webb, M.D. 1981b:47. *Parabolopona camphorae* Matsumura 1912b:288 (Japan: Kyushu, Kagoshima). Paraboloponinae. Palaearctic; Japan and Amani-Oshima.

Fdwardsiana Nast 1976a:166. Error for *Edwardsiana* Zachvatkin.

Fedotartessus Evans, F. 1981a:174. *F. popensis* n.sp. (New Guinea: Papua, Popondetta). Tartessinae. Oriental.

Ferganotettix Dubovskij 1966a:171. *F. elegans* n.sp. (USSR: Uzbekistan, Fergana Valley). Deltocephalinae, Athysanini. Palaearctic.
Subgenera:
Anaemotettix Korolevskaya 1980a:800. *F. (A.) klishinae* n.sp.
Asthenotettix Korolevskaya 1980a:801. *F. (A.) perarmata* n.sp.

Ferrariana Young 1977a:1029. *Tettigonia trivittata* Signoret 1854b:349 (Brazil). Cicadellinae, Cicadellini. Neotropical; Costa Rica and Panama, to Paraguay and Argentina.

Ficiana Ghauri 1963e:472. *F. pruthii* n.sp. (India: Coimbatore). Typhlocybinae, Empoascanini. Oriental.

Ficocyba Vidano 1960a:122 as subgenus of *Typhlocyba* Germar 1833a:180. *Typhlocyba ficaria* Horvath 1897b:636 (Italy). Typhlocybinae, Typhlocybini. Palaearctic.

Fieberelli 10:2659. Error for *Fieberiella* Signoret.

Fieberellia 10:2659. Error for *Fieberiella* Signoret.

Fieberia Signoret 1879a:52. *Selenocephalus florii* Stal 1864b:67 (Greece). 10:1702. Deltocephalinae, Fieberiellini. Palaearctic. Preoccupied by *Fieberia* Jakolev 1873, see *Fieberiella* Signoret 1880a:67.

Fieberialla 10:2659. Error for *Fieberiella* Signoret.

Fieberiella Signoret 1880a:67, replacement for *Fieberia* Signoret 1879a:52, not *Fieberia* Jakovlev 1873. *Selenocephalus florii* Stal 1864b:67 (Greece). 10:1702. Deltocephalinae, Fieberiellini. Palaearctic. [Nearctic].
Synonym:
Fieberia Signoret 1879a:52. *Selenocephalus florii* Stal 1864b:67.

FIEBERIELLINI 10:1701.

Fitchana Oman 1949a:133. *Acocephalus vitellinus* Fitch 1851a:57 (USA: New York). 10:2167. Deltocephalinae, Platymetopiini. Nearctic; northern latitudes.

Flammigeroidia Dlabola 1958c:56 as subgenus of *Erythroneura* Fitch 1851a:62. "*Erythroneura flammigera* Geoffr" = *Cicada flammigera* de Fourcroy 1785a:190 (France). Typhlocybinae, Erythroneurini. Palaearctic. Synonym of *Zygina* Fieber 1866a:509; Nast 1972a:304.

Flavitartessus Evans, F. 1981a:154. *Bythoscopus flavibasis* Walker 1870b:320 (Indonesia: Aru Island). Tartessinae. Oriental; New Guinea and New Britain.

Flexamia DeLong 1926d:22 as subgenus of *Deltocephalus* Burmeister 1838b:15. *Deltocephalus reflexus* Osborn & Ball 1897a:203 (USA: Iowa, Ames). 10:1748. Young & Beirne 1958a. Deltocephalinae, Deltocephalini. Nearctic.

Flexana DeLong & Freytag 1971a:321. *F. spinosa* n.sp. (Peru: Rio Santiago). Gyponinae, Gyponini. Neotropical.

Flexiamius 10:2659. Error for *Flexamia* DeLong.

Flexocerus Kuoh, C.L. & Fang 1985a:91. *F. flexureus* n.sp. (China: Guizhou). Idiocerinae. Oriental.

Floridonus Oman 1949a:114. *Thamnotettix taxodii* Sanders & DeLong 1923a:154 (USA: Florida). 10:1329. Deltocephalinae, Athysanini. Nearctic.

Folicana DeLong & Freytag 1972a:226. *F. nota* n.sp. (Brazil: Sao Paulo, Piracicaba). DeLong & Freytag 1972b. Gyponinae, Gyponini. Neotropical; northern South America.

Fonsecaiulus Young 1977a:760. *Tettigonia flavovittata* Stal 1859b:288 (Brazil: Rio de Janeiro). Cicadellinae, Cicadellini. Neotropical; Argentina, Bolivia, Brazil, Venezuela.

Forcipata DeLong & Caldwell 1936a:70. (Invalid, no type species designated).

Forcipata DeLong & Caldwell 1942e:49. *Forcipata loca* DeLong & Caldwell 1936a:71 (USA: Pennsylvania, Ohiopyle). 17:107. Typhlocybinae, Dikraneurini. Holarctic.

Formotettigella Ishihara 1965c:216. *F. shirozui* n.sp. (Formosa). Cicadellinae, Cicadellini. Oriental.

Foroa Linnavuori 1977a:10 & 35. *F. spiniloba* n.sp. (Ivory Coast: Foro-Foro). Penthimiinae. Ethiopian.

Foso Linnavuori & Al-Ne'amy 1983a:91. *F. longiceps* n.sp. (Ghana: near Foso). Selenocephalinae, Selenocephalini. Ethiopian; Guinean.

Frequenamia DeLong 1947a:63. *F. guerrera* n.sp. (Mexico: Guerrero, Pandancuarco). 10:2088. Deltocephalinae, Athysanini. Neotropical.
Synonyms:
Penebahita Linnavuori 1959b:166. *Deltocephalus venosulus* Berg 1879d:266.
Exobahita Linnavuori 1959b:174. *E. fallaciosa* n.sp.; Linnavuori & DeLong 1978b:118.

Freytagana DeLong 1975a:409. *F. gibsoni* n.sp. (Mexico: D.F.). Gyponinae, Gyponini. Neotropical.

Fridonus Oman 1949a:129. *F. concanus* n.sp. (USA: Texas, Concan). 10:1331. Deltocephalinae, Athysanini. Nearctic.

Frigartus Oman 1949a:150. *Athysanus frigidus* Ball 1899c:172 (USA: Colorado, Fort Collins). 10:270. Deltocephalinae, Athysanini. Nearctic; Great Plains, east of the Rocky Mtns.

Friscananus Ball 1936c:60. *Thamnotettix intricatus* Ball 1911a:198 (USA: California, San Francisco). 10:1280. Nielson 1957a, 1966a:333. Deltocephalinae, Athysanini. Nearctic. Synonym of *Colladonus* Ball 1936c:57.

FRISCANINA Anufriev 1978a:59, subtribe of PAGARONIINI.

Friscanus Oman 1938c:168. *Errhomenellus friscanus* Ball 1909b:182 (USA: California, San Francisco). 8:17. Oman & Musgrave 1975a:5; Anufriev 1978a:59. Cicadellinae, Pagaroniini. Nearctic; coastal habitats, southern California to northern Oregon.

Fruticidia Metcalf 1968a:1448, error for *Frutioidia* Zachvatkin 1946b:152. 17:1448.

Frutioidia Zachvatkin 1946b:152 as subgenus of *Zyginidia* Haupt 1929c:268. *Typhlocyba bisignata* Mulsant & Rey 1855a:241 (France: Provence). 17:1449. Dworakowska 1971b:110-113, 1979b:23-25. Typhlocybinae, Erythroneurini. Palaearctic and Oriental.
Subgenus:
Dworakowskellina Kocak 1981a:31, replacement for *Amicula* Dworakowska 1971b:111. *Erythroneura (Zygina) amicula* Linnavuori 1965a:16.

Fulvanus Linnavuori 1955a:110. *F. curvilinea* n.sp. (not stated). 10:1241. Deltocephalinae, Athysanini. Neotropical. Synonym of *Atanus* Oman 1938b:381; Linnavuori 1959b:296.

Fundarus Linnavuori & DeLong 1978f:55 as subgenus of *Scaphytopius* Ball 1931d:218. *S. (F.) vinculatus* n.sp. (Bolivia: Cochabamba). Deltocephalinae, Scaphytopiini. Neotropical.

Funkikonia Kato 1931e:438. *Ledra tuberculata* Kato 1929a:545 (Formosa). 4:58. Ledrinae, Petalocephalini. Oriental.

Furcatartessus Evans, F. 1981a:138. *F. waiwerensis* n.sp. (Australia: New South Wales, Waiwere, Narrabri). Tartessinae. Australian.

Furymela 12:42. Error for *Eurymela* Le Peletier & Serville.

Fusanus Linnavuori 1955b:123. *F. griseostriatus* n.sp. (Argentina: Mendoza). 10:1772. Linnavuori 1959b:135-137. Deltocephalinae, Deltocephalini. Neotropical.
Synonyms:
Metcalfiella Linnavuori 1955b:121 not *Metcalfiella* Goding 1929. *M. chacoensis* n.sp.
Zepama Metcalf 1967b:2241, replacement for *Metcalfiella* Linnavuori, not *Metcalfiella* Goding 1924. *Metcalfiella chacoensis* Linnavuori 1955b:122.

Fusigonalia Young 1977a:555. *Tettigonia lativittata* Fowler 1900d:281 (Panama: Bugaba). Cicadellinae, Cicadellini. Neotropical; southern Mexico, Panama, Central America, northern South America.

Fusiplata Ahmed, Manzoor 1969a:60. *F. quercicola* n.sp. (West Pakistan: Murree). Typhlocybinae, Dikraneurini. Palaearctic.

Futasujinoidella Kwon & Lee 1979c:54. *F. nobilis* n.sp. (Korea: Mt. Palgongsan). Deltocephalinae, Deltocephalini. Palaearctic; Korea.

Futasujinus Ishihara 1953b:47. *Deltocephalus candidus* Matsumura 1914a:208 (Japan). 10:1854. Deltocephalinae, Deltocephalini. Palaearctic; Japan, Korea, Soviet Maritime Territory.

Gabrita Walker 1858b:254. *G. annulivena* n.sp. (Brazil) = *Coelidia eburata* Walker 1851b:854. 11:28. Nielson 1983e:69-75. Coelidiinae, Gabritini. Neotropical.
Synonym:
Petalopoda Spangberg 1879a:18. *P. annulipes* n.sp.; Nielson 1983e:69.

GABRITINI Nielson 1983e:68, tribe in COELIDIINAE.

Galboa Distant 1909h:45. *G. typica* n.sp. (Seychelles: Mahe Island). 13:166. Hamilton, KGA 1980b:901. Macropsinae, Macropsini. Malagasian.

Galerius Distant 1918b:78. *G. indicatrix* n.sp. (South India). 10:1605. Deltocephalinae, Stenometopiini. Oriental. Synonym of *Doratulina* Melichar 1903b:198; Vilbaste 1965a:10.

Gamarex Hamilton, KGA 1975a:34 as subgenus of *Paraphlepsius* Baker 1897c:158. *Jassus fulvidorsum* Fitch 1851a:62 (USA: New York). Deltocephalinae, Athysanini. Nearctic.

Gambialoa Dworakowska 1972f:853. *Erythroneura gambiensis* Ross 1965a:265 (Gambia). Typhlocybinae, Erythroneurini. Ethiopian, Oriental; India.
Subgenus:
Nkasa Dworakowska 1974a:187. *G. (N.) ahmedi* n.sp.

Gannachrus Theron 1979a:85. *G. namaquanus* n.sp. (South Africa: Vanrhynsdorp). Agalliinae. Ethiopian; southern Africa.

Gannia Theron 1979a:77. *G. salsoli* n.sp. (South Africa: Oudtshoorn). Deltocephalinae. Ethiopian; Sudanese.

Garapita Oman 1938b:369. *G. garbosa* n.sp. (Argentina: Missiones, Loreto). 10:2104. Linnavuori 1959b:222-225. Deltocephalinae, Athysanini. Neotropical; Argentina, Brazil.
Subgenus:
Chlamydopita Linnavuori 1959b:225. *G. (C.) aurea* n.sp.

Gargaropsis Fowler 1896e:167. *G. innervis* n.sp. (Mexico: Guerrero). 15:97,98,223. Blocker 1979a:9,13,14. Iassinae. Neotropical.
Subgenus:
Curistuva Blocker 1979a:14. *Gargaropsis adibilis* Blocker 1975a:561.

Garlica Blocker 1976a:519. *G. hepneri* n.sp. (Panama: Pearl Island, San Jose). Iassinae. Neotropical; Brazil, Panama.

Gcaleka Naude 1926a:29. *G. laticephala* n.sp. (South Africa: Ceres). 8:218. Theron 1972a:207. Deltocephalinae. Ethiopian.

Gehundra Blocker 1976a:519. *G. tricosa* n.sp. (Peru: Estancia Naranjal, San Ramon, Junin). Iassinae. Neotropical; Brazil, Peru.

Geitogonalia Young 1977a:524. *Tettigonia quatuordecimmaculata* Taschenberg 1884a:447 (Brazil). Cicadellinae, Cicadellini. Neotropical.

Gelidanus Emeljanov 1966a:122 as subgenus of *Diplocolenus* Ribaut 1946b:82. *Cicada limbatella* Zetterstedt 1828a:522 (Lapland). Deltocephalinae, Paralimnini. Palaearctic. Synonym of *Erdianus* Ribaut 1952a:287; Knight 1974c:368.

Genatra Nielson 1983h:564. *G. spinosa* n.sp. (Brazil: Mt. Tumac Humac, Lunier River). Coelidiinae, Teruliini. Neotropical.

Gerganotettix Korolevskaya 1980a (English translation, Entomological Review 59(4)). Error for *Ferganotettix* Dubovskij 1966a.

Germaria de LaPorte 1832b:222. *G. cucullata* n.sp. (Unknown). 1:473. Young 1968a:26. Cicadellinae, Proconiini. Neotropical. Preoccupied by *Germaria* Robineau-Desvoidy 1830; synonym of *Proconia* LePeletier & Serville 1825a:610.

Gerostella Evans, J.W. 1954a:95. *Tettigonia billosa* Signoret 1860a:203 (Madagascar). 1:105. Cicadellinae, Cicadellini. Malagasian.

Gessius Distant 1908g:301. *G. verticalis* n.sp. (Burma: Ruby mines). 15:185. Selenocephalinae. Oriental.

Ghauriana Thapa 1985a:65. *G. pecularia* n.sp. (Nepal: Kathmandu, Bhandarkhal). Typhlocybinae, Empoascini. Oriental.

Ghamnotettix 10:2659. Error for *Thamnotettix* Zetterstedt.

Gicrantus Nielson 1982e:272. *Coelidia scutellaris* Linnavuori 1956a:33 (Bolivia). Coelidiinae, Coelidiini. Neotropical; Bolivia, Peru.

Giffardia Kirkaldy 1906c:336. *G. dolichocephala* n.sp. (Australia: Queensland, Cairns). 10:2075. Evans, J.W. 1966a:245. Deltocephalinae, Platymetopiini. Australian.

Giletiella Emelyanov 1962b:393. Error for *Gillettiella* Osborn 1930a.

Gillettiella Osborn 1930a:689. *Deltocephalus labiata* Gillette 1898b:26 (USA: Colorado). 10:1606. Deltocephalinae, Stenometopiini. Nearctic; central and southwest USA Great Plains.

Gindara Dworakowska 1980b:191. *G. temna* n.sp. (India: Uttar Pradesh, Mussoorie). Typhlocybinae, Erythroneurini. Oriental.

Giprus Oman 1949a:164. *Deltocephalus cinerosus* Van Duzee 1892d:305 (USA: California). 10:1655. Sawbridge 1975a. Deltocephalinae, Deltocephalini. Nearctic; USA: western states.

Gladionura Osborn 1930a:706. *Deltocephalus argenteolus* Uhler 1877a:473 (USA: Colorado). 10:1636. Deltocephalinae, Doraturini. Nearctic. Subgenus of *Athysanella* Baker 1898a:185; Ball & Beamer 1940a:9.

Gladioneura 10:2660. Error for *Gladionura* Osborn.

Glassocratus 9:120. Error for *Glossocratus* Fieber.

Gloridonus Ball 1936c:58. *Thamnotettix gloriosus* Ball 1910c:303 (USA: California, Tia Juana). 10:1282. Deltocephalinae, Athysanini. Nearctic.

Glossocratus Fieber 1866a:502. *G. floveolatus* n.sp. ("Seratov"). 9:8,120. Morrison 1973a:397, Linnavuori 1975a:9. Deltocephalinae, Hecalini. Palaearctic and Oriental.
Synonyms:
Ectomops Signoret 1879a:51. *E. chinensis* n.sp.
Ledrotypa Distant 1912d:442. *L. spatulata* n.sp.
Chelusa Signoret 1879a:51. *Acocephalus madagascariensis* Signoret 1880:51.

Glossonatus 9:120. Error for *Glossocratus* Fieber.
Glypona 3:221. Error for *Gypona* Germar.
Glyppocephalus 10:2661. Error for *Glyptocephalus* Edwards.

Glyptocephalus Edwards 1883b:148. *Athysanus canescens* Douglas & Scott 1873a:210 (England) = *Jassus (Athysanus) proceps* Kirschbaum 1868b:105. 10:1695. Deltocephalinae, Athysanini. Palaearctic. Synonym of *Rhytistylus* Fieber 1875a:404.

Gnathnodus 10:2661. Error for *Gnathodus* Fieber.
Gnathodes 10:2661. Error for *Gnathodus* Fieber.

Gnathodus Fieber 1866a:505. *Cicada punctata* Fabricius 1775a:687 (Europe). 10:2383. Deltocephalinae, Balcluthini. Palaearctic. Preoccupied by *Gnathodus* Pander 1856, see *Balclutha* Kirkaldy 1900b:243.

Gnatia Evans, J.W. 1941f:150. *G. angustata* n.sp. (Western Australia: Dedari, 40 miles W of Coolgardie). 16:8. Idiocerinae. Australian. Synonym of *Austrocerus* Evans, J.W. 1941e:37; Webb, M.D. 1983a:66.

Gnatodus 10:2663. Error for *Gnathodus* Fieber.

Gobicuellus Dlabola 1967d:223. *G. dzadagadus* n.sp. (Boreal Mongolia: Dzun-Mod). Deltocephalinae, Paralimnini. Palaearctic.
Synonym:
Acharista Emeljanov 1968a:149. *A. nudiventris* n.sp.; Nast 1984a:396.

Goblinaja Kramer 1965a:72. *G. otavaloa* n.sp. (Ecuador: Otavalo). Iassinae. Neotropical.

Goifa Dworakowska 1977i:18. *G. fasciata* n.sp. (Vietnam: Hanoi). Typhlocybinae, Empoascini. Oriental; India, Vietnam.

Goldeus Ribaut 1946b:85. *Deltocephalus harpago* Ribaut 1925a:194 (France). 10:1437. Deltocephalinae, Paralimnini. Palaearctic; Mediterranean subregion.
Synonym:
Muleyrechia Linnavuori 1956d:164. *Jassargus melillensis* Linnavuori 1955c:25.

GONIAGNATHINI 10:1899. Nast 1972a:321.

Goniagnathus Fieber 1866a:506. *Jassus brevis* Herrich-Schaeffer 1835a:71 (Germany). 10:1899. Linnavuori 1978c:486-497. Deltocephalinae, Goniagnathini. Palaearctic; Eurasian and northern Africa.
Subgenus:
Tamaricades Emeljanov 1962a:163. *Athysanus decoratus* Haupt 1917d:246.
Synonym:
Goniozygum Bergroth 1920a:29, replacement for *Goniognathus*, an emendation of *Goniagnathus* Fieber 1866a:506.

Goniagnothus 10:2663. Error for *Goniagnathus* Fieber 1866a:506.
Goniagnatus 10:2663. Error for *Goniagnathus* Fieber.
Goniognathus 10:2663. Emendation of *Goniagnathus* Fieber.
Goniagrathus 10:2663. Error for *Goniagnathus* Fieber.
Goniozygium 10:2663. Error for *Goniozygum* Bergroth.

Goniozygum Bergroth 1920a:29, replacement for *Goniognathus* of authors, not *Goniagnathus* Fieber 1866a:506. 10:1902, 2663. Deltocephalinae, Goniagnathini. Palaearctic. Synonym of *Goniagnathus* Fieber 1866a:506.

Gorgonalia Young 1977a:174. *Tettigonia sirena* Stål 1864a:76 (Mexico). Cicadellinae, Cicadellini. Neotropical; Mexico.

Goryphaeus 10:2663. Error for *Coryphaeus* Fieber.

Goska Dworakowska 1981j:245. *G. sessila* n.sp. (India: Tamil Nadu, near Shambaganur). Typhlocybinae, Erythroneurini. Oriental; India.

Graminella DeLong 1936a:218. *Thamnotettix aureovittatus* Sanders & DeLong 1920b:16 (Florida). 10:856. Kramer 1967a. Deltocephalinae, Deltocephalini. Nearctic, Neotropical; West Indies, Mexico to Brazil and Paraguay.

Graminiella 10:2664. Error for *Graminella* DeLong.

Grammacephalus Haupt 1929c:258. *Platymetopius pugio* Noualhier 1895:176 (Asia Minor). 10:2343. Deltocephalinae, Scaphoideini. Palaearctic; Mediterranean.

GRANULIDAE Hong 1980a:63, 69. Fossil.

Granulus Hong 1980a:64,69. *G. tongchuanensis* n.sp. Fossil: Middle Triassic, Tongchuan formation; China, Shalixi Province.

Graphacephala 1:681. Error for *Graphocephala* Van Duzee.
Graphlocephala 1:682. Error for *Graphocephala* Van Duzee.
Graphlocelphala 1:682. Error for *Graphocephala* Van Duzee.

Graphocephala Van Duzee 1916a:66. *Cicada coccinea* Forster 1771a:69 (North America). 1:366. Young 1977a:849. Cicadellinae, Cicadellini. Nearctic and Neotropical; as far south as Colombia and French Guiana. [Palaearctic].
Synonyms:
Neokolla Melichar 1926a:343. *Tettigonia hieroglyphica* Say 1830b:313.
Keonolla Oman 1949a:74. *Proconia confluens* Uhler 1861b:285.
Hordnia Oman 1949a:70. *Tettigonia circellata* Baker 1898c:285 = *Tettigonia atropunctata* Signoret 1854b:354.
Marathonia Oman 1949a:72. *Cicadella marathonensis* Olsen 1918a:3.

GRAPHOCEPHALINI 1:366.

Graphoceras 10:2664. Error for *Graphocraerus* Thomson.
Graphocoraeus 10:2664. Error for *Graphocraerus* Thomson.

Graphocraerus Thomson 1869a:57 as subgenus of *Jassus*. *Cicada ventralis* Fallen 1806a:18 (Sweden). 10:1707. Deltocephalinae, Athysanini. Palaearctic; Eurasia. [Nearctic].

Graphocraetus 10:2664. Error for *Graphocraerus* Thomson.
Graphocraeus 10:2664. Error for *Graphocraerus* Thomson.
Graphocresus 10:2664. Error for *Graphocraerus* Thomson.
Graphocreus 10:2664. Error for *Graphocraerus* Thomson.
Graphocroerus 10:2664. Error for *Graphocraerus* Thomson.

Graphogonalia Young 1977a:981. *G. evagorata* n.sp. (Mexico: Yucatan, Merida). Cicadellinae, Cicadellini. Neotropical; Mexico, Central America, Panama, Venezuela.

Grapocraerus 10:2664. Error for *Graphocraerus* Thomson.

Gratba Dworakowska 1982a:104. *G. astrata* n.sp. (India: Tamil Nadu, Conoor Sim's Park). Typhlocybinae, Typhlocybini. Oriental; south India.

Grathodus 10:2644. Error for *Gnathodus* Fieber.

Greceasca Thapa 1985a:67 as subgenus of *Empoasca* Walsh 1862a:149. *E. (G.) kapoori* (Nepal: Kathmandu, Godawari). Typhlocybinae, Empoascini. Oriental.

Gredzinskiya* Dworakowska 1972j:113. *G. lipcowa* n.sp. (Vietnam: Bao-Ha, S of Fan-Si-Pan). Typhlocybinae, Erythroneurini. Oriental. * Name erroneously credited to Jaczewski in Zoological Record 109:146.

Gressittella Evans, J.W. 1972a:194. *G. bella* n.sp. (New Guinea: Vogelkop, Bomberi). Penthimiinae. Oriental.

Gressittocerus Maldonado-Capriles 1985a:270. *G. elongatus* n.sp. (SE New Guinea: Papua, Cape Rodney). Idiocerinae. Oriental.

Grootonia Webb, M.D. 1983b:248. *G. mella* n.sp. (South Africa: Messina). Idiocerinae. Ethiopian; Botswana, Kenya, South Africa.

Grunchia Kramer 1963a:47. *Batrachomorphus (Stragania) grossus* Linnavuori 1957b:148 (Peru: Callanga). Iassinae. Neotropical; Bolivia, Ecuador, Peru.

Grypotes Fieber 1866a:503. *Jassus puncticollis* Herrich-Schaeffer 1834d:7 (Bavaria). 10:1916. Deltocephalinae, Grypotini. Palaearctic; Mediterranean and Canary Islands.
Synonym:
Protaenia Thomson 1869a:58. *Jassus pinetellus* Zetterstedt 1840a:1077 = *Jassus puncticollis* Herrich-Schaeffer 1834d:7.

GRYPOTINI 10:1916.

Guadlera DeLong 1980h:65 as subgenus of *Eutettis* (sic!), error for *Eutettix* Van Duzee 1892e:307. *E. (G.) discapa* n.sp. (Mexico: Guadalajara Rd., K 116). Deltocephalinae. Neotropical.

Guaporea Linnavuori & DeLong 1978b:123. *Eutettix laticeps* Osborn 1923c:54 (Brazil: Rio Guapore, below Rio S. Miguel). Deltocephalinae, Athysanini. Neotropical.

Gubela Distant 1910e:232. *G. bellicosa* n.sp. (South Africa: Natal). 5:88. Ulopinae, Ulopini. Ethiopian. Synonym of *Coloborrhis* Germar 1836a:72.

Gudwana Distant 1917c:189. *G. typica* n.sp. (Australia: New South Wales). 4:93. Ledrinae, Ledrini. Australian. Synonym of *Porcorhinus* Goding 1903a:38.

Guheswaria Thapa 1983a:175. *G. linguplata* n.sp. (Nepal: Kathmandu, Guheswari area). Typhlocybinae, Typhlocybini. Oriental.

Guinobata Mahmood 1967a:24 as subgenus of *Bakera* Mahmood 1976a:23. *B. (G.) nigroscuta* n.sp. (Philippines: Luzon, Albay, Guinobatan). Hongsaprug & Wilson 1985a. Typhlocybinae, Erythroneurini. Oriental.
Synonym:
Sandalla Mahmood 1967a:24. *S. nigroclypelli* n.sp.

Guliga Distant 1908g:326. *G. erebus* n.sp. (Burma: Ruby mines). 11:28. Coelidiinae, Thargiini. Oriental. Synonym of *Thagria* Melichar 1903b:176; Nielson 1977a:9.

Gullifera Webb, M.D. 1980a:846. *G. idas* n.sp. (Aldabra: West Island). Typhlocybinae, Dikraneurini. Malagasian.

Gunghuyana Distant 1910e:244. *G. gazana* n.sp. (Rhodesia). 10:542. Deltocephalinae. Ethiopian.

Gunhilda Distant 1918b:88. *G. noctua* n.sp. (South India: Nilgiri Hills). 14:3. Agalliinae, Agalliini. Oriental; southern India.

Gununga Melichar 1914b:129. *G. jacobsoni* n.sp. (Java). 1:253. Cicadellinae, Cicadellini. Oriental; Java, Malaysia, Sumatra, Vietnam.

Gurawa Distant 1908g:262. *G. vexillum* n.sp. (India: Sikkim). 8:217. Aphrodinae, Aphrodini. Oriental; Flores Island, Indonesia; India.

Gyphona 3:222. Error for *Gypona* Germar.

Gypnonana Zoological Record 109, Sec. 13f:146-7, 1972. Error for *Gyponana* Ball.

Gypona Germar 1821a:73. *Cercopis glauca* Fabricius 1803a:91 (South America). 3:13. DeLong & Freytag 1962a, 1964a, 1972a. Gyponinae, Gyponini. Neotropical, Nearctic.
Subgenera:
Marganalana Metcalf 1949b:277. *M. testacea* n.sp.
Obtusana DeLong & Freytag 1964a:29. *Gypona melanota* Spangberg 1878a:19.
Paragypona DeLong & Freytag 1964a:7. *Cercopis thoracica* Fabricius 1803a:91.
Elevanosa DeLong 1977g:37. *G. (E.) vertara* n.sp.
Carnoseta DeLong 1981a:510. *G. (C.) colomella* n.sp.

Gyponana Ball 1920a:25 as subgenus of *Gypona* Germar 1821a:73. *Tettigonia octolineata* Say 1825a:340 (USA: Missouri). 3:72, 77. DeLong & Freytag 1964a:80, 1972a:225, 1972c. Gyponinae, Gyponini. Nearctic; Neotropical to northern South America.
Subgenera:
Clovana DeLong & Freytag 1964a:118. *Gyponana omani* DeLong 1942d:55.
Zerana DeLong & Freytag 1964a:118. *G. (Z) apicata* n.sp.
Sternana DeLong & Freytag 1964a:119. *G. (S.) tenuis* n.sp.
Pandara DeLong & Freytag 1972c:160. *G. (P.) eleganta* n.sp.
Spinanella DeLong & Freytag 1972c:158. *G. (S.) rubralineata* n.sp.

GYPONIDAE 3:5.
GYPONINAE 3:13.
GYPONINI 3:13.

Habenia Dworakowska 1972c:477. *H. octobris* n.sp. (Ivory Coast: Lamto Forest). Typhlocybinae, Empoascini. Ethiopian.
Subgenus:
Atucla Dworakowska 1972c:477. *H. (A.) octava* n.sp.

Habralebra Young 1952b:33. *Protalebra nicaraguensis* Baker 1903d:6 (Nicaragua: San Marcos). 17:81. Typhlocybinae, Alebrini. Neotropical; West Indies, Central and South America.

Habrostis Dubovskij 1966a:163. *Thamnotettix suturalis* Melichar 1898a:64 (Turkestan). Deltocephalinae, Fieberiellini. Palaearctic.

Hackeriana Evans, J.W. 1936b:67. *H. huonensis* n.sp. (Tasmania: Huonville). 4:130. Evans, J.W. 1966a:128-129. Ledrinae, Thymbrini. Australian; Tasmania, southern and western Australia.

Hadralebra Young 1952b:19. *Dikraneura (Hyloidea) laticeps* Osborn 1928a:277 (Bolivia: Prov. del Sara). 17:19. Typhlocybinae, Alebrini. Neotropical.

Hadria Metcalf & Bruner 1936a:945. *H. convertibilis* n.sp. (Cuba: Pinar del Rio Prov., Sierra Rangel, Las Animas). 1:395. Young 1977a:965. Cicadellinae, Cicadellini. Neotropical; Cuba, Hispaniola.
Synonyms:
Lucumius Metcalf & Bruner 1936a:968. *L. triangularis* n.sp.
Arezzia Metcalf & Bruner 1936a:957. *A. maestralis* n.sp.

Hadroca Theron 1974b:161. *Euscelis ramosa* Naude 1926a:65 (South Africa: Stellenbosch). Deltocephalinae. Ethiopian.

Hajra Dworakowska 1981j:243. *H. iridescens* n.sp. (India: Karnataka, Nagarahole). Typhlocybinae, Erythroneurini. Oriental; south India.

Haldorellus Menezes 1973a:115 as subgenus of *Haldorus* Oman 1938b:373. *H. (H.) krameri* n.sp. (Brazil: Sao Paulo, Jundiai). Deltocephalinae, Deltocephalini. Neotropical.

Haldorus Oman 1938b:373 as subgenus of *Deltocephalus* Burmeister 1838b:15. *Thamnotettix venatus* Osborn 1924c:425 (Bolivia: Santa Cruz). 10:1878. Linnavuori 1959b:141. Deltocephalinae, Deltocephalini. Neotropical/Nearctic; southern USA, Panama, Bolivia, Brazil, Argentina.
Subgenera:
Parahaldorus Linnavuori 1959b:142. *H. (P.) truncatistylus* n.sp.
Eohaldorus Linnavuori 1959b:149. *Deltocephalus australis* DeLong 1926d:90.
Haldorellus Menezes 1973a:115. *H. (H.) krameri* n.sp.
Synonym:
Cumora Oman 1938b:374. *C. angulata* n.sp.

Hamana DeLong 1942d:85. *Gypona dictatoria* Gibson 1919a:91 (USA: Arizona). 3:123. DeLong & Freytag 1966c, 1972a:225. Gyponinae, Gyponini. Nearctic; southwest USA, northern Mexico.

Hameedia Ahmed, Manzoor 1972a:67. *H. erythrocephala* n.sp. (East Pakistan: Dacca). Typhlocybinae, Dikraneurini. Oriental. Synonym of *Uzeldikra* Dworakowska 1971d:585; Sharma 1978a:12.

Hamolidia Nielson 1982e:195. *H. hama* n.sp. (Colombia). Coelidiinae, Coelidiini. Neotropical; Bolivia, Colombia.

Handianus Ribaut 1942a:265. *Jassus procerus* Herrich-Schaeffer 1835c:10 (Germany). 10:412. Deltocephalinae, Athysanini. Palaearctic.
Subgenera:
Dlabolia Lang 1945a:62. *Deltocephalus ignoscus* Fieber 1869a:208.
Usuironus Ishihara 1953a:197. *Athysanus ogikubionis* Matsumura 1914a:188.
Pycnoides Emeljanov 1964f:32. *Jassus flavovarius* Herrich-Schaeffer 1835c:9.
Cyclopherus Emeljanov 1964f:35. *Handianus (C.) erotiae* n.sp.
Ephemerinus Emeljanov 1964f:35. *Handianus (E.) magnificus* n.sp.

Handlirschiana Metcalf & Wade 1966a:220, replacement for *Mesojassus* Handlirsch 1939a:145, not *Mesojassus* Tillyard 1916a:34. *Mesojassus pachyneurus* Handlirsch 1939a:146. Fossil: Jurassic, Germany.

Hangklipia Linnavuori 1972b:205 & 235. *Camptelasmus signatus* Linnavuori 1961:457 (South Africa: K. Hangklip, 14 miles south of Strand). Ledrinae, Petalcephalini. Ethiopian.

Hanshumba Young 1977a:260. *H. brasura* n.sp. (Brazil: Santa Catherina, Hansa Humboldt). Cicadellinae, Cicadellini. Neotropical.

Haranga Distant 1908g:248. *Penthimia orientalis* Walker 1851b:841 (northern India). 3:211. Linnavuori 1977a:55. Penthimiinae. Oriental, Ethiopian; equatorial Africa.

Haranthus Nielson 1975a:191. *H. pendiculus* n.sp. (New Guinea: Feramin). Coelidiinae, Tharrini. Oriental; New Guinea.

Harasupia Nielson 1979b:35. *Coelidia marginata* Stal 1864a:85 (Mexico: Salle). Coelidiinae, Teruliini. Neotropical and Nearctic; southern USA, Mexico, Costa Rica.

Harathanus Nielson 1979b:4. Error for *Haranthus* Nielson.

Hardiana Mahmood 1967a:14. *H. assamensis* n.sp. (India: Assam, 10 km N of Tinsukia). Typhlocybinae, Erythroneurini. Oriental. Synonym of *Thaia* Ghauri 1962a:253; Dworakowska 1970c:87 & 88.

Hardya Edwards 1922a:206. *Aphrodes melanopsis* Hardy 1850a:427 (England). 10:253. Deltocephalinae, Athysanini. Palaearctic, Nearctic.
  Subgenera:
  Eohardya Zachvatkin 1946b:171. *H. (E.) mira* n.sp.
  Mimohardya Zachvatkin 1946b:170. *H. (M.) heptneri* n.sp.
Hardyopsis Ribaut 1948b:218. *Hardya insularis* Lindberg 1948b:155 (Cyprus). 10:271. Deltocephalinae, Athysanini. Palaearctic. Synonym of *Eohardya* Zachvatkin 1946b:171; Linnavuori 1962a:47.
  [Note: Nast 1972a:375 list both *Eohardya* and *Hardyopsis* as synonyms of *Stenometopiellus* Haupt 1917d:250, but Dlabola 1981a:290 does not accept that synonymy. There appears to be agreement that *Hardyopsis* is a junior synonym of *Eohardya*].
Harmata Dworakowska 1976b:30. *Typhlocyba condensata* Jacobi 1941a:312 (Sunda: W. Flores, Badjawa). Typhlocybinae, Erythroneurini. Oriental.
Hatigoria Distant 1908g:258. *H. praeiens* n.sp. (Burma: Karen Hills). 2:10. Hylicinae, Sudrini. Oriental.
Hatralixia Webb, M.D. 1983a:88. *H. pallida* n.sp. (Australia: Northern Territory, Arnhem Land, Maningrida). Idiocerinae. Australian.
Hauptidia Dworakowska 1970k:620. *Typhlocyba distinguenda* Kirschbaum 1868b:183 (unknown). Typhlocybinae, Erythroneurini. Palearctic, Oriental; India.
  Subgenus:
  Melicharidia Dworakowska 1970k:624. *Erythroneura aridula* Linnavuori 1956b:138.
  Synonym:
  Kodaikanalia Ramakrishnan & Menon 1973a:35; Dworakowska 1981d:377.
Havelia Ahmed, Manzoor 1971d:277. *H. alba* n.sp. (West Pakistan: Abbottabad). Typhlocybinae, Erythroneurini. Oriental.
Hazaraneura Samad & Ahmed, Manzoor 1979a:249. *H. kaghanensis* n.sp. (Pakistan: Garhi Habibullah, N.W.F.P.). Typhlocybinae, Dikraneurini. Oriental.
Hclas 9:120. Error for *Hecalus* Stal.
Hebata DeLong 1931b:32 as subgenus of *Empoasca* Walsh 1862a:149. *Empoasca nigra* Gillette & Baker 1895a:108 (USA: Colorado, NW of North Park). 17:425. Typhlocybinae, Empoascini. Nearctic; western regions.
Hebecephalus DeLong 1926d:58 as subgenus of *Deltocephalus* Burmeister 1838b:15. *Deltocephalus signatifrons* Van Duzee 1892d:305 (USA: NW Colorado). 10:1854. Deltocephalinae, Deltocephalini. Nearctic.
Hebenarus DeLong 1944c:41. *H. pallidus* n.sp. (Mexico: D.F.). 10:2249, 2250. Deltocephalinae, Scaphytopiini. Nearctic. Synonym of *Scaphytopius* Ball 1931d:218.
Hebeta Beirne 1956a:56. Error for *Hebata* DeLong.
Hebexa Oman 1949a:162. *H. incognita* n.sp. (USA: Utah, Snyderville). 10:1572. Deltocephalinae, Deltocephalini. Nearctic; northern inter-mountain region.
Hecalapona DeLong & Freytag 1975c:547. *H. bexa* n.sp. (Panama: Darien Prov., Santa Fe). Gyponinae, Gyponini. Neotropical; Mexico to Brazil and Paraguay, also Grenada, West Indies.
  Subgenera:
  Nulapona DeLong & Freytag 1975c:562. *H. (N.) quina* n.sp.
  Carapona DeLong & Freytag 1975c:565. *H. (C.) vulta* n.sp.
HECALIDAE 9:5.
HECALINI 9:6.
Hecalocratus Evans, J.W. 1966a:135. *H. pallidus* n.sp. (Australia: Queensland, Cairo Station). Deltocephalinae, Hecalini. Australian.
Hecaloidea 10:2665. Error for *Hecaloidia* Osborn.
Hecaloidella Osborn 1934a:173. *H. nitida* n.sp. (Samoa: Upolu). 10:2352. Deltocephalinae, Scaphytopiini. Samoa.
Hecaloidia Osborn 1923c:28. *H. nervosa* n.sp. (Bolivia: Villa Bella). 10:1741. Linnavuori 1959b:182. Deltocephalinae, Athysanini. Neotropical; Bolivia, Brazil.
Hecalus Stal 1864b:65. *Petalocephala paykulli* Stal 1854b:252 (Senegal). 9:8. Linnavuori 1975a:19; Morrison 1973a:419-434. Deltocephalinae, Hecalini. All major land masses.
  Synonyms:
  Parabolocratus Fieber 1866a:502. *P. glaucescens* n.sp.
  Thomsonia Signoret 1879a:51. *Hecalus kirschbaumi* Stal 1870c:737 = *Acocephalus porrectus* Walker 1858c:362.
  Thomsoniella Signoret 1880a:52, emendation of *Thomsonia* Signoret 1879a:51.
  Columbanus Distant 1916a:224. *C. misranus* n.sp.
Hecalusu Ishihara 1965d. Error for *Hecalus* Stal.
Hecatus 9:121. Error for *Hecalus* Stal.
Hecullus Oman 1949a:31. *Hecalus bracteatus* Ball 1901a:4 (USA: Colorado, Rocky Ford). 9:22. Deltocephalinae, Hecalini. Nearctic; southwest Great Plains region.
Hegira Oman 1938b:383. *H. brunnea* n.sp. (Argentina, Missiones, Loreto). 10:160. Linnavuori 1959b:283-284. Deltocephalinae, Athysanini. Neotropical; Argentina, Brazil.
Helechara 1:682. Error for *Helochara* Fitch.
Heleochara 1:682. Error for *Helochara* Fitch.
Heliona Melichar 1903b:215. *H. constricta* n.sp. (Ceylon: Peradeniya). 17:611. Typhlocybinae, Helionini. Oriental; Sri Lanka, and Palaearctic.
Helionidea Lindberg 1960a:47. Error for *Helionidia* Zachvatkin.
Helionides Matsumura 1931b:79. *H. singularis* n.sp. (Japan: Honshu). 17:241. Typhlocybinae, Dikraneurini. Oriental; southern China, and Palaearctic; Japan.
Helionidia Zachvatkin 1946a:152. *H. biplagiata* n.sp. (Palestine). 17:1461. Dworakowska 1970e, 1971b. Typhlocybinae, Erythroneurini. Palaearctic, Ethiopian, Oriental.
HELIONINI 17:611.
Heliotettix Rodrigues 1968a:10. *H. fernandesi* n.sp. (Portugal: Bombarral, Estremadura). Deltocephalinae, Fieberiellini. Palaearctic; Morocco, Portugal.
Hellerina Dworakowska 1972f:849. *H. ptasia* n.sp. (Republic of the Congo: Elizabethville). Typhlocybinae, Zyginellini. Ethiopian; Equatorial Africa.
Helochara Fitch 1851a:56. *H. communis* n.sp. (USA: New York). 1:356. Young 1977a:601. Cicadellinae, Cicadellini. Nearctic.
Helocharina Melichar 1926a:343. *Tetigonia gayi* Spinola 1852a:285 (Chile). 1:365. Young 1977a:608. Cicadellinae, Cicadellini. Neotropical; Chile.
HELOCHARINI Metcalf 1965a:356, tribe of TETTIGELLINAE = CICADELLINAE.
Helochora 1:682. Error for *Helochara* Fitch.
Heloehara 1:682. Error for *Helochara* Fitch.
Hemipeltis Spinola 1850b:132. *H. chilensis* n.sp. (Chile). 4:108. Ledrinae, Petalocephalini. Neotropical; Chile.
Hemisudra Schmidt 1911a:228. *H. borneensis* n.sp. (Borneo). 2:7. Hylicidae, Sudrini. Oriental; Borneo.
Henchia 10:2665. Error for *Henschia* Lethierry.
Hengchunia Vilbaste 1969a:8. *Thamnotettix koshunensis* Matsumura 1914a:178 (Formosa: Heng-chun). Deltocephalinae, Deltocephalini. Oriental.
Henriana Emeljanov 1964c:412 as subgenus of *Cicadula* Zetterstedt 1840a:296. *Jassus frontalis* Herrich-Schaeffer 1835a:70 (Great Britain). Deltocephalinae, Cicadulini. Palaearctic.
Henribautia Young & Christian 1952a:96, in Young 1952a. *Typhlocyba nigricephala* Beamer 1943c:131 (USA: Texas, Orange County). 17:933. Typhlocybinae, Typhlocybini. Nearctic; Sonoran.
Henschia Lethierry 1892a:69. *H. seticauda* n.sp. (Hungary). 10:1744. Deltocephalinae, Paralimnini. Palaearctic; central Europe, south Russia, Kazakhstan, Mongolia.
Hensleyella Webb, M.D. 1983b:227. *H. ipoa* n.sp. (Tanzania: Tanganyika). Idiocerinae. Ethiopian.
Hephates Lindberg 1960a:47. Error for *Hephathus* Ribaut.
Hephathus Amyot 13:251. Invalid, not binominal.
Hephathus Ribaut 1952a:437. *Bythoscopus nanus* Herrich-Schaeffer 1835a:69 (Bavaria). 13:160. Hamilton, K.G.A. 1980b:904. Macropsinae, Macropsini. Palaearctic and Australian; possibly Oriental.
  Synonym:
  Asmaropsis Linnavuori 1978b:17. *A. troilos* n.sp.; Hamilton, KGA 1980b:904.
Hephatus Lindberg 1964a:67. Error for *Hephathus* Ribaut.
Hepneriana* Dworakowska 1972j:114. *H. wroblewskae* n.sp. (Java: Tjibulan, near Bogor). Typhlocybinae, Erythroneurini. Oriental. Synonym of *Mandera* Ahmed, Manzoor 1971b:190; Dworakowska and Viraktamath 1975a:529. * Name erroneously credited to Jaczewski in Zoological Record 109:148.

Heptathus Dlabola 1957b:300, 1961b:289. Error for *Hepthathus* Ribaut 1952.

Hesium Ribaut 1942a:262. *Cicada biguttata* Fallen 1806a:27 (Sweden) = *Hesium falleni* Metcalf 1955a:265, replacement for *Cicada biguttata* Fallen 1806, not *Cicada biguttata* Fabricius 1781. 10:303. Deltocephalinae, Athysanini. Palaearctic; Eurasia.

Hespenedra Kramer 1966a:492. *Thlasia chilensis* Spinola 1852a:277 (Chile: Valdivia, Chesque). Ledrinae. Neotropical.

Heterometopia Melichar 1925a:386 as subgenus of *Oncometopia* Stal 1869a:62. *O. (H.) reticulata* n.sp. (Peru). Preoccupied by *Heterometopia* Macquart 1844, see *Hyogonia* China. 1:584. Young 1968a:230. Cicadellinae, Proconiini. Neotropical. Synonym of *Hyogonia* China 1927d:283.

Heterostemma Melichar 1924a:227. *Tettigonia nervosa* Signoret 1855c:524 (Colombia). 1:493. Young 1968a:30. Cicadellinae, Proconiini. Neotropical. Synonym of *Diestostemma* Amyot & Serville 1843a:572.

Hikangia Nielson 1983e:9. *H. liberiensis* n.sp. (Liberia: Toppita). Coelidiinae, Hikangiini. Ethiopian; Cameroon, Liberia, Zaire.

HIKANGIINI Nielson 1983e:9, tribe in COELIDIINAE.

Hilaius Anufriev 1978a:58. Error for *Hylaius* Oman & Musgrave 1975a.

Hiltus Theron 1974b:148. *Chlorotettix africanus* Naude 1926a:76 (South Africa: Natal, Helton Road). Deltocephalinae, Paralimnini. Ethiopian; South Africa.

Hiratettix Matsumura 1931b:59. *Hiratettix arisanellus* Matsumura 1932a:102 (Formosa). 17:241. Typhlocybinae, Typhlocybini. Oriental; Burma, India, Taiwan.

Hirotettix 17:1503. Error for *Hiratettix* Matsumura.

Hishimonoides Ishihara 1965b:20. *H. sellatiformis* n.sp. (Japan: Honshu, Yamagata Pref., Shijo). Deltocephalinae, Opsiini. Oriental.

Hishimonus Ishihara 1953b:38. *Acocephalus discigutta* Walker 1857b:172 (Sarawak). 10:711. Deltocephalinae, Opsiini. Oriental.

Hishiomonus Zoological Record 96, Section 13:361, 1959. Error for *Hishimonus* Ishihara 1953b.

Histipagus Remane & Asche 1980a:85. *H. stipaphagus* n.sp. (Spain). Deltocephalinae, Paralimnini. Palaearctic.

Hododoecus Linnavuori & DeLong 1978c:208. Error for *Hodoedocus* Jacobi 1910b:126. 7:32.

Hodoedocus Jacobi 1910b:126. *H. acuminifrons* n.sp. (Tanzania: Tanganyika). 7:23. Linnavuori & DeLong 1978c:208. Deltocephalinae, Stenometopiini. Paleotropical: East Africa, Philippines, Formosa. Australian. [Neotropical; Guatemala?].
Synonyms:
Acurhinus Osborn 1920a:158. *Dorydium maculatum* Osborn 1909a:464.
Stenometopius Matsumura 1914a:217. *S. formosanus* n.sp.
Afrolimnus Evans, J.W. 1955a:38. *A. ribauti* n.sp.

Hodoedoecus 7:32. Error for *Hodoedocus* Jacobi.

Homa Distant 1908g:400. *H. insignis* n.sp. (Ceylon: Peradeniya). 17:440. Mahmood 1967a:44,45. Typhlocybinae, Typhlocybini. Oriental; Ceylon, Malaysia, Philippines.

Homalodisca Stal 1869a:63. *Cicada triquetra* Fabricius 1803a:63 (South America). 1:499. Young 1968a:193. Cicadellinae, Proconiini. Central and southern USA, Mexico, Central America, Trinidad and South America to Colombia, Ecuador and Brazil.

Homalodisa 1:682. Error for *Homalodisca* Stal.
Homalodissa 1:682. Error for *Homalodisca* Stal.

Homalogoniella Melichar 1925a:341. *Tettigonia pubescens* Signoret 1854d:721 (unknown). 1:523. Young 1977a:1098. Cicadellinae, Cicadellini (?). Neotropical.

Homogramma Emeljanov 1975a:386. *Melicharella proxima* Dlabola 1960c:8 (Iran: Iranshar). Adelungiinae, Adelungiini. Palaearctic. Preoccupied by *Homogramma* Guenee 1854, see *Emelyanogramma* Kocak 1981a:33.

Homolodisca 1:683. Error for *Homalodisca* Stal.

Homoscarta Melichar 1924a:199. *Tettigonia irregularis* Signoret 1855b:232 (Bolivia). 1:602. Young 1968a:46. Cicadellinae, Proconiini. Neotropical.

homsoniella Evans, J.W. 1966a:134. Error for *Thomsoniella* Signoret.

Hordnia Oman 1949a:70. *Tettigonia circellata* Baker 1898c:285 (USA: California) = *Tettigonia artopunctata* Signoret 1854b:354. 1:388. Young 1977a:849. Cicadellinae, Cicadellini. Nearctic, Neotropical. Synonym of *Graphocephala* Van Duzee 1916a:66; Young 1977a:849.

Horouta Knight, W.J. 1975a:205. *H. inconstans* n.sp. (New Zealand: Canterbury Prov., Mt.Cook Nat'l. Park, Hooker Valley, 3.2 km N of Hermitage, 1307 m). Deltocephalinae, Deltocephalini. Australian; New Zealand.

Hortensia Metcalf & Bruner 1936a:928. *Tettigonia similis* Walker 1851b:769 (North America). 1:430. Cicadellinae, Cicadellini. Nearctic/Neotropical; southeastern USA, Mexico, Central America and all South American countries except Chile and Uruguay.

Horvathiella Matsumura 1914a:234. *H. arisana* n.sp. (Formosa). 8:224. Cicadellinae, Anoterostemmini. Oriental. Preoccupied by *Horvathiella* Poppins 1912, see *Agrica* Strand 1942a:393.

Houtbayana Linnavuori 1961a:476. *H. orosioides* n.sp. (South Africa: Cape Prov., Cape Peninsula, Hout Bay, Skoorsteenkop). Deltocephalinae, Athysanini. Ethiopian; South Africa.

Huachia Linnavuori 1959b:191. *H. rugicollis* n.sp. (Bolivia: Huachi, Beni). Deltocephalinae, Athysanini. Neotropical; Bolivia, Brazil, Ecuador.

Huancabamba Linnavuori 1959b:204. *H. rotundiceps* n.sp. (Peru: Huancabamba). Deltocephalinae, Athysanini. Neotropical; Peru.

Huleria Ball 1902b:57. *H. quadripunctata* n.sp. (USA: California, Los Angeles County). 10:1084. Deltocephalinae, Cochlorhini. Nearctic; subsonoran region.

Humpatagallia Linnavuori & Viraktamath 1973a:485. *H. scutellaris* n.sp. (Guinea: Nimba). Agalliinae, Agalliini. Ethiopian; Guinea.

Hussa Distant 1918b:68. *H. insignis* n.sp. (South India: Kodaikanal). 10:2347. Barnett 1977a:494. Deltocephalinae, Scaphoideini. Oriental. Synonym of *Scaphoideus* Uhler 1889a:33.

Hussainiana Mahmood & Ahmed 1969a:87. *H. tripunctata* n.sp. (East Pakistan: Rangpur). Dworakowska 1971c:493. Oriental; Bangladesh.
[NOTE: See Dworakowska 1971c:493 for sub-family placement, not Typhlocybinae].

Hyalocerus Maldonado-Capriles 1977b:358. *H. aurantius* n.sp. (British Guiana: near Mazaruni Hd., Pakaraima Mts.). Idiocerinae. Neotropical; British Guiana, Peru.

Hyalojassus Evans, J.W. 1972b:648. *H. takensis* n.sp. (Thailand: Sam Ngow Tak). Iassinae, Hyalojassini. Oriental.

HYALOJASSINI Evans, J.W. 1972b:648, tribe in JASSINAE = IASSINAE.

Hybla McAtee 1932b:119. *H. maculata* n.sp. (Puerto Rico). 17:284. Typhlocybinae, Dikraneurini. Neotropical.

Hybos 17:1503. Error for *Kybos* Fieber.

Hybrasil Kirkaldy 1907d:41. *H. brani* n.sp. (Fiji). 10:2089. Linnavuori & Al-Ne'amy 1983a:22. Selenocephalinae, Bhatiini. Oceania.

Hydabricta Webb, M.D. 1983a:80. *H. streeta* n.sp. (Australia: Queensland, Windsor Tableland via Mt. Caroline). Idiocerinae. Australian.

Hylaius Oman & Musgrave 1975a:6 as subgenus of *Bathysmatophorus* Sahlberg 1871a:109. *Errhomenus oregonensis* Baker 1898:262 (USA: Oregon). Cicadellinae, Errhomenini. Nearctic; northwest California to central Washington, USA.

Hylica Stal 1863c:593. *H. paradoxa* n.sp. (Burma). 2:6. Hylicinae, Hylicini. Oriental; Borneo, Burma, India, Java.

HYLICIDAE 2:5.
HYLICINAE 2:5.
HYLICINI 2:5.

Hyloidea McAtee 1926b:162 as subgenus of *Dikraneura* Hardy 1850a:423. *D. (H.) depressa* n.sp. (Puerto Rico: Vega Alta). 17:224. Typhlocybinae, Dikraneurini. Neotropical, Nearctic. Subgenus of *Alconeura* Ball & DeLong 1925a:334.
Synonym:
Dikraneuroidea Lawson 1929d:307. *D. beameri* n.sp.

Hymetta McAtee 1919a:121. *Tettigonia trifasciata* Say 1825a:343 (USA: Missouri). 17:1466. Typhlocybinae, Erythroneurini. Nearctic.

Hyogonia China 1927d:283, replacement for *Heterometopia* Melichar 1925a:386, not *Heterometopia* Marquart 1844. *Oncometopia (H.) reticulata* Melichar 1925a:386 (Peru). 1:584. Young 1968a:230. Emmrich & Lauterer 1975a:309-314. Cicadellinae, Proconiini. Neotropical; Bolivia, Brazil, Peru, Venezuela.
Synonym:
Heterometopia Melichar 1925a:386, preoccupied.

Hypacostemma Linnavuori 1961a:465. *H. viridissima* n.sp. (South Africa: Cape Province, Tzitzikama Forest, Stormsrivierpiek). Deltocephalinae, Athysanini. Ethiopian; South Africa.

HYPACOSTEMMINI Linnavuori & Al-Ne'amy 1983a:54, tribe in SELENOCEPHALINAE.

Hypaulacia Amyot 15:223. Invalid, not binominal.

Hypericiella Dworakowska 1970j:559 as subgenus of *Zygina* Fieber 1866A:509. *Typhlocyba hyperici* Herrich-Schaeffer 1836a:4 (Germany). Typhlocybinae, Erythroneurini. Palaearctic.

Hyposadianus 10:2665. Error for *Hypospadianus* Ribaut.

Hypospadianus Ribaut 1942a:264. *Cicada torneella* Zetterstedt 1828a:528 (Lapland). 10:1242, 2665. Deltocephalinae, Athysanini. Palaearctic. Synonym of *Colladonus* Ball 1936c:57.

Hyposticta Amyot 17:1503. Invalid, not binominal.
Hypostilba Amyot 17:1503. Invalid, not binominal.

Ianeira Linnavuori 1969a:1152. *I. palaemon* n.sp. (Cameroon). Selenocephalinae, Ianeirini. Ethiopian; Guinean.

IANEIRINI Linnavuori 1978a:43, tribe of SELENOCEPHALINAE.

Iassargus Mitjaev 1968b:34. Error for *Jassargus*.

IASSIDAE 10:3, 11:5, 15:5.

IASSINAE 11:11, 15:19. Placed on the Official List of Family-Group Names in Zoology, ICZN Opinion 612, 1961.

IASSINI 15:20.

Iassomorphus Theron 1972a:202. *Bythoscopus drakensteini* Naude 1926a:18 (South Africa: Stellenbosch, Jonkershoek). Iassinae, Iassini. Ethiopian; Sudeanese subregion, south and northeast Africa, southwest Arabia.

Iassus Fabricius 1803a:85. *Cicada lanio* Linnaeus 1761a:242 (Sweden). 15:21. Iassinae, Iassini. Palaearctic and Africa.
Synonym:
Straganiassus Anufriev 1971g:87. *Stragania Matsumurai* Metcalf 1955a:266; Nast 1981a:258.
[NOTE: Placed on the Official List of Generic Names in Zoology, ICZN Opinion 612].

Ibadarrus Remane & Asche 1980a:73. *I. gracilior* n.sp. (Spain). Deltocephalinae, Paralimnini. Palaearctic; Spain.

Iberia Kirkaldy 1907d:40. *Stegelytra bolivari* Signoret 1880b:203 (Spain). 11:93. Stegelytrinae. Palaearctic; southern Europe, northern Africa.

Icaia Linnavuori 1973b:15. *I. gnathenion* n.sp. (Peru: Ica, Hda. Sta. Rosa). Deltocephalinae, Doraturini. Neotropical; Bolivia, Ecuador, Peru.

Icaiana Dworakowska 1981b:18 as subgenus of *Acia* McAtee 1934a:109. *A. (I.) tina* n.sp. (Zaire: Lubumbashi "Elizabethville"). Typhlocybinae, Empoascini. Ethiopian.

Ichthyobelus Melichar 1925a:360. *I. bellicosus* n.sp. (Bolivia: Coroico). 1:540. Young 1968a:135. Cicadellinae, Proconiini. Neotropical; Bolivia, Colombia, Ecuador, Peru.

Ideocerus 16:235. Error for *Idiocerus* Lewis.

Idia Fieber 1866a:509. *Typhlocyba scutellaris* Herrich-Schaffer 1838c:13 (Provence). 17:1439. Typhlocybinae, Typhlocybini. Palaearctic. Preoccupied by *Idia* Huebner 1813, see *Zyginidia* Haupt 1929c:268.

Idiocerella Evans, J.W. 1941e:39. *I. obscura* n.sp. (Australia: Kangaroo Island, Flinders Chase). 16:216. Webb, M.D. 1983a:93. Idiocerinae. Australian.

IDIOCERIDAE 16:5.

Idiocerinus Baker 1915c:320. *I. melichari* n.sp. (Philippine Islands: Luzon). 16:224. Idiocerinae. Oriental; Philippines.

Idioceris 16:235. Error for *Idiocerus* Lewis.

Idioceroides Matsumura 1912b:324. *I. tettigoniformis* n.sp. (Formosa). 16:217. Maldonado-Capriles 1976a:139. Agalliinae. Oriental.

Idiocerra 16:235. Error for *Idiocerus* Lewis.

Idiocerus Lewis 1834a:47. *I. stigmaticalis* n.sp. (Great Britain). 16:8. Hamilton, KGA 1980a:825. Idiocerinae. Holarctic, Malagasian; Ethiopia, Chile, South Africa.
Subgenera:
Liocratus Dubovskij 1966a:119. *Idiocerus chivensis* Kusnezov 1929a:309.
Populicerus Dlabola 1974b:62. *Cicada populi* Linnaeus 1861a:242.
Parocerus Vilbaste 1980b:32. *Idiocerus laurifoliae* Vilbaste 1965b:35.
Metidiocerus Ossiannilsson 1981a:318. *Idiocerus crassipes* J. Sahlberg 1871a:143.
Koreocerus Kwon 1985a:65. *Idiocerus koreanus* Matsumura 1915a:173.
Nabicerus Kwon 1985a:66. *Idiocerus fuscescens* Anufriev 1971c:677.
Bicernarus Kwon 1985a:66. *Idiocerus ishiyamae* Matsumura 1905a:67.
Podulmorinus Kwon 1985a:69. *Idiocerus vitticollis* Matsumura 1905a:69.
Pugnostilus Kwon 1985a:70. *Idiocerus latistylus* Vilbaste 1968a:69.
Synonyms:
Sahlbergotettix Zachvatkin 1953a:213. *Idiocerus salicicola* Flor 1861a:163.
Acericerus Dlabola 1974b:61. *Idiocerus rotundifrons* Kirschbaum 1868a:5.
Viridicerus Dlabola 1974b:62. *Bythoscopus ustulatus* Mulsant & Rey 1855a:217.
Tremulicerus Dlabola 1974b. "*Idiocerus fasciatus* Fieber 1868a:455 = *Cicara* (sic!) *tremulae* Estlund 1796a:129."
Stenidiocerus Ossiannilsson 1981a:348. *Bythoscopus poecilus* Herrich-Schaeffer 1835a:69.

Idiocoris Ishihara 1965d. Error for *Idiocerus* Lewis.

Idiodonus Ball 1936c:57. *Jassus kennicottii* Uhler 1864a:161 (USA: Maryland). 10:1284, 1304. DeLong 1984c. Deltocephalinae, Athysanini. Holarctic; Neotropical.
Synonyms:
Josanus DeLong 1938f:244. *Phlepsius josea* Ball 1900c:347.
Orolix Ribaut 1942a:267. *Cicada cruentata* Panzer 1799a:15.
Angulanus DeLong 1946b:30. *Idiodonus (Angulanus) incisurus* n.sp.
[NOTE: *Angulanus* is included as a synonym (rather than as a subgenus as originally described, and as treated by Metcalf 10:1304) because DeLong did not mention the name in his 1984c paper in which *incisurus*, the type species of *Angulanus*, was included in *Idiodonus* without reservation].

Idionannus Linnavuori 1956a:16. *I. elegantulus* n.sp. (Paraguay: S. Bernardino). Idiocerinae. Neotropical.

Idioscopus Baker 1915c:320. *Idiocerus clypealis* Lethierry 1889c:252 (India: West Bengal). 16:9. Maldonado-Capriles 1964a:90-100; Maldonado-Capriles 1971a:184. Idiocerinae. Oriental, Malagasian; Ethiopian.
Synonym:
Chunrocerus Zachvatkin 1946:154.

Idiotettix Osborn 1929a:465. *Thamnotettix magnificus* Osborn 1924c:424 (Bolivia: Quatro Ojos). 16:225. Linnavuori 1959b:27, Kramer 1963b. Neobalinae. Neotropical. Synonym of *Calliscarta* Stal 1869a:82.

Idliocerus 16:235. Error for *Idiocerus* Lewis.
Idocerus 16:235. Error for *Idiocerus* Lewis.
Idocierus 16:236. Error for *Idiocerus* Lewis.

Idona DeLong 1931b:50 as subgenus of *Empoasca* Walsh 1862a:149. *Empoasca minuenda* Ball 1921a:23 (USA: Florida, Miami). 17:202. Typhlocybinae, Dikraneurini. Nearctic.
[NOTE: Moznette's 1919a:46 use of a Ball manuscript name in connection with comments about "*Empoasca minuenda* Ball" was apparently considered by Metcalf 1968a:205, to have validated the name. Subsequent authors have not accepted that interpretation].

Idoneus 10:2665. Error for *Idiodonus* Ball.
Idriocerus 16:236. Error for *Idiocerus* Lewis.

Idyia Linnavuori 1960b:58 as subgenus of *Cicadulina* China 1926c:43 (Fiji: Viti Levu, Suva). *C. (I.) fijiensis* n.sp. Deltocephalinae, Macrostelini.

Iedidia Amyot 17:1503. Invalid, not binominal.

Ifeia Linnavuori & Al-Ne'amy 1983a:93. *Selenocephalus africanus* Stal 1854b:254 (Sierra Leone: Afzelius). Selenocephalinae, Selenocephalini. Ethiopian; Guinean.

Ifeneura Ghauri 1975b:489. *I. oliarana* n.sp. (Nigeria: Ibadan, Moor plantation). Typhlocybinae, Erythroneurini. Ethiopian.

Ifugoa Dworakowska & Pawar 1974a:587. *I. mikra* n.sp. (Philippines: Ifugo). Typhlocybinae, Empoascini. Oriental.

Igerna Kirkaldy 1903f:13, replacement for *Pachynus* Stal 1866a:127, not *Pachynus* Rafinesque 1815. *Bythoscopus (Oncopsis) bimaculicollis* Stal 1855a:100 (South Africa: Natal). 14:146. Agalliinae, Agalliini. Ethiopian.

Igutettix Matsumura 1932a:105. *I. pulverosus* n.sp. (Japan: Honshu). 17:242. Typhlocybinae, Dikraneurini. Palaearctic; Japan.

Ikomella Ishihara 1961a:253. *I. confersa* n.sp. (Thailand: Chieng Mai). Mukariinae, Mukariini. Oriental. Synonym of *Mukaria* Distant 1908g:269; Linnavuori 1979b:985.

Ilagia Kramer & DeLong 1968a:172. *I. zebula* n.sp. (Mexico: San Luis Potosi, Tamazunchale). Deltocephalinae, Athysanini. Neotropical.

Imbecilla Dworakowska 1970m:703. *Erythroneura lubiae* China 1931a:53 (Sudan). Typhlocybinae, Erythroneurini. Northern Ethiopian and southern Palaearctic.

Impoasca 17:1503. Error for *Empoasca* Walsh.

Imugina Mahmood 1967a:27. *I. robusta* n.sp. (Philippines: Luzon, N. Viscaya, Imugin). Typhlocybinae, Erythroneurini. Oriental.

Inazuma Ishihara 1953b:15. *Deltocephalus dorsalis* de Motschulsky 1859b:114. 10:1601. Deltocephalinae, Deltocephalini. Palaearctic. Subgenus of *Recilia* Edwards 1922a:206; Kwon & Lee 1979b:80.

Indianella Ramakrishnan & Menon 1973a:22. *I. indraprasthana* n.sp. (India: Delhi, I.A.R.I.). Typhlocybinae, Erythroneurini. Oriental. Synonym of *Accacidia* Dworakowska 1971b:113; Dworakowska & Sohi 1978a:39.

Indiocerus 16:236. Error for *Idiocerus* Lewis.

Indodikra Sharma, B. 1979a:5. Invalid, no type species designated.

Indoformosa Ramakrishnan & Ghauri 1979b:198. *Zygina indica* Datta 1969a:391 (India: Andra Pradesh, Agricultural Research Station). Typhlocybinae, Erythroneurini. Oriental. Synonym of *Empoascanara* Distant 1918b:94; Dworakowska 1980d:195.

Inemadara Ishihara 1953b:48. *Deltocephalus oryzae* Matsumura 1902a:390 (Japan). 10:1599. Deltocephalinae, Deltocephalini. Palaearctic. Synonym of *Togacephalus* Matsumura 1940d:38; Kwon & Lee 1979b:74, 76.

[NOTE: Nast 1972a:343 lists *Inemadara* Ishihara as a synonym of *Recilia* Edwards, but later authors (see index) have not recognized the synonymy].

Inemedara Claridge & Wilson 1981c:22, error for *Inemadara* Ishihara.

Infulatartessus Evans, F. 1981a:168. *I. nondugensis* n.sp. (New Guinea: Wau, 1200 m). Tartessinae. Oriental; New Guinea, Bismarck Archipelago.

Inghamia Evans, J.W. 1966e:248. *I. dayi* n.sp. (Australia: Queensland, Ingham). Deltocephalinae, Platymetopiini. Australian.

Inoclapis Nielson 1979b:131. *I. peruviensis* n.sp. (Peru: Colonia Perene, Rio Perene, 29 km NE of La Merced, Jamin). Coelidiinae, Teruliini. Neotropical.

Insulanus Linnavuori 1960a:303 as subgenus of *Deltocephalus* Burmeister 1838b:15. *Stirellus subviridis* Metcalf 1946d:125 (Guam). Deltocephalinae, Deltocephalini. Oriental; throughout Oceania.

Inuyana Young 1977a:778. *I. juninensis* n.sp. (Peru: Chanchamyo). Cicadellinae, Cicadellini. Neotropical; Peru and SW Brazil.

Ionia Ball 1933c:226. *I. triunata* n.sp. (USA: Arizona, Sabino Canyon near Tucson). 14:111. Agalliinae, Agalliini. Nearctic. Subgenus of *Aceratagallia* Kirkaldy 1907d:30.

Iowanus Ball 1918b:382 as subgenus of *Phlepsius* Fieber 1866a:514. *P. (I.) handlirschi* n.sp. (Mexico: Guerrero). 10:691. Deltocephalinae, Athysanini. Nearctic. Subgenus of *Texananus* Ball 1918b:381.

Ipelloides Evans, J.W. 1966a:34. *I. macleayi* n.sp. (South Australia). Eurymelinae, Ipoini. Australian.

Ipo Kirkaldy 1906c:464. *I. ambita* n.sp. (Australia: Queensland). 12:34. Evans, J.W. 1966a:52. Eurymelinae, Ipoini. Australian.

Ipocerus Baker 1915c:319. *I. kirkaldyi* n.sp. (Philippines: Palawan). 16:219. Idiocerinae. Oriental.

Ipocerus Evans, J.W. 1934a:165. *Ipo procurrens* Jacobi 1909a:342 (Southwestern Australia). 12:38, 42. Preoccupied by *Ipocerus* Baker 1915c:319, see *Bakeriana* Evans, J.W. Eurymelinae, Ipoini. Australian. Synonym of *Bakeriana* Evans, J.W. 1954a:129.

Ipoella Evans, J.W. 1934a:157. *I. fidelis* n.sp. (Australia: Queensland, Bunya Mountains). 12:31. Evans, J.W. 1966a:40. Eurymelinae, Ipoini. Australian.
Synonym:
Anipo Evans, J.W. 1934a:159. *Eurymela porriginosa* Signoret 1850c:512.

Ipoides Evans, J.W. 1934a:155. *I. hackeri* n.sp. (Australia: Queensland, Brisbane). 12:32. Evans, J.W. 1966a:36. Eurymelinae, Ipoini. Australian.

IPOINAE 12:28.

Ipolo Evans, J.W. 1966a:39. "*Ipoides davisi* Evans", lapsus for *Ipoella davisi* Evans, J.W. 1947f:227 (Northwest Australia, Walcott Inlet, Isdell River). Eurymelinae, Ipoini. Australian.

Iposa Evans, J.W. 1977a:86. *Anipo fusca* Evans, J.W. 1941f:143 (Western Australia: Perth). Eurymelinae, Ipoini. Australian.

Iposcopus Baker 1915c:319. *I. distanti* n.sp. (Philippines: Mindanao). 16:228. Idiocerinae. Oriental.

Iragua Melichar 1926a:342. *Tettigonia diversa* Signoret 1855a:49 (French Guiana). Young 1977a:399. 1:264. Cicadellinae, Cicadellini. Neotropical.
Synonyms:
Eugonia Melichar 1926a:342. *Tettigonia diversa* Signoret 1855a:49.
Eugonalia Evans, J.W. 1947a:161. *Tettigonia diversa* Signoret 1855a:49.

Iraqerus Ghauri 1972a:207. *I. neveosparsus* n.sp. (Iraq: Golle). Unassigned, formerly Coelidiinae; Nielson 1975a:11. Palaearctic.

Irenaella Linnavuori 1977a:47. *I. kalypso* n.sp. (Ivory Coast: Foro-Foro). Penthimiinae. Ethiopian.

Irenara Ramakrishnan & Ghauri 1979b:204. "*Thamnotettix limbata* Matsumura, 1907b:89", lapsus for *Typhlocyba (Zygina) limbata* Matsumura 1910e:120 (Japan). Typhlocybinae, Erythroneurini. Oriental. Synonym of *Empoascanara* Distant 1918b:94; Dworakowska 1980d:195.

Iriatartessus Evans, F. 1981a:181. *I. maai* n.sp. (New Guinea (SW): Vogelkop, Fak Fak, S. coast of Bomberai). Tartessinae. Australian.

Irinula Ribaut 1948c:58. *Cicadula erythrocephala* Ferrari 1882a:46. 10:2027. Deltocephalinae, Macrostelini. Palaearctic. Synonym of *Nesoclutha* Evans, J.W. 1947e:126; Vilbaste 1976a:28.

Isaca Walker 1857b:172. *I. bipars* n.sp. (Sarawak). 15:169. Selenocephalinae, Selenocephalini. Oriental.

Iseza Dworakowska 1981h:322. *I. setna* n.sp. (Zaire: Lubumbashi "Elizabethville"). Typhlocybinae, Erythroneurini. Ethiopian; Namibia, Nigeria, Sudan, Tanzania, Zaire.
Subgenus:
Tamaga Dworakowska 1981h:325. *I. (T.) dubia* n.sp.

Ishidaella Matsumura 1912a:41. *Tettigonia albomarginata* Signoret 1853b:347 (Australia). 1:463. Cicadellinae, Cicadellini. Australian: Australia and Tasmania.

Ishidella 1:683. Error for *Ishidaella* Matsumura.

Ishiharanara Ramakrishnan & Ghauri 1979b:208 as subgenus of *Sawaiinara* Ramakrishnan & Ghauri 1979b:208. *Empoascanara wasata* Dworakowska 1978a:156 (Malaysia: W. coast Penang I., Pulau). Typhlocybinae, Erythroneurini. Oriental. Synonym of *Empoascanara* Distant 1918b:94; Dworakowska 1980d:195.

Ishiharella Dworakowska 1970h:716. *Empoasca polyphemus* Matsumura 1931b:82 (Japan: Kyoto). Typhlocybinae, Empoascini. Palaearctic.

Isogonalia Young 1977a:819. *Tettigonia sexlineata* Signoret 1855d:792 (Guatemala). Cicadellinae, Cicadellini. Neotropical; Mexico, Central America and Panama.

Iturnoria Evans, J.W. 1954a:108. *I. insulana* n.sp. (Madagascar). 11:82. Unassigned, formerly Coelidiinae; Nielson 1975a:11. Malagasian.

Ivorycoasta Dworakowska 1972f:851. *I. pulchra* n.sp. (Sudan: Equatoria, Loka forest). Typhlocybinae, Erythroneurini. Ethiopian.

Jacobiasca Dworakowska 1972i:29, as subgenus of *Austroasca* Lower 1952a:202. "*Chlorita lybica* Berg. et Zan." presumably intended to be *Chlorita libyca* de Bergevin & Zanon 1922a:58 (Libya). Typhlocybinae, Empoascini. Palaearctic.

Jacobiella Dworakowska 1972c:481. "*Chlorita facialis* Jac. 1910" (1912b:69?). Typhlocybinae, Empoascini. Ethiopian; Equatorial Africa.

Jafar Kirkaldy 1903f:13, replacement for *Setabis* Stal 1866a:111, not *Setabis* Westwood 1851. *Gypona javeti* Signoret 1858a:342 (Calabar). 3:214. Linnavuori 1977a:53. Penthimiinae. Ethiopian; Guinean.
Synonym:
Auropenthimia Evans, J.W. 1947a:255. *A. aburensis* n.sp.

Jakarellus Webb, M.D. 1980a:857. "*Scaphoideus ineffectus* Baker 1924d:367" replacement for *Scaphoideus tessellatus* Distant 1917a:319 (Seychelles), not *Scaphoideus tesselatus* Osborn 1909a:465. Deltocephalinae, Platymetopiini. Ethiopian.

Jakrama Young 1977a:686. *Tettigonia servillei* Signoret 1853b:330 (Brazil). Cicadellinae, Cicadellini. Neotropical; northern South America.

Jalalia Ahmed, Manzoor 1970b:177. *J. colorata* n.sp. (West Pakistan: Abbottabad, Thai). Typhlocybinae, Erythroneurini. Oriental.

Jalorpa Nielson 1979b:71. *J. larseni* n.sp. (Brazil: Minas Gerais, Bello Horizonte). Coelidiinae, Teruliini. Neotropical.

Jamacerus Freytag 1969a:352. *J. farri* n.sp. (West Indies: Jamaica, Portland, Hardwar Gap). Idiocerinae. Neotropical.

Jamitettix Matsumura 1940d:40. *J. kotonis* n.sp. (Formosa). 15:181. Selenocephalinae, Bhatiini. Oriental; Formosa and western Micronesia.

Janastana Young 1977a:214. *Tettigonia distinguenda* Fowler 1900a:257 (Panama). Cicadellinae, Cicadellini. Neotropical; Colombia, Costa Rica, Ecuador, Mexico, Panama.

Janitettix Zoological Record 97, Section 13:331, 1960. Error for *Jamitettix* Matsumura 1940d:40.

Jannius Theron 1982b:25. *J. mecus* n.sp. (South Africa: Natal, Ladysmith). Deltocephalinae, Deltocephalini. Ethiopian; South Africa, southwest Africa.

Janus 11:149. Error for *Jassus* Germar.

Japanagallia Ishihara 1955a:215. *Agallia pteridis* Matsumura 1905a:68 (Japan). 14:60. Agalliinae, Agalliini. Palaearctic; Japan.

Japananus Ball 1931d:218. *Platymetopius hyalinus* Osborn 1900d:501 (USA: District of Columbia). 10:2244. Deltocephalinae, Scaphytopiini. Palaearctic. [Nearctic].

Japanus 10:2665. Error for *Japananus* Ball.

JASCOPIDAE Hamilton, KGA 1971a:943. Fossil.

Jascopus Hamilton, KGA 1971a:944. *J. notabilis* n.sp. Fossil: Upper Cretaceous amber; Canada, Manitoba, Cedar Lake. Evans, J.W. 1972c.

Jasius 11:149. Error for *Jassus* Stal.

JASSARGINI Emelyanov 1962b:393, tribe of DELTOCEPHALINAE.

Jassargus Zachvatkin 1933b:268. *Jassus (Deltocephalus) distinguendus* Flor 1861a:240 (Europe). 10:1201, 1218. Deltocephalinae, Paralimnini. Palaearctic.
Subgenera:
Arrailus Ribaut 1952a:255. *Deltocephalus flori* Fieber 1869a:210.
Aurkius Ribaut 1952a:257. *Deltocephalus repletus* Fieber 1869a:208.
Sayetus Ribaut 1952a:257. *Deltocephalus sursumflexus* Then 1902a:189.
Synonym:
Lausulus Ribaut 1946b:84. *Deltocephalus pseudocellaris* Flor 1861a:547.

JASSIDAE 1:16, 3:12, 10:1, 11:5, 15:10.
JASSINAE 10:8, 11:11, 15:20.
JASSINI 10:10, 11:16.
Jassius 11:149. Error for *Jassus* Stal.

Jassoidula Osborn 1934a:182. *J. straminea* n.sp. (Samoa: Tutuila). 11:82. Coelidiinae, Tharrini. Oceania. Synonym of *Tharra* Kirkaldy 1906c:342; Nielson 1975a:34.

Jassolidia Nielson 1982e:284. *Coelidia munda* Stal 1862b:51 (Brazil). Coelidiinae, Coelidiini. Neotropical.

Jassonirvana Baker 1923a:399. *J. lineata* n.sp. (Philippines: Luzon). 7:23. Nirvaninae, Nirvanini. Oriental.

Jassopsis Scudder 1890a:312. *J. evidens* n.sp. Metcalf & Wade 1966a:218. Fossil: Miocene; USA, Colorado.

Jassosqualus Kramer 1964b:122. *Carchariacephalus smithii* Baker 1897d:153 (Brazil: Rio de Janeiro). Nirvaninae. Neotropical.

Jassulus Evans, J.W. 1955a:30. *J. brunneus* n.sp. (Belgian Congo: Elizabethville). 15:115. Linnavuori & Quartau 1975a:10. Iassinae, Iassini. Ethiopian; Congolese.
Synonym:
Rhodulopa Linnavuori 1961a:456. *R. granulosa* n.sp.

Jassus Burmeister 1835a:104. 10:957.

Jassus Fallen 1806b:115. 15:223, 15:22. Unnecessary emendation of *Iassus* Fabricius.
[NOTE: Placed on the List of Rejected and Invalid Names in Zoology, ICZN Opinion 612].

Jassus Stal 1854b:255. 11:33.

Jasus 11:176. Error for *Jassus* Germar.

Javadikra Dworakowska 1971d:585. *J. marcowa* n.sp. (Viet-Nam: Lao Kay). Typhlocybinae, Dikraneurini. Oriental; India, Taiwan, Vietnam. Synonym of *Karachiota* Ahmed, Manzoor 1969a:58; Sohi & Dworakowska 1984a:165.

Jawigia Nielson 1979b:106. *Jassus angulatus* Spangberg 1878b:10 (Brazil: Sao Paulo). Coelidiinae, Teruliini. Neotropical; Brazil, Paraguay.

Jdiocerus 16:236. Error for *Idiocerus* Lewis.

Jenolidia Nielson 1982e:81. *J. jenniferae* n.sp. (Malaysia: Sabah, Sandakan Bay). Coelidiinae, Coelidiini. Oriental; Malaysia.

Jikradia Nielson 1979b:75. *Jassus olitorius* Say 1830b:310 ("Amer. Bor."). Coelidiinae, Teruliini. Nearctic and Neotropical to Central America and the Galapagos Islands.

Jilijapa Melichar 1925a:356. *J. armata* n.sp. (Cochin China). 1:537. Cicadellinae, Proconiini. Oriental.

Jilinga Ghauri 1974c:551. *Deltocephalus darjilingensis* Distant 1918b:82 (Darjeeling, E. Himalayas). Deltocephalinae, Deltocephalini. Oriental; India.

Jimara Dworakowska 1977h:337. *J. bifasciata* n.sp. (Zaire: Lubumbashi "Elizabethville"). Typhlocybinae, Dikraneurini. Ethiopian; tropical Africa.

Jiridlabolina Kocak 1981a:32, replacement for *Sagittifer* Dlabola 1961b:345, not *Sagittifer* Kuhl 1920. *Sagittifer optatus* Dlabola 1961b:346 (USSR: Tadzhikistan, Tigrowaja Balka). Deltocephalinae, Doraturini. Palaearctic. Synonym of *Aconura* Lethierry 1876a:9.

Jiutepeca Linnavuori & DeLong 1978e:49. *Dicyphonia nigrita* Ball 1937a:132 (Mexico: Mexico City to Acupulco Road, km 399). Deltocephalinae, Hecalini. Neotropical; Mexico.

Jivena Blocker 1976a:521. *J. ambita* n.sp. (Peru: Dep. Junin, San Ramon, Estancia Naranjal, 1000 m). Iassinae. Neotropical; Peru.

Jogocerus Viraktamath, C.A. 1979b:17. *J. freytagi* n.sp. (India: Karnataka, Jog Falls). Idiocerinae. Oriental.

Johanus Theron 1974b:160. *Euscelis cypraea* Naude 1926a:64 (South Africa: Stellenbosch, Jonkershoek). Deltocephalinae, Athysanini. Ethiopian.

Joruma McAtee 1924c:34. *J. pisca* n.sp. (USA: Maryland, Plummers Island). 17:599, 602. Typhlocybinae, Jorumini. Nearctic; Neotropical.
Subgenus:
Jorumidia Young 1952b:107. *Joruma curvata* Osborn 1928a:284.
Synonym:
Jorumella McAtee 1934a:109. *Joruma ascripta* McAtee 1926b:165.

Jorumella McAtee 1934a:109 as subgenus of *Joruma* McAtee 1924a:34. *Joruma ascripta* McAtee 1926b:165 (Costa Rica). 17:602. Typhlocybinae, Jorumini. Neotropical. Synonym of *Joruma* McAtee 1924c:34.

Jorumidea 17:1504. Error for *Jorumidia* Young.

Jorumidia Young 1952b:107 as subgenus of *Joruma* McAtee 1924c:34. *Joruma curvata* Osborn 1928a:284 (Brazil: Lagoa Feia). 17:606. Typhlocybinae, Jorumini. Neotropical.

JORUMINI 17:599.

Josanus DeLong 1938f:244 as subgenus of *Phlepsius* Fieber 1866a:514. *Phlepsius josea* Ball 1900c:347 (USA: Colorado). 10:1284. Deltocephalinae, Athysanini. Nearctic. Synonym of *Idiodonus* Ball 1936c:57.

Jozima Young 1977a:294. "The species identified and illustrated as *Tettigonia leucopa* Walker 1858b:217" (Peru: Rio Napo). Cicadellinae, Cicadellini. Neotropical.

Jubrinia Linnavuori 1962a:54. *J. distincta* n.sp. (Israel: Beit Jubrin). Deltocephalinae, Opsiini. Palaearctic and Ethiopian: Egypt, Israel, Ethiopia, Congo, South Africa.

Jukaruka Distant 1907g:190. *J. typica* n.sp. (Australia: Queensland). Evans, J.W. 1966a:99,100. Ledrinae, Petalocephalini. Australian.

Juliaca Melichar 1926a:344. *Microgoniella (Microscita) naevula* Melichar 1951a:110 (Colombia). 1:295. Young 1977a:443. Cicadellinae, Cicadellini. Neotropical; northern South America.
Synonym:
Microscita Melichar 1951a:93. *M. (M.) naevula* Melichar 1951a:110.

Julipopa Blocker 1979a:35. *J. liolova* n.sp. (Peru: Chancay, 4 mi. N. Lima). Iassinae. Neotropical; Chile, Peru.

Kaapia Theron 1983a:147. *K. darwini* n.sp. (South Africa: Cape Province, Cape Point). Deltocephalinae, Athysanini. Ethiopian; South Africa.

Kabakra Dworakowska 1979a:155. *K. augusta* n.sp. (Vietnam: Prov. Nghe-An, Quy-chau). Typhlocybinae, Erythroneurini. Oriental.

Kadrabia Dworakowska & Sohi 1978b:467. *K. viraktamathi* n.sp. (India: Karnataka, Jog Falls). Typhlocybinae, Typhlocybini. Oriental.

Kahaono Kirkaldy 1906c:358. *K. hanuala* n.sp. (Australia, Queensland, Brisbane). 17:106. Evans, J.W. 1966a:260-261. Dworakowska 1972l:195, 197. Typhlocybinae, Dikraneurini. Australian.

Kahono 17:1504. Error for *Kahaone* Kirkaldy.

Kahavalu Kirkaldy 1906c:371. *K. gemma* n.sp. (Australia, New South Wales, Sydney). 5:79. Ulopinae, Ulopini. Australian.

KAHAVALUIDAE 5:3.

Kaila Dworakowska 1974a:148. *K. synavei* n.sp. (Congo: Dimonika, Mayumbe). Typhlocybinae, Empoascini. Ethiopian; Congo and Aldabra Id.

Kalasha Distant 1908g:254. *K. nativa* n.sp. (India: Assam, Sadeya). 2:9. Hylicinae, Sudrini. Oriental; India.

Kalimorpha Nielson 1979b:304. *Coelidia flaviceps* Stal 1864a:85 (Mexico). Coelidiinae, Teruliini. Neotropical; Mexico and Central America.

Kalkandelenia Kocak 1981a:32, replacement for *Matuta* Emeljanov 1966a:99, not *Matuta* Grote 1874. *Tettigonia guttiger* Uhler 1896a:294 (Japan). Cicadellinae, Pagaroniini. Palaearctic. Synonym of *Pagaronia* Ball 1902a:19.

Kalkiana Sohi, Viraktamath & Dworakowska 1980a:279. *K. bambusa* n.sp. (India: Haryana, Kalka). Typhlocybinae, Dikraneurini. Oriental.

Kallebra McAtee 1926b:152 as subgenus of *Protalebra* Baker 1899b:402. *P. (K.) ninettae* n.sp. (Brazil: Chapada). 17:21. Typhlocybinae, Alebrini. Neotropical. Synonym of *Paralebra* McAtee 1926b:147; Young 1957c:145.

Kaltitartessus Evans, F. 1981a:135. *K. mouldsi* n.sp. (Australia: Queensland, Mt. Lewis, SW of Mossman). Tartessinae. Australian.

Kamaza Dworakowska 1977f:303. *K. reducta* n.sp. (India: Asarori Range, between Dehra Dun and Mohand). Typhlocybinae, Dikraneurini. Oriental; India.

Kana Distant 1908g:285. *K. thoracica* n.sp. (Ceylon: Pundaluoya). 7:5. Nirvaninae, Nirvanini. Oriental; Ceylon, India, SE Asia, Philippines.

Kanguza Dworakowska 1972k:121. *K. ibis* n.sp. (Java: Baluran, Beokol near Banjuwangi). Typhlocybinae, Erythroneurini. Oriental. Subgenus of *Empoascanara* Distant 1918b:94; Dworakowska 1980d:194.

Kanorba Oman 1938b:374. *K. reflexa* n.sp. (Brazil: Corumba). 10:2203. Linnavuori 1959b:94-96. Deltocephalinae, Deltocephalini. Neotropical; Argentina, Brazil.

Kansendria Kramer 1971b:206. *Polyamia kansiensis* Tuthill 1930a:46 (USA: Kansas, Saline Co.). Deltocephalinae, Deltocephalini. Nearctic.

Kanziko Linnavuori & Al-Ne'amy 1983a:56. *Phlepsius quadripunctatus* Melichar 1911a:109 (East Africa: Mont Nyro, "Afr. Or. Angl."). Selenocephalinae, Adamini. Ethiopian; East Africa.

Kapateira Young 1977a:805. *Oncometopia rosipennis* Osborn 1926b:172 (Colombia). Cicadellinae, Cicadellini. Neotropical; Bolivia, Colombia, Ecuador, Panama, Venezuela.

Kapsa Dworakowska 1972b:402. *Typhlocyba furcifrons* Jacobi 1941a:316 (E. Floreo, Geli Moetse, Keli Mutu ?). Typhlocybinae, Erythroneurini. Oriental; India, Indonesia, Vietnam.

Karachiota Ahmed, Manzoor 1969a:58. *K. azadirachtae* n.sp. (West Pakistan: Karachi). Typhlocybinae, Dikraneurini. Oriental; India and Pakistan.
Synonyms:
Javadikra Dworakowska 1971d:585; Sohi & Dworakowska 1984a:165.
Pusatettix Ramakrishnan & Menon 1971a:463. *P. bipunctatus* n.sp.

Karajassus Martynov 1927a:1352. *K. crassinervis* n.sp. Metcalf & Wade 1966a:218. Fossil.

Karasekia Melichar 1912c:117. *Wolfella lata* Melichar 1905a:209 (Tanganyika). 2:14. Linnavuori 1972b:202. Hylicinae, Sudrini. Ethiopian; East Africa.

Karataviella Bekker-Migdisova 1949a:25. *K. brachyptera* n.sp. Metcalf & Wade 1966a:209. Fossil: Mesozoic, Turkestan.

Kareskia 2:18. Error for *Karasekia* Melichar.

Karoseefa Webb, M.D. 1981b:70. *K. brevipenis* n.sp. (Borneo: Sarawak, foot of Mount Dulit). Paraboloponinae. Oriental; Borneo, Sabah and Sarawak.

Kartwa Distant 1908g:394. *K. mustelina* n.sp. (India: Calcutta). 8:209. Deltocephalinae, Chiasmusini. Oriental. Synonym of *Chiasmus* Mulsant & Rey 1855a:215.

Kasachstanicus Emeljanov 1962b:394, error for *Kazachstanicus* Dlabola.

Kashitettix Ishihara 1952e:60. *Eupteryx quercus* Matsumura 1916b:394 (Japan: Honshu). 17:777. Typhlocybinae, Eupterygini. Palaearctic. Synonym of *Aguriahana* Distant 1918b:105; Dworakowska 1972a:278.

Kasinella Evans, J.W. 1971a:44. *K. siamensis* n.sp. (Thailand: Makomswan). Unassigned, formerly Coelidiinae; Nielson 1975a:11. Oriental.

Kasunga Linnavuori 1979b:970. *K. theobromae* n.sp. (Ghana: Tafo). Nirvaninae, Nirvanini. Ethiopian; Guinean.

Kaszabianus Nast 1972a:499. Error for *Kaszabinus* Dlabola 1965c.

Kaszabinus Dlabola 1965c:115. *K. tridenticus* n.sp. (Mongolia: Zuun-Chara, 850 m). Deltocephalinae, Deltocephalini. Palaearctic; Mongolia, Tuva.
Synonym:
Sigista Emeljanov 1966a:125. *Deltocephalus burjata* Kusnezov 1929b:180.

Katipo Evans, J.W. 1934a:151. *Eurymeloides rubrivenosus* Kirkaldy 1906c:353 (Australia: New South Wales, Mittagong). 12:37. Evans, J.W. 1966a:44. Eurymelinae, Ipoini. Australian.

Kaukania* Dworakowska 1972j:112. *K. anser* n.sp. (Vietnam: Mouong- Xen, ca. 900 m). Typhlocybinae, Erythroneurini. Oriental. * Name erroneously credited to Jaczewski in Zoological Record 109:148.

Kazachstanicus Dlabola 1961b:312. *K. margaritae* n.sp. (USSR: Kazakhstan). Deltocephalinae, Deltocephalini. Palaearctic.

Keia Theron 1984c:317. *Aletta pulchra* Evans, J.W. 1955a:27 (South Africa). Coelidiinae (provisional). Ethiopian.

Keonolla Oman 1949a:74. *Proconia confluens* Uhler 1861b:285 (USA: Washington Territory). 1:278. Young 1977a:828. Cicadellinae, Cicadellini. Nearctic.

[NOTE: Young (l.c.) apparently considers *Keonolla* a synonym of *Graphocephala* Van Duzee 1916a:66 although there is but passing mention of the name on page 828].

Khamiria Dlabola 1979a:252. *K. mangrovecola* n.sp. ("S-Iran, Mangrove Formation"). Paraboloponinae. Palaearctic. Synonym of *Dryadomorpha* Kirkaldy 1906c:335; Webb, M.D. 1981b:49.

Khoduma Dworakowska 1972b:403. *K. jacobii* n.sp. (East Flores, Geli Moetoe). Typhlocybinae, Erythroneurini. Oriental. Synonym of *Arboridia* Zachvatkin 1946b:153; Dworakowska & Viraktamath 1975a:529.

Khyphocotis 4:141. Error for *Kyphocotis* Kirkaldy.

Kiamoncopsis Linnavuori 1978b:15. *K. quartaui* n.sp. (Angola: Cela). Macropsinae, Macropsini. Ethiopian. Subgenus of *Pediopsoides* Matsumura 1912b:305; Hamilton, KGA 1980b:899.

Kidraneuroidea Mahmood 1967a:11. *K. rubrovittata* n.sp. (Singapore). Typhlocybinae, Dikraneurini. Oriental; Malaysia.

Kidrella Young 1952b:54. *Dikraneura santana* Beamer 1936a:8 (USA: Arizona, Patagonia). 17:191. Typhlocybinae, Dikraneurini. Nearctic; Arizona.

Kiknchiella 1:683. Error for *Kikuchiella* Kato.

Kikraneura 17:1504. Error for *Dikraneura* Hardy.

Kikuchiella Kato 1932b:223. *K. gracilis* n.sp. (Manchuria). 1:429. Cicadellinae, Cicadellini. Palaearctic; Manchuria, Japan.

Kinonia Ball 1933c:224. *K. elongata* n.sp. (USA: Arizona). 10:1582. Deltocephalinae, Stenometopiini. Nearctic.

Kinnonia Vilbaste 1965a:5. Error for *Kinonia* Ball.

Kirkaldiella Osborn 1935b:13. *K. euphorbiae* n.sp. (USA: Hawaii, Molokai). 10:464. Deltocephalinae, Athysanini. Oceania; Hawaii.

Kirkaldykra Dworakowska 1971d:585. *Dikraneura apis* Dworakowska 1969b:248 (South India: Deveraya Druy). Typhlocybinae, Dikraneurini. Oriental; India.

Kirotettix 10:2667. Error for *Kirrotettix* Haupt.

Kirrotettix Haupt 1929c:255, replacement for *Cicadulina* Haupt 1927a:38, not *Cicadulina* China 1926c:43. *Cicadulina pallida* Haupt 1927a:39 (Palestine). 10:2479. Deltocephalinae, Macrostelini. Palaearctic. Synonym of *Cicadulella* China 1926c:43.

Kitara Anufriev 1971a:104. Error for *Kutara* Distant 1908g:308.

Kivulopa Linnavuori 1972a:129. *K. mwenga* n.sp. (Zaire: Kivu, Leleup, Terr. Mwenga, Laclungive, 2700 m). Ulopinae, Ulopini. Ethiopian; Tanzania, Zaire.

Knightipsis Dworakowska 1969e:440. *K. knighti* n.sp. (India: Punjab, Kumaun Hills). Typhlocybinae, Typhlocybini. Oriental; India.

Knullana DeLong 1941a:86. *Thamnotettix perexigua* Ball 1900c:339 (Mexico: Cuernavaca). 10:840. Deltocephalinae, Cicadulini. Neotropical; Mexico and southern Arizona.

Kodaikanalia Ramakrishnan & Menon 1973a:35. *K. primulae* n.sp. (India: Kodaikanal). Typhlocybinae, Erythroneurini. Oriental. Synonym of *Hauptidia* Dworakowska 1970k:620; Dworakowska 1981d:377.

Koebelea 4:141. Error for *Koebelia* Baker.

Koebelia Baker 1897a:176. *K. californica* n.sp. (USA: California). 4:121. Oman 1971b. Koebeliinae. Nearctic; western North America.

KOEBELIINAE 4:120.

KOEBELIINI 4:121.

Kogigonalia Young 1977a:207. *K. dietzi* n.sp. (Venezuela: Ptaritepui, 30 mi. N of Kavanayen, 1800 m). Cicadellinae, Cicadellini. Neotropical; Colombia, French Guiana, Peru, Venezuela.

Kolba 1:683. Error for *Kolla* Distant.

Kolella Evans, J.W. 1966a:148. *"Kolla pupula* Kirby" = *Tettigonia pupula* Kirby 1891a:169 (Ceylon: Punduloya). Cicadellinae, Cicadellini. Australian, Oriental: Sri Lanka.

Kolla Distant 1908g:223. *K. insignis* n.sp. (India: Kurseong). 1:435. Cicadellinae, Cicadellini. Oriental, Ethiopian, Australian.

Kopamerra Webb, M.D. 1983b:220. *Idiocerus haupti* Melichar 1908a:65-66 (Tanzania). Idiocerinae. Ethiopian; Afrotropical region and Madagascar.

Koperta Dworakowska 1972k:120. *Typhlocyba gemina* Jacobi 1941a:315 ("Lombok Plawangan"). Typhlocybinae, Erythroneurini. Oriental; India, Indonesia.

Korana Distant 1910e:241. *K. maculosa* n.sp. (South Africa: Natal). 15:207. Deltocephalinae. Linnavuori & Quartau 1975a:155. Ethiopian; South Africa and tropical Africa.

Koreanopsis Kwon & Lee 1979c:50. *K. koreana* n.sp. (Korea: Mt. Odaesan). Selenocephalinae, Bhatiini. Palaearctic.

Koreascleroracus Kwon & Lee 1979d:48. Invalid, *nomen nudum*.

Koreocerus Kwon 1985a:65 as subgenus of *Idiocerus* Lewis 1834a:47. *Idiocerus koreanus* Matsumura 1915a:173 (Korea). Idiocerinae. Palaearctic.

Korsigianus Nielson 1979b:319. *Daridna exoptata* Walker 1858c:320 (Peru). Coelidiinae, Teruliini. Neotropical.

Kosasia Distant 1910e:240. *K. typica* n.sp. (South Africa: Durban). 7:23. Linnavuori & DeLong 1978c:208. Nirvaninae, Nirvanini. Ethiopian region.
Synonym:
 Atritona Melichar 1914g:5. *A. paradoxa* n.sp.

Kosmiopelex Kirkaldy 1906c:334. *K. varicolor* n.sp. (Australia: Queensland, Bundaberg). 8:208. Evans, J.W. 1974a:172. Deltocephalinae, Chiasmusini. Australian. Synonym of *Chiasmus* Mulsant & Rey 1855a:215; Evans, J.W. 1974a:172.

Kosmiopelix 8:264. Error for *Kosmiopelex* Kirkaldy.

Kotwaria Dworakowska 1984a:13. *K. ramlyi* n.sp. (Malaysia: Penang I.). Typhlocybinae, Typhlocybini. Oriental.

Kramerana DeLong & Thambimuttu 1973b:165. *K. linnavuorii* n.sp. (Chile: Salta de Laja). Deltocephalinae, Athysanini. Neotropical; Chile.

Krameraxus Maldonado-Capriles 1968a:35. *K. leucornatus* n.sp. (Puerto Rico: El Verde Experimental Forest Station, near El Yunque). Deltocephalinae. Neotropical.

Krameriata Dworakowska 1977e:846. *K. hanae* n.sp. (Uganda: Ankole, Kichivamba). Typhlocybinae, Empoascini. Ethiopian; Uganda, Zaire.

Kramerolidia Nielson 1982e:236. *K. krameri* n.sp. (West Indies: Cuba, Pico Turquino). Coelidiinae, Coelidiini. Neotropical.

Kravilidius Nielson 1979b:13. *Jassus formosus* Spangberg 1878b:3 ("Brazilian"). Coelidiinae, Teruliini. Neotropical; Brazil.

Krisna Kirkaldy 1900b:243, replacement for *Siva* Spinola 1850b:127, not *Siva* Hodgson 1838. *Siva strigicollis* Spinola 1850b:128 (Madras). 15:197, 198. Linnavuori & Quartau 1975a:147. Iassinae, Krisnini. Oriental region (India to China and Japan); Ethiopian region (Guinean); Neotropical (Puerto Rico).

Krisnella Linnavuori & Al-Ne'amy 1983a:47 as subgenus of *Adama* Dlabola 1980e:89. *A. (K.) elongata* n.sp. (Zaire: Elizabethville). Selenocephalinae, Adamini. Ethiopian; southern and East African.

KRISNIDAE Ishihara 1961a:241.

KRISNINI 15:193.

Krocobella Kramer 1964b:118. *K. colotes* n.sp. (Brazil: Rio de Janeiro, Angra, Jussaral). Nirvaninae. Neotropical.

Krocodona Kramer 1964b:114. *K. sauridion* n.sp. (Honduras: La Fraqua). Nirvaninae. Neotropical.

Krocozzota Kramer 1964b:115. *K. languria* n.sp. (Panama: Canal Zone). Nirvaninae. Neotropical.

Kronos Distant 1917a:307. *K. typicus* n.sp. (Seychelles Islands). 3:217. Penthimiinae. Oriental.

Kropka Dworakowska 1970k:618. *K. vidanoi* n.sp. (Bulgaria: Rodopi, Plovdio Distr.). Typhlocybinae, Erythroneurini. Palaearctic; southern Europe and Iran.

Krosolus Nielson 1982e:37. *Jassus centroafricanus* Jacobi 1912a:38 (East Africa: Kwidjivi). Coelidiinae, Coelidiini. Ethiopian; Burundi, Ugandi, Zaire.

Kumba Linnavuori & Al-Ne'amy 1983a:85. *Selenocephalus armatissimus* Linnavuori 1969a:1165 (Republic of Guinea: Nimba). Selenocephalinae, Selenocephalini. Ethiopian; Guinean.

Kunasia Distant 1908g:339. *K. nivosa* n.sp. (Burma: Tenasserim, Myitta). 11:27. Unassigned, formerly Coelidiinae, Nielson 1975a:11. Oriental.

Kunzeana Oman 1949a:83. *Dikraneura kunzei* Gillette 1898a:721 (USA: Arizona). 17:194. Typhlocybinae, Dikraneurini. Nearctic; Sonoran. Neotropical; Cuba, Mexico to Panama.

Kunzella Young 1952b:65. *Dikraneura marginella* Baker 1925a:160 (USA: Florida, Miami). 17:282. Typhlocybinae, Dikraneurini. Nearctic; Florida. Neotropical; Cuba, Puerto Rico, Central America to Panama.

Kurotsuyanus Ishihara 1953b:4. *Tettigoniella sachalinensis* Oshanin 1912a:100 (Sahkalin). 1:401. Kwon 1983a:17. Cicadellinae, Pagaroniini. Palaearctic; Japan, Korean Peninsula, Sakhalin.

Kusala Dworakowska 1981a:317. *K. sagittata* n.sp. (India: Tamil Nadu, Shambaganur "Silver Cascade" 1800 m). Typhlocybinae, Erythroneurini. Oriental; India, Pakistan, Vietnam.

Kuscheliola Evans, J.W. 1957a:372. *K. reticulata* n.sp. (Chile: Juan Fernandes Is., Masatierra, Alto Ingles). Agalliinae, Agalliini. Neotropical; southern Chile.

Kutara Distant 1908g:308. *K. brunescens* Distant 1908g:308 (Ceylon). 15:184. Linnavuori & Al-Ne'amy 1983a:22. Selenocephalinae, Ianerini. Oriental; Ceylon, India, Borneo, Java. Palaearctic; USSR Maritime Territory.

Kuznetsovium Zachvatkin 1953c:231. *K. aristidae* n.sp. (USSR: Kazakhstan). Eupelicinae, Paradorydiini. Palaearctic. Synonym of *Paradorydium* Kirkaldy 1901f:339; Nast 1972a:230-231.

Kwempia Ahmed, Manzoor 1979a:34. *K. rugosa* n.sp. (Uganda). Typhlocybinae, Erythroneurini. Ethiopian.

Kyboasca Zachvatkin 1953c:228. *Chloria bipunctata* Oshanin 1871a:212 (Turkestan). Typhlocybinae, Empoascini. Palaearctic. [Nearctic].

Kybos Fieber 1866a:508. *Cicada smaragdula* Fallen 1806a:37 (Sweden). 17:364. Dworakowska 1973c, 1976a, 1977d. Ross 1963a. Typhlocybinae, Empoascini. Holarctic. Australia. Subgenus of *Empoasca* Walsh 1862a:149; or often accorded generic rank.

Kyphocotes 4:141. Error for *Kyphocotis* Kirkaldy.

Kyphocotis Kirkaldy 1906c:370. *K. tessellata* n.sp. (Australia: Queensland, Bundaberg). 4:118. Evans, J.W. 1966a:106-108. Ledrinae, Stenocotini. Australian.

Kyphoctella Evans, J.W. 1966a:111. *K. distorta* n.sp. (Australia: Northern Territory, Adelaide River). Ledrinae, Stenocotini. Australian.

Labocurtidia Nielson 1979b:46. *L. curtisi* n.sp. (Brazil: Chapada). Coelidiinae, Teruliini. Neotropical; Brazil, Argentina, Paraguay.

Laburrus Ribaut 1942a:268. *Athysanus limbatus* Ferrari 1882a:127 (Switzerland) = *Athysanus quadratus* Forel 1862a:3. 10:79. Deltocephalinae, Athysanini. Palaearctic; Eurasia.
Subgenus:
Esolanus Ribaut 1952a:87. *Athysanus pellax* Horvath 1903c:472.

Labururs Ishihara 1975d. Error for *Laburrus* Ribaut.

Ladoffa Young 1977a:339. *Tettigonia ignota* Walker 1851b:766 (Unknown). Cicadellinae, Cicadellini. Neotropical; southern Mexico, Central America, Panama, northern South America.

Ladya Theron 1982b:25. *L. longipennis* n.sp. (South Africa: Natal, Ladysmith). Deltocephalinae, Deltocephalini. Ethiopian.

Laevicephalus DeLong 1926d:64 as subgenus of *Deltocephalus* Burmeister 1838b:15. *Deltocephalus sylvestris* Osborn & Ball 1897a:213 (USA: Iowa, Ames). 10:1830. Deltocephalinae, Deltocephalini. Nearctic, Neotropical: Mexico northward to central Manitoba, Canada, and eastward to the Atlantic.

Laevisephalus 10:2667. Error for *Laevicephalus* DeLong.

Laevilidia Nielson 1979b:32. *L. vermiculata* n.sp. (Brazil: Minas Gerais, Diamantino). Coelidiinae, Teruliini. Neotropical.

LAEVIPEDES 5:1.

Lajolla Linnavuori 1959b:160. *L. perlata* n.sp. (Panama: La Jolla). Deltocephalinae, Athysanini. Neotropical.

Lametettix Vilbaste 1961a:45. Invalid.

Lamia Linnavuori 1960b:39. *L. cydippe* n.sp. (Fiji: Viti Levu, Lami). Linnavuori & Al-Ne'amy 1983a:22. Selenocephalinae, Bhatiini. Australian; Australia, Fiji.

Lampridius Distant 1918b:58. *L. spectabilis* n.sp. (Burma: Dawna Hills). 10:419. Viraktamath, S. & Viraktamath, C. A. 1980a:139-140. Deltocephalinae, Opsiini. Oriental; Burma, India.

Lamprotettix Ribaut 1952a:267. *Cicada octopunctata* von Schrank 1796a:211 (Germany) = *Cicada nitidula* Fabricius 1787a:273. 10:2052. Deltocephalinae, Athysanini. Palaearctic; western Europe, British Islands.

Lamtoana Dworakowska 1972f:853. *L. sordida* n.sp. (Ivory Coast, Lamto). Typhlocybinae, Erythroneurini. Ethiopian.

Laneola Young 1977a:1075. *Tettigonia rubicauda* Signoret 1854b:351 (Paraguay). Cicadellinae, Cicadellini. Neotropical; Brazil, Paraguay.

Lankasca Ghauri 1963e:470. *Empoasca centromaculata* Melichar 1903b:213 (Ceylon: Peradeniya). Typhlocybinae, Empoascini. Oriental; India, Sri Lanka.

Laokayana Dworakowska 1972i:27. *Empoasca bombaxia* Ghauri 1965a:683 (East Pakistan: Dacca). Typhlocybinae, Empoascini. Oriental. Synonym of *Amrasca* Ghauri 1967a:159; Dworakowska & Viraktamath 1975a:530.

Lareba Young 1957c:180. *Protalebra variata* Ruppel & DeLong 1953c:226 (Mexico: Guerrero, Cutzemala). Typhlocybinae, Alebrini. Neotropical; Mexico.

Largulara DeLong & Freytag 1972f:292 as subgenus of *Polana* DeLong 1942d:110. *P. (L.) fantasa* n.sp. (Peru: Napo R.). Gyponinae, Gyponini. Neotropical; Mexico, Peru.

Lascumbresa Linnavuori & Delong 1979b:132. *L. armata* n.sp. (Panama: Las Cumbres). Deltocephalinae, Athysanini. Neotropical.

Lasioscopus China 1926d:294. *Eurymeloides acmaeops* Jacobi 1909a:340 (unknown). 12:41. Evans, J.W. 1966a:78. Eurymelinae, Pogonoscopini. Australian; Australia.

Lataba DeLong & Triplehorn 1978a:180 as subgenus of *Curtara* DeLong & Freytag 1972a:231. *C. (L.) basala* n.sp. (Paraguay: Fiebriz). Gyponinae, Gyponini. Neotropical.

Latalus DeLong & Sleesman 1929a:87. *Amblycephalus sayii* Fitch 1851a:61 (USA: New York). 10:1777. Deltocephalinae, Deltocephalini. Nearctic; northern latitudes or high elevations.
Synonym:
Quontus Oman 1949a:170. *Deltocephalus misellus* Ball 1899b:191.

Lataponana DeLong 1977e:109 as subgenus of *Ponana* Ball 1920a:93. *P. (L.) ampa* n.sp. (Peru: Sinchona). Gyponinae, Gyponini. Neotropical.

Latenus DeLong & Knull 1971a:54 as subgenus of *Scaphoideus* Uhler 1889a:33. *Scaphoideus veterator* DeLong & Beery 1936a:334 (USA: Pennsylvania, Loyalsock). 10:2110. Barnett 1977a:545. Deltocephalinae, Scaphoideini. Nearctic.

Laterana DeLong 1936a:217 as subgenus of *Ballana* DeLong 1936a:217. *Thamnotettix dissimilata* Ball 1910c:306 (USA: California, Colfax). 10:820. Deltocephalinae, Athysanini. Nearctic. Synonym of *Ballana* DeLong 1936a:217.

Lausulus Ribaut 1946b:84. *Deltocephalus pseudocellaris* Flor 1861a:547 (Europe). 10:1201. Deltocephalinae, Paralimnini. Palaearctic. Synonym of *Jassargus* Zachvatkin 1933b:268.

Lautereria Young 1977a:1060. *Cicadella signatula* Osborn 1926b:200 (Bolivia: Prov. del Sara). Cicadellinae, Cicadellini. Neotropical; Bolivia, Brazil, Colombia, Peru.

Lautereriana Dworakowska 1974a:173. *L. nigroscutellata* n.sp. (Congo: Odzala). Typhlocybinae, Zyginellini. Ethiopian.

Latulus 10:2667. Error for *Latalus* DeLong & Sleesman.

Lavrushinia Cockerell 1925a:10. *L. elegantula* n.sp. Metcalf & Wade 1966a:209. Fossil: Tertiary, Siberia.

Lawsonellus Young 1957c:182. *Dikraneura attenuata* Osborn 1928a:269 (Bolivia: Prov. del Sara). Typhlocybinae, Alebrini. Neotropical.

Leavicephalus 10:2668. Error for *Laevicephalus* DeLong.

Lebaja Young 1977a:301. *L. mediana* n.sp. (Brazil: Minas Gerais, Vicosa). Cicadellinae, Cicadellini. Neotropical.

Lebora China 1927d:283, replacement for *Parametopia* Melichar 1925a:387, not *Parametopia* Reitter 1884. *Cicada orbona* Fabricius 1798a:520 ("America"). 1:585. Young 1968a:220. Cicadellinae, Proconiini. Neotropical. Isogenotypic synonym of *Oncometopia* Stal 1869a:62.

Lebradea Remane 1959a:386. *L. calamagrostidis* n.sp. (Germany: Lebrade). Deltocephalinae, Paralimnini. Holarctic; northern latitudes.

Lecacis Theron 1982b:23. *L. platypennis* n.sp. (South Africa: Natal, Ladysmith). Deltocephalinae, Deltocephalini. Ethiopian.

Lectotypella Dworakowska 1972k:118. *Zygina albisoma* Matsumura 1932a:108 (Formosa). Typhlocybinae, Erythroneurini. Oriental; Formosa, Japan, Indonesia.

Leda 4:141. Error for *Ledra* Fabricius.

Ledeira Dworakowska 1969f:488. *L. callosa* n.sp. (East Nepal: Taplejung Distr., Sangu, c. 6200 ft.). Typhlocybinae, Zyginellini. Oriental; India, Nepal.

Ledophora Heer 1853b:116. *L. producta* n.sp. Metcalf & Wade 1966a:209. Fossil: Tertiary, Croatia.
Ledra Fabricius 1803a:13. *Cicada aurita* Linnaeus 1758a:435 (Germany). 4:8. Ledrinae, Ledrini. Palaearctic, Oriental.
Synonym:
Ledraria Rafinesque 1815a:121.
Ledracephala Evans, J.W. 1947a:252. *Ledra brevifrons* Walker 1851b:825 (Australia). 4:100. Ledrinae, Petalocephalini. Australian. Synonym of *Rubria* Stal 1866a:104; Evans, J.W. 1966a:102.
Ledracorrhis Evans, J.W. 1959a:497. *L. rugosa* n.sp. (Madagascar). Ledrinae, Ledrini. Malagasian.
Ledracotis Evans, J.W. 1937b:162. *L. gunnensis* n.sp. (Australia: New South Wales, Gunning). 4:119. Evans, J.W. 1966a:111. Ledrinae, Stenocotini. Australian.
Ledramorpha 4:141. Error for *Ledromorpha* Stal.
Ledraprora Evans, J.W. 1936a:40. *L. insularis* n.sp. (Australia: Kangaroo Island). 4:132. Evans, J.W. 1966a:113. Ledrinae, Thymbrini. Australian.
Ledraria Rafinesque 1815a:121, replacement for *Ledra* Fabricius 1803a:13. *Cicada aurita* Linnaeus 1758a:435 (Germany). 4:9. Ledrinae, Ledrini. Palaearctic. Synonym of *Ledra* Fabricius 1803a:13.
Ledrella Evans, J.W. 1936a:40. *L. brunnea* n.sp. (Australia: Victoria, Kiata). 1:133. Evans, J.W. 1966a:113. Ledrinae, Thymbrini. Australian.
LEDRELLINI 4:124.
LEDRIDAE 4:1. Kirschbaum [1867] 1868b:14, type genus *Ledra* Fabricius 1803a:24. Placed in the Official List of Family-Group Names in Zoology, ICZN Opinion 647, 1962.
LEDRINAE 4:7.
LEDRINI 4:8.
Ledroides Dammerman 1910a:59. *L. reticuloides* n.sp. (Sumatra). 3:217. Penthimiinae. Oriental.
Ledromdrpha 4:141. Error for *Ledromorpha* Stal.
Ledromorpha Stal 1864b:68. *Fulgora planirostris* Donovan 1805a:1 (unknown). 4:97. Evans, J.W. 1966a:96. Ledrinae, Ledrini. Australian.
Ledropsella Evans, J.W. 1966a:101. *Platyledra monstrosa* Evans, J.W. 1939a:45 (Western Australia: King George's Sound). Ledrinae, Ledrini. Australian.
Ledropis Evans, J.W. 1966a:101. Error for *Ledropsis* White 1844a.
Ledropsis White 1844a:425. *L. cancroma* n.sp. (Hong Kong). 4:86. Evans, J.W. 1966a:99. Ledrinae, Petalocephalini. Oriental, Australian.
Synonym:
Scaphocephalus Matsumura 1905a:52. *Petalocephala discolor* Uhler 1896a:290.
Ledrotypa Distant 1912d:442. *L. spatulata* n.sp. (India: Himalayas). 9:7. Deltocephalinae, Hecalini. Oriental. Synonym of *Glossocratus* Fieber 1866a:502; Morrison 1973a:397.
Lemellus Oman 1949a:165. *Deltocephalus bimaculatus* Gillette & Baker 1895a:86 (USA: Colorado). 10:1863. Deltocephalinae, Deltocephalini. Nearctic; Rocky Mts., Cascade Mts.
Leofa Distant 1918b:86. *L. mysorensis* n.sp. (south India: Mysore). 8:225. Cicadellinae, Anoterostemmini. Oriental.
Lepropsis 4:141. Error for *Ledropsis* White.
Leptchloris 17:1504. Error for *Leptochloris* Amyot.
Leptochloris Amyot 17:1504. Invalid, not binominal.
Lerda 4:142. Error for *Ledra* Fabricius.
Lettigonia 1:683. Error for *Tettigonia* Olivier.
Leucopepla Kirkaldy 1907d:87. *Tettigonia bituberculata* Signoret 1855c:528 (Uruguay: Rio Negro). 1:469. Young 1968a:30. Cicadellinae, Proconiini. Neotropical. Synonym of *Diestostemma* Amyot & Serville 1843a:572; Young 1968a:30.
Leuconeura Ishihara 1978a:11. *L. ipomoeae* n.sp. (Japan). Typhlocybinae. Oriental.
Leucospilus Amyot 15:223. Invalid, not binominal.
Levantotettix Lindberg 1953a:110,114. *L. striatus* n.sp. (Palestine). Selenocephalinae, Selenocephalini. Palaearctic. Synonym of *Selenocephalus* Germar 1833a:180; Linnavuori 1962a:50.
Libengaia Linnavuori 1969a:1160. *Eutettix vermiculatus* Melichar 1912c:119 (Belgian Congo). Deltocephalinae, Opsiini. Ethiopian; Republic of Congo.
Lichtrea Dworakowska 1976b:37. *L. nigrescens* n.sp. (Cameroon: Mukonje). Typhlocybinae, Erythroneurini. Ethiopian; Cameroon. Palaearctic; Nepal.

Licolidia Nielson 1979b:96. *L. angusta* n.sp. (Guyana: Tanaka Crk.). Coelidiinae, Teruliini. Neotropical.
Licontinia Nielson 1979b:141. *Daridna introducens* Walker 1858a:108 ("Amaz."). Coelidiinae, Teruliini. Neotropical; Mexico, Central America, northern South America.
Liguropia Haupt 1930a:156. *L. menozzii* n.sp. (Italy) = *Notus juniperi* Lethierry 1876a:12. 17:99. Nast 1972a:254. Typhlocybinae, Dikraneurini. Palaearctic; Mediterranean.
Limassola Dlabola 1974d:305. Error for *Limassolla* Dlabola 1965a:663.
Limassolla Dlabola 1965e:663. *Zyginella pistaciae* Linnavuori 1962a:67 (Israel: Rehovot). Typhlocybinae, Zygellini. Palaearctic; Mediterranean, Japan, Okinawa. Oriental; India.
Synonym:
Pruthius Mahmood 1967a:33; Dworakowska 1969d:433.
Limbanus Oman 1949a:131. *Thamnotettix limbatus* Van Duzee 1890g:92 (USA: California). 10:943. Deltocephalinae, Athysanini. Nearctic.
Limentinus Distant 1917a:316. *L. aldabranus* n.sp. (Aldabra Islands). 11:30. Coelidiinae, Coelidiini. Malagasian; Seychelles, Madagascar.
Limmasolla Chou & Ma 1981a. Error for *Limassolla* Dlabola.
Limmotettix 10:2668. Error for *Limotettix* Sahlberg.
Limnotettix 10:2668. Error for *Limotettix* Sahlberg.
Limonattus Amyot 10:2668. Invalid, not binominal.
Limonettix 10:2668. Error for *Limotettix* Sahlberg.
Limonus Kwon & Lee 1979d:48. Invalid, *nomen nudum*.
Limotetlix 10:2668. Error for *Limotettix* Sahlberg.
Limotettice 10:2668. Error for *Limotettix* Sahlberg.
Limotettitx 10:2668. Error for *Limotettix* Sahlberg.
Limottettix 10:2669. Error for *Limotettix* Sahlberg.
Limotettix Sahlberg 1871a:224. *Cicada striola* Fallen 1806a:31 (Sweden). 10:128. Deltocephaline, Athysanini. Holarctic.
Subgenus:
Neodrylix Emeljanov 1966a:103. *Athysanus parallelus* Van Duzee 1891a:169.
Synonym:
Drylix Edwards 1922a:206. *Cicada striola* Fallen 1806a:31.
LIMOTETTIXINA 10:128.
Limpica Cheng, Y.J. 1980a:85. *L. forcata* n.sp. (Paraguay: NW Asuncion, 2 km E. Limpo). Deltocephalinae, Deltocephalini. Neotropical.
Linacephalus Evans, J.W. 1977a:93. *Paradorydium michaelseni* Jacobi 1909a:339 (Western Australia). Ulopinae, Cephalini. Australian.
Lindbergana Metcalf 1952a:229, replacement for *Nesotettix* Lindberg 1936a:6, not *Nesotettix* Holdhaus 1909. *Nesotettix freyi* Lindberg 1936a:6 (Canary Islands). 10:2461. Xestocephalinae, Xestocephalini. Ethiopian. Synonym of *Xestocephalus* Van Duzee 1892d:298; Linnavuori 1959b:36.
Lindbergina Dlabola 1958c:54 as subgenus of *Youngia* Dlabola 1958c:54. *Typhlocyba aurovittata* Douglas 1875c:76 (England). Typhlocybinae, Alebrini or Typhlocybini. Palaearctic.
Subgenus:
Youngiada Dlabola 1959b:195. *Typhlocyba loewi* Lethierry 1884b:131.
[NOTE: Metcalf 1968a:45 placed *Typhlocyba aurovittata* as a synonym of *Alebra albostriella* Fallen 1862a:54. European workers (e.g. Ribaut 1936b:113, Nast 1972a:284) include the species in Typhlocybini].
Linnatanus Menezes 1973b:135, replacement for *Tubulanus* Linnavuori 1955a:109, not *Tubulanus* Reinier 1804. *Tubulanus nitidus* Linnavuori 1955a:110 (Brazil). Deltocephalinae, Athysanini. Neotropical. Synonym of *Atanus* Oman 1938b:381; Linnavuori & DeLong 1977c:209.
Linnavuoriana Dlabola 1958c:54. *Cicada decempunctata* Fallen 1806a:41 (Sweden). Typhlocybinae, Typhlocybini. Palaearctic.
Subgenus:
Sharmana Dworakowska 1982a:123. *Agnesiella swargii* Sharma 1978a:12.

225

Linnavuoriella Evans, J.W. 1966a:134. *"Parabolocratus arcuatus* Motschulsky" = *Platymetopius arcuatus* Motschulsky 1859b:115 (India). Deltocephalinae, Hecalini. Oriental, Australian. Synonym of *Hecalus* Stal 1864b:65; Morrison 1973a:410.

Linnavuorina Kocak 1981a:32, replacement for *Carinifer* Linnavuori 1952b:185, not *Carinifer* Hamm 1881. *Carinifer maculiceps* Linnavuori 1952b:185 (Turkestan). Deltocephalinae, Doraturini. Palaearctic. Synonym of *Aconura* Lethierry 1876a:9; Vilbaste 1965a:9.

Liocratus Dubovskij 1966a:119 as subgenus of *Idiocerus* Lewis 1834a:47. *I. chivensis* Kusnezov 1929a:309 (Turkestan). Hamilton, KGA 1980a:825, 838. Idiocerinae. Palaearctic.

Liojassus Handlirsch 1939a:146. *L. affinis* n.sp. Metcalf & Wade 1966a:217. Fossil: Jurassic; Germany, Mecklenburg.

Lipata Dworakowska 1974a:145. *L. rosea* n.sp. (Congo: Odzala). Typhlocybinae, Empoascini. Ethiopian.

Lissoscarta Stal 1869a:69. *Cicada vespiformis* Fabricius 1803a:68 ("Amer. Merid."). 1:652. Young 1977a:149. Cicadellinae, Cicadellini. Neotropical; Brazil, British Guiana, Bolivia, Peru.

Listrophora Boulard 1971a:712. *L. evansi* n.sp. (Central African Republic: Boukoko). Linnavuori 1979a:650-652. Eupelicinae, Listrophorini. Ethiopian; Angola, Central African Republic.

LISTROPHORINI Boulard 1971a:712, tribe of EUPELICINAE.

Litura Knight, W.J. 1970b:179. *L. unda* n.sp. (Ceylon: Deniyaya). Deltocephalinae, Opsiini. Oriental; Borneo, Ceylon, Malaya, Sarawak. Ethiopian; Cameroon.

Livasca Dworakowska & Viraktamath, C.A. 1978a:540 as subgenus of *Empoasca* Walsh 1862a:149. *E. (L.) malliki* n.sp. (India: Karnataka, Jog Falls, 1700 m.). Typhlocybinae, Empoascini. Oriental.

Lodia Nielson 1982e:222. *L. glabrosa* n.sp. (Ecuador: Tena). Coelidiinae, Coelidiini. Neotropical; Ecuador, Peru.

Lodiana Nielson 1982e:86. *L. alata* n.sp. (China: Fukien, Shaowu, Tachulan, 1000 m.). Coelidiinae, Coelidiini. Oriental; widespread. Palaearctic; Japan.

Loepotettix Ribaut 1942a:264 as subgenus of *Thamnotettix* Zetterstedt 1838a:292. *Jassus (Thamnotettix) dilutior* Kirschbaum 1868b:92 (Europe). 10:812. Ossiannilsson 1983a:732. Deltocephalinae, Athysanini. Palaearctic.

Logata 1:683. Error for *Lojata* Strand.

Loipothea Linnavuori 1969a:1173. *L. elegans* n.sp. (Cameroon). Deltocephalinae. Ethiopian; Cameroon, Kenya, Republic of Congo, Tanzania.

Loja Schmidt 1932a:47. *L. ohausi* n.sp. (Ecuador: Loja). 1:658. Young 1968a:25. Cicadellinae, Proconiini. Neotropical. Preoccupied by *Loja* Giglio-Tos 1898. Synonym of *Lojata* Strand 1933a:122.

Lojanus Linnavuori 1959b:272. *L. wagneri* n.sp. (Ecuador: Loja). Deltocephalinae. Neotropical.

Lojata Strand 1933a:122, replacement for *Loja* Schmidt 1932a:45, not *Loja* Giglio-Tos 1898. *Loja ohausi* Schmidt 1932a:47 (Ecuador: Loja). 1:658. Young 1968a:25. Cicadellinae, Proconiini. Neotropical.

Loka Linnavuori & Al-Ne'amy 1983a:75. *Selenocephalus bacteriphora* de Bergevin 1926c:260 (French Congo). Selenocephalinae, Selenocephalini. Ethiopian; Sudanese.

Lokia Thapa 1984b:46. *L. celtiana* n.sp. (Nepal: Kathmandu, Bhandarkhal). Typhlocybinae, Erythroneurini. Oriental.

Lonatura Osborn & Ball 1898f:83. *L. catalina* n.sp. (USA: Iowa, Ames). 10:1476. Kramer 1967d. Deltocephalinae. Nearctic; grasslands of central, southeastern and southwestern North America. (Australia ?).

Lonenus DeLong 1939a:33 as subgenus of *Scaphoideus* Uhler 1889a:34. *Scaphoideus intricatus* Uhler 1889a:34 (USA: Virginia "New Market"). 10:2106. Barnett 1977a:544. Deltocephalinae, Scaphoideini. Nearctic.

Lonotura 10:2669. Error for *Lonatura* Osborn & Ball.

Loralia Evans, J.W. 1966a:233. *L. pulcherrima* n.sp. (Australia: New South Wales, Mt. Kosiusko, Wilson's Valley). Deltocephalinae, Deltocephalini. Australian.

Lorellana DeLong & Kolbe 1975a:9. *L. tropicana* n.sp. (Panama: Las Cumbres). Deltocephalinae, Deltocephalini. Neotropical; Panama.

Loreta Linnavuori 1959b:132. *L. ornaticeps* n.sp. (Venezuela: Amazonas, San Fernando de Atabapo). Deltocephalinae, Deltocephalini. Neotropical; Argentina, Brazil, Panama, Paraguay, Peru, Venezuela, West Indies.
Subgenus:
Bahitella Linnavuori 1959b:134. *L. (B.) lineaticollis* n.sp.

Lova 1:683. Error for *Loja* Schmidt.

Lowata Dworakowska 1977i:30. *L. bifasciata* n.sp. (Vietnam: Nghe-an, Quy-chau). Typhlocybinae, Zyginellini. Oriental.

Lualabanus Linnavuori 1975a:17. *L. curticeps* n.sp. (Zaire: "Paro Nat. Upemba, Kalumengongo"). Deltocephalinae, Hecalini. Ethiopian; savannah zones of eastern Zaire and the Sudan.

Lublinia Dworakowska 1970l:629. *Erythroneura (Zygina) gediensis* Linnavuori 1962a:72 (Israel: Ein Gedi). Typhlocybinae, Erythroneurini. Palaearctic; Mediterranean. Ethiopian; Ivory Coast, Sudan, South Africa.

Lucumius Metcalf & Bruner 1936a:968. *L. triangularis* n.sp. (Cuba: Camaguey). 1:393. Young 1977a:965. Cicadellinae, Cicadellini. Neotropical; Cuba. Synonym of *Hadria* Metcalf & Bruner 1936a:945; Young 1977a:965.

Luheria Osborn 1923c:31. *L. constricta* n.sp. (Brazil: Bahia, Barra). 8:226. Linnavuori 1959b:61. Deltocephalinae, Luheriini. Neotropical; Argentina, Brazil.

LUHERIINI Linnavuori 1959b:60, tribe of DELTOCEPHALINAE.

Lupola Nielson 1982e:43. *Coelidia maculinervis* Stal 1854b:254 (Sierra Leone: Afzelino). Coelidiinae, Coelidiini. Ethiopian; Cameroon, Ghana, Ivory Coast, Liberia, Nigeria, Port Guinea, Sierra Leone.

Lusitanocephalus Quartau 1970a:19 & 22. *L. sacarraoi* n.sp. (Portugal: Sintra, Estremadura). Deltocephalinae, Deltocephalini. Palaearctic.

Luteobalmus Maldonado-Capriles 1977b:359. *L. maculatus* n.sp. (British Guiana: New River). Idiocerinae. Neotropical.

Luvila Dworakowska 1974a:143. *L. ceglasta* n.sp. (Congo: Odzala). Typhlocybinae, Empoascini. Ethiopian; Cameroon, Central African Republic, Congo, Nigeria, Sudan, Zaire.
Synonym:
Erinwa Ghauri 1975b:483. *E. delicata* n.sp.

Luzoniana Metcalf 1953a:47, replacement for *Luzoniella* Melichar 1926a:341, not *Luzoniella* Karny 1926. *L. philippina* n.sp. (Philippines: Luzon). 1:252. Cicadellinae, Cicadellini. Oriental.

Luzoniella Melichar 1926a:341. Preoccupied by *Luzoniella* Karny 1926, see *Luzoniana* Metcalf 1953a:47.

Lycioides Oman 1949a:144. *Phlepsius (Dixianus) lycioides* Ball 1931g:88 (USA: Arizona, Tucson). 10:575. Deltocephalinae, Athysanini. Nearctic.

Lycisca Linnavuori 1979a:653. *L. umbrina* n.sp. (South Africa: Cape Province, Drakensbergen near Rhodes). Eupelicinae, Paradorydiini. Ethiopian. Preoccupied by *Lycisca* Spinola 1840, see *Afralycisca* Kocak 1981a:32.

Lystridea Baker 1898d:261. *L. conspersa* n.sp = *Bathysmatophorus uhleri* Baker 1898d:261 (USA: California, Dunsmuir). 8:32. Oman & Musgrave 1975a:8; Anufriev 1978a:58. Cicadelline, Errhomenini. Nearctic; mountain chapparal habitats of California, Oregon and Nevada.

Macednus Emeljanov 1962a:165. *M. marginatus* n.sp. (USSR: Maritime Territory). Deltocephalinae, Athysanini. Palaearctic. Synonym of *Albicostella* Ishihara 1953a:196; Vilbaste 1967a:44.

Macnedus Hamilton, KGA 1975a:497. Error for *Macednus* Emeljanov 1962a:165.

Macrasana DeLong & Hershberger 1947e:78, as subgenus of *Crassana* DeLong & Hershberger 1974e:76. *C. (M.) marginella* n.sp. (Mexico: Guererro, Chilpancingo). 10:466. Deltocephalinae, Athysanini. Neotropical.

Macrasteles 10:2669. Error for *Macrosteles* Fieber.

Macroceps Melichar 1902b:75. *M. ahngeri* n.sp. (Transcaspia). 14:164. Adelungiinae, Adelungiini. Palaearctic. Preoccupied by *Macroceps* Signoret 1879a, see *Bergevina* Evans, J.W. 1947a:189.

Macroceps Signoret 1879a:53. *M. fasciatus* Signoret 1880c:364 (Australia). 4:127. Evans, J.W. 1966a:116-117. Ledrinae, Thymbrini. Australian.
Synonym:

Macrometopius Horvath 1914a:660, replacement for *Macroceps* Signoret.
MACROCEPSINI 4:123.
Macroceratogonia Kirkaldy 1906c:323. *M. aurea* n.sp. (Australia: Queensland, Kuranda). 7:2. Nirvaninae, Macroceratogoniini. Australian; Australia, New Caledonia.
MACROCERATOGONIINA 7:2.
MACROCERATOGONIINI 7:2.
Macrocerus Evans, J.W. 1941e:39. *M. minutus* n.sp. (South Australia: Kangaroo Isl., Flinders Chase). 16:216. Preoccupied by *Macrocerus* de Motschulsky 1845, see *Zaletta* Metcalf 1952a:229. Idiocerinae. Australian.
Macrolestes Zoological Record 97, Section 13:331, 1960. Error for *Macrosteles* Fieber 1866.
Macrometopius Horvath 1914a:660, replacement for *Macroceps* Signoret 1879a:53, not *Macroceps* Melichar 1902. *Macroceps fasciatus* Signoret 1880a:364 (Australia). 4:127. Ledrinae, Thymbrini. Australian. Synonym of *Macroceps* Signoret 1879a:53.
Macropis 13:251. Error for *Macropsis* Lewis.
Macropsella Hamilton, KGA 1980b:901. *Macropsis saidora* Evans, J.W. 1971b:352 (NE New Guinea: 19 km SE Okapa, 1800 m). Macropsinae, Macropsini. Oriental; New Guinea, and northern Australian.
MACROPSIDAE 13:5. Placed on official list of Family-Group names in Zoology, ICZN Opinion 603.
Macropsidius Ribaut 1952a:436. *Pediopsis dispar* Fieber 1868a:459 (Saratov). 13:159. Hamilton, KGA 1980b:907. Macropsinae, Macropsini. Palaearctic, Nearctic.
  [NOTE: Hamilton, KGA 1980b:907 placed *Macropsidius* as a subgenus of *Macropsis* Lewis. Vilbaste 1980a:29 & 170 retained the taxon as genus. Which is the more recent publication is uncertain].
MACROPSIDAE Evans, J.W. 1938c:43, type genus *Macropsis* Lewis 1834a:49. Placed in the Official List of Family-Group Names in Zoology, ICZN Opinion 603, 1961.
MACROPSINAE 13:7.
MACROPSINI 13:7.
Macropsis Lewis 1834a:49. *Iassus prasinus* Boheman 1852b:123 (Sweden). 13:7. Linnavuori 1978b; Hamilton, K.G.A. 1980b:904. Macropsinae, Macropsini. Cosmopolitan, excepting Australian region?
Subgenera:
Parapediopsis Hamilton, KGA 1980b:905. *Macropsis benguetensis* Merino 1935a:321.
Macropsidius Ribaut 1952a:436. *Pediopsis dispar* Fieber 1868a:459. (See comment under *Macropsidius* Ribaut.)
Neomacropsis Hamilton, KGA 1980b:911. *Pediopsis basalis* Van Duzee 1889a:171.
Synonym:
Tsavopsis Linnavuori 1978b:14. *T. tuberculata* n.sp.
[NOTE: Acting under plenary powers the ICZN, Opinion 603, rejected all previous designations of type species and designated *Iassus prasina* Boheman 1852b:123 as the type species of *Macropsis* Lewis].
Macropus 13:252. Error for *Macropsis* Lewis.
Macrospsis 13:252. Error for *Macropsis* Lewis.
Macrosteles Fieber 1866a:504. *Cicada sexnotata* Fallen 1806a:34 (Sweden). 10:2486. Deltocephalinae, Macrostelini. Holarctic; and So. Africa.
Synonyms:
Acrostigmus Thomson 1869a:76. *Cicada sexnotata* Fallen 1806a:34.
Erotettix Haupt 1929c:255. *Thamnotettix cyane* Boheman 1845b:158.
MACROSTELINI 10:2467.
Macrotartessus Evans, F. 1981a:174. *M. straatmani* n.sp. (NE New Guinea: 6 km W of Green Riv.). Tartessinae. Oriental.
Macrostellaria 10:2667. Error for *Macrosteles* Fieber.
Macrostyles 10:2557. Error for *Macrosteles* Fieber.
Macroteles 10:2570. Error for *Macrosteles* Feiber.
Macsrosteles 10:2670. Error for *Macrosteles* Fieber.
Macugonalia Young 1977a:941. *Tettigonia sobrina* Stal 1862c:41 (Brazil: Rio de Janeiro). Cicadellinae, Cicadellini. Neotropical.

Macunolla Young 1977a:801. *Tettigonia ventralis* Signoret 1854a:21 (Colombia). Cicadellinae, Cicadellini. Neotropical; West Indies, Panama, and South America south to Argentina.
Macustus Ribaut 1942a:3, 268. *Cicada grisescens* Zetterstedt 1828a:530 (Lapland). 10:230. Deltocephalinae, Athysanini. Holarctic.
Macutella Evans, J.W. 1972a:183. *M. lutea* n.sp. (NW New Guinea, Waris, S. of Sukarnapura, 450 m). Penthimiinae. Oriental.
Maemichungella Kwon & Lee 1979d:48. Invalid, *nomen nudum*.
Magadella Evans, J.W. 1954a:126. *M. brunnea* n.sp. (Madagascar). 10:2243. Deltocephalinae, Platymetopiini. Malagasian.
MAGNENTIINI Linnavuori 1978a:34, tribe of NIONIINAE.
Magnentius Singh-Pruthi 1930a:6. *M. clavatus* n.sp. (India: Madras). 13:246. Linnavuori 1978a:34-35. Nioniinae, Magnentiini. Oriental; India. Ethiopian; Zaire.
Maguangua Melichar 1926a:341. *Tettigonia inconspicua* Walker 1870b:303 (New Guinea: Waigeo). 1:233. Cicadellinae, Cicadellini. Oriental.
Mahalana Distant 1918b:64. *M. fidelis* n.sp. (India: Coimbatore). 10:2204. Deltocephalinae. Oriental; India.
Mahellus Nielson 1982e:84. *Jassus determinatus* Distant 1917a:316 (Seychelles). Coelidiinae, Coelidiini. Oriental; India. Malagasian; Seychelles.
Mahmoba Dworakowska 1982a:148 as subgenus of *Shamala* Dworakowska 1980b:169. *S. (M.) prospecta* n.sp. (India: W. Bengal, Darjeeling). Typhlocybinae, Typhlocybini. Oriental.
Mahmoodia Dworakowska 1970a:693. *M. mirabilis* n.sp. (Vietnam: Distr. Nghe An, Phu Quy). Typhlocybinae, Typhlocybini. Oriental.
Mahmoodiana Ahmed, Manzoor & Waheed 1971a:116. *M. acuta* n.sp. (West Pakistan: Lahore). Typhlocybinae, Dikraneurini. Oriental. Synonym of *Togaritettix* Matsumura 1931b:71; Sohi & Dworakowska 1984a:166.
Maichewia Linnavuori & Al-Ne'amy 1983a:69. *M. ererensis* n.sp. (Ethiopia: Agheresalam). Selenocephalinae, Selenocephalini. Ethiopian; south and NE Africa, SW Arabia.
Maiestas Distant 1917a:312. *M. illustris* n.sp. (Seychelles). 10:2175. Deltocephalinae, Platymetopiini. Oriental.
Mainda Distant 1908g:229. *M. praeculta* n.sp. (NW India: Kumaon). 6:50. Cicadellinae, Evacanthini. Oriental; India.
Mairana DeLong 1984g:117 as subgenus of *Angubahita* DeLong 1982a:185. *A. (M.) nigrens* n.sp. (Bolivia: Santa Cruz, Mairana). Deltocephalinae, Athysanini. Neotropical.
Makilingana Mahmood 1967a:15. *M. tripunctata* n.sp. (Philippines: Luzon, Mt. Makiling). Typhlocybinae, Erythroneurini. Oriental.
Makilingia Baker 1914a:409. *M. nigra* n.sp. (Philippines: Luzon). 1:659. Young 1968a:16; 1977a:1106. Makilingiinae. Oriental.
MAKILINGIINI 1:659.
Malagasiella Evans, J.W. 1954a:108. *M. minima* n.sp. (Madagascar). 11:82. Unassigned, formerly Coelidiinae; Nielson 1975:11. Malagasian.
Malasiella Evans, J.W. 1954a:125. *M. coronopunctata* n.sp. (Madagascar). 10:2243. Deltocephalinae, Platymetopiini. Malagasian.
Maldonadora Webb, M.D. 1983b:228. *M. rixia* n.sp. (Nigeria: Udo FR, M W State). Idiocerinae. Ethiopian; Cameroon, Central Africa Republic, Nigeria.
Malendea Linnavuori & Al-Ne'amy 1983a:83. *M. pungens* n.sp. (Cameroon). Selenocephalinae, Selenocephalini. Ethiopian; Guinean.
Malichus Distant 1918b:24. *M. capitatus* n.sp. (Ceylon: Peradeniya). 3:213. Penthimiinae. Oriental.
Malicia Evans, J.W. 1954a:116. *M. maculata* n.sp. (Madagascar). 15:208. Acostemminae. Malagasian.
Malipo Evans, J.W. 1966a:54. *Ipo speciosa* Evans, J.W. 1941f:144 (Western Australia: Dedari). Eurymelinae, Ipoini. Australian.
Malissiana Evans, J.W. 1954a:100. *Tettigonia madagascariensis* Signoret 1855b:344, pl. 11, fig. 5. (Madagascar). 1:105. Cicadellinae, Cicadellini. Malagasian.
Malmaemichungia Kwon 1983a:21. *M. brachycephala* n.sp. (Korea: Prov. J B, Mt. Daedunsan). Cicadellinae, Errhomenini. Palaearctic.

Mandera Ahmed, Manzoor 1971b:190. *M. heterostyla* n.sp. (East Pakistan: Comilla). Typhlocybinae, Erythroneurini. Oriental; India, Malaysia, Pakistan.
Synonym:
Hepneriana Dworakowska 1972j:114. *H. wroblewskae* n.sp.; Dworakowska & Viraktamath, C.A. 1975a:529.

Mandola Dworakowska & Viraktamath, C.A. 1975a:525. *Zygina quadrinotata* Ahmed 1969c:171 (West Pakistan: Murree). Typhlocybinae, Erythroneurini. Oriental; northwest India, Pakistan.

Mangganeura Ghauri 1967a:164. *M. reticulata* n.sp. (Malaysia: Selangor). Hongsaprug & Wilson 1985a:180. Typhlocybinae, Erythroneurini. Oriental; SE Asia.

Manzoonara Ramakrishnan & Ghauri 1979b:205. *Erythroneura hazarensis* Ahmed 1970b:179 (West Pakistan: Thai). Typhlocybinae, Erythroneurini. Oriental. Synonym of *Empoascanara* Distant 1918b:94; Dworakowska 1980d:195.

Manzutus Oman 1949a:73. *Cicadella huachucana* Ball 1936a:19 (USA: Arizona, Huachuca Mts.). 1:286. Young 1977a:838. Cicadellinae, Cicadellini. Nearctic; Arizona, Mexico.

Mapochia Distant 1910e:231. *M. collaris* n.sp. (Kaffraria). 9:116. Linnavuori 1979a:669-672. Eupelicinae, Paradorydiini. Ethiopian.

Mapochiella Evans, J.W. 1966a:140. *M. rotundata* n.sp. (Australia: Queensland, Deception Bay). Eupelicinae, Paradorydiini. Australian; Queensland.

Maranata Blocker 1979a:26. *"Bythoscopus nigrofrons* Osborn" 1924b:389 = *Bythoscopus nigrifrons* Osborn 1924b:389 (Brazil: Chapada). Iassinae, Iassini. Neotropical.

Marathonia Oman 1949a:72. *Cicadella marathonensis* Olsen 1918a:3 (USA: Texas, Marathon). 1:285. Cicadellinae, Cicadellini. Nearctic. Synonym of *Graphocephala* Van Duzee 1916a:66; Young 1977a:849.

Marcapatiana Nielson 1979b:139. *M. emmrichi* n.sp. (Peru: Marcapata). Coelidiinae, Teruliini. Neotropical; Peru.

Mareba Distant 1908b:77. *M. erisia* n.sp. (Ecuador: Cachabe). 1:642. Young 1968a:60. Cicadellinae, Proconiini. Neotropical; Ecuador, French Guiana, Panama.
Synonyms:
Caripuna Melichar 1926a:318. *Tettigonia guerini* Signoret 1855a:51.
Anchura Melichar 1926a:344. *M. erisia* Distant 1908b:77, preoccupied.
Anchuralia Evans, J.W. 1947a:159. *M. erisia* Distant 1908b:77.

Mareja Melichar 1926a:343. *Tettigonia limbaticollis* Stal 1864a:75 (Mexico) = *Tettigonia urbana* Stal 1864a:74. 1:412. Young 1977a:977. Cicadellinae, Cicadellini. Neotropical; Mexico, Central America, Colombia, Venezuela.

Margalana DeLong & Freytag 1972a:218. Error for *Marganalana* Metcalf 1949b:277.

Margana DeLong 1942d:109. *Ponana marginifrons* var. *suilla* Ball 1935c:503 (USA: Arizona, Penal Mts. above Superior). 3:154. Gyponinae, Gyponini. Nearctic. Preoccupied by *Margana* Walker 1865, see *Marganana* DeLong 1948b:101.

Marganalana Metcalf 1949b:277. *M. testacea* n.sp. (British Guiana). 3:153. Gyponinae, Gyponini. Neotropical. Subgenus of *Gypona* Germar 1821a:73; DeLong & Freytag 1964a:7, 37.

Marganana DeLong 1948b:101, replacement for *Margana* DeLong 1942d:109, not *Margana* Walker 1865. *Ponana marginifrons* var. *suilla* Ball 1935c:503 (Pinal Mountains above Superior, Arizona). 3:153. DeLong & Freytag 1963a. Gyponinae, Gyponini. Nearctic.
Synonym:
Margana DeLong 1942d:109, preoccupied.
Subgenus:
Declivana Delong & Freytag 1963a:262. *M. (D.) equata* n.sp.

Maricaona Caldwell 1952a:61. *M. polyamia* n.sp. (Puerto Rico: Marico Insular Forest, Maricao-Santana Grande Road, km 15.6, 2450 ft). 10:1853. Deltocephalinae, Deltocephalini. Neotropical; Caribbean.

Marizella Schmidt 1928c:82. *M. polita* n.sp. (Ecuador: Jnez). 1:655. Young 1977a:104. Cicadellinae, Cicadellini. Neotropical. Synonym of *Dilobopterus* Signoret 1850b:284; Young 1977a:104.

Marolda Dworakowska 1977e:851 as subgenus of *Empoasca* Walsh 1862a:149. *E. (M.) testacea* n.sp. (Zaire: Haut-Vele, Yebo-Moto). Typhlocybinae, Empoascini. Ethiopian.

Maroopula Webb, M.D. 1983a:72. *M. trixia* n.sp. (Australia: mid Queensland). Idiocerinae. Australian.

Marquardtella Schmidt 1930b:122. *M. krusei* n.sp. (New Guinea). 15:20. Iassinae, Iassini. Oriental; New Guinea. Synonym of *Thalattoscopus* Kirkaldy 1905b:334; Knight, W.J. 1983a:29.

Marquesia Osborn 1934b:250. *M. atra* n.sp. (Marquesas Islands: Taupaooa, 2750 ft). 1:2486. Deltocephalinae, Macrostelini. Oceania. Preoccupied by *Marquesia* Malloch 1932, see *Marquesitettix* Metcalf 1967c:2486.

Marquesitettix Metcalf 1952a:229, replacement for *Marquesia* Osborn 1934b:250, not *Marquesia* Malloch 1932. *M. atra* n.sp. (Marquesas Isl: Tapuaooa, 2750 ft). 10:2486. Deltocephalinae, Macrostelini. Oceania; (Marquesas Islands).

Mascarenotettix Evans, J.W. 1954a:95. *Poeciloscarta retinens* 1951a:82 (Madagascar). 1:105. Cicadellinae, Cicadellini. Malagasian.

Mascoitanus Linnavuori & Heller 1961a:12 as subgenus of *Brasilanus* Linnavuori 1959b:287. *B. (M.) lateralis* n.sp. (Peru: "Hda. Mascoitania, Urwald"). Deltocephalinae, Athysanini. Neotropical.

Masiripius Dlabola 1981a:273. *Platymetopius zizyphi* de Bergevin 1922a:63 (Algeria). Deltocephalinae, Platymetopiini. Palaearctic; Algeria, Iran, Libya.

Matatua Knight, W.J. 1976a:85. *M. montivaga* n.sp. (New Zealand: Canterbury Prov., Mt. Cook N.P., Mt. Sebastopol, Red Lakes). Typhlocybinae, Dikraneurini. Australian; New Zealand.

Matsumurana Distant 1917a:317. *M. facialis* n.sp. (Seychelles). 10:2354. Xestocephalinae, Xestocephalini. Malagasian.

Matsumurasca Anufriev 1973b:438 & 440 as subgenus of *Empoasca* Walsh 1862a:149. *Empoasca diversa* Vilbaste 1968a:87 (USSR: Maritime Territory). Typhlocybinae, Empoascini. Palaearctic.

Matsumuratettix Metcalf 1952a:229, replacement for *Epitettix* Matsumura 1914:194 not *Epitettix* Hancock 1907. *Epitettix hiroglyphicus* n.sp. (Formosa). 10:878. Deltocephalinae, Athysanini. Oriental.
Synonym:
Epitettix Matsumura 1914a:194. *E. hiroglyphicus* n.sp.

Matsumurella Ishihara 1953a:200. *Jassus (Allygus) kogotensis* Matsumura 1914a:206 (Japan: Honshu). 10:1004. Deltocephalinae, Athysanini. Palaearctic; Manchurian subregion.
Subgenus:
Ochta Emeljanov 1977a:152. *M. (O.) minor* n.sp.
Synonym:
Shonenus Ishihara 1958a:232. *Jassus (Allygus) kogotensis* Matsumura 1914a:200.

Matsumurina Dworakowska 1972b:403. *Zygina kagina* Matsumura 1932a:112 (Formosa). Typhlocybinae, Erythroneurini. Palaearctic, Oriental; Formosa, Japan.

Mattogrossus Linnavuori 1959b:129 as subgenus of *Unerus* DeLong 1936a:219. *U. (M.) colonoides* n.sp. (Brazil: Chapada). Deltocephalinae, Deltocephalini. Neotropical.

Matuta Emeljanov 1966a:99. *Tettigonia guttiger* Uhler 1896a:294 (Japan). Cicadellinae, Pagaroniini. Palaearctic. Preoccupied by *Matuta* Grote 1874, see *Kalkandelenia* Kocak 1981a:32. Synonym of *Pagaronia* Ball 1902a:19; Anufriev 1970f:553.

Mavromoustaca Dlabola 1967e:37 as subgenus of *Osbornellus* Ball 1932a:17. *O. (M.) consanguineus* n.sp. (Cyprus). Deltocephalinae, Scaphoideini. Palaearctic.

Mavromoustacus Dlabola 1974a:68. Error for *Mavromoustaca* Dlabola 1967.

Maximianus Distant 1918b:79. *M. notatus* n.sp. (S. India: Kodiakanal). 10:1585. Deltocephalinae, Athysanini. Oriental; southern India.

Mblokoa Linnavuori 1972a:135. *M. angustiformis* n.sp. (Sudan: Bahr el Ghazal). Ulopinae, Ulopini. Ethiopian; Sudanian.

Mcateeana Christian 1952a:1130. *Empoa querci sexnotata* Van Duzee 1914a:57 (USA: California, Alpine). 17:933. Typhlocybinae, Typhlocybini. Nearctic; California.

Mecdaria Amyot 13:252. Invalid, not binominal.

Medlerola Young 1977a:436. *M. motiva* n.sp. (French Guiana: Maroni River). Cicadellinae, Cicadellini. Neotropical.

Megabahita Linnavuori & DeLong 1978b:120. *Eutettix irroratus* Osborn 1923c:51 (Bolivia: Santa Cruz). Deltocephalinae, Athysanini. Neotropical; Bolivia, Brazil, Panama, Paraguay.

Megabyzus Distant 1908g:294. *M. signandus* n.sp. (Ceylon: Amotapuro). 10:1695. Deltocephalinae, Athysanini. Oriental.

Megacoelidia Kramer & Linnavuori 1959a:55. *M. splendida* n.sp. (Brazil: Amazonas, Esperanza). Neocoelidiinae. Neotropical; Bolivia, Brazil.

Megadorus Linnavuori 1959b:231 as subgenus of *Mimodorus* Linnavuori 1959b:229. *M. (M.) pseudundatus* n.sp. (Brazil: Corumba). Deltocephalinae, Athysanini. Neotropical.

Megagallia Linnavuori 1954e:128. *Bythoscopus punctaticollis* Stal 1862e:54 (Brazil: Rio de Janeiro). 14:150. Agalliinae, Agalliini. Neotropical.

Megalidia Nielson 1982e:270. *M. elongata* n.sp. (Bolivia: Yungas de Palmar). Coelidiinae, Coelidiini. Neotropical.

Megalopenthimia Evans, J.W. 1954a:113. *M. flava* n.sp. (Madagascar). 3:216. Penthimiinae. Malagasian.

Megalophtalmus 5:98. Error for *Megophthalmus* Curtis.

Megalophthalmus 5:98. Error for *Megophthalmus* Curtis.

Megalopsius Emeljanov 1961a:123. *M. oshanini* n.sp. (USSR: Turkmenia: Farab). Deltocephalinae, Goniagnathini. Palaearctic.

Megalotettigella Ishihara 1953b:16. *Cicada ferruginea* Fabricius 1787a:269 (China). 1:233. Cicadellinae, Cicadellini. Palaearctic. Synonym of *Bothrogonia* Melichar 1926a:341.

Megaulon Theron 1975a:201. *Deltocephalus chlorellus* Naude 1926a:44 (South Africa: Cape Province, Viljoen's Pass). Deltocephalinae, Deltocephalini. Ethiopian.

Megipocerus Zachvatkin 1945a:3. *M. mordvilkoi* n.sp. (Ussuri Prov., Sutshan district, Fanza). 16:221. Idiocerinae. Palaearctic; USSR Maritime Territory.

MEGOPHTHALMINAE 5:7.

MEGOPHTHALMINI 5:13.

Megophthalmus Curtis 1833a:193. *M. bipunctatus* n.sp. (Great Britain) = *Cicada scanica* Fallen 1806b:113. 5:14. Megophthalminae. Palaearctic; Europe, northern Africa.
Synonym:
Paropia Germar 1833a:181. *Ulopa scanica* Germar 1833a:181.

Megopthalmus 5:98. Error for *Megophthalmus* Curtis.

Megothalmus 5:98. Error for *Megophthalmus* Curtis.

Megulopa Lindberg 1925a:106. *M. sahlbergorum* n.sp. (Egypt). 5:86. Linnavuori 1972a:131. Ulopinae, Ulopini. Ethiopian; Sudanian. Palaearctic; Mediterranean.

Meketia Dworakowska 1982a:107. *Camulus bellulus* Distant 1918b:98 (S. India: Kodiakanal). Typhlocybinae, Typhlocybini. Oriental.

Melanderus Amyot 6:61. Invalid, not binominal.

Melaneura Amyot 1:684. Invalid, not binominal.

Melanochloa Amyot 1:684. Invalid, not binominal.

Melanochlora 1:684. Error for *Melanochloa* Amyot, invalid.

Melicharella Semenow 1902a:353, replacement for *Ahngeria* Melichar 1902b:76, not *Ahngeria* Kokujew 1902. *M. (elicharella) planifrons* Oshanin 1906a:67 = *Ahngeria planifrons* Melichar 1902b:76 (Transcaspia). 14:162. Adelungiinae, Adelungiini. Palaearctic.

Melicharellaustroagallia Kameswara Rao 1976a:11. Invalid, *nomen nudum*.

MELICHARELLINAE 14:159.

MELICHARELLINI Nast 1972a:210.

Melicharidia Dworakowska 1970k:624 as subgenus of *Hauptidia* Dworakowska 1970k:620. *Erythroneura aridula* Linnavuori 1956b:138 (Israel: Jerusalem). Typhlocybinae, Erythroneurini. Palaearctic.

Melichariella Matsumura 1914a:236. *M. satsumensis* n.sp. (Japan). 10:547. Linnavuori & Al-Ne'amy 1983a:23. Selenocephalinae, Bhatiini. Oriental/Palaearctic; Formosa, Japan, Micronesia.

Melillaia Linnavuori 1971a:60. *Thamnotettix desbrochersi* Lethierry 1889a:318 (Algeria: Oran). Deltocephalinae, Athysanini. Ethiopian; Mediterranean.

Mellia Schmidt 1920m:127, preoccupied by *Mellia* Herrmannsen 1847, see *Melliola* Hedicke 1923a:72. *M. granulata* n.sp. (China). 2:17. Hylicinae, Melliolini. Oriental; China.

Melliala 2:18. Error for *Melliola* Hedicke.

MELLIINI 2:16.

Melliola Hedicke 1923a:72, replacement for *Mellia* Schmidt 1920m:127, not *Mellia* Herrmannsen 1847. *Mellia granulata* Schmidt 1920m:128 (China). 2:17. Hylicinae, Melliolini. Oriental; China.

MELLIOLINI Metcalf 1962a:16, tribe of HYLICINAE.

Melopopterus 17:1504. Error for *Molopopterus* Jacobi.

Memnonia Ball 1900a:66. *M. consobrina* n.sp. (USA: Colorado). 8:55. Deltocephalinae, Hecalini. Nearctic; central Great Plains and Sonoran subregion.

Mendera Ahmed, Manzoor 1972a:67. Error for *Mandera* Ahmed, Manzoor.

Mendozellus Linnavuori 1959b:117 as subgenus of *Amplicephalus* DeLong 1926d:83. *Spathifer dubius* Linnavuori 1955b:126 (Argentina: Mendoza). Deltocephalinae, Deltocephalini. Neotropical.
[NOTE: Raised to generic rank; Kramer 1971d:261 & 263].

Mendrausus Ribaut 1946b:84. *Deltocephalus chyzeri* Horvath 1897b:636 (Hungary). 10:1475. Deltocephalinae, Paralimnini. Palaearctic; Eurasia.

Mendreus Ribaut 1946b:83. *Deltocephalus serratus* Ribaut 1925a:189 (France). 10:1394. Deltocephalinae, Paralimnini. Palaearctic.

Menosoma Ball 1931a:4. *M. stonei* n.sp. (USA: Florida, Sanford). 10:530. Deltocephalinae, Athysanini. Neotropical/Nearctic; eastern and central North America, south in tropical regions to Brazil.

Meremra Dworakowska & Viraktamath 1979a:58. *M. puncta* n.sp. (India: Karnataka, Mudigere). Typhlocybinae, Erythroneurini. Oriental.

Meroleucocerus Maldonado-Capriles 1972b:631. *M. punctipennis* n.sp. (Viet Nam: Dalat, 6 km S). Idiocerinae. Oriental.

Mesadorus Linnavuori 1955b:120. *M. undatus* n.sp. (Brazil: Rio Grande do Sul). 10:932. Deltocephalinae, Athysanini. Neotropical.

Mesagallia Zachvatkin 1946b:157-158. *M. georgii* n.sp. (Turkey: Ankara). 14:108. Agalliinae, Agalliini. Palaearctic; Turkey, USSR: Kirghirzia.

Mesamia Ball 1907a:59 as subgenus of *Eutettix* Van Duzee 1892e:307. *E. (M.) nigridorsum* n.sp. (USA). 10:2075. Deltocephalinae, Athysanini. Nearctic; southern Canada to central Mexico.

MESAMIINI 10:2075.

Mesanua 10:2670. Error for *Mesamia* Ball.

Mesargus Melichar 1903b:175. *M. asperstus* n.sp. (Ceylon: Henaratgoda). 5:84. Ulopinae, Ulopini. Oriental; Ceylon, India.
Synonyms:
Moonia Distant 1908g:197. *M. sancita* n.sp.
Sitades Distant 1912e:608. *S. fasciatus* n.sp.
[NOTE: The generic concept exemplified by *Mesargus*, *Moonia* and *Sitades* is not clear. Linnavuori 1972a:128 placed *Moonia* as a subgenus of *Coloborrhis* Germar 1836a:72, but Vilbaste 1975a:230 made no mention or action in connection with the above synonymy. Davis, R.B. 1975a treated *Coloborrhis, Mesargus* and *Moonia* as separate genera].

Mesobana Melichar 1926a:322 as subgenus of *Abana* Distant 1908b:72. *Amblydisca pomposula* Jacobi 1905c:167 (Peru). 1:645. Cicadellinae, Proconiini. Neotropical. Synonym of *Abana* Distant; Young 1968a:148.

Mesodicus Fieber 1866a:501. *M. foveolatus* n.sp. ("Southern Europe," error ?). 4:42. Ledrinae, Xerophloeini. Nearctic, Neotropical. Synonym of *Xerophloea* Germar 1839a:190.

Mesodiscus 4:142. Error for *Mesodicus* Fieber.

Mesodorydium Melichar 1914g:4. *M. famelicum* n.sp. (Africa). 9:117. Linnavuori 1979a:655. Eupelicinae, Paradorydiini. Ethiopian. Synonym of *Paradorydium* Kirkaldy 1901f:339; Linnavuori 1979a:655.

Mesogonia Melichar 1926a:343. *Tettigonia ferrugatula* Breddin 1901h:107 (Ecuador). 1:267. Young 1977a:480. Cicadellinae, Cicadellini. Neotropical; Bolivia, Brazil, Peru, Ecuador.
Synonym:
Santarema Melichar 1926a:344. *Tettigonia ferrugatula* Breddin 1901h:107.

Mesoparopia Matsumura 1912a:27. *M. nitobei* n.sp. (Formosa). 5:77. Ulopinae, Ulopini. Oriental.

Mesojassoides Oman 1937a:38. *M. gigantea* n.sp. Metcalf & Wade 1966a:223. Fossil: Cretaceous; Colorado.

Mesojassus Tillyard 1916a:34. *M. ipsviciensis* n.sp. Metcalf & Wade 1966a:219. Fossil: Triassic; Australia, Queensland.

Mesojassus Handlirsch 1939a:145. *M. pachyneurus* n.sp. Preoccupied by *Mesojassus* Tillyard 1916a:34, see *Handlirschiana* Metcalf & Wade 1966a:220. Fossil: Jurassic; Germany, Mecklenburg.

Mesotettix Matsumura 1914a:195. *M. shokaensis* n.sp. (Formosa). 10:879. Deltocephalinae. Oriental.

Metacephalus DeLong & Martinson 1973f:225. *M. albocrux* n.sp. (Peru: Pucallpa, Loreto). [Salient characters for placement of the genus in higher category not given; probably Xestocephalinae, Portanini]. Neotropical.

Metagoldeus Remane & Asche 1980a:81. *M. simplicipenis* n.sp. (Morocco). Deltocephalinae, Paralimnini. Palaearctic; Mediterranean subregion.

Metalimnus Ribaut 1948c:59. *Deltocephalus formosus* Boheman 1845b:155 (Lapland). 10:1880. Deltocephalinae, Paralimnini. Northern Palaearctic.

Metapocirtus Costa 1834b:85. *M. listatus* n.sp. (Italy). 16:233. Idiocerinae. Palaearctic.

Metaporcitus 16:236. Error for *Metapocirtus* Costa.

Metascarta Melichar 1926a:341. *Tettigonia flavipes* Signoret 1854a:22 (Bolivia). 1:196. Young 1977a:1098. Cicadellinae, Cicadellini. Neotropical.

Metathysanus Dahl 1912a:439. *Athysanus obsoletus* Kirschbaum 1858a:7 (Prussia). 10:12,14,2670. Deltocephalinae, Athysanini. Palaearctic. Synonym of *Conosanus* Osborn & Ball 1902a:236.

Metcalfiella Linnavuori 1955b:121-122. *M. chacoensis* Linnavuori n.sp. (Paraguay: Chaco). 10:2241. Deltocephalinae, Deltocephalini.
Preoccupied by *Metcalfiella* Goding 1929, see *Zepama* Metcalf 1967c:2241. Synonym of *Fusanus* Linnavuori 1955b:123-124; Linnavuori 1959b:135.

Metidiocerus Ossiannilsson 1981a:318. *Idiocerus crassipes* J. Sahlberg 1871a:143 (Karelia). Idiocerinae, Idiocerini. Palaearctic. Subgenus of *Idiocerus* Lewis 1834a:47; Kwon 1985a:68.
Synonym:
Stenidiocerus Ossiannilsson 1981a:348; Kwon 1985a:68.

Metriosteles 10:2570. Error for *Macrosteles* Fieber.

Mexara Oman 1949a:176. *Amplicephalus atascasus* Ball 1936b:20 (USA: Arizona, Atascasa Mt., east of Ruby). 10:1824. Deltocephalinae, Deltocephalini. Nearctic.

Mexicananus DeLong 1944e:89. *M. levis* n.sp. (Mexico: Chiapas, Finca Vergel, Valley of Rio Huixtla). 10:710. Deltocephalinae, Athysanini. Neotropical.

Mexolidia Nielson 1983h:564. *M. terminata* n.sp. (Mexico: "M.B.194"). Coelidiinae, Teruliini. Neotropical.

Mfutila Dworakowska 1974a:225. *M. sanga* n.sp. (Congo: Odzala). Typhlocybinae, Typhlocybini. Ethiopian.

Mgenia Theron 1984c:321. *Coelidia fuscovaria* Stal 1855a:99 (Caffraria). Coelidiinae, (provisional). Ethiopian; South Africa.

Miarogonalia Young 1977a:728. *Tettigonia chevrolatii* Signoret 1855d:778 (Venezuela). Cicadellinae, Cicadellini. Neotropical; Colombia, Venezuela.

Micantulina Anufriev 1970d:262. *Cicadula micantula* Zetterstedt 1840a: 229 (Lapland). Typhlocybinae, Dikraneurini. Transpalaearctic.
Subgenus:
Anufrievola Kocak 1981a:32. *Typhlocyba stigmatipennis* Mulsant & Rey 1855a:245.

Michalowskiya Dworakowska 1972k:122. *M. lutea* n.sp. (China: Kiangsu, Hangchow). Typhlocybinae, Erythroneurini. Oriental; China, India.

Micrelloides Evans, J.W. 1973a:189. *M. molaris* n.sp. (Western Australia: Millstream area). Deltocephalinae, Hecalini. Australian.

Microgoniella Melichar 1926a:342. *Cicada pudica* Fabricius 1803a:65 ("Amer. merid."). 1:84,90. Young 1977a:410. Cicadellinae, Cicadellini. Neotropical; Costa Rica to southern Brazil.

Microledrella Evans, J.W. 1969a:745. *M. minuta* n.sp. (New Guinea: Wissel Lakes, Enarotadi, 1900 m). Ledrinae, Thymbrini. Oriental.

Microlopa Evans, J.W. 1966a:87. *M. minuta* n.sp. (Tasmania: Cradle Mt. near Waldheim). Ulopinae, Ulopini. Australian.

Micropsis 13:252. Error for *Macropsis* Lewis.

Microscita Melichar 1951a:93 as subgenus of *Microgoniella* Melichar 1926a:342. *Microgoniella (Microscita) naevula* n.sp. (Colombia). 1:90. Cicadellinae, Cicadellini. Neotropical. Synonym of *Juliaca* Melichar 1926a:344; Young 1977a:443.

Microtartessus Evans, F. 1981a:123. *Tartessus idyia* Kirkaldy 1907a:44 (Australia: Queensland, Nelson). Tartessinae. Australian.

Midoria Kato 1931e:439. *M. capitata* n.sp. (Formosa). 4:82. Ledrinae, Petalocephalini. Oriental; Formosa, Japan.

Mileeva Nast 1972a:252. Error for *Mileewa* Distant 1908g:238.

Mileewa Distant 1908g:238. *M. margheritae* n.sp. (Assam: Margherita). 1:297. Linnavuori 1979a:717. Cicadellinae, Mileewanini. Oriental; Ethiopian, and eastern Palaearctic.
Synonyms:
Bolinlila Distant 1910e:234. *B. rhodesiana* n.sp.
Tylozygoides Matsumura 1912a:42. *T. artemisae* n.sp. = *M. margheritae* Distant.
Faenius Distant 1918b:14. *F. lynchi* n.sp.
Augulus Distant 1918b:98. *A. typicus* n.sp.

MILEEWANINI 1:296.

Milotartessus Evans, F. 1981a:178. *Tartessus sananas* Distant 1912h:603 (New Guinea). Tartessinae. Oriental; New Guinea.

Mimallygus Ribaut 1948c:58. *Jassus lacteinervis* Kirschbaum 1868b:103 (Prussia). 10:1003. Deltocephalinae, Athysanini. Palaearctic; Europe.

Mimodorus Linnavuori 1959b:229. *M. diabolus* n.sp. (Argentina: Missiones, Loreto). Deltocephalinae, Athysanini. Neotropical; Argentina, Brazil, Paraguay.
Subgenus:
Megadorus Linnavuori 1959b:231. *M. (M.) pseudundatus* n.sp.

Mimodrylix Zachvatkin 1935a:108. *Athysanus capicola* Stal 1855a:99 (Kaffraria). 10:308. Deltocephalinae, Athysanini. Ethiopian, Oriental. Synonym of *Exitianus* Ball 1929a:5.

Mimohardya Zachvatkin 1946b:170 as subgenus of *Hardya* Edwards 1922a:206. *H. (M.) heptneri* n.sp. (Turkmenia, nr Kushka). 10:266. Deltocephalinae, Athysanini. Palaearctic.

Mimotelliae 10:2670. Error for *Mimotettix* Matsumura.

Mimotettix Matsumura 1914a:197. *M. kawamurae* n.sp. (Japan: Kyushu). 10:880. Deltocephalinae, Athysanini. Oriental; Formosa, southern Japan. Eastern Palaearctic; Korea.

Mindanaoa Mahmood 1967a:42. *M. orietalis* n.sp. (Philippines: Mindanao, Zamboanga). Typhlocybinae, Typhlocybini. Oriental.

Miochlorotettix Pierce, W.D. 1963a:73. Fossil: Miocene, (California, San Bernadino County, Calico Mtns). Invalid, no type species designated.

Miomesamia Pierce, W.D. 1963a:81. *M. juliae* n.sp., by monotypy. Fossil: Miocene, (California, San Bernadino County, Calico Mtns., SW 1/4 Section 24, R1E, T10 N).

Miraldus Lindberg 1960a:52. *Deltocephalus truncatus* Melichar 1902b:80 (Spain). Deltocephalinae, Deltocephalini. Palaearctic; Spain.

Mirzacha Amyot 10:2670. Invalid, not binominal.

Mirzayansus Dlabola 1979a:251. *M. denaicus* n.sp. (Iran: Fars, west slope of Dena-Gebirge). Deltocephalinae, Athysanini. Palaearctic.

Mitelloides Evans, J.W. 1939a:46. *M. moanensis* n.sp. (Australia: Torres Strait, Banks I., Moa). 4:126. Evans, J.W. 1966a:126. Ledrinae, Thymbrini. Australia and New Guinea.

Mitjaevia Dworakowska 1970n:763. *Erythroneura amseli* Dlabola 1961b:297 (Afghanistan: Herat, 900 m). Typhlocybinae, Erythroneurini. Palaearctic; Siberian subregion.

Mocoa Linnavuori & DeLong 1978d:236. *M. elegans* n.sp. (Colombia: 15 mi SW Mocoa, Narins, 1600 m). Deltocephalinae, Athysanini. Neotropical.

Mocuastrum Emeljanov 1972b:248 as subgenus of *Mocuellus* Ribaut 1946b:83. *Mocuellus bogdianus* Dlabola 1967d:221 (Mongolia: Bogd). Deltocephalinae, Paralimnini. Palaearctic.

Mocuellus Ribaut 1946b:83. *Deltocephalus collinus* Boheman 1850a:261 (Gotland). 10:1375,1377. Emeljanov 1966a:130,131; Ross & Hamilton, KGA 1970b. Deltocephalinae, Paralimninini. Holarctic.
Subgenera:
Erzaleus Ribaut 1952a:296. *Jassus (Deltocephalus) metrius* Flor 1861a:264.
Mocuola Emeljanov 1964f:48. *M. (M.) pulchellus* n.sp.
Falcitettix Linnavuori 1953d:58. *F. sibiricus* n.sp.
Mocuastrum Emeljanov 1972b:248. *Mocuellus bogdianus* Dlabola 1967d:221.
Promocuus Emeljanov 1972b:248. *Mocuellus hordei* Emeljanov 1964f:47.
Mocuola Emeljanov 1964f:48 as subgenus of *Mocuellus* Ribaut 1946b:83. *M. (M.) pulchellus* n.sp. (USSR: Kazakhstan). Deltocephalinae, Paralimnini. Palaearctic.
Mocustus Mitjaev 1968b:43. Error for *Macustus* Ribaut.
Mocydia Edwards 1922a:206. *Jassus croceus* Herrich-Schaeffer 1837a:7 (Germany). 10:794. Deltocephalinae, Athysanini. Palaearctic; European and Mediterranean subregion.
Mocydiopsis Ribaut 1939a:274. *Jassus attenuatus* Germar 1821a:91 (Prussia). 10:802. Deltocephalinae, Athysanini. Palaearctic; European and Mediterranean subregion.
Modderena Theron 1984c:323. *Equeefa albicosta* Naude 1926a:81 (South Africa). Coelidiinae, (provisional). Ethiopian; South Africa.
Mogangella Dlabola 1957a:51. *M. straminea* n.sp. (Anatolia). Deltocephalinae, Paralimnini. Palaearctic; European USSR, Turkey.
Mogangina Emeljanov 1962a:174. *M. bromi* n.sp. (USSR: Kazakhstan). Deltocephalinae, Paralimnini. Palaearctic; USSR: Altai Mts., Kazakhstan, Kirghizia.
Mogenola Blocker 1979a:39. *M. deserta* n.sp. (St. Kitts: Brimstone Mts.). Iassinae. Neotropical; Leeward Is., West Indies.
Mohunia Distant 1908g:272. *M. splendens* n.sp. (Burma: Tenasserim, Myitta). 7:16. Nirvaninae, Nirvanini. Oriental; Burma, Java, Thailand.
Molomea China 1927d:283, replacement for *Centrometopia* Melichar 1925a:399, not *Centrometopia* Ragonot 1887. *Tettigonia personata* Signoret 1854b:364 (Brazil: Sao Paulo). 1:591. Cicadellinae, Proconiini. Neotropical; Brazil, Ecuador, Paraguay, Peru.
Synonyms:
Centrometopia Melichar 1925a:399, preoccupied.
Centrometopides Strand 1928a:73, replacement for *Centrometopia* Melichar 1925.
Molopopterus Jacobi 1910b:133. *M. nigriplaga* n.sp. (Tanganyika). 17:779. Dworakowska 1973a. Typhlocybinae, Eupterygini. Ethiopian; East, West and South African subregions.
Momoria Blocker 1979a:48. *Stragania misella* Stal 1864a:85 (Mexico: Vera Cruz). Iassinae. Western Nearctic and Neotropical; south to Brazil and Paraguay, excepting Chile and Peru.
Mongoloiassus Dlabola 1965c:119. Error for *Mongolojassus* Zachvatkin.
Mongolojascus Vilbaste 1980b:74. Error for *Mongolojassus* Emeljanov.
Mongolojassus Zachvatkin 1953d:247. *Deltocephalus sibiricus* Horvath 1901a:272 (Siberia). Deltocephalinae, Paralimnini. Palaearctic; European and Siberian subregions.
Synonym:
Omskius Linnavuori 1954g:183. *Deltocephalus sibiricus* Horvath 1901a:272; Emeljanov 1962a:181.
Monobazus Distant 1908g:351. *M. himalayensis* n.sp. (N.E. Himalayas: Kurseong). 10:541. Deltocephalinae, Athysanini. Northern India.
Monosoma Cheng 1980a:101. Error for *Menosoma* Ball.
Monteithia Evans, J.W. 1968a:216. *M. anomala* n.sp. (NE New Guinea: Keglsugl, Mt. Wilhelm, 2500-2700 m). Ulopinae, Monteithiini. Oriental; New Guinea.
MONTEITHIINI Evans, J.W. 1968a:221, tribe of ULOPINAE.
Moonia Distant 1908g:197. *M. sancita* n.sp. (Northern India: Musoorie). 5:80. Linnavuori 1972a:128. Ulopinae, Ulopini. Oriental. Synonym of *Mesargus* Melichar 1903b:175; Vilbaste 1975a:230.
Moorada Ghauri 1975b:481. *M. ibadensis* n.sp. (Nigeria: Ibadan, Moor Plantation). Deltocephalinae, Athysanini. Ethiopian.

Mordania Dworakowska 1979c:313. *M. lutea* n.sp. (Nigeria: K State, N. Bussa). Typhlocybinae, Zyginellini. Ethiopian.
Morinda Emeljanov 1972d:106. *Euscelis sibiricus* Emeljanov 1962a:170 (Siberia). Deltocephalinae, Cicadulini. Palaearctic.
Mormoria Blocker 1979a:1. Error for *Momoria* Blocker 1979a:48.
Moskgha Deeming & Webb 1982a:486. *M.* laminata n.sp. (Nigeria: Zaria, Samaru). Deltocephalinae, Athysanini. Ethiopian.
Morvellina DeLong 1980h:69. Error for *Norvellina* Ball 1931a:2.
Motaga Dworakowska 1979b:12. *M. rokfa* n.sp. (Vietnam: Hanoi). Typhlocybinae, Erythroneurini. Oriental; India, Vietnam.
Motschulskyia Kirkaldy 1905a:226, replacement for *Conometopus* de Motschulsky 1863a:103, not *Conometopus* Fieber 1858. *Conometopus inspiratus* Motschulsky 1863a:104 (Ceylon). 17:232. Dworakowska 1969b, 1971d. Typhlocybinae, Dikraneurini. Oriental; Ceylon, India, Korea.
Subgenus:
Togaritettex Matsumura 1931b:71. *T. serratus* n.sp.; Dworakowska 1971d:579.
Synonym (of *Togaritettix*):
Mahmoodiana Ahmed & Waheed 1971a:116; Sohi & Dworakowska 1984a:166.
Motrchulskla 17:1504. Error for *Motschulskyia* Kirkaldy.
Motschulskia 17:1504. Error for *Motschulskyia* Kirkaldy.
Mucaria 7:32. Error for *Mukaria* Distant.
Mucidiopsis Pelov 1968a:161. Error for *Mocidiopsis* Ribaut.
Mucrometopia Melichar 1925a:403 as subgenus of *Oncometopia* Stal 1869a:62. *Tettigonia caudata* Walker 1851b:749 (Bolivia). 1:594. Young 1977a:156. Cicadellinae, Cicadellini. Neotropical; Bolivia, Peru, Colombia.
Mucydia Pelov 1968a:161. Error for *Mocydia* Edwards.
Muinocerus Ghauri 1985c:67. *M. qadirii* n.sp. (Malaysia: Sarawak, Semengoh, Engkabang plantation). Idiocerinae. Oriental.
Muirella Kirkaldy 1907d:79. *M. oxyomma* n.sp. (Fiji). 11:22. Nielson 1975a:34. Coelidiinae, Tharrini. Oceania. Synonym of *Tharra* Kirkaldy 1906c:324; Nielson 1975a:34.
Mukaria Distant 1908g:269. *M. penthimioides* n.sp. (Ceylon: Maskeliya). 7:30. Mukariinae, Mukariini. Oriental; Ceylon, Formosa, India, New Britain, New Guinea, Thailand.
Synonyms:
Parabolotettix Matsumura 1912b:280. *P. maculatus* n.sp.
Ikomella Ishihara 1961a:253. *I. confersa* n.sp.
Mukariana 7:33. Error for *Mukaria* Distant.
MUKARIINI 7:30.
Mukwana Distant 1908g:317. *M. introducta* n.sp. (Ceylon: Kurungala). 11:88. Coelidiinae, Thagriini. Oriental. Synonym of *Thagria* Melichar 1903b:176; Nielson 1977a:9.
Muleyrechia Linnavuori 1956d:164. *Jassargus melillensis* Linnavuori 1955c:25 (Spanish Morocco). Deltocephalinae, Paralimnini. Palaearctic; Mediterranean subregion. Synonym of *Goldeus* Ribaut 1946b:85; Dlabola 1974c:107.
Mulsantina Anufriev 1970f:262 as subgenus of *Micantulina* Anufriev 1970d:262. *Typhlocyba stignatipennis* Mulsant & Rey 1855a:245 (France). Typhlocybinae, Dikraneurini. Palaearctic. Preoccupied by *Mulsantina* Weise 1906, see *Anufrievola* Kocak 1981a:32.
Muluana Dworakowska 1979c:310. *M. jarka* n.sp. (Zaire: Lubumbashi, "Elizabethville"). Typhlocybinae, Zyginellini. Ethiopian.
Mulungaella Linnavuori 1977a:45 as subgenus of *Uzelina* Melichar 1903b:181. *U. (M.) chinchonae* n.sp. (Zaire: Kivu, Mulungu). Penthimiinae. Ethiopian.
Murata Melichar 1926a:344. (Invalid) 1:396.
Murreeana Ramakrishnan & Menon 1971a:458. *M. duttai* n.sp. (Pakistan: Muree Hills). Typhlocybinae, Dikraneurini. Oriental.
Murzacha 10:2671. Error for *Mirzacha* Amyot.
Musbrnoia Dworakowska 1972b:395. *Aidola fumistriga* Melichar 1914b:143 (Java). Typhlocybinae, Erythroneurini. Oriental; Java, Vietnam.
Musgraviella Evans, J.W. 1966a:187. *M. tasmaniensis* n.sp. (Tasmania: Tasman Peninsula). Idiocerinae. Australian.
Musosa Linnavuori 1977a:11 & 52. *M. aeaea* n.sp. (Zaire: Musosa). Penthimiinae. Ethiopian; Zaire.
Myerslopella Evans, J.W. 1977a:100. *M. taylori* n.sp. (Australia: Queensland, Mt. Lewis, 1060 m). Ulopinae, Myerslopiini. Australian.

Myerslopia Evans, J.W. 1947c:143. *M. magna* n.sp. (New Zealand: Prov. Wellington, Waimarino). 5:95. Knight 1973b:986-1006. Ulopinae, Myerslopiini. Australian; Tasmania, New Zealand, New South Wales.

MYERSLOPIINI Evans, J.W. 1957a:367, tribe of ULOPINAE.

Myiltana Mahmood 1973b. Error for *Myittana* Distant 1908g:340.

Myittana Distant 1908g:340. *M. dohertyi* n.sp. (Burma: Tenasserim, Myitta). 10:1893. Deltocephalinae, Paralimnini. Oriental.

Myogonia Melichar 1926a:332. *Tettigonia limpida* Signoret 1855c:512 (Brazil: Bahia). 1:651. Cicadellinae, Proconiini. Neotropical. Synonym of *Teletusa* Distant 1908b:78; Young 1968a:162.

Myrmecophryne Kirkaldy 1906c:461. *M. formiceticola* n.sp. (Australia: Queensland, Bundaberg). 10:2354. Evans, J.W. 1966a:255-257. Xestocephalinae, Xestocephalini. Australian.
[NOTE: Linnavuori 1960b:32 lists *Myrmecophryne* Kirkaldy as a synonym of *Xestocephalus* Van Duzee 1892d, but does not repeat that placement in his 1979b:930 list of synonyms of *Xestocephalus*. The synonymy suggested by Linnavuori 1960b has not been accepted by subsequent authors].

Myrmecoscopus Evans, J.W. 1966a:79. *M. minutus* n.sp. (Western Australia: Cannington). Eurymelinae, Pogonoscopini. Australian.

Mysolis Kirkaldy 1904c:279, replacement for *Norsia* Walker 1870b:326, not *Norsia* Walker 1867. *Norsia flavidorsum* Walker 1870b:326 ("Mysol"). 15:182. Selenocephalinae, Selenocephalini. Oriental.
Synonyms:
Norsia Walker 1870b:326. *N. flavidorsum* n.sp.
Norsiana Distant 1908f:148. *Norsia flavidorsum* Walker 1870a:326.

Mysticana DeLong & Freytag 1976a:3 as subgenus of *Curtara* DeLong & Freytag 1972a:231. *Gypona mystica* Spangberg 1878a:71 (Mexico). Gyponinae, Gyponini. Neotropical.

Nabicerus Kwon 1985a:66 as subgenus of *Idiocerus* Lewis 1834a:47. *Idiocerus fuscescens* Anufriev 1971b:677 (USSR: Primorsky Dist.). Idiocerinae. Palaearctic.

Nacia Amyot 15:223. Invalid, not binominal.

Nacolus Jacobi 1914a:381. *N. gavialis* n.sp. (Formosa). 2:11. Hylicinae, Sudrini. Oriental; China, Formosa, India, Japan.
Synonym:
Ahenobarbus Distant 1918b:28. *A. assamensis* n.sp.

Naevus Knight 1970b:174. *Hishimonus dorsiplaga* Heller & Linnavuori 1968a:9 (Ethiopia: Illubabor, Gore, 2007 m). Deltocephalinae, Opsiini. Ethiopian; Congo, Ethiopia, Nigeria. Malagasian, Aldabra.

Nahuelbuta Linnavuori & DeLong 1977c:207 as subgenus of *Carelmapu* Linnavuori 1959b:220. *C. (N.) aureonitens* n.sp. (Chile: Chepu', 19 km E Chiloe Is.). Deltocephalinae, Athysanini. Neotropical; Chile.

Nakaharanus Ishihara 1953a:192. *Eutettix nakaharae* Matsumura 1914a190 (Japan). 10:504. Deltocephalinae, Platymetopiini. Palaearctic; Japan, Korea.

Nakula Distant 1918b:39. *N. multicolor* n.sp. (Lower Burma: Mergui). 15:180. Selenocephalinae. Oriental.

Naltaca Young 1977a:904. *N. eremica* n.sp. (Mexico: Puebla, 3 mi. NW Petlalcingo). Cicadellinae, Cicadellini. Neotropical.

Namiocerus Ghauri 1985a:41. *Bythoscopus cephalotes* Walker 1857b:174 (Sarawak). Idiocerinae. Oriental.

Namsangia Distant 1908g:259. *N. garialis* n.sp. (Assam: Margherita). 1:659. Cicadellinae. Oriental; India, Assam, China, Yunnam Prov.

Nanctasus Linnavuori 1959b:99 as subgenus of *Amplicephalus* DeLong 1926a:83. *A. (N.) bolivicus* n.sp. (Bolivia). Deltocephalinae, Deltocephalini. Neotropical.

Nandara Dworakowska 1984a:17. *N. flavescens* n.sp. (Malaysia: Penang I.). Typhlocybinae, Erythroneurini. Oriental.

Nandidrug Distant 1918b:74. *N. speciosum* n.sp. (S. India: Nandrug). 10:2346. Deltocephalinae, Stenometopiini. Oriental. Synonym of *Doratulina* Melichar 1903b:198; Vilbaste 1965a:10.

Nanipoides Evans, J.W. 1966a:47. *Ipoides maculosa* Evans, J.W. 1939a:48 (South Australia: Frome Downs Station). Eurymelinae, Ipoini. Australian.

Nannicerus Maldonado-Capriles 1977c:605. *N. gracilis* n.sp. (British Guiana: Demarara). Idiocerinae. Neotropical.

Nannogonalia Young 1977a:745. *Tettigonia circumcincta* Signoret 1855d:775 ("Brasil Nirm."). Cicadellinae, Cicadellini. Neotropical; Brazil.

Nanopsis Freytag 1974a:606. *Jassus verticus* Say (sic!) = *Jassus verticis* Say 1830b:308 (USA: Missouri). Macropsinae, Macropsini. Nearctic. Subgenus of *Pediopsoides* Matsumura 1912b:305; Hamilton, K.G.A. 1980b:897.
[NOTE: Hamilton also states that *verticis* Say, as interpreted by Freytag, is misidentified].

Nanosius Dlabola 1974c:122. *Deltocephalus chloroticus* Melichar 1896a:233 ("Carniola", Yugoslavia). Deltocephalinae, Deltocephalini. Palaearctic; southern Europe.

Naphotettix 15:223. Error for *Nephotettix* Matsumura.

Napo Linnavuori & DeLong 1976a:34. *N. brazosellus* n.sp. (Peru: Napo River). Deltocephalinae, Athysanini. Neotropical.

Napochia Lindberg 1956a:1209. Error for *Mapochia* Distant.

Naracia Dworakowska 1981b:25 as subgenus of *Acia* McAtee 1934a:113. *A. (N.) assamensis* n.sp. (India: Assam, Mariani, Meleng Forest). Typhlocybinae, Empoascini. Oriental.

Naratettix Matsumura 1931b:73. *Erythria zonata* Matsumura 1915a:159 (Japan). 17:234. Dworakowska 1980a. Typhlocybinae, Dikraneurini. Palaearctic; Japan, Korea, Soviet Maritime Territory.

Narecho Jacobi 1910b:127. *N. pallioviridis* n.sp. (Tanganyika). 7:19. Nirvaninae, Nirvanini. Ethiopian; East Africa and adjacent Congo, and southern Africa.
Synonym:
Parabolitus Naude 1926a:37. *P. anceps* n.sp.; Theron 1972a:207.

Narta Dworakowska 1979c:311. *N. waska* n.sp. (Zaire: Elisabethville). Typhlocybinae, Zyginellini. Ethiopian.

Nastoides 15:223. Error for *Nasutoides* Ball.

Nasutoideus Ball 1931d:219. *Platymetopius nasutus* Van Duzee 1907a:64 (Jamaica). 10:2264. Deltocephalinae, Scaphytopiini. Neotropical. Synonym of *Scaphytopius (Cloanthanus)* Ball 1931d:219.

Nataretus Theron 1980a:285. *Typhlocyba tricolor* Walker 1851b:905 (Natal). Deltocephalinae, Athysanini. Ethiopian; South Africa.

Naudeus Theron 1982b:23. *Deltocephalus bivittatus* Naude 1926a:44 (South Africa: Natal, Cedara). Deltocephalinae, Deltocephalini. Ethiopian.

Navaia Linnavuori 1960b:52. *N. deiphobe* n.sp. (Fiji: Viti Levu, Navai). Deltocephalinae. Oceania: Fiji.

Ndua Linnavuori 1978a:35. *N. barbata* n.sp. (Guinea: Nimba). Nioniinae, Magnentiini. Ethiopian.

Nedangia Nielson 1982e:232. *N. carinata* n.sp. (Ecuador: Rio Verde). Coelidiinae, Coelidiini. Neotropical.

Negosiana Oman 1949a:45. *Gypona negotiosa* Gibson 1919a:97 (USA: south-eastern states). 3:66. DeLong & Freytag 1972a:223. Gyponinae, Gyponini. Nearctic; Canada to Mexico, east of Continental Divide.

Nehela White 1878a:473. *N. vulturina* n.sp. (St. Helena). 14:143. Agalliinae, Nehelini. Ethiopian; Saint Helena.

NEHELINAE 14:3.

Neiva Melichar 1925a:368. *N. rufipes* n.sp. (Ecuador). 1:558. Young 1977a:30. Cicadellinae, Cicadellini. Neotropical; Bolivia, Peru, Ecuador.
Synonym:
Acharista Melichar 1925a:368. *A. variolata* n.sp.; Young 1977a:30.

Nelidina DeLong 1953a:129 as subgenus of *Coelidiana* Oman 1938b:397. *C. (N.) defila* n.sp. (Peru: San Ramon). 11:111. Kramer 1964d:272. Neocoelidiinae. Neotropical.

Neoaliturus Distant 1918b:63, replacement for *Aliturus* Distant 1908g:398, not *Aliturus* Fairmaire 1902. *Aliturus gardineri* Distant 1908g:398 (Laccadive Archipelago, Minikoi). 10:1893. Oman 1970a. Deltocephalinae, Opsiini. Oriental, Palaearctic.
Synonyms:
Bothrognathus Bergroth 1920a:29, replacement for *Aliturus* Distant 1908g:398.
Alituralis Merino 1934a:77, replacement for *Aliturus* Distant 1908g:398.
Distomotettix Ribaut 1938a:97. *Jassus fenestratus* Herrich-Schaeffer 1834a:5.

Neoarthaldeus Kwon & Lee 1979d:48. Invalid, *nomen nudum*.

Neobala Oman 1938b:396. *Thamnotettix pallidus* Osborn 1923c:67 (Bolivia: Santa Cruz) = *Neobala boliviensis* Metcalf 1955a:265, replacement for *Thamnotettix pallidus* Osborn, not *Thamnotettix karrooensis* var *pallidus* Cogan 1916. 10:809. Linnavuori 1959b:23-25, Kramer 1963b:208-209. Neobalinae. Neotropical; Bolivia, Brazil, Venezuela.

NEOBALINAE Linnavuori 1959b:17.

Neobassareus Kocak 1981a:32, replacement for *Bassareus* Linnavuori 1979b:999, not *Bassareus* Haldeman 1849. *Pseudobalbillus villiersi* Linnavuori 1969a:1147 (Congo: Odzala). Mukariinae. Ethiopian.

Neobufonaria Kocak 1981a:32, replacement for *Bufonaria* Emeljanov 1963a:1581, not *Bufonaria* Schumacher 1817. *Bufonaria oshanini* Emeljanov 1963b:1581 (USSR: Tyan Shan). Ulopinae, Ulopini. Palaearctic.

Neocaelidia 11:177. Error for *Neocoelidia* Gillette & Baker.

Neocoelidea 11:177. Error for *Neocoelidia* Gillette & Baker.

Neocoelidia Gillette & Baker 1895a:103. *N. tumidifrons* n.sp. (USA: Colorado, Fort Collins). 11:96. Kramer 1964d:262. Neocoelidiinae. Nearctic/Neotropical; Canada to Costa Rica.
Synonyms:
Paracoelidia Baker 1898h:292. *P. tuberculata* n.sp.
Stenocoelidia DeLong 1953a:104. *S. virgata* n.sp.
Eurycoelidia DeLong 1953a:112. *Neocoelidia pulchella* Ball 1909c:168.

Neocoelidiana DeLong 1953a:125. *Neocoelidia obscura* Baker 1898h:289 (USA: Arizona). 11:106, 107. Kramer 1964d:273. Neocoelidiinae. Neotropical/Nearctic; western USA to Panama.

NEOCOELIDIINAE 11:95.
NEOCOELIDIINI 11:95.

Neocrassana Linnavuori 1959b:286. *N. undata* n.sp. (Brazil: Rio Grande). Deltocephalinae, Athysanini. Neotropical; Brazil, Panama.

Neodartellus Evans, J.W. 1955a:32. *N. maculatus* n.sp. (Belgian Congo: Masombwe). 3:210. Linnavuori 1977a:14. Penthimiinae. Ethiopian.

Neodartus Melichar 1903b:162. *N. acocephaloides* n.sp. (Ceylon). 3:205. Linnavuori 1977a:13. Penthimiinae. Oriental.
Synonyms:
Neovulturnus Evans, J.W. 1937a:150. *Vulturnus vanduzeei* Kirkaldy 1907d:83.
Vulturnellus Evans, J.W. 1966a:220. *V. shephardi* n.sp.
[NOTE: See comments under *Neovulturnus* Evans, J.W.].

Neodeltocephalus Linnavuori 1959b:132. *N. asper* n.sp. (Argentina: Prov. Missiones, Loreto). Deltocephalinae, Deltocephalini. Neotropical; Argentina, Brazil.

Neodikrella Young 1952b:66. *Dikraneura (Hyloidea) disconotata* Osborn 1928a:275 (Brazil: Bahia). 17:283. Typhlocybinae, Dikraneurini. Neotropical; Brazil.

Neodonus DeLong & Hershberger 1948b:159. *N. piperatus* n.sp. (Mexico: Michoacan, Morelia). 10:1237. Deltocephalinae, Platymetopiini. Neotropical; Mexico.

Neodaratura Linnavuori 1956a:10. Invalid.

Neodrylix Emeljanov 1966a:103 as subgenus of *Limotettix* 1871a:224. "*Thamnotettix parallelus* V.D." (sic!) = *Athysanus parallelus* Van Duzee 1891a:169 (Canada: Ontario near South Falls on Muskoka River); Vilbaste 1973b:206-207. Deltocephalinae, Athysanini. Holarctic.

Neohecalus Linnavuori 1975a:14, replacement for *Hecalus* of North American authors, not Stal 1864b:65. *Glossocratus lineatus* Uhler 1877a:463 (USA: New Jersey). Deltocephalinae, Hecalini. Nearctic; eastern and central USA and adjacent Canada.

Neohegira Linnavuori & DeLong 1978c:205. *N. breviceps* n.sp. (Bolivia: Cochabamba, Chapare, Alto Palmar, 1100 m). Deltocephalinae, Deltocephalini. Neotropical.

Neojoruma Young 1952b:107. *Joruma adusta* McAtee 1924c:35 (Brazil: Chapada). 17:607. Typhlocybinae, Jorumini. Neotropical.

NEOKILLINI (sic) Metcalf 1965a:265. Error for NEOKOLLINI.

Neokolla Melichar 1926a:343. *Tettigonia hieroglyphica* Say 1830b:313 (USA: Arkansas). 1:269. Young 1977a:828, 849, 852. Cicadellinae, Cicadellini. Nearctic. Synonym of *Graphocephala* Van Duzee 1916a:66.

NEOKOLLINI Metcalf 1965a:265, tribe of TETTIGELLINAE = CICADELLINAE.

Neolimnus Linnavuori 1953a:114. *Scaphoideus aegyptiacus* Matsumura 1908a:29 (Egypt). 10:2156. Deltocephalinae, Athysanini. Palaearctic; Mediterranean & Siberian subregions, Sudan, SW Africa.

Neomacropsis Hamilton, KGA 1980b:911 as subgenus of *Macropsis* Lewis 1834a:49. *Pediopsis basalis* Van Duzee 1889a:171 (USA: Montana). Macropsinae, Macropsini. Nearctic.

Neomesus Linnavuori 1959b:233. *Paramesus obtusiceps* Berg 1884a:37 (Uruguay). Deltocephalinae, Athysanini. Neotropical; Argentina, Brazil, Chile, Uruguay.

Neometopia Schroeder 1959a:44. *Oncometopia fowleri* Distant 1908b:67 (Mexico: Teapa, Tabasco). Cicadellinae, Proconiini. Neotropical. Synonym of *Egidemia* China 1927d:283; Young 1968a:176.

Neonirvana Oman 1936c:116. *N. hyalina* n.sp. (Costa Rica: San Pedro de Montes de Oca). 7:20. Nirvaninae, Nirvanini. Neotropical; Brazil, Costa Rica, Panama.

Neopapyrina Kocak 1981a:33, replacement for *Papyrina* Emeljanov 1962a: 162, not *Papyrina* Moench 1853. *Papyrina viridis* Emeljanov 1962a:163 (USSR: Uzbekistan). Deltocephalinae, Athysanini. Palaearctic. Synonym of *Scenergates* Emeljanov 1972d:106; Kocak 1983a:70.

Neopenthimia Evans, J.W. 1972a:193. *N. pulchra* n.sp. (New Britain: Vunabakan, 10 km E of Keravat). Penthimiinae. Oriental.

Neopetalocephala 4:142. Error for *Neotituria* Kato.

Neophlepsius Dubovskij 1966a:173. *N. concinnus* n.sp. (USSR: Tadzikistan). Deltocephalinae, Athysanini. Palaearctic. Preoccupied by *Neophlepsius* Linnavuori 1955b:118, see *Elphnesopius* Nast 1984a:396.

Neophlepsius Linnavuori 1955b:118. *Phlepsius gracilis* Osborn 1923c:57 (Bolivia: Santa Cruz). 10:546. Linnavuori 1959b:191-194. Deltocephalinae, Athysanini. Neotropical; Argentina, Bolivia, Brazil, Paraguay.
Subgenus:
Nesolanus Linnavuori 1959b:194. *N. (N.) disonymos* n.sp.
[NOTE: Linnavuori 1959b:192 lists *Phlepsius gracilis* Osborn 1923c as a junior synonym of *Phlepsius multifarius* Berg 1884b:117, and (on pages 193-4) points out that he had previously misidentified *gracilis* Osborn, the type species of *Neophlepsius*].

Neophotettix 10:2672. Error for *Nephotettix* Matsumura.

Neopolana DeLong, Wolda & Estribi 1983a:473. Error for *Neoponana*.

Neoponana DeLong & Freytag 1967a:43 as subgenus of *Ponana* Ball 1920a:93. *P. (N.) demela* n.sp. (Peru: Monson Valley, Tingo Maria). Gyponinae, Gyponini. Neotropical.

NEOPSINAE Linnavuori in Linnavuori & DeLong 1977c:174.

Neopsis Oman 1938b:353. *Pediopsis elegans* Van Duzee 1907a:58 (Jamaica: Mandeville). 13:167. Linnavuori 1978c:18. Macropsinae, Neopsini. Neotropical; Bolivia, Brazil, Colombia, Ecuador, Jamaica.

Neoslossonia Van Duzee 1909a:218. *N. atra* n.sp. (USA: Florida, Sanford) = *Dorycephalus putnamae* Osborn 1907a:163. 9:64. Oman 1985a:317. Eupelicinae, Dorycephalini. Nearctic; Florida, Georgia.

Neosteles 10:2672. Error for *Nesosteles* Kirkaldy.

Neotartessus Evans, F. 1981a:126. *Tartessus flavipes* Spangberg 1878c:4 (Northern Australia). Tartessinae. Australian.

Neotharra Nielson 1975a:32. *N. ventrospiculata* n.sp. (New Guinea: Waris, S. of Hollandia). Coelidiinae, Tharrini. Oriental.

Neotitulia 4:142. Error for *Neotituria* Kato.

Neotituria Kato 1932b:220. *Ledropsis kongosana* Matsumura 1915a:173 (Korea). 4:59. Ledrinae, Petalocephalini. Oriental, Palaearctic.

Neovulturnus Evans, J.W. 1937a:150. *Vulturnus vanduzeei* Kirkaldy 1907d:83 (Australia: New South Wales). 3:205, 223. Penthimiinae, Australian.

[NOTE: The name is listed by Metcalf 1962b:205 and Evans, J.W. 1966a:213 as a junior synonym of *Neodartus* Melichar 1903b:162. However, Evans, J.W. 1977a:113 uses *Neovulturnus* as a valid generic name, at the same time relegating *Vulturnellus* Evans, J.W. to synonymy under *Neodartus*].

Nephoptettix 10:2672. Error for *Nephotettix* Matsumura.

Nephoptettixs 10:2672. Error for *Nephotettix* Matsumura.

Nephoris Jacobi 1912a:39. *N. chalybaea* n.sp. (Belgian Congo). 10:418. Deltocephalinae, Athysanini. Ethiopian; tropical Africa.

Nephotetix 10:2672. Error for *Nephotettix* Matsumura.

Nephotettis 10:2672. Error for *Nephotettix* Matsumura.

Nephotettix Matsumura 1902a:356. *Selenocephalus cincticeps* Uhler 1896a:292 (Japan). 10:423. Linnavuori 1956a, Ishihara 1964a, Ghauri 1968a & 1971a. Deltocephalinae, Athysanini. Southern Palaearctic, Ethiopian, Malagasian, Oriental, Austro-Oriental, Australian regions and Micronesia.
[NOTE: Hamilton, K.G.A. 1983b:498, implies probable establishment in the Nearctic region, SE USA].

Nephotetttix 10:2672. Error for *Nephotettix* Matsumura.

Nephrotettix 10:2672. Error for *Nephotettix* Matsumura.

Nereius Linnavuori 1959b:218 as subgenus of *Osbornellus* Ball 1932a:17. *Osbornellus bimarginatus* DeLong 1923b:261 (Puerto Rico). Deltocephalinae, Scaphoideini. Neotropical; West Indies.

Nerminia Kocak 1981b:124, replacement for *Tubulanus* Linnavuori 1955a:109, not *Tubulanus* Reinier 1804. *T. nitidus* Linnavuori 1955a:109 (Brazil: Bahia). Deltocephalinae, Athysanini. Neotropical. Synonym of *Atanus* Oman 1938b:381; Linnavuori & DeLong 1977c:209.

Neruotettix 10:2672. Error for *Nephotettix* Matsumura.

Nesaloha Oman 1943b:33. *N. cantonis* n.sp. (Canton Island). 10:1077. Deltocephalinae, Opsiini. Oceania. Synonym of *Orosius* Distant 1918b:85.

Nesocerus Freytag & Knight 1966a:82. *N. spurus* n.sp. (Madagascar Est, Sambava). Idiocerinae. Malagasian.

Nesoclutha Evans, J.W. 1947e:126. *N. obscura* n.sp. (Australia: Victoria). 10:2442. Deltocephalinae, Macrostelini. Australian; Australia, New Zealand, Lord Howe I., Norfolk I. Ethiopian; Cape Verde Is., South Africa. Palaearctic; Mediterranean subregion.
Synonym:
Irinula Ribaut 1948c:58. *Cicada erythrocephala* Ferrari 1882a:46; Vilbaste 1976a:28.

Nesolanus Linnavuori 1959b:194 as subgenus of *Neophlepsius* Linnavuori 1955b:118. *N. (N.) disonymos* n.sp. (Brazil: Nova Teutonia). Deltocephalinae, Athysanini. Neotropical.

Nesolina Osborn 1935b:60. *N. lineata* n.sp. (Hawaii: Oahu, Diamond Head). 10:2478. Deltocephalinae, Macrostelini. Hawaiian Islands.

Nesoniella 10:2672. Error for *Nesoriella* Osborn.

Nesophrosyne Kirkaldy 1907h:160. *Eutettix perkinsi* Kirkaldy 1904a:178 (Hawaii: Oahu). 10:1675, 1678. Linnavuori 1960a:320- 322. Deltocephalinae, Opsiini. Oceania; Hawaiian Islands.
Subgenus:
Nesoreias Kirkaldy 1910e:573. *N. (N.) insularis* n.sp.

Nesophryne Kirkaldy 1907h:160. *N. filicicola* n.sp. (Hawaii: Kauai). 10:1673. Deltocephalinae, Athysanini. Oceania; Hawaii.

Nesophyla Osborn 1934b:251. *N. picta* n.sp. (Marquesas Islands.: Nukuhiva, Ooumi). 10:2484. Deltocephalinae, Macrostelini. Oceania.

Nesophyrosyne 10:2673. Error for *Nesophrosyne* Kirkaldy.

Nesopteryx Matsumura 1931b:60. *N. arisana* n.sp. (Formosa). 17:26. Typhlocybinae, Alebrini. Oriental, Palaearctic. Synonym of *Alebra* Fieber 1872a:14.

Nesoreias Kirkaldy 1910e:573 as subgenus of *Nesophrosyne* Kirkaldy 1907h:160. 10:1692. *N. (N.) insularis* n.sp. (Hawaii: Kilauea). Deltocephalinae, Athysanini. Oceania; Hawaii.

Nesorias 10:2673. Error for *Nesoreias* Kirkaldy.

Nesoriella Osborn 1934b:248. *N. maculata* n.sp. (Marquesas Islands: Hivasa, Mount Temetiu). 10:1670. Deltocephalinae, Athysanini. Oceania.

Nesosteles Kirkaldy 1906c:343. *N. hebe* n.sp. (Australia: Queensland). 10:2442. Deltocephalinae, Balcluthini. Cosmopolitan. Synonym of *Balclutha* Kirkaldy 1900b:243.

Nesostelus 10:2673. Error for *Nesosteles* Kirkaldy.

Nesotettix Lindberg 1936a:6. *N. freyi* n.sp. (Canary Islands). 10:2461. Preoccupied by *Nesotettix* Holdhaus 1909, see *Lindbergana* Metcalf 1952a:229. Xestocephalinae, Xestocephalini. Ethiopian. Synonym of *Xestocephalus* Van Duzee 1892d:298; Linnavuori 1959b:36.

Nesotettix 17:1505. Error for *Nesopteryx* Matsumura.

Nesothamnus Linnavuori 1959b:244. *N. sanguineus* n.sp. (Surinam: Paramaribo). Deltocephalinae, Athysanini. Neotropical; Brazil, Panama, Surinam.

Neurotettix Matsumura 1914a:192. *N. horishanus* n.sp. (Formosa: Horisha). 10:879. Deltocephalinae, Athysanini. Oriental/Oceania; Formosa and Marquesas Islands.

Newmaniana Evans, J.W. 1941f:152 (1942). *N. viridis* n.sp. (Western Australia). 11:124. Evans, F. 1981a:118. Tartessinae. Australian.
Synonyms:
Euprora Evans, J.W. 1938b:9, not *Euprora* Busck 1906. *E. mullensis* n.sp.
Stenotartessus Evans, J.W. 1947a:207. *Euprora mullensis* Evans, J.W. 1938b:10.

Ngangula Dworakowska 1974a:202 as subgenus of *Ntotila* Dworakowska 1974a:199. *N. (N.) mochidai* n.sp. (Congo: Odzala). Typhlocybinae, Erythroneurini. Ethiopian; Congo.

Ngoma Dworakowska 1974a:206. *N. rubromaculata* n.sp. (Congo: Odzala). Typhlocybinae, Erythroneurini. Ethiopian; Congo.

Ngombela Dworakowska 1974a:193. *N. meinanderi* n.sp. (Congo: Odzala). Typhlocybinae, Erythroneurini. Ethiopian; Congo.

Ngunga Dworakowska 1974a:189. *N. soosi* n.sp. (Congo: Odzala). Typhlocybinae, Erythroneurini. Ethiopian; Congo.

Nicolaus Lindberg 1958a:176. *N. xerophilus* n.sp. (Cape Verde Islands: Sao Nicolau, Ponta Chao Grande). Deltocephalinae. Ethiopian; Cape Verde Is.

Nielsonia Young 1977a:794. *N. praestigia* n.sp. (Panama: Cerro Campana). Cicadellinae, Cicadellini. Neotropical; Costa Rica, Ecuador, Panama.

Nielsoniella Linnavuori 1977a:11 & 50. *N. vitellina* n.sp. (Nigeria: K. St., New Bussa). Penthimiinae. Ethiopian.

Niema Dworakowska 1979b:3 as subgenus of *Thaia* Ghauri 1962a:253. *T. (N.) enica* n.sp. (Vietnam: Prov. Nghe-An, Quy-Chau). Typhlocybinae, Erythroneurini. Oriental.

Nigridonus Oman 1949a:125. *Eutettix (Mesamia) illumina* Ball 1909a:80 (USA: Arizona). 10:1275. Deltocephalinae, Athysanini. Nearctic.

Nigritartessus Evans, F. 1981a:137. *N. henriettensis* n.sp. (Australia: Queensland). Tartessinae. Australian.

Nihilana DeLong & Freytag 1972f:295 as subgenus of *Polana* DeLong 1942d:110. *P. (N.) pressa* n.sp. (Honduras: La Ceiba). Gyponinae, Gyponini. Neotropical.

Nikkotettix Matsumura 1931b:76. *N. galloisi* n.sp. (Japan: Honshu, Chuzenji). 17:268. Typhlocybinae, Empoascini. Palaearctic; Japan, USSR Maritime Territory.

Nionia Ball 1915a:165. *Goniagnathus* palmeri Van Duzee 1891a:171 (USA: North Carolina, Balsam). 13:247. Linnavuori 1978a:33-34. Nioniinae, Nioniini. Nearctic, Neotropical.

NIONIINI 13:246.

Nionis 13:252. Error for *Nionia* Ball.

Nirvana Kirkaldy 1900d:293. *N. pseudommatos* n.sp. (Ceylon). 7:7. Nirvaninae, Nirvanini. Oriental, Paleotropical. Australian.

NIRVANIDAE 7:1.

NIRVANINI 7:5

Nirvanoides Baker 1923a:396. *N. amboinensis* n.sp. (Amboina). 7:22. Nirvaninae, Nirvanini. Oriental; Moluccas, New Guinea.

Niravna 7:33. Error for *Nirvana* Kirkaldy.

Nirtana 7:33. Error for *Nirvana* Kirkaldy.

Nisitra Walker 1870b:327. *N. telifera* n.sp. ("Mysol"). 11:85. Coelidiinae, Tharrini. Oriental. Preoccupied by *Nisitra* Walker 1869, see *Nisitrana* Metcalf 1952a:229.

Nisitrana Metcalf 1952a:229, replacement for *Nisitra* Walker 1870b:327, not *Nisitra* Walker 1869. *Nisitra telifera* Walker 1870b:328 ("Mysol"). 11:85. Coelidiinae, Tharrini. Oriental. Synonym of *Tharra* Kirkaldy 1906c:324; Nielson 1975a:34.

Nkaanga Dworakowska 1974a:206. *N. lobata* n.sp. (Congo: Sibiti). Typhlocybinae, Typhlocybini. Ethiopian.

Nkasa Dworakowska 1974a:187 as subgenus of *Gambialoa* Dworakowska 1972f:853. *G. (N.) ahmedi* n.sp. (Congo: Dimonika, Mayumbe). Typhlocybinae, Erythroneurini. Ethiopian.

Nkima Dworakowska 1976b:28 as subgenus of *Optya* Dworakowska 1974a:141. *O. (N.) manua* n.sp. (Mt. Cameroon, Mannsguell top, 2250 m). Typhlocybinae, Empoascini. Ethiopian.

Nkonba Dworakowska 1974a:202. *N. youngi* n.sp. (Congo: Sibiti). Typhlocybinae, Erythroneurini. Ethiopian.

Nkumba Dworakowska 1974a:208. *N. wumba* n.sp. (Congo: Odzala). Typhlocybinae, Typhlocybini. Ethiopian.

Nlunga Dworakowska 1974a:191. *N. reeneni* n.sp. (Congo: Dimonika, Mayumbe). Typhlocybinae, Erythroneurini. Ethiopian. Subgenus of *Thaia* Ghauri 1962a:253; Dworakowska 1976b:50.

Nollia Hamilton, KGA 1983d:19. *Neopsis pallidus* Linnavuori & DeLong 1977c:175 (Chile: Chiloe Is.). Macropsinae, Neopsini. Neotropical.

Noritonus Nielson 1979b:164. *N. multus* n.sp. (Bolivia: Trinidad). Coelidiinae, Teruliini. Neotropical; Argentina, Bolivia, Brazil, Paraguay.

Norsia Walker 1870b:326. *N. flavidorsum* n.sp. ("Mysol"). 15:182. Preoccupied by *Norsia* Walker 1867, see *Mysolis* Kirkaldy 1904c:279. Selenocephalinae, Selenocephalini. Oriental.

Norsiana Distant 1908f:148, replacement for *Norsia* Walker 1870b:326, not *Norsia* Walker 1867. *Norsia flavidorsum* Walker 1870b:326 ("Mysol"). 15:182. Selenocephalinae, Selenocephalini. Oriental. Synonym of *Mysolis* Kirkaldy 1904c:279.

Nortoides Evans, J.W. 1972a:192. *N. bilobata* n.sp. (NE New Guinea: Kassam, 1350 m, 48 km E of Kainantu). Penthimiinae. Oriental.

Norva Emeljanov 1969a:1100. *N. anufrievi* n.sp. (USSR: Maritime Territory). Deltocephalinae, Platymetopiini. Palaearctic; Japan, Korea, USSR Maritime Territory.

Norvellina Ball 1931a:2. *Eutettix mildredae* Ball 1901d:45 (USA: Colorado, Manitou). 10:505. Deltocephalinae, Platymetopiini. Nearctic; southern Canada to Mexico.

Notocephalius Jacobi 1909a:339. *N. hartmeyeri* n.sp. (Western Australia). 5:47. Evans, J.W. 1977a:99. Ulopinae, Cephalelini. Australia and Tasmania.
Synonym:
Anacephaleus Evans, J.W. 1936a:43. *A. minutus* n.sp.

Notus Fieber 1866a:508. *Cicada flavipennis* Zetterstedt 1828a:525 (Lapland). 17:243. Typhlocybinae, Dikraneurini. Holarctic.
Synonym:
Curta DeLong & Caldwell 1937a:30. *Dikraneura (Curta) alta* n.sp.

Novolopa Evans, J.W. 1966a:88. *N. townsendi* n.sp. (New Zealand: South Island, Mt. Owen, 5000 ft). Ulopinae, Ulopini. Australian; New Zealand.

Novothymbris Evans, J.W. 1941c:162. *Diedrocephala zealandica* Myers 1023a:409 (New Zealand: Nelson, Dun Mtn.). 4:124. Ledrinae, Thymbrini. Australian; New Zealand, Chatham Islands.

Nsanga Dworakowska 1974a:189. *N. mahmoodi* n.sp. (Congo: Dimonika, Mayumbe). Typhlocybinae, Erythroneurini. Ethiopian.

Nsesa Dworakowska 1974a:233. *N. matteba* n.sp. (Congo: Dimonika, Mayumbe). Typhlocybinae, Erythroneurini. Ethiopian, Congo, Zaire.

Nsimbala Dworakowska 1974a:213. *N. tukula* n.sp. (Congo: Odzala). Typhlocybinae, Erythroneurini. Ethiopian; Congo, Zaire.

Ntanga Dworakowska 1974a:196. *N. ghaurii* n.sp. (Congo: Odzala). Typhlocybinae, Erythroneurini. Ethiopian.

Ntkkotettix 17:1505. Error for *Nikkotettix* Matsumura.

Ntotila Dworakowska 1974a:199. *N. ishiharai* n.sp. (Congo: Dimonika, Mayumbe). Typhlocybinae, Erythroneurini. Ethiopian.
Subgenus:
Ngangula Dworakowska 1974a:202. *N. (N.) mochidai* n.sp.

Nubelella Evans, J.W. 1972a:194. *N. leopardina* n.sp. (NE New Guinea: Mt. Kaindi, 2350 ft). Penthimiinae. Oriental.

Nubelloides Evans, J.W. 1972a:181. *N. albomaculata* n.sp. (NW New Guinea: Waris). Penthimiinae. Oriental.

Nudulidia Nielson 1982e:276. *N. multidentata* n.sp. (Ecuador: Huigra). Coelidiinae, Coelidiini. Neotropical.

Nulapona DeLong & Freytag 1975c:562 as subgenus of *Hecalapona* DeLong & Freytag 1975c:547. *H. (N.) guina* n.sp. (Peru: Sinchona). Gyponinae, Gyponini. Neotropical.

Nullamia DeLong 1970a:118. *N. fuegoensis* n.sp. (Chile: Magallanes Prov., Payne). Deltocephalinae, Deltocephalini. Neotropical; southern Chile.

Nullana DeLong 1976f:26. *N. huallaga* n.sp. (Peru: Huallaga Valley, Huanuco). Gyponinae, Gyponini. Neotropical; Brazil, Ecuador, "Guyana" (French Guiana?), Panama, Peru.

Nurenus Oman 1949a:127. *Idiodonus snowi* Ball 1937b:26 (USA: Arizona, Pinal Mts.). 10:1328. Deltocephalinae, Platymetopiini. Nearctic.

Nyndgama Webb, M.D. 1983a:81. *N. arowa* n.sp. (Australia: Queensland National Park). Idiocerinae. Australian.

Nzinga Dworakowska 1974a:197. *N. delongi* n.sp. (Congo: Dimonika, Mayumbe). Typhlocybinae, Erythroneurini. Ethiopian.

Obothrogonia Yang in Yang & Li 1980a:194 as subgenus of *Bothrogonia* Melichar (China) 1926a:341. *B. (O.) sinica* Yang & Li n.sp. (Zhejiang: Tienmushan, 920 m). Cicadellinae, Cicadellini. Palaearctic.

Obtusama Teson 1972a:59. Error for *Obtusana* DeLong & Freytag 1964a.

Obtusana DeLong & Freytag 1964a:29 as subgenus of *Gypona* Germar 1821a:73. *Gypona melanota* Spangberg 1878a:19 (eastern USA). Gyponinae, Gyponini. Neotropical, Nearctic.

Occinirvana Evans, J.W. 1941f:156. *O. eborea* n.sp. (Western Australia: Perth). 7:24. Evans, J.W. 1966a:151. Nirvaninae, Occinirvanini. Australian.

OCCINIRVANINI Evans, J.W. 1966a:151, tribe of NIRVANINAE.

Occiplanocephalus Evans, J.W. 1941f:146. *O. ravus* n.sp. (Western Australia: Dedari). 10:1891. Evans, J.W. 1966a:223. Deltocephalinae, Deltocephalini. Australian.

Oceanopona Linnavuori 1960a:299. *O. croceipennis* n.sp. (Ponape: Mt. Temwetemwensekir, 180 m). Paraboloponinae. Oceania; Eastern Caroline Islands, Ponape, Kusaie.

Ochrostacta Stal 1869a:61. *Tettigonia diadema* Burmeister 1835a:120 (Uruguay). 1:654. Young 1968a:145. Cicadellinae, Proconiini. Neotropical; Argentina, Brazil, Paraguay, Uruguay.
Synonym:
Andamarca Melichar 1926a:336. *Tettigonia physocephala* Signoret 1854d:720.

Ochrostaeta 1:684. Error for *Ochrostacta* Stal.

Ochta Emeljanov 1977a:152 as subgenus of *Matsumurella* Ishihara 1953a:200. *M. (O.) minor* n.sp. (Mongolia). Deltocephalinae, Athysanini. Palaearctic.

Ociepa Dworakowska 1977e:850 as subgenus of *Empoasca* Walsh 1862a:149. *E. (O.) medleri* n.sp. (Nigeria: NW State, Badeggi). Typhlocybinae, Empoascini. Ethiopian.

Oclasma Melichar 1905a:293. *O. degenerata* n.sp. (Tanganyika). 5:89. Ulopinae, Ulopini. Ethiopian. Synonym of *Coloborrhis* Germar 1836a:72.

Oconura 10:2673. Error for *Aconura* Lethierry.

Odmiella Linnavuori 1978d:464. *Stenomiella falcata* Linnavuori 1969a:1180-1181 (Congo: Odzala). Paraboloponinae. Ethiopian; Guinean.

Odomas Jacobi 1912a:32. *O. myops* n.sp. (Uganda). 5:37. Linnavuori 1972a, 1973a. Megophthalminae. Ethiopian; mountainous areas of eastern Africa, the Congo and Cameroons.

Odzalana Linnavuori 1969a:1154. *O. villiersi* n.sp. (Congo: Odzala). Deltocephalinae, Hecalini. Ethiopian; Congo.

Oeogonalia Young 1977a:220. *Tettigonia fossulata* Signoret 1855b:237 (Colombia). Cicadellinae, Cicadellini. Neotropical; Colombia, Brazil.

Ohausia Schmidt 1911b:299. *O. nigra* n.sp. (Ecuador: Loji). 3:125. Kramer 1966a:471,490. Ledrinae. Neotropical.

Okaundua Linnavuori 1969a:1177. *O. consita* n.sp. (SW Africa: Okaundua, near Okahandja). Deltocephalinae, Athysanini. Ethiopian; Ethiopia, SW Africa.

Okubasca Dworakowska 1982b:53 as subgenus of *Empoasca* Walsh 1862a:149. *Empoasca okubella* Matsumura 1931b:81 (Japan: Honshu, Okubo near Tokyo). Typhlocybinae, Empoascini. Palaearctic.

Ollarianus Ball 1936c:59. *Eutettix balli* Van Duzee 1907a:68 (Jamaica). 10:498. Deltocephalinae, Athysanini. Nearctic; Sonoran. Neotropical; Mexican, Antillean.

Olokemeja Ghauri 1975b:485. *O. kilima* n.sp. (Nigeria: Moor Plantation). Typhlocybinae, Empoascini. Ethiopian.

Olszewskia Dworakowska 1974a:242. *O. stodruga* n.sp. (Congo: Odzala). Typhlocybinae, Typhlocybini. Ethiopian.

Omagua Melichar 1924a:200. *Tettigonia fitchii* Signoret 1855b:230 ("Cayenne"). 1:522. Young 1968a:156. Cicadellinae, Proconiini. Neotropical; Brazil, British Guiana, French Guiana.

Omanana DeLong 1942a:293. *Athysanus litigiosus* Ball 1901d:51 (Mexico: Cuernavaca ?). 10:535. Deltocephalinae, Athysanini. Neotropical; Mexican subregion.

Omanella Merino 1936a:358. *O. barberi* n.sp. (Philippines: Mindanao, Zamboanga, Basilan). 15:170. Selenocephalinae, Selenocephalini. Oriental.

Omanesia Thapa 1983a:173. *O. reclinata* n.sp. (Nepal: Kathmandu, Bhandarkhal). Typhlocybinae, Eupterygini. Oriental.

Omaniella Ishihara 1953a:196. *O. flavopicta* n.sp. (Japan: Sapporo). 10:163. Deltocephalinae, Athysanini. Palaearctic. Synonym of *Ophiola* Edwards 1922a:206.

Omanolidia Nielson 1982e:294. *O. omani* n.sp. (Mexico: Guerrero, 3 km S., 1.7 km E. of Chilpancingo). Coelidiinae, Coelidiini. Neotropical; Colombia, El Salvador, Guatemala, Mexico.

Omaranus Distant 1918b:5. *O. typicus* n.sp. (India: Calcutta). 7:19. Nirvaninae, Nirvanini. Oriental.
Synonym:
Didius Distant 1918b:36. *D. sexualis* n.sp.

Omarus 7:33. Error for *Omaranus* Distant.

Omegalebra Young 1957c:199. *Protalebra vexillifera* Baker 1899b:404 (Brazil: Chapada). Typhlocybinae, Alebrini. Neotropical; Brazil, Argentina, Costa Rica, Panama, West Indies.

Omiya Dworakowska 1981b:40. *O. secunda* n.sp. (Nigeria: W. State, Ife-Ife). Typhlocybinae, Empoascini. Ethiopian; Nigeria. Oriental; Vietnam.

Omskius Linnavuori 1954g:183 as subgenus of *Arocephalus* Ribaut 1946b:85. *Deltocephalus sibiricus* Horvath 1901a:272 (Siberia). 10:1454. Emeljanov 1962a:181. Deltocephalinae, Paralimnini. Palaearctic. Synonym of *Mongolojassus* Zachvatkin 1953d:247.

Onblavia Nielson 1979b:309. *O. flavocapitata* n.sp. (Venezuela: San Esteban). Coelidiinae, Teruliini. Neotropical.

Onchopsis 13:252. Error for *Oncopsis* Burmeister.

Oncometapia 1:684. Error for *Oncometopia* Stal.

Oncometopia Stal 1869a:62, replacement for *Proconia* Amyot & Serville 1843, not *Proconia* Le Peletier & Serville 1825. *Cicada undata* Fabricius 1794a:32 ("Carolina"), preoccupied = *Cicada orbona* Fabricius 1798a:520. 1:559,566. Young 1968a:220. Cicadellinae, Proconiini. Neotropical; Mexican & Brazilian subregions. Nearctic; east, central and southern USA.
Subgenus:
Similitopia Schroder 1959a:45. *Oncometopia fuscipennis* Fowler 1899b:230.
Synonym:
Lebora China 1927d:283. *Cicada orbona* Fabricius 1798a:520.

ONCOPSIDAE 13:6.

Oncopis 13:252. Error for *Oncopsis* Burmeister.

Oncopsis Burmeister 1838a:10 as subgenus of *Bythoscopus* Germar 1833a:180. *Cicada flavicollis* Linnaeus 1761a:242 (Sweden). 13:168. Hamilton, KGA 1980b:887. Macropsinae, Macropsini. Holarctic, Oriental.
Subgenus:
Parasitades Singh-Pruthi 1936a:106. *P. baileyi* n.sp.
Synonym:
Zinneca Amyot & Serville 1843a:579. *Z. flavidorsum* n.sp.

Onega Distant 1908a:528. *O. avella* n.sp. (Ecuador: Baiza). 1:264. Young 1977a:285. Cicadellinae, Cicadellini. Neotropical; Bolivia, Colombia, Ecuador, Peru.

Oneratulus Vilbaste 1975a:233. *"Jassus?* (sic!) *curtulus* Motschulsky" 1863a:98 (Ceylon: Patannas Mount). Deltocephalinae. Oriental.

Oniella Matsumura 1912a:46. *O. leucocephala* n.sp. (Japan). 7:14. Cicadellinae, Evacanthini. Palaearctic; Japan, Kurile Islands. Oriental; China.

Oniroxis China 1925c:483. *O. kariana* n.sp. (Yunnan, China). 8:225. Cicadellinae, Anoterostemmini. Oriental; China, Yunnan Prov.

Onocopsis 13:252. Error for *Oncopsis* Burmeister.

Onukia Matsumura 1912a:44. *O. onukii* n.sp. (Japan). 6:48. Cicadellinae, Evacanthini. Oriental; China. Palaearctic; Manchurian subregion.
Synonym:
Apphia Distant 1918b:4. *A. burmanica* n.sp.

Onukiades Ishihara 1963b:3. *Euacanthus formosanus* Matsumura 1912a:40 (Formosa: Hoppo). Cicadellinae, Evacanthini. Oriental; Formosa.

Onukigallia Ishihara 1955a:215, 217. *Agallia onukii* Matsumura 1912b: 315 (Japan). 14:61. Agalliinae, Agalliini. Palaearctic; Japan, USSR Maritime Territory.

Onura Oman 1938b:387. *O. eburneola* n.sp. (Argentina: Missiones, Loreto). 10:839. Linnavuori 1959b:238. Deltocephalinae, Athysanini. Neotropical; Argentina, Brazil, Paraguay.

Ootacamundus Distant 1918b:57. *O. typicus* n.sp. (S. India: Ootacamund). 10:2353. Xestocephalinae, Xestocephalini. Oriental.

Opamata Dworakowska 1971g:652. *O. kweitniowa* n.sp. (Vietnam: Bao-Ha, S of Fan-Si-Pan). Typhlocybinae, Typhlocybini. Oriental; Thailand, Vietnam.

Ophiala 10:2673. Error for *Ophiola* Edwards.

Ophiola Edwards 1922a:206. *Cicada striatula* Fallen 1806a:31 (Sweden), homonym of *Cicada striatula* Fabricius 1794 = *Iassus corniculus* Marshall 1866l:119. 10:162. Ossiannilsson 1983a:752. Deltocephalinae, Athysanini. Palaearctic.
Synonyms:
Ophiolix Ribaut 1942a:264. *Thamnotettix paludosus* Boheman 1854a:34.
Omaniella Ishihara 1953a:196. *O. flavopicta* n.sp.

Ophiolix Ribaut 1942a:264 as subgenus of *Limotettix* Sahlberg 1871a:31. *Thamnotettix paludosus* Boheman 1845a:34, 1845b:158 (Sweden). 10:160. Ossiannilsson 1983a:753. Deltocephalinae, Athysanini. Palaearctic. Synonym of *Ophiola* Edwards 1922a206; Ossiannilsson 1983a:753.

Ophionotum Emeljanov 1964b:56. *Phlepsius pulcher* Melichar 1898a:65 (Turkestan). Deltocephalinae, Athysanini. Palaearctic.

Ophiuchus Distant 1918b:33. *O. princeps* n.sp. (India: Travancore). 7:21. Nirvaninae, Nirvanini. Oriental; India. (Australian?).

Ophuchus 7:33. Error for *Ophiuchus* Distant.

Opio Evans, J.W. 1934a:151, 165. *Bythoscopus multistriga* Walker 1858a:105 (Australia: New South Wales). 12:38. Eurymelinae, Ipoini. Australian.

OPIOINAE 12:28.

Opostigmus Amyot. 10:2673. Invalid, not binominal.

Opsianus Linnavuori 1960a:316. *Euscelis picturatus* Metcalf 1946d:123 (Guam). Deltocephalinae, Opsiini. Oceania; Micronesia.

OPSIINI Emeljanov 1962b:389, tribe of DELTOCEPHALINAE.

Opsius Fieber 1866a:505. *O. stactogalus* n.sp. (France). 10:448. Deltocephalinae, Opsiini. Palaearctic, (now cosmopolitan on *Tamarix*).

Optocerus Freytag 1969a:348. *O. dicranus* n.sp. (Panama: Davien Prov., Santa Fe). Idiocerinae. Neotropical; Mexico, Panama.

Optya Dworakowska 1974a:141. *O. odzalae* n.sp. (Congo: Odzala). Typhlocybinae, Empoascini. Ethiopian; Afrotropical.
Subgenus:
Nkima Dworakowska 1976b:28. *O. (N.) manua* n.sp.

Oragua Melichar 1926a:344. *Tettigonia nebulosa* Signoret 1854b:343 (Brazil: "Fernambuco"). 1:294. Young 1977a:612. Cicadellinae, Cicadellini. Neotropical; southern Mexico to southern Brazil and Argentina.

Orechona Melichar 1926a:345. *O. superba* n.sp. (Brazil). 1:27. Young 1977a:252. Cicadellinae, Cicadellini. Neotropical; Brazil, Ecuador.

Orectogonia Melichar 1926a:345. *Tettigonia sparsuta* Signoret 1855c:508 (Venezuela). 1:417. Young 1968a:257. Cicadellinae, Proconiini. Neotropical. Synonym of *Acrogonia* Stal 1869a:61; Young 1968a:257.

Orientalebra Dworakowska 1971c:497. *O. gembata* n.sp. (Vietnam: Bao-Ha, S. of Fan-Si-Pan). Typhlocybinae, Alebrini. Oriental; Malaysia, Vietnam.

Orientus DeLong 1938b:217. *Phlepsius ishidae* Matsumura 1902a:382 (Japan). 10:707. Deltocephalinae, Athysanini. Palaearctic; Manchurian subregion. [Nearctic].

Oriotura 10:2674. Error for *Driotura* Osborn & Ball.

Orocastus Oman 1949a:161. *Deltocephalus perpusillus* Ball & DeLong 1926a:214 (USA: Montana, Sydney). 10:1390. Deltocephalinae, Deltocephalini. Nearctic; northern Rocky Mountain subregion.

Orolis 10:2674. Error for *Orolix* Ribaut.

Orolix Ribaut 1942a:267. *Cicada cruentata* Panzer 1799a:15 (Austria). 10:1284. Deltocephalinae, Athysanini. Palaearctic. Synonym of *Idiodonus* Ball 1936c:57.

Orosius Distant 1918b:85. *O. albicinctus* n.sp. (S. India: Kodiakanal). 10:1670. Ghauri 1966a, Linnavuori 1960a:320. Deltocephalinae, Opsiini. Oriental, Australian, Oceania.
Synonym:
Nesaloha Oman 1943b:33. *N. cantonis* n.sp.; Ghauri 1966a:18.

Orsalebra Young 1952b:23. *O. robusta* n.sp. (Ecuador: Bolivar, Hacienda Talahua). 17:20. Typhlocybinae, Alebrini. Neotropical; Chile, Colombia, Ecuador.

Ortega Melichar 1924a:199. *Tettigonia truncatipennis* Signoret 1854d: 717 (Colombia). 1:604. Young 1977a:40. Cicadellinae, Cicadellini. Neotropical.

Orthojassus Jacobi 1914a:382. *O. philagroides* n.sp. (Formosa). 11:30. Coelidiinae, Thagriini. Oriental. Synonym of *Thagria* Melichar 1903b:176, Nielson 1977a:9.

Orucyba Ghauri 1975b:491. *O. mollis* n.sp. (Nigeria: Ibadan, Moor Plantation). Typhlocybinae, Zyginellini. Ethiopian. Synonym of *Tataka* Dworakowska 1974a:174; Dworakowska 1979c:306.

Osbornellus Ball 1932a:17. *Scaphoideus auronitens* Provancher 1889a:276 (Canada: Quebec). 10:2177. Linnavuori 1959b:209-219. Deltocephalinae, Scaphoideini. Nearctic, Neotropical, Palaearctic.
Subgenera:
Sorbonellus Linnavuori 1959b:210. *Osbornellus infuscatus* Linnavuori 1955a:102.
Nereius Linnavuori 1959b:218. *Scaphoideus bimarginatus* DeLong 1923b:261.
Mavromoustaca Dlabola 1967e:37. *O. (M.) consanguineus* n.sp.

Osbornitettix Metcalf 1952a:229, replacement for *Calotettix* Osborn 1934b:247, not *Calotettix* Bruner 1908. *Calotettix metrosideri* Osborn 1934b:247 (Marquesas Islands). 10:2339. Paraboloponinae. Oceania; Marquesas Islands. Synonym of *Dryadomorpha* Kirkaldy 1906c:335; Webb, M.D. 1981b:49.

Osbornulus Young 1957c:164. *Dikraneura quadrifasciata* Osborn 1928a:272 (Bolivia: Prov. del Sara). Typhlocybinae, Alebrini. Neotropical; Bolivia, Brazil.

Osella Evans, J.W. 1972a:191. *O. anomala* n.sp. (NE New Guinea: Torricelli Mts., 750 m). Penthimiinae. Oriental.

Ossana Distant 1914m:518. *O. bicolor* n.sp. (Nigeria: Lagoo, 70 miles E., near Oni Clearg). 15:117. Iassinae, Iassini. Ethiopian. Synonym of *Batracomorphus* Lewis 1834a:51; Linnavuori & Quartau 1975a:30.

Ossiannilssonia Young & Christian 1952a:97 in Young 1952a. *Typhlocyba berenice* McAtee 1926a:35 (USA: District of Columbia). 17:954. Preoccupied by *Ossiannilssonia* Lambers 1952, see *Ossiannilssonola* Christian 1953a:1107,1132. Typhlocybinae, Typhlocybini. Nearctic.

Ossiannilssonola Christian 1953a:1107, replacement for *Ossiannilssonia* Young & Christian 1952a:97, not *Ossiannilssonia* Lambers 1952. *Typhlocyba berenice* McAtee 1926a:35. (USA: District of Columbia). 17:954. Typhlocybinae, Typhlocybini. Nearctic; widespread. Palaearctic; European subregion.

Ossuaria Dworakowska 1979b:18. *O. agara* n.sp. (Vietnam: Nghe-An, Quy-Chau). Typhlocybinae, Erythroneurini. Oriental.

Otbatara Dworakowska 1984a:16. *O. parvula* n.sp. (Malaysia: Kuala Trangganu). Typhlocybinae, Erythroneurini. Oriental.

Oxoometopia 1:685. Error for *Oncometopia* Stal.

Oxycoryphia Melichar 1926a:344. *Tettigonia spatulata* Signoret 1854d: 722 (Colombia). 1:414. Cicadellinae, Cicadellini. Neotropical. Synonym of *Platygonia* Melichar 1925a:340; Young 1977a:420.

Oxygonalia Evans, J.W. 1947a:163, replacement for *Oxygonia* Melichar 1926a:344, not *Oxygonia* Mannerheim 1837. *Ciccus ignifer* Walker 1851b:804 (Colombia). 1:412. Young 1977a:420. Synonym of *Platygonia* Melichar 1925a:340; Young 1977a:420. Cicadellinae, Cicadellini. Neotropical.

Oxygonia Melichar 1926a:344. *Ciccus ignifer* Walker 1851b:804 (Colombia). 1:412. Preoccupied by *Oxygonia* Mannerheim 1837, see *Oxygonalia* Evans, J.W. 1947a. Synonym of *Platygonia* Melichar 1925a:340; Young 1977a:420.

Oxytettigella Metcalf 1952a:229, replacement for *Oxytettix* Ribaut 1942a:263, not *Oxytettix* Rehn 1929. *Jassus (Thamnotettix) viridinervis* Kirschbaum 1868b:97 (Italy). 10:335. Deltocephalinae, Athysanini. Palaearctic; Mediterranean subregion.

Oxytettix Ribaut 1942a:263. *Jassus (Thamnotettix) viridinervis* Kirschbaum 1868b:97 (Italy). 10:335. Preoccupied by *Oxytettix* Rehn 1929, see *Oxytettigella* Metcalf 1952a:229.

Ozias Jacobi 1912a:40. *O. pedestris* n.sp. (Belgian Congo). 5:79. Linnavuori 1972a:148. Agallinae. Ethiopian, tropical Africa. Pacechia Amyot 13:252. Invalid, not binomial.

Pachitea Melichar 1926a:344. *Tettigoniella habenula* Jacobi 1905c:183 (Peru). 1:411. Young 1977a:328. Cicadellinae, Cicadellini. Neotropical; Bolivia, Colombia, Ecuador, Peru, Venezuela.

Pachodus Linnavuori 1961a:474. *P. filigranus* n.sp. (S. Rhodesia; Victoria Falls). Deltocephalinae, Athysanini. Ethiopian.

Pachyledra Schumacher 1912b:248. *P. kamerunensis* n.sp. (Cameroons). 4:83. Ledrinae, Petalocephalini. Ethiopian. Synonym of *Petalocephala* Stal 1854b:251; Linnavuori 1972b:214.

Pachymetopius Matsumura 1914a:214. *P. decoratus* n.sp. (Formosa). 16:232. Maldonado-Capriles 1975a:306. Coelidiinae, unassigned. Oriental.

Pachynopsis 15:224. Error for *Pachyopsis* Uhler.

Pachynus Stal 1866a:127. *Bythoscopus (Oncopsis) bimaculicollis* Stal 1855a:100 (South Africa, Natal). 14:146. Preoccupied by *Pachynus* Rafinesque 1815, see *Igerna* Kirkaldy 1903f:13.

Pachyopsis Uhler 1877a:466. *P. laetus* n.sp. (USA: Colorado). 15:93. Blocker 1979a:11. Iassinae, Iassini. Nearctic, Neotropical; Western USA to Venezuela.
Synonym:
Straganiopsis Baker 1903e:10. *Macropsis idioceroides* Baker 1900a:55.

Pachytettix Linnavuori 1959b:244. *P. sagittarius* n.sp. (Peru: Madre de Dios). Deltocephalinae, Athysanini. Neotropical; Bolivia, Peru.

Pactana Linnavuori 1960a:269. *P. elegantula* n.sp. (Carolina Islands: Yap, Mt. Madaade). Nirvaninae, Nirvanini. Micronesia; Caroline Islands.

Paganalia Distant 1917a:314. *P. virescens* n.sp. (Seychelles Islands). 10:2175. Paraboloponinae. Malagasian. Synonym of *Dryadomorpha* Kirkaldy 1906c:335; Webb, M.D. 1981b:49.

Pagaronia Ball 1902a:19. *P. tredecimpunctata* n.sp. (USA: California). 8:13. Anufriev 1970c, 1971d, 1978a; Oman & Musgrave 1975a:4; Kwon & Lee 1978a; Kwon 1983a. Cicadellinae, Pagaroniini. Palaearctic; Manchurian subregion. Nearctic; USA: California, Oregon.
Subgenus:
Parapagaronia Kwon & Lee 1978a:8. *P. jenjouristi* Anufriev 1970g:556.
Synonyms:
Matuta Emeljanov 1966a:99. *Tettigonia guttiger* Uhler 1896a:294.
Kalkandelenia Kocak 1981a:32. *Tettigonia guttiger* Uhler 1896a:294.

PAGARONIINI Anufriev 1978a:58; tribe of CICADELLINAE.

Paivanana Distant 1918b:95. *P. indra* n.sp. (South India: Kodiakanal). 10:1672. Deltocephalinae, Stenometopiini. Oriental. Synonym of *Doratulina* Melichar 1903b:198; Vilbaste 1965a:10.

Pakeasta Ahmed, Manzoor 1971b:188. *P. notata* n.sp. (East Pakistan: Dacca). Typhlocybinae, Erythroneurini. Oriental. Synonym of *Diomma* de Motschulsky 1863a:102; Dworakowska 1981c:367.

Palicus Stal 1866a:120. *Coelidia lineoligera* Stal 1855a:98 (Natal). 11:17,178. Theron 1984c:313. Coelidiinae. Ethiopian. Preoccupied by *Palicus* Philippi 1838. Synonym of *Aletta* Metcalf 1952a:229.

Palingonalia Young 1977a:427. *Tettigonia bigutta* Signoret 1854a:6 (Brazil: Bahia). Cicadellinae, Cicadellini. Neotropical; Brazil, French Guiana.

Paluda DeLong 1937b:233. *Thamnotettix placidus* Osborn 1905a:536 (USA: New York), homonym of *Thamnotettix placidus* Horvath 1897 = *Thamnotettix mellus* Sanders & DeLong 1917a:91. 10:840. Deltocephalinae, Athysanini. Holarctic.
Subgenus:
Rhopalopyx Ribaut 1939a:267. *Jassus preyssleri* Herrich-Schaeffer 1838c:7.

Palus DeLong & Sleesman 1929a:85 as subgenus of *Flexamia* DeLong 1926d:22. "*Flexamia (P.) delector*" = *Deltocephalus delector* Sanders & DeLong 1919a:233 (USA: Pennsylvania). 10:2036, 2674. Emeljanov 1966a:110-115. Deltocephalinae, Paralimnini. Nearctic. Preoccupied by *Palus* Palmer 1928; synonym of *Cosmotettix* Ribaut 1942a:267.

Pamplona Melichar 1926a:343. *Tettigonia feralis* Fowler 1899e:254 (Panama: "V. de Chiriqui", 2-300 ft.). 1:401. Young 1977a:430. Cicadellinae, Cicadellini. Neotropical; Bolivia, Brazil, Colombia, Ecuador, Panama, Peru.

Pamplonoidea Young 1977a:429. *P. yalea* n.sp. (Brazil: Rio de Janeiro). Cicadellinae, Cicadellini. Neotropical.

Pandacerus Webb, M.D. 1983b:240. *Idiocerus sinuatus* Webb, M.D. 1976a:302 (South Africa: East Cape Province, Katberg). Idiocerinae. Ethiopian; Ethiopia, South Africa, Seychelles.

Pandara DeLong & Freytag 1972c:160 as subgenus of *Gyponana* Ball 1920a:85. *G. (P.) eleganta* n.sp. (Bolivia: Dept. Santa Cruz, Prov. San Estaban, Mururina, 40 km N. of Santa Cruz, 1120 ft. el.). Gyponinae, Gyponini. Neotropical.

Panimius Amyot 15:224. Invalid, not binominal.

Panolidia Nielson 1979b:33. *P. melanota* n.sp. (Panama: Pacora). Coelidiinae, Teruliini. Neotropical.

Pantallus Emeljanov 1961a:127. *Deltocephalus alboniger* Lethierry 1889b:81 (Inkutsk). Deltocephalinae, Paralimnini. Palaearctic; Manchurian subregion.

Pantalus Dlabola 1967d:225. Error for *Pantallus* Emeljanov.

Pantanarendra Ramakrishnan & Ghauri 1971b:199. *Zygina fumigata f. sonani* Matsumura 1931b:66 (Formosa). Typhlocybinae, Erythroneurini. Oriental. Synonym of *Empoascanara* Distant 1918b:94; Dworakowska 1980d:195.

Panthimia 3:224. Error for *Penthimia* Germar.

Paolia Lower 1952a:208 as subgenus of *Austroasca* Lower 1952a:202. *Empoasca bancrofti* Evans, J.W. 1938c:40 (Australia: Queensland). 17:618. Typhlocybinae, Empoascini. Australian.
[NOTE: Raised to generic status by Vilbaste 1968a:80].

Paolicia Dworakowska 1981f:592, replacement for *Paoliella* Dworakowska 1981b:4, not *Paoliella* Theobald 1928. *Empoasca brevis* Dworakowska 1972c:483 (former German East Africa: Tanganyika ?). Typhlocybinae, Empoascini. Ethiopian. Subgenus of *Acia* McAtee 1934a:109; Dworakowska 1982a:6.

Paoliella Dworakowska 1981b:4 as subgenus of *Acia* McAtee 1934a:113. *Empoasca brevis* Dworakowska 1972c:483 (former German East Africa: Tanganyika ?). Typhlocybinae, Empoascini. Ethiopian. Preoccupied by *Paoliella* Theobald 1928, see *Paolicia* Dworakowska 1981f:592.

Papallacta Schmidt 1932a:47. *P. haenschi* n.sp. (Ecuador). 1:658. Cicadellinae, Proconiini. Neotropical. Synonym of *Splonia* Signoret 1891a:467; Young 1968a:141.

Papyrina Emeljanov 1962a:162. *P. viridis* n.sp. (USSR: Uzbekistan). Deltocephalinae, Athysanini. Palaearctic. Preoccupied by *Papyrina* Moench 1853, see *Scenergates* Emeljanov 1972d:106.

Parabahita Linnavuori 1959b:177. *P. vezenyii* n.sp. (Argentina: Tucuman). Deltocephalinae, Athysanini. Neotropical; tropical zones of South America and Panama.

Parablocratus 9:121. Error for *Parabolocratus* Fieber.
Parablotettix 7:33. Error for *Parabolotettix* Matsumura.
Parabolacratis 9:121. Error for *Parabolocratus* Fieber.
Parabolacratus 9:121. Error for *Parabolocratus* Fieber.

Parabolitus Naude 1926a:37. *P. anceps* n.sp. (South Africa: Jonkershoek). 9:24. Nirvaninae, Nirvanini. Ethiopian. Synonym of *Narecho* Jacobi 1910b:127; Theron 1972a:207.

Parabolocratalis Evans, J.W. 1955a:10. *P. viridis* n.sp. (Belgian Congo: Kiamokota-Kiwakishi). 9:49. Linnavuori 1975a:14-16. Deltocephalinae, Hecalini. Ethiopian; Sudan to Transvaal.

Parabolocrates 9:121. Error for *Parabolocratus* Fieber.

Parabolocratus Fieber 1868a:502. *P. glaucescens* n.sp. (Malaga). 9:24. Deltocephalinae, Hecalini. Synonym of *Hecalus* Stal 1864b:65; Linnavuori 1975a:19.

Paraboloiratus 9:121. Error for *Parabolocratus* Fieber.

Parabolopona Matsumura 1912b:228. *Parabolocratus guttatus* Uhler 1896a:291 (Japan). 15:138. Webb, M.D. 1981b:41-47. Paraboloponinae. Oriental, Palaearctic; Asia as far south as Nepal, and to the Philippines.

PARABOLOPONIDAE Ishihara 1953b:5, 20.

Parabolotettix Matsumura 1912b:280. *P. maculatus* n.sp. (Formosa). 7:30. Mukariinae, Murakiini. Oriental. Synonym of *Mukaria* Distant 1908g:269.

Paracanthus Anufriev 1978a:63 as subgenus of *Evacanthus* Le Peletier & Serville 1825a:612. *Evacanthus ogumae* Matsumura 1911b:21. (Sakhalin). Cicadellinae, Evacanthini. Palaearctic.

Paracatua Melichar 1926a:340. *Tettigonia rubrolimbata* Signoret 1854d:718 (Colombia: Bogata). 1:232. Young 1977a:35. Cicadellinae, Cicadellini. Neotropical; Colombia, Venezuela.

Paracarinolidia Nielson 1979b:26. *Coelidia guttulata* Stal 1862e:51 (Brazil: Rio de Janeiro). Coelidiinae, Teruliini. Neotropical; Brazil, Ecuador.

Paracephaleus Evans, J.W. 1942c:49. *P. montanus* n.sp. (Tasmania: Hobart, Mt. Wellington). 5:47. Knight 1973b:981-986. Ulopinae, Cephalelini. Australian; Australia, New Zealand, Tasmania.

Paracephalus Evans, J.W. 1966a:93, error for *Paracephaleus* Evans, J.W.

Parachurnoides Maldonado-Capriles 1977a:317. *P. wandae* n.sp. (British Guiana: Katari Sources). Idiocerinae. Neotropical.

Paracicadula Osborn 1934c:260. *P. coniceps* n.sp. ("Hatutaa", Hatutu, Marquesas Islands). 7:17. Nirvaninae, Nirvanini. Oceania; Marquesas Islands.

Paracoelida 11:178. Error for *Paracoelidea* Baker.

Paracoelidea Baker 1898h:292. *P.tuberculata* n.sp. (USA: Massachusetts). 11:103. Neocoelidiinae. Nearctic. Synonym of *Neocoelidia* Gillette & Baker 1895a:103; Kramer 1964d:262.

Paracoelidia 11:178. Error for *Paracoelidea* Baker.

Paracostemma Linnavuori & Al-Ne'amy 1983a:31 as subgenus of *Adama* Dlabola 1980e:89. *Selenocephalus variabilis* Evans, J.W. 1955a:34 (Belgian Congo: Kaziba). Selenocephalinae, Adamini. Ethiopian.

Paracrocampsa Young 1968a:128. *Amblydisca amida* Distant 1908b:70 (Ecuador: Cachabe). Cicadellinae, Proconiini. Neotropical; Colombia, Ecuador.

Paracyba Vilbaste 1968a:96. *Zygina akashiensis* Takahashi 1928a:442 (Japan: Kyushu). Typhlocybinae, Typhlocybini. Palaearctic; Manchurian subregion.

Paradirydium 8:264 and 9:121. Error for *Paradorydium* Kirkaldy.

PARADORYDIINI 9:104.

Paradorydium Kirkaldy 1901f:339, replacement for *Dorydium* Burmeister 1839a:3, not *Dorydium* Burmeister 1835. *Dorydium lanceolatum* Burmeister 1839a:4 (Sicily) = *Jassus paradoxus* Herrich-Schaeffer 1837a:6 (Bavaria). 9:106. Linnavuori 1979a:655-669. Eupelicinae, Paradorydiini. Warmer parts of Palaearctic, Ethiopian, and Australian regions.
Subgenus:

Penedorydium Linnavuori 1979a:667. *P. (P.) perforatum* n.sp.
Synonyms:
Carphosoma Royer 1907a:29, replacement for *Dorydium* Burmeister 1839a:3.
Deltodorydium Kirkaldy 1907d:73. *Paradorydium brighami* n.sp.
Mesodorydium Melichar 1914g:4. *M. famelicum* n.sp.
Bumizana Distant 1918b:32. *B. elongata* n.sp.
Semenovium Kusnetzov 1929a:312. *S. ferganae* n.sp.
Kuznetsovium Zachvatkin 1953c:231. *K. artistidae* n.sp.
Paradrabescus Kuoh, C. L. 1985a:379, 382. *P. testaceus* n.sp. (China: Yunnan, Cangyuan Country, 1010 m.). Iassinae, Drabescini. Palaearctic; China; Yunnan.
Parafieberiella Dlabola 1974a:69. *P. olivacea* n.sp. (SW Iran: Faro, Kazerun). Deltocephalinae, Fieberielliini. Palaearctic; Mediterranean subregion.
Paragallia 14:172. Error for *Peragallia* Ribaut.
Paraganus Linnavuori 1955a:111. *P. spinosus* n.sp. (Paraguay: Toldo Cue). 10:1241. Deltocephalinae, Athysanini. Neotropical.
Paragomia 8:264. Error for *Pagaronia* Ball.
Paragonalia Evans, J.W. 1947a:163, replacement for *Paragonia* Melichar 1926a:342, not *Paragonia* Huebner 1823. *Tettigonia sanguinicollis* Latreille 1811a:191 (Cuba or Guiana). 1:256. Young 1977a:1098. Cicadellinae, Cicadellini. Neotropical; Colombia, Cuba, Guiana, Surinam.
Paragonia 8:264. Error for *Pagaronia* Ball.
Paragonia Melichar 1926a:342. *Tettigonia sanguinicollis* Latreille 1811a:191 (Cuba or Guiana). 1:256. Preoccupied by *Paragonia* Huebner 1823, see *Paragonalia* Evans, J.W. 1947a:163.
Paragygrus Emeljanov 1964c:423 as subgenus of *Paralimnus* Matsumura 1902a:386. *P. (P.) major* n.sp. (Kazakhstan). Nast 1972a:407. Deltocephalinae, Paralimnini. Palaearctic.
Paragypona DeLong & Freytag 1964a:7 as subgenus of *Gypona* Germar 1821a:73. *Cercopis thoracica* Fabricius 1803a:91 (South America). Gyponinae, Gyponini. Neotropical.
Parahaldorus Linnavuori 1959b:142 as subgenus of *Haldorus* Oman 1949a:73. *H. (P.) truncatistylus* n.sp. (Brazil: Nova Teutonia). Deltocephalinae, Deltocephalini. Neotropical.
Paraidioscopus Maldonado-Capriles 1964a:98. *Idioscopus tagalicus* Baker 1915c:338 (Philippines: Luzon). Idiocerinae. Oriental; Philippines, Luzon and Palawan.
Paralaevicephalus Ishihara 1953b:14. *Deltocephalus nigrofemoratus* Matsumura 1902a:399 (Japan: Honshu). 10:1307. Deltocephalinae, Paralimnini. Palaearctic; Japan, Korea. Oriental: Taiwan.
Paralebra McAtee 1926b:147,151 as subgenus of *Protalebra* Baker 1899b:402. *Protalebra similis* Baker 1899b:402 (Brazil: Corumba). 17:22. Typhlocybinae, Alebrini. Neotropical.
Synonyms:
Plagalebra McAtee 1926b:147, 150. *Protalebra singularis* Baker 1899b:402.
Kallebra McAtee 1926b:152. *Protalebra (K.) ninettae* n.sp.
[NOTE: Both Young 1952a:28 and Metcalf 1968a:22,23 consider *Paralebra* and *Plagalebra* McAtee as synonyms. Young, as first reviser, suppressed *Plagalebra*; Metcalf, presumably on page priority grounds, placed *Paralebra* as the junior synonym].
Paralidia Nielson 1982e:234. *Coelidia plaumanni* Linnavuori 1956a:32 (Brazil: Nova Teutonia). Coelidiinae, Coelidiini. Neotropical.
Paralimnellus Emeljanov 1972d:107. *Paralimnus cingulatus* Dlabola 1960b:2 (Iraq). Deltocephalinae, Paralimnini. Palaearctic; Iran, Iraq, Tadzhikistan, Uzbekistan.
PARALIMNINI [as PARALIMNUSARIA] 10:1864.
Paralimnoidella Kwon & Lee 1979c:60. *P. elegans* n.sp. (Korea: Mt. Sobaeksan). Deltocephalinae, Paralimnini. Palaearctic.
Paralimnus Matsumura 1902a:386. *P. fallaciosus* n.sp. (Jeso Island). 10:1864. Nast 1972a:407. Deltocephalinae, Paralimnini. Palaearctic, Oriental: widespread.
Subgenera:
Anthocallis Emeljanov 1946c:423. *Paralimnus minor* Kusnezov 1929a:319; Nast 1972a:407.
Bubulcus Dlabola 1961b:320. *Paralimnus cingulatus* Dlabola 1960b:2; Nast 1972a:407.

Paragygrus Emeljanov 1964c:423. *Paralimnus major* n.sp.; Nast 1972a:407.
Synonym:
Diemoides Evans, J.W. 1938b:13. *D. smithtoniensis* n.sp.
Parallaxis McAtee 1926b:153. *P. vacillans* n.sp. (Brazil: Chapada). 17:101. Linnavuori 1954f. Typhlocybinae, Dikraneurini. Neotropical; Brazilian and Mexican subregions.
Parallygus Melichar 1903b:179. *P. divaricatus* n.sp. (Ceylon). 10:548. Deltocephalinae, Athysanini. Oriental; Ceylon, Formosa, India.
Paramacroceps de Bergevin 1926a:30. *P. fouquei* n.sp. (Algeria). 14:159. Adelungiinae, Adelungiini. Palaearctic; Mediterranean subregion.
Paramecus 10:2675. Error for *Paramesus* Fieber.
Parameeus 10:2675. Error for *Paramesus* Fieber.
Paramelia Evans, J.W. 1954a:124. *P. typica* n.sp. (Madagascar). 10:2242. Deltocephalinae, Platymetopiini. Malagasian.
Paramenus 10:2675. Error for *Paramesus* Fieber.
Paramerus 10:2675. Error for *Paramesus* Fieber.
Paramesanus Dlabola 1979b:135. *P. wittmeri* n.sp. (Saudi Arabia). Deltocephalinae, Paralimnini. Palaearctic.
Paramesius 10:2675. Error for *Paramesus* Fieber.
Paramesodes Ishihara 1953b:14, 45. *Athysanus albinervosus* Matsumura 1902a:372 (Japan: Kyushu). 10:2088. Deltocephalinae, Athysanini. Palaearctic, Oriental.
Synonym:
Coexitianus Dlabola 1960a:252. *Athysanus albinervosus* Matsumura 1902a:372.
Paramesus Fieber 1866a:506. *Athysanus obtusifrons* Stal 1853a:175 (Sweden). 10:2061. Deltocephalinae, Paralimnini. Palaearctic; European, Manchurian and Mediterranean subregions. [Nearctic; northeast Canada].
Synonym:
Dochmocarus Thomson 1869a:65. *Cicada nervosa* Fallen 1826a:39 = *Athysanus obtusifrons* Stal 1853a:175.
Parametopia Melichar 1925a:387 as subgenus of *Oncometopia* Stal 1869a:62. *Cicada orbona* Fabricius 1798a:520 (America). 1:585. Cicadellinae, Proconiini. Neotropical. Preoccupied by *Parametopia* Reitter 1884, see *Lebora* China 1927d:283. Synonym of *Oncometopia* Stal 1869a:62.
Paranastus Emeljanov 1972d:107. *Paramesus reticulatus* Horvath 1897b:628 (Hungary). Deltocephalinae, Paralimnini. Palaearctic. Preoccupied by *Paranastus* Roewer 1943, see *Parapotes* Emeljanov 1975a:390.
Parandanus Linnavuori & DeLong 1976a:34. *P. ornatus* n.sp. (Peru: Sinchona near Tingo Maria). Deltocephalinae, Deltocephalini. Neotropical; Bolivia, Peru.
Paranoplus Linnavuori 1961a:473. *P. maculatus* n.sp. (South Africa: Cape Prov., Tzitzikama Forest, Stormsrivierpiek). Deltocephalinae, Athysanini. Ethiopian. Synonym of *Dagama* Distant 1910e:243; Theron 1980b:293.
Paraonukia Ishihara 1963b:5. *P. keitonis* n.sp. (Formosa: Keito). Cicadellinae, Evacanthini. Oriental.
Parapagaronia Kwon & Lee 1978a:8 as subgenus of *Pagaronia* Ball 1902a:19. *Pagaronia jenjouristi* Anufriev 1970f:556 (Japan: Tokyo, Takao-San). Cicadellinae, Pagaroniini. Palaearctic.
Parapediopsis Hamilton, KGA 1980b:905 as subgenus of *Macropsis* Lewis 1834a:49. *Macropsis benguetensis* Merino 1936a:321 (Philippines: Luzon, Baguio). Macropsinae, Macropsini. Oriental, Australian.
Parapetalocephala Kato 1931e:435. *P. montana* n.sp. (Formosa). 4:64. Ledrinae, Petalocephalini. Oriental; Formosa. Palaearctic; Japan, Korea.
Paraphelpsius 10:2675. Error for *Paraphlepsius* Baker.
Paraphlepsius Baker 1897c:158. *P. ramosus* n.sp. (USA: New York, Ithaca). 10:578. Hamilton, KGA 1975a:18. Deltocephalinae, Athysanini. Nearctic, Neotropical; southern Canada south to the Carribean and Guatemala.
Subgenera:
Strephonius Hamilton, KGA 1975a:23. *Phlepsius tigrinus* Ball 1909a:80.
Sabix Hamilton, KGA 1975a:29. *Jassus irroratus* Say 1830b:308.
Gamarex Hamilton, KGA 1975a:34. *Jassus fulvidorsum* Fitch 1851a:62.

Paraphysius Hamilton, KGA 1975a:68. *Phlepsius lascivius* Ball 1900b:200.
[NOTE: See comment under *Phlepsius*].

Paraphlesius 10:2675. Error for *Paraphlepsius* Baker.

Parapholis Uhler 1877a:461. *P. peltata* n.sp. (USA: Massachusetts). 4:43. Ledrinae, Xerophloeini. Nearctic. Synonym of *Xerophloea* Germar 1839a:190.

Paraphrodes Linnavuori 1979a:681. *Aphrodes flavigera* Evans, J.W. 1955a:15 (Belgian Congo: Kiamokota-Kiwakishi). Aphrodinae, Paraphrodini. Ethiopian; widespread in tropical Africa.

PARAPHRODINI Linnavuori 1979a:680; tribe of APHRODINAE.

Paraphysius Hamilton, KGA 1975a:68 as subgenus of *Paraphlepsius* Baker 1897c:158. *Phlepsius lascivius* Ball 1900b:200 (USA: Colorado, Fort Collins). Deltocephalinae, Athysanini. Nearctic.

Parapotes Emeljanov 1975a:390, replacement for *Paranastus* Emeljanov 1972d:107, not *Paranastus* Roewer 1943. *Paramesus reticulatus* Horvath 1897b:628 (Hungary). Deltocephalinae, Paralimnini. Palaearctic; European subregion.

Parapulopa 5:99. Error for *Paropulopa* Fieber.

Parargus Emeljanov 1961a:127. *P. kerzhneri* n.sp. (Kazakhstan: Karaganda region, 40 km south of Zhan-Ark). Deltocephalinae, Paralimnini. Palaearctic; Siberian subregion.

Parasitades Singh-Pruthi 1936a:106. *P. baileyi* n.sp. (India: Sikkim). 13:165. Macropsinae, Macropsini. Oriental. Subgenus of *Oncopsis* Burmeister 1838a:10; Hamilton, K.G.A. 1980b:890.

Parasudra Schmidt 1909d:263. *P. sumatrana* n.sp. (Sumatra). 2:8. Hylicinae, Sudrini. Oriental.

Paratanus Young 1957a:14. *Atanus exitiosus* Beamer 1943a:178 (Argentina: Rio Negro Valley). Deltocephalinae, Athysanini. Neotropical; Argentina, Brazil, Chile, Ecuador, Paraguay, Venezuela.

Paraterulia Nielson 1979b:108. *Terulia magna* Baker 1898h:292 (Brazil: Para). Coelidiinae, Teruliini. Neotropical; Brazil, Guyana, Peru, Venezuela.

Parathaia Kuoh, C. L. 1982a:398, 403. *P. bimaculata* n.sp. (China: Jiangxi, Lushan, 1100 m). Typhlocybinae, Erythroneurini. Palaearctic; Manchurian subregion.

Parathona Melichar 1926a:330. *Cicada lyncea* Fabricius 1787a:269 (French Guiana) = *Cicada cayennensis* Gmelin 1798a:2105, replacement for *Cicada lyncea* Fabricius 1787a:269, not *Cicada lyncea* Fabricius 1775. 1:648. Young 1977a:303. Cicadellinae, Cicadellini. Neotropical; Argentina, Bolivia, Brazil, Colombia, Panama, Trinidad, Venezuela.

Paratubana Young 1977a:239. *Tettigonia vittifacies* Signoret 1855d:774 (unknown). Cicadellinae, Cicadellini. Neotropical; Brazil, Argentina.

Paratyphlocyba Ahmed, Manzoor 1985c:98. *P. hazarensis* n.sp. (Pakistan: Abbottabad). Typhlocybinae, Typhlocybini. Oriental.

Paraulacizes Young 1968a:93. *Cicada irrorata* Fabricius 1794a:33 (USA: "Carolina"). Cicadellinae, Proconiini. Neotropical, Nearctic; southeast and central USA, Mexico, Central America and Panama.

Parazyginella Chou, I. & Zhang 1985a:295, 299. *P. lingtianensis* n.sp. (China: Guangxi Prov., Lingtruan, Lingtian). Typhlocybinae, Zyginellini. Palaearctic.

Parcana DeLong & Freytag 1974a:193 as subgenus of *Acuera* DeLong & Freytag 1972a:229. *A. (P.) ultima* n.sp. (Mexico: Chiapas). Gyponinae, Gyponini. Neotropical.

Paremesus 10:2675. Error for *Paramesus* Fieber.

Parinaeota Melichar 1926a:341. *P. bakeri* Young 1977a:79. (Peru: Marcapata). 1:232. Cicadellinae, Cicadellini. Neotropical.

Paririana 3:225. Error for *Prairiana* Ball.

Parmesus 10:2675. Error for *Paramesus* Fieber.

Parocerus Vilbaste 1980b:32. *Idiocerus laurifoliae* Vilbaste 1965b:35 (USSR: Kazakhstan, Altai Mts.). Idiocerinae. Palaearctic. Subgenus of *Idiocerus* Lewis 1834a:47; Kwon 1985a:64.

Parohinka Webb, M.D. 1981b:57. *Muirella longiseta* Melichar 1914b: 3-135 (Java). Paraboloponinae. Oriental; NE India to Philippines and southeast Asia, south to New Guinea and Pacific Islands.

Paromenia Melichar 1926a:342. *Tettigonia rufa* Walker 1851b:742 (Venezuela). 1:257. Young 1977a:263. Cicadellinae, Cicadellini. Neotropical; Mexico and Guatemala to Bolivia and SE Brazil.
Synonyms:
Alocha Melichar 1926a:342; Young 1977a:263.
Scarisana Metcalf 1949b:277; Young 1977a:263.

Paropia Germar 1833a:181. *Ulopa scanica* Germar 1833a:181 = *Cicada scanica* Fallen 1806b:113 (Sweden). 5:14. Megophthalminae. Palaearctic. Synonym of *Megophthalmus* Curtis 1833a:193.

PAROPIINAE 5:7.

PAROPIINI 5:13.

Paropulopa Fieber 1866a:500. *P. lineata* n.sp. (France). 5:33. Megophthalminae. Palaearctic.

Parunculus Emeljanov 1962a:176. *Deltocephalus pantherinus* Kusnezov 1929b:183 (Transbaikalia). Deltocephalinae, Deltocephalini. Palaearctic.
[NOTE: Emeljanov 1964f:44 points out that his 1962a:176 concept of *Deltocephalus pantherinus* Kusnezov 1929b:183, named as type species of *Parunculus*, was incorrect, and referred *D. pantherinus* to the genus *Rosenus*. This case should be referred to the Commission under the provisions of ICZN Article 70(b)].

Parvulana DeLong & Freytag 1972f:289 as subgenus of *Polana* DeLong 1942d:110. *P. (P.) alata* n.sp. (Brazil: "Museo La Plata"). Gyponinae, Gyponini. Neotropical.

Pasadenus Ball 1936c:60. *Thamnotettix pasadena* Ball 1914a:212 (USA: California, Pasadena). 10:944. Nielson 1969a. Deltocephalinae, Athysanini. Nearctic; California.

Pasara Dworakowska 1981e:180. *P. minuta* n.sp. (India: Tamil Nadu, Silver Cascade near Shambaganur). Typhlocybinae, Erythroneurini. Oriental.

Pasaremus Oman 1949a:160. *Deltocephalus concentricus* Van Duzee 1894f:208 (USA: NW Colorado). 10:2060. Deltocephalinae, Cicadulini. Nearctic; Rocky Mountain subregion.

Pascoepus Webb, M.D. 1983a:50. *Idiocerus hyleorais* Kirkaldy 1907d:34 (Australia: Queensland, Bundaberg). Idiocerinae. Australian.

Pataganus Linnauori 1956a:9. Invalid, *nomen nudum*.

Patalimnus 10:2675. Error for *Paralimnus* Matsumura.

Paternus Distant 1918b:71. *P. pusanus* n.sp. (India: Bihar, Pusa). 10:2349. Deltocephalinae, Stenometopiini. Oriental. Synonym of *Doratulina* Melichar 1903b:198; Vilbaste 1965a:10.

Paulianiana Evans. J.W. 1954a:91. *P. dracula* n.sp. (Madagascar). 5:79. Linnavuori 1972a:143. Ulopinae, Myerslopiini. Malagasian.

Paulomanus Young 1952b:108. *P. cecropiae* n.sp. (Panama: Canal Zone, Summit). 17:607. Typhlocybinae, Jorumini. Neotropical.

Pauripo Evans, J.W. 1934a:151. *P. insularis* n.sp. (South Australia: Kangaroo Island). 12:37. Eurymelinae, Eurmelini. Australian.

Pauroeurymela Evans, J.W. 1933a:79. *Eurymela amplicincta* Walker 1858a:84 (Australia: New South Wales). 12:25. Eurymelinae, Eurymelini. Australian; Australia.

Pawiloma Young 1977a:729. *Tettigonia amoena* Walker 1851b:759 (Venezuela). Cicadellinae, Cicadellini. Neotropical; Panama to Brazil, Paraguay and Argentina.

Pazu Oman 1949a:171. *Hebecephalus balli* Beamer 1936d:252 (USA: Arizona, Cochise). 10:1812. Deltocephalinae, Deltocephalini. Nearctic; Sonoran.

Peayanus Nielson 1979b:65. *P. peayi* n.sp. (Peru: Callanga). Coelidiinae, Teruliini. Neotropical; Bolivia, Peru.

Peconus Oman 1949a:171. *Hebecephalus scriptanus* Oman 1934a:77 (USA: Arizona, Mustang Mt.). 10:1824. Deltocephalinae, Deltocephalini. Nearctic; Sonoran.

Pectinapyga Osborn 1930a:691. *P. texana* n.sp. (USA: Texas, Brownsville). 10:1623. Deltocephalinae, Doraturini. Nearctic. Synonym of *Athysanella (Amphipyga)* Osborn 1930a:691.

Pectinopyga 10:2675. Error for *Pectinapyga* Osborn.

Pedarium Emeljanov 1961a:125. *P. ruderale* n.sp. (USSR: Kazakhstan, Chu, Dzhambul region). Deltocephalinae, Opsiini. Palaearctic.

Pedematia Amyot 15:224. Invalid, not binominal.

Pediaspis 13:252. Error for *Pediopsis* Burmeister.

Pedopsis 13:259. Error for *Pediopsis* Burmeister.

Pedionis Hamilton, K.G.A. 1980b:891. *Pediopsis garuda* Distant 1916a:239 (India: Madras). Macropsinae, Macropsini. Oriental, Oceania, Australian.

Subgenus:
Thyia Hamilton, K.G.A. 1980b:894. *Macropsis thyia* Kirkaldy 1907d:36.

Pediopsis Burmeister 1838a:11 as subgenus of *Bythoscopus* Germar 1833a:180. *Jassus tiliae* Germar 1831a:pl. 14 (Germany). 13:8, 252. Hamilton, K.G.A. 1980b:902. Macropsinae, Macropsini. Holarctic, Australian, Oriental.

Pediopsoides Matsumura 1912b:305. *P. formosanus* n.sp. (Formosa). 15:22, 36, 224. Hamilton, K.G.A. 1980b:895. Macropsinae, Macropsini. Oriental, Palaearctic, Nearctic, Ethiopian.

Subgenera:
Sispocnis Anufriev 1967a:174. *Bythoscopus juglans* Matsumura 1912b:304.
Nanopsis Freytag 1974a:605. *Jassus verticis* Say 1830b:308.
Kiamoncopsis Linnavuori 1978b:15. *K. quartaui* n.sp.
Celopsis Hamilton, K.G.A. 1980b:896. *Macropsis dapitana* Merino 1936a:324.

Pedioscopus Kirkaldy 1906c:349. *P. philenor* n.sp. (Australia: Queensland, Cairns). 16:217. Idiocerinae. Oriental; Papuan. Australian; Australia. Oceania; Fiji, Samoa, Micronesia.

Pedumella Ball & Beamer 1940a:65 as subgenus of *Athysanella* Baker 1898a:185. *A. (P.) spatulata* n.sp. (USA: Arizona, St. John). 10:1649. Deltocephalinae, Doraturini. Nearctic.

Pegogonia Young 1977a:722. *Cicada rufipes* Fabricius 1803a:68 ("Amer. merid."). Cicadellinae, Cicadellini. Neotropical; northern South America.

Peitouellus Vilbaste 1969a:7. *Thamnotettix hokutonis* Matsumura 1914a:181 (Formosa: Pei-tou). Deltocephalinae, Deltocephalini. Oriental.

Peltocephala 4:143. Error for *Petalocephala* Stal.

Peltocheirus Walker 1858b:247. *Tettigonia bigibbosa* Signoret 1855c:510 (French Guiana). 1:649. Young 1968a:118. Cicadellinae, Proconiini. Neotropical; northern South America.

Synonym:
Cymbalopus Kirkaldy 1907d:88. *Tettigonia bigibbosa* Signoret 1855c:510.

Peltophlyctis Amyot 15:224. Invalid, not binominal.

Peltospila Amyot 17:1506. Invalid, not binominal.

Pemoasca Mahmood 1967a:43. *P. brunneipunctata* n.sp. (Philippines: Luzon, Mt. Makiling). Typhlocybinae, Typhlocybini. Oriental.

Penaia Freytag & DeLong 1971a:541 as subgenus of *Chilenana* DeLong & Freytag 1967b:106. *Chilenana flexa* DeLong & Freytag 1967b:112 (Chile: Chiloe Prov., Toi-goi). Gyponinae, Gyponini. Neotropical.

Penangiana Mahmood 1967a:25. *P. pallida* n.sp. (Malaya: Penang Is.). Typhlocybinae, Empoascini. Oriental.

Pendarus Ball 1927d:262 as subgenus of *Phlepsius* Fieber 1866a:514. *Phlepsius slossoni (slossonae)* Ball 1905b:209 (USA: Florida, Biscayne Bay). 10:578. Hamilton, K.G.A. 1975a:5. Deltocephalinae, Athysanini. Nearctic; Allegheny subregion.

Subgenus:
Remadosus Ball 1929a:2. *Athysanus magnus* Osborn & Ball 1897a:225.

Penebahita Linnavuori 1959b:166 as subgenus of *Bahita* Oman 1938b:379. *Deltocephalus venosulus* Berg 1879d:266 (Argentina: Buenos Aires). Deltocephalinae, Athysanini. Neotropical. Synonym of *Frequenamia* DeLong 1947a:63; Linnavuori & DeLong 1978b:115.

Penedorydium Linnavuori 1979a:667 as subgenus of *Paradorydium* Kirkaldy 1901f:339. *P. (P.) perforatum* n.sp. (Zaire: Lubumbashi). Eupelicinae, Paradorydiini. Ethiopian.

Penehuleria Beamer 1934a:43. *P. acuticephala* n.sp. (USA: California, Mint Canyon). 10:1088. Deltocephalinae, Cochlorhinini. Nearctic.

Penestirellus Beamer & Tuthill 1934a:21. *P. catalinus* n.sp. (USA: Arizona, Sabino Canyon). 10:115. Linnavuori 1959b:319. Deltocephalinae, Stenometopiini. Nearctic. Synonym of *Stirellus* Osborn & Ball 1902a:250; Linnavuori 1959a:319.

Penestragania Beamer & Lawson 1945a:50 as subgenus of *Stragania* Stal 1862e:49. *Pachyopsis robustus* Uhler 1877a:467 (USA: New Mexico). 15:97,99. Iassinae, Iassini. Nearctic and Neotropical; Mexican, Antillean and Brazilian subregions. [NOTE: Raised to generic status by Blocker 1979a:40].

Pentastigmops Amyot 17:1506. Invalid, not binominal.

Penthemia 3:225. Error for *Penthimia* Germar.

Penthimia Germar 1821a:46. *Cercopis atra* Fabricius 1794a:50 (Europe) = *Cicada nigra* Goeze 1778a:161. 3:160. Penthimiinae. Cosmopolitan.

Penthimidia Haglund 1899a:55. *P. eximia* n.sp. (Cameroon). 3:205. Linnavuori 1977a:57. Penthimiinae. Ethiopian; Guinean.

Penthimiella Evans, J.W. 1972a:187. *P. amboinensis* n.sp. (Amboina). Penthimiinae. Oriental; Molucca Islands.

PENTHIMIINAE 3:157.

Penthimiola Linnavuori 1959b:54 = *Penthimiola* Lindberg 1958a:208, *nomen nudum*. *Penthimiola fasciolata* Lindberg 1958a:208 (Cape Verde Islands: Sao Tiago, Ribeira da Boa Entrada). Linnavuori 1977a:37. Penthimiinae. Ethiopian and Malagasian regions.

[NOTE: Questionable record from Neotropical region; Argentina].

Penthimiolla Webb, M.D. 1980a:830, 833. Error for *Penthimiola* Linnavuori.

Penthimiopsis Evans, J.W. 1972a:188. *P. parva* n.sp. (Bismarck Archipelago: Manus, Rossum, 6 km SE of Lorengau). Penthimiinae. Oriental; New Guinea and Bismarck Archipelago.

Penthopitta Amyot 10:2676. Invalid, not binominal.

Penthotaenia Amyot 8:264. Invalid, not binominal.

Pentimia 3:225. Error for *Penthimia* Germar.

Pentoffia Kramer 1964b:119. *P. nivata* n.sp. (Colombia: Cali Valle). Nirvaninae, Nirvanini. Neotropical.

Pentria Evans, J.W. 1972a:192. *P. albobrunnea* n.sp. (NE New Guinea: Kassam, 1350 m). Penthimiinae. Australian.

Peragalia Logvinenko 1957b:66. Error for *Peragallia* Ribaut.

Peragallia Ribaut 1948c:59. *Bythoscopus sinuatus* Mulsant & Rey 1855a:222 (Provence). 14:80. Agalliinae, Agalliini. Palaearctic. Synonym of *Austroagallia* Evans, J.W. 1935c:70; LeQuesne 1964a:3.

Peranoa DeLong 1980c:217 as subgenus of *Ponana* Ball 1920a:93. *P. (P.) perusana* n.sp. (Peru: Cusco, Santa Isabel). Gyponinae, Gyponini. Neotropical.

Perophlepsius Heller & Linnavuori 1968a:4; invalid, *nomen nudum*.

Perotettix Ribaut 1942a:260. *Thamnotettix pictus* Lethierry 1880a:64 (France). 10:221. Deltocephalinae, Athysanini. Palaearctic; European subregion.

Perubahita Linnavuori & DeLong 1978b:113. *P. longifal* n.sp. (Peru: Sinchona). Deltocephalinae, Athysanini. Neotropical; Peru.

Perubala Linnavuori 1959b:20. *P. fasciata* n.sp. (Peru). Neobalinae. Neotropical; Bolivia, Peru.

Perugrampta Kramer 1965a:69. *P. curcoensis* n.sp. (Peru: Cusco, Santa Isabel). Nirvaninae, Nirvanini. Neotropical; Peru.

Perulidia Nielson 1979b:103. *P. pilosa* n.sp. (Bolivia: Las Juntas). Coelidiinae, Teruliini. Neotropical; Bolivia, Brazil, Peru.

Petalacephala 4:143. Error for *Petalocephala* Stal.

Petalocephala Stal 1854b:251. *P. bohemani* n.sp. (Java). 4:65. Linnavuori 1972b:214-224. Ledrinae, Petalocephalini. Oriental, Ethiopian, Palaearctic.

Synonym:
Pachyledra Schumacher 1912b:248. *P. kamerunensis* n.sp.

PETALOCEPHALINI Metcalf 1962c:58; tribe of LEDRINAE.

Petalocephaloides Kato 1931e:436. *P. laticapitata* n.sp. (Formosa). 4:83. Ledrinae, Petalocephalini. Oriental; Japan, Taiwan.

Petalocera 4:143. Error for *Petalocephala* Stal.

Petalopoda Spangberg 1879a:18. *P. annulipes* n.sp. (French Guiana). 11:29. Nielson 1983e:69. Coelidiinae, Gabritini. Neotropical. Synonym of *Gabrita* Walker 1858b:254; Nielson 1983e:69.

Petatocephala 4:143. Error for *Petalocephala* Stal.

Pettya Kirkaldy 1906c:343. *P. anemolua* n.sp. (Australia: Queensland, Cairns). 10:2462. Typhlocybinae, Erythroneurini. Australian.
Synonym:
Eutamborina Evans, J.W. 1942a:27. *E. punctata* n.sp.

Peyerimhoffiola de Bergevin 1928b:274. *P. galeata* n.sp. (Algeria). 14:165. Adelungiinae, Peyerimhoffiolini. Palaearctic; Algeria, Iraq, Tunisia.

PEYERIMHOFFIOLINI Al-Ne'amy & Linnavuori 1982a:111; tribe of ADELUNGIINAE.

Phaeida Emeljanov 1962a:172. *P. tesquorum* n.sp. (USSR: Kazakhstan). Deltocephalinae, Athysanini. Palaearctic.

Phelepsius 10:2676. Error for *Phlepsius* Fieber.

Phepsius 10:2676. Error for *Phlepsius* Fieber.

Phera Stal 1864a:77. *P. tiarata* n.sp. (Mexico: Vera Cruz) = *Tettigonia centrolineata* Signoret 1855b:239. 1:495. Young 1968a:184. Cicadellinae, Proconiini. Neotropical; south Texas and Mexico to southeast Brazil.
Synonym:
Capinota Melichar 1926a:319. *C. fowleri* n.sp.

PHEREURHININAE Kramer 1976a:117.

Phereurhinus Jacobi 1905c:168. *P. batillus* n.sp. ("Peru N Rioja"). 1:413. Kramer 1976a:124. Phereurhininae. Neotropical; Brazil, Peru.

Pherodes Fowler 1899b:225. *P. flammeicolor* n.sp. (Mexico: Tabasco, Teapa). 1:419. Young 1968a:257. Cicadellinae, Proconiini. Neotropical. Synonym of *Acrogonia* Stal 1869a:67; Young 1968a:257.

Philaia Dlabola 1952a:48. *P. jassargiforma* n.sp. (Slovakia). 10:1242. Deltocephalinae, Paralimnini. Palaearctic; Siberian subregion.

Philaja Mitjaev 1968b:33, 40. Error for *Philaia* Dlabola.

Philippocerus Maldonado-Capriles 1972b:627. *P. albipennis* n.sp. (Philippines: Palawan, 3 km NE Tinabog). Idiocerinae. Oriental.

Philipposcopus Maldonado-Capriles 1972a:542. *Pedioscopus maquilingensis* Baker 1915c:331 (Philippines: Luzon, Mt. Maguiling). Idiocerinae. Oriental.

Philotartessus Evans, F. 1981a:183. *Bythoscopus dimidiatus* Walker 1870b:319 (Dorey I = Manokwari I, West Irian). Tartessinae. Oriental: New Guinea, Solomon Islands.

Phlagothamnus Zoological Record 98, Sec. 13:370, 1961. Error for *Phlogothamnus* Ishihara 1961a.

Phlebiastes Emeljanov 1961a:128. *P. elymi* n.sp. (USSR: Kazakhstan, 40 km south of Zhan-Ark, Karaganda region). Deltocephalinae, Paralimnini. Palaearctic; USSR, Kazakhstan, Kirghizia.

Phlegsius 10:2676. Error for *Phlepsius* Fieber.

Phlepisus 10:2676. Error for *Phlepsius* Fieber.

Phlepsanus Oman 1949a:140. *Phlepsius nigrifrons* Ball 1905b:210 (USA: Arizona, Hot Springs). 10:655. Deltocephalinae, Athysanini. Nearctic; Rocky Mountain subregion.

Phlepsias 10:2676. Error for *Phlepsius* Fieber.

Phlepsidius Emeljanov 1961a:124. *P. desertorum* n.sp. (USSR: Kazakhstan, 40 km south of Zhan-Ark, Karaganda region). Deltocephalinae, Athysanini. Palaearctic: USSR, Kazakhstan, Uzbekistan, Tadzhikistan.

PHLEPSIINI 10:336.

Phlepsinus 10:2676. Error for *Phlepsius* Fieber.

Phlepsius Fieber 1866a:514. *P. maculatus* n.sp. (Germany) = *Jassus intricatus* Herrich-Schaeffer 1838b:5. 10:549. Deltocephalinae, Athysanini. Palaearctic; Mediterranean subregion.
[NOTE: Pierce's 1963a:80 fossil record for the Nearctic region applies to *Paraphlepsius*, a Nearctic taxon].

Phlepsis 10:2677. Error for *Phlepsius* Fieber.

Phlepsobahita Linnavuori 1959b:195. *P. irrorata* n.sp. (Argentina: Tucuman). Deltocephalinae, Athysanini. Neotropical.

Phlepsopsius Dlabola 1979b:132. *P. arabicus* n.sp. (Saudi Arabia). Deltocephalinae, Athysanini. Palaearctic.

Phlepsus 10:2677. Error for *Phlepsius* Fieber.

Phlogis Linnavuori 1979a:684. *P. mirabilis* n.sp. (Cameroon). Phlogisinae. Ethiopian.

PHLOGISINAE Linnavuori 1979a:683.

Phlogospila Amyot 17:1506. Invalid, not binominal.

Phlogotaenia Amyot 17:1506. Invalid, not binominal.

Phlogotettix Ribaut 1942a:262. *Jassus cyclops* Mulsant & Rey 1855a:227 (France). 10:382. Deltocephalinae, Platymetopiini. Palaearctic; European, Siberian, and Manchurian subregions.

Phlogothamnus Ishihara 1961a:248. *P. maculiceps* n.sp. (Thailand: Doi Inthanon). Deltocephalinae, Platymetopiini. Oriental; Cambodia, Thailand.

Phlogottettix Dlabola 1981a:263. Error for *Phlogotettix* Ribaut.

Pholetaera Zetterstedt 1840a:288. *Cercopis rustica* Fallen 1826a:23 (Sweden). 8:59. Aphrodinae, Aphrodini. Palaearctic. Synonym of *Aphrodes* Curtis 1829a:193.

Pholetera 8:265. Error for *Pholetaera* Zetterstedt.

Phrymomorphus 10:2678. Error for *Phrynomorphus* Curtis.

PHRYNOMORPHINAE 10:8.

PHRYNOMORPHINI 10:10.

Phrynomorphus Curtis 1833a:194. *P. nitidus* n.sp. (England). 10:12. Deltocephalinae, Athysanini. Palaearctic. Synonym of *Euscelis* Brulle 1832a:109.

Phrynomorphyes 10:2678. Error for *Phrynophyes* Kirkaldy.

Phrynophyes Kirkaldy 1906c:327. *P. phrynophyes* n.sp. (Australia: Queensland, Bundaberg). Deltocephalinae, Stenometopiini. [Australian?]. Synonym of *Doratulina* Melichar 1903b:198; Vilbaste 1965a:10.
[NOTE: Evans, J.W. 1966a:231 treats *Phrynophyes* as distinct genus, but (1977a:119) apparently accepts the Vilbaste synonymy].

Phrynophryes 10:2678. Error for *Phrynophyes* Kirkaldy.

Phrynormorphus 10:2678. Error for *Phrynomorphus* Curtis.

Phycotettix Haupt 1929c:260. *Thamnotettix paryphantus* Lethierry 1878d:xxviii (Algeria). 10:1742. Deltocephalinae, Fieberiellini. Palaearctic; Mediterranean subregion.

Phytotartessus Evans, F. 1981a:163. *P. transversus* n.sp. (New Guinea: Irian, Bodem, 11 km SE Oeberfaren). Tartessinae. Oriental; New Guinea.

Piaraneura 17:1506. Error for *Dicraneura* Puton.

Pibrochoides Haupt 1929c:252. *Tettigonia rugicollis* Signoret 1855c:525 (Mexico). 1:494,686. Young 1968a:30. Cicadellinae, Proconiini. Neotropical. Synonym of *Diestostemma* Amyot & Serville 1843a:572.

Picchuia Linnavori & Delong 1979a:51. *P. pungens* n.sp. (Peru: Machu Picchu). Deltocephalinae, Deltocephalini. Neotropical.

Picchusteles Linnavuori & Delong 1976a:37. *P. inca* n.sp. (Peru: Machu Picchu). Deltocephalinae, Macrostelini. Neotropical.

Piezauchenia Spinola 1850a:58. *P. aphrophoroides* n.sp. (Chile). 15:187. Linnavuori 1959b:12. Ledrinae, Xerophloeini. Neotropical.

Pilosana Nielson 1983e:30. *Jassus gratiosus* Spangberg 1879a:25 (Mexico: Salle). Coelidiinae, Youngolidiini. Neotropical; Mexico, West Indies, Central America, Panama, Brazil, Ecuador, Guyana, Venezuela.

Pingellus Evans, J.W. 1966a:230. *P. nigroflavus* n.sp. (Australia: Queensland, Lammington National Park). Deltocephalinae, Deltocephalini. Australian.

Pinumius Ribaut 1946b:84. *Deltocephalus areatus* Stal 1858e:193 (Irkutsk). 10:1424. Deltocephalinae, Deltocephalini. Holarctic; northern latitudes.

Piorella Evans, J.W. 1972a:178. *P. gressetti* n.sp. (NE New Guinea: Mt. Piora, 3200 m). Penthimiinae. Oriental.

Pisacha Distant 1908g:230. *P. primitiva* n.sp. (Burma: Ruby Mines). 1:462. Cicadellinae, Cicadellini. Oriental. Preoccupied by *Pisacha* Distant 1906, see *Pisachoides* Distant 1914i:333.

Pisachoides Distant 1914i:333, replacement for *Pisacha* Distant 1908g:230, not *Pisacha* Distant 1906. *Pisacha primitiva* Distant 1908g:231 (Burma: Ruby Mines). 1:462. Cicadellinae, Cicadellini. Oriental.

Pithyotettix Ribaut 1942a:261. *Cicada abietina* Fallen 1806a:28 (Sweden). 10:223. Deltocephalinae, Athysanini. Palaearctic; European and Siberian subregions.
Subgenus:

Abietotettix Mitjaev 1965a:1260. *P. (A.) sibiricus* n.sp.

Placidellus Evans, J.W. 1971a:43. *P. ishiharei* n.sp. (Thailand: San Ngow Tak). Unassigned, formerly Coelidiinae; Nielson 1975a:11. Oriental.

Placidus Distant 1908g:341. *P. hornei* n.sp. (India: NW Provinces). Unassigned, formerly Coelidiinae; Nielson 1975a:11. Oriental; India. Palaearctic; Afghanistan.

Placotettix Ribaut 1942a:262. *Jassus (Thamnotettix) taeniatifrons* Kirschbaum 1868b:89 (Italy). 10:384. Deltocephalinae, Athysanini. Palaearctic; European and Mediterranean subregions.

Plagalebra McAtee 1926b:147,150 as subgenus of *Protalebra* Baker 1899b:402. *Protalebra singularis* Baker 1899b:402 (Brazil: Chapada). 17:22. Typhlocybinae, Alebrini. Neotropical. Synonym of *Paralebra* McAtee 1926b:147; Young 1952a:28.

Planaphrodes Hamilton, K.G.A. 1975d:1012. "*Cicada tricincta* Curtis" = *(Acucephalus) tricinctus* Curtis 1836a:pl.620 (England). Aphrodinae, Aphrodini. Palaearctic, Nearctic (adventive, Newfoundland).

Planicephalus Linnavuori 1954e:143 as subgenus of *Deltocephalus* Burmeister 1838b:15. *Jassus (Deltocephalus) flavicosta* Stal 1862a:53 (Brazil: Rio de Janeiro). 10:1091, 2679. Kramer 1971d:255. Deltocephalinae, Deltocephalini. Nearctic; Alleghany subregion. Neotropical; Mexican, Carribean and Brazilian subregions.

Planolidia Nielson 1982e:309. *P. arida* n.sp. (Panama: Bugaba). Coelidiinae, Coelidiini. Neotropical; Costa Rica, Panama.

Plapigella Nielson 1979b:134. *Jassus elegans* Spangberg 1878b:6 ("Rio Negro"). Coelidiinae, Teruliini. Neotropical; Bolivia, Peru, Venezuela.

Platentomus Theron 1980a:287. *Jassus (Deltocephalus) sobrinus* Stal 1859b:294 (South Africa: Cape of Good Hope). Deltocephalinae, Deltocephalini. Ethiopian.

Platyacina Emeljanov 1964f:19 as subgenus of *Aconura* Lethierry 1876a:LXXXV. *Aconura depressa* n.sp. (USSR: Kazakhstan). Deltocephalinae, Doraturini. Palaearctic.

Platycyba Matsumura 1932a:103. *P. bistriata* n.sp. (Okinawa). 17:1047. Typhlocybinae, Zyginellini. Palaearctic; Japan, Ryukyu Islands.

Platyetopius 10:2679. Error for *Platymetopius* Burmeister.

Platyeurymela Evans, J.W. 1933a:78. *Eurymela semifascia* Walker 1851b:643 ("New Holland", Australia). 12:10. Eurymelinae, Eurymelini. Australian; Australia, Tasmania.

Platygonia Melichar 1925a:340. *Tettigonia praestantior* Fowler 1899e:254 (Panama: Bugaba). 1:521. Young 1977a:420. Cicadellinae, Cicadellini. Neotropical; Brazil, Colombia, Costa Rica, Ecuador, Panama, Peru.
Synonyms:
Oxycoryphia Melichar 1926a:344. *Tettigonia spatulata* Signoret 1854d:722.
Oxygonalia Evans, J.W. 1947a:163. *Ciccus ignifer* Walker 1851b:804.

Platyhynna Berg 1884a:26. *Epiclines bdellostoma* Berg 1879d:235 ("Misiones", Argentina). 4:96. Ledrinae, Petalocephalini. Neotropical; Argentina, Brazil.

PLATYJASSINI 15:208.

Platyjassus Evans, J.W. 1954a:113. *P. viridis* n.sp. (Madagascar). 15:208. Linnavuori & Quartau 1975a:157. Iassinae, Platyjassini. Malagasian.

Platyledra Evans, J.W. 1936a:39. *P. hirsuta* n.sp. (South Australia: Ooldea). 4:99. Ledrinae, Petalocephalini. Australian.

PLATYMETOPIINI 10:2209.

Platymethopius 10:2679. Error for *Platymetopius* Burmeister.
Platymetobius 10:2679. Error for *Platymetopius* Burmeister.

Platymetopius Burmeister 1838b:16 as subgenus of *Jassus. Cicada vittata* Fabricius 1775a:684 (Europe). 10:2209. Deltocephalinae, Platymetopiini. Palaearctic; widespread.
Subgenus:
Quernus Dlabola 1974a:57. *Platymetopius signoreti* Metcalf 1967c:2230.
Synonym:
Eremitopius Lindberg 1927b:29. *E. albus* n.sp.

Platymetopuis 10:2679. Error for *Platymetopius* Burmeister.
Platymetopus 10:2679. Error for *Platymetopius* Burmeister.
Platymoides 10:2679. Error for *Platymoideus* Ball.

Platymoideus Ball 1931d:219. *Platymetopius trilineatus* Ball 1916b:204 (USA: California, Santa Margarita). 10:2264. Deltocephalinae, Scaphytopiini. Nearctic. Synonym of *Scaphytopius (Cloanthanus)* Ball 1931d:219.

Platymstopius 10:2680. Error for *Platymetopius* Burmeister.

Platypona DeLong 1982e:140. *P. sinverda* n.sp. (Peru: Sinchona). Gyponinae, Gyponini. Neotropical.

Platyproctus Lindberg 1925a:112. *P. tessellatus* n.sp. (Transcaspia). 14:160. Adelungiinae, Adelungiini. Palaearctic; Mediterranean subregion.

Platyrectus Datta 1973i:435. Error for *Platyretus* Melichar.

Platyretus Melichar 1903b:174. *P. marginatus* n.sp. (Ceylon: Peradeniya). 10:2089. Deltocephalinae, Athysanini. Oriental.

Platyscopus Evans, J.W. 1941f:145. *P. badius* n.sp. (Western Australia: Dendari). 3:204. Penthimiinae. Australian.

Platytetticis Strand 1942a:393, replacement for *Platytettix* Matsumura not *Platytettix* Hancock 1906. *Motschulskyia pulchra* Matsumura 1916b:397 (Japan: Honshu). 17:1046. Typhlocybinae, Erythroneurini. Palaearctic/Oriental; Japan, India, Taiwan. Synonym of *Diomma* de Motschulsky 1863a:102; Dworakowska 1981c:367.
Synonym:
Platytettix Matsumura 1932a:104, preoccupied.

Platytettix Matsumura 1932a:104. *Motschukskyia pulchra* Matsumura 1916b:397 (Japan: Honshu). 17:1046, 1506. Typhlocybinae, Erythroneurini. Palaearctic, Oriental. Preoccupied by *Platytettix* Hancock 1906, see *Platytetticis* Strand. Synonym of *Diomma* de Motschulsky 1863a:102; Dworakowska 1981c:367.

Platytettix Vilbaste 1961a:45. *P. viridis* n.sp. (USSR: Uzbekistan). Deltocephalinae, Athysanini. Palaearctic. Preoccupied by *Platytettix* Hancock 1906, see *Scenergates* Emeljanov 1972d:106.

Playmetopius 10:2680. Error for *Platymetopius* Burmeister.

Pleargus Emeljanov 1964c:430. *Deltocephalus pygmaeus* Horvath 1897b:634 (Hungary). Deltocephalinae, Deltocephalini. Palaearctic; European and Siberian subregions.

Pleopardus Emeljanov 1975a:386. *Platyproctus rubiginosis* Mitjaev 1969a:366 (USSR: Kazakhstan). Adelungiinae, Adelungiini. Palaearctic.

Plepsius 10:2680. Error for *Phlepsius* Fieber.

Plerogonalia Young 1977a:526. *Tettigoniella rudicula* Jacobi 1905c:184 ("Peru S; Marcapata"). Cicadellinae, Cicadellini. Neotropical; Bolivia, Ecuador, Peru.

Plesiommata Provancher 1889a:263. *P. biundulata* n.sp. (Canada: Quebec) = *Tettigonia tripunctata* Fitch 1851a:55. 1:458. Young 1977a:595. Cicadellinae, Cicadellini. Nearctic, Neotropical.
Synonym:
Provancherana Hamilton, K.G.A. 1976a:36. *Tettigonia tripunctata* Fitch 1851a:55 (USA: New York).
[NOTE: Hamilton, K.G.A. 1976a:36 referred *Plesiommata* Provancher to the Cercopoidea; Young 1977a:59 rejected that assignment and placed *Provancherana* Hamilton, K.G.A. 1976a:36 as a junior synonym of *Plesiommata* (Cicadellidae)].

Plessiommata 1:686. Error for *Plesiommata* Provancher.

Plexitartessus Evans, F. 1981a:151. *Tartessus pulchellus* Spangberg 1878c:10 (Australia: Queensland, Cape York Peninsula). Tartessinae. Australian, Oriental; Australia and New Guinea.

Plumerella DeLong 1942b:200, 1:686. Error for *Plummerella* DeLong.

Plummerella DeLong 1942b:200. *P. alpina* n.sp. (Mexico: Distrito Federal). 1:343,344. Young 1977a:902. Cicadellinae, Cicadellini. Nearctic, Neotropical.
[NOTE: Two spellings "*Plumerella*" and "*Plummerella*" appear in DeLong's 1942b paper. DeLong 1944b:68, first reviser, used the spelling "*Plummerella*," which conforms to the stated patronymic intent in the 1942b publication].

Plumosa Sohi 1977a:355. *P. emarginata* n.sp. (India: Dehra Dun). Typhlocybinae, Erythroneurini. Oriental.

Pochara 1:686. Error for *Poochara* Stal.

Podiopsis 13:259. Error for *Pediopsis* Burmeister.

Podulmorinus Kwon 1985a:69 as subgenus of *Idiocerus* Lewis 1834a:47. *Idiocerus vitticollis* Matsumura 1905a:69 (Japan). Idiocerinae. Palaearctic.

Poecilocarda Metcalf 1952a:228, replacement for *Poeciloscarta* Melichar 1926a:342, not *Poeciloscarta* Stal 1869a:73. *Tettigonia binaria* Signoret 1860a:204 (Madagascar). 1:95. Linnavuori 1979a:694-705. Cicadellinae, Cicadellini. Malagasian and Ethiopian; widespread in Africa.

Poeciloscata 1:686. Error for *Poeciloscarta* Stal.

Poeciloscarta Stal 1869a:73 as subgenus of *Tettigonia*. *Cicada cardinalis* Fabricius 1803a:71 ("Amer. merid."). 1:58. Young 1977a:197. Cicadellinae, Cicadellini. Neotropical; Bolivia, Brazil, Colombia, the Guianas, Peru.

POGONOSCOPINAE 12:39.

Pogonoscopus China 1924c:529. *P. myrmex* n.sp. (Western Australia: Perth). 12:40. Evans, J.W. 1966a:75-77. Eurymelinae, Pogonoscopini. Australian; Australia.

Polana DeLong 1942d:110. *Gypona quadrinotata* Spangberg 1878a:56 (USA: Georgia & Texas). 3:155. DeLong & Freytag 1972a:226, 1972f. Gyponinae, Gyponini. Nearctic/Neotropical; southern USA, Central America, Antillean subregion and much of South America.
Subgenera:
Angusana DeLong & Freytag 1972f:254. *Gypona exornata* Fowler 1903b:315.
Bohemanella DeLong & Freytag 1972f:242. *Gypona bohemani* Stal 1864a:81.
Bulbusana DeLong & Freytag 1972f:256. *P. (B.) plumea* n.sp.
Declivella DeLong & Freytag 1972f:241. *P. (D.) danesa* n.sp.
Largulara DeLong & Freytag 1972f:292. *P. (L.) fantasa* n.sp.
Nihilana DeLong & Freytag 1972f:295. *P. (N.) pressa* n.sp.
Polanana DeLong & Freytag 1972f:309. *Gypona venosa* Stal 1854b:252.
Polanella DeLong & Freytag 1972f:314. *P. (P.) cupida* n.sp.
Parvulana DeLong & Freytag 1972f:289. *P. (P.) alata* n.sp.
Striapona DeLong 1979g:187. *P. (S.) desela* n.sp.
Validapona DeLong 1979g:188. *P. (V.) lamina* n.sp.

Polanana DeLong & Freytag 1972f:309 as subgenus of *Polana* DeLong 1942d:110. *Gypona venosa* Stal 1854b:252 (Brazil: Minas Gerais). Gyponinae, Gyponini. Neotropical.

Polanella DeLong & Freytag 1972f:314 as subgenus of *Polana* DeLong 1942a:110. *P. (P.) cupida* n.sp. ("Stieglmayr, Rio Grande do Sul"). Gyponinae, Gyponini. Neotropical.

Poliona Emeljanov 1972d:105. *Phlepsius microcephalus* Kusnezov 1929a:315 (Turkestan). Deltocephalinae, Athysanini. Palaearctic; Mediterranean subregion.

Polisanella Melichar 1926a:343. *P. bakeri* Metcalf 1953a:50 (Philippines: Luzon, Mount Polis). 1:265. Cicadellinae, Cicadellini. Oriental.

Polluxia Dworakowska 1974a:168. *P. vultuosa* n.sp. (Congo: Odzala). Typhlocybinae, Zyginellini. Ethiopian.

Polyamia DeLong 1926d:62 as subgenus of *Deltocephalus* Burmeister 1838b:15. *Deltocephalus weedi* Van Duzee 1892d:306 (USA: Mississippi). 10:1788. Deltocephalinae, Deltocephalini. Nearctic and Neotropical; widespread.

Polyania 10:2680. Error for *Polyamia* DeLong.

Polymia 10:2680. Error for *Polyamia* DeLong.

Polynia Zachvatkin 1953c:229 as subgenus of *Kyboasca* Zachvatkin 1953c:228. *Chlorita vittata* Lethierry 1884c:65 (Saratov). Typhlocybinae, Empoascini. Palaearctic. Synonym of *Austroasca* Lower 1952a:202; Dworakowska 1970h:712.

Ponana Ball 1920a:93 as subgenus of *Gypona* Germar 1821a:73. *Gypona scarlatina* Fitch 1851a:57 (USA: New York). 3:125, 132. DeLong & Freytag 1967a, 1972a:225. Gyponinae, Gyponini. Neotropical, Nearctic; from eastern Canada S and SW to Colombia and Peru.
Subgenera:
Bulbana DeLong 1942d:107. *B. pura* n.sp.
Neopanana DeLong & Freytag 1967a. *P. (N.) demela* n.sp.
Lataponana DeLong 1977e:109. *P. (L.) ampa* n.sp.
Proxaponana DeLong 1977e:111. *P. (P.) pertenua* n.sp.
Peranoa DeLong 1980c:217. *P. (P.) perusana* n.sp.

Ponanella DeLong & Freytag 1969b:303. *P. ena* n.sp. (Mexico: Oaxaca, Chichicazapa). Gyponinae, Gyponini. Neotropical; tropics of Mexico, Central and South America.

Poniagnathus Dubovskij 1978a:44. Error for *Goniagnathus* Fieber.

Ponona 3:225. Error for *Ponana* Ball.

Pontimia 3:225. Error for *Penthimia* Germar.

Poochara Stal 1869a:77 as subgenus of *Tettigonia*. *Tettigonia farinosa* Fabricius 1803a:70 (Sumatra). 1:259. Cicadellinae, Cicadellini. Oriental; Indo-Malayan subregion.

Populicerus Dlabola 1974b:62. *Cicada populi* Linnaeus 1761a:242 (Sweden). Idiocerinae. Palaearctic. Subgenus of *Idiocerus* Lewis 1834a:47; Hamilton, K.G.A. 1980a:825 & 838; Kwon 1985a:67.

Poralimnus 10:2680. Error for *Paralimnus* Matsumura.

Porcorhinus Goding 1903a:38. *P. mastersi* n.sp. (Australia: New South Wales). 4:93. Evans 1966a:97. Ledrinae, Ledrini. Australian.
Synonym:
Gudwana Distant 1917c:189. *G. typica* n.sp.

PORTANINI Linnavuori 1959b:45; tribe of XESTOCEPHALINAE.

Portanus Ball 1932a:18. *Scaphoideus stigmosus* Uhler 1895a:77 (St. Vincent Island). 10:2202. Kramer 1964e. Xestocephalinae, Portanini. Neotropical; Brazilian and Antillean subregions.

Postumus Distant 1918b:84. *P. fascialis* n.sp. (South India: Chikkaballapura). 8:224. Deltocephalinae, Chiasmusini. Oriental. Synonym of *Chiasmus* Mulsant & Rey 1855a:215; Evans, J.W. 1974a:172.

Poyamia 10:2681. Error for *Polyamia* DeLong.

Pradhanasundra Ramakrishnan & Ghauri 1979b:201. *Empoascanara capreola* Dworakowska 1978a:154 (Nigeria: W State, Ile-Ife). Typhlocybinae, Erythroneurini. Ethiopian. Synonym of *Empoascanara* Distant 1918b:94; Dworakowska 1980d:195.

Praganus Dlabola 1949a:2. *Deltocephalus hofferi* Dlabola 1947a:19 (Central Bohemia: Devin, near Prague). 10:1200. Deltocephalinae, Deltocephalini. Palaearctic; southern Eurasia.

Prainiana 3:225. Error for *Prairiana* Ball.

Prairiana Ball 1920a:90 as subgenus of *Gypona* Germar 1821a:73. *Gypona cinerea* Uhler 1877a:460 (USA: Colorado, Manitou). 3:55. Gyponinae, Gyponini. Nearctic and Neotropical: widespread.

Prasutagus Distant 1918b:53. *P. pulchellus* n.sp. (South India: Nandidrug). 10:418. Deltocephalinae, Athysanini. Oriental.

Pratura Theron 1982b:21. *P. graminea* n.sp. (South Africa: Natal, Ladysmith). Deltocephalinae, Deltocephalini. Ethiopian.

Pravistylus Theron 1975a:195. *Deltocephalus eductus* Naude 1926a:43 ("Petrusburg, O.F.S.", Orange Free State). Deltocephalinae, Deltocephalini. Ethiopian; South Africa.

Preconia 1:687. Error for *Proconia* LePeletier & Serville.

Premanus DeLong 1944g:98. *P. hebatus* n.sp. (Mexico: San Luis Potosis, Tamazunchale). 10:547. Deltocephalinae, Athysanini. Neotropical.

Prescottia Ball 1932a:16. *Scaphoideus lobatus* Van Duzee 1894f:211 (USA: New York, Lancaster). 10:2198. Deltocephalinae, Scaphoideini. Nearctic; eastern USA and adjacent Canada, also Sonoran.

Preta Distant 1908g:234. *Signoretia gratiosa* Melichar 1903b:160 (Ceylon: Peradeniya). 6:57. Cicadellinae, Evacanthini. Oriental; Ceylonese and Indo-Malayan subregions.

Pretioscopus Webb, M.D. 1983b:243. *Idioscopus flavosignatus* Webb 1976a:302 (Nigeria: SE State, Oban Rest House). Idiocerinae. Ethiopian; tropical Africa.

Priariana 3:225. Error for *Prairiana* Ball.

Procama Young 1968a:83. *Amblydisca fluctuosa* Fowler 1898a:211 ("Caldera, 1200 ft"). Cicadellinae, Proconiini. Neotropical; Panama.

Procandea Young 1968a:69. *Tettigonia corticata* Signoret 1855b:226 (Peru). Cicadellinae, Proconiini. Neotropical; Amazonas, Bolivia, Ecuador, Peru.

Procephaleus Evans, J.W. 1936a:43. *P. bulbosa* n.sp. (Western Australia: Carlisle). 5:39. Ulopinae, Cephalelini. Australian.

Proceps Mulsant & Rey 1855a:237. *P. acicularis* n.sp. (France). 10:2336. Deltocephalinae, Platymetopiini. Palaearctic; Mediterranean subregion.

Proconia Le Peletier & Serville 1825a:610. *Cicada cristata* Fabricius 1803a:5, 62 (Surinam), preoccupied = *Proconia esmeraldae* Melichar 1924a:203. 1:472. Young 1968a:26. Cicadellinae, Proconiini. Neotropical; northern and central South America.
Synonyms:
Germaria LaPorte 1832b:222. *G. cuculata* n.sp. = *Cicada marmorata* Fabricius 1803a:61.
Zyzza Kirkaldy 1900b:243, replacement for *Germaria* LaPorte, preoccupied. *Germaria cuculata* LaPorte 1832b:223.
Eustollia Goding 1925a:105. *Cicada jubata* Stoll 1781a:30 = *Cicada marmorata* Fabricius 1803a:61.

PROCONIIDAE 1:8.
PROCONIINAE 1:466.
PROCONIINI 1:472.

Proconobola Young 1968a:64. *Amblydisca callidula* Jacobi 1905c:65 (Peru S: Marcapata). Cicadellinae, Proconiini. Neotropical; Bolivia, Ecuador, Peru.

Proconopera Young 1968a:113. *Amblydisca pullula* Jacobi 1905c:166 (Peru S: Marcapata). Cicadellinae, Proconiini. Neotropical; Bolivia, Colombia, Ecuador, Peru.

Proconosama Young 1968a:106. *Aulacizes alalia* Distant 1908b:76 (Bolivia: Toungas de la Paz). Cicadellinae, Proconiini. Neotropical; Bolivia, Colombia, Ecuador, Peru, Venezuella.

Prodesmia Amyot 8:265. Invalid, not binominal.

Proekes Theron 1975a:204. *Deltocephalus cephaleus* Naude 1926a:44 (South Africa: Viljoen's Pass, Cape Province). Deltocephalinae, Stenometopiini. Ethiopian.

Proeonia 1:687. Error for *Proconia* LePeletier & Serville.

Promecopsis Dumeril 1806. Suppressed by the ICZN for purposes of the Law of Priority but not for the Law of Homonymy, Opinion 605, 1961.

Promocuus Emeljanov 1972b:248 as subgenus of *Mocuellus* Ribaut 1946b:83. *Mocuellus hordei* Emeljanov 1964f:47 (Mongolia). Deltocephalinae, Paralimnini. Palaearctic.

Propetes Walker 1851b:797. *P. compressa* n.sp. ("Para", Brazil). 1:516. Young 1968a:204. Cicadellinae, Proconiini. Neotropical; Mexico to the Guianas and Ecuador.

Proranus Spinola 1850b:122. *P. ghilianii* n.sp. ("Para", Brazil). 4:94. Kramer 1966a:481-488. Ledrinae, Petalocephalini. Neotropical; Bolivia, Brazil, Colombia, Panama, Paraguay, Peru, Venezuela.

Proskura Dworakowska 1981j:225. *P. depressa* n.sp. (India: Madhya Pradesh, Choral, 35 km S of Indore). Typhlocybinae, Erythroneurini. Oriental; India.

Prosopoxys Jacobi 1917a:546. *P. thoracicus* n.sp. (Madagascar). 15:115. Iassinae, Iassini. Malagasian.

Prostictops Amyot 15:224. Invalid, not binominal.

Prostigmoderus Amyot 8:265. Invalid, not binominal.

Protaenia Amyot 10:2681. Invalid, not binominal.

Protaenia Thomson 1869a:58 as subgenus of *Jassus*. *Jassus pinetellus* Zetterstedt 1840a:1077 (Lapland) = *Jassus puncticollis* Herrich-Schaeffer 1834d:7. 10:1917. Deltocephalinae, Grypotini. Palaearctic. Synonym of *Grypotes* Fieber 1866a:503.

Protalebra Baker 1899b:402. *Alebra curvilinea* Gillette 1898a:710 (Brazil: Chapada). 17:84. Typhlocybinae, Alebrini. Neotropical; Argentina, Brazil, Virgin Islands.

Protalebrella Young 1952b:38. *Protalebra brasiliensis* Baker 1899b:405 (Brazil: Minas Gerais). 17:92. Typhlocybinae, Alebrini. Neotropical, Nearctic; southern USA, Antillean, Brazilian, and Mexican subregions.

Protartessus Evans, F. 1981a:138. *Tartessus spinosus* Evans, J.W. 1936b:55 (Western Australia: Carnac I.). Tartessinae. Australian.

Protelebra 17:1507. Error for *Protalebra* Baker.

Protochlorotettix Pierce, W.D. 1963a:78. *P. calico* n.sp. by monotypy. Fossil: Miocene, California, San Bernardino County, Calico Mtns. NE 1/4 Section 19, R2W T10N. Nearctic.

Protoenia 10:2681. Error for *Protaenia* Thomson.

Protolebra 17:1507. Error for *Protalebra* Baker.

Protonesis Spinola 1850b:125. *P. delegorguei* n.sp. (South Africa: Port Natal). 11:83. Linnavuori 1970a:183-184. Unassigned, formerly Coelidiinae; Nielson 1975a:11. Ethiopian.

Protranus DeLong 1980h:68 as subgenus of *Scaphytopius* Ball 1931d:218. *S. (P.) abutus* n.sp. (Brazil: Caraguatatatuba). Deltocephalinae, Scaphytopiini. Neotropical.

Provancherana Hamilton, K.G.A. 1976a:36, replacement for *Plesiommata* of authors, not Provancher 1889a:263, with type species *Tettigonia tripunctata* Fitch 1851a:55 (USA: New York), and characteristics as described by Oman 1949a:67. Cicadellinae, Cicadellini. Nearctic, Neotropical. Synonym of *Plesiommata* Provancher 1889a:263; Young 1977a:595.

Proxaponana DeLong 1977e:111 as subgenus of *Ponana* Ball 1920a:93. *P. (P.) pertenua* n.sp. (Peru: Sinchona). Gyponinae, Gyponini. Neotropical.

Proxima DeLong & Freytag 1975a:111. *P. ocellata* n.sp. (Brazil: E. Santo, Tijuco Presto). Gyponinae, Gyponini. Neotropical.

Pruthiana Izzard 1955a:186. *P. sexnotata* n.sp. (South India). Nirvaninae, Nirvanini. Oriental.

Pruthiorosius Ghauri 1963f:559. *Orosius maculatus* Pruthi 1930a:67 (Central India). Deltocephalinae, Opsiini. Oriental.

Pruthius Mahmood 1967a:33. *P. aureatus* n.sp. (Singapore). Typhlocybinae, Typhlocybini. Oriental. Synonym of *Limassola* Dlabola 1965e:663; Dworakowska 1969d:433.

Psammootettix 10:2681. Error for *Psammotettix* Haupt.

Psammotettis Servadei 1972a:20. Error for *Psammotettix* Haupt.

Psammotettix Haupt 1929c:262. *Athysanus maritimus* Perris 1857a:172. 10:1486. Deltocephalinae, Paralimnini. Holarctic; widespread.
Synonym:
Ribautiellus Zachvatkin 1933b:268. *Cicada striata* Linnaeus 1758a:437.

Psamotettix 10:2681. Error for *Psammotettix* Haupt.

Psegmatus Fieber 1875a:402. *P. lethierryi* n.sp. ("Georgia", USSR). 9:50. Deltocephalinae, Hecalini. Palaearctic; Southern Russia, Caucasus.

Pseudaconura Linnavuori 1952b:182. *P. luxorensis* n.sp. (Egypt: Luxor). 10:1598. Deltocephalinae, Stenometopiini. Ethiopian. Synonym of *Doratulina* Melichar 1903b:198; Nast 1972a:354.

Pseudalaca Linnavuori 1959b:236. *Agallia multipunctata* Osborn 1923b:10 (Paraguay). Deltocephalinae. Neotropical.

Pseudaligia Kramer & DeLong 1968a:171. *P. nigropunctata* n.sp. (Mexico: Guerrero, Iguala). Deltocephalinae, Athysanini. Neotropical.

Pseudaraldus Bonfils 1981a:304. *P. giustinai* n.sp. (Corsica). Deltocephalinae, Paralimnini. Palaearctic; Mediterranean subregion.

Pseudobalbillus Jacobi 1912a:37. *P. protrudens* n.sp. (Belgian Congo). 7:3. Nirvaninae, Macroceratogoniini. Ethiopian; Guinean.

Pseudocephalelus Linnavuori 1969b:214. *Cephalelus bleusei* Puton 1898a:172 (Algeria). Deltocephalinae, Athysanini. Palaearctic; Algeria, Morocco.

Pseudoidioscopus Maldonado-Capriles 1977b:362. *P. bipunctatus* n.sp. (British Guiana: New River). Idiocerinae. Neotropical.

Pseudolausulus Wagner & Franz 1961a:150. *Deltocephalus laciniatus* Then 1896a:189 (Salzburg). Deltocephalinae, Paralimnini. Palaearctic.

Pseudolausus Servadei 1968a:168. Error for *Pseudolausulus* Wagner & Franz.

Pseudometopia Schmidt 1928c:74. *P. appendiculata* n.sp. (Colombia). 1:559. Young 1968a:99. Cicadellinae, Proconiini. Neotropical; Trinidad Island, northern South America to Paraguay, Brazil and Argentina.

Pseudonirvana Baker 1923a:386. *P. sandakanensis* n.sp. (Borneo). 7:26. Nirvaninae, Nirvanini. Oriental. Synonym of *Sophonia* Walker 1870b:327.

Pseudophera Melichar 1925a:332. *Proconia atra* Walker 1851b:789 (Honduras). 1:514. Young 1968a:206. Cicadellinae, Proconiini. Neotropical; southern Mexico to Colombia and Ecuador.

Pseudophlepsius Zachvatkin 1924a:128. *Phlepsius binotatus* Signoret 1880b:189 (Saratov). 10:706. Deltocephalinae, Athysanini. Palaearctic; Siberian subregion.

Pseudoscarta Melichar 1926a:341. *Cicada fastuosa* Fabricius 1803a:70 ("Amer. merid."). 1:206. Young 1977a:104. Cicadellinae, Cicadellini. Neotropical. Synonym of *Dilobopterus* Signoret 1850b:284; Young 1977a:104.

Pseudosubhimalus Ghauri 1974c:553. *Ophiola bicolor* Pruthi 1936a:123 (India: Punjab). Deltocephalinae, Deltocephalini. Oriental; India, Pakistan.

Pseudosudra Schmidt 1920l:118. *Sudra borneensis* Schmidt 1909d:265 (North Borneo). 2:9. Hylicinae, Sudrini. Oriental.

Pseudotettix 10:2681. Error for *Speudotettix* Ribaut.

Pseudothaia Kuoh, C.L. 1982a:402 & 404. *P. striata* n.sp. (Hainan Island). Typhlocybinae, Erythroneurini. Oriental.

Pseupalus Remane & Asche 1980a:88. *P. graecanarus* n.sp. (Canary Islands). Deltocephalinae, Paralimnini. Ethiopian.

Pseutettix DeLong 1967d:210. *P. mexicana* n.sp. (Mexico: Morelos, Cuernavaca). Deltocephalinae, Athysanini. Neotropical; Mexico.

Psibala Kramer 1963b:206. *P. empusa* n.sp. (Bolivia: Huachi Beni). Neobalinae. Neotropical.

Psimmythimas Amyot 11:178. Invalid, not binominal.

Psymmitimas 11:178. Error for *Psimmythimas* Amyot, invalid.

Pteropyx Haupt 1927a:27. *P. hyalinus* n.sp. (Palestine). 10:2484. Deltocephalinae, Paralimnini. Palaearctic; Mediterranean subregion.

Pugla Distant 1908g:318. *P. sigillaris* n.sp. (Burma: Tenasserim, Myitta). 10:2351. Deltocephalinae, Platymetopiini. Oriental.

Pugnostilus Kwon 1985a:70 as subgenus of *Idiocerus* Lewis 1834a:47. *Idiocerus latistylus* Vilbaste 1968a:69 (USSR: Maritime Territory). Idiocerinae. Palaearctic.

Punahuana Young 1977a:317. *Cicadella brunneatula* Osborn 1926b:185 (Bolivia). Cicadellinae, Cicadellini. Neotropical; Bolivia, Peru.

Punctigerella Vilbaste 1968a:112 as subgenus of *Erythroneura* Fitch 1851a:62. *E. (P.) lamellaris* n.sp. (USSR: Maritime Territory). Typhlocybinae, Erythroneurini. Palaearctic.

Purpuranus Zachvatkin 1933b:262. *Platymetopius rubrostriatus* Horvath 1907a:317 (Caucasus). 10:2205. Deltocephalinae, Scaphytopiini. Palaearctic. Synonym of *Stymphalus* Stal 1866a:121; Nast 1972a:345.

Pusaneura Ramakrishnan & Menon 1971a:459. *P. signata* n.sp. (India: Delhi, I.A.R.I.). Typhlocybinae, Dikraneurini. Oriental. Synonym of *Uzeldikra* Dworakowska 1971d:585; Sharma 1978a:12.

Pusatettix Ramakrishnan & Menon 1971a:463. *P. bipunctaus* n.sp. (India: Delhi, I.A.R.I). Typhlocybinae, Dikraneurini. Oriental. Synonym of *Javadikra* Dworakowska 1971d:585; Dworakowska 1977a:597, which is a junior synonym of *Karachiota* Ahmed, Manzoor 1969a:58; Sohi & Dworakowska 1984a:165.

Putoniessa Kirkaldy 1907d:50. *P. dignissima* n.sp. (Australia: New South Wales). 4:129. Evans, J.W. 1966a:118-123. Ledrinae, Thymbrini. Australian.

Putoniessiella Evans 1969a:746. *P. sagitta* n.sp. (Western Australia: Tammin). Ledrinae, Thymbrini. Australian.

Pycnoides Emeljanov 1964f:32 as subgenus of *Handianus* Ribaut 1942a:265. *Jassus flavovarius* Herrich-Schaeffer 1835c:9 (Germany). Deltocephalinae, Athysanini. Palaearctic.

Pygometopia Schroeder 1960a:319 as subgenus of *Oncometopia* Stal 1869a:62. *Oncometopia (Mucrometopia) infuscata* Melichar 1925a:404 (Peru). Cicadellinae, Cicadellini. Neotropical; northern South America. Synonym of *Teleogonia* Melichar 1924a:198; Young 1977a:158.

Pygotettix Matsumura 1940d:42. *P. formosanus* n.sp. (Formosa). 10:389. Deltocephalinae, Athysanini. Oriental.

Pyramidotettix Matsumura 1931b:59. *Conometopus citri* Matsumura 1907c:113 (Japan). 17:229. Dworakowska 1970h:707. Typhlocybinae, Dikraneurini. Palaearctic. Synonym of *Zyginella* Loew 1885a:346; Dworakowska 1970h:707.

Pyramidotettitx 17:1507. Error for *Pyramidotettix* Matsumura.

Pyrotaenia Amyot 17:1507. Invalid, not binominal.

PYTHAMNINAE 6:45.

Pythamus Melichar 1903b:161. *P. dealbatus* n.sp. (Ceylon). 6:46. Cicadellinae, Evacanthini. Oriental; Borneo, Ceylon, Philippines, Malaysia.

Pythonirvana Baker 1923a:397. *P. muiri* n.sp. (Borneo). 7:22. Nirvaninae, Nirvanini. Oriental; Borneo, Ambon (Ambonia).

Qadria Mahmood 1967a:18. *Empoasca rubronotata* Distant 1918b:92 (South India: Kodiakanal). Typhlocybinae, Erythroneurini. Oriental; Bangladesh, India, Pakistan, Vietnam.
Synonym:
Spatulostylus Ramakrishnan & Menon 1973a:16. *S. variegatus* n.sp.

Quadria Dworakowska 1977f:293. Error for *Qadria* Mahmood 1967a:18.

Quartasca Dworakowska 1972i:27 as subgenus of *Laokayana* Dworakowska 1972i:27. *L. (Q.) czerwcowa* n.sp. (Vietnam: Bao-Ha, S of Fan-Si-Pan). Typhlocybinae, Empoascini. Oriental. Subgenus of *Amrasca* Ghauri 1967a:159; Dworakowska 1977i:16.

Quartauropa Webb, M.D. 1983b:252. *Idioscopus nigrocellus* Webb, M.D. 1976a:330 (Angola: Tumdavla). Idiocerinae. Ethiopian; Angola.

Quartausius Dlabola 1974c:123. *Q. dalmatinus* n.sp. (Yugoslavia: Srebrenica). Deltocephalinae, Paralimnini. Palaearctic.

Quaziptus Kramer 1965b:29. *Q. chapini* Kramer n.sp. (Columbia: Cundinamarca, Guasca, 3300 m). Deltocephalinae, Deltocephalini. Neotropical.

Quercinirvana Ahmed, Manzoor & Mahmood 1970a:260. *Q. longicephala* n.sp. (W. Pakistan: Abbotabad). Nirvaninae, Nirvanini. Oriental; Pakistan, Bangladesh.

Quernus Dlabola 1974a:57 as subgenus of *Platymetopius* Burmeister 1838b:16. *Platymetopius signoreti* Metcalf 1967c:2230 (Southern Europe) = *Platymetopius viridinervis* Signoret 1880b:192, not Kirschbaum 1868. Deltocephalinae, Platymetopiini. Palaearctic.

Quichira Young 1968a:233. *Q. tegminis* Young n.sp. (Panama: Volcan de Chiriqui, 8000 ft). Cicadellinae, Proconiini. Neotropical.

Quilopsis Webb, M.D. 1983a:65. *Q. koebelei* n.sp. (West Australia: Yanchep). Idiocerinae. Australian.

Quontus Oman 1949a:170. *Deltocephalus misellus* Ball 1899b:191 (USA: Colorado). 10:1777. Deltocephalinae, Deltocephalini. Nearctic. Synonym of *Latalus* DeLong & Sleesman 1929a:87.

Raabeina* Dworakowska 1972j:113. *R. fuscofasciata* n.sp. (Vietnam: Bao-Ha, S of Fan-Si-Pan). Typhlocybinae, Erythroneurini. Oriental. * Name erroneously credited to Jaczewski in Zoological Record 109:155.

Rabela Young 1952b:21. *Protalebra tabebuiae* Dozier 1927b:260 (Puerto Rico: Rio Piedras). 17:20. Typhlocybinae, Alebrini. Neotropical; Antillian subregion and Brazil.

Rabiana Mahmood 1967a:36. *R. delicatula* n.sp. (Philippines: Luzon, Mt. Makiling). Typhlocybinae, Typhlocybini. Oriental.

Racinolidia Nielson 1983h:569. *R. amazonensis* n.sp. (Brazil: Amazon, Fonteboa). Coelidiinae, Teruliini. Neotropical.

Radhades Distant 1912e:606. *R. crassus* n.sp. (Burma: Tenasserim). 5:87. Ulopinae, Ulopini. Oriental.

Ragia Theron 1973a:29. *Cicadula flavoalbida* Naude 1926a:83. (South Africa: Petrusburg, O.F.S.). Deltocephalinae, Deltocephalini. Ethiopian.

Ramakrishnania Dworakowska 1974a:171. *R. ushae* n.sp. (Congo: Sibiti). Typhlocybinae, Zyginellini. Ethiopian.

Ramania* Dworakowska 1972j:107. *R. javaica* n.sp. (Java: Tjibulan, near Bogor). Typhlocybinae, Erythroneurini. Oriental. * Name erroneously credited to Jaczewski in Zoological Record 109:155.

Ramosulus Young 1977a:438. *Cicadella corrugipennis* Osborn 1926a:204 (Bolivia). Cicadellinae, Cicadellini. Neotropical; Bolivia, Brazil, Ecuador, Peru.

Ramsisia Einyu & Ahmed, Manzoor 1979a:304. *R. ankolensis* n.sp. (Uganda). Typhlocybinae, Dikraneurini. Ethiopian.

Ranbara Dworakowska 1983a:116. *R. variabilis* n.sp. (Thailand: Chiang Dao, Chiang Mai). Typhlocybinae, Erythroneurini. Oriental.

Raphirinus 1:687. Error for *Raphirhinus* de LaPorte.

Raphirhinus de LaPorte 1832d:413. *Fulgora adscendens* Fabricius 1787a:260 (Cayenne) = *Cicada phosphorea* Linnaeus 1758a:434. 1:528. Young 1968a:157. Cicadellinae, Proconiini. Neotropical; northern South America and Trinidad Island.

Ratbura Mahmood 1967a:17. *Empoasca nagpurensis* Distant 1918b:93 (India: Nagpur). Typhlocybinae, Erythroneurini. Oriental. Synonym of *Empoascanara* Distant 1918b:94; Dworakowska 1980d:193-194.

Ratburella Ramakrishnan & Menon 1973a:20. *R. unipunctata* n.sp. (India: Delhi, I.A.R.I.). Typhlocybinae, Erythroneurini. Oriental.
Subgenus:
Burara Dworakowska 1980b:184. *R. (B.) maxima* n.sp.

Ratjalia Dworakowska 1981j:239. *R. plova* n.sp. (India: Karnataka, 18- 25 km E. of Mudigere). Typhlocybinae, Erythroneurini. Oriental.

Raunoia Dworakowska & Lauterer 1975a:37. *R. luciphila* n.sp. (Guinea: Guekedou). Typhlocybinae, Zyginellini. Ethiopian. Synonym of *Sundara* Ramakrishnan & Menon 1972b:186; Dworakowska 1977i:32.

Raventazonia. Error for *Reventazonia* Linnavuori 1959b:138, Zoological Record 97, Section 13:333, 1960.

Rawania Ghauri 1963e:465. *Empoasca petasata* Melichar 1903b:215 (Ceylon: Kandy). Typhlocybinae, Empoascini. Oriental.

Readionia Young 1952b:61 as subgenus of *Dikrella* Oman 1949a:83. *Dikraneura readionis* Lawson 1930e:39 (USA: Arizona, Pima Co.). 17:281. Typhlocybinae, Dikraneurini. Nearctic.

Recilia Edwards 1922a:206. *Jassus (Deltocephalus) coronifer* Marshall 1866c:222 (England). 10:934. Kramer 1962b, Ghauri 1980b. Deltocephalinae, Deltocephalini. Palaearctic, Ethiopian, Oriental.
Subgenera:
Togacephalus Matsumura 1940d:38. *Deltocephalus distinctus* de Motschulsky 1859b:112.
Inazuma Ishihara 1953b:15. *Deltocephalus dorsalis* de Motschulsky 1859b:114.
Synonym (of *Togacephalus*):
Inemadara Ishihara 1953b:48. *Deltocephalus oryzae* Matsumura 1902a:390.

Redaprata Blocker 1979a:64. *R. culisona* n.sp. (Peru: E. side of Carpish Mts., 2800 m, 40 mi SW Tingo Maria). Iassinae. Neotropical.

Refrolix Theron 1984b:227. *R. ruensis* n.sp. (South Africa: Villiersdorp). Deltocephalinae, Athysanini. Ethiopian; South Africa.

Regalana DeLong & Freytag 1975b:121. *R. corona* n.sp. (Panama: Canal Zone, Alhajuelo). Gyponinae, Gyponini. Neotropical; Brazil, Panama.

Relaba Young 1957c:161. *R. williamsi* n.sp. (Ecuador: Tena). Typhlocybinae, Alebrini. Neotropical.

Relipo Evans, J.W. 1977a:88. *R. oenpellensis* n.sp. (Australia: Northern Territory, Birraduk Creek, 18 km E by N of Oenpelli). Eurymelinae, Ipoini. Australian.

Remadosus Ball 1929a:2. *Athysanus magnus* Osborn & Ball 1897a:225 (USA: Iowa, Ames). 10:444. Deltocephalinae, Athysanini. Nearctic. Subgenus of *Pendarus* Ball 1927d:262; Hamilton, K.G.A. 1975a:7.

Remarana DeLong & Freytag 1976a:54 as subgenus of *Curtara* DeLong & Freytag 1972a:231. *C. (R.) remara* n.sp. (Peru: Huanuco, Huallaga River valley). Gyponinae, Gyponini. Neotropical.

Remmia Vilbaste 1968a:91. *R. orbigera* n.sp. (USSR: Maritime Territory). Typhlocybinae, Zyginellini. Palaearctic. Synonym of *Zyginella* Loew 1885a:346; Dworakowska 1970h:707.

Remoya Webb, M.D. 1983b:254. *Idioscopus aldabraensis* Webb, M.D. 1976a:318 (Aldabra: S. Island, Anse Cedre). Idiocerinae. Ethiopian.

Renonus DeLong 1959a:325. *R. rubraviridis* n.sp. (Mexico: Guererro, Iguala). Deltocephalinae, Athysanini. Neotropical.

Renosteria Theron 1974b:147. *Chlorotettix spadix* Naude 1926a:57 (South Africa: Cape Province, Ceres). Deltocephalinae, Athysanini. Ethiopian.

Resomus Amyot 14:172. Invalid, not binominal.

Reticana DeLong & Freytag 1964a:121. *Gypona lineata* Burmeister 1839b:4 (Brazil). Gyponinae, Gyponini. Neotropical; Argentina, Brazil, Colombia.

Reticopsis Hamilton, K.G.A. 1980b:885. *Pediopsis nubila* Van Duzee 1890d:37 (USA: California). Macropsinae, Macropsini. Nearctic; southern California.

Retusana DeLong & Freytag 1976a:57 as subgenus of *Curtara* DeLong & Freytag 1972a:231. *C. (R.) retusa* n.sp. (French Guiana: Guyane). Gyponinae, Gyponini. Neotropical.

Retusanus DeLong 1945g:135. *R. punctatus* n.sp. (Mexico: Guerrero, Iguala). 10:1657. Deltocephalinae, Athysanini. Neotropical.

Reuplemmeles Evans, J.W. 1966a:209. *Reuteriella hobartensis* Evans, J.W. 1938b:11 (Tasmania: Hobart). Iassinae, Reuplemmelini. Australian; Australia, Tasmania, New Guinea.

REUPLEMMELINI Evans, J.W. 1966a:209, tribe of JASSINAE = IASSINAE.

Reuteria Signoret 1879a:51. *R. flavescens* Signoret 1880a:46 (Tasmania). 15:135. Deltocephalinae, Hecalini. Australian. Preoccupied by *Reuteria* Puton 1875, see *Reuteriella* Signoret 1879b:267.

Reuteriella Signoret 1879b:267, replacement for *Reuteria* Signoret 1879a:51, not *Reuteria* Puton 1875. *Reuteria flavescens* Signoret 1880a:46 (Tasmania). 15:135. Deltocephalinae, Hecalini. Australian; Tasmania.

REUTERIELLINI 15:134.

Reventazonia Linnavuori 1959b:138. *R. atrifrons* n.sp. (Costa Rica: Farm Hamburg, Reventazon). Deltocephalinae, Deltocephalini. Nearctic- Neotropical; Brazil, Costa Rica. USA: central & southeast.

Rhabdotalebra Young 1952b:36. *Protalebra octolineata* Baker 1903d:7 (Nicaragua: San Marcos or Managua). 17:80. Typhlocybinae, Alebrini. Neotropical; Antillean and Brazilian subregions.

Rhaphidorhimus 1:687. Error for *Raphirhinus* de LaPorte.
Rhaphidorhinus 1:687. Error for *Raphirhinus* de LaPorte.
Rhaphidorhynchus 1:687. Error for *Raphirhinus* de LaPorte.
Rhaphirhinus 1:687. Error for *Raphirhinus* de LaPorte.
Rhaphirinus 1:687. Error for *Raphirhinus* de LaPorte.
Rhaphirrhinus 1:687. Error for *Raphirhinus* de LaPorte.
Rhaphorhinus 1:687. Error for *Raphirhinus* de LaPorte.

Rhinocerotis Theron 1977a:115. *Empoasca exilis* Naude 1926a:91 (South Africa: Ellensburg, Stellenbosch). Typhlocybinae, Empoascini. Ethiopian.

Rhoananus Dlabola 1949a:2 as subgenus of *Sorhoanus* Ribaut 1946b:85. *Deltocephalus hypochlorus* Fieber 1869a:215 (Austria). 10:1474. Deltocephalinae, Paralimnini. Palaearctic; European subregion.

Rhobala Kramer 1963b:204. *R. lemur* n.sp. (Bolivia: Huachi Beni). Neobalinae. Neotropical; Bolivia, Brazil.

Rhodulopa Linnavuori 1961a:456. *R. granulosa* n.sp. (S. Rhodesia: Victoria Falls). Iassinae, Iassini. Ethiopian. Synonym of *Jassulus* Evans, J.W. 1955a:30; Linnavuori & Quartau 1975a:10.

Rhogosana Osborn 1938a:14. *R. rugulosa* n.sp. (French Guiana). 3:55. DeLong & Freytag 1972a:223. Gyponinae, Gyponini. Neotropical; Bolivia, Brazil, the Guianas, Venezuela.

Rhombopsana Metcalf 1952a:229, replacement for *Rhombopsis* Haupt 1927a:22, not *Rhombopsis* Gardner 1916. *Rhombopsis virens* Haupt 1927a:23 (Palestine). 10:2345. Paraboloponinae. Palaearctic. Synonym of *Dryadomorpha* Kirkaldy 1906c:335; Webb, M.D. 1981b:49.

Rhombopris 10:2681. Error for *Rhombopsis* Haupt.

Rhombopsis Haupt 1927a:22. *R. virens* n.sp. (Palestine). 10:2345, 2681. Preoccupied by *Rhombopsis* Gardner 1916, see *Rhombopsana* Metcalf 1952a. Paraboloponinae. Palaearctic. Synonym of *Dryadomorpha* Kirkaldy 1906c:335; Webb, M.D. 1981b:49.

Rhopalogonia Melichar 1926a:341. *Tettigonia scita* Walker 1821a:63 (Venezuela). 1:255. Young 1977a:255. Cicadellinae, Cicadellini. Neotropical.

Rhopalopyx Ribaut 1939a:267. *Jassus preyssleri* Herrich-Schaeffer 1838c:7 (Bohemia). 10:397. Deltocephalinae, Athysanini. Palaearctic. Subgenus of *Paluda* DeLong 1937b:233; Emeljanov 1962a:164.

Rhothidus Stal 1865b:157. *Ledra navicula* Walker 1851b:826 (Australia: New South Wales). 4:135. Ledrinae, Thymbrini. Australian. Synonym of *Rhotidus* Walker 1862a:318.

Rhotidoides Evans, J.W. 1936b:59. *R. norfolkensis* n.sp. (Tasmania: New Norfolk). 4:126. Evans, J.W. 1966a:123-124. Ledrinae, Thymbrini. Australian; Australia, Tasmania.

Rhotidus Walker 1862a:318. *R. cuneatus* n.sp. (Australia: Queensland). 4:134. Evans, J.W. 1966a:127-128. Ledrinae, Thymbrini. Australian; Australia, New Guinea, Tasmania.
Synonym:
Rhothidus Stal 1865b:157. *Ledra navicula* Walker 1851b:826.

Rhusia Theron 1977a:111. *Erythroneura maculicosta* Naude 1926a:99 (South Africa: Ceres). Dworakowska 1981h:327-329. Typhlocybinae, Erythroneurini. Ethiopian; central and South Africa.

Rhusopus Webb, M.D. 1983b:250. *Idiocerus cuneiformis* Naude 1926a:16 (South Africa: Cape of Good Hope). Idiocerinae. Ethiopian.

Rhutelorbus Webb, M.D. 1981b:56. *R. merinoi* n.sp. (Malaya: Kuala Lumpur). Paraboloponinae. Oriental; Philippines, Malaya, Borneo.

Rhytidodus Fieber 1872a:8 as subgenus of *Idiocerus* Lewis 1834a:47. *Idiocerus germari* Fieber 1868a:451 (Europe) = *Cicada decimaquarta* von Schrank 1776a:76. 16:209. Idiocerinae. Palaearctic; (adventive widely elsewhere).

[NOTE: Metcalf 1966d:209-210 lists *Rhytidodus* Fieber as invalid. Nast 1972a:215 credits the name to "Fieber 1868," presumably Fieber 1868a, and states that the type species is "*Idiocerus germari* Fieber 1868 by subsequent designation." Dlabola 1965b:71 cites "Zachvatkin, 1953 hatte aber schon nach dem einzigen Art-Typus die Gattungsmerkmale bei *I. decimusquartus* Schrank ausreichend klar erkannt und diese hohere Einheit wurde deswegen von ihm und von neuem monotypisch errichtet."].

Rhystistylus 10:2682. Error for *Rhytistylus* Fieber.

Rhytistylus Fieber 1875a:404. *Jassus (Athysanus) proceps* Kirschbaum 1868b:105 (Prussia). 10:1695. Deltocephalinae, Athysanini. Palaearctic; European subregions.
Synonyms:
Edwardsiastes Kirkaldy 1900b:243. *Jassus (Athysanus) proceps* Kirschbaum 1868b:105.
Glyptocephalus Edwards 1883b:148. *Athysanus canescens* Douglas & Scott 1873a:210 = *Jassus proceps* Kirschbaum 1868b:105.

Ribautanus Dlabola 1980c:76 as subgenus of *Diplocolenus* Ribaut 1946b:82. *Diplocolenus convenarum* Ribaut 1946b:85 (France). Deltocephalinae, Paralimnini. Palaearctic.

Ribautiana Zachvatkin 1947a:113. *Cicada ulmi* Linnaeus 1758a:439 (Sweden). 17:907. Typhlocybinae, Typhlocybini. Holarctic.

Ribautiellus Zachvatkin 1933b:268. *Cicada striata* Linnaeus 1758a:437 (Europe). 10:1486. Deltocephalinae, Paralimnini. Holarctic. Synonym of *Psammotettix* Haupt 1929c:262.

Ribaytiana Lindberg 1960a:47. Error for *Ribautiana* Zachvatkin.

Rikana Nielson 1983e:27. *R. larseni* n.sp. (Peru: Tingo Maria). Coelidiinae, Youngolidiini. Neotropical; Guyana, Peru.

Rinconada Linnavuori & DeLong 1977c:203. *R. simplex* n.sp. (Chile: Santiago, Rinconada Maipes). Deltocephalinae, Athysanini. Neotropical.

Rineda Linnavuori & DeLong 1978b:129. *R. cornuta* n.sp. (Panama: Darien Prov., Santa Fe). Deltocephalinae, Athysanini. Neotropical.

Romanius Emeljanov 1966a:119 as subgenus of *Adarrus* Ribaut 1946b:83. *Adarrus servadeinus* Dlabola 1958a:12 (Italy). Deltocephalinae, Paralimnini. Palaearctic.

Ropalopyx Mitjaev 1968b:43. Error for *Rhopalopyx* Ribaut.
Rophalogonia 1:688. Error for *Rhopalogonia* Melichar.
Rosemus Nast 1982a:340. Error for *Rosenus* Oman.

Rosenus Oman 1949a:170. *Deltocephalus cruciatus* Osborn & Ball 1898f:77 (USA: Iowa, Ames). 10:1809. Deltocephalinae, Paralimnini. Holarctic.
Synonym:
Arctotettix Linnavuori 1952b:185. *Deltocephalus abiskcoensis* Lindberg 1926b:112.

Rosopaella Webb, M.D. 1983a:6. *Idiocerus kirkaldyi* Evans, J.W. 1935c:79 (Australia: New South Wales, Leura). Idiocerinae. Australian; Tasmania, New South Wales, Victoria, West Australia, Queensland.

Rotifunkia China 1926b:672. *Paropia guttifera* Walker 1851b:845 (unknown). 16:227. Idiocerinae. Ethiopian; Ethiopia, Sierra Leone.

Rotigonalia Young 1977a:521. *Cyclogonia rudicula* var. *concedula* Melichar 1926a:349 (Bolivia: Cochabamba). Cicadellinae, Cicadellini. Neotropical; Bolivia, Brazil, Costa Rica, Ecuador, French Guiana, Peru.

Rotundicerus Maldonado-Capriles 1977a:323. *R. luteus* n.sp. (British Guiana: Katari). Idiocerinae. Neotropical.

Roxasella Merino 1936a:358. *R. camusi* n.sp. (Philippines: Luzon, Laguna Prov., Los Banos). 15:170. Selenocephalinae, Selenocephalini. Oriental.

Ruandopsis Linnavuori 1978b:15. *R. kayovea* n.sp. (Ruanda: Kayove Terr. Kisenyi, 2000 m). Macropsinae, Macropsini. Ethiopian; Ruanda, Zaire. Oriental; New Guinea; northern Australian.

Rubacea DeLong 1977c:89 as subgenus of *Tenuacia* DeLong. *T. (R.) rubera* n.sp. (Peru: Sinchona). Gyponinae, Gyponini. Neotropical.

Rubria Stal 1865b:158 as subgenus of *Petalocephala* Stal 1854b:251. *P. (R.) sanguinosa* n.sp. (North Australia). 4:102. Evans, J.W. 1966a:102. Ledrinae, Petalocephalini. Australian; Australia, Tasmania.
Synonym:
Ledracephala Evans, J.W. 1947a:252. *Ledra brevifrons* Walker 1851b:825.

Rugosa 3:226. Error for *Rugosana* DeLong.

Rugosana DeLong 1942d:64 as subgenus of *Gyponana* Ball 1920a:25. *Gypona rugosa* Spangberg 1878a:6 (Mexico). 3:114. Gyponinae, Gyponini. Nearctic; eastern, southern & southwestern USA. Neotropical; Mexico and Guatemala.

Rugosella Freytag & DeLong 1971a:543 as subgenus of *Chilenana* DeLong & Freytag 1967b:106. *C. (R.) obrienorum* n.sp. (Chile: Chiloe Island, 9 km N Castro). Gyponinae, Gyponini. Neotropical.

Ruppeliana Young 1977a:747. *Tettigonia signiceps* Stal 1862e:39 (Brazil: Rio de Janeiro). Cicadellinae, Cicadellini. Neotropical.

Saavedra Linnavuori & DeLong 1978b:122. *Mesamia fasciata* Osborn 1923c:47 (Brazil: Rio Guapore below Rio S. Miguel). Deltocephalinae, Athysanini. Neotropical; Argentina, Brazil.

Sabelanus Ribaut 1959a:404 as subgenus of *Diplocolenus* Ribaut 1946b:82. *Diplocolenus nasti* Wagner 1939a:168 (Carpathia). Deltocephalinae, Paralimnini. Palaearctic. Synonym of *Erdianus* Ribaut 1952a:287; Knight 1974c:368.

Sabima Distant 1908g:324. *S. prima* n.sp. (India: Assam, Margherita). 11:20. Coelidiinae, Thagriini. Oriental. Synonym of *Thagria* Melichar 1903b:176; Nielson 1977a:9.

Sabimamorpha Schumacher 1915b:123. *S. speciosissima* n.sp. (Formosa). 11:21. Unassigned, formerly Coelidiinae; Nielson 1975a:11, 1977a:3. Oriental.

Sabimoides Evans, J.W. 1947a:254. *S. cardamomi* n.sp. (Malaysia). 11:94. Coelidiinae, Thagriini. Oriental. Synonym of *Thagria* Melichar 1903b:176; Nielson 1977a:10.

Sabix Hamilton, KGA 1975a:29 as subgenus of *Paraphlepsius* Baker 1897c:158. *Jassus irroratus* Say 1830b:308 (USA: Indiana). Deltocephalinae, Athysanini. Nearctic.

Sabourasca Ramakrishnan & Menon 1972b:183. *S. peculiaris* n.sp. (India: Bihar, Sabour). Typhlocybinae, Empoascini. Oriental. Synonym of *Empoasca* Walsh 1862a:149; Dworakowska 1977f:283.

Sacapome Schumacher 1915b:127. *S. formosana* n.sp. (Formosa). 17:1045. Typhlocybinae, Typhlocybini. Oriental.

Sagatus Ribaut 1948c:57. *Cicada punctifrons* Fallen 1826a:42 (Sweden). 10:2630. Deltocephalinae, Macrostelini. Palaearctic; European subregion.

Sagittifer Dlabola 1961b:345. *S. optatus* n.sp. (USSR: Tadzikhistan). Preoccupied by *Sagittifer* Kuhl 1820, see *Jiridlabolina* Kocak 1981a:32. Deltocephalinae, Doraturini. Palaearctic. Synonym of *Aconura* Lethierry 1876a:9; Nast 1972a:349.

Sahlbergotettix Zachvatkin 1953b:213. *Idiocerus salicicola* Flor 1861a:163 (Livonia). Idiocerinae. Palaearctic. Synonym of *Idiocerus* Lewis 1834a:47; Hamilton, K.G.A. 1980a:825.

Sailerana Young 1977a:308. *Tettigonia solitaris* Signoret 1853b:346 (Brazil: Para). Cicadellinae, Cicadellini. Neotropical; Brazil, Colombia, French and British Guiana, Peru.

Sajda Dworakowska 1981j:244. *S. flava* n.sp. (India: Karnataka, Nandi Hills). Typhlocybinae, Erythroneurini. Oriental.

Sajitettix 17:1507. Error for *Sujitettix* Matsumura.
Salka Dworakowska 1972e:778. *Zygina nigricans* Matsumura 1932a:114 (Formosa). Typhlocybinae, Erythroneurini. Oriental; India, Taiwan, Vietnam.
Salsolibia Theron 1979a:83. *S. knersi* n.sp. (South Africa: Vanrhynsdorp). Deltocephalinae, Athysanini. Ethiopian.
Salsolicola Theron 1979a:80. *S. plana* n.sp. (South Africa: Oudtshoorn). Deltocephalinae, Athysanini. Ethiopian; South Africa.
Salvina Melichar 1926a:344. *Tettigonia dorsisignata* Fowler 1900d:282 (Panama: Volcan de Chiriqui, 4-6000 ft). 11:113. Kramer 1964d:270. Neocoelidiinae. Neotropical.
Samuraba Linnavuori 1961a:480. *S. elegans* n.sp. (South Africa: Cape Province, Kokstad). Deltocephalinae, Deltocephalini. Ethiopian.
Sanachus Amyot 10:2682. Invalid, not binominal.
Sanatana Dworakowska 1984a:16. *S. malaica* n.sp. (Malaysia: Penang I.). Typhlocybinae, Erythroneurini. Oriental.
Sanctahelenia Dlabola 1976a:278. *S. synavei* n.sp. (St. Helena). Deltocephalinae, Athysanini. Ethiopian.
Sanctanus Ball 1932a:10. *Jassus sanctus* Say 1830b:307 (USA: Indiana). 10:2091, 2101. Deltocephalinae, Deltocephalini. Nearctic and Neotropical; eastern and central USA to Brazil.
Subgenus:
Cruciatanus DeLong & Hershberger 1946a:208. *Scaphoideus cruciatus* Osborn 1911c:253.
Sanctonus Linsley & Usinger 1966a:138. Error for *Sanctanus* Ball.
Sandalla Mahmood 1967a:24. *S. nigroclypelli* n.sp. (N. Borneo: Sabah, Sandakan). Typhlocybinae, Erythroneurini. Oriental. Synonym of *Bakera (Guinobata)* Mahmood 1967a:24; Hongsaprug & Wilson 1985a:179.
Sandanella Mahmood 1967a:21. *S. bakeri* n.sp. (No. Borneo: Sabah, Sandakan). Typhlocybinae, Erythroneurini. Oriental.
SANDERSELLINI 11:94.
Sandersellus DeLong 1945d:414. *S. carinatus* n.sp. (Peru: Sinchona). 11:95. Nielson 1975a:20. Coelidiinae, Sandersellini. Neotropical; Bolivia, Peru.
Synonym:
Cixidocoelidia Linnavuori 1956a:34. *C. truncatipennis* n.sp.
Sandia Theron 1982b:29. *S. brevis* n.sp. (South Africa: Natal, Ashburton). Typhlocybinae, Erythroneurini. Ethiopian.
Sanestebania Linnavuori & DeLong 1978b:130. *S. rotundiceps* n.sp. (Bolivia: San Esteban, 49 km N Santa Cruz). Deltocephalinae, Athysanini. Neotropical.
Sanluisia Linnavuori 1959b:94. *S. atricapilla* n.sp. (Paraguay: San Luis). Deltocephalinae, Deltocephalini. Neotropical.
Sannella Dworakowska 1982a:141. *S. otiosa* n.sp. (Nepal: Kathmandu Valley, 8 km W Nagarkot). Typhlocybinae, Typhlocybini. Oriental.
Sansalvadoria Schroeder 1959a:48. *S. bimaculata* n.sp. (El Salvador). Cicadellinae, Proconiini. Neotropical. Synonym of *Acrogonia* Stal 1869a:67; Young 1968a:257.
Santarema Melichar 1926a:344. *Tettigonia ferrugatula* Breddin 1901h:107 (Ecuador). 1:424. Cicadellinae, Cicadellini. Neotropical. Synonym of *Mesogonia* Melichar 1926a:343; Young 1977a:480.
Sanuca DeLong 1980h:66. *S. badia* n.sp. (Mexico: Guererro, Iguala). Deltocephalinae, Athysanini. Neotropical.
Sapingia Nielson 1979b:117. *Coelidia adspersa* Stal 1854b:254 (Uruguay: Montevideo). Coelidiinae, Teruliini. Neotropical; Argentina, Peru, Bolivia, Brazil, Paraguay, Uruguay.
Sapoba Linnavuori & Al-Ne'amy 1983a:96. *Selenocephalus flagellifer* Linnavuori 1969a:1168 (Cameroon). Selenocephalinae, Selenocephalini. Ethiopian; Guinean.
Sapporoa Dworakowska 1972e:775 as subgenus of *Alnella* Anufriev. *A. (S.) watanabei* n.sp. (Japan: Kyushu, Kagoshima). Typhlocybinae, Erythroneurini. Palaearctic. Synonym of *Alnella* Anufriev 1971b:109, which is a subgenus of *Alnetoidia* Dlabola 1958c:55; Dworakowska 1979b:30.
Saranella Young 1952b:68. *Dikraneura (Hyloidea) micronotata* Osborn 1928a:278 (Bolivia: Prov. del Sara). 17:285. Typhlocybinae, Dikraneurini. Neotropical.

Sarascarta Young 1952b:66. *Dikraneura (Hyloidea) fulva* Osborn 1928a:277 (Bolivia: Prov. del Sara). 17:284. Typhlocybinae, Dikraneurini. Neotropical; Argentina, Bolivia, Puerto Rico.
Sarbazius Dlabola 1977b:257 as subgenus of *Neolimnus* Linnavuori 1953a:114. *N. (S.) superlaminatus* n.sp. (Iran: Sarbaz). Deltocephalinae, Athysanini. Palaearctic.
Sardius Ribaut 1946b:82. *Iassus (Deltocephalus) argus* Marshall 1866c:221 (England). 10:1387. Deltocephaline, Athysanini. Palaearctic.
Sardiopsis Mitjaev 1971a:135, 153. *Sardius kasachstanicus* Mitjaev 1967c:1206 (USSR: Kazakhstan). Deltocephalinae, Athysanini. Palaearctic; Kazakhstan.
Sarejuia Ghauri 1974c:556. *Chikkaballapura quinquemaculata* Distant 1918b:108 (India: United Provinces, Kumaon). Typhlocybinae, Tyhphlocybini. Oriental. Synonym of *Draberiella* Dworakowska 1971d:647, a subgenus of *Agnesiella* Dworakowska 1970c:211; Sharma & Malhotra 1981a:41.
Sarpestus Spangberg 1878c:10. *S. specularis* n.sp. (New Guinea: Mysol). 11:144. Evans, F. 1981a:179. Tartessinae. Oriental.
Satsumanus Ishihara 1953a:193. *Eutettix satsumae* Matsumura 1914a:191 (Japan: Kyushu). 10:504. Deltocephalinae, Athysanini. Oriental; southern Japan, China, Fiji, Samoa, Caroline and Cook Islands.
Saudallygus Dlabola 1979b:136. *S. curvatus* n.sp. (Saudi Arabia). Deltocephalinae. Palaearctic.
Savanicus Dlabola 1977b:258. *S. sirik* n.sp. (Iran: 16 km northerly from Jask). Deltocephalinae, Paralimnini. Palaearctic.
Savitara Dworakowska 1984a:1. *S. albida* n.sp. (Malaysia: Penang I.). Typhlocybinae, Typhlocybini. Oriental.
Sawainara Ramakrishnan & Ghauri 1979b:208. *Empoascanara lutea* Dworakowska 1977f:298 (India: Asarori Range between Dehra Dun and Mohand). Typhlocybinae, Erythroneurini. Oriental. Synonym of *Empoascanara* Distant 1918b:94; Dworakowska 1980d:195.
Sayara Ramakrishnan & Ghauri 1979b:209 as subgenus of *Sawainara* Ramakrishnan & Ghauri 1979b:208. *Empoascanara mana* Dworakowska & Pawar 1974a:583 (Philippines: Ifugo). Typhlocybinae, Erythroneurini. Oriental. Synonym of *Empoascanara* Distant 1918b:94; Dworakowska 1980d:195.
Sayetus Ribaut 1952a:257 as subgenus of *Jassargus* Zachvatkin 1933b:268. *Deltocephalus sursumflexus* Then 1902a:189 (Styria, Carinthia, Austria). 10:1233. Deltocephalinae, Paralimnini. Palaearctic.
Scaphaideus 10:2682 and 10:2685. Error for *Scaphoideus* Uhler.
Scaphetus Evans, J.W. 1966a:237. *S. brunneus* n.sp. (New Zealand: Bay of Islands). Deltocephalinae, Deltocephalini. Australian.
Scaphhoideus 10:2682. Error for *Scaphoideus* Uhler.
Scaphitopius Suehiro 1961a. Error for *Scaphytopius* Ball 1931d.
Scaphocephalus Matsumura 1905a:52. *Petalocephalus discolor* Uhler 1896a:290 (Japan). 4:86. Ledrinae, Petalocephalini. Oriental. Synonym of *Ledropsis* White 1844a:425.
Scaphoidella Vilbaste 1968a:133. *S. arboricola* n.sp. (USSR Maritime Territory). Deltocephalinae, Athysanini. Palaearctic; Siberia.
SCAPHOIDEINI 10:2075.
Scaphoidens 10:2682. Error for *Scaphoideus* Uhler.
Scaphoides Lodos 1982a:136. Error for *Scaphoideus* Uhler.
Scaphoideus Uhler 1889a:33. *Jassus immistus* Say 1830b:306 (USA: Indiana or Missouri). 10:2105, 2151. Barnett, D.E. 1977a. Deltocephalinae, Scaphoideini. Cosmopolitan.
Subgenera:
Lonenus DeLong 1939a:33. *Scaphoideus intricatus* Uhler 1889a:34.
Latenus DeLong & Knull 1971a:54. *Scaphoideus veterator* DeLong & Beery 1936a:334.
Synonyms:
Hussa Distant 1918b:68. *H. insignis* n.sp.
Bolanus Distant 1918b:89. *B. baeticus* n.sp.
Angenus DeLong & Knull 1971a:54. *Jassus immistus* Say 1830b:306.
Scaphoidius 10:2684. Error for *Scaphoideus* Uhler.
Scaphoidophyes Kirkaldy 1906e:154 as subgenus of *Scaphoideus* Uhler 1889a:33. *S. (S.) annae* n.sp. (Ivory Coast). 10:2150. Barnett, D.E. & Freytag 1976a. Deltocephalinae, Scaphoideini. Ethiopian.

Scaphoidophytes Barnett, D.E. & Freytag 1976a, error for *Scaphoidophyes* Kirkaldy.
Scaphoidula Osborn 1923c:41. *S. cingulatus* n.sp. (Bolivia: Prov. del Sara). 10:2203. Linnavuori 1959b:227-229. Deltocephalinae. Neotropical.
Scaphoidulina Osborn 1934b:245. *S. obliqua* n.sp. (Marquesas Islands: Uapou, Teavaituhai, Paaumea side). 10:2468. Deltocephalinae, Macrostelini. Oceania; Marquesas Islands.
Scaphoidus 10:2684. Error for *Scaphoideus* Uhler.
Scaphoridens Zoological Record 107, Sec. 13:692. Error for *Scaphoideus* Uhler 1889a.
Scaphotettix Matsumura 1914a:227. *S. viridis* n.sp. (Formosa). 10:494. Deltocephalinae, Athysanini. Oriental.
Scaphytoceps Dlabola 1957b:287. *S. melleus* n.sp. (Afghanistan: Nuristan, Bashgutal 1100 m). Deltocephalinae, Scaphytopiini. Palaearctic.
SCAPHYTOPIINI 10:2243.
Scaphytopius Ball 1931d:218. *Platymetopius elegans* Van Duzee 1890g:94 (USA: California). 10:2249, 2259. Deltocephalinae, Scaphytopiini. Nearctic/Neotropical; widespread; southern Canada to Brazil and the Galapagos. [Hawaiian Islands].
  Subgenera:
    Cloanthanus Ball 1931d:219. *Platymetopius angustatus* Osborn 1905a:518.
    Convelinus Ball 1931d:220. *Platymetopius nigricollis* Ball 1916b:205.
    Protranus DeLong 1980h:68. *S. (P.) abutus* n.sp.
  Synonyms:
    Deltopinus Ball 1931d:218. *Platymetopius nigriviridis* Ball 1909c:163.
    Nasutoideus Ball 1931d:219. *Platymetopius nasutus* Van Duzee 1907a:64.
    Platymoideus Ball 1931d:219. *Platymetopius trilineatus* Ball 1916b:204.
    Turneus DeLong 1944a:168. *T. serrellus* n.sp.
    Hebenarus DeLong 1944c:41. *H. pallidus* n.sp.
Scapidonus Nielson 1983h:566. *S. horridus* n.sp. (Brazil: Bahia, Iquassu). Coelidiinae, Teruliini. Neotropical.
Scapoideus 10:2684. Error for *Scaphoideus* Uhler.
SCARIDAE 3:5.
Scaries 3:226. Error for *Scaris* LePeletier & Serville.
SCARINI 3:13.
Scaris LePeletier & Serville 1825a:609. *Iassus ferrugineus* Fabricius 1803a:86 (Colombia). 3:69. DeLong & Freytag 1972a:221; Freytag & DeLong 1982a. Gyponinae, Gyponini. Neotropical; Central and South America, from Costa Rica to Brazil and Uruguay.
  Synonyms:
    Darma Walker 1858a:102. *D. bipunctata* n.sp.
    Clinonaria Metcalf 1949b:277. *C. bicolor* n.sp.
Scarisana Metcalf 1949b:277. *S. variabilis* n.sp. (British Guiana). 3:66. Cicadellinae, Cicadellini. Neotropical. Synonym of *Paromenia* Melichar 1926a:342; Young 1977a:263.
Scaroidana Osborn 1938a:49. *S. flavida* n.sp. (Bolivia). 3:55. DeLong & Freytag 1972a:218. Iassinae. Neotropical; Bolivia, Brazil, Panama, Paraguay, Peru.
Scaropsia Blocker 1979a:10. *S. trombida* n.sp. (Argentina: Tucuman). Iassinae. Neotropical.
Scenergates Emeljanov 1972d:106, replacement for *Papyrina* Emeljanov 1962a:162, not *Papyrina* Moench 1853. *Papyrina viridis* Emeljanov 1962a:163 (USSR: Uzbekistan). Deltocephalinae, Athysanini. Palaearctic.
  Synonyms:
    Platytettix Vilbaste 1961a:45, preoccupied. *P. viridis* n.sp.
    Papyrina Emeljanov 1962a:162. *P. viridis* n.sp.
    Neopapyrina Kocak 1981a:33. *Papyrina viridis* Emeljanov 1962a:163.
Schildola Young 1977a:680. *S. opaca* n.sp. (Costa Rica: Tucurrique). Cicadellinae, Cicadellini. Neotropical; Costa Rica, Ecuador, Peru, Panama, Colombia.
Scinda DeLong & Ruppel 1951a:95 as subgenus of *Cicadella* Dumeril 1806a:266 (suppressed name). *C. (S.) scarlatina* n.sp. (Mexico: Morelos, Lagunas de Cimpaola). 17:778. Typhlocybinae, Eupterygini. Neotropical; Mexico.

Schistogonalia Young 1977a:812. *S. prava* n.sp. (Costa Rica: Irazu). Cicadellinae, Cicadellini. Neotropical; Costa Rica, Panama.
Schizandrasca Anufriev 1972a:36. *Alebroides ussurica* Vilbaste 1968a:73 (USSR Maritime Territory). Typhlocybinae, Empoascini. Palaearctic; Manchurian subregion.
Schleroracus 10:2684. Error for *Scleroracus* Van Duzee.
Scleroracus Van Duzee 1894c:136. *Athysanus anthracinus* n.sp. (USA: Iowa). 10:163. Deltocephalinae, Athysanini. Holarctic. [NOTE: As indicated (Oman 1947a:204-205), Van Duzee's 1894c:136 association of the generic names *Conogonus* and *Scleroracus* with his description of *Athysanus anthracinus* validated them as of that date in accord with ICZN Opinions 4, 24, 53 and others, as well as then existing and current Codes. Metcalf 1952a:231, 1967a:163 erroneously attributed *Scleroracus* to Oman, dating from 1949].
Scophoideus 10:2685. Error for *Scaphoideus* Uhler.
Scopogonalia Young 1977a:529. *Tettigonia subolivacea* Stal 1862e:42 (Brazil). Cicadellinae, Cicadellini. Neotropical; Argentina, Colombia, Bolivia, Brazil, Paraguay, Peru, Venezuela.
Scoposcartula Young 1977a:669. *Tettigonia oculata* Signoret 1853b:344 (Venezuela: La Guaria). Cicadellinae, Cicadellini. Neotropical; Argentina, Bolivia, Brazil, British Guiana, Costa Rica, Panama, Paraguay.
Secopennis DeLong & Sleesman 1929a:85 as subgenus of *Flexamia* DeLong 1926d:22. *Deltocephalus slossonae* Ball 1905a:119 (USA: Florida, Biscayne Bay). 10:1773. Deltocephalinae, Deltocephalini. Nearctic; southeastern USA.
Sectoculus Morrison 1973a:395. *S. carinatus* n.sp. (Laos: Vientiane Prov., Ban Van Eue). Deltocephalinae, Hecalini. Oriental.
Segonalia Young 1977a:999. *S. steinbachi* n.sp. (Bolivia: Santa Cruz). Cicadellinae, Cicadellini. Neotropical.
Selachina Emeljanov 1962c:769. *Carchariacephalus apicalis* Matsumura 1908a:43 (Algeria). Deltocephalinae, Athysanini. Palaearctic.
Selanocephalus 15:224. Error for *Selenocephalus* Germar.
SELENOCEPHALINI 15:136.
Selenocephalus Germar 1833a:180. *Jassus obsoletus* Germar 1817b:281 (Dalmatia) = *Cicada plana* Turton 1802a:597. 15:142. Selenocephalinae, Selenocephalini. Palaearctic; subtropical & warm temperature areas. Oriental and Oceania (?); Linnavuori & Al Ne'amy 1983a:69.
  Synonym:
    Levantotettix Lindberg 1953a:110. *L. striatus* n.sp.
SELENOMORPHINI Evans, J.W. 1974a:166, tribe of JASSINAE = IASSINAE.
Selenomorphus Evans, J.W. 1974a:166. *S. nigrovenatus* n.sp. (New Caledonia: Col d'Amieu, 650 m). Iassinae, Selenomorphini. Australian.
Selenonephalus 15:225. Error for *Selenocephalus* Germar.
Selenophalus 15:225. Error for *Selenocephalus* Germar.
Selenophares 15:225. Error for *Selenocephalus* Germar.
Selenophores 15:225. Error for *Selenocephalus* Germar.
Selenophorus 15:225. Error for *Selenocephalus* Germar.
Selenopsis Spinola 1850a:59. *S. subaptera* n.sp. (South Africa: Natal). Unassigned, formerly Coelidiinae; Nielson 1975a:11. Ethiopian; South Africa. (Para?).
Selvitsa Young 1977a:659. *Tettigonia humeralis* Signoret 1853b:369 (Brazil: "San Paolo"). Cicadellinae, Cicadellini. Neotropical; Bolivia, Brazil, Ecuador, Panama, Venezuela.
Semenovium Kusnezov 1929a:312. *S. ferganae* n.sp. (Turkestan). 9:75. Eupelicinae, Paradorydiini. Palaearctic. Subgenus of *Paradorydium* Kirkaldy 1901f:339; Emeljanov 1964c:380.
Sempia Dworakowska 1970m:703. *Erythroneura (Zygina) capreola* Linnavuori 1964a:340 (Sinai: Wadi Feiran). Typhlocybinae, Erythroneurini. Palaearctic.
Sequoiatettix Bliven 1955a:3. *Colladonus eurekae* Bliven 1954a:117 (USA: California, Eureka). 10:1312. Nielson 1966a:333. Deltocephalinae, Athysanini. Nearctic. Synonym of *Colladonus* Ball 1936c:57.
Seriana Dworakowska 1971e:345. *S. frater* n.sp. (Java: Tjibulan, near Bogar). Typhlocybinae, Erythroneurini. Oriental; Ceylon and Indo-Malayan subregions.

Serpa Distant 1908a:529. *Tettigonia plumbea* Walker 1851b:754 (Ecuador). 1:256. Young 1977a:1094. Cicadellinae, Cicadellini. Neotropical; Colombia, Ecuador.

Serratulus Mahmood 1967a:46. *S. recurvatus* n.sp. (Philippines: Mindanao, Zamboanga). Typhlocybinae, Typhlocybini. Oriental.

Serratus Linnavuori 1959b:250. *S. clypealis* n.sp. (Argentina: Tucuman). Deltocephalinae, Athysanini. Neotropical.

Serridonus Linnavuori 1959b:181. *S. longistylus* n.sp. (Argentina: Formosa). Deltocephalinae, Athysanini. Neotropical; Argentina, Paraguay.

Sestrelicola Remane & Asche 1980a:91. *S. garrafer* n.sp. (Portugal). Deltocephalinae, Paralimnini. Palaearctic.

Setabis Stal 1866a:111. *Gypona javeti* Signoret 1858a:342 (Calabar). 3:214. Penthimiinae. Ethiopian. Preoccupied by *Setabis* Westwood 1851, see *Jafar* Kirkaldy 1903f:13.

Shaddai Distant 1918b:15. *S. typicus* n.sp. (India: Darjeeling). 17:26. Dworakowska 1971c:494. Typhlocybinae, Alebrini. Oriental; India, China, Vietnam.
Synonym:
Sinalebra Zachvatkin 1936a:16. *S. hummeli* n.sp.

Shamala Dworakowska 1980b:169. *S. mikra* n.sp. (India: Assam, Burnihat). Typhlocybinae, Typhlocybini. Oriental; India, Nepal.
Subgenus:
Mahmoba Dworakowska 1982a:148. *S. (M.) prospecta* n.sp.

Sharmana Dworakowska 1982a:123 as subgenus of *Linnavuoriana* Dlabola 1958c:54. *Agnesiella swaraji* Sharma 1978a:12 (India: Jammu). Typhlocybinae, Typhlocybini. Palaearctic.

Shirazia Dlabola 1977b:248. *S. eminens* n.sp. (Iran: 30 km easterly from Kazerun). Deltocephalinae, Platymetopiini. Palaearctic. Preoccupied by *Shirazia* Amsel 1954, see *Burakia* Kocak 1981a:32.

Shonenus Ishihara 1958a:232, unnecessary replacement for *Matsumurella* Ishihara 1953a:200. *Jassus (Allygus) kogotensis* Matsumura 1914a:206 (Japan: Honshu). Deltocephalinae, Athysanini. Palaearctic. Synonym of *Matsumurella* Ishihara 1953a.

Shurabella Bekker-Migdisova 1949a:28. *S. lepyroniopsis* n.sp Metcalf & Wade 1966a:210. Fossil: Mesozoic, Leninabad.

Sibovia China 1927d:283. *Tettigonia sagata* Signoret 1854a:27 (Mexico). 1:30. Young 1977a:696. Cicadellinae, Cicadellini. Neotropical; Mexico to southern Brazil, Antilles subregion. Nearctic; southern USA.
Synonyms:
Ceratagonia Melichar 1926a:350. *Tettigonia recta* Fowler 1900a:264.
Entogonia Melichar 1926a:360. *Tettigonia* Signoret 1854a:27.
Ceratagoniella Metcalf 1952a:228, replacement for *Ceratagonia* Melichar 1926a:350.

Sichaea Stal 1866a:106. *Acocephalus missellus* Stal 1855a:98 ("Caffraria"). 5:84. Linnavuori 1972a:148. Ledrinae. Ethiopian; southern Africa.

Sichafa Zoological Record 109, Sec. 13f:155, 1972. Error for *Sichaea* Stal 1866a.

Sichea 5:99. Error for *Sichaea* Stal.

Sicistella Emeljanov 1972d:111. *Orocastus aridus* Emeljanov 1964f:46 (USSR: Kazakhstan). Deltocephaline, Paralimnini. Palaearctic; Kazakhstan, Turkmenia.

Sidelloides Evans, J.W. 1972a:180. *S. histrica* n.sp. (NE New Guinea: Toricelli Mts., Koiniri). Penthimiinae. Oriental.

Siderojassus Evans, J.W. 1972b:658. *S. kaindii* n.sp. (NE New Guinea: Mt. Kaindi). Iassinae, Reuplemmelini. Oriental.

Sigista Emeljanov 1966a:125. *Deltocephalus burjata* Kusnezov 1929b:180 (Transbaikalia). Deltocephalinae, Paralimnini. Palaearctic; Siberian and Manchurian subregions. Synonym of *Kaszabinus* Dlabola 1965c:115; Nast 1972a:424.

Signoretia Stal 1859b:289. *Thamnotettix malaya* Stal 1855b:192 (Malacca). 6:53. Linnavuori 1978a:36-42. Signoretiinae, Signoretiini. Oriental and Ethiopian; widespread in warm regions.

SIGNORETIINAE 6:52.

Similitopia Schroder 1959a:45. *Oncometopia fuscipennis* Fowler 1899b:230 (Mexico: Guerrero, Chilpancingo). Cicadellinae, Proconiini. Neotropical. Subgenus of *Oncometopia* Stal 1869a:62; Young 1968a:230.

Sinalebra Zachvatkin 1936a:16. *S. hummeli* n.sp. (China). 17:97. Dworakowska 1971c:494. Typhlocybinae, Alebrini. Oriental. Synonym of *Shaddai* Distant; Dworakowska 1977i:9.

Sincholata DeLong 1982f:477. *S. dicera* n.sp. (Peru: Sinchono). Deltocephalinae, Scaphoideini. Neotropical.

Sinchonoa Linnavuori & DeLong 1978b:132. *S. pilosa* n.sp. (Peru: Sinchono). Deltocephalinae, Athysanini. Neotropical

Sinchora DeLong 1979a:229 as subgenus of *Curtara* DeLong & Freytag 1972a:231. *C. (S.) regela* n.sp. (Brazil: Minas Gerais). Gyponinae, Gyponini. Neotropical.

Singapora Mahmood 1967a:20. *S. nigropunctata* n.sp. (Singapore). Typhlocybinae, Erythroneurini. Oriental; widespread. Palaearctic; Manchurian subregion.
Synonym:
Erythroneuropsis Ramakrishnan & Menon 1973a:37. *E. indicus* n.sp.

Singhardina Mahmood 1967a:32. *S. robusta* n.sp. (Singapore). Typhlocybinae, Typhlocybini. Oriental. Subgenus of *Eurhadina* Haupt 1929b:1075; Dworakowska 1969g:80.

Sirosoma McAtee 1933a:545. *S. hiaticula* n.sp. (North Borneo). 17:1474. Typhlocybinae, Erythroneurini. Oriental; Borneo, Philippines, Singapore.

Sisimitalia Young 1977a:756. *Tettigonia vulnerata* Signoret 1855d:782 (Guatemala). Cicadellinae, Cicadellini. Neotropical; S. Mexico, Costa Rica, Guatemala, Honduras, Nicaragua, Panama, Colombia.

Sispocnis Anufriev 1967a:174 as subgenus of *Oncopsis* Burmeister 1838a:10. *Bythoscopus juglans* Matsumura 1912b:304 (Japan: Hokkaido). Macropsinae, Macropsini. Palaearctic. Subgenus of *Pediopsoides* Matsumura 1912b:305; Hamilton, K.G.A. 1980b:897.

Sitades Distant 1912e:608. *S. fasciatus* n.sp. (India: Bengal). 5:88. Ulopinae, Ulopini. Oriental. Synonym of *Mesargus* Melichar 1903b:175; Vilbaste 1975a:230.

Siva Spinola 1850b:127. *S. strigicollis* n.sp. (Madras). 15:197. Preoccupied by *Siva* Hodgson 1838, see *Krisna* Kirkaldy 1900b:243. Iassinae, Krisnini. Oriental. Synonym of *Krisna* Kirkaldy 1900b:243.

Ska Dworakowska 1976b:31. *S. argentea* n.sp. (E. Flores: "Geli Moetoe" ? Keli Mutu). Typhlocybinae, Erythroneurini. Ethiopian.

Smaris 3:226. Error for *Scaris* LePeletier & Serville.

Smicrocotis Kirkaldy 1906c:370. *S. obscura* n.sp. (Australia: New South Wales, Sydney). 4:116. Evans, J.W. 1966a:108-110. Ledrinae, Stenocotini. Australian; Australia.

Smicrotis 4:144. Error for *Smicrocotis* Kirkaldy.

Sobara Oman 1949a:134. *Eutettix palliolata* Ball 1902a:13 (USA: Texas). 10:2172. Deltocephalinae, Scaphoideini. Nearctic; southwest USA.

Sobrala Dworakowska 1977i:12. *S. trnava* n.sp. (Vietnam: Prov. Lao-cai, Sa-Pa, 1650 m). Typhlocybinae, Alebrini. Oriental.

Socistella Logvinenko 1984a:27. Error for *Sicistella* Emeljanov.

Sohinara Ramakrishnan & Ghauri 1979b:205. *Empoascanara falcata* Dworakowska 1971e:341 (Vietnam: Prov. Ninh-Binh, Cuo-phuong). Typhlocybinae, Erythroneurini. Oriental. Synonym of *Empoascanara* Distant 1918b:94; Dworakowska 1980d:195.

Soibanga Distant 1908g:236. *Tettigonia bella* Walker 1851b:778 (North India). 1:91. Cicadellinae, Cicadellini. Oriental.

Solanasca Ghauri 1974a:425. *Empoasca solana* DeLong 1931b:50 (USA: Louisiana, Baton Rouge). Typhlocybinae, Empoascini. Nearctic; seasonal in northern latitudes. Neotropical; Mexico to northern South America, also Antilles subregion.

Soleatus DeLong 1971a:54. *Prescottia bicalcea* DeLong 1941c:181 (Mexico: Chiapas, Finca Vergel). Deltocephalinae, Scaphoideini. Neotropical.

Solenocephalus 15:225. Error for *Selenocephalus* Germar.

Solenopyx Ribaut 1939a:273. *Cicada sulphurella* Zetterstedt 1828a:534 (Sweden). 10:1930. Deltocephalinae, Cicadulini. Palaearctic. Synonym of *Elymana* DeLong 1936a:218.

Soleroracus Mitjaev 1971a:135. Error for *Scleroracus* Van Duzee 1894c.

Sonesimia Young 1977a:1084. *Tettigonia grossa* Signoret 1854a:25 (uncertain, see comment). Cicadellinae, Cicadellini. Neotropical.

[NOTE: Young & Beier 1963a:569, 570 record a specimen with labels *"grossa;* det. Signoret." and "Caracas; Coll. Signoret.", but do not indicate a lectotype designated. Young 1977a:1085 refers to an illustration prepared from lectotype, yet does not record Venezuela as a locality for the genus. Metcalf 1965a:290 does not list a locality attributed to Signoret].

Sonronius Dorst 1937a:9. *Cicadula dahlbomi* Zetterstedt 1840a:297 (Lapland) = *Cicadula maculipes* Zetterstedt 1840a:297. 10:2617. Deltocephalinae, Macrostelini. Holarctic.

Soortana Distant 1908g:319. *S. simulata* n.sp. (Ceylon: Kandy). 11:89. Coelidiinae, Thagriini. Oriental. Synonym of *Thagria* Melichar 1903b:176; Nielson 1977a:9.

Soosiulus Young 1977a:369. *Poeciloscarta fabricii* Metcalf 1965a:65 (French Guiana). Cicadellinae, Cicadellini. Neotropical.

Sophonia Walker 1870b:327. *S. rufitelum* n.sp. (New Guinea: Mysol). 7:26. Nirvaninae, Nirvanini. Oriental; widespread.
Synonym:
Pseudonirvana Baker 1923a:386. *P. sandakanensis* n.sp.

Soracte Kirkaldy 1907d:55. *S. apollonos* n.sp. (Australia: Queensland, Cairns). 10:1748. Evans, J.W. 1966a:225. Deltocephalinae, Platymetopiini. Australian; Australia.

Soractellus Evans, J.W. 1966a:225. *S. brunneus* n.sp. (Australia: Queensland, Ingham). Deltocephalinae, Deltocephalini. Australian.

Sorbonellus Linnavuori 1959b:210 as subgenus of *Osbornellus* Ball 1932a:17. *Osbornellus infuscatus* Linnavuori 1955a:102 (Brazil: Rio Grande de Sul). Deltocephalinae, Scaphoideini. Neotropical.

Sordana DeLong 1976c:92. *Gypona sordida* Stal 1854b:252 (Brazil: Minas Gerais). Gyponinae, Gyponini. Neotropical.

Sordanus Zoological Record 103, Sec. 13:397, 1966. Error for *Sorhoanus* Ribaut 1946b.

Sorhoanus Ribaut 1946b:85. *Cicada assimilis* Fallen 1806a:22 (Sweden). 10:1454, 1465. Deltocephalinae, Paralimnini. Palaearctic; European and Siberian subregions. (? Nearctic).
Subgenus:
Rhoananus Dlabola 1949a:2. *Deltocephalus hypochlorus* Fieber 1869a:215.

Soronius Wagner & Franz 1961a:140. Error for *Sonronius* Dorst 1937a.

Sorrhoanus Dworakowska 1969a:52. Error for *Sorhoanus* Ribaut 1946b.

Sotanus Ribaut 1942a:265. *Athysanus thenii* Loew 1885a:350 (southern Tyrol). 10:391. Deltocephalinae, Athysanini. Palaearctic; central Europe. Ethiopian; Tanganyika.

Spanbergia 9:121. Error for *Spangbergiella* Signoret.
Spanbergiella 9:121. Error for *Spangbergiella* Signoret.
Spangbergia 9:122. Error for *Spangbergiella* Signoret.

Spangbergiella Signoret 1879b:273. *Glossocratus vulneratus* Uhler 1877a:464 (USA: Texas). 9:51. Linnavuori 1959b:64. Deltocephalinae, Hecalini. Nearctic; southern USA. Neotropical; Antillean subregion, and from Mexico to Brazil and Uruguay.
Synonym:
Bergiella Baker 1897c:157. *Parabolocratus uruguayensis* Berg 1884a:36.

Spangebergiella 9:122. Error for *Spangbergiella* Signoret.

Spanigorlus Nielson 1979b:322. *S. braziliensis* n.sp. (Brazil: Nova Teutonia). Coelidiinae, Teruliini. Neotropical.

Spanotartessus Evans, F. 1981a:128. "*Tartessus obscurus* Evans" 1936b:54 (Australia: New South Wales, Leura) = *Tartessus evansi* Metcalf 1955a:266. Tartessinae. Australian.

Spartopyge Young & Beirne 1958a:48. *Flexamia mexicana* DeLong & Hershberger 1948a:136 (Mexico: Guerrero, Iguala). Deltocephalinae, Deltocephalini. Neotropical.

Spathanus DeLong 1945a:158. *Athysanus acuminatus* Baker 1896c:25 (USA: New Mexico). 10:538. Ellsbury & Nielson 1978a. Deltocephalinae, Athysanini. Nearctic.

Spathifer Linnavuori 1955b:125. *S. fuscatus* n.sp. (Paraguay). 10:1880. Linnavuori 1959b:140. Deltocephalinae, Deltocephalini. Neotropical.

Spatulostylus Ramakrishnan & Menon 1973a:16. *S. variegatus* n.sp. (India: Delhi, I.A.R.I.). Typhlocybinae, Erythroneurini. Oriental. Synonym of *Qadria* Mahmood 1967a:18; Dworakowska 1977f:293.

Speudotettix Ribaut 1942a:261. *Cicada subfuscula* Fallen 1806a:30 (Sweden). 10:239. Deltocephalinae, Athysanini. Palaearctic; European, Mediterranean and Siberian subregions.

Sphaeropogonia Breddin 1901f:100. *S. aureatula* n.sp. (Ecuador). 1:657. Young 1977a:257. Cicadellinae, Cicadellini. Neotropical; Bolivia, Peru, Ecuador.

Sphinctogonia Breddin 1901a:128. *Tettigonia guttivitta* Walker 1870b:301 (Celebes). 1:56. Cicadellinae, Cicadellini. Oriental; Indo-Malayan subregion.

Spinanella DeLong & Freytag 1972c:158 as subgenus of *Gyponana* Ball 1920a:85. *G. (S.) rubralineata* n.sp. (Venezuela: Coiripito). Gyponinae, Gyponini. Neotropical.

Spinolidia Nielson 1982e:286. *Jassus flavifrons* Osborn 1924c:448 (Bolivia: Corioco, Yungas). Coelidiinae, Coelidiini. Neotropical; Bolivia, Brazil, Colombia, Ecuador, Peru.

Spinulana DeLong 1967a:20. *S. varigata* n.sp. (Mexico: Guerrero, Iguala). Deltocephalinae, Athysanini. Neotropical.

Splonia Signoret 1891a:467. *S. acutalis* n.sp. ("Senegal", error; Young 1968a:144). 1:539. Young 1968a:141. Cicadellinae, Proconiini. Neotropical; Venezuela, Ecuador.
Synonyms:
Syringophora Kirkaldy 1907d:87. *Raphirhinus brevis* Walker 1851b:807.
Papallacta Schmidt 1932a:47. *P. haenschi* n.sp.

Sprundigia Nielson 1979b:24. *S. plana* n.sp. (Guyana: Essequibo R., Moraballi Crk.). Coelidiinae, Teruliini. Neotropical.

Stacla Dworakowska 1969e:439 as subgenus of *Eupteryx* Curtis 1829a:192. *E. (S.) cristagalli* n.sp. (East Nepal: Tapelung Distr., Sangu, c. 6200 ft). Dworakowska 1978b:707-709. Typhlocybinae, Eupterygini. Oriental.

Stactogala Amyot 10:2685. Invalid, not binominal.
Stactopeltus Amyot 17:1507. Invalid, not binominal.

Stalinabada Dlabola 1961b:344. *S. paraconurae* n.sp. (USSR: Tadzhikistan). Deltocephalinae, Doraturini. Palaearctic. Synonym of *Aconura* Lethierry 1876a9; Nast 1972a:349.

Stalolidia Nielson 1979b:146. *Coelidia cingulata* Stal 1862e:50 (Brazil). Coelidiinae, Teruliini. Neotropical; Argentina, Bolivia, Brazil, Paraguay, Peru.

Stegelythra Dlabola 1963c:328. Error for *Stegelytra* Mulsant & Rey.

Stegelytra Mulsant & Rey 1855a:224. *S. alticeps* n.sp. (Provence). 11:90. Stegelytrinae. Palaearctic; Mediterranean subregion.

STEGELYTRINI 11:90.

Stehlikiana Young 1977a:56. *Tettigoniella halticula* Jacobi 1905c:176 (Peru S: Marcapata). Cicadellinae, Cicadellini. Neotropical; Bolivia, Brazil, Colombia, Ecuador, Honduras, Venezuela, Peru.

Stehliksia Dworakowska 1972f:851. *Zygina scutata* Melichar 1905a:304 (Tanganyika). Typhlocybinae, Erythroneurini. Ethiopian; Kenya, Tanzania.

Stellena Theron 1973a:33. *Cicadula nigrifrons* Naude 1926a:86 (South Africa: Stellenbosch, Jonkershoek) = *Cicadula capensis* Metcalf 1955a:266. Deltocephalinae, Macrostelini. Ethiopian.

Stenagallia Evans, J.W. 1957a:370. *S. sagittaria* n.sp. (Chile: Juan Fernandes I., Masatierro, Picacho Central). Agalliinae, Agalliini. Neotropical.

Stenalsella Evans, J.W. 1966a:117. *S. testacea* n.sp. (Australia: Queensland, Mt. Glorious). Ledrinae, Thymbrini. Australian.

Stenidiocerus Ossiannilsson 1981a:348. *Bythoscopus poecilus* Herrich-Schaeffer 1835a:69 (Germany). Idiocerinae. Palaearctic. Synonym of *Metidiocerus* Ossiannilsson 1981a:318; Kwon 1985a:68.

Stenipo Evans, J.W. 1934a:151. *S. swani* n.sp. (Western Australia: Rottnest Island). 12:33. Evans, J.W. 1966a:51. Eurymelinae, Ipoini. Australian; western and southern Australia.

Stenocladus 15:226. Error for *Bythoscopus* Germar.

Stenocoelidia DeLong 1953a:104. *S. virgata* n.sp. (Mexico: Morelos, Cuernavaca). 11:114. Neocoelidiinae. Neotropical, Nearctic. Synonym of *Neocoelidia* Gillette & Baker 1895a:103; Kramer 1964d:262.

STENOCOTINAE 4:110.

Stenocotis Stal 1854b:254. *S. subvittata* n.sp. (Australia: Van Dieman's Land) = *Ledra depressa* Walker 1851b:817. 4:111. Evans, J.W. 1966a:104-106. Ledrinae, Stenocotini. Australia, all states; Tasmania.

Stenogiffardia Evans, J.W. 1977a:116. *S. elongata* n.sp. (Australia: NW Queensland, Richmond). Deltocephalinae, Platymetopiini. Australian.

Stenoledra Evans, J.W. 1954a:93. *S. decorsei* n.sp. 4:110. (Madagascar). Ledrinae, Petalocephalini. Malagasian.

Stenometohardya Dlabola 1981a:292. *S. veriviva* n.sp. (Iran). Deltocephalinae, Athysanini. Palaearctic.

Stenometopiellus Haupt 1917d:250. *S. sigillatus* n.sp. (Turkmenia). 10:1024. Deltocephalinae, Athysanini. Palaearctic; European, Mediterranean, Siberian subregions. (? Nearctic).
Synonym:
Diplocolenoidea Linnavuori 1953d:58. *D. turkestanica* n.sp.

STENOMETOPIINI 7:28.

Stenometopius Matsumura 1914a:217. *S. formosanus* n.sp. (Formosa). 7:29. Synonym of *Hodoedocus* Jacobi 1910b:126; Linnavuori & DeLong 1978c:208. Deltocephalinae, Stenometopiini. Oriental.

Stenometoupius Linnavuori & DeLong 1978c:208. Error for *Stenometopius* Matsumura 1914a:217.

Stenomiella Evans, J.W. 1955a:39. *S. viridis* n.sp. (Belgian Congo: Lusinga). 10:1891. Linnavuori 1978d:462. Paraboloponinae. Ethiopian; Congo.

Stenomisella Evans, J.W. 1954a:123. *S. nigrovenella* n.sp. (Madagascar). 10:78. Deltocephalinae, Athysanini. Malagasian.

Stenopiellus Lindberg 1960b:62. Error for *Stenometopiellus* Haupt.

Stenopsoides Evans, J.W. 1941f:153. *S. turneri* n.sp. (Western Australia: Dedari). 13:166. Macropsinae, Macropsini. Australian.

Stenortor 7:34. Error for *Stenotortor* Baker.

Stenoscopus Evans, J.W. 1934a:166. *S. drummondi* n.sp. (Western Australia: Beverly). 13:167. Hamilton, K.G.A. 1980b:887. Macropsinae, Macropsini. Australian.

Stenotartessus Evans, J.W. 1947a:207, replacement for *Euprora* Evans, J.W. 1938b:9, not *Euprora* Busck 1906. *Euprora mullensis* Evans, J.W. 1938b:10 (Western Australia: Mullewa). 11:144. Evans, F. 1981a:118. Tartessinae. Australian. Synonym of *Newmaniana* Evans, J.W. 1941F:152 (1942); Evans, F. 1981a:118.

Stenotortor Baker 1923a:377. *S. inocarpi* n.sp. (Singapore Island). 7:3. Nirvaninae, Macroceratogoniini. Oriental.

Stephanolla Young 1977a:1024. *Tettigonia rufoapicata* Fowler 1900d:286 (Panama: Bugaba). Cicadellinae, Cicadellini. Neotropical; Costa Rica, Colombia, Ecuador, Guatemala, Nicaragua, Panama.

Sternana DeLong & Freytag 1964a:119 as subgenus of *Gyponana* Ball 1920a:85. *G. (S.) tenuis* n.sp. (Mexico: Michoacan, Zitacuara). Gyponinae, Gyponini. Neotropical.

Stiapona DeLong 1979g:187. Error for *Striapona* DeLong, l.c.

Stictocoris Thomson 1869a:51 as subgenus of *Jassus*. *Cicada lineata* Fabricius 1787a:270 (Saxony), preoccupied by *Cicada lineata* Linnaeus 1758 = *Cicada hybneri* Gmelin 1789a:2107. 10:947. Deltocephalinae, Athysanini. Palaearctic; all subregions.

Stictorcoris 10:2686. Error for *Stictocoris* Thomson.
Stictoris 10:2686. Error for *Stictocoris* Thomson.

Stictoscarta Stal 1869a:61. *Tettigonia sulcicollis* Germar 1821a:62 (Brazil: Sao Paulo). 1:600. Young 1968a:53. Cicadellinae, Proconiini. Neotropical; Argentina, British Guiana, Brazil, Colombia, Trinidad.

Stigmocratus Amyot 10:2686. Invalid, not binominal.

STIRELLINI 10:11. Emeljanov 1966b:609, tribe of DELTOCEPHALINAE.

Stirellus Osborn & Ball 1902a:250 as subgenus of *Athysanus* Burmeister 1838b:14. *Athysanus bicolor* Van Duzee 1892a:114 (USA: Kansas and Missouri). 10:103. Deltocephalinae, Stenometopiini. Nearctic, Neotropical.
Synonym:
Penestirellus Beamer & Tuthill 1934a:21. *P. catalinus* n.sp.

[NOTE: Emeljanov 1962b:393 (page 238, English translation) listed 18 names (*Anemochrea* Kirkaldy, *Anemolua* Kirkaldy, *Arya* Distant, *Bella* Pruthi, *Bituitus* Distant, *Campbellinella* Distant, *Giletiella* (sic) Osborn, *Kinonia* Ball, *Nandidrug* Distant, *Paivanana* Distant, *Paternus* Distant, *Penestirellus* Beamer & Tuthill, *Phrynophyes* Kirkaldy, *Pseudaconura* Linnavuori, *Sunda* Pruthi, *Trebellius* Distant, *Umesaona* Ishihara, and *Volusenus* Distant, plus *Aconura* of authors, (not Lethierry) as new synonyms of *Stirellus* Osborn & Ball. These synonymies have not been generally accepted. Vilbaste 1965a:10 lists 12 of the names (*Anemochrea, Arya, Bella, Bituitus, Campbellinella, Nandidrug, Paivanana, Paternus, Phrynophyes, Sunda, Umesaona & Volusenus*) as synonyms of *Doratulina* Melichar, along with *Aconura* of authors, (not Lethierry), and considers *Pseudaconura* Linnavuori a valid genus].

Stonasla White 1878a:472. *S. undulata* n.sp. (St. Helena). 14:4. Agalliinae, Agalliini. Malagasian; Saint Helena, Seychelles.

Stoneana DeLong 1943b:448. *S. marthae* n.sp. (Mexico: Cuernavaca, Acapulco Rd.). 10:1747. Deltocephalinae, Athysanini. Neotropical.

Stragania Stal 1862e:49 as subgenus of *Gypona* Germar 1821a:73. *S. ornatula* n.sp. (Brazil). 15:97. Blocker 1979a:29. Iassinae, Iassini. Neotropical; Brazil, Colombia, Costa Rica.

Straganiassus Anufriev 1971g:87. *Stragania matsumurai* Metcalf 1955a:266 (Japan) = *Macropsis dorsalis* Matsumura 1912b:300, not *Macropsis dorsalis* Provancher 1889. Iassinae, Iassini. Palaearctic. Synonym of *Iassus* Fabricius 1803a:85; Viraktamath, C.A. 1979a:98.

Straganiopsis Baker 1903e:10. *Macropsis idioceroides* Baker 1900a:55 (USA: New Mexico). 15:94. Blocker 1979a:11. Iassinae, Iassini. Nearctic. Synonym of *Pachyopsis* Uhler 1877a:466.

Stragnia 15:226. Error for *Stragania* Stal.
Strangania 15:226. Error for *Stragania* Stal.

Strephonius Hamilton, K.G.A. 1975a:23 as subgenus of *Paraphlepsius* 1897c:158. *Phlepsius tigrinus* Ball 1909a:80 (USA: District of Columbia). Deltocephalinae, Athysanini. Nearctic.

Streptanulus Emeljanov 1962a:171 as subgenus of *Streptanus* Ribaut 1942a:261. *Streptanus josifovi* Dlabola 1957c:117 (Bulgaria). Deltocephalinae, Athysanini. Palaearctic.

Streptanus Ribaut 1942a:261. *Cicada sordida* Zetterstedt 1828a:531 (Lapland). 10:272. Deltocephalinae, Athysanini. Palaearctic; all subregions.
Subgenus:
Streptanulus Emeljanov 1962a:171.

Streptopyx Linnavuori 1958a:36. *S. tamaninii* n.sp. (Italy: Trentino). Deltocephalinae, Athysanini. Palaearctic; Italy, Yugoslavia.

Striapona DeLong 1979g:187 as subgenus of *Polana* DeLong 1942d:110. *P. (S.) desela* n.sp. (Peru: Iquitos). Gyponinae, Gyponini. Neotropical.

Striatellus Zoological Record 109, Sec. 13f:156, 1972. Error for *Stirellus* Osborn & Ball 1902a.

Strictogonia Melichar 1926a:358. *Cicada unifasciata* Fabricius 1803a:72 ("Amer. Merid."). 1:57. Cicadellinae, Cicadellini. Neotropical; Brazil, French Guiana.

Stroggylocephalus Flor 1861a:210 as subgenus of *Acocephalus* Burmeister 1835a:111. *Cicada agrestis* Fallen 1806a:23 (Sweden). 8:34. Ossiannilsson 1981a:379-383. Aphrodinae, Aphrodini. Palaearctic; Nearctic; Oriental; widespread, adventive to Nearctic (?).

Strongilocephalus Logvinenko 1957b:65. Error for *Stroggylocephalus* Flor.

Stronglylocephalus Kirschbaum 8:266. Error for *Stroggylocephalus* Kirschbaum.

Strogylocephalus 8:266. Error for *Stroggylocephalus* Flor.
Strongylocephalus 8:266. Error for *Stroggylocephalus* Flor.

Strongylocephalus Kirschbaum 1868b:73. Error for *Stroggylocephalus* Flor 1861a:210.

Strongylomma Spinola 1850b:135. *S. caffra* n.sp. ("South Africa: Natal", error). 16:6. Linnavuori 1970a:184-185. Idiocerinae. South America; Webb, M.D. 1983b:212.

Strorgylocephua Cantoreanu 1965d:141. Error for *Stroggylocephalus* Flor.

Stymphalella Evans, J.W. 1954a:126. *S. minuta* n.sp. (Madagascar). 10:2242. Deltocephalinae, Platymetopiini. Malagasian.

Stymphalus Stal 1866a:121. *Platymetopius rubrolineatus* Stal 1855a:99 (South Africa: Natal). 10:2104. Linnavuori 1978c:479. Deltocephalinae, Scaphytopiini. Ethiopian; Palaearctic; Oriental; widespread.
Synonyms:
Varta Distant 1908g:320. *V. rubrofasciata* n.sp.
Purpuranus Zachvatkin 1933b:262. *Platymetopius rubrostriatus* Horvath 1907a:317.

Suarezia Linnavuori & DeLong 1978b:121. *Eutettix reflexus* Osborn 1923c:55 (Bolivia: Puerto Suarez). Deltocephalinae, Athysanini. Neotropical.

Subbanara Ramakrishnan & Ghauri 1979b:201. *Empoascanara linnavuorii* Dworakowska 1972k:121 (Java: Baluran, Beokol near Banjuwangi). Typhlocybinae, Erythroneurini. Oriental. Synonym of *Empoascanara* Distant 1918b:94; Dworakowska 1980d:195.

Subhimalus Ghauri 1971c:113. *S. fuscus* n.sp. (India: Simla). Deltocephalinae, Deltocephalini. Oriental; India.

Subrasaca Young 1977a:472. *Tettigonia ignicolor* Signoret 1854a:8 (Brazil). Cicadellinae, Cicadellini. Neotropical; Argentina, Brazil, ? Paraguay.

Sudanoiassus Linnavuori & Quartau 1975a:130 as subgenus of *Batracomorphus* Lewis 1834a:51. *B. (S.) kivuensis* n.sp. (Zaire: Kivu, Rutshuru). Iassinae, Iassini. Ethiopian.

Sudra Distant 1908g:257. *S. notanda* n.sp. (Burma: Karen Hills). 2:7. Hylicinae, Sudrini. Oriental; Borneo, Burma, Indochina, Sumatra, Thailand.

SUDRINI 2:7.

Sujitettix Matsumura 1931b:76. *S. ferrugineus* n.sp. (Japan). 17:268. Typhlocybinae, Typhlocybini. Palaearctic; Oriental, Japan, China, India. Synonym of *Apheliona* Kirkaldy 1907d:67; Dworakowska 1970a:691; 1977d:611.

Sulamicerus Dlabola 1974b:62. *Idiocerus stali* Fieber 1868a:453 (Rhodes). Idiocerinae. Palaearctic; Mediterranean subregion.

Sulcana DeLong & Freytag 1966b:309. *S. brevis* n.sp. (Brazil: Chapada). Gyponinae, Gyponini. Neotropical.

Sunda Singh-Pruthi 1936a:112. *S. ribeiroi* n.sp. (India: Bengal). 10:2352. Deltocephalinae, Stenometopiini. Oriental. Synonym of *Doratulina* Melichar 1903g:198; Vilbaste 1965a:10.

Sundapteryx Dworakowska 1970h:708. "*Chlorita biguttula* Ishida," presumably an error for *Chlorita biguttula* Shiraki 1912a:96 (Formosa). Typhlocybinae, Empoascini. Oriental. Synonym of *Amrasca* Ghauri 1967a:159; Dworakowska & Viraktamath 1975a:530.

Sundara Ramakrishnan & Menon 1972b:186. *S. quadrimaculata* n.sp. (India: Delhi, I.A.R.I.). Typhlocybinae, Zyginellini. Oriental, widespread. Ethiopian; central Africa.
Synonym:
Raunoia Dworakowska & Lauterer 1975a:37. *R. luciphila* n.sp.

Svanetia Schengeliya & Dlabola 1964a:659. *S. kobachidzeina* n.sp. (USSR: Georgia). Deltocephalinae, Athysanini. Palaearctic. Preoccupied by *Svanetia* Hesse 1926, see *Transcaucasica* Kocak 1981a:32.

Swarajnara Ramakrishnan & Ghauri 1979b:200. *Ratbura unipunctata* Mahmood 1967:17 (Thailand: Ratburi). Typhlocybinae, Erythroneurini. Oriental. Synonym of *Empoascanara* Distant 1918b:94; Dworakowska 1980d:195.

Sylhetia Ahmed, Manzoor 1972a:67,70. *S. punctata* n.sp. (East Pakistan: Mymensingh). Typhlocybinae, Zyginellini. Oriental.

Symphyrrhoea Emeljanov 1961a:123. Invalid, *nomen nudum*.

Symphypiga 14:172. Error for *Symphypyga* Haupt.

Symphypya 14:172. Error for *Symphypyga* Haupt.

Symphypyga Haupt 1917d:239. *S. obsoleta* n.sp. (Uzbekistan). 14:166. Adelungiinae, Adelungiini. Palaearctic; Mediterranean subregion.

Syncharina Young 1977a:603. *Tettigonia punctatissima* Signoret 1854a:16 (Colombia). Cicadellinae, Cicadellini. Neotropical; Argentina, Bolivia, Brazil, Chile, Colombia.

Synogonia Melichar 1926a:344. *Tettigonia nasuta* Fowler 1900:291 (Guatemala: Vera Paz). Kramer 1976c:38. 1:410. Nirvaninae, Nirvanini. Neotropical; Guatemala, Honduras.

SYNOPHROPSINI 10:1746.

Synophropsis Haupt 1926b:308. *S. wagneri* n.sp. (Dalmatia) = *Thamnotettix lauri* Horvath 1897b:631. 10:1746. Deltocephalinae, Platymetopiini. Palaearctic; Mediterranean subregion.

Synphropis 10:2686. Error for *Synophropsis* Haupt.

Syringius Emeljanov 1966a:101 as subgenus of *Allygus* Fieber 1875a:410. *Allygus syrinx* Dlabola 1961b:330 (USSR: Uzbekistan). Deltocephalinae, Athysanini. Palaearctic.

Syringophora Kirkaldy 1907b:87. *Raphirhinus brevis* Walker 1851b:807 (Venezuela). 1:538. Cicadellinae, Proconiini. Neotropical. Synonym of *Splonia* Signoret 1891a:467; Young 1968a:141.

Szymczakowskia Dworakowska 1974a:238. *S. linnavuorii* n.sp. (Congo: Odzala). Typhlocybinae, Erythroneurini. Ethiopian; Cameroon, Congo.

Taberinha Linnavuori & Heller 1961a:9. Error for *Taperinha* Linnavuori.

Tacora Melichar 1926a:359. *Tettigonia dilecta* Walker 1851b:747 (Brazil: "Ega"). Young 1977a:313. Cicadellinae, Cicadellini. Neotropical.

Taeniocerus Dlabola 1974b:64. *Bythoscopus ocularis* Mulsant & Rey 1855a:220 (Provence). Idiocerinae. Palaearctic. Preoccupied by *Taeniocerus* Kaup 1871, see *Bugraia* Kocak 1981a:32.

Tafalka Dworakowska 1979c:309. *T. bosa* n.sp. (Cameroon: Dimako). Typhlocybinae, Zyginellini. Ethiopian; Cameroon, Central African Republic, Zaire.

Tahara Nielson 1977a:7. *T. quadrispiculata* n.sp. (NW New Guinea: Waris, S. of Hollandia). Coelidiinae, Thagriini. Oriental.

Taharana Nielson 1982e:50. *Coelidia sparsa* Stal 1854b:254 (Philippines). Coelidiinae, Coelidiini. Oriental; widespread.

Tahuampa Young 1977a:46. *Tettigonia septemfasciata* Signoret 1853b:332 (Venezuela: Laguaria). Cicadellinae, Cicadellini. Neotropical; Peru, Venezuela.

Tahura Melichar 1926a:343. *T. fowleri* Kramer 1976c:39 (Peru: Callanga). 1:410. Nirvaninae, Nirvanini. Neotropical.

Takagiana Dworakowska 1974a:179. *T. crinita* n.sp. (Congo: Odzala). Typhlocybinae, Zyginellini. Ethiopian.

Takagiella Vilbaste 1969a:4. *Deltocephalus tezuyae* Matsumura 1902a:391 (Japan: Akashi, Takasago). Deltocephalinae, Deltocephalini. Oriental/Palaearctic; Japan, Korea, Malaya.

Takama Dworakowska & Viraktamath 1975a:523. *T. magna* n.sp. (India: Mysore, Bangalore). Typhlocybinae, Erythroneurini. Oriental.

Tamaga Dworakowska 1981h:325 as subgenus of *Iseza* Dworakowska 1981h:322. *I. (T.) dubia* n.sp. (Nigeria: N W State, Badeggi). Typhlocybinae, Erythroneurini. Ethiopian.

Tamaricades Emeljanov 1962a:163 as subgenus of *Goniagnathus* Fieber 1866a:506. *Athysanus decoratus* Haupt 1917d:246 (Turkestan). Linnavuori 1978c:497. Deltocephalinae, Goniagnathini. Palaearctic.

Tamaricella Zachvatkin 1946a:153 as subgenus of *Helionidia* Zachvatkin 1946a:152. *Zygina jaxartensis* Oshanin 1871a:213 (Turkestan). 17:1463. Dworakowska 1971b:103-110. Typhlocybinae, Erythroneurini. Palaearctic; Mediterranean and Siberian subregions, Canary Islands.

Tambila Distant 1908g:247. *T. greeni* n.sp. (Ceylon, Kandy). 3:210. Penthimiinae. Oriental; Ceylon, India.

Tambilia 3:227. Error for *Tambila* Distant.

Tamnotettix 10:2686. Error for *Thamnotettix* Zetterstedt.

Tantogonalia Young 1977a:1071. *T. praecalva* n.sp. (Colombia: Antioquia, Palmira). Cicadellinae, Cicadellini. Neotropical; Colombia, Paraguay.

Tantulidia Nielson 1979a:654. *Coelidia rufifrons* Walker 1851b:854 (Honduras). Coelidiinae, Tinobregmini. Neotropical; Mexico to Ecuador.

Tapajosa Melichar 1924a:241. *Tettigonia fulvopunctata* Signoret 1854c:484 (Brazil: Bahia). 1:511. Young 1968a:234. Cicadellinae, Proconiini. Neotropical; Argentina, Bolivia, Brazil, Ecuador, Venezuela.

Taperinha Linnavuori 1959b:188. *T. bifurcata* n.sp. (Brazil: Chapada). Deltocephalinae, Athysanini. Neotropical.

Tapetia Emeljanov 1964b:52. *T. lydiae* n.sp. (Turkmenia). Deltocephalinae, Athysanini. Palaearctic.

[NOTE: There are two spellings given for the type species; *lydiae* for the designated taxon, and *lidiae* for the new species described].

Tartessella Evans, J.W. 1936b:56 (1937). *T. attenuata* n.sp. (Western Australia: Mullewa). 11:142. Evans, F. 1981a:120. Tartessinae. Australian.

TARTESSINAE 11:123.

Tartessoides Evans, J.W. 1936b:55 (1937). *T. griseus* n.sp. (Australia: between Everard and Warburton Ranges). 11:142. Evans, F. 1981a:118. Tartessinae. Australian.

Tartessops Evans, F. 1981a:155. *Bythoscopus colligatus* Walker 1870b: 319 (New Guinea). Tartessinae. Oriental; New Guinea.

Tartesus 11:179. Error for *Tartessus* Stal.

Tartessus Stal 1865b:156. *Bythoscopus malayus* Stal 1859b:290 (Malacca). 11:124. Evans, F. 1981a:173. Tartessinae. Oriental/Australian.

Taslopa Evans, J.W. 1942a:29. *T. montana* n.sp. (Tasmania: Hobart, Mt. Wellington). 5:78. Ulopinae, Ulopini. Australian; Australia, Tasmania.

Tasmanotettix Evans, J.W. 1938b:12. *T. maculata* n.sp. (Tasmania). 10:102. Deltocephalinae. Australian. Synonym of *Thamnophryne* Kirkaldy 1907d:61; Evans, J.W. 1966a:232.

Tasnimocerus Ghauri 1975c:287. *T. clypeatus* n.sp. (India: Simla, Mashobra). Idiocerinae. Oriental.

Tataka Dworakowska 1974a:174. *T. boulardi* n.sp. (Congo: Odzala). Typhlocybinae, Zyginellini. Ethiopian.
Synonym:
Orucyba Ghauri 1975b:491. *O. mollis* n.sp.

Taurotettix Haupt 1929c:259. *Thamnotettix beckeri* Fieber 1872a:11 (Southern Russia). 10:1739. Deltocephalinae, Fieberiellini. Palaearctic; European and Mediterranean subregions.

Tautocerus Anufriev 1971a:98. *T. dworakowskae* Anufriev n.sp. (Korea: Phyonjang, Moranbong). Idiocerinae. Oriental.

Tautoneura Anufriev 1969b:186. *T. tricolor* n.sp. (USSR Maritime Territory, Jaovlevka). Typhlocybinae, Erythroneurini. Palaearctic; Siberian and Manchurian subregions. Oriental; India.

Tbilisica Dlabola 1958b:333. *T. denticulata* n.sp. (USSR: Tbilisi). Deltocephalinae, Athysanini. Palaearctic; Mediterranean subregion.

Teitogonia 1:688. Error for *Tettigonia* Olivier.

Teleogonia Melichar 1925a:361. *Tettigonia fusca* Walker 1851b:741 (Colombia). 1:540. Young 1977a:158. Cicadellinae, Cicadellini. Neotropical; Bolivia, Colombia, Ecuador, French Guiana, Paraguay, Peru, Venezuela.
Synonym:
Pygometopia Schroeder 1960a:319. *Oncometopia (Mucrometopia) infuscata* Melichar 1925a:404.

Teletusa Distant 1908b:78. *T. paraguayensis* n.sp. (Paraguay: San Bernardino). 1:523. Young 1968a:162. Cicadellinae, Proconiini. Neotropical; Amazonas to Bolivia, to Argentina and Paraguay.
Synonym:
Myogonia Melichar 1926a:332. *Tettigonia limpida* Signoret 1855c:512.

Telligonia 1:688. Error for *Tettigonia* Olivier.

Telopetulcus Evans, J.W. 1972b:660. *T. dubius* n.sp. (NW New Guinea: Waris, S. of Hollandia). Acostemminae. Oriental.

Telusus Oman 1949a:171. *Deltocephalus blandus* Gillette 1898b:26 (USA: Colorado). 10:1811. Deltocephalinae, Deltocephalini. Nearctic.

Temnopsis Signoret 1879b:267. 9:22. *Nomen nudum*.

Tenarus 10:2686. Error for *Tenuarus* DeLong.

Tengirhinus Ishihara 1953b:17. *T. tengu* n.sp. (Japan: Shikoku). 6:52. Cicadellinae, Evacanthini. Palaearctic.

Tenobregmus 11:179. Error for *Tinobregmus* Van Duzee.

Tenuacia DeLong 1977c:88. *T. macera* n.sp. (Panama: Darien Province, Santa Fe). Gyponinae, Gyponini. Neotropical; Costa Rica, Panama, Peru.
Subgenus:
Rubacea DeLong 1977c:89. *T. (R.) rubera* n.sp.

Tenuarus DeLong 1944c:52 as subgenus of *Hebenarus* DeLong 1944c:41. *H. (T.) spinosa* n.sp. (Mexico: Chiapas, Vergel, valley of the Rio Huixtla). 10:2249. Deltocephalinae, Scaphytopiini. Neotropical.

Tenucephalus DeLong 1944i:236. *T. marginellus* n.sp. (Mexico: Guerrero, Iguala). 9:60. Deltocephalinae, Hecalini. Neotropical; Mexico, Panama, Peru, Bolivia.

Tenuisanus DeLong 1944j:73. *T. costatus* n.sp. (Mexico: Guerrero, Iguala). 10:1747. Deltocephalinae, Athysanini. Neotropical.

Tenuitartessus Evans, F. 1981a:132. *Tartessus blundellensis* Evans, J.W. 1936b:55 (Australia: A.C.T., Blundello). Tartessinae. Australian.

Terulia Stal 1862e:50. *T. ferruginea* n.sp. (Brazil). 11:25. Nielson 1979b:19-24. Coelidiinae, Teruliini. Neotropical; Brazil, Panama, Colombia, Nicaragua, Peru, Venezuela.

TERULIINI Nielson 1979b:10, tribe of COELIDIINAE.

TETARTOSTYLINI 10:1929.

Tetartostylus Wagner 1951a:39. *Athysanus pellucidus* Melichar 1896a:258 (Italy), a primary homonym. 10:1929. Deltocephalinae, Tetarostylini. Ethiopian; South Africa. Palaearctic; southern Europe, northern Africa.
[NOTE: See Nast 1985a:147-148].

Tetelloides Evans, J.W. 1955a:19. *T. rubigella* n.sp. (Belgian Congo: Lusinga, 1700 m). 1:104. Cicadellinae, Cicadellini. Ethiopian. Synonym of *Baramapulana* Distant 1910e:235; Linnavuori 1979a:688.

Tetigonia Blanchard 1852 (in Gay, C. 1852):282, a junior homonym of *Tettigonia* Linnaeus 1758. Placed in the Official Index of Rejected and Invalid Generic Names in Zoology, ICZN Opinion 299, 1954.

Tetigonia Fourcroy 1785a:193, a junior homonym of *Tettigonia* Linnaeus 1758. Placed in the Official Index of Rejected and Invalid Generic Names in Zoology, ICZN Opinion 299, 1954.

Tetigonia Geoffroy 1762a:429. *Cicada viridis* Linnaeus 1758a:438 (Europe). 1:107. Cicadellinae, Cicadellini. Palaearctic. Synonym of *Cicadella* Latrielle 1817a:406. ICZN 1963, Opinion 647. Placed in the Official Index of Rejected and Invalid Generic Names in Zoology, ICZN Opinion 299, 1954.

Tetigoniella 1:718. Error for *Tettigoniella* Jacobi.

TETIGONIIDAE 1:10.

TETIGONIINAE 1:25.

TETIGONIINI 1:26.

Tetitgonia 1:688. Error for *Tettigonia* Olivier.

Tetragonia 1:688. Error for *Tretogonia* Melichar.

Tetrastactus Amyot 10:2686. Invalid, not binominal.

Tetratostylus Zoological Record 98, Sec. 13:370. Error for *Tetartostylus* Wagner 1951a.

Tetrogonia 1:688. Error for *Tretogonia* Melichar.

Tettegonia 1:688. Error for *Tettigonia* Olivier.

Tettigdnia 1:688. Error for *Tettigonia* Olivier.

Tettigella China & Fennah 1945a:711. *Cicada viridis* Linnaeus 1758a:438 (Europe). 1:106. Young 1968a:15. Cicadellinae, Cicadellini. Palaearctic. Synonym of *Cicadella* Latreille 1817a:406. ICZN Opinion 647, 1963.

TETTIGELLIDAE 1:7.

TETTIGELLINAE 1:25

TETTIGELLINI 1:26.

Tettigonia Fabricius, J.C. 1775a:678, a junior homonym of *Tettigonia* Linnaeus 1758a. Placed in the Official Index of Rejected and Invalid Generic Names in Zoology, ICZN Opinion 299, 1954.

Tettigonia Olivier 1789a:24, replacement for *Tetigonia* Geoffroy 1762a:429. *Cicada viridis* Linnaeus 1758a:438 (Europe). Cicadellinae, Cicadellini. Palaearctic.

Tettigoniella Jacobi 1904a:778. *Cicada viridis* Linnaeus 1758a:438 (Europe). 1:107. Cicadellinae, Cicadellini. Palaearctic. Synonym of *Cicadella* Latreille 1817a:406. ICZN 1963, Opinion 647.

TETTIGONIELLIDAE 1:12.

TETTIGONIELLINAE 1:25.

TETTIGONIELLINI 1:26.

TETTIGONIIDAE 1:7.

TETTIGONIINAE 1:25.

TETTIGONIINI 1:26.

Tettignoia 1:689. Error for *Tettigonia* Olivier.

Tettigoma 1:689. Error for *Tettigonia* Olivier.

Tettigonella 1:689. Error for *Tettigoniella* Jacobi.

Tettingonia 1:725. Error for *Tettigonia* Olivier.

Tettisama Young 1977a:785. *Tettigonia bisellata* Signoret 1862a:586 (Peru). Cicadellinae, Cicadellini. Neotropical; northern South America to Paraguay and Argentina.

Tettiselva Young 1977a:101. *T. pedia* n.sp. (Peru: Rio Santiago). Cicadellinae, Cicadellini. Neotropical; Ecuador, Peru.

Tettogoniella 1:725. Error for *Tettigoniella* Jacobi.

Texacanus 10:2686. Error for *Texananus* Ball.

Texananus Ball 1918b:381 as subgenus of *Phlepsius* Fieber 1866a:514. *P. (T.) mexicanus* n.sp. (Mexico). 10:658, 663. Deltocephalinae, Athysanini. Nearctic/Neotropical; Mexican & Antillean subregions.
Subgenera:
    Iowanus Ball 1918b:382. *Phlepsius (Iowanus) handlirschi* Ball 1918b:383.
    Aridanus DeLong & Hershberger 1949a:173. *Phlepsius aerolatus* Baker 1898f:30.
Synonym:
    Zioninus Ball 1918b:388. *Phlepsius extremus* Ball 1901a:10.

Texanus 10:2686. Error for *Texananus* Ball.

Teyasteles Linnavuori 1969a:1185. *Cicadula divisifrons* Naude 1926a:83 (South Africa: Natal, Cedara). Deltocephalinae, Macrostelini. Ethiopian.

Thagria Melichar 1903b:176. *T. fasciata* n.sp. (Ceylon). 11:88. Nielson 1977a:9. Coelidiinae, Thagriini. Oriental; Indo-Malaya, Fiji.
Synonyms:
    Mukwana Distant 1908g:317. *M. introducta* n.sp.
    Soortana Distant 1908g:319. *S. simulata* n.sp.
    Dharmma Distant 1908g:323. *D. projecta* n.sp.
    Sabima Distant 1908g:324. *S. prima* n.sp.
    Guliga Distant 1908g:326. *G. erebus* n.sp.
    Orthojassus Jacobi 1914a:382. *O. philagroides* n.sp.
    Sabimoides Evans, J.W. 1947a:254. *S. cardamomi* n.sp.

THAGRIINI 11:87.

Thaia Ghauri 1962a:253. *T. oryzivora* n.sp. (Thailand: Bangkok). Dworakowska 1970c. Typhlocybinae, Erythroneurini. Oriental, Ethiopian.
Subgenera:
    Nlunga Dworakowska 1974a:191. *N. reenei* n.sp; Dworakowska 1976b:50.
    Niema Dworakowska 1979b:3. *T. (N.) enica* n.sp.
Synonym:
    Hardiana Mahmood 1967a:14. *H. assamensis* n.sp.; Dworakowska 1970c:88.

Thailocyba Mahmood 1967a:35. *T. elbeli* n.sp. (Thailand: Loei, Dansai, Na Haeo, Ban Na Muang). Typhlocybinae, Typhlocybini. Oriental.

Thailus Mahmood 1967a:14. *T. nigroscutellatus* n.sp. (Thailand: Loei, Dansai, Ban Na Muang). Typhlocybinae, Erythroneurini. Oriental.

Thalasia 4:145. Error for *Thlasia* Germar.

Thalattoscopus Kirkaldy 1905b:334. *T. dryas* n.sp. (New Britain). 15:20. 1975a:30. Iassinae, Iassini. Oriental.
Synonym:
    Marquardtella Schmidt 1930b:122. *M. krusei* n.sp.

Thammotettix 10:2686. Error for *Thamnotettix* Zetterstedt.

Thamnocettix 10:2686. Error for *Thamnotettix* Zetterstedt.

Thamnophryne Kirkaldy 1907d:61. *T. nysias* n.sp. (Australia: New South Wales, Mittagong). 10:793. Deltocephalinae. Australian; Australia, Tasmania.
Synonym:
    Tasmanotettix Evans, J.W. 1938b:12. *T. maculata* n.sp.

Thamnotelia 10:2687. Error for *Thamnotettix* Zetterstedt.
Thamnoteltix 10:2687. Error for *Thamnotettix* Zetterstedt.
Thamnotethix 10:2687. Error for *Thamnotettix* Zetterstedt.
Thamnotetix 10:2687. Error for *Thamnotettix* Zetterstedt.
Thamnotetlix 10:2687. Error for *Thamnotettix* Zetterstedt.
Thamnotetrix 10:2687. Error for *Thamnotettix* Zetterstedt.
Thamnotettex 10:2687. Error for *Thamnotettix* Zetterstedt.
Thamnotettiex 10:2687. Error for *Thamnotettix* Zetterstedt.
Thamnotettis 10:2687. Error for *Thamnotettix* Zetterstedt.

Thamnotettix Zetterstedt 1838a:292. *Cicada prasina* Fallen 1806a:27 (Sweden), not *Cicada prasina* Fabricius 1794 = *Cicada confinis* Zetterstedt 1828a:527. 10:716. Ossiannilsson 1983a:732. Deltocephalinae, Athysanini. Palaearctic, Nearctic.
Subgenus:
    Loepotettix Ribaut 1942a:264. *Jassus (Thamnotettix) dilutior* Kirschbaum 1868b:92.
Synonym:
    Thamnus Fieber 1866a:505. *Cicada confinis* Zetterstedt 1828a:527.

THAMNOTETTIXINI 10:715.

Thamnotettrix 10:2687. Error for *Thamnotettix* Zetterstedt.
Thamnotlttix 10:2687. Error for *Thamnotettix* Zetterstedt.
Thamnottetix 10:2688. Error for *Thamnotettix* Zetterstedt.
Thamnottettix 10:2688. Error for *Thamnotettix* Zetterstedt.
Thamnottix 10:2688. Error for *Thamnotettix* Zetterstedt.

Thamnus Fieber 1866a:505. *Cicada confinis* Zetterstedt 1828a:527 (Scandinavia). 10:717. Deltocephalinae, Athysanini. Palaearctic. Synonym of *Thamnotettix* Zetterstedt 1838a:292.

Thamonotettix 10:2688. Error for *Thamnotettix* Zetterstedt.

Thamotettix 10:2688. Error for *Thamnotettix* Zetterstedt.

Thampoa Mahmood 1967a:33. *T. dansaiensis* n.sp. (Thailand: Loei, Dansai, Na Haeo, Ban Na Muang). Typhlocybinae, Typhlocybini. Oriental.

Thannotettix 10:2688. Error for *Thamnotettix* Zetterstedt.
Thannotettix 10:2688. Error for *Thamnotettix* Zetterstedt.

Tharra Kirkaldy 1906c:324. *T. labena* n.sp. (Australia: Queensland, Kuranda). 11:23. Nielson 1975a:34. Coelidiinae, Tharrini. Australian, Oriental, Oceania.
Synonyms:
    Muirella Kirkaldy 1907d:79. *M. oxyomma* n.sp.
    Nisitra Walker 1870b:327. *N. telifera* n.sp. = *Coelidia tiarata* Stal 1865b:159.
    Nisitrana Metcalf 1952a:229. *Nisitra telifera* Walker 1870b:328.
    Jassoidula Osborn 1934a:182. *J. straminea* n.sp.

THARRINI Nielson 1975a:131, tribe of COELIDIINAE.

Thatuna Oman 1938c:176. *T. gilletti* n.sp. (USA: Idaho, Moscow). 8:31. Oman & Musgrave 1975a:8; Anufriev 1978a:58. Cicadellinae, Errhomenini. Nearctic; northern Idaho, adjacent areas of Montana, Washington, British Columbia.

Thaumaptoscopus 3:227. Error for *Thaumatoscopus* Kirkaldy.

Thaumastus Stal 1864b:67. *Ledra marmorata* Blanchard 1840a:194 (Madagascar). 4:109. Ledrinae, Petalocephalini. Malagasian. Preoccupied by *Thaumastus* Martens 1860, see *Betsileonas* Kirkaldy 1903f:13.

Thaumatopoides Evans, J.W. 1947a:255. *T. caffra* n.sp. (Southwest Africa). 3:212. Linnavuori 1977a:11. Penthimiinae. Ethiopian; South Africa, southwest Africa.

THAUMATOSCOPINAE 3:217.

Thaumatoscopus Kirkaldy 1906c:462. *T. galeatus* n.sp. (Australia: Queensland, Cairns). 3:218. Penthimiinae. Australian; Australia. Oriental; Philippines.

Theasca Dworakowska 1972c:483. *T. rubrata* Dworakowska n.sp. (Zaire: Elizabethville). Typhlocybinae, Empoascini. Ethiopian; Zaire, Tanganyika.

Theronopus Webb, M.D. 1983b:232. *Idiocerus angulatus* Webb, M.D. 1975a:181 (Kenya: Wajir). Idiocerinae. Ethiopian; Afrotropical, widespread.

Thlasia Germar 1836a:71. *T. brunnipennis* n.sp. (South Africa: Cape of Good Hope). 4:101. Linnavuori 1972b:232-234. Ledrinae, Petalocephalini. Ethiopian; South Africa.

Thnunatettix 10:2688. Error for *Thamnotettix* Zetterstedt.

Thomnotettix 10:2688. Error for *Thamnotettix* Zetterstedt.

Thompsoniella 10:2688. Error for *Thomsoniella* Signoret.

Thomsonia Signoret 1879a:51. *Hecalus kirschbaumii* Stal 1870c:737 (Philippines) = *Acocephalus porrectus* Walker 1858b:363. 10:1719. Morrison 1973a:411. Deltocephalinae, Hecalini. Oriental. Synonym of *Hecalus* Stal 1964b:65.
[NOTE: Signoret 1880a:52 emended *Thomsonia* to *Thomsoniella* apparently because he considered it preoccupied by *Thompsonia* Kossmann 1872. See footnote by Distant 1908g:273].

Thomsoniella Signoret 1880a:52, emendation of *Thomsonia* Signoret 1879a:51. *Hecalus kirschbaumii* Stal 1870c:737 (Philippines) = *Acocephalus porrectus* Walker 1858b:362. 10:1719. Morrison 1973a:410. Deltocephalinae, Hecalini. Oriental. Synonym of *Hecalus* Stal 1864b:65; Morrison 1973a:410.

Thyia Hamilton, K.G.A. 1980b:894 as subgenus of *Pedionis* Hamilton, K.G.A. 1980b:891. *Macropsis thyia* Kirkaldy 1907d:36 (Australia: Queensland, Kuranda). Macropsinae, Macropsini. Australian.

Thymbrella Evans, J.W. 1969a:747. *T. tamminensis* n.sp. (Western Australia: Tammin). Ledrinae, Thymbrini. Australian.

THYMBRINI 4:123.

Thymbris Kirkaldy 1907d:49. *T. inachis* n.sp. (Australia: Queensland). 4:131. Ledrinae, Thymbrini. Australian.

Thyphlocyba 17:1507. Error for *Typhlocyba* Germar.

Tiaja Oman 1941a:207. *Paropulopa californica* Ball 1909b:184 (USA: California, Salinas). 5:35. Sawbridge 1975b. Megophthalminae. Nearctic; Californian subregion.

Tialidia Nielson 1982e:20. *T. congoensis* n.sp. (Uganda). Coelidiinae, Coelidiini. Ethiopian; Nigeria, Uganda, Zaire.

Tiaratus Emeljanov 1961a:129. *T. caricis* n.sp. (USSR: Kazakhstan, 40 km south of Zhan-Ark, Karaganda region). Deltocephalinae, Paralimnini. Palaearctic; Siberian subregion.

Tichocoelidia Kramer 1962a:104. *T. clarkei* n.sp. (Colombia: Cundinam-arca, Rio Sumapaz Gorge, E of Melgar). Neocoelidiinae. Neotropical.

Tideltellus Kramer 1971d:263. *Deltocephalus marinus* Osborn & Metcalf 1920a:110 (USA: North Carolina, Wrightsville Beach). Deltocephalinae, Deltocephalini. Nearctic; southeast USA.

Tigriculus Dlabola 1961b:323. *T. cinnamocolorata* n.sp. (USSR: Tadzikhistan). Deltocephalinae, Deltocephalini. Palaearctic.

Timeus 10:2689. Error for *Tumeus* DeLong.

Tinderella Webb, M.D. 1983a:91. *T. maondica* n.sp. (Australia: Queensland, Tamborine Mt.). Idiocerinae. Australian.

Tingolix Linnavuori & DeLong 1978d:234. *T. piperatus* n.sp. (Peru: Monson Valley, Tingo Maria). Deltocephalinae, Athysanini. Neotropical.

Tingopyx Linnavuori & DeLong 1978d:228. *T. ramosus* n.sp. (Peru: Monson Valley, Tingo Maria). Deltocephalinae, Athysanini. Neotropical.

TINOBREGMINI 11:12.

Tinobregmus Van Duzee 1894f:213. *T. vittatus* n.sp. (USA: Florida). 11:12. Nielson 1975a:16. Coelidiinae, Tinobregmini. Nearctic; central and southern USA. Neotropical; Antillean and Mexican subregions.

Tinocripus Nielson 1982e:212. *T. gladius* n.sp. (Peru: Torentoy Canyon, 2000-2200 m). Coelidiinae, Coelidiini. Neotropical; Peru.

Tiphlocyba 17:1507. Error for *Typhlocyba* Germar.

Tipuana Melichar 1926a:344. *T. expallida* Young 1977a:339 (Peru: Cuzco, Hacienda Maria). 1:396. Cicadellinae, Cicadellini. Neotropical; Brazil, French Guiana, Peru.

Titia Stal 1866a:105. *Acocephalus punctiger* Stal 1855a:98 (So. Africa: Cape of Good Hope). 8:219. Ledrinae, Petalocephalini. Ethiopian. Preoccupied by *Titia* Meigen 1800, see *Titiella* Bergroth 1920a:29.

Titiella Bergroth 1920a:29, replacement for *Titia* Stal 1866a:105, not *Titia* Meigen 1800. *Acocephalus punctiger* Stal 1855a:98 (South Africa: Cape of Good Hope). 8:219. Linnavuori 1972b:238-239. Ledrinae, Petalocephalini. Ethiopian.

Tituria Stal 1865b:158 as subgenus of *Petalocephala* Stal 1854b:251. *P. (T.) nigromarginata* n.sp. (Malacca). 4:59. Linnavuori 1972b:224-232. Ledrinae, Petalocephalini. Oriental; widespread. Ethiopian.

Tlagonalia Young 1977a:1037. *Tettigonia nigroguttata* Signoret 1855d:772 (Mexico). Cicadellinae, Cicadellini. Neotropical; southern Mexico to Guatemala.

Tlasia 4:145. Error for *Thlasia* Germar.

Toba Schmidt 1911b:301. *T. fasciculata* n.sp. (Sumatra). 11:21. Unassigned, formerly Coelidiinae; Nielson 1975a:11. Oriental.

Tocephalus Evans, J.W. 1966a:134. Error for *Acocephalus* Burmeister.

Togacephalus Matsumura 1940d:38. *Deltocephalus distinctus* de Motschulsky 1859b:112 (Ceylon). 10:1485. Deltocephalinae, Deltocephalini. Oriental. Subgenus of *Recilia* Edwards 1922a:206; Kwon & Lee 1979b:74.

Togaricrania Matsumura 1931b:72. *T. rubrovitta* n.sp. (Japan: Honshu?, or Hong Kong Islands). 17:233. Typhlocybinae, Dikraneurini. Oriental; China, Japan.

Togaritettix Matsumura 1931b:71. *T. serratus* n.sp. (Japan). 17:98. Typhlocybinae, Dikraneurini. Palaearctic. Subgenus of *Motschulskyia* Kirkaldy 1905a:226; Dworakowska 1971d:579. Synonym:
Mahmoodiana Ahmed, Manzoor & Waheed 1971a:116. *M. acuta* n.sp.

Tolasella Evans, J.W. 1972a:176. *T. festa* n.sp. (NE New Guinea: Goroka-Kabebe, 1800 m). Penthimiinae. Oriental; New Guinea.

Toldoanus Linnavuori 1954e:144. *Deltocephalus marginellus* Osborn 1923c:41 (Bolivia: Prov. del Sara). 10:1891. Linnavuori 1959b:138. Deltocephalinae, Deltocephalini. Neotropical; Bolivia, Argentina, Brazil, Paraguay.

Tolua Melichar 1926a:298. *Aulacizes multiguttata* Stal 1864a:80 (Mexico). 1:619. Cicadellinae, Proconiini. Neotropical. Synonym of *Amblydisca* Stal 1869a:61; Young 1968a:109.

Tomaloides Evans, J.W. 1972a:177. *T. sheperdi* n.sp. (Australia: New South Wales, Broken Hill). Penthimiinae. Australian.

Tomopennis Maldonado-Capriles 1984a:97. *T. tumidus* n.sp. (Guyana). Idiocerinae. Neotropical.

Tongdotettix Kwon 1980b:3. *T. youngeouni* n.sp. (Southern Korea: Gyeongsangnamdo Prov., Tongdoso Temple). Deltocephalinae, Platymetopiini. Palaearctic.

Torenadoga Blocker 1979a:32. *T. youngi* n.sp. (Venezuela: E. Merida, Acequias, 8000 ft). Iassinae. Neotropical.

Toroa Ahmed, Manzoor 1979a:41. *T. einyui* n.sp. (Uganda). Typhlocybinae, Erythroneurini. Ethiopian.

Toropsis Hamilton, K.G.A. 1980b:886. *Oncopsis balli* Kirkaldy 1907d:38 (Australia: Queensland, Bungaberg). Macropsinae, Macropsini. Australian.

Torresabela Young 1977a:790. *Tettigonia fairmairei* Signoret 1853c:685 (Brazil). Cicadellinae, Cicadellini. Neotropical; Argentina, Bolivia, Brazil, Ecuador, Paraguay, Venezuela.

Tortigonalia Young 1977a:1053. *T. torta* n.sp. (Peru: Monson Valley, Tingo Maria). Cicadellinae, Cicadellini. Neotropical; Bolivia, Brazil, Ecuador, Peru.

Tortor Kirkaldy 1907d:42. *T. daulais* n.sp. (Australia: Queensland, Bundaberg). 7:14. Evans, J.W. 1966a:153. Nirvaninae, Nirvanini. Australian.
Synonym:
Austronirvana Evans, J.W. 1941e:41. *A. flavus* n.sp.

Tortotettix Theron 1982b:27. *T. dispar* n.sp. (South Africa: Natal, Ladysmith). Deltocephalinae, Deltocephalini. Ethiopian.

Tortusana DeLong & Freytag 1974a:191 as subgenus of *Acuera* DeLong & Freytag 1972a:229. *A. (T.) angera* n.sp. (Peru: Tingo Maria). Gyponinae, Gyponini. Neotropical.

Tozzita Kramer 1964d:267. *T. ips* n.sp. (Bolivia: Riberalta). Neocoelidiinae. Neotropical; Bolivia, Brazil.

Trachygonalia Young 1977a:694. *Tettigonia germari* Signoret 1853b:359 (Brazil). Cicadellinae, Cicadellini. Neotropical.

Tphlocyba 17:1507. Error for *Typhlocyba* Germar.

Tpplocyba 17:1508. Error for *Typhlocyba* Germar.

Tragua 1:725. Error for *Iragua* Melichar.

Traiguma Distant 1908g:261. *T. nasuta* n.sp. (Southern India). 2:11. Hylicinae, Hylicini. Oriental.

Transcaucasica Kocak 1981a:32, replacement for *Svanetia* Schengeliya & Dlabola 1964a:659, not *Svanetia* Hesse 1926. *Svanetia kobachidzeina* Schengeliya & Dlabola 1964a:660 (USSR: Georgia). Deltocephalinae, Athysanini. Palaearctic.

Trebellius Distant 1918b:52. *T. albifrons* n.sp. (India: United Provinces, Allahabad District). 10:115. Deltocephalinae, Stenometopiini. Oriental. Synonym of *Doratulina* Melichar 1903b:198; Nast 1972a:354.

Trebellus Emeljanov 1968b:250. Error for *Trebellius* Distant 1918b.

Tremulicerus Dlabola 1974b:63. *Idiocerus balcanicus* Horvath 1903a:24 (Serbia). Idiocerinae. Palaearctic. Synonym of *Idiocerus* Lewis 1834a:47; Hamilton, K.G.A. 1980a:825.

Tretogonia Melichar 1924a:198. *Tettigonia pruinosa* Walker 1851b:743 (Colombia) preoccupied = *Tretagonia notatifrons* Melichar 1926a:274. 1:598. Young 1968a:166. Cicadellinae, Proconiini. Neotropical; Panama to Argentina, Paraguay, and Uruguay.

Triassojassus Tillyard 1919a:866. *T. proavitus* n.sp. Metcalf & Wade 1966a:225. Fossil: Triassic; Australia, Queensland.
Trichogonia Breddin 1901c:75. *T. ardentula* n.sp. (Ecuador: Banos). 1:646. Young 1977a:49. Cicadellinae, Cicadellini. Neotropical; Ecuador to Bolivia.
Triquetolidia Nielson 1982e:262. *T. youngi* n.sp. (Paraguay: Jabaty). Coelidiinae, Coelidiini. Neotropical; Brazil, Paraguay, Guyana, Cuba, Surinam.
Triviotartessus Evans, F. 1981a:175. *Tartessus trivialis* Spangberg 1878c:5 ("Mysol"). Tartessinae. Oriental; New Guinea.
Trocnada Walker 1858a:103. *T. dorsigera* n.sp. (Australia: New South Wales, Sydney). 15:133. Iassinae, Trocnadini. Australian.
  Synonym:
    Abelterus Stal 1865b:157. *A. incarnatus* n.sp.
Trocnadella Singh-Pruthi 1930a:18. *T. shillongensis* n.sp. (India: Assam, Shillong). 15:116. Iassinae, Trocnadini. Oriental; India, Formosa, Nepal.
TROCNADINI 15:138.
Tropicanus DeLong 1944d:87 as subgenus of *Phlepsius* Fieber 1866a:514. *Allygus costomaculatus* Van Duzee 1894f:207 (USA: Texas). 10:569. Deltocephalinae, Athysanini. Neotropical; Antillean and Mexican subregions. Nearctic; Sonoran.
  Subgenus:
    Cabimanus Linnavuori 1959b:204. *Tropicanus singularis* DeLong 1944d:92.
Trypanalebra Young 1952b:27. *Protalebra maculata* Baker 1903d:6 (Nicaragua: San Marcos or Managua). 17:22. Typhlocybinae, Alebrini. Neotropical; Antillean subregion and Central America.
Tsavopsis Linnavuori 1978b:14. *T. tuberculata* n.sp. (Kenya: Tsavo Park, Kitani Lodge). Macropsinae, Macropsini. Ethiopian. Synonym of *Macropsis* Lewis 1834a:49; Hamilton, K.G.A. 1980b:904.
Tuakamara Webb, M.D. 1980a:859. *T. aphila* n.sp. (Aldabra: South Island, Takamaka Grove). Deltocephalinae, Platymetopiini. Malagasian.
Tuberana DeLong & Freytag 1971a:317. *T. tubera* n.sp. (Peru: Tingo Maria). Gyponinae, Gyponini. Neotropical; Peru.
Tubiga Young 1977a:657. *T. debilis* n.sp. (Colombia: Choco Dept., Camp Curiche). Cicadellinae, Cicadellini. Neotropical; Colombia, Peru.
Tubulanus Linnavuori 1955a:109. *T. nitidus* n.sp. (Brazil: Bahia). 10:1241. Deltocephalinae, Athysanini. Neotropical. Preoccupied by *Tubulanus* Reinier 1804. Synonym of *Atanus* Oman 1938b:381; Linnavuori & DeLong 1977c:209.
Tumeus DeLong 1944a:168. *T. serrellus* n.sp. (Mexico: Chiapas, Vergel). 10:2249. Deltocephalinae, Scaphytopiini. Nearctic. Synonym of *Scaphytopius* Ball 1931b:218.
Tumocerus Evans, J.W. 1941f:149. *T. varius* n.sp. (Western Australia: Dendari, 40 mi W of Coolgardie). 16:7. Idiocerinae. Australian.
Tumupasa Linnavuori 1959b:140. *T. harpago* n.sp. (Bolivia: Tumapasa). Deltocephalinae, Deltocephalini. Neotropical.
Tungurahuala Kramer 1965a:68. *T. basiliscus* n.sp. (Ecuador: Banos, Mt. Tungurahua, 2500 m). Nirvaninae. Neotropical.
Turitia Schumacher 1912b:248. *T. uniformis* n.sp. ("Cameroons"). 4:58. Ledrinae, Petalocephalini. Ethiopian.
Turrutus Ribaut 1946b:84. *Jassus (Deltocephalus) socialis* Flor 1861a:242 (Livonia). 10:1415. Deltocephalinae, Deltocephalini. Palaearctic; European and Mediterranean subregions.
Tuzinka Dworakowska & Viraktamath 1979a:56. *T. acuta* n.sp. (India: Karnataka, Mudigere). Typhlocybinae, Erythroneurini. Oriental.
Twingingia 10:2689. Error for *Twiningia* Ball.
Twiningia Ball 1931c:93. *Scaphoideus blandus* Ball 1901a:7 (USA: Colorado, Rifle). 10:2160. Deltocephalinae, Scaphoideini. Nearctic; Rocky Mountain subregion.
Tylissus Stal 1870c:739. *T. nitens* n.sp. (Philippines). 15:192. Selenocephalinae, Selenocephalini. Oriental.
Tylozygoides Matsumura 1912a:42. *T. artemisae* n.sp. (Formosa) = *Mileewa margheritae* Distant 1908g:238. 1:298. Cicadellinae, Mileewanini. Oriental. Synonym of *Mileewa* Distant 1908g:238.

Tylozygus Fieber 1865a:44. *Tettigonia nigrolineata* Herrich-Schaeffer 1838c:17 ("Germany" error ?) = *Tettigonia bifida* Say 1830b:313. 1:447. Young 1977a:585. Cicadellinae, Cicadellini. Neotropical/Nearctic; eastern North America, Antillean subregion, Central America and northern South America.
Typhlcoyba 17:1508. Error for *Typhlocyba* Germar.
Typhlecyba 17:1508. Error for *Typhlocyba* Germar.
Typhlicyba 17:1508. Error for *Typhlocyba* Germar.
Typhlocbya 17:1508. Error for *Typhlocyba* Germar.
Typhlochyba 17:1508. Error for *Typhlocyba* Germar.
Typhlociba 17:1508. Error for *Typhlocyba* Germar.
Typhlocpba 17:1509. Error for *Typhlocyba* Germar.
Typhlocyba Germar 1833a:180. *Cicada quercus* Fabricius 1977a:298 (Europe). 17:781. Placed in the Official List of Generic Names in Zoology, ICZN Opinion 605, 1961. Typhlocybinae, Typhlocybini. Holarctic. [Australian?].
  Subgenera:
    Empoa Fitch 1851a:63. *E. querci* n.sp.
    Anomia Fieber 1866a:509. *Cicada quercus* Fabricius 1777a:298.
    Eupterycyba Dlabola 1958c:55. *Typhlocyba jucunda* Herrich-Schaeffer 1837a:16.
    Ficocyba Vidano 1960a:112. *Typhlocyba ficaria* Horvath 1897b:636.
  Synonym (of *Empoa*):
    Empoides Vilbaste 1968a:93. *E. rubellus* n.sp.
Typhlocybella Baker 1903b:3. *T. minima* n.sp. (Nicaragua: Managua). 17:191. Typhlocybinae, Dikraneurini. Neotropical; Antillean subregion, Central America, Panama, Colombia.
TYPHLOCYBIDAE 17:1. Kirschbaum 1868b:16, type genus *Typhlocyba* Germar 1833a:180. Placed in the Official List of Family-Group Names in Zoology, ICZN Opinion 605, 1961.
TYPHLOCYBINAE 17:100.
TYPHLOCYBINI 17:780.
Typhlocyca 17:1508. Error for *Typhlocyba* Germar.
Typhlocyda 17:1508. Error for *Typhlocyba* Germar.
Typhlopcyba 17:1509. Error for *Typhlocyba* Germar.
Typhlophora 17:1509. Error for *Typhlocyba* Germar.
Typhloryba 17:1509. Error for *Typhlocyba* Germar.
Typhloyba 17:1509. Error for *Typhlocyba* Germar.
Typhocyba 17:1509. Error for *Typhlocyba* Germar.
Typholocyba 17:1509. Error for *Typhlocyba* Germar.
Typhrocyba 17:1509. Error for *Typhlocyba* Germar.
Typhylocyba 17:1509. Error for *Typhlocyba* Germar.
Typhyocyba 17:1509. Error for *Typhlocyba* Germar.
Typlocyba 17:1509. Error for *Typhlocyba* Germar.
Typthlocyba 17:1509. Error for *Typhlocyba* Germar.
Tzitzikamaia Linnavuori 1961a:469. *T. silvicola* n.sp. (South Africa: Cape Prov., Tzitzikama Forest, Storms River mouth). Deltocephalinae, Athysanini. Ethiopian; South Africa.
Uhlariella 10:2689. Error for *Uhleriella* Ball.
Uhlerella 10:2689. Error for *Uhleriella* Ball.
Uhleriella Ball 1902b:54. *Deltocephalus coquilletti* Van Duzee 1890b:95 (USA: California) = *Cochlorhinus unispinosus* Beamer 1940a:54. 10:1079. Deltocephalinae, Cochlorhinini. Nearctic. Synonym of *Cochlorhinus* Uhler 1876a:358.
Uhlieriella 10:2690. Error for *Uhleriella* Ball.
Ujna Distant 1908g:239. *U. delicatula* n.sp. (Ceylon: Peradeniya). 1:296. Cicadellinae, Mileewanini. Oriental; Ceylon, Burma, Philippines. Malagasian; Madagascar, Seychelles Islands.
Ullopa 5:100. Error for *Ulopa* Fieber.
Ulopa Fallen 1814a:19. *U. obtecta* n.sp. (Sweden) = *Cercopis reticulata* Fabricius 1794a:57. 5:51. Linnavuori 1972a:128-129. Ulopinae, Ulopini. Palaearctic; European and Mediterranean subregions. Oriental; India, central provinces. Ethiopian; Tanganyika. Malagasian; Madagascar.
Ulopella Poisson 1938a:13. *U. termiticola* n.sp. (French Guinea: Conakry). 5:80. Linnavuori 1972a:132-133. Ulopinae, Ulopini. Ethiopian; tropical Africa.
ULOPIDAE 5:1.
ULOPINAE 5:50.
Uloprora Evans, J.W. 1939a:44. *U. risdonensis* n.sp. (Tasmania: Risdon). 5:77. Ulopinae, Ulopini. Australian; Tasmania, Australia.

Ulozena Melichar 1926a:345. *Tettigonia lineatocollis* Signoret 1854d:728 (Madagascar). 1:419. Cicadellinae, Cicadellini. Malagasian; Madagascar (Italian Somaliland ?).

Umesaona Ishihara 1961a:246. *U. asiatica* n.sp. (Cambodia: Kompong Cham). Deltocephalinae, Stenometopiini. Oriental. Synonym of *Doratulina* Melichar 1903b:198; Vilbaste 1965a:10.

Umesdena Vilbaste 1965a:10. Error for *Umesaona* Ishihara ?

Unerus DeLong 1936a:219 as subgenus of *Deltocephalus* Burmeister 1838b:15. *Thamnotettix colonus* Uhler = *Deltocephalus colonus* Uhler 1895a:80 (St. Vincent Island). 10:857. Linnavuori 1959b:126. Deltocephalinae, Deltocephalini. Neotropical/Nearctic; Southern USA and Antilles subregion to Brazil.
Subgenus:
Mattogrossus Linnavuori 1959b:129. *U. (M). colonoides* n.sp.

Unguitartessus Evans, F. 1981a:124. *U. cairnsensis* n.sp. (Australia: Queensland, Cairns). Tartessinae. Australian.

Unitra Dworakowska 1974a:152. *U. bufonis* n.sp. (Congo: Odzala). Typhlocybinae, Empoascini. Ethiopian; Congo.

Unoka Lawson 1928a:456. *Athysanus ornatus* Gillette 1898b:29 (USA: Colorado), preoccupied = *Unoka gillettei* Metcalf 1955a:266. 10:1659. Deltocephalinae, Deltocephalini. Nearctic; central Great Plains and Rocky Mountain subregion.

Uperogonalia Young 1977a:54. *U. reducta* n.sp. (Peru: Tarma). Cicadellinae, Cicadellini. Neotropical.

Upsicella Maldonado-Capriles 1972a:545 as subgenus of *Angusticella* Maldonado-Capriles 1972a:545. *Pedioscopus similis* Baker 1915c:331 (Philippines: Mindanao, Dapitan). Idiocerinae. Oriental.

Urganus Dlabola 1965c:113. *U. paradarrinus* n.sp. (Mongolia: "Central aimak, So von somon Bajancogt 1600 m"). Deltocephalinae, Deltocephalini. Palaearctic; Siberian and Manchurian subregions.

Urmila Dworakowska 1981j:230. *U. tripunctata* n.sp. (India: Karnataka, Nandi Hills). Typhlocybinae, Erythroneurini. Oriental.

Usanus DeLong 1947b:110. *U. stonei* n.sp. (Mexico: Guerrero, Iguala). 10:1025. Deltocephalinae, Athysanini. Neotropical.

Ushamenona Malhotra & Sharma 1974a:245 as subgenus of *Apheliona* Kirkaldy 1907d:67. *Sujitettix aryavartha* Ramakrishnan & Menon 1972a:114 (India: Delhi I.A.R.I.). Typhlocybinae, Empoascini. Oriental. Synonym of *Aphelonia* Kirkaldy 1907d:67; Dworakowska 1977d:611.

Usharia Dworakowska 1977i:18. *U. mata* n.sp. (Vietnam: Prov. Lao-cai, Sa-pa, 1650 m). Typhlocybinae, Empoascini. Oriental; India, Malaysia, Nepal, Vietnam.

Ussuriasca Anufriev 1972e:36. *Empoasca olivacea* Anufriev 1969b:173 (USSR Maritime Territory, Barabach-Levada). Typhlocybinae, Empoascini. Palaearctic.

Usuironus Ishihara 1953a:197. *Athysanus ogikubonis* Matsumura 1914a:188 (Japan: Honshu). 10:390. Deltocephalinae, Athysanini. Palaearctic. Subgenus of *Handianus* Ribaut 1942a:265; Emeljanov 1964b:32.

Uzeldikra Dworakowska 1971d:585. *Empoasca citrina* Melichar 1903b:213 (Ceylon: Peradeniya). Typhlocybinae, Dikraneurini. Oriental; Ceylon, India, Malaysia.
Synonyms:
Pusaneura Ramakrishnan & Menon 1971a:459. *P. signata* n.sp.
Hameedia Ahmed, Manzoor 1972a:67. *H. erythrocephala* n.sp.

Uzelina Melichar 1903b:181. *U. laticeps* n.sp. (Ceylon: Bundarawella). 3:204. Linnavuori 1977a:41. Penthimiinae. Oriental; Ceylon. Ethiopian; Ethiopia, South Africa, Tanzania, Zaire.
Subgenus:
Mulungaella Linnavuori 1977a:45. *U. (M.) cinchonae* n.sp.

Validapona DeLong 1979g:188 as subgenus of *Polana* DeLong 1942d:110. *P. (V.) lamina* n.sp. (Colombia: Choco Dept., Camp Teresita). Gyponinae, Gyponini. Neotropical.

Vangama Distant 1908g:260. *V. steneosaura* n.sp. (India: NW Provinces, Kuamon, Bhim Tal). 6:44. Cicadellinae, Evacanthini. Oriental.

Varicopsella Hamilton, K.G.A. 1980b:900. *Macropsis breakeyi* Merino 1936a:320 (Philippines: Mindanao). Macropsinae, Macropsini. Oriental; Borneo, India, Philippines.

Varta Distant 1908g:320. *V. rubrofasciata* n.sp. (India: Bengal). 10:2351. Linnavuori 1978d:479. Deltocephalinae, Scaphytopiini. Oriental. Synonym of *Stymphalus* Stal 1866a:121; Linnavuori 1978d:479.

Vecaulis Theron 1975a:200. *Deltocephalus attenuatus* Naude 1926a:44 (South Africa: Cape Province, Viljoen's Pass). Deltocephalinae, Deltocephalini. Ethiopian.

Velu Ghauri 1963a:467. *V. caricae* n.sp. (India: Poona, Velu). Typhlocybinae, Empoascini. Oriental.

Verdanulus Emeljanov 1966a:122 as subgenus of *Diplocolenus* Ribaut 1946b:82. *Jassus nigrifrons* Kirschbaum 1868b:139 (Austria). Deltocephalinae, Paralimnini. Palaearctic. Synonym of *Diplocolenus* Ribaut 1946b:82; Knight 1974c:367.

Verdanus Oman 1949a:165. *Deltocephalus evansi* Ashmead 1904a:132 (Alaska). 10:1357. Emeljanov 1966a:122, Knight 1974c:368. Deltocephalinae, Paralimnini. Holarctic. Subgenus of *Diplocolenus* Ribaut 1946b:82.

Vermara Dworakowska 1980b:174. *V. neglecta* n.sp. (India: West Bengal, Darjeeling). Typhlocybinae, Erythroneurini. Oriental.

Vernobia Nielson 1979b:17. *V. johnsoni* n.sp. (Peru: Yurac, 108 km E. of Tingo Maria). Coelidiinae, Teruliini. Neotropical.

Versigonalia Young 1977a:1088. *Tettigonia ruficauda* Walker 1851b:763 (Brazil). Cicadellinae, Cicadellini. Neotropical; southern Brazil, northern Argentina.

Vertanus Hepner 1946a:87 as subgenus of *Scaphytopius* Ball 1931d:218. *S. (V.) ulcus* n.sp. (USA: Texas, Hidalgo). 10:2248. Deltocephalinae, Scaphytopiini. Neotropical. Synonym of *Ascius* DeLong 1943f:250.

Vertigella Evans, J.W. 1972a:182. *V. kaindensis* n.sp. (SW New Guinea: Mt. Kaindi, 2350 m). Penthimiinae. Oriental; New Guinea.

Vicosa Linnavuori & DeLong 1978c:202. *V. bicornis* n.sp. (Brazil: Vicosa). Deltocephalinae, Deltocephalini. Neotropical.

Vietnara Ramakrishnan & Ghauri 1979b:210. *Typhlocyba maculifrons* de Motschulsky 1863a:103 (Ceylon). Typhlocybinae, Erythroneurini. Oriental. Synonym of *Empoascanara* Distant 1918b:106; Dworakowska 1980d:195.

Vidanoana Young 1977a:1079. *Tettigonia flavomaculata* Blanchard in Spinola 1852a:282 (Chile). Cicadellinae, Cicadellini. Neotropical.

Vilargus Theron 1975a:198. *Deltocephalus pumilicans* Naude 1926a:44 (South Africa: Cape Province, Viljoen's Pass, near Elgin). Deltocephalinae, Deltocephalini. Ethiopian.

Vilbasteana Anufriev 1970d:261. *"Dicraneura" oculata* Lindberg 1929b:10 (USSR Maritime Territory). Typhlocybinae, Dikraneurini. Palaearctic; Siberian subregion.

Virganana DeLong & Thambimuttu 1973b:167. *V. herbida* n.sp. (Chile: Malleco Prov., Angol). Deltocephalinae, Athysanini. Neotropical.

Viridasca Ramakrishnan & Menon 1972b:185. *V. albomaculata* n.sp. (India: Delhi, I.A.R.I.). Typhlocybinae, Empoascini. Oriental. Synonym of *Eremochlorita* Zachvatkin 1946b:150; Dworakowska 1977f:285.

Viridicerus Dlabola 1974b:62. *Bythoscopus ustulatus* Mulsant & Rey 1855a:217 (Provence). Idiocerinae. Palaearctic. Synonym of *Idiocerus* Lewis 1834a:37; Hamilton, K.G.A. 1980a:825.

Viridomarus Distant 1918b:69. *V. capitatus* n.sp. (South India: Chikkaballapura). 10:2348. Deltocephalinae, Stenometopiini. Oriental; India.

Viriosana DeLong 1936a:217 as subgenus of *Ballana* DeLong 1936a:217. *Thamnotettix viriosa* Ball 1910a:266 (USA: California). 10:820. Deltocephalinae, Athysanini. Nearctic. Synonym of *Ballana* DeLong 1936a:217.

Volusenus Distant 1918b:72. *V. lahorensis* n.sp. (India: Punjab). 10:1585. Deltocephalinae, Stenometopiini. Oriental. Synonym of *Doratulina* Melichar 1903b:198; Vilbaste 1965a:10.

Vulturnellus Evans, J.W. 1966a:220. *V. shephardi* n.sp. (Australia: New South Wales, Broken Hill). Penthimiinae. Australian. Synonym of *Neodartus* Melichar 1903b:162; Evans, J.W. 1977a:113.

Vulturnus Kirkaldy 1906c:463. *V. vulturnus* n.sp. (Australia: Queensland, Cairns). 3:214. Penthimiinae. Australian.

Wadkufia Linnavuori 1965a:13. *W. elegans* n.sp. (Libya: Wadi-el-Kub). Stegelytrinae. Palaearctic.

Wadkupfia Nielson 1975a:11. Error for *Wadkufia* Linnavuori.

Wagneriala Anufriev 1970h:635, replacement for *Wagneriana* Anufriev 1970f:262, not *Wagneriana* McCook 1904. *Notus minimus* Sahlberg 1871a:168 (Scandinavia). Typhlocybinae, Dikraneurini. Palaearctic; European and Siberian subregions.

Wagneriana Anufriev 1970d:262. "*Notus minutus* J. Sahlberg 1871" = *Notus minimus* Sahlberg 1871a:168 (Scandinavia). Preoccupied by *Wagneriana* McCook 1904, see *Wagneriala* Anufriev 1970h:635. Typhlocybinae, Dikraneurini. Palaearctic.

Wagneripteryx Dlabola 1958c:53. *Cicadula gemari* Zetterstedt 1840a:301 (Lapland). Typhlocybinae, Typhlocybini. Palaearctic. Synonym of *Aguriahana* Distant 1918b:105; Dworakowska 1972a:278.

Wagneriunia Dworakowska 1969c:381. *W. thecata* n.sp. (East Nepal: Taplejung District, between Sangu and Tamrang, 2500 ft). Typhlocybinae, Typhlocybini. Oriental.

Wakaya Linnavuori 1960b:43. *W. obtusiceps* n.sp. (Fiji: Wakaya Island). Deltocephalinae, Deltocephalini. Oceania; Fiji.

Wania Liu 1939c:297. *W. membracioidea* n.sp. (China: Anhwei). 2:13. Hylicinae, Sudrini. Oriental. Synonym of *Balala* Distant 1908g:250.

Wanritettix Vilbaste 1969a:12. *Thamnotettix wanrianus* Matsumura 1914a:177 (Formosa: Wanri). Deltocephalinae, Athysanini. Oriental.

Warodia Dworakowska 1970f:215. *Typhlocyba hoso* Matsumura 1932a:97 (Japan: Honshu). Typhlocybinae, Typhlocybini. Palaearctic.

Watanabella Vilbaste 1969a:10. *Thamnotettix montivagus* Baker 1924d:367, replacement for *Thamnotettix montanus* Matsumura 1914a:182 (Japan). Deltocephalinae, Athysanini. Palaearctic; Manchurian subregion.

Watara Dworakowska 1977f:295. *Typhlocyba sudra* Distant 1908g:412 (India: Calcutta). Typhlocybinae, Erythroneurini. Oriental.

Webaskola Blocker 1979a:66. *Macropsis smithii* Baker 1900a:56 (Brazil: Chapada). Iassinae. Neotropical; Brazil, Peru.

Webbanara Ramakrishnan & Ghauri 1979b:210. *Typhlocyba fumigata* Melichar 1903b:217 (Ceylon: Pattipola). Typhlocybinae, Erythroneurini. Oriental. Synonym of *Empoascanara* Distant 1918b:94; Dworakowska 1980d:195.

Wemba Dworakowska 1974a:159. *W. ossiannilssoni* n.sp. (Congo: Mbila, (Mts. du Chaillu). Typhlocybinae, Empoascini. Ethiopian; Congo, Kenya, Liberia. Malagasian; Madagascar.

Westindica Ramakrishnan & Ghauri 1979b:199 as subgenus of *Indoformosa* Ramakrishnan & Ghauri 1979b:198. *Empoascanara stilleri* Dworakowska 1978a:157 (Malaysia: W. coast of Penang I., Pulau). Typhlocybinae, Erythroneurini. Oriental. Synonym of *Empoascanara* Distant 1918b:94; Dworakowska 1980d:195.

Wiata Dworakowska 1972:850. *W. pallida* n.sp. (Ivory Coast: Lamto). Dworakowska 1981i. Typhlocybinae, Zyginellini. Ethiopian.

Willeiana Young 1977a:167. *Tettigoniella vallonia* Distant 1908a:522 (Peru). Cicadellinae, Cicadellini. Neotropical; Bolivia, Costa Rica, Peru.

Williamsiana Goding 1926a:103. *W. ferruginosa* n.sp. (Ecuador). 1:485. Young 1968a:67. Cicadellinae, Proconiini. Neotropical. Synonym of *Zyzzogeton* Breddin 1902a:176.

Wiloatma Webb, M.D. 1983a:76. *W. caprilesi* n.sp. (Australia: Queensland, Bamboo Ck., near Miallo, N. of Mossman). Idiocerinae. Australian.

Witera Dworakowska 1981j:245. *W. semilunaris* n.sp. (India: Karnataka, 13 km N. of Mulbagala). Typhlocybinae, Erythroneurini. Oriental.

Woldana DeLong 1981d:207. *W. campana* n.sp. (Panama: Cerro Campana). Gyponinae, Gyponini. Neotropical.

Wolfella Spinola 1850b:120. *W. caternaultii* n.sp. (Guinea). 2:15. Kramer 1965c:167-180; Linnavuori 1972b:197-202. Hylicinae, Hylicini. Ethiopian; equatorial central and west Africa.

Wolffella 2:18. Error for *Wolfella* Spinola.

Woodella Evans, J.W. 1966a:87. *W. wanungarae* n.sp. (Australia: Queensland, McPherson Ranges, Mt. Wanungara). Ulopinae, Ulopini. Australian.

Wutingia Melichar 1926a:372. *W. nigronervosa* n.sp. (China: Hopeh Prov., "Pekin"). 1:41. Cicadellinae, Cicadellini. Oriental.

Xanthaphala Amyot 17:1509. Invalid, not binominal.

Xanthochrea Amyot 15:226. Invalid, not binominal.

Xantholius Sherborn 15:226. Invalid, not binominal.

Xantholues Amyot 15:226. Invalid, not binominal.

Xanthopala 17:1510. Error for *Xanthaphala* Amyot, not binomianl.

Xedreota Kramer 1966a:495. *Xerophloea tuberculata* Osborn 1938a:16 (British Guiana: Bartica). Ledrinae, Xerophloeini. Neotropical; Brazil, British Guiana, Guyane.

Xenocoelidia Kramer 1959a:30. *X. youngi* n.sp. (Colombia: Buenaventura). Neocoelidiinae. Neotropical; Colombia.

Xenogonalia Young 1977a:219. *Cicadella longicornis* Osborn 1926b:183 (Brazil). Cicadellinae, Cicadellini. Neotropical; Brazil, Peru.

Xerochlorita Zachvatkin 1953c:229 as subgenus of *Chlorita* Fieber 1866a:508. *Chlorita prasina* Fieber 1884a:62 (Southern Russia). Typhlocybinae, Empoascini. Palaearctic.

Xerophaloea 4:145. Error for *Xerophloea* Germar.

Xerophlaeum 4:145. Error for *Xerophloea* Germar.

Xerophlaea 4:145. Error for *Xerophloea* Germar.

Xerophlea 4:145. Error for *Xerophloea* Germar.

Xerophloca 4:145. Error for *Xerophloea* Germar.

Xerophloea Germar 1839a:190. *X. grisea* n.sp. (Brazil) = *Cercopis viridis* Fabricius 1794a:50. 4:42. Ledrinae, Xerophloeini. Western Hemisphere; widespread from southern Canada to Argentina.
Synonyms:
Mesodicus Fieber 1866a:501. *M. foveolatus* n.sp.
Parapholis Uhler 1877a:461. *P. peltata* n.sp.

XEROPHLOEINI 4:42.

Xeroplaea 4:146. Error for *Xerophloea* Germar.

Xerphloea 4:146. Error for *Xerophloea* Germar.

XESTOCEPHALINI 10:2353.

Xestocephalus Van Duzee 1892d:298. *Xestocephalus pulicarius* Van Duzee 1894f:215 (USA: New York, Lancaster). 10:2355. Linnavuori 1959b:36. Xestocephalinae, Xestocephalini. Cosmopolitan.
Synonym:
Lindbergana Metcalf 1952a:229. *Nesotettix freyi* Lindberg 1936a:6.

Xestpcephalus 10:2690. Error for *Xestocephalus* Van Duzee.

Xestrocephalus 10:2690. Error for *Xestocephalus* Van Duzee.

Xextocephalus 10:2690. Error for *Xestocephalus* Van Duzee.

Xiqilliba Kramer 1964d:268. *X. bellator* n.sp. (Brazil: Itaituba). Neocoelidiinae. Neotropical.

Xyphon Hamilton, K.G.A. 1985a:83. *Diedrocephala flaviceps* Riley 1880f:78 (USA: Texas). Cicadellinae, Cicadellini. Nearctic; southeastern and south-central USA, and Antillean subregion.

Yachandra Webb, M.D. 1983b:230. *Idiocerus projectus* Webb, M.D. 1975a:173 (Southwest Africa: Aus). Idiocerinae. Ethiopian; South Africa, southwest Africa.

Yakunopona Ishihara 1954f:12. *Y. yakushimensis* n.sp. (Japan: Yakushima, south of Kyushu). 15:140. Paraboloponinae. Oriental. Synonym of *Dryadomorpha* Kirkaldy 1906c:335; Webb, M.D. 1981b:49.

Yamotettix 10:2690. Error for *Yamatotettix* Matsumura.

Yamatatettix Lee, C.E. & Kwon 1979b:912. Error for *Yamatotettix* Matsumura 1914a.

Yamatotettix Matsumura 1914a:183. *Y. falvovittatus* n.sp. (Japan: Shikoku, Iyo, Imogo Valley). 10:877. Deltocephalinae, Macrostelini. Oriental/Palaearctic; Japan, Korea, Thailand.

Yanocephalus Ishihara 1953b:48. *Deltocephalus yanonis* Matsumura 1902a:400 (Japan). 10:1598. Deltocephalinae, Deltocephalini. Palaearctic; Manchurian subregion.

Yaoundea Linnavuori 1979b:964. *Y. semela* n.sp. (Cameroon). Nirvaninae, Nirvanini. Ethiopian; Cameroon, Liberia, Nigeria, Zaire.

Yasumatsuus Ishihara 1971a:18. *Kolla mimica* Distant 1908g:225 (India: Bengal, Calcutta) = *Tettigonia quinquenotata* Stal 1870c:734. Cicadellinae, Cicadellini. Oriental. Synonym of *Cofana* Melichar 1926a:345; Young 1979a:1; Linnavuori 1979a:705.

Yisiona Kuoh, C.L. 1981b:206 & 217. *Y. ziheina* n.sp. (China: Quinhai-Xizang Plateau, Zogang, 3700 m). Typhlocybinae, Typhlocybini. Palaearctic.

Yochlia Nielson 1979b:57. *Y. brevis* n.sp. (French Guiana). Coelidiinae, Teruliini. Neotropical; Brazil, French Guiana, Guyana.

Yotala Melichar 1925a:336. *Y. boliviana* n.sp. (Bolivia: Coroico). 1:517. Young 1968a:122. Cicadellinae, Proconiini. Neotropical; Bolivia, Brazil.

Youngama Ahmed, Manzoor 1969d:313. *Y. spinistyla* n.sp. (West Pakistan: Murree). Typhlocybinae, Typhlocybini. Oriental. Synonym of *Aguriahana* Distant 1918b:105; Dworakowska 1972a:278.

Youngia Dlabola 1958c:54. *Typhlocyba loewi* Lethierry 1884d:131 (Austria). Preoccupied by *Youngia* Jones & Kirby 1886, see *Youngiada* Dlabola 1959b:195. Typhlocybini or Alebrini; see remarks under *Lindbergina* Dlabola 1958c:54. Palaearctic.

Youngiada Dlabola 1959b:195, replacement for *Youngia* Dlabola 1958c:54, not *Youngia* Jones & Kirby 1886. *Typhlocyba loewi* Lethierry 1884d:131 (Austria). LeQuesne 1977a:87. Typhlocybinae, Typhlocybini or Alebrini; see remarks under *Lindbergina* Dlabola 1958c:54. Palaearctic. Subgenus of *Lindbergina* Dlabola 1958c:54.

Youngolidia Nielson 1983e:51. *Y. latula* n.sp. (Bolivia). Coelidiinae, Youngolidiini. Neotropical; Bolivia, Colombia, Peru, Venezuela.

YOUNGOLIDIINI Nielson 1983e:18, tribe of COELIDIINAE.

Yunga Melichar 1924a:208. *Aulacizes coriacea* Stal 1864a:80 (Mexico). 1:468. Young 1968a:57. Cicadellinae, Proconiini. Neotropical; Colombia, Costa Rica, Mexico, Panama.

Yungasia Linnavuori 1959b:207. *Y. digitata* n.sp. (Bolivia: Nor Yungas, Caranavi). Deltocephalinae, Athysanini. Neotropical; Bolivia, Peru, Brazil, Ecuador.

Yuraca Linnavuori & DeLong 1978d:231. *Y. flavomarginata* n.sp. (Peru: Yurac, 67 miles E. of Tingo Maria). Deltocephalinae, Athysanini. Neotropical.

Zabrosa Oman 1949a:128. *Thamnotettix amazonensis* Osborn 1923c:65 (Brazil: Minas Gerais). 10:1330. Linnavuori 1959b:246-247. Deltocephalinae, Athysanini. Neotropical; Argentina, Bolivia, Brazil, Paraguay. Nearctic; USA, Texas.

Zaletta Metcalf 1952a:229, replacement for *Macrocerus* Evans, J.W. 1941e:39, not *Macrocerus* de Motschulsky 1945. *Macrocerus minutus* Evans, J.W. 1941e:39 (South Australia, Kangaroo I., Flinders Chase). 16:216. Idiocerinae. Australian; Australia.
Synonym:
Macrocerus Evans, J.W. 1941e:39. *M. minutus* n.sp.

Zandeana Linnavuori & Al-Ne'amy 1983a:Abstract. *Nomen nudum.*
[NOTE: Apparently an error for *Zinjella* Linnavuori & Al-Ne'amy].

Zanjoneura Ghauri 1974d:638. *Z. kenyana* n.sp. (Kenya: Nairobi, National Agricultural Laboratories). Typhlocybinae, Erythroneurini. Ethiopian.

Zapycna Emeljanov 1968a:147. *Z. eremita* n.sp. (Mongolia: 8 km SSW of Nomgona). Deltocephalinae, Opsiini. Palaearctic.

Zaruma Melichar 1926a:347. *Z. vexata* n.sp. (Ecuador). 1:27. Young 1977a:93. Cicadellinae, Cicadellini. Neotropical; Bolivia, Peru, Ecuador.

Zelenius Emeljanov 1966a:129. *Laevicephalus orientalis* DeLong & Davidson 1935b:167 (USA: Pennsylvania). Deltocephalinae, Deltocephalini. Palaearctic; Nearctic; widespread.

Zelopsis Evans, J.W. 1966a:168. *Z. nothofagi* n.sp. (New Zealand: Nelson Prov., Aniseed Valley). Macropsinae, Macropsini. Australian.

Zennica 16:236. Error for *Zinneca* Amyot & Serville.

Zepama Metcalf 1967b:2241, replacement for *Metcalfiella* Linnavuori 1955b:121, not *Metcalfiella* Goding 1929. *Metcalfiella chacoensis* Linnavuori 1955b:122 (Paraguay: Chaco). 10:2241. Linnavuori 1959b:135. Deltocephalinae, Deltocephalini. Neotropical. Synonym of *Fusanus* Linnavuori 1955b:123.

Zerana DeLong & Freytag 1964a:118 as subgenus of *Gyponana* Ball 1920a:85. *G. (Z.) apicata* n.sp. (Peru: Sinchona). Gyponinae, Gyponini. Neotropical.

Zercanus Dlabola 1965d:439. *Z. rubroocellatus* n.sp. (Jordan: Zerka, S. Amman). Deltocephalinae, Athysanini. Palaearctic.

Zestocephalus 10:2690. Error for *Xestocephalus* Van Duzee.

Ziczacella Anufriev 1970c:697 as subgenus of *Erythroneura* Fitch 1851a:62. *Erythroneura heptapotamica* Kusnezov 1928b:316 (Russia). Dworakowska 1970n:760, 1981d:371-377. Typhlocybinae, Erythroneurini. Palaearctic; Siberian and Manchurian subregions. Oriental; China, Vietnam.

Zigina 17:1510. Error for *Zygina* Fieber.

Zigyna 17:1510. Error for *Zygina* Fieber.

Zigzacella Dworakowska 1970n:760. Error for *Ziczacella* Anufriev 1970c:697.

Zilkaria Menezes 1974a:108. *Z. assymetrica* n.sp. (Brazil: Matto Grasso). Deltocephalinae, Athysanini. Neotropical.

Zinga Dworakowska 1972b:398. *Z. novembris* n.sp. (New Guinea: Papua, Kokoda, 1200 ft). Typhlocybinae, Erythroneurini. Oriental.

Zinislopa Webb, M.D. 1983a:89. *Z. troopa* n.sp. (Australia: Queensland, Mulgrave R., 4 mi. W. of Gordondale). Idiocerinae. Australian.

Zinjella Linnavuori & Al-Ne'amy 1983a:29 as subgenus of *Adama* Dlabola 1980e:89. *Distantia maculithorax* Jacobi 1910b:126 (Kenya: Kilimanjaro). Selenocephalinae, Adamini. Ethiopian.

Zinneca Amyot & Serville 1843a:579. *Z. flavidorsum* n.sp. (North America). 16:210. Macropsinae, Macropsini. Nearctic. Synonym of *Oncopsis* Burmeister 1838a:10; Hamilton, K.G.A. 1980b:887.

Zinnevia Amyot 15:226. Invalid, not binominal.

Zioninus Ball 1918b:388 as subgenus of *Phlepsius* Fieber 1866a:514. *Phlepsius extremus* Ball 1901a:10 (USA: Colorado). 10:658. Deltocephalinae, Athysanini. Nearctic. Synonym of *Texananus* Ball 1918b:381.

Zizyphoides Distant 1918b:73. *Z. indicus* n.sp. (S. India: Chikkaballapura). 10:2350. Paraboloponinae. Oriental. Synonym of *Dryadomorpha* Kirkaldy 1906c:335; Webb, M.D. 1981b:49.

Zonana DeLong & Freytag 1963b:263. *Z. flamma* n.sp. (Panama: Canal Zone, Paraiso). Gyponinae, Gyponini. Neotropical, Costa Rica, Panama.

Zonocyba Vilbaste 1982a:12. *Typhlocyba bifasciata* Boheman 1852a:79 (Germany: Hamburg). Typhlocybinae, Typhlocybini. Palaearctic; European subregion.

Zorka Dworakowska 1970f:216. *Z. ariadnae* n.sp. (Vietnam: Bao Ha, S. of Fan-Si-Pan). Typhlocybinae, Typhlocybini. Oriental; China, Vietnam.

Zubara Al-Ne'amy & Linnavuori 1982a:113. *Z. lycii* n.sp. (Iraq: near Zubar). Adelungiinae, Achrini. Palaearctic; Iraq.

Zuzza Metcalf 1965a:729 (index). Error for *Zyzza* Kirkaldy 1900b:243.

Zygima 17:1510. Error for *Zygina* Fieber.

Zygina Fieber 1866a:509. *Typhlocyba nivea* Mulsant & Rey 1855a:246 (Provence). 17:1329. Dworakowska 1970j. Typhlocybinae, Erythroneurini. Holarctic and Oriental, India.
Subgenus:
Hypericiella Dworakowska 1970i:559. *Typhlocyba hyperici* Herrich-Schaeffer 1836a:4.
Synonym:
Flammigeroidia Dlabola 1958c:56. "*Erythroneura flammigera* Geoffr."; Nast 1972a:304.

ZYGINAE 17:100.

Zyginella Loew 1855a:346. *Z. pulchra* n.sp. (Austria). 17:1475. Dworakowska 1969d, 1970h:707, 1979c. Typhlocybinae, Zyginellini. Palaearctic; Ethiopian; Oriental; widespread.
Synonyms:
Pyramidotettix Matsumura 1931b:59. *Conometopus citri* Matsumura 1907c:113.
Remmia Vilbaste 1968a:91. *R. orbigera* n.sp.

ZYGINELLINI Dworakowska 1977i:24, tribe of TYPHLOCYBINAE.

Zyginia 17:1510. Error for *Zygina* Fieber.

Zyginidea 17:1510. Error for *Zyginidia* Haupt.

Zyginidia Haupt 1929c:268, replacement for *Idia* Fieber, not *Idia* Huebner 1813. *Typhlocyba scutellaris* Herrich-Schaeffer 1838c:13 (Provence). 17:1438. Dworakowska 19701. Typhlocybinae, Typhlocybini. Palaearctic; European and Mediterranean subregions. Oriental; West Pakistan.
Synonym:
Idia Fieber 1966a:509. *Typhlocyba scutellaris* Herrich-Schaeffer 1838c:13.

Zyginidis 17:1510. Error for *Zyginidia* Haupt.

Zyginoides Matsumura 1931b:59. *Eupteryx taiwanus* Shiraki 1912a:95 (Formosa). 17:231. Dworakowska 1972g, 1981c. Typhlocybinae, Dikraneurini. Oriental: Formosa, India, Pakistan, New Guinea, Sunda Is. Ethiopian; Sudan. Synonym of *Diomma* Motschulsky 1863a:102; Dworakowska 1981c:367.

Zyginopsis Ramakrishnan & Menon 1973a:29. *Z. abundans* n.sp. (India: Delhi, I.A.R.I.). Typhlocybinae, Erythroneurini. Oriental.

Zygnia 17:1510. Error for *Zygina* Fieber.

Zyzyphoides Sharma, B. 1977a:62. Error for *Zizyphoides* Distant.

Zyzza Kirkaldy 1900b:243, replacement for *Germaria* LaPorte 1832b:222, not *Germaria* Robineau-Desvoidy 1830. *Germaria cucullata* de LaPorte 1832b:223 (unknown). 1:473. Young 1968a:26. Cicadellinae, Proconiini. Neotropical. Synonym of *Proconia* LePeletier & Serville 1825a:610.

Zyzzogedon 1:726. Error for *Zyzzogeton* Breddin.

Zyzzogeton Breddin 1902a:178. *Z. haenschi* n.sp. (Ecuador: Balzapamba). 1:485. Young 1968a:67. Cicadellinae, Proconiini. Neotropical; Colombia, Ecuador.

Synonym:

Williamsiana Goding 1926a:103. *W. ferruginosa* n.sp. = *Zyzzogeton haenschi* Breddin; Young 1968a:67.

INDEX

The index relates only to papers published between 1956-1985, as listed in the bibliography. It is divided into subject matter categories and subcategories, using the author/year/letter code system. References may be indexed under more than one category or subcategory, depending on content.

Except for the taxonomic index, single authors are listed with initials for all given names (e.g. Schvester, D. 1962a) whilst dual authors are listed without initials (e.g. Schulz & Meijer 1978a). In the case of more than two authors, only the first author's surname is given, followed by "et al." (e.g. Reynolds et al. 1957a) except for those papers where the senior author(s) and date are identical. In the latter case the surname of the second author and, if necessary, subsequent authors, is also given (e.g. Reynolds, Fukuto et al. 1960a; Reynolds, Stern et al. 1960a. Lamey, Surin & Leewangh 1967a; Lamey, Surin, Disthaporn et al. 1967a). In the taxonomic index, initials are omitted for single authors except where necessary to distinguish identical names (e.g. Evans, F.; Evans, J.W.). In the case of multiple authors, all surnames are given.

Entries in the subject index category biogeography, subcategory faunas, refer to papers providing in depth treatment of multi-generic faunas of a major geographic region, e.g. Evans 1966a, Knight 1975a, Linnavuori 1978a and 1979a, Vilbaste 1978a and Webb 1981b. These publications may be cited also in the taxonomic index and under individual genera in the generic check-list, if appropriate. However, many taxonomic papers may also contain distributional and faunal data. Since these papers may be listed only in the taxonomic index, both the biogeography category and the taxonomic index should be consulted by those interested in distribution and faunas.

In the taxonomic index the genera (but not subgenera) are arranged alphabetically. For each genus, the references are listed under two headings - "New species" and "Records". The latter refers to those references containing information that elaborates the understanding of the particular genus, such as keys to species and genera, generic revisions, distribution, seasonal occurrence, host associations, etc., each annotated when appropriate as a guide. For a taxon variably treated by authors as subgenus or genus, references to "New species" and "Records" are indexed under its own name when treated as a genus but under both its own name and that of its parent genus when treated as a subgenus.

For those references in which there is a temporary hiatus regarding the application of a name not involving synonymy, e.g. Ishihara 1958a for *Eupteryx* vs *Cicadella*, the reference is included under both genera in the taxonomic index. This procedure is not applied to ordinary synonyms since each has a separate entry in the taxonomic index for those references in which it is used as a "valid" generic name.

SUBJECT INDEX

GENERAL

Allen, R.T. 1983a; Anonymous 1964a, 1978c, 1985f; Asahina et al. 1959a; Barnes, J.K. 1984a; Claridge & Wilson 1982a; Cobben, R.H. 1978a, 1978b; Dantzig et al. 1964a; Deitz, L.L. 1979a; DeLong, D.M. 1984b; Dlabola, J. 1984b; Dobreanu & Manolache 1969a; Evans, J.W. 1956b, 1956c, 1957c, 1958a, 1958b, 1959b, 1960b, 1961c; Evans et al. 1970a; Fennah, R.G. 1983a; Fletcher, M.J. 1982a, 1984a; Kisimoto, R. 1973a; Knight et al. 1983a; Kukalova-Peck, J. 1978a; Lee, C.E. 1979a; Le Quesne, W.J. 1969a, 1969b, 1983a; Lodos & Onder 1984a; Logvinenko, V.N. 1956a, 1957a, 1957d, 1957e, 1962d, 1967d; Mahmood, S.H. 1973a; Mani, M.S. 1974a; Matsuda, R. 1963a, 1965a; Mound, L.A. 1983a; Muller, H.J. 1956a, 1960b; Nault & Rodriguez 1985a; Nielson, M.W. 1985a; Quartau, J.A. 1984c; Rodriguez, J.G. 1985a; Ross, H.H. 1956b, 1957c, 1958b, 1958c, 1962a, 1965b, 1970a; Russell, L.M. 1973a; Samsinak & Dlabola 1980a; Snodgrass, R.E. 1957a; Theron, J.G. 1974c, 1985a; Viraktamath, C.A. 1983a; Wagner, W. 1968a; Wagner & Franz 1961a; Weidner, H. 1977a; Young, D.A. 1958b, 1967a; Zahradnik, J. 1984a.

REVIEWS & COMPREHENSIVE

Avidov, Z. 1961a, 1969a; Chou et al. 1953a; Chu & Hirashima 1981a; Claridge, M.F. 1985a; Clausen, C.P. 1978a; Danilevskii, A.S. 1965a; Dantsig et al. 1964a; Day & Bennetts 1954a; DeLong, D.M. 1971b; Frankel & Hawkes 1975a; Fritzche et al. 1972a; Gyrisco, G.G. 1958a; Gyrisco et al. 1978a; Hainzelin, E. 1982a; Hamilton, K.G.A. 1985d; Hansen, H.P. 1961a; Harris, K.F. 1981b, 1984a; Harris & Maramorosch 1980a, 1982a; Heinze, K.G. 1959a; Horber, E. 1972a; Houk & Griffiths 1980a; Ishihara, T. 1966b; Jeppson & Carman 1960a; Joplin, C.E. 1974a; Kim & Lamey 1973a; Kiritani, K. 1979a; Kuc, J. 1966a; Ling, K.C. 1968a, 1969a, 1969b, 1972a, 1979a; Ling, Tiongco & Cabunagan 1983a; Maramorosch, K. 1963b, 1968a, 1969a, 1973a, 1979a, 1980a, 1981a; Maramorosch et al. 1970a, 1975a; Maramorosch & Harris 1979a, 1981a; Maramorosch & Jensen 1963a; Maramorosch & Raychandhuri 1981a; Markham, P.G. 1983a; Misra, B.C. 1980a; Mitsuhashi, J. 1965g; Nasu, S. 1967a, 1969b; Ossiannilsson, F. 1966a; Painter, R.H. 1958a; Pathak, M.D. 1967a, 1968a; Pearson, E.O. 1958a; Pholboon, P. 1965a; Razvyazkina, G.M. 1955a; Schneider, F. 1962a; Sinha, R.C. 1968b; Smith, K.M. 1957a, 1958a; Smith & Brierley 1956a; Soper, R.S. 1985a; Suenaga & Nakatsuka 1958a; Turnipseed & Kogan 1976a; Whitcomb, R.F. 1981a; Whitcomb & Black 1982a; Whitcomb & Tully 1979a; Yen, D.F. 1973a;

BIBLIOGRAPHIES

Anufriev & Osychnyuk 1985a; Chiu, S.C. 1958a; Deitz, L.L. 1979a; Gyrisco et al. 1978a; Hainzelin, E. 1982a; Hamilton, K.G.A. 1976a; Metcalf, Z.P. 1964a; Metcalf & Wade 1963a; Meyerdirk et al. 1983a; Nast, J. 1979a; Ryan et al. 1984a; Webster, J.A. 1975a.

CATALOGUES

Anonymous 1968e; Burnside, V.W. 1971a; Casale, A. 1981a; Drosopoulos, S. 1980a; Gaedike, H. 1971a; Georghiou, G.P. 1957a; Knutson & Murphy 1984a; Lauterer & Schroder 1970a; Lee, C.E. 1979a; Lee & Kwon 1979b; LeQuesne, W.J. 1964c; Mannheims, B. 1965a; McCabe & Johnson 1980a; Metcalf, Z.P. 1956a, 1962a, 1962b, 1962c, 1962d, 1963a, 1963b, 1963c, 1963d, 1964a, 1964b, 1965a, 1965b, 1966a, 1966b, 1966c, 1966d, 1967a, 1967b, 1967c, 1968a; Metcalf & Wade 1963a, 1966a; Nast, J. 1972a, 1976b, 1979a, 1982a; Oman, P.W. 1976a; Rikhter, V.A. 1968a; Salmon, M.A. 1954b; Schiemenz, H. 1970a; Servadei, A. 1967a; Subba Rao, B.R. 1983a; Vidano, C. 1968a; Wade, V. 1966a; Wagner, W. 1968b; Wagner & Franz 1961a; Wilson, M.R. 1981b.

BEHAVIOUR

Communication
Arai, Y. 1977a; Claridge, M.F. 1983a, 1985a, 1985b; Claridge & Howse 1968a; Claridge & Reynolds 1973a; Den Hollander, J. 1984a; Inoue, H. 1982a; Kullenberg & Wallin 1963a; Kumar & Sazena 1985a; Mebes, H.-D. 1974a; Michelsen et al. 1982a; Moore, T.E. 1961a; Purcell & Loher 1976a; Saxena, K.N. 1981a, 1985a; Saxena & Kumar 1980a, 1984a, 1984b; Schowalter, T.D. 1985a; Shaw, K.C. 1976a; Shaw et al. 1973a; Smith, J.W. 1970a; Strubing, H. 1958a, 1963a, 1965a, 1966a, 1967a, 1976a, 1978a, 1983a; Traue, D. 1978a, 1981a; Vargo, A. 1970a.

Feeding
Alivizatos, A.S. 1982a; Andow, D.A. 1984a; Andrzejewska, L. 1966a; Auclair et al. 1982a; Backus, E.A. 1984a, 1985a; Backus & McLean 1982a, 1983a, 1984a, 1985a; Basilio & Heinrichs 1981a; Carle & Moutous 1966a; Chang, V. 1974a, 1975a; Chang, V.C.S. 1978a; Chu & Liou 1981a; Claridge et al. 1983a; Crane, P.S. 1970a; Ghandhi, J.R. 1980a, 1980b, 1982a; Gunthart & Gunthart 1981a, 1983a; Gunthart & Wanner 1981a; Harris, K.F. 1981a; Harris et al. 1981a; Heinrichs & Rapusas 1984a, 1984b; Herman & Maramorosch 1977a; Hill, M.G. 1976a, 1982a; Hollebone et al. 1966a; Hori, Y. 1967a; Ishizaki, T. 1980a; Kawabe, S. 1978a, 1979a; Kawabe & McLean 1978a, 1980a; Kawabe et al. 1981a; Khan & Saxena 1984b, 1984c, 1985a, 1985b; Kido & Stafford 1965a; Kono et al. 1975a, 1982a; Koyama, K. 1971a, 1973a; Kurata & Sogawa 1976a; Lin & Hou 1980a, 1981a; Malabuyoc & Heinrichs 1981a; Matolcsy et al. 1968a; McClanahan, R.J. 1963a; McClure, M.S. 1980a; McCrae, A.W.R. 1975a; Moreau & Boulay 1967a; Munk, R. 1968a, 1968b; Naito, A. 1976a, 1976b, 1977a, 1977b; Naito & Masaki 1967a, 1967b; Newton et al. 1970a; Ogane et al. 1979a; Orenski, Murray et al. 1965a; Orenski et al. 1962a; Oya, S. 1980a; Oya & Sato 1981a; Patrick, C.R. 1970a; Pollard, D.G. 1968a, 1969a, 1971a; Raman et al. 1979a; Rao & Anjaneyula 1980a; Raven, J.A. 1983a; Saxena, K.N. 1954b; Saxena & Saxena 1974a, 1975a, 1975b; Saxena & Khan 1984a, 1985a; Sekido & Sogawa 1976a; Sogawa, K. 1967a, 1973a; Tallamy & Denno 1979a; Tonks, N.V. 1960a; Triplehorn et al. 1984a.

Movement
Adler, P.H. 1982a; Akino, K. 1969a; Alverson et al. 1977a; Andow & Kiritani 1984a; Capinera & Walmsley 1978a; Dysart, R.J. 1962a; Gifford & Trahan 1966a; Heady & Nault 1985a; Hepner et al. 1970a; Kakiya & Kiritani 1972a; Kieckhefer & Medler 1966a; Ling & Carbonell 1975a; Miah & Ling 1978a; Mohan & Janartharan 1985a, 1985b; Moustafa et al. 1985a; Oliveira & Arauja 1979a; Patange et al. 1981a; Perfect & Cook 1982a; Perkes, R.R. 1970a; Peterson et al. 1969a; Pienkowski & Medler 1966a; Purcell & Suslow 1982a; Reling, D. 1983a; Saroja, R. 1981a; Sathiyandandam & Subramanian 1982a; Saxena & Saxena 1974a, 1975a, 1975b; Srinivasan et al. 1981a; Stuben, M. 1973a; Waloff, N. 1984a.

Reproduction
Ameresekere, R.V.W.E. 1970a; Arai, Y. 1977a; Carlson, O.V. 1967a; Carlson & Hibbs 1962a, 1970a; Chen, C.C. 1969a; Claridge & Reynolds 1972a; Claridge et al. 1977a, 1978d; Claridge & Wilson 1978d; Cruz, Y.P. 1974a; Dobreanu & Manolache 1969a; Godoy, C. 1985a; Gomez & van Schoonhoven 1977a; Gunthart & Gunthart 1983a; Heady et al. 1985a; Hsu & Banttari 1979a; Ichikawa, T. 1979a; Inoue, H. 1982a, 1983a; Kennedy, G.G. 1971a; Khan & Agarwal 1984a; Kieckhefer & Medler 1964a; Kitajima & Landim 1972a; Kooner et al. 1978a; Kumar & Sazena 1978a, 1985a; Kunze, L. 1959a; Maillet, P.L. 1959b; Maramorosch, K. 1974a; Matsuzaki, M. 1975a; McClure, M.S. 1980a; Menezes, M. de 1978a; Miller & Hibbs 1963a; Nielson & Morgan 1982a; Nielson & Toles 1970a; O'Keeffe, L.W. 1965a; Oya, S. 1978a; Perkes, R.R. 1970a; Pollard & Yonce 1965a; Prestidge, R.A. 1982a; Purcell & Loher 1976a; Raatikainen et al. 1976a; Rajendra, Singh 1978a; Ramalho & Ramos 1979a; Rao & Anjaneyulu 1983a; Rapusas et al. 1985a; Reissig & Kamm 1975a; Saxena & Basit 1982b; Saxena & Kumar 1980a, 1984a, 1984b; Sekido & Sogawa 1976a; Seyedoleslami & Croft 1980a; Simmons et al. 1985a; Simonet & Pienkowski 1977a; Simonet, Pienkowski & Wolf 1979a; Stiling, P.D. 1980b; Thompson, P. 1978a; Tsai & Anwar 1977a; Wen & Lee 1980a;

Secretion
Basden, R. 1966a; Chu & Liou 1981a; Sogawa, K. 1967a, 1967b.

Excretion
Chu & Liou 1981a; Malabuyoc & Heinrichs 1981a; Schefer-Immel, V. 1957a; Smith & Littau 1960a; Vidano & Arzone 1984a.

Social
Lamborn, W.A. 1914a; Lavigne, R. 1966a; Leech, H.B. 1966a.

General
Ahmed et al. 1977a; Allen, A.A. 1978a; Budh, D.S. 1975a; Chandy, K.C. 1957a; Claridge & Wilson 1978b; Dabrowski, Z.T. 1985a; Khan & Saxena 1985c; Knull & Knull 1960a; Leath & Byers 1977a; Patrick, C.R. 1970a; Raine, J. 1960a; Saxena, K.N. 1979a, 1981a; Schuler, L. 1956a; Tonks, N.V. 1960a; Turner & Pollard 1959a; Valley & Wheeler 1985a; Wheeler & Walley 1980a; Whittaker, J.B. 1969a;

BIOGEOGRAPHY

Distribution
Aguilar, F.P.G. 1964a; Allen, R.T. 1983a; Alvarenga, M. 1962a; Anjaneyulu & Chakraborti 1977a; Anonymous 1957b, 1959a, 1961a, 1961b, 1963b, 1966a, 1968b, 1968f, 1970a, 1971b, 1974b, 1974c, 1975a, 1976b, 1976c, 1978b, 1980a, 1982a; Anufriev, G.A. 1971h; Arai, Y. 1978a; Attard, G. 1985a; Badmin, J.S. 1979a, 1981a, 1985a; Bharadwaj & Yadav 1980a; Barnes, P. 1954a; Barnett, D.E. 1975a; Beardsley, J.W. 1956a, 1961a, 1964a, 1965a, 1966a, 1967a, 1969a, 1969b; Beardsley & Funaski 1976a; Begley & Butler 1980a; Bianchi, F.A. 1955a; Bindra et al. 1970a; Blocker, H.D. 1971a; Blocker & Wesley 1985a; Bonfils & Delplanque 1971a; Bonfils & Della Giustina 1978a, 1978b; Bonfils & Lauriant 1975a; Bonfils et al. 1974a; Carnegie, A.J.M. 1976a; Chen, C.C. 1972a; Chen & Sogawa 1969a; Chong, M. 1965a; Claridge & Wilson 1978e; Cruz, Y.P. 1974a; Cunningham, H.B. 1964a; D'Aguilar & Della Giustina 1974a; Davis, B.N.K. 1983a, 1984a; Della Giustina & Meusnier 1982a; DeLong, D.M. 1983d; Dlabola, J. 1959b, 1963d, 1964d, 1965a, 1970a, 1971a, 1971b, 1974b, 1974d, 1974e, 1977c, 1977d, 1980c; Dolling, W.R. 1980a; Dubovskij, G.K. 1965b, 1966a, 1968a, 1970a, 1978a, 1980a; Dubovskij & Turgunov 1971a; D'Urso, V. 1981b; Emeljanov, A.F. 1967a, 1972c; Emelyanov et al. 1968a; Emmrich, R. 1980a; Evans, J.W. 1957c, 1958b, 1959b, 1963b, 1965a, 1966a, 1968a, 1971b, 1971c, 1972a, 1977a, 1982a; Freytag, P.H. 1971b; Gamez et al. 1979a; Ghauri, M.S.K. 1963d, 1968a, 1971a; Gill & Oman 1982a; Gonzalez B., J.E. 1959a; Gourlay, E.S. 1964a; Gowda et al. 1983a; Gu & Ito 1981a; Gunthart, H. 1971c, 1984b; Hamilton, K.G.A. 1983b, 1985c; Hardee et al. 1962a; Hoebeke & Wheeler 1983a; Hoffman & Taboada 1965a; Hulden, L. 1984b; Iida, T. 1965a; Kalkandelen, A. 1985a; Kalkandelen & Fox 1968a; Karaman, S. 1966a; Kartal, V. 1982a, 1983a; Kato & Wakamatsu 1978a; Kawase, E. 1971a; Khaire & Bhapkar 1971a; Knight, W.J. 1968a, 1974c, 1976b, 1981a, 1983a, 1983b; Kuoh, C.L. 1964a; Linnavuori, R. 1959b, 1960a, 1960b; Linnavuori & DeLong 1977c; Ma, N. 1982a, 1983a; Maillet, P.L. 1956a; Mason & Yonke 1971a; McClure, M.S. 1980b; Mitjaev, I.D. 1974a, 1974b, 1975b, 1979c, 1979e; Miyamoto & Miyatake 1963a; Morris, M.G. 1972a; Moshida et al. 1979a; Musil, M. 1956a, 1958c, 1960a, 1963a; Neparidze & Dekanoidze 1971a; Nielson, M.W. 1975a, 1977a, 1979b, 1979d, 1982e, 1983e; Nielson & Morgan 1982a; Niemczyk & Guyer 1963a; Novak & Wagner 1962a; Novoa & Alayo 1985a; Okali, I. 1963a, 1964a, 1968a; Oman, P.W. 1967a; Oman & Krombein 1968a; Prior, R.N.B. 1966a; Quartau, J.A. 1982d; Rao, V.P. 1965a; Remane, R. 1961d, 1968a, 1983a, 1984a; Rose, D.J.W. 1983a; Ross, H.H. 1968a; Ross & Hamilton 1970a; Salmon, M.A. 1959a; Saringer, G. 1961a; Schroder, H. 1959a; Servadei, A. 1957a, 1958a, 1960a, 1967a, 1968a, 1969a, 1971a, 1972a; Siwi & Roechan 1983a; Sohi, A.S.

1977b; Soto et al. 1976a; Subba Rao et al. 1968a; Synave, H. 1961a, 1966a, 1966b, 1966c, 1976a; Synave & Dlabola 1976a; Taboada & Hoffman 1965a; Teraguchi et al. 1981a; Vidano, C. 1965a; Vidano & Arzone 1985a, 1985b; Vilbaste, J. 1983a; Wagner & Franz 1961a; Yonce, C.E. 1983a; Yoshioka, E. 1965a.

Regional and Local Faunas

Afghanistan: Dlabola, J. 1957b, 1964a, 1971a, 1972a.

Afrotropical: Dabrowski & Okoth 1985a; Dlabola, J. 1964b; Einyu & Ahmed 1979a, 1980a, 1982a, 1983a; Ghauri, M.S.K. 1968a, 1974d; Linnavuori, R. 1972a, 1972b, 1975a, 1977a, 1978a, 1978b, 1978d, 1979a, 1979b; Linnavuori & Al-Ne'amy 1983a; Linnavuori & Quartau 1975a; Quartau, J.A. 1978b, 1980a; Van Stalle, J. 1982a; Villiers, A. 1956a; Webb, M.D. 1975a, 1976a, 1981a, 1983b.

Albania: Csiki, E. 1940a; Dlabola, J. 1964c.

Aldabra Islands: Webb, M.D. 1980a.

Arctic/Sub-Arctic: Lindberg, H. 1958b.

Argentina: Marino de Remes Lenicov, A.M. 1982a.

Australia: Evans, J.W. 1956c, 1959b, 1960b, 1966a, 1977a; Evans et al. 1970a; Kitching et al. 1973a; Knight, W.J. 1976b; Webb, M.D. 1983a; Woodward et al. 1974a.

Austria: Wagner & Franz 1961a; Nicolaus, M. 1957a.

Azores: Lindberg, H. 1960c; Quartau, J.A. 1979a, 1980b, 1982c.

Bangladesh: Ahmed & Samad 1972b; Mahmood & Ahmed 1969a.

Barro Colorado Island: Wolda, H. 1982a.

Bolivia: Linnavuori & DeLong 1976a, 1977b.

Brazil: Alvarenga, M. 1962a; Bergmann, Moret et al. 1984a; Bergmann, Ramiro et al. 1984a; Menezes, M. de 1972a; Penny & Arias 1982a; Piza, S. 1968a; Zanol & Menezes 1982a.

Bulgaria: Dirimanov & Karisanov 1964a, 1965a; Pelov, V. 1968a.

Canada: Beirne, B.P. 1956a; Downes, W. 1957a; George, J.A. 1959a; Hamilton, K.G.A. 1970a, 1972a, 1972b, 1972d, 1985c; Phillips, J.H.H. 1951a; Scudder, G.G.E. 1961a; Varty, I.W. 1963a.

Canary Islands: Lindberg, H. 1960d; Lindberg & Wagner 1965a; Remane, R. 1984a.

Cape Verde Islands: Lindberg, H. 1958a, 1958b.

Central America: Linnavuori & DeLong 1977b, 1978c, 1979b, 1979c.

Chatham Island: Knight, W.J. 1976b.

Chile: Linnavuori & Delong 1977c, 1979e.

China: Anufriev, G.A. 1979a; Chou & Ma 1981a; Dubovskij & Turgunov 1971a; Kuoh, C.L. 1981b, 1985b; Kuoh & Fang 1984a; Ma, N. 1982a, 1983a; Popov, Y.A. 1963a.

Cocos-Keeling Islands: Gibson-Hill, G.A. 1950a.

Colombia: Anonymous 1968e; Linnavuori, R. 1968b.

Congo: Linnavuori, R. 1969a.

Cook Islands: Eyles & Linnavuori 1974a.

Crete: Asche, M. 1980b.

Cuba: Linnavuori, R. 1973c.

Cyprus: Georghiou, G.P. 1957a.

Czechoslovakia: Dlabola, J. 1956a, 1957c, 1957d, 1958a, 1961a, 1965a, 1977d; Jansky, V. 1983a, 1984a, 1985a, 1985b; Lauterer, P. 1957a, 1958a, 1978a, 1980a, 1983b, 1984a; Lauterer & Okali 1974a; Musil, M. 1956a, 1958a, 1958c, 1958d, 1959a, 1960a, 1961b, 1963a; Okali, I. 1959a, 1960a, 1963a, 1964a.

Danger Island: Linnavuori, R. 1956c.

Denmark: Kristensen, N.P. 1965a, 1965b; Ossiannilsson, F. 1978a, 1981a, 1983a; Trolle, L. 1966a, 1968a, 1974a.

Egypt: Hamad et al. 1981a; Linnavuori, R. 1964a, 1969b; Mohamed Hamed, K. 1964a.

Ethiopia: Heller & Linnavuori 1968a.

Europe: Anufriev, G.A. 1979e; Dantsig et al. 1964a; Dlabola, J. 1956b, 1958a, 1961a, 1965e, 1979a; Emeljanov, A.F. 1964c, 1967b; Vidano, C. 1959a, 1965b; Vilbaste, J. 1973b, 1976a, 1982a; Wilson, M.R. 1981a.

Fiji: Linnavuori, R. 1960b.

Finland: Albrecht, A. 1977a; Hulden, L. 1983a, 1984b; Hulden & Heikinheimo 1984a; Kontkanen, P. 1960a; Linnavuori, R. 1969c; Raatikainen, M. 1968a.

France: Bailly-maitre, A. 1955a; Bonfils & Della Giustina 1978a, 1978b; Bonfils & Lauriant 1975a; Bonfils & Schvester 1960a; D'Aguilar & Della Giustina 1974a; Della Giustina, W. 1977a, 1983a; Maillet, P.L. 1956a; Ribaut, H. 1959a, 1959b; Tribout, P. 1956a; Villiers, A. 1971a.

French Antilles: Bonfils & Delplanque 1971a.

Galapagos Islands: Linsley & Usinger 1966a.

Germany: Emmrich, R. 1966b, 1975b; Haupt et al. 1969a; Nicolaus, M. 1957a; Remane, R. 1961a, 1961d, 1962a; Sander, F.W. 1984a; Schiemenz, H. 1964a, 1965a, 1969a, 1970a, 1971a.

Greece: Asche, M. 1980a; Drosopoulos, S. 1980a; Wagner, W. 1959c.

Hungary: Orosz, A. 1979a, 1981a; Saringer, G. 1958a, 1958b; Soos, A. 1956a.

India: Anjaneyulu & Chakraborti 1977a; Bharadqaj & Yadav 1980a; Datta & Pramanik 1977a; Ghauri, M.S.K. 1980b; Ghosh, M. 1974a; Ishihara, T. 1979a; Kameswara Rao, P. 1976a; Kumar, A.R.V. 1983a; Mahmood, S.H. 1980a; Maldonado-Capriles, J. 1964a, 1965a; Malhotra & Sharma 1977b; Ramachandra Rao, K. 1973a; Rao, K.R. 1980a, 1981a; Ruppel, R.F. 1969a; Santok Singh et al. 1955a; Sharma, B. 1977b; Sohi, A.S. 1977b; Sohi & Dworakowska 1981a, 1984a; Verma, K.L. 1976a; Viraktamath, C.A. 1981a; Viraktamath & Dworakowska 1979a; Viraktamath & Sohi 1980a; Wesley, C.S. 1983a.

Indonesia: Hokyo et al. 1977a; Mochida et al. 1979a; Rees, C.J.C. 1983a.

Iran: Dlabola, J. 1960c, 1971a, 1971b, 1974a, 1977b, 1979a, 1981a, 1984a; Dlabola & Heller 1962a; Mirzayans, H. 1976a.

Iraq: Shalaby et al. 1968a.

Israel: Klein & Raccah 1984a; Klein et al. 1982a; Linnavuori, R. 1962a.

Italy: D'Urso, V. 1980b, 1981a, 1981b; D'Urso et al. 1984a, 1984b; Linnavuori, R. 1959a, 1959c; Servadei, A. 1957a, 1958a, 1960a, 1967a, 1968a, 1969a, 1971a, 1972a; Vidano, C. 1959b, 1964b, 1981a; Wagner, W. 1959a.

Japan: Arai, Y. 1978a; Asahina et al. 1959a; Hori, Y. 1969a, 1982a; Ishihara, T. 1956a, 1957a, 1958a, 1959a, 1961b, 1965d, 1968a, 1983a; Miyamoto & Miyatake 1963a; Mochida, O. 1973a.

Jordan: Dlabola, J. 1965d.

Juan Fernandez Island: Evans, J.W. 1957a.

Kermadec Islands: Knight, W.J. 1976b.

Korea: Chang & Choe 1982a; Choe, K.R. 1985a; Kwon, Y.J. 1983a; Kwon & Lee 1978b, 1978c, 1979a, 1979c, 1979d, 1979g, 1979h; Lee, C.E. 1971a, 1979a; Lee & Kwon 1976a, 1979b.

Kurile Islands: Anufriev, G.A. 1968d, 1970c, 1977b; Ishihara, T. 1966a.

Lapland: Hulden & Albrecht 1984a.

Lebanon: Dlabola, J. 1965d; Abdul-Nour, H. 1984a, 1985a.

Lord Howe Island: Knight, W.J. 1976b.

Madeira: Lindberg, H. 1960c, 1961a; Quartau, J.A. 1981f; Remane, R. 1968a, 1984a.

Malaysia: Carnegie, A.J.M. 1976a; Hokyo et al. 1977a; Kathirithamby, J. 1978a.

Mauritius: Mamet, J. 1957a.

Mediterranean: Dlabola, J. 1984a; Linnavuori, R. 1958a, 1965a; Vilbaste, J. 1976a.

Mexico: Linnavuori & DeLong 1978e; Ramos-Elorduy de Conconi, J. 1972c; Ruppel, R.F. 1958a.

Micronesia: Linnavuori, R. 1960a, 1975c.

Mongolia: Dlabola, J. 1965c, 1966a, 1967a, 1967b, 1967c, 1967d, 1968a, 1968b, 1970b; Emeljanov, A.F. 1977a; Emeljanov et al. 1968a.

Morocco: Lindberg, H. 1956b, 1963a, 1964a; Linnavuori, R. 1956d.

Nearctic: Greene, J.F. 1971a; Hamilton, K.G.A. 1975d, 1980a, 1980b, 1982a, 1983b, 1983d; Hamilton & Ross 1972a, 1975a; Kramer, J.P. 1964a, 1966a.

Neotropical: Kramer, J.P. 1964a, 1964b, 1966a, 1976a, 1976c; Linnavuori, R. 1956a, 1957a, 1959a, 1965b, 1975b, 1978c; Linnavuori & DeLong 1977a, 1978b, 1978c, 1978d, 1979a, 1979b, 1979d; Maldonado-Capriles, J. 1957a, 1984a; Menezes, M. de 1973b, 1974a.

Nepal: Thapa, V.K. 1983a, 1984a, 1984b, 1985a.

Netherlands: Cobben & Gravestein 1958a; Cobben & Rozeboom 1978a; Gravestein, W.H. 1953a, 1965a; Ulenberg et al. 1983a.
New Guinea: Evans, J.W. 1982a.
New World: Kramer, J.P. 1964a, 1966a.
New Zealand: Dumbleton, L.J. 1964a, 1967a; Evans, J.W. 1963b, 1966a, 1977a; Gourlay, E.S. 1964a; Knight, W.J. 1973a, 1973b, 1974a, 1974b, 1975a, 1976a, 1976b.
Nicaragua: Estrada, R.F.A. 1960a.
Niger: Lindberg, H. 1956a.
Nigeria: Deeming, J.C. 1981a.
Niue Island: Eyles & Linnavuori 1974a.
Norfolk Island: Knight, W.J. 1976b.
Norway: Hulden, L. 1982a; Ossiannilsson, F. 1962a, 1974a; Schulz, C.A. 1976a.
Oriental: Mahmood, S.H. 1962a, 1967a, 1968a, 1973b, 1975a; Sohi, A.S. 1983a; Viraktamath, C.A. [undated].
Pakistan: Ahmed, M. 1971a, 1972b, 1985c; Dlabola, J. 1971a; Ghauri, M.S.K. 1965a, 1972c; Mahmood, S.H. 1973b, 1975a, 1980a.
Palaearctic: Nast, J. 1972a, 1979a, 1982a.
Panama: Wolda, H. 1977a, 1980a, 1980b.
Peru: Aguilar, F.P.G. 1964a; Dourojeanni & Marc 1965a; Escalante et al. 1981a; Linnavuori, R. 1973b, 1975d; Linnavuori & DeLong 1976a; Linnavuori & Heller 1961a.
Philippines: Mahmood, S.H. 1967a; Maldonado-Capriles, J. 1964a.
Phoenix Island: Linnavuori, R. 1956c.
Poland: Chudzika, E. 1981a, 1982a; Gajewski, A. 1961a; Gebicki et al. 1982a; Jasinska, J. 1980a; Nast, J. 1973a, 1976a, 1976b; Smreczynski, S. 1955a.
Portugal: Cardoso, A.M. 1974a; Lindberg, H. 1960a, 1962a; Quartau, J.A. 1981e; Quartau & Rodrigues 1969a.
Rennell Island: Knight, W.J. 1982a.
Rio De Oro: Lindberg, H. 1956b.
Romania: Cantoreanu, M. 1959a, 1961a, 1963a, 1963b, 1965a, 1965b, 1965d, 1968b, 1969a, 1969b, 1971b, 1971c, 1975a; Dobreanu & Manolache 1969a.
Salvage Islands: Quartau, J.A. 1984a.
Saudi Arabia: Dlabola, J. 1979b, 1980e.
Scandinavia: Ossiannilsson, F. 1976a, 1978a, 1981a, 1983a.
SE Asia: Claridge & Wilson 1981c; Ishihara, T. 1961a; Knight, W.J. 1981a, 1983a; Lee & Kwon 1977b; Webb, M.D. 1981b.
Selvagens Islands: Quartau, J.A. 1981f.
South Africa: Linnavuori, R. 1961a; Theron, J.G. 1970b, 1972a, 1973a, 1973b, 1974a, 1975a, 1976a, 1977a, 1978a, 1980a, 1980b, 1982b, 1983a, 1984c, 1985a; Van Rensburg, G.D.J. 1983a.
Spain: Gomez-Menor, J. 1956a; Linnavuori, R. 1956d; Medina et al. 1981a, 1982a; Morris, M.G. 1978a; Strubing, H. 1981a.
Spanish Sahara: Lindberg, H. 1965a.
St. Helena: Dlabola, J. 1976a; Synave, H. 1976a; Synave & Dlabola 1976a.
Sudan: Dlabola, J. 1964b.
Sweden: Gyllensvard, N. 1963a, 1965a, 1969a; Ossiannilsson, F. 1955a, 1961b.
Switzerland: Baggiolini, Canevascini, Caccia et al. 1968a; Gunthart, H. 1971a, 1971b, 1974a, 1984a.
Syria: Dlabola, J. 1965d.
Taiwan: Anonymous 1963a; Chen & Sogawa 1969a; Ishihara, T. 1965c; Lin, K.J. 1979a.
Thailand: Hongsaprug, W. 1983a, 1983b; Kawase, E. 1971a; Kitbamroong & Freytag 1978a; Mahmood, S.H. 1967a.
Tromelin Island, Mascarene Basin: China, W.E. 1955b.
Tunisia: Linnavuori, R. 1971a.
Turkey: Dlabola, J. 1957a, 1971a, 1971b, 1979a; Kalkandelen, A. 1974a, 1980a; Kartal, V. 1982a, 1983a; Kocak, A.O. 1981a, 1983a; Lodos, N. 1982a; Lodos & Kalkandelen 1981a, 1982a, 1982b, 1983a, 1983b, 1983c, 1984a, 1984b, 1984c, 1984d, 1985a, 1985b, 1985c; Lodos & Onder 1984a.
United Kingdom: Allen, A.A. 1962a, 1963a, 1963b, 1963c, 1964a, 1965a, 1978a, 1982a, 1985a; Andrewes, C.H. 1977a; Badmin, J.S. 1970a, 1979a, 1981a, 1985a; Bennett, S.H. 1959a; Davis, B.N.K. 1983a, 1984a; Dolling, W.R. 1980a; Flint, J.H. 1963a, 1976a; Kathirithamby, J. 1974a; Le Quesne, W.J. 1959a, 1961a, 1961b, 1964a, 1964c, 1965a, 1965b, 1965c, 1968b, 1969a, 1969b, 1969c, 1974a, 1976a, 1977a, 1983c, 1984a; Le Quesne & Morris 1971a; Le Quesne & Payne 1981a; Morris, M.G. 1966a; Nelson, J.M. 1971a; Payne, K. 1981c; Prior, R.N.B. 1966a; Richards, O.W. 1964a; Salmon, M.A. 1954b, 1959a; Steel & Woodroffe 1969a; Whalley, P.E.S. 1955a; Whittaker, J.B. 1964a, 1965a, 1969a; Woodroffe, G.E. 1967a, 1968a, 1971a, 1971b, 1972a, 1972b.
USA: Anonymous 1957b, 1959a, 1961a, 1975a; Beardsley, J.W. 1956a, 1961a, 1964a, 1965a, 1966a, 1969a, 1969b, 1976a; Beardsley & Funasaki 1976a; Begley & Butler 1980a; Beirne, B.P. 1956a; Bianchi, F.A. 1955a; Blocker & Reed 1976a; Blocker et al. 1971a; Bonnefil, L. 1969a; Chong, M. 1965a; Cunningham, H.B. 1964a; Cwikla & Blocker 1981b; Douglas et al. 1966a; Freytag, P.H. 1971b, 1976a; Fullaway, D.T. 1956a; Hoffman & Taboada 1960a, 1961a; Horning & Barr 1970a; Knowlton, G.F. 1955a, 1955b, 1955c; Maclean, D.B. 1984a; Mason & Yonke 1970a, 1971a; Mead, F.W. 1957a, 1965a, 1966a, 1981a; Musgrave, C.A. 1979a; Namba, R. 1956b; Napompeth & Nishida 1971a; Russell, L.M. 1973a; Stafford & Summer 1963a; Suehiro, A. 1961a; Taboada, O. 1959a, 1964a, 1979a, 1982a; Taboada & Burger 1967a; Taboada et al. 1975a; Wolfe, H.R. 1955c; Yonce, C.E. 1983a; Yoshioka, E. 1965a.
USSR: Abdurakhimov & Dubovsky 1970a; Anufriev, G.A. 1971h, 1978a, 1981c; Arakelian et al. 1984a; Danka, L. 1959a, 1961a, 1961b, 1964a; Dantsig et al. 1964a; Davletshina et al. 1981a; Davletshina & Radzivilovskaya 1965a; Dlabola, J. 1958b, 1961b; Dubovskij, G.K. 1964b, 1966a, 1967a, 1968a, 1970a, 1970b, 1978a, 1980a; Dubovskij & Karimov 1970a; Dubovskij & Tursunkhodzhaev 1970a; Emeljanov, A.F. 1964c, 1964e, 1967b, 1969b; Kholmuminov & Dubovskij 1979a; Kirejtshuk, A.G. 1977a; Korolevskaya, L.L. 1968b, 1974a, 1975a, 1976a, 1977a, 1978a, 1978b, 1979b, 1980a; Lindberg, H. 1960b; Logvinenko, V.N. 1956a, 1957a, 1957b, 1957c, 1957d, 1957e, 1959a, 1959b, 1959c, 1960b, 1961c, 1962d, 1962e, 1963a, 1963b, 1964b, 1965a, 1966a, 1967d, 1969a, 1984a; Mitjaev, I.D. 1963b, 1968b, 1971a, 1971b, 1974a, 1974b, 1975b, 1979a, 1979d, 1979f; Pogvinenko, V.M. 1962a; Popov, Y.A. 1963a; Schengeliya, E.S. 1956a, 1961a, 1964a, 1966a; Shengeliya & Dlabola 1964b; Talitskii & Logvinenko 1965a, 1966a; Tereshko, L.I. 1965a; Varzimska, R. 1983a; Vilbaste, J. 1955a, 1958a, 1958b, 1959b, 1962b, 1964a, 1965b, 1968a, 1969b, 1973b, 1974a, 1979a, 1980a, 1980b, 1983a; Zachvatkin, A.A. 1948a, 1948b, 1953a, 1953b, 1953c, 1953d, 1953e.
Yugoslavia: Dlabola & Jankovic 1981a; Jancovik, L. 1966a, 1971a; Kiauta, B. 1962a; Novak & Wagner 1962a; Papovic, R.M. 1977a; Tanasijevic, N. 1966a.

BIOLOGY

Life Histories

General: Chiswell, J.R. 1964a; Srivastava & Bisaria 1981a; Srivastava & Misra 1981a; Sun, S.C. 1963a.
Aceratagallia curvata: Nielson & Toles 1968a.
Acinopterus angulatus: Nielson & Toles 1968a.
Agallia constricta: Mitsuhashi & Maramorosch 1963a, 1963b.
Agalliopsis novella: Mitsuhashi & Maramorosch 1963a, 1963b.
Alebra albostriella: Mazzone, P. 1976a.
Alnetoidea alneti: Mazzone, P. 1976a.
Amrasca splendens: Das et al. 1976a.
Amritodes atkinsoni: Patel et al. 1975a.
Aphrodes bicinctus: Saringer, G. 1961a.
Austroasca lybica: Monsef, A. 1981a.
Carneocephala nuda: Nielson & Toles 1970a.
Carneocephala triguttata: Nielson & Toles 1970a.
Cassianeura cassiae: Iqbal Singh 1983a.
Cicadella viridis: Chu & Teng 1950a.
Cicadulina bipunctella zeae: Ammar, El-D. 1977a.
Cicadulina chinai: Ammar, El-D. 1975b.
Cicadulina triangula: Dabrowski, Z.T. 1985a.
Circulifer tenellus: Bindra & Sohi 1969a; Cook, W.C. 1967a.
Coloborrhis corticina camerunensis: Mbondji, P.M. 1982b, 1983a.
Colladonus clitellarius: George & Davidson 1959a.
Colladonus geminatus: Nielson, M.W. 1968a.
Colladonus mendicus: Rice & Jones 1974a.
Cuerna arida: Nielson & May 1975a.

Cuerna balli: Nielson & May 1975a.
Cuerna costalis: Turner & Pollard 1959b.
Dalbulus elimatus: Barnes, P. 1954a.
Dalbulus maidis: Barnes, P. 1954a; Davis, R. 1966a; Mitsuhashi & Maramorosch 1963a, 1963b; Pitre, H.N. 1970a, 1970b.
Deltocephalus sonorus: Stoner & Gustin 1968a.
Dikrella scinda: Ramos-Elorduy de Conconi, J. 1972b.
Draeculacephala antica: Mason & Yonke 1971b.
Draeculacephala delongi: Mason & Yonke 1971b.
Draeculacephala mollipes: Bridges & Pass 1967a, 1970a; Mason & Yonke 1971b.
Draeculacephala portola: Mason & Yonke 1971b.
Empoasca dolichi: Parh & Taylor 1981a.
Empoasca fabae: Hogg, D.B. 1985a; Medler et al. 1966a.
Empoasca kerri: Butt et al. 1981a.
Empoasca kraemeri: Wilde et al. 1976a.
Empoasca signata: Naheed & Ahmed 1980b.
Empoasca vitis: Moutous & Fos 1976a.
Endria inimica: Coupe & Shulz 1968a, 1968b.
Erythroneura aclys: Richter, F.T. 1970a.
Erythroneura alneti: Cantoreanu, M. 1964a.
Erythroneura imeretina: Dekanoidze, G.I. 1962a.
Erythroneura lawsoni: McClure, M.S. 1974a.
Eupteryx atropunctata: Hoebeke & Wheeler 1983a.
Eupteryx notata: Payne, K. 1981a.
Exitianus exitiosus: Gustin & Stoner 1973a.
Fieberiella florii: Swenson, K.G. 1974a.
Forcipata loca: MacRae & Yonke 1984a.
Graminella nigrifrons: Stoner & Gustin 1967a.
Graminella sonora: Stoner & Gustin 1968a.
Graphocephala coccinea: Wheeler & Walley 1980a.
Graphocephala versuta: Turner & Pollard 1959b.
Hishimonus phycitis: Bindra & Singh 1969a.
Homalodisca coagulata: Turner & Pollard 1959b.
Homalodisca insolita: Turner & Pollard 1959b.
Idiocerus atkinsoni: Alam & Mandal 1964a.
Macropsis fuscula: Fluiter et al. 1956a; Fluiter & van der Meer 1958a; Tonks, N.V. 1960a.
Macrosteles fascifrons: Mitsuhashi & Maramorosch 1963a, 1963b; Mitsuhashi et al. 1965a.
Macrosteles sexnotatus: Becker, M. 1974a.
Marathonia nigrifascia: Ramos-Elorduy de Conconi, J. 1970a; Serrano-Limon, G. 1970a; Serrano-Limon & Ramos Elorduy de Conconi 1972a.
Nephotettix bipunctatus: Alam & Islam 1959a.
Nephotettix cincticeps: Hokyo, N. 1972a; Hokyo & Kuno 1977a; Mitsuhashi & Kono 1975a; Mochida, O. 1970a; Murai & Kiritani 1970a; Sameshima & Nagai 1962a.
Nephotettix nigropictus: Dhawan & Sajjan 1976a.
Nephotettix virescens: Anjaneyulu, A. 1980a; Chakravarti et al. 1979a; Fachrudin 1980a.
Ollarianus balli: Dabek, A.J. 1982a.
Oncometopia alpha: Nielson et al. 1975a.
Oncometopia undata: Turner & Pollard 1959b.
Oncopsis alni: Cantoreanu, M. 1965c.
Orosius albicinctus: Bindra & Singh 1970a; Sundararaju & Jayaraj 1977a.
Paraphlepsius irroratus: Chiykowski, L.N. 1985a.
Paraulacizes irroratus: Mason & Yonke 1971b.
Ribautiana tenerrima: Raine, J. 1960a.
Scaphytopius acutus: Palmiter et al. 1960a.
Scaphytopius acutus acutus: Meyer, J.R. 1984a.
Scaphytopius cinerus: Meyer, J.R. 1984a.
Scaphytopius delongi: Swenson, K.G. 1971a.
Scaphytopius frontalis: Meyer, J.R. 1984a.
Scaphytopius magdalensis: Meyer, J.R. 1984a.
Scaphytopius nitridus: Nielson & Morgan 1982a.
Scaphytopius verecundus: Meyer, J.R. 1984a.
Typhlocyba ishidai: Tsugawa et al. 1966a.
Typhlocyba prunicola: Mulla, M.S. 1957a.
Typhlocyba quercus: Mulla, M.S. 1957a.
Zygina binotata: Malik et al. 1979a.
Zygina rubronotata: Ahmed & Ahmed 1980a.
Zyginidia pullula: Vidano & Arzone 1985a.
Zyginidia quyumi: Jabbar & Ahmed 1975a.

Bionomics
Abu Yaman, I.K. 1967a; Abul-Nasr et al. 1969a; Adler, P.H. 1982a; Ahmed, Manzoor 1973a; Ahmed & Ahmed 1980a; Ahmed & Samad 1972a; Ahmed, Baluch, Ahmed & Shaukat 1980a; Ahmed, Ahmed et al. 1981a; Aitchison, C.W. 1978a; Akingbohungbe, A.E. 1983a; Akino, K. 1969a; Alam & Mandal 1964a; Alam, S. 1971a, 1977a; Alam & Islam 1959a; Ammar, El-Desouky 1975b, 1977a; Ammar & Farrag 1980a; Ammar & Hosny 1969a; Ammar et al. 1978a, 1978b, 1981a, 1983a; Ananthanarayanan & Abraham 1956a; Anderson, B.L. 1984a; Andow, D.A. 1984a; Andrzejewska, L. 1960a, 1964a, 1966b, Anonymous 1982b; Apple & Falter 1962a; Arzone, A. 1972a; Atwal et al. 1969a; Bae, T.U. 1985a; Bairyamova, V. 1982a; Balasubramanian, G. 1979a; Ball, J.C. 1979a; Baloch & Soomro 1980a; Barbosa, A.J. da S. 1954a, 1954b; Barnes, P. 1954a; Basu et al. 1976a; Becker, M. 1974a, 1979a; Betsch, W.D. 1978a; Bhalani & Patel 1981a; Bindra & Deol 1972a; Bindra & Kaur 1969a; Bindra & Singh 1969a, 1970a; Bindra & Sohi 1969a, 1970a; Bonnefil, L. 1969a, 1972a; Boulard, M. 1978a; Boyd & Pitre 1968a, 1969a; Brar, J.S. 1974a; Brar & Singh 1978a; Bridges & Pass 1967a, 1970a; Brown et al. 1955a, 1957a, 1959a, 1960a; Butt & Ahmed 1981a; Butt et al. 1981a; Byers & Yung 1979a; Cantoreanu, M. 1964a, 1965c, 1965d, 1969a; Carlson & Hibbs 1962a; Cattaneo & Arzone 1983a; Caudwell et al. 1969a, 1970a; Chaboussou, F. 1971a; Chakravarti et al. 1979a; Chandy, K.C. 1957a; Chattopadhyay & Mukhopadhyay 1977a; Chaudhary et al. 1980a; Chen & Ko 1978a; Cheng & Pathak 1971a; Chettanachit & Disthaporn 1982a; Chiswell, J.R. 1964a; Chiykowski, L.N. 1970a, 1985a; Chiykowski & Hamilton 1985a; Chu et al. 1961a, 1981a; Claridge, M.F. 1980a; Claridge & Nixon 1981a; Claridge & Reynolds 1972a; Claridge & Wilson 1976a, 1978a, 1981a; Cobben, R.H. 1956a, 1956b; Cobben & Rozeboom 1978a; Combs, R.L. 1967a; Coupe & Shulz 1968a, 1968b; Dabek, A.J. 1979a, 1982a; Dabrowski, Z.T. 1985a; Dabrowski & Okoth 1985a; Damsteegt, V.D. 1984a; Das et al. 1969a; Davis, R. 1966a; Davletishina & Radzirilovskaya 1965a; Davoodi, Z. 1980a; Decker & Cunningham 1967a, 1968a; Decker et al. 1971a; Decker & Maddox 1967a; Deitz et al. 1976a; Dekanoidze, G.I. 1962a; DeLong, D.M. 1970b, 1971b; Deol & Bindra 1978a; Dhawan & Sajjan 1976a; Diaz-Chavez, A.J. 1969a; Di Martino, E. 1956a; Dirimanov & Karisanov 1964a, 1964c, 1965a; Dlabola, J. 1957d; Dookia & Poonia 1981a; Douglass & Halleck 1957a, 1958a; Downes, W. 1957a; Drosopoulos, S. 1984a; Duffield, C.A.W. 1963a; Durant, J.A. 1968b, 1973a; Einyu & Ahmed 1977a; El-Kady et al. 1974a, 1974b; Elliott, D.R. 1981a; El-Nahal et al. 1979a, 1981a; El- Saadany & Abdel-Fattah 1980a; Emelyanov, A.F. 1964e; Escobedo et al. 1985a; Eskafi & Schoonhoven 1981a; Evans, D.E. 1966a; Fachrudin 1980a; Falk, J.H. 1982a; Fernando, H.E. 1959a; Fleischer et al. 1982a, 1983a; Flinn, P.W. 1981a; Flinn & Hower 1984a; Flock et al. 1962a; Flores & Aguirre 1972a; Fluiter & van der Meer 1958a; Frediani, D. 1956b; Gaborjanyi & Saringer 1967a; Garcia et al. 1979a; Genung & Mead 1969a; Giorgadze, D.C. 1968a; Glass et al. 1966a; Goodman & Toms 1956a; Greene, G.L. 1967a; Gromadzka, J. 1970a; Gu & Ito 1982a; Gunthart, H. 1977a; Gunthart & Gunthart 1981a; Gustin & Stoner 1968a; Habib et al. 1972a; Hagel & Landis 1967a; Halteren, P. van 1970a; Hegab, A.M. 1985a; Hegab et al. 1980a, 1984a; Helaly et al. 1985a; Henriquez, M.R. 1971a; Herms, D.A. 1984a; Hernandez, J.C. 1982a; Heyer et al. 1984a, 1984b; Hill, M.G. 1976a, 1982a; Hirao & Inoue 1978b; Hiremath, S.C. 1979a; Hoebeke & Wheeler 1983a; Hogg, D.B. 1985a; Hokyo, N. 1971a, 1972a, 1976a; Hokyo & Kuno 1977a; Hokyo et al. 1976a; Hopkins & Johnson 1984a; Hori, Y. 1967a; Hosny & El- Dessouki 1967a, 1969a; Hsieh, C.Y. 1975a; Hsieh & Dyck 1975a; Hsu & Banttari 1979a; Huang & Lo 1964a; Hummelen & Soenarjo 1977a; Ichikawa & Kiritani 1973a; Inoue, H. 1966a; Inoue & Hirao 1980a; Iqbal, Singh 1983a; Ito, Y. 1982a; Ito et al. 1983a; Ito & Johraku 1982a; Jabbar, A. 1984a; Jabbar & Ahmed 1974a, 1976a; Jabbar, Ahmed & Samad 1977a, 1980a; Jabbar et al. 1982a; Jayaraj & Basheer 1964a, 1964b; Jayaraj & Venugopal 1964a; Jensen, D.D. 1958a, 1971b; Jensen & Chapman 1979a; Jhoraku, T. 1966a; Jhoraku & Kato 1974a; Jhoraku et al. 1983a; John & Ghosh

1981a; Joyce, R.J.V. 1961a; Kabir & Choudhury 1975a; Kakiya & Kiritani 1972a; Kaloostian, G.H. 1956a; Karim & Pathak 1979a; Katayama, E. 1975a; Kaur et al. 1971a; Khan & Saxena 1985c; Kidokoro, T. 1979a, 1980a; Kieckhefer, R.W. 1962a; Kim & Hyun 1979a; Kira et al. 1978a; Kiritani et al. 1970a; Kirollos & Hibbs 1971a; Kisimoto, R. 1959b, 1959c; Kitajima & Landim 1972a; Kittur et al. 1985a; Klein & Raccah 1980a; Knowlton, G.F. 1962a; Koesnang & Rao 1980a; Kohno & Hashimoto 1978a; Kooner & Deol 1982a; Kooner et al. 1978a, 1980a; Korolevskaya, L.L. 1971a, 1971b; Kouskolekas, C.A. 1964a; Kouskolekas & Decker 1966a; Kozhevnikova, A.G. 1975a; Kuno, E. 1968a, 1973a, 1980a, 1984a; Kuno & Hokyo 1970a, 1976a; Kuwahara, M. 1974a; Lacy, G.H. 1982a; Lal, R. 1946a; Lauterer & Bures 1984a; Leite Filho & Ramalho 1979a; Le Quesne, W.J. 1964b; Liang, W.F. 1981a; Lim, G.S. 1969a; Ling & Miah 1980a; Ling & Tiongco 1975a, 1977a, 1979a; Linskii, V.G. 1980a; Macphee, A.W. 1979a; Madden, L.V. 1985a; Mahmood et al. 1964a; Maillet, P.L. 1959b, 1962a; Malabuyoc & Heinrichs 1981a; Malhotra & Sharma 1977c; Malik et al. 1979a; Mal'kooskiy, M.P. 1956a; Mani & Jayaraj 1976a; Manjunath & Urs 1979a; Martinez, M.R. 1975a; Martinez & Ramos-Elorduy de Concini 1978a; Massee, A.M. 1953a; Mathur & Chaturvedi 1980a; Maung, M.M. 1985a; Mbondji, P.M. 1982b, 1983a; McCarthy, H.R. 1956a; McClure, M.S. 1974a, 1980a, 1980b; Medina et al. 1981a, 1982a; Medler et al. 1966a; Meyerdirk & Hessein 1985a; Miller & De Lyzer 1960a; Miller & Hibbs 1963a; Misra & Israel 1968b; Misra & Reddy 1985a; Mitjaev, I.D. 1963b, 1975b; Miyahara et al. 1983a; Miyai et al. 1978a; Mochida & Kuno 1962a; Moffitt, H.R. 1968a; Moffitt & Reynolds 1972a; Moore, L. 1973a; Moreau, J.P. 1963a; Moreau & Leclant 1973a; Morris, M.G. 1966a; Mukhopadhyay et al. 1979a; Mulla, M.S. 1957a; Muller, H.J. 1955a, 1955b, 1956b, 1956c, 1957a, 1960b, 1960c, 1961a, 1962a, 1964a, 1965a, 1974a, 1976a, 1979a, 1980a, 1981a, 1981b, 1983a, 1984a; Murai & Kiritani 1970a; Murata et al. 1965a; Musgrave, C.A. 1974a; Musil, M. 1958a, 1960a; Naba, K. 1983a; Nachiappan & Baskaran 1984a; Nagaich, B.B. 1975a; Naheed & Ahmed 1980b; Naheed et al. 1981a; Nakamura et al. 1967a; Narayanasamy et al. 1979a; Nasu, S. 1963a; Nasu & Nakasuka 1973a; Nielson, M.W. 1949a, 1968a, 1983j; Nielson & Freytag 1976a; Nielson & May 1975a; Nielson et al. 1975a; Nielson & Morgan 1982a; Nielson & Toles 1968a; Niemczyk & Guyer 1963a; Nikolova, V. 1969a; Nishida et al. 1976a; Nixon, G.A. 1984a; Noda et al. 1968a; Nowacka, W. 1964a, 1966a, 1978a, 1978b, 1982a; O'Keefe, L.W. 1965a; Okoth et al. 1985a; Oliveira et al. 1981a; Oman, P.W. 1971a, 1972a; Orita, S. 1971a; Otake, A. 1966a; Oya, S. 1978a; Page, F.D. 1979a, 1983a; Palis et al. 1984a; Palmiter & Adams 1957a; Pant & Gupta 1984a; Parh, I.A. 1982a; Patel et al. 1975a, 1975b; Perfect & Cook 1982a; Pienkowski, R.L. 1970a; Pitre, H.N. 1967a, 1968a; Pitre & Hepner 1967a; Pizzamiglio, M.A. 1979a; Pollard, H.N. 1962a, 1965a, 1965b; Pollard & Kaloostian 1961a; Pomazkov, Y.I. 1966a; Pophaly et al. 1979a; Port, G.R. 1981a; Prestidge, R.A. 1982a, 1982b; Purcell, A.H. 1976a; Purcell & Suslow 1984a; Raatikainen & Vasarainen 1973a; Rajendra Singh 1978a, 1982a; Ramakrishnan & Ghauri 1979a; Ramalho & Ramos 1979a; Ramalho & Albuqueque 1979a; Ramos-Elorduy de Conconi, J. 1970a; Raney & Yeargan 1977a; Rasmy & Hassib 1974a; Razvyazkina, G.M. 1960a; Reddy & Garg 1983a; Reddy & Mishra 1983a; Reddy et al. 1983a; Rees, C.J.E. 1983a; Reissig & Kamm 1975a; Remane, R. 1958a; Rice & Jones 1974a; Richter, F.T. 1970a; Ricou, G. 1960a; Rizk & Ahmed 1981a; Rose, D.J.W. 1972b, 1973a, 1974a; Ruan, Y.L. 1985a; Saini, R.S. 1967a, 1967b; Sajian & Dhawan 1977a; Sam & Chelliah 1984a; Samad & Ahmed 1979a, 1979b; Sameshima & Nagar 1962a; Sasaba, R. 1974a; Sasaba & Kiritani 1971a; Sato & Sogawa 1981a; Saxena & Saxena 1971a; Schefer-Immel, V. 1957a; Schiemenz, H. 1969a; Schuler, L. 1956a; Schvester, Moutous, Bonfils et al. 1962a; Senapati & Hohanty 1980a; Senapati & Khan 1978a; Serrano-Limon, G. 1970a; Serrano-Limon & Ramos Elorduy de Conconi 1972a; Sharma et al. 1981a; Shukla & Anjaneyulu 1982c; Sidhu & Dhawan 1978a; Simmons et al. 1985a; Simmons, Pass et al. 1984a; Simmons, Yeargan et al. 1984a; Simonet, D.E. 1978a; Simonet & Pienkowski 1977a,
1977b, 1979a, 1979b, 1980a; Simonet et al. 1979a; Siva Shankra Sastry, K.S. 1958a; Soejitno et al. 1974a; Sohi & Dhaliwal 1985a; Song, Y.H. 1978a; Sriharan & Garg 1975a; Srinavasan et al. 1981a; Srivastava & Bisaria 1981a; Srivastava & Misra 1981a; Staples et al. 1970a; Stathopoulos, D.G. 1967a; Stevens, M.M. 1985a; Stewart, A.J.A. 1981a; Stiling, P.D. 1980a, 1980b; Stoner & Gustin 1967a, 1968a; Strong & Rawlins 1957a; Sugimoto, A. 1969a, 1977a, 1981a; Sun, S.C. 1963a; Sundararaju & Jayaraj 1977a; Swenson, K.G. 1971a, 1971b, 1974a; Swenson & Kamm 1975a; Tay, E.B. 1972a; Tereshko, L.I. 1965a; Theron, J.G. 1979a, 1981a; Thomas, P.E. 1972a; Thompson, P. 1978a; Timian & Alm 1973a; Tonks, N.V. 1960a, 1963a; Tsai & Anwar 1977a; Tsugawa et al. 1966a; Turner & Pollard 1959a; Valle, R.R. 1985a, 1985b; Valle & Kuno 1984a; Valley & Wheeler 1985a; Van Rensburg, G.D.J. 1982a, 1982b; Varty, I.W. 1963a, 1964a, 1967a, 1967b; Vidano, C. 1959c, 1960a, 1960b, 1960c, 1961b, 1961d, 1961e, 1962a, 1964b, 1981a; Vidano & Arzone 1983a; Viggiani, G. 1971a, 1973a; Viraktamath & Viraktamath 1981a, 1982a; Waite, G.K. 1976a, 1977a; Wakibe et al. 1983a; Waloff, N. 1983a; Walter, S. 1978a; Weaver, C.R. 1959a; Webster et al. 1968b; Wei & Brooks 1979a; Wen & Lee 1978a; Whalley, P.E.S. 1971a; Wheeler & Walley 1980a, 1980b; Wu & Ruan 1982a; Yonce, C.E. 1983a; Yusof, O.H. 1982a; Zachvatkin, A.A. 1953a.

Ecology

Abdel-Salam, F. 1967a; Abdul-Nour, H. 1984b, 1985a; Ahmed, M. 1973a, 1983a; Ahmed et al. 1977a; Ahmed & Ahmed 1980a; Aitchison, C.W. 1978a, 1984a; Akingbohungbe, A.E. 1983a; Akino, K. 1969a; Altieri et al. 1977a, 1978a; Alverson et al. 1977a, 1980a; Alverson, All & Kuhn 1980a; Ammar et al. 1981a; Andow, D.A. 1984a; Andrzejewska, L. 1960a, 1960b, 1960c, 1962a, 1964a, 1965a, 1966b, 1971a, 1978a, 1981a, 1984a; Arzone, A. 1972a; Arzone & Vidano 1984a; Bae, T.U. 1985a; Baluch et al. 1980a; Barnes, P. 1954a; Batra, S.W.T. 1979a; Becker, M. 1974a; Blocker et al. 1972a; Blocker et al. 1971a; Blocker & Reed 1976a; Bonfils & Della Giustina 1978b; Brown & Eads 1965a, 1969a; Butt & Ahmed 1981a; Butt et al. 1981a; Cameron, G.N. 1972a; Cancelado & Yonke 1970a; Cantoreanu, M. 1969a, 1971b, 1973a; Chakravarthy et al. 1985a; Chang & Oka 1984a; Chen, C.C. 1970a; Chen & Ko 1978a; Chen, Lieu et al. 1978a; Cheng & Birch 1978a; Chhabra et al. 1976a; Chu et al. 1961a; Claridge, D.W. 1982a; Claridge et al. 1983a; Claridge & Wilson 1976a, 1978a, 1978b, 1978c, 1978d, 1981a, 1981b, 1982a, 1982b; Cobben & Rozeboom 1978a; Cook, W.C. 1967a; Coupe & Shulz 1968a; Cunningham, H.B. 1967a; Davletshina et al. 1981a; DeLong, D.M. 1965a, 1966a, 1970b; Denno, R.F. 1977a, 1980a; Diaz-Chavez, A.J. 1969a; Dobreanu & Manolache 1969a; Dookia & Poonia 1981a; Douglass & Hallock 1958a; Drosopoulos, S. 1984a; Dubovskij, G.K. 1965a, 1965b; Dubovskij & Turgunov 1978a; Durant, J.A. 1973a; D'Urso et al. 1984b; Einyu & Ahmed 1977a; Elliott, D.R. 1981a; Emmrich, R. 1966b; Gebicki, C. 1979a, 1983a; Goel, S.C. 1978a; Gunthart, H. 1977a, 1984b; Gunthart & Thaler 1981a; Guppy, J.C. 1958a; Gyorffy & Pollak 1983a; Harakly & Assem 1978a; Hawkins & Cross 1982a; Henriquez, M.R. 1971a; Hewitt & Burleson 1976a; Hill, M.G. 1976a; Ho & Chen 1968a; Hokkanan & Raatikainen, M. 1977a; Hokyo et al. 1977a; Hosny & El-Dessouki 1967a; Huff, F.A. 1963a; Jayaraj & Basheer 1964a; Jayaraj & Venugopal 1964a; Jensen, Stafford et al. 1965a; Jhoraku & Kato 1974a; Jhoraku et al. 1983a; Joyce, R.J.V. 1961a; Jurisoo, V. 1964a; Kelly & Klostermeyer 1984a; Kieckhefer, R.W. 1962a; Kieckhefer & Medler 1964a; Kira et al. 1978a; Kiritani et al. 1970a; Klein & Raccah 1980a; Klimaszewski, Wojciechowski, Czylok et al. 1980a; Klimaszewski, Wojciechowski, Gebicki et al. 1980a; Kobayashi et al. 1973a, 1974a; Kobayashi, T. 1961a; Koponen, S. 1978a; Koppany & Wolcsanszky 1956a; Koppany, T. 1969a; Korolevskaya, L.L. 1968b, 1974a; Koshihara, T. 1972a; Kouskolekas, C.A. 1964a; Krishnaiah & Ramachander 1979a; Kristensen, N.P. 1965a; Kuno, E. 1968a, 1973a, 1980a; Kuno & Hokyo 1970a; Leeuwangh & Leuamsang 1967a; Lehmann, W. 1973b, 1973c; Leite Filho & Ramalho 1979a; Le Quesne, W.J. 1972a; Lin & Liu 1984a; Lindberg, H. 1958a; Logvinenko, V.N. 1957d, 1957e, 1959b,

1961c, 1962b, 1964a, 1964b; Madden, L.V. 1985a; Mahmood et al. 1964a; Maillet, P.L. 1959a, 1962a; Malhotra & Sharma 1977c; Malik et al. 1979a; Mal'kooskiy, M.P. 1956a; Mani & Jayaraj 1976a; Manley, G.V. 1976a; Martinez, M.R. 1975a; Martinez & Ramos-Elorduy de Concini 1978a; Mathur & Chaturvedi 1980a; Maung, M.M. 1985a; McClure, M.S. 1974a, 1980b, 1982a; McClure & Price 1975a, 1976a; Meade, A.B. 1962a; Meade & Peterson 1964a, 1967a; Medler, J.T. 1966a; Meyer & Colvin 1985a; Meyerdirk & Hessein 1985a; Meyerdirk & Oldfield 1985a; Miller, L.A. 1956a; Miller & De Lyzer 1960a; Miller, R.L. 1962a; Miller & Hibbs 1963a; Misra & Reddy 1985a; Mitjaev, I.D. 1967b, 1970b, 1971c, 1971d, 1973b, 1974b, 1979e; Mochida & Kuno 1962a; Mochida et al. 1979a; Mohammad et al. 1980a; Moore, L. 1973a; Moravskaja, A.S. 1956a; Morris, M.G. 1971a, 1972a, 1974a, 1974b, 1978a, 1978b, 1982a, 1984a, 1984b; Muller, H.J. 1955a, 1955b, 1956b, 1956c, 1957a, 1960b, 1960c, 1961a, 1962a, 1964a, 1965a, 1974a, 1976a, 1979a, 1980a, 1981a, 1981b, 1983a, 1984a; Murata et al. 1965a; Naba, K. 1983a; Nagel, H.G. 1973a; Naheed & Ahmed 1974a; Nasu & Nakasuka 1973a; Nault et al. 1983a; Nault & Madden 1985a; Neparidze, N.N. 1969a; Neparidze & Dekanoidze 1971a; Nielson & Bleak 1963a; Nielson & Kaloostian 1956a; Nikolova, V. 1969a; Nishida et al. 1976a; Noda et al. 1968a; Noon, Z.B. 1962a; Obretel, R. 1969a; Ohmart et al. 1983a; Okada, T. 1971a; O'Keeffe, L.W. 1965a; Okada, T. 1971a; Orita, S. 1971a; Otake, A. 1966a; Oya, S. 1979a; Oya & Sato 1973a; Page, F.D. 1979a, 1983a; Parh, I.A. 1983a; Pathak, M.D. 1968a; Pathak, P.K. 1983a; Payne, K. 1981b, 1981c; Perkes, R.R. 1970a; Peterson, A.G. 1973a; Peterson et al. 1969a; Pienkowski, R.L. 1962a, 1970a; Pienkowski & Medler 1962a, 1963a, 1964a, 1966a; Pimentel & Wheeler 1973b; Pitre, H.N. 1968b; Pitre et al. 1967a; Pollard & Kaloostian 1961a; Pollard et al. 1959a; Popov, Y.A. 1963a; Port, G.R. 1981a; Poston & Ogunlana 1973a; Poston & Pedigo 1975a; Pradhan, S. 1964a; Prestidge, R.A. 1982b; Prestidge & McNeill 1982a, 1983a, 1983b; Purcell, A.H. 1985a; Purcell & Frazier 1985a; Quartau, J.A. 1981g; Quisenberry et al. 1979a; Raatikainen, M. 1971a; Raatikainen & Vasarainen 1971a, 1973a, 1976a; Rack, K. 1984a; Rajamohan et al. 1974a; Ramos- Elorduy de Conconi, J. 1973a; Rao & Anjaneyulu 1982a, 1983a; Rasmy & Hassib 1974a; Raupp & Denno 1979a; Raven, J.A. 1983a; Reddy et al. 1983a; Rees, C.J.C. 1983a; Remane, R. 1958a; Ricou & Duval 1969a; Room & Wardhaugh 1977a; Rose, D.J.W. 1972a, 1972b, 1973b; Ross, H.H. 1957c, 1958c, 1970a; Ruppel, R.F. 1974a; Saini, R.S. 1967a, 1967b; Sam & Chelliah 1984b; Sander, F.W. 1984a; Sasaba & Kiritani 1971a; Sasamoto et al. 1968a; Sato & Sogawa 1981a; Savage, J.M. 1958a; Saxena et al. 1974a; Schefer-Immel, V. 1957a; Schiemenz, H. 1964a, 1965a, 1969a, 1970a, 1971a; Schoonhoven et al. 1978a, 1981a; Schowalter, T.D. 1985a; Schvester, Moutous, Bonfils et al. 1962a; Schwarz, R. 1959a; Schwemmler, W. 1978a, 1984a; Senapati & Hohanty 1980a; Senapati & Khan 1978a; Servadei, A. 1968a; Seyedoleslami, H. 1978a; Seyedoleslami & Croft 1980a; Sharma & Badan 1985b; Sharonova, M.W. 1969a; Simonet & Pienkowski 1979b; Singh, K.M. & Singh, R.N. 1978a; Singh, R.N. & Singh, K.M. 1978a; Singh et al. 1975a; Siva Shankara Sastry, K.S. 1958a; Smith, D. 1962a; Srinivasan et al. 1981a; Stathopoulos, D.G. 1967a; Stewart, A.J.A. 1983a; Stiling, P.D. 1980c; Stilling, D. 1978a; Strubing, H. 1956b, 1958a, 1963a, 1965a, 1966a, 1970a, 1976a, 1978a; Suenaga, H. 1962a; Suenaga & Nakatsuka 1958a; Swenson & Kamm 1975a; Takai et al. 1972a, 1965a; Tallamy & Denno 1979a; Tandon et al. 1983a; Tashev, D.G. 1968a; Tay, E.B. 1972a; Taylor & Kamara 1974a; Teraguchi et al. 1981a; Thresh, J.M. 1980a; Timmer et al. 1982a; Tormala, T. 1977a; Tormala & Vanninen 1983a; Troxclair & Boethel 1984a; Upadhyay, R.K. 1984a; Upadhyay et al. 1981a; Valle, R.R. 1985a; Van Halteren, P. 1979a; Vaszai-Vinag, E. 1984a; Vidano, C. 1961b, 1961c, 1965a, 1966b; Vidano & Arzone 1985b; Vilbaste, J. 1955a; Vincent, J.J. 1975a; Wagner, W. 1958a; Waite, G.K. 1976a; Waloff, N. 1978a, 1979a, 1980a, 1981a, 1984a; Waloff & Solomon 1973a; Waloff & Thompson 1980a; Way, M.O. 1983a; Way et al. 1982a, 1983a, 1984a; Weaver, C.R. 1959a; Wen & Lee 1978a, 1985a; Whalley, P.E.S. 1955a, 1971a; Wheeler, A.G. 1971a, 1974a, 1974b, 1977a; Wheeler & Walley 1980b; Whitcomb, R.F. 1958a, 1973a; Whitcomb et al. 1973a, 1972a; Whitmore et al. 1981a; Whittaker, J.B. 1964a, 1965a, 1967a, 1969a, 1984a; Whittaker & Warrington 1985a; Wilde, W.H.A. 1962a, 1962b; Williams, D.W. 1982a, 1984a; Wilson et al. 1973a, 1979a; Witsack, W. 1985a; Wolda, H. 1977a, 1979a, 1979b, 1980a, 1980b, 1982a, 1983a; Wonn, L. 1956a; Yang & Pan 1979a; Yano et al. 1981a, 1984a, 1985a; Yonce, C.E. 1983a; Yoshimeki, M. 1966a; Yusof, O.H. 1982a; Zachvatkin, A.A. 1953a; Zalom, F.G. 1981a; Zhizhilashvili, T.I. 1954a.

Mutualistic Micro-organisms

Brooks, M.A. 1985a; Chang & Musgrave 1972a, 1975a, 1975b; Houk & Griffiths 1980a; Kaiser, B. 1980a; Khan, A.M. 1975a, 1976a, 1977a; Korner, H.K. 1969a, 1969b, 1972a, 1974a, 1976a, 1978a; Laporte, M. 1966a; Louis & Laporte 1969a; Louis & Nicolas 1976a; Louis et al. 1976a; Mahdihassan, S. 1957a, 1970a, 1976a, 1978a; Mitsuhashi & Kono 1975a; Mitsuhashi et al. 1965a; Muller, H.J. 1962a; Muller, J. 1969a; Noda, H. 1980a; Noda & Mittler 1983a; Orenski, Mitsuhashi et al. 1965a; Rack, K. 1984a; Sander, K. 1968a; Schwemmler, W. 1978a, 1984a; Schwemmler et al. 1971a; Wei, L.-Y. 1978a; Wei & Brooks 1978a.

Dispersal

Chakravarthy & Rao 1985a; Chiykowski, L.N. 1958a; Chiykowski & Chapman 1965a; Cook, W.C. 1967a; Coudriet & Tuttle 1963a; Cunningham, H.B. 1967a; Decker, G.C. 1959a; Douglass, J.R. 1954a; Drake & Chapman 1965a; Dysart, R.J. 1962a; Ghauri, M.S.K. 1983b; Glick, P.A. 1960a; Gressitt & Nakata 1958a; Hagel & Hampton 1970a; Hagel et al. 1973a; Harrell & Hoizapfel 1966a; Harrison, R.G. 1980a; Hawkins & Cross 1982a; Henderson, C.F. 1955a; Hopkins & Johnson 1984a; Hosny & El-Dessouki 1969a; Huff, F.A. 1963a; Ito & Miyashita 1961a, 1965a; Jabbar, Ahmed & Ahmed 1977a; Johnson, C.G. 1969a; Joomaye, A. 1976a; Joyce, C.R. 1955a, 1962a, 1963a; Kuno & Hokyo 1976a; McClure, M.S. 1980a, 1980b; Medler, J.T. 1957a, 1960b; Miah & Ling 1978a; Miyashita et al. 1964a; Nichiporich, W. 1965a; Peterson et al. 1969a; Pienkowski, R.L. 1970a; Pienkowski & Medler 1963a, 1964a; Pitre et al. 1967a; Pollard et al. 1959a; Poston & Pedigo 1975a; Purcell & Frazier 1985a; Purcell & Suslow 1982a; Raatikainen, M. 1972a; Ramalho, F.S. 1978a; Reddy & Garg 1983a; Reddy & Mishra 1983a; Rose, D.J.W. 1972a, 1972b, 1973c; Ross, H.H. 1959c, 1968a; Schneider, F. 1962a; Schulz & Meijer 1978a; Taimr & Dlabola 1965a; Taylor, R.A.J. 1985a; Vidano & Arzone 1984c; Wallis, R.L. 1962a; Waloff, N. 1973a; Yoshimeki, M. 1966a

All Others

Inoue, H. 1983a; Misra, B.C. 1980a; Mound, L.A. 1983a; Tribout, P. 1956a.

FOSSILS

Bekker-Migdisova, E.E. 1962a, 1967a; Bode, A. 1953a; Elias, S.A. 1982a; Evans, J.W. 1956a, 1956b, 1958c, 1961b, 1961c, 1963c, 1964a, 1971d; Hamilton, K.G.A. 1971a; Hong, You-Chong 1980a; Kukalova-Peck, J. 1978a; Pierce, W.D. 1963a; Ross, H.H. 1965b; Statz, G. 1950a; Tasch & Riek 1969a; Wootton, R.J. 1981a.

GENETICS

Bhattacharya, A.K. 1972a, 1973a, 1973b, 1973c, 1973d, 1973e, 1974a, 1975a, 1975b, 1980a; Cruz, Y.P. 1974a; Den Hollander, J. 1982a; Gallun, R.L. 1972a; Halkka, O. 1959a, 1960a, 1960b, 1962a, 1965a; Halkka & Heinonen 1964a, 1966a; John & Claridge 1974a; Kawai, A. 1977a; Kitajima & Landim 1972a; Manna & Bhattacharya 1973a; Meinhardt, H. 1981a; Menezes & Coelho 1982a; Mitsuhashi, J. 1966a, 1967b; Nielson & Toles 1970a; Norment et al. 1972a; Parida & Dalua 1981a, 1981b; Saitoh et al. 1970a; Sander, K. 1959a, 1960a, 1962a, 1967a, 1968a; Saxena & Barrion 1985a; Sharma & Gupta 1980a, 1980b, 1980c; Stewart, A.J.A. 1983a; Tokumitsu & Maramorosch 1967a, 1968a; Whitten, M.J. 1965a, 1968a; Whitten & Taylor 1969a.

GALLS
Maramorosch et al. 1961a; Mitjaev, I.D. 1968c.

HOST PLANTS
General: Ahmed, M. 1970c, 1970e, 1971b, 1971c, 1984a; Akingbohungbe, A.E. 1983a; Allen, A.A. 1963c; Ammar & Hosny 1969a; Anjaneyulu, Shukla et al. 1982a; Arzone, A. 1972a; Batra & Jangyani 1960a; Beardsley, J.W. 1956a, 1961a, 1966a, 1969a, 1969b; Bhatnagar, V.S. 1974a; Bindra & Sohi 1968a; Blisard, T.J. 1931a; Bos, L. 1981a; Boyd & Pitre 1969a; Brar & Singh 1981a; Claridge et al. 1977a; Claridge & Wilson 1978c; Cobben, R.H. 1956b, 1978b; Cook, W.C. 1967a; Dhawan & Sajjan 1977a; Douglass & Hallock 1956a, 1957a; Downes, W. 1957a; Dubovskij, G.K. 1968a; Einyu & Ahmed 1977a; El-Nahal et al. 1979a; Emeljanov, A.F. 1964e, 1967a; Freytag, P.H. 1965b; Fullaway, D.T. 1956a; Gajewskaja, N.S. 1965a; Gallun, R.L. 1972a; Gambrell & Gilmer 1956a; Ghandhi, J.R. 1980a, 1980b, 1982a; Ghauri, M.S.K. 1974c; Ghosh & Mukhopadhyay 1978a; Giunchi, P. 1952-53a; Goble, H.W. 1967a; Greene, G.L. 1967a; Guagliumi, P. 1965a, 1965b; Gullyev, A. 1965a; Gunthart, H. 1984a; Hagel et al. 1973a; Hallock & Douglass 1956a; Hawkins & Cross 1982a; Helson, G.A.H. 1950a; Hoebeke & Wheeler 1983a; Hongsaprug & Wilson 1985a; Huber & Osmun 1966a; Hudon et al. 1980a; Ishihara, T. 1965b, 1967a, 1982a; Jayaraj, S. 1976a; Kamal, El Din M.H. 1964a; Khan et al. 1964a; Kim, J. 1963a; Klein & Raccah 1980a; Knight, W.J. 1966b; Kooner & Deol 1982a; Langlitz, H.O. 1964a; Lehker, G.E. 1957a; Lehmann & Skadow 1970a; Lim, G.S. 1976a; Lindberg, H. 1958a; Lodos & Kalkandelen 1981a, 1982a, 1982b, 1983a, 1983b, 1983c, 1984a, 1984b, 1984c, 1984d, 1985a, 1985b, 1985c; Logvinenko, V.N. 1959b, 1962c, 1973a; Lomakina, L.G. 1963a, 1967a; MacNay, C.G. 1961c; Mahmood et al. 1964a; Mallamaire, A. 1954a; McClanahan, R.J. 1963a; McClure, M.S. 1980a; Menezes, M. de 1978a; Mitjaev, I.D. 1971d; Nielson & Morgan 1982a; Oldfield & Kaloostian 1979a; Payne, K. 1981a; Peterson, A.G. 1973a; Pholboon, P. 1965a; Pitre, H.N. 1967a; Purcell, A.H. 1976a; Quartau, J.A. 1980a; Rajendra Singh 1982a; Rao & Anjaneyulu 1977a, 1978a; Rao & John 1974a; Saxena & Basit 1982a; Saxena et al. 1974a; Saxena & Saxena 1974a, 1975a, 1975b; Shiomi & Sugiura 1984a; Sohi & Dhaliwal 1985a; Stiling, P.D. 1980b; Sundararaju & Jayaraj 1977a; Swenson, K.G. 1974a; Tashev, D.G. 1968a; Thomas, P.E. 1972a; Ting & Ong 1974a; Tsai & Anwar 1977a; Van Rensburg, G.D.J. 1982b; Vidano, C. 1981a; Vidano & Arzone 1978a; Viswanathan & Kalode 1981a; Wagner & Franz 1961a; Wakibe et al. 1983a; Wallis, R.L. 1960a, 1962b; Wilbur, D.A. 1954a.

Trees: Ahmed, M. 1971a; Anufriev & Zhil'tsova 1982a; Attard, G. 1985a; Claridge & Nixon 1981a; Claridge & Reynolds 1972a; Claridge & Wilson 1978e, 1981a; Gebicki, C. 1983a; Ghauri, M.S.K. 1975c; Gibson, L.P. 1973a; Gunthart & Gunthart 1983a; Hamilton, K.G.A. 1985d; Hepner, L.W. 1969a; Herms, D.A. 1984a; Iqbal, S. 1983a; Ivliev & Kononov 1966a; Klimaszewski, Wojciechowski, Czylok et al. 1980a; Koponen, S. 1978a; Lehmann, W. 1973c; Le Quesne, W.J. 1964e; Lezhava, V.V. 1953a; Loos, C.A. 1965a; Markelova, Y.M. 1962a; Matesova, G.I. 1960a; McClure, M.S. 1975a; McClure & Price 1975a, 1976a; Schefer-Immel, V. 1957a; Semede, C.M.B. 1961a; Varty, I.W. 1963a, 1964a, 1967a, 1967b; Vidano & Arzone 1981a; Viggiani, G. 1971a, 1973a; Vilbaste, J. 1979a.

Herbs: Alghali & Domingo 1982a; Bigornia, A.E. 1963a; Gentsch, B.J. 1982a; Klimaszewski, Wojciechowski, Gebicki et al. 1980a; Lamp, Morris et al. 1984a.

Anacardiaceae: Ghauri, M.S.K. 1967a; Glass et al. 1966a; Hongsaprug, W. 1984a; Hongsaprug & Wilson 1985a; Viraktamath, C.A. 1973a.

Aquifoliaceae: Ross, H.H. 1953b.

Cannabidaceae: Lehmann & Schmidt 1969a.

Caricaceae: Ghauri, M.S.K. 1974a; Ivancheva et al. 1967a.

Chenopodiaceae: Blaine et al. 1982a; Goeden & Ricker 1968a; Hori, Y. 1967a; Jabbar, A. 1984a; Kheyri, M. 1969a; Theron, J.G. 1979a.

Combretaceae: Knight, W.J. 1973c.

Compositae: Batra, S.W.T. 1979a; Frick & Hawkes 1970a; Ingram, B.F. 1969a; Theron, J.G. 1984a, 1984b.

Convolvulaceae: Habeck, D.H. 1960a; Ishihara, T. 1978a.

Cucurbitaceae: Howe & Rhodes 1976a.

Cyperaceae: Dabek, A.J. 1979a.

Elaeagnaceae: Emeljanov, A.F. 1964a.

Ericaceae: Gunthart, H. 1971c; Hopkins & Johnson 1984a; MacNay, C.G. 1961a; Wheeler & Walley 1980a.

Euphorbiaceae: Jayaraj, S. 1966e, 1968a.

Fagaceae: Mendes, M.A. 1959a.

Gramineae: Ammar et al. 1983a; Athwal et al. 1971a; Avesi & Khush 1984a; Choi & Lee 1976b; Chu & Liou 1981a; Chu et al. 1981a; Coupe & Schulz 1968a; Cwikla & Blocker 1981b; Denno, R.F. 1977a, 1980a; Gamez, R. 1983b; Gamez & Leon 1983a, 1985a; Garg & Sethi 1980a; Genung & Mead 1969a; Ghauri, M.S.K. 1961a, 1962a, 1974b, 1981a; Gromadzka, J. 1970b; Gyorffy & Pollak 1983a; Hardee et al. 1963a; Harizanov, A. 1970a; Hashmi et al. 1983a; Hawkins, J.A. 1979a; Hawkins et al. 1979a; Hill, M.G. 1976a, 1982a; Hinckley, A.D. 1963a; Hunang & Shichen 1981a; Ishihara, T. 1976a; Israel & Misra 1968a; Jabbar & Ahmed 1974a, 1975a, 1976a, 1980a; Jabbar, Ahmed & Ahmed 1977a; Jabbar, Ahmed & Samad 1977a, 1980a; Janjua, N.A. 1957a; Kalandadze et al. 1954a; Kalode, M.B. 1983a; Kathirithamby, J. 1981a; Kelly & Klostermeyer 1984a; Kerr, T.W. 1957b; Khaire & Bhapkar 1971a; Kharizanov, A. 1970a; Khurana et al. 1974a; King, T.H. 1968a; Kira et al. 1978a; Klimaszewski, Wojciechowski, Gebicki et al. 1980a; Kobayashi et al. 1973a; Kramer & Whitcomb 1968a; Kuoh, C.L. 1983a; Lange & Grigarick 1970a; Leeuwangh, J. 1968a; Leeuwangh & Leuamsang 1967a; Lodos, N. 1981a; MacRae & Yonke 1984a; Mishra et al. 1973a; Nault, L.R. 1985a; Nault & DeLong 1980a; Nault et al. 1983a; Patel, R.K. 1976a; Pathak, P.K. 1983a; Pitre et al. 1966a; Prestidge & McNeill 1983a; Sogawa & Saito 1981a, 1983a; Sohi et al. 1980a; Watts, J.G. 1963a; Whitcomb, R.F. 1958a; Whitcomb, Coan et al. 1973a; Whitmore et al. 1981a; Wilson et al. 1973a.

Guttiferae: Johansson, S. 1962a.

Juncaceae: Ali, A.M. 1978a.

Labiatae: Gould, G.E. 1960a.

Leguminosae: Beccari, F. 1970a; Bergman, B.H.H. 1956a; Betsch, W.D. 1978a; Bonnefil, L. 1972a; Ghauri, M.S.K. 1975a, 1978a, 1985b; Gist et al. 1968a; Gunthart & Wanner 1981a; Guppy, J.C. 1958a; Gyrisco, G.G. 1958a; Gyrisco et al. 1978a; Hagel, G.T. 1969a; Harakly & Assem 1978a; Helaly et al. 1985a; Herms, D.A. 1984a; Hewitt & Burleson 1976a; Jayaraj & Seshadri 1967a; Joplin, C.E. 1974a; Kerr, T.W. 1957a; Kerr & Stuckey 1956a; Kincade et al. 1970a; Koblet-Gunthardt, M. 1975a; Korytkowski & Torres 1966a; Kouskolekas, C.A. 1964a; Kuklarni & Saoji 1961a; Lal, S.S. 1985a; Landis & Hagel 1969a; Luckmann, W.H. 1971a; Mahmood et al. 1969a; Malik et al. 1979a; Quisenberry et al. 1983a; Tugwell et al. 1973a; Valley & Wheeler 1985a; Ward et al. 1977a; Wheeler & Walley 1980b.

Malvaceae: Allen, A.A. 1982a; Annappan et al. 1965a; Ghauri, M.S.K. 1964c; Hassanein et al. 1971a; Klein, M. 1984a; Matthews & Tunstall 1967a.

Moraceae: Ahmed & Malik 1972a; Ghauri, M.S.K. 1963e.

Pistaciaceae: Alkan, B. 1957a.

Polygonaceae: Garneau, A. 1984a.

Rosaceae: Ahmed & Waheed 1971a; Klein, M. 1984a; Le Quesne, W.J. 1964e; MacNay, C.G. 1955a, 1957a; Woodroffe G.E. 1971c; Wagner, W. 1964a.

Rubiaceae: Ghauri, M.S.K. 1964d.

Rutaceae: Ahmed, M. 1970d; Ghauri, M.S.K. 1963c; Hermoso de Mendoza & Medina 1979a; Hoelscher, C.E. 1966a; Jeppson & Carman 1960a; Kohno & Nagahama 1970a.

Scrophulariaceae: Lauterer, P. 1980b; Sankaran & Rao 1966a.

Solanaceae: Choudhury et al. 1983a; Gromadzka, J. 1970a; Harakly, F.A. 1974a; Herrera A., J.M. 1963a; Thomas & Boll 1977a; Thomas & Martin 1971a; Young & Cannon 1973a.

Sterculiaceae: Ghauri, M.S.K. 1969a; Lodos, N. 1969a, 1969b.

Tamaricaceae: Mahadevan & Rangarajan 1975a.

Theaceae: Ghauri, M.S.K. 1964a.

Umbelliferae: Davletishina & Radzivilovskaya 1965a.

Urticaceae: Ahmed, M. 1971d; Le Quesne, W.J. 1972a; Stiling, P.D. 1980c; Stilling, D. 1978a.

Vitidaceae: Ghauri, M.S.K. 1963a; Gunthart & Gunthart 1967a; Harizanov, A. 1969a; Iren, Z. 1976a; Kharizanov, A. 1969a; Klein, M. 1984a; Vidano, C. 1959c, 1962c.

INSECT COLLECTIONS

Anonymous 1965b (Illinois Natural History Survey); Anufriev, G.A. 1979a (Jacobi); Capco, S.R. 1960a (Baker); Carrillo et al. 1966a (National Institute of Agricultural Research, Mexico); Casale, A. 1981a (Spinola); Datta & Dhar 1984a (general); DeLong, D.M. 1977b (Osborn), 1979f (Stal, Spangberg); Dlabola, J. 1960a (Horvath), 1963c (Haupt); Emmrich, R. 1973a (Staatlichen Museum fur Tierkunde, Dresden); Freytag, P.H. 1971a (general); Gaedike, H. 1971a (Deutschen Entomologischen Institutes); Gibson & Carrillo 1959a (Oficina de Estudios Especiales, Mexico); Hamilton, K.G.A. 1976a (Provancher); Kamal, El Din M.H. 1964a (Ministry of Agriculture, Egypt); Lauterer & Schroder 1970a (Moravian Museum); Lee et al. 1976a (National Science Museum, Korea); Linnavuori, R. 1956a (Hungarian National Museum, European museums), 1956e (Stal, Spangberg); Mannheims, B. 1965a (Zoologischen Forschungsinstitut und Museum A. Koenig); McCabe & Johnson 1980a (New York State Museum); Mohamed Hamed, K. el Din 1964a (Ministry of Agriculture, Egypt); Schoenfeld, P. 1984a (Jacobi); Sohi, A.S. 1983a (Merino, Schumacher); Somadikarta et al. 1966a (Museum Zoologicum Bogoriense); Tanner & Harris 1969a (Brigham Young University); Theron, J.G. 1980a (Stal); Torres, B.A. 1960a (Berg); Vidano & Arzone 1976b (Spinola); Vilbaste, J. 1969a (Matsumura), 1973a (Flor), 1975a (Motschulsky), 1976a (Matsumura); Viraktamath, C.A. 1973b (Matsumura), 1979a (Matsumura), 1980c (Pruthi), 1981b (Distant); Wagner, W. 1968b (Zoologischen Staatsinstitutes und Zoologischen Museums Hamburg); Webb, D.W. 1980a (Illinois Natural History Survey); Webb, M.D. 1979a (Rambur); Weidner & Wagner 1968a (Zoologischen Staatsinstitutes und Zoologischen Museums Hamburg); Young, D.A. 1957b (Osborn), 1963a (Naturhistoriska Riksmuseet Stockholm), 1964a (Signoret), 1965b (Breddin), 1965c (British Museum), 1965d (Fabricius), 1974a (Museum National d'Histoire Naturelle, Paris); Young & Beier 1963a (Natural History Museum Vienna); Young & Lauterer 1964a (Jacobi), 1966a (Moravian Museum); Young & Nast 1963a (Schmidt); Young & Soos 1964a (Hungarian Natural History Museum); Zack, R.S. 1984a (James).

HOST PLANT RESISTANCE

Host Plants

General: Cantello & Sanford 1984a; Frankel & Hawkes 1975a; Guevara, C.J. 1966a; Horber, E. 1972a; Howe & Manglitz 1959a; Jayaraj, S. 1966a, 1976a; Kooner & Deol 1982a; Kuc, J. 1966a; Luginbill, P. 1969a; Painter, R.H. 1958a, 1968a; Saxena, K.N. 1969a; Schalk & Radcliffe 1976a; Sleesman, J.P. 1970a; Sprague & Dahms 1972a; Tingey, W.M. 1981a, 1985a, 1985b; Wakibe et al. 1983a; Webster, J.A. 1975a.

Alfalfa: Barnes et al. 1970a; Dudley et al. 1963a; Farrar & Woddworth 1939a; Hanson, C.H. 1969a; Hanson et al. 1963a, 1964a, 1965a, 1972a; Horber, E. 1976a, 1976b; Jarvis & Kehr 1964a, 1966a; Kehr, W.R. 1970a, 1970b; Kehr et al. 1967a, 1968a; Kehr & Manglitz 1984a; Kindler & Kehr 1970a, 1970b, 1974a; Kindler et al. 1973a; Moore, G.D. 1971b; Newton & Barnes 1965a; Nielson & Lehman 1980a; Nielson & Schonhorst 1965a; Pimentel & Wheeler 1973a; Roof et al. 1976a; Schillinger et al. 1964a; Soper et al. 1984a; Sorenson & Horber 1974a; Taylor, N.L. 1955a, 1956a; Webster et al. 1968a, 1968b; Wilson, M.C. 1958a; Wilson & Davis 1958a, 1962a; Wilson et al. 1955a; Wressell, H.B. 1960a.

Cacao: Bruneau de Mire & Lotode 1974a.

Carrot: Schultz & Chapman 1979a.

Castor: Dorairaj et al. 1963a; Jayaraj, S. 1966d, 1967a, 1967e, 1968a, 1969a; Jayaraj & Seshadri 1966a; Saini & Chhabra 1968a; Varisai & Doriraj 1967a.

Cotton: Agarwal et al. 1978a; Agarwal & Krishnananda 1976a; Ambekar & Kalbhor 1981a; Annappan, R.S. 1960a; Annappan et al. 1965a; Balasubramanian et al. 1977a; Evans, D.E. 1965a; Gulab et al. 1983a; Hosny & El-Dessouki 1968a; Joshi & Rao 1959a; Kamel & Farouk 1965a; Khan & Agarwal 1981a; Krishnananda & Agarwal 1979a, 1980a; Mathur & Bhandari 1978a; Mavi & Sidhu 1982a; Mehtre et al. 1981a; Pandya & Patel 1964a; Pollard & Saunders 1956a; Premusekar 1985a; Reed, W. 1974a; Sharma et al. 1981a; Sidhu & Dhawan 1980a; Singh et al. 1972a; Tidke & Sane 1962a.

Cranberry: Changler et al. 1947a; Wilcox, R.B. 1951a.

Eggplant (Brinjal): Bindra & Mahal 1981a; Mote, U.N. 1982a.

Flax: Martin et al. 1961a.

Forage: Fiori & Dolan 1981a; Horber, E. 1974a, 1974b; Manglitz & Gorz 1972a; Manglitz et al. 1976a; Manglitz & Jarvis 1966a; Quisenberry et al. 1983a; Quisenberry & Yonke 1981a, 1981b, 1981c.

Grapevine: Schvester et al. 1967a; Tracy, R.K. 1977a.

Groundnuts: Amin, P.W. 1983a; Campbell & Emery 1971a; Campbell et al. 1975a, 1976a; Smith et al. 1985a.

Legumes: Anonymous 1985c; Assa, A.D. 1976a; Chalfant, R.B. 1965a; Chhabra & Kooner 1981a; Cruz, C. 1976a; Enkerlin & Medina 1979a; Eskafi & Van Schoonhoven 1978a, 1981a; Galwey & Evans 1982a; Galwey et al. 1985a; Garcia et al. 1981a; Gates, D. 1945a; Guevara, C.J. 1957a; Gui, H.L. 1945a; Hohmann et al. 1980a; Karel & Malinga 1980a; Leon, M.J.R.de 1978a; Lyman & Cardona 1982a; Lyman et al. 1982a; Marfo, K.O. 1985a; Medina, R.M. 1974a; Miah & Husain 1981a; Moraes & Oliveira 1981a; Nangju et al. 1979a; Rachie et al. 1976a; Raheja, A.K. 1976a; Ram et al. 1984a; Raman, K.V. 1977a, 1978a; Regupathy et al. 1975a; Sagar & Mehta 1982a; Schoonhoven et al. 1985a; Simmons, Pass et al. 1984a; Simmons, Yeargan et al. 1984a; Singh, S.R. 1976a, 1977a; Singh et al. 1976a; Taylor, W.E. 1976a; Tissot, A.N. 1932a; Turnipseed & Sullivan 1976a; Wells et al. 1984a; Wolfenbarger, D.A. 1961a; Wolfenbarger & Sleesman 1959a, 1961a, 1961b, 1961c, 1963a.

Lettuce: Zalom, F.G. 1981a.

Maize (Corn): All et al. 1977a; Collins & Pitre 1969a, 1969b; Durant, J.A. 1968b; Hao & Pitre 1970a; Knoke et al. 1977a; Scott & Rosenkranz 1981a.

Mango: Nachiappan & Baskaran 1984a.

Melon: Kennedy et al. 1975a.

Okra (Bhendi): Kishore et al. 1983a; Mahal & Balraj 1982a, 1982b, 1982c; Mote, U.N. 1980a; Regupathy & Jayaraj 1972a, 1974a; Teli & Dalaya 1981a; Uthamasamy, S. 1969a, 1980a, 1985a; Uthamasamy & Subramaniam 1985a; Uthamasamy et al. 1973a, 1976a.

Ornamentals: El-Kifel et al. 1976a.

Potato: Abdel-Salam et al. 1972a; Gardner et al. 1945a; Klashorst & Tingey 1979a; Lauer & Radcliffe 1967a; Martinez & Ramos-Elorduy de Concini 1978a; Maughan, F.B. 1937a; Peterson, A.G. 1950a; Radcliffe & Lauer 1967a, 1968a; Sanford, L.L. 1973a, 1982a; Sanford et al. 1972a, 1984a; Sanford & Ladd 1979a; Sanford & Sleesman 1970a, 1974a, 1974b, 1975a; Schalk et al. 1975a; Sleesman, J.P. 1940a, 1945a; Stevenson, F.J. 1956a; Tingey & Plaisted 1976a.

Rice: Ando & Kishino 1981a; Anonymous 1985d, 1985e; Athwal & Pathak 1972a; Athwal et al. 1971a; Auclair et al. 1982a; Cabunagan et al. 1984a, 1985a; Chandramohan & Chelliah 1984a; Chang et al. 1975a; Chattopadhyay & Mukhopadhyay 1975a; Chelliah et al. 1981a; Chen, C.C. 1975a, 1979a; Chen, L.C. 1975a; Cheng & Pathak 1972a; Chiu, Lin et al. 1968a; Choi, S.Y. 1975a; Choi et al. 1976a; Choi, Song & Park 1973a; Choi, Song, Park et al. 1973a; Chou & Cheng 1971a; Dahal & Hibino 1985a; Dutt & Biswas 1979a; Feng, Y.X. 1981a; Fujimura & Somasundaram 1984a; Fukamachi, S. 1976a; Hashioka, Y. 1952a; Heinrichs, E.A. 1984a; Heinrichs, Medrano & Rapusas 1985a; Heinrichs, Medrano, Rapusas et al. 1985a; Heinrichs, Medrano, Sunio et al. 1982a; Heinrichs & Rapusas 1983a, 1984a, 1985a, 1985b; Hibino et al. 1983a; Imbe, T. 1981a; Karim, A.N.M.R. 1978a; Karim & Pathak 1982a; Kawabe, S. 1979a; Khan & Saxena 1985a, 1985b; Khush, G.S. 1973a, 1977a, 1980a; Khush & Beachell 1972a; Kim, K.C. 1978a; Kishino & Ando 1978a, 1979a; Kobayashi et al. 1983a; Koesnang & Rao 1980a; Koshihara, T. 1971a; Krishnaiah, K. 1975a; Krishnaiah & Bari 1977a; Lin et al. 1984a; Ling & Miah 1980a; Malabuyoc & Heinrichs

1981a; Maung, M.M. 1985a; Misra et al. 1983a; Mitra & John 1973a; Morinaka & Sakurai 1970a; Mukhopadhyay & Saha 1981a; Muniyappa & Raju 1981a; Muniyappa & Ramakrishnan 1976b; Nanda et al. 1976a; Oya & Sato 1980a, 1981a; Panda et al. 1984a; Park & Lee 1976a; Pathak, M.D. [undated], 1969a, 1971a, 1977a; Pathak et al. 1969a; Pathak & Saxena 1980a; Rahman & Hibino 1985b; Rapusas et al. 1985a; Rapusas & Heinrichs 1981a, 1982a, 1982b, 1985a; Razzaque & Heinrichs 1985a, 1985b, 1985c, 1985d; Sastry & Rao 1976a; Sato & Sogawa 1981a; Seetharaman & John 1981a; Sekar & Cheliah 1983a; Sekizawa & Ogawa 1980a; Shumiya et al. 1984a; Sidhwi & Khush 1984a; Singh, K.G. 1979a; Siwi, B.H. 1983a; Siwi & Khush 1977a; Siwi et al. 1978a; Sogawa & Saito 1981a, 1983a; Sonku, U. 1973a; Sridhar et al. 1978a; Srisamudh & Rejesus 1972a; Srivastava & Bisaria 1981a; Takita & Habibuddin 1985a; Vaithilingam et al. 1979a; Velusamy et al. 1975a; Venkatanarayanan, D. 1971a; Veronica & Kalode 1983a; Viswanathan & Kalode 1981a, 1984a; Wilde & Apostol 1983a; Yein et al. 1980a.

Sesamum: Regupathy & Jayaraj 1973b.

Sorghum: Tomeu & Moseley 1972a.

Sugarcane: Stevenson et al. 1972a.

Sunflower: Deshmukh & Akhare 1979a; Rogers, C.E. 1981a; Sethi et al. 1978a.

Tomato: Thomas & Martin 1971a; Young & Cannon 1973a.

Mechanisms of Resistance

General: Dahlman, D.L. 1965a; Dahlman & Hibbs 1967a; Dahlman et al. 1981a; Roof et al. 1974a; Saxena & Basit 1982b; Tingey, W.M. 1985a; Webster, J.A. 1975a.

Alfalfa: Horber et al. 1974a; Newton & Barnes 1965a; Roof, M.E. 1974a; Roof et al. 1976a; Shade et al. 1979a; Simonet, Pienkowski et al. 1979a; Taylor, N.L. 1956a.

Castor: Jayaraj, S. 1964a, 1965a, 1966b, 1966e, 1967b, 1968b.

Cotton: Agarwal & Krishnananda 1976a; Ambekar & Kalbhor 1981a; Balasubramanian & Gopalan 1978a, 1979a, 1981a; Bhat et al. 1981a; Evans, D.E. 1965a; Khan & Agarwal 1981a; Knight, R.L. 1952a; Mehtre et al. 1981a; Premusekar, 1985a; Sharma & Agarwal 1983a, 1983b; Sikka et al. 1966a; Tidke & Sane 1962a; Yadava et al. 1967a.

Eggplant (Brinjal): Srinivasan & Chelliah 1980a, 1981a; Subbaratnam et al. 1983a.

Forage: Manglitz et al. 1976a; Pieters, A.J. 1929a.

Groundnuts: Campbell & Emery 1972a;

Legumes: Bernard & Singh 1968a; Broersma et al. 1972a; Chhabra et al. 1981a, 1984a; Galwey, N.W. 1983a; Hartwig & Edwards 1970a; Lee, Y.I. 1983a; Pillemer & Tingey 1976a, 1978a; Raman et al. 1980a; Robbins et al. 1979a; Singh et al. 1971a; Turnipseed, S.G. 1977a; Wilde & Van Schoonhoven 1976a; Wolfenbarger & Sleesman 1961a.

Okra (Bhendi): Mahal & Balraj 1982a, 1982b, 1982c; Uthamasamy, S. 1985a; Uthamasamy et al. 1971a; Uthamasamy, Jayaraj et al. 1972a; Uthamasamy, Subramaniam et al. 1972a, 1973a, 1976a.

Potato: Hibbs et al. 1964a, 1964b; Orgell et al. 1959a; Raman et al. 1979a; Sanford et al. 1972a; Tingey & Gibson 1978a; Tingey & Laubengayer 1981a; Tingey et al. 1978a; Tingey & Sinden 1982a.

Rice: Athwal & Pathak 1972a; Athwal et al. 1971a; Auclair et al. 1982a; Avesi & Khush 1984a; Dutt & Biswas 1979a; Karim, A.N.M.R. 1978a; Kawabe, S. 1978a, 1985a; Malabuyoc & Heinrichs 1981a; Sidhwi & Khush 1984a; Sogawa & Saito 1983a; Venkatanarayanan, D. 1971a.

NATURAL ENEMIES

Parasites

Abdul-Nour, H. 1970a, 1971a, 1976a; Albrecht, A. 1980a; Ali, A.M. 1978a; Annecke, D.P. 1965a; Arcidiacono, S. 1965a; Arzone, A. 1974a, 1974b, 1974c; Badmin, J.S. 1970a; Bakkendorf, O. 1971a; Baldridge & Blocker 1980a; Baltazar, C.R. 1981a; Barrett & Westdal 1961a; Barrett et al. 1965a; Barrion & Litsinger 1983a; Bazlul Huw, S. 1985a; Bentur & Kalode 1980a; Bergman, B.H.H. 1957a; Blanchard, E.E. 1966a; Brooks, J.C. 1979a; Cantoreanu, M. 1971a; Carle, P. 1967a; Chandra, G. 1978a, 1980a, 1980b, 1980c; Chang, Y.D. 1980a; Chu & Hirashima 1981a; Claridge & Reynolds 1972a; Clausen, C.P. 1978a; Cobben, R.H. 1956a; Currado, I. 1983a; Currado & Olmi 1979a; Doutt, R.L. 1961a; Doutt & Nakata 1965a, 1973a; Doutt et al. 1966a; Emmrich, R. 1966a; Fiori, A. 1983a; Flock et al. 1962a; Freytag, P.H. 1976c, 1985a; Freytag & Back 1977a; George, J.A. 1959b; Ghai & Ahmed 1975a; Gomez & van Schoonhoven 1977a; Gourlay, E.S. 1964a; Greathead, D.J. 1983a; Hardy, D.E. 1971a; Henderson, C.F. 1955a; Hincks, W.K. 1963a; Hirashima, Y. 1981a; Hirashima et al. 1979a; Hirashima & Kifune 1978a, 1985a; Hoelscher, C.E. 1966a; Hower & Davis 1984a; Hsieh, C.Y. 1975a; Huffacker et al. 1954a; Hulden, L. 1984a; Huq, S. 1984a, 1985a; Jervis, M.A. 1978a, 1980a, 1980b, 1980c; Kaloostian, G.H. 1956b; Kapoor & Grewal 1984a; Kathirithamby, J. 1977a, 1979a; Kido et al. 1983a; Kifune, T. 1983a; Kim, J.B. 1984a; Kim & Kim 1984a; Kiritani et al. 1971a; Kiritani & Kono 1975a; Koizumi, K. 1959a, 1960a; Kumar et al. 1983a; Lauterer, P. 1981b, 1983a; Lin, K.S. 1974a; Lindberg, H. 1950a; Maillet, P.L. 1960a; Manjunath et al. 1978a; Marino de Remes Lenicov & Teson 1975a; Marletto & Maggiora 1983a; Marshall & Morgan 1956a; Maung, M.M. 1985a; Mbondji, P.M. 1984a; McKenzie & Beirne 1972a; Meyer & Bruyn 1984a; Meyerdirk & Hessein 1985a; Misra & Krishna 1982a; Misra et al. 1984a; Mitjaev, I.D. 1963a, 1963d; Miura, T. 1976a, 1976b, 1978a, 1979a; Miura et al. 1979a, 1981a; Moczar, L. 1979a; Mulla, M.S. 1956a; Muller, H.J. 1960a; Nair, K.K. 1983a; Napompeth & Tanagsnakod 1976a; Nayak & Srivastava 1979a; Nishida et al. 1976a; Olmi, M. 1975a; Orita, S. 1969a, 1972a; Otake, A. 1967a; Parker, H.L. 1967a; Pena & Shepard 1985a; Pierce, H.D. 1972a; Pizzamiglio, M.A. 1979a; Quezada, J.R. 1979a; Raatikainen, M. 1972a; Raatikainen & Heikinheimo 1974a; Rao, B.R.S. 1983a; Rao & Reddy 1982a; Remane & Schulz 1973a; Richards, O.W. 1972a; Sahad, K.A. 1982a, 1982b, 1982c; Sahad & Hirashima 1984a; Santa Cruz, S. 1965a; Sasaba & Kiritani 1972a; Schauff, M.E. 1981a; Seyedoleslami, H. 1978a; Seyedoleslami & Croft 1980a; Shibang, Q. 1983a; Shimada, K. 1972a; Singh & Parshad 1967a; Sperka & Freytag 1975a; Stiling, P.D. 1980b; Strubing, H. 1956a; Subba Rao, B.R. 1966a, 1983a, 1984a; Subba Rao et al. 1965a, 1968a; Taguchi, H. 1975a; Valentine, E.W. 1963a; Vidano, C. 1962b; Viggiani, G. 1981a, 1985a; Vungsilabutr, P. 1978a; Walker, I. 1979a; Waloff, N. 1974a, 1975a, 1980a, 1981a, 1982a; Wen & Lee 1980a; Whalley, P.E.S. 1956a, 1970a; Williams, D.W. 1984a; Wolfe, H.R. 1958b; Yang, O. 1955a; Yang et al. 1982a; Yano, K. 1979a; Yano et al. 1984a, 1985a; Yashiro, N. 1979a; Yasumatsu et al. 1975a; Yen, D.F. 1973a; Ylonen & Raatikainen 1984a; Zhong, L. 1983a.

Predators

Aitchison, C.W. 1984a; Alkhzishvili, T.V. 1953a; Arcidiacono, S. 1965a; Barrion & Litsinger 1981a, 1982a; Brooks, J.C. 1979a; Chang & Oka 1984a; Chen et al. 1974a; Chiu & Cheng 1976a; Chiu et al. 1974a; Chu & Hirashima 1981a; Chu, Ho et al. 1975a, 1976a; Chu, Lin et al. 1976a, 1976b, 1977a; Chu & Reid 1982a; Clausen, C.P. 1978a; Davidson & Landis 1938a; Deseo, V.K. 1959a; Esipenko, P.A. 1973a; Evans, H.E. 1968a; Flinn et al. 1985a; Garg & Sethi 1983a; Ghorpade, K.D. 1979a; Glen, D.M. 1975a; Greathead, D.J. 1983a; Gurdip & Sandhu 1976a; Hespenheide & Rubke 1977a; Hirashima, Y. 1981a; Hirashima et al. 1979a; Howell & Pienkowski 1971a; Hsieh, C.Y. 1975a; Hsieh & Dyck 1975a; Hurd & Eisenberg 1984a; Janvier, H. 1956a; Kajak et al. 1972a; Kang & Kiritani 1978a; Kawahara & Kiritani 1975a; Kido et al. 1983a; Kiritani & Kakiya 1975a; Kiritani & Kawahara 1973a; Kiritani, Kawahara et al. 1972a; Kiritani & Kono 1975a; Kiritani et al. 1971a; Krishnasamy et al. 1984a; Lauterer, P. 1981a; Lauterer & Bures 1984a; Lavallee & Shaw 1969a; Lim & Jusho 1979a; Lin, K.S. 1974a; Macphee, A.W. 1979a; Manjunath et al. 1978a; Manley, G.V. 1976a, 1977a; Marshall & Morgan 1956a; Martinez & Pienkowski 1982a, 1983a; Maung, M.M. 1985a; Misra et al. 1984a; Miyai et al. 1978a; Muir, R.C. 1966a; Nair, K.K. 1983a; Nakamura, K. 1977a; Napompeth & Tanagsnakod 1976a; O'Brien & Kurczewski 1982a; Okuma et al. 1978a; Pawar, A.D. 1975a; Perfect et al. 1983a; Pophaly et al. 1979a; Prior, R.N.B. 1972a; Raatikainen, M. 1972a; Raney & Yeargan 1977a; Rao, B.R.S. 1983a; Rao & Reddy 1982a; Rao et al. 1981a; Rensner et al.

1983a; Reyes & Gariel 1975a; Samal & Misra 1983a; Sasaba & Kiritani 1974a; Schmutterer, H. 1974a; Shibang, Q. 1983a; Somchoudhury, A.K. 1981a; Sorenson et al. 1976a; Subba Rao, B.R. 1983a; Subba Rao et al. 1965a, 1968a; Suzuki & Kiritani 1974a; Tanaka & Ito 1982a; Tandon & Lal 1983a; Tseng et al. 1976a; Waloff, N. 1980a; Wei et al. 1984a; Wheeler, A.G. 1977a; Whitaker, J.O. 1972a; Whittaker & Warrington 1985a; Wilson & Oliver 1969a; Yadava & Shaw 1968a; Yano et al. 1981a; Yasumatsu et al. 1975a; Yen, D.F. 1973a; Zhong, L. 1983a; Zhou & Tang 1982a, 1984a; Zhu & Zheng 1984a.

Pathogens

Aguda, Litsinger et al. 1984a; Balasubramanian & Mariappan 1983a; Balazy et al. 1980a; Ben-Ze'ev & Kenneth 1981a; Brooks, J.C. 1979a; Delius et al. 1984a; Devanesan et al. 1979a; Folliot & Maillet 1967a; Jensen, D.D. 1962a, 1968a, 1969b; Jensen et al. 1967a, 1972a; Kenneth & Olmert 1975a; Kenneth et al. 1971a; Kido et al. 1983a; Kiritani & Kono 1975a; Kuruvilla et al. 1980a; Li, H.K. 1985a; Marshall & Morgan 1956a; Mathai et al. 1979a; Mitjaev, I.D. 1963a, 1963d; Nayak & Srivastava 1979a; Rao, P.S. 1975a; Soper, R.S. 1985a; Srivastava et al. 1965a; Turian, G. 1960a; Whitcomb, Shapiro et al. 1966a; Yalovitasyn, M.V. 1962a; Yen & Tsai 1970a; Zang & Luo 1976a.

Scavengers

Samsinak & Dlabola 1980a.

NOMENCLATURE

Akingbohungbe, A.E. 1983a; Barrion & Litsinger 1981b; China, W.E. 1957a; Chopra, N.P. 1973a; Deitz, L.L. 1979a; Ghauri, M.S.K. 1983a; Heller, F.R. 1961a; Holthuis, L.B. 1960a; Kapoor & Sohi 1972a; Kocak, A.O. 1981a, 1981b, 1981c, 1983a; Laffoon, J.L. 1960a; Le Quesne, W.J. 1983b; Ling, K.C. 1973a; Oman, P.W. 1970a;

PEST STATUS

Crops

General: Abraham et al. 1970a; Ahmed, Naheed, Ahmed & Baluch 1980a; Ahmed & Ahmed 1980b; Akingbohungbe, A.E. 1983a; Ammar & Hosny 1969a; Anonymous undated, 1928a, 1957b, 1958a, 1960a, 1961b, 1961c, 1964a, 1966a, 1968b, 1968c, 1968e, 1968f, 1971b, 1972a, 1972b, 1973a, 1974b, 1974c, 1974d, 1975b, 1976b, 1976c, 1976d, 1978b, 1980a, 1981a, 1982a, 1984a, 1984b; Antova, Y.K. 1958a; Arakelian et al. 1984a; Armitage, H.M. 1955a; Arzone, A. 1972a; Asena, N. 1970a; Avidov, Z. 1961a, 1969a; Bairyamova, V. 1970a, 1970b, 1977a, 1982a; Batiashvili & Dekanoidze 1967a; Batra & Jangyani 1960a; Beirne, B.P. 1972a; Bertels & Baucke 1966a; Boulard, M. 1969a; Brown et al. 1959a; Chou et al. 1953a; Costa, A.S. 1965a; David, W.A.L. 1958a; DeLong, D.M. 1965a; Dobretsova, V.N. 1935a; Dogger, J.R. 1958a; El-Nahal et al. 1981a; Emeljanov, A.F. 1972a, 1972e; Fletcher, M.J. 1984a; Foster, J.A. 1982a; Gaborjanyi & Nagy 1972a; Ghauri, M.S.K. 1974c, 1979a; Ghauri & Onder 1980a; Giunchi, P. 1952-53a; Guagliumi, P. 1965a, 1965b; Gullyev, A. 1965a; Hall, W.J. 1956a; Harizanov, A. 1968a; Hoebeke, E.R. 1980a; Hongsaprug & Wilson 1985a; Huang & Lo 1964a; Huber & Osmun 1966a; Hudon et al. 1980a; Ishihara, T. 1965b, 1966b, 1967a, 1982a; Khan et al. 1964a; Kido, H. 1980a; Kim, J. 1963a; Knight, W.J. 1966b; Knowlton, G.F. 1958a; Kuoh, C.L. 1966a; Kwon, Y.J. 1985a; Langlitz, H.O. 1964a; Lattin & Oman 1983a; Lehker, G.E. 1957a; Lodos & Kalkandelen 1981a, 1982a, 1982b, 1983a, 1983b, 1983c, 1984a, 1984b, 1984c, 1984d, 1985a, 1985b, 1985c; Logvinenko, V.N. 1962c, 1973a; MacNay, C.G. 1950a, 1961b, 1961c; Mallamaire, A. 1954a; Meyer & Osmun 1970a; Miller, L.A. 1956b; Milliron, H.E. 1958a; Mitjaev, I.D. 1962b, 1963a; Monty, J. 1977a; Moretti, M.F.D. 1956a; Munro, J.A. 1954a; Musil, M. 1960a; Nowacka, W. 1978a; Nowacka & Adamska-Wilczek 1972a; Osmun, J.V. 1957a, 1958a, 1959a; Pedigo, L.P. 1972a; Peterson, G.D. 1957a; Pitre & Hepner 1967a; Pradhan, S. 1964a; Rivnay, E. 1962a; Rose, D.J.W. 1983a; Ruppel & DeLong 1956b; Schlottfeldt, C.S. 1944a; Shands, W.A. 1964a; Sohi, A.S. 1977b, 1977c; Squire, F.A. 1972a; Sridhar et al. 1972a; Stafford & Summer 1963a; Suehiro, A. 1961a; Szent-Ivany & Barrett 1960a; Theron, J.G. 1980c, 1981a; Tonks, N.V. 1963a; Usman, S. 1953a; Velikan et al. 1980a; Vidano, C. 1981a; Wan, M.T.K. 1969a; Whellan, J.A. 1964a; Wilson, K.J. 1964a; Wilson, M.R. 1985a; Young, D.A. 1956a; Zanol, K.M.R. 1980a.

Agave: Halffter, G. 1957a.

Alfalfa: Armbrust & Ruesink 1973a; Blair, B.D. 1975a; Blair & Niemczyk 1969a; Byers & Bierlein 1984a; Byers & Hower 1976a, 1976b; Christensen, C. 1982a; Faris et al. 1981a; Flinn, P.W. 1981a; Flinn & Hower 1984a; Gentsch, B.J. 1982a; Gist et al. 1968a; Kehr et al. 1970a; Kindler, Ogden et al. 1968a; Kindler et al. 1973a; Kouskolekas, C.A. 1964a; Medler, J.T. 1958c; Mohammad Zaky, S.H.F. 1982a; Muka, A.A. 1975a; Nielson & Currie 1962a; Nowacka, W. 1966a; Pimentel & Wheeler 1973b; Poston & Ogunlana 1973a; Smith & Medler 1959a; Smith & Franklin 1961a; Smith & Ellis 1982a, 1983a; Tanasijevic, N. 1964a; Weaver, C.R. 1954a; Wheeler, A.G. 1971a, 1974b; Wilson et al. 1955a, 1979a; Womack, C.L. 1984a; Zajac & Wilson 1983a.

Apple: Bennett, S.H. 1959a; Markelova, E.M. 1961a; Markelova, Y.M. 1962a; Matesova, G.I. 1960a; Vincent, J.J. 1975a; Yang, C. 1965a.

Apricot: Bonfils et al. 1974a.

Avocado: Wysoki & Izhar 1978a.

Bamboo: Yang, C. 1984a.

Banana: Nickel, J.L. 1962a.

Blueberry: Hopkins & Johnson 1984a.

Blackberry: Williams, D.W. 1982a, 1984a; Woodroffe, G.E. 1971c.

Cacao: Anonymous 1968a, 1969b; Boulard, M. 1969b, 1975a; Bruneau de Mire & Lotode 1974a; Ceballos, B.I. 1961a; Fernando, H.E. 1959a; Ghauri, M.S.K. 1969a; Lodos, N. 1969b; Salas & Hansen 1963a.

Cashew: Babu et al. 1983a.

Cereals: Ahmed & Jabbar 1972a, 1977a; Ahmed et al. 1973a; Anonymous 1971a, 1971d; Arzone & Vidano 1984a; Bogavac & Antonijevic 1964a; Chatelain, L. 1956a; Dlabola, J. 1960d; Dubovskij, G.K. 1964a; Gromadzka, J. 1970b; Harizanov, A. 1970a; Hashmi et al. 1983a; Kalandadze et al. 1954a; Kharizanov, A. 1970a; Khurana et al. 1974a; Merzheevskaia, O.I. 1955a; Nowacka, W. 1982a; Painter, R.H. 1955a; Patel, R.K. 1976a; Prabhakar et al. 1981a; Raatikainen, M. 1971a, 1972a; Raatikainen & Vasarainen 1971a, 1976a; Rawat & Sahu 1969a; Shurovenkov, Y.B. 1964a; Squire, F.A. 1972b; Tanasijevic, N. 1965a; Vidano & Arzone 1984c, 1985b.

Cherry: Evenhuis, H.H. 1955a; Fos, A. 1977a; Phillips, J.H.H. 1951a; Taboada et al. 1975a; Wilde, W.H.A. 1962b.

Citrus: Annecke, D.P. 1965a; Annecke & Mynhardt 1968a; Di Martino, E. 1956a; Ghauri, M.S.K. 1963c; Hermoso de Mendoza & Medina 1979a; Jeppson & Carman 1960a; Lee et al. 1981a; Silveira Neto et al. 1983a; Velimirovic, V. 1980a.

Coconut: Eskafi, F.M. 1982a; Ghauri, M.S.K. 1980c; Zelazny & Pacumbaba 1982a.

Coffee: Ghauri, M.S.K. 1964d.

Cotton: Abul-Nasr et al. 1969a; Afzal & Ghani 1953a; Alkhazishvili, T.V. 1953a; Askari & Hussain 1977a; Bailey, J.C. 1982a; Balasubramanian, G. 1979a; Baloch & Soomro 1980a; Bhat et al. 1984a; Bigi, F. 1953a; Butani, D.K. 1967a, 1975a, 1976a; Cadou, J. 1970a; Chakravarthy et al. 1985a; Couillaud & Daeschner 1971a; Deeming, J.C. 1981a; Dubovskij, G.K. 1964c; Duviard, D. 1973a; Evans, D.E. 1966a; Ghauri, M.S.K. 1963d, 1964c; Hassanein et al. 1971a; Legal, J. 1961a; Matthews & Tunstall 1967a; Moffitt, H.R. 1968a; Moffitt & Reynolds 1972a; Monsef, A. 1981a; Parsons, F.S. 1956a; Pearson, E.O. 1958a; Proctor, J.H. 1958a, 1961a; Rao & Rao 1984a; Ripper & George 1965a; Room & Wardhaugh 1977a; Senapati & Hohanty 1980a; Sidhu & Dhawan 1976a, 1980a, 1981a; Sohi, A.S. 1976b, 1983a; Sohi, G.S. 1964a; Subramanian, T.R. 1957a; Sweeney, R.C.H. 1961a; Szymkowski & Yepex 1963a; Zhou et al. 1981a.

Cranberry: Chandler et al. 1947a; MacNay, C.G. 1961a.

Cucurbits: Howe & Rhodes 1976a; Korolevskaya, L.L. 1973a; Rizk & Ahmed 1981a.

Eggplant (Brinjal): Ahmed, M. 1982b, 1985b; Baluch et al. 1980a; Kisha, J.S.A. 1981a.

Fig: Ghauri, M.S.K. 1963e.

Forage: App, B.A. 1960a; Beccari, F. 1970a; Blocker et al. 1972a; Childers & Dickson 1980a; Deseo, V.K. 1959a; Dubovskij, G.K. 1960a, 1960b, 1963a, 1964a; Genung, W.G. 1956a; Hagel, G.T. 1969a; Hardee et al. 1963a; Hawkins, J.A. 1979a; Hawkins et al. 1979a; Kerr, T.W. 1957a, 1957b; Kerr & Stuckey 1956a; Kira et al. 1978a; Landis & Hagel 1969a; Manglitz & Jarvis 1966a; Nagel, H.G. 1973a; Niemczyk & Guyer 1963a; Nowacka, W. 1982a; Obrtel, R. 1969a; Painter, R.H. 1955a; Quisenberry & Yonke 1981a, 1981b; Ram & Gupta 1976a; Ricou & Duval 1969a; Romankow, W. 1963a; Sachan, J.N. 1980a; Tanasijevic, N. 1963a; Waite, G.K. 1976b; Wallace, C.R. 1967a; Watts, J.G. 1963a.

Fruit: Ahmed, M. 1978a, 1982a, 1983a; Ahmed & Naheed 1982a; Ahmed, Naheed & Shafiq 1980a; Ahmed & Samad 1980a; Ahmed, Samad & Naheed 1981a; Andison, H. 1954a; Athwal et al. 1980a; Batiashvili & Dekanoidze 1967b; Bergmann, Moreti et al. 1984a; Dubovskij, G.K. 1965a, 1965b; Escalante et al. 1981a; Gambrell & Gilmer 1956a; Goble, H.W. 1967a; Hamilton, K.G.A. 1985d; Lehmann, W. 1973b, 1973c; MacNay, C.G. 1955a, 1957a; Massee, A.M. 1953a; Mitjaev, I.D. 1962a; Newcomer, E.J. 1966a; Nielson, M.W. 1949a; Nielson & Kaloostian 1956a; Samad & Ahmed 1979a, 1979b; Smol'yannikov, V.V. 1980a; Velikan et al. 1984a; Vereshchagina & Vereshchagina 1969a; Wadhi & Batra 1964a.

Grapevine: Abu Yaman, I.K. 1967a; Anonymous 1980b; Arakelian & Actvatsatryan 1977a; Batiashvili & Dekanoidze 1967b; Boller & Baggiolini 1970a; Boller et al. 1970a; Bonfils & Schvester 1960a; Bournier, A. 1977a; Carle, P. 1965a; Carle & Moutous 1966a; Gartel, W. 1965a; Ghauri, M.S.K. 1963a; Gunthart & Gunthart 1967a, 1967b; Harizanov, A. 1969a; Iren, Z. 1976a; Jensen et al. 1969a; Jubb et al. 1983a; Kharizanov, A. 1969a; Kido et al. 1984a; Monge Cassillas, J. 1981a; Nachev, P. 1960a; Schruft, G. 1962a; Schvester, D. 1972a; Schvester, Moutous, Bonfils et al. 1962a; Taschenberg, E.F. 1973a; Theron, J.G. 1982a; Velimirovic, V. 1966a; Vidano, C. 1958a, 1962a, 1962c, 1966b; Vidano & Arzone 1983a; Williams, D.W. 1982a, 1984a.

Groundnuts: Amin, P.W. 1982a; Bastos Cruz et al. 1962a; Bergman, B.H.H. 1956a; Ellis, C.R. 1984a; Ellis & Roy 1980a; Kuklarni & Saoji 1961a; Mercer, P.C. 1977a; Passlow, T. 1969a; Rose, D.J.W. 1962b.

Hazel: Viggiani, G. 1971a, 1973a.

Hop: Lehmann & Schmidt 1969a; Mitjaev, I.D. 1968d.

Legumes: Altieri et al. 1978a; Anonymous 1957c, 1983a; Ballantyne, B. 1971a; Blickenstaff & Huggans 1962a; Broersma, D.B. 1968a; Butt & Ahmed 1981a; Butt et al. 1981a; Cavalcante et al. 1975a; Cress & Wells 1976b; Deitz et al. 1976a; Dhuri & Singh 1983a; Dubovskij, G.K. 1962a; Eckenrode & Ditman 1963a; Garcia et al. 1979a; Hagel & Hampton 1970a; Hammad, S.M. 1978a; Helaly et al. 1985a; Joplin, C.E. 1974a; Kincade et al. 1970a; Korytkowski & Torres 1966a; Lal, S.S. 1985a; Leite Filho & Ramalho 1979a; Luckmann, W.H. 1971a; Miranda Colin, S. 1971a; Moraes et al. 1980b; Moraes & Ramalho 1980a; Murguido & Beltran 1983a; Murguido & Ruiz 1982a; Ogunlana & Pedigo 1974a, 1974b; Oliveira et al. 1981a; Parh, I.A. 1983a, 1983c; Pedigo, L.P. 1974a; Peter Ooi, A.C. 1973a; Peterson, A.G. 1971a; Poston & Ogunlana 1973a; Rachie et al. 1976a; Raney & Yeargan 1977a; Rangaiah & Sehgal 1984a; Ruppel & Idrobo 1962a; Sales, F.M. 1979a; Schoonhoven & Cardona 1980a; Sharma et al. 1968a; Singh et al. 1975a, 1976a; Singh, K.M. & Singh, R.N. 1978a; Singh, R.N. & Singh, K.M. 1978a; Srivastava et al. 1972a; Tugwell et al. 1973a; Turnipseed & Kogan 1976a; Vidano, C. 1960b; Whitfield & Ellis 1977a; Wilson & Genung 1956a; Yadav & Yadav 1983a.

Maize (Corn): Ahmed & Jabbar 1972a, 1977a; Altieri et al. 1978a; Bogavac, M. 1968a; Bushing & Burton 1974a; Bushing et al. 1975a; Douglas et al. 1966a; Dubovskij, G.K. 1962c; Durant, J.A. 1968a, 1968b; Durant & Hepner 1969a; Elerdashvili & Dekanoidze 1961a; Emeljanov, A.F. 1960a; Escalante et al. 1981a; Estrada, R.F.A. 1960a; Ghauri, M.S.K. 1961a, 1975b, 1981a; Guthrie, E.J. 1978a; Lodos, N. 1981a; Nault, L.R. 1984b; Painter, R.H. 1955a; Pitre & Hepner 1967a; Rajagopal & Channabasavanna 1975a; Twine, P.H. 1971a; Vaszai-Virag, E. 1984a; Vidano & Arzone 1984b.

Mango: Ahmed, Baluch, Ahmed & Shaukat 1980a; Ahmed, Ahmed et al. 1981a; Ghauri, M.S.K. 1967a; Hongsaprug, W. 1984a; Hongsaprug & Wilson 1985a; Tandon & Lal 1983a; Viraktamath & Viraktamath 1985a.

Mint: Ozer, M. 1964a.

Miscellaneous: Ahmed & Jabbar 1970a; Ahmed & Ahmed 1980a; Baloch & Khan 1977a; Bertels, A. 1962a; Bigornia, A.E. 1963a; Blaine et al. 1982a; Ghauri, M.S.K. 1975a, 1985b; Malik et al. 1979a; Mitjaev, I.D. 1960a; Moiseeva, N.V. 1963a; Moore, K.M. 1965a; Moreau & Leclant 1973a; Moutous & Fos 1975a; Pinto & Frommer 1980a; Theron, J.G. 1978a; Vidano & Arzone 1976a, 1978a.

Okra: Ahmed, M. 1983c; Ahmed & Waheed 1984a; Dhamdhere et al. 1985a; Rawat & Sahu 1975a; Srinivasan & Krishnakumar 1983a; Srinivasan et al. 1981a.

Ornamentals: Albouy, J. 1966a; Andison, H. 1954a; Bohm, O. 1963a; El-Dine & Rizkallah 1971a; Goble, H.W. 1967a; Hamilton, K.G.A. 1985d; Lomakina, L.G. 1963a, 1967a; Nikolova, V. 1969a; Sharonova, M.V. 1969a; Vereshchagina, V.V. 1962a.

Papaw: Ghauri, M.S.K. 1974a.

Peach: Nielson, M.W. 1979d; Purcell & Suslow 1984a; Taboada et al. 1975a.

Pear: Ghauri, M.S.K. 1975c; Purcell & Suslow 1984a.

Pecan: Wysoki & Izhar 1978a.

Pistachio: Alkan, B. 1957a; Ozer, M. 1958a; Wagner, W. 1958c; Wagner & Duzgunes 1960a;

Plum: Natchev, P. 1965a.

Pomegranate: Dubovskij, G.K. 1962b.

Potato: Abdel-Salam et al. 1972a; Ahmed, M. 1984b; Ahmed, Ahmed et al. 1984a; Arestegui, P.A. 1976a; Cancelado et al. 1976a; Choudhury et al. 1983a; Cress & Wells 1976a; El-Saadany & Abdel-Fattah 1980a; Escalante et al. 1981a; Gromadzka, J. 1970a; Herrera, A.J.M. 1963a; Hofmaster, R.N. 1959b, 1959c; Martinez, M.R. 1975a; Martinez & Ramos-Elorduy de Concini 1978a; Maughan, F.B. 1937a; McCarthy, H.R. 1956a; Nagaich, B.B. 1975a; Prasad, S.K. 1957a; Radcliffe, E.B. 1982a; Rossiter, P.D. 1975a; Schultz, J.T. 1976a.

Raspberry: Mitjaev, I.D. 1960b; Reitzel, J. 1964a, 1971a; Woodroffe, G.E. 1971c.

Rhubarb: Garneau, A. 1984a.

Rice: Ahmed et al. 1970a; Ahmed & Samad 1972a; Alam, M.Z. 1967a; Alam, S. 1971a, 1977a; Alam & Alam 1979a; Alam & Islam 1959a; Alghali & Domingo 1982a; Anonymous 1965a, 1966b, 1967a, 1967b, 1968d, 1970c, 1970d, 1971c, 1972c, 1976a, 1985b; Banerjee, S.K. undated; Banerjee, S.N. 1964a; Bergmann, Ramiro et al. 1984a; Bhalla & Pawar 1975a; Birat, R.B.S. 1963a; Butani & Jotwani 1976a; Cendana & Calora 1967a; Cheng, C.H. 1976a, 1979a; Claridge, M.F. 1980a; Claridge & Wilson 1981c; Dean, G.J.W. 1976a; Descamps, M. 1956a; Dinther, J.B.M. van 1963a; Diwakara, M.C. 1975a; Dubovskij, G.K. 1962d; Dyck et al. 1981a; Fernando et al. 1954a; Fujimura & Somasundaram 1984a; Garg & Sethi 1980a; Ghauri, M.S.K. 1962a, 1974b; Ghosh & John 1979a; Grigarick, A.A. 1984a; Hinckley, A.D. 1963a; Hunang & Shichen 1981a; Israel & Misra 1968a; Janjua, N.A. 1957a; Kabir & Choudhury 1975a; Kalode, M.B. 1983a; Kathirithamby, J. 1981a; Kim & Lamey 1973a; King, T.H. 1968a; Kisimoto, R. 1984a; Krishnasamy et al. 1984a; Kuno, E. 1984a; Kuoh, C.L. 1964a, 1983a; Lange & Grigarick 1970a; Leeuwangh, J. 1968a; MacQuillan, M.J. 1975a; Manjunath & Urs 1979a; Mishra, U.S. 1977a; Misra, B.C. 1980a; Misra & Israel 1968a, 1970a; Misra et al. 1984a; Mitra & John 1973a; Mokrotovarov, S. 1965a; Nasu, S. 1967a; Ngoan, N.D. 1971a; Nguyen, C.T. 1982a; Nishida et al. 1976a; Olmi, M. 1968a; Otake & Hokyo 1976a; Ou, S.H. 1972a; Ou & Ling 1966a; Ou & Rivera 1969a; Paik, W.H. 1967a; Pathak, M.D. 1967a, 1968a; Pawar & Bhalla 1967a; Rao & Halteren 1976a; Rapusas et al. 1985a; Shagir et al. 1983a; Singh, K.G. [undated]; Singh et al. 1977a; Siva Shankara Sastry, K.S. 1958a; Soehardjan, M. 1971a, 1973a; Sohi, A.S. 1983a; Son, B.I. 1973a, 1973b; Soto et al. 1976a; Soto & Siddiqui 1978a; Suenaga & Nakatsuka 1958a; Suhardjan et al. 1973a; Tarafder et al. 1980a; Trinh, T.T. 1980a; Upadhyay, R.K.

1984a; Van Dinther, J.B.M. 1963a; Van Halteren, P. 1979a; Van Halteren & Sama 1973a; Wan, M.T.K. 1972a; Way et al. 1982a, 1984a; Wilson, M.R. 1983a; Wilson & Claridge 1985a; Yunus & Rothschild 1967a; Yusof, O.H. 1982a.

Safflower: David & Janagarajan 1969a; Selim, A.A. 1977a.

Sesame: Abraham et al. 1973a; Nath, D.K. 1975a; Ozer, M. 1964a.

Strawberry: Mitjaev, I.D. 1960b; Taksdal, G. 1977a.

Sugarbeet: Chrzanowski, J. 1955a; Douglass, J.R. 1954a; Douglass et al. 1956a; Flock & Deal 1959a; Fullerton, D. 1964a; Hills & Brubaker 1968a; Hills et al. 1963a; Kheyri, M. 1969a; Meyerdirk & Hessein 1985a; Minoranskiy & Logvinenko 1968a; Reynolds et al. 1967a; Zverezomb-Zubovskii, E.V. 1957a.

Sugarcane: Agarwal & Siddiqi 1964a; Ananthanarayanan & Abraham 1955a; Anonymous 1976e; Bharadwaj & Yadav 1980a; Buzacott, J.H. 1953a; Carnegie, A.J.M. 1976a; Costilla et al. 1971a; Garcia et al. 1979a; Ishihara, T. 1976a; Kira et al. 1978a; Rao, V.P. 1965a; Sastry, K.S.S. 1957a, 1957b; Tripathi et al. 1978a.

Sunflower: Deshmukh & Akhare 1979a; Ingram, B.F. 1969a; Mohammad et al. 1980a; Nowacka & Bielejewski 1978a; Rajamohan et al. 1974a; Sattar et al. 1984a; Sethi et al. 1978a; Silveira-Guido & Carbonell Bruhn 1965a.

Sweet Potato: Ishihara, T. 1978a.

Tea: Ghauri, M.S.K. 1964a.

Trees: Aibasov, Kh.A. 1974a; Brown & Eads 1965a, 1969a; De Silva, M.D. 1961a; Furniss & Carolin 1977a; Ghauri, M.S.K. 1978a; Herms, D.A. 1984a; Ivliev & Kononov 1966a; Lauterer, P. 1980b; Mahmood et al. 1969a; Mendes, M.A. 1959a; Mitjaev, I.D. 1970a; Ossowski, L.L.J. 1957a; Razvyazkina, G.M. 1955a; Schindler, U. 1960a; Sharma & Malhotra 1981a; Tsai & Mead 1982a; Vidano, C. 1961d; Vidano & Arzone 1981a.

Truck crops: El-Kady et al. 1974b; Habib et al. 1975a, 1975b, 1976a; Harakly, F.A. 1974a; Khristova & Loginova 1975a.

Vegetables: Ahmed, M. 1978a, 1982a, 1983a, 1985a; Ahmed, Baluch, Ahmed & Ahmed 1980a; Ahmed & Naheed 1982a; Ahmed & Samad 1980a; Ahmed, Samad & Naheed 1981a; Apple & Falter 1962a; Butani & Jotwani 1983a; Dubovskij & Tursunkhodzhaev 1970a; Goble, H.W. 1967a; Greene, G.L. 1967a; Korolevskaya, L.L. 1971a, 1971b, 1973a; Lall, B.S. 1964a; Nowacka & Zoltanska 1974a; Ozer, M. 1964a; Prasad, S.K. 1957a; Samad & Ahmed 1979a, 1979b; Wressell, H.B. 1971a.

Walnut: Natchev, P. 1965a.

Human Nuisance

McCrae, A.W.R. 1975a.

Non-Pathogenic Damage

Ahmed & Naheed 1982a; Ahmed, Naheed, Ahmed & Baluch 1980a; Ahmed & Waheed 1984a; Ahmed & Ahmed 1980b; Ahmed, Ahmed, Baluch & Naheed 1984a; Aibasov, Kh.A. 1974a; Alam & Alam 1979a; Andrzejewska, L. 1966a; Askari & Hussain 1977a; Atwal et al. 1980a; Baggiolini, M. 1968a; Baggiolini et al. 1972a; Baggiolini, Canevascini, Tencalla et al. 1968a; Bailly-Maitre, A. 1955a; Barnes & Newton 1963a; Bianchi, F.A. 1955a; Bigornia, A.E. 1963a; Bushing & Burton 1974a; Bushing et al. 1975a; Chenon, R.D.de 1979a; Choudbury et al. 1983a; Ellis, C.R. 1984a; Faris et al. 1981a; Gardner & Cannon 1972a; Genung, W.G. 1956a; Hower, A.A. 1979a, 1979b; Hower & Byers 1977a; Hower & Muka 1975a; Jabbar & Ahmed 1980a; Jackson et al. 1965a; James & Granovsky 1927a; Jayaraj, S. 1966c, 1966d, 1967c, 1967d; Jayaraj & Seshadri 1966a; Jenkins & Smith 1977a; Kantack et al. 1960a; Kisha, J.S.A. 1981a; Kohno & Nagahama 1970a; Kouskolekas & Decker 1968a; Ladd, T.L. 1963a; Ladd & Rawlins 1965a; Lodos, N. 1969a; Loos, C.A. 1965a; Manglitz & Jarvis 1966a; Matthews & Tunstall 1967a; Maughan, F.B. 1937a; Mitjaev, I.D. 1968c, 1970a; Mohammad Zaky, S.H.F. 1982a; Mohammad et al. 1980a; Molina Valero & Viana 1970a; Moore, G.D. 1968a, 1971a; Moutous & Fos 1973a, 1975a; Murguido Beltran 1983a; Murisier & Jelmini 1985a; Naba, K. 1981a, 1982a, 1983a; N achiappan & Baskaran 1984a; Nakasuji & Nomura 1968a; Newton et al. 1970a; Nickel, J.L. 1962a; Nielson & Bleak 1963a; Ogunlana & Pedigo 1974a, 1974b; Parh, I.A. 1983b,

1983c; Pedigo, L.P. 1972a, 1974a; Peterson, G.D. 1957a; Pienkowski, R.L. 1970a; Pollard, D.G. 1968a; Prasad, S.K. 1957a, 1961a; Quisenberry & Yonke 1981c; Raman et al. 1978a; Rao & Anjaneyulu 1980a; Renard et al. 1982a; Ruppel & Idrobo 1962a; Rygg, T. 1981a; Salas & Hansen 1963a; Sankaran & Rao 1966a; Schindler, U. 1960a; Schmalscheidt, W. 1985a; Schoonhoven et al. 1978a; Seshadri & Seshu 1956a; Shurovenkov, Y.B. 1964a; Srivastava et al. 1972a; Takagi, K. 1981a; Theron, J.G. 1974a; Touzeau, J. 1968a; Tripathi et al. 1978a; Uchida & Ushiyama 1969a; Usman, S. 1953a; Van Dinther, J.B.M. 1963a; Vidano, C. 1959c, 1962c, 1963a, 1963b, 1963c, 1966a; Viggiani, G. 1971a; Walgenbach & Wyman 1985a; Weaver, C.R. 1954a; Whittaker, J.B. 1984a; Whittaker & Warrington 1984a, 1985a; Womack, C.L. 1984a; Yang, C. 1965a, 1984a; Zaky, S.J. 1981a.

Pest Management and Control

General: Ahmed, Manzoor 1973a; Allen et al. 1972a; Altieri et al. 1977a; Ameresekere, R.V.W.E. 1970a; Andrzejewska, L. 1978a; Anonymous 1958a, 1958b, 1959b; Asena, N. 1970a; Atkinson et al. 1956a; Baltazar, C.R. 1981a; Basu et al. 1982a; Bentur & Kalode 1980a; Broadbent, L. 1957a, 1969a; Brown et al. 1955a, 1957a, 1959a, 1960a; Burgaud & Cessac 1962a; Carle, P. 1964a, 1964b; Chapman, R.F. 1973a; Chiykowski, L.N. 1958a; Choi & Lee 1976a; Chu, Ho et al. 1976a; Chu, Lin et al. 1976a; Colomes, M. 1965a; David, W.A.L. 1958a; Deay, H.O. 1961a; Dirimanov & Karisanov 1964b; Ditman et al. 1957a; Duffus, J.E. 1983a; Eto et al. 1966a; Fluiter & van der Meer 1958a; Forsythe & Gyrisco 1961a; Freitag & Smith 1969a; Gaborjanyi & Saringer 1967a; Gunathilagaraj & Jayaraj 1979a; Hamilton, C.C. 1959a; Hamilton, C.D. 1957a; Hanna, M.A. 1970a; Harris & Maramorosch 1982a; Hashmi & Nizam-ud-Din 1978a; Haynes et al. 1957a; Hirata & Sogawa 1976a; Hofmaster, R.N. 1963a; Hower & Davis 1984a; Huang & Lo 1964a; Ionica & Grigorescu 1978a; Jensen, J.O. 1983a; Kaloostian, Oldfield, Gumph et al. 1979a; Kandoria & Haracharan 1984a; Kantack & Berndt 1972a; Kazano, H. 1983a; Kazano et al. 1969a, 1969b, 1970a, 1975a, 1978a, 1983a; Kiritani & Kawahara 1973a; Koyama, T. 1971a; Kurosu, Y. 1972a; Lacy et al. 1979a; Lim, G.S. 1969a; Lindsten, K. 1979a; Lo & Huang 1963a; Luginbill, P. 1969a; Maramorosch, K. 1982a; Martinez & Pienkowski 1983a; Matolcsy et al. 1968a; McBride, D.K. 1972a; Megenasa, T. 1971a; Meleshko, R. 1965a; Miller, L.A. 1960a; Napompeth & Tanagsnakod 1976a; Nielson, M.W. 1979d; Ouyang et al. 1984a; Perrin & Gibson 1985a; Rawlins & Glidden 1969a; Reynolds et al. 1957a; Reynolds, Stern et al. 1960a; Richter, F.T. 1970a; Riddell, J.A. 1982a; Ripper, W.E. 1956a; Rivnay, E. 1962a; Rose, D.J.W. 1983a; Saini & Cutkomp 1967a; Saint-Aubin et al. 1965a, 1966a; Saxena, K.N. 1981a; Saxena & Khan 1984a; Schultz, G.A. 1979a; Shibang, Q. 1983a; Singh, M.P. et al. 1984a; Sinha, R.C. 1979a; Sleesman, J.P. 1957a, 1958a; Stevenson & Pree 1985a; Taylor, J.H. 1971a; Tingey, W.M. 1981a; Urbauer & Pruess 1973a; Vora et al. 1985a; Wan, M.T.K. 1969a; Welbourn, W.C. 1983a; Westdal et al. 1959a, 1961a; Whitcomb et al. 1983a; Wolfe, H.R. 1956a, 1958a; Wu & Ruan 1982a; Ying, S.H. 1982a.

Alfalfa: Anonymous 1974a, 1978a; Armbrust & Ruesink 1973a; Blair, B.D. 1975a; Byers & Hower 1976a, 1976b; Byers et al. 1977a; Cuperus, G.W. 1982a; Cuperus & Radcliffe 1983a; Cuperus et al. 1983a; Davey & Manson 1958a; Edwards, C.R. 1974a; Fenton, F.A. 1959a; Forsythe et al. 1962a; Hill et al. 1969a; Huber & Giese 1973a; Huddleston & Gyrisco 1961a; Isenhour, D.J. 1985a; Judge et al. 1970a; Kehr et al. 1974a, 1975a; Kindler, Manglitz et al. 1968a; Koinzan & Pruess 1975a; Lamp, Barney et al. 1984a; Lamp et al. 1985a; Medler, J.T. 1956a; Medler & Brooks 1957a; Onstad et al. 1984a; Passlow & Waite 1969a; Pienkowski, R.L. 1970a; Pienkowski & Medler 1962a; Pruess et al. 1977a; Shaw & Zeiner 1964a; Steinhauer et al. 1962a; Waite, G.K. 1976b, 1978a; Walstrom, R.J. 1961a; Weaver, J.E. 1984a; Wilson, M.C. 1984a; Wilson et al. 1968a, 1979a; Wilson & Cleveland 1955a.

Apple: Abdel-Salam, F. 1967a; Asquith & Hull 1973a; Chiswell, J.R. 1964a; Clancy & McAlister 1956a; Gambrell & Gilmer 1960a; Garman, P. 1959a; Hamilton & Cleveland 1958a; Herne et al. 1973a; Markelova, Y.M. 1968a; Muir, R.C.

1966a; Nordlander, G. 1977a; Teulon & Penman 1984a; Trammel, K. 1974a; Tsugawa et al. 1966a.

Black currant: Pomazkov, Y.I. 1966a.

Cacao: Ceballos, B.I. 1961a; Fernando, H.E. 1959a.

Carrots: Eckenrode, C.J. 1973a; Fisher et al. 1980a; Henne, R.C. 1970a.

Cashew: Babu et al. 1983a.

Castor: Deshmukh et al. 1979a; Govindan et al. 1979a; Patel & Patel 1967a.

Celery: Freitag et al. 1962a.

Cereals: All et al. 1981a; Chatelain, L. 1956a; Chiang, H.C. 1977a; Choi, S.Y. 1973a; Chu et al. 1961a; Corella et al. 1969a; Grigorov, S. 1967a; Gyrisco & Armbrust 1966a; Mzira, C.N. 1984b; Radulescu & Munteanu 1970a; Rawat & Sahu 1969a; Rawat et al. 1980a; Ruppel & Janes 1970a; Westdal & Richardson 1972a.

Citrus: Ishiguro et al. 1971a; Uchida & Ushiyama 1969a.

Cotton: Agarwal et al. 1979a, 1983a; Ahmad & Shafl 1966a; Angelini & Vandamme 1965a; Atwal & Singh 1969a; Avtar Singh & Butani 1963a; Bailey, J.C. 1982a; Banerjee & Katiyar 1984a; Banerjee et al. 1977a; Barbosa, A.J. da S. 1954a, 1954b; Beingolea, G.O.D. 1957a, 1958a; Borle et al. 1980a; Butani, D.K. 1967a, 1975a, 1976a; Butani et al. 1977a; Butani & Sahni 1966a, 1966b; Butani & Surjit Singh 1965a; Chiang, H.C. 1977a; Couillaud & Aubertin 1970a; Couillaud & Daeschner 1971a; Dahiya & Singh 1982a; Dargan et al. 1968a; Deal, A.S. 1956a; Dilbagh & Singh 1978a; Gameel, O.I. 1974a; Gupta & Kavadia 1984a; Gurdip Singh et al. 1974a; Hanna, A.D. 1969a; Harcharan & Mann 1979a; Hassan et al. 1960a, 1975a; Hassanein et al. 1970a; Javaid, I. 1979a; Jayaraj & Venugopal 1964a; Jayaswal & Saini 1982a; Joginder Singh et al. 1982a; Joyce, R.J.V. 1955a, 1961a; Kandoria & Haracharan 1980a, 1982a, 1982b; Kareem et al. 1977a; Khan et al. 1971a; Krishen et al. 1966a; Lazarevic, B.M. 1970a; Legal, J. 1961a; Lindley, C.D. 1972a; Mabbett, T. 1980a; Moiz & Naqui 1971a; Parencia, C.R. 1968a; Patel et al. 1957a, 1961a, 1985a; Peswani et al. 1979a; Proctor, J.H. 1958a, 1961a; Rajendra & Teotia 1978a; Rao & Agarwal 1981a; Rao et al. 1984a; Rasmy & Hassib 1974a; Regupathy & Subramaniam 1980a; Reynolds & Deal 1956a; Ripper & George 1965a; Sandhu et al. 1978a; Sandhu & Harcharan 1975a, 1979a, 1980a; Sardar Singh et al. 1958a; Sidhu & Dhawan 1976a; Simons, J.N. 1958a; Tao, C.H. 1958a, 1960a; Thimmaiah, G. 1977a; Thimmaiah et al. 1980a; Toms & Goodman 1957a; Tunstall & Matthews 1966a; Van Der Laan, P.A. 1961a; Visvanathan & Abdul-Kareem 1983a; Wright, W.E. 1970a; Yein, B.R. 1981a.

Cranberry: Chandler et al. 1947a.

Cucurbits: Pareek & Noor 1980a; Shaheen, A.H. 1976a; Shaheen et al. 1973a.

Eggplant (Brinjal): Bindra, Khatri, Sohi, Deol et al. 1973a; Deol et al. 1978a; Deshpande et al. 1974a; Kisha, J.S.A. 1978a; Mohan et al. 1980a; Mote, U.N. 1978a, 1981a; Narayanasamy, Raghunathan et al. 1979a; Shah, Purohit et al. 1984a; Sohi et al. 1974a; Subbaratnam & Butani 1982a, 1984a; Tewari & Moorthy 1983a; Veeraval & Baskaran 1976a.

Forage: Byers & Jung 1979a; Cancelado & Yonke 1970a; Falk, J.H. 1982a; Grossmann, R.E. 1957a; Gyrisco, G.G. 1958a; Gyrisco & Armbrust 1966a; Hawkins, J.A. 1979a; Hawkins et al. 1979a; Kelly & Klostermeyer 1984a; Kerr & Stuckey 1956a; Manglitz et al. 1976a; Ruppel, R.F. 1975b; Ruppel & Janes 1970a; Sachan, J.N. 1980a; Tormala, T. 1977a.

Fruit: Downing, R.S. 1962a; Gambrell & Gilmer 1956a; Mal'kooskiy, M.P. 1956a; Raine & Tonks 1960a; Vereshchagina & Vereshchagina 1969a.

Grapevine: Aliniazee et al. 1971a; Carle, P. 1965b; Carle & Schvester 1964a; Caudwell, Schvester et al. 1972a, 1973a; Chiba, M. 1970a; Dekanoidze, G.I. 1962a; Flaherty et al. 1978a; Flores & Aguirre 1972a; Jensen, F.L. 1969a; Jensen, Lynn et al. 1965a; Jensen et al. 1961a; Lynn et al. 1965a; Malbrunot & Francois 1965a; Moutous & Carle 1965a; Moutous & Fos 1971a; Moutous et al. 1977a; Purcell, A.H. 1979c; Schvester, D. 1962a, 1970a; Schvester, Moutous et al. 1962b, 1963a; Sohi et al. 1975a; Stafford et al. 1960a; Stafford & Kido 1969a; Taschenberg, E.F. 1957a, 1973a.

Groundnuts: Arant, F.S. 1954a; Howe & Miller 1954a; Jenkins & Smith 1977a; Patel & Vora 1981a; Saboo & Puri 1978a.

Guava: Wen & Lee 1985a.

Legumes: Agyen-Sampong, M. 1976a; Altieri et al. 1977a; Anonymous 1957c, 1960c; Assa, A.D. 1976a; Benavides, M. 1955a; Bindra & Sagar 1976a; Bortoli & Giacomini 1981a; Bowyer & Atherton 1972a; Brett & Brubaker 1956a; Brown et al. 1959a, 1960a; Buchmeier & Edwards 1979a; Cardona et al. 1981a; Castro, G.D. 1971a; Cavalcante et al. 1975a; Chalfant, R.B. 1965a; Choi, S.Y. 1973a; Cisneros & Fausto 1959a; Cruz, C. 1975a, 1981a; Diaz, C.G. 1971a, 1973a; Ditman et al. 1954a; Ditman & Wiley 1958a; Eckenrode, C.J. 1981a; Eckenrode & Ditman 1963a; Eskafi & Van Schoonhoven 1981a; Ezueh, M.I. 1976a; Gatoria & Harcharan 1984a; Gonzalez B., J.E. 1959a, 1960a; Guyer et al. 1960a; Hammad, S.M. 1978a; Hernandez, J.C. 1982a; Hodjat, S.H. 1970a, 1972a; Hohmann et al. 1980a; Judge et al. 1970a; Lal, S.S. 1985a; Litsinger & Ruhendi 1984a; Mancia et al. 1973a; Martel, C.G. 1958a; Mayse, M.A. 1978a; Mishra & Saxens 1983a; Moraes et al. 1980a, 1980b; Murguido, C.A. 1982a, 1983a, 1983b; Murguido & Izquierdo 1982a; Nangju et al. 1979a; Peay, W.E. 1950a; Peay & Oliver 1964a; Peterson, A.G. 1971a; Petty & Bigger 1966a; Probst & Everly 1957a; Raheja, A.K. 1976a; Ratcliffe et al. 1960a; Regupathy et al. 1975a; Ruhendi & Litsenger 1982a; Sarmiento & Cisneroa 1966a; Satyavir 1983a; Schwartz et al. 1961a; Sherman & Tamashiro 1957a; Sherman et al. 1954a; Singh, S.R. 1976a; Srivastava, O.S. 1973a; Taylor, W.E. 1976a; Troxlair & Boethel 1984a; Turnipseed, S.G. 1972a; Turnipseed & Sullivan 1976a; Verma & Pant 1979a; Vyas & Saxena 1981a, 1983a; Wells et al. 1984a; Wolfenbarger, D.O. 1963a; Wressell & Miller 1960a; Yein, B.R. 1982a.

Lettuce: Hoffman, J.R. 1952a; Rawlins & Gonzales 1966a; Richardson & Westdal 1963a, 1964a; Strong & Rawlins 1958a, 1959a, 1959b; Thompson, L.S. 1967a; Thompson & Rawlins 1961a; Zalom, F.G. 1981a.

Maize (Corn): Ahmed & Jabbar 1974a; Akram & Yunis 1972a; All et al. 1977a, 1981a; Bhirud & Pitre 1972a, 1972b, 1972c; Brown et al. 1959a, Chiang, H.C. 1977a; Drinkwater et al. 1979a; Hernandez, J.C. 1982a; Kuhn et al. 1975a; Lindley, C.D. 1972a; Pitre, H.N. 1968a, 1968c; Rains & Christensen 1983a; Rose, D.J.W. 1973d; Rossel, H.W. 1984a; Ruppel & Janes 1970a; Sajian et al. 1982a; Sandhu et al. 1977a; Schwartz et al. 1961a; Tapia & Saenz 1971a; Van Rensburg & Walters 1978a.

Mango: Alam, M.Z. 1964a; Bindra et al. 1971a; Dakshinamurthy, A. 1984a; De & Dutta 1955a; Gandhale et al. 1975a; Hiremath, S.C. 1979a; Latif & Qayyum 1950a; Maheswariah, B.M. 1957a; Palo & Garcia 1935a; Patel & Hadli 1953a; Roy & Ram 1952a; Sarma et al. 1981a; Shah, Jhala et al. 1983a, 1984a; Shukla & Prasad 1984a; Sunder & Ali 1961a; Tandon & Lal 1979a; Thontadarya et al. 1978a; Uppal & Wagle 1944a; Wagle, P.V. 1934a; Wen & Lee 1978a; Yazdani & Mehto 1980a.

Mint: Gould, G.E. 1960a.

Miscellaneous: Baloch & Khan 1977a; Kaushik et al. 1978a; Li, H.K. 1982a; Raju, A.K. et al. 1983a; Rao et al. 1983a.

Mulberry: Chanoki et al. 1978a; Kariappa, B.K. 1978a.

Okra (Bhendi): Babu & Azam 1982a; Butani & Verma 1976a; Darshan et al. 1982a; Dewan et al. 1967a; Dhamdhere, Deole et al. 1981a; Dhamdhere, Singh et al. 1981a; Dixit et al. 1977a; Easwaramoorthy et al. 1976a; Faleiro et al. 1982a; Gayen, A.K. 1975a; Gulab & Chopra 1979a; Gupta & Dhari 1978a; Hasabe & Moholkar 1981a; Jat, N.R. 1981a; Jotwani et al. 1967a; Krishnaiah et al. 1976a; Mohan et al. 1983a; Nair et al. 1977a; Patel et al. 1980a; Rai et al. 1980a; Rawat & Jakhmola 1977a; Rawat & Sahu 1975a; Regupathy & Jayaraj 1972a, 1973a, 1974a; Sarma & Rao 1979a; Satpathy & Mishra 1969a; Sidhu & Simwat 1975a; Srinivasan & Krishnakumar 1983a; Uthamasamy & Balasubramanian 1978a.

Onion: Miyahara et al. 1983a.

Ornamentals: Anonymous 1956a; Barteneva, R.D. 1963a; Dirimanov & Karisanov 1964c; Freitag et al. 1954a; Groen & Slogteren 1974a; Morallo-Rejesus & Eroles 1980a; Sinha & Petersen 1972a.

Peach: Kaloostian & Pollard 1962a; Lacy, G.H. 1982a; McClure et al. 1982a; Mowry & Whalon 1984a; Schaefers & Brann 1974a; Sun, S.C. 1963a.

Pistachio: Tokmakoglu & Celik 1972a.

Plum: Pelet et al. 1968a.

Potato: Abdel-Salam et al. 1971a; Ahmed, Baluch et al. 1981a; Awate et al. 1978a; Awate & Pokharkar 1978a; Bacon, O.G. 1960a; Cancelado & Radcliffe 1979a, 1979b; Cancelado et al. 1976a; Ditman et al. 1961a; El-Saadany et al. 1976a; Fernando & Manickavasagar 1958a; Gerhardt & Turley 1961a; Graham et al. 1967a; Guevara, C.J. 1949a; Hofmaster, R.N. 1958a, 1959a; Hofmaster & Dunton 1961a; Hofmaster et al. 1967a, 1968a; Hofmaster & Waterfield 1965a; Knoke, J.K. 1962a; Libby & Hartberg 1972a; Linn et al. 1948a; Mavi & Harcharan 1975a; Misra & Lal 1981a, 1985a; Moore, D.H. 1959a; Patterson, R.S. 1962a; Patterson & Rawlins 1964a; Peterson, A.G. 1958a; Prasad, S.K. 1960a; Rioux, G. 1963a; Rossiter, P.D. 1975a, 1977a; Rygg, T. 1981a; Schwartz et al. 1961a; Sleesman, J.P. 1956a; Sleesman & Hedden 1958a; Trojanowski, H. 1963a; Walgenbach, J.F. 1982a; Walgenbach & Wyman 1984a, 1985a; Wressell & Driscoll 1964a.

Raspberry: Pomazkov, Y.I. 1966a; Reitzel, J. 1964a, 1971a; Van der Meer & Fluiter 1962a, 1970a.

Rice: Abe & Okamoto 1975a; Alam, M.Z. 1967a; Ananthanarayanan & Abraham 1956a; Anonymous 1977a, 1978d, 1980c; Arizono et al. 1971a; Bae & Pathak 1969a; Bang & Kae 1963a; Basilio & Heinrichs 1981a, 1981b, 1982a; Baskaran et al. 1976a; Bergonia, H.T. 1978a; Bhaktavatsalam & Anjaneyulu 1984a; Bowling, C.C. 1961a, 1970a; Burton et al. 1980a; Butani & Jotwani 1976a; Chand, P. 1984a; Chandra, G. 1980c; Chandramohan & Kumaraswami 1978a; Chen, C.C. 1978a; Cheng, C.H. 1976a, 1979a; Chiang, H.C. 1977a; Chiu & Cheng 1976a; Choi, S.Y. 1973a, 1976a; Choi et al. 1975a; Chowdhury & Alam 1979a; Chu, Ho et al. 1975a, 1976a; Chu, Lin et al. 1976a, 1976b, 1977a; Dahiphale et al. 1979a; Diwakara, M.C. 1975a; Dyck et al. 1981a; Endo & Masuda 1979a; Fabellar et al. 1981a; Fabellar & Mochida 1985a; Feng et al. 1979a; Fernando et al. 1954a; Fukuda, H. 1966a; Garg & Sethi 1982a; Ghosh et al. 1979a; Grigarick, A.A. 1984a; Gunathilagaraj & Jayaraj 1977a; Hama, H. 1975b, 1976a, 1977a; Hashizume, B. 1964a; Heinrichs, E.A. 1979a; Heinrichs & Arceo 1978a; Heinrichs et al. 1984a; Heyde et al. 1984a, 1985a; Hirai et al. 1968a; Hokyo, N. 1976a; Horimoto et al. 1976a; Hosoda & Fujiwara 1977a; Hsieh, C.Y. 1976a; Inoue, H. 1985a; Ishiguro & Saito 1970a; Ishiguro et al. 1971a; Ishihara & Edwards 1973a; Ishii, M. 1973a; Israel et al. 1968a; Iwata, T. 1981a; Iwaya & Kollmer 1975a; Jhoraku, T. 1966a; John, V.T. 1966a; Joshi, G. 1983a; Kamiwada et al. 1976a; Kato et al. 1979a; Kawahara et al. 1971a; Kazano et al. 1969a, 1978a, 1983a; King, T.H. 1968a; Kiritani, K. 1976a, 1977a, 1979a, 1983a; Kiritani, Inoue et al. 1972a; Kiritani et al. 1971a; Kiritani & Kono 1975a; Kittur et al. 1985a; Kobayashi, T. 1961a; Kojima & Ishizuka 1960a; Kono et al. 1975a, 1976a, 1982a; Ku & Wang 1978b; Kuno, E. 1984a; Kurata & Sogawa 1976a; Kyomura & Takahashi 1979a; Lim, G.S. 1972b, 1973a; Lindley, C.D. 1972a; Liu, Chen et al. 1983a; Liu & Chang 1979a, 1981a; Liu & Cheng 1980a; Maeda & Moriya 1972a; Maiti et al. 1980a; Mani & Jayaraj 1976b, 1976c; Mariappan et al. 1982a, 1983a; Mariappan & Saxena 1983a, 1983b, 1984a; Mishra & Shankar 1980a; Mishra & Kaushik 1976a; Mitra et al. 1970a; Miyahara et al. 1976a; Mochida & Valencia 1984a; Moriya, S. 1978a; Moriya & Maeda 1975a; Nagai et al. 1983a; Nakamura et al. 1971a; Nakasuga & Higuchi 1972a; Narayanasamy, Baskaran et al. 1976a; Okamoto, D. 1970a; Ozaki, Kassai et al. 1984a; Patange et al. 1981a; Pathak, M.D. 1966a, 1977a; Pathak, P.K. 1983a; Pathak et al. 1967a, 1974a; Pillai et al. 1983a; Rahman et al. 1985a; Rao & Anjaneyulu 1979b, 1980b; Ruan, Y.L. 1985a; Ryu et al. 1977a; Sakai et al. 1967a; Sakurai & Morinaka 1970a; Sam & Chelliah 1984c; Sanchez, F.F. 1977a; Sasaba & Kiritani 1975a; Sasaba et al. 1973a; Sasamoto et al. 1968a; Satapathy & Anjaneyulu 1982a, 1982b, 1982c, 1983a, 1985a; Satari, G. 1983a; Saxena & Khan 1985a; Shagir & Halteren 1976a; Shibuya, M. 1956a; Shimada & Araki 1976a; Shukla & Anjaneyulu 1980a, 1981a, 1981b, 1981c, 1982a, 1982b; Singh, S.R. 1967a; Sriharan & Garg 1976a; Srinivasan, S. 1980a, 1980b; Subramanian & Balasubramanian 1976a; Sun et al. 1975a; Takahashi, Y. 1979a; Takahashi & Kiritani 1973a; Takahashi et al. 1977a; Tarafder et al. 1980a; Thomas & John 1980a; Thontadarya & Devaiah 1975a; Tsuji & Fujita 1978a; Valencia et al. 1983a; Venkataraman et al. 1971a, 1973a; Wan, M.T.K. 1972a; Way, M.O. 1983a; Way et al. 1984a; Wilkins et al. 1984a; Wouters, L.J.A. 1963a; Yasumatsu et al. 1975a; Young, V.L. 1961a; Zhong, L. 1983a; Zhu & Zheng 1984a.

Sesame: Abraham et al. 1977a; Bindra, Khatri, Sohi & Deol 1973a; Venkatarao & Shanmugam 1983a.

Strawberry: Thompson et al. 1973a.

Sugarbeet: Ameresekere & Georghiou 1971a; Dorst, H.E. 1960a; Duffus, J.E. 1983a; Finkner & Scott 1972a; Flock, R.A. 1977a; Gibson & Fallini 1963a; Hallock & Deen 1956a; Harries & Valcarce 1957a; Hills et al. 1960a, 1964a; Jabbar, A. 1984a; Landis et al. 1970a; Reynolds, Fukuto et al. 1960a; Ritenour et al. 1970a; Wilson, H.L. 1955a, 1961a.

Sugarcane: Bhatia, G.N. 1972a; Costillo et al. 1971a; Pan, Y.S. 1981a.

Sunflower: Balasubramanian & Chelliah 1985a; Deshmukh et al. 1980a; Silveira-Guido & Carbonell Bruhn 1965a.

Tea: Chen, H.T. et al. 1978a.

Tobacco: Paddick & French 1964a, 1968a, 1972a.

Tomato: Agarwal & Kushwaha 1979a; Osmelak, J.A. 1984a.

Trees: Anonymous 1972d; Cathey et al. 1975a; Tashiro, H. 1973a.

Truck crops: Bronson & Rust 1951a; Khristova & Loginova 1975a.

Vegetables: Deay et al. 1959a; Granett & Reed 1960a; Greene, G.L. 1967a.

Pesticide Resistance

Anonymous 1970b, 1970c; Asakawa & Kazano 1976a; Chen, Chiang et al. 1978a; Eto et al. 1966a; Genung, W.G. 1957a; Gupta & Kavadia 1984a; Hama, H. 1975a, 1975b, 1976a, 1977a, 1980a, 1980b, 1983a, 1984a; Hama & Iwata 1971a, 1972a, 1973a, 1973b, 1978a; Hama et al. 1977a, 1979a, 1980a; Hama & Yamasaki 1981a; Hamilton & Fahey 1954a; Hanna, M.A. 1970a; Hayashi & Hayakawa 1962a; Hosoda & Fujiwara 1977a; Inoue, H. 1985a; Iwata, T. 1981a; Iwata & Hama 1971a, 1971b, 1972a, 1977a, 1981a; Jensen et al. 1961a; Kao et al. 1981a, 1982a; Kojima, Ishizuka et al. 1963a; Kojima, Kitakata et al. 1963a; Ku & Wang 1978a, 1981a; Kyomura & Takahashi 1979a; Lee & Yoo 1975a; Mitra et al. 1978a, 1982a; Miyata & Saito 1976a, 1978a, 1982a; Miyata, Saito et al. 1980a, 1981a; Miyata, Sakai et al. 1981a; Moriya & Maeda 1975a, 1976a; Motoyama et al. 1984a; Nagata, T. 1983a; Nagata & Mochida 1984a; Nakasuga & Higuchi 1972a; Ozaki, K. 1966a, 1983a; Ozaki & Kassai 1984a; Ozaki & Kurosu 1967a; Ozaki, Kassai et al. 1984a; Ozaki et al. 1966a; Ozaki, Saski et al. 1984a; Saito & Miyata 1982a; Shim, J.W. 1978a; Shimada, K. 1975a; Takahashi, Y. 1979a; Takahashi & Kiritani 1973a; Takahashi et al. 1977a; Trammel, K. 1974a; Tsuji & Fujita 1978a; Voss, G. 1983a; Wilson, H.L. 1961a; Yamamoto et al. 1977a, 1978a, 1983a; Yoshioka et al. 1975a.

PHYSIOLOGY

Anderson, B.L. 1984a; Babu, T.H. 1975a; Basden, R. 1966a; Berlin, L.C. 1962a; Berlin & Hibbs 1963a; Bhalani & Patel 1981a; Bhole & Srivastava 1979a; Brooks, M.A. 1985a; Carle, P. 1965b; Ghandhi, J.R. 1978a; Gouranton, J. 1967a, 1968a; Gouranton & Maillet 1966a; Gromadzka, J. 1971a; Gunthart, M. 1975a; Gustin, R.D. 1974a; Hama, H. 1975b, 1976a, 1977a, 1978a, 1980a, 1980b, 1983a, 1984a; Hama & Iwata 1971a, 1972a, 1973a, 1978a; Hama et al. 1977a, 1979a, 1980a; Hama & Yamasaki 1981a; Harris, K.F. 1981a; Hou & Brooks 1977a, 1978a; Hulbert & Schaller 1972a; Ishizaki, T. 1980b; Iwata & Hama 1971b, 1972a; Kamm & Swenson 1972a; Kazano, H. 1983a; Kazano et al. 1978a; Kirollos & Hibbs 1971a; Kisimoto, R. 1959a; Koblet-Gunthardt, M. 1975a; Lin & Hou 1980a, 1981a; Lindsay & Marshall 1981a; Lomakina et al. 1963a; Malabuyoc & Heinrichs 1981a; Manglitz et al. 1976a; Miles, P.W. 1972a; Moriya & Maeda 1976a; Munk, R. 1968a, 1968b; Noda et al. 1973a; Noguchi et al. 1968a; Norment et al. 1972a; Nuorteva, P. 1956a; Oman,

P.W. 1967a; Orenski et al. 1962a; Oya, S. 1980a; Pant & Gupta 1984a; Prestidge, R.A. 1982a; Sander, K. 1959a, 1960a, 1962a, 1967a, 1968a; Saxena, K.N. 1954b, 1979a; Smith & Littau 1960a, 1960b; Sogawa, K. 1967a, 1967b, 1968a, 1968b, 1973a; Sugiyama et al. 1971a; Tokumitsu & Maramorosch 1967a; Vidano & Arzone 1984a; Wei, L.Y. 1978a; Wei & Brooks 1979a; Yamamoto et al. 1977a, 1983a; Yoshimeki, M. 1966a.

PHYLOGENY

Blocker, H.D. 1979c, 1983a; Cobben, R.H. 1965a; Cunningham, H.B. 1962a; Cunningham et al. 1965a; Davis, R.B. 1970a, 1975a; Dlabola, J. 1958c; Emeljanov, A.F. 1977b; Emeljanov & Kuznetsova 1983a; Evans, J.W. 1957b, 1958a, 1960a, 1962a, 1963a, 1964a, 1969a, 1971a, 1972a, 1972c; 1975a; Halkka, O. 1960b; Hamilton, K.G.A. 1970b, 1972c, 1975b, 1983c, 1983d; Inoue, H. 1983a; Inoue et al. 1979a; Ishihara, T. 1959b; Johnson, D.H. 1970a; Knight, W.J. 1974c; Kramer, J.P. 1961b; Kukalova-Peck, J. 1978a; Lee & Kwon 1979a; Ling, K.C. 1968b; Mahmood & Ahmed 1968a; Mani, M.S. 1974a; Matsuda, R. 1963a, 1965a; Moore & Ross 1957a; Nault, L.R. 1985a; Nault & DeLong 1980a; Qadri, M.A.H. 1967a; Quartau, J.A. 1983c, 1984c, 1985b, 1985c; Ramakrishnan & Ghauri 1979a; Remane, R. 1984a; Remane & Koch 1977a; Remane & Schulz 1977a; Ross, H.H. 1953b, 1957b, 1958a, 1958b, 1963a, 1965a, 1965b, 1968a; Ross et al. 1965a; Ross & Hamilton 1970a; Tribout, P. 1956a; Triplehorn & Nault 1985a; Vilbaste, J. 1975b; Wagner, W. 1968a.

STRUCTURE

Morphology

Ahmed & Lodhi 1970a; Ammar, El-Desouky 1978a; Anufriev & Khokhlova 1981a; Arora & Singh 1962a; Bailly-maitre, A. 1955a; Barnett, D.E. 1977b; Becker, M. 1979a, 1981a; Bednarczyk, J. 1983a; Berlin, L.C. 1962a; Berlin & Hibbs 1963a; Bhattacharya, A.K. 1973e; Blocker, H.D. 1984a; Blocker & Triplehorn 1985a; Boulard, M. 1978a; Brar & Singh 1981a; Capel-Williams, G. 1978a; Carle & Amargier 1965a; Chen, J. 1964a; Chudzika, E. 1980a; Claridge & Nixon 1981a; Cobben, R.H. 1978b; Cunningham & Ross 1965a; Cwickla, P.S. 1984a; Cwikla & Freytag 1983a; Davoodi, Z. 1980a; DeLong, D.M. 1984b; Den Hollander, J. 1984a; Dirimanov & Karisanov 1964b, 1964c; Dlabola, J. 1956c; Dobreanu & Manolache 1969a; D'Urso & Ippolito 1984a; El-Kady et al. 1974a; Emeljanov, A.F. 1977b, 1981a; Emeljanov & Kusnetzova 1983a; Evans, J.W. 1957b, 1968a, 1969a, 1972a, 1973b, 1975a, 1975b; Forbes & MacCarthy 1969a; Forbes & Raine 1973a; Frediani, D. 1955a, 1956a, 1956b, 1958a; Gunthart, H. 1978a; Hamilton, K.G.A. 1970b, 1972c, 1975c, 1983c, 1983d; Helms, T.J. 1967a, 1968a; Hill, B.G. 1969a; Hoebeke & Wheeler 1983a; Hulbert & Schaller 1972a; Iwata, T. 1972a; Jabbar & Ahmed 1975a; Jansson, A.M. 1973a; Kathirithamby, J. 1974b, 1974c, 1976a, 1979a; Kawai, A. 1977a; Khokhlova & Anufriev 1981a; Kisimoto, R. 1959a, 1959c; Kukalova-Peck, J. 1978a; Lee & Kwon 1979a; Le Quesne, W.J. 1965d, 1978a; Le Quesne & Woodroffe 1976a; Liang, W.F. 1981a; MacRae & Yonke 1984a; Mahdihassan, S. 1970a; Maldonado-Capriles, J. 1985b; Malhotra & Sharma 1977c; Matsuda, R. 1963a, 1965a; Mayse, M.A. 1981a; Mazzone, P. 1976a; Mbondji, P.M. 1982a, 1982b, 1983a; Mishra, R.K. 1979a; Mochida, O. 1970a; Mukharji, S.P. 1959a, 1962a; Muller, H.J. 1960a; Munk, R. 1967a; Musgrave, C.A. 1974a; Nowacka, W. 1978b; Olmi, M. 1976a; Ossiannilsson, F. 1978a; Page, F.D. 1979a; Parh, I.A. 1982a; Patrick, C.R. 1970a; Pollard, D.G. 1972a; Pollard, H.N. 1962a, 1965b; Pollard & Yonce 1965a; Quartau, J.A. 1978a, 1980a; Ramachandra Rao, K. 1969a, 1973b; Ramakrishnan, U. 1983c; Ramakrishnan & Ghauri 1979a; Remane & Schulz 1977a; Ruppell & Romero 1972a; Sato & Sogawa 1981a; Savinov, A.B. 1983a, 1984a; Sawai Singh, G. 1971a; Saxena, K.N. 1954a; Saxena et al. 1985a; Shaw & Carlson 1979a; Shcherbakov, D.Ye. 1981a, 1982a; Simonet & Pienkowski 1980a; Siwi, S.S. 1979a, 1985a; Siwi & Roechan 1985a; Smith & Georghiou 1972a; Snodgrass, R.E. 1957a; Sogawa, K. 1965a; Sohi & Kapoor 1973a; Sohi & Sandhu 1971a; Srivastava, B.K. 1958a; Strubing, H. 1978a, 1981a; Thapa & Sohi 1984a; Theron, J.G. 1970b; Tribout, P. 1956a; Valley & Wheeler 1985a; Vidano, C. 1961c; Vilbaste, J. 1975b; Wagner, W. 1958a; Walter, S. 1978a; Ylonen & Raatikainen 1984a; Zabel, U. 1978a, 1980a.

Anatomy

Ahmad, I. 1983a; Ammar, El-Desouky 1985a; Arora & Singh 1962a; Backus, E.A. 1985a; Backus & McLean 1982a, 1983a; Balasubramamian, A. 1972a; Becker, M. 1979a; Bednarczyk, J. 1983a; Bharadwaj et al. 1966a; Bhola & Srivastava 1979a; Brar & Singh 1978a; Capel-Williams, G. 1978a; Carle & Amargier 1965a; Dobreanu & Manolache 1969a; D'Urso & Ippolito 1984a; Emeljanov & Kuznetsova 1983a; Gil-Fernandez & Black 1965a; Gouranton, J. 1967a, 1968b, 1968c; Gouranton & Maillet 1966a; Granados, R.R. 1969a; Granados, Hirumi et al. 1967a; Gunthart, H. 1977a; Gunthart, M. 1975a; Helms, T.J. 1967a, 1968a, 1968b; Hemmati, K. 1979a; Howse & Claridge 1970a; Katayama, E. 1975a; Koblet- Gunthardt, M. 1975a; Kunze, L. 1959a; Littau, V.C. 1960a; Matsuda, R. 1963a, 1965a; Mishra, R.K. 1979a; Mukharji, S.P. 1959a, 1962a; Munk, R. 1967a, 1967b; Nuorteva, P. 1956a; Prigent, J.P. 1961a, 1962a; Raine & Forbes 1971a; Remane & Schulz 1973a; Savinov, A.B. 1984a; Saxena, K.N. 1954a, 1955a; Shukla, S.P. 1961a; Sogawa, K. 1965a; Tribout, P. 1956a.

Ultrastructure

Ameresekere et al. 1971a; Ammar El-Desouky 1985a, 1985b; Balasubramamian, A. 1972a; Carle & Amargier 1965a; Cobben, R.H. 1965a; Da Cruz & Kitajima 1972a; Forbes & Raine 1973a; Giannotti, J. 1969b; Giannotti & Devauchelle 1969a; Gouranton, J. 1967a, 1968b, 1968c; Gouranton & Maillet 1966a; Granados, R.R. 1969a, 1972a; Granados & Meehan 1973a; Granados, Ward et al. 1967a, 1968a; Gunthart, H. 1977a; Hamon, C. 1969a, 1969b; Hamon & Folliot 1969a; Heady & Nault 1984a, 1985b; Hemmati, K. 1979a; Hirumi et al. 1967a; Hirumi & Maramorosch 1963b, 1969a; Houk & Griffiths 1980a; Howse & Claridge 1970a; Jackle & Schmidt 1978a; Kaiser, B. 1980a; Khan, A.M. 1975a, 1976a, 1977a; Kitajima & Landim 1972a; Korner, H.K. 1969a, 1969b, 1972a, 1974a; Littau, V.C. 1960a; Littau & Maromorosch 1956a, 1958a, 1960a; Lomakina et al. 1963a; Louis & Nicolas 1976a; Louis et al. 1976a; Mahdihassan, S. 1978a; Maillet & Folliot 1967a, 1967b, 1968a; Maillet et al. 1968a; Maramorosch et al. 1968a; Matsuzaki, M. 1975a; Mukharji, S.P. 1959a, 1962a; Munk, R. 1967a, 1967b, 1968c; Prigent, J.P. 1961a, 1962a; Raine & Forbes 1971a; Raine et al. 1976a; Saxena, K.N. 1955a; Schwemmler & Kemner 1984a; Schwemmler et al. 1971a; Shikata, E. 1979a; Silvere & Georgadze 1978a; Smith & Littau 1960b; Sogawa, K. 1965a; Tiivel, T. 1984a; Vogel, O. 1978a, 1982a, 1982b, 1983a, 1983b.

Development

Ammar, El-Desouky 1978a; Becker, M. 1979a; Capel-Williams, G. 1978a; Heinrichs & Rapusas 1984a; Helms, T.J. 1967a, 1968a; Hou & Brooks 1977a; Hulbert & Schaller 1972a; Katayama, E. 1975a; Kathirithamby, J. 1974c; Khan & Saxena 1985b; Kitajima & Landim 1972a; Korner, H.K. 1969a, 1969b; MacRae & Yonke 1984a; Malhotra & Sharma 1977c; Matsuzaki, M. 1975a; Mazzone, P. 1976a; Muller, H.J. 1960b, 1965a, 1973a, 1983a; Sander, K. 1959a, 1960a, 1962a, 1967a; Vogel, O. 1978a, 1982a, 1982b, 1983a, 1983b; Zabel, U. 1978a, 1980a.

TECHNIQUES

Collecting and Sampling

Alverson et al. 1977a; Ammar El-Desouky 1975a; Ammar & Farrag 1980a; Ammar et al. 1978a; Anonymous 1960b; Audras, G. 1959a; Beardsley, J.W. 1958a; Beirne, B.P. 1955a; Blocker & Reed 1976a; Bram et al. 1965a; Capinera & Walmsley 1978a; Chakravarthy & Rao 1985a; Chapman & Kingborn 1955a; Cherry et al. 1977a; Chhabra et al. 1976a; Coineau, Y. 1962a; Deay, H.O. 1961a; DeLong & Davidson 1936a; Diniz, M. de Assuncao 1964a; Faleiro et al. 1982a; Fenton & Howell 1957a; Fleischer & Allen 1982a; Fleischer et al. 1982a, 1983a; Fox, J.W. 1963a; Fraval, A. 1968a; Frost, S.W. 1956a, 1957a; Glick, P.A. 1957a, 1960a; Gressitt & Gressitt 1962a; Gressitt & Nakata 1958a; Gressitt et al. 1961a; Gu & Ito 1982b; Harakly & Shalaby 1981a; Harrell & Hoizapfel 1966a; Helm et al. 1980a; Imbe, T. 1981a; Johnson et al. 1957a; Kohno & Hashimoto 1978a; Krishnaiah & Ramachander 1979a; Lehmann, W. 1973b; Lim & Jusho 1979a; Lippold et al. 1977a; Luna et al. 1983a; Mabbett et al. 1984a; Mayse et al. 1978a; Medler, J.T. 1966a; Meyer & Colvin 1985a; Meyerdirk & Oldfield 1985a; Mitsuhashi, J. 1970a; Mohan & Janartharan 1985a, 1985b; Moustafa et al. 1985a; Muramatsu et al. 1970a; Nakamura et al. 1967a; Oliveira & Arauja 1979a; Parencia, C.R. 1968a; Payne, K. 1981b; Perfect et al. 1983a; Perron & Crete 1968a; Pruess et al. 1977a; Pruess & Whitmore 1976a; Purcell & Elkinton 1980a; Ramalho, F.S. 1978a; Ramalho & Albuquerque 1979a; Reling, D. 1983a; Reling & Taylor 1984a; Ross, H.H. 1956b, 1962a; Saroja, R. 1981a; Sasamoto et el. 1968a; Sathiyandandam & Subramanian 1982a; Saugstad et al. 1967a; Schafer & Taubert 1984a; Schuler, L. 1954a; Schwarz, R. 1959a; Shepard, M. 1984a; Shepard et al. 1985a; Simonet & Pienkowski 1977a, 1977b, 1979a, Simonet et al. 1978a; Simonet, Pienkowski, Martinez et al. 1979a; Smith & Ellis 1982a; Southwood & Pleasance 1962a; Sriharan & Garg 1975a; Stuben, M. 1973a; Sturani, M. 1948a; Takai et al. 1965a, 1972a; Theron, J.G. 1985b; Tormala, T. 1982a, 1983a; Townes, H. 1962a; Tsai & Mead 1982a; Twinn, D.C. 1964a; Upadhyay et al. 1981a; Vincent, J.J. 1975a; Walgenbach et al. 1985a; Wilde, W.H.A. 1962a; Yoshimoto & Gressitt 1960a, 1963a; Yoshimoto et al. 1962a; Zhou et al. 1981a; Zile Singh & Vaishampayan 1976a.

Disease Transmission

Allen & Donndelinger 1982a; Alverson, All & Bush 1980a; Alverson, All & Kuhn 1980a; Anderson et al. 1974a; Anjaneyulu, Singh et al. 1982a; Archer et al. 1982a; Auger & Shalla 1975a; Bantarri & Zeyen 1968a; Bolton et al. 1984a; Cabauatan & Ling 1979a; Caudwell et al. 1974a, 1977a; Caudwell & Larrue 1977a; Chaduneli & Chkheidze 1974a; Chen, S.X. 1985a; Chiu & Black 1967a; Chiu et al. 1966a; Chiykowski, L.N. 1983a; Clark & Whitcomb 1983a; Conner & Banttari 1979a; Daniels et al. 1973a; Davis et al. 1970a; Fraval, A. 1968a; Gamez & Black 1967a; Gao, D.M. 1983a, 1985a; Giannotti, J. 1972a; Giannotti et al. 1969a; Giannotti, Vago & Duthoit 1968a; Grylls & Waterford 1976a; Hafidi et al. 1979a; Herman & Maramorosch 1977a; Hibino & Cabuatan 1985a; Hirumi & Maramorosch 1963b, 1968a; Holdeman & McCartney 1965a; Hsieh & Roan 1967a; Hsu & Black 1974a; Kimura, I. 1980a, 1982a, 1985a; Kimura & Black 1971a, 1972a; Kiritani & Nakasuji 1977a; Kirkpatrick et al. 1985a; Kleinhempel et al. 1974a; Kunkel, L.O. 1954a; Lee, P.E. 1963a; Lee & Chiykowski 1963a; Lee, P.W. 1963a; Lehmann, W. 1982a; Ling et al. 1979a; Liu & Black 1978a; Long & Timian 1971a; Magyarosy, A.C. 1980a; Magyarosy & Sylvester 1979a; Maramorosch, K. 1955a, 1955b, 1956a, 1956c, 1960b, 1965a, 1965b; Markham et al. 1983a; Mitsuhashi, J. 1967a; Mitsuhashi & Maramorosch 1964a; Mumford, D.L. 1982a; Nagaraj & Black 1961a; Nagaraj et al. 1961a; Nakasuji & Kiritani 1972a; Nakasuji et al. 1975a, 1985a; Nasu, Jensen & Richardson 1974a; Nasu, Jensen, Richardson et al. 1974a, 1974b; Omura et al. 1984a; Omura, Saito et al. 1982a; Purcell & Finlay 1979b; Raju & Nyland 1981a; Raju et al. 1979a, 1984a; Rana et al. 1975a; Reddy, D.V.R. 1966a; Reddy, H.R. 1975a; Richardson & Westdal 1967a; Saillard et al. 1980a; Seki & Onizuka 1964a; Sherman & Maramorosch 1977a; Sinha, R.C. 1968c, 1969a, 1974a; Sinha & Black 1962a; Sinha & Reddy 1964a; Sinha et al. 1964a; Smith et al. 1981a; Timian & Alm 1973a; Toyoda et al. 1965a; Van Rensburg, G.D.J. 1979a; Whitcomb, R.F. 1964a, 1983a; Whitcomb & Coan 1982a; Yoshii et al. 1961a.

Museum Preservation

Barnett, D.E. 1976a; Beamer, R.H. 1940a; Beirne, B.P. 1955a; DeLong & Davidson 1936a; Diniz, M. de Assuncao 1964a; Gurney et al. 1964a; Ossiannilsson, F. 1957a, 1958b; Quartau, J.A. 1984b; Ross, H.H. 1956b, 1962a; Sabrosky, C.W. 1966a, 1971a; Schuler, L. 1954a; Theron, J.G. 1985c.

Tissue Culture

Black, L.M. 1969a, 1979a; Brooks, M.A. 1985a; Hirumi, H. 1965a; Hirumi & Maramorosch 1963a, 1964a, 1964b, 1964c; Hollebone et al. 1966a; Hsu et al. 1983a; Kimura, I. 1980a, 1982a, 1985a; Kimura & Black 1971a, 1972a; Lehmann, W. 1982a; Maramorosch, K. 1965b, 1979a; Maramorosch, Mitsuhashi et al. 1965a; Martinez-Lopez & Black 1974a; McIntosh et al. 1973a; Mitsuhashi, T. 1964b, 1965a, 1965b, 1965c, 1965d, 1965f, 1965g, 1965h, 1966b, 1967a, 1967b, 1969a, 1975a; Mitsuhashi & Maramorosch 1964b; Richardson & Jensen 1971a; Senboku & Shikata 1980a; Tokumitsu & Maramorosch 1967a, 1968a; Vago & Flandre 1963a.

Rearing

Aguda, Centina et al. 1984a; Bindra & Sohi 1969a; Brooks, M.A. 1985a; Caudwell & Larrue 1977a; Chandra, G. 1978a, 1980b; Chiykowski, L.N. 1985a; Dahlman, D.L. 1963a; Fraval, A. 1968a; Genyte & Staniulis 1976a; George & Davidson 1959a; Ghosh et al. 1968a; Gu & Ito 1982a; Hou, R.F. 1976a; Hou & Brooks 1975a; Hou & Lin 1979a; Huq, S. 1984a; Jensen & Chapman 1979a; Kieckhefer & Medler 1960a; Kim et al. 1985a; Koyama, K. 1973b; Medrano et al. 1984a; Mitsuhashi, J. 1964a, 1965a, 1965e, 1974a, 1979a; Mitsuhashi & Maramorosch 1963a, 1963b; Moreau, J.P. 1963a; Moutous & Fos 1976a; Palmiter et al. 1960a; Razzaque et al. 1985a; Ricou, G. 1960a; Soejitno et al. 1974a; Strong & Rawlins 1957a; Sugimoto, A. 1969a, 1977a, 1981a; Wei & Brooks 1979a.

Others

Ameresekere, R.V.W.E. 1970a; Ameresekere & Georghiou 1971a; Anderson et al. 1974a; Anonymous 1970b; Bekker-Migdisova, E.E. 1967a; Bindra & Sohi 1970a; Birch & Fleischer 1984a; Chakravarthy & Rao 1985a; Chandra, G. 1978a; Dillon, L.S. 1954a; Forsythe & Gyrisco 1961a; Gaddoura & Venkatraman 1967a; Gunthart, H. 1979a; Halkka, O. 1965a; Hanson, C.H. 1969a; Hanson et al. 1972a; Harcharan & Mann 1979a; Hay & Meyer 1961a; Hokyo, N. 1971a; Horber, E. 1974b; Horimoto et al. 1976a; Jansson, A.M. 1973a; Kamm & Ritcher 1972a; Kawabe, S. 1978a; Kawabe et al. 1980a, 1981a; Kawabe & McLean 1978a, 1980a; Kazano et al. 1975a; Kehr, W.R. 1970a, 1970b; Kehr & Manglitz 1984a; Khan & Saxena 1984a, 1984b, 1984c; Kiritani, Kawahara et al. 1972a; Kishino & Ando 1978a; Knight, W.J. 1965a; Knutson, L. 1976a; Koppany, T. 1969a; Koyama, K. 1972a; Le Quesne, W.J. 1984a; Leston, D. 1955a; Lin & Liu 1984a; Madden, L.V. 1985a; Maramorosch, K. 1956a; Maramorosch & Jernberg 1964a; Mavi & Sidhu 1982a; Medrano et al. 1984a; Miah & Husain 1983a; Mitsuhashi, J. 1974a, 1979a; Mitsuhashi & Maramorosch 1963b; Miyai et al. 1978a; Miyata et al. 1980a; Morinaka & Sakurai 1970a; Moriya & Maeda 1975a; Murguido & Ruiz 1982a; Naba, K. 1982a; Nagaich, B.B. 1975a; Nagaraj & Black 1961a; Naito, A. 1964a, 1965a; Nakamura, K. 1977a; Palis et al. 1984a; Pierce, H.D. 1976a; Pollard, D.G. 1971a; Prior, R.N.B. 1964a; Proeseler & Eisbein 1974a; Quartau, J.A. 1978b, 1981c, 1982a, 1982b, 1982e, 1983a, 1983b, 1983d; Quartau & Davies 1983a, 1984a, 1985a; Rao & John 1971a; Razzaque, Heinrichs & Rapusas 1985a; Roof et al. 1972a; Sabrosky, C.W. 1971a; Saini & Chhabra 1968a; Sasaba, R. 1974a; Sasaba & Kiritani 1975a; Sasaba et al. 1973a; Schillinger et al. 1964a; Smith et al. 1985a; Taimr & Dlabola 1965a; Takai et al. 1965a, 1972a; Theron, J.G. 1985c; Tingey, W.M. 1985b; Van Rensburg & Walters 1977a; Vogel, O. 1982b, 1983a; Wilson & Davis 1962a.

VECTORS OF PLANT PATHOGENS

General

Abeygunawardena, D.V.W. 1969a; Abeygunawardena et al. 1971a; Agarkov, V.A. 1964a; Aguiero et al. 1979a; Akingbohungbe, A.E. 1983a; Albouy, J. 1966a; Albouy et al. 1967a; Amen, C.R. 1953a; Anonymous 1957a, 1967c, 1969a, 1981b, 1985g; Ballantyne, B. 1971a; Bantarri, E.E. 1966a; Belli, G. 1974a; Bennett, C.W. 1971a; Bernhard et al. 1977a; Bhirud & Pitre 1972a, 1972b, 1972c; Bindra, O.S. 1973a; Bindra, Khatri & Sohi 1973a; Black, L.M. 1962a; Bos, L.

1981a; Bowyer & Atherton 1971a; Brcak, J. 1979a; Bremer & Raatikainen 1975a; Brooks, M.A. 1985a; Cadman, C.H. 1961a; Carter, W. 1961a, 1962a, 1973a; Catara, A. 1983a; Cattaneo & Arzone 1983a; Chaudhuri, R.P. 1955a; Chenon, R.D.de 1979a; Chiu, R.J. 1982a; Chiykowski, L.N. 1964a, 1965c, 1973a, 1974a, 1974b; Chiykowski & Wolynetz 1981a; Cohen, G. 1961a; Conti, M. 1981a, 1983a, 1985a; Converse et al. 1982a; Crall & Stover 1957a; Cropley, R. 1960a; Damsteegt, V.D. 1980a, 1983a; D'Arcy & Nault 1982a; David & Alexander 1984a; Davis et al. 1978a; Davis, R.E. 1974a; Day & Bennetts 1954a; Delgadillo-Sanchez, F. 1984a; Derrick & Newsom 1984a; Dlabola, J. 1960d; Dollet, M. 1980a; Eastman et al. 1984a; Evenhuis, H.H. 1955b; Fajemisin et al. 1976a, 1977a; Fisher et al. 1973a; Flock, R.A. 1977a; Fluiter, H.J.de 1958a; Fluiter et al. 1955a; Foster, J.A. 1982a; Fraval, A. 1968a; French & Feliciano 1982a; Fritzche et al. 1972a; Fullerton, D. 1964a; Furuta, T. 1977a; Grylls, N.E. 1979a; Hanson et al. 1964a; Harris, K.F. 1980a; Harris, K.F., ed. 1984a; Harris & Maramorosch 1980a, 1982a; Harris et al. 1981a; Hoffman & Taboada 1965a; Ishihara, T. 1969a; Janson & Ellet 1963a; Julia, J.F. 1979a; Kaloostian et al. 1976a; Kiritani, K. 1983a; Lim, G.S. 1972b; Lindsten, K. 1985a; Maramorosch, K. 1954a, 1956b, 1959b, 1959c, 1963b, 1980a; Maramorosch & Harris 1979a, 1981a; Maramorosch & Oman 1966a; Maramorosch et al. 1974a; Martinez Lopez et al. 1974a; Mercer, P.C. 1977a; Miller, P.R. 1966a; Morris, D.O. 1954a; Muller, H.J. 1979a; Nault & Knoke 1981a; Nielson, M.W. 1968a, 1979c; Otake, A. 1983a; Ou, S.H. 1972a; Purcell, A.H. 1982a, 1982b; Radulescu & Munteanu 1970a; Raven, J.A. 1983a; Reyes, G.M. 1957a; Sales, F.M. 1979a; Schafers & Brann 1974a; Schieber & Costillo 1960a, 1962a; Schneider, C.L. 1959a; Schvester, D. 1965a; Serrano, F.B. 1957a; Sinha, R.C. 1963a, 1981a; Smith & Brierly 1956a; Smith, K.M. 1957a, 1958a; Swincer, D.E. 1984a; Szekessy, V. 1969a; Taboada & Burger 1967a; Taboada & Hoffman 1965a; Thung & Hadiwidjaja 1957a; Trinh, T.T. 1980a; Trujillo et al. 1974a; Vacke, J. 1966a; Vidano, C. 1959c, 1965c; Vidano & Arzone 1984c; Viennot-Bourgin, G. 1981a; Vlasor & Rodina 1969a; Volk, J. 1958a, 1967a; Yonce, C.E. 1983a.

Bacteria

Adlerz & Hopkins 1978a, 1979a; Auger et al. 1974a, 1974b; Belli & Osler 1975a; Briansky et al. 1982a, 1983a; Crall & Stover 1957a; D'Arcy & Nault 1982a; Davis et al. 1978a, 1981a; Davis & Whitcomb 1971a; Duncan & Genereux 1960a; French & Feliciano 1982a; Goheen et al. 1973a; Greber & Gowanlock 1979a; Grylls, N.E. 1954a, 1955a; Grylls et al. 1974a; Hewitt et al. 1958a; Hopkins, D.L. 1977a; Hopkins & Adlerz 1980a; Hopkins et al. 1973a, 1978a; Hopkins & Mollenhauer 1973a; Kalkandelen & Fox 1968a; Kaloostian & Pollard 1962a; Kaloostian et al. 1962a; Knight, K.G. 1961a; Lee et al. 1981a; Maillet, P.L. 1970a, 1970b; Maramorosch, K. 1955a, Maramorosch et al. 1975a; McCoy et al. 1978a, 1983a; Mircetich et al. 1976a; Nielson & Gill 1984a; Nyland et al. 1973a, 1981a; Purcell, A.H. 1974a, 1975a, 1979b, 1980a, 1981a, 1985a; Purcell & Finlay 1979a; Purcell et al. 1979a; Raine et al. 1976a; Raju, Goheen et al. 1983a; Stoner, W.N. 1953a, 1958a; Timmer & Lee 1985a; Tracy, R.K. 1977a; Turner & Pollard 1956a, 1959a, 1959b; Vidano, C. 1963b; Whitcomb & Black 1969a; Windsor & Black 1973a; Yonce, C.E. 1983a.

Epidemiology

Alverson, D.R. 1979a; Autrey & Ricaud 1983a; Belli et al. 1975a; Bennett, C.W. 1967b; Bindra, Khatri, Sohi, Deol et al. 1973a; Caudwell, Kuszala et al. 1972a; Chen et al. 1980a; Chiykowski, L.N. 1981a; Clark, R.L. 1968a; Desmidts et al. 1973a; Duffus, J.E. 1983a; Gamez & Saavedra 1985a; Ishijima, T. 1969a; Kiritani, K. 1981a; Lim, G.S. 1972a; Ling, K.C. 1975a; Ling, Tiongco & Fores 1983a; Maramorosch & Harris 1981a; Mukhopadhyay & Chowdhury 1973a; Muniyappa & Ramakrishnan 1976a; Nakasuji, F. 1974a; Nakasuji & Kiritani 1977a; Nakasuji et al. 1975a, 1985a; Nyland et al. 1981a; Osler et al. 1973a; Ossiannilsson, F. 1966a; Peterson, G.W. 1984a; Pitre & Boyd 1970a; Purcell, A.H. 1975a; Rao & John 1974a; Rose, D.J.W. 1974a, 1978a; Saavedra, F. 1982a; Savio & Conti 1983a; Schultz, G.A. 1979a; Shinkai, A. 1960a; Thresh, J.M. 1974a, 1980a, 1980b; Timmer & Lee 1985a; Vidano & Arzone 1983a.

Mycoplasma-like Organisms (MLO)

General: Alivizatos, A.S. 1982a, 1983a; Archer et al. 1982a; Baker et al. 1983a; Behncken, G.M. 1984a; Belli et al. 1972a; Belli & Osler 1975a; Bindra, Khatri & Sohi 1973a; Blattny, C. Jr 1963a; Bove et al. 1979a; Bove & Saillard 1979a; Bovey, B. 1957a, 1958a, 1958b; Bowyer, J.W. 1972a; Brcak, J. 1954a; Calavan et al. 1974a; Calavan & Oldfield 1979a; Catara, A. 1984a; Cattaneo & Arzone 1983a; Caudwell, A. 1977a; Caudwell et al. 1969a, 1970a; Caudwell, Larrue & Kuszala 1978a; Charbonneau et al. 1979a; Chiykowski, L.N. 1958a, 1964a, 1967b, 1969a, 1973a, 1974a, 1977a, 1979a; Chiykowski & Sinha 1969a, 1970a; Clark, T.B. 1982a; Cousin & Moreau 1966a; Cousin et al. 1965a; Dabek, A.J. 1982a; Daniels, M.J. 1979a, 1979b; D'Arcy & Nault 1982a; Davis, R.E. 1979a; Davis & Whitcomb 1971a; Davis et al. 1968a, 1970a; Doi et al. 1967a; El-Bolok, M.M. 1981a; Fisher et al. 1973a; Fluiter & van der Meer 1956a; Frederiksen, R.A. 1962a, 1964a; Freitag, J.H. 1958a, 1963a, 1964a, 1967a; Freitag & Smith 1969a; Gaborjanyi & Nagy 1972a; Galvez & Miah 1969a; Ghosh, S.K. 1981a; Giannotti, J. 1972a; Giannotti, Devauchelle et al. 1968a; Giannotti et al. 1969a, 1970a; Giannotti, Vago, Devauchelle et al. 1968a; Gold, R.E. 1974a, 1979a; Gold & Sylvester 1982a; Gouranton & Maillet 1973a; Granados, R.R. 1965a; Granados & Chapman 1968a, 1968b; Hagel & Landis 1967a; Hagel et al. 1973a; Harder & Westdal 1971a; Hayflick & Arai 1973a; Heinze & Kunze 1955a; Hemmati, K. 1979a; Hemmati & McLean 1980a; Herman & Maramorosch 1977a; Hibben et al. 1973a; Hill, S.A. 1980a; Hirumi & Maramorosch 1963b, 1969a, 1969b; Horvath, J. 1974a; Hsu & Banttari 1979a; Hull, R. 1972a; Iwaki, M. 1979a; Jensen, D.D. 1968a, 1969a, 1971b, 1972a; Jensen & Nasu 1970a; Jensen & Richardson 1972a; Jensen, J.O. 1983a; Jenser et al. 1981a; Jones et al. 1977a; Kaloostian & Pierce 1972a; Kaloostian, Oldfield, Pierce, Calavan, Granett, Rana, Gumpf et al. 1975a; Kirkpatrick et al. 1985a; Kitajima & Costa 1972a; Kleinhempel et al. 1974a; Kochman & Ksiazek 1967a; Kunkel, L.O. 1954a, 1957a; Lacy et al. 1979a; Lastra & Trujillo 1976a; Lee et al. 1973a, 1973b; Lee, P.E. 1962a; Lee & Chiykowski 1963b; Lee & Jensen 1963a; Lee & Robinson 1958a; Lehmann, W. 1971a, 1973a; Lehmann & Skadow 1971a; Ling, Tiongco & Cabunagan 1983a; Littau & Maramorosch 1956a, 1958a, 1960a; Liu, Gumpf et al. 1983a, 1983b; Maillet et al. 1968a; Maramorosch, K. 1955b, 1956c, 1958a, 1959a, 1962a, 1962b, 1973a, 1981a, 1982a; Maramorosch & Kondo 1978a; Maramorosch et al. 1968a, 1970a, 1975a; Maramorosch & Raychandhuri 1981a; Markham, P.G. 1983a; Markham et al. 1977a, 1983a; Markham & Pinner 1984a; Markham & Townsend 1974a, 1977a, 1979a; Martinez, M.R. 1975a; Marwitz et al. 1984a; McCoy, R.E. 1979a; Miyahara et al. 1982a; Muniyappa & Veeresh 1980a; Musil, M. 1958b, 1959b, 1960b, 1964a, 1964c, 1965a, 1966a; Musil & Valenta 1958a; Nasu, Jensen & Richardson 1970a; Nasu, Jensen, Richardson et al. 1974b; Nault, L.R. 1983a, 1983b, 1984a; Nault et al. 1979a; Nichiporich, W. 1965a; Nielson, M.W. 1979d; Nour, M.A. 1962a; Ogunlana & Pedigo 1984a; O'Hayer et al. 1983a; Oldfield, G.N. 1980a; Oldfield et al. 1978a; Onishchenko, A.N. 1984a; Peterson & Saini 1964a; Ploaie, P.G. 1981a; Pollini et al. 1984a; Posnette & Ellenberger 1963a; Purcell, A.H. 1979a, 1985a; Purcell, Raju et al. 1980a, 1981a; Purcell, Richardson et al. 1981a; Raine & Forbes 1969a, 1971a; Raine et al. 1976a; Raju et al. 1984a; Razvyazkina, G.M. 1959a, 1959b; Rice & Jones 1972a; Russo et al. 1976a; Sackston, W.E. 1959a; Saglio & Whitcomb 1979a; Saillard et al. 1980a, 1984a; Samyn et al. 1982a; Savio & Conti 1983a; Savulescu & Ploaie 1962a; Sen-Sarma, P.K. 1984a; Shikata, E. 1979b; Shikata et al. 1969a; Shinkai, A. 1975a; Shiomi & Choi 1983a; Shiomi & Sugiura 1984a, 1984b; Sinha, R.C. 1979a, 1983a, 1984a; Smith et al. 1981a; Stanarius et al. 1976a, 1980a; Stoner, W.N. 1965a; Swenson, K.G. 1971b; Taboada, O. 1979a; Tsai, J.H. 1979a; Vago & Giannotti 1972a; Valenta, V. 1958a; Valenta et al. 1961a; Vanderveken, J. 1968a; Varney, E.H. 1977a; Vasudeva & Sahambi 1959a; Vidano, C. 1965c, 1972a; Vignault et al. 1980a; Wakibe & Miyahara 1984a; Wallis, R.L. 1960a; Westdal et al. 1959a, 1961a;

Whitcomb, R.F. 1972a, 1972b, 1981a, 1983a; Whitcomb & Black 1982a; Whitcomb & Bov 1983a; Whitcomb & Davis 1970a, 1970b; Whitcomb, Jensen et al. 1966a; Whitcomb & Tully 1979a; Williamson & Whitcomb 1974a; Wolfe, H.R. 1955a, 1955b; Zaman, M.Q. 1980a.

Acholeplasma sp.: Eden-Green et al. 1985a.

Alfalfa: Bowyer, J.W. 1974a; Hanson et al. 1963a; James & Granovsky 1927a; Raine, J. 1967a; Weaver, C.R. 1954a.

Alfalfa Yellow Mosaic: Manglitz & Kreitlow 1960a.

Aster: Maramorosch et al. 1962a; Shikata et al. 1968a.

Aster Yellows: Bantarri, E.E. 1966a; Bantarri & Moore 1960a; Charbonneau et al. 1979a; Chiykowski, L.N. 1958a, 1962b, 1963a, 1965b, 1967c, 1973a, 1977a, 1977b, 1979a; Chiykowski & Sinha 1969a; Chiykowski & Wolynetz 1981a; Davis et al. 1968a, 1970a; Doi et al. 1967a; El-Bolok, M.M. 1981a; Esau et al. 1976a; Frederiksen, R.A. 1964b; Frederiksen & Christensen 1957a; Freitag, J.H. 1956a, 1958a, 1962a, 1962b, 1963a, 1964a, 1967a; Freitag et al. 1954a; Freitag & Smith 1969a; George & Richardson 1957a; Gill et al. 1969a; Granados, R.R. 1965a; Granados & Chapman 1968a, 1968b; Groen & Slogteren 1974a; Hagel & Landis 1967a, 1974a; Hayflick & Arai 1973a; Hemmati, K. 1979a; Hemmati & McLean 1980a; Henne, R.C. 1970a; Herman & Maramorosch 1977a; Hiruki & Chen 1978a; Hirumi & Maramorosch 1963b, 1969a, 1969b; Hoffman, J.R. 1952a; Hsu & Banttari 1979a; Jensen, J.O. 1983a; Kochman & Ksiazek 1967a; Kunkel, L.O. 1957a; Lee, P.E. 1962a; Lee & Chiykowski 1963b; Lee & Robinson 1958a; Littau & Maramorosch 1956a, 1958a, 1960a; Maillet et al. 1968a; Maramorosch, K. 1955b, 1956c, 1958a, 1959a, 1962a, 1962b; Maramorosch & Kondo 1978a; Maramorosch et al. 1962a; Martin et al. 1961a; Mitsuhashi & Maramorosch 1964a; Muratomaa, A. 1967a, 1969a; Nichiporich, W. 1965a; Orenski, Murray et al. 1965a; Peterson & Saini 1964a; Raju et al. 1984a; Sackston, W.E. 1959a; Schultz & Chapman 1979a; Shikata et al. 1968a; Shinkai, A. 1975a; Shiomi & Sugiura 1984b; Sinha, R.C. 1983a; Slogteren & Muller 1972b; Smith et al. 1981a; Strong & Rawlins 1958a; Swenson, K.G. 1971b; Wallis, R.L. 1960a; Westdal et al. 1959a, 1961a; Westdal & Richardson 1972a; Whitcomb, R.F. 1972b; Whitcomb & Davis 1970b; Zalom, F.G. 1981a.

Barley Yellow Dwarf: Gill et al. 1969a.

Bean Yellow Mosaic: Manglitz & Kreitlow 1960a.

Blueberry: Blattny, C. Jr 1963a; Blattny & Blattny 1970a; Varney, E.H. 1977a, 1981a.

Blueberry Stunt: Varney, E.H. 1981a.

Brinjal Little Leaf: Bindra, Khatri, Sohi, Deol et al. 1973a; Sarkar & Kulshreshtha 1978a; Shantha & Lakshmanan 1984a; Srinivasan, K. 1978a; Srinivasan & Chelliah 1980a, 1981a.

Brittle Root Disease: Fletcher et al. 1981a.

Bupleurum falcatum: Shiomi & Choi 1983a.

Carrot: Hagel & Landis 1974a; Henne, R.C. 1970a; Mitsuhashi & Maramorosch 1964a; Schultz & Chapman 1979a.

Celery: Chiykowski, L.N. 1962b, 1963a, 1977b; George & Richardson 1957a; Jensen, D.D. 1955a, 1956b, 1957b; Nasu, Jensen & Richardson 1974a; Nasu, Jensen, Richardson et al. 1974a.

Cereals: Bantarri & Moore 1960a, 1962a; Bantarri & Zeyen 1970a; Chiykowski, L.N. 1963a, 1965b, 1967c; Chiykowski & Wolynetz 1981a; Freitag, J.H. 1962b; Gill et al. 1969a; Harder & Westdal 1971a; Munteanu et al. 1983a; Westdal & Richardson 1966a, 1972a.

Cherry: Gonot & Purcell 1981a; Jensen & Thomas 1954a, 1955a; Nielson & Jones 1954a; Suslow & Purcell 1982a; Taboada et al. 1975a; Wilde, W.H.A. 1960a; Wilks & Welsh 1964a.

Cherry Buckskin: Gonot & Purcell 1981a; Jensen, D.D. 1956b; Jensen & Thomas 1954a, 1955a.

Chokeberry: Peterson, G.W. 1984a; Rosenberger & Jones 1978a.

Citrus: Calavan et al. 1979a; Catara, A. 1984a; Gumpf & Calavan 1981a; Hafidi et al. 1979a; Kaloostian, Oldfield, Pierce, Calavan, Granett, Rana & Gumpf 1975a; Kaloostian, Oldfield, Gumph et al. 1979a; Kaloostian, Oldfield, Pierce et al. 1979a; Kaloostian & Pierce 1972a; Markham et al. 1974a; Oldfield & Kaloostian 1979a; Oldfield et al. 1976a; Oldfield, Kaloostian, Pierce, Calavan et al. 1977a.

Citrus Little Leaf: Markham et al. 1974a.

Clover: Bos & Grancini 1965a; Bovey, B. 1957a; Chiykowski, L.N. 1961a, 1962a, 1962c, 1965a, 1965c, 1967a, 1969a, 1973b, 1974b, 1975a, 1976a; Chiykowski & Sinha 1970a; Cousin, M.T. 1968a; Cousin & Moreau 1966a; Cousin et al. 1965a, 1968a; Evenhuis, H.H. 1958a, 1958b; Giannotti, J. 1969a, 1969b; Giannotti & Devauchelle 1969a; Giannotti, Vago & Duthoit 1968a; Gourret et al. 1973a; Lee & Chiykowski 1963a; Maillet, P.L. 1970a, 1970b; Maillet & Gouranton 1971a; Manglitz & Kreitlow 1960a; Moreau et al. 1968a; Musil, M. 1960b, 1961a, 1962a, 1966b; Ploaie, P.G. 1967a; Posnette & Ellenberger 1963a; Raine, J. 1967a; Razvyazkina, G.M. 1959b; Sinha & Paliwal 1970a; Vanderveken, J. 1964a, 1965a; Webb & Schultz 1955a; Windsor & Black 1973a.

Clover Club Leaf: Windsor & Black 1973a.

Clover Dwarf: Musil, M. 1962a, 1966b.

Clover Phyllody: Chiykowski, L.N. 1961a, 1962a, 1962c, 1965a, 1967a, 1973b, 1974b, 1975a; Chiykowski & Craig 1978a; Cousin, M.T. 1968a; Cousin et al. 1968a; Evenhuis, H.H. 1958a, 1958b; Giannotti, J. 1969a, 1969b; Giannotti & Devauchelle 1969a; Giannotti, Vago & Duthoit 1968a; Gouranton & Maillet 1973a; Gourret et al. 1973a; Lee & Chiykowski 1963a; Maillet, P.L. 1970a, 1970b; Maillet & Gouranton 1971a; Maillet et al. 1968a; Musil, M. 1960b, 1961a; Ploaie, P.G. 1967a; Raine, J. 1967a; Sinha & Paliwal 1970a; Vanderveken, J. 1965a.

Clover Yellow Edge: Chiykowski, L.N. 1976a.

Coconut: Dabek, A.J. 1982b; Eden-Green, S.J. 1979a; Eden-Green et al. 1985a; Ghauri, M.S.K. 1980c; Heinze, K.G. 1972a; Tsai & Anwar 1977a.

Corn: Ancalmo & Davis 1961a; Bradfute et al. 1981a; Chen & Liao 1975a; Choudhury & Rosenkranz 1973a; Collins & Pitre 1969a, 1969b; Davis, R.E. 1974a, 1974b, 1977a; Davis et al. 1972a; Davis & Worley 1973a; Granados, R.R. 1969a; Granados, Granados et al. 1968a; Granados, Gustin et al. 1968a; Granados & Maramorosch 1967a; Granados et al. 1966a, 1966b; Granados, Maramorosch et al. 1966a, 1966b, 1968a; Granados & Whitcomb 1971a; Hao & Pitre 1970a; Kitajima & Costa 1972a; Kramer, J.P. 1964f, 1967a; Kramer & Whitcomb 1968a; Lastra & Trujillo 1976a; Madden & Nault 1983a; Madden et al. 1984a; Malaguti & Ordosgoitty 1969a; Maramorosch, K. 1957a, 1958c, 1959a, 1963a, 1963c; Maramorosch, Orenski et al. 1965a; Nault, L.R. 1980a, 1983a, 1983b, 1984a; Nault & Bradfute 1977a, 1979a; Nault et al. 1979a; Orenski, Murray et al. 1965a; Pitre, H.N. 1966a, 1968a, 1968b; Pitre et al. 1967a; Ramirez et al. 1975a; Rosenkranz, E. 1969a; Schieber & Costillo 1960a, 1962a; Shikata et al. 1968a; Stoner, W.N. 1965a; Stoner & Gustin 1967a; Stoner & Ullstrup 1964a; Uribe & Valencia 1970a; Williamson & Whitcomb 1975a; Wolanski & Maramorosch 1979a.

Corn Stunt: Alivizatos, A.S. 1983a; Ancalmo & Davis 1961a; Bradfute et al. 1981a; Chen & Liao 1975a; Choudhury & Rosenkranz 1973a; Collins & Pitre 1969a, 1969b; Davis, R.E. 1974a, 1974b, 1977a; Davis et al. 1972a; Davis & Worley 1973a; Granados, R.R. 1969a; Granados, Granados et al. 1968a; Granados, Gustin et al. 1968a; Granados & Maramorosch 1967a; Granados et al. 1966a, 1966b; Granados, Maramorosch et al. 1966a, 1966b, 1968a; Granados & Whitcomb 1971a; Hao & Pitre 1970a; Kramer, J.P. 1964f, 1967a; Kramer & Whitcomb 1968a; Madden & Nault 1983a; Madden et al. 1984a; Malaguti & Ordosgoitty 1969a; Maramorosch, K. 1957a, 1958c, 1959a, 1960b, 1963a, 1963c; Maramorosch, Orenski et al. 1965a; Markham et al. 1977a; Nault, L.R. 1980a; Nault & Bradfute 1977a, 1979a; Pitre, H.N. 1966a, 1968a, 1968b; Pitre et al. 1967a; Ramirez et al. 1975a; Rosenkranz, E. 1969a; Schieber & Costillo 1960a, 1962a; Shikata et al. 1968a; Stoner, W.N. 1965a; Stoner & Gustin 1967a; Stoner & Ullstrup 1964a; Uribe & Valencia 1970a; Williamson & Whitcomb 1975a; Wolanski & Maramorosch 1979a.

Cotton: Desmidts et al. 1973a; Laboucheix et al. 1972a, 1973a; Moreau, J.P. 1969a; Rassel & Desmidts 1976a.

Cotton Phyllody: Desmidts et al. 1973a; Rassel & Desmidts 1976a.

Cucumber: Freitag, J.H. 1956a.

Cucurbits: Chou et al. 1975a; Freitag, J.H. 1956a.

Eggplant (Brinjal): Bindra, Khatri, Sohi, Deol et al. 1973a; Sarkar & Kulshreshtha 1978a; Shantha & Lakshmanan 1984a; Srinivasan, K. 1978a; Srinivasan & Chelliah 1980a, 1981a.
Flavescence Doree: Baggiolini, Canevascini, Caccia et al. 1968a; Belli et al. 1973a; Carle & Moutous 1967a; Caudwell, A. 1981a; Caudwell et al. 1969a, 1970a, 1971a; Caudwell, Kuszala et al. 1972a; Caudwell, Larrue, Moutous et al. 1978a; Gartel, W. 1965a; Moutous, G. 1977a; Osler et al. 1975a; Schvester, D. 1965a, 1966a; Schvester, Carle et al. 1961a, 1963a, 1963b, 1967a, 1970a; Schvester, Moutous & Carle 1962a; Vidano, C. 1964a.
Flax: Bantarri & Moore 1962a; Frederiksen, R.A. 1964a, 1964b; Frederiksen & Christensen 1957a; Frederiksen & Goth 1959a; Martin et al. 1961a.
Flax Crinkle Disease: Frederiksen & Goth 1959a.
Floral Virescence of Cotton: Laboucheix et al. 1972a, 1973a; Moreau, J.P. 1969a.
Gladioli: Freitag et al. 1954a; Groen & Slogteren 1974a; Slogteren & Muller 1972b.
Grapevine: Baggiolini, Canevascini, Caccia et al. 1968a; Belli et al. 1973a; Belli & Osler 1975a; Carle & Moutous 1967a; Caudwell, A. 1981a; Caudwell et al. 1969a, 1970a, 1971a; Caudwell, Larrue, Moutous et al. 1978a; Gartel, W. 1965a; Moutous, G. 1977a; Osler et al. 1975a; Schvester, D. 1965a, 1966a; Schvester, Carle et al. 1961a, 1963a, 1963b, 1967a, 1970a; Schvester, Moutous & Carle 1962a; Vidano, C. 1963b, 1964a, 1965c.
Grasses: Bantarri, E.E. 1966a; Muratomaa, A. 1967a, 1969a.
Green Petal of Clover: Posnette & Ellenberger 1963a.
Green Petal Disease of Primula: Stevens & Spurdon 1972a.
Green Petal of Rape: Horvath, J. 1969a.
Groundnuts: Bergman, B.H.H. 1956b.
Hairy Sprout of Potato: Khanna, V.M. 1973a.
Herbs: Gaborjanyi & Nagy 1972a; Jensen, D.D. 1971a.
Horseradish: Duffus et al. 1982a; Fletcher et al. 1981a.
Horseradish Bitter Root: Duffus et al. 1982a.
Hyacinths: Slogteren & Muller 1972a, 1972b.
Legumes: Bowyer, J.W. 1974a; Bowyer & Atherton 1971b, 1972a; Caudwell, Kuszala et al. 1972a; Freitag, J.H. 1962a; Hiruki & Chen 1978a; Iwaki, M. 1979a; Jackson & Zettler 1983a; McCoy et al. 1983a; Murayama, D. 1966a; Licha-Baquero, M. 1979a; Thung & Hadiwidjaja 1957a.
Legume Little Leaf: Bowyer, J.W. 1974a; Bowyer & Atherton 1971b, 1972a; Jackson & Zettler 1983a.
Lethal Yellowing: Dabek, A.J. 1982b; Eden-Green, S.J. 1979a; Fisher et al. 1973a; Ghauri, M.S.K. 1980c; Heinze, K.G. 1972a; Tsai & Anwar 1977a.
Lettuce: Hoffman, J.R. 1952a; Koga et al. 1982a; Novak, J.B. 1961a; Strong & Rawlins 1958a; Zalom, F.G. 1981a.
Little Cherry Disease: Nielson & Jones 1954a; Wilde, W.H.A. 1960a; Wilks & Welsh 1964a.
Little Leaf: Behncken, G.M. 1984a.
Machismo Disease of Sorghum: Granada, G.A. 1979a.
Maize Bushy Stunt: Nault, L.R. 1980a.
Marginal Flavescence of Potato: Nagaich & Giri 1971a; Nagaich et al. 1974a.
Mosaic I: Bergman, B.H.H. 1956b.
Mulberry: Faan et al. 1964a; Giorgadze & Tulashvili 1973a; Ishiie & Matsuno 1971a, 1971b; Ishiie et al. 1967a; Ishijima, T. 1969a, 1971a; Ishijima & Ishiie 1981a; Ishizaki, T. 1980a, 1980b; Tahama, Y. 1963a, 1964a, 1968a.
Mulberry Curly Leaf: Giorgadze & Tulashvili 1973a.
Mulberry Dwarf: Doi et al. 1967a; Faan et al. 1964a; Ishiie & Matsuno 1971a, 1971b; Ishiie et al. 1967a; Ishijima, T. 1969a, 1971a; Ishijima & Ishiie 1981a; Ishizaki, T. 1980a, 1980b; Tahama, Y. 1963a, 1964a, 1968a.
Oat Blue Dwarf: Bantarri & Moore 1962a; Bantarri & Zehen 1970a; Hsu & Banttari 1979a.
Onion: Koga et al. 1982a; Miyahara et al. 1982a, 1983a.
Onion Biwadama: Koga et al. 1982a.
Onion Yellow Dwarf: Kochman & Ksiazek 1967a.
Ornamentals: Baker et al. 1983a; Lehmann, W. 1969a; Stanarius et al. 1976a; Tsai et al. 1972a.
Padi Jantan of Rice: Singh et al. 1970a; Soon & Guan 1968a.
Papaya: Haque & Parasram 1973a; Story & Halliwell 1969a.

Papaya Bunchy Top: Haque & Parasram 1973a; Story & Halliwell 1969a.
Peach: Frazier & Jensen 1970a; Jensen, D.D. 1953a, 1957a, 1957b, 1958a; Jensen et al. 1952a, 1967a; Jensen & Thomas 1954a; Lacy, G.H. 1982a; McClure, M.S. 1980a, 1980b, 1982a; McClure et al. 1982a; Nasu, Jensen & Richardson 1970a; Palmiter & Adams 1957a; Purcell & Suslow 1984a; Rosenberger & Jones 1978a; Sinha & Chiykowski 1980a; Taboada et al. 1975a.
Peach X-Disease: McClure, M.S. 1980a, 1980b.
Peach Yellow Leaf Roll: Jensen, D.D. 1953a, 1955a, 1956b, 1957a, 1957b, 1958a; Jensen et al. 1952a; Nyland et al. 1981a; Purcell & Suslow 1984a.
Pear: Nyland et al. 1981a; Purcell & Suslow 1984a.
Periwinkle: Dabek, A.J. 1982c; Maramorosch, K. 1960b; Oldfield, Kaloostian, Pierce, Granett et al. 1977a; Rassel & Desmidts 1976a.
Periwinkle Phyllody: Dabek, A.J. 1982b, 1982c.
Periwinkle Virescence: Oldfield, Kaloostian, Pierce, Granett et al. 1977a.
Phyllody: Chiykowski, L.N. 1967b; Giannotti, Devauchelle et al. 1968a; Giannotti et al. 1970a; Lehmann, W. 1969a; Nour, M.A. 1962a; Vasudeva & Sahambi 1959a.
Pierce's Disease: Vidano, C. 1963b.
Potato: Fukushi et al. 1955a; Hagel & Landis 1974a; Harding & Teakle 1985a; Khanna, V.M. 1973a; Martinez, M.R. 1975a; Martinez & Ramos-Elorduy de Concini 1978a; Nagaich & Giri 1971a, 1973a; Nagaich et al. 1974a; Norris, D.O. 1954a; Raine, J. 1967a; Sahtiyanci, S. 1966a; Sekiyama & Fukushi 1955a; Shatrughna Singh & Nagaich 1979a; Webb & Schultz 1955a.
Potato Purple Top Wilt: Harding & Teakle 1985a; Martinez & Ramos- Elorduy de Concini 1978a; Norris, D.O. 1954a; Shiomi & Sugiura 1984b; Webb & Schultz 1955a.
Primulas: Marwitz et al. 1984a; Stevens & Spurdon 1972a.
Purple Top Roll of Potato: Nagaich & Giri 1973a; Nagaich et al. 1974a; Shatrughna Singh & Nagaich 1979a.
Rape: Horvath, J. 1969a; Lehmann & Skadow 1971a.
Raspberry: Converse et al. 1982a; Fluiter & van der Meer 1956a; Keldysh & Pomazkov 1967a.
Raspberry Dwarf: Keldysh & Pomazkov 1967a.
Rhynchosia: Dabek, A.J. 1983a.
Rhynchosia Little Leaf: Dabek, A.J. 1983a.
Rice: Arjunan et al. 1985a; Basu et al. 1974a; Chen, C.C. 1970b; Chen & Ko 1978a; Ching-Chung Chen 1970a; Chiu, Lin et al. 1968a; Faan et al. 1983a; Galvez & Miah 1969a; Govindu et al. 1968a; Hashioka, Y. 1952a; Hirao & Inoue 1978b; Iida & Shinkai 1950a; John, V.T. 1966a; Ling, Tiongco & Cabunagan 1983a; Mathew & Abraham 1977a; Mitra & John 1973a; Morinaka & Sakurai 1970a; Muniyappa & Raju 1981a; Muniyappa & Ramakrishnan 1976a, 1976b, 1980a; Muniyappa & Viraktamath 1981a; Nasu et al. 1967a; Ouchi & Suenaga 1963a, 1963b, 1968a; Raychaudhuri et al. 1967a, 1973a; Rivera et al. 1963a; Saito et al. 1976a; Sameshima & Nagai 1962a; Shikata, E. 1979b; Shikata et al. 1968a; Shinkai, A. 1959a, 1960a; Singh, K.G. 1971b; Singh et al. 1970a; Soon & Guan 1968a; Takahashi, Y. 1963a, 1963b; Takahashi & Sekiya 1962a; Wathanakul, L. 1964a.
Rice Brown Spot: Hashioka, Y. 1952a.
Rice Gall Dwarf: Faan et al. 1983a.
Rice Orange Leaf: Rivera et al. 1963a; Saito et al. 1976a; Singh, K.G. 1971b; Wathanakul, L. 1964a.
Rice Yellow Dwarf: Arjunan et al. 1985a; Basu et al. 1974a; Chen, C.C. 1970b; Chen & Ko 1978a; Ching-Chung Chen 1970a; Chiu, Lin et al. 1968a; Govindu et al. 1968a; Hashioka, Y. 1952a; Hirao & Inoue 1978b; Iida & Shinkai 1950a; John, V.T. 1966a; Mathew & Abraham 1977a; Mitra & John 1973a; Morinaka & Sakurai 1970a; Muniyappa & Raju 1981a; Muniyappa & Ramakrishnan 1976a, 1976b, 1980a; Muniyappa & Viraktamath 1981a; Nasu et al. 1967a; Ouchi & Suenaga 1963a, 1963b, 1968a; Raychaudhuri et al. 1967a, 1973a; Sameshima & Nagai 1962a; Shikata et al. 1968a; Shinkai, A. 1959a, 1960a; Takahashi, Y. 1963a, 1963b; Takahashi & Sekiya 1962a.
Rubus: Klein et al. 1976a; Van Der Meer & Fluiter 1970a.
Rubus Stunt: Jenser et al. 1981a; Van Der Meer & Fluiter 1970a.
Safflower: Klein, M. 1970a; Raccah & Klein 1982a.

Safflower Phyllody: Klein, M. 1970a; Raccah & Klein 1982a.
Sandal Spike: Sen-Sarma, P.K. 1982a; Sivaramakrishnan & Sen-Sarma 1978a; Varmah, J.C. 1981a.
Sesamum: Abraham et al. 1977a; Anonymous 1967c; Kooner et al. 1978a; Murugesan et al. 1973a; Regupathy & Jayaraj 1973b; Sahambi, H.S. 1970a; Sundararaju & Jayaraj 1977a.
Sesamum Phyllody: Abraham et al. 1977a; Anonymous 1967c; Kooner et al. 1978a; Murugesan et al. 1973a; Regupathy & Jayaraj 1973b; Sahambi, H.S. 1970a; Sundararaju & Jayaraj 1977a.
Sickle Hare's Ear Yellows: Shiomi & Choi 1983a.
Sida: Laboucheix et al. 1973a.
Solanaceae: Freitag, J.H. 1962a; Marchoux et al. 1970a.
Sorghum: Granada, G.A. 1979a.
Spinach: Esau et al. 1976a.
Spiroplasma citri: Archer et al. 1982a; Bove et al. 1979a; Calavan et al. 1979a; Fletcher et al. 1981a; Hafidi et al. 1979a; Kaloostian, Oldfield, Pierce et al. 1979a; Liu, Gumpf et al. 1983a, 1983b; Markham & Townsend 1974a; Meyerdirk et al. 1983a; Ogunlana & Pedigo 1984a; O'Hayer et al. 1983a; Oldfield, G.N. 1980a; Oldfield & Kaloostian 1979a; Oldfield et al. 1978a; Oldfield, Kaloostian, Pierce, Calavan et al. 1977a; Russo et al. 1976a; Saillard et al. 1980a.
Stolbur Disease: Abu Yaman, I.K. 1971a; Brcak, J. 1954a; Gaborjanyi & Saringer 1967a; Giannotti, Vago & Duthoit 1968a; Maillet et al. 1968a; Musil, M. 1958b, 1959b, 1960b, 1966a; Musil & Valenta 1958a; Sahtiyanci, S. 1966a; Savulescu & Ploaie 1962a; Valenta, V. 1958a.
Stone Fruits: Gilmer et al. 1966a; Jensen, D.D. 1962a.
Strawberry: Bovey, B. 1958a; Chiykowski, L.N. 1962c; Chiykowski & Craig 1978a; Frazier & Jensen 1970a; Frazier & Posnette 1956a, 1957a; Greber & Gowanlock 1979a; Hill, S.A. 1980a; Misiga et al. 1960a; Posnette & Ellenberger 1963a; Shiomi & Sugiura 1983a; Thompson, L.S. 1968a.
Strawberry Green Petal: Chiykowski, L.N. 1962c; Chiykowski & Craig 1978a; Frazier & Posnette 1956a, 1957a; Misiga et al. 1960a; Posnette & Ellenberger 1963a; Thompson, L.S. 1968a.
Stubborn Disease: Bove et al. 1979a; Calavan et al. 1974a, 1979a; Gumpf & Calavan 1981a; Hafidi et al. 1979a; Kaloostian, Oldfield, Pierce, Calavan, Granett, Rana & Gumpf 1975a; Kaloostian, Oldfield, Gumph et al. 1979a; Kaloostian, Oldfield, Pierce, Calavan, Granett, Rana, Gumpf et al. 1975a; Lee et al. 1973a, 1973b; Oldfield & Kaloostian 1979a; Oldfield et al. 1976a, 1978a.
Sugarbeet: Meyerdirk et al. 1983a.
Sugarcane: Chen, C.T. 1973a; Lee & Chen 1972a; Matsumoto et al. 1968a, 1968b; Shikata et al. 1968a.
Sugarcane White Leaf: Chen, C.T. 1973a; Lee & Chen 1972a; Matsumoto et al. 1968a, 1968b; Shikata et al. 1968a.
Sweet Potato: Jackson & Zettler 1983a; Murayama, D. 1966a; Shinkai, A. 1964a, 1968a; Tsai et al. 1972a; Yang & Chou 1982a.
Tomato: Abu Yaman, I.K. 1971a; Bowyer, J.W. 1974a; Giannotti, Vago & Duthoit 1968a.
Tomato Big Bud: Bowyer, J.W. 1974a; Norris, D.O. 1954a.
Trees: Ghosh, S.K. 1981a; Hibben et al. 1973a; Jin et al. 1981a; Kim, C.J. 1966a; Seliskar & Wilson 1981a; Sen-Sarma, P.K. 1982a, 1984a; Sivaramakrishnan & Sen-Sarma 1978a; Varmah, J.C. 1981a; Zhang et al. 1983a.
Vegetables: Novak, J.B. 1961a.
Western X-disease: Frazier & Jensen 1970a; Gilmer et al. 1966a; Gilmer & McEwen 1958a; Gold, R.E. 1974a, 1979a; Gold & Sylvester 1982a; Gonot & Purcell 1981a; Jensen, D.D. 1962a, 1968a, 1969a, 1971a, 1971b, 1972a; Jensen et al. 1967a; Jensen & Richardson 1972a; Kirkpatrick et al. 1985a; Lacy, G.H. 1982a; Lacy et al. 1979a; Lee & Jensen 1963a; McClure, M.S. 1980b, 1982a; McClure et al. 1982a; Nasu, Jensen & Richardson 1970a, 1974a; Nasu, Jensen, Richardson et al. 1974a, 1974b; Nielson, M.W. 1979d; Nielson & Jones 1954a; Nyland et al. 1981a; Palmiter & Adams 1957a; Peterson, G.W. 1984a; Purcell, A.H. 1979a; Purcell, Raju et al. 1980a, 1981a; Purcell, Richardson et al. 1981a; Raju et al. 1984a; Rice & Jones 1972a; Rosenberger & Jones 1978a; Sinha & Chiykowski 1980a; Suslow & Purcell 1982a; Taboada, O. 1979a; Taboada et al. 1975a; Whitcomb, R.F. 1972b; Whitcomb, Jensen et al. 1966a; Wolfe, H.R. 1955a, 1955b.
Wheat Striate Mosaic: Westdal & Richardson 1966a.
Wheat Yellow Dwarf: Munteanu et al. 1983a.
Witches' Broom: Bergman, B.H.H. 1956b; Blattny & Blattny 1970a; Bos & Grancini 1965a; Bowyer, J.W. 1974a; Chou et al. 1975a; Converse et al. 1982a; Doi et al. 1967a; Fukushi et al. 1955a; Jackson & Zettler 1983a; Jin et al. 1981a; Kim, C.J. 1966a; Klein et al. 1976a; Licha-Baquero, M. 1979a; McCoy et al. 1983a; Murayama, D. 1966a; Nagaich et al. 1974a; Nour, M.A. 1962a; Raine, J. 1967a; Sekiyama & Fukushi 1955a; Shatrughna Singh & Nagaich 1979a; Shinkai, A. 1964a, 1968a, 1975a; Shiomi & Sugiura 1983a; Thung & Hadiwidjaja 1957a; Tsai et al. 1972a; Yang & Chou 1982a; Zhang et al. 1983a.
Yellows Diseases: Bowyer, J.W. 1972a; Caudwell, A. 1977a; Caudwell, Larrue & Kuszala 1978a; Chiykowski, L.N. 1965c, 1974a; Davis & Whitcomb 1971a; Giannotti et al. 1969a; Giannotti, Vago, Devauchelle et al. 1968a; Giannotti, Vago & Duthoit 1968a; Greber & Gowanlock 1979a; Hanson et al. 1963a; James & Granovsky 1927a; Koga et al. 1982a; Kunkel, L.O. 1954a; Marchoux et al. 1970a; Miyahara et al. 1983a; Moreau et al. 1968a; Musil, M. 1964a, 1964c, 1965a; Novak, J.B. 1961a; Seliskar & Wilson 1981a; Shikata et al. 1969a; Slogteren & Muller 1972a, 1972b; Valenta, V. 1958a; Valenta et al. 1961a; Weaver, C.R. 1954a.

Viruses

General: Abeygunawardena, D.V.W. 1969a; Abeygunawardena et al. 1971a; Agarkov, V.A. 1972a; All et al. 1981a; Amen, C.R. 1953a; Belli, G. 1974a; Bennett, C.W. 1971a; Bindra, Khatri & Sohi 1973a; Black, L.M. 1953a, 1958a, 1958b, 1962a, 1979a; Black et al. 1958a; Bos, L. 1981a; Bovey, B. 1957a, 1958a, 1958b; Bradfute et al. 1981a; Brakke et al. 1953a; Bremer & Raatikainen 1975a; Cadman, C.H. 1961a; Catara, A. 1984a; Caudwell, A. 1977a; Chaudhuri, R.P. 1955a; Chen, C.C. 1978a; Chiu et al. 1966a; Chiykowski, L.N. 1964a; Conti, M. 1981a, 1985a; Coudriet & Tuttle 1963a; Cousin & Moreau 1966a; Cousin et al. 1965a; Cropley, R. 1960a; D'Arcy & Nault 1982a; David & Alexander 1984a; Day & Bennetts 1954a; Evenhuis, H.H. 1955b; Fajemisin & Shoyinka 1977a; Fluiter, H.J. de 1958a; Fluiter et al. 1955a; Forbes & MacCarthy 1969a; Frazier, N.W. 1965a; Frederiksen, R.A. 1962a, 1964a; Fritzsche et al. 1972a; Gaborjanyi & Nagy 1972a; Gaborjanyi & Saringer 1967a; Galvez & Miah 1969a; Gamez, R. 1969a; Gamez & Black 1967a, 1968a; Gamez & Leon 1985a; Gamez & Saavedra 1985a; Gao, D.M. 1983a, 1985a; Ghosh, S.K. 1985a; Gilmer & McEwen 1958a; Gondran, J. 1967a; Granados, R.R. 1969b, 1972a; Greber, R.S. 1983a, 1984a, 1984b; Grylls, N.E. 1954a, 1961a; Grylls & Waterford 1976a; Hainzelin, E. 1982a; Hansen, H.P. 1961a; Harizanov, A. 1968a; Harris, K.F. 1979a, 1981b, 1981c, 1983a; Harrison, B.D. 1981a; Hatta & Francki 1982a; Heinrichs, E.A. 1979a; Heinze, K.G. 1959a; Heskova et al. 1961a; Hibben et al. 1973a; Hill, S.A. 1980a; Hills et al. 1963a; Hirumi & Maramorosch 1968a; Hirumi et al. 1967a; Hoffman & Taboada 1962a; Holdeman & McCartney 1965a; Hoppe, W. 1976a; Hsu & Banttari 1979a; Hsu & Black 1974a; Hsu et al. 1977a, 1983a; Igarashi et al. 1982a; Iida, T. 1965a; Ishihara, T. 1965a; Ishihara & Nasu 1966a; Ishii, M. 1973a; Ivancheva et al. 1967a; Iwaki, M. 1979a; Jensen, D.D. 1956a, 1959a, 1959b, 1963a, 1969b; Jensen et al. 1972a; Keshavamurthy & Yaraguntaiah 1969a; Kimura & Black 1971a, 1972a, 1972b; Kiritani, K. 1981a; Knoke et al. 1977a; Kochman & Ksiazek 1967a; Krczal, H. 1960a; Lapierre, H. 1980a; Lastra & Trujillo 1976a; Lehmann, W. 1982a; Lehmann & Skadow 1970a; Lherault, P. 1968a; Lim, G.S. 1969a; Lindsten, K. 1979a; Lindsten et al. 1970a; Ling, K.C. 1969a, 1969b, 1972a, 1973a, 1979a; Ling, Tiongco & Cabunagan 1983a; Liu & Black 1978a; Liu et al. 1973a; Lockhart et al. 1985a; Maillet & Folliot 1967a, 1967b; Maramorosch, K. 1954a, 1956b, 1958a, 1959a, 1959b, 1960a, 1963b, 1964a, 1964b, 1965a, 1965b, 1967a, 1968a, 1969a, 1969b, 1970a; Maramorosch et al. 1961a, 1977a; Maramorosch & Shikata 1965a; Maramorosch, Shikata & Granados 1969a; Maramorosch, Shikata, Hirumi et al. 1969a; Maramorosch & Jensen 1963a; Maramorosch & Kondo 1978a; Maramorosch, Mitsuhashi et al. 1965a; Mariappan et al. 1984a; Markham et al. 1984b; Mathur & Chaturvedi 1980a; Mathur & Singh 1975a; McClure, M.S.

1980a; Mercer, P.C. 1977a; Merr, F.A. 1981a; Miah, M.S.A. 1971a; Miller, P.R. 1954a, 1966a; Mitsuhashi, J. 1965b; Muratomaa & Valenta 1968a; Murthy et al. 1975a; Musil, M. 1958b, 1959a, 1960a, 1960b, 1962b, 1962c, 1965a; Musil & Valenta 1958a; Nagaich, B.B. 1960a; Nagaraj & Black 1961a, 1962a; Nagaraju et al. 1984a; Nakasuji & Nomura 1968a; Nasu, S. 1963a, 1967a, 1969b; Nault, L.R. 1983a, 1983b, 1984a, 1984b; Nault et al. 1979a, 1982a; Nault & Knoke 1981a; Nielson, M.W. 1962c, 1968b; Novak, J.B. 1961a; Nowacka & Hoppe 1969a; Ofori & Francki 1985a; Okoth et al. 1985a; Oldfield, G.N. 1980a; Oman, P.W. 1969a; Otake, A. 1983a; Ou & Ling 1966a; Ou & Rivera 1969a; Pirone et al. 1972a; Pridantseva, E.A. 1972a; Raychaudhuri et al. 1969a; Razvyazkina, G.M. 1962a; Razvyazkina & Pridantseva 1968a; Reddy, D.B. 1968a; Ricaud & Felix 1976a; Richardson & Raabe 1956a; Revilla, M.V.A. 1965a; Rocha-Pena, M. 1981a; Rossel et al. 1984a; Saito, Y. 1977a, 1977b; Seetharaman & John 1981a; Selsky, M.I. 1961a; Shikata, E. 1979b; Shikata & Maramorosch 1965c, 1965d, 1967a, 1969a; Shikata et al. 1964a, 1966a; Shinkai, A. 1962a, 1965a; Sinha, R.C. 1963a, 1968b, 1973a, 1973b; Slykhuis, J.T. 1951a, 1961a; Smith & Brierley 1956a; Smith, K.M. 1957a, 1958a; Stoner, W.N. 1965a; Streissle et al. 1968a; Sulochana, C.B. 1984a; Szekessy, V. 1969a; Thomas, P.E. 1972a; Timian, R.G. 1960a; Trujillo et al. 1974a; Vacke, J. 1966a; Varney, E.H. 1977a; Vidano, C. 1972a; Vlasov & Rodina 1969a; Volk, J. 1958a, 1967a; Wathanakul & Weerapat 1969a; Wegorek, W. 1967a; Wenzil, H. 1959a; Whellan, J.A. 1964a; Whitcomb, R.F. 1969a, 1972a, 1972b; Whitcomb & Black 1969a; Whitcomb & Davis 1970a; Yokayama & Sakai 1975a; Yoshii, H. 1959a.

Axonopus Chlorotic Streak: Van Velsen, R.J. 1967a.

Bajra Streak: Seth et al. 1972a.

Barley Stripe Mosaic: Chiko, A.W. 1973a.

Barley Yellow Dwarf: Gill et al. 1969a.

Bermuda Grass Etched Line Virus: Lockhart et al. 1985a.

Blueberry: Varney, E.H. 1977a.

Cantaloupe: Hills et al. 1961a, 1964a; Hills & Taylor 1954a.

Cereals: Agarkov, V.A. 1972a; All et al. 1981a; Bantarri & Moore 1962a; Bantarri & Zeyen 1968a, 1970a; Bremer & Raatikainen 1975a; Chiko, A.W. 1973a; Conti, M. 1981a; Galvez et al. 1963a; Gill et al. 1969a; Greber, R.S. 1979a, 1983a, 1984a; Grylls, N.E. 1961a, 1963a, 1963b; Hamilton, R.E. 1964a; Harder & Westdal 1971a; Heskova et al. 1961a; Hoppe, W. 1972a, 1974a, 1976a; Iida, T. 1965a; Kuklarni et al. 1980a; Lapierre, H. 1980a; Lee, P.E. 1963a; Lee, P.W. 1963a; Lee & Bell 1963a; Lindsten et al. 1970a; Lomakina et al. 1963a; Long & Timian 1971a; Maramorosch et al. 1977a; Mayhew & Flock 1981a; Mzira, C.N. 1984b; Nowacka & Hoppe 1969a; Paliwal, Y.C. 1968a; Polyakova, G.P. 1972a; Pridantseva, E.A. 1972a; Radulescu & Munteanu 1970a; Razvyazkina & Polyakova 1967a, 1967b; Razvyazkina & Pridantseva 1968a; Seth et al. 1972a; Sinha, R.C. 1968c, 1970a; Slykhuis, J.T. 1951a, 1953a, 1961a, 1963a; Slykhuis & Sherwood 1964a; Sulochana, C.B. 1984a; Timian, R.G. 1960a, 1978a, 1985a; Vacke & Hoppe 1975a; Vlasov & Rodina 1969a; Vlasov et al. 1971a; Westdal & Richardson 1966a; Zeyen & Banttari 1972a.

Cereal Chlorotic Mottle: Greber, R.S. 1979a; Greber & Gowanlock 1979b.

Cereal Enation Disease: Harder & Westdal 1971a.

Chilo Iridescent Virus: Jensen et al. 1972a.

Chloris Striate Mosaic: Grylls & Waterford 1976a.

Citrus: Catara, A. 1984a.

Clover: Bovey, B. 1957a; Cousin & Moreau 1966a; Cousin et al. 1965a; Gondran, J. 1967a; Grylls, N.E. 1955a; Grylls et al. 1974a; Krczal, H. 1960a; Lherault, P. 1968a; Selsky & Black 1961a; Whitcomb, R.F. 1972b.

Coconut: Zelazny & Pacumbaba 1982a.

Coconut Cadang Cadang: Zelazny & Pacumbaba 1982a.

Corn Mosaic: McEwen & Kawanishi 1967a.

Cucumber Mosaic: Hills et al. 1961a.

Enanismo of Small Grains: Galvez et al. 1963a.

Fiji Disease: Hatta & Francki 1982a.

Flax: Bantarri & Moore 1962a; Frederiksen, R.A. 1964a.

Giallume Disease of Rice: Belli et al. 1974a.

Grasses: Conti, M. 1981a; Ghosh & John 1979a; Greber, R.S. 1979a, 1983a, 1984a; Grylls, N.E. 1961a, 1963a; Knoke et al. 1983a; Mishra et al. 1973a; Muratomaa & Valenta 1968a; Revilla, M.V.A. 1965a; Van Velsen, R.J. 1967a.

Grassy Stunt: Palmer & Rao 1981a.

Groundnuts: Mercer, P.C. 1977a.

Herbs: Gaborjanyi & Nagy 1972a; Jensen, D.D. 1956a.

Legumes: Abney et al. 1976a; Caudwell, A. 1977a; Iwaki, M. 1979a; Jayaraj & Seshadri 1967a; Lo, T.C. 1966a; Regupathy et al. 1975a; Wegorek, W. 1967a.

Lettuce: Novak, J.B. 1961a.

Maize (Corn): All et al. 1977a, 1981a; Alverson, D.R. 1979a; Anonymous 1985a; Boccardo et al. 1980a; Bradfute et al. 1972a, 1981a, 1985a; Choudhury & Rosenkranz 1983a; Dabrowski, Z.T. 1985a; Damsteegt, V.D. 1980a, 1983a, 1984a; Delgadillo-Sanchez, F. 1984a; Escobedo et al. 1985a; Espinoza & Gamez 1980a; Fajemisin et al. 1976a; Fajemisin & Shoyinka 1977a; Gamez, R. 1969a, 1975a, 1977a, 1980a, 1980b, 1983a, 1983b; Gamez et al. 1979a, 1981a; Gamez & Leon 1983a, 1985a; Gonzalez & Gamez 1974a; Gordon & Nault 1977a; Graham, C.L. 1979a; Granados, R.R. 1969b; Greber, R.S. 1984b; Grylls, N.E. 1975a; Guthrie, E.J. 1978a; Hainzelin, E. 1982a; Harris & Childress 1980a; Holdeman & McCartney 1965a; Kitajima & Gamez 1977a, 1983a; Knoke et al. 1983a; Lastra & Trujillo 1976a; Maramorosch, K. 1959b; Maramorosch et al. 1961a; Markham et al. 1984a, 1984b; McEwen & Kawanishi 1967a; Milne & Lovisolo 1977a; Mzira, C.N. 1984a; Nault, L.R. 1983a, 1983b, 1984a, 1984b; Nault et al. 1973a, 1979a, 1980a, 1982a; Nault & Knoke 1981a; Ofori & Francki 1983a, 1984a, 1985a; Paniagua & Gamez 1976a; Pirone et al. 1972a; Pitre, H.N. 1966a; Ricaud & Felix 1976a; Rico de Cujia & Martinez-Lopez 1977a; Rocha-Pena, M. 1981a; Rose, D.J.W. 1962a, 1973c; Rossel, H.W. 1984a; Rossel et al. 1980a, 1984a; Saavedra, F. 1982a; Scott & Rosenkranz 1981a; Soto, P.E. 1978a; Stoner, W.N. 1965a; Storey et al. 1966a; Trujillo et al. 1974a; Van Rensburg, G.D.J. 1981a; Van Rensburg & Kuhn 1977a; Wolanski & Maramorosch 1979a.

Maize Chlorotic Dwarf: All et al. 1977a; Alverson, D.R. 1979a; Choudhury & Rosenkranz 1983a; Gordon & Nault 1977a; Harris & Childress 1980a; Scott & Rosenkranz 1981a.

Maize Dwarf: Nault et al. 1973a.

Maize Dwarfing Disease: Bradfute et al. 1972a.

Maize Dwarf Mosaic: All et al. 1977a; Alverson, D.R. 1979a; Bradfute et al. 1985a; Knoke et al. 1983a; Scott & Rosenkranz 1981a.

Maize Mottle: Rossel et al. 1980a.

Maize Mottle Chlorotic Stunt: Rossel, H.W. 1984a.

Maize Rayado Fino: Bradfute et al. 1985a; Delgadillo-Sanchez, F. 1984a; Escobedo et al. 1985a; Espinoza & Gamez 1980a; Gamez, R. 1975a, 1977a, 1980a, 1980b, 1983a, 1983b; Gamez et al. 1979a, 1981a; Gamez & Leon 1983a; Gonzalez & Gamez 1974a; Kitajima & Gamez 1977a, 1983a; Lockhart et al. 1985a; Nault et al. 1980a; Paniagua & Gamez 1976a; Rico de Cujia & Martinez-Lopez 1977a; Saavedra, F. 1982a; Wolanski & Maramorosch 1979a.

Maize Rough Dwarf: Milne & Lovisolo 1977a.

Maize Streak: Anonymous 1985a; Dabrowski, Z.T. 1985a; Damsteegt, V.D. 1980a, 1983a, 1984a; Fajemisin et al. 1976a; Fajemisin & Shoyinka 1977a; Graham, C.L. 1979a; Guthrie, E.J. 1978a; Markham et al. 1984a; Mzira, C.N. 1984a, 1984b; Rose, D.J.W. 1962a, 1973c; Soto, P.E. 1978a; Storey et al. 1966a; Van Rensburg, G.D.J. 1981a; Van Rensburg & Kuhn 1977a.

Maize Wallaby Ear: Boccardo et al. 1980a; Grylls, N.E. 1975a; Ofori & Francki 1983a, 1984a.

Mulberry: Giorgadze, D.C. 1968a.

Mulberry Disease: Giorgadze, D.C. 1968a.

Mustard: Mathur & Singh 1975a.

Oat Blue Dwarf: Bantarri & Moore 1962a; Bantarri & Zeyen 1968a, 1970a; Hsu & Banttari 1979a; Lockhart et al. 1985a; Long & Timian 1971a; Timian, R.G. 1985a; Zeyen & Banttari 1972a.

Okra: Regupathy & Jayaraj 1972a.

Onion Yellow Dwarf: Kochman & Ksiazek 1967a.

Papaya: Ivancheva et al. 1967a.

Potato: Amen, C.R. 1953a; Chiu et al. 1970a; Evenhuis, H.H. 1955b; Falk et al. 1981a; Khanna, V.M. 1973a; Larson, R.H. 1945a; Raymer, W.B. 1956a; Schumann et al. 1980a.
Potato Hairy Sprout: Khanna, V.M. 1973a.
Potato Late-Breaking Virus: Raymer, W.B. 1956a.
Potato Spindle Tuber: Schumann et al. 1980a.
Potato Yellow Dwarf: Chiu et al. 1970a; Falk et al. 1981a; Hsu & Black 1974a; Larson, R.H. 1945a; Liu & Black 1978a; Nagaich, B.B. 1960a.
Ragged Stunt: Palmer & Rao 1981a; Panda et al. 1984a.
Ragi Disease: Keshavamurthy & Yaraguntaiah 1969a.
Ragi Mosaic: Murthy et al. 1975a.
Raji Streak: Nagaraju et al. 1984a.
Raspberry: Cadman, C.H. 1961a.
Rhododendron: Viennot-Bourgin 1981a.
Rhododendron Bud Blast: Viennot-Bourgin 1981a.
Rice: Abeygunawardena, D.V.W. 1969a; Abeygunawardena et al. 1971a; Aguiero et al. 1979a; Anjaneyulu, A. 1975a, 1975b, 1980a; Anjaneyulu & Chakraborti 1977a; Anjaneyulu & John 1972a; Anjaneyulu, Shukla et al. 1981a, 1982a; Anjaneyulu, Singh et al. 1982a; Anjaneyulu et al. 1980a; Bae & Pathak 1969a; Basu et al. 1974a, 1976a; Belli et al. 1974a; Cabauatan & Hibino 1984a; Cabunagan et al. 1984a, 1985a; Chen, C.C. 1978a; Chen & Chiu 1980a; Chen et al. 1980a; Chen & Shikata 1972a; Chettanachit & Disthaporn 1982a; Chiu, R.J. 1982a; Chiu & Jean 1967a; Chiu, Jean et al. 1968a; Chiu et al. 1965a; Dahal & Hibino 1985b; Faan et al. 1983a; Fajardo et al. 1964a; Furuta, T. 1977a; Galvez & Miah 1969a; Galvez et al. 1971a; Gao, D.M. 1985a; Ghosh & John 1979a; Ghosh et al. 1979a; Gopalkrishnan et al. 1974a; Govindu et al. 1968a; Halteren et al. 1974a; Hasanuddin & Ling 1980a, 1980b; Hashizume, B. 1964a; Heinrichs, E.A. 1979a; Heinrichs & Rapusas 1983a, 1984a, 1984b; Hibino, H. 1980a, 1983a, 1983b; Hibino & Cabuatan 1985a; Hibino, Roechan et al. 1978a; Hibino, Saleh et al. 1978a; Hibino et al. 1979a, 1983a; Hino et al. 1974a; Hirai et al. 1968a; Hirao & Inoue 1978a, 1979c, 1980a; Hirao et al. 1974a; Hsieh, S.P.Y. 1969a, 1969b; Hsieh et al. 1970a; Hsieh & Liao 1974a; Hsieh & Roan 1967a; Inoue, H. 1977a, 1978a, 1979a; Inoue & Hirao 1980a, 1981a; Inoue & Omura 1982a; Inoue et al. 1980a; Ishii, M. 1973a; Ishii & Ono 1966a; Iwasaki et al. 1978a, 1979a; John, V.T. 1965a, 1966a; John et al. 1979a; John & Ghosh 1981a; Kamiwada et al. 1976a; Kannaiyan et al. 1978a; Khan & Saxena 1985c; Kimura, I. 1962a, 1976a, 1980a, 1982a, 1985a; Kimura & Fukushi 1960a; Kimura, T. 1975a; Kimura et al. 1975a; Kiritani, K. 1981a; Kiritani & Nakasuji 1977a; Kiritani & Sasaba 1978a; Kitagawa & Shikata 1969a, 1973a, 1974a; Koesnang & Rao 1980a; Kono et al. 1976a; Lamey, Surin & Leewangh 1967a; Lamey, Surin, Disthaporn et al. 1967a; Lim, G.S. 1969a, 1972a; Lin et al. 1984a; Ling, K.C. 1966a, 1968b, 1968c, 1969b, 1970a, 1972a, 1973a, 1975a, 1979a; Ling & Carbonell 1975a; Ling et al. 1979a, 1981a; Ling & Miah 1980a; Ling & Palomar 1966a; Ling & Tiongco 1975a, 1977a, 1979a; Ling, Tiongco & Cabunagan 1983a; Ling, Tiongco & Fores 1983a; Lippold et al. 1970a; Maramorosch, K. 1969b; Maramorosch et al. 1961a; Mariappan et al. 1982a, 1983a, 1984a; Mariappan & Saxena 1983a, 1983b, 1984a; Mathur & Chaturvedi 1980a; Miah, M.S.A. 1971a; Mishra et al. 1973a, 1976a; Mitra & John 1973a; Morinaka et al. 1980a, 1982a; Mukhopadhyay & Chowdhury 1970a, 1973a; Mukhopadhyay et al. 1978a; Nanda et al. 1976a; Narayanasamy, P. 1972a, 1972b; Nasu, S. 1963a, 1967a, 1969b; Nishi et al. 1975a; Nuque & Miah 1969a; Omura et al. 1980a, 1980b; Omura, Inoue et al. 1982a; Omura, Kimura et al. 1982a; Omura, Saito et al. 1982a; Ou & Ling 1966a; Ou & Rivera 1969a; Ou et al. 1965a; Palmer & Rao 1981a; Palomar & Ling 1966a; Panda et al. 1984a; Putta et al. 1980a, 1982a; Rahman & Hibino 1985a, 1985b; Rao & Anjaneyulu 1977a, 1978a, 1979a, 1980b, 1982a; Rao & John 1974a; Raychaudhuri et al. 1967b, 1969a, 1973a; Reddy, D.B. 1968a; Reyes, G.M. 1957a; Rivera et al. 1969a, 1972a; Rivera & Ou 1965a, 1967a; Saito, Y. 1977a, 1977b; Saito et al. 1975a, 1978a, 1981a; Satomi et al. 1975a; Seetharaman & John 1981a; Serrano, F.B. 1957a; Shikata, E. 1979b; Shinkai, A. 1956a, 1960a, 1962a, 1965a, 1977a; Shukla & Anjaneyulu 1981c, 1982a, 1982b; Singh, K.G. 1971a, 1979a; Siwi & Roechan 1983a; Sogawa, K. 1976a; Sonku, U. 1973a; Sridhar et al. 1978a; Suenaga, H. 1962a; Ting, W.P. 1971a; Ting & Ong 1974a; Ting & Paramsothy 1970a; Toyoda et al. 1964a, 1965a; Van Halteren & Sama 1973a; Wathanakul, L. 1964a; Wathanakul & Weerapat 1969a; Xie & Lin 1983a; Xie et al. 1981a, 1982a, 1984a; Yokayama & Sakai 1975a; Yokayama et al. 1974a; Yoshii, H. 1959b; Yoshii et al. 1961a.
Rice Black-Streaked Dwarf: Kitagawa & Shikata 1969a, 1973a, 1974a.
Rice Blast: Singh, K.G. 1979a.
Rice Bunchy Stunt: Xie & Lin 1983a; Xie et al. 1984a.
Rice Chlorotic Streak: Anjaneyulu et al. 1980a.
Rice Dwarf: Gao, D.M. 1985a; Hashizume, B. 1964a; Ishii & Ono 1966a; Kimura, I. 1962a, 1976a, 1980a, 1982a, 1985a; Kimura & Fukushi 1960a; Kimura, T. 1975a; Kiritani & Nakasuji 1977a; Kiritani & Sasaba 1978a; Kono et al. 1976a; Omura, Inoue et al. 1982a; Reyes, G.M. 1957a; Shinkai, A. 1956a, 1960a; Suenaga, H. 1962a; Toyoda et al. 1964a, 1965a; Xie et al. 1981a, 1982a.
Rice Gall Dwarf: Faan et al. 1983a; Inoue & Omura 1982a; Morinaka et al. 1980a, 1982a; Omura et al. 1980a, 1980b; Omura, Kimura et al. 1982a; Putta et al. 1980a, 1982a.
Rice Penyakit Merah: Lim, G.S. 1972a; Ou et al. 1965a; Singh, K.G. 1971a, 1979a; Ting, W.P. 1971a; Ting & Ong 1974a; Ting & Paramsothy 1970a.
Rice Ragged Stunt: Aguiero et al. 1979a.
Rice Stripe: Hirai et al. 1968a; Sonku, U. 1973a.
Rice Stunt: Serrano, F.B. 1957a; Yoshii, H. 1959b; Yoshii et al. 1961a.
Rice Transitory Yellowing: Chen & Chiu 1980a; Chen et al. 1980a; Chen & Shikata 1972a; Chiu, R.J. 1982a; Chiu & Jean 1967a; Chiu, Jean et al. 1968a; Chiu et al. 1965a; Gao, D.M. 1985a; Hsieh, S.P.Y. 1969a, 1969b; Hsieh et al. 1970a; Hsieh & Liao 1974a; Hsieh & Roan 1967a; Inoue, H. 1979a; Inoue et al. 1980a; Saito et al. 1978a.
Rice Tungro: Anjaneyulu, A. 1975a, 1975b, 1980a; Anjaneyulu & Chakraborti 1977a; Anjaneyulu & John 1972a; Anjaneyulu, Shukla et al. 1981a, 1982a; Anjaneyulu, Singh et al. 1982a; Bae & Pathak 1969a; Basu et al. 1974a, 1976a; Cabauatan & Hibino 1984a; Cabunagan et al. 1984a, 1985a; Chettanachit & Disthaporn 1982a; Dahal & Hibino 1985b; Fajardo et al. 1964a; Galvez et al. 1971a; Ghosh & John 1979a; Ghosh et al. 1979a; Gopalkrishnan et al. 1974a; Govindu et al. 1968a; Halteren et al. 1974a; Hasanuddin & Ling 1980a, 1980b; Heinrichs & Rapusas 1983a, 1984a, 1984b; Hibino, H. 1980a, 1983a, 1983b; Hibino & Cabuatan 1985a; Hibino, Roechan et al. 1978a; Hibino, Saleh et al. 1978a; Hibino et al. 1979a, 1983a; John, V.T. 1965a, 1966a; John et al. 1979a; John & Ghosh 1981a; Kannaiyan et al. 1978a; Khan & Saxena 1985c; Koesnang & Rao 1980a; Lamey, Surin & Leewangh 1967a; Lin et al. 1984a; Ling, K.C. 1966a, 1968b, 1968c, 1970a, 1975a; Ling & Carbonell 1975a; Ling et al. 1979a, 1981a; Ling & Miah 1980a; Ling & Palomar 1966a; Ling & Tiongco 1975a, 1977a, 1979a; Ling, Tiongco & Fores 1983a; Lippold et al. 1970a; Mariappan et al. 1982a, 1983a; Mariappan & Saxena 1983a, 1983b, 1984a; Mishra et al. 1973a, 1976a; Mitra & John 1973a; Mukhopadhyay & Chowdhury 1970a, 1973a; Mukhopadhyay et al. 1978a; Nanda et al. 1976a; Narayanasamy, P. 1972a, 1972b; Nuque & Miah 1969a; Omura, Saito et al. 1982a; Palmer & Rao 1981a; Palomar & Ling 1966a; Panda et al. 1984a; Rahman & Hibino 1985a, 1985b; Rao & Anjaneyulu 1977a, 1978a, 1979a, 1980b, 1982a; Rao & John 1974a; Raychaudhuri et al. 1967b, 1973a; Rivera et al. 1969a, 1972a; Rivera & Ou 1965a, 1967a; Saito, Y. 1977a; Saito et al. 1975a, 1981a; Shukla & Anjaneyulu 1981c, 1982a, 1982b; Siwi & Roechan 1983a; Sogawa, K. 1976a; Sridhar et al. 1978a; Van Halteren & Sama 1973a; Wathanakul, L. 1964a.
Rice Waika: Furuta, T. 1977a; Hibino, H. 1983b; Hirao & Inoue 1978a, 1979c, 1980a; Hirao et al. 1974a; Inoue, H. 1977a, 1978a; Inoue & Hirao 1980a, 1981a; Iwasaki et al. 1978a, 1979a; Kamiwada et al. 1976a; Kimura et al. 1975a; Nishi et al. 1975a; Saito, Y. 1977a; Satomi et al. 1975a; Shinkai, A. 1977a; Yokayama et al. 1974a.
Rice Yellow-Orange Leaf: Hino et al. 1974a; Lamey, Surin, Disthaporn et al. 1967a.
Rugose Leaf Curl: Grylls, N.E. 1954a, 1955a; Grylls et al. 1974a.
Sorghum Stunt Mosaic: Mayhew & Flock 1981a.
Soybean Mosaic: Abney et al. 1976a.

Soybean Rosette: Lo, T.C. 1966a.
Spinach: Heskova et al. 1961a; Richardson & Raabe 1956a.
Sterility Mosaic: Jayaraj & Seshadri 1967a.
Strawberry: Bovey, B. 1958a; Cropley, R. 1960a; Frazier, N.W. 1975a; Greber & Gowanlock 1979b; Hill, S.A. 1980a; Krczal, H. 1960a; Lherault, P. 1968a.
Strawberry Pallidosis: Frazier, N.W. 1975a.
Sugarbeet: Bennett, C.W. 1953a, 1962a, 1962b, 1967a, 1971a; Bennett & Costa 1949a; Bennett & Tanrisever 1957a, 1958a; Duffus & Gold 1973a; Duffus et al. 1982a; Fullerton, D. 1964a; Gibson & Oliver 1970a; Giddings, N.J. 1954a; Gracia & Feldman 1972a; Hills & Brubaker 1968a; Hills et al. 1963a; Izadpanah & Shepherd 1973a; Kaur et al. 1971a; Kheyri, M. 1969a; Landis et al. 1970a; Magyarosy, A.C. 1978a, 1980a; Magyarosy & Duffus 1977a; Magyarosy & Sylvester 1979a; Meyerdirk et al. 1983a; Mumford & Peay 1970a; Schneider, C.L. 1959a; Simons, J.N. 1962a; Staples et al. 1970a; Thomas, P.E. 1972a.
Sugarbeet Curly Top: Bennett, C.W. 1953a, 1962a, 1962b, 1967a, 1971a; Bennett & Costa 1949a; Bennett & Tanrisever 1957a, 1958a; Duffus & Gold 1973a; Duffus et al. 1982a; Fullerton, D. 1964a; Gardner & Cannon 1972a; Gibson & Oliver 1970a; Giddings, N.J. 1954a; Gracia & Feldman 1972a; Heggestad & Moore 1959a; Hills, O.A. 1958a; Hills & Brubaker 1968a; Hills et al. 1961a, 1964a; Hills & Taylor 1954a; Izadpanah & Shepherd 1973a; Kaur et al. 1971a; Keener, P.D. 1956a; Kheyri, M. 1969a; Landis et al. 1970a; Magyarosy, A.C. 1978a, 1980a; Magyarosy & Duffus 1977a; Magyarosy & Sylvester 1979a; Meyerdirk et al. 1983a; Mumford & Peay 1970a; Schneider, C.L. 1959a; Simons, J.N. 1962a; Thomas, P.E. 1972a; Thomas & Boll 1977a; Thomas & Martin 1971a; Young & Cannon 1973a.
Sugarbeet Western Yellows: Landis et al. 1970a.
Sugarbeet Yellow Vein: Staples et al. 1970a.
Sugarcane: Ammar et al. 1982a; David & Alexander 1984a.
Sugarcane Streak: Ammar et al. 1982a.
Sweet Potato: Habeck, D.H. 1960a.
Sweet Potato Internal Cork Virus: Habeck, D.H. 1960a.
Tobacco: Bennett & Costa 1949a; Heggestad & Moore 1959a; Helson, G.A.H. 1950a; Olivares & San Juan 1966a; Ossiannilsson, F. 1958a; Schmutterer, H. 1956a; Selsky, M.I. 1961a.
Tobacco Club Root: Selsky, M.I. 1961a.
Tobacco Curly Top: Bennett & Costa 1949a.
Tobacco Leaf Curl: Olivares & San Juan 1966a.
Tobacco Mosaic: Ossiannilsson, F. 1958a; Schmutterer, H. 1956a.
Tobacco Yellow Dwarf: Helson, G.A.H. 1950a.
Tomato: Bennett & Costa 1949a; Gardner & Cannon 1972a; Thomas & Boll 1977a; Thomas & Martin 1971a; Young & Cannon 1973a.
Tomato Curly Top: Bennett & Costa 1949a.
Trees: Hibben et al. 1973a; Jensen, D.D. 1956a; Merr, F.A. 1981a.
Vegetables: Keener, P.D. 1956a; Novak, J.B. 1961a.
Water Melon: Hills, O.A. 1958a.
Wheat Mosaic: Hoppe, W. 1974a; Lomakina et al. 1963a; Polyakova, G.P. 1972a; Razvyazkina & Polyakova 1967a, 1967b; Vacke & Hoppe 1975a; Vlasov et al. 1971a.
Wheat Streak: Kuklarni et al. 1980a.
Wheat Streak Mosaic: Hoppe, W. 1972a.
Wheat Striate Mosaic: Grylls, N.E. 1963a, 1963b; Hamilton, R.E. 1964a; Lee, P.E. 1963a; Lee, P.W. 1963a; Lee & Bell 1963a; Paliwal, Y.C. 1968a; Sinha, R.C. 1968c, 1970a; Slykhuis, J.T. 1953a, 1963a; Slykhuis & Sherwood 1964a; Timian, R.G. 1978a; Westdal & Richardson 1966a.
Wheat Yellow Stunt: Radulescu & Munteanu 1970a.
Wound Tumor: Black, L.M. 1958a; Black et al. 1958a; Brakke et al. 1953a; Chiu et al. 1966a; Gamez & Black 1968a; Hirumi & Maramorosch 1968a; Kimura & Black 1971a, 1972a, 1972b; Liu & Black 1978a; Liu et al. 1973a; Maramorosch, K. 1959a; Maramorosch & Shikata 1965a; Nagaich, B.B. 1960a; Nagaich & Black 1961a; Selsky, M.I. 1961a; Selsky & Black 1961a; Shikata & Maramorosch 1965c, 1965d, 1967a; Shikata et al. 1964a, 1966a; Streissle et al. 1968a; Whitcomb, R.F. 1972b; Whitcomb & Black 1969a.
Yellow Mosaic: Regupathy et al. 1975a.

Yellow Vein Mosaic: Regupathy & Jayaraj 1972a.

PATHOGEN/VECTOR INTERACTIONS

Amin, P.W. 1977a, 1977b; Amin & Jensen 1971a, 1971b; Bantarri & Zeyen 1976a, 1979a; Black, L.M. 1959a, 1984a; Black & Brakke 1952a; Chen, C.T. 1979a; Chen et al. 1975a; Chettanachit & Disthaporn 1982a; Childress, S.A. 1980a, 1981a; Chiykowski, L.N. 1965b, 1967a, 1967b, 1967c, 1973b, 1976a, 1977a; Chiykowski & Craig 1978a; Chiykowski & Sinha 1969a, 1980a; Choudhury & Rosenkranz 1973a, 1983a; Clark, T.B. 1982a; Clark & Whitcomb 1983a; Cohen, G. 1961a; Conner & Banttari 1979a; Damsteegt, V.D. 1984a; Daniels, M.J. 1979a, 1979b; Daniels et al. 1973a; Davis et al. 1968a; Delius et al. 1984a; Devanesan et al. 1979a; Duffus & Gold 1973a, 1982a; Eden-Green et al. 1985a; El-Bolok, M.M. 1981a; Esau et al. 1976a; Espinoza & Gamez 1980a; Folliot & Maillet 1967a; Frazier & Jensen 1970a; Freitag, J.H. 1956a, 1958a, 1963a, 1964a, 1967a, 1969a; Fukushi, T. 1969a; Fukushi & Shikata 1963a; Gamez, R. 1983b; Gamez & Leon 1983a, 1985a; Gamez et al. 1981a; Giannotti, J. 1969b; Giannotti & Devauchelle 1969a; Giannotti et al. 1970a; Giorgadze & Tulashvili 1973a; Godoy, C. 1985a; Gouranton & Maillet 1973a; Granados, Hirumi et al. 1967a; Granados & Meehan 1973a, 1975a; Granados, Ward et al. 1967a, 1968a; Harris, K.F. 1979a, 1983a; Harris & Childress 1980a; Harrison, B.D. 1981a; Hemmati, K. 1979a; Hibino, H. 1983a; Hirumi et al. 1967a; Hirumi & Maramorosch 1963b; Hsieh, S.P.Y. 1969a; Hsu & Black 1974a; Hsu et al. 1977a; Hsu & Banttari 1979a; Ishii & Ono 1966a; Ishizaki, T. 1980a, 1980b; Jensen, D.D. 1953a, 1958a, 1959a, 1959b, 1962a, 1963a, 1968a, 1969b, 1971b; Jensen et al. 1967a, 1972a; Jensen & Nasu 1970a; Jones et al. 1977a; Kimura, I. 1980a, 1982a, 1985a; Kimura & Black 1972a; Kitagawa & Shikata 1974a; Kitajima & Costa 1972a; Kitajima & Gamez 1977a, 1983a; Kooner et al. 1978a; Kunkel, L.O. 1954a, 1957a; Ling, K.C. 1966a, 1968c, 1969a; Ling & Tiongco 1975a, 1977a, 1979a; Littau & Maramorosch 1956a, 1958a, 1960a; Liu et al. 1983b; Lomakina et al. 1963a; Madden & Nault 1983a; Madden et al. 1984a; Magyarosy & Duffus 1977a; Magyarosy & Sylvester 1979a; Maillet, P.L. 1970a, 1970b; Maillet & Folliot 1967a, 1967b, 1968a; Maillet & Gouranton 1971a; Maillet et al. 1968a; Maramorosch, K. 1956c, 1958b, 1958c, 1959c, 1962a, 1964a, 1964b, 1965b, 1967a, 1968a, 1969a, 1969b, 1970a, 1980a; Maramorosch et al. 1968a, 1970a, 1975a; Maramorosch & Jensen 1963a; Maramorosch, Mitsuhashi et al. 1965a; Maramorosch & Shikata 1965a; Maramorosch, Shikata & Granados 1969a; Maramorosch, Shikata, Hirumi et al. 1969a; Markham, P.G. 1983a; Markham & Pinner 1984a; Mitsuhashi, J. 1965b, 1965d, 1965f, 1965h, 1967a, 1969a; Mitsuhashi & Nasu 1967a; Miyahara et al. 1976a; Moreau & Boulay 1967a; Mukhopadhyay et al. 1978a; Mumford, D.L. 1982a; Murisier & Jelmini 1985a; Musil, M. 1962c, 1964a, 1964b, 1964c, 1966a, 1966b; Nagaich, B.B. 1960a, 1975a; Najaraj & Black 1962a; Nagaraju, S.V. 1981a; Nakasuji, F. 1974a; Nakasuji & Kiritani 1970a, 1971a, 1972a, 1977a; Narayanasamy, P. 1972a, 1972b; Nasu, S. 1965a, 1969a; Nasu et al. 1967a, 1970a, 1974a, 1974b; Nasu, Kono et al. 1974a; Nault, L.R. 1980a, 1983a, 1983b; Nault & Bradfute 1979a; Nault et al. 1973a, 1980a, 1982a, 1984a; Niazi et al. 1985a; Ofori & Francki 1984a, 1985a; Oman, P.W. 1969a; Omura et al. 1982a, 1984a; Onishchenko, A.N. 1984a; Orenski, S.W. 1964a; Osler et al. 1973a; Ou et al. 1965a; Paliwal, Y.C. 1968a; Palomar & Ling 1966a; Paniagua & Gamez 1976a; Peters, D. 1971a; Peterson & Saini 1964a; Pirone et al. 1972a; Pitre, H.N. 1966a; Purcell, A.H. 1978a, 1981a, 1982a, 1982b, 1985a; Purcell & Finlay 1979a, 1979b; Purcell, Richardson et al. 1981a; Raatikainen et al. 1976a; Rahman & Hibino 1985a; Raine & Forbes 1969a, 1971a; Raine et al. 1976a; Raju & Nyland 1981a; Raju et al. 1979a, 1984a; Rana et al. 1975a; Rao & Anjaneyulu 1982a; Reddy & Black 1966a, 1972a; Regupathy & Jayaraj 1972a, 1973b; Renard et al. 1982a; Rico de Cujia & Martinez-Lopez 1977a; Rivera, C. 1981a; Rivera et al. 1981a; Saglio & Whitcomb 1979a; Saito, Y. 1977a, 1977b; Schvester, D. 1962a; Seki & Onizuka 1964a; Selsky & Black 1961a; Shaskolskaya, D.D. 1962a; Sherman & Maramorosch 1977a; Shikata, E. 1966a,

1979a, 1979b; Shikata & Maramorosch 1965a, 1965b, 1965c, 1965d, 1967a, 1967b, 1969a; Shikata et al. 1964a, 1966a, 1968a; Shinkai, A. 1956a, 1958a, 1959a, 1960a, 1960b, 1962a, 1964a; Shiomi & Sugiura 1984b; Shukla & Anjaneyulu 1981c, 1982a, 1982b; Silvere & Giorgadze 1978a; Sinha, R.C. 1963a, 1964a, 1965a, 1965b, 1967a, 1968a, 1969a, 1981a; Sinha & Black 1962a, 1963a, 1965a, 1967a; Sinha & Chiykowski 1966a, 1967a, 1967b, 1967c, 1968a, 1969a, 1980a; Sinha & Paliwal 1970a; Sinha & Peterson 1972a; Sinha & Reddy 1964a; Sinha et al. 1964a; Sinha & Shelley 1965a; Slykhuis & Sherwood 1964a; Smith et al. 1981a; Spendlove et al. 1967a; Srinivasan, K. 1978a; Srinivasan & Chelliah 1980a, 1981a; Stanarius et al. 1976a, 1980a; Staples et al. 1970a; Takahashi, Y. 1963a, 1963b; Takahashi & Sekiya 1962a; Tantera & Roechan 1978a; Thomas & Boll 1977a; Thomas & Martin 1971a; Timian & Alm 1973a; Ting & Paramsothy 1970a; Townsend et al. 1977a; Tsai et al. 1972a; Vago & Flandre 1963a; Van Rensburg, G.D.J. 1981a; Vidano, C. 1967a; Volk, J. 1958a, 1967a; Whitcomb, R.F. 1964a, 1968a, 1969a; Whitcomb & Black 1959a, 1961a; Whitcomb & Bov 1983a; Whitcomb et al. 1966a, 1966b, 1967a, 1968a, 1968b, 1968c, 1974a, 1983a; Whitcomb & Coan 1982a; Whitcomb & Davis 1970a, 1970b; Whitcomb & Jensen 1968a; Whitcomb & Richardson 1966a; Whitcomb & Tully 1973a, 1973b; Whitcomb, Tully et al. 1973a; Whitcomb & Williamson 1975a, 1979a; Yoshii, H. 1959b.

VARIATION

Polymorphism

Blocker, H.D. 1967a, 1984a; Blocker & Triplehorn 1985a; Chudzicka, E. 1980a; Claridge, M.F. 1981a; Claridge & Nixon 1981a; Cwikla, P.S. 1980a; DeLong, D.M. 1984b; Dlabola, J. 1956c; Dworakowska et al. 1978a; Frediani, D. 1955a, 1956a; Ghosh, A. 1980a; Gunthart, H. 1974a, 1977a, 1978a, 1979a; Hamilton. K.G.A. 1972c, 1982a; Harrison, R.G. 1980a; Iwata, T. 1972a; Kathirithamby. J. 1976a; Knight, W.J. 1967a, 1968a; LeQuesne, W.J. 1965d, 1983a; Mishra, R.K. 1979a; Mitjaev, I.D. 1973b; Muller, H.J. 1956b, 1957a, 1962b, 1979a, 1979b, 1981a; Musgrave, C.A. 1975a; Nielson & Freytag 1976a; Olmi, M. 1976a; Oman, P.W. 1970a, 1985a; Ramachandra Rao, K. 1973b, 1973c; Remane & Koch 1977a; Richards, O.W. 1961a; Rose, D.J.W. 1972b; Ruppel, R.F. 1965a; Sawbridge, J.R. 1975a; Simonet & Pienkowski 1980a; Stewart, A.J.A. 1981a; Strubing, H. 1978b; Strubing & Hasse 1974a; Vidano, C. 1961b, 1961c; Wagner, W. 1958b, 1968a.

Colour Polymorphism

Claridge, M.F. 1981a; Dworakowska et al. 1978a; Ghosh, A. 1980a; Gunthart, H. 1978a; Hamilton, K.G.A. 1983a; Kawai, A. 1977a; Khokhlova & Anufriev 1981a; Linnavuori & DeLong 1978e; Maldonado- Capriles, J. 1985b; Muller, H.J. 1956b, 1957a, 1974a; Sogawa & Saito 1983a; Stewart, A.J.A. 1981a; Stiling, P.D. 1980a.

Wing Polymorphism

Becker, M. 1981a; Evans, J.W. 1957a, 1966a, 1968a; Harrison, R.G. 1980a; Knight, W.J. 1973b; Kramer, J.P. 1967d, 1976b; LeQuesne, W.J. 1964d; Linnavuori, R. 1972a; Linnavuori & DeLong 1977a, 1977c; Oman, P.W. 1985a; Oman & Musgrave 1975a; Taylor, R.A.J. 1985a; Theron, J.G. 1978b, 1982b; Waloff, N. 1973a, 1980a, 1983a; Wilson & Claridge 1985a.

Acoustic Variation

Claridge, M.F. 1985b; Strubing, H. 1976a.

Geographical Variation

Anufriev & Zhil'tsova 1982a; Claridge, M.F. 1983a, 1985b; Dlabola, J. 1980c; Ellsbury & Nielson 1978a; Khokhlova & Anufriev 1981a; Knight, W.J. 1983b; LeQuesne, W.J. 1965d, 1978a, 1983a; LeQuesne & Woodroffe 1976a; Maldonado Capriles, J. 1985b; Muller, H.J. 1981a; Musgrave, C.A. 1975a; Oman, P.W. 1970a; Ramakrishnan, U. 1983c; Remane & Koch 1977a; Ruppel, R.F. 1965a, 1969a; Sato & Sogawa 1981a; Saxena et al. 1985a; Sogawa & Saito 1983a; Stewart, A.J.A. 1981a; Stiling, P.D. 1980a; Triplehorn & Nault 1985a; Wagner, W. 1968a.

Seasonal Variation (Photoperiod)

Claridge & Wilson 1978c; Danilevskii, A.S. 1965a; Hulbert & Schaller 1972a; Kisimoto, R. 1959c; LeQuesne, W.J. 1983a; Muller, H.J. 1955b, 1955c, 1956b, 1956c, 1957a, 1959a, 1960b, 1962b, 1964a, 1965a, 1973a; 1979a, 1983a, 1984a; Nielson & Freytag 1976a; Song, Y.H. 1978a; Strubing, H. 1976a, 1980a, 1981a, 1984a; Strubing & Hasse 1974a; Vidano, C. 1961b; Wagner, W. 1968a; Whitcomb et al. 1972a.

Parasitism

Kathirithamby, J. 1974b, 1976a, 1979a; LeQuesne, W.J. 1983a; Muller, H.J. 1960a; Olmi, M. 1976a; Oman, P.W. 1985a; Remane & Schulz 1973a; Sperka & Freytag 1975a; Stiling, P.D. 1980a; Vidano, C. 1962b; Yano et al. 1985a; Ylonen & Raatikainen 1984a.

Genetic Variation

Halkka, O. 1965a; John & Claridge 1974a; Khokhlova & Anufriev 1981a; Manna & Bhattacharya 1973a; Whitten & Taylor 1969a.

Sexual Dimorphism

Anufriev, G.A. 1971a; Dworakowska, I. 1967a; Kwon, Y.J. 1983a; Linnavuori & Delong 1978a, 1978e; Nielson, M.W. 1975a, 1977a, 1979a, 1982e; Oman, P.W. 1971a; Oman & Musgrave 1975a; Ossiannilsson, F. 1981a, 1983a; Ruppel et al. 1972a; Theron, J.G. 1982a, 1984a; Webb, M.D. 1979a.

Hybridisation

Inoue et al. 1979a; Ling, K.C. 1968b; Nielson & Morgan 1982a; Nielson & Toles 1970a; Ramakrishnan & Ghauri 1979a; Ross, H.H. 1958a.

Abnormalities

Mitjaev, I.D. 1973b.

TAXONOMIC INDEX

Aaka

New species: **Dworakowska** 1972e.

Abana

New species: **Young** 1968a.

Records: **Young** 1965c (lectotype designation); 1968a (n. sp. syn.). **Young & Soos** 1964a (notes on type material). **Young & Lauterer** 1966a (lectotype designation). **Young & Nast** 1973a (notes on type specimen).

Abdistragania

New species: **Blocker** 1979b.

Record: **Blocker** 1979b (n. subgenus proposed).

Abietotettix

New species: **Mitjaev** 1965a.

Record: **Hamilton, K. G. A.** 1975b (APHRODINAE, DELTOCEPHALINI, PLATYMETOPIINA).

Abimwa

New species: **Linnavuori** 1978a.

Record: **Linnavuori** 1978a (n. comb. from *Phlepsius*; key to species). **Linnavuori & Al-Ne'amy** 1983a (catalogued; SELENOCEPHALINAE, IANERINI).

Abrabra

New species: **Dworakowska** 1976b.

Record: **Dworakowska** 1977i (Vietnam).

Abrela

Record: **Young** 1957c (n. comb. from *Diceratalebra*).

Absheta

New species: **Blocker** 1979a.

Record: **Blocker** 1979a (n. comb. from *Macropsis*).

Acacimenus

Records: **Dlabola** 1984a (Iran). **Nast** 1982a (catalogued; DELTOCEPHALINAE).

Acacioiassus

New species: **Linnavuori & Quartau** 1975a.

Records: **Linnavuori & Quartau** 1975a (n. comb. from *Jassus* and *Macropsis*; n. sp. syn.; key to species).

Acastroma
  New species: **Linnavuori** 1969a.
  Record: **Linnavuori & Al-Ne'amy** 1983a (structural characters).
Accacidia
  New species: **Dworakowska** 1971b, 1974a, 1979a. **Einyu & Ahmed** 1983a.
  Records: **Dworakowska** 1971b (n. comb. from *Erythroneura*). **Dworakowska & Sohi** 1978a (= *Indianella*, n. syn.; n. sp. syn.). **Theron** 1977a (n. comb. from *Erythroneura*).
Aceratagallia
  Records: **Beirne** 1956a (Canada; key to species). **Davis, R. B.** 1975a (morphology; phyletic inferences). **Datta & Dhar** 1984a (descriptive notes, subgenus *Ionia*). **Hamilton, K. G. A.** 1972d (Canada: Manitoba); 1976a (identity of Provancher species). **Kramer** 1964a (in key to New World AGALLIINAE). **Linnavuori** 1973c (Cuba). **Nielson** 1968b (characterization of vector species).
Acericerus
  New species: **Dlabola** 1974a.
  Records: **Dlabola** 1974b (n. comb. from *Idiocerus*). **Lodos & Kalkandelen** 1981a (Turkey). **Nast** 1982a (catalogued; IDIOCERINAE).
Achaetica
  New species: **Dlabola** 1965e. **Emeljanov** 1959a, 1962a, 1964f, 1964c, 1979a. **Mitjaev** 1971a. **Vilbaste** 1980b.
  Records: **Emeljanov** 1962b (EUSCELINAE, OPSIINI, ACHAETICINA); 1964c (European USSR; key to species); 1974f (key to species); 1977a (Mongolia). **Hamilton, K. G. A.** 1975b (APHRODINAE, APHRODINI, ACHAETICINA). **Mitjaev** 1971a (Uzbekistan); 1971b (as *Achetica* and *Achactica*; southern Kazakhstan).
Acharis
  New species: **Kwon & Lee** 1979b.
  Records: **Dworakowska** 1973d (Korea). **Emeljanov** 1966a (n. comb. from *Deltocephalus*); 1977a (Mongolia). **Hamilton, K. G. A.** 1975b (APHRODINAE, DELTOCEPHALINI, DELTOCEPHALINA). **Lee, C. E.** 1979a (Korea, redescribed). **Vilbaste** 1968a (coastal regions of the USSR); 1980b (Tuva).
Acharista Emeljanov
  New species: **Emeljanov** 1968a.
  Records: **Hamilton, K. G. A.** 1975b (APHRODINAE, DELTOCEPHALINI, DELTOCEPHALINA). **Nast** 1984a (= *Gobicuellus*, n. syn.).
Acharista Melichar
  Records: **Young** 1977a (= *Neiva*, n. syn.). **Young & Lauterer** 1966a (lectotype designation). **Young & Soos** 1964a (notes on type material).
Achrus
  New species: **Dlabola** 1961b, 1968a, 1981a. **Emeljanov** 1964f, 1975a.
  Records: **Anufriev** 1978a (USSR: Maritime Territory). **Davis, R. B.** 1975a (morphology; phyletic inferences). **Dlabola** 1960a (n. comb. from *Idiocerus*); 1968b (Mongolia). **Dubovskij** 1978a (Uzbekistan, Zarafshan Valley); 1980a (USSR: western Turkemia). **Emeljanov** 1975a (in key to genera of ADELUNGIINI); 1977a (Mongolia). **Korolevskaya** 1975a (Tadzhikistan). **Lindberg** 1960b (Soviet Armenia). **Mitjaev** 1968b (eastern Kazakhstan); 1971a (Kazakhstan; in key to genera of MELICHARELLINAE); 1971b (southern Kazakhstan).
Acia
  New species: **Dworakowska** 1981b. **Ahmed, Manzoor** 1979a.
  Records: **Dworakowska** 1979b (Taiwan); 1981b (n. subgenera defined; n. comb. from *Empoasca*); 1984a (*Paolicia*, replacement for *Paoliella*, subgenus). **Knight, W. J.** 1982a (Rennell Island).
Acinopterus
  New species: **Linnavuori & DeLong** 1977d (new name).
  Records: **Beirne** 1956a (Canada). **Hamilton, K. G. A.** 1975b (APHRODINAE, DELTOCEPHALINI, PLATYMETOPIINA). **Linnavuori** 1956a (Colombia); 1959b (ACINOPTERINI; n. comb. from *Deltocephalus*; key to species); 1973c (Antillean subregion). **Linnavuori & DeLong** 1977d (review of Mexican and neotropical species). **Nielson** 1968b (characterization of vector species)

Acocephalus
  Records: **Kocak** 1981a (n. comb. to *Aphrodes*). **Vilbaste** 1973a (notes on Flor's material); 1976a (notes on Matsumura's material).
Acojassus
  New species: **Evans, J. W.** 1972b.
  Record: **Knight, W. J.** 1983b (n. comb. to *Batracomorphus*).
Aconura
  New species: **Dlabola** 1957b, 1984a. **Emeljanov** 1964c, 1964f, 1972b. **Lindberg** 1956a. **Logvinenko** 1975a. **Vilbaste** 1962b.
  Records: **Datta** 1972c, 1972h (descriptive notes). **Dlabola** 1957b (Turkey); 1958a (Cyprus); 1960a (notes on Horvath's material); 1961b (central Asia); 1965c, 1966a, 1967b, 1967c, 1967d, 1968a, 1970b (Mongolia); 1971a, 1972a (Afghanistan); 1981a (Iran). **Dubovskij** 1966a, 1972h (Uzbekistan); 1978a (Uzbekistan, Zarafshan Valley); 1980a (USSR: western Turkmenia). **Dworakowska** 1969a (Mongolia). **Emeljanov** 1964c (European USSR; key to species); 1977a (Mongolia). **Ghauri** 1963f (n. comb. from *Arya, Paternus, Sunda*). **Hamilton, K. G. A.** 1975b ((APHRODINAE, APHRODINI, DORATURINA). **Ishihara** 1958a (Japan: north Honshu). **Jankovic** 1966a (Serbia). **Kalkandelen** 1974a ((Turkey; in key to genera). **Kholmuminov & Dlabola** 1979a (USSR: Golodnostep plain). **Kwon** 1980a (n. comb. to *Changwania*). **Lindberg** 1958a (Cape Verde Islands); 1965a (Spanish Sahara). **Linnavuori** 1960a (Micronesia: Bonin Islands); 1961a (Southern Rhodesia); 1962a (Israel); 1964a (Egypt); 1965a (Mediterranean). **Lodos & Kalkandelen** 1985c (Turkey). **Mitjaev** 1968b (eastern Kazakhstan); 1971a (Kazakhstan; key to species). **Servadei** 1958a (Italy). **Theron** 1975a (n. comb. to *Aconurella* and *Stirellus*). **Vilbaste** 1965a (genus redefined); 1980b (Tuva; key to species).
Aconurella
  New species: **Anufriev** 1972b. **Dlabola** 1957b. **Emeljanov** 1964d. **Ghauri** 1964c, 1981a. **Theron** 1982b. **Webb, M. D.** 1980a.
  Records: **Anufriev** 1978a (USSR: Maritime Territory). **Bonfils & Della Giustina** 1978a (Corsica). **Dlabola** 1956a (Czechoslovakia); 1957a (Turkey); 1958b (Caucasus); 1960c (Iran); 1961b (central Asia); 1963c (notes on Haupt's material); 1964b (Sudan); 1965c, 1966a, 1967b, 1967c, 1967d, 1968b, 1970b (Mongolia); 1971a (Iran); 1972a (Afghanistan); 1977c (Mediterranean); 1981a (Iran). **Dubovskij** 1966a (Uzbekistan); 1978a (Uzbekistan, Zarafshan Valley). **Dworakowska** 1969a (Mongolia). **Emeljanov** 1964c, 1977a (Mongolia). **Ghauri** 1981a (key to Ethiopian species). **Hamilton, K. G. A.** (APHRODINAE, APHRODINI, DORATURINA). **Heller & Linnavuori** 1968a (Aethiopia: Konso). **Jankovic** 1966a (Serbia). **Kholmuminov & Dlabola** 1979a (USSR: Golodnostep plain). **Korolevskaya** 1973a (Tadzhikistan). **Lee, C. E.** 1979a (Korea, redescribed). **Lee, C. E. & Kwon** 1977b (Korea). **Lindberg** 1958a (Cape Verde Islands); 1961a (Madeira Islands). **Linnavuori** 1956d (Spanish Morocco); 1962a (Israel); 1964a (Egypt); 1965a (Mediterranean). **Lodos & Kalkandelen** 1985c (Turkey). **Mitjaev** 1968b, 1971a (Kazakhstan; key to species). **Sharma, B.** 1977b (India: Jammu). **Servadei** 1968a (Italy). **Theron** 1975a (n. comb. from *Aconura*). **Vilbaste** 1965a (list of included species); 1965b (Altai Mts.); 1968a (coastal regions of the USSR); 1976a (n. comb. from *Thamnotettix*); 1980b (Tuva).
Aconurina
  Record: **Hamilton, K. G. A.** 1975b (APHRODINAE, APHRODINI, DORATURINA).
Aconuromimus
  Record: **Evans, J. W.** 1966a (= *Euleimonios*, n. syn.).
Acostemana
  Record: **Evans, J. W.** 1972b (to ACOSTEMMINAE).
Acostemma
  New species: **Evans, J. W.** 1959a.
Acostemmella
  Records: **Evans, J. W.** 1972b (to ACOSTEMMINAE). **Linnavuori & Quartau** 1975a (= *Acropona*; ACROPININAE).

Acrobelus
New species: **Young** 1968a.
Record: **Young** 1968a (key to species)

Acrocampsa
New species: **Young** 1968a.
Records: **Young** 1968a (n. sp. syn.). **Young & Lauterer** 1966a (lectotype designations).

Acrogonia
New species: **Metcalf** 1965a (new name).
Record: **Young** 1968a (n. gen. syn.; n. sp. syn.; key to species).

Acropona
Records: **Evans, J. W.** 1972b (to ACOSTEMMINAE). **Linnavuori & Quartau** 1975a (=*Acostemma*).

Acrulogonia
New species: **Young** 1977a.
Record: **Young** 1977a (n. comb. from *Poeciloscarta* and *Tettigoniella*; key to species).

Acuera
New species: **DeLong & Freytag** 1974a. **DeLong & Wolda** 1982b.
Records: **DeLong** 1977b (n. comb. from *Ponana*). **DeLong & Freytag** 1972a (in key to genera of GYPONINAE); 1974a (key to subgenera and species).

Acunasus
New species: **DeLong** 1980h.
Record: **Cwikla & Blocker** 1981a (comparative notes).

Acuponana
New species: **DeLong & Bush** 1971a. **DeLong & Freytag** 1970a. **DeLong & Wolda** 1983a.
Records: **DeLong & Freytag** 1970a (key to species); 1972a (in key to genera of GYPONINAE).

Acurhinus
Records: **Linnavuori** 1958b (=*Flexamia*). **Linnavuori & DeLong** 1978c (=*Hodoedocus*).

Acusana
New species: **DeLong & Freytag** 1966a.
Records: **DeLong & Freytag** 1966a (key to species); 1972a (in key to genera of GYPONINAE). **Linnavuori** 1973c (Cuba).

Adama
New species: **Linnavuori & Al-Ne'amy** 1983a.
Record: **Linnavuori & Al-Ne'amy** 1983a (keys to subgenera and species).

Adarrus
New species: **Cobben** 1979a. **Dlabola** 1958a, 1961c, 1980a, 1980b. **D'Urso** 1984a. **Emeljanov** 1966a, 1972b. **Gunthart, H.** 1985a. **Heller** 1975a. **Logvinenko** 1966b. **Mitjaev** 1980d. **Remane & Asche** 1980a.
Records: **Anufriev** 1977b (Kurile Islands). **Bonfils & Della Giustina** 1978a (Corsica). **Della Giustina** 1983a (France). **Dlabola** 1958a (Yugoslavia); 1967b (Mongolia). **Dubovskij** (Uzbekistan). **Dworakowska** 1973d (n. comb. to *Kaszabinus*). **Emeljanov** 1964c (European USSR; key to species); 1977a (n. comb. to *Anargella*; Mongolia). **Hamilton, K. G. A.** 1975b (APHRODINAE, DELTOCEPHALINI, DELTOCEPHALINA). **Jankovic** 1971a (Serbia). **Lee, C. E.** 1979a (Korea, redescribed). **Lee, C. E. & Kwon** 1977b (Korea). **LeQuesne** 1969c (Britain). **Lindberg** 1960a, 1962a (Portugal). **Mitjaev** 1968b (Eastern Kazakhstan); 1971a (Kazakhstan). **Nast** 1955a (Poland); 1972a (catalogued); 1976a (Poland: Pieniny Mts.) 1976b (Poland, catalogued); 1981a (Italy). **Ossiannilsson** 1983a (Sweden). **Quartau & Rodrigues** 1969a (Portugal). **Remane & Asche** 1980a (subgenera defined). **Servadei** 1959a, 1960a, 1968a, 1969a, 1971a, 1972a (Italy). **Vilbaste** 1965b (Altai Mts.); 1980b (Tuva).

Adelungia
New species: **Dlabola** 1984a. **Mitjaev** 1971a.
Records: **Al-Ne'amy & Linnavuori** 1982a (structural characteristics). **Davis, R. B.** 1975a (morphology; phyletic inferences). **Dlabola** 1961b (central Asia); 1981a, 1984a (Iran). **Dubovskij** 1978a (Uzbekistan, Zarafshan Valley; 1980a (USSR: western Turkmenia). **Emeljanov** 1975a (composition and characteristics of the tribe ADELUNGIINI). **Korolevskaya** 1975a (Tadzhikistan). **Mitjaev** 1971a (key to species; Kazakhstan).

Adoratura
New species: **Dlabola** 1958b.

Records: **Dlabola** 1961b (central Asia). **Hamilton, K. G. A.** 1975b (APHRODINAE, APHRODINI, DORATURINA). **Lindberg** 1960a (Soviet Armenia).

Aequcephalus
New species: **DeLong & Thambimuttu** 1973b.
Records: **Cwikla & Blocker** 1981a (comparative notes). **Linnavuori & DeLong** 1977c (Chile).

Aethiopulopa
New species: **Linnavuori** 1972a. **Van Stalle** 1983c.
Record: **Linnavuori** 1972a (descriptions; key to species). **Van Stalle** 1982a (tropical Africa).

Aeturnus
Record: **Nielson** 1975a (transferred from COELIDIINAE to DELTOCEPHALINAE).

Aflexia
Record: **Hamilton, K. G. A.** 1975b (APHRODINAE, DELTOCEPHALINI, DELTOCEPHALINA).

Afrakeura
New species: **Einyu & Ahmed** 1979a.
Record: **Ahmed, Manzoor** 1983b (remarks about venation).

Afrakra
New species: **Dworakowska** 1979e.

Afralebra
Records: **Dworakowska** 1971c (n. comb. from *Empoasca*; n. sp. syn.).

Afrasca
Record: **Dworakowska & Lauterer** 1975a (n. comb. from *Empoasca*).

Afrascius
New species: **Linnavuori** 1969a, 1978d.
Record: **Linnavuori** 1978d (key to species).

Africoelidia
New species: **Nielson** 1982e.
Record: **Nielson** 1982e (key to species).

Afridonus
New species: **Nielson** 1983e.
Record: **Nielson** 1983e (key to species).

Afroccidens
New species: **Dworakowska** 1972i, 1974a. **Ghauri** 1969a, 1975b.
Records: **Dworakowska** 1972c (Ivory Coast & Zaire); 1974a (Congo); 1976b (Cameroon Mtn.); 1977e (tropical Africa).

Afroiassus
New species: **Linnavuori & Quartau** 1975a.
Record: **Linnavuori & Quartau** 1975a (n. comb. from *Iassus*; key to species).

Afroideus
New species: **Linnavuori** 1961a.

Afroindica
Records: **Dworakowska** 1980d (=*Empoascanara*, n. syn.). **Ramakrishnan & Ghauri** 1979b (in key to genera of Oriental ERYTHRONEURINI).

Afrokana
New species: **Heller** 1972a.
Record: **Linnavuori** 1979b (redescription; in key to genera of NIRVANINI).

Afrolimnus
Records: **Linnavuori** 1979b (=*Hodoedocus*; DELTOCEPHALINAE, STENOMETOPIINI). **Linnavuori & DeLong** 1978c (=*Hodoedocus*; DELTOCEPHALINAE, STIRELLINI).

Afronirvana
Records: **Linnavuori** 1979b (in key to African genera; key to the *Afronirvana* group of genera and species; NIRVANINI).

Afrorubria
New species: **Linnavuori** 1972b.
Records: **Linnavuori** 1972b. (n. comb. from *Camptelasmus, Petalocephala, Rubria*; key to species). **Theron** 1976b, 1980a (redescription of species).

Afrosteles
Record: **Theron** 1975b (n. comb. from *Dalbulus*).

Afrosus
Records: **Heller & Linnavuori** 1968a (South Africa). **Theron** 1975b (Neotropical occurrence record questioned).

Agalita
New species: **Evans, J. W.** 1957a. **Linnavuori & DeLong** 1977c.
Records: **Kramer** 1964a (=*Agalliana*). **Linnavuori & DeLong** 1977c (removed from synonymy with *Agalliana*).

Agallia

New species: **Cwikla & DeLong** 1985a. **Dlabola** 1957b, 1958a, 1961b, 1965d, 1981a, 1984a. **Dubovskij** 1970b. **Dutra** 1966a, 1967a, 1969a, 1970a, 1971a, 1972a (new name), 1974a, 1976a, 1977a. **Emeljanov** 1964f. **Evans, J. W.** 1957a. **Kameswara Rao & Ramakrishnan** 1978b. **Kramer** 1964a, 1976c. **Lindberg** 1960a. **Linnavuori** 1956a, 1956d, 1960a, 1961a, 1965b, 1971a. **Linnavuori & DeLong** 1977c, 1979c, 1979d. **Linnavuori & Heller** 1961a. **Logvinenko** 1983a. **Mitjaev** 1975a. **Oman** 1971c. **Quartau** 1971a. **Sawai Singh** 1969a. **Villiers** 1956a. **Viraktamath, C. A.** 1980b.

Records: **Anufriev** 1970e (n. comb. to *Dryodurgades*). **Beirne** 1956a (Canada). **Bonfils & Della Giustina** 1978a (Corsica). **Cantoreanu** 1961a (Romania). **Davis, R. B.** 1975a (morphology; phyletic inferences). **Dlabola** 1956a (Czechoslovakia); 1956b (Hungary); 1957b (Afghanistan); 1959a (macropterous morph of *A. brachyptera*); 1960a (notes on Horvath's material); 1961b (central Asia); 1965c (Mongolia); 1965d (Jordan); 1974d (Morocco); 1981a (Iran). **Dubovskij** 1978a (Uzbekistan, Zarafshan Valley). **Dworakowska** 1969a, 1973d (Mongolia). **Emeljanov** 1964c (European USSR; key to species); 1977a (Mongolia). **Gravestein** 1965a (The Netherlands: Terschelling I.) **Hamilton, K. G. A.** 1972d (Canada: Manitoba); 1976a (notes on Provancher's material). **Heller & Linnavuori** 1968a (n. sp. syn.). **Ishihara** 1961a (Thailand). **Jankovic** 1966a (Serbia). **Kameswara Rao & Ramakrishnan** 1978b (key to Indian species). **Kholmuminov & Dlabola** 1979a (USSR: Golodnostep plain). **Korolevskaya** 1973a (Tadzhikistan). **Kramer** 1964a (=*Alloproctus*; n. comb. from *Alloproctus*; in key to New World genera of AGALLIINAE). **Lauterer** 1957a, 1984a (Czechoslovakia). **LeQuesne** 1965c (Britain). **LeQuesne & Woodroffe** 1976a (intraspecific variation in male genitalia). **Lindberg** 1956a (Spanish Sahara); 1960a, 1962a ((Portugal); 1963a (Morocco); 1958a (Cape Verde Islands); 1960a (Portugal). **Linnavuori** 1956d (Spanish Morocco); 1956e (notes on type specimens); 1962a (Israel); 1964a (Egypt); 1973c (Cuba); 1975d (Peru). **Linsley & Usinger** 1966a (Galapagos Islands). **Lodos** 1982 (Turkey). **Lodos & Kalkandelen** 1981a (Turkey). **Mitjaev** 1968b (eastern Kazakhstan); 1971a (Kazakhstan). **Logvinenko** 1957b (USSR: Ukraine); 1963b (USSR: Carpathians); 1975a (n. comb. to *Mesagallia*). **Nast** 1972a (catalogued); 1973a (Poland); 1976a (Poland: Pieniny Mts.); 1976b (Poland; catalogued). **Ossiannilsson** 1981a (Fennoscandia & Denmark; key to species). **Quartau & Rodrigues** 1969a (Portugal). **Servadei** 1957a, 1960a, 1968a (Italy). **Theron** 1970b (redescription of Cogan's species). **Vilbaste** 1958a (Estonia); 1973a (notes on Flor's material); 1979a (Vooremaa hardwood/spruce forest). **Viraktamath, C. A.** 1973b (notes on Matsumura's material); 1980b (Juan Fernandez Islands; key to species). **Wagner** 1959c (Greece). **Zachvatkin** 1953b (central Russia; Oka district).

Agalliana

Records: **Davis, R. B.** 1975a (morphology; phyletic inferences). **Kramer** 1964a (synopsis). **Linnavuori** 1956a (Argentina). **Linnavuori & DeLong** 1977c (Chile; characters differentiating *Agalita* and *Agalliana*). **Marino de Remes Lenicov** 1982a(Argentina; key to species). **Nielson** 1968b (characterization of vector species).

Agalliopsis

New species: **Cwikla & DeLong** 1985a. **Kramer** 1960a, 1964a, 1976c. **Linnavuori** 1956a, 1965b. **Linnavuori & DeLong** 1979c, 1979d. **Oman** 1970b, 1970c.

Records: **Beirne** 1956a (Canada; key to species). **Davis, R. B.** 1975a (morphology; phyletic inferences). **Hamilton, K. G. A.** (Canada: Manitoba). **Kramer** 1960a (male of *A. inscripta* described); 1964a (synopsis; in key to genera of New World AGALLIINAE); 1965a (n. sp. syn.). **Linnavuori** 1956a (Brazil, Ecuador). **Linsley & Usinger** 1966a (Galapagos Islands). **Nielson** 1968b (characterization of vector species). **Oman** 1970b (revision of *A. novella* complex).

Agalliota

New species: **Dutra & Egler** 1982a. **Linnavuori** 1956a.

Records: **Davis, R. B.** (morphology; phyletic inferences; does not =*Igerna*). **Kramer** 1964a (synopsis; n. sp. syn.). **Linnavuori & Heller** 1961a (Peru). **Theron** 1980a (not a synonym of *Igerna*).

Agapelus

Records: **Dlabola** 1967b (Afghanistan). **Emeljanov** 1964c (European USSR; key to species); 1966a (=*Palus*). **Hamilton, K. G. A.** 1975b (APHRODINAE, DELTOCEPHALINI, DELTOCEPHALINA).

Agelina

Record: **Linnavuori** 1959b (redescribed; DELTOCEPHALINAE, MACROSTELINI).

Agellus

Record: **Hamilton, K. G. A.** 1975b (APHRODINAE, DELTOCEPHALINI, MACROSTELINI).

Aglaenita

Record: **Young** 1977a (identity discussed, probably =*Onega* Distant).

Aglena

Records: **Bonfils & Della Giustina** 1978a (Corsica). **Cantoreanu** 1963b (Romania). **Dlabola** 1964c (Albania); 1981a (Iran). **Emeljanov** 1964c (European USSR; key to species). **Hamilton, K. G. A.** 1975b (APHRODINAE, DELTOCEPHALINI, DELTOCEPHALINA). **Kalkandelen** 1974a (Turkey). **Linnavuori** 1956d (Spain); 1962a (Israel). **Logvinenko** 1957b (USSR: Ukraine). **Servadei** 1957a, 1968a (Italy).

Aglenita = *Aglaenita*.

Agnesiella

New species: **Chou, I. & Ma** 1981a. **Dworakowska** 1977i, 1982a. **Sharma, B.** 1978a, 1979a. **Sharma, B. & Malhotra** 1981a. **Thapa** 1984a.

Records: **Ahmed, Manzoor** 1985c (Pakistan). **Dworakowska** 1970f (n. comb. from *Typhlocyba*); 1971g (new subgenus proposed; n. comb. from *Chikkaballapura*); 1977i (n. comb. from *Typhlocyba*); 1982a (n. comb. to *Linnavuoriana*). **Sharma, B.** 1979a (India: Nagaland, Kohima). **Sohi & Dworakowska** 1984a (Indian species listed). **Thapa** 1984a (Nepal; host records).

Agrosoma

New species: **Medler** 1960a. **Young** 1977a.

Records: **Medler** 1960a (key to species). **Ramos-Elorduy** 1972c (Mexico: Vera Cruz).

Aguahua

Record: **Young** 1977a (n. comb. from *Tettigonia*).

Aguana

New species: **Young** 1977a.

Record: **Young** 1977a (n. sp. syn.; key to species).

Aguatala

New species: **Young** 1977a.

Agudus

New species: **Cheng, Y. J.** 1980a. **DeLong & Linnavuori** 1978b. **Linnavuori** 1959b.

Records: **Cheng, Y. J.** 1980a (Paraguay). **Linnavuori** 1959b (redescribed; n. comb. from *Platymetopius*; key to species).

Aguriahana

New species: **Chou, I. & Ma** 1981a. **Dworakowska** 1972a, 1982a. **Samad & Ahmed** 1979a.

Records: **Ahmed, Manzoor** 1985c (Pakistan). **Anufriev** 1977b (Kurile Islands); 1978a (key to species). **Dworakowska** 1972a (redescribed; n. gen. syn; n. comb. from *Asymmetropteryx, Cicadella, Eupteroidea, Eupteryx, Eurhadina, Evansioma, Kashitettix, Typhlocyba, Wagneripteryx, Youngama;* species groups defined; checklist of species; key to species); 1977i (Vietnam); 1980b (India); 1982a (n. comb. from *Typhlocyba*). **Hamilton, K. G. A.** 1983b (species adventive to Nearctic). **Kirejtshuk** 1977a (USSR: Kharkov region). **Lee, C. E.** 1979a (Korea, redescribed). **LeQuesne & Payne** 1981a (Britain; key to species). **Lodos & Kalkandelen** 1984a (Turkey). **Ma, N.** 1982a (China). **Nast** 1976a (Poland: Pieniny Mts.); 1976b (Poland; catalogued). **Ossiannilsson** 1981a (Fennoscandia & Denmark; key to species). **Sohi & Dworakowska** 1984a (list of Indian species). **Thapa** 1983a (Nepal). **Vilbaste** 1973a (notes on Flor's material); 1980b (Tuva). **Wilson, M. R.** 1978a (descriptions and key characters for 5th instars).

Ahimia
New species: **Dworakowska** 1979c.

Ahmedra
New species: **Dworakowska & Viraktamath** 1979a.
Records: **Dworakowska** 1980b (India). **Dworakowska & Viraktamath** 1979a (n. comb. from *Basilana*). **Sohi & Dworakowska** 1984a (list of Indian species).

Aidola
Record: **Dworakowska** 1972b (= *Basilana*, n. gen. syn.).

Aindrahamia
New species: **Linnavuori** 1965a.
Record: **Hamilton, K. G. A.** 1975b (APHRODINAE, DELTOCEPHALINI, PLATYMETOPIINA).

Airosius
Records: **Hamilton, K. G. A.** 1975b (APHRODINAE, DELTOCEPHALINI, DELTOCEPHALINA).

Airosus
Records: **Emeljanov** 1964c (European USSR). **Hamilton, K. G. A.** 1975b (APHRODINAE, DELTOCEPHALINI, DELTOCEPHALINA). **Ossiannilsson** 1983a (subgenus of *Cosmotettix*).

Ajika
New species: **Dworakowska** 1979b.

Akotettix
New species: **Dworakowska** 1974a. **Ramakrishnan & Menon** 1972a.
Records: **Dworakowska** 1970a (lectotype designation; = *Banosa*, n. syn.); 1974a (new subgenus added). **Dworakowska & Sohi** 1978b (subgenus of *Dialecticopteryx*; n. comb. to *Dialecticopteryx*). **Sohi & Dworakowska** 1984a (list of Indian species).

Alaca
Record: **Linnavuori** 1959b (redescribed).

Aladozoa
New species: **Linnavuori** 1969a.

Alanus
Records: **Cwikla & Blocker** 1981a (comparative notes). **Young** 1957a (= *Atanus*, n. syn.).

Alapona
New species: **DeLong** 1980c.

Alapus
Record: **Hamilton, K. G. A.** 1975b (APHRODINAE, DELTOCEPHALINI, DELTOCEPHALINA).

Albera
New species: **Young** 1957c.
Record: **Young** 1957c (n. comb. from *Protalebra*).

Albicostella
New species: **Dlabola** 1970b. **Kwon & Lee** 1979b. **Vilbaste** 1967a.
Records: **Anufriev** 1977b (Kurile Islands); 1978a (USSR: Maritime Territory). **Emeljanov** 1977a (Mongolia; n. sp. syn.). **Hamilton, K. G. A.** 1975b (APHRODINAE, DELTOCEPHALINI, ATHYSANINA). **Kwon & Lee** 1979b (Korea). **Lee, C. E.** 1979a (Korea, redescribed). **Lee, C. E. & Kwon** 1977b (Korea). **Vilbaste** 1967a (key to males); 1968a (coastal regions of the USSR).

Alconeura
New species: **Ruppel** 1966a.
Record: **Young** 1957b (redescription of type material).

Alebra
New species: **Anufriev** 1969b. **Dworakowska** 1968d. **Mitjaev** 1963a. **Zachvatkin** 1948a.
Records: **Anufriev** 1978a (USSR: Maritime Territory; key to species). **Beirne** 1956a (Canada: southern British Columbia and southern Ontario). **Dlabola** 1957a (Turkey); 1958b (Caucasus); 1958c (in key to Palaearctic genera of TYPLOCYBINAE); 1965d (Jordan); 1971a (Turkey); 1977c (Mediterranean); 1981a (Iran). **Drosopoulos** 1984a (Greece; *albostirella* complex). **Dworakowska** 1971c (species groups defined). **Emeljanov** 1964c (European USSR). **Gunthart, H.** 1971a (Switzerland). **Gyllensvard** 1969a (Sweden). **Kirejtshuk** 1977a (USSR: Karkov region). **Jankovic** 1966a (Serbia). **Jasinska** 1980a (Poland: Bledowska Wilderness). **Lee, C. E.** 1979a (Korea, redescribed). **LeQuesne** 1976a (key to British species). **LeQuesne & Payne** 1981a (Britain). **Lindberg** 1961a (Madeira); 1962a (Portugal). **Linnavuori** 1959b (*A. dorsalis* Gillette, *Incertae sedis*; Brazil); 1962a (Israel); 1965a (Turkey). **Lodos** 1982a (Turkey). **Lodos & Kalkandelen** 1983b (Turkey). **Logvinenko** 1957b (USSR: Ukraine). **Mitjaev** 1963b (Kazakhstan); 1968a (eastern Kazakhstan); 1971a (Kazakhstan; in key to genera; key to species). **Nast** 1972a (catalogued); 1976a (Poland: Pieniny Mts.); 1976b (Poland; catalogued). **Ossiannilsson** 1981 (Fennoscandia & Denmark; key to species). **Quartau** 1979a (Azores). **Quartau & Rodrigues** 1969a (Portugal). **Servadei** 1957a, 1960a, 1968a, 1971a (Italy). **Trolle** 1966a (Denmark). **Varty** 1967a (Canada; host association). **Viggiani** 1971a (Italy). **Vilbaste** 1959b (Estonia); 1968a (coastal regions of the USSR); 1973a (notes on Flor's material); 1980b (Tuva). **Wilson, M. R.** (descriptions and key characters for fith instars). **Young** 1957c (n. sp. syn.; key to species). **Zachvatkin** 1953b (central Russia; Oka district).

Alebroides
New species: **Datta & Ghosh** 1973a. **Dworakowska** 1976b, 1977i, 1980b, 1981e. **Linnavuori** 1960a. **Sohi & Dworakowska** 1979a. **Vilbaste** 1968a.
Records: **Ahmed, Manzoor** 1985c (n. comb. from *Empoasca*). **Anufriev** 1969b, 1978a (USSR: Maritime Territory). **Dworakowska** 1979i (n. sp. syn.; n. comb. from *Empoasca* and *Paolia*); 1971f (n. comb. from *Typhlocyba*; corrections); 1976b (data for lectotype specimens); 1977d (n. comb. from *Empoasca*, *Empoanara* & *Typlocyba*; n. sp. syn; records and notes); 1977i (Vietnam); 1979a (n. sp. syn.); 1981e (India & Nepal); 1982a (Japan; Korea). **Dworakowska, Nagaich & Singh** 1978a (notes on color forms). **Lee, C. E.** 1979a (Korea, redescribed). **LeQuesne & Payne** 1981a (Britain). **Linnavuori** 1960a (Micronesia: Bonin Ids., Eastern Caroline Ids.; key to species). **Sohi & Dworakowska** 1979a (key to species from India & Tibet); 1984a (list of Indian species).

Alemaia
New species: **Heller & Linnavuori** 1968a.

Aletta
Records: **Nielson** 1975a (removed from COELIDIINAE; unassigned). **Theron** 1976a (*albicosta* Naude returned to *Equeefa*); 1984c (redescribed; provisionally COELIDIINAE; n. comb. to *Keia*).

Algothyma
Record: **Young** 1977a (identity unknown).

Aligia
New species: **Kramer & DeLong** 1968a.
Records: **Beirne** 1956a (Canada: southern British Columbia). **Hamilton, K. G. A.** 1975b (APHRODINAE, DELTOCEPHALINI, PLATYMETOPIINA).

Alishania
Records: **Hamilton, K. G. A.** 1975b (as *Atishania*; APHRODINAE, unplaced). **Vilbaste** 1969a (n. comb. from *Thamnotettix*).

Alladanus
New species: **DeLong & Harlan** 1968a.
Records: **Cwikla & Blocker** 1981a (comparative notes). **DeLong & Harlan** 1968a (in key to Mexican genera of the *Eutettix* complex; key to species).

Allectus
Records: **Vilbaste** 1965a (= *Doratulina*). **Viraktamath, S. & Viraktamath, C. A.** (male genital structures described, illustrated).

Allogonia
New species: **Young** 1977a.
Record: **Young** 1977a (n. comb. from *Tettigonia*; n. sp. syn.; key to species).

Alloproctus
Records: **Evans, J. W.** 1957a (transferred from JASSINAE to AGALLIINAE). **Kramer** 1964a (= *Agallia*, n. syn.; n. comb. to *Agallia*). **Linnavuori & DeLong** 1977c (= *Agallia*).

Allotapes
New species: **Emeljanov** 1964b.
Record: **Hamilton, K. G. A.** 1975b (APHRODINAE; unplaced).

Allygianus
Record: **Hamilton, K. G. A.** 1975b (APHRODINAE, DELTOCEPHALINI, PLATYMETOPIINA).

Allygidius
New species: **Dlabola** 1980a. **Logvinenko** 1975a, 1979a.

Records: **Bonfils & Della Giustina** 1978a (Corsica). **Cantoreanu** 1968b (Danube river delta); 1971c (Romania). **Dlabola** 1957a (Turkey); 1958b (Caucasus); 1960a (n. sp. syn.); 1964c (Albania); 1971b (Turkey); 1977c (Mediterranean); 1981a (Iran). **Emeljanov** 1964c (European USSR; key to species). **Hamilton, K. G. A.** 1975b (APHRODINAE, DELTOCEPHALINI, PLATYMETOPIINA). **ICZN** 1985a (placed on Official List of Generic Names in Zoology). **Jankovic** 1966a (Serbia). **Kalkandelen** 1974a (Turkey). **LeQuesne** 1969c (Britain). **Lindberg** 1960b (Soviet Armenia). **Linnavuori** 1959a (Mt. Sibillini); 1959c (Mt. Picentini); 1962a (Israel). **Logvinenko** 1957b (USSR: Ukraine). **Mitjaev** 1968b (eastern Kazakhstan); 1971a (Kazakhstan). **Nast** 1955a (Poland); 1972a (catalogued); 1976a (Poland: Pieniny Mts.); 1976b (Poland; catalogued). **Ossiannilsson** 1983a (Fennoscandia & Denmark). **Servadei** 1957a, 1960a, 1968a, 1971a, 1972a (Italy). **Vilbaste** (Altai Mts.).

Allygiella
Record: **Hamilton, K. G. A.** 1975b (APHRODINAE, DELTOCEPHALINI, PLATYMETOPIINA).

Allygus
New species: **Dlabola** 1961b. **Emeljanov** 1966a, 1979a. **Linnavuori** 1965a. **Mitjaev** 1971a.
Records: **Beirne** 1956a (Canada: southern British Columbia). **Bonfils & Della Giustina** 1978a (Corsica). **Cantoreanu** 1965a (Romania); 1968b (Danube river delta). **Della Giustina** 1983a (France). **Dlabola** 1960a (notes on *A. theryi* Horvath); 1964c (Albania); 1971b (Turkey); 1981a (Iran). **Downes** 1957a (Canada: British Columbia, adventive). **Dubovskij** 1966a (Uzbekistan: Fergana Valley). **Emeljanov** 1964c (European USSR; key to species). **Hamilton, K. G. A.** 1975b (APHRODINAE, DELTOCEPHALINI, PLATYMETOPIINA); 1983b (adventive to Nearctic). **Jankovic** 1966a, 1971a (Serbia). **Jansky** 1985a (Slovakia). **Kartal** 1982a (Turkey). **LeQuesne** 1969c (Britain). **Lindberg** 1960a ((Portugal); 1963a (Morocco). **Linnavuori** 1959c (Mt. Picentini); 1962a (Israel); 1965a (Spanish Morocco & Tunisia). **Logvinenko** 1957b, 1984a (USSR: Ukraine). **Mitjaev** 1971a (Kazakhstan; key to species). **Nast** 1955a (Poland); 1972a (catalogued); 1976a (Poland: Pieniny Mts.); 1976b (Poland; catalogued). **Ossiannilsson** 1974a (Norway); 1982a (lectotype designations); 1983a (Fennoscandia & Denmark: key to species); 1983b (proposed designation of type species). **Quartau & Rodrigues** 1969a (Portugal). **Servadei** 1957a, 1960a, 1968a, 1971a, 1972a (Italy). **Vilbaste** 1973a (Latvia; notes on Flor's material); 1976a (sp. syn. noted); 1979a (Vooremaa hardwood/spruce forest). **Wagner** 1959c (Greece). **Zachvatkin** 1953b (central Russia; Oka district).

Almunisna
New species: **Dworakowska** 1969c.

Alnella
New species: **Anufriev** 1971a. **Dworakowska** 1972e.
Records: **Anufriev** 1977b (Kurile Islands); 1978a (USSR: Maritime Teriitory; key to species). **Dworakowska** 1972e (new subgenus proposed); 1979b (subgenus of *Alnetoidia*; *Sapporoa* a junior synonym).

Alnetoidia
New species: **Anufriev** 1971a, 1972c. **Chou, I. & Ma** 1981a. **Dworakowska** 1972e, 1979b, 1980b.
Records: **Anufriev** 1969b (Soviet Maritime Territory; key to species); 1972c (notes on Matsumura's material; n. sp. syn.; key to species); 1977b (Kurile Islands); 1978a (USSR: Maritime Territory; key to species). **Bonfils & Della Giustina** 1978a (Corsica). **Cantoreanu** 1971c (Romania). **Claridge, M. F. & Wilson, M. R.** 1981 (host associations, Britain). **Dlabola** 1958b (Caucasus); 1958c (in key to genera of Palaearctic TYPHLOCYBONAE); 1981a (Iran). **Dworakowska** 1972e (n. comb. from *Zygina*); 1979b (*Alnella* a subgenus; n. gen. syn.). **Emeljanov** 1964c (European USSR). **Gunthart, H.** 1971a (Switzerland). **Jankovic** 1966a (Serbia). **Jasinska** 1980a (Poland: Bledowska Wilderness). **Nast** 1972a (catalogued); 1976a (Poland: Pieniny Mts.); 1976b (Poland; catalogued). **Ossiannilsson** 1974a (Norway); 1981a (Fennoscandia & Denmark; in key to genera). **Sohi & Dworakowska** 1984a (list of Indian species).

Viggiani 1971a (Italy). **Vilbaste** 1968a (coastal areas of the USSR; n. comb. from *Zygina*); 1973a (Estonia, Latvia; notes on Flor's material); 1980b (Tuva).

Alobaldia
Records: **Anufriev** 1978a (USSR: Maritime Territory; in key to genera of DELTOCEPHALINI). **Choe** 1985a (Korea). **Emeljanov** 1972d (n. comb. from *Thamnotettix*; n. sp. syn.). **Hamilton, K. G. A.** (APHRODINAE, DELTOCEPHALINI, DELTOCEPHALINA). **Kwon & Lee** 1979b (Korea). **Lee, C. E.** 1979a (Korea, redescribed). **Nast** 1982a (catalogued; DELTOCEPHALINAE).

Alocephalus
Record: **Evans, J. W.** 1977a (n. comb. from *Cephalelus*).

Alocha
Record: **Young** 1977a (= *Paromenia*, n. syn.).

Alocoelidia
Record: **Nielson** 1975b (removed from COELIDIINAE: unassigned).

Alodeltocephalus
New species: **Evans, J. W.** 1966a.
Records: **Evans, J. W.** 1966a (n. comb. from *Deltocephalus* and *Phrynomorphus*; n. sp. syn.); 1977a (type locality of *A. longuinquus*; error in identity of type species noted). **Knight, W. J.** 1975a (designated type species misidentified; redescriptions provided).

Aloeurymela
New species: **Evans, J. W.** 1965b, 1966a.

Aloipo
Record: **Evans, J. W.** (n. comb. from *Ipoides*).

Alopenthimia
New species: **Evans, J. W.** 1972a.

Aloplemmeles
New species: **Evans, J. W.** 1966a.

Alosarpestus
New species: **Evans, F.** 1981a.

Alospangbergia
New species: **Evans, J. W.** 1973a.

Alotartessella
Record: **Evans, F.** 1981a (n. comb. from *Tartessella*).

Alotartessus
Record: **Evans, F.** 1981a (n. comb. from *Tartessus*).

Aloxestocephalus
New species: **Evans, J. W.** 1973a.

Alseis
Record: **Evans, J. W.** 1966a (synopsis).

Altaiotettix
New species: **Vilbaste** 1965b.
Records: **Dlabola** 1967b, 1967d, 1968b, 1970b (Mongolia). **Dworakowska** 1969a (Mongolia). **Hamilton, K. G. A.** 1975b (APHRODINAE, DELTOCEPHALINI, DELTOCEPHALINA). **Vilbaste** 1980b (Tuva; in key to genera).

Amahuaka
New species: **Linnavuori & DeLong** 1977a. **Young** 1965a.
Records: **Ahmed, Manzoor** 1983b (remarks on wing venation). **Linnavuori & DeLong** 1977a (genus reviewed; species groups defined). **Young** 1965a (key to species).

Amalfia
Record: **Young** 1977a (position uncertain; structural characters noted).

Ambara
New species: **Dworakowska** 1981j.
Record: **Sohi & Dworakowska** 1984a (Indian species listed).

Amberbakia
Record: **Evans, J. W.** 1972a (location of type specimen).

Ambigonalia
Record: **Young** 1977a (n. comb. from *Tettigonia*).

Amblydisca
New species: **Young** 1968a.
Records: **Young** 1968a (n. gen. syn.; n. comb. from *Aulacizes*; n. sp. syn.; key to species). **Young & Lauterer** 1964a, 1966a, (lectotype designations). **Young & Nast** 1963a (lectotype designations). **Young & Soos** 1964a (notes on type specimen).

Amblyscarta
New species: **Young** 1977a.

Records: **Young** 1965c (lectotype designations); 1977a (n. comb. from *Sphaeropogonia* & *Tettigonia*; n. sp. syn.; key to species; list of nominal taxa).

Amblyscartidia
New species: **Young** 1977a.
Record: **Young** 1977a (n. comb. from *Tettigonia*; n. sp. syn.; key to species; list of nominal taxa).

Amblysellus
New species: **DeLong & Hamilton** 1974a. **Kramer** 1971a.
Records: **Beirne** 1956a (Canada). **DeLong & Hamilton** 1974a (key to species). **Hamilton, K. G. A.** 1975b (APHRODINAE, DELTOCEPHALINI, DELTOCEPHALINA); 1976a (sp. syn. of Provancher species). **Kramer** 1971a (n. comb. from *Deltocephalus*; key to species).

Amblytelinus
Record: **Hamilton, K. G. A.** 1975b (APHRODINAE, DELTOCEPHALINI, ATHYSANINA).

Amimenus
Records: **Anufriev** 1978a (USSR: Maritime Territory; in key to genera). **Hamilton, K. G. A.** 1975b (APHRODINAE, DELTOCEPHALINI, PLATYMETOPIINA). **Kwon & Lee** 1979b (Korea). **Lee, C. E.** 1979a (Korea, redescribed). **Lee, C. E. & Kwon** 1977b (Korea).

Amphigonalia
New species: **Nielson & Gill** 1984a.
Records: **Nielson & Gill** 1984a (key to species). **Young** 1977a (n. comb. from *Neokolla* and *Tettigonia*; n. sp. syn.; key to species).

Amphipyga
Record: **Hamilton, K. G. A.** 1975b (APHRODINAE, DELTOCEPHALINI, DORATURINA).

Amplicephalus
New species: **Cheng, Y. J.** 1980a. **DeLong** 1984d. **Dlabola** 1967g. **Linnavuori** 1959b, 1973d. **Linnavuori & DeLong** 1976a, 1977c, 1979a.
Records: **Anufriev** 1978a (USSR: Maritime Territory; in key to genera). **Beirne** 1956a (Canada). **Cwikla & Blocker** 1981a (comparative notes). **Hamilton, K. G. A.** 1972d (Canada: Manitoba); 1975b (APHRODINAE, DELTOCEPHALINI, DELTOCEPHALINA); 1976a (specific synonymy of Provancher species); 1983a (transboreal taxon). **Kramer** 1971b (redefinition; key to Nearctic species). **Lee, C. E.** 1979a (Korea, redescribed). **Linnavuori** 1956a (Neotropical species); 1959b (subgenera *Cruciatanus*, *Endria*, *Mendozellus*, *Nanctasus*, *Sanctanus*; n. comb. from *Athysanus*, *Deltocephalus*, *Sanctanus*, *Scaphoideus*, *Spathifer* and *Thamnotettix*; n. sp. syn.; key to subgenera; key to species). **Linnavuori & DeLong** 1977c (key to Chilean species); 1978d (descriptive notes for *A. discalis*). **Linnavuori & Heller** 1961a (Peru). **Vilbaste** 1980a (Kamchatka); 1980b (Tuva).

Amrasca
New species: **Ahmed, Manzoor, Samad & Naheed** 1981a. **Dworakowska** 1977i. **Einyu & Ahmed** 1980a. **Ghauri** 1967a, 1983a. **Ramakrishnan & Menon** 1972a. **Sohi** 1977a.
Records: **Ahmed, Manzoor** 1985c (Pakistan). **Dworakowska** 1976b (notes on Melichar's type material); 1977f (northern India); 1977i (*Quartasca* transferred from *Laokayana*, to subgeneric status); 1979a (China, India); 1980b (India); 1982b (range of *A. biguttula*). **Dworakowska & Viraktamath** 1975a (n. gen. syn.). **Ghauri** 1967a (n. comb. from *Empoasca*; key to species); 1983a (nomenclature). **Knight, W. J.** 1982a (Rennell Island). **Sharma, B.** 1977b (India: Jammu). **Sohi & Dworakowska** 1984a (list of Indian species). **Thapa** 1983a, 1985a (Nepal: Kathmandu Valley).

Amritodus
New species: **Ahmed, S. S., Naheed & Ahmed** 1980a. **Viraktamath, C. A.** 1976b.
Records: **Anufriev** 1970b (n. comb. from *Idiocerus*). **Dlabola** 1974b (in key to genera of IDIOCERINAE). **Sharma** 1977b (India: Jammu). **Viraktamath, C. A.** 1976b (key to species).

Amurta
New species: **Dworakowska** 1977i, 1982a.

Amylidia
New species: **Nielson** 1983h.

Anacephaleus
Records: **Evans, J. W.** 1966a. (=*Cephalelus*). **Evans, J. W.** 1977a (removed from synonymy with *Cephalelus*).

Anaceratagallia
New species: **Dlabola** 1957a, 1957b, 1958a, 1960b, 1960c, 1967a, 1984a. **Dubovskij** 1966a, 1984a. **Mitjaev** 1967a, 1969a, 1969b, 1971a. **Vilbaste** 1959a.
Records: **Anufriev** 1978a (USSR: Maritime Territory; in key to genera). **Cantoreanu** 1968a (Danube river delta); 1971c (Romania). **Cardosa** 1974a (Portugal). **Dlabola** 1956b (Hungary); 1957a (Turkey); 1958b (Afghanistan); 1958b (Caucasus); 1961b (central Asia); 1964c (Albania); 1965c (Mongolia); 1965d (Jordan); 1967b, 1967c, 1967d, 1968a (Mongolia); 1971b (Turkey); 1971b (Switzerland); 1972a (Afghanistan); 1981a (Iran). **Dubovskij** 1966a (Uzbekistan; in key to genera); 1978a (Uzbekistan, Zarafshan Valley); 1980a (USSR: Turkemania). **Jankovic** 1966a, 1971a (Serbia). **Lindberg** 1960a (Portugal); 1960b (Soviet Armenia). **Lodos & Kalkandelen** 1981a (Turkey). **Logvinenko** 1975a (n. comb. to *Mesagallia*); 1984a (USSR: Ukraine). **Mitjaev** 1967a, 1968b (eastern Kazakhstan); 1971a (Kazakhstan; key to species). **Nast** 1972a (catalogued); 1976a (Poland: Pieniny Mts.); 1976b (Poland; catalogued). **Ossiannilsson** 1981a (=*Agallia*). **Quartau** 1979a, 1980b (Azores). **Quartau & Rodrigues** 1969a (Portugal). **Remane** 1961d (Germany). **Servadei** 1960a, 1968a, 1971a (Italy). **Vilbaste** 1962b (eastern part of the Caspian lowlands); 1964a (Auwiesen Estlands); 1965b (Altai Mts.); 1973a (notes on Flor's material); 1980b (Tuva).

Anacornutipo
Records: **Davis, R. B.** 1975a (morphology; phyletic inferences). **Evans, J. W.** 1966a (reviewed).

Anacotis
Records: **Evans, J. W.** 1966a (reviewed).

Anacrocampsa
New species: **Young** 1968a.
Record: **Young** 1968a (n. comb. from *Amblydisca* & *Tettigonia*; key to species).

Anacuerna
Record: **Young** 1968a (n. comb. from *Cuerna*).

Anaemotettix
Record: **Nast** 1982a (catalogued; DELTOCEPHALINAE).

Anaka
New species: **Dworakowska & Viraktamath** 1975a.
Record: **Sohi & Doworakowska** 1984a (list of Indian species).

Anareia
New species: **Vilbaste** 1965b.
Records: **Dlabola** 1967b, 1967d, 1968a, 1968b (Mongolia). **Hamilton, K. G. A.** (APHRODINAE, DELTOCEPHALINI, DELTOCEPHALINA). **Vilbaste** 1980b (Tuva; in key to genera).

Anargella
Records: **Emeljanov** 1977a (Mongolia; n. comb. from *Adarrus*). **Nast** 1982a (catalogued; DELTOCEPHALINAE).

Anchura
Record: **Young** 1968a (=*Mareba*, n. syn.).

Anchuralia
Record: **Young** 1968a (=*Mareba*, n. syn.).

Ancudana
New species: **DeLong & Martinson** 1974b.
Record: **Cwikla & Blocker** 1981a (comparative notes).

Andamarca
Record: **Young** 1968a (=*Ochrostacta*, n. syn.).

Andanus
New species: **Linnavuori** 1959b.
Record: **Linnavuori & Heller** 1961a (Peru).

Andrabia
New species: **Ahmed, Manzoor** 1970d. **Dworakowska & Sohi** 1978b.
Records: **Ahmed, Manzoor** 1983b (remarks about male style shape); 1985c (Pakistan).

Anemochrea
Records: **Evans, J. W.** 1966a (synopsis; DELTOCEPHALINAE, DELTOCEPHALINI); 1977a (=*Aconura* ?). **Vilbaste** 1965a (relation to *Aconura* questioned).

Anemolua
  Record: Evans, J. W. 1966a (synopsis; DELTOCEPHALINAE, PLATYMETOPIINI).

Aneono
  New species: Dworakowska 1972l. Evans, J. W. 1966a.
  Record: Evans, J. W. 1966a (synopsis; n. comb. from *Empoa*).

Angolaia
  New species: Linnavuori & Al-Ne'amy 1983a.

Angubahita
  New species: DeLong 1982a, 1984g.
  Record: DeLong 1984g (new subgenus proposed).

Angucephala
  New species: DeLong & Freytag 1975a.

Angusticella
  Record: Maldonado-Capriles 1976a (generic characters tabulated).

Anidiocerus
  New species: Maldonado-Capriles 1976a.
  Record: Maldonado-Capriles 1976a (generic characters tabulated).

Anipo
  Record: Davis, R. B. 1975a (morphology; phyletic inferences).

Ankosus
  Record: Oman & Musgrave 1975a (n. comb. from *Errhomus*; in key to genera of ERRHOMENINI).

Annidion
  Record: Linnavuori 1975a (identity unknown, original description quoted).

Anomia
  Record: Beirne 1956a (= *Typhlocyba*).

Anomiana
  Record: Hamilton, K. G. A. 1975b (APHRODINAE, DELTOCEPHALINI, MACROSTELINA).

Anoplotettix
  New species: Dlabola 1965d, 1970a, 1971b, 1974c, 1981a, 1984a. Emeljanov 1962a. Logvinenko 1975a. Remane 1966a. Wagner 1959c.
  Records: Bonfils & Della Giustina 1978a (Corsica). Dlabola 1958a (Italy); 1964c (Albania); 1974c (synopsis); 1981a (Iran). Emeljanov 1962a (notes on *A. loewi*); 1964c (European USSR; key to species). Hamilton, K. G. A. 1975b (APHRODINAE, DELTOCEPHALINI, PLATYMETOPIINA). Kartal 1982a (Turkey). Linnavuori 1959a (Mt. Sibillini). Logvinenko 1959a (USSR: Transcarpathian region); 1963b (USSR: Carpathians). Quartau & Rodrigues 1969a (Portugal). Servadei 1957a, 1960a, 1968a, 1969a, 1971a, 1972a (Italy). Wagner 1959c (key to species).

Anoscopus
  Records: Hamilton, K. G. A. 1975b (APHRODINAE, APHRODINI, APHRODINA); 1975d (synopsis; n. sp. syn.; key to species); 1976a (synonymy of Provancher species); 1983b (transboreal taxa). Jasinska 1980a (Poland: Beldowska Wilderness). Nast 1976a (Poland: Pieniny Mts.). Ossiannilsson 1981a (key to species of Fennoscandia & Denmark). Vilbaste 1980b (Tuva).

Anoterostemma
  Records: Datta & Dhar 1984a (descriptive notes). Dlabola 1957a (Turkey); 1981a (Iran). Emeljanov 1964c (European USSR). Hamilton, K. G. A. 1975b (APHRODINAE, APHRODINI, ANOTEROSTEMMINA). Lodos & Kalkandelen 1983a (Turkey; CICADELLINAE). Mitjaev 1968a (easterm Kazakhstan); 1971a (Kazakhstan; in key to genera). Vilbaste 1962b (eastern part of the Caspian lowlands); 1980b (Tuva).

Anthocallis
  Record: Hamilton, K. G. A. (APHRODINAE, DELTOCEPHALINI, DELTOCEPHALINA).

Antoniellus
  New species: Linnavuori 1959b.
  Records: Cheng, Y. J. 1980a (Paraguay). Cwikla & Blocker 1981a (comparative notes). Linnavuori 1959b (DELTOCEPHALINAE, EUSCELINI). Linnavuori & DeLong 1977c (in key to the *Yungasia* group of genera).

Anufrievia
  New species: Dworakowska 1970n, 1976b, 1977f. Dworakowska & Viraktamath 1978a.
  Records: Dworakowska 1970n (n. comb. from *Zygina*). Lee, C. E. 1979a (Korea, redescribed). Sohi & Dworakowska 1984a (list of Indian species). Thapa 1984b (Nepal).

Anufrieviella
  Record: Nast 1981a (n. comb. from *Straganiassus*).

Apetiocellata
  New species: Dworakowska 1980b. Sohi 1977a.
  Records: Sohi 1977a (DIKRANEURINI). Sohi & Dworakowska 1984a (list of Indian species).

Aphanalebra
  Record: Young 1957c (reviewed; in key to genera of ALEBRINI).

Apheliona
  New species: Dworakowska & Sohi 1978b. Thapa 1985a.
  Records: Anufriev 1972a (notes about male genitalia). Dworakowska 1970a (= *Sujitettix*, n. gen. syn.); 1972g (Java); 1977d (= *Chikkaballapura*, = *Ushamenona*; n. gen. syn.; n. sp. syn.); 1980b (India); 1982b (Japan). Malhotra & Sharma, B. 1974a (n. comb. from *Sujitettix*; n. subgenus added; key to subgenera and species); 1977c (immature stages). Sharma, B. 1977b (India: Jammu). Sohi & Dworakowska 1984a (list of Indian species).

Aphrodes
  New species: Bliven 1957a. Dlabola 1960a, 1965c, 1971b. Cantoreanu 1968a. Dubovskij 1966a. Emeljanov 1964c, 1972b. Kuoh, C. L. 1981b. Linnavuori 1979a. Logvinenko 1965a, 1966b, 1967b, 1968a, 1968b, 1971a, 1983a. Metcalf 1963c (new name). Mitjaev 1967a, 1979b. Rodrigues 1968a. Saringer, Gy. 1959a. Tshmir 1977a. Vilbaste 1965b. Zachvatkin 1948b.
  Records: Anufriev 1977b (Kurile Islands); 1978a (USSR: Maritime Territory; key to species). Beirne 1956a (Canada; key to species). Bonfils & Della Giustina 1978a (Corsica). Cantoreanu 1965d (Romania); 1968b (Danube river delta); 1971c (Romania). Cardosa 1974a (Portugal). Dlabola 1956a (Czechoslovakia); 1957a (Turkey); 1957b (Afghanistan); 1958b (Caucasus); 1961b (central Asia); 1964c (Albania); 1965c (Mongolia); 1965d (Jordan); 1966a, 1967b (Mongolia); 1967e (Bohemia); 1970b (Mongolia); 1971b (Iran); 1972a (Afghanistan); 1977c (Mediterranean); 1981, 1984a (Iran); Dubovskij 1966a (Uzbekistan; key to species); 1978a (Uzbekistan, Zarafshan Valley). Duffield 1963a (synopsis of British species; key to species). Emeljanov 1964c (European USSR; key to species); 1977a (Mongolia; n. sp. syn.). Emmrich 1980a (systematic relationships, species level). Gravestein 1965a (The Netherlands: Terschelling Island). Gyllensvard 1963a (Sweden). Hamilton, K. G. A. 1972d (Canada: Manitoba; SCARINAE); 1975b (APHRODINAE, APHRODINI); 1975d (reviewed; key to species of APHRODINA); 1983b (adventive taxa in Nearctic). Ishihara 1958a (Japan); 1966a (Kurile Islands). Jankovic 1966a, 1971a (Serbia). Jasinska 1980a (Poland: Bledowska Wilderness). Kholmuminov & Dlabola 1979a (USSR: Golodnostep plain). Kocak 1981b (n. sp. syn.). Korolevskaya 1975a (Tadzhikistan). Lauterer 1957a (Czechoslovakia). LeQuesne 1964a (n. sp. syn.); 1965c (Britain). Lindberg 1960a (Portugal); 1960b (Soviet Armenia); 1960c (Azores); 1961a (Madeira); 1963a (Morocco). Linnavuori 1956d (Spain & Spanish Morocco); 1959a (Mt. Sibillini); 1959c (Mt. Picentini); 1962a (Israel); 1979a (*A. prominens* Walker referred to HECALINI, unassigned). Lodos 1982a (Turkey). Logvinenko 1957b, 1984a (USSR: Ukraine). Mitjaev 1968b (eastern Kazakhstan); 1971a (Kazakhstan; key to species); 1979b (key to species). Nast 1972a (catalogued); 1976a (Poland: Pieniny Mts.; revised sp. syn.); 1976b (Poland, catalogued). Nielson 1968b (characterization of vector species). Ossiannilsson 1981a (key to species of Fennoscandia & Denmark). Quartau 1979a, 1980b (Azores). Quartau & Rodrigues 1969a (Portugal). Saringer, Gy. 1958a (Hungary; revision and supplement). Saringer, Gyula 1958a (Hungary; review). Schulz 1976a (Norway). Servadei 1957a, 1958a, 1960a, 1968a, 1969a, 1971a, 1972a (Italy). Vilbaste 1958a, 1959b (Estonia); 1964a (Auwiesen Estlands); 1965b (Altai Mts.; n. sp. syn.); 1968a (coastal region of the USSR); 1973a (identity of Flor's material; 1976a (Matsumura's species transferred from *Acocephalus*); 1979a (Vooremaa hardwood/spruce forest);

1980a (Kamchatka); 1980b (Tuva). **Woodroffe** 1967a (Britain). **Zachvatkin** 1953b (central Russia; Oka district).

Aplanatus
New species: **Cheng, Y. J.** 1980a.
Record: **Cwikla & Blocker** 1981a (comparative notes).

Aplanus
Record: **Hamilton, K. G. A.** 1975b (APHRODINAE, DELTOCEPHALINI, PLATYMETOPIINA).

Apogonalia
New species: **Costa Lima** 1963a.
Record: **Costa Lima** 1963a (n. comb. from *Tettigonia*).

Apphia
Record: **Young** 1965c (lectotype designations).

Apulia
New species: **Young** 1977a.
Records: **Young** 1965c (lectotype designations); 1977a (n. comb. from *Tettigonia*; n. sp. syn.).

Apulina
Record: **Young** 1977a (=*Apulia*).

Arahura
New species: **Knight, W. J.** 1975a.

Araldus
Record: **Hamilton, K. G. A.** 1975b (APHRODINAE, DELTOCEPHALINI, DELTOCEPHALINA).

Arapona
New species: **DeLong** 1979b.

Arawa
New species: **Knight, W. J.** 1975a.
Records: **Knight, W. J.** 1975a (n. comb. from *Deltocephalus*; key to New Zealand species); 1976b (Chatham Island).

Arbela
Record: **Anufriev** 1972a (n. comb. from *Alebra*).

Arbelana
Record: **Anufriev** 1975a (=*Arbela*).

Arboridia
New species: **Diabola** 1971a. **Dworakowska** 1977f, 1979b, 1980b. **Dworakowska & Viraktamath** 1978a. **Kirejtshuk** 1975a. **Korolevskaya** 1979a. **Sohi & Sandhu** 1971a. **Vilbaste** 1980b.
Records: **Ahmed, Manzoor** 1985c (Pakistan; n. comb. from *Erythroneura*). **Anufriev** 1977b (Kurile Islands); 1978a (USSR: Maritime Territory; key to species). **Bonfils & Della Giustina** 1978a (Corsica). **Choe** 1985a (Korea). **Della Giustina** 1983a (France). **Diabola** 1958c (in key to Palaearctic genera of TYPHLOCYBINAE); 1971a (Turkey); 1972a (Afghanistan); 1974d (Europe) 1981a (Iran). **Dworakowska** 1980g (n. comb. from *Erythroneura*, *Typhlocyba* and *Zyginidia*; n. sp. syn.); 1977b (probably east Asiatic affinities); 1981e (India). **Dworakowska & Viraktamath** 1975a (n. gen. syn.; n. comb. from *Erythroneura* and *Khoduma*); 1978a (n. comb. from *Erythroneura* and *Typhlocyba*). **Kartal** 1983a (Turkey). **Kirejtshuk** 1977a (USSR: Kharkov region). **Lauterer** 1980a (Czechoslovakia). **Korolevskaya** 1978a (Tadzhikistan). **Lee, C. E.** 1979a (Korea, redescribed). **Lee, C. E. & Kwon** 1977b (Korea). **LeQuesne & Payne** 1981a (Britain). **Lodos** 1982a (Turkey). **Lodos & Kalkandelen** 1984c (Turkey). **Nast** 1976b (Poland; catalogued). **Ossiannilsson** 1981a (Fennoscandia & Denmark). **Sharma, B.** 1984a (India: J. & K., Kishtwar). **Sohi & Dworakowska** 1984a (list of Indian species). **Sohi & Sandhu** 1971a (n. subgenus proposed). **Vilbaste** 1973a (identity of Flor's material); 1979a (Vooremaa hardwood/spruce forest); 1980b (Tuva).

Arctotettix
Record: **Hamilton, K. G. A.** 1975b (APHRODINAE, DELTOCEPHALINI, DELTOCEPHALINA).

Arenua
Record: **Anufriev** 1972a (n. comb. from *Dikraneura*).

Arezzia
New species: **Diabola & Novoa** 1976b.
Record: **Diabola & Novoa** 1976b (n. comb. from *Hadria* and *Hortensia*; key to species).

Argaterma
Record: **Synave** 1976a (key to species).

Argyrilla
Records: **Emeljanov** 1972a (n. comb. from *Athysanus*). **Hamilton, K. G. A.** 1975b (APHRODINAE, APHRODINI; unplaced). **Nast** 1982a (catalogued; DELTOCEPHALINAE)

Aricanus
New species: **Linnavuori** 1956a, 1959b.
Record: **Linnavuori & DeLong** 1977c (Chile; additional characters, revised characterization of aedeagal structure).

Ariellus
Records: **Emeljanov** 1964c (European USSR). **Hamilton, K. G. A.** 1975b (APHRODINAE, DELTOCEPHALINI, DELTOCEPHALINA). **Ossiannilsson** 1983a (Fennoscandia & Denmark).

Arocephalus
New species: **Diabola** 1967a, 1971b. **D'Urso** 1978a. **Emeljanov** 1962a, 1964c, 1964f. **Mitjaev** 1969c. **Remane & Asche** 1980a. **Servadei** 1972a. **Vilbaste** 1965b.
Records: **Bonfils & Della Giustina** 1978a (Corsica). **Cantoreanu** 1965d (Romania); 1971c (Romania). **Cardosa** 1974a (Portugal). **Diabola** 1957b (Afghanistan); 1961b (USSR: Transcaucasus); 1967d, 1968b, 1970b (Mongolia); 1981a (Iran). **Dubovskij** 1966a (Uzbekistan). **D'Urso** 1980b (Sicily). **Emeljanov** 1964c (European USSR; key to species); 1977a (Mongolia). **Gravestein** 1965a (The Netherlands: Terschelling I.) **Hamilton, K. G. A.** 1975b (APHRODINAE, DELTOCEPHALINI, DELTOCEPHALINA). **Jasinska** 1980a (Poland: Bledowska Wilderness). **Kalkandelen** 1974a (Turkey). **Korolevskaya** 1979b (Tadzhikistan). **Lindberg** 1960a (Portugal). **LeQuesne** 1969c (Britain). **Lindberg** 1960a (Portugal). **Linnavuori** 1956d (Spain). **Logvinenko** 1957b (USSR: Ukraine). **Mitjaev** 1968a (eastern Kazakhstan); 1971a (Kazakhstan; key to species). **Nast** 1972a (catalogued); 1976a (Poland: Pieniny Mts.); 1976b (Poland; catalogued). **Ossiannilsson** 1983a (Fennoscandia & Denmark; key to species). **Quartau & Rodrigues** 1969a (Portugal). **Servadei** 1957a, 1960a, 1968a, 1971a, 1972a, 1977a (Italy; key to species). **Vilbaste** 1964a (Auwiesen Estlands); 1965b (Altai Mts.); 1973a (identity of Flor's material); 1980b (Tuva).

Arrailus
Record: **Emeljanov** 1964c (European USSR). **Hamilton, K. G. A.** 1975b (APHRODINAE, DELTOCEPHALINI, DELTOCEPHALINA).

Arrugada
New species: **Linnavuori & DeLong** 1978a.
Record: **Linnavuori & DeLong** 1978a (redescription of subfamily; phyletic inferences).

Arthaldeus
New species: **Emeljanov** 1966a. **Remane** 1960a.
Records: **Beirne** 1956a (Canada: southeastern regions). **Bonfils & Della Giustina** 1978a (Corsica). **Cantoreanu** 1965d (Romania). **Dubovskij** 1966a (Uzbekistan). **Dworakowska** 1969a (Mongolia). **Emeljanov** 1964c (European USSR; key to species); 1977a (Mongolia). **Gravestein** 1965a (The Netherlands: Terschelling I.). **Hamilton, K. G. A.** 1975b (APHRODINAE, DELTOCEPHALINI, DELTOCEPHALINA). **Jasinska** 1980a (Poland: Bledowska Wilderness). **LeQuesne** 1969c (Britain). **Logvinenko** 1957b (USSR: Ukraine). **Mitjaev** 1968b (eastern Kazakhstan); 1971a (Kazakhstan). **Nast** 1972a (catalogued); 1976a (Poland: Pieniny Mts.); 1976b (Poland; catalogued). **Ossiannilsson** 1974a (Norway); 1983a (Fennoscandia & Denmark; key to species). **Remane** 1960a (key to species). **Servadei** 1968a, 1971a (Italy). **Vilbaste** 1962b (eastern part of the Caspian lowlands); 1964a (Auwiesen Estlands); 1965b (Altai Mts.); 1973a (identity of Flor's material); 1980b (Tuva).

Artianus
Records: **Cantoreanu** 1961a (Romania); 1971c (Romania). **Della Giustina** 1983a (France). **Diabola** 1957a (Turkey); 1958b (Caucasus); 1961b (central Asia); 1964c (Albania). **Dubovskij** 1966a (Uzbekistan; in key to genera). **Emeljanov** 1964c (European USSR; key to species). **Hamilton, K. G. A.** 1975b (APHRODINAE, DELTOCEPHALINI, ATHYSANINA). **Jankovic** 1966a (Serbia). **Kalkandelen** 1974a (Turkey). **Lindberg** 1960b (Soviet Armenia).

Logvinenko 1957b (USSR: Ukraine). **Mitjaev** 1971a (Kazakhstan; in key to genera). **Nast** 1972a (catalogued); 1973a (Poland); 1976a (Poland: Pieniny Mts.); 1976b (Poland; catalogued). **Ribaut** 1959a (France). **Servadei** 1957a, 1960a, 1968a, 1971a (Italy).

Articoelidia
New species: **Nielson** 1979b, 1983b.
Record: Nielson 1979b (n. comb. from *Coelidia*; key to species).

Artucephalus
Record: **Cwikla & Blocker** 1981a (comparative notes).

Aruena
Record: **Anufriev** 1978a (USSR: Maritime Territory; in key to DIKRANEURINI).

Arundanus
Record: **Hamilton, K. G. A.** 1975b (APHRODINAE, DELTOPCEPHALINI, DELTOCEPHALINA).

Arya
Records: **Datta** 1973c (descriptive notes). **Ramachandra Rao** 1967a (India: Calcutta, Kalyani); 1967b (illustration); 1973a (India: Kalyani). **Vilbaste** 1965a (= *Doratulina*).

Ascius
Records: **Cwikla & Blocker** 1981a (comparative notes). **Hamilton, K. G. A.** 1975b (APHRODINAE, DELTOCEPHALINI, PLATYMETOPIINA).

Asialebra
Record: **Dworakowska** 1971c (n. comb. from *Alebra*).

Asianidia
New species: **Korolevskaya** 1976b.
Records: **Dlabola** 1972a (Afghanistan); 1981a (Iran). **Dubovskij** 1978a (Uzbekistan, Zarafshan Valley). **Dworakowska** 1970m (n. comb. from *Erythroneura*); 1970n (n. comb. from *Erythroneura* and *Zygina*). **Korolevskaya** 1974a, 1976a (Tadzhikistan; key to species).

Asiotoxum
Records: **Emeljanov** 1964b (n. comb. from *Phlepsius*). **Hamilton, K. G. A.** (APHRODINAE; unplaced).

Asmaropsis
New species: **Linnavuori** 1978b.

Aspilodora
Record: **Young** 1977a (comments on characters; placement uncertain).

Assina
New species: **Dworakowska** 1979b.

Assiuta
New species: **Emeljanov** 1975a. **Linnavuori** 1969b.
Record: **Emeljanov** 1975a (in key to genera of ADELUNGIINI; food plants).

Astenogonia
Record: **Young** 1968a (= *Acrogonia*, n. syn.).

Asthenotettix
Record: **Nast** 1982a (catalogued; DELTOCEPHALINAE).

Asymmetrasca
Records: **Dlabola** 1958c (in key to Palaearctic genera of TYPHLOCYBINAE); 1965d (Jordan); 1971a (Pakistan); 1981a (Iran). **Dworakowska** 1968d (Korea). **Lodos & Kalkandelen** 1983c (Turkey). **Nast** 1972a (catalogued; = *Empoasca*). **Servadei** 1968a, 1971a (Italy).

Asymmetropteryx
Record: **Emeljanov** 1964c (European USSR).

Atanus
New species: **DeLong** 1978a, 1982f. **Cheng, Y. J.** 1980a. **Linnavuori** 1959b. **Linnavuori & DeLong** 1976b, 1977c, 1979b. **Linnavuori & Heller** 1961a.
Records: **Cheng, Y. J.** 1980a (Paraguay). **DeLong** 1978a (key to species). **Hamilton, K. G. A.** 1975b (APHRODINAE, DELTOCEPHALINI, PLATYMETOPIINA). **Linnavuori** 1956a (n. sp. syn.); 1959b (n. gen. syn.; n. comb. from *Atanus, Fulvanus* and *Athysanus*; key to species). **Linnavuori & DeLong** 1976a (Bolivia and Peru); 1977b (notes on type of *A. lobatus*); 1978c (in key to genera). **Young** 1957a (= *Alanus*, n. syn.).

Athysanella
New species: **Beirne** 1955a. **Blocker** 1967b, 1984a, 1985a. **Blocker & Wesley** 1985a. **Emeljanov** 1970a. **Johnson, D. H. & Blocker** 1979a. **Wesley & Blocker** 1985a.
Records: **Beirne** 1956a (Canada; key to subgenera and species). **Blocker** 1984a (structural variation). **Emeljanov** 1977a (Mongolia). **Hamilton, K. G. A.** 1972d, (Canada: Manitoba); 1975b (APHRODINAE, APHRODINI, DORATURINA). **Scudder** 1961a (Canada: British Columbia). **Wesley & Blocker** 1985a (n. subgenera proposed; key to species).

Athysanopsis
Records: **Anufriev** 1978a (USSR: Maritime Territory). **Dlabola** 1967b (Mongolia). **Dworakowska** 1973d (Korea). **Hamilton, K. G. A.** 1975b (APHRODINAE; unplaced). **Ishihara** 1958b (Japan). **Lee, C. E.** 1979a (Korea, redescribed). **Lee, C. E. & Kwon** 1977b (Korea). **Linnavuori & Al-Ne'amy** 1983a (SELENOCEPHALINAE, BHATIINI). **Vilbaste** 1968a (coastal regions of the USSR).

Athysanus
New species: **Emeljanov** 1964f.
Records: **Anufriev** 1977b (Kurile Islands); 1978a (USSR: Maritime Territory). **Badmin** 1981a (Britain: Kent). **Bonfils & Della Giustina** 1978a (Corsica). **Cantoreanu** 1971c (Romania). **Datta** 1972a (descriptive notes). **Dlabola** 1965c, 1967c, 1970b (Mongolia). **Dubovskij** 1966a (Uzbekistan; in key to genera). **Emeljanov** 1964c (European USSR; key to species); 1977a (Mongolia). **Evans, J. W.** 1966a (*A. negatus* F. B. White, *incertae sedis*). **Gravestein** 1965a (The Netherlands: Terschelling I.) **Hamilton, K. G. A.** 1975b (APHRODINAE, DELTOCEPHALINI, ATHYSANINA). **Jankovic** 1966a (Serbia). **LeQuesne** 1965b (Isle of Wight); 1969c (in key to genera of ATHYSANINI of Britain). **Logvinenko** 1957b (USSR: Ukraine). **Mitjaev** 1968b (eastern Kazakhstan); 1971a (Kazakhstan, key to two species). **Nast** 1972a (catalogued); 1976b (Poland; catalogued). **Ossiannilsson** 1983a (Fennoscandia & Denmark; key to species). **Ramachandra Rao** 1973a (as subgenus of *Phrynomorphus*). **Salmon** 1959a (Britain). **Servadei** 1957a, 1958a, 1968a, 1969a (Italy). **Theron** 1970b (Cogan's species redescribed). **Vilbaste** 1958a (Estonia); 1964a (Auwiesen Estlands); 1965b (Altai Mts.); 1968a (coastal regions of the USSR); 1973a (identity of Flor's material); 1976a (notes on Matsumura's material); 1980b (Tuva). **Zachvatkin** 1953b (central Russia; Oka district).

Atkinsoniella
Records: **Datta** 1980a (descriptive notes). **Young** 1965c (lectotype designation).

Atlantisia
New species: **Dlabola** 1976a.

Atractotypus
Record: **Hamilton, K. G. A.** 1975b (APHRODINAE, EUPELICINI).

Attenuipyga
New species: **Oman** 1985a.
Records: **Emeljanov** 1966a (generic status; in key to genera of DORYCEPHALINAE). **Oman** 1985a (genus redefined; key to species).

Augulus
Record: **Young** 1965c (lectotype designation).

Aulacizes
Records: **Schroder** 1959a (diagnosis). **Young** 1963a, 1965c (lectotype designations); 1968a (n. sp. syn.). **Young & Lauterer** 1966a (lectotype designations). **Young & Nast** 1963a (notes on type specimens). **Young & Soos** 1964a (lectotype designation).

Auridius
New species: **Hamilton, K. G. A. & Ross** 1972a.
Records: **Beirne** 1956a (Canada: western Provinces; key to species). **Hamilton, K. G. A.** 1972d (Canada: Manitoba); 1975b (APHRODINAE, DELTOCEPHALINI, DELTOCEPHALINA).

Aurkius
Record: **Hamilton, K. G. A.** 1975b (APHRODINAE, DELTOCEPHALINI, DELTOCEPHALINA).

Aurogonalia
New species: **Young** 1977a.

Auropenthimia
Record: **Linnavuori** 1977a (= *Jafar*, n. syn.).

Australoscopus
Record: **Evans, J. W.** 1966a (reviewed).

Austroagallia
New species: **Bindra, Singh, Sohi & Gill** 1973a. **Linnavuori** 1969b. **Kameswara Rao, Ramakrishnan & Ghai** 1979b. **Viraktamath, C. A.** 1972a. **Viraktamath, S. & Viraktamath, C. A.** 1981a.
Records: **Bonfils & Della Giustina** 1978a (Corsica). **Davis, R. B.** 1975a (morphology, phyletic inferences). **Dlabola** 1971a, 1972a, (Afghanistan); 1981a (Iran). **Dubovskij** 1980a (USSR: western Turkmenia). **Evans, J. W.** 1966a (Australia; n. sp. syn.); 1971b (Palaearctic taxon, adventive to Australia and adjacent regions); 1982a (biogeography). **Lauterer** 1984a (Moravia). **LeQuesne** 1964a (=*Peragallia*, n. syn.); 1965c (Britain). **Linnavuori** 1964a (Egypt); 1965a (Libya, Tunisia, Spanish Morocco); 1969a (Congo). **Lodos & Kalkandelen** 1981a (Turkey). **Nielson** 1968b (characterization of vector species). **Servadei** 1968a, 1971a (Italy). **Sharma, B.** 1977b (India: Jammu). **Viraktamath, C. A.** 1972a (n. comb. from *Agallia*; generic characters); 1973a (n. comb. from *Agallia*). **Viraktamath, C. A. & Sohi** 1980a (review of species known from India).

Austroagalloides
New species: **Evans, J. W.** 1977a.
Records: **Davis, R. B.** 1975a (morphology; phyletic inferences). **Evans, J. W.** 1966a (reviewed).

Austroasca
New species: **Ahmed, Manzoor, Samad & Waheed** 1981a. **Dworakowska** 1970h, 1971f. **Kirejtshuk** 1975a. **Mitjaev** 1979a.
Records: **Ahmed, Manzoor** 1985c (Pakistan; n. comb. to *Jacobiasca*). **Choe** 1985a (Korea). **Dworakowska** 1970h (n. comb. from *Chlorita, Empoasca* and *Kyboasca*); 1971f (n. comb. from *Empoasca* and *Typhlocyba*); 1972g (Ceylon); 1972i (n. subgenus proposed; n. sp. syn.); 1973b (many Palaearctic records); 1976b (notes on Jacobi's type material); 1982a (China, Japan, Mongolia). **Emeljanov** 1977a (Mongolia). **Evans, J. W.** 1966a (=*Empoasca*). **Kirejtshuk** 1977a (USSR: Karkov region). **Korolevskaya** 1977a (central Asia). **Lee, C. E.** 1979a (Korea, redescribed). **Lee, C. E. & Kwon** 1977b (Korea). **Lodos** 1982a (Turkey). **Lodos & Kalkandelen** 1973c (Turkey). **Thapa** 1985a (as *Astroasca*; Nepal: Kathmandu Valley). **Vilbaste** 1968a (coastal regions of the USSR); 1980b (Tuva).

Austrocerus
New species: **Webb, M. D.** 1983a.
Records: **Evans, J. W.** 1966a (reviewed). **Webb, M. D.** 1983a (=*Gnatia*, n. syn.; n. comb. from *Gnatia, Idiocerus* and *Tumocerus*; key to species).

Austrolopa
Records: **Davis, R. B.** 1975a (morphology; phyletic inferences). **Evans, J. W.** 1966a (synopsis; n. sp. syn.).

Austronirvana
Record: **Evans, J. W.** 1966a (=*Tortor*, n. syn.).

Austrotartessus
New species: **Evans, F.** 1981a.
Record: **Evans, F.** 1981a (n. comb. from *Tartessus*).

Awasha
New species: **Heller & Linnavuori** 1968a.

Ayubiana
New species: **Ahmed, Manzoor** 1969a.
Record: **Ahmed, Manzoor** 1985c (Pakistan).

Aztrania
New species: **Blocker** 1979a.

Baaora
New species: **Dworakowska** 1981g.
Records: **Dworakowska** 1981g (n. comb. from *Typhlocyba*). **Sharma, B.** 1984a (India: J. & K., Kishtwar). **Sohi & Dworakowska** 1984a (list of Indian species).

Babacephala
New species: **Ishihara** 1958b.

Backhoffella
Records: **Young** 1977a (redescribed). **Young & Nast** 1963a (lectotype designation).

Badylessa
New species: **Dworakowska** 1981f.

Baguoidea
New species: **Mahmood** 1967a.

Records: **Dworakowska** 1973b (n. comb. from *Empoasca*); 1984a (Malaysia).

Bahita
New species: **Cheng, Y. J.** 1980a. **Linnavuori** 1959b. **Linnavuori & DeLong** 1978b. **Linnavuori & Heller** 1961a.
Records: **Cheng, Y. J.** 1980a (Paraguay). **Linnavuori** 1956a (Brazil, Ecuador); 1959b (n. comb. from *Deltocephalus, Eutettix* and *Phlepsius*; n. sp. syn.; key to subgenera and species). **Linnavuori & DeLong** 1978b (in key to genera of the *Bahita* group).

Baileyus
Records: **Datta** 1972h (descriptive notes). **Hamilton, K. G. A.** 1975d (APHRODINAE, EUPELICINI).

Bakera
New species: **Mahmood** 1967a. **Hongsaprug & Wilson** 1985a.
Records: **Dworakowska** 1970a (n. comb. from *Typhlocyba*; lectotype selection; ERYTHRONEURINI); 1976b (Philippines). **Hongsaprug & Wilson** (additional generic and specific characters).

Bakeriana
New species: **Evans, J. W.** 1966a.
Records: **Davis, R. B.** 1975a (morphology; phyletic inferences). **Evans, J. W.** 1966a (synopsis).

Bakshia
New species: **Dworakowska** 1977f, 1981j.
Record: **Sohi & Dworakowska** 1984a (list of Indian species).

Balacha
Record: **Young** 1977a (key to species).

Balala
Record: **Datta** 1980a (descriptive notes).

Balanda
New species: **Dworakowska** 1979b.

Balbillus
New species: **Linnavuori** 1979b.

Balcanocerus
New species: **Dlabola** 1974a.
Records: **Dlabola** 1974b (n. comb. from *Idiocerus*); 1977c (Mediterranean); 1981a (Iran). **Emeljanov** 1977a (Mongolia). **Hamilton, K. G. A.** 1980a (n. comb. from *Idiocerus*; list of Nearctic species). **Lodos** 1982a (as *Balkanocerus*; Turkey). **Lodos & Kalkandelen** 1981a (Turkey). **Nast** 1976b (Poland; catalogued); 1982 (catalogued).

Balclutha
New species: **Blocker** 1967a, 1968a, 1981a. **Blocker & Nixon** 1978a. **Dlabola** 1961b, 1967b. **Dubovskij** 1970b. **Dubovskij** 1970b. **Ghauri** 1971c. **Kuoh, C. L.** 1981b. **Lindberg** 1958a. **Linnavuori** 1959b, 1960a. **Mitjaev** 1971a. **Namba** 1956a. **Ossiannilsson** 1961a. **Sharma, B. & Badan** 1985a. **Theron** 1973a. **Vilbaste** 1968a. **Webb, M.D.** 1980a.
Records: **Beirne** 1956a (Canada; key to genera of BALCLUTHINI; key to species). **Blocker** 1967a (systematics of western hemisphere taxa; n. comb. from *Agellus, Eugnathodus* and *Nesosteles*; n. sp. syn.); 1971a (Bahama Islands, Newfoundland, Panama); 1981a (Panama). **Bonfils & Della Giustina** 1978a (Corsica). **Cantoreanu** 1963a, 1965b, 1965d, 1971c (Romania). **Cheng, Y. J.** 1980a (Paraguay; list of species). **Dlabola** 1957a, 1957b, (Afghanistan); 1958b (Caucasus); 1959a (southern Germany, Bohemia); 1959b (southern Europe); 1961b (central Asia); 1961c (Italy); 1963c (Haupt material); 1964b (Sudan); 1965c (Mongolia); 1965d (Jordan); 1966a, 1967c, 1967d, 1968a, 1968b, 1970b (Mongolia); 1971a (Afghanistan, Pakistan, Turkey); 1972a (Afghanistan); 1977c (Mediterranean); 1981a (Iran). **Dubovskij** 1966a, (Uzbekistan; key to species); 1978a (Uzbekistan, Zarafshan Valley); 1980a (USSR: western Turkmenia). **Emeljanov** 1964c (European USSR; key to species); 1977a (Mongolia). **Evans, J. W.** 1966a (n. gen. syn.; n. comb. from *Eusceloscopus*; n. sp. syn.); 1977a (sp. syn.); 1974a (New Caledonia). **Ghauri** 1963f (n. comb. from *Eugnathodus*). **Hamilton, K. G. A.** 1972b, 1972d, (Canada: Manitoba); 1975b (APHRODINAE, DELTOCEPHALINI, MACROSTELINA); 1976a (identity of *B. rosea*); 1983b (transboreal species). **Heller & Linnavuori** 1968a (east Africa & Sudan). **Ishihara** 1958a (Japan); 1961a (Thailand); 1965c (Formosa). **Jankovic** 1966a (Serbia). **Jasinska** 1980a

(Poland: Bledowska Wilderness). **Kalkandelen** 1974a (Turkey). **Kholmuminov & Dlabola** 1979a (USSR: Golodnostep Plain). **Knight, W. J.** 1982a (Rennell Island). **Kwon & Lee** 1979b (Korea). **Lee, C. E.** 1979a (Korea, redescribed). **LeQuesne** 1969c (Britain). **Lindberg** 1958a (Cape Verde Islands); 1960c (Azores); 1961a (Madeira); 1962a (Portugal). **Linnavuori** 1956a (Chile, Colombia, Trinidad); 1959b (n. comb. from *Eugnathodus* and *Nesosteles*; n. sp. syn.; key to Neotropical species); 1960a (Micronesia; key to species); 1961a (South Africa; n. comb. from *Cicadula*); 1962a (Israel); 1964a (Egypt); 1965a (Spanish Morocco); 1969a (Congo); 1975c (Micronesia). **Linnavuori & DeLong** 1977c (Chile). **Linsley & Usinger** 1966a (Galapagos Islands). **Lodos** 1982a (Turkey). **Lodos & Kalkandelen** 1985b (Turkey). **Logvinenko** 1957b, 1984a (USSR: Ukraine). **Mitjaev** 1968a (eastern Kazakhstan); 1971a (Kazakhstan). **Namba** 1956a (Hawaii; key to species). **Nast** 1972a (catalogued); 1976a (Poland: Pieniny Mts.); 1976b (Poland; catalogued). **Ossiannilsson** 1961b (Sweden); 1974a (Norway); 1983a (Fennoscandia & Denmark). **Quartau** 1979a, 1980b (Azores). **Schulz** 1976a (Norway). **Servadei** 1957a, 1958a, 1960a, 1968a, 1969a, 1972a (Italy). **Sharma, B.** 1977b (India: Jammu). **Sharma, B. & Badan** 1985a (India: Jammu). **Theron** 1970b (specific synonymies discussed); 1973a (identity of Naude species; n. sp. syn.). **Triplehorn & Nault** 1985a (tribal placement discussed). **Vilbaste** 1959b (Estonia); 1962b (eastern part of the Caspian lowlands); 1964a (Auwiesen Estlands); 1965b (Altai Mts.); 1968a (coastal regions of the USSR); 1973a (notes on Flor's species); 1976a (Matsumura's species described as *Gnathodus*; sp. syn. noted); 1979a (Vooremaa hardwood/spruce forest); 1980a (Kamchatka); 1980b (Tuva). **Webb, M. D.** 1980a (Aldabra).

Balcluthina
New species: **Datta** 1972a (descriptive notes).

Baldriga
New species: **Blocker** 1969a, 1982a.
Record: **Blocker** 1979a (n. comb. from *Gargaropsis*).

Baldulus
New species: **Kramer & Whitcomb** 1968a.
Records: **Hamilton, K. G. A.** 1975b (APHRODINAE, DELTOCEPHALINI, MACROSTELINA). **Triplehorn & Nault** 1985a (generic affinities; phyletic inferences).

Baleja
New species: **Young** 1977a.
Record: **Young** 1977a (n. comb. from *Aulacizes* and *Tettigonia*; n. sp. syn.; key to species).

Balera
New species: **Ruppel** 1959a. **Young** 1957c.
Records: **Young** 1957b (n. comb. from *Empoasca*); 1957c (key to species).

Ballana
New species: **DeLong** 1964a.
Records: **Beirne** 1956a (Canada). **Hamilton, K. G. A.** 1975b (APHRODINAE, DELTOCEPHALINI, ATHYSANINA).

Balocerus
New species: **Freytag & Morrison** 1972a, 1973a. **Webb, M. D.** 1983a.
Record: **Webb, M. D.** 1983a (Australia; n. comb. from *Idiocerus*; key to species).

Balocha
New species: **Kameswara Rao & Ramakrishnan** 1979a. **Kameswara Rao, Ramakrishnan & Ghai** 1979a. **Maldonado-Capriles** 1961a, 1968a, 1970a. **Webb, M. D.** 1983a.
Records: **Kameswara Rao & Ramakrishnan** 1979a (key to species); 1983a (redescription of *B. bicolor*). **Kameswara Rao, Ramakrishnan & Ghai** 1979a (key to known species). **Maldonado-Capriles** 1961a (genus redescribed; n. comb. from *Idiocerinus*; n. comb. to *Pedioscopus*; key to species); 1964a (n. comb. from *Idiocerus*); 1968b (keys to two species and subspecies); 1970a (list of verified species and their geographic distribution). **Sharma, B.** 1977b (India: Jammu). **Viraktamath, C. A.** 1976b (key to Indian species). **Webb, M. D.** (n. comb. from *Idiocerus*; key to Australian species).

Balocharella
New species: **Webb, M. D.** 1983a.
Record: **Webb, M. D.** 1983a (key to species).

Baluba
New species: **Nielson** 1979b.
Record: **Nielson** 1979b (key to species).

Bambusana
New species: **Anufriev** 1969a.
Records: **Anufriev** 1969a (n. comb. from *Thamnotettix*); 1977b (Kurile Islands). **Hamilton, K. G. A.** 1975a (APHRODINAE, DELTOCEPHALINI, CICADULINA).

Bampurius
New species: **Dlabola** 1977a, 1984a.
Records: **Dlabola** 1977b (n. comb. from *Neolimnus*); 1981a (Iran). **Nast** 1982a (catalogued; DELTOCEPHALINAE).

Bandara
New species: **DeLong** 1980h.
Records: **Beirne** 1956a (Canada; key to species). **Hamilton, K. G. A.** 1975b (APHRODINAE, DELTOCEPHALINI, PLATYMETOPIINA). **Linnavuori** 1956a (Trinidad); 1959b (Colombia, Haiti, Panama, Trinidad).

Bandaromimus
New species: **Linnavuori** 1959b. **Linnavuori & Heller** 1961a.

Bannalgaechungia
New species: **Kwon** 1983a.

Banosa
New species: **Mahmood** 1967a.
Record: **Dworakowska** 1970a (=*Akotettix*, n. syn.).

Banus
Record: **Datta & Dhar** 1984a (descriptive notes).

Baramapulana
New species: **Linnavuori** 1979a.
Records: **Linnavuori** 1979a (=*Tetelloides*, n. syn.; n. comb. from *Tetelloides*; key to species). **Young** 1965c (lectotype designation).

Barbinolla
Record: **Young** 1977a (n. comb. from *Microgoniella* and *Tettigonia*; key to species).

Bardana
New species: **DeLong** 1980h.

Bardera
Record: **Linnavuori & Al-Ne'amy** (n. comb. from *Phlepsius*).

Barela
New species: **Young** 1957c.
Record: **Young** 1957c (n. comb. from *Protalebra*; n. sp. syn.; key to species).

Barodecus
New species: **Nielson** 1979b.

Baroma
Record: **Linnavuori** 1959b (redescribed).

Bascarrhinus
New species: **Kramer** 1966a.
Record: **Kramer** 1966a (in key to genera of New World LEDRINAE; key to species).

Basilana
New species: **Mahmood** 1967a.
Records: **Dworakowska** 1972b (=*Aidola*, n. syn.). **Mahmood** 1967a (n. comb. from *Typhlocyba*).

Bassareus
New species: **Linnavuori** 1979b.
Record: **Linnavuori** 1979b (n. comb. from *Pseudobalbillus*; key to species).

Basuaneura
New species: **Ramakrishnan & Menon** 1971a.
Record: **Dworakowska** 1977a (=*Erythria*, n. syn.).

Basutoia
New species: **Linnavuori** 1961a.

Batarius
Record: **Hamilton, K. G. A.** 1975b (APHRODINAE, DELTOCEPHALINI, DELTOCEPHALINA).

Bathysmatophorus
New species: **Anufriev** 1971a, 1977b. **Ishihara** 1957a. **Kwon** 1983a. **Mitjaev** 1967a.
Records: **Anufriev** 1971a, 1978a (USSR: Far East and Maritime Territory; key to species). **Anufriev & Emeljanov** 1968a (n. sp. syn.). **Emeljanov** 1964c (European USSR); 1977a

(Mongolia). **Emmrich** 1974a (n. sp. syn.). **Gyllensvard** 1963a (Sweden). **Ishihara** 1958a (listed; EVACANTHIDAE); 1968a (Japan; ERRHOMENELLIDAE). **Kwon** 1983a (Korea; CICADELLINAE, ERRHOMENINI). **Lee, C. E.** 1979a (Korea, redescribed). **Lee, C. E. & Kwon** 1977b (Korea). **Mitjaev** 1968b (eastern Kazakhstan); 1971a (Kazakhstan). **Oman & Musgrave** 1975a (n. comb. from *Errhomus*; subgenus for Nearctic species). **Ossiannilsson** 1981a (CICADELLINAE, ERRHOMENINI; in key to genera). **Vilbaste** 1968a (USSR: coastal regions); 1969b (Taimyr); 1980b (Tuva).

Batracomorphus
New species: **Anufriev** 1971b, 1981a (new name). **Blote** 1964a. **Dlabola** 1964a, 1979a. **Evans, J. W.** 1966a, 1972b. **Dubovskij** 1970b. **Ghauri** 1964d. **Heller & Linnavuori** 1968a. **Kameswara Rao & Ramakrishnan** 1980a. **Knight, W. J.** 1983b. **Lindberg** 1958a. **Linnavuori** 1957b, 1960a, 1969a, 1971a. **Linnavuori & Heller** 1961a. **Linnavuori & Quartau** 1975a. **Quartau** 1968a, 1981a. **Webb, M. D.** 1980a.

Records: **Anufriev** 1971b (USSR: Primorsky District; key to species); 1977b (Kurile Islands); 1978a (USSR: Maritime Territory); 1979a (northeast China; notes on Jacobi's material). **Chang, Y. D. & Choe** 1982a (Korea). **Choe** 1985a (Korea). **Dlabola** 1957a (Turkey); 1958a (Cyprus); 1958b (Caucasus); 1960c (Iran); 1961b (central Asia); 1964a (Afghanistan); 1964b (Sudan); 1964c (Albania); 1965d (Jordan); 1967b, 1967c, 1970b (Mongolia); 1972a (Afghanistan); 1981 (Iran). **Dubovskij** 1966a (Uzbekistan); 1978a (Uzbekistan, Zarafshan Valley). **Emeljanov** 1964c (European USSR); 1977a (Mongolia). **Evans, J. W.** 1972b (review; JASSINAE, JASSINI); 1974a (n. comb. from *Bythoscopus* and *Selenocephalus*; New Caledonia); 1966a (n. comb. from *Bythoscopus* and *Eurinoscopus*); 1982a (biogeography). **Eyles & Linnavuori** 1974a (Niue Island). **Heller & Linnavuori** 1968a (IASSINAE). **Ishihara** 1961a (Thailand; n. comb. from *Pachyopsis*; JASSIDAE); 1966a (Kuriles and Japan; JASSIDAE). **Kameswara Rao & Ramakrishnan** 1980a (key to Indian species). **Kholmuminov & Dlabola** 1979a (USSR: Golodnostep Plain). **Knight** 1974b (key to New Zealand species; JASSINAE); 1976b (Norflok Id.; Kermadec Id.; JASSINAE); 1983b (n. gen. syn.; n. sp. syn.; n. comb. from *Acojassus, Bythoscopus, Edijassus, Eurinoscopus, Idiocerus, Iassus, Macropisis, Pachyopsis* and *Stragania*; key to Australian and Eastern Oriental species). **Lauterer** 1978a (Moravia, Slovakia). **Lee, C. E.** 1979a (Korea, redescribed). **Lee, C. E. & Kwon** 1977b (Korea; IASSINAE). **LeQuesne** 1965c (Britain). **Lindberg** 1958a (Cape Verde Islands; JASSIDAE, JASSINAE); 1960b (Soviet Armenia). **Linnavuori** 1957b (n. comb. from *Stragania*; key to species; IASSINAE); 1958a (Israel species; n. sp. syn.); 1960a (Micronesia; n. comb. from *Bythoscopus*; key to species; IASSINAE); 1960b (Fiji; n. comb. from *Eurinoscopus*; IASSINAE); 1961a (South Africa; n. comb. from *Bythoscopus*; IASSINAE); 1962a (Israel); 1964a (Egypt); 1965a (Libya); 1969a (Congo); 1975c (Micronesia). **Linnavuori & Quartau** 1975a (definition of subgenera; key to species). **Lodos & Kalkandelen** 1981a (Turkey). **Logvinenko** 1957b (USSR: Ukraine). **Mitjaev** 1968b (eastern Kazakhstan); 1971a (Kazakhstan; key to species). **Nast** 1972a (catalogued); 1976a (Poland; catalogued). **Ossiannilsson** 1981a (Fennoscandia & Denmark; IASSINAE). **Quartau** 1968a (Cape Verde Island; key to species); 1981a (redescriptions of African species; key to Ethiopian species). **Servadei** 1960a (Italy). **Theron** 1972a (n. comb. from *Eurinoscopus*; redescription of Naude's species); 1980a (Stal's *B. subolivaceus*). **Vilbaste** 1962b (eastern part of the Caspian lowlands); 1965b (Altai Mts.); 1968a (coastal regions of the USSR); 1980b (Tuva). **Viraktamath, C. A.** 1979a (n. comb. from *Iassus*). **Zachvatkin** 1953b (central Russia; Oka district).

Baya
New species: **Dworakowska** 1972b.

Beamerana
New species: **Ruppel** 1975a.
Record: **Ruppel** 1975a (phylogeny).

Beamerulus
New species: **Young** 1957c.
Record: **Young** 1957c (key to species).

Begonalia
Record: **Young** 1977a (n. comb. from *Cicadella* and *Tettigoniella*; n. sp. syn.; key to species).

Beirneola
New species: **Young** 1977a.
Record: **Young** 1977a (n. comb. from *Entogonia* and *Tettigonia*; n. sp. syn.; key to species).

Belaunus
Record: **Hamilton, K. G. A.** 1975b (APHRODINAE, DELTOCEPHALINI, DELTOCEPHALINA).

Bella
Records: **Datta** 1972h (descriptive notes). **Vilbaste** 1965a (=*Doratulina*, n. syn.).

Benala
New species: **Linnavuori & DeLong** 1979b.
Records: **Kramer** 1963b (in key to genera of NEOBALINAE). **Linnavuori** 1959b (redescribed; in key to genera of NEOBALIINAE [sic]).

Benglebra
New species: **Mahmood & Ahmed** 1969a.
Records: **Dworakowska** 1971c (placement in TYPHLOCYBINAE questioned; EUSCELINAE suggested). **Mahmood & Ahmed** 1969a (in key to TYPHLOCYBINAE, ALEBRINI of East Pakistan).

Bengueta
New species: **Mahmood** 1967a.

Beniledra
New species: **Linnavuori** 1972b.

Bergallia
New species: **Linnavuori** 1956a, 1973b, 1973d, 1975d. **Linnavuori & DeLong** 1977c. **Marino de Remes Lenicov** 1982a.
Records: **Davis, R. B.** 1975a (morphology; phyletic inferences). **Kramer** 1964a (in key to New World genera of AGALLIINAE; n. comb. from *Agalita*). **Linnavuori** 1975d (Peru). **Linnavuori & DeLong** 1977c (n. comb. from *Agallia*; n. sp. syn.). **Marino de Remes Lenicov** 1982a (key to species).

Bergevina
Records: **Davis, R. B.** 1975a (morphology; phyletic inferences). **Dlabola** 1961b (central Asia).

Bergolix
New species: **Linnavuori** 1959b.
Records: **Linnavuori** 1959b (n. comb. from *Athysanus*; key to species).

Bertawolia
New species: **Blocker** 1979a.

Betarmonia
Records: **Young** 1977a (identity unknown).

Bharata
Records: **Young** 1965c (lectotype designation). **Young & Lauterer** 1966a (lectotype designation).

Bharinka
New species: **Webb, M. D.** 1983a.

Bharoopra
New species: **Webb, M. D.** 1983a.

Bhatia
Records: **Ishihara** 1961a (Thailand; DELTOCEPHALIDAE). **Hamilton, K. G. A.** 1975b (APHRODINAE; unassigned). **Linnavuori** 1960a (Micronesia: Bonin Islands; DELTOCEPHALINAE, EUSCELINI). **Linnavuori & Al-Ne'amy** 1983a (SELENOCEPHALINAE, BHATIINI; in key to genera).

Bhooria
Record: **Young** 1965c (lectotype designation).

Biadorus
New species: **Nielson** 1979b.
Record: **Nielson** 1979b (n. comb. from *Coelidia*; key to species).

Biluscelis
New species: **Dlabola** 1980d.
Record: **Nast** 1982a (catalogued; DELTOCEPHALINAE).

Bilusius
New species: **Logvinenko** 1974a.

Record: **Hamilton, K. G. A.** 1975b (APHRODINAE, DELTOCEPHALINI, ATHYSANINA). **Logvinenko** 1957b (USSR: Ukraine; key to species).
Bituitus
Record: **Vilbaste** 1965a (=*Doratulina*, n. syn.).
Biza
New species: **Kramer** 1962a, 1967b.
Records: **Kramer** 1962a (key to species); 1964d (in key to genera of NEOCEOLIDIINAE).
Blarea
New species: **Young** 1957c.
Record: **Young** 1957c (in key to genera of ALEBRINI).
Bloemia
Record: **Theron** 1974b (n. comb. from *Euscelis*).
Bobacella
New species: **Emeljanov** 1962a.
Records: **Dlabola** 1960a (n. comb. from *Driotura*; n. sp. syn.); 1965c, 1967b, 1970b (Mongolia). **Emeljanov** 1962a (EUSCELINAE); 1964c (European USSR); 1977a (Mongolia). **Hamilton, K. G. A.** 1975b (APHRODINAE, DELTOCEPHALINI, ATHYSANINA). **Vilbaste** 1965b (Altai Mts.); 1980b (Tuva).
Bolanusoides
Records: **Dworakowska** 1982a (India). **Sohi & Dworakowska** 1984a (list of Indian species). **Thapa & Sohi** 1984a (*B. heros* redescribed).
Bolarga
New species: **Linnavuori** 1959b.
Records: **DeLong** 1982c (n. comb. to *Daltonia*). **Linnavuori** 1959b (redescribed; key to species).
Bolidiana
New species: **Nielson** 1979b.
Record: **Nielson** 1979b (n. comb. from *Coelidia*: n. sp. syn.; key to species).
Bolivaia
New species: **DeLong** 1982f. **Linnavuori & DeLong** 1979a, 1979b.
Boliviella
New species: **DeLong** 1969a.
Bolotheta
New species: **Kramer** 1963c.
Records: **Cwikla & Blocker** 1981a (comparative notes). **Kramer** 1963c (n. comb. from *Neocoelidia*; key to species).
Bonamus
New species: **Linnavuori & Heller** 1961a.
Records: **Linnavuori** 1957a (redescribed; HECALINAE); 1959b (DELTOCEPHALINAE, HECALINI).
Bonaspeia
New species: **Linnavuori** 1961a. **Theron** 1974b.
Records: **Linnavuori** 1961a (n. comb. from *Euscelis*). **Theron** 1974b (n. comb. from *Euscelis*).
Bonneyana
Record: **Hamilton, K. G. A.** 1975b (APHRODINAE, DELTOCEPHALINI, PLATYMETOPIINA).
Bordesia
New species: **Lindberg** 1956b.
Records: **Lindberg** 1956b (Spanish Sahara). **Linnavuori** 1964a (Egypt); 1975a (in key to genera; DELTOCEPHALINAE, HECALINI).
Borditartessus
New species: **Evans, F.** 1981a.
Record: **Evans, F.** 1981a (n. comb. from *Tartessus*).
Boreotettix
New species: **Emeljanov** 1966a.
Records: **Anufriev** 1978a (USSR: Maritime Territory; in key to genera of PARALIMNINI). **Dlabola** 1967d (Mongolia). **Emeljanov** 1966a (n. comb. from *Sorhoanus*). **Hamilton, K. G. A.** 1975b (APHRODINAE, DELTOCEPHALINI, DELTOCEPHALINA). **Mitjaev** 1971a (Kazakhstan). **Ossiannilsson** 1983a (Fennoscandia & Denmark). **Vilbaste** 1980a (Kamchatka); 1980b (Tuva; n. sp. syn.).
Borogonalia
New species: **Young** 1977a.
Record: **Young** 1977a (n. comb. from *Tettigonia* and *Tettigoniella*; n. sp. syn.).
Borulla
New species: **Dworakowska, Sohi & Viraktamath** 1980a.

Record: **Sohi & Dworakowska** 1984a (list of Indian species).
Bothrogonia
New species: **Capco** 1959a. **Ishihara** 1962a. **Li, Z. Z.** 1983a. **Yang, C. K. & Li** 1980a.
Records: **Capco** 1959a (key to species). **Datta & Pramanik** 1977a (India: A. P., Subansiri Division). **Dlabola** 1981a (Iran). **Dworakowska** 1973d (Korea). **Ishihara** 1958a (Japan: north Honshu); 1962a (synopsis); 1965c (Formosa); 1983a (notes about type locality). **Kwon** 1983a (Korea). **Lee, C. E.** 1979a (Korea, redescribed). **Lee, C. E. & Kwon** 1977b (Korea). **Yang, C. K. & Li** 1980a (new subgenera defined).
Boulardus
New species: **Nielson** 1983e.
Record: **Nielson** 1983e (key to species).
Bousaada
New species: **Linnavuori** 1971a.
Records: **Dlabola** 1980d (Spain). **Hamilton, K. G. A.** 1975b (APHRODINAE, DELTOCEPHALINI, PLATYMETOPIINA.) **Vilbaste** 1976a (=*Epicephalius*, n. syn.; n. sp. syn.).
Brachydella
Record: **Hamilton, K. G. A.** 1975b (APHRODINAE, APHRODINI, DORATURINA).
Brachylope
New species: **Logvinenko** 1969a.
Record: **Hamilton, K. G. A.** 1975b (APHRODINAE, unassigned).
Brachylorus
New species: **Maldonado-Capriles** 1972b.
Brachypterona
New species: **Quartau** 1981d.
Records: **Hamilton, K. G. A.** 1975b (APHRODINAE, DELTOCEPHALINI, ATHYSANINA). **Lindberg & Wagner** 1965a (Canary Islands). **Quartau** 1981d (generic affinities).
Brasa
Records: **Davis, R. B.** (morphology; phyletic inferences). **Kramer** 1964a (in key to New World genera of AGALLIINAE).
Brasilanus
New species: **Linnavuori** 1959b. **Linnavuori & Heller** 1961a.
Records: **Cwikla & Blocker** 1981a (comparative notes). **Linnavuori & Heller** 1961a (EUSCELINAE, EUSCELINI; new subgenera proposed).
Brasura
New species: **Nielson** 1982e.
Record: **Nielson** 1982e (key to species).
Brazosa
New species: **Linnavuori & DeLong** 1976a. **Linnavuori & Heller** 1961a.
Records: **Linnavuori** 1956a (Peru); 1959b (redescribed). **Linnavuori & DeLong** 1976a (key to species.
Brenda
Record: **Davis, R. B.** 1975a (morphology; phyletic inferences).
Brevolidia
New species: **Nielson** 1982e.
Record: **Nielson** 1982e (n. comb. from *Aletta*; key to species).
Brincadorus
Record: **Linnavuori** 1959b (redescribed; DELTOCEPHALINAE, EUSCELINI).
Britimnathista
New species: **Dworakowska** 1969f.
Record: **Sohi & Dworakowska** 1984a (list of Indian species).
Brunerella
Record: **Young** 1957c (n. comb. from *Protalebra*; in key to genera of ALEBRINI; key to species).
Brunotartessus
New species: **Evans, F.** 1981a.
Record: **Evans, F.** 1981a (n. comb. from *Tartessus*).
Bubacua
Record: **Young** 1977a (n. comb. from *Hortensia*).
Bubulcus
Record: **Hamilton, K. G. A.** 1975b (APHRODINAE, DELTOCEPHALINI, DELTOCEPHALINA).
Bucephalogonia
Record: **Young** 1977a (n. sp. syn.).

Bufonaria
New species: Emeljanov 1963a.

Bulotartessus
New species: Evans, F. 1981a.

Bumizana
New species: Morrison 1973a. Viraktamath, C. A. 1976a.
Records: Morrison 1973a (in key to genera of Oriental HECALINAE; key to species). Theron 1982b (=*Paradorydium*, n. syn.). Viraktamath, C. A. 1976a (key to Indian species).

Bundabrilla
New species: Webb, M. D. 1983a.

Bunyipia
Records: Dworakowska 1981c (to subgenus of *Diomma*). Vilbaste 1975a (=*Diomma s. str.*).

Busonia
New species: Maldonado-Capriles 1977d.
Record: Maldonado-Capriles (n. comb. to *Busoniomimus*; key to species).

Busoniomimus
New species: Viraktamath, C. A. & Murphy 1980a. Viraktamath, S. & Viraktamath, C. A. 1985a. Webb, M. D. 1983a.
Records: Maldonado-Capriles 1977d (n. comb. from *Busonia*). Viraktamath, C. A. & Murphy 1980a (n. comb. from *Amritodus*). Webb, M. D. 1983a (n. comb. from *Idioscopus*; key to species).

Byphlocyta
New species: Ahmed, Manzoor 1971c. Samad & Ahmed 1979a.
Records: Ahmed, Manzoor 1985c (Pakistan). Dworakowska 1982a (n. comb. to *Linnavuoriana*).

Bythonia
New species: Linnavuori 1959b.
Record: Linnavuori 1959b (redescribed; BYTHONIINAE; key to species).

Bythoscopus
Records: Datta 1972b (descriptive notes). Hamilton, K. G. A. 1980b (lectotype designation). Knight 1983b (n. comb. to *Batracomorphus*). Zachvatkin 1953b (central Russia; Oka district).

Cabrellus
New species: Emeljanov 1979a.
Records: Emeljanov 1979a (raised to generic status). Hamilton, K. G. A. (APHRODINAE, DELTOCEPHALINI, DELTOCEPHALINA).

Cabrulus
Records: Beirne 1956a (Canada: Alberta and Saskatchewan). Hamilton, K. G. A. 1975b (APHRODINAE, DELTOCEPHALINI, DELTOCEPHALINA). Ross & Hamilton 1972a (as subgenus of *Orocastus*).

Caelidioides
Record: Nielson 1975a (removed from COELIDIINAE; unassigned).

Caffretus
Record: Lindberg 1958a (Cape Verde Islands; JASSIDAE, EUSCELINAE).

Caffrolix
New species: Linnavuori 1961a.
Records: Linnavuori 1961a (n. comb. from *Athysanus* and *Thamnotettix*; EUSCELINAE, EUSCELINI). Theron 1974b (n. comb. from *Euscelis*); 1980a (n. sp. syn.; EUSCELINAE); 1983a (n. comb. from *Euscelis*).

Cahya
New species: Linnavuori & DeLong 1979b. Menezes 1973b (new name).
Record: Menezes 1973b (n. comb. from *Thamnotettix*).

Caladonus
Records: Hamilton, K. G. A. 1975b (APHRODINAE, DELTOCEPHALINI, PLATYMETOPIINA). Nielson & Kaloostian 1956a (USA: Utah).

Calamotettix
New species: Dlabola 1961b. Emeljanov 1959a, 1962a. Vilbaste 1980b.
Records: Anufriev 1978a (USSR: Maritime Territory; in key to genera of PARALIMNINI). Dlabola 1968a (Mongolia). Dworakowska 1973a (Mongolia). Emeljanov 1977a (Mongolia). Hamilton, K. G. A. 1975b (APHRODINAE, DELTOCEPHALINI, DELTOCEPHALINA). Lauterer 1980a (Slovakia). Mitjaev 1971a (Kazakhstan).

Calanana
Records: Beirne 1956a (Canada: southern British Columbia). Hamilton, K. G. A. 1975b (APHRODINAE, DELTOCEPHALINI, CICADULINA).

Caldwelliola
New species: Young 1977a.
Record: Young 1977a (n. comb. from *Tettigonia*; key to species).

Calidanus
Record: Hamilton, K. G. A. 1975b (APHRODINAE, DELTOCEPHALINI, DELTOCEPHALINA).

Calliscarta
New species: Kramer 1963b. Linnavuori & Heller 1961a.
Records: Kramer 1963b (in key to genera of NEOBALINAE). Linnavuori 1959b (=*Idiotettix*, n. syn.; n. comb. from *Idiotettix* and *Thamnotettix*; key to species). Linnavuori & DeLong 1978c (descriptive note). Linnavuori & Heller 1961a (Peru; descriptive note).

Callistrophia
New species: Mitjaev 1971a.
Records: Anufriev 1978a (USSR: Maritime Territory; in key to ATHYSANINI). Anufriev & Emeljanov 1968a (n. sp. syn.). Dlabola 1965c, 1967b, 1967c, 1970b (Mongolia). Emeljanov 1964c (European USSR). Hamilton, K. G. A. 1975b (APHRODINAE, DELTOCEPHALINI, CICADULINA). Lee, C. E. 1979a (Korea, redescribed). Lee, C. E. & Kwon 1977b (Korea). Mitjaev 1968b (eastern Kazakhstan); 1971a (Kazakhstan; key to two species); 1971b (southern Kazakhstan). Vilbaste 1965b (Altai Mts.); 1968a (coastal regions of the USSR); 1980a (Kamchatka); 1980b (Tuva).

Calodia
New species: Nielson 1982e.
Record: Nielson 1982e (n. comb. from *Coelidia*; key to species).

Caloduferna
New species: Webb, M. D. 1980a.
Record: Webb, M. D. 1980a (DELTOCEPHALINAE, JASSARGINI).

Calonia
Record: Hamilton, K. G. A. 1975b (APHRODINAE, DELTOCEPHALINI, COCHLORHININI).

Calotartessus
Record: Evans, F. 1981a (n. comb. from *Tartessus*).

Calotettix
New species: Eyles & Linnavuori 1974a.
Records: Eyles & Linnavuori 1974a (Cook Island; recescribed; PARABOLOPONINAE). Webb, M. D. 1981b (=*Dryadomorpha*).

Camaija
New species: Young 1977a.
Record: Young 1977a (n. comb. from *Tettigonia*; key to species).

Campbellinella
Records: Evans, J. W. 1966a (n. comb. from *Phrynomorphus*). Vilbaste 1965a (=*Doratulina*, n. syn.).

Campecha
Record: Young 1977a (redescribed).

Camptelasmus
New species: Linnavuori 1961a.
Record: Linnavuori 1972b (in key to genera of LEDRINAE; taxonomic position obscure). Theron 1976a (n. comb. to *Titiella*).

Camulus
Record: Sharma, B. 1979a (India: Nagaland, Kohima).

Canariotettix
Record: Hamilton, K. G. A. 1975b (APHRODINAE, DELTOCEPHALINI, DELTOCEPHALINA).

Candulifera
New species: Webb, M. D. 1983a.

Capeolix
New species: Linnavuori 1961a.

Caphodellus
New species: Linnavuori & DeLong 1976a.

Caphodus
New species: Linnavuori & DeLong 1977b.
Records: Linnavuori 1956a (Paraguay); 1959b (redescribed).

Capinota
Record: **Young** 1968a (=*Phera*, n. syn.).
Caplopa
Records: **Linnavuori** 1961a (South Africa: Cape Province; ULOPINAE); 1972a (n. comb. from *Ulopa*; correction of specific record). **Theron** 1980a (redescription of *C. sordida*).
Capoideus
Record: **Theron** 1974b (n. comb. from *Scaphoideus*).
Caragonalia
Record: **Young** 1977a (n. comb. from *Tettigonia*; key to species).
Caranavia
New species: **Linnavuori** 1959b. **Linnavuori & Heller** 1961a.
Record: **Linnavuori & DeLong** 1978c (in key to genera of the *Atanus* group).
Carchariacephalus
New species: **Evans, J. W.** 1974a.
Records: **Emeljanov** 1962c (n. comb. to *Selachina*). **Evans, J. W.** 1974a (NIRVANINAE, NIRVANINI; comparative notes). **Kramer** 1964b (n. comb. to *Jassosqualus*). **Vilbaste** 1976a (notes on Matsumura's material).
Cardioscarta
Records: **Schroder** 1959a (diagnosis). **Young** 1977a (n. comb. from *Cicada* and *Tettigonia*; n. sp. syn.; key to species). **Young & Lauterer** 1966a (lectotype designations). **Young & Soos** 1964a (lectotype designation).
Carelmapu
New species: **Linnavuori** 1959b.
Records: **Linnavuori & DeLong** 1977b (in key to genera of the *Osbornellus* group); 1977c (new subgenus proposed).
Cariancha
Record: **Linnavuori** 1959b (redescribed).
Caribovia
New species: **Young** 1977a.
Records: **Ghauri** 1980c (*C. intensa* redescribed). **Young** 1977a (n. comb. from *Cicadella, Entogonia, Hortensia* and *Tettigonia*; n. sp. syn.; key to species).
Carinifer
New species: **Dlabola** 1961b.
Records: **Dlabola** 1960a (n. comb. from *Aconura*); 1960c (Iran); 1961b (central Asia); 1963c (n. comb. from *Thamnotettix*; n. sp. syn.). **Hamilton, K. G. A.** 1975b (APHRODINAE, APHRODINI, DORATURINA). **Linnavuori** 1962a (Israel). **Vilbaste** 1965a (=*Aconura*).
Carinolidia
New species: **Nielson** 1979b.
Caripuna
Record: **Young** 1968a (=*Mareba*, n. syn.).
Carneocephala
Records: **Alayo & Novoa** 1985a (Cuba). **Beardsley** 1967a (Hawaii). **Datta** 1980a (descriptive notes). **Hamilton, K. G. A.** 1985a (redefined; n. comb. to *Xyphon*). **Linnavuori** 1973a (Cuba); 1979b (Africa: Ivory Coast, Upper Volta; adventive). **Mead** 1965a (USA: Florida). **Nielson** 1968b (characterization of vector species). **Nielson & Toles** 1970a (interspecific hybridization). **Schroder** 1959a (diagnosis). **Young** 1977a (discussion of species concepts).
Carsonus
Records: **Anufriev** 1978a (CICADELLINAE, ERRHOMENINI). **Hamilton, K. G. A.** 1975d (APHRODINAE, ERRHOMENINI). **Oman & Musgrave** 1975a (CICADELLINAE, ERRHOMENINI).
Caruya
New species: **Linnavuori & DeLong** 1978d.
Record: **Cwikla & Blocker** 1981a (comparative notes).
Carvaka
New species: **Evans, J. W.** 1966a.
Record: **Ross** 1968a (n. comb. to *Hybrasil*; n. sp. syn.).
Cassianeura
New species: **Dworakowska** 1980b, 1984a. **Ramakrishnan & Menon** 1973a.
Records: **Ahmed, Manzoor** 1985c (Pakistan). **Dworakowska** 1977f (India: Dehra Dun); 1980b (south India). **Dworakowska & Viraktamath** 1975a (n. comb. from *Erythroneura*). **Ramakrishnan & Menon** 1974a (in key to genera of ERYTHRONEURINI). **Sohi** 1976a (n. comb. from *Erythroneura*; n. sp. syn.). **Sohi & Dworakowska** 1984a (list of Indian species).

Castoriella
New species: **Dworakowska** 1974a.
Record: **Dworakowska** 1979c (TYPHLOCYBINAE, ZYGINELLINI; list of species).
Catagonalia
Record: **Young** 1977a (n. comb. from *Cicadella*; n. sp. syn.; key to species).
Catorthorrhinus
Records: **Young** 1965c (lectotype designation); 1968a (redescribed).
Cazenus
Records: **Beirne** 1956a (Canada: northern & northwestern regions). **Hamilton, K. G. A.** 1972a (Canada: Manitoba); 1975b (APHRODINAE, DELTOCEPHALINI, DELTOCEPHALINA).
Cechenotettix
New species: **Dlabola** 1974d. **Lindberg** 1964a.
Records: **Bonfils & Della Giustina** 1978a (Corsica). **Lindberg** 1965a (Spanish Sahara). **Linnavuori** 1956d (Spanish Morocco; CICADELLINAE, SYNOPHROPSINI); 1965a (Tunisia). **Quartau & Rodrigues** 1969a (Portugal). **Vilbaste** 1976a (notes on Matsumura's material; lectotype designation).
Cedarotettix
Records: **Theron** 1975a (n. comb. from *Deltocephalus*); 1982a (supplemental descriptive notes).
Celsanus
Record: **Linnavuori** 1956a (Brazil).
Centrometopia
Records: **Lauterer & Schroder** 1970a (lectotype designation). **Schroder** 1959a (n. comb. to *Molomea*). **Young** 1968a (=*Molomea*). **Young & Nast** 1963a (notes on type specimens).
Centrometopoides
Record: **Young** 1968a (=*Molomea*).
Cephalelus
Records: **Davis, R. B.** 1975a (morphology; phyletic inferences). **Evans, J. W.** 1966a (n. comb. from *Paradorydium*; n. sp. syn.); 1977a (in key to genera of CEPHALELINI; removal of genera from synonymy with *Cephalelus*). **Linnavuori** 1961a (South Africa: Cape Province); 1965a (Spanish Morocco); 1972a (in key to genera of ULOPINAE; key to species).
Cephalius
New species: **Villiers** 1956a.
Record: **Hamilton, K. G. A.** 1975b (APHRODINAE, PARADORYDIINI).
Ceratagallia
Records: **Davis, R. B.** 1975a (morphology; phyletic inferences). **Kramer** 1964a (in key to New World genera of AGALLIINAE).
Ceratogonia
Record: **Young & Lauterer** 1966a (lectotype designation).
Cerneura
New species: **Ghauri** 1978a.
Cerrillus
Records: **Linnavuori** 1957a, 1959b (in key to Neotropical genera of HECALINAE); 1975b (DELTOCEPHALINAE, CERRILLINI).
Cerus
New species: **Theron** 1984b.
Records: **Theron** 1975a (n. comb. from *Deltocephalus*); 1984b (key to species).
Cestius
New species: **Sohi** 1972a. **Viraktamath, C. A.** 1978a.
Records: **Hamilton, K. G. A.** 1975b (APHRODINAE, DELTOCEPHALINI, PLATYMETOPIINA). **Knight, W. J.** 1973c (affinities and differences *vis a vis Hishimonus*). **Mahmood** 1975b (genus distinct from *Hishimonus* and *Hishimonoides*). **Sawai Singh** 1971b (n. comb. from *Hishimonus* and *Hishimonoides*.
Chacotettix
New species: **Linnavuori** 1959b.
Record: **Linnavuori** 1959b (revised status, subgenus of *Chlorotettix*).
Changwhania
New species: **Kwon** 1980a.

Records: **Kwon** 1980a (n. comb. from *Aconura*; subtribe DELTOCEPHALINA). **Nast** 1982a (catalogued; DELTOCEPHALINAE).

Chaparea
New species: **Linnavuori** 1959b.
Records: **Linnavuori** 1959b (n. comb. from *Thamnotettix*; key to species).

Chelidinus
New species: **Dlabola** 1980d. **Emeljanov** 1962a.
Records: **Anufriev** 1978a (USSR: Maritime Territory; in key to PARALOMNINI). **Dlabola** 1967b, 1967d (Mongolia). **Dworakowska** 1973d (Korea). **Emeljanov** 1964c (European USSR); 1977a (Mongolia). **Hamilton, K. G. A.** 1975b (APHRODINAE, DELTOCEPHALINI, DELTOCEPHALINA). **Lee, C. E.** 1979a (Korea, redescribed). **Logvinenko** 1984a (USSR: Ukraine). **Mitjaev** 1968a (eastern Kazakhstan); 1971a (Kazakhstan). **Vilbaste** 1968a (coastal regions of the USSR); 1980b (Tuva).

Chiapasa
Record: **Young** 1968a (= *Diestostemma*, n. syn.).

Chiasmus
New species: **Sawai Singh** 1969a. **Theron** 1982b.
Records: **Bonfils & Della Giustina** 1978a (Corsica). **Datta** 1972e, 1972h (descriptive notes). **Dlabola** 1957a (Turkey); 1957b (Afghanistan); 1958b (Caucasus); 1961b (central Asia); 1961c (Romania); 1964a (Egypt); 1964c (Albania); 1965c (Mongolia); 1972a (Afghanistan); 1981 (Iran). **Dubovskij** 1966a (Uzbekistan); 1978a (Uzbekistan, Zarafshan Valley). **Emeljanov** 1964c (European USSR). **Evans, J. W.** 1974a (n. gen. syn.; n. sp. syn.; New Caledonia; DELTOCEPHALINAE). **Hamilton, K. G. A.** 1975b, 1975d (APHRODINAE, EUPELICINI). **Jankovic** 1966a (Serbia). **Kalkandelen** 1974a (Turkey). **Kholmuminov & Dlabola** 1979a (USSR: Golodnostep Plain). **Lindberg** 1960a (Portugal); 1965a (Spanish Sahara). **Linnavuori** 1961a (n. comb. from *Postumus*); 1962a (Israel); 1964a (Egypt); 1965a (Spanish Morocco). **Lodos & Kalkandelen** 1985c (Turkey). **Quartau & Rodrigues** 1969a (Portugal). **Servadei** 1968a (Italy). **Sharma, B.** 1977b (India: Jammu). **Vilbaste** 1976a (specific synonymy noted).

Chibala
New species: **Linnavuori & DeLong** 1977c.

Chichahua
New species: **Young** 1977a.
Record: **Young** 1977a (key to species).

Chigallia
New species: **Linnavuori & DeLong** 1977c.

Chileanoscopus
New species: **Freytag & Morrison** 1967a.
Records: **Freytag & Morrison** 1967a (n. comb. from *Idiocerus*; key to species). **Linnavuori & DeLong** 1977c (Chile).

Chilelana
New species: **DeLong** 1969a.
Records: **Linnavuori & DeLong** 1977c (Chile, sand dunes & sandy beaches). **Nielson** 1975a (redescribed; COELIDIINAE, TINOBREGMINI; Chile and Bolivia).

Chilenana
New species: **DeLong & Freytag** 1967b. **Freytag & DeLong** 1971a.
Records: **DeLong & Freytag** 1967b (key to species); 1972a (in key to genera of GYPONINAE). **Freytag & DeLong** 1971a (new subgenera proposed). **Linnavuori & DeLong** 1977c (Chile).

Chinaella
New species: **Evans, J. W.** 1966a.
Record: **Evans, J. W.** 1966a (synopsis).

Chinaia
New species: **Kramer** 1958a. **Linnavuori** 1956a, 1965b.
Records: **Kramer** 1959a (n. comb. from *Coelidia*; *Neocoelidia punctata* transferred from NEOCOELIDIINAE to DELTOCEPHALINAE, tribe unassigned); 1964d (in key to genera of NEOCOELIDIINAE); 1967b (n. sp. syn.).

Chinchinota
New species: **Kramer** 1967b.

Chlidochrus
New species: **Emeljanov** 1962a.

Records: **Hamilton, K. G. A.** 1975b (APHRODINAE, APHRODINI, ACHAETICINA). **Mitjaev** 1971a (Kazakhstan; key to species).

Chloothea
New species: **Emeljanov** 1959a, 1964d.
Records: **Dlabola** 1967b, 1967c, 1968b, 1970b (Mongolia). **Dworakowska** 1969a (Mongolia). **Emeljanov** 1964c (European USSR); 1966a (supplemental and comparative notes); 1977b (Mongolia). **Hamilton, K. G. A.** 1975b (APHRODINAE, DELTOCEPHALINI, DELTOCEPHALINA). **Mitjaev** 1968a (eastern Kazakhstan); 1971a (Kazakhstan). **Vilbaste** 1965b (Altai Mts.); 1980b (Tuva).

Chloriona
New species: **Dlabola** 1960b.

Chlorita
New species: **Ahmed, Manzoor** 1979a. **Dlabola** 1958b, 1959b, 1961b, 1963d, 1965c, 1967a, 1967e, 1971a, 1975a. **Dubovskij** 1970b. **Dworakowska** 1968b, 1968d, 1970h, 1977b, 1977d, 1977f, 1981e. **Einyu & Ahmed** 1980a. **Emeljanov** 1964c, 1977a. **Kirejtshuk** 1975a. **Korolevskaya** 1978a. **Linnavuori** 1962a, 1964a, 1965a. **Logvinenko** 1966b. **Mitjaev** 1963b, 1969a. **Sharma, B.** 1984a. **Vidano** 1964b. **Vilbaste** 1961b, 1965b, 1980b. **Wagner** 1959b. **Zachvatkin** 1953c, 1953d.
Records: **Cantoreanu** 1963b, 1971c (Romania). **Della Giustina** 1983a (France). **Dlabola** 1958b (Caucasus); 1958c (in key to genera); 1959b (notes on zoogeography); 1961c (southern Europe); 1965c, 1966a, 1967b, 1967c, 1967d, 1968b, 1970b (Mongolia); 1981a (Iran). **Dubovskij** 1966a (Uzbekistan); 1978a (Uzbekistan, Zarafshan Valley;). **Dworakowska** 1969a (Mongolia; n. sp. syn.); 1970i, 1977b (n. sp. syn.); 1977d (notes on Vidano's species); 1980b (northern India); 1981e (India & Nepal); 1982a (Korea & Japan). **Dworakowska & Sohi** 1978a (n. comb. from *Viridasca*). **Emeljanov** 1964c (European USSR; key to species); 1977a (Mongolia). **Ishihara** 1958a (Japan: north Honshu); 1961a (Thailand). **Jankovic** 1966a (Serbia). **Kirejtshuk** 1977a (USSR: Karkov region). **Korolevskaya** 1977a (central Asia). **Lee, C. E.** 1979a (Korea, redescribed). **LeQuesne** 1961a (Britain); 1964a (revised sp. syn.). **LeQuesne & Payne** 1981a (Britain). **Linnavuori** 1964a (Egypt). **Lodos** 1982a (Turkey). **Lodos & Kalkandelen** 1973c (Turkey). **Logvinenko** 1957b, 1984a (USSR: Ukraine). **Mahmood** 1967a (= *Empoasca*). **Mitjaev** 1963b, 1963c, 1968b, 1971a (Kazakhstan; key to species). **Nast** 1955a (Poland); 1972a (catalogued); 1976b (Poland; catalogued). **Ossiannilsson** 1981a (Fennoscandia & Denmark; key to species). **Servadei** 1960a, 1968a, 1969a, 1971a (Italy). **Sohi & Dworakowska** 1984a (list of Indian species). **Thapa** 1985a (Nepal). **Vilbaste** 1962b (eastern part of the Caspian lowlands); 1964a (Auwiesen Estlands); 1965b (Altai Mts.); 1980b (Tuva). **Wagner** 1959b (key to members of *viridula* species complex). **Zachvatkin** 1953c (n. comb. to *Kyboasca*).

Chloroasca
New species: **Anufriev** 1972a.

Chlorogonalia
New species: **Young** 1977a.
Record: **Young** 1977a (n. comb. from *Tettigonia*; key to species).

Chloronana
New species: **DeLong** 1980a. **DeLong & Freytag** 1964b.
Record: **DeLong & Freytag** 1972a (in key to genera of GYPONINAE).

Chloropelix
New species: **Lindberg** 1958a.
Records: **Hamilton, K. G. A.** 1975b (APHRODINAE, PARADORYDIINI). **Heller & Linnavuori** 1968a (Ethiopia; PARADORYDIINAE). **Lindberg** 1961a (Madeira); 1963a (Morocco); 1965a (Spanish Morocco). **Lindberg & Wagner** 1965a (Canary Islands). **Linnavuori** 1962a (Israel); 1979a (DORYCEPHALINAE, PARADORYGIINI).

Chlorotettix
New species: **Cheng, Y. J.** 1980a. **Cwikla & Freytag** 1982a. **DeLong** 1959a, 1980a, 1982h. **DeLong & Linnavuori** 1978b, 1979a. **DeLong & Martinson** 1974a. **Linnavuori** 1959b, 1973b, 1973c, 1973d. **Linnavuori & DeLong** 1979b.

Records: **Beirne** 1956a (Canada; key to species). **Cheng, Y. J.** (Paraguay; in key to genera of DELTOCEPHALINI). **Datta & Dhar** 1984a (descriptive notes). **Hamilton, K. G. A.** 1972d (Canada: Manitoba); 1975b (APHRODINAE, DELTOCEPHALINI, ATHYSANINA). **Linnavuori** 1956a (Argentina, Colombia, Trinidad); 1959b (key to neotropical species); 1973c (Cuba). **Nielson** 1968b (characterization of vector species). **Theron** 1974b (Naude's species; n. comb. to *Hiltus* and *Renosteria*).

Chonosina
New species: **Linnavuori & DeLong** 1978b.
Record: **Cwikla & Blocker** 1981 (comparative notes).

Chromagallia
Records: **Davis, R. B.** 1975a (morphology; phyletic inferences). **Kramer** 1964a (key to species)

Chromogonia
Record: **Young** 1977a (status uncertain).

Chroocacus
New species: **Emeljanov** 1962a.
Record: **Emeljanov** 1964c (European USSR). **Hamilton, K. G. A.** 1975b (APHRODINAE, unassigned).

Chudania
New species: **Heller** 1972a.

Chunra
New species: **Webb, M. D.** 1983b.
Records: **Webb, M. D.** 1983a (comparative notes; in key to Australian genera of IDIOCERINAE); 1983b (key to Afrotropical species).

Chunrocerus
Records: **Cantoreanu** 1963a (Romania). **Dlabola** 1971b (Turkey).

Chunroides
New species: **Maldonado-Capriles** 1975b.
Records: **Freytag** 1971a (n. comb. from *Brachybelus*). **Maldonado-Capriles** 1975b (redescribed; key to species); 1977a (key characters).

Cibra
New species: **Young** 1977a.

Cicada
New species: **Metcalf** 1965a (new name).
Records: **Young** 1965d, 1974a (lectotype designations).

Cicadella
New species: **Ossiannilsson** 1981a.
Records: **Anufriev** 1977b (Kurile Islands); 1978a (USSR: Maritime Territory). **Bonfils & Della Giustina** 1978a (Corsica). **Cantoreanu** 1965d (*sensu* Dumeril, Romania). **Choe** 1985a (Korea). **Dlabola** 1957d (Turkey); 1958b (Caucasus); 1960c (Iran); 1961b (central Asia); 1964b (Sudan); 1964c (Albania); 1966a, 1967b, 1967c, 1968b, 1970b (Mongolia); 1971a (Iran); 1972a (Afghanistan); 1981a, 1984a (Iran); **Dubovskij** 1966a (Uzbekistan); 1978a (Uzbekistan, Zarafshan Valley); 1980a (USSR: western Turkmenia). **D'Urso** 1980b (Sicily). **Emeljanov** 1964c (European USSR); 1977b (Mongolia). **Evans, J. W.** 1966a (*sensu* Dumeril 1806; = *Eupteryx*). ICZN Opinion 647, 1963 (*Cicadella* Latreille validated under Plenary Powers; *Cicadella* Dumeril 1806 suppressed). **Ishihara** 1958a (*sensu* Dumeril 1806; Japan); 1965c (*sensu* Dumeril; north Honshu). **Jankovic** 1966a, 1971a (Serbia). **Jansky** 1983a (Slovakia). **Jasinska** 1980a (Poland: Bledowska Wilderness). **Korolevskaya** 1975a (Tadzhikistan). **Kwon** 1983a (Korea). **Lee, C. E.** 1979a (Korea, redescribed). **Lee, C. E. & Kwon** 1977b (*sensu* Dumeril 1806; Korea). **LeQuesne** 1965a (Britain). **Lindberg** 1958a (for a species of *Cofana*). **Linnavuori** 1959c (Mt. Picentini); 1960b (Fiji, for a species of *Cofana*); 1961a (South Africa, for a species of *Cofana*); 1962a (Israel); 1965a (Turkey). **Lodos** 1982a (Turkey). **Lodos & Kalkandelen** 1983a (Turkey). **Logvinenko** 1957b (USSR: Ukraine). **Mitjaev** 1968a (eastern Kazakhstan); 1971a (Kazakhstan). **Nast** 1972a (catalogued); 1976a (Poland: Pieniny Mts.); 1976b (Poland; catalogued). **Ossiannilsson** 1981a (Fennoscandia & Denmark; key to species). **Servadei** 1960a (Italy). **Viggiani** 1973a (Italy). **Vilbaste** 1962b (Eastern part of the Caspian lowlands); 1964a (Auwiesen Estlands); 1965b (Altai Mts.); 1968a (USSR: Primorsky District); 1963a (Estonia & Latvia; notes on Flor's material); 1979a (Vooremaa hardwood/spruce forest); 1980b (Tuva). **Young** 1968a (discussion of status;); 1977a (systematics of related genera). **Zachvatkin** 1953b (central Russia; Oka district).

Cicadula
New species: **Dubovskij** 1966a. **Hamilton, K. G. A.** 1972b, 1976a. **Salmon** 1954a. **Vilbaste** 1965b, 1969b.
Records: **Anufriev** 1977b (Kurile Islands); 1978a (USSR: Maritime Territory; key to species). **Beirne** 1956a (Canada; key to species). **Cantoreanu** 1971c (Romania). **Datta** 1971i, 1973l (descriptive notes). **Dlabola** 1957a (Turkey); 1957b (Afghanistan); 1959b (southern Europe); 1961b (central Aisa); 1961c (Austria); 1964c (Albania); 1964d (Europe); 1967b, 1967d, 1970b (Mongolia); 1971a (Pakistan); 1977c (Mediterranean); 1981a (Iran). **Dubovskij** 1966a (Uzbekistan; key to species); 1970b (Uzbekistan); 1978a (Uzbekistan, Zarafshan Valley). **Dworakowska** 1973d (Mongolia). **Emeljanov** 1964c (European USSR; key to species; 1977a (Mongolia). **Hamilton, K. G. A.** 1972d (Canada: Manitoba); 1975b (APHRODINAE, DELTOCEPHALINI, CICADULINA). ICZN 1961 (*Cicadula sulphurella* placed on Official List of Specific Names in Zoology). **Ishihara** 1966a (Kurile Islands). **Jansky** 1983a (Slovakia). **Jasinska** 1980a (Poland: Bledowska Wilderness). **Kalkandelen** 1974a (Turkey; key to subgenera and species). **Korolevskaya** 1973a, 1979b (Tadzhikistan). **Lee, C. E.** 1979a (Korea, redescribed). **Lee, C. E. & Kwon** 1977b (Korea). **LeQuesne** 1964a (comparative notes; remarks on specific synonymy); 1969c (Britain); 1983c (*C. flori* in Britain). **Linnavuori** 1955c (*C. persimilis* removed from synonymy); 1962a (Israel). **Logvinenko** 1957b, 1957c, 1962e, 1984a (USSR: Ukraine). **Mitjaev** 1968a (eastern Kazakhstan); 1971a (Kazakhstan; key to species). **Nast** 1972a (catalogued); 1976b (Poland: Pieniny Mts.); 1976b (Poland; catalogued). **Ossiannilsson** 1955a (comments on specific synonymy); 1974a (Norway); 1983a (Fennoscandia & Denmark; key to species). **Quartau & Rodrigues** 1969a (Portugal). **Servadei** 1968a, 1969a, 1971a, 1972a (Italy). **Sharma, B.** 1977b (India: Jammu). **Theron** 1970b (identity of *C. longiforma* Cogan). **Trolle** 1966a (Denmark). **Vilbaste** 1958a (Estonia); 1964a (Auwiesen Estlands); 1965b (Altai Mts.); 1968a (coastal regions of the USSR); 1973a (notes on Flor's material); 1976a Matsumura's species described as *Cicadula*; n. comb. to *Nesoclutha*; n. comb. from *Thamnotettix*; n. sp. syn.); 1980a (Kamchatka); 1980b (Tuva). **Wagner** 1967a (affinities of *Paluda*, *Elymana*, *Cicadula* discussed). **Zachvatkin** 1953b (central Russia; Oka district).

Cicadulella
Records: **Hamilton, K. G. A.** 1975b (APHRODINAE, DELTOCEPHALINI, MACROSTELINA). **Lindberg** 1958a (Cape Verde Islands). **Linnavuori** 1962a (Israel); 1964a (Egypt).

Cicadulina
New species: **Fennah** 1959a. **Ghauri** 1961a, 1964b, 1971b. **Lindberg** 1958a. **Linnavuori** 1960b. **Ruppel** 1965a. **Ruppel & DeLong** 1956a. **Van Rensburg** 1983a.
Records: **Dlabola** 1956a (Czechoslovakia); 1957a (Turkey); 1960c (Iran); 1965d (Jordan); 1977c (Mediterranean); 1981a (Iran). **El-Desouky** 1978a (diagnostic characteristics of species). **Evans, J. W.** 1966a (n. comb. from *Limotettix*; n. sp. syn.); 1974a (New Caledonia). **Fennah** 1959a (key to African species). **Ghauri** 1972e (India). **Hamilton, K. G. A.** 1975b (APHRODINAE, DELTOCEPHALINI, MACROSTELINA). **Heller & Linnavuori** 1968a (n. sp. syn.). **Kitching, Grylls & Waterford** 1973a (revised synonymies; morphological data; interbreeding trials; *capitata* Kirkaldy returned to *Limotettix*). **Lee, C. E.** 1979a (Korea, redescribed). **Lindberg** 1961a (Madeira). **Linnavuori** 1959b (Neoptropical species); 1960a (Micronesia: Bonin Islands, Caroline Atolls, Yap); 1960b (Fiji; new subgenus proposed); 1961a (central and southern Africa); 1962a (Israel); 1964a (Egypt); 1965a (Libya); 1975c (Micronesia: Bonin Islands, Mariana, Palau). **Lodos** 1982a (Turkey). **Lodos & Kalkandelen** 1985b (Turkey). **Nielson** 1968b (characterization of vector species). **Ruppel** 1965a (key to species; intraspecific and geographic variation); 1969a (inter-population variation). **Sharma, B.** 1977b (India: Jammu). **Van Rensburg** 1983a (neotype designation;

key to African species; occurrence records for all species). **Vilbaste 1976a** (n. sp. syn.). **Webb, M. D. 1980a** (Aldabra)

Cicciana
Record: **Young 1968a** (key to species).

Ciccus
Records: **Young 1965c** (lectotype designation); **1968a** (key to species).

Ciminius
New species: **Young 1977a**.
Records: **Marino de Remes Lenicov & Teson 1983a** (lectotype designation for *platensis* (Berg)). **Young 1977a** (n. comb. from *Acocephalus, Cicadella* and *Tettigonia*; key to species).

Cinerogonalia
Record: **Young 1977a** (n. comb. from *Cicadella*).

Circulifer
New species: **Lindberg 1958a. Linnavuori 1961a**.
Records: **Beardsley 1961a, 1965a** (Hawaii); **1966a** (Hawaii; Nihoa Island, French Frigate Shoal, Kure Atoll). **Bindra & Sohi 1970a** (hybridization experiments). **Bindra, Singh & Sohi 1970a** (n. sp. syn.; Indian taxa). **Bindra, Singh, Sohi & Gill 1973a** (Northwest India; hybridization experiments). **Cantoreanu 1959b, 1965b, 1965d** (Romania); **1968b** (Danube river delta). **Chong 1965a** (Hawaii). **Dlabola 1957a** (Turkey; n. comb. from *Thamnotettix*; n. sp. syn.); **1958a** (southern Europe); **1960c** (Iran); **1961b** (central Asia); **1968a, 1968b** (Mongolia). **Hamilton, K. G. A. 1975b** (APHRODINAE, DELTOCEPHALINI, PLATYMETOPIINA). **Klein & Racca 1984a** (Israel). **Klein, Raccah & Oman 1982a** (Israel). **LeQuesne 1969b** (Channel Islands). **Lindberg 1960a** (Portugal); **1960b** (Soviet Armenia); **1961a** (Madeira); **1962a** (Portugal); **1963a** (Morocco); **1965a** (Spanish Sahara). **Linnavuori 1955c** (n. sp. syn.); **1956d** (Spain, Spanish Morocco); **1961a** (n. comb. from *Thamnotettix*); **1962a** (Israel); **1964a** (Egypt; n. sp. syn.). **Lodos & Kalkandelen 1985a** (Turkey). **Logvinenko 1957b** (USSR: Ukraine). **Meyerdirk, Oldfeld & Hessein 1983a** (bibliography of *C. tenellus*). **Quartau 1984a** (Small Salvage Island). **Quartau & Rodrigues 1969a** (Portugal). **Servadei 1957a, 1958a, 1960a, 1968a, 1969a, 1971a** (Italy). **Suehiro 1961a** (Hawaii, Kauai). **Yoshioka, E. 1965a** (Hawaii, Hawaii, Kawaihae).

Citorus
New species: **Linnavuori 1977a**.
Records: **Linnavuori 1961a** (South Africa; EUSCELINAE, PENTHIMIINI); **1977a** (n. comb. from *Penthimia* and *Selenocephalus*; key to species; PENTHIMIINAE). **Theron 1980a** (n. gen. syn.; South Africa; PENTHIMIINAE).

Citripo
Record: **Evans, J. W. 1966a** (synopsis).

Ciudadrea
Records: **Dworakowska 1970k** (n. comb. from *Zygina*); **1971b** (n. comb. from *Erythroneura*).

Cixidocoelidia
New species: **Linnavuori 1956a**.
Record: **Nielson 1975a** (=*Sandersellus*, n. syn.; COELIDIINAE, SANDERSELLINI).

Cleptochiton
New species: **Emeljanov 1959a**.
Records: **Hamilton, K. G. A. 1975b** (APHRODINAE, DELTOCEPHALINI, DELTOCEPHALINA). **Logvinenko 1984a** (USSR: Ukraine).

Clinonana
New species: **Kramer 1966a**.
Record: **Kramer 1966a** (n. comb. from *Gypona*; n. sp. syn.; in key to New World LEDRINAE; key to species)

Clinonaria
New species: **DeLong & Freytag 1969a. Teson 1971a** (as *Clinoria*).
Record: **DeLong & Freytag 1969a** (n. comb. from *Darma, Gypona, Ponana* and *Scaris*; n. sp. syn.).

Clinonella
Record: **DeLong & Freytag 1972a** (in key to genera of GYPONINAE).

Cloanthanus
Record: **Hamilton, K. G. A. 1975b** (APHRODINAE, DELTOCEPHALINI, PLATYMETOPIINA).

Clorindaia
New species: **Linnavuori 1975b**.
Records: **Cwikla & Blocker 1981a** (comparative notes). **Linnavuori & DeLong 1977c** (in key to genera of the *Faltala* group; DELTOCEPHALINAE, ATHYSANINI).

Clydacha
New species: **Kramer 1976a**.
Record: **Kramer 1976a** (key to species; PHEREURHININAE).

Clypeolidia
Record: **Nielson 1982e** (n. comb. from *Coelidia*).

Cochanga
New species: **Nielson 1979b**.

Cochlorhinus
New species: **Bliven 1955c**.
Records: **Datta & Ghosh 1974a** (descriptive notes). **Hamilton, K. G. A. 1975b** (APHRODINAE, DELTOCEPHALINI, COCHLORHININA). **Linnavuori & DeLong 1977c** (Chile, adventive ?; DELTOCEPHALINAE, COCHLORHININI).

Cocoelidia
Record: **Kramer 1964d** (n. comb. from *Neocoelidiana*; raised to generic rank; in key to genera of NEOCOELIDIINAE).

Codilia
New species: **Nielson 1982e**.
Record: **Nielson 1982e** (key to species).

Coelana
New species: **Kramer 1964d**.
Record: **Kramer 1964d** (raised to generic status; n. comb. from *Coelidiana*; key to species).

Coelella
Record: **Kramer 1964d** (raised to generic rank; n. comb. from *Coelidiana*).

Coelestinus
New species: **Emeljanov 1964c**.
Records: **Anufriev 1978a** (USSR: Maritime Territory; in key to genera of PARALIMNINI). **Dlabola 1967b** (as subgenus of *Palus*; Afghanistan). **Dworakowska 1969a** (Mongolia). **Emeljanov 1966a** (generic status; n. comb. from *Palus*); **1977a** (Mongolia). **Hamilton, K. G. A. 1975b** (APHRODINAE, DELTOCEPHALINI, DELTOCEPHALINA). **Mitjaev 1971a** (Kazakhstan). **Vilbaste 1980b** (Tuva).

Coelidia
New species: **Linnavuori 1956a. Metcalf 1964b** (new name). **Nielson 1982e, 1983i**.
Records: **Beirne 1956a** (Canada). **Heller & Linnavuori 1968a** (Ethiopia). **Ishihara 1965c** (Formosa). **Kramer 1959a** (n. comb. from *Chinaia*). **Linnavuori 1956a** (Neotropical COELIDIINAE); **1960a** (Micronesia); **1960b** (Fiji); **1975c** (Micronesia; n. sp. syn.). **Linsley & Usinger 1966a** (Galapagos Islands). **Nielson 1968b** (characterization of vector species); **1982e** (genus redefined; key to species); **1983i** (revised key to species; specific homonymy). **Ramos Elorduy 1972c** (Mexico; catalogued). **Servadei 1969a** (?Coelidia, Italy).

Coelidiana
New species: **DeLong & Kolbe 1975a. Kramer 1967b. Linnavuori 1965b. Linnavuori & Heller 1961a**.
Records: **Kramer 1964d** (n. gen. syn.; n. comb. from *Neocoelidia*; in key to genera of NEOCOELIDIINAE); **1967b** (key to species). **Linnavuori 1956a** (Neotropical species); **1965b** (DELTOCEPHALINAE, NEOCOELIDIINI).

Coelidroma
New species: **Kramer 1967b**.

Coelogypona
New species: **DeLong & Freytag 1966b**.
Record: **DeLong & Freytag 1972a** (in key to genera of GYPONINAE).

Coelopola
Record: **Young 1968a** (=*Ciccus*).

Coexitianus
New species: **Dlabola 1959c**.
Record: **Dlabola 1960a** (n. comb. from *Athysanus*).

Cofana
New species: **Young 1979a**.

Records: **Ghauri** 1980a (Papua New Guinea). **Lindberg** 1958a (as *Cicadella*; Cape Verde Islands). **Linnavuori** 1960b (as *Cicadella*; Fiji); 1961a (as *Cicadella*; South Africa); 1979a (n. comb. from *Poecilocarda*; key to species). **Young** 1979a (genus redefined; n. gen. syn.; n. comb. from *Poecilocarda, Tettigonia* and *Tettigoniella*; n. sp. syn.).

Coganoa
New species: **Dworakowska** 1967b, 1979a.

Coganus
Record: **Theron** 1978b (n. comb. from *Deltocephalus*).

Colimona
Records: **Ramos Elorduy** 1972c (Mexico; catalogued). **Schroder** 1959a (diagnosis). **Young** 1977a (= *Apogonalia*, n. syn.).

Colladonus
New species: **Linnavuori** 1959b. **Nielson** 1957a, 1962a.
Records: **Anufriev** 1977b (Kurile Islands); 1978a (USSR: Maritime Territory). **Beirne** 1956a (Canada; key to species). **Bliven** 1958a (hybrid origin for *C. flavocapitatus* ?); 1963a (*C. uhleri* removed from synonymy). **Cantoreanu** 1971c (as *Coladonus*; Romania). **Dlabola** 1966a, 1967b, 1967c, 1967d, 1968a, 1968b, 1970b (Mongolia). **Emeljanov** 1964c (European USSR); 1977a (Mongolia). **Hamilton, K. G. A.** 1972d (Canada: Manitoba); 1975b (APHRODINAE, DELTOCEPHALINI, PLATYMETOPIINA); 1983b (transboreal species). **Kartal** 1983a (Turkey). **Lauterer** 1980a (Moravia and Slovakia). **Lee, C. E.** 1979a (Korea, redescribed). **Lee, C. E. & Kwon** 1977b (Korea). **LeQuesne** 1969c (Britain). **Linnavuori** 1959b (n. comb. from *Idiodonus*; key to species). **Logvinenko** 1957b (USSR: Ukraine). **Nast** 1972a (catalogued); 1976b (Poland; catalogued). **Nielson** 1957a (redescribed; key to species); 1966a (n. gen. syn.; Provancher taxon removed from the genus, unassigned); 1968b (characterization of vector species). **Ossiannilsson** 1983a (Fennoscandia & Denmark). **Scudder** 1961a (Canada: British Columbia). **Vilbaste** 1962b (eastern part of the Caspian lowlands); 1965b (Altai Mts.); 1968a (coastal regions of the USSR); 1969b (Taimyr); 1979a (Vooremaa hardwood/spruce forest); 1980a (Kamchatka); 1980b (Tuva).

Coloana
New species: **Dworakowska** 1971e, 1981a. **Sohi** 1977a.
Record: **Sohi & Dworakowska** 1984a (list of Indian species).

Coloborrhis
New species: **Evans, J. W.** 1959a. **Linnavuori** 1972a. **Mbondji** 1982b (and 1983a, a duplicate publication).
Records: **Davis, R. B.** 1975a (morphology; phyletic inferences). **Emeljanov** 1963a (n. comb. from *Ulopa*). **Evans, J. W.** 1959a (remarks on evolution and speciation; key to Mascarene species). **Linnavuori** 1961a (central, east and South Africa); 1972a (in key to genera; key to species; ULOPINAE, ULOPINI). **Theron** 1980a (all parts of southern Africa). **Van Stalle** 1982a (tropical Africa).

Colobotettix
Records: **Della Giustina** 1983a (France). **Hamilton, K. G. A.** 1975b (APHRODINAE, DELTOCEPHALINI, PLATYMETOPIINA). **Nast** 1955a (Poland); 1972 (catalogued); 1976b (Poland; catalogued). **Ossiannilsson** 1983a (east Fennoscandia; DELTOCEPHALINAE, ATHYSANINI). **Ribaut** 1959a (France).

Cololedra
New species: **Evans, J. W.** 1969a.

Columbonirvana
New species: **Linnavuori** 1959b.
Record: **Kramer** 1964b (in key to Neotropical genera of NIRVANINAE).

Comanopa
New species: **Blocker** 1979a, 1982a.
Record: **Blocker** 1979a (n. comb. from *Stragania*).

Comayagua
New species: **Linnavuori & DeLong** 1978c.
Record: **Cwikla & Blocker** 1981a (comparative notes).

Commellus
Records: **Beirne** 1956a (Canada; key to species). **Hamilton, K. G. A.** 1972d (Canada: Manitoba).

Conala
New species: **Kramer** 1963b.

Records: **Kramer** 1963b (in key to genera of Neotropical NEOBALINAE; key to species). **Linnavuori** 1959b (redescribed).

Conbalia
New species: **Nielson** 1979b, 1981a.
Records: **Nielson** 1979b (key to species); 1981a (n. comb. from *Coelidia*; revised key to species).

Concavifer
New species: **Dlabola** 1960c. **Emeljanov** 1962b. **Kartal** 1982a.
Records: **Dlabola** 1961b (USSR: Tadzhikistan); 1981a, 1984a (Iran). **Emeljanov** 1977a (Mongolia). **Hamilton, K. G. A.** 1975b (APHRODINAE, DELTOCEPHALINI, PLATYMETOPIINA). **Linnavuori** 1962a (Israel).

Concepciona
New species: **Linnavuori & DeLong** 1977c.
Record: **Cwikla & Blocker** 1981a (comparative notes).

Condylotes
New species: **Dubovskij** 1966a. **Emeljanov** 1959a. **Mitjaev** 1967c, 1969a, 1969c.
Records: **Dubovskij** 1978a (Uzbekistan, Zarafshan Valley); 1980a (USSR: western Turkmenia). **Emeljanov** 1964c (European USSR). **Hamilton, K. G. A.** 1975b (APHRODINAE, DELTOCEPHALINI, ATHYSANINA). **Mitjaev** 1968a (eastern Kazakhstan); 1971a (Kazakhstan; in key to genera of DELTOCEPHALINAE; key to species).

Confucius
New species: **Linnavuori** 1972b.
Record: **Linnavuori** 1972b (in key to genera of LEDRINAE; n. comb. from *Ledropsis* and *Petalocephala*; n. sp. syn; key to species).

Coniferadonus
Record: **Bliven** 1958a (comparative notes); 1963a (comments on specific synonymy). **Nielson** 1966a (= *Colladonus*, n. syn).

Conlopa
New species: **Evans, J. W.** 1971c.
Record: **Linnavuori** 1972a (in key to genera of African ULOPINI; Zaire).

Conodonus
Record: **Hamilton, K. G. A.** 1975b (APHRODINAE, DELTOCEPHALINI, PLATYMETOPIINA).

Conogonia
Records: **Young** 1965b (lectotype designation). **Young & Nast** 1963a (notes on type specimens).

Conogonus
Record: **Hamilton, K. G. A.** 1975b (APHRODINAE, DELTOCEPHALINI, ATHYSANINA).

Conojassus (error for *Coriojassus*)
Record: **Linnavuori & Quartau** 1975a (IASSINAE, HYALOJASSINI).

Conometopus
Record: **Vilbaste** 1975a (as *Conometopius*; notes on de Motschulsky's material)

Conosanus
Records: **Bonfils & Della Giustina** 1978a (Corsica). **Dlabola** 1977c (Mediterranean). **Hamilton, K. G. A.** 1975b (APHRODINAE, DELTOCEPHALINI, ATHYSANINA). **Nast** 1972a (catalogued); 1976a (Poland: Pieniny Mts.); 1976b (Poland; catalogued). **Ossiannilsson** 1983a (Denmark and southern Sweden; DELTOCEPHALINAE, ATHYSANINI). **Quartau** 1979a (Azores). **Quartau & Rodrigues** 1969a (Portugal).

Consepusa
New species: **Linnavuori & DeLong** 1977c.
Records: **Cwikla & Blocker** 1981 (comparative notes). **Linnavuori & DeLong** 1977b (in key to Neotropical genera of the *Osbornellus* group).

Convelinus
Records: **Hamilton, K. G. A.** 1975b (APHRODINAE, DELTOCEPHALINI, PLATYMETOPIINA). **Linnavuori** 1956a (Colombia).

Conversana
New species: **DeLong** 1967e.
Record: **Cwikla & Blocker** 1981a (comparative notes).

Copididonus
New species: **Linnavuori** 1975d. **Zanol & Sakakibara** 1985a.

Records: **Linnavuori** 1956a (Argentina & Paraguay); 1959b (n. comb. from *Athysanus* and *Thamnotettix*; n. sp. syn.; key to species); 1975d (descriptive notes). **Linnavuori & DeLong** 1978d (in key to Neotropical genera of the *Copididonus* group). **Linnavuori & Heller** 1961a (Peru). **Zanol & Sakikabara** 1985a (synopsis).

Corilidia
New species: **Nielson** 1982a.
Record: **Nielson** 1982a (in key to genera of TINOBREGMINI).

Coriojassus
New species: **Evans, J. W.** 1972b.
Record: **Linnavuori & Quartau** 1975a (as *Conojassus*; IASSINAE, HYALOJASSINI).

Cornutipo
New species: **Evans, J. W.** 1969b, 1970a.
Records: **Davis, R. B.** 1975a (morphology; phyletic inferences). **Evans, J. W.** 1966a (synopsis); 1977a (=*Cornutipoides*, n. gen. syn.).

Cornutipoides
Records: **Davis, R. B.** 1975a (morpholgy; phyletic inferences). **Evans, J. W.** 1966a (synopsis); 1977a (=*Cornutipo*).

Coronigonalia
New species: **Young** 1977a.
Record: **Young** 1977a (n. comb. from *Tettigonia*; key to species).

Coronigoniella
New species: **Young** 1977a.
Record: **Young** 1977a (n. comb. from *Cicada* and *Graphocephala*; n. sp. syn.; key to species).

Coronophtus
New species: **Van Stalle** 1983b.

Cortona
Record: **Linnavuori** 1959b (redescribed).

Corupiana
New species: **Nielson** 1979b.

Corymbonotus
New species: **Maldonado-Capriles** 1977b.
Record: **Maldonado-Capriles** 1977b (in key to Idiocerine genera from Guyana).

Coryphaelus
Records: **Hamilton, K. G. A.** 1975b (APHRODINAE, DELTOCEPHALINI, ATHYSANINA). **Logvinenko** 1957b, 1957c (USSR: Ukraine). **Nast** 1972a (catalogued); 1973a (Poland); 1976a (Poland: Pieniny Mts.); 1976b (Poland; catalogued). **Ossiannilsson** 1983a (Sweden; DELTOCEPHALINAE, CORYPHAELINI).

Cosmotettix
New species: **Ossiannilsson** 1976a. **Vilbaste** 1980a.
Records: **Anufriev** 1977b (Kurile Islands); 1978a (USSR: Maritime Territory). **Dworakowska** 1973d (Mongolia). **Emeljanov** 1977a (Mongolia). **Hamilton, K. G. A.** 1975b (APHRODINAE, DELTOCEPHALINI, DELTOCEPHALINA). **Jasinska** 1980a (Poland: Bledowska Wilderness). **Lauterer** 1980a, 1984a (Moravia). **Logvinenko** 1984a (USSR: Ukraine). **Nast** 1972a (catalogued); 1976b (Poland; catalogued). **Ossiannilsson** 1976a (key to species of Fennoscandia); 1983a (Fennoscandia & Denmark; key to species). **Vilbaste** 1980a (Kamchatka); 1980b (Tuva).

Costamia
Record: **Cwikla & Blocker** 1981a (comparative notes).

Costanana
New species: **DeLong & Freytag** 1972a, 1972e. **DeLong & Wolda** 1983a.
Records: **DeLong & Freytag** 1972a (in key to genera of GYPONINAE); 1972e (n. comb. from *Gypona*; key to species).

Coulinus
New species: **Emeljanov** 1966c.
Records: **Beirne** 1956a (Canada; key to species). **Emeljanov** 1964c (European USSR). **Hamilton, K. G. A.** 1972d (Canada: Manitoba); 1975b (APHRODINAE, DELTOCEPHALINI, ATHYSANINA). **Vilbaste** 1969b (Taimyr).

Cozadanus
New species: **DeLong & Harlan** 1968a.
Records: **Cwikla & Blocker** 1981a (comparative notes). **DeLong & Harlan** 1968a (key to species).

Crassana
Record: **Hamilton, K. G. A.** 1975b (APHRODINAE, DELTOCEPHALINI, PLATYMETOPIINA).

Crassinolanus
New species; **Nielson** 1982e.

Crepluvia
New species: **Nielson** 1979b, 1983d.
Records: **Nielson** 1979b (n. comb. from *Coelidia*; key to species); 1983d (revised key to species).

Cribrus
Records: **Hamilton, K. G. A.** 1975b (APHRODINAE, DELTOCEPHALINI, DELTOCEPHALINA). **Ross & Hamilton** 1972b (n. comb. from *Deltocephalus*).

Crinolidia
New species: **Nielson** 1982e.

Crinorus
New species: **Nielson** 1982e.

Crossogonalia
Record: **Young** 1977a (n. comb. from *Tettigonia*).

Crumbana
Record: **Hamilton, K. G. A.** 1975b (APHRODINAE, DELTOCEPHALINI, DELTOCEPHALINA).

Cruziella
New species: **Linnavuori & DeLong** 1979a.
Record: **Cwikla & Blocker** 1981a (comparative notes).

Ctenurella
New species: **Vilbaste** 1968a.
Records: **Anufriev** 1978a (USSR: Maritime Territory; in key to genera of DELTOCEPHALINI). **Hamilton, K. G. A.** 1975b (APHRODINAE, unplaced). **Lee, C. E.** 1979a (Korea, redescribed).

Cubnara
New species: **Dworakowska** 1981e.
Records: **Dworakowska** 1979a (n. comb. from *Typhlocyba*). **Sohi & Dworakowska** 1984a (list of Indian species).

Cubrasa
Record: **young** 1977a (n. comb. from *Poeciloscarta*).

Cuerna
New species: **Hamilton, K. G. A.** 1970a. **Nielson** 1965a.
Records: **Beirne** 1956a (Canada; key to species). **Datta** 1980a (descriptive notes). **Hamilton, K. G. A.** 1970d (Canada: Manitoba) **Nielson** 1965a (key to species); 1968b (characterization of vector species). **Schroder** 1961a (Argentina, Paraguay). **Young** 1968a (synopsis).

Cuitlana
New species: **Young** 1977a.
Record: **Young** 1977a (n. comb. from *Tettigonia*; key to species).

Culumana
New species: **DeLong** 1979c, 1984a. **DeLong & Freytag** 1972a, 1972d.
Records: **DeLong** 1984a (key to species). **DeLong & Freytag** 1972a (in key to genera of GYPONINAE); 1972d (key to species).

Cumbrenanus
New species: **DeLong & Cwikla** 1984a.

Cumora
Record: **Linnavuori** 1959b (=*Haldorus*, n. syn.).

Curtara
New species: **DeLong** 1977a, 1979a, 1980e, 1983a, 1984h. **DeLong & Foster** 1982b. **DeLong & Freytag** 1972a, 1976a. **DeLong & Triplehorn** 1978a, 1979a. **DeLong & Wolda** 1978a, 1982b, 1984a.
Records: **DeLong** 1977b (n. comb. from *Gypona* and *Prairiana*; n. sp. syn.); 1979a (new subgenus proposed); 1979f (n. comb. from *Gypona*). **DeLong & Freytag** 1972a (in key to genera of GYPONINAE); 1976a (six subgenera recognized; keys to subgenera and species). **Linnavuori & DeLong** 1977c (Chile).

Cyanidius
New species: **Dubovskij** 1966a.
Records: **Dubvskij** 1966a (in key to genera of EUSCELINI); 1970b (Uzbekistan); 1978a (Uzbekistan, Zarafshan Valley); 1980a (USSR: western Turkmenia). **Emeljanov** 1964f (n. comb. from *Euscelidius*). **Hamilton, K. G. A.** 1975b (APHRODINAE; *incertae sedis*). **Korolevskaya** 1979b (Tadzhikistan). **Mitjaev** 1971a (Kazakhstan; key to species).

Cyclogonia
New species: **Young** 1977a.
Records: **Young** 1977a (n. comb. from *Cicadella*, *Tettigonia* and *Tettigoniella*; n. sp. syn.; key to species). **Young & Lauterer** 1966a (lectotype designations).

Cyclopherus
Record: **Hamilton, K. G. A.** 1975b (APHRODINAE, DELTOCEPHALINI, ATHYSANINA).

Cymbalopus
Record: **Young** 1968a (= *Peltocheirus*).

Cyperana
Records: **Emeljanov** 1964c (European USSR). **Hamilton, K. G. A.** 1975b (APHRODINAE, DELTOCEPHALINI, CICADULINA). **Ossiannilsson** 1983a (subgenus of *Cicadula*).

Cyrta
Record: **Nielson** 1975a (removed from COELIDIINAE; unassigned).

Cyrtodisca
Records: **Schroder** 1959a (diagnosis). **Young** 1968a (redescribed). **Young & Soos** 1964a (lectotype designation).

Dagama
New species: **Theron** 1980b.
Records: **Linnavuori & Quartau** 1975a (n. comb. from *Bythoscopus*). **Theron** 1980b (= *Paranoplus*; n. comb. from *Paranoplus*; n. sp. syn).

Dalbulus
New species: **Linnavuori** 1959b. **Triplehorn & Nault** 1985a.
Records: **Linnavuori** 1959b (new subgenus proposed; key to species). **Nielson** 1968a (characterization of vector species). **Theron** 1975b (n. comb. to *Afrosteles*). **Triplehorn & Nault** 1985a (key to species; phyletic inferences).

Daltonia
Records: **DeLong** 1982c (n. comb. from *Bolarga*). **Hamilton, K. G. A.** 1975b (APHRODINAE, DELTOCEPHALINI, DELTOCEPHALINA).

Daluana
New species: **Ramakrishnan** 1982a.

Dalus
Record: **Dubovskij** 1980a (USSR: western Turkmenia). **Emeljanov** 1975a (n. comb. from *Symphypyga*; phyletic inferences).

Dampfiana
Record: **Cwikla & Blocker** 1981a (comparative notes).

Dananea
New species: **Linnavuori** 1972a.

Danbara
Record: **Hamilton, K. G. A.** 1975b (APHRODINAE, DELTOCEPHALINI, PLATYMETOPIINA).

Dapitana
New species: **Dworakowska** 1972i. **Mahmood** 1967a.

Dardania
Records: **Evans, J. W.** 1972b (to ACOSTEMMINAE). **Linnavuori & Quartau** 1975a (Madagascar; in key to genera of ACROPONINAE). **Nielson** 1975a (transferred from COELIDIINAE to STEGELYTRINAE).

Daridna
Record: **Nielson** 1982e (removed from synonymy).

Dariena
New species: **Linnavuori & DeLong** 1977b.
Record: **Cwikla & Blocker** 1981a (comparative notes).

Dasmeusa
New species: **Young** 1977a.
Record: **Young** 1965d (lectotype designation); 1977a (key to species).

Dattasca
Records: **Dworakowska** 1979a (n. comb. from *Alebroides*). **Sohi & Dworakowska** 1984a (list of Indian species).

Davisonia
Records: **Beirne** 1956a (Canada; key to species). **Hamilton, K. G. A.** 1972d (Canada: Manitoba); 1975b (APHRODINAE, DELTOCEPHALINI, MACROSTELINA). **Moore, T. E. & Ross** 1957a (as subgenus of *Macrosteles*). **Zachvatkin** 1953b (central Russia; Oka district).

Davmata
New species: **Dworakowska** 1979b.

Dayoungia
New species: **Kramer** 1976a.
Record: **Kramer** 1976a (in key to genera of PHEREURHININAE; key to species).

Dayus
New species: **Dworakowska** 1971f. **Dworakowska & Viraktamath** 1978a. **Mahmood** 1967a.
Records: **Dworakowska** 1971f (n. comb. from *Cicadula*, *Empoasca* and *Homa*; n. sp. syn.); 1980b, 1981e (southern India); 1982b (Japan, Okinawa); 1984a (Singapore). **Eyles & Linnavuori** 1974a (Niue Island). **Lee, C. E.** 1979a (Korea, redescribed). **Mahmood** 1967a (in key to genera of Oriental TYPHLOCYBINI). **Sohi & Dworakowska** 1984a (list of Indian species).

Dechacona
Record: **Young** 1968a (n. comb. from *Tettigonia*).

Declivara
New species: **DeLong & Freytag** 1971a.
Record: **DeLong & Freytag** 1972a (in key to genera of GYPONINAE).

Decua
Record: **Young** 1977a (key to species).

Delopa
New species: **Evans, J. W.** 1971c. **Van Stalle** 1983a.
Record: **Linnavuori** 1972a (in key to genera of African ULOPINAE; redescribed).

Deltanus
Record: **Hamilton, K. G. A.** 1975b (APHRODINAE, DELTOCEPHALINI, DELTOCEPHALINA).

Deltazotus
Record: **Hamilton, K. G. A.** 1975b (APHRODINAE, DELTOCEPHALINI, DELTOCEPHALINA). **Kramer** 1971d (n. comb. from *Deltocephalus*).

Deltocephalus
New species: **DeLong** 1980h, 1984f. **Dlabola** 1964b. **Evans, J. W.** 1966a. **Hamilton, K. G. A. & Ross** 1975a. **Heller & Linnavuori** 1969a. **Knight, W. J.** 1976b. **Kramer** 1963c, 1965b, 1971e. **Lindberg** 1958a. **Linnavuori** 1960a, 1961a, 1962. **Metcalf** 1967b (new name). **Sawai Singh** 1969a. **Zachvatkin** 1953e.
Records: **Anufriev** 1977b (Kurile Islands). **Beardsley** 1966a (Hawaii; Nihoa Island; French Frigate Shoal; Kure Atoll). **Beirne** 1956a (Canada; key to species). **Cantoreanu** 1965d (Romania); 1968b (Danube river delta); 1971c (Romania). **Datta** 1972b (descriptive notes). **Datta & Dhar** 1984a (descriptive notes). **Della Giustina** 1983a (France). **DeLong** 1966a (Alaska; Muir Inlet). **Dlabola** 1957a (Turkey); 1958a (Caucasus); 1960c (Iran); 1961b (central Asia); 1964c (Albania); 1965c, 1967c, 1968a, 1968b, 1970b (Mongolia). **Dubovskij** 1966a (Uzbekistan); 1978a (Uzbekistan, Zarafshan Valley). **Emeljanov** 1964c (European USSR); 1977a (Mongolia). **Evans, J. W.** 1966a (n. comb. from *Divitiacus* and *Phrynomorphus*; n. sp. syn.). **Hamilton, K. G. A.** 1972a (Canada: Manitoba); 1975b (APHRODINAE, DELTOCEPHALINI, DELTOCEPHALINA). **Heller** 1961a (n. sp. syn.). **Heller & Linnavuori** 1968a (Ethiopia). **Ishihara** 1961a (Cambodia); 1966a (Kurile Islands). **Jankovic** 1966a (Serbia). **Kalkandelen** 1974a (Turkey). **Knight, W. J.** 1975a (*Recilia* as subgenus); 1976b (Kermadec Islands); 1982a (Rennell Island). **Korolevskaya** 1978b (central Asia). **Kramer** 1962b (comparative notes on *Deltocephalus* and *Recilia*; n. comb. to *Recilia*); 1967d (n. comb. from *Lonatura*); 1971a (n. comb. to *Amblysellus*); 1971c (n. sp. syn.; reinstated synonymy; key to Nearctic species).. **Lee, C. E.** 1979a (Korea, redescribed). **Lee, C. E. & Kwon** 1977b (Korea) **LeQuesne** 1969c (in key to genera of British DELTOCEPHALINI; key to species); 1964a (n. sp. syn.). **Lindberg** 1956a (Niger River floodplain, Dogo); 1960b (Soviet Armenia). **Linnavuori** 1956a (Neotropical region); 1959b (new subgenus proposed); 1960a (new subgenus proposed; n. comb. from *Stirellus*); 1960b (key to subgenera; remarks on distribution of the genus and the subgeneric segregates); 1961a (n. comb. from *Thamnotettix*; South Africa); 1962a (Israel); 1964a (Egypt; *Recilia* as subgenus). **Linnavuori & Heller** 1961a (Peru; EUSCELINAE, EUSCELINI). **Linsley & Usinger** 1966a (Galapagos Islands). **Lodos & Kalkandelen** 1985c (Turkey).

Logvinenko 1957b (USSR: Ukraine); 1963b (USSR: Carpathians; *Recilia* as subgenus). **Mitjaev** 1968b (eastern Kazakhstan); 1971a (Kazakhstan; in key to genera of DELTOCEPHALINAE). **Nast** 1972a (catalogued); 1973a (Poland); 1976a (Poland: Pieniny Mts.); 1976b (Poland; catalogued). **Ossiannilsson** 1974a (Norway); 1983a (Fennoscandia & Denmark). **Quartau & Rodrigues** 1969a (Portugal). **Ross & Hamilton** 1972b (n. sp. syn.). **Servadei** 1957a, 1958a, 1960a, 1968a, 1969a, 1971a, 1972a (Italy). **Sharma, B.** 1977b (India: Jammu). **Suehiro** 1961a (Hawaii, Kauai). **Theron** 1970b (redescription of Cogan species); 1975a (n. comb. of Naude species to *Cerus, Ebarrius, Cedarotettix, Pravistylus, Vecaulis, Vilargus, Elginus, Megaulon* and *Proekes*). **Vilbaste** 1959b (Estonia); 1962b (eastern part of Caspian lowlands); 1964a (Auwiesen Estlands); 1965b (Altai Mts.); 1973a (identity of Flor's material); 1975a (identity of de Motschulsky's species; n. sp. syn.); 1976a (Matsumura species transferred to *Diplocolenus, Pantallus, Psammotettix* and *Turrutus*; specific synonymy noted); 1980a (Kamchatka); 1980b (Tuva). **Viraktamath, S. & Viraktamath, C. A.** 1980a (*Divitiacus* removed from synonymy with *Deltocephalus*). **Zachvatkin** 1953e (USSR: shores of Lake Ksubsugul).

Deltocoelidia
New species: **Kramer** 1961a.
Record: **Kramer** 1964d (in key to genera of NEOCOELIDIINAE).

Deltodorydium
Record: **Evans, J. W.** 1966a (= *Paradorydium*, n. syn.).

Deltolidia
Record: **Nielson** 1982e (n. comb from *Coelidia*; in key to genera of COELIDIINI).

Deltopinus
Record: **Hamilton, K. G. A.** 1975b (APHRODINAE, DELTOCEPHALINI, PLATYMETOPIINA).

Deltorhynchus
Record: **Cwikla & Blocker** 1981a (comparative notes).

Depanana
Records: **Young** 1965c (lectotype designation); 1968a (n. comb. from *Amblydisca*; n. sp. syn.).

Depanisca
Record: **Young** 1964a (lectotype designation); 1968a (n. comb. from *Amblydisca* and *Tettigonia*).

Derakandra
New species: **Blocker** 1979a.

Derogonia
Record: **Young** 1977a (identity uncertain).

Derriblocera
New species: **Nielson** 1983h.

Desamera
Record: **Young** 1968a (n. comb. from *Cicada* and *Coelopola*; n. sp. syn.).

Deselvana
Record: **Young** 1968a (n. comb. from *Amblydisca, Aulacizes, Ciccus, Proconia, Teletusa* and *Tettigonia*; n. sp. syn.).

Desertana
New species: **DeLong & Martinson** 1973b. **Linnavuori & DeLong** 1977c.
Records: **Cwikla & Blocker** 1981a (comparative notes). **Linnavuori & DeLong** 1977c (key to species).

Destria
Records: **Hamilton, K. G. A.** 1972d (Canada: Manitoba); 1975b (APHRODINAE, DELTOCEPHALINI, DELTOCEPHALINA). **Kramer** 1967d (n. comb. from *Lonatura*); 1976b (key to species).

Dharmma
Record: **Nielson** 1977a (n. gen. syn.).

Dhongariva
New species: **Webb, M. D.** 1983a.
Record: **Webb, M. D.** 1983a (in key to genera of Australian IDIOCERINAE).

Diacra
New species: **Diabola** 1961b, 1968b, 1984a. **Emeljanov** 1961a, 1969a.
Records: **Dubovskij** 1978a (Uzbekistan, Zarafshan Valley); 1980a (USSR: western Turkmenia). **Emeljanov** 1977a (Mongolia; n. comb. to *Zapycna*; n. sp. syn.). **Hamilton, K. G. A.** 1975b (APHRODINAE, APHRODINI, ACHAETICINA). **Korolevskaya** 1978b (central Asia). **Mitjaev** 1971a (Kazakhstan; in key to genera of DELTOCEPHALINAE).

Dialecticopteryx
New species: **Dworakowska** 1980b.
Records: **Dworakowska** 1970a (Australia); 1976b (n. comb. from *Typhlocyba*).

Diceratalebra
New species: **Young** 1957c (new name).
Record: **Young** 1957c (n. comb. from *Protalebra*; n. sp. syn.).

Dichometopia
Records: **Young** 1968a (= *Egidemia*). **Young & Nast** 1963a (notes on type specimens).

Dichrophelps
New species: **Young** 1968a.
Records: **Young** 1968a (redescribed; key to species; n. sp. syn.). **Young & Lauterer** 1966a (lectotype designation). **Young & Nast** 1963a (notes on type specimens).

Dicodia
New species: **Nielson** 1982e.
Record: **Nielson** 1982e (n. comb. from *Coelidia*; in key to genera of COELIDIINI; key to species).

Dicolecia
New species: **Nielson** 1982e.
Record: **Nielson** 1982e (in key to genera of COELIDIINI).

Dicrallygus
Record: **Hamilton, K. G. A.** 1975b (APHRODINAE, DELTOCEPHALINI, PLATYMETOPIINA).

Dicraneurula
New species: **Vilbaste** 1968a.
Records: **Anufriev** 1969b (Soviet Maritime Territory); 1978a (USSR: Maritime Territory; n. comb. from *Dikraneura*; n. sp. syn.). **Dworakowska** 1972k (Korea). **Lee, C. E.** 1979a (Korea, redescribed). **Vilbaste** 1968a (coastal regions of the USSR).

Dicranoneura
New species: **Beirne** 1955a.
Records: **Diabola** 1958c (in key to genera of Palaearctic TYPHLOCYBINAE); 1967b, 1967c, 1967d, 1968a, 1968b (Mongolia). **Dubovskij** 1966a (Uzbekistan). **Emeljanov** 1964c (European USSR; key to species). **Logvinenko** 1957b (USSR: Ukraine). **Servadei** 1957a (Italy). **Trolle** 1966a (Denmark). **Vilbaste** 1964a (Auwiesen Estlands); 1965b (Altai Mts.).

Dictyodisca
Record: **Young** 1968a (redescribed)

Dicyphonia
Records: **Hamilton, K. G. A.** 1975b, 1975d (APHRODINAE, HECALINI). **Linnavuori** 1959b (DELTOCEPHALINAE, HECALINI). **Linnavuori & DeLong** 1978e (n. comb. to *Jiutepeca*).

Didius
Record: **Datta** 1972e (descriptive notes).

Diedrocephala
New species: **Sakikabara & Cavichioli** 1982a. **Young** 1977a.
Records: **Young** 1965c (lectotype designation); 1977a (key to species).

Diemoides
Record: **Evans, J. W.** 1966a (= *Paralimnus*, n. syn).

Diestostemma
New species: **Young** 1968a.
Records: **Young** 1965c (lectotype designation); 1968a (n. gen. syn.; n. sp. syn.; n. comb. from *Heterostemma* and *Leucopepla*; key to species); 1974a (lectotype designation). **Young & Lauterer** 1966a (lectotype designation). **Young & Nast** 1963a (notes on type material). **Young & Soos** 1964a (notes on type material).

Diglenita
Records: **Linnavuori** 1970a (Spinola specimen not found); 1979a (n. comb. from *Aphrodes*).

Dikraneura
New species: **Ahmed, Manzoor** 1969a. **Anufriev** 1969b. **Datta & Ghosh** 1973b, 1974a. **Dworakowska** 1969b. **Dworakowska, Singh & Bagaich** 1979a. **Dworakowska & Sohi** 1978a. **Knight** 1965b, 1968a. **Linnavuori** 1962a. **Metcalf** 1968a (new name). **Vilbaste** 1968a.

Records: **Ahmed, Manzoor** 1985c (Pakistan). **Anufriev** 1978a (USSR: Maritime Territory; in key to genera of DIKRANEURINI). **Beirne** 1956a (as *Dicraneura*; Canada; key to species). **Cantoreanu** 1971c (Romania). **Dlabola** 1957a (Turkey); 1958b (Caucasus); 1958c (in key to genera of TYPHLOCYBINAE); 1961a (central Asia); 1961c (Romania); 1964c (Albania); 1966a, 1967b, 1967c (Mongolia); 1967e (Israel, Switzerland); 1970b (Mongolia); 1971b (Turkey). **Dubovskij** 1966a (Uzbekistan). **Dworakowska** 1968a (Mongolia); 1971d (USSR: Irkutsk); 1972k (Korea). **Emeljanov** 1964c (European USSR; key to species); 1977a (Mongolia). **Evans, J. W.** 1966a (Australia; synopsis). **Gunthart, H.** 1971a (as *Dicraneura*, host, season). **Gyllensvard** 1963a (Sweden). **Hamilton, K. G. A.** 1976a (notes on Provancher's species); 1983b (notes on erroneous records). **Jankovic** 1966a, 1977a (as *Dicraneura*; Serbia). **Knight, W. J.** 1965b (redescription of *D. micantula*); 1968a (genus redefined; n. sp. syn.; key to species). **Lauterer** 1957a (Czechoslovakia). **LeQuesne & Payne** 1981a (Britain). **Lindberg** 1960b (Soviet Armenia). **Lodos & Kalkandelen** 1983b (Turkey). **Mahmood** 1967a (supplemental notes). Mitjaev 1963b (Kazakhstan); 1968b (as *Dicraneura*; Kazakhstan); 1971a (Kazakhstan; key to species). **Nast** 1972a (catalogued); 1976a (Poland: Pieniny Mts.); 1976b (Poland; catalogued). **Ossiannilsson** 1955a (male genital structures); 1981a (Fennoscandia & Denmark). **Remane** 1961d (comparative notes). **Servadei** 1960a, 1968a, 1971a, 1972a (Italy). **Vidano** 1965a (systematics). **Vilbaste** 1959b (Estonia); 1968a (coastal regions of the USSR); 1980b (Tuva).

Dikrella
New species: **Ruppel** 1966a. **Young** 1956a.
Records: **Beirne** 1956a (Canada; key to species). **Ramos Elorduy** 1972c (Mexico). **Varty** 1967a (Canada). **Young** 1957b (notes on type material).

Dikrellidia
Record: **Young** 1957b (n. comb. from *Dikraneura*).

Dilobopterus
New species: **Young** 1977a.
Records: **Young** 1965c (lectotype designation); 1977a (n. gen. syn.; n. comb. from *Amblyscarta*, *Cicadella*, *Marizella*, *Tettigoniella* and *Tettigonia*; n. sp. syn.; key to species). **Young & Beier** 1963a (type specimens in the Signoret collection).

Dio
Records: **Datta** 1972e (descriptive notes). **Ghauri** 1974c (= *Pusatettix*, n. syn.; n. sp. syn.).

Diomma
New species: **Dworakowska** 1981c (redescribed and reviewed); 1984a (Malaysia). **Sohi & Dworakowska** 1984a (list of Indian species). **Vilbaste** 1975a (identity of *D. ochracea*).

Diplocolenoidea
Records: **Hamilton, K. G. A.** 1975b (APHRODINAE, DELTOCEPHALINI, ATHYSANINA). **Nast** 1972a (catalogued; = *Stenometopiellus*, n. syn.).

Diplocolenus
New species: **Anufriev** 1970e, 1971a, 1971c. **Dlabola** 1965c, 1980c. **Emeljanov** 1962a, 1964c, 1966a. **Kalkandelen** 1972a. **Knight** 1974c. **Linnavuori** 1959a. **Logvinenko** 1966b, 1971a. **Ribaut** 1959b. **Ross & Hamilton** 1970a, 1972a. **Servadei** 1960a. **Vilbaste** 1965b, 1980b.
Records: **Anufriev** 1970e (USSR: south of the Primor'ya Territory); 1978a (USSR: Maritime Territory); 1981b (key to Palaearctic species). **Beirne** 1956a (Canada). **Cantoreanu** 1965d (Romania); 1971c (Romania). **Della Giustina** 1983a (France). **Dlabola** 1957a (Turkey); 1958b (Caucasus); 1961b (Siberia, Uzbekistan); 1964c (Albania); 1965c, 1966a, 1967b, 1967c, 1967d, 1968a, 1968b (Mongolia); 1970a (Palaearctic); 1970a (Mongolia); 1980c (review of Palaearctic species; n. subgenus; zoogeography). **Emeljanov** 1964c (European USSR; key to species); 1966a (new subgenera; n. gen. syn.; key to subgenera); 1977a (Mongolia; n. sp. syn.). **Hamilton, K. G. A.** 1972d (Canada: Manitoba); 1975b (APHRODINAE, DELTOCEPHALINA, DELTOCEPHALINA); 1983b (notes on Nearctic and Palaearctic species). **Ishihara** 1966a (Kurile Islands). **Kalkandelen** 1974a (Turkey; key to subgenera and species); 1980a (description, female of *D. bekiri*). **Knight, W. J.** 1974c (n. gen. syn.; n. sp. syn.; n. status of specific names; key to subgenera; key to species). **Lee, C. E.** 1979a (Korea, redescribed). **Lee, C. E. & Kwon** 1977b (Korea). **LeQuesne** 1969c (Britain). **Logvinenko** 1957b, 1984a (USSR: Ukraine). **Mitjaev** 1968a (eastern Kazakhstan); 1971a (Kazakhstan; key to 8 species). **Nast** 1972a (catalogued); 1973a (Poland); 1976a (Poland: Pieniny Mts.); 1976b (Poland; catalogued); 1977a (identity of *D. sudeticus*). **Olmi** 1976a (morphological variation). **Ossiannilsson** 1974a (Norway); 1983a (Fennoscandia & Denmark). **Ribaut** 1959a (France). **Ross & Hamilton** 1970a (suggested phylogeny). **Schulz** 1976a (Norway). **Servadei** 1958a, 1968a, 1969a, 1971a, 1972a (Italy). **Vilbaste** 1959b (Estonia); 1962b (eastern part of the Caspian lowlands); 1965b (Altai Mts.); 1968a (coastal regions of the USSR); 1969a (notes on Matsumura's material); 1973a (notes on Flor's material); 1976a (transfer of species from *Deltocephalus* noted); 1980b (Tuva). **Ylonen & Raatikainen** 1984a (variation in male genital structures).

Distantasca
New species: **Dworakowska** 1972i.
Records: **Dworakowska** 1972i (n. comb. from *Empoasca*: n. sp. syn.). **Sharma, B.** 1977b (India: Jammu).

Distantessus
Record: **Evans, F.** 1981a (n. comb. from *Tartessus*).

Distantia
New species: **Linnavuori** 1961, 1969a. **Linnavuori & Al-Ne'amy** 1983a.
Records: **Campbell, M. B. S. C. & Davies** 1985a (species redescribed). **Heller & Linnavuori** 1968a (Ethiopia: Awash). **Linnavuori & Al-Ne'amy** 1983a (n. comb. from *Selenocephalus*; key to species). **Linnavuori & Quartau** 1975a (n. comb. from *Siva* to "*Distantiella*," error for *Distantia*?). **Theron** 1974b (n. comb. from *Selenocephalus*).

Distomotettix
Records: **Dlabola** 1958b (Caucasus); 1960c (Iran; n. comb. from *Circulifer* and *Thamnotettix*); 1961b (central Asia). **Hamilton, K. G. A.** 1975b (APHRODINAE, DELTOCEPHALINI, PLATYMETOPIINA). **Lindberg** 1960b (Soviet Armenia). **Logvinenko** 1957b (USSR: Ukraine). **Zachvatkin** 1953b (central Russia; Oka district).

Divitiacus
Records: **Evans, J. W.** 1966a (n. sp. syn.). **Viraktamath, S. & Viraktamath, C. A.** (genus redefined).

Divus
Records: **Datta** 1972e (descriptive notes).

Dixianus
Record: **Hamilton, K. G. A.** 1975b (APHRODINAE, DELTOCEPHALINI, PLATYMETOPIINA).

Dlabolaia
New species: **Linnavuori** 1959b.
Record: **Hamilton, K. G. A.** 1975b (APHRODINAE, DELTOCEPHALINI, ATHYSANINA). **Theron** 1980a (= *Citorus*).

Dlabolia
New species: **Logvinenko** 1957b. 1957c (USSR: Ukraine).

Dlabolaiana
New species: **Dworakowska** 1974a.
Record: **Dworakowska** 1979c (Zaire).

Dlabolaracus
Records: **Dlabola** 1972a (Afghanistan). **Hamilton, K. G. A.** 1975b (APHRODINAE, DELTOCEPHALINI, CICADULINA). **Remane** 1961b (n. comb. from *Mocydiopsis*).

Docalidia
New species: **Nielson** 1979b, 1982f, 1982g, 1982h.
Record: **Nielson** 1979b (n. comb. from *Coelidia* and *Jassus*; n. sp. syn.; key to species).

Docotettix
New species: **Dlabola** 1977b.
Records: **Dlabola** 1958a (Cyprus). **Hamilton, K. G. A.** 1975b (APHRODINAE, DELTOCEPHALINI, PLATYMETOPIINA).

Doda
Record: **Nielson** 1975a (removed from COELIDIINAE; unassigned).

Doleranus
New species: **DeLong & Cwikla** 1985a. **Linnavuori** 1959b.
Records: **Hamilton, K. G. A.** 1975b (APHRODINAE, DELTOCEPHALINI, ATHYSANINA). **Linnavuori** 1959b (n. comb. from *Chlorotettix*; n. sp. syn.; key to species).

Dolichopscerus
New species: **Maldonado-Capriles** 1985a.
Record: **Maldonado-Capriles** 1985a (comparative notes).

Doliotettix
Records: **Anufriev** 1978a (USSR: Maritime Territory; in key to genera of ATHYSANINI). **Beirne** 1956a (Alaska). **Emeljanov** 1964c (European USSR). **Hamilton, K. G. A.** 1975b (APHRODINAE, DELTOCEPHALINI, ATHYSANINA); 1983b (transboreal taxon). **Lauterer** 1984a (Czechoslovakia). **Nast** 1972a (catalogued); 1976b (Poland; catalogued). **Ossiannilsson** 1974a (Norway); 1983a (Fennoscandia & Denmark). **Schulz** 1976a (Norway). **Vilbaste** 1965b (Altai Mts.); 1969b (Taimyr); 1979a (Vooremaa hardwood/spruce forest); 1980a (Kamchatka); 1980b (Tuva).

Dolyobius
New species: **Linnavuori** 1959b.

Domelia
New species: **Ahmed, Manzoor & Waheed** 1971a.
Record: **Ahmed, Manzoor** 1985c (Pakistan). **Dworakowska** 1978b (=*Eupteryx*, subgenus *Stacla*); 1982a (=*Eupteryx* (*Stacla*)).

Donleva
New species: **Blocker** 1979a, 1982a.
Records: **Blocker** 1979a (n. comb. from *Batracomorphus*; in key to genera of Western Hemisphere IASSINAE).

Dorada
Record: **Linnavuori** 1959b (not DELTOCEPHALINAE, probably GYPONINAE).

Doratulina
New species: **Kwon & Lee** 1979e. **Viraktamath, C. A.** 1976a.
Records: **Choe** 1985a (Korea). **Dlabola** 1981a (Iran). **Dubovskij** 1978a (Uzbekistan, Zarafshan Valley). **Emelvajov** 1968b (=*Stirellus*). **Hamilton, K. G. A.** 1975b (APHRODINAE, STIRELLINI). **Kartal** 1982a (Turkey). **Korolevskaya** 1978b (central Asia). **Kwon & Lee** 1979e (Korea; key to species). **Lee, C. E.** 1979a (Korea, redescribed). **Lee, C. E. & Kwon** 1977b (Korea). **Servadei** 1968a, 1969a (Italy). **Vilbaste** 1965a (generic synonymy and list of species). **Viraktamath, C. A.** 1976a (new subgenus proposed). **Viraktamath, S. & Viraktamath, C. A.** 1980a (type species redescribed).

Doratura
New species: **Dlabola** 1959c, 1961b, 1967a, 1981a. **D'Urso** 1983a. **Dworakowska** 1967a. **Emeljanov** 1966a. **Logvinenko** 1961a, 1975a. **Mitjaev** 1971a. **Vilbaste** 1961b.
Records: **Anufriev** 1978a (USSR: Maritime Territory; key to species). **Beirne** 1956a (Canada; adventive). **Cantoreanu** 1963a, 1965d (Romania); 1971c (Romania). **Dlabola** 1956b (Europe; n. sp. syn.); 1957a (Turkey); 1958b (Caucasus); 1961b (central Asia); 1964c (Albania); 1965c, 1967b, 1967d, 1968a, 1968b, 1970b (Mongolia); 1981a (Iran). **Dubvoskij** 1966a (Uzbekistan; key to species); 1978a (Uzbekistan, Zarafshan Valley). **D'Urso** 1980b (Eurasia). **Dworakowska** 1968c (key to 7 Polish species); 1969a (Mongolia); 1973d (n. sp. syn.). **Emeljanov** 1964c (European USSR; key to species); 1977a (Mongolia). **Hamilton, K. G. A.** 1972d (Canada: Manitoba); 1975b (APHRODINAE, APHRODINI, DORATURINA); 1983b (transboreal taxon). **Jankovic** 1966a (Serbia). **Jasinska** 1980a (Poland: Bledowska Wilderness). **Kalkandelen** 1974a (key to Turkish species). **Karaman** 1966a (Macedonia). **Korolevskaya** 1978b (central Asia). **Lauterer** 1957a (Czechoslovakia); 1984a (Moravia, Slovakia). **Lee, C. E.** 1979a (Korea, redescribed). **LeQuesne** 1964a (n. sp. syn.); 1969c (Britain). **Lindberg** 1960a (Portugal); 1960b (Soviet Armenia); 1962a (Portugal). **Linnavuori** 1959a (Mt. Sibellini); 1959c (Mt. Picentini); 1962a (Israel); 1965a (Mediterranean). **Lodos & Kalkandelen** 1985c (Turkey). **Logvinenko** 1957b (USSR: Ukraine); 1961a (synopsis). **Mitjaev** 1968a (eastern Kazakhstan); 1971a (Kazakhstan; key to 7 species-level taxa). **Nast** 1955a (Poland); 1972a (catalogued); 1976a (Poland: Pieniny Mts.); 1976b (Poland; catalogued). **Ossiannilsson** 1983a (key to species of Fennoscandia & Denmark). **Quartau & Rodrigues** 1969a (Portugal). **Servadei** 1957a, 1960a, 1968a, 1969a, 1971a, 1972a (Italy). **Shulz** 1976a (Norway). **Vilbaste** 1962b (eastern part of the Caspian lowlands); 1964a (Auweisen Estlands); 1965b (Altai Mts.); 1968a (coastal regions of the USSR); 1969b (Estonia); 1973a (notes on Flor's material); 1980b (Tuva).

Doraturopsis
Records: **Dlabola** 1981a (Iran). **Dubovskij** 1966a (Uzbekistan); 1978a (Uzbekistan, Zarafshan Valley). **Emeljanov** 1964c (European USSR); 1977a (Mongolia). **Hamilton, K. G. A.** 1975b (APHRODINAE, APHRODINI, DORATURINA). **Kalkandelen** 1974a (Turkey). **Korolevskaya** 1978b (central Asia). **Lodos & Kalkandelen** 1985c (Turkey). **Logvinenko** 1962e (USSR: Ukraine). **Mitjaev** 1968b (eastern Kazakhstan); 1971a (Kazakhstan; in key to genera of DELTOCEPHALINAE; key to species). **Vilbaste** 1980b (Tuva).

Dorrotartessus
New species: **Evans, F.** 1981a.

Dorycara
New species: **Oman** 1985a.
Records: **Emeljanov** 1966a (in key to genera of DORYCEPHALINAE). **Oman** 1985a (genus redefined; key to Nearctic genera of DORYCEPHALINAE; n. comb. from *Dorycephalus*; key to species).

Dorycephalus
New species: **Emeljanov** 1964d.
Records: **Beirne** 1956a (Canada). **Dlabola** 1965c, 1967b, 1967c, 1968b, 1970b (Mongolia). **Emeljanov** 1966a (in key to genera of DORYCEPHALINAE); 1977a (Mongolia). **Hamilton, K. G. A.** 1972d (Canada: Manitoba). **Lodos & Kalkandelen** 1981a (Turkey). **Logvinenko** 1962e (USSR: Ukraine). **Mitjaev** 1968b (eastern Kazakhstan); 1971a (Kazakhstan; in key to genera of DORYCEPHALINAE). **Oman** 1985a (n. comb. to *Dorycara*). **Vilbaste** 1962b (eastern part of the Caspian lowlands); 1980b (Tuva).

Dorycnia
New species: **Dworakowska** 1979b.
Record: **Dworakowska** 1972b (n. comb. from *Typhlocyba*).

Dorydiella
Record: **Hamilton, K. G. A.** 1975b (APHRODINAE, DELTOCEPHALINI, PLATYMETOPIINA).

Draberiella
Record: **Sharma, B. & Malhotra** 1981a (n. gen. syn.).

Drabescoides
Record: **Kwon & Lee** 1979c (n. comb. from *Drabescus* and *Kutara*; n. sp. syn.). **Lee, C. E.** 1979a (Korea, redescribed).

Drabescus
New species: **Anufriev** 1971a. **Evans, J. W.** 1966a, 1972a. **Kuoh, C. L.** 1985a. **Kwon & Lee** 1979c. **Linnavuori** 1960a. **Vilbaste** 1968a.
Records: **Anufriev** 1971a (Soviet Far East; key to species); 1977b (Kurile Islands); 1978a (USSR: Maritime Territory; key to species); 1979a (n. comb. from *Selenocephalus*; n. sp. syn.). **Chang, Y. D. & Choe** 1982a (Korea). **Choe** 1985a (Korea). **Evans, J. W.** 1972a (descriptive notes). **Hamilton, K. G. A.** 1975b (APHRODINAE, SELENOCEPHALINI). **Ishihara** 1958a (Japan: north Honshu). **Lee, C. E. & Kwon** 1977b (Korea). **Linnavuori** 1960a (Micronesia: Kusaie, Ponape; key to species); 1969a (Congo); 1978a (descriptive notes). **Linnavuori & Al-Ne'amy** 1983a (to SELENOCEPHALINAE, DRABESCINI). **Vilbaste** 1968a (coastal regions of the USSR).

Draeculacephala
New species: **DeLong** 1967c. **Hamilton, K. G. A.** 1967a, 1985a. **Young & Davidson** 1959a.
Records: **Beirne** 1956a (Canada; key to species). **Hamilton, K. G. A.** 1972d (Canada: Manitoba); 1985a (genus redefined; key to species). **Napompeth & Nishida** 1971a (Hawaii; diagnostic characters). **Nielson** 1968b (characterization of vector species). **Schroder** 1959a (diagnosis). **Young** 1977a (n. sp. syn.). **Young & Davidson** 1959a (key to species).

Dragonana
New species: **DeLong & Freytag** 1964c.

Records: **DeLong & Freytag** 1964c (key to species); 1972a (in key to genera of GYPONINAE).

Drakensbergena
New species: **Linnavuori** 1961a, 1979a.
Record: **Linnavuori** 1979a (key to species).

Dremuela
New species: **Evans, J. W.** 1966a.

Drionia
Record: **Hamilton, K. G. A.** 1975b (APHRODINAE, DELTOCEPHALINI, CHLORHININA).

Driotura
Records: **Beirne** 1956a (Canada). **Dlabola** 1960a (notes on Horvath's material). **Hamilton, K. G. A.** 1972d (Canada: Manitoba); 1975b (APHRODINAE, APHRODINI, DORATURINA). **Logvinenko** 1957b (USSR: Ukraine). **Scudder** 1961a (Canada: British Columbia).

Drordana
New species: **Nielson** 1983e.
Record: **Nielson** 1983e (n. comb. from *Coelidia*).

Dryadomorpha
New species: **Webb, M. D.** 1981a, 1981b.
Records: **Evans, J. W.** 1966a (synopsis). **Linnavuori** 1960b (Fiji). **Webb, M. D.** 1981a (key to African species); 1981b (n. gen. syn.; n. sp. syn.; n. comb. from *Khamiria*; key to species).

Drylix
Records: **Hamilton, K. G. A.** 1975b (APHRODINAE, DELTOCEPHALINI, ATHYSANINA). **Lee, C. E. & Kwon** 1977b (Korea).

Dryodurgades
New species: **Dlabola** 1957a. **Vilbaste** 1968a. **Wagner** 1963b.
Records: **Anufriev** 1970e (USSR: south of the Primor'ya Territory; n. comb. from *Agallia*; key to species); 1978a (USSR: Maritime Territory; key to species). **Bonfils & Della Giustina** 1978a (Corsica). **Della Giustina** 1983a (France). **Dlabola** 1958a (Yugoslavia); 1964c (Albania); 1981a (Iran). **Emeljanov** 1964c (European USSR). **Jansky** 1984a (Slovakia). **Kartal** 1982a (Turkey). **Lee, C. E.** 1979a (Korea, redescribed). **Lee, C. E. & Kwon** 1977b (Korea). **Lindberg** 1960b (Soviet Armenia). **Linnavuori** 1965a (Mediterranean). **Lodos & Kalkandelen** 1981a (Turkey). **Mitjaev** 1971a (Kazakhstan). **Quartau & Rodrigues** 1969a (Portugal). **Servadei** 1960a, 1968a, 1969a, 1971a. **Vilbaste** 1965b (Altai Mts.); 1968a (coastal regions of the USSR). **Wagner** 1963b (revision of the European taxa; key).

Duanjina
New species: **Kuoh, C. L.** 1981b.

Duatartessus
New species: **Evans, F.** 1981a.
Record: **Evans, F.** 1981a (n. comb. from *Bythoscopus*).

Dudanus
New species: **Dlabola** 1956a. **Emeljanov** 1964d.
Records: **Emeljanov** 1964c (European USSR); 1977a (Mongolia). **Hamilton, K. G. A.** 1975b (APHRODINAE, DELTOCEPHALINI, PLATYMETOPIINA). **Mitjaev** 1971a (Kazakhstan; in key to genera of DELTOCEPHALINAE). **Vilbaste** 1980b (Tuva).

Dumorpha
New species: **DeLong** 1983e. **DeLong & Freytag** 1975e.
Record: **DeLong** 1983e (key to species).

Durgades
New species: **Kameswara Rao & Ramakrishnan** 1978c.
Records: **Dlabola** 1956a (Czechoslovakia). **Datta** 1973c (descriptive notes). **Viraktamath, C. A.** 1973b (n. comb. from *Agallia*).

Durgula
New species: **Emeljanov** 1964f.
Record: **Dubovskij** 1978a (Uzbekistan, Zarafshan Valley). **Mitjaev** 1971a (in key to genera of AGALLIINAE).

Dussana
Record: **Datta** 1972e (descriptive notes).

Duttaella
New species: **Ramakrishnan & Menon** 1971a.

Dwightia
New species: **Linnavuori & Al-Ne'amy** 1983a.

Dworakowskaia
New species: **Chou, I. & Zhang** 1985a.

Dziwneono
New species: **Dworakowska** 1972l.

Ebarrius
New species: **Nast** 1977a.
Records: **Bonfils & Della Giustina** 1978a (Corsica). **Dlabola** 1957a (Turkey); 1958b (Caucasus); 1961b (central Asia); 1964c (Albania); 1967b, 1968a, 1968b (Mongolia). **Dworakowska** 1973d (Mongolia). **Emeljanov** 1964c (European USSR); 1977a (Mongolia). **Hamilton, K. G. A.** 1975b (APHRODINAE, DELTOCEPHALINI, DELTOCEPHALINA). **Jankovic** 1966a (Serbia). **Jasinska** 1980a (Poland: Bledowska Wilderness). **Kalkandelen** 1974a (Turkey). **LeQuesne** 1964a, 1969c (Britain). **Linnavuori** 1959a (Mt. Sibillini). **Logvinenko** 1963a (USSR: Ukraine). **Nast** 1955a (Poland); 1972a (catalogued); 1976a (Poland: Pieniny Mts.); 1976b (Poland; catalogued); 1977a (synopsis). **Ossiannilsson** 1974a (Norway); 1983a (Fennoscandia & Denmark). **Servadei** 1971a, 1972a (Italy). **Theron** 1975a (n. comb. from *Deltocephalus*). **Vilbaste** 1965b (Altai Mts.); 1980a (Kamchatka).

Ectomops
Record: **Morrison** 1973a (= *Glossocratus*, n. syn.).

Ectopiocephalus
Records: **Evans, J. W.** 1966a (n. comb. from *Scaris*; n. sp. syn.; in key to genera of PENTHIMIINAE); 1972a (in key to genera of Australian and New Guinea PENTHIMIINAE).

Ectypus
Record: **Young** 1977a (tribal placement and identity uncertain).

Ederranus
Records: **Anufriev** 1978a (USSR: Maritime Territory; in key to genera of DELTOCEPHALINAE, ATHYSANINI). **Hamilton, K. G. A.** 1975b (APHRODINAE, DELTOCEPHALINI, ATHYSANINA). **Ossiannilsson** 1983a (key to species of Fennoscandia and Denmark). **Vilbaste** 1968a (coastal regions of the USSR; n. sp. syn.); 1980b (Tuva).

Edijassus
New species: **Evans, J. W.** 1972b.
Record: **Knight, W. J.** 1983b (n. comb. to *Batracomorphus*).

Edwardsiana
New species: **Ahmed, Manzoor & Naheed** 1981a. **Anufriev** 1968b, 1975a. **Arzone** 1975a. **Dlabola** 1967a, 1967b, 1967d, 1967f, 1971b, 1974a. **Dubovskij** 1966a. **Dworakowska** 1968d, 1971g. **Jancovic** 1977a and 1978a (duplicate descriptions of the same species). **Kirejtshuk** 1975a, 1977a. **Lauterer** 1958a. **Logvinenko** 1966b. **Mitjaev** 1963b, 1963c, 1968a, 1979b. **Vidano** 1961a, 1961d. **Vilbaste** 1968a, 1980b. **Zachvatkin** 1948a.
Records: **Ahmed, Manzoor** 1985a (Pakistan). **Anufriev** 1975a (northern latitudes); 1977b (Kurile Islands); 1978a (USSR: Maritime Territory). **Beirne** 1956a (Canada; key to species). **Cantoreanu** 1961a, 1963b, 1969b, 1971c (Romania). **Choe** 1985a (Korea). **Claridge, M. F. & Wilson** 1978a (new records for Britain). **Della Giustina** 1983a (France). **Dlabola** 1958a (Czechoslovakia, Yugoslavia); 1958b (Caucasus); 1959a (specific synonymy indicated); 1959b (Austria); 1960c (Iran); 1961b (central Asia); 1961c (southern Europe); 1965e (Cyprus); 1967b, 1967d (Mongolia); 1971a (Iran, Turkey); 1977c (Mediterranean); 1981a (Iran). **Dubovskij** 1966a (Uzbekistan; key to species); 1978a (Uzbekistan, Zarafshan Valley); 1980a (USSR: western Turkmenia). **Dworakowska** 1968d, 1971g (n. comb. from *Typhlocyba*); 1982a (synopsis; n. comb. from *Typhlocyba*; n. sp. syn.). **Emeljanov** 1964c (European USSR); 1977a (Mongolia). **Gunthart, H.** 1971a (Switzerland). **Hamilton, K. G. A.** 1983a (as subgenus of *Typhlocyba*). **Jankovic** 1966a, 1971a (Serbia). **Jasinska** 1980a (Poland: Bledowska Wilderness). **Kirejtshuk** 1977a (USSR: Kharkov region). **Korolevskaya** 1974a, 1979b (Tadzhikistan). **Lauterer** 1958a, 1980a, 1983b, 1984a (Czechoslovakia). **Lee, C. E.** 1979a (Korea, redescribed). **Lee, C. E. & Kwon** 1977b (Korea). **LeQuesne & Payne** 1981a (Britain). **Lodos & Kalkandelen** 1984a (Turkey). **Logvinenko** 1984a (USSR: Ukraine). **Mitjaev** 1963b (Kazakhstan); 1968b (eastern Kazakhstan); 1971a, 1979b (Kazakhstan; keys to species). **Nast** 1972a (catalogued); 1976a (Poland: Pieniny Mts.); 1976b (Poland; catalogued). **Ossiannilsson** 1981a (Fennoscandia & Denmark). **Servadei**

1960a, 1968a, 1969a (Italy). **Trolle** 1968a (Denmark). **Varty** 1967a (Canada). **Viggiani** 1971a (figures; Italy). **Vilbaste** 1964a (Auwiesen Estlands); 1965b (Altai Mts.); 1968a (coastal regions of the USSR); 1973a (notes on Flor's material); 1980a (Kamchatka; sp. syn. discussed); 1980b (Tuva). **Wilson, M. R.** 1978a (descriptions and key characters for 5th instars.

Edwardsiastes
Record: **Hamilton, K. G. A.** 1975b (APHRODINAE, DELTOCEPHALINI, ATHYSANINA).

Egenus
New species: **DeLong & Linnavuori** 1978b. **Linnavuori** 1957a.
Records: **Linnavuori** 1956a (Brazil); 1957a (in key to genera of HECALINAE; redescription); 1959b (DELTOCEPHALINAE, HECALINI).

Egidemia
New species: **Schroder** 1972a. **Young** 1968a.
Records: **Schroder** 1959a (raised to generic rank; n. comb. from *Oncometopia*); 1960b (synopsis). **Young** 1968a (n. gen. syn.; n. comb. from *Oncometopia* and *Proconia*; n. sp. syn.; key to species).

Ehagua
Record: **Young** 1977a (n. comb. from *Cicada* and *Cicadella*; n. sp. syn.; key to species).

Elabra
New species: **Young** 1957c.
Records: **Young** 1957b (identity of Osborn's material); 1957c (synopsis; n. comb. from *Protalebra*; key to species).

Elbelus
New species: **Chou, I. & Ma** 1981a. **Dworakowska** 1972j. **Mahmood** 1967a.
Records: **Dworakowska** 1972j, 1979b (Vietnam). **Mahmood** 1967a (in key to genera of Oriental TYPHLOCYBINAE). **Sharma, B.** 1978a (India: Nagaland, Kohima; summary of geographic distribution). **Sohi & Dworakowska** 1984a (list of Indian species).

Elburzia
New species: **Dlabola** 1974a.
Records: **Dlabola** 1981a, 1984a (Iran). **Hamilton, K. G. A.** 1975b (APHRODINAE; unassigned). **Nast** 1972a (catalogued; DELTOCEPHALINAE).

Eldama
New species: **Dworakowska** 1972b.

Eldarbala
New species: **Young** 1977a.

Elginus
Record: **Theron** 1975a (n. comb. from *Deltocephalus*).

Elphnesopius
Record: **Nast** 1984a (n. comb. from *Neophlepsius* Dubovskij 1966a).

Elrabonia
New species: **Linnavuori** 1959b.

Elymana
New species: **Dworakowska** 1968a. **Hamilton, K. G. A.**, *in* Chiykowski & Hamilton 1985a. **Mitjaev** 1969c. **Vilbaste** 1966a.
Records: **Anufriev** 1977b (Kurile Islands); 1978a (USSR: Maritime Territory). **Anufriev & Emeljanov** 1968a (Soviet Far East). **Beirne** 1956a (Canada; key to species). **Chiykowski & Hamilton** 1985a (n. sp. syn.; key to world species). **Dlabola** 1957a (Turkey); 1965c, 1967b, 1967c, 1967d, 1968a, 1968b, 1970b (Mongolia). **Dubovskij** 1966a (Uzbekistan; key to species). **Dworakowska** 1968a (key to Palaearctic species); 1973d (Korea). **Emeljanov** 1964c (European USSR; key to species); 1966a (n. comb. from *Thammotettix*); 1977a (Mongolia). **Gyllensvard** 1963a, 1969a (Sweden). **Hamilton, K. G. A.** 1972d (Canada: Manitoba); 1975b (APHRODINAE, DELTOCEPHALINI, ATHYSANINA); 1983b (taxon adventive to Nearctic). **Jankovic** 1966a, 1971a (Serbia). **Jasinska** 1980a (Poland: Bledowska Wilderness). **Lauterer** 1984a (Czechoslovakia). **Lee, C. E.** 1979a (Korea, redescribed). **Lee, C. E. & Kwon** 1977b (Korea). **LeQuesne** 1969c (Britain). **Lindberg** 1960a (Portugal). **Linnavuori** 1959c (Mt. Picentini). **Logvinenko** 1957b, 1963a (USSR: Ukraine). **Mitjaev** 1968b (eastern Kazakhstan); 1971a (Kazakhstan; key to 2 species). **Nast** 1972a (catalogued); 1976a (Poland: Pieniny Mts.); 1976b (Poland; catalogued). **Ossiannilsson** 1983a (key to species of Fennoscandia & Denmark). **Schulz** 1976a (Norway). **Servadei** 1972a (Italy). **Vilbaste** 1964a (Auwiesen Estlands); 1965b (Altai Mts.); 1968a (coastal regions of the USSR); 1973a (identity of Flor's material); 1979a (Vooremaa hardwood/spruce forest); 1980b (Tuva). **Wagner** 1967a (affinities with *Cicadula* and *Paluda* noted).

Emelyanoviana
Records: **Anufriev** 1970d (n. comb. from *Typhlocyba*). **Bonfils & Della Giustina** 1978a (Corsica). **Dlabola** 1977c (Mediterranean). **Dworakowska** 1971d (Hungary, Romania, Yugoslavia). **Kirejtshuk** 1977a (USSR: Kharkov region). **LeQuesne & Payne** 1981a (Britain). **Nast** 1972a (catalogued); 1976a (Poland: Pieniny Mts.); 1976b (Poland; catalogued). **Ossiannilsson** 1981a (Fennoscandia & Denmark).

Emeljanovianus
Records: **Hamilton, K. G. A.** 1975b (APHRODINAE, DELTOCEPHALINI, DELTOCEPHALINA). **Vilbaste** 1980b (Tuva; generic rank; n. sp. syn.).

Empoa
New species: **Anufriev** 1979d. **Dworakowska** 1977b.
Records: **Beirne** 1956a (Canada; key to species). **Dworakowska** 1971g (*Empoides* to subgeneric status under *Empoa*; n. comb. from *Edwardsiana*, *Empoides* and *Typhlocyba*; n. sp. syn.); 1973d (Mongolia); 1982a (summary of Asiatic records). **Hamilton, K. G. A.** 1983a (as subgenus of *Typhlocyba*). **Lee, C. E. & Kwon** 1977b (Korea). **Varty** 1967a (Canada).

Empoanara
New species: **Ahmad [Ahmed], Manzoor** 1969b. **Ramakrishnan & Menon** 1972a.
Records: **Ahmed, Manzoor** 1985c (Pakistan; n. comb. to *Alebroides*). **Dworakowska** 1980b (notes on type specimens). **Mahmood** 1967a (n. comb. from *Empoasca* and *Typhlocyba*). **Sohi & Dworakowska** 1984a (list of Indian species).

Empoasca
New species: **Ahmed, Manzoor** 1979a. **Ahmed, Manzoor & Samad** 1972b. **Ahmed, Manzoor, Samad & Naheed** 1981a. **Anufriev** 1969b, 1970g, 1973b. **Bergman** 1956a. **Cunningham & Ross** 1975b. **Datta & Ghosh** 1973b. **DeLong** 1982g. **DeLong & Guevara C.** 1954b. **DeLong & Liles** 1956a. **Dlabola** 1957a, 1957b, 1967a. **Dworakowska** 1968d, 1971f, 1972c, 1972f, 1972g, 1972h, 1973b (new name), 1974a, 1976b, 1977b (new name), 1977e, 1977g, 1977i, 1980b, 1981e, 1981f, 1982b, 1984a. **Dworakowska & Lauterer** 1975a. **Dworakowska & Pawar** 1974a. **Dworakowska & Sohi** 1978b. **Dworakowska & Trolle** 1976a. **Dworakowska & Viraktamath** 1978a, 1979a. **Einyu & Ahmed** 1980a. **Evans, J. W** 1966a. **Eyles & Linnavuori** 1974a. **Ghauri** 1964a, 1965a, 1979a. **Hamilton, K. G. A.** 1982b. **Haque & Parasram** 1973a. **Heller & Linnavuori** 1968a. **Kuoh, C. L.** 1981b. **Langlitz** 1964a. **Lindberg** 1958a, 1960b. **Linnavuori** 1960a, 1960b, 1965a, 1975c. **Linnavuori & DeLong** 1977c. **Logvinenko** 1980a, 1980b, 1981c. **Mahmood, Ahmed & Aslam** 1969a. **Metcalf** 1968a (new name). **Mitjaev** 1963b, 1971a, 1980c. **Naheed & Ahmed** 1980a. **Ross** 1959a, 1959b, 1963a, 1965a. **Ross & Cunningham** 1960a. **Ross & Moore** 1957a. **Ruppel & DeLong** 1956b. **Ruppel & Romero** 1972a. **Sharma, B.** 1984a. **Sohi** 1977a. **Thapa** 1985a. **Theron** 1974a. **Torres** 1960b. **Vilbaste** 1961a, 1965b, 1966a, 1968a. **Webb, M. D.** 1980a. **Young** 1956a.
Records: **Ahmed, Manzoor** 1985c (Pakistan). **Anufriev** 1969b (USSR: Maritime Territory); 1970g (identity of *E. kontkaneni*); 1973b (new subgenus proposed; key to species of the Soviet Maritime Territory); 1977b (Kurile Islands). **Beardsley** 1966a (Hawaii; Nihoa and Necker Islands; Kure Atoll); 1969a, 1969b (Hawaii; Oahu; *E. solana* complex). **Beirne** 1956a (Canada; key to species). **Bianchi** 1955a (Hawaii; Oahu; *E. solana*). **Bonfils & Della Giustana** 1978a (Corsica). **Bonnefil** 1969a (comparative notes). **Cantoreanu** 1959a, 1965d, 1971c (Romania). **Cardosa** 1974a (Portugal). **Cunningham** 1962a (*E. flavescens* in North America); 1967a (migration of *E. fabae*). **Cunningham & Ross** 1965a (specific characters in females); 1965b (keys to species complexes). **Datta** 1973a (descriptive notes). **Davis, B. N. K.**

1983a, 1984a (London, U. K.). **Della Giustina** 1983a (France). **Dlabola** 1956b (southern Europe); 1957a (Turkey); 1957b (Afghanistan); 1958b (Caucasus); 1958c (in key to genera of Palaearctic TYPHLOCYBINAE); 1960c (Iran); 1961b (central Asia); 1961c (Austria); 1964a (Afghanistan); 1964c (Albania); 1965c (Mongolia); 1965d (Jordan); 1966a, 1967b, 1967c, 1968a, 1968b (Mongolia); 1977b (Iran); 1977d (Czechoslovakia); 1977c (Mediterranean). **Dubovskij** 1978a (Uzbekistan, Zarafshan Valley). **Dworakowska** 1968b (Mongolia); 1968d (Korea); 1970h (n. comb. to *Austroasca*); 1971f (n. comb. to *Dayus* & *Austroasca*; n. sp. syn.); 1972c (Zaire); 1972h (China; notes on Matsumura type specimens); 1973b (n. comb. to *Babuoidea*; comments on species identity and possible synonymy); 1974a (Congo); 1976a (*Kybos* as subgenus); 1976b (Oriental and Ethiopian taxa); 1976c (discussion of species groups); 1977b (n. sp. syn.); 1977e (Ethiopian fauna; new subgenera proposed); 1977f (India; n. comb. from *Sabourasca*; n. comb. to *Empoascanara*); 1979a (notes on Singh-Pruthi's type material); 1980b (India); 1982b (new subgenus proposed; n. comb. to *Schizandrasca*; n. sp. syn.). **Dworakowska & Sohi** 1978a, 1978b (India). **Dworakowska & Trolle** 1976a (Kenya). **Emeljanov** 1964c (European USSR; key to species); 1977a (Mongolia). **Evans, J. W.** 1966a (n. gen. syn.; n. comb. from *Dialecticopteryx* and *Austroasca*; n. sp. syn.). **Ghauri** 1963d (distinctive features of major pest species); 1979a (lectotype designation; n. sp. syn.); 1983a (nomenclature). **Gunthart, H.** 1971a (Switzerland). **Gyllensvard** 1965a, 1969a (Sweden). **Hamilton, K. G. A.** 1976a (identity of Provancher species); 1983b (transboreal taxa). **Ishihara** 1961a (Thailand). **Jankovic** 1966a (Serbia). **Jasinska** 1980a (Poland: Bledowska Wilderness). **Kholomuminov & Dlabola** 1979a (USSR: Golodnostep Plain). **Kirejtshuk** 1977a (USSR: Kharkov region). **Korolevskaya** 1974a (Tadzhikistan); 1977a (central Asia). **Langlitz** 1964a (Peru; pest species). **Lauterer** 1980a (Moravia). **Lee, C. E.** 1979a (Korea, redescribed). **Lee, C. E. & Kwon** 1977b (Korea). **LeQuesne** 1961a (*Kybos* spp. of Britain). **LeQuesne & Payne** 1981a (Britain). **Lindberg** 1960a (Portugal); 1960b (Soviet Armenia). **Linnavuori** 1960a (Micronesia; key to species); 1960b (Fiji; n. comb. from *Cicadula*; key to species); 1962a (Israel); 1964 (Egypt); 1965a (Libya); 1975c (n. comb. to *Sundapteryx*; Micronesia). **Lodos** 1982a (Turkey). **Lodos & Kalkandelen** 1983c (Turkey). **Logvinenko** 1957b, 1984a (USSR: Ukraine). **Mahmood** 1967d (comparative notes). **Mitjaev** 1963b (Kazakhstan); 1968a (eastern Kazakhstan); 1971a (Kazakhstan; key to species). **Nast** 1967a (Poland: Pieniny Mts.); 1967b (Poland; catalogued); 1972a (catalogued). **Nielson** 1968b (characterization of vector species). **Okada** 1971a (Japan: Kyushu). **Ossiannilsson** 1956a (species differentiation); 1981a (key to species of Fennoscandia & Denmark). **Quartau & Rodrigues** 1969a (Portugal). **Ramos-Elorduy** 1972c (Mexico). **Ross** 1959a (key to species of the *fabae* complex; phyletic inferences); 1959b (Neotropical members of the *fabae* complex); 1963a (subgenus *Kybos*; key to Nearctic species). **Ross & Cunningham** 1960a (*solana* complex; key to species). **Ross, Decker & Cunningham** 1965a (evolutionary hypothesis and phyletic inferences). **Ross & Moore** 1957a (*fabae* complex; key to species). **Ruppel & DeLong** 1956b (Colombia; pests in highland crops). **Servadei** 1957a, 1960a, 1968a, 1971a (Italy). **Sharma, B.** 1984a (India: J. & K., Kishtwar). **Sohi** 1976a (northern India). **Sohi & Dworakowska** 1984a (list of Indian species). **Thapa** 1985a (subgenera defined). **Theron** 1977a (n. comb. to *Rhinocerotis*). **Vidano** 1985a (key to species on *Vitis* in Italy). **Viggiani** 1971a (Italy). **Vilbaste** 1959b (Estonia); 1962b (eastern part of the Caspian lowlands); 1964a (Auwiesen Estlands); 1965b (Altai Mts.); 1966a (Primorje-Gebiet); 1968a (coastal regions of the USSR); 1973a (notes on Flor's material); 1979a (Vooremaa hardwood/spruce forest); 1980b (Tuva). **Wilson, M. R.** 1978a (descriptions and key characters of fifth instars). **Young** 1957b (redescription of Osborn's Neotropical species; n. comb. from *Erythroneura*). **Zachvatkin** 1953c (Astrakhan: Trans-Volga sand areas); 1953d (central Russia; Oka district).

Empoascanara
New species: **Ahmed, Manzoor** 1979a. **Dworakowska** 1971e, 1972e, 1972k, 1976b, 1977f, 1978a, 1979b, 1979f, 1980b, 1980c, 1980d (new name). **Dworakowska & Pawar** 1974a. **Dworakowska & Trolle** 1976a. **Einyu & Ahmed** 1982a. **Hongsaprug** 1983a. **Thapa** 1984b.
Records: **Ahmed, Manzoor** 1985c (Pakistan). **Anufriev** 1977b (Kurile Islands). **Chang, Y. D. & Choe** 1985a (Korea). **Datta** 1973a (descriptive notes). **Dworakowska** 1971e (n. comb. from *Typhlocyba*); 1972e (n. comb. from *Thamnotettix, Typhlocyba* and *Zygina*); 1977f (n. comb. from *Empoasca* and *Zyginidia*); 1978a (taxonomic features *vis a vis Seriana*); 1979a (n. comb. from *Zygina*; n. sp. syn.); 1979b (new subgenus proposed; n. sp. syn.); 1979f (Asia); 1980c (Australia and the Orient); 1980d (systematics; n. gen. syn; n. comb. from *Zygina*); 1984a (Malaysia & Singapore). **Dworakowska & Trolle** 1976a (Kenya). **Dworakowska & Viraktamath** 1975a (n. comb. from *Erythroneura* and *Zygina*; n. sp. syn.); 1978a (India). **Einyu & Ahmed** 1982a (Uganda; key to species). **Hongsaprug** 1983a (Thailand). **Lee, C. E.** 1979a (Korea, redescribed). **Lee, C. E. & Kwon** (Korea). **Mahmood** 1967a (comparative notes). **Ramakrishnan & Ghauri** 1979b (in key to genera of the *Empoascanara* complex). **Sohi** 1976a (n. comb. from *Zygina*). **Sohi & Dworakowska** 1984a (list of Indian species). **Thapa** 1984b (Nepal). **Vilbaste** 1975a (notes on two of de Motschulsky's species).

Empoides
New species: **Dworakowska** 1970f. **Vilbaste** 1968a.
Records: **Anufriev** 1971a (n. comb. from *Edwardsiana*; n. sp. syn.); 1977b (Kurile Islands); 1978a (USSR: Maritime Territory). **Choe** 1985a (Korea). **Emeljanov** 1977a (Mongolia). **Lee, C. E.** 1979a (Korea, redescribed).

Enantiocephalus
Records: **Bonfils & Della Giustina** 1978a (Corsica). **Cantoreanu** 1959a (Romania). **Della Giustina** 1983a (France). **Dubovskij** 1966a (Uzbekistan; in key to genera of JASSARGINI). **Emeljanov** 1964c (European USSR). **Hamilton, K. G. A.** 1975b (APHRODINAE, DELTOCEPHALINI, DELTOCEPHALINA). **Lauterer** 1957a (Czechoslovakia). **Logvinenko** 1957b (USSR: Ukraine). **Mitjaev** 1968b (eastern Kazakhstan); 1971a (Kazakhstan; in key to genera of DELTOCEPHALINAE). **Vilbaste** 1965b (Altai Mts.); 1980b (Tuva).

Endoxoneura
Record: **Young** 1957b (n. comb. from *Dikraneura*).

Endria
Records: **Beirne** 1956a (Canada). **Emeljanov** 1964c (European USSR). **Hamilton, K. G. A.** 1972d (Canada: Manitoba); 1975b (APHRODINAE, DELTOCEPHALINI, DELTOCEPHALINA). **Kramer** 1967d (n. comb. from *Lonatura*). **Lauterer** 1980a (Moravia; alary polymorphism discussed). **Lee, C. E. & Kwon** 1977b (Korea). **Linnavuori** 1959b (as subgenus of *Amplicephalus*). **Nielson** 1968b (characterization of vector species). **Ossiannilsson** 1983a (Fennoscandia & Denmark). **Remane** 1961a (Palaearctic; adventive). **Vilbaste** 1969a (n. comb. from *Deltocephalus*; n. sp. syn.); 1980a (Kamchatka); 1980b (Tuva; =*Amplicephalus*).

Entogonia
Record: **Young & Nast** 1963a (lectotype designations).

Eohardya
Records: **Dlabola** 1964c (Albania); 1971a (as subgenus of *Hardya*, new status; =*Hardyopsis*, n. syn.); 1977c (Mediterranean); 1981a (Iran). **Hamilton, K. G. A.** 1975b (APHRODINAE, DELTOCEPHALINI, ATHYSANINI). **Kalkandelen** 1974a (Turkey; as subgenus; =*Hardyopsis*). **Linnavuori** 1962a (Israel; as subgenus of *Hardya*; = *Hardyopsis*, n. syn.).

Eovulturnops
Record: **Linnavuori** 1959b (redescribed; DELTOCEPHALINAE, PENTHIMIINI).

Ephelodes
New species: **Emeljanov** 1972d.
Records: **Hamilton, K. G. A.** 1975b (APHRODINAE, DELTOCEPHALINI, PLATYMETOPIINA). **Nast** 1982a (catalogued; DELTOCEPHALINAE).

Ephemerinus
Record: **Hamilton, K. G. A.** (APHRODINAE, DELTOCEPHALINI, ATHYSANINA).

Epiacanthus
New species: **Anufriev** 1976a. **Hori** 1982a.
Records: **Anufriev** 1976a (n. comb. from *Deltocephalus*; key to species); 1977b (Kurile Islands; CICADELLINAE); 1978a (USSR: Maritime Territory; CICADELLINAE, PAGARONIINI). **Ishihara** 1958a (Japan: north Honshu; ERRHOMENELLIDAE); 1966a (Kurile Islands). **Kwon** 1983a (Korea; CICADELLINAE, PAGARONIINI). **Lee, C. E.** 1979a (Korea, redescribed). **Lee, C. E. & Kwon** 1977b (Korea). **Vilbaste** 1968a (coastal regions of the USSR); 1975a (n. comb. from *Deltocephalus*; n. sp. syn.).

Epicephalius
Records: **Hamilton, K. G. A.** 1975b (APHRODINAE, PARADORYDIINI). **Vilbaste** 1976a (as *Epicephalus*; n. gen. syn.; n. sp. syn.).

Epiclines
Record: **Linnavuori** 1959b (to XEROPHLOEINAE).

Epignoma
New species: **Dworakowska** 1972i, 1974a, 1977c. **Einyu & Ahmed** 1980a. **Lauterer** 1973a.
Records: **Dworakowska** 1972i (n. comb. from *Zygina*); 1977c (=*Olokemeja*; reviewed).

Epipsychidion
New species: **Evans, J. W.** 1969a.
Record: **Evans, J. W.** 1966a (synopsis).

Epitettix
Record: **Ishihara** 1967a (redescribed).

Equeefa
Records: **Linnavuori** 1961a (South Africa). **Nielson** 1975a (removed from COELIDIINAE; unassigned). **Theron** 1984c (redescribed; provisionally retained in COELIDIINAE).

Eralba
Record: **Young** 1957c (n. comb. from *Protalebra*).

Erdianus
Record: **Hamilton, K. G. A.** 1975b (APHRODINAE, DELTOCEPHALINI, DELTOCEPHALINA).

Eremochlorita
New species: **Dlabola** 1961b, 1964b. **Mitjaev** 1980b. **Zachvatkin** 1953c.
Records: **Dlabola** 1958b (Caucasus); 1971a (Turkey, Iran); 1972a (Afghanistan); 1981a (Iran). **Dubovskij** 1966a (Uzbekistan; in key to genera of TYPHLOCYBINAE); 1978a (Uzbekistan, Zarafshan Valley); 1980a (USSR: western Turkmenia). **Dworakowska** 1977f (as subgenus of *Chlorita*). **Emeljanov** 1964c (European USSR); 1977a (Mongolia). **Kholomuminov & Dlabola** 1979a (USSR: Golodnostep Plain). **Kirejtshuk** 1975a (as subgenus of *Chlorita*; eastern Ukraine); 1977a (USSR: Kharkov region). **Korolevskaya** 1978a (Tadzhikistan; revised comb., revised sp. syn.). **Lodos & Kalkandelen** 1983c (Turkey). **Mitjaev** 1963b (Kazakhstan; n. sp. syn.); 1968b (eastern Kazakhstan); 1971a (Kazakhstan; in key to genera of TYPHLOCYBINAE); 1980b (systematics; key to species groups and species). **Vilbaste** 1962b (eastern part of the Caspian lowlands).

Eremophlepsius
New species: **Dlabola** 1961b.
Records: **Dubovskij** 1966a (Uzbekistan; in key to genera of EUSCELINAE); 1978a (Uzbekistan, Zarafshan Valley); 1980a (USSR: western Turkmenia). **Emeljanov** 1969b (n. comb. from *Pseudophlepsius*); 1977a (Mongolia). **Hamilton, K. G. A.** 1975b (APHRODINAE, DELTOCEPHALINI, PLATYMETOPIINA). **Kholomuminov & Dlabola** 1979a (USSR: Golodnostep Plain). **Korolevskaya** 1978b (central Asia). **Lindberg** 1960b (Soviet Armenia). **Mitjaev** 1968a (eastern Kazakhstan); 1971a (Kazakhstan; in key to genera of DELTOCEPHALINAE).

Ericotettix
New species: **Lindberg** 1960a.
Records: **Hamilton, K. G. A.** 1975b (APHRODINAE, DELTOCEPHALINI, PLATYMETOPIINA). **Lindberg** 1960a (Portugal). **Quartau & Rodrigues** 1969a (Portugal). **Rodrigues** 1968a (Portugal). **Vilbaste** 1976a (revised sp. syn.; lectotype designation for *E. albovarius*).

Erinwa
New species: **Ghauri** 1975b.

Erotettix
Records: **Cantoreanu** 1968b (Danube river delta). **Emeljanov** 1964c (European USSR). **Hamilton, K. G. A.** 1975b (APHRODINAE, DELTOCEPHALINI, MACROSTELINA). **LeQuesne** 1969c (Britain).

Errastunus
Records: **Anufriev** 1978a (USSR: Maritime Territory; in key to genera of PARALIMINI). **Beirne** 1956a (Canada). **Cantoreanu** 1971c (Romania). **Dlabola** 1961b (central Asia); 1965c, 1967b, 1967c, 1967d, 1968b (Mongolia). **Emeljanov** 1964c (European USSR). **Hamilton, K. G. A.** 1975b (APHRODINAE, DELTOCEPHALINI, DELTOCEPHALINA). **Jasinska** 1980a (Poland: Bledowska Wilderness). **LeQuesne** 1969c (Britain). **Logvinenko** 1957b (USSR: Ukraine). **Mitjaev** 1968b (eastern Kazakhstan); 1971a (Kazakhstan; in key to genera of DELTOCEPHALINAE). **Nast** 1972a (catalogued); 1976a (Poland: Pieniny Mts.); 1976b (Poland; catalogued). **Ossiannilsson** 1983a (Fennoscandia & Denmark). **Remane** 1961d (notes on *E. antennalis*). **Servadei** 1968a, 1971a (Italy). **Vilbaste** 1964a (Auwiesen Estlands); 1965b (Altai Mts.); 1969a (n. sp. syn.); 1969b (Taimyr); 1980a (Kamchatka); 1980b (Tuva).

Errhomenellus
Records: **Cantoreanu** 1971c (Romania). **Dlabola** 1958a (Czechoslovakia). **Servadei** 1968a (Italy). **Young** 1968a, 1977a (not CICADELLINAE).

Errhomenus
Records: **Cobben** 1978a (The Netherlands). **Emeljanov** 1964c (European USSR). **Lauterer** 1957a (Czechoslovakia); 1983b (discussion of possible sp. syn.). **Nast** 1955a (Poland); 1972a (catalogued); 1976a (Poland: Pieniny Mts.); 1976b (Poland; catalogued).

Errhomus
New species: **Bliven** 1957a.
Records: **Beirne** 1956a (Canada: southern British Columbia). **Oman & Musgrave** 1975a (redefined; n. comb. to *Ankosus* and *Bathysmatophorus*).

Eryapus
Records: **Evans, J. W.** 1972b (to ACOSTEMMINAE). **Linnavuori & Al-Ne'amy** 1983a (comparative note).

Erythria
New species: **Dlabola** 1977a, 1984a. **Dlabola & Jankovic** 1981a. **Dworakowska** 1977a, 1979e. **Vidano** 1959b.
Records: **Ahmed, Manzoor** 1985c (Pakistan). **Cantoreanu** 1971c (Romania). **Dlabola** 1958a (Yugoslavia); 1958c (in key to genera of Palaearctic TYPHLOCYBINAE); 1977c (Mediterranean); 1981a (Iran). **Dworakowska** 1971d (*Erythridea* a subgenus; n. comb. from *Erythridea*); 1977a (systematics discussed; n. gen. syn.; n. comb. from *Dio* and *Basuaneura*); 1979a (n. sp. syn.; n. comb. from *Dikraneura*); 1979e (n. sp. syn.). **Emeljanov** 1964c (European USSR; key to species). **Gunthart, H.** 1971a (Switzerland). **Jankovic** 1966a (Serbia). **LeQuesne & Payne** 1981a (Britain). **Lodos & Kalkandelen** 1983b (Turkey). **Logvinenko** 1984a (USSR: Ukraine). **Nast** 1972a (catalogued); 1976a (Poland: Pieniny Mts.); 1976b (Poland; catalogued). **Ossiannilsson** 1981a (Fennoscandia & Denmark). **Schulz** 1976a (Norway). **Servadei** 1968a, 1969a, 1972a (Italy). **Sohi & Dworakowska** 1984a (list of Indian species). **Vidano** 1959b (systematics; in key to DIKRANEURINI; key to species); 1965a (synopsis of European taxa).

Erythridea
New species: **Vidano** 1959b.
Records: **Dworakowska** 1971d (as subgenus of *Erythria*; n. comb. to *Erythria*). **Servadei** 1968a, 1969a (Italy). **Vidano** 1959b (in key to DIKRANEURINI; key to species); 1965a (synopsis of European taxa).

Erythrogonia
New species: **Medler** 1963a. **Young** 1977a.

Records: **Medler** 1963a (redescription; key to species). **Schroder** 1959a (diagnosis). **Young** 1977a (lectotype designations; n. comb. from *Tettigonia*; n. sp. syn.; generic groups defined).

Erythroneura
New species: **Ahmed, Manzoor** 1970a, 1970b, 1971b, 1985c. **Anufriev** 1969b, 1969c, 1971e. **Dlabola** 1957a, 1957b, 1957c, 1961b, 1963a, 1965d, 1968a. **Dworakowska** 1970d, 1972g. **Emeljanov** 1964f. **Gunthart, H.** 1971b. **Heller & Linnavuori** 1968a. **Hepner** 1966a, 1966b, 1966c, 1966d, 1967a, 1967b, 1967c, 1972a, 1972b, 1972c, 1973a, 1975a, 1976a, 1976b, 1976c, 1976d, 1977a, 1977b, 1977c, 1978a. **Ishihara** 1958a, 1965c. **Linnavuori** 1956b, 1960a, 1960b, 1962a, 1964a, 1965a. **Mitjaev** 1969b, 1971a. **Ribaut** 1959b. **Richards & Varty** 1963a. **Ross** 1956a, 1957a. **Samad & Ahmed** 1979b. **Sohi & Kapoor** 1973b, 1974a. **Vilbaste** 1968a.

Records: **Ahmed, Manzoor** 1972a (notes on color similarities); 1983b (remarks on shape of male style); 1985c (n. comb. to *Arboridia*). **Anufriev** 1969b (Soviet Maritime Territory); 1969c (new subgenus proposed); 1971a (n. sp. syn.); 1971e (key to species of subgenus *Punctigerella*). **Beirne** 1956a (Canada; key to subgenera and species). **Cantoreanu** 1964a, 1965a, 1965c, 1965d (Romania). **Dlabola** 1956a (southern Europe); 1957a (Turkey); 1958a (Cyprus); 1958b (Caucasus); 1959b (southern Europe); 1961b (Dagestan, Tadzhikistan); 1961c (Romania); 1964a (Afghanistan); 1964c (Albania); 1965d (Jordan); 1967b, 1967d (Mongolia); 1972a (Afghanistan). **Dworakowska** 1968b (Mongolia); 1970d (new subgenus proposed; n. comb. from *Zygina*; n. sp. syn.); 1970g (n. comb. to *Arboridia*; n. sp. syn.); 1970j (n. comb. to *Zygina*); 1970k (n. comb. to *Hauptidia*). **Emeljanov** 1964c (European USSR; key to species). **Evans, J. W.** 1966a (Australia; synopsis). **Gerard** 1972a (n. comb. to *Zygina*). **Gunthart, H.** 1971a (Switzerland). **Heller & Linnavuori** 1968a (African species). **Ishihara** 1958a (Japan: north Honshu); 1961a (Thailand); 1965c (Formosa). **Jankovic** 1966a, 1971a (Serbia). **Lauterer** 1957a (Czechoslovakia). **Lee, C. E.** 1979a (Korea, redescribed). **Lee, C. E. & Kwon** 1977b (Korea). **LeQuesne** 1964a (Britain). **LeQuesne & Payne** 1981a (Britain). **Lin, K. E.** 1971a (Taiwan). **Lindberg** 1960b (Soviet Armenia); 1961a (Madeira); 1962a (Portugal). **Lindberg & Wagner** 1965a (Canary Islands). **Linnavuori** 1960b (Fiji; n. sp. syn.; key to species); 1961a (South Africa); 1962a (Israel); 1964a (Egypt). **Logvinenko** 1957b (USSR: Ukraine); 1959a (USSR: Transcarpathian region). **Mahmood** 1967a (n. comb. from *Empoasca*). **Mitjaev** 1963b, 1963c (Kazakhstan); 1968b (eastern Kazakhstan); 1971a (Kazakhstan; key to nine species); 1971b (USSR: southern Kazakhstan). **Nast** 1955a (Poland); 1972a (catalogued). **Remane** 1962a (northwest Germany). **Ross** 1958a (phylogeny suggested). **Servadei** 1957a, 1968a, 1971a (Italy). **Sohi** 1976a (n. comb. to *Cassianeura*); 1977a (northwestern India). **Thapa** 1984b (Nepal). **Theron** 1977a (n. comb. to *Accacidia, Molopopterus, Helionidia* and *Rhusia*). **Trolle** 1966a (Denmark). **Varty** 1964a, 1967b (Canada). **Vidano** 1959a (Italy). **Viggiani** 1971a (Italy). **Vilbaste** 1958a (Estonia); 1962b (eastern part of the Caspian lowlands); 1964a (Auwiesen Estlands); 1965b (Altai Mts.); 1968a (coastal regions of the USSR). **Young** 1957b (comment on the distribution of *E. elegantula*).

Erythroneuropsis
New species: **Ramakrishnan & Menon** 1973a.

Record: **Ramakrishnan & Menon** 1974a (India; in key to genera of ERYTHRONEURINI; key to species).

Erzaleus
Records: **Hamilton, K. G. A.** 1975b (APHRODINAE, DELTOCEPHALINI, DELTOCEPHALINA). **Mitjaev** 1968b (eastern Kazakhstan). **Vilbaste** 1965b (Altai Mts.).

Esolanus
Record: **Hamilton, K. G. A.** 1975b (APHRODINAE, DELTOCEPHALINI, ATHYSANINA).

Euacanthella
New species: **Evans, J. W.** 1966a.

Records: **Evans, J. W.** 1966a (n. sp. syn.); 1974a (n. sp. syn.; revised sp. syn.; EUACANTHELLINAE). **Knight, W. J.** 1974b (New Zealand; n. sp. syn.; APHRODINAE).

Euacanthus
Records: **Emeljanov** 1964c (European USSR). **Jankovic** 1971a (Serbia). **Lindberg** 1960b (Soviet Armenia). **Servadei** 1960a (Italy). **Vilbaste** 1973a (notes on Flor's material). **Young** 1963a (lectotype designation). **Zachvatkin** 1953b (central Russia; Oka district).

Eualebra
Record: **Young** 1957b (notes on Osborn's types).

Eugnathodus
Records: **Datta** 1972b, 1973d (descriptive notes). **Hamilton, K. G. A.** 1975b (APHRODINAE, DELTOCEPHALINI, MACROSTELINA).

Eugonalia
Record: **Young** 1977a (= *Iragua*, n. syn.).

Euleimonios
Record: **Evans, J. W.** 1966a (n. gen. syn.; n. sp. syn.).

Eulonus
Record: **Hamilton, K. G. A.** 1975b (APHRODINAE, DELTOCEPHALINI, COCHLORHININA).

Eupelix
Records: **Bonfils & Della Giustina** 1978a (Corsica). **Cantoreanu** 1965d (Romania); 1968b (Danube river delta); 1971c (Romania). **Dlabola** 1957a (Turkey); 1957b (Afghanistan); 1958b (Caucasus); 1961b (central Asia); 1964c (Albania); 1965d (Jordan); 1967b (Mongolia); 1972a (Afghanistan); 1977c (Mediterranean); 1981a (Iran). **Dubovskij** 1966a (Uzbekistan; HECALINAE, HECALINI); 1978a (Uzbekistan, Zarafshan Valley); 1980a (USSR: western Turkmenia). **Dworakowska** 1969a (Mongolia). **Emeljanov** 1964c (European USSR); 1977a (Mongolia). **Gravestein** 1965a (The Netherlands: Terschelling I.). **Hamilton, K. G. A.** 1975b (APHRODINAE, EUPELICINI). **Jankovic** 1966a (Serbia). **Jasinska** 1980a (Poland: Bledowska Wilderness). **Kholomuminov & Dlabola** 1979a (USSR: Golodnostep Plain). **LeQuesne** 1965c (Britain). **Lindberg** 1960a (Portugal); 1960b (Soviet Armenia); 1963a (Morocco). **Linnavuori** 1956d (Spain; EUPELICINAE). **Lodos & Kalkandelen** 1981a (Turkey). **Logvinenko** 1957b (USSR: Ukraine). **Mitjaev** 1971a (Kazakhstan; in key to DORYCEPHALINAE). **Nast** 1972a (catalogued; DORYCEPHALINAE); 1976a (Poland: Pieniny Mts.). **Ossiannilsson** 1974a (Norway); 1981a (Fennoscandia & Denmark). **Quartau & Rodrigues** 1969a (Portugal). **Servadei** 1957a, 1960a, 1968a, 1971a, 1972a (Italy). **Vilbaste** 1959b (Estonia); 1973a (notes on Flor's material); 1980b (Tuva).

Eupenthimia
New species: **Evans, J. W.** 1972a.

Eupterella
Record: **Mahmood** 1967a (in key to genera of TYPHLOCYBINI).

Eupteroidea
Records: **Anufriev** 1969b (Soviet Maritime Territory). **Beirne** 1956a (Canada: southern Ontario; in key to genera of TYPHLOCYBINI). **Dworakowska** 1967b (Mongolia); 1972a (= *Aguriahana*, n. syn.). **Emeljanov** 1964c (European USSR). **Jankovic** 1966a (Serbia). **Kristensen** 1965b (Denmark). **Vilbaste** 1968a (n. comb. from *Eupteryx*; n. gen. syn.; n. sp. syn.).

Eupterycyba
Records: **Jankovic** 1966a (Serbia). **Kirejtshuk** 1977a (as *Euptericyba*; USSR: Kharkov region). **LeQuesne & Payne** 1981a (Britain). **Lodos & Kalkandelen** 1984a (Turkey). **Ossiannilsson** 1981a (Fennoscandia & Denmark). **Wilson, M. R.** 1978a (description and key characters of 5th instars).

Eupteryx
New species: **Ahmed, Manzoor** 1969b. **Christian** 1956a. **Dlabola** 1957a, 1963d, 1965c, 1967a, 1967e, 1974a, 1974c, 1981a. **Dworakowska** 1969e, 1970b, 1971a, 1972d, 1976b, 1978b, 1980b. **Emeljanov** 1964c. **Heller & Linnavuori** 1968a. **Linnavuori** 1962a, 1965a, 1968a, 1969a. **Logvinenko** 1966b, 1967a, 1978a. **Metcalf** 1968a (new name). **Meusnier** 1982a. **Mitjaev** 1971a. **Samad & Ahmed** 1979a. **Sawai Singh** 1969a. **Zachvatkin** 1948b.

Records: **Ahmed, Manzoor** 1985c (Pakistan). **Allen, A. A.** 1965a, 1982a (England). **Anufriev** 1978a (USSR: Maritime Territory; key to species). **Anufriev & Emeljanov** 1968a (Soviet Far East). **Beirne** 1956a (Canada; key to species). **Bonfils & Della Giustina** 1978a (Corsica). **Cantoreanu** 1963b, 1965b, 1971c (Romania). **Christian** 1956a (key to species). **Della Giustina** 1981a (description of male); 1983a (France). **Dlabola** 1956a (Czechoslovakia); 1957a (Turkey); 1957b (Afghanistan); 1958a (southern Europe); 1958b (Caucasus); 1958c (in key to Palaearctic TYPHLOCYBINAE); 1959b (Italy); 1960c (Iran); 1961b (central Asia); 1961c (Romania); 1964a (Afghanistan); 1964c (Albania); 1965d (Jordan); 1967b, 1967c, 1967d, 1970b (Mongolia); 1971a (Afghanistan, Crete, Turkey); 1972a (Afghanistan); 1977c (Mediterranean); 1981a (Iran); 1984a (Canary Islands). **Dubovskij** 1966a (Uzbekistan; key to species); 1978a (Uzbekistan, Zarafshan Valley). **Dumbleton** 1967a (as *Cicadella*; *E. melissae* in New Zealand). **Dworakowska** 1968d (China, Korea); 1970b (synopsis; key to species); 1971 (synopsis of north African taxa); 1972d, 1977b (n. sp. syn.); 1978b (suggested gen. syn.); 1982a (synopsis; n. sp. syn.). **Emeljanov** 1977a (Mongolia). **Evans, J. W.** 1966a (Australia ?). **Flint** 1976a (Britain: Kent). **Gravestein** 1965a (The Netherlands: Terschelling I.). **Gunthart, H.** 1971a (Switzerland). **Gyllensvard** 1969a (Sweden). **Hamilton, K. G. A.** 1983b (taxa adventive to Nearctic). **Hoebeke** 1980a (Nearctic; adventive). **Hoebeke & Wheeler** 1983a (as *Cicadella*; Nearactic; adventive). **Ishihara** 1958a (Japan: north Honshu); 1965c (as *Cicadella*; Formosa). **Jankovic** 1966a, 1971a (Serbia). **Jansky** 1985a (Slovakia). **Kartal** 1983a (Turkey). **Kholomuminov & Dlabola** 1979a (USSR: Golodnostep Plain). **Kirejtshuk** 1977a (USSR: Kharkov region). **Knight, W. J.** 1976a (New Zealand; in key to genera of TYPHLOCYBINAE). **Korolevskaya** 1978a (Tadzhikistan). **Lauterer** 1957a (Czechoslovakia); 1983b (Moravia); 1984a (Moravia, Slovakia). **Lee, C. E.** 1979a (Korea, redescribed). **LeQuesne** 1974a (Britain). **LeQuesne & Payne** 1981a (Britain). **LeQuesne & Woodroffe** 1976a (intraspecific variation in genitalic structures). **Lindberg** 1960a (Portugal); 1961a (Madeira); 1962a (Portugal). **Linnavuori** 1959c (Mt. Picentini); 1962a (Israel); 1964a (Egypt). **Lodos & Kalkandelen** 1984b (Turkey). **Logvinenko** 1957b (USSR: Ukraine). **Mahmood** 1967a (in key to genera of TYPHLOCYBINAE). **Mitjaev** 1963b (Kazakhstan); 1968b (eastern Kazakhstan); 1971a (Kazakhstan; key to species). **Nast** 1955a (Poland); 1972a (catalogued); 1967a (Poland: Pieniny Mts.). **Ossiannilsson** 1974a (Norway); 1981a (Fennoscandia & Denmark; key to species). **Quartau** 1979a, 1980b (Azores). **Quartau & Rodrigues** 1969a (Portugal). **Remane** 1961d (habitat notes); 1962a (northwest Germany). **Schulz** 1967a (Norway). **Servadei** 1957a, 1960a, 1968a, 1971a, 1972a (Italy). **Sharma, B.** 1984a (India: J. & K., Kishtwar). **Sohi** 1977a (northwestern India). **Sohi & Dworakowska** 1984a (list of Indian species). **Stewart** 1981a (nymphal polymorphism). **Stiling** 1980a (color polymorphism in nymphs). **U. S. Department of Agriculture** 1980a (North American records for Old World species). **Vilbaste** 1959b (Estonia); 1964a (Auwiesen Estlands); 1966a (Primorje-Gebiet); 1968a (coastal regions of the USSR); 1973a (notes on Flor's material); 1976a (notes on Matsumura's species described as *Typhlocyba*; n. sp. syn.); 1979a (Vooremaa hardwood/spruce forest); 1980b (Tuva). **Zachvatkin** 1953b (central Russia; Oka district).

Euragallia

New species: **Cwikla & DeLong** 1985a. **Kramer** 1964a, 1976c.

Records: **Davis, R. B.** 1975a (morphology; phyletic inferences). **Kramer** 1964a (in key to New World genera of AGALLINAE). **Linnavuori** 1956a (Brazil, Paraguay, Surinam, Trinidad). **Linnavuori & DeLong** 1979c (Central America & Panama); 1979d (male genitalia described).

Eurhadina

New species: **Ahmed, Manzoor** 1969d. **Anufriev** 1969b. **Dlabola** 1967a. **Dworakowska** 1967b, 1969g, 1971g, 1978b, 1981g. **Linnavuori** 1962a. **Logvinenko** 1978a. **Zachvatkin** 1948a.

Records: **Ahmed, Manzoor** 1985c (Pakistan). **Anufriev** 1977b (Kurile Islands); 1978a (USSR: Maritime Territory). **Claridge, M. F. & Wilson, M. R.** 1978a (Britain). **Dlabola** 1958a (Caucasus); 1959a (sp. syn. noted); 1967b (Mongolia). **Dworakowska** 1967b ((Mongolia; key to species); 1969g (checklist of Palaearctic and Oriental species; *Singhardina* as subgenus, n. status; key to species); 1971g (n. sp. syn.); 1982a (synopsis); 1984a (listed). **Emeljanov** 1964c (European USSR; key to species). **Gunthart, H.** 1971a (Switzerland). **Hamilton, K. G. A.** 1983b (species adventive to Nearctic). **Jankovic** 1966a, 1971a (Serbia). **Jansky** 1984a (Slovakia). **Kartal** 1983a (Turkey). **Kirejtshuk** 1977a (USSR: Kharkov region). **Lee, C. E.** 1979a (Korea, redescribed). **Lee, C. E. & Kwon** 1977b (Korea). **LeQuesne & Payne** 1981a (Britain). **Logvinenko** 1957b (USSR: Ukraine); 1959a (USSR: Transcarpathian region). **Mitjaev** 1968b (eastern Kazakhstan); 1971a (Kazakhstan; in key to genera of TYPHLOCYBINAE). **Nast** 1972a (catalogued); 1976a (Poland: Pieniny Mts.). **Ossiannilsson** 1981a (Fennoscandia & Denmark; key to species). **Sharma, B.** 1977b (India: Jammu); 1984a (India: J. & K., Kishtwar). **Sohi & Dworakowska** 1984a (list of Indian species). **Vilbaste** 1968a (coastal regions of the USSR); 1973a (notes on Flor's material). **Wilson, M. R.** 1978a (description and key characters of 5th instars; key to nymphs). **Woodroffe** 1971b (Britain). **Zachvatkin** 1953b (central Russia; Oka district).

Eurinoscopus

Record: **Knight, W. J.** 1983b (n. comb. to *Batracomorphus*).

Euronirvanella

New species: **Evans, J. W.** 1966a.

Eurymela

New species: **Evans, J. W.** 1969b. **Metcalf** 1965b (new name).

Records: **Davis, R. B.** 1975a (morphology; phyletic inferences). **Evans, J. W.** 1966a (in key to genera of EURYMELINI; revised specific synonymy; key to species). **Stevens, M. M.** 1985a (lectotype designations).

Eurymelella

Record: **Evans, J. W.** 1966a (transferred from EURYMELINI to IPOINI).

Eurymelessa

Records: **Davis, R. B.** 1975a (morphology; phyletic inferences). **Evans, J. W.** 1966a (synopsis).

Eurymelita

Records: **Davis, R. B.** 1975a (morphology; phyletic inferences). **Evans, J. W.** (synopsis; n. sp. syn.).

Eurymeloides

New species: **Evans, J. W.** 1966a.

Records: **Davis, R. B.** (morphology; phyletic inferences). **Evans, J. W.** 1966a (synopsis; n. sp. syn.; key to species).

Eurymelops

Records: **Davis, R. B.** 1975a (morphology; phyletic inferences). **Evans, J. W.** 1966a (synopsis; key to species).

Eurypella

Record: **Evans, J. W.** 1966a (synopsis; n. comb. from *Bakeriola*).

Eusallya

New species: **Evans, J. W.** 1972a.

Eusama

Record: **Hamilton, K. G. A.** 1975b (APHRODINAE, DELTOCEPHALINI, PLATYMETOPIINA).

Euscelidius

New species: **Dlabola** 1957a, 1957b. **Emeljanov** 1962a. **Logvinenko** 1977a.

Records: **Beirne** 1956a (Canada: southern British Columbia). **Bonfils & Della Giustina** 1978a (Corsica). **Cantoreanu** 1971c (Romania). **Cardosa** 1974a (Portugal). **Dlabola** 1957a (Turkey); 1958a (southern Europe); 1958b (Caucasus); 1960c (Iran); 1961b (central Asia); 1965d (Jordan); 1972a (Afghanistan); 1981a (Iran). **Dubovskij** 1966a (Uzbekistan; in key to genera of EUSCELINAE). **Emeljanov** 1964c (European USSR; key to species). **Hamilton, K. G. A.** 1972d (Canada: Manitoba); 1975b (APHRODINAE, DELTOCEPHALINI, ATHYSANINA); 1983b (adventive to Nearctic). **Jankovic** 1966a, 1971a (Serbia). **Kalkandelen** 1974a (Turkey; key to species). **LeQuesne** 1969c (Britain). **Lindberg** 1960a (Portugal); 1960b (Soviet Armenia); 1960c (Azores); 1963a (Morocco). **Linnavuori** 1956d (Spanish Morocco); 1962a (Israel); 1965a (Turkey). **Logvinenko**

1957b (USSR: Ukraine); 1977a (Ukraine; key to species). **Mitjaev** 1968a (eastern Kazakhstan); 1971a (Kazakhstan; key to species). **Nast** 1972a (catalogued); 1976b (Poland; catalogued). **Nielson** 1968b (characterization of vector species). **Ossiannilsson** 1983a (Fennoscandia & Denmark). **Quartau** 1979a (Azores). **Quartau & Rodrigues** 1969a (Portugal). **Servadei** 1968a (Italy). **Vilbaste** 1962b (eastern part of the Caspian lowlands); 1965b (Altai Mts.).

Euscelis
New species: **Dlabola** 1971b. **Emeljanov** 1962a. **Pierce, W. D.** 1963a (fossil). **Strubing** 1980a, 1984a. **Vilbaste** 1980b.

Records: **Anufriev** 1977b (Kurile Islands). **Anufriev & Emeljanov** 1968a (Soviet Far East). **Beirne** 1956a (Canada: southwest British Columbia). **Bonfils & Della Giustina** 1978a (Corsica). **Cantoreanu** 1965d (Romania); 1968b (Danube river delta); 1969b, 1971c (Romania). **Cardosa** 1974a (Portugal). **Della Giustina** 1983a (France). **Dlabola** 1957a (Turkey); 1958b (Caucasus); 1960c (Iran); 1961b (central Asia); 1964c (Albania); 1965d (Jordan); 1967b, 1967c (Mongolia); 1971b (Turkey); 1972a (Afghanistan); 1974d (Mediterranean); 1977a (Iran); 1977c (Mediterranean); 1981a, 1984a (Iran). **Dubovskij** 1966a (Uzbekistan; in key to genera of EUSCELINAE); 1978a (Uzbekistan, Zarafshan Valley); 1980a (USSR: western Turkmenia). **Emeljanov** 1964c (European USSR; key to species); 1977a (Mongolia). **Gravestein** 1965a (The Netherlands: Terschelling I.). **Hamilton, K. G. A.** 1975b (APHRODINAE, DELTOCEPHALINI, ATHYSANINA); 1983b (adventive to Nearctic). **Jansky** 1983a (Slovakia). **Jasinska** 1980a (Poland: Bledowska Wilderness). **Kalkandelen** 1974a (Turkey; key to species). **Lauterer** 1957a (Czechoslovakia). **LeQuesne** 1959a (synonymy of British species). **Lindberg** 1960a (Portugal); 1960b (Soviet Armenia); 1961a (Madeira); 1962a (Portugal). **Linnavuori** 1956d (Spain & Spanish Morocco); 1959c (Mt. Picentini); 1962a (Israel); 1965a (Turkey). **Lodos** 1982a (Turkey). **Logvinenko** 1957b, 1957c (USSR: Ukraine). **Mitjaev** 1971a (Kazakhstan; in key to genera of DELTOCEPHALINAE). **Nast** 1955a (Poland); 1972a (catalogued); 1976a (Poland: Pieniny Mts.); 1976b (Poland; catalogued). **Nielson** 1968b (characterization of vector species). **Ossiannilsson** 1983a (Fennoscandia & Denmark). **Remane** 1967a (phylogeny). **Servadei** 1957a, 1958a, 1960a, 1968a, 1969a, 1972a (Italy). **Theron** 1974b (n. comb. to *Caffrolix, Bonaspeia, Bloemia, Hadroca, Johanus, Capoideus* and *Tzitzikamaia*). **Trolle** 1968a (Denmark). **Vilbaste** 1958a (Estonia); 1965b (Altai Mts.); 1973a (notes on Flor's material); 1980a (Kamchatka); 1980b (Tuva). **Wagner** 1959c (Greece). **Woodroffe** 1968a (Britain).

Eusceloidia
Record: **Linnavuori** 1958b (redescribed; key to species).

Eusora
Record: **Hamilton, K. G. A.** 1975b (APHRODINAE: unassigned).

Eustollia
Record: **Young** 1968a (=*Proconia*).

Eutambourina
Record: **Evans, J. W.** 1966a (=*Pettya*, n. syn.).

Eutandra
New species: **Webb, M. D.** 1983a.
Record: **Webb, M. D.** 1983a (in key to genera of Australian IDIOCERINAE).

Eutartessus
New species: **Evans, F.** 1981a.

Eutettix
New species: **DeLong & Harlan** 1968a.
Records: **Datta** 1973b (descriptive notes). **DeLong & Harlan** 1968a (key to Mexican species). **Hamilton, K. G. A.** 1975b (APHRODINAE, DELTOCEPHALINI, PLATYMETOPIINA). **Linnavuori** 1958b (*E. mimicus* Osborn, *incertae sedis*; Bolivia). **Ramos Elorduy** 1976c (Mexico).

Evacanthus
New species: **Anufriev** 1970c. **Emeljanov** 1966a (as *Euacanthus*). **Hamilton, K. G. A.** 1983b. **Kuoh, C. L.** 1981a, 1981b. **Li, Z. Z.** 1985b.

Records: **Anufriev** 1977b (Kurile Islands); 1978a (USSR: Maritime Territory). **Beirne** 1956a (Canada and Alaska). **Bonfils & Della Giustina** 1978a (Corsica). **Cantoreanu** 1965d, 1971c (Romania). **Datta** 1973b (as *Euacanthus*; descriptive notes). **Datta & Pramanik** 1977a (India: A. P., Subansiri Division). **DeLong** 1966a (Alaska, Muir Inlet). **Dlabola** 1957a (Turkey); 1957b (Afghanistan); 1958b (Caucasus); 1961b (central Asia); 1961c (Romania); 1964c (Albania); 1967b, 1967c, 1967d, 1968a, 1968b, 1970b (Mongolia); 1972a (Afghanistan). **Dubovskij** 1966a (Uzbekistan); 1978a (Uzbekistan, Zarafshan Valley); 1980a (USSR: western Turkmenia). **Hamilton, K. G. A.** 1972d (Canada: Manitoba); 1983b (taxa adventive to Nearctic). **Ishihara** 1958a (Japan: north Honshu); 1965c (Formosa); 1966a (Kurile Islands). **Jankovic** 1966a (Serbia). **Kwon** 1983a (Korea; CICADELLINAE, EVACANTHINI; key to Korean species). **Lee, C. E.** 1979a (Korea, redescribed). **Lee, C. E. & Kwon** 1977b (Korea). **LeQuesne** 1965c (Britain). **Lindberg** 1960a (as *Euacanthus*; Portugal). **Linnavuori** 1959a (Mt. Sibillini); 1959c (Mt. Picentini). **Lodos & Kalkandelen** 1983a (Turkey). **Logvinenko** 1957b (USSR: Ukraine). **Mitjaev** 1968b (eastern Kazakhstan); 1971a (Kazakhstan; key to species). **Nast** 1972a (catalogued); 1976a (Poland: Pieniny Mts.); 1976b (Poland; catalogued). **Ossiannilsson** 1974a (Norway); 1981a (Fennoscandia & Denmark; key to species). **Schulz** 1976a (Norway). **Servadei** 1968a, 1969a, 1971a, 1972a (Italy). **Vilbaste** 1959b (Estonia); 1964a (Auwiesen Estlands); 1965b (Altai Mts.); 1968a (coastal regions of the USSR); 1973a (notes on Flor's material); 1979a (Vooremaa hardwood/spruce forest); 1980a (Kamchatka); 1980b (Tuva).

Evanirvana
New species: **Hill, B. G.** 1973a.

Evansiola
New species: **Evans, J. W.** 1957a, 1968a.
Record: **Linnavuori & Delong** 1977c (Chile: Juan Fernandez Islands; EVANSIOLINAE).

Evansioma
New species: **Ahmed, Manzoor** 1969d.
Record: **Ahmed, Manzoor** 1985c (Pakistan).

Evansolidia
New species: **Nielson** 1982e.
Record: **Nielson** 1982e (n. comb from IASSUS; key to species).

Evinus
New species: **Dlabola** 1977b.
Record: **Nast** 1982a (catalogued; HECALINAE).

Excavanus
Record: **Cwikla & Blocker** 1981a (comparative notes).

Excultanus
Records: **Hamilton, K. G. A.** 1975b (APHRODINAE, DELTOCEPHALINI, PLATYMETOPIINA). **Linnavuori** 1959b (generic status; n. comb. from *Phlepsius*; key to Neotropical species). **Nielson** 1968b (characterization of vector species).

Exitianiellus
New species: **Evans, J. W.** 1966a.

Exitianus
New species: **Ghauri** 1972b, 1974b. **Linnavuori** 1959b. **Ross** 1968a. **Villiers** 1956a.
Records: **Beirne** 1956a (Canada). **Bonfils & Della Giustina** 1978a (Corsica). **Cardosa** 1974a (Portugal). **Cheng, Y. J.** (Paraguay). **Dlabola** 1957a (Turkey); 1957b (Afghanistan); 1958a (southern Europe); 1960c (Iran); 1961b (central Asia); 1963c (notes on Haupt's material); 1964a (Afghanistan); 1964b (Sudan); 1964c (Albania); 1965d (Jordan); 1971a (Afghanistan, Turkey, Pakistan); 1972a (Afghanistan); 1977c (Mediterranean); 1981a (Iran). **Dubovskij** 1978a (Uzbekistan, Zarafshan Valley). **Evans, J. W.** 1966a (Australia; n. comb. from *Nephotettix*; n. sp. syn.); 1974a (New Caledonia). **Eyles & Linnavuori** 1974a (Niue Island, Cook Islands). **Hamilton, K. G. A.** 1972d (Canada: Manitoba); 1975b (APHRODINAE, DELTOCEPHALINI, DORATURINA). **Heller & Linnavuori** 1968a (Ethiopia). **Kalkandelen** 1974a (Turkey). **Knight, W. J.** 1976b (Lord Howe, Norfolk & Kermadec Islands); 1982a (Rennell Island). **Kwon & Lee** 1979b (n. sp. syn.). **Lee, C. E.** 1979a (Korea, redescribed). **Lindberg** 1956a

(Niger River floodplain; Dogo, Mopti, Saredena); 1958a (Cape Verde Islands); 1960a (Portugal); 1961a (Madeira); 1963a (Morocco). **Lindberg & Wagner** 1965a (Canary Islands). **Linnavuori** 1956a (Argentina, Brazil, Colombia, Paraguay, Uruguay); 1956c (Phoenix & Danger Islands); 1956d (Spanish Morocco); 1959b (key to Neotropical species); 1960a (Micronesia, widespread; n. comb. from *Mimodrylix*); 1960b (Fiji); 1961a (South Africa; n. comb. from *Athysanus* and *Phrynomorphus*); 1962a (Israel); 1964a (Egypt); 1965a (Tunisia); 1973c (Cuba); 1975c (n. sp. syn.; Micronesia). **Linnavuori & DeLong** 1976a (Peru); 1977c (Chile); 1978d (n. sp. syn.). **Linsley & Usinger** 1966a (Galapagos Islands). **Quartau** 1984a (Azores). **Quartau & Rodrigues** 1969a (Portugal). **Ross** 1968a (n. comb. from *Athysanus, Euscelis* and *Phrynomorphus*; n. comb. to *Hybrasil*; species groups defined; n. sp. syn.). **Servadei** 1960a, 1968a (Italy). **Sohi & Kapoor** 1973b (India: Coimbatore). **Sharma, B.** 1977b (India: Jammu). **Suehiro** 1961a (Hawaii; Kauai). **Theron** 1980a (note on *capicola* Stal); 1982 (South Africa). **Vilbaste** 1975a (placement of *Jassus fusconervosus* confirmed); 1976a (transfer of species from *Athysanus* noted). **Webb, M. D.** 1980a (Aldabra).

Exogonia
Record: **Young** 1977a (n. comb. from *Cardioscarta* and *Tettigonia*; n. sp. syn.; key to species).

Exolidia
Records: **Kramer** 1963b (in key to genera of NEOBALINAE). **Linnavuori** 1959b (redescribed).

Extrusanus
Records: **Beirne** 1956a (Canada). **Hamilton, K. G. A.** 1972d (Canada: Manitoba); 1975b (APHRODINAE, DELTOCEPHALINI, ATHYSANINA).

Ezrana
New species: **Evans, J. W.** 1969a.

Faenius
Record: **Young** 1965c (lectotype designation).

Fagocyba
New species: **Arzone** 1976a. **Lauterer** 1983b.
Records: **Arzone** 1976a (key to species). **Cantoreanu** 1965a (Romania). **Dlabola** 1958b (Caucasus); 1958c (in key to genera of Palaearctic TYPHLOCYBINAE); 1961b (central Asia); 1977c (Mediterranean); 1981a (Iran). **Dworakowska** 1982a (Iran). **Emeljanov** 1964c (European USSR). **Gunthart, H.** 1971a (Switzerland). **Hamilton, K. G. A.** 1983b (adventive to Nearctic). **Jankovic** 1971a (Serbia). **Lauterer** 1983b (new subgenus proposed). **LeQuesne & Payne** 1981a (Britain). **Linnavuori** 1959c (Mt. Picentini). **Nast** 1972a (catalogued); 1976a (Poland: Pieniny Mts.); 1976b (Poland; catalogued). **Ossiannilsson** 1981a (Fennoscandia & Denmark; key to species). **Servadei** 1960a, 1968a (Italy). **Trolle** 1966a (Denmark). **Viggiani** 1971 (Italy). **Vilbaste** 1959b (Estonia). **Wilson, M. R.** (descriptions and key characters of 5th instars).

Faiga
New species: **Dworakowska** 1980b.
Record: **Sohi & Dworakowska** 1984a (list of Indian species).

Falcitettix
New species: **Vilbaste** 1965b, 1980b.
Records: **Anufriev** 1978a (USSR: Maritime Territory; in key to genera of PARALIMNINI). **Hamilton, K. G. A.** 1975b (APHRODINAE, DELTOCEPHALINI, DELTOCEPHALINA). **Mitjaev** 1968b (eastern Kazakhstan). **Vilbaste** 1965b (Altai Mts.); 1968a (coastal regions of the USSR); 1980b (Tuva).

Faltala
New species: **Cheng, Y. J.** 1980a.
Records: **Linnavuori** 1959b (redescribed). **Linnavuori & DeLong** 1977c (in key to genera of the *Faltala* group; ATHYSANINI).

Farynala
New species: **Ahmed, Manzoor & Naheed** 1981a. **Dworakowska** 1970f, 1977i, 1982a. **Sharma, B.** 1977a.
Records: **Ahmed, Manzoor** 1985c (Pakistan). **Dworakowska** 1980b (India: Simla); 1982a (Vietnam). **Sharma, B.** 1984a (India: J. & K., Kishtwar). **Sohi & Dworakowska** 1984a (list of Indian species).

Favintiga
Record: **Webb, M. D.** 1981b (n. comb. from *Parabolopona*; in key to genera of PARABOLOPONINAE).

Fedotartessus
New species: **Evans, F.** 1981a.

Ferganotettix
New species: **Dubovskij** 1966a. **Korolevskaya** 1980a.
Records: **Hamilton, K. G. A.** 1975b (APHRODINAE; unassigned). **Korolevskaya** 1980a (new subgenera proposed; key to subgenera and species).

Ferrariana
Record: **Young** 1977a (n. comb. from *Graphocephala* and *Tettigonia*; key to species).

Ficiana
New species: **Ghauri** 1963e.
Records: **Mahmood** 1967a (in key to genera of TYPHLOCYBINI). **Sohi & Dworakowska** 1984a (list of Indian species).

Ficocyba
Records: **Bonfils & Della Giustina** 1978a (Corsica). **Dworakowska** 1982a (Israel). **Lodos & Kalkandelen** 1984a (Turkey).

Fieberiella
New species: **Dlabola** 1965a, 1984a. **Emeljanov** 1964c. **Linnavuori** 1962a. **Wagner** 1963a, 1963b.
Records: **Beirne** 1956a (Canada: southern Ontario; adventive). **Dlabola** 1957a (Turkey); 1958b (Caucasus); 1961b (central Asia); 1964c (Albania); 1965a (key to species); 1965d (Lebanon; FIEBERIELLINI); 1971b (Turkey); 1981a (Iran). **Dubovskij** 1966a (Uzbekistan; FIEBERIELLINI); 1978a (Uzbekistan, Zarafshan Valley). **Emeljanov** 1964c (European USSR; key to species). **Hamilton, K. G. A.** 1975b (APHRODINAE, DELTOCEPHALINI, PLATYMETOPIINA); 1983b (adventive to the Nearctic). **Jankovic** 1971a (Serbia). **Kalkandelen** 1974a (Turkey). **Korolevskaya** 1978b (central Asia). **Linnavuori** 1959c (Mt. Picentini); 1965a (Turkey). **Linnavuori & Al-Ne'amy** 1983a (remarks about the tribe FIEBERIELLINI). **Lodos** 1982a (Turkey). **Logvinenko** 1957b (USSR: Ukraine). **Mitjaev** 1971a (Kazakhstan; in key to genera of DELTOCEPHALINAE). **Nast** 1955a (Poland); 1972a (catalogued); 1973b (Poland); 1976b (Poland; catalogued). **Nielson** 1968b (characterization of vector species). **Scudder** 1961a (Canada: British Columbia). **Servadei** 1960a, 1968a, 1969a, 1971a. **Wagner** 1963b (revision of European taxa).

Fitchana
Record: **Hamilton, K. G. A.** 1975b (APHRODINAE, DELTOCEPHALINI, PLATYMETOPIINA).

Flammigeroidia
New species: **Dlabola** 1968a.
Records: **Dlabola** 1958c (in key to genera of Palaearctic TYPHLOCYBINAE); 1970b (Mongolia); 1971a (Iran, Turkey); 1972a (Afghanistan). **Gunthart, H.** 1979a (n. sp. syn.; experimental evidence). **Vilbaste** 1968a (coastal regions of the USSR); 1973a (notes on Flor's material); 1976a (identity of Matsumura's species described as *Typhlocyba*).

Flavitartessus
Record: **Evans, F.** 1981a (n. comb from *Bythoscopus*).

Flexamia
New species: **Hamilton, K. G. A. & Ross** 1975a. **Ross & Cooley** 1969a. **Young & Beirne** 1958a.
Records: **Beirne** 1956a (Canada; key to species). **Hamilton, K. G. A.** 1972d (Canada: Manitoba); 1975b (APHRODINAE, DELTOCEPHALINI, DELTOCEPHALINA). **Linnavuori** 1959a (=*Acurhinus*). **Young & Beirne** 1958a (key to species).

Flexana
New species: **DeLong & Freytag** 1971a.
Record: **DeLong & Freytag** 1972a (in key to genera of GYPONINAE).

Flexocerus
New species: **Kuoh, C. L. & Fang** 1985a.

Floridonus
Record: **Hamilton, K. G. A.** 1975b (APHRODINAE, DELTOCEPHALINI, PLATYMETOPIINA).

Folicana
New species: **DeLong & Foster** 1982b. **DeLong & Freytag** 1972a, 1972b. **Freytag** 1979a.
Records: **DeLong & Freytag** 1972a (in key to genera of GYPONINAE); 1972b (key to species).

Fonsecaiulus
New species: **Young** 1977a.
Record: **Young** 1977a (n. comb. from *Cicadella, Entogonia* and *Tettigonia*; key to species).

Forcipata
New species: **Anufriev** 1969b. **Vidano** 1965b.
Records: **Anufriev** 1977b (Kurile Islands); 1978a (USSR: Maritime Territory). **Beirne** 1956a (Canada; as syn. of *Dicranoneura*). **Della Giustina** 1983a (France). **Dlabola** 1970b (Mongolia). **D'Urso** 1980b (Sicily). **Dworakowska** 1968d, 1972k (Korea). **Emeljanov** 1977a (Mongolia). **Jasinska** 1980a (Poland: Bledowska Wilderness). **Kirejtshuk** 1977a (USSR: Kharkov region). **Kocak** 1983a (comments about date validated). **Lee, C. E.** 1979a (Korea, redescribed). **LeQuesne & Payne** 1981a (Britain). **Nast** 1972a (catalogued); 1976a (Poland: Pieniny Mts.); 1976b (Poland; catalogued). **Ossiannilsson** 1974a (Norway); 1981a (Fennoscandia & Denmark; key to species). **Schulz** 1976a (Norway). **Servadei** 1968a (Italy). **Vidano** 1965a, 1965b (Palaearctic taxa; key to species). **Vilbaste** 1968a (coastal regions of the USSR); 1969b (Taimyr); 1973a (notes on Flor's material); 1979a (Vooremaa hardwood/spruce forest); 1980a (Kamchatka); 1980b (Tuva)

Formotettigella
New species: **Ishihara** 1965c.

Foroa
New species: **Linnavuori** 1977a.

Foso
New species: **Linnavuori & Al-Ne'amy** 1983a
Record: **Linnavuori & Al-Ne'amy** 1983a (n. comb. from *Selenocephalus*).

Frequenamia
New species: **DeLong** 1982a, 1984g. **Kramer & DeLong** 1968a. **Linnavuori & DeLong** 1978b.
Records: **Cwikla & Blocker** 1981a (comparative note). **Linnavuori & DeLong** 1978b (n. gen. syn.; n. comb. from *Bahita, Exobahita, Penebahita*); 1978d (n. comb. from *Bahita*).

Freytagana
New species: **DeLong** 1975b.

Fridonus
Record: **Hamilton, K. G. A.** 1975b (APHRODINAE, DELTOCEPHALINI, PLATYMETOPIINA).

Frigartus
Records: **Beirne** 1956a (Canada). **Hamilton, K. G. A.** 1972d (Canada: Manitoba); 1975b (APHRODINAE, DELTOCEPHALINI, ATHYSANINA).

Friscananus
Record: **Hamilton, K. G. A.** 1975b (APHRODINAE, DELTOCEPHALINI, PLATYMETOPIINA).

Friscanus
Records: **Nielson** 1968b (characterization of vector species). **Oman & Musgrave** 1975a (redefined).

Frutioidia
New species: **Dworakowska** 1971b, 1976b, 1977f, 1979b, 1980b. **Korolevskaya** 1978a (Tadzhikistan). **Logvinenko** 1981c. **Webb, M. D.** 1980a.

Fulgora
Record: **Young** 1965d (lectotype designation).

Fulvanus
Record: **Linnavuori** 1959b (=*Atanus*, n. syn.; n. sp. syn.).

Furcatartessus
New species: **Evans, F.** 1981a.

Fusanus
New species: **Cheng, Y. J.** 1980a. **Linnavuori** 1959b.
Records: **Linnavuori** 1956a (Argentina); 1959b (key to species).

Fusigonalia
New species: **Young** 1977a.
Record: **Young** 1977a (n. comb. from *Tettigonia*).

Fusiplata
New species: **Ahmed, Manzoor** 1969a, 1985c.
Records: **Ahmed, Manzoor** 1969a (key to species); 1985c (Pakistan).

Futasujinoidella
New species: **Kwon & Lee** 1979c.
Records: **Lee, C.E.** 1979a (Korea, redescribed). **Nast** 1982a (catalogued; DELTOCEPHALINAE).

Futasujinus
New species: **Dlabola** 1967a. **Emeljanov** 1966a. **Kwon & Lee** 1979c. **Vilbaste** 1966a.
Records: **Anufriev** 1977b (Kurile Islands); 1978a (USSR: Maritime Territory). **Dlabola** 1967b (Mongolia). **Dworakowska** 1973d (Korea). **Emeljanov** 1966a (n. comb. from *Deltocephalus*; n. sp. syn.); 1977a (Mongolia). **Hamilton, K. G. A.** 1975b (APHRODINAE, DELTOCEPHALINI, DELTOCEPHALINA). **Ishihara** 1958a (Japan: north Honshu). **Lee, C. E.** 1979a (Korea, redescribed). **Lee, C. E. & Kwon** 1977b (Korea). **Vilbaste** 1967a (east Asiatic region; key to males); 1968a (coastal regions of the USSR).

Galboa
New species: **Hamilton, K. G. A.** 1980b (comparative notes).

Galerius
Records: **Datta** 1973a (descriptive notes). **Vilbaste** 1965a (=*Doratulina*, n. syn).

Gambialoa
New species: **Ahmed, Manzoor** 1979a. **Dworakowska** 1974a, 1979a, 1979b, 1980b, 1981e, 1981h. **Dworakowska & Trolle** 1976a. **Einyu & Ahmed** 1982a.
Records: **Ahmed, Manzoor** 1985c (Pakistan). **Dworakowska** 1972f (n. comb. from *Erythroneura*); 1974a (new subgenus proposed); 1979a (n. comb. from *Zygina* and *Zyginopsis*; n. sp. syn.). **Dworakowska & Trolle** 1976a (n. comb. from *Zygina*). **Einyu & Ahmed** 1982a (key to species). **Sohi & Dworakowska** 1984a (list of Indian species).

Gannachris
New species: **Theron** 1979a.
Records: **Al-Ne'amy & Linnavuori** 1982a (n. comb. from *Agallia*; transferred to AGALLIINAE). **Theron** 1979a (ADELUNGIINAE, ACHRINI).

Gannia
New species: **Linnavuori & Al-Ne'amy** 1983a. **Theron** 1979a.
Records: **Linnavuori & Al-Ne'amy** 1983a (key to species). **Theron** 1979a (EUSCELINAE).

Garapita
New species: **Linnavuori** 1959b.
Record: **Linnavuori** 1959b (new subgenus proposed; key to subgenera and species).

Gargaropsis
New species: **Blocker** 1975a.
Records: **Blocker** 1975a (key to species); 1979a (new subgenus proposed; in key to genera). **Kramer** 1963a (in key to genera).

Garlica
New species: **Blocker** 1976a, 1979a, 1982a.
Record: **Blocker** 1979a (in key to genera).

Gcaleka
Records: **Linnavuori** 1961a (South Africa: Cape Province); 1975a (EUSCELINI). **Theron** 1972a (resdescribed; referred to EUSCELINI).

Gehundra
New species: **Blocker** 1976a.
Record: **Blokjer** 1979a (n. comb. from *Bythoscopus* and *Stragania*).

Geitogonalia
Record: **Young** 1977a (n. comb. from *Cardioscarta* and *Tettigonia*; n. sp. syn.).

Gelidanus
Record: **Hamilton, K. G. A.** 1975b (APHRODINAE, DELTOCEPHALINI, DELTOCEPHALINA).

Genatra
New species: **Nielson** 1983h.

Germaria
Records: **Young** 1965c (lectotype designation); 1968a (=*Proconia*).

Ghauriana
New species: **Thapa** 1985a.

Gicrantus
Record: **Nielson** 1982e (n. comb. from *Coelidia*).

Giffardia
Record: **Evans, J. W.** 1966a (synopsis; DELTOCEPHALINAE, PLATYMETOPIINI).

Gillettiella
Records: **Emeljanov** 1968b (=*Stirellus*). **Hamilton, K. G. A.** 1975b (APHRODINAE, STIRELLINI). **Vilbaste** 1965a (comments *vis a vis Stirellus*).

Gindara
New species: **Dworakowska** 1980b.
Record: **Sohi & Dworakowska** 1984a (list of Indian species).

Giprus
New species: **Sawbridge** 1975a.
Records: **Hamilton, K. G. A.** 1975b (APHRODINAE, DELTOCEPHALINI, DELTOCEPHALINA). **Sawbridge** 1975a (DELTOCEPHALINAE; key to species).

Gladionura
Records: **Hamilton, K. G. A.** 1975b (APHRODINAE, DELTOCEPHALINI, DORATURNINA).

Gloridonus
Record: **Hamilton, K. G. A.** 1975b (APHRODINAE: unassigned).

Glossocratus
New species: **Kwon & Lee** 1979a. **Linnavuori** 1975a. **Morrison** 1973a.
Records: **Anufriev** 1978a (USSR: Maritime Territory; in key to genera of HECALINAE). **Dlabola** 1966a, 1967b, 1967c, 1968a, 1968b (Mongolia); 1972a (Afghanistan). **Emeljanov** 1964c (European USSR); 1977a (Mongolia). **Hamilton, K. G. A.** 1975b (APHRODINAE, HECALINI). **Kwon & Lee** 1979a (Korea; in key to genera of HECALINI). **Lee, C. E.** 1979a (Korea, redescribed). **Linnavuori** 1961a (South Africa; n. sp. syn.); 1969a (Congo); 1975a (n. gen. syn.; n. comb. from *Hecalus*; n. sp. syn.; key to species of the Ethiopian region). **Mitjaev** 1971a (Kazakhstan). **Morrison** 1973a (n. gen. syn.; n. comb. from *Ectomops*, *Hecalus* and *Ledrotypa*; n. sp. syn.; key to Oriental species). **Vilbaste** 1980b (Tuva).

Gnathodus
Records: **Vilbaste** 1976a (notes on Matsumura's species). **Webb, M. D.** 1980a (Aldabra; revised sp. syn.)

Gnatia
Records: **Evans, J. W.** 1966a (synopsis). **Webb, M. D.** 1983a (=*Austrocerus*, n. syn.; n. comb. to *Austroçerus*).

Gobicuellus
New species: **Dlabola** 1967a, 1968b.
Records: **Emeljanov** 1977a (Mongolia; n. sp. syn.) **Hamilton, K. G. A.** 1975b (APHRODINAE, DELTOCEPHALINI, DELTOCEPHALINA). **Nast** 1972a (catalogued); 1984a (=*Acharista*, n. syn.).

Goblinaja
New species: **Kramer** 1965a.
Record: **Blocker** 1979a (in key to genera).

Goifa
New species: **Dworakowska** 1971i, 1980b.
Record: **Sohi & Dworakowska** 1984a (list of Indian species).

Goldeus
New species: **Dlabola** 1974c, 1977a. **Quartau** 1972a. **Remane & Asche** 1980a.
Records: **Della Giustina** 1983a (France). **Dlabola** 1974c (=*Muleyrechia*, n. syn.; n. comb. from *Muleyrechia*; Portugal, Spain & Spanish Morocco). **Hamilton, K. G. A.** 1975b (APHRODINAE, DELTOCEPHALINI, DELTOCEPHALINA). **Quartau & Rodrigues** 1969a (Portugal).

Goniagnathus
New species: **Dlabola** 1957a, 1961b. **Emeljanov** 1962a. **Linnavuori** 1978d. **Webb, M. D.** 1980a.
Records: **Anufriev** 1978a (USSR: Maritime Territory; DELTOCEPHALINAE, GONIAGNATHINI). **Bonfils & Della Giustina** 1978a (Corsica). **Datta** 1973e (descriptive notes). **Dlabola** 1957a (Turkey); 1957b (Afghanistan); 1958b (Caucasus); 1961b (central Asia); 1964b (Afghanistan); 1964c (Albania); 1965c (Mongolia) 1965d (Jordan); 1967b, 1968a, 1968b (Mongolia); 1971a (Iran); 1972a (Afghanistan); 1981a (Iran). **Dubovskij** 1966a (Uzbekistan; in key to genera of EUSCELINAE, GONIAGNATHINI); 1978a (Uzbekistan, Zarafshan Valley); 1980a (USSR: western Turkmenia). **Emeljanov** 1962a (new subgenus proposed); 1964c (European USSR; key to species); 1977a (Mongolia). **Hamilton, K. G. A.** 1975b (APHRODINAE, SELENOCEPHALINI). **Kalkandelen** 1974a (Turkey; key to species). **Korolevskaya** 1978b (central Asia). **Kwon & Lee** 1979b (Korea; n. sp. syn.). **Lee, C. E.** 1979a (Korea, redescribed). **Lindberg** 1956b (Morocco); 1958a (Cape Verde Islands); 1960b (Soviet Armenia); 1962a (Portugal); 1963a (Morocco). **Linnavuori** 1956b (Israel); 1956f (Spanish Morocco); 1959c (Mt. Picentini); 1962a (Israel; n. comb. from *Athysanus*); 1964a (Egypt); 1965a (Tunisia, Turkey); 1978d (Ethiopian region; n. comb. from *Athysanus*; key to species). **Lodos & Kalkandelen** 1985a (Turkey). **Logvinenko** 1957b, 1962e, 1984a (USSR: Ukraine). **Mitjaev** 1968b (eastern Kazakhstan); 1971 (Kazakhstan; key to six species). **Nast** 1972a (catalogued); 1973a (Poland); 1976b (Poland; catalogued)). **Quartau & Rodrigues** 1969a (Portugal). **Servadei** 1957a, 1960a, 1968a, 1971a (Italy). **Sharma, B.** 1977b (India: Jammu) **Theron** 1980a (supplemental notes). **Vilbaste** 1962b (eastern part of the Caspian lowlands); 1965b (Altai Mts.); 1968a (coastal regions of the USSR); 1980b (Tuva).

Gorgonalia
Record: **Young** 1977a (n. comb. from *Cicadella* and *Tettigoniella*; n. sp. syn.).

Goska
New species: **Dworakowska** 1981j.
Record: **Sohi & Dworakowska** 1984a (list of Indian species).

Graminella
New species: **Kramer** 1965b. **Linnavuori** 1959b. **Linnavuori & DeLong** 1979a. **Menezes** 1974a.
Records: **Beardsley** 1966a (as *Deltocephalus*; Nihoa Island, French Frigate Shoal; Kure Atoll). **Beirne** 1956a (Canada). **Cheng, Y. J.** 1980a (Paraguay). **Hamilton, K. G. A.** 1972d (Canada: Manitoba); 1975b (APHRODINAE, DELTOCEPHALINI, DELTOCEPHALINA). **Kramer** 1967a (key to species). **Linnavuori** 1956a (Argentina, Brazil, Colombia, Trinidad); 1973c (Cuba). **Linnavuori & Heller** 1961a (Peru). **Taboada** 1972a. (Michigan).

Grammacephalus
New species: **Dlabola** 1984a. **Linnavuori** 1978d. **Viraktamath** 1981a.
Records: **Dlabola** 1960c (Iran); 1965d (Jordan); 1981a (Iran). **Hamilton, K. G. A.** 1975b (APHRODINAE, DELTOCEPHALINI, PLATYMETOPIINA). **Linnavuori** 1962a (Israel); 1964a (Afghanistan); 1978d (Ethiopian region; n. comb. from *Platymetopius*). **Viraktamath** 1981c (n. comb. from *Platymetopius*; key to Indian species).

Granulus (fossil)
New species: **Hong** 1980a.

Graphocephala
New species: **Hamilton, K. G. A.** 1985b. **Young** 1977a.
Records: **Badmin** 1979a (England: Kent). **Beirne** 1956a (Canada). **D'Aguilar & Della Giustina** 1974a (France). **Datta** 1980a (descriptive notes). **Della Giustina** 1983a (France). **Downes** 1957a (Canada: British Columbia; adventive). **Hamilton, K. G. A.** 1983b (adventive to Palaearctic); 1985b (key to members of the *coccinea* complex). **LeQuesne** 1965c (Britain). **Nielson** 1968b (characterization of vector species). **Ramos Elorduy** (Mexico). **Schroder** 1959a (diagnosis). **Young** 1977a (n. gen. syn.; n. comb. from *Cicadella*, *Keonolla*, *Neokolla*, *Poeciloscarta* and *Tettigonia*; n. sp. syn.; key to males).

Graphocraerus
Records: **Anufriev** 1978a (USSR: Maritime Territory). **Cantoreanu** 1971c (Romania). **Dlabola** 1957a (Turkey); 1958b (Caucasus); 1964c (Albania); 1981a (Iran). **D'Urso** 1980b (Italy). **Emeljanov** 1964c (European USSR); 1977a (Mongolia). **Hamilton, K. G. A.** 1975b (APHRODINAE, DELTOCEPHALINI, DELTOCEPHALINA); 1983b (eastern Canada; adventive from Palaearctic). **Jasinska** 1980a (Poland: Bledowska Wilderness). **LeQuesne** 1969c (Britain). **Linnavuori** 1959a (Mt. Sibillini). **Logvinenko**

1957b (USSR: Ukraine). **Mitjaev** 1968b (eastern Kazakhstan); 1971a (Kazakhstan; in key to genera of DELTOCEPHALINAE). **Nast** 1972 (catalogued); 1976a (Poland: Pieniny Mts.); 1976b (Poland; catalogued). **Ossiannilsson** 1983a (Fennoscandia & Denmark). **Schulz** 1976a (Norway). **Servadei** 1968a, 1971a (Italy). **Vilbaste** 1962b (eastern part of the Caspian lowlands); 1964a (Auwiesen Estlands); 1968a (coastal regions of the USSR); 1973a (notes on Flor's material); 1979a (Vooremaa hardwood/spruce forest); 1980b (Tuva).

Graphogonalia
New species: **Young** 1977a.
Record: **Young** 1977a (n. comb. from *Tettigonia*; key to species).

Gratba
New species: **Dworakowska** 1982a.
Record: **Sohi & Dworakowska** 1984a (list of Indian species).

Gredzinskiya
New species: **Dworakowska** 1972j.

Gressittella
New species: **Evans, J. W.** 1972a.

Gressittocerus
New species: **Maldonado-Capriles** 1985a.
Record: **Maldonado-Capriles** 1985a (comparative notes).

Grootonia
New species: **Webb, M. D.** 1983b.
Record: **Webb, M. D.** 1983b (key to species).

Grunchia
New species: **Blocker** 1975b, 1979a, 1982a.
Records: **Blocker** 1975b (key to species); 1979a (list of species). **Kramer** 1963a (n. comb. from *Batracomorphus*; in key to New World genera of IASSINAE).

Grypotes
Records: **Dlabola** 1957a (Turkey). **Emeljanov** 1964c (European USSR; key to species). **Hamilton, K. G. A.** 1975b (APHRODINAE, APHRODINI, ANOTEROSTEMMINA). **Jankovic** 1966a (Serbia). **Kalkandelen** 1974a (Turkey). **LeQuesne** 1969c (Britain). **Lindberg** 1962a (Portugal); 1963a (Morocco). **Linnavuori** 1956d (Spanish Morocco); 1962a (Israel); 1965a (Libya, Tunisia, Turkey). **Lodos & Kalkandelen** 1985a (Turkey). **Logvinenko** 1957b (USSR: Ukraine). **Nast** 1972a (catalogued); 1976b (Poland; catalogued). **Ossiannilsson** 1983a (Fennoscandia & Denmark; DELTOCEPHALINAE). **Quartau & Rodrigues** 1969a (Portugal). **Rodrigues** 1968a (Portugal). **Servadei** 1957a, 1960a (Italy). **Zachvatkin** 1953b (central Russia; Oka district).

Guaporea
New species: **Linnavuori & DeLong** 1978d.
Records: **Cwikla & Blocker** 1981a (comparative notes). **Linnavuori & DeLong** 1978b (n. comb. from *Eutettix*; in key to genera of the Bahita group).

Guheswaria
New species: **Thapa** 1983a.

Guliga
Record: **Nielson** 1977a (= *Thagria*, n. syn.).

Gullifera
New species: **Webb, M. D.** 1980a.

Gurawa
Records: **Datta** 1973a (descriptive notes). **Datta & Dhar** 1984a (descriptive notes). **Hamilton, K. G. A.** 1975d (APHRODINAE, EUPELICINI).

Gypona
New species: **DeLong** 1977f, 1979e, 1980f, 1980g, 1981a, 1982d, 1983c. **DeLong & Foster** 1981a, 1982b. **DeLong & Freytag** 1962a, 1964a, 1975d. **DeLong & Kolbe** 1974a, 1975c. **DeLong & Linnavuori** 1977a. **DeLong & Martinson** 1972a. **DeLong & Triplehorn** 1978a, 1979a. **DeLong & Wolda** 1984b. **Linnavuori & DeLong** 1977c. **Metcalf** 1962b (new name). **Teson** 1972a, 1972b.
Records: **Beirne** 1956a (Canada). **DeLong** 1977b (notes on Osborn's species; n. comb. to *Bahapona, Curtara, Hecalapona* and *Prairiana*); 1979f (notes on Spangberg and Stal types; n. comb. to *Curtara, Polana, Ponana* and *Prairiana*). **DeLong & Freytag** 1962a (n. sp. syn.); 1964a (synopsis; new subgenera proposed; n. sp. syn.; keys to subgenera and species); 1966a (n. comb. from *Hamana*); 1972a (in key to genera of GYPONINAE). **Freytag & Cwikla** 1982a (Dominica). **Freytag & DeLong** 1968a (n. comb. to *Clinonaria*). **Hamilton, K. G. A.** 1972d (Canada: Manitoba). **Linnavuori & DeLong** 1977c (Chile).

Gyponana
New species: **Bliven** 1958a. **DeLong** 1983f, 1984h. **DeLong & Freytag** 1964a, 1972c. **DeLong & Wolda** 1972a, 1984b. **Freytag & DeLong** 1975a. **Hamilton, K. G. A.** 1972b, 1982a. **Ramos Elorduy** 1972a.
Records: **Beirne** 1956a (Canada; key to species). **DeLong** 1984b (comments on male genital structures as criteria for species). **DeLong & Freytag** 1964a (synopsis; new subgenera proposed; keys to subgenera and species). **Hamilton, K. G. A.** 1972d (Canada: Manitoba); 1976a (identity of *G. quebecensis* (Provancher)); 1982a (review of the nominotypical subgenus; n. sp. syn.). **Nielson** 1968a (characterization of vector species). **Ramos Elorduy** 1972a (Mexico; key to species); 1972c (Mexico: Vera Cruz region).

Habenia
New species: **Dworakowska** 1972c.

Habralebra
New species: **Young** 1957c.
Record: **Young** 1957c (key to species).

Habrostis
Record: **Dubovskij** 1966a (n. comb. from *Thamnotettix*).

Hackeriana
New species: **Evans, J. W.** 1969a.
Record: **Evans, J. W.** 1966a (synopsis; n. sp. syn.).

Hadralebra
New species: **Ruppel** 1959a.
Records: **Ruppel** 1959a (discussion of subfamily placement). **Young** 1957c (discussion of subfamily placement).

Hadria
New species: **Dlabola & Novoa** 1976a.
Records: **Dlabola & Novoa** (revision of the genus; n. comb. to *Arezzia*; morphological notes). **Young** 1977a (n. gen. syn.; n. comb. from *Arezzia, Cicadella* and *Graphocephala*; n. sp. syn.; key to species).

Hadroca
Record: **Theron** 1974b (n. comb. from *Euscelis*).

Hajra
New species: **Dworakowska** 1981j.
Record: **Sohi & Dworakowska** 1984a (list of Indian species).

Haldorus
New species: **Cheng, Y. J.** 1980a. **Linnavuori** 1959b. **Linnavuori & DeLong** 1979a, 1979b. **Menezes** 1973a, 1974a.
Records: **Cwikla & Blocker** 1981a (comparative notes). **Hamilton, K. G. A.** 1975b (APHRODINAE, DELTOCEPHALINI, DELTOCEPHALINA). **Linnavuori** 1956a (Argentina, Brazil, Colombia); 1959b (= *Cumora*, n. syn.; n. comb. from *Cumora, Deltocephalus, Thamnotettix*; n. sp. syn.; new subgenera proposed). **Menezes** 1973a (new subgenus proposed).

Hamana
New species: **DeLong & Freytag** 1966c.
Records: **DeLong & Freytag** 1966c (synopsis; revised comb.; key to species); 1972a (in key to genera of GYPONINAE).

Hameedia
New species: **Ahmed, Manzoor** 1972a.
Record: **Sharma, B.** 1978a (= *Uzeldikra*, n. syn.).

Hamolidia
New species: **Nielson** 1982e.
Record: **Nielson** 1982e (key to species).

Handianus
New species: **Dlabola** 1959a, 1960b, 1961b, 1963d, 1971a, 1981a. **Emeljanov** 1964c, 1964f. **Linnavuori** 1962a. **Logvinenko** 1967c, 1975a. **Mitjaev** 1969a, 1975a. **Vilbaste** 1980b. **Zachvatkin** 1948a.
Records: **Anufriev** 1978a (USSR: Maritime Territory); 1979a (notes on Jacobi's species from China). **Cantoreanu** 1968b (Danube river delta); 1971c (Romania). **Dlabola** 1957a (Turkey); 1958b (Caucasus); 1961b (central Asia); 1968a, 1968b, (Mongolia); 1981a (Iran). **Dubovskij** 1966a (Uzbekistan; in key to genera of DELTOCEPHALINI); 1978a (Uzbekistan, Zarafshan Valley); 1980a (USSR: western Turkmenia). **Dworakowska** 1973d (Korea). **Emeljanov** 1964c (European USSR; key to species); 1964f

(new subgenera proposed; key to subgenera and species); 1977a (Mongolia). **Hamilton, K. G. A.** 1975b (APHRODINAE, DELTOCEPHALINI, ATHYSANINA). Ishihara 1961a (Thailand, identity questioned). **Jankovic** 1966a, 1971a (Serbia). **Kalkandelen** 1974a (Turkey; key to species). **Korolevskaya** 1979b (Tadzhikistan). **Lauterer** 1983b (Moravia). **Lee C. E.** 1979a (Korea, redescribed). **Lee, C. E. & Kwon** 1977b (Korea). **Lindberg** 1960a (Portugal). **Linnavuori** 1959a (Mt. Sibillini). **Logvinenko** 1957b, 1957c, 1984a (USSR: Ukraine). **Mitjaev** 1968b (eastern Kazakhstan); 1971a (Kazakhstan; key to 13 species). **Nast** 1972a (catalogued); 1973a (Poland); 1976a (Poland: Pieniny Mts.); 1976b (Poland; catalogued). **Servadei** 1960a, 1969a (Italy). **Vilbaste** 1962b (eastern part of the Caspian lowlands); 1965b (Altai Mts.); 1968a (coastal regions of the USSR); 1973a (notes on Flor's material); 1980b (Tuva; key to species).

Hangklipia
Records: **Linnavuori** 1972b (n. comb. from *Camptelasmus;* key to species). **Theron** 1976a (n. comb. to *Titiella*).

Hanshumba
New species: **Young** 1977a.

Haranga
New species: **Linnavuori** 1977a.

Haranthus
New species: **Nielson** 1975a.

Harasupia
New species: **Nielson** 1979b, 1983b.
Record: **Nielson** 1979b (in key to genera of TERULIINI; n. comb. from *Coelidia*; key to species).

Hardiana
New species: **Mahmood** 1967a. **Ramakrishnan & Menon** 1974a.
Records: **Mahmood** 1967a (in key to genera of Oriental TYPHLOCYBINAE). **Ramakrishnan & Menon** 1974a (in key to genera of ERYTHRONEURINI). **Sohi** 1976a (n. comb. to *Thaia*; synonymy with *Thaia* noted).

Hardya
New species: **Dlabola** 1968a. **Vilbaste** 1969b.
Records: **Beirne** 1956a (Canada; key to species). **Della Giustina** 1983a (France). **Dlabola** 1957a (Turkey); 1957b (Afghanistan); 1958b (Caucasus); 1961b (central Asia); 1964c (Albania); 1967b, 1970b (Mongolia); 1971b (Switzerland); 1972a (Afghanistan): 1981a (Iran). **Dubovskij** 1966a (Uzbekistan); 1978a (Uzbekistan, Zarafshan Valley). **Dworakowska** 1969a (redescription; n. sp. syn.). **Emeljanov** 1964c (European USSR); 1977a (Mongolia). **Hamilton, K. G. A.** 1975b (APHRODINAE, DELTOCEPHALINI, ATHYSANINA); 1983b (transboreal taxon). **Jankovic** 1966a (Serbia). **Jasinska** 1980a (Poland: Bledowska Wilderness). **Kalkandelen** 1974a (Turkey; key to subgenera and species); 1975a (generic synonymy; key to species). **Korolevskaya** 1979b (Tadzhikistan). **LeQuesne** 1969c (Britain). **Lindberg** 1960a (Portugal); 1960b (Soviet Armenia). **Linnavuori** 1962a (*Eohardya* to generic status); 1965a (Spanish Morocco, Turkey). **Logvinenko** 1957b (USSR: Ukraine). **Mitjaev** 1968b (eastern Kazakhstan); 1971a (Kazakhstan; key to 3 species). **Nast** 1972a (catalogued); 1973a (Poland); 1976a (Poland: Pieniny Mts.); 1976b (Poland; catalogued). **Ossiannilsson** 1973a (Fennoscandia). **Remane** 1961d (Germany). **Servadei** 1958a, 1968a, 1971a (Italy). **Vilbaste** 1965b (Altai Mts.); 1969b (Taimyr); 1980a (Kamchatka); 1980b (Tuva).

Hardyopsis
Records: **Cantoreanu** 1961a (Romania). **Dlabola** 1956b (southern Europe); 1958b (Caucasus); 1961b (central Asia); 1970a (Slovakia); 1971a (Iran; =*Eohardya*). **Hamilton, K. G. A.** 1975b (APHRODINAE, DELTOCEPHALINI, ATHYSANINA). **Jankovic** 1966a (Serbia). **Lindberg** 1960b (Soviet Armenia). **Linnavuori** 1962a (Israel; =*Eohardya*; n. comb. to *Eohardya*; n. sp. syn.; additional sp. syn. suggested). **Servadei** 1971a (Italy).

Harmata
Record: **Dworakowska** 1976b (n. comb. from *Typhlocyba*).

Hatralixia
New species: **Webb, M. D.** 1983a.

Hauptidia
New species: **Dlabola** 1979a.
Records: **Bonfils & Della Giustina** 1978a (Corsica). **Dlabola** 1981a (Iran). **Dworakowska** 1970k (n. comb. from *Erythroneura, Typhlocyba, Zyginidia,* and *Zygina*; n. sp. syn.); 1977b (n. comb. from *Erythroneura*); 1981d (=*Kodiakanalia*; n. comb. from *Kodiakanalia* and *Zygina*). **LeQuesne & Payne** 1981a (Britain). **Lodos & Kalkandelen** 1984c (Turkey). **Logvinenko** 1984a (USSR: Ukraine). **Nast** 1972a (catalogued); 1976a (Poland: Pieniny Mts.); 1976b (Poland; catalogued). **Ossiannilsson** 1981a (Fennoscandia & Denmark). **Quartau** 1981e (Portugal). **Sohi & Dworakowska** 1984a (list of Indian species).

Havelia
New species: **Ahmed, Manzoor** 1971d.
Records: **Ahmed, Manzoor** 1983b (remarks on male style shape); 1985c (Pakistan).

Hazaraneura
New species: **Samad & Ahmed** 1979a.
Record: **Ahmed, Manzoor** 1985c (Pakistan).

Hebecephalus
New species: **Dlabola** 1965c. **Emeljanov** 1972d, 1976a. **Hamilton, K. G. A. & Ross** 1972a. **Vilbaste** 1965b.
Records: **Beirne** 1956a (Canada; key to species). **Dlabola** 1966a, 1968a, 1970b (Mongolia). **Hamilton, K. G. A.** 1972d (Canada: Manitoba); 1975b (APHRODINAE, DELTOCEPHALINI, DELTOCEPHALINA). **Vilbaste** 1965b (Altai Mts.); 1980b (Tuva).

Hebenarus
Record: **Hamilton, K. G. A.** 1975b (APHRODINAE, DELTOCEPHALINI, PLATYMETOPIINA)

Hebexa
Record: **Hamilton, K. G. A.** 1975b (APHRODINAE, DELTOCEPHALINI, DELTOCEPHALINA).

Hecalapona
New species: **DeLong** 1976b, 1977d, 1981b. **DeLong & Foster** 1982b. **DeLong & Freytag** 1975c. **DeLong & Triplehorn** 1979a. **DeLong & Wolda** 1984b.
Records: **DeLong** 1977b (n. comb. from *Gypona*). **DeLong & Freytag** 1975c (n. comb. from *Gypona*; keys to subgenera and species).

Hecalocratus
New species: **Evans, J. W.** 1966a.

Hecaloidia
Records: **Linnavuori** 1959b (synopsis; DELTOCEPHALINAE, EUSCELINI). **Linnavuori & DeLong** 1978b (in key to genera of DELTOCEPHALINAE).

Hecalus
New species: **Dlabola** 1963c. **Emeljanov** 1964c. **Ishihara** 1961a. **Kwon & Lee** 1979a. **Linnavuori** 1975a. **Linnavuori & DeLong** 1977c. **Morrison** 1973a. **Villiers** 1956a.
Records: **Anufriev** 1978a (USSR: Maritime Territory). **Anufriev & Emeljanov** 1968b (Soviet Far East; n. sp. syn.). **Beirne** 1956a (Canada: southern Manitoba). **Choe** 1985a (Korea). **Dlabola** 1960a (notes on Horvath's material); 1963c (notes on Haupt's material); 1964a (Afghanistan); 1966a, 1967b, 1967c, 1968a (Mongolia); 1972a (Afghanistan); 1974d (Mediterranean; n. sp. syn.); 1981a (Iran). **Dubovskij** 1978a (Uzbekistan, Zarafshan Valley). **Dworakowska** 1973d (Mongolia). **Emeljanov** 1977a (Mongolia). **Hamilton, K. G. A.** 1972d (Canada: Manitoba); 1975b (APHRODINAE, HECALINI). **Heller & Linnavuori** 1968a (Ethiopia). **Kholmuminov & Dlabola** 1979a (USSR: Golodnostep Plain). **Kwon & Lee** 1979a (Korea; key to species). **Lee C. E.** 1979a (Korea, redescribed). **Lee, C. E. & Kwon** 1977b (Korea). **Lindberg** 1956a (Niger River flood plain; Dogo, Mopti). **Linnavuori** 1961a (n. gen. syn.; n. comb. from *Parabolocratus*); 1962a (Israel; n. comb. from *Parabolocratus*); 1964a (Egypt); 1965a (Turkey); 1973c (Cuba); 1975a (Ethiopian region; n. comb. from *Parabolocratus*; key to species). **Linnavuori & DeLong** 1978e (Chile, Mexico). **Lodos & Kalkandelen** 1981a (Turkey). **Mitjaev** 1971a (Kazakhstan; HECALINAE). **Morrison** 1973a (n. gen. syn.; n. comb. from *Columbanus, Ledrotypa* and *Parabolocratus*; n. sp. syn.; key to species). **Sharma, B.** 1977b (India: Jammu). **Theron** 1982a (South

Africa). **Vilbaste** 1962b (eastern part of the Caspian lowlands); 1968a (coastal regions of the USSR); 1975a (identity of *Platymetopius lineolatus* de Motschulsky); 1980b (Tuva). **Webb, M. D.** 1980a (Aldabra).

Hecullus
Record: **Hamilton, K. G. A.** 1975b (APHRODINAE, HECALINI).

Hegira
Records: **Linnavuori** 1956a (Brazil); 1959b (redescribed).

Heliona
New species: **Dlabola** 1960c. **Lindberg** 1958a.
Records: **Dlabola** 1964b (Sudan); 1965d (Jordan). **Dworakowska** 1970e (=*Acia*, n. syn.). **Linnavuori** 1962a (Israel); 1964a (Egypt).

Helionidia
New species: **Ahmed, Manzoor** 1970b, 1971b. **Ahmed, Manzoor & Khokhar** 1971a. **Ahmed, Manzoor & Samad** 1972b. **Dlabola** 1961b, 1967d, 1974d. **Dworakowska** 1970e, 1971b, 1980b, 1981e, 1981h. **Dworakowska & Viraktamath** 1975a. **Einyu & Ahmed** 1983a. **Logvinenko** 1965a. **Mitjaev** 1969a. **Vilbaste** 1961b. **Webb, M. D.** 1980a. **Zachvatkin** 1953c.
Records: **Ahmed, Manzoor** 1985c (Pakistan). **Dlabola** 1957a (Turkey); 1958c (in key to genera of Palaearctic TYPHLOCYBINAE); 1960c (Iran); 1961b (central Asia); 1963c (notes on Haupt's material); 1964a (Afghanistan); 1964b (Sudan); 1965d (Jordan); 1968a, 1968b (Mongolia); 1971a (Afghanistan, Pakistan); 1972a (Afghanistan); 1981a (Iran). **Dworakowska** 1970e (n. comb. from *Heliona*); 1971b (*Tamaricella* to generic rank. n. comb. to *Tamaricella*); 1972f (Congo, Tanganyika); 1977b (n. sp. syn.); 1977f (n. sp. syn.). **Dworakowska & Viraktamath** 1975a (n. comb. to *Mitjaevia*). **Emeljanov** 1964c (European USSR; key to species). **Lindberg** 1960a (Portugal). **Linnavuori** 1962a (Israel; n. sp. syn.); 1964a (Egypt); 1965a (Greece, Turkey). **Lodos & Kalkandelen** 1984c (Turkey). **Mitjaev** 1963b (Kazakhstan); 1968b (eastern Kazakhstan); 1971a (Kazakhstan; key to 6 species); 1971b (southern Kazakhstan). **Servadei** 1960a (Italy). **Sharma, B.** 1977a (northwest India). **Sohi & Dworakowska** 1984a (list of Indian species). **Theron** 1977a (n. comb. from *Erythroneura*). **Vilbaste** 1962b (eastern part of the Caspian lowlands).

Heliotettix
New species: **Rodrigues** 1968a.
Records: **Hamilton, K. G. A.** 1975b (APHRODINAE, DELTOCEPHALINI, PLATYMETOPIINA). **Nast** 1982a (catalogued; DELTOCEPHALINAE). **Quartau & Rodrigues** 1969a (Portugal). **Vilbaste** 1976a (transfer of species from *Thamnotettix* noted; n. sp. syn.).

Hellerina
Record: **Dworakowska** 1979c (Ethiopia, Tanzania, Zaire).

Helochara
Records: **Beirne** 1956a (Canada). **Datta** 1980a (descriptive notes). **Hamilton, K. G. A.** 1972d (Canada: Manitoba). **Nielson** 1968b (characterization of vector species). **Young** 1977a (n. sp. syn.).

Helocharina
New species: **Young** 1977a.
Record: **Young** 1977a (key to species).

Hemipeltis
Records: **Kramer** 1966a (summary of available information; position uncertain; LEDRINAE). **Linnavuori & DeLong** 1977c (Chile; Spinola records; LEDRINAE).

Hengchunia
Records: **Hamilton, K. G. A.** 1975b (APHRODINAE, DELTOCEPHALINI, DELTOCEPHALINA). **Vilbaste** 1969a (n. comb. from *Thamnotettix*).

Henschia
Records: **Dlabola** 1956a (Czechoslovakia); 1958a (southern Europe); 1967b, 1967c, 1967d, 1968b (Mongolia). **Dworakowska** 1969a (Mongolia). **Emeljanov** 1964c (European USSR);1977a (Mongolia). **Hamilton, K. G. A.** 1975b (APHRODINAE, DELTOCEPHALINI, DELTOCEPHALINA). **Logvinenko** 1962e (USSR: Ukraine). **Mitjaev** 1968a (eastern Kazakhstan); 1971a (Kazakhstan; in key to genera of DELTOCEPHALINAE).

Vilbaste 1962b (eastern part of the Caspian lowlands); 1965b (Altai Mts.); 1980b (Tuva).

Hensleyella
New species: **Webb, M. D.** 1983b.

Hephathus
New species: **Dlabola** 1957b, 1961b. **Mitjaev** 1967a. **Vilbaste** 1966a.
Records: **Anufriev & Emeljanov** 1968a (Soviet Far East). **Bonfils & Della Giustina** 1978a (Corsica). **Cantoreanu** 1971c (Romania). **Dlabola** 1957a (Turkey); 1958b (Caucasus); 1961b (central Asia); 1964c (Albania); 1965d (Jordan); 1971a, 1972a (Afghanistan). **Dubovskij** 1978a (Uzbekistan, Zarafshan Valley). **Emeljanov** 1964c (European USSR). **Hamilton, K. G. A.** 1980b (n. gen. syn.; n. comb. from *Asmaropsis* and *Macropsis*). **Kholmuminov & Dlabola** 1979a (USSR: Golodnostep Plain). **Korolevskaya** 1975a (Tadzhikistan). **LeQuesne** 1961b, 1965c (Britain). **Lindberg** 1960a (as *Hephates*; Portugal); 1964a (as *Hephatus*; Morocco). **Lodos & Kalkandelen** 1981a (Turkey). **Logvinenko** 1957b (USSR: Ukraine). **Mitjaev** 1968a (eastern Kazakhstan); 1971a (Kazakhstan; key to 3 species). **Nast** 1972a (catalogued); 1976b (Poland; catalogued). **Ossiannilsson** 1983a (Fennoscandia & Denmark). **Quartau & Rodrigues** 1969a (Portugal). **Servadei** 1960a, 1968a (Italy). **Vilbaste** 1959b (Estonia); 1968a (coastal regions of the USSR).

Hepneriana
New species: **Dworakowska** 1972j.
Record: **Dworakowska & Viraktamath** 1975a (=*Mandera*, n. syn.).

Hesium
Records: **Dlabola** 1958b (Caucasus); 1960a (notes on Horvath's material); 1964c (Albania). **Emeljanov** 1964c (European USSR). **Hamilton, K. G. A.** 1975b (APHRODINAE, DELTOCEPHALINI, PLATYMETOPIINA). **Nast** 1972a (catalogued); 1976a (Poland: Pieniny Mts.); 1976b (Poland; catalogued). **Ossiannilsson** 1983a (Fennoscandia & Denmark). **Schulz** 1976a (Norway). **Servadei** 1968a, 1972a (Italy).

Hespenedra
Records: **Kramer** 1966a (in key to genera of New World LEDRINAE; n. comb. from *Thlasia*). **Linnavuori & DeLong** 1977c (Chile).

Heterometopia
Records: **Schroder** 1959b (n. comb. to *Hyogonia*). **Young** 1968a (=*Hyogonia*).

Heterostemma
Records: **Young** 1968a (=*Diestostemma*, n. syn.). **Young & Lauterer** 1966a (lectotype designation).

Hikangia
New species: **Nielson** 1983e.
Record: **Nielson** 1983e (key to species).

Hiltus
New species: **Theron** 1974b.
Record: **Theron** 1974b (n. comb. from *Chlorotettix*).

Hiratettix
New species: **Dworakowska** 1982a.
Record: **Dworakowska** 1982a (redescribed; to TYPHLOCYBINI).

Hishimonoides
New species: **Anufriev** 1970a. **Ishihara** 1965b. **Kuoh, C. L.** 1976a. **Mahmood** 1975b.
Records: **Emeljanov** 1977a (Mongolia). **Hamilton, K. G. A.** 1975b (APHRODINAE, DELTOCEPHALINI, PLATYMETOPIINA). **Mahmood** 1975b (comparative notes). **Sawai Singh** 1971a, 1971b (=*Cestius*).

Hishimonus
New species: **Emeljanov** 1969a. **Heller & Linnavuori** 1968a. **Ishihara** 1972a (new name). **Knight, W. J.** 1970a, 1973c. **Kuoh, C. L.** 1976a. **Linnavuori** 1969a. **Okada** 1978b.
Records: **Anufriev** 1978a (USSR: Maritime Territory). **Arai** 1978a (Japan, "an unrecorded species"). **Choe** 1985a (Korea). **Dlabola** 1957b, 1972a (Afghanistan). **Dworakowska** 1973d (Korea). **Evans, J. W.** 1966a (n. comb. from *Eutettix*). **Hamilton, K. G. A.** 1975b (APHRODINAE, DELTOCEPHALINI, PLATYMETOPIINA). **Ishihara** 1958a (DELTOCEPHALIDAE); 1959b (revised sp. syn.);

1961a (DELTOCEPHALIDAE); 1963a (key to species); 1967a (n. comb. from *Eutettix*); 1972a (removed from synonymy). **Knight, W. J.** 1970a (revision and summary; n. sp. syn.; key to species). **Lee C. E.** 1979a (Korea, redescribed). **Lee, C. E. & Kwon** 1977b (Korea). **Lindberg** 1958a (Cape Verde Islands). **Linnavuori** 1960b (Fiji; key to species). **Mahmood** 1975b (comparative notes). **Nielson** 1968b (characterization of vector species). **Sawai Singh** 1971a (= *Cestius*; n. comb. to *Cestius*); 1971b (= *Cestius*).

Histipagus
New species: **Remane & Asche** 1980a.
Record: **Nast** 1982a (catalogued; DELTOCEPHALINAE).

Hodoedocus
New species: **Linnavuori** 1979b.
Records: **Linnavuori** 1979b (as *Hododoecus*; = *Stenometopius*, = *Afrolimnus*, n. syn.; n. comb. from *Afrolimnus*; STENOMETOPIINI). **Linnavuori & DeLong** 1978c (as *Hododoecus*; = *Acurhinus*, n. syn.).

Homa
New species: **Dworakowska** 1984a.
Record: **Mahmood** 1967a (comparative note).

Homalodisca
New species: **Metcalf** 1965a (new name). **Schroder** 1957a. **Young** 1968a.
Records: **Nielson** 1968b (characterization of vector species). **Pollard, Turner & Kaloostian** 1959a (USA: southeast). **Ramos Elorduy** 1972c (Mexico). **Schroder** 1957a (n. sp. syn.). **Young** 1958a (USA; key to species); 1968a (genus redescribed; n. sp. syn.; key to species). **Young & Lauterer** 1966a (lectotype designation). **Young & Nast** 1963a (notes on type material).

Homalogoniella
Record: **Young** 1977a (subfamily placement uncertain).

Homogramma
Records: **Emeljanov** 1975a (n. comb. from *Melicharella*). **Kocak** 1981a (replacement name, *Emelyanogramma*). **Nast** 1982a (catalogued; AGALLIINAE).

Homoscarta
Records: **Young** 1968a (n. sp. syn.; key to species). **Young & Nast** 1963a (notes on type specimen).

Hordnia
Records: **Nielson** 1968b (characterization of vector species). **Young** 1977a (= *Graphocephala*, n. syn.).

Horouta
New species: **Knight, W. J.** 1975a.

Hortensia
New species: **Young** 1977a.
Records: **Schroder** 1959a (diagnosis). **Young** 1977a (n. comb. from *Tettigonia*; n. sp. syn.; key to species).

Houtbayana
New species: **Linnavuori** 1961a.

Huachia
New species: **Linnavuori** 1959a. **Linnavuori & DeLong** 1978c, 1978d.
Records: **Linnavuori & DeLong** 1978b (in key to genera of the *Bahita* group); 1978c (comparative notes).

Huancabamba
New species: **Linnavuori** 1959b.
Record: **Linnavuori & DeLong** 1977c (in key to genera of the *Yungasia* group; DELTOCEPHALINAE, ATHYSANINI).

Huleria
Record: **Hamilton, K. G. A.** 1975b (APHRODINAE, DELTOCEPHALINI, COCHLORHININA).

Humpatagallia
New species: **Linnavuori & Viraktamath** 1973a.
Record: **Linnavuori & Viraktamath** 1973a (n. comb. from *Nehela*; key to species).

Hussa
Record: **Sharma, B.** 1977a (India: Jammu).

Hussainiana
New species: **Mahmood & Ahmed** 1969a.
Record: **Dworakowska** 1971c (placement in TYPHLOCYBINAE questioned; probably CICADELLINAE).

Hyalocerus
New species: **Maldonado-Capriles** 1977b.

Record: **Maldonado-Capriles** 1977b (in key to four Neotropical genera).

Hyalojassus
New species: **Evans, J. W.** 1972b.
Record: **Linnavuori & Quartau** 1975a (tribal position discussed, related to IASSINI).

Hybrasil
Records: **Linnavuori** 1960b (redescribed). **Linnavuori & Al-Ne'amy** 1983a (in key to genera of BHATINI, SELENOCEPHALINAE). **Ross** 1968a (n. comb. from *Exitianus*).

Hydrabricta
New species: **Webb, M. D.** 1983a.

Hymetta
Record: **Beirne** 1956a (Canada; in key to genera of ERYTHRONEURINI).

Hyogonia
New species: **Emmrich & Lauterer** 1975a.
Records: **Emmrich & Lauterer** 1975a (redescription of type species). **Schroder** 1959a (generic rank; n. comb. from *Heterometopia*). **Young** 1968a (redescribed.

Hypacostemma
New species: **Linnavuori** 1961a.
Record: **Linnavuori & Al-Ne'amy** 1983a (SELENOCEPHALINAE, HYPACOSTEMMINI).

Hypospadianus
Record: **Hamilton, K. G. A.** 1975b (APHRODINAE, DELTOCEPHALINI, PLATYMETOPIINA).

Ianeira
New species: **Linnavuori** 1969a, 1978a.
Records: **Linnavuori** 1978a (in key to genera of IANEIRINI, DRABESCINAE). **Linnavuori & Al-Ne'amy** 1983a (IANEIRINI, SELENOCEPHALINAE).

Iassomorphus
New species: **Linnavuori & Quartau** 1975a. **Quartau** 1981b, 1985a. **Webb, M. D.** 1980a.
Record: **Linnavuori & Quartau** 1975a (n. comb. from *Batracomorphus* and *Eurinoscopus*; key to species).

Iassus
New species: **Dlabola** 1965c. **Ghosh, M.** 1974a. **Linnavuori** 1969a. **Metcalf** 1966c (new name). **Orosz** 1979a.
Records: **Anufriev** 1978a (USSR: Maritime Territory; key to species). **Badmin** 1985a (Britain). **Bonfils & Della Giustina** 1978a (Corsica). **Cantoreanu** 1968b (Danube river delta). **Dlabola** 1957a (Turkey); 1964c (Albania); 1967d (Mongolia); 1981a (Iran). **Heller & Linnavuori** 1968a (Ethiopia). **ICZN** 1961b (Placed on Official List of Generic Names in Zoology, Opinion 612). **Knight, W. J.** 1983b (n. comb. to *Batracomorphus*). **Lauterer** 1957a (Czechoslovakia); 1984a (Moravia, Slovakia; *I. mirabilis* redescribed). **Lee, C. E.** 1979a (Korea, redescribed). **Linnavuori** 1957b (characterized); 1965a (Turkey). **Lodos & Kalkandelen** 1981a (Turkey). **Nast** 1972a (catalogued); 1976b (Poland; catalogued); 1981a (redescription of *I. mirabilis*). **Orosz** 1979a (key to European species). **Ossiannilsson** 1981a (Fennoscandia & Denmark). **Sharma, B.** 1977b (India: Jammu). **Viraktamath, C. A.** (n. comb. from *Batracomorphus*; n. comb. to *Batracomorphus*). **Wilson, M. R.** 1981a (Britain; key to species). **Young** 1965c (lectotype designation).

Ibadarrus
New species: **Remane & Asche** 1980a.
Record: **Nast** 1982a (catalogued; DELTOCEPHALINAE).

Iberia
Record: **Nielson** 1975a (transferred from COELIDIINAE to STEGELYTRINAE).

Icaia
New species: **Blocker** 1983a. **DeLong** 1983b. **Linnavuori** 1973b.
Records: **Blocker** 1983a (key to species; checklist; suggested phylogeny). **Cwikla & Blocker** 1981a (comparative notes).

Ichthyobelus
New species: **Kramer** 1976c. **Young** 1968a.
Records: **Young** 1968a (key to species). **Young & Lauterer** 1966a (lectotype designation).

Idiocerella
Records: **Evans, J. W.** 1966a (synopsis). **Webb, M. D.** 1983a (redescribed; identity uncertain).

Idiocerinus
Record: **Dlabola** 1974b (n. comb. to *Sulamicerus*; not Palaearctic).

Idioceroides
Record: **Maldonado-Capriles** 1976a (transferred from IDIOCERINAE to AGALLIINAE).

Idiocerus
New species: **Anufriev** 1970c, 1971b, 1971d, 1978a. **Dlabola** 1964a, 1965d, 1967a. **Dubovskij** 1966a, 1970b. **Emeljanov** 1972b. **Evans, J. W.** 1959a. **Freytag** 1962a, 1965a, 1967a, 1975a. **Freytag & Cwikla** 1984a. **Freytag & Knight** 1966a. **Hamilton, K. G. A.** 1980a, 1985c. **Heller** 1969a. **Heller & Linnavuori** 1968a. **Ishihara** 1956a. **Kwon** 1985a. **Linnavuori** 1956a, 1961a. **Metcalf** 1966d (new name). **Mitjaev** 1967a. **Nast** 1984a (new name). **Vilbaste** 1965b, 1968a. **Wagner** 1958c. **Wagner & Duzgunes** 1960a. **Webb, M. D.** 1975a.

Records: **Allen, A. A.** 1963a, 1964a, 1978a, 1985a (Britain). **Anufriev** 1968a (characters and status of *Rhytidodus*); 1971a (Soviet Far East; comparative notes); 1971b (Primorsky District); 1978a (USSR: Maritime Territory; n. sp. syn; key to species). **Anufriev & Emeljanov** 1968a (Soviet Far East). **Beirne** 1956a (Canada; key to species). **Bliven** 1963a, 1966a (n. sp. syn.). **Bonfils & Della Giustina** 1978a (Corsica). **Datta** 1972j, 1973l (descriptive notes). **Cantoreanu** 1961a, 1971c (Romania). **Dlabola** 1957a (Turkey); 1957b (Afghanistan); 1958a (Czechoslovakia, southern Europe); 1960a (notes on Horvath's material); 1961b (central Asia); 1964c (Albania); 1965c (Mongolia); 1965d (Jordan); 1966a, 1967b, 1967c, 1968a, 1968b, 1970b (Mongolia); 1971b (Turkey); 1972a (Afghanistan); 1974b (n. comb. to *Balcanocerus, Liocratus, Populicerus, Sahlbergotettix, Sulamicerus, Taeniocerus, Tremulicerus* and *Viridicerus*); 1977c (Mediterranean); 1981a (Iran). **Downes** 1957a (Canada: British Columbia). **Dubovskij** 1978a (Uzbekistan, Zarafshan Valley). **Dumbleton** 1967a (New Zealand). **Emeljanov** 1964c (European USSR); 1977a (Mongolia). **Evans, J. W.** 1966a (synopsis of Australian fauna); 1974a (New Caledonia; n. comb. from *Nehela*, tentative). **Freytag** 1965a (lectotype designations; n. sp. syn.; key to species; suggested phylogeny); 1976a (USA: Kentucky). **Freytag & Knight** 1966a (Madagascar; n. comb. to *Nesocerus*; key to species). **Hamilton, K. G. A.** 1976a (lectotype designation for Provancher species); 1980a (n. gen. syn.; n. sp. syn.; key to Nearctic genera and subgenera; key to species); 1983b (taxa adventive to the Nearctic). **Ishihara** 1956a (Japan; key to species); 1958a (Japan: north Honshu); 1961a (Thailand); 1966a (Kurile Islands). **Jankovic** 1966a, 1971a (Serbia). **Jasinska** 1980a (Poland: Bledowska Wilderness). **Kameswara Rao** 1976a (revised synonymy; n. comb. to *Idioscopus*). **Kameswara Rao & Ramakrishnan** 1979c (n. comb. to *Idioscopus*). **Knight, W. J.** 1974b (Palaearctic species adventive to New Zealand); 1983b (n. comb. to *Batracomorphus*). **Kwon** 1985a (Korea; new subgenera proposed; changes in status of generic-level names; n. sp. syn.; key to species). **Lauterer** 1957a (Czechoslovakia). **Lee C. E.** 1979a (Korea, redescribed). **Lee, C. E. & Kwon** 1977b (Korea). **LeQuesne** 1964a, 1965c (Britain). **Lindberg** 1960a, 1962a (Portugal); 1964a (Morocco). **Linnavuori** 1956a (Brazil); 1959c (Mt. Picentini); 1961a (South Africa); 1962a (Israel); 1965a (Libya). **Lodos & Kalkandelen** 1981a (Turkey). **Logvinenko** 1957b (USSR: Ukraine); 1959a (USSR: Transcarpathian region). **Maldonado-Capriles** 1971a (n. gen. syn.; comparative notes about *Balcanocerus, Idiocerus* and *Idioscopus*). **Mitjaev** 1968a (eastern Kazakhstan); 1971a (Kazakhstan; key to 21 species). **Nast** 1955a (Poland); 1972a (catalogued); 1976a (Poland: Pieniny Mts.); 1976b (Poland; catalogued); 1984a (replacement name for *I. pallidus* Mitjaev). **Ossiannilsson** 1981a (Fennoscandia & Denmark). **Quartau & Rodrigues** 1969a (Portugal). **Servadei** 1957a, 1960a, 1968a, 1971a, 1972a (Italy). **Theron** 1976a (n. comb. to *Idioscopus*). **Vilbaste** 1962b (eastern part of the Caspian lowlands); 1974a (Auwiesen Estlands); 1965b (Altai Mts.); 1958a (coastal regions of the USSR); 1973a (notes on Flor's material); 1975a (discussion of identity of "*Idiocerus ? subopacus* de Motschulsky; not an *Idiocerus*); 1976a (n. sp. syn.); 1980a (Kamchatka). **Webb, M. D.** 1975a (Ethiopian region; key to species); 1979a (n. sp. syn.; *Jassus corixoides* Rambur n. syn. of *I. decimaquartus*; lectotype designation for Rambur species). **Zachvatkin** 1953b (n. comb. to *Sahlbergotettix*).

Idiodonus
New species: **DeLong** 1983d.
Records: **Anufriev** 1977b (Kurile Islands); 1978a (USSR: Maritime Territory). **Beirne** 1956a (Canada; key to species). **DeLong** 1984c (revised key to species). **Dlabola** 1967b, 1967c, 1968b, 1970b (Mongolia). **Emeljanov** 1964c (European USSR); 1977a (Mongolia). **Hamilton, K. G. A.** 1975b (APHRODINAE, DELTOCEPHALINI, PLATYMETOPIINA); 1976a (notes on Provancher species). **Jankovic** 1971a (Serbia). **Lee C. E.** 1979a (Korea, redescribed). **Lee, C. E. & Kwon** 1977b (Korea). **Mitjaev** 1968b (eastern Kazakhstan); 1971a (Kazakhstan; in key to genera of DELTOCEPHALINAE). **Nast** 1972a (catalogued); 1976a (Poland: Pieniny Mts.); 1976b (Poland; catalogued). **Nielson** 1968b (characterization of vector species). **Quartau & Rodrigues** 1969a (Portugal). **Servadei** 1960a, 1972a (Italy). **Vilbaste** 1959b (Estonia); 1964a (Auwiesen Estlands); 1965b (Altai Mts.); 1968a (coastal regions of the USSR); 1973a (notes on Flor's material); 1980a (Kamchatka); 1980b (Tuva).

Idionannus
New species: **Linnavuori** 1956a.

Idioscopus
New species: **Ahmed, S. S., Naheed, R. & Ahmed** 1980a. **Freytag & Knight** 1966a. **Kuoh, C. L. & Fang** 1985b. **Maldonado-Capriles** 1974a. **Viraktamath, C. A.** 1976b, 1979b, 1979c, 1980c. **Viraktamath, S. & Viraktamath, C. A.** 1985a. **Webb, M. D.** 1976a.
Records: **Dlabola** 1974b (in key to Palaearctic genera of IDIOCERINAE); 1984a (Iran and Mediterranean region). **Freytag & Knight** 1966a (key to species from Madagascar). **Hongsaprug** 1984a (Thailand). **Kameswara Rao** 1976a (n. comb. from *Idiocerus*). **Kameswara Rao & Ramakrishnan** 1979c (n. comb. from *Idiocerus*). **Maldonado-Capriles** 1964a (key to species); 1971a (n. gen. syn.); 1973a (n. comb. from *Idiocerus*); 1974a (list of species retained in *Idioscopus*, and species transferred to *Amritodus* and *Paraidioscopus*); 1985b (revised key to species). **Ramachandra Rao** 1973a (India: Poona). **Sharma, B.** 1977b (India: Jammu). **Theron** 1976a (n. comb. from *Idiocerus*). **Viraktamath, C. A.** (redescription of Singh Pruthi's species). **Viraktamath C. A. & Murphy** (lectotype designation; notes on specific synonymy and species differentiation). **Webb, M. D.** 1976a (Ethiopian region; species groups defined; n. comb. from *Idiocerus*; key to species); 1980a (Aldabra).

Idiotettix
Records: **Linnavuori** 1956a (Brazil, Paraguay, Peru); 1959b (= *Calliscarta*, n. syn.).

Idona
New species: **Cwikla & Freytag** 1982a.
Record: **Mead** 1957a (USA: Florida).

Ifeia
New species: **Linnavuori & Al-Ne'amy** 1983a.
Record: **Linnavuori & Al-Ne'amy** 1983a (n. comb. from *Selenocephalus*; key to species).

Ifeneura
New species: **Ghauri** 1975b.

Ifugoa
New species: **Dworakowska** 1980b. **Dworakowska & Pawar** 1974a.
Record: **Sohi & Dworakowska** 1984a (list of Indian species).

Igerna
Records: **Davis, R. B.** 1975a (morphology; phyletic inferences; n. comb. from *Nehela*). **Theron** 1980a (notes on *I. bimaculicollis*).

Ikomella
New species: **Ishihara** 1961a.
Record: **Linnavuori** 1979b (= *Mukaria*, n. syn.).

Ilagia
New species: **Kramer & DeLong** 1968a.
Record: **Cwikla & Blocker** 1981a (comparative notes).

Imbecilla
New species: **Ahmed, Manzoor 1979a. Dworakowska 1970m, 1972f, 1974a, 1981h. Dworakowska & Lauterer 1975a. Einyu & Ahmed 1983a.**
Records: **Dworakowska** 1970m (n. comb. from *Erythroneura*; n. sp. syn.). **Lodos & Kalkandelen** 1984c (Turkey).

Imugina
New species: **Mahmood 1967a.**

Inazuma
Records: **Hamilton, K. G. A.** 1975b (APHRODINAE, DELTOCEPHALINI, DELTOCEPHALINA). **Ishihara** 1961a (Thailand). **Kwon & Lee** 1979b (Korea; as subgenus of *Recilia*). **Lee, C. E. & Kwon** 1977b (Korea). **Nielson** 1968b (=*Recilia*, n. syn.).

Indianella
New species: **Ramakrishnan & Menon 1973a.**
Records: **Ramakrishnan & Menon** 1973a (key to species); 1974a (in key to genera).

Indodikra
Record: **Sharma, B.** 1979a (India: Nagaland, Kohima).

Indoformosa
Records: **Dworakowska** 1980d (=*Empoascanara*, n. syn.; discussion of taxonomy of *Empoascanara*). **Ramakrishnan & Ghauri** 1979b (in key to genera of the *Empoascanara* complex).

Inemadara
Records: **Hamilton, K. G. A.** 1975b (APHRODINAE, DELTOCEPHALINI, DELTOCEPHALINA). **Lee, C. E. & Kwon** 1977b (Korea). **Linnavuori** 1960a (Ceylon; western Micronesia). **Nast** 1972a (catalogued; =*Recilia*).

Infulatartessus
New species: **Evans, F. 1981a.**

Inghamia
New species: **Evans, J. W. 1966a.**

Inoclapis
New species: **Nielson 1979b.**

Inuyana
New species: **Young 1977a.**
Record: **Young** 1977a (n. comb. from *Cicadella*; key to species).

Ionia
Record: **Davis, R. B.** 1975a (morphology; phyletic inferences).

Iowanus
Record: **Hamilton, K. G. A.** 1975b (APHRODINAE, DELTOCEPHALINI, PLATYMETOPIINA).

Ipelloides
New species: **Evans, J. W. 1966a.**

Ipo
New species: **Evans, J. W. 1966a, 1969b.**
Records: **Davis, R. B.** 1975a (morphology; phyletic inferences). **Evans, J. W.** 1966a (synopsis); 1969a, 1977a (additional geographic records).

Ipoella
New species: **Evans, J. W. 1966a, 1973a, 1977a.**
Records: **Davis, R. B.** 1975a (morphology; phyletic inferences). **Evans, J. W.** 1966a (synopsis; n. comb. from *Anipo*; n. sp. syn.).

Ipoides
New species: **Evans, J. W. 1966a, 1973a.**
Records: **Evans, J. W.** 1966a (n. comb. from *Ipo*; n. sp. syn.); 1974a (New Caledonia).

Ipolo
Record: **Evans, J. W.** 1966a (n. comb. from *Ipoella*).

Iposa
New species: **Evans, J. W. 1977a.**

Iragua
New species: **Young 1977a.**
Record: **Young** 1977a (n. gen. syn.; n. comb. from *Diedrocephala* and *Tettigonia*; key to species).

Iraqerus
New species: **Ghauri 1972a.**
Record: **Nielson** 1975a (removed from COELIDIINAE; unassigned).

Irenaella
New species: **Linnavuori 1977a. Webb, M. D. 1980a.**

Irenara
Records: **Dworakowska** 1980d (=*Empoascanara*; n. syn.). **Ramakrishnan & Ghauri** 1979b (n. comb. from *Empoascanara* and *Thamnotettix*; in key to genera of the *Empoascanara* complex).

Iriatartessus
New species: **Evans, F. 1981a.**

Irinula
New species: **Lindberg 1958a. Linnavuori 1960a.**
Records: **Dlabola** 1957a (Turkey); 1959b (southern Europe); 1977c (Mediterranean); 1981a (Iran). **Hamilton, K. G. A.** 1975b (APHRODINAE, DELTOCEPHALINI, MACROSTELINA). **Jankovic** 1966a (Serbia). **Kalkandelen** 1974a (Turkey). **Lindberg** 1961a (Madeira). **Linnavuori** 1960a (Micronesia: southwest Caroline Islands, Tobi; key to species); 1961a (South Africa: Cape Province).

Iseza
New species: **Dworakowska 1981h.**

Ishidaella
Records: **Ishihara** 1958a (Japan: north Honshu); 1966a (Kurile Islands). **Vilbaste** 1968a (coastal regions of the USSR).

Ishiharanara
Records: **Dworakowska** 1980d (=*Empoascanara*, n. syn.). **Ramakrishnan & Ghauri** 1979b (in key to genera of the *Empoascanara* complex).

Ishiharella
New species: **Dworakowska 1982b.**
Records: **Dworakowska** 1970h (n. comb. from *Empoasca*; lectotype designated); 1982b (Japan: Hokkaido, Honshu, Shikoku).

Isogonalia
New species: **Young 1977a.**
Record: **Young** 1977a (n. comb. from *Tettigonia*; key to species).

Iturnoria
Record: **Nielson** 1975a (removed from COELIDIINAE; unassigned).

Ivorycoasta
New species: **Dworakowska 1972f.**

Jacobiasca
New species: **Ahmed, Manzoor 1979a. Dworakowska 1977g, 1977i, 1984a. Einyu & Ahmed 1980a.**
Records: **Ahmed, Manzoor** 1985c (Pakistan; n. comb. from *Austroasca*). **Dworakowska** 1976b (generic status; Taiwan); 1977d (n. comb. from *Amrasca*; n. sp. syn.); 1977e (Africa: Guinea, Nigeria, Rhodesia, Sudan, Zaire); 1979a (n. comb. from *Empoasca*); 1980b (southern India); 1982b (Bonin Islands, India, Japan, Vietnam); 1984a (Malaysia). **Dworakowska & Sohi** 1978b (India). **Sohi & Dworakowska** 1984a (list of Indian species).

Jacobiella
New species: **Webb, M. D. 1980a.**
Records: **Dworakowska** 1972c (n. comb. from *Chlorita*). **Dworakowska & Trolle** 1976a (Africa: Kenya, Nigeria, Sudan, Tanganyika, Zaire).

Jafar
Record: **Linnavuori** 1977a (n. gen. syn.; n. sp. syn.).

Jakarellus
Record: **Webb, M. D.** 1980a (n. comb. from *Scaphoideus*).

Jakrama
New species: **Young 1977a.**
Record: **Young** 1977a (n. comb. from *Tettigonia*; key to species).

Jalalia
New species: **Ahmed, Manzoor 1970b.**
Record: **Ahmed, Manzoor** 1983b (remarks about male stlye shape).

Jalorpa
New species: **Nielson 1979b.**

Jamacerus
New species: **Freytag 1969a, 1970a.**

Jamitettix
New species: **Linnavuori 1960a.**
Records: **Linnavuori** 1960a (generic affinities; key to species); 1975c (Micronesia; Guam). **Linnavuori & Al-Ne'amy** 1983a (BHATIINI, SELENOCEPHALINAE).

Janastana
Records: **Young** 1977a (n. comb. from *Cardioscarta* and *Tettigonia*; n. sp. syn.; key to species).

Jannius
New species: **Theron** 1982b.

Japanagallia
Records: **Anufriev** 1978a (USSR Maritime Territory). **Dworakowska** 1973d (Korea). **Ishihara** 1958a (Japan; descriptive notes). **Lee C. E.** 1979a (Korea, redescribed). **Lee, C. E. & Kwon** 1977b (Korea). **Viraktamath, C. A.** 1973b (n. comb. from *Agallia*).

Japananus
New species: **Bonfils** 1981a.
Records: **Anufriev** 1979a (USSR Maritime Territory). **Cantoreanu** 1971c (Romania). **Della Giustina** 1983a (France). **Emeljanov** 1964c (European USSR). **George** 1959a (Canada). **Hamilton, K. G. A.** 1975b (APHRODINAE, DELTOCEPHALINI, PLATYMETOPIINA); 1983a (adventive to Nearctic region). **Ishihara** 1968a (Japan). **Kwon & Lee** 1979b (Korea). **Lauterer** 1980a (Czechoslovakia); 1984a (Moravia; adventive). **Lee, C. E.** 1979a (Korea, redescribed). **Vilbaste** 1968a (coastal regions of the USSR).

Jascopus (fossil)
New species: **Hamilton, K. G. A.** 1971a.
Records: **Evans, J. W.** 1972c (discussion of taxonomic placement). **Hamilton, K. G. A.** 1971a (JASCOPIDAE).

Jassargus
New species: **Dlabola** 1958b. **D'Urso** 1980a, 1982a. **Emeljanov** 1966a. **Lindberg** 1960a, 1963b. **Linnavuori** 1955c. **Logvinenko** 1961b, 1963b, 1965a, 1966b, 1983a. **Mitjaev** 1967c. **Sawai Singh** 1969a. **Vilbaste** 1965b. **Wagner** 1958a.
Records: **Anufriev** 1977b (Kurile Islands); 1978a (USSR Maritime Territory). **Cantoreanu** 1959a, 1963a, 1965d, 1971c (Romania). **Della Giustina** 1983a (France). **Dlabola** 1957a (Turkey); 1961b (central Asia); 1964c (Albania); 1967b, 1967b, 1967c, 1970b (Mongolia). **D'Urso** 1980b (Corsica); 1982a (comparative notes). **Emeljanov** 1964c (European USSR; key to species); 1972e (USSR); 1977a (Mongolia). **Gyllensvard** 1969a (Sweden). **Hamilton, K. G. A.** 1975b (APHRODINAE, DELTOCEPHALINI, DELTOCEPHALINA). **Ishihara** 1961a (Thailand; n. comb. from *Deltocephalus*). **Jankovic** 1971a (Serbia). **Jansky** 1984a, 1985a (Slovakia). **Jasinska** 1980a (Poland, Bledowska Wilderness). **Lee, C. E.** 1979a (Korea, redescribed). **Lee, C. E. & Kwon** 1977b (Korea). **LeQuesne** 1964a, 1969c (Britain). **Lindberg** 1960a (Portugal). **Logvinenko** 1957b, 1963a (USSR: Ukraine). **Mitjaev** 1968b (as *Iassargus*; eastern Kazakhstan); 1971a (Kazakhstan; key to three species). **Nast** 1972a (catalogued); 1976a (Poland: Pieniny Mts.); 1976b (Poland; catalogued). **Ossiannilsson** 1974a (Norway); 1983a (Fennoscandia & Denmark; systematics; key to species). **Quartau & Rodrigues** 1969a (Portugal). **Ribaut** 1959a (France). **Schulz** 1976a (Norway). **Servadei** 1957a, 1958a, 1960a, 1968a, 1969a, 1971a, 1972a (Italy). **Vilbaste** 1958a, 1959b (Estonia); 1964a (Auwiesen Estlands); 1965b (Altai Mts.); 1969a (n. sp. syn.); 1973a (lectotype designation) 1980a (Kamchatka); 1980b (Tuva).

Jassoidula
Record: **Nielson** 1975a (= *Tharra*, n. syn.).

Jassolidia
Record: **Nielson** 1982e (n. comb. from *Jassus*).

Jassosqualus
Record: **Kramer** 1964b (n. comb. from *Carchariacephalus*).

Jassus
New species: **Anufriev** 1971a, 1977b. **Lindberg** 1958a.
Records: **Anufriev** 1977b (Kurile Islands). **Anufriev & Emeljanov** 1968a (Soviet Far East). **Datta** 1973e, 1973m (descriptive notes). **Emeljanov** 1964c (European USSR; key to species); 1977a (Mongolia). **ICZN** 1961b (placed on the Official List of Rejected and Invalid Names in Zoology, Opinion 612). **Lindberg** 1962a (Portugal). **Logvinenko** 1957b (USSR: Ukraine). **Servadei** 1968a, 1971a (Italy). **Vilbaste** 1973a (notes on Flor's material). **Webb, M. D.** 1979a (n. comb. to *Idiocerus*).

Javadikra
New species: **Dworakowska** 1971d.
Records: **Dworakowska** 1976b (Taiwan). **Dworakowska & Viraktamath** 1978a (southern India).

Jawigia
Record: **Nielson** 1979b (n. comb. from *Coelidia*; n. sp. syn.).

Jenolidia
New species: **Nielson** 1982e.
Record: **Nielson** 1982e (key to species).

Jikradia
New species: **Nielson** 1979b.
Record: **Nielson** 1979b (n. comb. from *Coelidia*; new status for species level taxa; key to species).

Jilijapa
Record: **Young** 1974a (lectotype designation).

Jilinga
Record: **Ghauri** 1974c (n. comb. from *Deltocephalus*).

Jimara
New species: **Dworakowska** 1977h.
Record: **Ahmed, Manzoor** 1983b (remarks about shape of male style.

Jiutepeca
New species: **Linnavuori & DeLong** 1978e.
Records: **Cwikla & Blocker** 1981a (comparative notes). **Linnavuori & DeLong** 1978e (n. comb. from *Dicyphonia*; systematics).

Jivena
New species: **Blocker** 1976a.
Record: **Blocker** 1979a (in key to New World genera of IASSINAE).

Jogocerus
New species: **Viraktamath, C. A.** 1979b.

Johanus
Record: **Theron** 1975b (n. comb. from *Euscelis*).

Joruma
Records: **Mahmood** 1967a (comparative notes). **Young** 1957b (notes on Osborn's species).

Josanus
Record: **Hamilton, K. G. A.** (APHRODINAE, DELTOCEPHALINI, PLATYMETOPIINA).

Jozima
Record: **Young** 1977a (conditional identification of type species).

Jubrinia
New species: **Heller & Linnavuori** 1968a. **Linnavuori** 1962a, 1969a. **Theron** 1971a.
Records: **Dlabola** 1977c (Mediterranean). **Hamilton, K. G. A.** 1975b (APHRODINAE, DELTOCEPHALINI, PLATYMETOPIINA). **Linnavuori** 1964a (Egypt).

Jukaruka
New species: **Evans, J. W.** 1966a.
Record: **Evans, J. W.** 1966a (synopsis).

Juliaca
New species: **Young** 1977a.
Record: **Young** 1977a (n. gen. syn.; n. sp. syn.; n. comb. from *Cardioscarta*, *Cicadella*, *Microgoniella*, *Tettigonia*, and *Tettigoniella*; key to species).

Julipopa
New species: **Blocker** 1979a.
Record: **Blocker** 1979a (n. comb. from *Batracomorphus*; in key to New World genera of IASSINAE).

Kaapia
New species: **Theron** 1983a.

Kabakra
New species: **Dworakowska** 1979a.

Kadrabia
New species: **Dworakowska & Sohi** 1978b.
Record: **Sohi & Dworakowska** 1984a (list of Indian species).

Kahaono
New species: **Dworakowska** 1972l. **Evans, J. W.** 1966a.
Record: **Evans, J. W.** 1966a (synopsis).

Kahavalu
Record: **Evans, J. W.** 1966a (synopsis).

Kaila
New species: **Dworakowska** 1974a, 1981f. **Webb, M. D.** 1980a.

Kalimorpha
New species: **Nielson** 1979b.
Record: **Nielson** 1979b (key to species).

Kalkiana
New species: **Sohi, Viraktamath & Dworakowska** 1980a.
Record: **Sohi & Dworakowska** 1984a (list of Indian species).

Kallebra
  Record: **Young** 1957c (n. gen. syn.; n. comb. to *Paralebra*).
Kaltitartessus
  New species: **Evans, F.** 1981a.
Kamaza
  New species: **Dworakowska** 1977f.
  Records: **Dworakowska** 1980b (India). **Dworakowska & Sohi** 1978b (India). **Sohi & Dworakowska** 1984a (list of Indian species).
Kanguza
  New species: **Dworakowska** 1972k.
  Records: **Dworakowska** 1980d (subgenus of *Empoascanara*, n. status). **Ramakrishnan & Ghauri** 1979b (in key to genera of the *Empoascanara* complex).
Kanorba
  New species: **Linnavuori** 1959b.
  Record: **Linnavuori** 1959b (key to species).
Kansendria
  Records: **Hamilton, K. G. A.** 1975b (APHRODINAE, DELTOCEPHALINI, DELTOCEPHALINA). **Kramer** 1971b (n. comb. from *Amplicephalus*).
Kanziko
  Record: **Linnavuori & Al-Ne'amy** 1983a (n. comb. from *Phlepsius*).
Kapateira
  Records: **Young** 1977a (n. comb. from *Oncometopia*).
Kapsa
  New species: **Dworakowska** 1979b, 1980b, 1981e, 1981j. **Dworakowska, Nagaich & Singh** 1978a. **Dworakowska & Sohi** 1978b.
  Records: **Ahmed, Manzoor** 1985c (Pakistan). **Dworakowska** 1972b (n. comb. from *Typhlocyba*); 1976b (notes on lectotype specimens). **Dworakowska, Nagaich & Singh** (n. comb. from *Erythroneura*). **Sharma, B.** 1984a (India: J. & K., Kishtwar). **Sohi & Dworakowska** 1984a (list of Indian species).
Karachiota
  New species: **Ahmed, Manzoor** 1969b.
  Record: **Ahmed, Manzoor** 1985c (Pakistan). **Sohi & Dworakowska** 1984a (list of Indian species).
Karasekia
  Record: **Linnavuori** 1972b (key to species).
Karoseefa
  New species: **Webb, M. D.** 1981b.
  Record: **Webb, M. D.** 1981b (key to species).
Kartwa
  Record: **Hamilton, K. G. A.** 1975d (APHRODINAE, EUPELICINI).
Kasinella
  New species: **Evans, J. W.** 1971a.
  Record: **Nielson** 1975a (removed from COELIDIINAE; unassigned).
Kasunga
  New species: **Linnavuori** 1979b.
  Record: **Linnavuori** 1979b (n. comb. from *Kosasia*).
Kaszabinus
  New species: **Diabola** 1965c.
  Records: **Diabola** 1967c, 1968b, 1970b (Mongolia). **Dworakowska** 1973d (Mongolia; n. comb. from *Adarrus*). **Emeljanov** 1977a (Mongolia; n. comb. from *Deltocephalus*; n. sp. syn.). **Hamilton, K. G. A.** 1975b (APHRODINAE, DELTOCEPHALINI, DELTOCEPHALINA). **Vilbaste** 1980b (Tuva).
Katipo
  Records: **Davis, R. B.** 1975a (morphology; phyletic inferences).
Kaukania
  New species: **Dworakowska** 1972j.
  Records: **Dworakowska** 1967b (Upper Laos); 1979b (Vietnam).
Kazachstanicus
  New species: **Diabola** 1961b, 1965e. **Dworakowska** 1973d.
  Records: **Diabola** 1965c, 1967b, 1970b (Mongolia). **Emeljanov** (as *Kasachstnicus*) 1964c (European USSR; key to species); 1972e (Kazakhstan); 1977a (Mongolia). **Hamilton, K. G. A.** 1975b (APHRODINAE, DELTOCEPHALINI, DELTOCEPHALINA). **Logvinenko** 1984a (USSR: Ukraine). **Mitjaev** 1968b (eastern Kazakhstan); 1971a (as *Kasachstanicus*; Kazakhstan; in key to genera of DELTOCEPHALINAE). **Vilbaste** 1962b (eastern part of the Caspian lowlands); 1965B (Altai Mts.); 1980b (Tuva).
Keia
  New species: **Theron** 1984c.
  Record: **Theron** 1984c (n. comb. from *Aletta*).
Keonolla
  New species: **DeLong & Currie** 1960a.
  Records: **Beirne** 1956a (Canada: British Columbia). **Nielson** 1968b (characterization of vector species). **Young** 1977a (=*Graphocephala*, n. syn.).
Khamiria
  New species: **Diabola** 1979a.
  Records: **Nast** 1982a (catalogued; DELTOCEPHALINAE). **Webb, M. D.** 1981b (=*Dryadomorpha*, n. syn.).
Khoduma
  New species: **Dworakowska** 1972b.
  Record: **Dworakowska & Viraktamath** 1975a (=*Arboridia*, n. syn.).
Kiamoncopsis
  New species: **Linnavuori** 1978b.
  Records: **Hamilton, K. G. A.** 1980b (subgenus of *Pediopsoides*). **Linnavuori** 1978b (key to species).
Kidraneuroidea
  New species: **Mahmood** 1967a.
  Record: **Dworakowska** 1984a (Singapore).
Kinonia
  Records: **Hamilton, K. G. A.** 1975b (APHRODINAE, STIRELLINI). **Vilbaste** 1965a (as *Kinnonia*; generic affinities).
Kirkaldykra
  Records: **Dworakowska** 1971d (n. comb. from *Dikraneura*); 1980b (southern India). **Sohi & Dworakowska** 1984a (list of Indian species).
Kivulopa
  New species: **Linnavuori** 1972a. **Van Stalle** 1982a, 1983c.
  Record: **Linnavuori** 1972a (n. comb. from *Ulopa*; in key to genera of African ULOPINAE). **Van Stalle** 1983c (east Africa).
Knightipsis
  New species: **Dworakowska** 1969e.
  Records: **Dworakowska** 1980b, 1982a (India). **Sharma, B.** 1977b (India: Jammu).
Knullana
  New species: **DeLong & Martinson** 1973a.
  Records: **Cwikla & Blocker** 1981a (comparative notes). **Hamilton, K. G. A.** 1975b (APHRODINAE, DELTOCEPHALINI, PLATYMETOPIINA).
Kodaikanalia
  New species: **Ramakrishnan & Menon** 1973a.
  Records: **Dworakowska** 1981d (=*Hauptidia*, n. syn.; n. comb. to *Hauptidia*). **Ramakrishnan & Menon** 1974a (in key to genera of ERYTHRONEURINI).
Koebelia
  New species: **Oman** 1971b.
  Records: **Beirne** 1956a (Canada: British Columbia). **Evans, J. W.** 1969a (discussion of taxonomic placement). **Kramer** 1966a (removed from LEDRINAE to KOEBELIINAE). **Oman** 1971b (review of subfamily).
Kogigonalia
  New species: **Young** 1977a.
  Record: **Young** 1977a (n. comb. from *Cardioscarta* and *Tacora*; key to species).
Kolella
  Record: **Evans, J. W.** 1966a (n. comb. from *Kolla* and *Tettigonia*; n. sp. syn.).
Kolla
  Records: **Anufriev** 1977b (Kurile Islands); 1978a (USSR: Maritime Territory); 1979a (n. sp. syn.). **Datta** 1973e, 1973m (descriptive notes). **Datta & Dhar** 1984a (descriptive notes). **Datta & Pramanik** 1977a (India: A. P., Subsanari Division). **Emeljanov** 1977a (Mongolia). **Ghosh, M.** 1974a (redescription of *K. maculifrons*). **Evans, J. W.** 1966a (n. comb. from *Cicadella* and *Tettigoniella*; n. sp. syn.); 1974a (New Caledonia records not of *Kolla*; unassigned); 1982a (Oriental origin presumed). **Heller & Linnavuori** 1968a (tropical Africa). **Ishihara** 1961a (Thailand); 1971a (revised

sp. syn.). **Kwon** 1983a (in key to genera of CICADELLINI; n. sp. syn.). **Lee, C. E.** 1979a (Korea, redescribed). **Lee, C. E. & Kwon** 1977b (Korea). **Linnavuori** 1969a (Congo); 1979a (*K. semipellucida* Jacobi "apparently...*Mileewa*"). **Ramachandra Rao** 1973a (India: Poona). **Sharma, B.** 1977b (India: Jammu). **Vilbaste** 1975a (confirms placement of *Jassus latruncularis* de Motschulsky). **Young** 1965c (lectotype designation). **Young & Nast** 1963a (type specimen information).

Kopamerra
New species: **Webb, M. D.** 1983b.
Record: **Webb, M. D.** 1983b (n. comb. from *Idiocerus*).

Koperta
New species: **Dworakowska** 1981e, 1981j.
Records: **Dworakowska** 1972g (n. comb. from *Typhlocyba*); 1976b (lectotype specimens); 1981j (India). **Sohi & Dworakowska** 1984a (list of Indian species).

Korana
Records: **Linnavuori** 1969a (Congo). **Linnavuori & Quartau** 1975a (transferred from IASSINAE, KRISNINI to DELTOCEPHALINAE, EUSCELINI). **Theron** 1980a (redescription of *K. rorulenta* (Stal)).

Koreanopsis
New species: **Kwon & Lee** 1979c.
Records: **Lee, C. E.** 1979a (Korea, redescribed). **Nast** 1982a (catalogued; DRABESCINAE).

Korsigianus
Record: **Nielson** 1979b (n. comb. from *Coelidia*).

Kosasia
New species: **Linnavuori** 1969a, 1979b.
Records: **Linnavuori** 1979b (n. comb. from *Kasunga*; key to species). **Linnavuori & DeLong** 1978c (removed from synonymy with *Hodoedocus*; NIRVANINAE). **Theron** 1982b (redescribed; NIRVANINAE).

Kosmiopelex
Records: **Evans, J. W.** 1966a (n. sp. syn.; APHRODINI); 1974a (=*Chiasmus*, n. syn; DELTOCEPHALINAE). **Hamilton, K. G. A.** 1975d (APHRODINAE, EUPELICINI).

Kotwaria
New species: **Dworakowska** 1984a.

Kramerana
New species: **DeLong & Thambimuttu** 1973b.
Records: **Cwikla & Blocker** 1981a (comparative notes). **Linnavuori & DeLong** 1977a (in key to genera of the *Faltala* group).

Krameraxis
New species: **Maldonado-Capriles** 1968a.
Record: **Cwikla & Blocker** 1981a (comparative notes).

Krameriata
New species: **Dworakowska** 1977e.

Kramerolidia
Record: **Nielson** 1982e (key to species).

Kravilidius
Record: **Nielson** 1979b (n. comb. from *Coelidia*).

Krisna
New species: **DeLong** 1982c. **Linnavuori** 1969a. **Linnavuori & Quartau** 1975a.
Records: **Datta** 1973e, 1973m (descriptive notes). **Ishihara** 1961a (Thailand). **Linnavuori** 1959b (key to Neotropical species); 1969a (Congo). **Linnavuori & Quartau** 1975a (IASSINAE, KRISNINI; species transferred to "*Distantiella*" (error for *Distantia*), SELENOCEPHALINAE; and to *Hypocostemma*, DELTOCEPHALINAE, EUSCELINI). **Linnavuori & DeLong** 1977a (IASSINAE, KRISNINI). **Sharma, B.** 1977b (India: Jammu).

Krocobella
New species: **Kramer** 1964b.

Krocodona
New species: **Kramer** 1964b.

Krocozzota
New species: **Kramer** 1964b.

Kropka
New species: **Dworakowska** 1970k.
Records: **Dlabola** 1971a (Iran, Turkey); 1977c (Mediterranean); 1981a (Iran). **Dworakowska** 1970k (n. comb. from *Erythroneura* and *Zygina*); 1981d (n. sp. syn.; comments on composition of the genus). **Lodos & Kalkandelen** 1984c (Turkey).

Krosolus
Record: **Nielson** 1982e (n. comb. from *Jassus*).

Kunasia
Record: **Nielson** 1975a (removed from COELIDIINAE; unassigned).

Kunzeana
Records: **Beardsley** 1956a (Hawaii; Oahu). **Mahmood** 1967a (morphology). **Namba** 1956b (Hawaii; Oahu). **Young** 1957b (species descriptions).

Kurotsuyanus
Record: **Kwon** 1983a (Korea).

Kusala
New species: **Dworakowska** 1981a.
Records: **Dworakowska** 1981a (n. comb. from *Erythroneura* and *Diomma*). **Sohi & Dworakowska** 1984a (list of Indian species).

Kuscheliola
New species: **Evans, J. W.** 1957a.
Records: **Kramer** 1964a (in key to New World genera of AGALLIINAE). **Linnavuori & DeLong** 1977c (Chile).

Kutara
New species: **Heller & Linnavuori** 1968a.
Records: **Heller & Linnavuori** 1968a (EUSCELINAE, EUSCELINI). **Linnavuori** 1978a (DRABESCINAE, DRABESCINI; in key to genera of DRABESCINI; African records erroneous). **Linnavuori & Al-Ne'amy** 1983a (SELENOCEPHALINAE, BHATIINI; in key to genera of BHATIINI). **Vilbaste** 1968a (USSR: coastal regions).

Kuznetsovium
New species: **Zachvatkin** 1953c.
Record: **Logvinenko** 1957b (USSR: Ukraine).

Kwempia
New species: **Ahmed, Manzoor** 1979a.

Kyboasca
New species: **Anufriev** 1979c. **Dlabola** 1967a. **Dubovskij** 1966a. **Kuoh, C. L.** 1981b. **Mitjaev** 1963c, 1971a.
Records: **Anufriev** 1969b, 1978a (USSR: Maritime Territory); 1979c (USSR: European zone). **Della Giustina** 1983a (France). **Dlabola** 1958c (Caucasus); 1958c (in key to genera of Palaearctic TYPHLOCYBINAE); 1960c (Iran); 1961b (central Asia); 1965c, 1967b (Mongolia); 1970a (Bohemia); 1970b (Mongolia); 1971a (Iran, Turkey); 1972a (Afghanistan); 1981a (Iran). **Dubovskij** 1978a (Uzbekistan, Zarafshan Valley); 1980a (USSR: western Turkmenia). **Dworakowska** 1967b (USSR: Maritime Territory); 1968b, 1968d (many Palaearctic sites); 1970h (n. comb. from *Empoasca*); 1973b (comments on species criteria); 1976b (China: Peking); 1977b (n. sp. syn.); 1982b (China, Japan, Korea). **Emeljanov** 1964c (European USSR; key to species); 1977a (Mongolia). **Hamilton, K. G. A.** 1983b (as subgenus of *Empoasca*; adventive to Nearctic). **Jankovic** 1966a (Serbia). **Kholmuminov & Dlabola** 1979a (USSR: Golodnostep Plain). **Kirejtshuk** 1977a (USSR: Kharkov region). **Korolevskaya** 1974a (Tadzhikistan); 1977a (central Asia). **Lauterer** 1984a (Czechoslovakia). **Lee, C. E.** 1979a (Korea, redescribed). **LeQuesne** 1961a (Britain). **LeQuesne & Payne** 1981a (Britain). **Lodos** 1982a (Turkey). **Lodos & Kalkandelen** 1983a (Turkey). **Mitjaev** 1963b (Kazakhstan); 1968b (eastern Kazakhstan); 1971a (Kazakhstan; key to eight species). **Nast** 1972a (catalogued); 1976b (Poland: Pieniny Mts.; catalogued). **Ossiannilsson** 1981a (Fennoscandia & Denmark). **Servadei** 1968a (Italy). **Vilbaste** 1965b (Altai Mts.); 1980b (Tuva). **Zachvatkin** 1953c (n. comb. from *Chlorita*).

Kybos
New species: **Dlabola** 1958b, 1963d, 1967a. **Dworakowska** 1968d, 1973c, 1976a. **Gerard** 1972a. **Hamilton, K. G. A.** 1972a. **Kuoh, C. L.** 1981b. **Logvinenko** 1980a, 1980b. **Mitjaev** 1963b, 1968a, 1971a, 1980c. **Ramakrishnan & Menon** 1972a. **Vilbaste** 1961b, 1980b. **Wagner** 1959a. **Zachvatkin** 1953a, 1953d.
Records: **Anufriev** 1978a (USSR: Maritime Territory; key to species). **Cantoreanu** 1971c (Romania). **Claridge, M. F. & Wilson, M. R.** (host associations). **Dlabola** 1958a (Czechoslovakia and southern Europe); 1958c (in key to

genera of Palaearctic TYPHLOCYBINAE); 1959a (Bohemia); 1961b (central Asia); 1964d (Bohemia); 1967b, 1967d, (Mongolia); 1971a (Turkey); 1972a (Afghanistan); 1981a (Iran). **Dubovskij** 1966a (Uzbekistan; key to species); 1970b (Uzbekistan); 1978a (Uzbekistan, Zarafshan Valley). **Dworakowska** 1967a (Mongolia); 1973c (n. comb. from *Empoasca*; synopsis of Palaearctic fauna); 1976a (subgenus of *Empoasca*: n. sp. syn.; species groups defined; synopsis of Palaearctic fauna); 1977d (notes on type specimens); 1982b (China, Japan, Korea). **Emeljanov** 1964c (European USSR; key to species); 1977a (Mongolia). **Gunthart, H.** 1971a (Switzerland). **Gyllensvard** 1963a (Sweden). **Hamilton, K. G. A.** 1972a (as subgenus of *Empoasca*; key to species of southern interior of British Columbia, Canada). **Jankovic** 1966a, 1971a (Serbia). **Jasinska** 1980a (Poland: Bledowska Wilderness). **Kirejtshuk** 1977a (USSR: Karkov region). **Knight, W. J.** 1976a (New Zealand; in key to genera of TYPHLOCYBINAE). **Korolevskaya** 1974a (Tadzhikistan). **Lee, C. E.** 1979a (Korea, redescribed). **Lee, C. E. & Kwon** 1977b (Korea). **LeQuesne** 1961a (Britain). **LeQuesne & Payne** 1981a (Britain). **Logvinenko** 1957b (USSR: Ukraine). **Mahmood** 1967a (as synonym of *Empoasca*). **Mitjaev** 1963b (Kazakhstan); 1968b (eastern Kazakhstan); 1971a (Kazakhstan; key to 11 species). **Nast** 1972a (catalogued); 1976a (Poland: Pieniny Mts.); 1976b (Poland; catalogued). **Ossiannilsson** 1974a (Norway); 1981a (Fennoscandia & Denmark; as synonym of *Empoasca*). **Schulz** 1976a (Norway). **Servadei** 1960a, 1968a, 1969a, 1971a, 1972a (Italy). **Trolle** 1966a (Denmark). **Vilbaste** 1958a, 1958b (Estonia); 1962b (eastern part of the Caspian lowlands); 1965b (Altai Mts.); 1968a (coastal regions of the USSR); 1973a (notes on Flor's material); 1980a (Kamchatka); 1980b (Tuva). **Wagner** 1959a (Italy; key to 2 species). **Wilson, M. R.** 1978a (descriptions and key characters of 5th instars). **Zachvatkin** 1953a, 1953b (central Russia; Oka district).

Kyphocotis
  Record: **Evans, J. W.** 1966a (n. comb. from *Stenocotis*; n. sp. syn.; key to species).

Kyphoctella
  New species: **Evans, J. W.** 1966a.

Labocurtidia
  New species: **Nielson** 1979b, 1983a.
  Records: **Nielson** 1979b (key to species); 1983a (revised key to species).

Laburrus
  New species: **Emeljanov** 1962a. **Mitjaev** 1971a. **Vilbaste** 1961a, 1965b, 1966a.
  Records: **Anufriev** 1977b (Kurile Islands); 1978a (USSR: Maritime Territory). **Choe** 1985 (Korea). **Dlabola** 1957a (Afghanistan); 1958a (Czechoslovakia & southern Europe); 1958b (Caucasus); 1961b (central Asia); 1965c, 1967b, 1967c, 1967d, 1968a, 1968b, 1970b (Mongolia); 1972a (Afghanistan); 1981a (Iran). **Dubovskij** 1966a, 1970b (Uzbekistan); 1980a (USSR: western Turkmenia). **Dworakowska** 1973d (Korea, Mongolia). **Emeljanov** 1964c (European USSR; key to species); 1977a (Mongolia). **Hamilton, K. G. A.** 1975b (APHRODINAE, DELTOCEPHALINI, ATHYSANINA). **Ishihara** 1958a (Japan: north Honshu). **Kalkandelen** 1974a (Turkey; key to subgenera and species). **Kholmuminov & Dlabola** 1979a (USSR: Golodnostep Plain). **Korolevskaya** 1979b (Tadzhikistan). **Lee, C. E.** 1979a (Korea, redescribed). **Lee, C. E. & Kwon** 1977b (Korea). **Lindberg** 1960b (Soviet Armenia). **Logvinenko** 1957b, 1957c (USSR: Ukraine). **Mitjaev** 1968a (eastern Kazakhstan); 1971a (Kazakhstan); 1971b (southern Kazakhstan). **Nast** 1972a (catalogued); 1976b (Poland; catalogued). **Ossiannilsson** 1983a (Fennoscandia & Denmark). **Servadei** 1958a, 1968a, 1971a (Italy). **Vilbaste** 1965b (Altai Mts.); 1968a (USSR: coastal regions); 1973a (notes on Flor's material); 1980b (Tuva).

Ladoffa
  Record: **Young** 1977a (n. comb. from *Tettigonia*; key to species).

Ladya
  New species: **Theron** 1982b.

Laevicephalus
  New species: **Hamilton, K. G. A.** 1972b. **Hamilton, K. G. A. & Ross** 1975a. **Ross & Hamilton** 1972b.
  Records: **Beirne** 1956a (Canada; key to species). **Hamilton, K. G. A.** 1972d (Canada: Manitoba); 1975b (APHRODINAE, DELTOCEPHALINI, DELTOCEPHALINA). **Ross & Hamilton** 1972b (genus redefined; n. comb. from *Orocastus*; n. comb. to *Cribrus*; n. sp. syn.; species groups defined).

Laevilidia
  New species: **Nielson** 1979b.

Lajolla
  New species: **Linnavuori** 1959b.
  Record: **Linnavuori & DeLong** 1978b (in key to genera of the *Bahita* group).

Lamia
  New species: **Evans, J. W.** 1966a. **Linnavuori** 1960b.
  Record: **Linnavuori & Al-Ne'amy** 1983a (SELENOCEPHALINAE, BHATIINI).

Lampridius
  Record: **Viraktamath, S. & Viraktamath, C. A.** 1980a (genus redefined).

Lamprotettix
  Records: **Cantoreanu** 1971c (Romania). **Dlabola** 1961c (Romania); 1977c (Mediterranean). **Emeljanov** 1964c (European USSR). **Hamilton, K. G. A.** 1975b (APHRODINAE, DELTOCEPHALINI, PLATYMETOPIINA). **Jansky** 1983a (Slovakia). **LeQuesne** 1969c (Britain; DELTOCEPHALINAE, ATHYSANINI). **Nast** 1972a, 1976b (catalogued; DELTOCEPHALINAE, ATHYSANINI). **Ossiannilsson** 1983a (Fennoscandia & Denmark; DELTOCEPHALINAE, ATHYSANINI). **Servadei** 1957a, 1960a (Italy). **Vilbaste** 1973a (notes on Flor's material).

Lamtoana
  New species: **Ahmed, Manzoor** 1979a. **Dworakowska** 1982f. **Einyu & Ahmed** 1983a.

Laneola
  Records: **Young** 1977a (n. comb. from *Tettigonia*). **Young & Beier** 1963a (lectotype designation).

Lankasca
  New species: **Dworakowska** 1980b. **Ramakrishnan & Menon** 1972a.
  Records: **Dworakowska & Viraktamath** 1978a (India). **Ghauri** 1963e (n. comb. from *Empoasca*). **Mahmood** 1967a (in key to genera of Oriental TYPHLOCYBINAE). **Sohi & Dworakowska** 1984a (list of Indian species).

Laokayana
  New species: **Dworakowska** 1972i.
  Records: **Dworakowska** 1972g (China: Canton); 1972i (n. comb. from *Empoasca*; new subgenus proposed). **Dworakowska & Viraktamath** 1975a (=*Amrasca*, n. syn.).

Lareba
  Record: **Young** 1957c (n. comb. from *Protalebra*).

Lascumbresa
  New species: **Linnavuori & DeLong** 1979b.
  Record: **Cwikla & Blocker** 1981a (comparative notes).

Lasioscopus
  Records: **Davis, R. B.** 1975a (morphology; phyletic inferences). **Evans, J. W.** 1966a (synopsis).

Latalus
  New species: **Ross & Hamilton** 1972a.
  Records: **Beirne** 1956a (Canada; key to species). **Hamilton, K. G. A.** 1972d (Canada: Manitoba); 1975b (APHRODINAE, DELTOCEPHALINI, DELTOCEPHALINA); 1983b (Peru). **Ross & Hamilton** 1972b (n. comb. to *Deltocephalus*; n. sp. syn.)

Laterana
  Record: **Hamilton, K. G. A.** 1975b (APHRODINAE, DELTOCEPHALINI, ATHYSANINA).

Lausulus
  Record: **Hamilton, K. G. A.** 1975b (APHRODINAE, DELTOCEPHALINI, DELTOCEPHALINA).

Lautereria
  New species: **Young** 1977a.
  Record: **Young** 1977a (n. comb. from *Cardioscarta* and *Cicadella*; key to species).

Lautereriana
  New species: **Dworakowska** 1974a.

Lawsonellus
  Record: **Young** 1975c (n. comb. from *Elabra*).

Lebaja
New species: **Young** 1977a.

Lebora
Records: **Schroder** 1959a (generic rank; n. comb. from *Parametopia* and *Oncometopia*). **Young** 1968a (= *Oncometopia*).

Lebradea
New species: **Dlabola** 1967a. **Ossiannilsson** 1976a. **Remane** 1959a.
Records: **Anufriev** 1977b (Kurile Islands); 1978a (USSR: Maritime Territory). **Dlabola** 1967b; 1967c (Mongolia). **Dworakowska** 1969a (Mongolia). **Emeljanov** 1966a (n. comb. from *Sorhoanus*); 1977a (Mongolia). **Hamilton, K. G. A.** 1975b (APHRODINAE, DELTOCEPHALINI, DELTOCEPHALINA). **Lee, C. E.** 1979a (Korea, redescription). **Lee, C. E. & Kwon** 1977b (Korea). **Ossiannilsson** 1976a (key to Palaearctic species); 1983a (Fennoscandia & Denmark; key to species; DELTOCEPHALINAE, PARALIMNINI). **Vilbaste** 1968a (USSR: coastal regions); 1969b (Taimyr); 1980a (Kamchatka); 1980b (Tuva).

Lecacis
New species: **Theron** 1982h.

Lectotypella
New species: **Dworakowska** 1976b, 1979b, 1981e.
Record: **Dworakowska** 1972k (n. comb. from *Zygina*; n. sp. syn.).

Ledeira
New species: **Dworakowska** 1969f.
Records: **Dworakowska** 1979c (India). **Sohi & Dworakowska** 1984a (list of Indian species).

Ledra
Records: **Anufriev** 1978a (USSR: Maritime Territory). **Bonfils & Della Giustina** 1978a (Corsica). **Choe** 1985a (Korea). **Datta & Dhar** 1984a (descriptive notes). **Dlabola** 1977c (Mediterranean). **Emeljanov** 1964c (European USSR). **Ishihara** 1958a (Japan: north Honshu). **Jankovic** 1966a (Serbia). **Kramer** 1966a (n. comb. to *Zyzzogeton*). **Kwon & Lee** 1978b (male genital structures described). **Lee, C. E.** 1979a (Korea, redescription). **Lee, C. E. & Kwon** 1977b (Korea). **LeQuesne** 1965c (Britain). **Lodos** 1982a (Turkey). **Lodos & Kalkandelen** 1981a (Turkey). **Nast** 1972a, 1976b (catalogued). **Ossiannilsson** 1981a (Fennoscandia & Denmark). **Servadei** 1960a, 1968a, 1971a (Italy). **Sharma, B.** 1977b (India: Jammu). **Viggiani** 1974a (Italy). **Vilbaste** 1968a (USSR: coastal regions).

Ledracorrhis
New species: **Evans, J. W.** 1959a.

Ledracotis
Record: **Evans, J. W.** 1966a (synopsis).

Ledraprora
Record: **Evans, J. W.** 1966a (synopsis).

Ledrella
Records: **Evans, J. W.** 1966a (synopsis); 1982a (phyletic inferences).

Ledroides
Record: **Evans, J. W.** 1972a (descriptive note; PENTHIMIINAE).

Ledromorpha
Record: **Evans, J. W.** 1966a (synopsis; n. sp. syn.).

Ledropsella
Record: **Evans, J. W.** 1966a (n. comb. from *Platyledra*).

Ledropsis
Records: **Anufriev** 1978a (USSR: Maritime Territory; in key to genera of LEDRINAE). **Evans, J. W.** 1966a (synopsis; n. sp. syn.). **Kwon & Lee** 1978b (Korea; synopsis). **Lee, C. E.** 1979a (Korea, redescription). **Lee, C. E. & Kwon** 1977b (Korea). **Vilbaste** 1968a (USSR: coastal regions).

Ledrotypa
Record: **Morrison** 1973a (= *Glossocratus*, n. syn.).

Lemellus
Records: **Beirne** 1956a (Canada; western regions). **Hamilton, K. G. A.** 1972d (Canada: Manitoba); 1975b (APHRODINAE, DELTOCEPHALINI, DELTOCEPHALINA).

Leofa
Records: **Datta** 1973f, 1973n (descriptive notes). **Datta & Dhar** 1984a (descriptive notes).

Leuconeura
Record: **Nast** 1982a (catalogued; TYPHLOCYBINAE).

Leucopepla
Records: **Young** 1968a (= *Diestostemma*, n. syn.). **Young & Lauterer** 1966a (lectotype designation). **Young & Nast** 1963a (notes on type specimens).

Levantotettix
New species: **Lindberg** 1953a.
Record: **Linnavuori** 1962a (= *Selenocephalus*, n. syn.).

Libengaia
New species: **Linnavuori** 1969a.
Record: **Linnavuori** 1969a (n. comb. from *Eutettix*).

Lichtrea
New species: **Dworakowska** 1976b, 1981e.

Licolidia
New species: **Nielson** 1979b.

Licontinia
New species: **Nielson** 1979b.
Record: **Nielson** 1979b (n. comb. from *Terulia*; n. sp. syn.; key to species).

Liguropia
Records: **Bonfils & Della Giustina** 1978a (Corsica). **Dlabola** 1958c (in key to genera of Palaearctic DIKRANEURINI); 1967e (Cyprus). **Linnavuori** 1965a (Turkey). **Lodos & Kalkandelen** 1983b (Turkey). **Logvinenko** 1984a (USSR: Ukraine). **Servadei** 1968a (Italy). **Vidano** 1965a (Mediterranean region).

Limassolla
New species: **Chou, I. & Ma** (as *Limmasolla*). **Chou, I. & Zhang** 1985a. **Dworakowska** 1969d, 1972g, 1977i. **Dworakowska & Lauterer** 1975a. **Sohi** 1976a.
Records: **Chou, I. & Zhang** 1985a (key to Chinese species). **Dlabola** 1974d (Egypt). **Dworakowska** 1969d (= *Pruthius*, n. syn.; redescribed; n. comb. from *Pruthius* and *Zygina*); 1972g (n. sp. syn.; ZYGINELLINI; known species listed). **Dworakowska & Lauterer** 1975a (Taiwan). **Sohi** 1976a (n. comb. from *Pruthius*).

Limbanus
Record: **Hamilton, K. G. A.** 1975b (APHRODINAE, DELTOCEPHALINI, ATHYSANINA).

Limentinus
Records: **Nielson** 1982e (n. comb. from *Coelidia*; n. sp. syn.). **Webb, M. D.** 1980a (Aldabra).

Limotettix
New species: **Dlabola** 1967c. **Dubovskij** 1966a. **Emeljanov** 1962a, 1964c, 1966a. **Evans, J. W.** 1966a. **Knight, W. J.** 1975a. **Kuoh, C. L.** 1981b (as *Limmotettix*). **Vilbaste** 1965b, 1966a, 1968a, 1973b, 1980b.
Records: **Anufriev** 1977b (Kurile Islands); 1978a (USSR: Maritime Territory); 1979a (notes on Jacobi species). **Badmin** 1970a (Britain). **Beirne** 1956a (Canada; key to species). **Cantoreanu** 1965d (Romania); 1968b (Danube river delta); 1971c (Romania). **Choe** 1985a (Korea). **Dlabola** 1957a (Turkey); 1957b (Afghanistan); 1958b (Caucasus); 1961b (central Asia); 1963c (notes on Haupt's material); 1964c (Albania); 1966a, 1967b, 1967c, 1967d, 1970b (Mongolia); 1972a (Afghanistan); 1981a (Iran). **Dubovskij** 1966a, 1970b (Uzbekistan); 1978a (Uzbekistan, Zarafshan Valley). **Emeljanov** 1964c (European USSR; key to species); 1966a (new subgenus proposed; key to Palaearctic species); 1977a (Mongolia). **Hamilton, K. G. A.** 1972d (Canada: Manitoba); 1975b (APHRODINAE, DELTOCEPHALINI, ATHYSANINA); 1983b (transboreal taxon). **Ishihara** 1966a (Kurile Islands). **Kalkandelen** 1974a (Turkey). **Kitching, Grylls & Waterford** 1973a (redescriptions; interbreeding trials; *capitata* Kirkaldy returned from *Cicadulina*). **Knight, W. J.** 1975a (New Zealand; n. comb. from *Cicadula*); 1976b (Chatham Island). **Korolevskaya** 1979b (Tadzhikistan). **Lee, C. E. & Kwon** 1979b (Korea, redescribed). **LeQuesne** 1969c (Britain). **Lindberg** 1960a (Portugal). **Linnavuori** 1959a (Mt. Sibillini); 1959b (West Indies). **Logvinenko** 1957b, 1984a (USSR: Ukraine). **Mitjaev** 1968b (eastern Kazakhstan); 1971a (Kazakhstan; key to species). **Nast** 1972a (catalogued);

1976a (Poland: Pieniny Mts.); 1976b (Poland; catalogued); 1985a (n. comb. to *Tetartostylus*). **Ossiannilsson** 1983a (Fennoscandia & Denmark; key to species). **Remane** 1962a (northwest Germany). **Servadei** 1958a, 1968a (Italy). **Vilbaste** 1958a, 1959b (Estonia); 1962b (eastern part of the Caspian lowlands); 1964a (Auwiesen Estlands); 1965b (Altai Mts.); 1968a (USSR: coastal regions); 1973a (notes on Flor's material); 1973b (synopsis; Nearctic species referred to subgenus *Neodrylix*; key to four Palaearctic species); 1980a (Kamchatka); 1980b (Tuva). **Zachvatkin** 1953b (central Russia; Oka district).

Limpica
New species: **Cheng, Y. J.** 1980a.
Record: **Cwikla & Blocker** 1981a (comparative notes).

Linacephalus
Record: **Evans, J. W.** 1977a (n. comb. from *Cephalelus*).

Lindbergana
Records: **Hamilton, K. G. A.** 1975b (APHRODINAE, XESTOCEPHALINI). **Linnavuori** 1959b (=*Xestocephalus*, n. syn.; (XESTOCEPHALINAE, XESTOCEPHALINI).

Lindbergina
New species: **Asche** 1980b. **LeQuesne** 1977a.
Records: **Dlabola** 1958c (in key to genera of Palaearctic TYPHLOCYBINAE). **Dworakowska** 1982a (n. comb. from *Youngiada*). **Kocak** 1983a (senior synonym of *Youngaiada*). **Lauterer** 1984a (Moravia). **LeQuesne** 1977a (n. comb. from *Typhlocyba*; n. sp. syn.). **LeQuesne & Payne** 1981a (Britain).

Linnatanus
Record: **Cwikla & Blocker** 1981a (comparative notes).

Linnavuoriana
New species: **Dlabola** 1961b. **Dubovskij** 1966a. **Dworakowska** 1968b, 1971g, 1982a. **Mitjaev** 1963b.
Records: **Anufriev** 1977b (Kurile Islands); 1978a (USSR: Maritime Territory). **Dlabola** 1958a (Czechoslovakia); 1958c (in key to genera of Palaearctic TYPHLOCYBINAE); 1959b (Austria); 1966a, 1967b, 1968a, 1968b (Mongolia); 1971a, 1972a (Afghanistan); 1981a (Iran). **Dubovskij** 1966a, 1970b (Uzbekistan); 1978a (Uzbekistan, Zarafshan Valley). **Dworakowska** 1971g, (n. sp. syn.); 1982a (new subgenus proposed; n. comb. from *Agnesiella* and *Byphlocyta*; summary of included taxa). **Emeljanov** 1964c (European USSR; key to species); 1977a (Mongolia). **Gunthart, H.** 1971a (Switzerland). **Jankovic** 1971a (Serbia). **Kirejtshuk** 1977a (USSR: Karkov region). **Korolevskaya** 1974a (Tadzhikistan). **Lee, C. E.** 1979a (Korea, redescribed). **LeQuesne & Payne** 1981a (Britain). **Linnavuori** 1965a (Turkey). **Mitjaev** 1963b (Kazakhstan); 1967b (southeastern Kazakhstan; sp. transferred from *Typhlocyba*); 1968b (eastern Kazakhstan); 1971a (Kazakhstan; key to 3 species). **Nast** 1972a (catalogued); 1976a (Poland: Pieniny Mts.); 1976b (Poland; catalogued). **Ossiannilsson** 1974a (Norway); 1981a (Fennoscandia & Denmark; key to species). **Quartau & Rodrigues** 1969a (Portugal). **Servadei** 1972a (Italy). **Sharma, B.** 1984a (India: J. & K., Kishtwar). **Sohi & Dworakowska** 1984a (list of Indian species). **Vilbaste** 1965b (Altai Mts.); 1968a (USSR: coastal regions); 1973a (notes on Flor's material); 1979a (Vooremaa hardwood/spruce forest); 1980b (Tuva). **Wilson, M. R.** (descriptions and key characters for 5th instars).

Linnavuoriella
New species: **Linnavuori** 1975a.
Records: **Evans, J. W.** 1966a (n. comb. from *Parabolocratus*). **Linnavuori** 1969a (Congo); 1975a (key to Ethiopian species); 1975c (n. comb. from *Parabolocratus*; Micronesia). **Morrison** 1973a (=*Hecalus*, n. syn.).

Liocratus
Records: **Dlabola** 1974b (generic status; in key to Palaearctic IDIOCERINAE). **Hamilton, K. G. A.** 1980a (subgenus of *Idiocerus*).

Lipata
New species: **Dworakowska** 1974a.

Lissocarta
New species: **Young** 1977a.
Record: **Young** 1977a (key to species).

Listrophora
New species: **Boulard** 1971a. **Linnavuori** 1979a.

Record: **Linnavuori** 1979a (key to species).

Litura
New species: **Knight, W. J.** 1970b. **Malhotra & Sharma, B.** 1977a.
Record: **Knight, W. J.** 1970b (n. comb. from *Hishimonus*).

Livasca
Record: **Sohi & Dworakowska** 1984a (subgenus of *Empoasca*).

Lodia
New species: **Nielson** 1982e.

Lodiana
New species: **Nielson** 1982e.
Record: **Nielson** 1982e (n. comb. from *Coelidia* and *Jassus*; n. sp. syn.; revised sp. syn.; key to species).

Loepotettix
Records: **Cardosa** 1974a (Portugal). **Hamilton, K. G. A.** 1975b (APHRODINAE, DELTOCEPHALINI, ATHYSANINA). **Nielson** 1968b (generic status; characterization of vector species). **Ossiannilsson** 1983a (as subgenus of *Thamnotettix*; Fennoscandia & Denmark).

Loipothea
New species: **Heller & Linnavuori** 1968a. **Linnavuori** 1969a.

Loja
Record: **Young & Nast** 1963a (notes on holotype specimen).

Lojanus
New species: **Linnavuori** 1959b.

Lojata
Record: **Young** 1968a (redescribed).

Loka
New species: **Linnavuori & Al-Ne'amy** 1983a.
Record: **Linnavuori & Al-Ne'amy** 1983a (n. comb. from *Distantia, Kutara* and *Selenocephalus*; key to species).

Lokia
New species: **Thapa** 1984b.

Lonatura
New species: **Beirne** 1955a. **Kramer** 1967d, 1976b.
Records: **Beirne** 1956a (Canada; key to species). **Evans, J. W.** 1966a (synopsis). **Hamilton, K. G. A.** 1972d (Canada: Manitoba); 1975b (APHRODINAE, DELTOCEPHALINI, DELTOCEPHALINA). **Kramer** 1967d (n. comb. to *Deltocephalus, Destria* and *Endria*; key to species).

Lonenus
Record: **Hamilton, K. G. A.** 1975b (APHRODINAE, DELTOCEPHALINI, PLATYMETOPIINA).

Loralia
New species: **Evans, J. W.** 1966a.

Lorellana
New species: **DeLong & Kolbe** 1975a.
Record: **Cwikla & Blocker** 1981a (comparative notes).

Loreta
New species: **DeLong** 1983b. **Linnavuori** 1959b. **Linnavuori & DeLong** 1978c, 1979a.
Record: **Linnavuori** 1959b (new subgenus proposed; key to subgenera and species).

Lowata
New species: **Dworakowska** 1977i.
Record: **Dworakowska** 1979c (ZYGINELLINI).

Lualabanus
New species: **Linnavuori** 1975a.
Record: **Linnavuori** 1975a (key to species).

Lublinia
New species: **Dworakowska** 1970l, 1972f.
Record: **Dworakowska** 1970l (n. comb. from *Erythroneura*).

Lucumius
Record: **Young** 1977a (=*Hadria*, n. syn.).

Luheria
Record: **Linnavuori** 1959b (redescribed; DELTOCEPHALINAE, LUHERIINI).

Lupola
New species: **Nielson** 1982e.
Record: **Nielson** 1982e (n. comb. from *Coelidia*; key to species).

Lusitanocephalus
New species: **Quartau** 1970a.
Records: **Hamilton, K. G. A.** 1975b (APHRODINAE, DELTOCEPHALINI, DELTOCEPHALINA). **Nast** 1982a (catalogued; DELTOCEPHALINAE).

Luteobalmus
New species: **Maldonado-Capriles** 1977b.

Luvila
New species: **Dworakowska** 1974a, 1977e.
Record: **Dworakowska** 1976b (Cameroon).

Lycioides
Record: **Hamilton, K. G. A.** 1975b (APHRODINAE, DELTOCEPHALINI, PLATYMETOPIINA).

Lycisca
New species: **Linnavuori** 1979a.

Lystridea
New species: **Kramer** 1967c.
Records: **Kramer** 1967c (key to species). **Oman & Musgrave** 1975a (redefined).

Macednus
New species: **Emeljanov** 1962a.
Records: **Hamilton, K. G. A.** 1975b (as *Macnedus*; APHRODINAE; unassigned). **Nast** 1972a (catalogued; DELTOCEPHALINAE). **Vilbaste** 1967a (=*Albicostella*, n. syn.).

Macroceps
New species: **Evans, J. W.** 1966a, 1969a.
Records: **Evans, J. W.** 1966a (synopsis). **Lindberg** 1960b (Soviet Armenia). **Mitjaev** 1971a (in key to genera of MELICHARELLINAE).

Macroceratogonia
Records: **Evans, J. W.** 1966a (redescribed); 1974a (n. sp. syn.).

Macropsella
New species: **Hamilton, K. G. A.** 1980b.
Record: **Hamilton, K. G. A.** 1980b (n. comb. from *Macropsis*).

Macropsidius
New species: **Dlabola** 1961b, 1963b, 1964c, 1965c, 1967a, 1974a, 1975a. **Emeljanov** 1964c, 1964f, 1972b. **Linnavuori** 1971a. **Logvinenko** 1965a, 1981b. **Mitjaev** 1967a, 1971a, 1973a. **Ribaut** 1959b. **Vilbaste** 1965b, 1980b.
Records: **Anufriev** 1978a (USSR: Maritime Territory). **Della Giustina** 1983a (France). **Dlabola** 1956a (Czechoslovakia); 1956b (Hungary); 1957a (Turkey); 1957b (Afghanistan); 1958b (Caucasus); 1961c (Italy); 1963b (in key to genera of European MACROPSINAE); 1964a (Afghanistan); 1967b, 1967c, 1967d (Mongolia); 1972a (Afghanistan); 1975a (n. sp. syn.); 1981a (Iran). **Dubovskij** 1978a (Uzbekistan, Zarafshan Valley); 1980a (USSR: Ukraine). **Emeljanov** 1964c (European USSR; key to species); 1977a (Mongolia). **Hamilton, K. G. A.** 1980b (as subgenus of *Macropsis*). **Kholmuninov & Dlabola** 1979a (USSR: Golodnostep Plain). **Korolevskaya** 1975a (Tadzhikistan). **Lee, C. E.** 1979a (Korea, redescribed). **Linnavuori** 1956d (Spain); 1965a (Tunisia). **Lodos & Kalkandelen** 1981a (Turkey). **Logvinenko** 1957b (USSR: Ukraine); 1981b (Caucasus; key to nine species). **Mitjaev** 1968b (eastern Kazakhstan); 1971a (Kazakhstan; key to eight species); 1973a (Kazakhstan). **Nast** 1972a (catalogued); 1976b (Poland; catalogued). **Servadei** 1968a, 1969a (Italy). **Vilbaste** 1962b (eastern part of the Caspian lowlands); 1965b (Altai Mts.); 1980b (Tuva).

Macropsis
New species: **Anufriev** 1971a. **Anufriev & Zhilysova** 1982a. **Dlabola** 1957b, 1958b, 1961b, 1963d, 1967e. **Dubovskij** 1966a. **Emeljanov** 1964a. 1964b. 1964f. **Evans, J. W.** 1971b, 1974a. **Hamilton, K. G. A.** 1972b, 1980b, 1983d. **Ishihara** 1961a. **Kameswara Rao & Ramakrishnan** 1979d. **Korolevskaya** 1963a. **Kuoh, C. L.** 1981a, 1981b. **Lindberg** 1958a. **Linnavuori** 1969a, 1978b. **Mitjaev** 1971a. **Nast** 1981a. **Vilbaste** 1965b, 1968a, 1980b. **Viraktamath** 1980a, 1981b. **Wagner** 1964a.
Records: **Anufriev** 1977b (Kurile Islands); 1978a (USSR: Maritime Territory; key to species); 1981a (n. sp. syn.; infraspecific variation). **Beirne** 1956a (Canada; key to species). **Bonfils & Della Giustina** 1978a (Corsica). **Cantoreanu** 1968b (Danube river delta). **Cobben** 1978a (discussion of species concepts). **Della Giustina** 1983a (France). **Dlabola** 1957a (Turkey); 1957b (Afghanistan); 1958b (Caucasus); 1959a (Bohemia); 1961b (central Asia); 1961c (Romania); 1964a (Afghanistan); 1964c (Albania); 1965d (Jordan); 1967b, 1967d, 1968b (Mongolia); 1972a (Afghanistan); 1974a (Iran); 1977c (Mediterranean); 1981a (Iran). **Downes** 1957a (Canada: British Columbia). **Dubovskij** 1966a (Uzbekistan; key to species); 1970b (Uzbekistan); 1978a (Uzbekistan, Zarafshan Valley); 1980a (USSR: western Turkmenia). **D'Urso** 1980b (Cyprus). **Emeljanov** 1964c (European USSR; key to species); 1977a (Mongolia). **Evans, J. W.** 1966a (synopsis; n. comb. from *Oncopsis*). **Freytag** 1976a (USA: Kentucky). **Hamilton, K. G. A.** 1972b, 1972d (Canada: Manitoba); 1976a (identity of Provancher species; n. sp. syn.); 1980b (n. gen. syn.; n. comb. from *Tsavopsis*); 1983b (species common to Palaearctic and Nearctic); 1983d (key to New World species; specific synonymies clarified). ICZN 1961a (designation of type species under Plenary Powers, Opinion 603). **Ishihara** 1966a (Kurile Islands). **Jankovic** 1966a, 1971a (Serbia). **Jasinska** 1980a (Poland: Bledowska Wilderness). **Kameswara Rao & Ramakrishnan** 1979d (India; key to species). **Khokhlova & Anufriev** 1981a (population variation in color and patterns). **Korolevskaya** 1975a (Tadzhikistan). **Lauterer** 1984a (Moravia and Slovakia). **Lee C. E.** 1979a (Korea, redescribed). **Lee, C. E. & Kwon** 1977b (Korea). **LeQuesne** 1961b, 1964b, 1965c (Britain). **Lindberg** 1960a (Portugal); 1960b (Soviet Armenia); 1964a (Morocco). **Lindberg & Wagner** 1965a (Canary Islands). **Linnavuori** 1959c (Mt. Picentini); 1962a (Israel); 1965a (Turkey); 1978b (Ethiopian region; species groups defined; key to species). **Lodos** 1982a (Turkey). **Lodos & Kalkandelen** 1981a (Turkey). **Logvinenko** 1957b (USSR: Ukraine). **Mitjaev** 1968b, 1971a (Kazakhstan; key to 28 species level taxa). **Nast** 1955a (Poland); 1972a (catalogued); 1976a (Poland: Pieniny Mts.); 1976b (Poland; catalogued). **Nielson** 1968b (characterization of vector species). **Ossiannilsson** 1974a (Norway); 1981a (Fennoscandia & Denmark; key to species). **Quartau & Rodrigues** 1969a (Portugal). **Schulz** 1976a (Norway). **Scudder** 1961a (Canada: British Columbia). **Servadei** 1957a, 1960a, 1968a, 1971a, 1972a (Italy). **Theron** 1970b (South Africa; notes on type material of *capensis* (Cogan)); 1980a (n. sp. syn.). **Valley & Wheeler** 1985a (illustrations of immature stages). **Vilbaste** 1958a (Estonia); 1962b (eastern part of the Caspian lowlands); 1968a (coastal regions of the USSR); 1973a (notes on Flor's material); 1980a (Kamchatka); 1980b (Tuva). **Viraktamath, C. A.** (India; key to species); 1981a (redescriptions). **Wagner** 1960a (Appeal to ICZN for designation of type species); 1964a (The Netherlands; key to species inhabiting ROSACEAE). **Zachvatkin** 1953b (central Russia; Oka district).

Macrosteles
New species: **Anufriev** 1968c. **Dlabola** 1962c, 1970b, 1974c. **Dubovskij** 1966a, 1970c. **Hamilton, K. G. A.** 1972b, 1983b. **Kuoh, C. L.** 1981b. **LeQuesne** 1968a. **Lindberg** 1963a. **Mitjaev** 1969a. **Moore, T. E. & Ross** 1957a. **Razvyaskina** 1957a (new name). **Vilbaste** 1965b, 1968a, 1980a, 1980b.
Records: **Anufriev** 1968c (Soviet Far East; key to species); 1977b (Kurile Islands); 1978a (USSR: Maritime Territory; key to species). **Anufriev & Emeljanov** 1968a (n. sp. syn.). **Beirne** 1956a (Canada; key to species). **Bonfils & Della Giustina** 1978a (Corsica. **Cantoreanu** 1959a, 1961a (Romania); 1968b (Danube river delta); 1971c (Romania). **Della Giustina** 1983a (France). **DeLong** 1966a, 1970b (Alaska; tidal flat habitat). **Dlabola** 1956b (Hungary, Romania); 1957a (Turkey); 1957b (Afghanistan); 1958a (Yugoslavia); 1958b (Caucasus); 1960c (Iran); 1961b (central Asia); 1964a (Afghanistan); 1964c (Albania); 1965c (Mongolia); 1965d (Jordan); 1966a, 1967b, 1967c, 1967d, 1968a, 1968b (Mongolia); 1970a (Bohemia); 1970b (Mongolia); 1971a (Iran); 1972a (Afghanistan); 1974d (Egypt); 1977c (Mediterranean); 1981a, 1984a (Iran). **Dubovskij** 1966a, 1970b (Uzbekistan); 1978a (Uzbekistan, Zarafshan Valley); 1980a (USSR: western Turkmenia). **D'Urso** 1980b (Cyprus). **Emeljanov** 1964c (European USSR; key to species); 1977a (Mongolia). **Gajewski** 1961a (Poland; key to species; sp. syn. noted). **Gravestein** 1965a (The Netherlands: Terschelling I.). **Hamilton, K. G. A.** 1972d (Canada: Manitoba); 1975b (APHRODINAE, DELTOCEPHALINI, MACROSTELINA); 1983b (transboreal taxon). **Heller & Linnavuori** 1968a (Ethiopia). **Ishihara** 1958a (Japan: north Honshu); 1966a (Kurile Islands). **Jankovic** 1966a (Serbia). **Jansson** 1973a (Sweden;

SEM scrutiny of cuticular topography). **Jasinska** 1980a (Poland: Bledowska Wilderness). **Kalkandelen** 1974a (Turkey; key to species). **Kholmuninov & Dlabola** 1979a (USSR: Golodnostep Plain). **Knight, W. J.** 1975a (New Zealand; adventive?). **Korolevskaya** 1971b, 1973a (Tadzhikistan); 1978b (central Asia). **Lauterer** 1957a (Czechoslovakia); 1980a (Slovakia); 1984a (Bohemia, Moravia, Slovakia). **Lee, C. E.** 1979a (Korea, redescribed). **LeQuesne** 1959a (Britain); 1968a (lectotype designation for *M. sexnotata*). **Lindberg** 1960a (Portugal); 1960b (Soviet Armenia); 1960c (Azores); 1961a (Madeira); 1962a (Portugal); 1965a (Spanish Sahara). **Lindberg & Wagner** 1965a (Canary Islands). **Linnavuori** 1956d (Spain & Spanish Morocco); 1959c (Mt. Picentini); 1959b (West Indies); 1961a (South Africa); 1962a (Israel); 1964a (Egypt). **Linnavuori & DeLong** 1977c (Chile; adventive). **Lodos** 1982a (Turkey). **Lodos & Kalkandelen** 1985b (Turkey). **Logvinenko** 1957b, 1957c, 1963a (USSR: Ukraine). **Mitjaev** 1968b (eastern Kazakhstan); 1971a (Kazakhstan; key to 11 species). **Moore, T. E. & Ross** 1957a (USA: Illinois; key to species). **Nast** 1955a (Poland); 1972a (catalogued); 1976a (Poland: Pieniny Mts.); 1976b (Poland; catalogued). **Nielson** 1968b (characterization of vector species). **Okali** 1963a (Slovakia); 1968a (middle and central Europe). **Ossiannilsson** 1974a (Norway); 1983a (Fennoscandia & Denmark; key to species). **Prior** 1966a (Britain). **Quartau** 1979a, 1980b (Azores). **Quartau & Rodrigues** 1969a (Portugal). **Raatikainen** 1968a (Finland). **Razvyaskina** 1957a, 1960a (descriptions). **Remane** 1962a (discussion of sp. syn.). **Ribaut** 1959a (France). **Servadei** 1957a, 1958a, 1960a, 1968a, 1971a, 1972a (Italy). **Sharma, B.** 1977b (India: Jammu). **Schulz** 1976a (Norway). **Theron** 1980a (identity of *Jassus dilectus* Stal). **Trolle** 1966a, 1968a (Denmark). **Vilbaste** 1958a, 1959b (Estonia); 1962b (eastern part of the Caspian lowlands); 1964a (Auwiesen Estlands); 1968a (USSR: coastal regions); 1969b (identity of Matsumura's material); 1973a (Flor's material); 1979a (Vooremaa hardwood/spruce forest); 1980a (Kamchatka); 1980b (Tuva). **Wagner** 1959c (Greece). **Woodroffe** 1971a (Britain). **Zachvatkin** 1953b (central Russia; Oka district).

Macrotartessus
New species: **Evans, F.** 1981a.

Macugonalia
New species: **Young** 1977a.
Record: **Young** (n. comb. from *Cicadella, Tettigella* and *Tettigonia*; n. sp. syn.; key to species).

Macunolla
New species: **Young** 1977a.
Record: **Young** 1977a (n. comb. from *Tettigonia*; n. sp. syn.).

Macustus
Records: **Anufriev** 1978a (USSR: Maritime Territory). **Beirne** 1956a (Canada). **Cantoreanu** 1971c (Romania). **Dlabola** 1967b, 1967d, 1970b (Mongolia). **Emeljanov** 1964c (European USSR); 1977a (Mongolia). **Gravestein** 1965a (The Netherlands: Terschelling I.). **Hamilton, K. G. A.** 1972d (Canada: Manitoba); 1975b (APHRODINAE, DELTOCEPHALINI, ATHYSANINA). **Jasinska** 1980a (Poland: Bledowska Wilderness). **LeQuesne** 1969c (Britain). **Mitjaev** 1971a (Kazakhstan). **Nast** 1972a (catalogued); 1976a (Poland: Pieniny Mts.); 1976b (Poland; catalogued). **Ossiannilsson** 1974a (Norway); 1983a (Fennoscandia & Denmark). **Schulz** 1976a (Norway). **Servadei** 1968a (Italy). **Vilbaste** 1964a (Auwiesen Estlands); 1965b (Altai Mts.); 1973a (notes on Flor's material); 1980a (Kamchatka); 1980b (Tuva).

Macutella
New species: **Evans, J. W.** 1972a.

Magnentius
New species: **Linnavuori** 1978a.
Records: **Datta** 1973n (descriptive notes). **Evans, J. W.** 1971b (PENTHIMIINAE); 1972a (comparative notes; PENTHIMIINAE). **Linnavuori** 1978a (NIONIINAE, MAGNENTIINI).

Maguangua
Record: **Young** 1965c (lectotype designation).

Mahalana
Record: **Webb, M. D.** 1981a (n. comb. from *Zizyphoides*; n. sp. syn.).

Mahellus
New species: **Nielson** 1982e.
Record: **Nielson** 1982e (n. comb. from *Coelidia*; key to species).

Mahmoodia
New species: **Dworakowska** 1970a.
Records: **Dworakowska** 1970a (n. comb. from *Typhlocyba*); 1976b (Indonesia); 1984a (Malaysia).

Mahmoodiana
New species: **Ahmed, Manzoor & Waheed** 1971a.
Records: **Ahmed, Manzoor** 1972a (in key to genera of East Pakistan TYPHLOCYBINAE); 1985c (Pakistan; n. comb. to *Togaritettix*).

Maichewia
New species: **Linnavuori & Al-Ne'amy** 1983a.
Record: **Linnavuori & Al-Ne'amy** 1983a (key to species).

Makilingana
New species: **Mahmood** 1967a.

Makilingia
Records: **Linnavuori** 1979a (CICADELLINAE, MAKILINGIINI). **Young** 1968a (MAKILINGIINAE); 1977a (not CICADELLINAE). **Young & Lauterer** 1966a (lectotype designations).

Malagasiella
Record: **Nielson** 1975a (removed from COELIDIINAE; unassigned).

Maldonadora
New species: **Webb, M. D.** 1983b.

Malendea
New species: **Linnavuori & Al-Ne'amy** 1983a.

Malichus
Record: **Evans, J. W.** 1972a (PENTHIMIINAE).

Malicia
Record: **Evans, J. W.** 1972b (to ACOSTEMNINAE).

Malipo
New species: **Evans, J. W.** 1966a.
Record: **Evans, J. W.** 1966a (n. comb. from *Ipo*).

Malmaemichungia
New species: **Kwon** 1983a.
Record: **Kwon** 1983a (CICADELLINAE, ERRHOMENINI, MALMAEMICHUNGIINA; in key to genera of ERRHOMENINI).

Mandera
New species: **Ahmed, Manzoor** 1971b. **Dworakowska** 1980b, 1981e, 1984a. **Sohi** 1977a.
Records: **Dworakowska** 1977f (northern India). **Sohi & Dworakowska** 1984a (list of Indian species).

Mandola
New species: **Sohi** 1977a.
Records: **Ahmed, Manzoor** 1985c (Pakistan). **Dworakowska & Viraktamath** 1975a (n. comb. from *Zygina*). **Sohi & Dworakowska** 1984a (list of Indian species).

Mangganeura
New species: **Ghauri** 1967a.
Record: **Hongsaprug & Wilson** 1985a (additional generic characters).

Manzoonara
Record: **Dworakowska** 1980d (=*Empoascanara*, n. syn.).

Manzutus
Records: **Young** 1968a (comments on tribal characteristics); 1977a (n. comb. from *Tettigonia*; n. sp. syn.; key to species).

Mapochia
New species: **Linnavuori** 1979a.
Record: **Lindberg** 1956a (as *Napochia*; Niger River flood plain; Dogo). **Linnavuori** 1979a (key to species).

Mapochiella
New species: **Evans, J. W.** 1966a.

Marathonia
Record: **Young** 1977a (=*Graphocephala*, n. syn.).

Marcapatiana
New species: **Nielson** 1979b.

Mareba
New species: **Young** 1968a.
Records: **Young** 1965c (lectotype designation); 1968a (n. gen. syn.; n. comb. from *Tettigonia*; n. sp. syn.; key to species).

Mareja
New species: **Young** 1977a.
Records: **Ramos Elorduy** 1972c (Mexico; catalogued). Schroder 1959a (diagnosis). **Young** 1977a (n. comb. from *Tettigonia*; key to species).

Marganana
New species: **DeLong & Freytag** 1963a.
Records: **DeLong & Freytag** 1963a (subgenera defined); 1972a (in key to genera of GYPONINAE).

Maricaona
New species: **Linnavuori** 1959b.
Records: **DeLong & Martinson** 1973g (redescription). Linnavuori 1959b (key to species).

Marizella
Records: **Young** 1977a (= *Dilobopterus*, n. syn.). **Young & Nast** 1963a (notes on type specimens).

Maroopula
New species: **Webb, M. D.** 1983a.

Masiripius
Records: **Dlabola** 1981a, 1984a (Iran).

Matatua
Record: **Knight, W. J.** 1976a (n. comb. from *Dikraneura*; lectotype designation).

Matsumurasca
Record: Nast 1982a (catalogued; TYPHLOCYBINAE).

Matsumuratettix
Record: **Hamilton, K. G. A.** 1975b (APHRODINAE, DELTOCEPHALINI, DELTOCEPHALINA).

Matsumurella
New species: **Anufriev** 1971d. **Emeljanov** 1962a, 1972b, 1977a.
Records: **Anufriev** 1971d (n. sp. syn.; key to species; =*Shonenus*, n. syn); 1978a (USSR: Maritime Territory); 1979a (northeast China). **Chang, Y. D. & Choe** 1982a (Korea; Mt. Gyeryong). **Emeljanov** 1977a (Mongolia). **Hamilton, K. G. A.** 1975b (APHRODINAE, DELTOCEPHALINI, ATHYSANINA). Ishihara 1958a (Japan: north Honshu). **Kown & Lee** 1979b (Korea). **Lee C. E.** 1979a (Korea, redescribed). **Lee, C. E. & Kwon** 1977b (Korea).

Matsumurina
Record: **Dworakowska** 1972b (n. comb. from *Zygina*).

Mattogrossus
New species: **Linnavuori** 1959b.

Matuta
Records: **Anufriev** 1970f (=*Pagaronia*, n. syn.). **Emeljanov** 1966a (n. comb. from *Epiacanthus*).

Mblokoa
New species: **Linnavuori** 1972a.
Record: **Van Stalle** 1983c (Zaire).

Medlerola
New species: **Young** 1977a.

Megabahita
New species: **DeLong** 1982a, 1984g. **Linnavuori & DeLong** 1978b.
Records: **Cwikla & Blocker** 1981a (comparative notes). **Linnavuori & DeLong** 1978b (n. comb. from *Eutettix*; in key to genera of the *Bahita* group).

Megacoelidia
New species: **Kramer & Linnavuori** 1959a.
Record: **Kramer** 1964d (redescribed; key to species).

Megagallia
Record: **Kramer** 1964a (in key to New World genera of AGALLIINAE).

Megalidia
New species: **Nielson** 1982e.

Megalotettigella
Record: **Sharma, B.** 1977b (India: Jammu).

Megalopsius
New species: **Emeljanov** 1961a.
Records: **Dlabola** 1981a, 1984a (Iran). **Hamilton, K. G. A.** 1975b (APHRODINAE, SELENOCEPHALINI).

Megaulon
Record: **Theron** 1975a (n. comb. from *Deltocephalus*).

Megipocerus
Records: **Anufriev** 1978a (USSR: Maritime Territory; in key to genera of IDIOCERINAE). **Dlabola** 1974b (in key to genera of Palaearctic IDIOCERINAE).

Megophthalmus
Records: **Bonfils & Della Giustina** 1978a (Corsica). Cantoreanu 1968b (Danube river delta). **Dlabola** 1957a (Turkey); 1964c (Albania); 1981a (Iran). **Emeljanov** 1964c (European USSR). Jankovic 1971a (Serbia). **Lindberg** 1960a (Portugal). **Linnavuori** 1959a (Mt. Sibillini); 1959c (Mt. Picentini); 1962a (Israel). **Lodos & Kalkandelen** 1981a (Turkey). **Logvinenko** 1957b, 1984a (USSR: Ukraine). **Nast** 1972a (catalogued); 1976a (Poland: Pieniny Mts.); 1976b (Poland; catalogued). **Ossiannilsson** 1981a (Fennoscandia & Denmark). **Quartau & Rodrigues** 1969a (Portugal). **Servadei** 1960a, 1968a, 1971a, 1972a (Italy). **Vilbaste** 1958a (Estonia).

Megulopa
New species: **Van Stalle** 1983a.
Records: **Davis, R. B.** 1975a (morphology; phyletic inferences). Lindberg 1956a (male genitalia figured; Niger River flood plain; Dogo); 1963a (Morocco). **Linnavuori** 1962a (Israel); 1972a (in key to African genera of ULOPINAE).

Meketia
Records: **Dworakowska** 1982a (n. comb. from *Carnulus*). **Sohi & Dworakowska** 1984a (list of Indian species).

Melicharella
New species: **Dlabola** 1960a, 1960c. **Dubovskij** 1968b. **Emeljanov** 1966a, 1975a. **Mitjaev** 1971a.
Records: **Davis, R. B.** 1975a (morphology; phyletic inferences). **Dlabola** 1981a (Iran). **Dubovskij** 1978a (Uzbekistan, Zarafshan Valley); 1980a (USSR: western Turkmenia). **Emeljanov** 1964c (European USSR); 1975a (n. comb. to *Homogramma*; in key to genera of ADELGUNIINI); 1977a (Mongolia). **Korolevskaya** 1975a (Tadzhikistan). **Lindberg** 1960b (Soviet Armenia; figures); 1963a (Morocco); 1965a (Spanish Sahara). **Mitjaev** 1968b (eastern Kazakhstan); 1971a (Kazakhstan; key to 7 species).

Melichariella
Records: **Lindberg & Wagner** 1965a (Canary Islands). **Linnavuori** 1975c (Micronesia: Bonin Islands). **Linnavuori & Al-Ne'amy** 1983a (in key to genera of BHATIINI, SELENOCEPHALINAE; list of known species). **Nast** 1972a (catalogued; =*Bhatia*).

Melillaia
Records: **Hamilton, K. G. A.** 1975b (APHRODINAE; unassigned). **Linnavuori** 1971a (transfer of species from *Thamnotettix* noted).

Memnonia
New species: **Linnavuori & DeLong** 1978e.
Records: **Hamilton, K. G. A.** 1975b (APHRODINAE, HECALINI). **Linnavuori** 1975a (EUSCELINI ?). **Linnavuori & DeLong** 1978e (DELTOCEPHALINAE, HECALINI).

Mendozellus
New species: **Linnavuori** 1959b. **DeLong** 1982c. **DeLong & Cwikla** 1985a. **Cheng, Y. J.** 1980a. **Linnavuori & DeLong** 1977c.
Records: **Hamilton, K. G. A.** 1975b (APHRODINAE, DELTOCEPHALINI, DELTOCEPHALINA). **Kramer** 1971d (n. comb. from *Deltocephalus*).

Mendrausus
Records: **Dlabola** 1967b (Mongolia). **Dworakowska** 1969a (Mongolia). **Emeljanov** 1964c (European USSR); 1977a (Mongolia). **Hamilton, K. G. A.** 1975b (APHRODINAE, DELTOCEPHALINI, DELTOCEPHALINA). **Jankovic** 1966a, 1971a (Serbia). **Logvinenko** 1957b (USSR: Ukraine). **Mitjaev** 1968b (eastern Kazakhstan); 1971a (Kazakhstan). **Nast** 1955a (Poland); 1972a (catalogued); 1976b (Poland; catalogued). **Ossiannilsson** 1983a (Fennoscandia & Denmark). **Vilbaste** 1965a (Altai Mts.); 1980b (Tuva).

Mendreus
Record: **Hamilton, K. G. A.** 1975b (APHRODINAE, DELTOCEPHALINI, DELTOCEPHALINA).

Menosoma
New species: **Cheng, Y. J.** 1980a. **Linnavuori** 1959b. **Linnavuori & DeLong** 1978b.
Records: **Cheng, Y. J.** 1980a (Paraguay; in key to genera of EUSCELINAE). **Hamilton, K. G. A.** 1975b (APHRODINAE, DELTOCEPHALINI, PLATYMETOPIINA). **Linnavuori** 1956a (Brazil,

Colombia); 1959b (key to Neotropical species). **Linnavuori & DeLong** 1978b (in key to the *Bahita* group of genera).

Meremra
New species: **Dworakowska & Viraktamath** 1979a.
Record: **Sohi & Dworakowska** 1984a (list of Indian species).

Meroleucocerus
New species: **Maldonado-Capriles** 1972b.

Mesadorus
Records: **Cheng, Y. J.** 1980a (Paraguay). **Linnavuori** 1959b (comparative notes).

Mesagallia
New species: **Logvinenko** 1975b.
Records: **Lodos & Kalkandelen** 1981a (Turkey). **Logvinenko** 1975b (n. comb. from *Agallia*; n. sp. syn.).

Mesamia
New species: **DeLong** 1980h, 1984g.
Records: **Beirne** 1956a (Canada; key to species). **Hamilton, K. G. A.** 1972d (Canada: Manitoba); 1975b (APHRODINAE, DELTOCEPHALINI, PLATYMETOPIINA). **Linsley & Usinger** 1966a (Galapagos Islands).

Mesargus
Records: **Davis, R. B.** 1975a (morphology; phyletic inferences). **Vilbaste** 1975a (= *Moonia,* = *Sitades*; generic and specific synonymy discussed).

Mesobana
Record: **Young** 1968a (= *Abana*).

Mesodorydium
Record: **Linnavuori** 1979a (= *Paradorydium*, n. syn.).

Mesogonia
New species: **Young** 1977a.
Record: **Young** 1977a (n. comb. from *Cicadella, Graphocephala, Kolla,* and *Tettigoniella*; n. sp. syn.; key to species).

Metacephalus
New species: **DeLong & Martinson** 1973f.

Metagoldeus
New species: **Bonfils** 1981a. **Remane & Asche** 1980a.
Records: **Della Giustina** 1983a (France). **Dlabola** 1982a (catalogued; DELTOCEPHALINAE).

Metalimnus
New species: **Emeljanov** 1966a.
Records: **Anufriev** 1977b (Kurile Islands); 1978a (USSR: Maritime Territory). **Cantoreanu** 1968b (Danube river delta). **Dlabola** 1976a (Saint Helena). **Dworakowska** 1973d (Mongolia). **Emeljanov** 1964c (European USSR; key to species); 1977a (Mongolia). **Hamilton, K. G. A.** 1975b (APHRODINAE, DELTOCEPHALINI, DELTOCEPHALINA). **Kwon & Lee** 1979b (Korea). **Lee, C. E.** 1979a (Korea, redescribed). **Lee, C. E. & Kwon** 1977b (Korea). **LeQuesne** 1969c (Britain). **Logvinenko** 1957b (USSR: Ukraine). **Nast** 1972a (catalogued); 1976a (Poland: Pieniny Mts.); 1976b (Poland; catalogued). **Ossiannilsson** 1983a (Fennoscandia & Denmark; key to species). **Remane** 1962a (northwest Germany). **Servadei** 1968a (Italy). **Vilbaste** 1964a (Auwiesen Estlands); 1965b (Altai Mts.); 1968a (USSR: coastal regions); 1973a (notes on Flor's material); 1980b (Tuva).

Metapocirtus
Record: **Servadei** 1969a (as *?Metaporcitus*; Italy).

Metascarta
Records: **Ramos Elorduy** 1972c (Mexico; catalogued). **Young** 1977a (identity unknown).

Metcalfiella
Records: **Linnavuori** 1956a (Argentina); 1959b (name invalid).

Metidiocerus
Record: **Ossiannilsson** 1981a (Fennoscandia & Denmark; key to species).

Mexara
Record: **Hamilton, K. G. A.** 1975b (APHRODINAE, DELTOCEPHALINI, DELTOCEPHALINA).

Mexolidia
New species: **Nielson** 1983h.

Mfutila
New species: **Dworakowska** 1974a.

Mgenia
New species: **Theron** 1984c.
Record: **Theron** 1984c (n. comb. from *Coelidia* and *Equeefa*).

Miarogonalia
Record: **Young** 1977a (n. comb. from *Apogonalia*).

Micantulina
Records: **Anufriev** 1970d (new subgenus proposed; n. comb. from *Cicadula, Dikraneura* and *Typhlocyba*). **Bonfils & Della Giustina** 1978a (Corsica). **Dlabola** 1971a (Iran, Turkey); 1977b (Iran); 1977c (Mediterranean); 1981a (Iran). **Emeljanov** 1977a (Mongolia). **Kirejtshuk** 1977a (USSR: Kharkov region). **Korolevskaya** 1974a (Tadzhikistan). **Lauterer** 1972a (n. sp. syn.). **Nast** 1972a (catalogued); 1976b (Poland; catalogued). **Ossiannilsson** 1981a (Fennoscandia & Denmark; key to species). **Vilbaste** 1979a (Vooremaa hardwood/spruce forest); 1980a (Kamchatka); 1980b (Tuva).

Michalowskiya
New species: **Dworakowska** 1972k, 1980b.
Record: **Sohi & Dworakowska** 1984a (list of Indian species).

Micrelloides
New species: **Evans, J. W.** 1973a.

Microgoniella
New species: **Young** 1977a.
Records: **Young** 1974a (lectotype designation); 1977a (n. comb. from *Cicadella* and *Erythrogonia*; n. sp. syn.; key to species). **Young & Soos** 1964a (lectotype designation).

Microledrella
New species: **Evans, J. W.** 1969a.
Record: **Evans, J. W.** 1982a (generic affinites; biogeography).

Microlopa
New species: **Evans, J. W.** 1966a.

Microscita
Record: **Young** 1977a (= *Juliaca*, n. syn.).

Microtartessus
New species: **Evans, F.** 1981a.
Record: **Evans, F.** 1981a (n. comb. from *Tartessus*).

Mileewa
New species: **Anufriev** 1971e. **Heller & Linnavuori** 1968a. **Linnavuori** 1969a, 1979a.
Records: **Anufriev** 1978a (USSR: Maritime Territory; CICADELLINAE, MILEEWANINI; key to 2 species). **Datta & Dhar** 1984a (descriptive notes). **Dworakowska** 1973d (Korea). **Evans, J. W.** 1966a (CICADELLINAE, MILEEWANINI). **Ishihara** 1965a (Formosa). **Kwon** 1983a (Korea; CICADELLINAE, MILEEWANINI). **Lee, C. E.** 1979a (Korea, redescribed). **Lee, C. E. & Kwon** (Korea; catalogued). **Linnavuori** 1961a (South Africa: Cape Province); 1979a (CICADELLINAE, MILEEWANINI; key to African species). **Linnavuori & DeLong** 1977a (comparative notes). **Mahmood** 1967a (subfamily placement discussed). **Vilbaste** 1968a (USSR: coastal regions). **Young** 1977a (not CICADELLINAE). **Young & Lauterer** 1966a (lectotype designation).

Milotartessus
Record: **Evans, F.** 1981a (n. comb. from *Tartessus*).

Mimallygus
New species: **Emeljanov** 1964f.
Records: **Dlabola** 1964c (southern Europe). **Emeljanov** 1977a (Mongolia) **Hamilton, K. G. A.** 1975b (APHRODINAE, DELTOCEPHALINI, PLATYMETOPIINA). **Vilbaste** 1980b (Tuva).

Mimodorus
New species: **Linnavuori** 1959b.
Record: **Linnavuori** 1959b (n. comb. from *Mesadorus*; keys to 2 subgenera and species).

Mimodrylix
Records: **Hamilton, K. G. A.** 1975b (APHRODINAE, APHRODINI, DORATURINA). **Ishihara** 1961a (Thailand; DELTOCEPHALIDAE).

Mimohardya
Record: **Hamilton, K. G. A.** 1975b (APHRODINAE, DELTOCEPHALINI, ATHYSANINA).

Mimotettix
New species: **Kwon & Lee** 1979b.
Records: **Hamilton, K. G. A.** 1975b (APHRODINAE, DELTOCEPHALINI, DELTOCEPHALINA). **Lee, C. E.** 1979a (Korea, redescribed).

Mindanaoa
New species: **Mahmood** 1967a.

Miraldus
New species: **Lindberg** 1960a.
Records: **Hamilton, K. G. A.** 1975b (APHRODINAE, DELTOCEPHALINI, DELTOCEPHALINA). **Lindberg** 1960a (n. comb. from *Deltocephalus*; Portugal). **Quartau & Rodrigues** 1969a (Portugal).

Mirzayansus
New species: **Dlabola** 1979a.
Record: **Nast** 1982a (catalogued; DELTOCEPHALINAE).

Mitelloides
Record: **Evans, J. W.** 1966a (synopsis).

Mitjaevia
New species: **Dworakowska** 1979b, 1980b. **Korolevskaya** 1976a.
Records: **Ahmed, Manzoor** 1985c (Pakistan). **Dlabola** 1972a (Afghanistan, central Asia). **Dworakowska** 1970n (n. comb. from *Erythroneura*); 1980b (n. comb. from *Helionidia*). **Dworakowska & Viraktamath** 1975a (n. comb. from *Helionidia*). **Korolevskaya** 1974a (Tadzhikistan); 1976a (n. comb. from *Erythroneura*; key to species). **Sohi & Dworakowska** 1984a (list of Indian species).

Mocoa
New species: **Linnavuori & DeLong** 1978d.
Record: **Cwikla & Blocker** 1981a (comparative notes).

Mocuastrum
Records: **Nast** 1982a (catalogued; DELTOCEPHALINAE). **Vilbaste** 1980b (Tuva).

Mocuellus
New species: **Dlabola** 1958b, 1965c, 1967a, 1967c, 1967d. **Emeljanov** 1962a, 1964f, 1966a, 1972b, 1979a. **Kalkandelen** 1972a. **Linnavuori** 1964a. **Logvinenko** 1960a, 1971a, 1975a. **Mitjaev** 1967c. **Ross & Hamilton** 1970b. **Vilbaste** 1961b, 1980b.
Records: **Anufriev** 1978a (USSR: Maritime Territory; in key to genera of PARALIMNINI). **Beirne** 1956a (Canada). **Cantoreanu** 1965d, 1971c (Romania). **Dlabola** 1957a (Turkey); 1958b (Caucasus); 1959b (southern Europe); 1961b (central Asia); 1965c, 1966a, 1967b, 1967c, 1967d, 1968a, 1968b, 1970b (Mongolia). **Dubovskij** 1966a, 1970b (Uzbekistan); 1978a (Uzbekistan, Zarafshan Valley). **Dworakowska** 1969a, 1973d (Mongolia). **Emeljanov** 1962a (n. comb. from *Deltocephalus*; n. sp. syn.); 1964c (European USSR; key to species); 1977a (Mongolia). **Gyllensvard** 1965a (Sweden). **Hamilton, K. G. A.** 1972d (Canada: Manitoba); 1975b (APHRODINAE, DELTOCEPHALINI, DELTOCEPHALINA). **Jankovic** 1966a (Serbia). **Jasinska** 1980a (Poland: Bledowska Wilderness). **Kalkandelen** 1974a (Turkey; key to species); 1984a (key to species). **Korolevskaya** 1979b (Tadzhikistan). **LeQuesne** 1969c (Britain). **Lindberg** 1961a (Madeira). **Logvinenko** 1957b, 1960a (USSR: Ukraine; key to 4 species). **Mitjaev** 1968a (eastern Kazakhstan); 1971a (Kazakhstan; key to 13 species). **Nast** 1955a (Poland); 1972a (catalogued); 1976a (Poland: Pieniny Mts.). **Ossiannilsson** 1983a (Fennoscandia & Denmark; key to subgenera and species). **Ross & Hamilton** 1970b (key to Nearctic species). **Vilbaste** 1962b (eastern part of the Caspian lowlands); 1965b (Altai Mts.); 1980b (Tuva).

Mocuola
Record: **Hamilton, K. G. A.** 1975b (APHRODINAE, DELTOCEPHALINI, DELTOCEPHALINA).

Mocydia
Records: **Bonfils & Della Giustina** 1978a (Corsica). **Dlabola** 1961b (central Asia); 1964c (Albania); 1965c (Mongolia); 1981a (Iran and Mediterranean). **Emeljanov** 1964c (European USSR). **Hamilton, K. G. A.** 1975b (APHRODINAE, DELTOCEPHALINI, CICADULINA). **Kalkandelen** 1974a (Turkey). **LeQuesne** 1969c (Britain). **Lindberg** 1960b (Soviet Armenia). **Logvinenko** 1957b, 1957c (USSR: Ukraine). **Muller, H. J.** 1981a (larval polymorphism). **Nast** 1972a, 1976b (catalogued). **Quartau & Rodrigues** 1969a (Portugal). **Servadei** 1957a, 1960a, 1968a (Italy).

Mocydiopsis
New species: **Dlabola** 1957b. **Remane** 1961b.
Records: **Bonfils & Della Giustina** 1978a (Corsica). **Cantoreanu** 1971c (Romania). **Della Giustina** 1983a (France). **Dlabola** 1981a (Iran). **Dubovskij** 1966a (Uzbekistan). **Emeljanov** 1964c (European USSR). **Gravestein** 1965a (The Netherlands: Terschelling I.). **Gyllensvard** 1963a (Sweden). **Hamilton, K. G. A.** 1975b (APHRODINAE, DELTOCEPHALINI, CICADULINA). **Jankovic** 1966a, 1971a (Serbia). **Kalkandelen** 1974a (Turkey). **Korolevskaya** 1979b (Tadzhikistan). **LeQuesne** 1969c (Britain). **Logvinenko** 1984a (USSR: Ukraine). **Nast** 1972a (catalogued); 1973a (Poland); 1976b (Poland; catalogued) **Ossiannilsson** 1961b (Sweden); 1983a (Fennoscandia & Denmark; key to species). **Quartau & Rodrigues** 1969a (Portugal). **Remane** 1961a (n. comb. to *Dlabolaracus*). **Servadei** 1957a, 1960a (Italy). **Vilbaste** 1976a (n. comb. from *Thamnotettix*; n. sp. syn.; lectotype designation).

Modderena
New species: **Theron** 1984c.
Record: **Theron** 1984c (n. comb. from *Aletta*).

Mogangella
New species: **Dlabola** 1957a, 1958b. **Remane & Asche** 1980a.
Records: **Dlabola** 1981a (Iran). **Emeljanov** 1964c (European USSR; key to species); 1977a (Mongolia). **Hamilton, K. G. A.** 1975b (APHRODINAE, DELTOCEPHALINI, DELTOCEPHALINA). **Kalkandelen** 1974a (redescription; Turkey). **Kocak** 1983a (considers *Mogangella* Dlabola 1957a invalid, synonym of *Mogangella* Kalkandelen). **Mitjaev** 1969b (Tien Shan & Karatau). **Vilbaste** 1962b (eastern part of the Caspian lowlands); 1980b (Tuva).

Mogangina
New species: **Emeljanov** 1962a.
Records: **Emeljanov** 1964c (European USSR). **Hamilton, K. G. A.** 1975b (APHRODINAE, DELTOCEPHALINI, DELTOCEPHALINA). **Mitjaev** 1969b (Tien Shan & Karatau); 1971a (Kazakhstan; key to species). **Vilbaste** 1965b (Altai Mts.).

Mogenola
New species: **Blocker** 1979a.

Mohunia
New species: **Ishihara** 1961a.

Molomea
New species: **Schroder** 1959a, 1960b, 1962a. **Young** 1968a.
Records: **Schroder** 1959a (generic rank; n. comb. from *Centrometopia*, *Parametopia* and *Tettigonia*); 1960b, 1962a (n. sp. syn.). **Young** 1968a (n. comb. from *Oncometopia* and *Tettigonia*; n. sp. syn.; key to species).

Molopopterus
New species: **Ahmed, Manzoor** 1979a. **Dworakowska** 1973a, 1974a, 1981h. **Einyu & Ahmed** 1983a. **Theron** 1978a.
Records: **Dworakowska** 1973a (n. comb. from *Erythroneura*; notes on Jacobi's type specimen); 1981h (Nigeria). **Theron** 1977a (n. comb. from *Erythroneura*); 1978a (n. comb. from *Typhlocyba*; redescription of the genus; review of southern African species).

Momoria
New species: **Blocker** 1979a, 1982a.

Mongolojassus
New species: **Dlabola** 1965c. **Emeljanov** 1962a, 1964d, 1972b, 1976a. **Mitjaev** 1969c.
Records: **Della Giustina** 1977a (French Alps); 1983a (France). **Dlabola** 1965c, 1967b, 1970b (Mongolia). **Dworakowska** 1969a (Mongolia). **Emeljanov** 1962a (= *Omskius*, n. syn.); 1964c (European USSR); 1964f (Kazakhstan); 1977a (Mongolia). **Hamilton, K. G. A.** 1975b (APHRODINAE, DELTOCEPHALINI, DELTOCEPHALINA). **Ossiannilsson** 1983a (east Fennoscandia; rare). **Vilbaste** 1965b (Altai Mts.); 1980b (Tuva).

Monteithia
New species: **Evans, J. W.** 1968a.

Moonia
New species: **Kameswara Rao & Ramakrishnan** 1978a, 1979b.
Records: **Davis, R. B.** 1975a (morphology; phyletic inferences). **Datta** 1973f (descriptive notes). **Dlabola** 1972a (n. comb. from *Ulopa*). **Kameswara Rao** 1976a (notes). **Kameswara Rao & Ramakrishnan** 1978a (synopsis; key to species). **Linnavuori** 1972a (as subgenus of *Coloborrhis*). **Vilbaste** 1975a (generic synonymy discussed).

Moorada
New species: **Ghauri** 1975b.
Mordania
New species: **Dworakowska** 1979c.
Morinda
Records: **Emeljanov** 1972d (n. comb. from *Euscelis*); 1977a (Mongolia). **Hamilton, K. G. A.** 1975b (APHRODINAE, DELTOCEPHALINI, CICADULINA). **Nast** 1982a (catalogued; DELTOCEPHALINAE). **Vilbaste** 1980a (Kamchatka).
Motaga
New species: **Dworakowska** 1979b, 1980b.
Record: **Sohi & Dworakowska** 1984a (list of Indian species).
Motschulskyia
New species: **Dworakowska** 1971d.
Records: **Dworakowska** 1969b (redescription of *M. inspirata*); 1971d (*Togaritettix* as subgenus; n. comb. from *Togaritettix*); 1976b (Taiwan); 1980b (India). **Lee, C. E.** 1979a (Korea, redescribed). **Sohi & Dworakowska** 1984a (list of Indian species). **Vilbaste** 1975a (notes of type specimen of *M. inspirata*).
Mucrometopia
Records: **Schroder** 1962a (generic rank; n. sp. syn.). **Young** 1977a (redescribed).
Muinocerus
New species: **Ghauri** 1985c.
Muirella
New species: **Linnavuori** 1960b.
Records: **Linnavuori** 1960b (key to species from Fiji). **Nielson** 1975a (= *Tharra*, n. syn.).
Mukaria
New species: **Evans, J. W.** 1973a. **Kuoh, C. L. & Kuoh, J. L.** 1983a.
Records: **Linnavuori** 1979b (= Ikomella, n. syn.; MUKARIINAE). **Sharma, B.** 1977b (India: Jammu).
Mukwana
Record: **Nielson** 1977a (= *Thagria*, n. syn.).
Muleyrechia
Records: **Dlabola** 1974c (= *Goldeus*, n. syn.). **Hamilton, K. G. A.** 1975b (APHRODINAE, DELTOCEPHALINI, DELTOCEPHALINA). **Linnavuori** 1956d (n. comb. from *Jassargus*).
Muluana
New species: **Dworakowska** 1979c.
Murreeana
New species: **Ramakrishnan & Menon** 1971a.
Musbrnoia
New species: **Dworakowska** 1979b.
Record: **Dworakowska** 1972b (n. comb. from *Aidola*).
Musgraviella
New species: **Evans, J. W.** 1966a. **Webb, M. D.** 1983a.
Record: **Webb, M. D.** 1983a (key to species).
Musosa
New species: **Linnavuori** 1977a.
Myerslopia
New species: **Knight, W. J.** 1973b.
Records: **Evans, J. W.** 1966a (synopsis; ULOPINAE, MYERSLOPIINI). **Knight, W. J.** 1973b (redescribed; ULOPINAE, MYERSLOPIINI). **Linnavuori** 1972a (characters summarized; MYERSLOPIINAE).
Myerslopella
New species: **Evans, J. W.** 1977a.
Myogonia
Records: **Young** 1968a (= *Teletusa*, n. syn.). **Young & Nast** 1963a (notes on type specimens).
Myrmecophryne
Records: **Davis, R. B.** 1975a (morphology; phyletic inferences). **Evans, J. W.** 1966a (synopsis).
Myrmecoscopus
New species: **Evans, J. W.** 1966a.
Naevus
New species: **Knight, W. J.** 1970b.
Records: **Knight, W. J.** 1970b (n. comb. from *Hishimonus*). **Webb, M. D.** 1980a (Aldabra).
Nakaharanus
New species: **Kwon & Lee** 1979c.

Records: **Hamilton, K. G. A.** 1975b (APHRODINAE, DELTOCEPHALINI, PLATYMETOPIINA). **Ishihara** 1963a (comparative notes). **Lee, C. E.** 1979a (Korea, redescribed).
Naltaca
New species: **Young** 1977a.
Namiocerus
Record: **Ghauri** 1985a (n. comb. from *Bythoscopus*).
Namsangia
New species: **Yang, C.** 1984a.
Record: **Young** 1965c (lectotype designation).
Nanctasus
New species: **Linnavuori** 1959b.
Nandara
New species: **Dworakowska** 1984a.
Nandidrug
Records: **Datta** 1973g (descriptive notes). **Sharma, B.** 1977b (India: Jammu). **Vilbaste** 1965a (= *Doratulina*, n. syn.).
Nanipoides
Record: **Evans, J. W.** 1966a (n. comb. from *Ipoides*).
Nannicerus
New species: **Maldonado-Capriles** 1977c.
Nannogonalia
Record: **Young** 1977a (n. comb. from *Tettigonia*).
Nanopsis
Records: **Freytag** 1974a (n. comb. from *Oncopsis*); 1976a (USA: Kentucky). **Hamilton, K. G. A.** 1983d (as subgenus of *Pediopsoides*).
Nanosius
Record: **Nast** 1982a (catalogued; DELTOCEPHALINAE).
Napo
New species: **Linnavuori & DeLong** 1976a.
Record: **Cwikla & Blocker** 1981a (comparative notes).
Naratettix
New species: **Dworakowska** 1972k, 1980a.
Records: **Anufriev** 1977b (Kurile Islands); 1978a (USSR: Maritime Territory). **Anufriev & Emeljanov** 1968a (Soviet Far East). **Dworakowska** 1968d (Korea); 1980a (n. sp. syn.; changes in status of Matsumura's species). **Lee, C. E.** 1979a (Korea, redescribed).
Narecho
New species: **Heller & Linnavuori** 1968a. **Linnavuori** 1979b. **Theron** 1970a, 1971a.
Records: **Linnavuori** 1979b (redescribed; in key to genera of African NIRVANINI; key to species). **Theron** 1972a (n. gen. syn.; n. comb. from *Parabolitus*).
Narta
New species: **Dworakowska** 1979c.
Nasutoideus
Record: **Hamilton, K. G. A.** 1975b (APHRODINAE, DELTOCEPHALINI, PLATYMETOPIINA).
Nataretus
Record: **Theron** 1980a (n. comb. from *Platyretus* and *Typhlocyba*).
Naudeus
Record: **Theron** 1982b (n. comb. from *Deltocephalus* and *Ebarrius*).
Navaia
New species: **Linnavuori** 1960b.
Record: **Linnavuori** 1960b (n. comb. from *Limotettix*).
Ndua
New species: **Linnavuori** 1978a.
Nedangia
New species: **Nielson** 1982e.
Negosiana
Records: **Beirne** 1956a (Canada: southern Manitoba). **DeLong & Freytag** 1972a (in key to genera of GYPONINAE).
Nehela
Record: **Davis, R. B.** 1975a (morphology; phyletic inferences; n. comb. from *Igerna*). **Synave & Dlabola** 1976a (St. Helena; characterized).
Neiva
New species: **Young** 1977a.
Records: **Young** 1977a (n. gen. syn.; n. comb. from *Acharista*; n. sp. syn.). **Young & Soos** 1964a (notes on type material).
Nelidina
New species: **Kramer** 1964d, 1967b.

Records: **Kramer** 1964d (generic status; in key to genera of NEOCOELIDIINAE; key to species); 1967b (key to species).

Neoaliturus
New species: **Dubovskij** 1966a. **Mitjaev** 1971a, 1975a.
Records: **Anufriev** 1978a (USSR: Maritime Territory; in key to some genera of OPSIINI). **Bonfils & Della Giustina** 1978a (Corsica). **Dlabola** 1964a (Afghanistan); 1964b (Sudan); 1964c (Albania); 1965c, 1967d (Mongolia); 1967e (Mongolia; n. sp. syn.); 1968a, 1968b, 1970b (Mongolia); 1971a (Afghanistan, Turkey); 1977c (Mediterranean); 1981a (Iran). **Dubovskij** 1966a (Uzbekistan; key to 3 species); 1978a (Uzbekistan, Zarafshan Valley); 1980a (USSR: western Turkmenia). **Emeljanov** 1964c (European USSR; key to species); 1977a (Mongolia). **Hamilton, K. G. A.** 1975b (APHRODINAE, DELTOCEPHALINI, PLATYMETOPIINA); 1983b (adventive to Nearctic). **Jankovic** 1966a (Serbia). **Jasinska** 1980a (Poland: Bledowska Wilderness). **Kalkandelen** 1974a (Turkey; key to species). **Kholmuminov & Dlabola** 1979a (USSR: Golodnostep Plain). **Korolevskaya** 1973a (Tadzhikistan); 1978b (central Asia). **Linnavuori** 1962a (Israel; n. gen. syn.; n. sp. syn.); 1964a (Egypt; n. sp. syn.); 1965a (Libya, Tunisia). **Lodos & Kalkandelen** 1985a (Turkey; comments on *Circulifer vis a vis Neoaliturus*). **Logvinenko** 1984a (USSR: Ukraine). **Mitjaev** 1968b (eastern Kazakhstan); 1971a (Kazakhstan; in key to genera of DELTOCEPHALINAE; key to 9 species). **Nast** 1972a (catalogued). **Oman** 1970a (characterization of *Circulifer* and *Neoaliturus*). **Theron** 1970b (redescription of Cogan's species); 1974b (n. comb. from *Thamnotettix*). **Vilbaste** 1972b (eastern part of the Caspian lowlands); 1965b (Altai Mts.); 1968a (coastal regions of the USSR); 1973a (notes on Flor's material); 1976a (notes on Matsumura's material; transfer of species from *Thamnotettix* noted; n. sp. syn.); 1980b (Tuva).

Neobala
New species: **Kramer** 1963b. **Linnavuori & DeLong** 1979b.
Records: **Kramer** 1963b (in key to genera of NEOBALINAE). **Linnavuori** 1959b (redescribed; key to species).

Neocoelidia
New species: **Kramer** 1967b.
Records: **Beirne** 1956a (Canada: southern Quebec, southern Saskatchewan). **Kramer** 1964d (n. gen. syn.; n. comb. from *Paracoelidia* and *Stenocoelidia*; n. sp. syn.). **Linnavuori** 1965b (Costa Rica).

Neocoelidiana
New species: **DeLong & Kolbe** 1975b.
Record: **Kramer** 1964d (redescribed; in key to genera of NEOCOELIDIINAE).

Neocrassana
New species: **Linnavuori** 1959b.

Neodartellus
Record: **Linnavuori** 1977a (Zaire)

Neodartus
Records: **Datta** 1973g (descriptive notes). **Evans, J. W.** 1966a (=*Neovulturnus*) [NOTE: New combinations recorded are in Metcalf 1962b]; 1972a (=*Neovulturnellus*, n. syn.); 1977a (n. comb. to *Vulturnus*; n. comb. from *Vulturnellus*). **Linnavuori** 1977a (redescribed). **Sharma, B.** 1977b (India: Jammu).

Neodeltocephalus
New species: **Linnavuori** 1959b.
Record: **Linnavuori & DeLong** 1978c (in key to 4 related Neotropical genera of DELTOCEPHALINI).

Neodonus
Record: **Hamilton, K. G. A.** 1975b (APHRODINAE, DELTOCEPHALINI, PLATYMETOPIINA).

Neodrylix
Record: **Hamilton, K. G. A.** 1975b (APHRODINAE, DELTOCEPHALINI, ATHYSANINA).

Neohecalus
Record: **Linnavuori** 1975a (n. comb. from *Hecalus*).

Neohegira
New species: **Linnavuori & DeLong** 1978c.
Record: **Cwikla & Blocker** 1981a (comparative notes).

Neokolla
New species: **DeLong & Currie** 1959a.

Records: **Beirne** 1956a (Canada). **DeLong & Currie** 1960a (synopsis). **Nielson** 1968b (characterization of vector species). **Young** 1977a (*Graphocephala*, n. syn.).

Neolimnus
New species: **Dlabola** 1964a.
Records: **Dlabola** 1960c (Iran); 1964b (Afghanistan); 1965d (Jordan); 1971a (Pakistan); 1972a (Afghanistan); 1977b, 1981a (Iran). **Heller & Linnavuori** 1968a (Ethiopia). **Hamilton, K. G. A.** 1975b (APHRODINAE, DELTOCEPHALINI, PLATYMETOPIINA). **Linnavuori** 1961a (southwest Africa); 1962a (Israel); 1964a (Egypt).

Neomesus
Records: **Linnavuori** 1959b (Uruguay; n. comb. from *Paramesus*). **Linnavuori & DeLong** 1977c (Chile; redescribed). **Menezes** 1975a (n. comb. from *Paramesus*; *N. obtusiceps* redescribed).

Neometopia
Records: **Schroder** 1959a (n. comb. from *Oncometopia*); 1960b (Brazil, Peru). **Young** 1968a (=*Egidemia*, n. syn.).

Neonirvana
Records: **Kramer** 1964b (in key to Neotropical genera of NIRVANINI). **Linnavuori** 1959b (redescribed).

Neopenthimia
New species: **Evans, J. W.** 1972a.

Neophlepsius Dubovskij
New species: **Dubovskij** 1966a. **Mitjaev** 1975a.
Record: **Nast** 1984a (name preoccupied, see *Elphnesopsius*).

Neophlepsius Linnavuori
New species: **Cheng, Y. J.** 1980a. **Linnavuori** 1959b. **Linnavuori & DeLong** 1979b. **Menezes** 1973b.
Records: **Linnavuori** 1956a (Brazil, Paraguay); 1959b (new subgenus proposed; n. sp. syn.; key to subgenera and species).

Neopsis
New species: **Hamilton, K. G. A.** 1983d. **Linnavuori** 1965b. **Linnavuori & DeLong** 1977c.
Records: **Hamilton, K. G. A.** 1983d (comparative notes; key to species). **Linnavuori** 1956a (Brazil); 1958c (NEOPSINAE). **Linnavuori & DeLong** 1977c (comparative notes).

Neoslossonia
Records: **Emeljanov** 1966a (in key to genera of DORYCEPHALINAE). **Oman** 1985a (redefined).

Neotartessus
Record: **Evans, F.** 1981a (n. comb. from *Tartessus*).

Neotharra
New species: **Nielson** 1975a.

Neotituria
Records: **Anufriev** 1978a (USSR: Maritime Territory). **Choe** 1985a (Korea). **Kwon & Lee** 1978b (Korea). **Lee, C. E.** 1979a (Korea, redescibed). **Lee, C. E. & Kwon** 1977b, 1979b (Korea).

Neovulturnus
Records: **Evans, J. W.** 1972a, 1977a (n. comb. from *Vulturnus*).

Nephotettix
New species: **Emeljanov** 1969b. **Ghauri** 1968a, 1971a. **Ishihara** 1964a. **Ishihara & Kawase** 1968a. **Kuoh, C. L.** 1981a. **Linnavuori** 1960a. **Mahmood & Aziz** 1979a.
Records: **Anufriev** 1978a (USSR: Maritime Territory). **Choe** 1985a (Korea). **Datta** 1973h (descriptive notes) **Evans, J. W.** 1966a (Australia; synopsis). **Ghauri** 1968a (African and Malagasian species); 1971a (n. comb. from *Selenocephalus*; n. sp. syn.). **Hamilton, K. G. A.** 1975b (APHRODINAE, APHRODINI, DORATURINA); 1983b (Nearctic record). **Heller & Linnavuori** 1968a (Ethiopia). **Hongsaprug** 1983b (Thailand). **Inoue** 1982a, 1983a (species concepts). **Ishihara** 1958a (Japan: north Honshu); 1961a (southeast Asia); 1964a (synopsis); 1965c (Formosa). **Kathirithamby** 1978a (Malaysia: Krian District). **Kawase** 1971a (Thailand). **Lee, C. E.** 1979a (Korea, redescribed). **Lee, C. E. & Kwon** 1977b (Korea). **Lindberg** 1956a (Niger River flood plain; Dogo, Mopti); 1963a (Morocco). **Linnavuori** 1956b (sp. syn. revised); 1960a (Micronesia); 1961a (southwest Africa); 1962a (Israel); 1964a (Egypt); 1983a (key to Asian species); 1975c (Micronesia: Guam, Palau, Yap). **Mahmood & Aziz** 1979a (Pakistan & Bangladesh). **Nielson** 1968b (characterization of vector species). **Ramachandra Rao** 1973a (India: Poona). **Ramakrishnan & Ghauri** 1979a

(Malaysia; hybridization). **Sharma, B.** 1977b (India: Jammu). **Vilbaste** 1975a (identity of Motschulsky's material).

Nesocerus
New species: **Freytag & Cwikla** 1984a. **Freytag & Knight** 1966a.
Record: **Freytag & Knight** 1966a (n. comb. from *Idiocerus*).

Nesoclutha
New species: **Logvinenko** 1977b.
Records: **Evans, J. W.** 1966a (n. comb. from *Eusceloscopus*; n. sp. syn.). **Knight, W. J.** 1975a (New Zealand); 1976b (Lord Howe Island, Norfolk Island). **Logvinenko** 1977b (= *Irinula*). **Nast** 1982a (catalogued; DELTOCEPHALINAE). **Nielson** 1968b (characterization of vector species). **Theron** 1982b (n. sp. syn.) **Vilbaste** 1976a (= *Irinula*, n. syn; n. comb. from *Cicadula*).

Nesophrosyne
New species: **Ishihara** 1965b. **Linnavuori** 1960a.
Records: **Dlabola** 1964b (Sudan). **Hamilton, K. G. A.** 1975b (APHRODINAE, DELTOCEPHALINI, PLATYMETOPIINA). **Heller & Linnavuori** 1968a (Ethiopia). **Ishihara** 1963a (n. comb. from *Eutettix*); 1965c (Formosa). **Lee, C. E. & Kwon** 1977b (Korea). **Lindberg** 1958a (Cape Verde Islands); 1961a (Madeira); 1963a (sp. syn. noted; Morocco). **Linnavuori** 1956c (Phoenix and Danger Islands); 1956d (Spanish Morocco); 1960a (Micronesia; n. comb. from *Orosius*); 1960b (Fiji; *Orosius* a subgenus with *Nesaloha* as synonym; *Kirkaldiella* probably subgenus); 1962a (Israel); 1964a (Egypt). **Nielson** 1968b (characterization of vector species).

Nesosteles
Records: **Beardsley** 1966a (Hawaii; Nihoa & Necker Islands). **Beirne** 1956a (Canada). **Hamilton, K. G. A.** 1975b (APHRODINAE, DELTOCEPHALINI, MACROSTELINA). **Ishihara** 1961a (Thailand).

Nesotettix
Record: **Lindberg** 1960d (Canary Islands). **Linnavuori** 1959b (= *Xestocephalus*, n. syn.).

Nesothamnus
New species: **Linnavuori** 1959b. **DeLong** 1982f.

Neurotettix
Records: **Hamilton, K. G. A.** 1975b (APHRODINAE, DELTOCEPHALINI, PLATYMETOPIINA). **Ishihara** 1963a (redescribed); 1965c (Formosa).

Newmaniana
Record: **Evans, F.** 1981a (n. gen. syn.; n. comb. from *Stenotartessus*).

Ngoma
New species: **Dworakowska** 1974a.

Ngombela
New species: **Dworakowska** 1974a.

Ngunga
New species: **Dworakowska** 1974a.

Nicolaus
New species: **Lindberg** 1958a.

Nielsonia
New species: **Young** 1977a.

Nielsoniella
New species: **Linnavuori** 1977a.

Nigridonus
Record: **Hamilton, K. G. A.** 1975b (APHRODINAE, DELTOCEPHALINI, PLATYMETOPIINA).

Nigritartessus
New species: **Evans, F.** 1981a.

Nikkotettix
Records: **Anufriev** 1969b (n. comb. from *Dikraneura*); 1972a (descriptive notes). **Dworakowska** 1982b (notes on holotype).

Nionia
New species: **Linnavuori** 1956a.
Records: **Datta & Dhar** 1984a (descriptive notes). **Evans, J. W.** 1971b (comments on systematics). **Linnavuori** 1956a (synopsis of Neotropical NIONIINAE; key to species); 1959b (characterization of NIONIINAE).)

Nirvana
New species: **Ishihara** 1958a.
Records: **Choe** 1985a (Korea). **Datta** 1973h (descriptive notes). **Evans, J. W.** (Australia; synopsis; in key to genera of NIRVANINI). **Ishihara** 1961a (Thailand). **Lee, C. E.** 1979a (Korea, redescribed). **Lee, C. E. & Kwon** 1977b (Korea). **Sharma, B.** 1977b (India: Jammu).

Nisitra
Record: **Nielson** 1975a (= *Tharra*, n. syn.).

Nisitrana
Record: **Nielson** 1975a (= *Tharra*, n. syn.).

Nkaanga
New species: **Dworakowska** 1974a.

Nkonba
New species: **Dworakowska** 1974a.

Nkumba
New species: **Dworakowska** 1974a.

Nlunga
New species: **Dworakowska** 1974a.

Nollia
New species: **Hamilton, K. G. A.** 1983d.
Record: **Hamilton, K. G. A.** 1983d (n. comb. from *Neopsis*).

Noritonus
New species: **Nielson** 1979b, 1983g.
Records: **Nielson** 1979b, 1983g (key to species).

Nortoides
New species: **Evans, J. W.** 1972a.

Norva
New species: **Anufriev** 1970a. **Emeljanov** 1969a.
Records: **Anufriev** 1978a (USSR Maritime Territory; in key to genera of OPSIINI). **Dworakowska** 1973d (Korea). **Hamilton, K. G. A.** 1975b (APHRODINAE, DELTOCEPHALINI, PLATYMETOPIINA). **Kwon & Lee** 1979b (Mt. Soelaksan). **Lee, C. E.** 1979a (Korea, redescribed).

Norvellina
New species: **DeLong** 1980h (as *Morvellina*). **Kramer & DeLong** 1969a.
Records: **Beirne** 1956a (Canada; key to species). **Hamilton, K. G. A.** 1975b (APHRODINAE, DELTOCEPHALINI, PLATYMETOPIINA).

Notocephalius
Records: **Evans, J. W.** 1966a (= *Cephalelus*); 1977a (removed from synonymy with *Cephalelus*; n. comb. from *Anacephaleus*; n. sp. syn.).

Notus
New species: **Anufriev** 1979b.
Records: **Anufriev** 1970e (USSR: south of the Primor'ya Territory); 1977b (Kurile Islands); 1978a (USSR: Maritime Territory; in key to genera of DIKRANEURINI; key to species). **Beirne** 1956a (Alaska, Canada: Northwest Territory and northern Manitoba). **Dlabola** 1958c (in key to Palaearctic genera of TYPHLOCYBINAE); 1961b (central Asia); 1967b, 1967d (Mongolia). **Dubovskij** 1966a (Uzbekistan); 1978a (Uzbekistan, Zarafshan Valley). **Dworakowska** 1971d (Yugoslavia). **Emeljanov** 1964c (European USSR); 1977a (Mongolia). **Hamilton, K. G. A.** 1983b (transboreal taxon). **Jankovic** 1971a (Serbia). **Kirejtshuk** 1977a (USSR: Karkov region). **Korolevskaya** 1977a (central Asia). **LeQuesne & Payne** 1981a (Britain). **Lodos & Kalkandelen** 1983b (Turkey). **Mitjaev** 1963b (Kazakhstan); 1968b (eastern Kazakhstan); 1971a (Kazakhstan; in key to genera of TYPHLOCYBINAE). **Nast** 1972a (catalogued); 1976a (Poland: Pieniny Mts.); 1976b (Poland; catalogued). **Ossiannilsson** 1981a (Fennoscandia & Denmark). **Servadei** 1958a, 1968a, 1969a (Italy). **Vidano** 1965a (Italy, and north, west and central Europe). **Vilbaste** 1964a (Auwiesen Estlands); 1965b (Altai Mts.); 1968a (coastal regions of the USSR); 1969b (Taimyr); 1980a (Kamchatka); 1980b (Tuva).

Novolopa
New species: **Evans, J. W.** 1966a. **Knight, W. J.** 1973b.
Record: **Knight, W. J.** 1973b (redescribed; key to species).

Novothymbris
New species: **Knight, W. J.** 1974a.
Records: **Evans, J. W.** 1966a (synopsis); 1969a (comparative notes). **Knight, W. J.** 1974a (redescribed; species groups defined; key to species); 1976b (Chatham Island).

Nsanga
New species: **Dworakowska** 1974a.
Nsesa
New species: **Dworakowska** 1974a, 1981h.
Nsimbala
New species: **Dworakowska** 1974a, 1981h.
Record: **Dworakowska** 1981h (Zaire).
Ntanga
New species: **Dworakowska** 1974a.
Ntotila
New species: **Dworakowska** 1974a.
Nubelella
New species: **Evans, J. W.** 1972a.
Nubelloides
New species: **Evans, J. W.** 1972a.
Nudulidia
New species: **Nielson** 1982e.
Nullamia
New species: **DeLong** 1970a. **Linnavuori & DeLong** 1977c.
Record: **Cwikla & Blocker** 1981a (comparative notes).
Nullana
New species: **DeLong** 1976f. **DeLong & Martinson** 1980a.
Nurenus
Record: **Hamilton, K. G. A.** 1975b (APHRODINAE, DELTOCEPHALINI, PLATYMETOPIINA).
Nyndgama
New species: **Webb, M. D.** 1983a.
Nzinga
New species: **Dworakowska** 1974a.
Occinirvana
Record: **Evans, J. W.** 1966a (NIRVANINAE, OCCINIRVANINI).
Oceanopona
New species: **Linnavuori** 1960a.
Records: **Evans, J. W.** 1966a (DELTOCEPHALINAE, PLATYMETOPIINI). **Linnavuori** 1960a (DELTOCEPHALINAE, PARABOLOPONINI). **Webb, M. D.** 1981b (comparative notes; in key to genera of PARABOLOPONINAE).
Ochtrostacta
Record: **Young** 1968a (= *Andamarca*, n. syn.; key to species).
Ochta
Records: **Nast** 1982a (catalogued; DELTOCEPHALINAE).
Odmiella
Records: **Linnavuori** 1978d (n. comb. from *Stenomiella*). **Webb, M. D.** 1981b (in key to genera of PARABOLOPONINAE).
Odomas
New species: **Linnavuori** 1972a, 1973a. **Van Stalle** 1983a, 1983b.
Records: **Davis, R. B.** 1975a (morphology; phyletic inferences). **Linnavuori** 1972a (key to species). **Van Stalle** 1982a, 1983a (systematics and distribution); 1983a (Mount Cameroon).
Odzalana
New species: **Linnavuori** 1969a, 1975a.
Record: **Linnavuori** 1975a (key to species).
Oeogonalia
Record: **Young** 1977a (n. comb. from *Tettigonia*).
Ohausia
Record: **Kramer** 1966a (comparative notes).
Okaundua
New species: **Heller & Linnavuori** 1968a. **Linnavuori** 1969a.
Record: **Linnavuori** 1969a (DELTOCEPHALINAE, EUSCELINI).
Ollarianus
New species: **DeLong** 1980h. **Linnavuori** 1959b. **Linnavuori & DeLong** 1977b.
Records: **Hamilton, K. G. A.** 1975b (APHRODINAE, DELTOCEPHALINI, PLATYMETOPIINA). **Linnavuori** 1956a (Colombia); 1959b (key to species). **Linnavuori & DeLong** 1978c (in key to genera of the *Atanus* group).
Olokemeja
New species: **Emeljanov Ghauri** 1975b.
Olszewskia
New species: **Dworakowska** 1974a.
Omagua
Record: **Young** 1968a (redefined).

Omanana
New species: **DeLong** 1980h.
Records: **Cwikla & Blocker** 1981a (comparative notes). **Hamilton, K. G. A.** 1975b (APHRODINAE, DELTOCEPHALINI, PLATYMETOPIINA).
Omanesia
New species: **Thapa** 1983a.
Omaniella
Record: **Hamilton, K. G. A.** 1975b (APHRODINAE, DELTOCEPHALINI, ATHYSANINA).
Omanolidia
New species: **Nielson** 1982e.
Record: **Nielson** 1982e (n. comb. from *Coelidia* and *Tettigella*; key to species).
Omegalebra
New species: **Young** 1957c.
Record: **Young** 1957c (n. comb. from *Protalebra*; key to species).
Omiya
New species: **Dworakowska** 1981b.
Omskius
Record: **Hamilton, K. G. A.** 1975b (APHRODINAE, DELTOCEPHALINI, DELTOCEPHALINA).
Onblavia
New species: **Nielson** 1979b.
Oncometopia
New species: **Emmrich** 1975a, 1984a. **Schroder** 1959a, 1960a, 1960b, 1962a, 1975a.
Records: **Emmrich** 1975a (n. sp. syn.; synopsis); 1984a (species groups defined). **Lauterer & Schroder** 1970a (notes on type material). **Nielson** 1968b (characterization of vector species). **Ramos Elorduy** 1972c (Mexico). **Schroder** 1959a (n. comb. to *Lebora*); 1960b (subgenera defined; lectotype designation). **Young** 1965c (lectotype designation); 1968a (*Similitopia* as subgenus, new status, n. sp. syn.).
Oncopsis
New species: **Anufriev** 1967a, 1971e. **Dlabola** 1963d. **Hamilton, K. G. A.** 1980b, 1983d. **Korolevskaya** 1984a. **Kuoh, C. L.** 1981b. **Lauterer & Anufriev** 1969a.
Records: **Anufriev** 1967a (new subgenus proposed; key to subgenera and species); 1970c (key to Palaearctic species); 1977b (Kurile Islands); 1978a (USSR: Maritime Territory; in key to genera of MACROPSINAE; key to species). **Beirne** 1956a (Canada; key to species). **Bliven** 1966a (remarks on the identity of *Jassus verticis* Say). **Cantoreanu** 1965c, 1971c (Romania). **Dlabola** 1958b (Caucasus); 1965c, 1967b, 1967d, 1968a, 1968b, 1970b (Mongolia); 1971b (Turkey); 1977c (Mediterranean); 1981a (Iran). **Dubovskij** 1966a (Uzbekistan). **D'Urso** 1980b (Italy); 1981a (Sicily). **Emeljanov** 1964c (European USSR; key to species); 1977a (Mongolia). **Evans, J. W.** 1971b (questionable placement of Australian taxa). **Freytag** 1976a (USA: Kentucky). **Gravestein** 1965a (The Netherlands: Terschelling I.). **Gyllensvard** 1965a (Sweden). **Hamilton, K. G. A.** 1976a (identity of Provancher material); 1980b (n. gen. syn.; *Parasitades* as subgenus; n. comb. from *Parasitades*; n. sp. syn.); 1983b (transboreal taxa); 1983d (in key to New World genera of MACROPSINAE; key to species). **Hori** 1969a (Japan: Shikoku). **Ishihara** 1966a (Kurile Islands). **Jankovic** 1966a, 1971a (Serbia). **Jasinska** 1980a (Poland: Bledowska Wilderness). **John, B. & Clardige** 1974a (chromosome variation). **Lauterer & Anufriev** 1976a (China and Far East; lectotype designation). **Lee, C. E.** 1979a (Korea, redescribed). **Lee, C. E. & Kwon** 1977b (Korea). **LeQuesne** 1961c, 1964a, 1965c (Britain). **Lindberg** 1960a (Portugal). **Linnavuori** 1959c (Mt. Picentini). **Linnavuori & DeLong** 1977c (South American taxa of uncertain identity). **Lodos & Kalkandelen** 1981a (Turkey). **Logvinenko** 1957b, 1984a (USSR: Ukraine). **Mitjaev** 1968b (eastern Kazakhstan); 1971a (Kazakhstan; key to 3 species). **Nast** 1972a (catalogued); 1976a (Poland: Pieniny Mts.); 1976b (Poland; catalogued). **Nixon** 1984a (biotaxonomy of the *O. flavicollis* complex). **Ossiannilsson** 1974a (Norway); 1981a (Fennoscandia & Denmark; key to species). **Schulz** 1976a (Norway). **Servadei** 1957a, 1960a, 1968a, 1972a (Italy). **Vilbaste** 1958a, 1959b (Estonia); 1965b (Altai Mts.); 1968a (coastal regions of the USSR); 1973a (notes on Flor's

material); 1979a (Vooremaa hardwood/spruce forest); 1980a (Kamchatka); 1980b (Tuva). **Zachvatkin** 1953b (central Russia; Oka district).

Onega
New species: **Young** 1977a.
Records: **Young** 1965c (lectotype designation); 1977a (n. comb. from *Tettigonia*; key to species).

Oneratulus
Record: **Vilbaste** 1975a (n. comb. from *Deltocephalus* and *Jassus*; n. sp. syn.; DELTOCEPHALINAE).

Oniella
Records: **Anufriev** 1977b (Kurile Islands). **Ishihara** 1958a (Japan: north Honshu).

Onukia
Records: **Anufriev** 1978a (USSR: Maritime Territory); **Ishihara** 1958a (Japan: north Honshu); 1963b (redescribed; EVACANTHIDAE; key to species); 1965c (Formosa). **Kwon** 1983a (Korea; in key to genera of EVACANTHINI). **Lee, C. E.** 1979a (Korea, redescribed). **Vilbaste** 1968a (coastal regions of the USSR).

Onukiades
Records: **Ishihara** 1963b (n. comb. from *Evacanthus*); 1965c (Formosa).

Onukigallia
Records: **Anufriev** 1977b (Kurile Islands); 1978a (USSR: Maritime Territory). **Dworakowska** 1973d (Korea). **Ishihara** 1958a (Japan: north Honshu). **Lee, C. E.** 1979a (Korea, redescribed). **Lee, C. E. & Kwon** 1977b (Korea). **Vilbaste** 1968a (USSR: Maritime Territory) **Viraktamath, C. A.** 1973b (n. comb. from *Agallia*).

Onura
Record: **Linnavuori** 1959b (n. comb. from *Agudus*; key to species).

Opamata
New species: **Dworakowska** 1971g, 1977i, 1982a.

Ophiola
Records: **Hamilton, K. G. A.** 1975b (APHRODINAE, DELTOCEPHALINI, ATHYSANINA); 1983b (transboreal taxon). **Lauterer** 1984a (Slovakia). **Ossiannilsson** 1983a (Fennoscandia & Denmark; =*Scleroracus*). **Zachvatkin** 1953b (central Russia; Oka district).

Ophiolix
Records: **Hamilton, K. G. A.** 1975b (APHRODINAE, DELTOCEPHALINI, ATHYSANINA). **Ossiannilsson** 1983a (as subgenus of *Ophiola*). **Vilbaste** 1965b (Altai Mts.); 1973a (notes on Flor's material); 1980b (Tuva).

Ophionotum
Records: **Dubovskij** 1966a (Uzbekistan); 1978a (Uzbekistan, Zarafshan Valley); 1980a (USSR: western Turkmenia). **Emeljanov** 1964b (n. comb. from *Phlepsius*). **Hamilton, K. G. A.** 1975b (APHRODINAE; unassigned).

Ophiuchus
New species: **Evans, J. W.** 1973b.
Record: **Evans, J. W.** 1966a (questionable placement for an Australian species).

Opio
Records: **Davis, R. B.** 1975a (morphology; phyletic inferences). **Evans, J. W.** 1966a (synopsis).

Opsianus
Record: **Linnavuori** 1960a (n. comb. from *Euscelis*).

Opsius
New species: **Della Giustina** 1981a. **Dlabola** 1960b, 1960c, 1965e. **Emeljanov** 1964c. **Lindberg** 1958a, 1963a. **Vilbaste** 1961a.
Records: **Allen, A. A.** 1963a, 1963b (Britain). **Beirne** 1956a (Canada). **Bonfils & Della Giustina** 1978a (Corsica). **Cantoreanu** 1968b (Danube river delta). **Dlabola** 1957a (Turkey); 1957b (Afghanistan); 1958a (southern Europe); 1958b (Caucasus); 1960c (Iran); 1961b (central Asia); 1964a (Afghanistan); 1964b (Sudan); 1968a, 1968b (Mongolia); 1971a (Turkey); 1972a (Afghanistan); 1981a, 1984a (Iran). **Dubovskij** 1966a (Uzbekistan); 1978a (Uzbekistan, Zarafshan Valley; 1980a (USSR: western Turkmenia). **Emeljanov** 1964c (European USSR; key to species); 1977a (Mongolia). **Hamilton, K. G. A.** 1975b (APHRODINAE, DELTOCEPHALINI, PLATYMETOPIINA); 1983b (Nearctic, adventive). **Heller & Linnavuori** 1968a (Ethiopia). **Jankovic** 1971a (Serbia). **Kalkandelen** 1974a (Turkey; key to species). **Kholmuminov & Dlabola** 1979a (USSR: Golodnostep Plain). **Korolevskaya** 1978b (central Asia). **LeQuesne** 1969c (Britain). **Lindberg** 1960a (Portugal); 1960b (Soviet Armenia); 1961a (Madeira); 1963a (Morocco). **Linnavuori** 1956d (Spanish Morocco); 1959b (Panama: Canal Zone); 1961a (South Africa; n. comb. from *Thamnotettix*); 1962a (Israel); 1964a (Egypt); 1965a (Mediterranean); 1973b (Cuba). **Linnavuori & DeLong** 1977c (Chile). **Lodos & Kalkandelen** 1985a (Turkey). **Logvinenko** 1957b (USSR: Ukraine). **Mitjaev** 1968a (eastern Kazakhstan); 1971a (Kazakhstan; key to 4 species). **Nast** 1972a (catalogued); 1976a (Poland: Pieniny Mts.); 1976b (Poland; catalogued). **Ossiannilsson** 1983a (Fennoscandia & Denmark). **Quartau** 1979a, 1980b (Azores). **Quartau & Rodrigues** 1969a (Portugal). **Servadei** 1957a, 1960a, 1968a, 1971a (Italy). **Theron** 1980a (characterization of Stal's *O. glaucovirens*). **Vilbaste** 1962 (eastern part of the Caspian lowlands). **Wagner** 1959c (Greece).

Optocerus
New species: **Freytag** 1969a.
Record: **Maldonado-Capriles** 1977a (in key to genera related to *Chunroides*).

Optya
New species: **Dworakowska** 1974a, 1976b, 1977e. **Dworakowska & Viraktamath** 1978a. **Webb, M. D.** 1980a.
Record: **Sohi & Dworakowska** 1984a (list of Indian species).

Oragua
New species: **Young** 1977a.
Record: **Young** 1977a (n. comb. from *Cicadella*, *Tettigella*, and *Tettigoniella*; key to species).

Orechona
Records: **Young** 1977a (redescribed). **Young & Lauterer** 1966a (lectotype designation).

Orectogonia
Record: **Young** 1968a (=*Acrogonia*, n. syn.).

Orientalebra
New species: **Dworakowska** 1971c, 1984a.

Orientus
Records: **Anufriev** 1978a (USSR: Maritime Territory). **George** 1959a (Canada). **Hamilton, K. G. A.** 1975b (APHRODINAE, DELTOCEPHALINI, PLATYMETOPIINA); 1983b (Nearctic, adventive). **Ishihara** 1958a (Japan: north Honshu); 1968a (Japan). **Lee, C. E.** 1979a (Korea, redescribed). **Lee, C. E. & Kwon** 1977b (Korea). **Vilbaste** 1968a (coastal regions of the USSR).

Orocastus
New species: **Emeljanov** 1964f. **Ross & Hamilton** 1972a.
Records: **Beirne** 1956a (Canada). **Emeljanov** 1972d (n. comb. to *Sicistella*). **Hamilton, K. G. A.** 1972d (Canada: Manitoba); 1975b (APHRODINAE, DELTOCEPHALINI, DELTOCEPHALINA). **Hamilton, K. G. A. & Ross** 1972a (n. sp. syn.). **Ross & Hamilton** 1972b (n. comb. to *Laevicephalus*).

Orolix
Record: **Hamilton, K. G. A.** 1975b (APHRODINAE, DELTOCEPHALINI, PLATYMETOPIINA).

Orosius
New species: **Dlabola** 1979a. **Ghauri** 1966a.
Records: **Datta** 1973n (descriptive notes). **Dlabola** 1957a (Turkey); 1971a (Pakistan); 1981a (Iran). **Evans, J. W.** 1974a (New Caledonia). **Eyles & Linnavuori** 1974a (Niue Island). **Ghauri** 1966a (n. sp. syn.; key to species). **Hamilton, K. G. A.** 1975b (APHRODINAE, DELTOCEPHALINI, PLATYMETOPIINA). **Kwon & Lee** 1979b (Korea; n. sp. syn.). **Lee, C. E.** 1979a (Korea, redescribed). **Linnavuori** 1975c (n. sp. syn.). **Lodos & Kalkandelen** 1985a (Turkey). **Nielson** 1968b (characterization of vector species). **Sharma, B.** 1977b (India: Jammu). **Webb, M. D.** 1980a (Astove Atoll).

Orsalebra
New species: **Kramer** 1965a. **Linnavuori & DeLong** 1977c. **Ruppel** 1959a.
Records: **Kramer** 1965a (key to species). **Young** 1957c (in key to genera of ALEBRINI).

Ortega
Record: **Young** 1977a (redescribed).
Orthojassus
Record: **Nielson** 1977a (= *Thagria*, n. syn.).
Orucyba
New species: **Ghauri** 1975b.
Record: **Dworakowska** 1979c (= *Tataka*, n. syn.; n. sp. syn.).
Osbornellus
New species: **DeLong** 1976a, 1982h. **DeLong & Martinson** 1976a, 1976b. **Dlabola** 1974a, 1976e, 1976g, 1984a. **Cheng, Y. J.** 1980a. **Ghauri** 1980c. **Linnavuori** 1959b. **Linnavuori & Heller** 1961a.
Records: **Beirne** 1956a (Canada; key to species). **Dlabola** 1977c (Mediterranean); 1981a (Iran). **Hamilton, K. G. A.** 1972d (Canada: Manitoba); 1975b (APHRODINAE, DELTOCEPHALINI, PLATYMETOPIINA); 1976a (identity of Provancher species). **Kartal** 1982a (changes in status of species-level taxa). **Linnavuori** 1956a (Argentina, Brazil, Paraguay); 1959b (new subgenus proposed; n. comb. from *Deltocephalus*; key to subgenera and species). **Linnavuori & DeLong** 1977b (in key to 4 related Neotropical genera). **Nielson** 1968b (characterization of vector species). **Ramos Elorduy** 1972c (Mexico).
Osbornulus
Record: **Young** 1957c (n. comb. from *Dikraneura*).
Osella
New species: **Evans, J. W.** 1972a.
Ossana
Record: **Linnavuori & Quartau** 1975a (= *Batracomorphus*; n. syn.).
Ossiannilssonola
New species: **Hamilton, K. G. A.** 1980b.
Records: **Beirne** 1956a (Canada; key to species). **Dlabola** 1958c (in key to genera of Palaearctic TYPHLOCYBINAE). **Emeljanov** 1964c (European USSR). **Jankovic** 1966a (Serbia). **LeQuesne & Payne** 1981a (Britain). **Nast** 1972a (catalogued); 1976a (Poland: Pieniny Mts.); 1976b (Poland; catalogued). **Ossiannilsson** 1981a (Palaearctic occurrence records).
Ossuaria
New species: **Dworakowska** 1979b.
Otbatara
New species: **Dworakowska** 1984a.
Oxycoryphia
Record: **Young** 1977a (= *Platygonia*, n. syn.).
Oxytettigella
Record: **Hamilton, K. G. A.** 1975b (APHRODINAE; unassigned).
Oxytettix
Record: **Linnavuori** 1956d (Spain). **Servadei** 1960a (Italy).
Ozias
Records: **Davis, R. B.** 1975a (morphology; phyletic inferences; ULOPINAE). **Linnavuori** 1972a (transferred from ULOPINAE to AGALLIINAE).
Pachitea
New species: **Young** 1977a.
Record: **Young** 1977a (n. comb. from *Tettigonia*; key to species).
Pachodus
New species: **Linnavuori** 1961a.
Pachymetopius
Record: **Maldonado-Capriles** 1975a (transferred from IDIOCERINAE to COELIDIINAE; comparative notes).
Pachynus
Record: **Linnavuori** 1961a (South Africa: Cape Province).
Pachyopsis
New species: **Blocker** 1979a, 1982a. **Kramer** 1963a. **Linnavuori** 1957b, 1965b.
Records: **Blocker** 1979a (in key to New World genera of IASSINAE). **Knight, W. J.** 1983b (n. comb. from *Batracomorphus*). **Kramer** 1973a (key to species).
Pachytettix
New species: **Linnavuori** 1959b. **Linnavuori & DeLong** 1976a.
Pactana
New species: **Linnavuori** 1960a.
Record: **Linnavuori** 1975c (Micronesia: Palau, Yap).
Paganalia
New species: **Linnavuori** 1978d.

Records: **Linnavuori** 1978d (n. gen. syn.; n. sp. syn.; key to species). **Webb, M. D.** 1980a (*Aldabra*; n. sp. syn.; lectotype designation); 1981a (= *Dryadomorpha*, n. syn.).
Pagaronia
New species: **Anufriev** 1970f, 1971c, 1971e. **Bliven** 1958a. **Choe** 1980a. **Hori** 1982a. **Kwon** 1981b. **Kwon & Lee** 1978a, 1980a, 1980b. **Okada** 1976a, 1978a.
Records: **Anufriev** 1970f (lectotype designation for *Tettigonia guttiger* Uhler); 1977b (Kamchatka); 1978a (USSR: Maritime Territory). **Datta & Ghosh** 1974a (descriptive notes). **Kwon** 1983a (in key to genera of Korean PAGARONIINI). **Kwon & Lee** 1978a (new subgenus proposed). **Lee, C. E.** 1979a (Korea, redescribed). **Nast** 1982a (catalogued; CICADELLINAE, PAGARONIINI). **Nielson** 1968b (characterization of vector species). **Oman & Musgrave** 1975a (redefined).
Paivanana
Records: **Datta** 1973a (descriptive notes). **Vilbaste** 1965a (= *Doratulina*, n. syn.).
Pakeasta
New species: **Ahmed, Manzoor** 1971b.
Record: **Dworakowska & Lauterer** 1975a (= *Zyginoides*, n. syn.).
Palicus
Records: **Linnavuori** 1961a (South Africa: Natal). **Nielson** 1975a (removed from COELIDIINAE; unassigned). **Theron** 1984c (returned to COELIDIINAE; provisional).
Palingonalia
Record: **Young** 1977a (redescribed).
Paluda
New species: **Anufriev** 1971a. **Dlabola** 1967b. **Emeljanov** 1962, 1964f. **Mitjaev** 1968a, 1969b. **Kalkandelen** 1972a.
Records: **Anufriev** 1977b (Kurile Islands); 1978a (USSR Maritime Territory). **Beirne** 1956a (Canada). **Bonfils & Della Giustina** 1978a (Corsica). **Della Giustina** 1983a (France) **Dlabola** 1967b, 1970b (Mongolia). **D'Urso** 1980b (Italy). **Dworakowska** 1969a (Mongolia). **Emeljanov** 1962a (= *Rhopalopyx*, n. syn.); 1964c (European USSR; key to species); 1977a (Mongolia). **Hamilton, K. G. A.** 1972d (Canada: Manitoba); 1975b (APHRODINAE, DELTOCEPHALINI, CICADULINA). **Kalkandelen** 1974a (key to species). **LeQuesne** 1969c (Britain). **Mitjaev** 1968b (eastern Kazakhstan); 1971a (Kazakhstan; key to 6 species). **Nast** 1972a (catalogued); 1976a (Poland: Pieniny Mts.); 1976b (Poland; catalogued). **Ossiannilsson** 1983a (Fennoscandia & Denmark). **Schulz** 1976a (Norway). **Trolle** 1968a (Denmark). **Vilbaste** 1973a (notes on Flor's material); 1980a (Kamchatka); 1980b (Tuva). **Wagner** 1967a (species groups defined; n. sp. syn.).
Palus
New species: **Emeljanov** 1966a. **Hamilton, K. G. A. & Ross** 1975a. **Linnavuori** 1959b. **Vilbaste** 1965b.
Records: **Beirne** 1956a (Canada). **Dlabola** 1964d (Bohemia); 1967b, 1967d (Mongolia). **Dubovskij** 1966a (Uzbekistan). **Dworakowska** 1969a (Mongolia). **Emeljanov** 1962a (new subgenus proposed); 1964c (European USSR; key to species); 1966a (= *Agapelus*, n. syn.; new subgenus proposed; key to subgenera and species). **Hamilton, K. G. A.** 1972d (Canada: Manitoba); 1975b (APHRODINAE, DELTOCEPHALINI, DELTOCEPHALINA). **Hamilton, K. G. A. & Ross** 1975a (key to species). **Jansky** 1983a (Slovakia). **LeQuesne** 1959a (Britain; key to species); 1969c (Britain). **Linnavuori** 1959b (new subgenera proposed; key to Neotropical subgenera and species). **Logvinenko** 1963a (USSR: Ukraine). **Mitjaev** 1968b (eastern Kazakhstan); 1971a (Kazakhstan; key to 2 species). **Nast** 1955a (n. comb. from *Deltocephalus*); 1972a (catalogued; preoccupied; = *Cosmotettix*, n. gen. syn.). **Remane** 1961c (n. comb. from *Deltocephalus*; key to species); 1962a (northwest Germany). **Vilbaste** 1958a (Estonia); 1964a (Auwiesen Estlands); 1965b (Altai Mts.); 1973a (notes on Flor's material).
Pamplona
New species: **Young** 1977a.
Record: **Young** 1977a (n. comb. from *Tettigonia*; key to species).
Pamplonoidea
New species: **Young** 1977a.

Pandacerus
New species: **Webb, M. D.** 1983b.
Record: **Webb, M. D.** 1983b (n. comb. from *Idioscopus*; key to species).

Panolidia
New species: **Nielson** 1979b.

Pantallus
Records: **Anufriev** 1978a (USSR: Maritime Territory). **Dlabola** 1965c, 1967b, 1967c, 1968a, 1968b, 1970b (Mongolia). **Dworakowska** 1973d (Mongolia). **Emeljanov** 1961a (n. comb. from *Deltocephalus*); 1964c (European USSR); 1977a (Mongolia). **Hamilton, K. G. A.** 1975b (APHRODINAE, DELTOCEPHALINI, DELTOCEPHALINA). **Lauterer** 1978a (Slovakia). **Mitjaev** 1971a (Kazakhstan; in key to genera of DELTOCEPHALINAE). **Vilbaste** 1965b (Altai Mts.); 1976a (identity of Matsumura's species); 1980b (Tuva).

Pantanarendra
Records: **Dworakowska** 1980d (=*Empoascanara*, n. syn.). **Ramakrishnan & Ghauri** 1979b (n. comb. from *Zygina* and *Empoascanara*).

Paolia
New species: **Ahmed, Manzoor & Samad** 1972b.

Paolicia
Record: **Dworakowska** 1984a (Malaysia: Penang I.)

Paoliella
Record: **Dworakowska** 1984a (name preoccupied, see *Paolicia*).

Papallacta
Records: **Young** 1968a (=*Splonia*, n. syn.). **Young & Nast** 1963a (notes on type specimens).

Papyrina
New species: **Emeljanov** 1962a.
Records: **Dubovskij** 1966a (Uzbekistan). **Mitjaev** 1971a (Kazakhstan).

Parabahita
New species: **DeLong** 1980a, 1982a. **Linnavuori** 1956a, 1959b. **Linnavuori & DeLong** 1978b, 1978c, 1978d. **Linnavuori & Heller** 1961a.

Parabolitus
New species: **Linnavuori** 1961a.
Record: **Theron** 1972a (=*Narecho*, n. syn.; n. sp. syn.; *P. foliaticeps* Linnavuori incorrectly placed).

Parabolocratalis
New species: **Linnavuori** 1975a.
Record: **Linnavuori** 1975a (key to species).

Parabolocratus
New species: **Capco** 1959b. **Linnavuori** 1960a. **Metcalf** 1973d (new name). **Sawai Singh** 1969a. **Villiers** 1956a.
Records: **Beirne** 1956a (Canada; key to species). **Capco** 1959b (Philippines; key to species). **Datta** 1972g (descriptive notes). **Dlabola** 1957a (Turkey); 1957b (Afghanistan); 1958b (Caucasus); 1960a (Iran); 1961b (central Asia); 1964a (Afghanistan); 1965d (Jordan). **Dubovskij** (Uzbekistan). **Hamilton, K. G. A.** 1972d (Canada: Manitoba); 1975b (APHRODINAE, HECALINI). **Ishihara** 1959a (synopsis); 1961a (Thailand). **Jankovic** 1966a (Serbia). **Lindberg** 1956a (Niger River flood plain; Dogo, Mopti); 1960b (Soviet Armenia); 1963a (Morocco). **Linnavuori** 1956d (Spanish Morocco); 1975a (=*Hecalus*; n. comb. to *Hecalus* and *Linnavuoriella*; n. sp. syn.). **Logvinenko** 1962e (USSR: Ukraine). **Ramachandra Rao** 1973a (India: Poona). **Servadei** 1960a (Italy). **Sharma, B.** 1977b (India: Jammu).

Parabolopona
New species: **Webb, M. D.** 1981b.
Records: **Hamilton, K. G. A.** 1975b (APHRODINAE, PARABOLOPONINI). **Lee, C. E.** 1979a (Korea, redescribed). **Lee, C. E. & Kwon** 1977b (Korea). **Webb, M. D.** 1981a (lectotype designation; key to species).

Paracanthus
Record: **Nast** 1982a (catalogued; CICADELLINAE).

Paracatua
Record: **Young** 1977a (n. comb. from *Scaris* and *Tettigonia*; key to species).

Paracarinolidia
New species: **Nielson** 1979b.
Record: **Nielson** 1979b (n. comb. from *Coelidia*; key to species).

Paracephaleus
New species: **Evans, J. W.** 1966a. **Knight, W. J.** 1973b.
Records: **Davis, R. B.** 1975a (morphology; phyletic inferences). **Evans, J. W.** 1966a (removed from synonymy with *Notocephalius*; n. comb. from *Cephalelus*; n. sp. syn.); 1977a (n. comb. from *Cephalelus*). **Knight, W. J.** 1973b (n. sp. syn.).

Parachunroides
New species: **Maldonado-Capriles** 1977a.
Record: **Maldonado-Capriles** 1977a (in key to 4 related genera).

Paracrocampsa
Record: **Young** 1968a (n. comb. from *Amblydisca, Aulacizes* and *Tettigonia*; key to species).

Paracyba
New species: **Dworakowska** 1977i.
Records: **Anufriev** 1977b (Kurile Islands). **Dworakowska** 1982a (Japan, Vietnam; discussion of specific synonymy). **Lee, C. E.** 1979a (Korea, redescribed). **Lee, C. E. & Kwon** 1977b (Korea). **Vilbaste** 1968a (coastal regions of the USSR).

Paradorydium
New species: **Dlabola** 1961b. **Evans, J. W.** 1966a, 1977a. **Knight, W. J.** 1973a. **Linnavuori** 1962a, 1964a, 1979a. **Theron** 1982b.
Records: **Dlabola** 1957a (Turkey); 1957b (Afghanistan); 1957d (Czechoslovakia); 1958b (southern Europe); 1960c (Iran); 1971a, 1972a (Afghanistan); 1981a (Iran). **Dubovskij** 1966a (Uzbekistan); 1978a (Uzbekistan, Zarafshan Valley); 1980a (USSR: western Turkmenia). **Emeljanov** 1964c (European USSR; key to species). **Evans, J. W.** 1966a (=*Deltodorydium*, n. syn.; n. comb. from *Deltodorydium*; n. sp. syn.). **Hamilton, K. G. A.** 1975b (APHRODINAE, PARADORYDIINI). **Knight, W. J.** 1973a (New Zealand; key to species). **Lindberg** 1958a (Cape Verde Islands); 1960b (Soviet Armenia); 1962a (Portugal). **Linnavuori** 1961a (South Africa: Cape Province, Basutoland); 1979a (=*Mesodorydium*, n. syn.; n. comb. from *Mesodorydium*; new subgenus proposed; key to species). **Lodos & Kalkandelen** 1981a (Turkey). **Mitjaev** 1968b (eastern Kazakhstan); 1971a (Kazakhstan; key to 2 species). **Quartau & Rodrigues** 1969a (Portugal). **Servadei** 1957a, 1960a (Italy). **Theron** 1976a (redescription of Naude species; n. sp. syn.); 1982b (=*Bumizana*, n. syn.).

Paradrabescus
New species: **Kuoh, C. L.** 1985a.
Record: **Kuoh, C. L.** 1985a (comparative notes).

Parafieberiella
New species: **Dlabola** 1974a.
Records: **Dlabola** 1981a (Iran). **Hamilton, K. G. A.** 1975b (APHRODINAE, DELTOCEPHALINI, PLATYMETOPIINA). **Nast** 1982a (catalogued; DELTOCEPHALINAE).

Paraganus
Record: **Linnavuori** 1959b (redescribed; comparative notes).

Paragonalia
Record: **Young** 1977a (identity obscure).

Paragygrus
New species: **Mitjaev** 1980d.
Records: **Anufriev** 1978a (USSR: Maritime Territory). **Choe** 1985a (Korea). **Emeljanov** 1964c (European USSR). **Hamilton, K. G. A.** 1975b (APHRODINAE, DELTOCEPHALINI, DELTOCEPHALINA). **Mitjaev** 1980d (key to 4 species). **Vilbaste** 1980b (Tuva).

Paraidioscopus
New species: **Viraktamath, C. A.** 1973a.
Records: **Maldonado-Capriles** 1964a (n. comb. from *Idioscopus*). **Viraktamath, C. A.** 1973a (key to species).

Paralaevicephalus
Records: **Hamilton, K. G. A.** 1975b (APHRODINAE, DELTOCEPHALINI, DELTOCEPHALINA). **Ishihara** 1958a (Japan: north Honshu). **Lee, C. E.** 1979a (Korea, redescribed). **Lee, C. E. & Kwon** 1977b (Korea).

Paralebra
New species: **Young** 1957c.
Record: **Young** 1957c (=*Kallebra*, n. syn.; key to species).

Paralidia
Record: **Nielson** 1982e (n. comb. from *Coelidia*).

Paralimnellus
  Records: **Emeljanov** 1972d (n. comb. from *Paralimnus*). **Hamilton, K. G. A.** 1975b (APHRODINAE, DELTOCEPHALINI, DELTOCEPHALINA). **Nast** 1982a (catalogued; DELTOCEPHALINAE).

Paralimnoidella
  New species: **Kwon & Lee** 1979c.
  Record: **Lee, C. E.** 1979a (Korea, redescribed).

Paralimnus
  New species: **Bonfils** 1981a. **Dlabola** 1960b, 1961b, 1967c, 1967d. **Dubovskij** 1966a, 1970c. **Emeljanov** 1964c, 1964f, 1972b. **Linnavuori** 1961a, 1964a. **Mitjaev** 1967c, 1969a, 1971a, 1975a. **Zachvatkin** 1953c.
  Records: **Anufriev & Emeljanov** 1968a (Soviet Far East). **Cantoreanu** 1965b (Romania); 1968b (Danube river delta). **Datta** 1973i, 1973l (descriptive notes). **Della Giustina** 1983a (France). **Dlabola** 1968a, 1968b (Mongolia). **Dubovskij** 1966a (Uzbekistan; key to species); 1980a (USSR: western Turkmenia). **Dworakowska** 1969a (Mongolia). **Emeljanov** 1964c (European USSR; key to species); 1972d (n. comb. to *Paralimnellus*); 1977a (Mongolia). **Evans, J. W.** 1966a (=*Diemoides*, n. syn.; n. comb. from *Diemoides*). **Gravestein** 1965a (The Netherlands: Terschelling I.). **Hamilton, K. G. A.** 1975b (APHRODINAE, DELTOCEPHALINI, DELTOCEPHALINA). **Heller** 1961a (n. sp. syn.). **Jankovic** 1966a (Serbia). **Kalkandelen** 1974a (Turkey). **Kholmuminov & Dlabola** 1979a (USSR: Golodnostep Plain). **Korolevskaya** 1973a, 1979a (Tadzhikistan). **Lee, C. E.** 1979a (Korea, redescribed). **Lee, C. E. & Kwon** 1977b (Korea). **LeQuesne** 1969c (Britain). **Logvinenko** 1957b, 1957c (USSR: Ukraine). **Mitjaev** 1968b (eastern Kazakhstan); 1971a (Kazakhstan; key to 13 species); 1971b (southern Kazakhstan). **Nast** 1972a (catalogued); 1976b (Poland; catalogued); 1982a (catalogued; DELTOCEPHALINAE). **Ossiannilsson** 1983a (Fennoscandia & Denmark; key to species). **Ramachandra Rao** 1969a (description of female). **Servadei** 1968a (Italy). **Sharma, B.** 1977b (India: Jammu). **Vilbaste** 1962b (eastern part of the Caspian lowlands); 1965b (Altai Mts.); 1980b (Tuva)

Parallaxis
  Records: **Linnavuori** 1956a (Argentina, Colombia, Costa Rica, Paraguay). **Ruppel** 1966a (Colombia). **Young** 1957b (notes on Osborn's type material).

Paramacroceps
  New species: **Dubovskij** 1966a, 1968b.
  Records: **Emeljanov** 1975a (Sahara; ADELUNGIINI). **Linnavuori** 1969b (n. comb. from *Platyproctus*; in key to north African genera of MELICHARELLINI). **Mitjaev** 1971a (Kazakhstan; in key to genera of MELICHARELLINAE).

Paramesanus
  Record: **Nast** 1982a (catalogued; DELTOCEPHALINAE).

Paramesodes
  New species: **Mahmood & Meher** 1983a. **Wilson, M. R.** 1983b.
  Records: **Dlabola** 1984a (Iran). **Dubovskij** 1966a (Uzbekistan). **Hamilton, K. G. A.** 1975b (APHRODINAE; unassigned). **Korolevskaya** 1979b (Tadzhikistan). **Kwon & Lee** 1979b (Korea; n. sp. syn.). **Lee, C. E.** 1979a (Korea, redescribed). **Linnavuori** 1962a (Israel; =*Coexitianus*, n. syn.). **Servadei** 1968a (Italy). **Wilson, M. R.** 1983b (redescribed; lectotype designation; n. sp. syn.; key to species).

Paramesus
  Records: **Anufriev** 1978a (USSR: Maritime Territory). **Beirne** 1956a (Canada). **Cantoreanu** 1968b (Danube river delta); 1971c (Romania). **Dlabola** 1957a (Turkey); 1957b (Afghanistan); 1958b (Caucasus); 1961b (central Asia); 1964c (Albania); 1967c (Mongolia); 1970a (Bohemia); 1972a (Afghanistan); 1981a (Iran). **Dubovskij** 1966a (Uzbekistan); 1978a (Uzbekistan, Zarafshan Valley). **Dworakowska** 1973d (Korea). **Emeljanov** 1964c (European USSR; key to species); 1972d (n. comb. to *Paranastus*); 1977a (Mongolia). **Gravestein** 1965a (The Netherlands: Terschelling I.). **Hamilton, K. G. A.** 1975b (APHRODINAE, DELTOCEPHALINI, DELTOCEPHALINA); 1983b (transboreal taxon). **Jankovic** 1971a (Serbia). **Kalkandelen** 1974a (Turkey). **Lee, C. E.** 1979a (Korea, redescribed). **Lee,** **C. E. & Kwon** 1977b (Korea). **LeQuesne** 1969c (Britain). **Lindberg** 1956a (Niger River flood plain; Tomara). **Linnavuori** 1962a (Israel). **Logvinenko** 1957b, 1957c (USSR: Ukraine). **Menezes** 1975a (n. comb. to *Neomesus*; species redescribed, including male sex). **Mitjaev** 1968b (eastern Kazakhstan); 1971a (Kazakhstan; key to 2 species). **Nast** 1955a (Poland); 1972a (catalogued); 1976b (Poland; catalogued). **Ossiannilsson** 1983a (Fennoscandia & Denmark). **Quartau & Rodrigues** 1969a (Portugal). **Servadei** 1957a, 1968a (Italy). **Vilbaste** 1962b (eastern part of the Caspian lowlands); 1980b (Tuva).

Parametopia
  Records: **Schroder** 1959a (n. comb. to *Lebora* and *Molomea*).

Paranastus
  Records: **Emeljanov** 1972d (n. comb. from *Paramesus*). **Hamilton, K. G. A.** 1975b (APHRODINAE, DELTOCEPHALINI, DELTOCEPHALINA). **Nast** 1982a (catalogued; DELTOCEPHALINAE).

Parandanus
  New species: **Linnavuori & DeLong** 1976a, 1979b.
  Record: **Cwikla & Blocker** 1981a (comparative notes).

Paranoplus
  New species: **Linnavuori** 1961a.

Paraonukia
  New species: **Ishihara** 1963b.

Parapagaronia
  Record: **Nast** 1982a (catalogued; CICADELLINAE).

Parapetalocephala
  New species: **Kwon & Lee** 1978b.
  Record: **Nast** 1982a (catalogued; LEDRINAE).

Paraphlepsius
  New species: **DeLong** 1982f. **DeLong & Linnavuori** 1978b. **Cwikla** 1980a. **Hamilton, K. G. A.** 1972b, 1975a.
  Records: **Hamilton, K. G. A.** 1972d (Canada: Manitoba); 1975a (subgenera and species groups defined; new subgenera proposed; n. comb. from *Phlepsius*; n. sp. syn.); 1975b (APHRODINAE, DELTOCEPHALINI, PLATYMETOPIINA). **Linnavuori** 1959b (key to Neotropical species); 1973c (Cuba). **Nielson** 1968b (characterization of vector species). **Scudder** 1971a (Canada: British Columbia).

Parapotes
  Records: **Nast** 1976b (Poland; catalogued); 1982a (catalogued; DELTOCEPHALINAE).

Parargus
  New species: **Emeljanov** 1961a, 1964f. **Mitjaev** 1969c. **Vilbaste** 1965b.
  Records: **Emeljanov** 1964c (European USSR). **Hamilton, K. G. A.** 1975b (APHRODINAE, DELTOCEPHALINI, DELTOCEPHALINA). **Mitjaev** 1968b (eastern Kazakhstan); 1971a (Kazakhstan; key to 2 species) **Vilbaste** 1980b (Tuva).

Parasitades
  Records: **Hamilton, K. G. A.** 1980a (subgenus of *Oncopsis*, n. status). **Viraktamath, C. A.** 1980a (comparative notes).

Paratanus
  New species: **Cheng, Y. J.** 1980a. **DeLong & Cwikla** 1985a. **Linnavuori** 1959b. **Linnavuori & DeLong** 1977c. **Young** 1957a.
  Records: **Linnavuori** 1959b (n. comb. from *Alaca*; key to species); 1975d (Peru). **Nielson** 1968b (characterization of vector species). **Young** 1957a (n. comb. from *Atanus*; key to species).

Paraterulia
  New species: **Nielson** 1979b.
  Record: **Nielson** 1979b (n. comb. from *Terulia*; key to species).

Parathaia
  New species: **Kuoh, C. L.** 1982a.

Parathona
  Record: **Young** 1977a (n. comb. from *Tettigonia*; n. sp. syn.; key to species).

Paratubana
  Record: **Young** 1977a (n. comb. from *Tettigonia*; key to species).

Paratyphlocyba
  New species: **Ahmed, Manzoor** 1985c.

Paraulacizes
  Record: **Young** 1968a (n. comb. from *Amblydisca, Aulacizes, Cicada, Oncometopia* and *Proconia*; n. sp. syn.).
Parazyginella
  New species: **Chou, I. & Zhang** 1985a.
Parinaeota
  New species: **Young** 1977a.
Parocerus
  Records: **Nast** 1982a (catalogued; IDIOCERINAE). **Vilbaste** 1980b (n. comb. from *Idiocerus*; key to species).
Parohinka
  New species: **Webb, M. D.** 1981b.
  Record: **Webb, M. D.** 1981b (n. comb. from *Dryaromorpha* and *Muirella*; key to species).
Paromenia
  New species: **Young** 1977a.
  Record: **Young** 1977a (n. gen. syn.; n. comb. from *Aulacizes* and *Tettigonia*; n. sp. syn.; key to species).
Paropia
  Record: **Wagner** 1959c (Greece).
Paropulopa
  Record: **Davis, R. B.** 1975a (morphology; phyletic inferences).
Parunculus
  New species: **Emeljanov** 1964f.
  Records: **Emeljanov** 1962a (n. comb. from *Deltocephalus*; 1977a (Mongolia). **Hamilton, K. G. A.** 1975b (APHRODINAE, DELTOCEPHALINI, DELTOCEPHALINA). **Mitjaev** 1971a (Kazakhstan; key to 2 species). **Vilbaste** 1965b (Altai Mts.); 1980b (Tuva).
Pasadenus
  Records: **Hamilton, K. G. A.** 1975b (APHRODINAE, DELTOCEPHALINI, PLATYMETOPIINA). **Nielson** 1969a (redefinition; key to species).
Pasara
  New species: **Dworakowska** 1981e.
  Record: **Sohi & Dworakowska** 1984a (list of Indian species).
Pasaremus
  Records: **Beirne** 1956a (Canada). **Hamilton, K. G. A.** 1975b (APHRODINAE, DELTOCEPHALINI, DELTOCEPHALINA).
Pascoepus
  New species: **Webb, M. D.** 1983a.
  Record: **Webb, M. D.** 1983a (n. comb. from *Idiocerus*; key to species).
Paternus
  Records: **Datta** 1973n (descriptive notes). **Vilbaste** 1965a (=*Doratulina*, n. syn.).
Pauripo
  Records: **Davis, R. B.** 1975a (morphology; phyletic inferences). **Evans, J. W.** 1966a (transferred from IPOINI to EURYMELINI).
Pauroeurymela
  Record: **Davis, R. B.** 1975a (morphology; phyletic inferences).
Pawiloma
  New species: **Young** 1977a.
  Record: **Young** 1977a (n. comb. from *Cardioscarta, Erythroneura* and *Tettigonia*; n. sp. syn.; key to species).
Pazu
  Record: **Hamilton, K. G. A.** 1975b (APHRODINAE, DELTOCEPHALINI, DELTOCEPHALINA).
Peayanus
  New species: **Nielson** 1979b, 1983f.
  Record: **Nielson** 1979b (n. comb. from *Coelidia*; key to species).
Peconus
  Record: **Hamilton, K. G. A.** 1975b (APHRODINAE, DELTOCEPHALINI, DELTOCEPHALINA).
Pectinapyga
  Record: **Hamilton, K. G. A.** 1975b (APHRODINAE, APHRODINI, DORATURINA).
Pedarium
  New species: **Emeljanov** 1961a.
  Records: **Diabola** 1984a (Iran). **Hamilton, K. G. A.** 1975b (APHRODINAE, DELTOCEPHALINI, ATHYSANINA). **Korolevskaya** 1978b (central Asia). **Mitjaev** 1971a (Kazakhstan; in key to genera of DELTOCEPHALINAE).

Pedionis
  New species: **Hamilton, K. G. A.** 1980b. **Viraktamath, C. A.** 1981b.
  Records: **Hamilton, K. G. A.** 1980b (n. comb. from *Macropsis*; new subgenera proposed). **Nast** 1982a (catalogued; MACROPSINAE).
Pediopsis
  New species: **Anufriev** 1971a.
  Records: **Anufriev** 1978a (USSR: Maritime Territory). **Emeljanov** 1964c (European USSR). **Hamilton, K. G. A.** 1980b (n. comb. from *Macropsis*; n. comb. to *Pedionis*; lectotype designation); 1983b (Nearctic region, adventive); 1983d (in key to genera of New World MACROPSINI). **Jankovic** 1966a (Serbia). **Lee, C. E.** 1979a (Korea, redescribed). **LeQuesne** 1965c (Britain). **Logvinenko** 1957b (USSR: Ukraine). **Nast** 1972a (catalogued); 1976a (Poland: Pieniny Mts.); 1976b (Poland; catalogued). **Ossiannilsson** 1981a (Fennoscandia & Denmark). **Theron** 1970b (redescription of *P. capensis*). **Vilbaste** 1973a (notes on Flor's material); 1975a (notes on Motschulsky's material).
Pediopsoides
  New species: **Viraktamath, C. A.** 1981b.
  Records: **Hamilton, K. G. A.** 1980b (new subgenus proposed; *Kiamoncopsis, Nanopsis* and *Syspocnis* as subgenera, new status); 1983d (subgenus *Nanopsis sensu* Hamilton 1980b for New World species).
Pedioscopus
  New species: **Linnavuori** 1960a, 1960b. **Maldonado-Capriles** 1972a.
  Records: **Evans, J. W.** 1966a (n. sp. syn.). **Linnavuori** 1960b (n. comb. from *Idiocerus*). **Maldonado-Capriles** 1961a (n. comb. from *Balocha*); 1968b (comparative notes); 1972a (list of incorrectly assigned species; key to species). **Webb, M. D.** 1983a (in key to genera of Australian IDIOCERINAE).
Pedumella
  Record: **Hamilton, K. G. A.** 1975b (APHRODINAE, APHRODINI, DORATURINA).
Pegogonia
  Record: **Young** 1977a (n. comb. from *Cicada* and *Tettigonia*).
Peitouellus
  Record: **Hamilton, K. G. A.** 1975b (APHRODINAE, DELTOCEPHALINI, DELTOCEPHALINA). **Vilbaste** 1969a (n. comb. from *Thamnotettix*).
Peltocheirus
  New species: **Young** 1968a.
  Records: **Young** 1968a (key to species). **Young & Lauterer** 1966a (lectotype designation).
Pemoasca
  New species: **Mahmood** 1967a.
  Record: **Mahmood** 1967a (in key to genera of Oriental TYPHLOCYBINAE).
Penangiana
  New species: **Mahmood** 1967a.
  Record: **Mahmood** 1967a (in key to genera of Oriental TYPHLOCYBINAE).
Pendarus
  New species: **Hamilton, K. G. A.** 1975a.
  Records: **Hamilton, K. G. A.** 1975a (subgeneric and species groups defined; n. comb. from *Paraphlepsius, Phlepsius* and *Remadosus*; n. sp. syn.); 1975b (APHRODINAE, DELTOCEPHALINI, PLATYMETOPIINA).
Penehuleria
  Records: **Datta & Dhar** 1984a (descriptive notes). **Hamilton, K. G. A.** 1975b (APHRODINAE, DELTOCEPHALINI, COCHLORHININA).
Penestirellus
  Records: **Linnavuori** 1959b (=*Stirellus*, n. syn.). **Hamilton, K. G. A.** 1975b (APHRODINAE, STIRELLINI).
Penestragania
  New species: **Blocker** 1979a, 1979b, 1982a.
  Records: **Blocker** 1979a (generic status; n. comb. from *Batracomorphus* and *Pachyopsis*; key to Neotropical species); 1979b (new subgenus proposed).
Penthimia
  New species: **Dlabola** 1958b, 1961b. **Kwon & Lee** 1978c. **Linnavuori** 1977a. **Logvinenko** 1983a.

Records: **Anufriev** 1978a (USSR: Maritime Territory). **Beirne** 1956a (Canada). **Bonfils & Della Giustina** 1978a (Corsica). **Cantoreanu** 1968b (Danube river delta); 1971c (Romania). **Datta** 1973i (descriptive notes). **Datta & Dhar** 1984a (descriptive notes). **Dlabola** 1957a (Turkey); 1957b (Afghanistan); 1960c (Iran); 1961b (central Asia); 1964c (Albania); 1971b (Turkey) 1972a (Afghanistan); 1981a (Iran). **D'Urso** 1980b (Italy). **Emeljanov** 1964c (European USSR). **Evans, J. W.** 1972a (n. comb. to *Vulturnus*). **Hamilton, K. G. A.** 1972d (Canada: Manitoba); 1976a (identity of Provancher material). **Ishihara** 1958a (Japan: north Honshu). **Jankovic** 1966a (Serbia). **Kwon & Lee** 1978c (Korea). **Lee, C. E.** 1979b (Korea, redescribed). **Lee, C. E. & Kwon** 1977b (Korea). **Lindberg** 1960a (Portugal); 1961a (Madeira). **Linnavuori** 1959c (Mt. Picentini); 1977a (definition of species groups; key to species). **Lodos & Kalkandelen** 1981a (Turkey). **Logvinenko** 1957b (USSR: Ukraine). **Nast** 1955a (Poland); 1972a (catalogued); 1976b (Poland; catalogued). **Servadei** 1957a, 1968a, 1979a (Italy). **Sharma, B.** 1977b (India: Jammu). **Theron** 1976a (n. comb. to *Thaumatopoides*); 1980 (South Africa; correction of record for *P. vinula*). **Vilbaste** 1968a (USSR: coastal regions); 1973a (notes on Flor's material).

Penthimidia
Records: **Linnavuori** 1977a (redescribed). **Linnavuori & Al-Ne'amy** 1983a (comparative notes).

Penthimiella
New species: **Evans, J. W.** 1972a.

Penthimiola
New species: **Lindberg** 1958a. **Linnavuori** 1977a.
Records: **Linnavuori** 1959b (Argentina ?); 1977a (in key to genera of African PENTHIMIINAE; key to species). **Theron** 1980a (correction of Argentina record). **Webb, M. D.** 1980a (Aldabra).

Penthimiopsis
New species: **Evans, J. W.** 1972a.

Pentoffia
New species: **Kramer** 1964b.

Pentria
New species: **Evans, J. W.** 1972a.

Peragallia
New species: **Dlabola** 1964a, 1964b. **Lindberg** 1958a. **Linnavuori** 1960b, 1962a. **Vilbaste** 1961a.
Records: **Dlabola** 1957a (Turkey); 1958b (Caucasus); 1960c (Iran); 1961b (central Asia); 1964a (Afghanistan); 1964b (Sudan); 1964c (Albania). **Dubovskij** 1966a (Uzbekistan). **Emeljanov** 1964c (European USSR). **Heller & Linnavuori** 1968a (Ethiopia). **LeQuesne** 1964a (= *Austroagallia*, n. syn.). **Lindberg** 1960a (Portugal); 1960b (Soviet Armenia); 1960d (Canary Islands); 1962a (Portugal); 1963a (Morocco); 1965a (Spanish Sahara). **Linnavuori** 1956d (Spanish Morocco); 1961a (South Africa; n. comb. from *Agallia*); 1962a (Israel). **Logvinenko** 1957b (USSR: Ukraine). **Nielson** 1968b (= *Austroagallia*). **Quartau & Rodrigues** 1969a (Portugal). **Servadei** 1960a (Italy). **Webb, M. D.** 1980a (Aldabra).

Perotettix
Records: **Anufriev** 1978a (USSR: Maritime Territory). **Cantoreanu** 1971c (Romania). **Hamilton, K. G. A.** 1975b (APHRODINAE, DELTOCEPHALINI, ATHYSANINA). **Jankovic** 1966a (Serbia). **Nast** 1955a (Poland); 1972a (catalogued); 1976b (Poland; catalogued). **Ossiannilsson** 1983a (Fennoscandia & Denmark). **Servadei** 1968a (Italy).

Perubahita
New species: **DeLong** 1982a. **Linnavuori & DeLong** 1978b.
Record: **Cwikla & Blocker** 1981a (comparative notes).

Perubala
New species: **Kramer** 1963b. **Linnavuori** 1959b, 1965b. **Linnavuori & DeLong** 1976a.
Records: **Kramer** 1963b (in key to genera of Neotropical NEOBALINAE). **Linnavuori** 1959b (key to species). **Linnavuori & DeLong** 1977c (in key to genera of NEOBALINAE).

Perugrampta
New species: **Kramer** 1965a. **Linnavuori & DeLong** 1978f.

Perulidia
New species: **Nielson** 1979b, 1983c.

Records: **Nielson** 1979b (key to species); 1983c (revised key to species).

Petalocephala
New species: **Dlabola** 1957b, 1981a. **Evans, J. W.** 1969a. **Kuoh, C. L.** 1984a. **Linnavuori** 1972b.
Records: **Anufriev** 1978a (USSR: Maritime Territory); 1979a (notes on Jacobi's material). **Datta** 1972a (descriptive notes). **Datta & Pramanik** 1977a (India: A. P., Subansiri Division). **Dlabola** 1972a (Afghanistan). **Evans, J. W.** 1969a (Australia and New Guinea). **Kwon & Lee** 1978b (Korea). **Lee, C. E.** 1979a (Korea, redescribed). **Lee, C. E. & Kwon** 1977b (Korea). **Lindberg** 1960d (Canary Islands). **Linnavuori** 1969a (Congo); 1972b (= *Pachyledra*, n. syn.; n. comb. from *Pachyledra*; n. sp. syn.; key to species). **Ramachandra Rao** 1973a (India: Poona). **Theron** 1976a (identity of Naude species); 1980a (notes on the type of *P. wahlbergi*). **Vilbaste** 1968a (USSR: coastal regions).

Pettya
Record: **Evans, J. W.** 1966a (= *Eutambourina*, n. syn.; n. comb. from *Eutambourina*).

Peyerimhoffiola
New species: **Al-Ne'amy & Linnavuori** 1982a.
Records: **Davis, R. B.** 1975a (morphology; phyletic inferences). **Emeljanov** 1975a (in key to genera of ADELUNGIINI). **Al-Ne'amy & Linnavuori** 1982a (redefined, (comparative notes; ADELUNGIINAE, PEYERIMHOFFIOLINI).

Phaeida
New species: **Emeljanov** 1962a.
Record: **Hamilton, K. G. A.** 1975b (APHRODINAE; unassigned).

Phera
New species: **Young** 1968a.
Records: **Schroder** 1959a (diagnosis). **Young** 1963a, 1965c (lectotype designations); 1968a (= *Capinota*, n. syn.; n. comb. from *Capinota* and *Homalodisca*; n. sp. syn.; key to species). **Young & Nast** 1963a (notes on type material).

Phereurhinus
New species: **Kramer** 1976a.
Record: **Kramer** 1976a (key to species).

Pherodes
Records: **Young** 1965c (lectotype designation); 1968a (= *Acrogonia*, n. syn.).

Philaia
Records: **Anufriev** 1978a (USSR: Maritime Territory). **Dlabola** 1965c, 1967b, 1967c, 1967d, 1968a 1968b (Mongolia); 1974d (Mediterranean). **Dworakowska** 1973d (Korea, Mongolia). **Emeljanov** 1964c (European USSR); 1977a (Mongolia). **Hamilton, K. G. A.** 1975b (APHRODINAE, DELTOCEPHALINI, DELTOCEPHALINA). **Lee, C. E.** 1979a (Korea, redescribed). **Logvinenko** 1963a (USSR: Ukraine). **Mitjaev** 1968b (eastern Kazakhstan); 1971a (Kazakhstan). **Vilbaste** 1980b (Tuva).

Philipposcerus
New species: **Maldonado-Capriles** 1972b.

Philotartessus
New species: **Evans, F.** 1981a.
Records: **Evans, F.** 1981a. (n. comb. from *Tartessus*; n. sp. syn.) **Knight, W. J.** 1982a (Rennell Island).

Phlebiastes
New species: **Emeljanov** 1961a, 1969b, 1972b (new name). **Mitjaev** 1969c.
Records: **Emeljanov** 1969b (n. comb. from *Arocephalus* and *Parargus*); 1977a (Mongolia; *Parargus* as subgenus; n. comb. from *Parargus*). **Hamilton, K. G. A.** 1975b (APHRODINAE, DELTOCEPHALINI, DELTOCEPHALINA). **Korolevskaya** 1979b (Tadzhikistan). **Mitjaev** 1971a (Kazakhstan; key to 2 species)

Phlepsanus
Record: **Hamilton, K. G. A.** 1975b (APHRODINAE, DELTOCEPHALINI, PLATYMETOPIINA).

Phlepsidius
New species: **Dubovskij** 1966a. **Emeljanov** 1961a, 1964f, 1979a. **Korolevskaya** 1961a, 1980a, 1981a. **Mitjaev** 1969b, 1971a.
Records: **Emeljanov** 1964c (European USSR). **Hamilton, K. G. A.** 1975b (APHRODINAE; unassigned). **Mitjaev** 1971a (Kazakhstan; key to 6 species).

Phlepsius
New species: **Dlabola** 1961b, 1974c, 1979a, 1984a. **Korolevskaya** 1961a. **Pierce, W. D.** 1963a (fossil). **Wagner** 1963b.
Records: **Bonfils & Della Giustina** 1978a (Corsica). **Datta** 1973i (descriptive notes). **Della Giustina** 1983a (France). **Dlabola** 1957b (Afghanistan); 1960c (Iran); 1961b (central Asia); 1964a (Afghanistan); 1964c (Albania); 1965d (Jordan); 1972a (Afghanistan); 1981a, 1984a (Iran). **Dubovskij** 1966a (Uzbekistan); 1978a (Uzbekistan, Zarafshan Valley); 1980a (USSR: western Turkmenia). **Emeljanov** 1964c (European USSR; key to species); 1977a (Mongolia). **Hamilton, K. G. A.** 1975b (APHRODINAE, SELENOCEPHALINI). **Heller & Linnavuori** 1968a (Ethiopia). **Kalkandelen** 1974a (Turkey; key to species). **Korolevskaya** 1973a, 1979b (Tadzhikistan). **Lindberg** 1960b (Soviet Armenia); 1962a (Portugal). **Linnavuori** 1961a (southern Africa: Cape Province; North and South Rhodesia); 1962a (Israel); 1965a (Mediterranean). **Logvinenko** 1957b (USSR: Ukraine). **Mitjaev** 1971a (Kazakhstan; key to 2 species). **Quartau & Rodrigues** 1969a (Portugal). **Servadei** 1957a, 1960a, 1968a, 1971a (Italy). **Wagner** 1959c (Greece); 1963b (revision of European species).

Phlepsobahita
New species: **Linnavuori** 1959b.

Phlepsopsius
Record: **Nast** 1982a (catalogued; DELTOCEPHALINAE).

Phlogis
New species: **Linnavuori** 1979a.

Phlogotettix
Records: **Anufriev** 1978a (USSR: Maritime Territory). **Choe** 1985a (Korea). **Dlabola** 1977c (Mediterranean); 1981a (Iran). **Dworakowska** 1973d (Korea). **Emeljanov** 1964c (European USSR). **Hamilton, K. G. A.** 1975b (APHRODINAE, DELTOCEPHALINI, PLATYMETOPIINA). **Jankovic** 1966a (Serbia). **Ishihara** 1958a (Japan: north Honshu). **Lauterer** 1984a (Moravia). **Lee, C. E.** 1979a (Korea, redescribed). **Lee, C. E. & Kwon** 1977b (Korea). **Linnavuori** 1962a (Israel). **Servadei** 1968a (Italy). **Vilbaste** 1968a (USSR: coastal regions).

Phlogothamnus
New species: **Ishihara** 1961a. **Linnavuori** 1969a.

Phrynomorphus
Records: **Hamilton, K. G. A.** 1975b (APHRODINAE, DELTOCEPHALINI, ATHYSANINA). **Ramachandra Rao** 1973a (*Athysanus* as a subgenus).

Phrynophyes
Records: **Evans, J. W.** 1966a (Australia; synopsis); 1977a (=*Aconura*; erroneous interpretation of Vilbaste's 1965a actions). **Vilbaste** 1965a (=*Doratulina*, n. syn.).

Phycotettix
New species: **Linnavuori** 1971a.
Records: **Dlabola** 1960a (notes on Horvath's material). **Hamilton, K. G. A.** 1975b (APHRODINAE; unassigned). **Lindberg** 1963a (Morocco). **Linnavuori** 1965a (Mediterranean). **Nast** 1982a (catalogued). **Servadei** 1969a (Italy). **Webb, M. D.** 1979a (n. comb. from *Tettigonia*; lectotype designation; n. sp. syn.).

Phytotartessus
New species: **Evans, F.** 1981a.
Record: **Evans, F.** 1981a. (n. comb. from *Bythoscopus*).

Picchuia
New species: **Linnavuori & DeLong** 1979a.
Record: **Cwikla & Blocker** 1981a (comparative notes).

Picchusteles
New species: **Linnavuori & DeLong** 1976a.
Record: **Cwikla & Blocker** 1981a (comparative notes).

Piezauchenia
Record: **Linnavuori** 1959b (? to XEROPHLOEINAE).

Pilosana
New species: **Nielson** 1983e.
Record: **Nielson** 1983e (key to species).

Pingellus
New species: **Evans, J. W.** 1966a.

Pinumius
New species: **Dlabola** 1965c. **Emeljanov** 1966a, 1972d. **Kuoh, C. L.** 1981a.
Records: **Dlabola** 1956b (Eurasia); 1961b (central Asia); 1965c, 1967b, 1967c, 1967d, 1968a, 1968b, 1970b (Mongolia). **Dworakowska** 1969a (Mongolia). **Emeljanov** 1964c (European USSR); 1977a (Mongolia) **Hamilton, K. G. A.** 1975b (APHRODINAE, DELTOCEPHALINI, DELTOCEPHALINA); 1983b (transboreal taxon). **Lauterer** 1983b (Moravia). **Logvinenko** 1963a (USSR: Ukraine). **Mitjaev** 1968b (eastern Kazakhstan); 1971a (Kazakhstan). **Nast** 1955a (Poland); 1972a (catalogued); 1976b (Poland; catalogued). **Ossiannilsson** 1983a (Fennoscandia & Denmark). **Vilbaste** 1965b (Altai Mts.); 1980b (Tuva).

Piorella
New species: **Evans, J. W.** 1972a.

Pisacha
Record: **Young** 1965c (lectotype designation).

Pithyotettix
New species: **Anufriev** 1971a, 1975a. **Emeljanov** 1966a, 1972d. **Mitjaev** 1965a. **Vilbaste** 1965b.
Records: **Anufriev** 1971a (subgenera defined); 1975a (key to species); 1977b (Kurile Islands) 1978a (USSR: Maritime Territory; key to species). **Cantoreanu** 1971c (Romania). **Dlabola** 1967b, 1967c (Mongolia). **Emeljanov** 1964c (European USSR; key to species); 1977a (Mongolia). **Hamilton, K. G. A.** 1975b (APHRODINAE, DELTOCEPHALINI, ATHYSANINA). **Mitjaev** 1965a (new subgenus proposed); 1968b (eastern Kazakhstan); 1971a (Kazakhstan). **Nast** 1972a (catalogued); 1976a (Poland: Pieniny Mts.); 1976b (Poland; catalogued). **Ossiannilsson** 1983a (Fennoscandia & Denmark). **Servadei** 1968a (Italy). **Vilbaste** 1965b (Altai Mts.); 1979a (Vooremaa hardwood/spruce forest); 1980b (Tuva).

Placidellus
New species: **Evans, J. W.** 1971a.
Record: **Nielson** 1975a (removed from COELIDIINAE; unassigned).

Placidus
New species: **Dlabola** 1957b.
Records: **Dlabola** 1972a (Afghanistan). **Evans, J. W.** 1971a (discussion of subfamily placement). **Nielson** 1975a (removed from COELIDIINAE; unassigned).

Placotettix
New species: **Dlabola** 1979a. **Linnavuori** 1965a. **Metcalf** 1967a (new name).
Records: **Allen, A. A.** 1962a (Britain: Kent). **Bonfils & Della Giustina** 1978a (Corsica). **Dlabola** 1958a (southern Europe); 1963c (identity of Haupt's material). **Hamilton, K. G. A.** 1975b (APHRODINAE, DELTOCEPHALINI, PLATYMETOPIINA). **LeQuesne** 1969c (Britain). **Quartau & Rodrigues** 1969a (Portugal). **Servadei** 1960a, 1968a (Italy).

Planaphrodes
New species: **Choe** 1981a.
Records: **Hamilton, K. G. A.** 1975d (n. comb. from *Aphrodes*; n. sp. syn.); 1983a (transboreal and adventive taxa).. **Lee, C. E.** 1979a (Korea, redescribed). **Lee, C. E. & Kwon** 1977b (Korea) **Nast** 1972a (catalogued); 1976a (Poland: Pieniny Mts.); 1982a (catalogued; APHRODINAE). **Ossiannilsson** 1981a (Fennoscandia & Denmark; key to species).

Planicephalus
New species: **Cheng, Y. J.** 1980a. **Linnavuori & DeLong** 1978c.
Records: **Hamilton, K. G. A.** 1975b (APHRODINAE, DELTOCEPHALINI, DELTOCEPHALINA). **Kramer** 1971d (generic status; n. comb. from *Deltocephalus*). **Linnavuori** 1959b (as subgenus of *Deltocephalus*; n. sp. syn.).

Planolidia
New species: **Nielson** 1982e.

Plapigella
New species: **Nielson** 1979b, 1983b.
Record: **Nielson** 1979b (n. comb. from *Coelidia*; key to species).

Platentomus
Record: **Theron** 1980a (n. comb. from *Thamnotettix*).

Platyacina
Record: **Hamilton, K. G. A.** 1975b (APHRODINAE, APHRODINI, DORATURINA).

Platycyba
    Record: **Dworakowska** 1979c (ZYGINELLINI; systematic position uncertain).
Platyeurymela
    Records: **Davis, R. B.** 1975a (morphology; phyletic inferences). **Evans, J. W.** 1966a (n. sp. syn.).
Platygonia
    New species: **Young** 1977a.
    Record: **Young** 1977a (n. gen. syn.; n. sp. syn.; n. comb. from *Ciccus* and *Tettigonia*; key to species).
Platyhynna
    Record: **Kramer** 1966a (redescription and comparative notes).
Platyjassus
    New species: **Evans, J. W.** 1959a.
    Records: **Evans, J. W.** 1959a (key to species). **Linnavuori & Quartau** 1975a (PLATYJASSINI characterized).
Platyledra
    New species: **Evans, J. W.** 1969a.
    Record: **Evans, J. W.** 1966a (n. comb. from *Ledropis* [sic]).
Platymetopius
    New species: **Dlabola** 1957b, 1958b, 1960b, 1961a, 1961b, 1964a, 1965c, 1967a, 1967e, 1971b, 1974a, 1980a, 1981a. **Dubovskij** 1966a, 1970b, 1970c, 1979a. **Emeljanov** 1964c, 1964f. **Evans, J. W.** 1966a. **Lindberg** 1958a, 1960b. **Linnavuori** 1962a, 1965a. **Logvinenko** 1969a. **Metcalf** 1967c (new name). **Mitjaev** 1968a, 1969b, 1980a. **Vilbaste** 1961a. **Wagner** 1959a.
    Records: **Cantoreanu** 1961a, 1963a (Romania). **Datta** 1973c (descriptive notes). **Dlabola** 1957a (Turkey); 1957b (Afghanistan); 1958a (southern Europe); 1958b (Caucasus); 1960a (notes on Horvath's material); 1960c (Iran); 1961a (central Europe; key to species); 1961b (central Asia); 1964a (Afghanistan); 1964c (Albania); 1965c (Mongolia); 1965d (Jordan); 1967b, 1967c (Mongolia); 1967e (southern Europe); 1968a, 1968b, 1970b (Mongolia); 1972a (Afghanistan); 1974a (Iran; new subgenus proposed); 1977c (Mediterranean); 1981a (Iran). **Dlabola & Heller** 1962a (Iran). **Dubovskij** 1966a (Uzbekistan); 1978a (Uzbekistan, Zarafshan Valley); 1980a (USSR: western Turkmenia). **Dworakowska** 1973d (Korea). **Emeljanov** 1964c (European USSR; key to species); 1977a (Mongolia). **Hamilton, K. G. A.** 1975b (APHRODINAE, DELTOCEPHALINI, PLATYMETOPIINA). **Jasinska** 1980a (Poland: Bledowska Wilderness). **Kalkandelen** 1974a (Turkey; key to species). **Korolevskaya** 1978b (Tadzhikistan). **Lee, C. E. & Kwon** 1977b (Korea). **LeQuesne** 1969c (Britain). **Linnavuori** 1956d (Spanish Morocco); 1962a (Israel); 1965a (Mediterranean). **Logvinenko** 1957b (USSR: Ukraine). **Mitjaev** 1968b (eastern Kazakhstan); 1971a (Kazakhstan; key to 10 species). **Nast** 1972a (catalogued); 1973a (Poland); 1976b (Poland; catalogued). **Ossiannilsson** 1983a (Fennoscandia & Denmark; key to species; ATHYSANINI). **Quartau & Rodrigues** 1969a (Portugal). **Servadei** 1957a, 1958a, 1960a, 1968a, 1971a (Italy). **Vilbaste** 1965b (Altai Mts.); 1968a (USSR: coastal regions); 1973a (notes on Flor's material); 1980b (Tuva). **Wagner** 1959a (Italy; description of *P. ferrari*; key to 3 species; species concepts).
Platymoideus
    Record: **Hamilton, K. G. A.** 1975b (APHRODINAE, DELTOCEPHALINI, PLATYMETOPIINA).
Platypona
    New species: **DeLong** 1982e.
Platyproctus
    New species: **Dlabola** 1960c. **Dubovskij** 1966a. **Linnavuori** 1962a, 1969b. **Mitjaev** 1969a, 1971a.
    Records: **Davis, R. B.** 1975a (morphology; phyletic inferences). **Dlabola** 1961b (central Asia); 1968b (Mongolia); 1972a (Afghanistan); 1981a (Iran). **Dubovskij** 1978a (Uzbekistan, Zarafshan Valley); 1980a (USSR: western Turkmenia). **Emeljanov** 1964c (European USSR); 1975a (in key to genera of ADELUNGIINI); 1977a (Mongolia). **Jankovic** 1966a, 1971a (Serbia). **Korolevskaya** 1975a (Tadzhikistan). **Lindberg** 1960b (Soviet Armenia; figures). **Linnavuori** 1962a (Israel); 1969b (n. comb. to *Paramacroceps*; key to northern African species). **Mitjaev** 1968b (eastern Kazakhstan); 1971a (Kazakhstan; key to 3 species).

Platyretus
    New species: **Heller & Linnavuori** 1968a.
    Records: **Datta** 1973i (as *Platyrectus*; descriptive note). **Ishihara** 1961a (Thailand).
Platyscopus
    New species: **Evans, J. W.** 1972a.
    Records: **Evans, J. W.** 1966a (n. sp. syn.); 1972a (in key to genera of Australia and New Guinea PENTHIMIINAE).
Platytetticis
    Record: **Lee, C. E.** 1979a (Korea, redescribed).
Platytettix Matsumura (TYPHLOCYBINAE)
    New species: **Ramakrishnan & Menon** 1974a.
    Record: **Ramakrishnan & Menon** (key to species).
Platytettix Vilbaste (DELTOCEPHALINAE)
    New species: **Vilbaste** 1961a.
Pleargus
    Records: **Emeljanov** 1964c (European USSR); 1977a (Mongolia). **Hamilton, K. G. A.** 1975b (APHRODINAE, DELTOCEPHALINI, DELTOCEPHALINA). **Vilbaste** 1965b (Altai Mts.); 1980b (Tuva).
Pleopardus
    Records: **Emeljanov** 1975a (in key to genera of ADELUNGIINI). **Nast** 1982a (catalogued; AGALLIINI).
Plerogonalia
    Record: **Young** 1977a (n. comb. from *Tettigoniella*; n. sp. syn.).
Plesiommata
    New species: **Young** 1977a.
    Records: **Beirne** 1956a (Canada). **Hamilton, K. G. A.** 1976a (=*Aphrophora*, CEROCOPIDAE). **Ishihara** 1971a (comparative note). **Schroder** 1959a (diagnosis). **Young** 1977a (=*Provancherana*, n. syn.; key to species); 1979a (comparative notes).
Plexitartessus
    New species: **Evans, F.** 1981a.
    Record: **Evans, F.** 1981a (n. comb. from *Tartessus*).
Plummerella
    Record: **Young** 1977a (key to species).
Plumosa
    New species: **Sohi** 1977a.
Poecilocarda
    New species: **Linnavuori** 1961a, 1969a.
    Records: **Boulard** 1969b (Central African Republic). **Heller & Linnavuori** 1968a (Ethiopia). **Linnavuori** 1969a (Congo); 1979a (n. sp. syn.; key to African species).
Poeciloscarta
    New species: **Metcalf** 1965a (new name).
    Records: **Ramos Elorduy** 1972c (Mexico). **Young** 1974a (lectotype designation); 1977a (redescribed). **Young & Lauterer** 1966a (lectotype designations).
Pogonoscopus
    Records: **Davis, R. B.** 1975a (morphology; phyletic inferences). **Evans, J. W.** 1966a (n. sp. syn.).
Polana
    New species: **DeLong** 1976g, 1979d, 1979e, 1980d, 1984e. **DeLong & Foster** 1982a, 1982b. **DeLong & Freytag** 1972f. **DeLong & Triplehorn** 1979a. **DeLong & Wolda** 1978a, 1982b, 1984a. **Freytag & Cwikla** 1982a.
    Records: **DeLong** 1976g (n. comb. from *Gypona*); 1979g (new subgenera proposed). **DeLong & Freytag** 1972a (in key to genera of GYPONINAE); 1972f (keys to subgenera and species). **Freytag & Cwikla** 1982a (key to species from Dominica).
Poliona
    New species: **Emeljanov** 1979a.
    Records: **Emeljanov** 1972d (n. comb. from *Phlepsius*). **Hamilton, K. G. A.** 1975b (APHRODINAE, DELTOCEPHALINI, PLATYMETOPIINA). **Nast** 1982a (catalogued; DELTOCEPHALINAE).
Polluxia
    New species: **Dworakowska** 1974a.
Polyamia
    New species: **DeLong** 1984f. **DeLong & Linnavuori** 1979a. **DeLong & Martinson** 1973c. **DeLong & Thambimuttu** 1973a. **Kramer** 1963c, 1965b. **Linnavuori & DeLong** 1978c.
    Records: **Beirne** 1956a (Canada). **DeLong & Thambimuttu** 1973a (n. comb. from *Deltocephalus*). **Hamilton, K. G. A.** 1975a (APHRODINAE, DELTOCEPHALINI,

DELTOCEPHALINA). **Kramer** 1973c (n. comb. from *Maricaona*).

Ponana
New species: **DeLong** 1977e, 1980c, 1981c. **DeLong & Freytag** 1967a. **DeLong & Kolbe** 1974b. **DeLong & Martinson** 1973d, 1980a. **DeLong, Wolda & Estribi** 1983a.
Records: **Beirne** 1956a (Canada; key to species). **DeLong** 1979f (Spangberg's species). **DeLong & Freytag** 1967a (key to subgenera and species); 1972a (in key to genera of GYPONINAE). **DeLong, Wolda & Estribi** 1983a (key to Panamanian species). **Hamilton, K. G. A.** 1972d (Canada: Manitoba); 1976a (identity of Provancher species; lectotype designation). **Ramos Elorduy** 1972c (Mexico).

Ponanella
New species: **DeLong** 1976d. **DeLong & Bush** 1971a. **DeLong & Freytag** 1969b.
Records: **DeLong & Freytag** 1969b (key to species); 1972a (in key to genera of GYPONINAE).

Poochara
Record: **Datta** 1980a (descriptive notes).

Populicerus
New species: **Dlabola** 1977b.
Records: **Anufriev** 1977b (Kurile Islands). **Dlabola** 1974b (n. comb. from *Idiocerus*); 1981a (Iran). **Emeljanov** 1977a (Mongolia) **Hamilton, K. G. A.** 1980a (as subgenus of *Idiocerus*, n. status). **Jasinska** 1980a (Poland: Bledowska Wilderness). **Lodos & Kalkandelen** 1981a (Turkey). **Nast** 1976a (Poland: Pieniny Mts.); 1982a (catalogued; IDIOCERINAE). **Ossiannilsson** 1981a (Fennoscandia & Denmark; key to species). **Vilbaste** 1980b (Tuva).

Porcorhinus
Record: **Evans, J. W.** 1966a (synopsis).

Portanus
New species: **DeLong** 1976e, 1980b, 1982b. **DeLong & Linnavuori** 1978b. **Kramer** 1961a, 1964e. **Linnavuori** 1959b. **Linnavuori & DeLong** 1979b.
Records: **Davis, R. B.** 1975a (morphology; phyletic inferences). **DeLong & Linnavuori** 1978b (n. comb. from *Scaphoideus*; XESTOCEPHALINAE). **Hamilton, K. G. A.** 1975b (APHRODINAE, SELENOCEPHALINI). **Kramer** 1974e (key to species; XESTOCEPHALINAE). **Linnavuori** 1956a (Peru); 1959b (XESTOCEPHALINAE, PORTANINI; key to species). **Linnavuori & Al-Ne'amy** 1983a (comments on systematics of PORTANINI).

Postumus
Records: **Evans, J. W.** 1974a (=*Chiasmus*, n. syn.). **Linnavuori** 1961a (n. comb. to *Chiasmus*).

Pradhanasundra
Records: **Dworakowska** 1980d (=*Empoascanara*, n. syn.). **Ramakrishnan & Ghauri** 1979b (n. comb. from *Empoascanara*).

Praganus
New species: **Mitjaev** 1975a.
Records: **Dlabola** 1956b (Hungary). **Emeljanov** 1964c (European USSR); 1977a (Mongolia). **Hamilton, K. G. A.** 1975b (APHRODINAE, DELTOCEPHALINI, DELTOCEPHALINA). **Mitjaev** 1968b (eastern Kazakhstan); 1971a (Kazakhstan). **Vilbaste** 1980b (Tuva).

Prairiana
New species: **Metcalf** 1962b (new name).
Records: **Beirne** 1956a (Canada). **DeLong** 1977b, 1979f (n. comb. from *Gypona*). **DeLong & Freytag** 1972a (in key to genera of GYPONINAE). **Hamilton, K. G. A.** 1972d (Canada: Manitoba).

Pratura
New species: **Theron** 1982b.

Pravistylus
Record: **Theron** 1975a (n. comb. from *Deltocephalus*).

Prescottia
New species: **DeLong** 1971a.
Records: **Beirne** 1956a (Canada: southeast region). **Hamilton, K. G. A.** 1975b (APHRODINAE, DELTOCEPHALINI, PLATYMETOPIINA).

Preta
Records: **Anufriev** 1971f (Ceylon; EVACANTHINAE, SIGNORETINI). **Ishihara** 1961a (Thailand).

Pretioscopus
New species: **Webb, M. D.** 1983b.
Record: **Webb, M. D.** 1983b (n. comb. from *Idioscopus*; key to species).

Procama
Record: **Young** 1968a (n. comb. from *Amblydisca*).

Procandea
New species: **Young** 1968a.
Record: **Young** 1968a (n. comb. from *Amblydisca, Stictocarta* and *Tettigonia*; key to species).

Procephaleus
Records: **Evans, J. W.** 1966a (=*Cephalelus*); 1977a (removed from synonymy with *Cephalelus*).

Proceps
Records: **Bonfils & Della Giustina** 1978a (Corsica). **Dlabola** 1957a (Turkey); 1958b (Caucasus); 1961b (central Asia); 1965d (Jordan); 1977c (Mediterranean); 1984a (Iran). **Emeljanov** 1964c (European USSR). **Hamilton, K. G. A.** 1975b (APHRODINAE, DELTOCEPHALINI, PLATYMETOPIINA). **Servadei** 1971a (Italy).

Proconia
New species: **Metcalf** 1965a (new name).
Records: **Young** 1965c (lectotype designation); 1968a (n. sp. syn.; key to species); 1974a (lectotype designation). **Young & Lauterer** 1966a (lectotype designation). **Young & Nast** 1963a (notes on type specimens).

Proconobola
Record: **Young** 1968a (n. comb. from *Amblydisca* and *Yunga*; key to species).

Proconopera
Record: **Young** 1968a (n. comb. from *Amblydisca*; n. sp. syn.; key to species).

Proconosama
New species: **Young** 1968a.
Record: **Young** 1968a (n. comb. from *Aulacizes* and *Tettigonia*; n. sp. syn.).

Proekes
Record: **Theron** 1975a (n. comb. from *Aconura* and *Deltocephalus*; n. sp. syn.).

Promecopsis
Record: **ICZN** 1961 (suppressed for purposes of the Law of Priority, but not for the Law of Homonymy, Opinion 605). **Wagner** 1960b (plea for suppression).

Promocuus
Record: **Nast** 1982a (catalogued; DELTOCEPHALINAE).

Propetes
Records: **Young** 1965c (lectotype designation); 1968a (n. sp. syn.). **Young & Lauterer** 1966a (lectotype designation). **Young & Nast** 1963a (notes on type material).

Proranus
New species: **Kramer** 1966a.
Record: **Kramer** 1966a (key to species).

Proskura
New species: **Dworakowska** 1981j.
Record: **Sohi & Dworakowska** 1984a (list of Indian species).

Protalebra
New species: **Young** 1957c.
Records: **Joyce, C. R.** 1962a, 1963a (Hawaii; Oahu). **Kim, J.** 1963a (Hawaii). **Ramos Elorduy** 1972c (Mexico). **Young** 1957b (descriptions of Osborn's type specimens); 1957c (in key to genera of ALEBRINI; key to species).

Protalebrella
New species: **Young** 1957c.
Record: **Young** 1957c (n. sp. syn.; key to species).

Protartessus
New species: **Evans, F.** 1981a.
Record: **Evans, F.** 1981a (n. comb. from *Bythoscopus* and *Tartessus*).

Protonesis
Records: **Linnavuori** 1970a (COELIDIINAE). **Nielson** 1975a (removed from COELIDIINAE; unassigned).

Provancherana
Records: **Hamilton, K. G. A.** 1976a (n. comb. from *Plesiommata*). **Young** 1977a (=*Plesiommata*, n. syn.).

Proxima
New species: **DeLong & Freytag** 1975a.

Pruthiana
New species: **Izzard** 1955a.

Pruthiorosius
Record: **Ghauri** 1963f (n. comb. from *Orosius*).

Pruthius
New species: **Mahmood** 1967a. **Ramakrishnan & Menon** 1972a.

Records: **Mahmood** 1967a (in key to genera of Oriental TYPHLOCYBINAE). **Ramakrishnan & Menon** 1972a (key to Indian species). **Sohi** 1976a (n. comb. to *Limassolla*). **Sohi & Kapoor** 1973b (India: Uttar Pradesh).

Psammotettix
New species: **Anufriev** 1976b. **DeLong** 1973a. **Dlabola** 1960b, 1961b, 1965c, 1966a, 1971a, 1979a. **Dubovskij** 1966a. **Emeljanov** 1962a, 1964f, 1966a, 1972b. **Greene** 1971a. **Korolevskaya** 1968a. **Kuoh, C. L.** 1981b. **Lindberg** 1958a, 1960a. **Linnavuori** 1965a (new name). **Logvinenko** 1961d, 1965a, 1966b, 1971a, 1977b, 1978a. **Mitjaev** 1969c, 1971a. **Orosz** 1981a. **Ossiannilsson** 1974a. **Remane** 1965a. **Ross & Hamilton** 1972a. **Sawai Singh** 1969a. **Vilbaste** 1960a, 1980a, 1980b. **Wagner** 1959a. **Zachvatkin** 1948b.

Records: **Anufriev** 1976b (n. sp. syn.); 1977b (Kurile Islands); 1978a (USSR: Maritime Territory). **Beirne** 1956a (Canada; key to species). **Bonfils & Della Giustina** 1978a (Corsica). **Cantoreanu** 1968b (Danube river delta); 1971c (Romania). **Della Giustina** 1983a (French Alps.). **DeLong** 1966a (Alaska; Muir Inlet). **Dlabola** 1956b (Czechoslovakia, Germany); 1957a (Turkey); 1957b (Afghanistan); 1958a (southern Europe); 1958b (Caucasus); 1959a (identity of *P. koeleriae* Zachvatkin); 1960a (n. comb. from *Deltocephalus*; n. sp. syn.); 1960c (Albania); 1961b (central Asia); 1961c (Romania); 1964a (Afghanistan); 1964c (Mongolia); 1965d (Jordan); 1967b, 1967c, 1967d, 1968a, 1968b, 1970a (Mongolia); 1971b (Italy); 1972a (Afghanistan); 1977c (Mediterranean); 1981a (Iran). **Dubovskij** 1966a (Uzbekistan); 1978a (Uzbekistan, Zarafshan Valley); 1980a (USSR: western Turkmenia). **D'Urso** 1980b (Italy). **Emeljanov** 1964c (European USSR; key to species); 1977a (Mongolia). **Gravestein** 1965a (The Netherlands). **Greene** 1971a (key to Nearctic species). **Gyllensvard** 1965a (Sweden). **Habib, El-Kady & Herakly** 1976a (n. sp. syn.). **Hamilton, K. G. A.** 1975b (APHRODINAE, DELTOCEPHALINI, DELTOCEPHALINA); 1983b (transboreal taxon). **Ishihara** 1958a (Japan: north Honshu). **Jankovic** 1971a (Serbia). **Jansky** 1985a (Slovakia). **Jasinska** 1980a (Poland, Bledowska Wilderness). **Kalkandelen** 1974a (Turkey; key to species). **Kholomuminov & Dlabola** 1979a (USSR: Golodnostep Plain). **Korolevskaya** 1963a, 1979b (Tadzhikistan).. **Lee, C. E.** 1979a (Korea, redescribed). **Lee, C. E. & Kwon** 1977b (Korea). **LeQuesne** 1959a (Britain; key to species); 1964a, 1969c (Britain). **Lindberg** 1958c (Arctic and Subarctic zones); 1960a (Portugal); 1961a (Madeira); 1962a (Portugal); 1963a (Morocco); 1965a (Spanish Sahara). **Linnavuori** 1956d (Spanish Morocco); 1959a (Mt. Sibillini); 1962a (Israel); 1964a (Egypt); 1965a (Libya, Tunisia). **Lodos** 1982a (Turkey). **Logvinenko** 1957b, 1963a, 1984a (USSR: Ukraine). **Mitjaev** 1968b (eastern Kazakhstan); 1971a (Kazakhstan; key to 17 species). **Moravskaja** 1956a (systematics). **Nast** 1955a (Poland); 1972a (catalogued); 1976a (Poland: Pieniny Mts.); 1976b (Poland; catalogued). **Nielson** 1968b (characterization of vector species). **Ossiannilsson** 1974a (Norway; notes on systematic position of new species); 1983a (Fennoscandia & Denmark; key to species). **Quartau** 1979a (Azores). **Remane** 1961d (identity of *P. notatus* discussed); 1962a (northwest Germany); 1965a (n. sp. syn.). **Ribaut** 1959a (France). **Schulz** 1976a (Norway). **Servadei** 1957a, 1958a, 1960a, 1968a, 1969a, 1971a, 1972a (Italy). **Vilbaste** 1958a, 1959b (Estonia); 1962b (eastern part of the Caspian lowlands); 1964a (Auwiesen Estlands); 1965b (Altai Mts.); 1968a (USSR: coastal regions); 1969a (notes on Matsumura's material); 1969b (Taimyr); 1973a (notes on Flor's material); 1975a (notes on Motschulsky's material); 1976a (transfer of species from *Deltocephalus* noted); 1979a (Vooremaa hardwood/spruce forest); 1980a (Kamchatka); 1980b (Tuva). **Wagner** 1958b (morphological variation).

Psegmatus
Record: **Hamilton, K. G. A.** 1975b (APHRODINAE, HECALINI).

Pseudaconura
Records: **Linnavuori** 1962a (Israel). **Nast** 1972a (catalogued; =*Doratulina*, n. syn.).

Pseudalaca
Record: **Linnavuori** 1959b (n. comb. from *Agallia*).

Pseudaligia
New species: **Kramer & DeLong** 1968a.
Record: **Cwikla & Blocker** 1981a (comparative notes).

Pseudaraldus
New species: **Bonfils** 1981a.

Pseudobalbillus
New species: **Linnavuori** 1969a, 1979b.
Record: **Linnavuori** 1979b (key to species).

Pseudocephalelus
Records: **Hamilton, K. G. A.** 1975b (APHRODINAE; unassigned). **Linnavuori** 1969b (n. comb. from *Cephalelus*).

Pseudoidioscopus
New species: **Maldonado-Capriles** 1977b.
Record: **Maldonado-Capriles** 1977b (key to species).

Pseudolausulus
Record: **Hamilton, K. G. A.** 1975b (APHRODINAE, DELTOCEPHALINI, DELTOCEPHALINA). **Servadei** 1968a (as *Pseudolausus*; Italy).

Pseudometopia
New species: **Young** 1968a.
Records: **Young** 1968a (n. comb. from *Aulacizes* and *Tettigonia*; n. sp. syn.; key to species). **Young & Nast** 1963a (notes on types).

Pseudonirvana
New species: **Evans, J. W.** 1966a, 1973a. **Kuoh, C. L.** 1973a. **Kuoh, C. L. & Kuoh, J. L.** 1983b.
Record: **Evans, J. W.** 1977a (=*Sophonia*).

Pseudophera
New species: **Kramer** 1976c. **Young** 1968a.
Records: **Young** 1968a (n. sp. syn.; key to species). **Young & Nast** 1963a (notes on type material).

Pseudophlepsius
New species: **Dubovskij** 1966a.
Records: **Dlabola** 1960a (identity of Horvath's material); 1961b (central Asia); 1964a (Afghanistan); 1968a, 1968b (Mongolia); 1972a (Afghanistan); 1981a (Iran). **Dubovskij** 1966a (Uzbekistan); 1978a (Uzbekistan, Zarafshan Valley). **Emeljanov** 1964c (European USSR). **Hamilton, K. G. A.** 1975b (APHRODINAE, DELTOCEPHALINI, PLATYMETOPIINA). **Logvinenko** 1959b, 1962e (USSR: Ukraine). **Mitjaev** 1971a (Kazakhstan). **Vilbaste** 1980b (Tuva). **Zachvatkin** 1953c (Astrakhan, Trans-Volga sand areas).

Pseudoscarta
Record: **Young** 1977a (=*Dilobopterus*, n. syn.).

Pseudosubhimalus
Record: **Ghauri** 1974c (n. comb. from *Ophiola*).

Pseudothaia
New species: **Kuoh, C. L.** 1982a.

Pseupalus
New species: **Remane & Asche** 1980a.
Record: **Nast** 1982a (catalogued; DELTOCEPHALINAE).

Pseutettix
New species: **DeLong** 1967d.
Record: **Cwikla & Blocker** 1981a (comparative notes).

Psibala
New species: **Kramer** 1963b.
Records: **Kramer** 1963b (in key to genera of NEOBALINAE). **Linnavuori & DeLong** 1977c (in key to genera of the *Neobala* group).

Pteropyx
New species: **Emeljanov** 1972d.
Records: **Emeljanov** 1972d (JASSARGINI). **Hamilton, K. G. A.** 1975b (APHRODINAE, DELTOCEPHALINI, MACROSTELINA). **Linnavuori** 1962a (Israel).

Punahuana
New species: **Young** 1977a.
Record: **Young** 1977a (n. comb. from *Cicadella*; key to species).

Punctigerella
Records: **Anufriev** 1969b, 1978a (USSR: Maritime Territory; generic status). **Lee, C. E.** 1979a (Korea, redescribed).

Purpuranus
Record: **Lee, C. E. & Kwon** 1977b (Korea). **Servadei** 1968a (Italy).

Pusaneura
New species: **Ramakrishnan & Menon** 1971a.
Record: **Sharma, B.** 1978a (= *Uzeldikra*, n. syn.).

Pusatettix
New species: **Ramakrishnan & Menon** 1971a.

Putoniessa
New species: **Evans, J. W.** 1966a, 1969a.
Record: **Evans, J. W.** 1966a (n. sp. syn.).

Putoniessiella
New species: **Evans, J. W.** 1969a.

Pycnoides
Record: **Hamilton, K. G. A.** 1975b (APHRODINAE, DELTOCEPHALINI, ATHYSANINA).

Pygometopia
Record: **Young** 1977a (= *Teleogonia*, n. syn.).

Pyramidotettix
New species: **Yang, C.** 1965a.

Qadria
New species: **Dworakowska** 1981e. **Dworakowska & Sohi** 1978a (as *Quadria*).
Records: **Ahmed, Manzoor** 1983b (remarks on shape of male style); 1985c (Pakistan). **Dworakowska** 1977f (as *Quadria*; n. gen. syn.; n. comb. from *Spatulostylus*); 1979a (n. comb. from *Spatulostylus* and *Zygina*); 1979b (as *Quadria*; Vietnam); 1980b (as *Quadria*; India). **Dworakowska & Viraktamath** 1978a (India). **Mahmood** 1967a (in key to genera of Oriental ERYTHRONEURINI; n. comb. from *Empoasca*). **Sohi & Dworakowska** 1984a (list of Indian species).

Quartauropa
Record: **Webb, M. D.** 1983b (n. comb. from *Idioscopus*).

Quartausius
New species: **Dlabola** 1974c.
Record: **Nast** 1982a (catalogued; DELTOCEPHALINAE).

Quaziptus
New species: **Kramer** 1965b.
Record: **Cwikla & Blocker** 1981a (comparative notes).

Quercinirvana
New species: **Ahmed, Manzoor & Mahmood** 1970a.

Quernus
Record: **Nast** 1982a (catalogued; DELTOCEPHALINAE).

Quichira
New species: **Young** 1968a.

Quilopsis
New species: **Webb, M. D.** 1983a.

Quontus
Record: **Hamilton, K. G. A.** 1975b (APHRODINAE, DELTOCEPHALINI, DELTOCEPHALINA).

Raabeina
New species: **Dworakowska** 1972j, 1979b.

Rabela
New species: **Young** 1957c.
Records: **Mead** 1957a, 1981a (USA: Florida). **Young** 1957c (key to species).

Rabiana
New species: **Mahmood** 1967a.

Racinolidia
New species: **Nielson** 1983h.

Ragia
Records: **Theron** 1973a (n. comb. from *Macrosteles*); 1982b (comparison with *Paramesanus*).

Ramakrishnania
New species: **Dworakowska** 1974a.

Ramania
New species: **Dworakowska** 1972j.

Ramosulus
New species: **Young** 1977a.
Record: **Young** 1977a (n. comb. from *Cardioscarta* and *Cicadella*; key to species).

Ramsisia
New species: **Einyu & Ahmed** 1979a.
Record: **Ahmed, Manzoor** 1983b (remarks on wing venation).

Ranbara
New species: **Dworakowska** 1983a.

Raphirhinus
Records: **Young** 1965a (lectotype designation); 1968a (n. sp. syn.). **Young & Lauterer** 1966a (lectotype designation).

Ratbura
New species: **Mahmood** 1967a.
Records: **Mahmood** 1967a (n. comb. from *Empoasca*; in key to genera of Oriental ERYTHRONEURINI). **Ramakrishnan & Ghauri** 1979b (in key to genera).

Ratburella
New species: **Dworakowska** 1977f, 1980b. **Ramakrishnan & Menon** 1973a.
Records: **Dworakowska** 1980b (new subgenus proposed). **Dworakowska & Sohi** 1978b (India: Punjab). **Sohi & Dworakowska** 1984a (list of Indian species).

Ratjalia
New species: **Dworakowska** 1981j.

Raunoia
New species: **Dworakowska & Lauterer** 1975a.
Record: **Dworakowska & Lauterer** 1975a (n. comb. from *Zyginella*).

Rawania
Record: **Dworakowska & Viraktamath** 1975a (Burma). **Ghauri** 1963e (n. comb. from *Empoasca*). **Mahmood** 1967a (in key to genera of Oriental TYPHLOCYBINI).

Recilia
New species: **Dlabola** 1965d. **Kramer** 1962b. **Kwon & Lee** 1979b. **Linnavuori** 1969a.
Records: **Anufriev** 1977b (Kurile Islands); 1978a (USSR: Maritime Territory). **Bonfils & Della Giustina** 1978a (Corsica). **Choe** 1985a (Korea). **Dlabola** 1977c (Mediterranean). **Dubovskij** 1966a (Uzbekistan); 1980a (USSR: western Turkmenia). **Dworakowska** 1973d (Korea). **Emeljanov** 1964c (European USSR). **Ghauri** 1980b (n. comb. from *Thamnotettix*). **Gravestein** 1965a (The Netherlands: Terschelling I.). **Hamilton, K. G. A.** 1975b (APHRODINAE, DELTOCEPHALINI, DELTOCEPHALINA). **Jankovic** 1971a (Serbia). **Kalkandelen** 1974a (Turkey; key to species). **Kramer** 1962b (Liberia; key to species). **Kholmuminov & Dlabola** 1979a (USSR: Golodnostep Plain). **Kwon & Lee** 1979b (*Togacephalus* and *Inazuma* as subgenera; n. comb. from *Togacephalus*). **Lee, C. E.** 1979a (Korea, redescribed). **LeQuesne** 1969c (Britain). **Linnavuori** 1969a (Congo); 1975c (Micronesia). **Lodos & Kalkandelen** 1985c (Turkey). **Logvinenko** 1963b (as subgenus of *Deltocephalus*; USSR: Ukraine). **Mitjaev** 1968b (eastern Kazakhstan); 1971a (Kazakhstan; key to 2 species). **Nast** 1972a (catalogued); 1976b (Poland; catalogued). **Nielson** 1968b (characterization of vector species; *Inazuma* a junior syn.). **Servadei** 1968a (Italy). **Theron** 1974b (n. comb. from *Thamnotettix*). **Vilbaste** 1965b (Altai Mts.); 1968a (USSR: coastal regions); 1975a (placement for *Deltocephalus dorsalis* Motschulsky); 1980b (Tuva).

Redaprata
New species: **Blocker** 1979a.

Refrolix
New species: **Theron** 1984b.

Regalana
New species: **DeLong & Freytag** 1975b.

Relaba
New species: **Young** 1957c.

Relipo
New species: **Evans, J. W.** 1977a.

Remadosus
Records: **Hamilton, K. G. A.** 1975a (as subgenus of *Pendarus*, n. status); 1975b (APHRODINAE, DELTOCEPHALINI, PLATYMETOPIINA).

Remmia
New species: **Vilbaste** 1968a.

Remoyia
Record: **Webb, M. D.** 1983b (n. comb. from *Idioscopus*).

Renonus
New species: **DeLong** 1959a.
Record: **Cwikla & Blocker** 1981a (comparative notes).

Renosteria
New species: **Theron** 1984a.
Records: **Theron** 1974b (n. comb. from *Tetartostylus*); 1984a (key to species).
Reticana
New species: **DeLong** 1980a.
Records: **DeLong & Freytag** 1964a (n. comb. from *Gypona*); 1972a (in key to genera of GYPONINAE).
Reticopsis
New species: **Hamilton, K. G. A.** 1983d.
Records: **Hamilton, K. G. A.** 1980b (n. comb. from *Pediopsis*); 1983d (key to species).
Retusanus
Record: **Cwikla & Blocker** 1981a (comparative notes).
Reuplemmeles
New species: **Evans, J. W.** 1972b.
Records: **Evans, J. W.** 1966a (n. comb. from *Reuteriella*). **Linnavuori & Quartau** 1975a (redefinition of REUPLEMMELINI).
Reuteriella
Record: **Evans, J. W.** 1966a (n. comb. from *Reuplemmeles*; HECALINI).
Reventazonia
New species: **Linnavuori** 1959b. **Linnavuori & DeLong** 1979a, 1979b.
Records: **Hamilton, K. G. A.** 1975b (APHRODINAE, DELTOCEPHALINI, DELTOCEPHALINA). **Kramer** 1971b (n. comb. from *Amplicephalus*).
Rhabdotalebra
New species: **Young** 1957c.
Record: **Young** 1957c (n. comb. from *Protalebra*; key to species).
Rhinocerotis
Record: **Theron** 1977a (n. comb. from *Empoasca*).
Rhoananus
Records: **Anufriev** 1978a (USSR: Maritime Territory). **Della Giustina** 1983a (France). **Dubovskij** 1966a (Uzbekistan); 1978a (Uzbekistan, Zarafshan Valley); 1980a (USSR: western Turkmenia). **Emeljanov** 1964c (European USSR). **Hamilton, K. G. A.** 1975b (APHRODINAE, DELTOCEPHALINI, DELTOCEPHALINA). **Jankovic** 1966a, 1971a (Serbia). **Kalkandelen** 1974a (Turkey). **Mitjaev** 1968a (eastern Kazakhstan); 1971a (Kazakhstan). **Nast** 1972a (catalogued); 1976b (Poland; catalogued). **Vilbaste** 1968a (USSR: coastal regions); 1980b (Tuva).
Rhobala
New species: **Kramer** 1963b.
Records: **Kramer** 1963b (in key to genera of NEOBALINAE). **Linnavuori & DeLong** 1977C (in key to genera of the *Neobala* group).
Rhodulopa
New species: **Linnavuori** 1961a.
Rhogosana
New species: **DeLong** 1975b, 1981c.
Record: **DeLong & Freytag** 1972a (in key to genera of GYPONINAE).
Rhombopsana
Records: **Dlabola** 1972a (Afghanistan, Iran, Israel). **Linnavuori** 1962a (Israel; EUSCELINAE, EUSCELINI); 1978d (=*Paganalia*, n. syn.; PARABOLOPINAE). **Webb, M. D.** 1981b (=*Dryadomorpha*; PARABOLONINAE).
Rhombopsis
Records: **Datta** 1973n (descriptive notes). **Dlabola** 1960c (Iran); 1974a (Afghanistan). **Linnavuori** 1978d (=*Paganalia*, n. syn.). **Webb, M. D.** 1980a (lectotype designation).
Rhopalogonia
Record: **Young** 1977a (redescribed).
Rhopalopyx
New species: **LeQuesne** 1964a. **Vilbaste** 1965a.
Records: **Anufriev** 1978a (USSR Maritime Territory). **Dlabola** 1957a (Turkey); 1958b (Caucasus); 1959b (Slovakia); 1960a (notes on Horvath's material); 1961b (central Asia); 1964d (Bohemia, Bulgaria, Romania, Slovakia); 1967c (Mongolia). **Emeljanov** 1964c (European USSR). **Jankovic** 1966a (Serbia). **LeQuesne** 1959a (Britain; key to species). **Logvinenko** 1957b, 1957c (USSR: Ukraine). **Mitjaev** 1968b (eastern Kazakhstan). **Nast** 1955a (Poland); 1972a (catalogued); 1976b (Poland; catalogued). **Ossiannilsson** 1983a (Fennoscandia & Denmark; key to species). **Quartau & Rodrigues** 1969a (Portugal). **Ribaut** 1959a (France). **Vilbaste** 1962a (genital structures illustrated); 1964a (Auwiesen Estlands); 1965b (Altai Mts.).
Rhothidus
Record: **Evans, J. W.** 1966a (n. comb. to *Thymbris*).
Rhotidoides
New species: **Evans, J. W.** 1966a.
Records: **Evans, J. W.** 1966a (n. comb. from *Bythoscopus* and *Thymbris*); 1969a (New Guinea).
Rhotidus
Record: **Evans, J. W.** 1966a (n. sp. syn.).
Rhusia
New species: **Dworakowska** 1981h.
Records: **Dworakowska** 1981h (n. comb. from *Typhlocyba*). **Theron** 1977a (n. comb. from *Erythroneura*).
Rhusopus
New species: **Webb, M. D.** 1983b.
Record: **Webb, M. D.** 1983b (key to species).
Rhutelorbus
New species: **Webb, M. D.** 1981b.
Rhytidodus
New species: **Anufriev** 1968a. **Dlabola** 1965b, 1968b, 1970a, 1974a. **Korolevskaya** 1964a. **Logvinenko** 1983a. **Mitjaev** 1970a, 1971a. **Vilbaste** 1980b.
Records: **Anufriev** 1968a (synopsis; key to species); 1978a (USSR: Maritime Territory). **Dlabola** 1960a (identity of Horvath's material); 1961a (central Asia); 1965b (Sudan; key to species); 1974a (Iran); 1974b (in key to genera of Palaearctic IDIOCERINAE; n. comb. from *Idiocerus*); 1981a (Iran). **Dubovskij** 1966a (Uzbekistan); 1978a (Uzbekistan, Zarafshan Valley). **Emeljanov** 1977a (Mongolia). **Hamilton, K. G. A.** 1980a (in key to genera of Nearctic IDIOCERINI); 1983b (adventive to Nearctic). **Korolevskaya** 1975a (Tadzhikistan). **Kwon** 1985a (in key to genera of Korean IDIOCERINAE). **Lodos** 1982a (Turkey). **Lodos & Kalkandelen** 1981a (Turkey). **Mitjaev** 1968b (eastern Kazakhstan); 1971a (Kazakhstan; key to 10 species). **Nast** 1972a (catalogued); 1976b (Poland; catalogued). **Ossiannilsson** 1981a (Fennoscandia & Denmark). **Servadei** 1968a, 1971a (Italy). **Vilbaste** 1980b (Tuva). **Webb, M. D.** 1983a (Australia: New South Wales, adventive; in key to genera of IDIOCERINAE). **Zachvatkin** 1953b (central Russia; Oka district).
Rhytistylus Fieber 1872a
New species: **Asche** 1980a.
Records: **Gravestein** 1965a (The Netherlands: Terschelling I.). **Hamilton, K. G. A.** 1975b (APHRODINAE, DELTOCEPHALINI, PLATYMETOPIINA).
Rhytistylus Fieber 1875a
Record: **Hamilton, K. G. A.** 1975b (APHRODINAE, DELTOCEPHALINI, ATHYSANINA).
Ribautanus
Record: **Nast** 1982a (catalogued; DELTOCEPHALINAE).
Ribautiana
New species: **Sharma, B.** 1984a. **Zachvatkin** 1948b.
Records: **Beirne** 1956a (Canada; key to species). **Bonfils & Della Giustina** 1978a (Corsica). **Dlabola** 1958a (Austria); 1958b (Caucasus); 1958c (in key to genera of Palaearctic TYPHLOCYBINAE); 1959a (n. sp. syn.); 1959b (southern Europe); 1961b (central Asia); 1961c (Bohemia); 1964c (Albania); 1971a (Turkey); 1977b, 1981a (Iran). **Dumbleton** 1964a (New Zealand, adventive). **Emeljanov** 1964c (European USSR; key to species). **Gunthart, H.** 1971a (Switzerland). **Hamilton, K. G. A.** 1983b (Nearctic, adventive). **Jankovic** 1966a (Serbia). **Kirejtshuk** 1977a (USSR: Kharkov region). **Knight, W. J.** 1976a (New Zealand). **Lauterer** 1958a (Czechoslovakia). **LeQuesne & Payne** 1981a (Britain). **Lindberg** 1960a (Portugal). **Linnavuori** 1962a (Israel). **Lodos & Kalkandelen** 1984a (Turkey). **Logvinenko** 1963a (USSR: Ukraine). **Nast** 1972a (catalogued); 1976a (Poland: Pieniny Mts.); 1976b (Poland; catalogued). **Ossiannilsson** 1974a (Norway); 1981a (Fennoscandia & Denmark). **Quartau** 1979a (Azores). **Servadei** 1968a, 1972a (Italy). **Sharma, B.** 1984a (northern

India). **Varty** 1967a (Canada). **Vilbaste** 1983a (notes on Flor's material). **Wilson, M. R.** 1978a (keys to 5th instars).
Ribautiellus
Record: **Hamilton, K. G. A.** 1975b (APHRODINAE, DELTOCEPHALINI, DELTOCEPHALINA).
Rikana
New species: **Nielson** 1983e.
Rinconada
New species: **Linnavuori & DeLong** 1977c.
Records: **Cwikla & Blocker** 1981a (comparative notes). **Linnavuori & DeLong** 1977c (in key to related genera).
Rineda
New species: **Linnavuori & DeLong** 1978b.
Record: **Cwikla & Blocker** 1981a (comparative notes).
Romanius
Record: **Hamilton, K. G. A.** 1975b (APHRODINAE, DELTOCEPHALINI, DELTOCEPHALINA).
Rosenus
New species: **Emeljanov** 1972b. **Hamilton, K. G. A. & Ross** 1975a. **Vilbaste** 1965b, 1980b.
Records: **Anufriev** 1978a (USSR: Maritime Territory). **Beirne** 1956a (Canada; key to species). **Dlabola** 1965c, 1967b, 1967c, 1967d, 1968b, 1970b (Mongolia). **Dworakowska** 1973d (Mongolia). **Emeljanov** 1964c (European USSR); 1964f (n. comb. from *Deltocephalus*); 1966a (n. comb. from *Sorhoanus*); 1977a (Mongolia). **Hamilton, K. G. A.** 1972d (Canada: Manitoba); 1975b (APHRODINAE, DELTOCEPHALINI, DELTOCEPHALINA). **Hamilton, K. G. A. & Ross** 1975a (key to Nearctic species). **Korolevskaya** 1979b (Tadzhikistan). **Nast** 1972a (catalogued). **Ossiannilsson** 1983a (Fennoscandia & Denmark). **Vilbaste** 1969b (Taimyr); 1980a (Kamchatka); 1980b (Tuva; n. sp. syn.).
Rosopaella
New species: **Webb, M. D.** 1983a.
Record: **Webb, M. D.** 1983a (n. comb. from *Idiocerus*; key to species).
Rotifunkia
New species: **Maldonado-Capriles** 1971b.
Records: **Maldonado-Capriles** 1971b (description of male; key to species). **Webb, M. D.** 1983b (key to species).
Rotigonalia
Record: **Young** 1977a (n. comb. from *Cicadella* and *Cyclogonia*; n. sp. syn.; key to species).
Rotundicerus
New species: **Maldonado-Capriles** 1977a.
Ruandopsis
New species: **Linnavuori** 1978b.
Records: **Hamilton, K. G. A.** 1980b (in key to genera and subgenera of MACROPSINI; n. sp. syn.). **Linnavuori** 1978b (key to species).
Rubria
New species: **Evans, J. W.** 1966a, 1969a.
Record: **Evans, J. W.** 1966a (n. comb. from *Ledracephala*; n. sp. syn.).
Rugosana
New species: **DeLong & Freytag** 1964a, 1973a.
Records: **DeLong & Freytag** 1964a (redescription; key to species); 1972a (in key to genera of GYPONINAE).
Ruppeliana
New species: **Young** 1977a.
Record: **Young** 1977a (n. comb. from *Tettigonia*).
Saavedra
Records: **Cwikla & Blocker** 1981a (comparative notes). **Linnavuori & DeLong** 1978b (n. comb. from *Menosoma*).
Sabelanus
Record: **Hamilton, K. G. A.** 1975b (APHRODINAE, DELTOCEPHALINI, DELTOCEPHALINA).
Sabima
Record: **Nielson** 1977a (= *Thagria*, n. syn.).
Sabimamorpha
Record: **Nielson** 1975a (removed from COELIDIINAE; unassigned).
Sabourasca
New species: **Ramakrishnan & Menon** 1972b.
Record: **Sohi & Dworakowska** 1984a (list of Indian species).

Sagatus
Records: **Cantoreanu** 1971c (Romania). **Dlabola** 1964a (Afghanistan). **Emeljanov** 1964c (European USSR). **Hamilton, K. G. A.** 1975b (APHRODINAE, DELTOCEPHALINI, MACROSTELINA). **Jasinska** 1980a (Poland: Bledowska Wilderness). **LeQuesne** 1969c (Britain). **Linnavuori** 1959b (in key to genera of MACROSTELINI); 1959c (Mt. Picentini). **Lodos & Kalkandelen** 1985b (Turkey). **Mitjaev** 1971a (Kazakhstan). **Moore, T. E. & Ross** (= *Davisonia*, n. syn.). **Nast** 1972a (catalogued); 1976a (Poland: Pieniny Mts.); 1976b (Poland; catalogued). **Okali** 1968a (central Europe, Slovakia). **Ossiannilsson** 1983a (Fennoscandia & Denmark). **Servadei** 1968a, 1971a (Italy). **Vilbaste** 1964a (Auwiesen Estlands); 1965b (Altai Mts.); 1973a (notes on Flor's material); 1980b (Tuva).
Sagittifer
New species: **Dlabola** 1961b. **Mitjaev** 1971a.
Records: **Dubovskij** 1966a (Uzbekistan). **Hamilton, K. G. A.** 1975b (APHRODINAE; unassigned). **Mitjaev** 1971a (Kazakhstan; key to 2 species).
Sahlbergotettix
New species: **Dubovskij** 1966a.
Records: **Anufriev** 1978a (USSR: Maritime Territory). **Dlabola** 1961a (Slovakia); 1967b (Mongolia); 1972a (Afghanistan); 1974b (n. comb. from *Idiocerus*; in key to genera of Palaearctic IDIOCERINAE). **Dubovskij** 1978a (Uzbekistan, Zarafshan Valley). **Jasinska** 1980a (Poland: Bledowska Wilderness). **Korolevskaya** 1975a (Tadzhikistan). **Logvinenko** 1984a (USSR: Ukraine). **Mitjaev** 1968a (eastern Kazakhstan); 1971a (Kazakhstan). **Ossiannilsson** 1981a (Fennoscandia & Denmark). **Vilbaste** 1973e (notes on Flor's material); 1980b (Tuva). **Zachvatkin** 1953b (n. comb. from *Idiocerus*).
Sailerana
New species: **Young** 1977a.
Record: **Young** 1977a (n. comb. from *Cardioscarta*, *Strictigonia* and *Tettigonia*; n. sp. syn.; key to species).
Sajda
New species: **Dworakowska** 1981j.
Record: **Sohi & Dworakowska** 1984a (list of Indian species).
Salka
New species: **Dworakowska** 1976b, 1979b, 1980b, 1981j.
Records: **Dworakowska** 1972e (n. comb. from *Zygina*); 1979a (n. comb. from *Empoasca*). **Sohi & Dworakowska** 1984a (list of Indian species).
Salsolibia
New species: **Theron** 1979a.
Salsolicola
New species: **Theron** 1979a.
Salvina
Record: **Kramer** 1964d (n. comb. from *Tettigonia*).
Samuraba
New species: **Linnavuori** 1961a.
Sanatana
New species: **Dworakowska** 1984a.
Sanctahelenia
New species: **Dlabola** 1976a.
Sanctanus
New species: **Kramer** 1963c.
Records: **Hamilton, K. G. A.** 1975b (APHRODINAE, DELTOCEPHALINI, DELTOCEPHALINA). **Kramer** 1963c (reinstated combination from *Amplicephalus*). **Linnavuori** 1956a (Colombia); 1959b (as subgenus of *Amplicephalus*).
Sandalla
New species: **Mahmood** 1967a.
Records: **Hongsaprug & Wilson** 1985a (= *Bakera* (*Guinobata*), n. syn.). **Mahmood** 1967a (in key to genera of BAKERINI).
Sandanella
New species: **Dworakowska** 1972k, 1981a. **Mahmood** 1967a.
Record: **Mahmood** 1967a (in key to genera of Oriental ERYTHRONEURINI).
Sandersellus
New species: **Nielson** 1975a, 1981b.

Record: **Nielson** 1975a (n. comb. from *Cixidocoelidia*; n. gen. syn.; key to species).

Sandia
New species: **Theron** 1982a.

Sanestebania
New species: **DeLong** 1984g. **Linnavuori & DeLong** 1978b.
Record: **Cwikla & Blocker** 1981a (comparative notes).

Sanluisia
New species: **Linnavuori** 1959b.

Sannella
New species: **Dworakowska** 1982a.
Records: **Dworakowska** 1982a (n. comb. from *Agnesiella* and *Blyphlocyta*). **Sohi & Dworakowska** 1984a (list of Indian species).

Sansalvadoria
New species: **Schroder** 1959a.

Santarema
Record: **Young** 1977a (= *Mesogonia*, n. syn.).

Sanuca
New species: **DeLong** 1980h.

Sapingia
New species: **Nielson** 1979b, 1983b.
Record: **Nielson** 1979b (n. comb. from *Coelidia*; key to species).

Sapoba
New species: **Linnavuori & Al-Ne'amy** 1983a.
Record: **Linnavuori & Al-Ne'amy** 1983a (key to species).

Sapporoa
New species: **Dworakowska** 1972e.
Records: **Dworakowska** 1979b (= *Alnetoidia* (*Alenlla*), n. syn.). **Nast** 1982a (catalogued; TYPHLOCYBINAE).

Sarascarta
Record: **Young** 1957b (redescribed).

Sarbazius
Record: **Nast** 1982a (catalogued; DELTOCEPHALINAE).

Sardiopsis
New species: **Mitjaev** 1971a.
Records: **Hamilton, K. G. A.** 1975b (APHRODINAE; unassigned). **Mitjaev** 1971b (southern Kazakhstan).

Sardius
New species: **Mitjaev** 1967c.
Records: **Bonfils & Della Giustina** 1978a (Corsica). **Dlabola** 1965d (Jordan). **Hamilton, K. G. A.** 1975b (APHRODINAE, DELTOCEPHALINI, ATHYSANINI). **LeQuesne** 1969c (Britain). **Nast** 1972a (catalogued); 1965b (Poland; catalogued).

Sarejuia
New species: **Ahmed, Manzoor** 1985c. **Ghauri** 1974c.
Record: **Sharma, B. & Malhotra** 1981a (= *Draberiella*, n. syn.).

Sarpestus
Record: **Evans, F.** 1981a (n. comb. from *Bythoscopus*; n. sp. syn.).

Satsumanus
New species: **Linnavuori** 1960a, 1960b.
Records: **Eyles & Linnavuori** 1974a (Cook Islands). **Hamilton, K. G. A.** 1975b (APHRODINAE, DELTOCEPHALINI, PLATYMETOPIINA). **Ishihara** 1963a (list of species). **Knight, W. J.** 1982a (Rennell Island).

Saudallygus
Record: **Nast** 1982a (catalogued; DELTOCEPHALINAE).

Savanicus
New species: **Dlabola** 1977b.
Record: **Nast** 1982a (catalogued; DELTOCEPHALINAE).

Savitara
New species: **Dworakowska** 1984a.

Sawainara
Records: **Dworakowska** 1980d (= *Empoascanara*, n. syn.). **Ramakrishnan & Ghauri** 1979b (n. comb. from *Empoascanara*).

Sayara
Record: **Dworakowska** 1980d (= *Empoascanara*, n. syn.).

Sayetus
Record: **Hamilton, K. G. A.** (APHRODINAE, DELTOCEPHALINI, DELTOCEPHALINA).

Scaphetus
New species: **Evans, J. W.** 1966a. **Knight, W. J.** 1975a.

Scaphoidella
New species: **Anufriev** 1977a. **Vilbaste** 1968a.
Record: **Anufriev** 1978a (USSR: Maritime Territory; key to species).

Scaphoideus
New species: **Barnett** 1977a, 1979a. **DeLong & Knull** 1971a. **Freytag** 1976b. **Ghosh, M.** 1974a. **Heller & Linnavuori** 1968a. **Kitbamroong & Freytag** 1978a. **Kwon & Lee** 1978d. **Linnavuori** 1969a. **Metcalf** 1967c (new name). **Vilbaste** 1968a.
Records: **Anufriev** 1970e (USSR: south of the Primor'ya Territory); 1978a (USSR: Maritime Territory). **Beirne** 1956a (Canada; key to species). **Barnett** 1975a (USA: Kentucky; n. sp. syn.); 1977a (n. gen. syn.; n. sp. syn.; key to subgenera and species). **Barnett & Freytag** 1976a (n. comb. to *Scaphoidophytes*). **Bonfils & Della Giustina** 1978a (Corsica). **Cantoreanu** 1965d (Romania). **Choe** 1985a (Korea). **Datta** 1973j (descriptive notes). **Della Giustina** 1983a (France). **Evans, J. W.** 1966a (Australia; generic placement questioned; n. sp. syn.); 1974a (New Caledonia). **Hamilton, K. G. A.** 1972d (Canada: Manitoba); 1975b (APHRODINAE, DELTOCEPHALINI, PLATYMETOPIINA); 1983b (transboreal taxon). **Ishihara** 1958a (Japan: north Honshu). **Kitbamroong & Freytag** 1978a (Thailand; key to species). **Kwon & Lee** 1979b (n. sp. syn.). **Lee, C. E. & Kwon** 1977b (Korea). **Lodos** 1982a (as *Scaphodes*; Turkey). **Nielson** 1968b (characterization of vector species). **Okada** 1977a (redescription of Matsumura species). **Sharma, B.** 1977b (India: Jammu). **Theron** 1974b (n. comb. to *Capoideus*). **Webb, M. D.** 1980a (lectotype designation).

Scaphoidophyes
New species: **Barnett** 1980a. **Barnett & Freytag** 1976a (as *Scaphoidophytes*).
Records: **Barnett** 1980a (key to species). **Barnett & Freytag** 1976a (as *Scaphoidophytes*; n. comb. from *Scaphoideus*).

Scaphoidula
New species: **Menezes** 1973b.
Record: **Linnavuori** 1959b (redescription; key to species).

Scaphytoceps
New species: **Dlabola** 1957b.
Records: **Dlabola** 1972f (Afghanistan). **Hamilton, K. G. A.** 1957b (APHRODINAE, DELTOCEPHALINI, PLATYMETOPIINA).

Scaphytopius
New species: **Cheng, Y. J.** 1980a. **Cwikla & Freytag** 1982a. **DeLong** 1980h. **DeLong & Linnavuori** 1978c, 1978f. **Hepner** 1946a. **Linnavuori** 1959b. **Linnavuori & DeLong** 1978c, 1978f, 1979b. **Linnavuori & Heller** 1961a. **Musgrave** 1975a.
Records: **Beardsley** 1956a, 1964a (Hawaii; Oahu); 1966a (Hawaii; Nihoa Island). **Beirne** 1956a (Canada; key to species). **Cwikla & Blocker** 1981a (comparative notes). **Hamilton, K. G. A.** 1972d (Canada: Manitoba); 1975b (APHRODINAE, DELTOCEPHALINI, PLATYMETOPIINA); 1976a (notes on Provancher material). **Hepner** 1946a (new subgenus proposed). **Joyce, C. R.** 1955a (Hawaii; Oahu). **Linnavuori** 1959b (n. comb. from *Platymetopius*; key to subgenera and species); 1973c (Cuba). **Linnavuori & DeLong** 1977b (Bolivia). **Linsley & Usinger** 1966a (Galapagos Islands). **Musgrave** 1975a (key to the *acutus* complex); 1979a (USA: Florida). **Nielon** 1968b (characterization of vector species). **Ramos Elorduy** 1972c (Mexico). **Suehiro** 1961a (as *Scaphitopius*; Hawaii; Kauai).

Scapidonus
New species: **Nielson** 1983h.

Scaris
New species: **Freytag & DeLong** 1982e.
Records: **DeLong & Freytag** 1972a (= *Clinonaria*, n. syn.). **Young** 1965c (lectotype designation).

Scarisana
Records: **DeLong & Freytag** 1972a (transferred from GYPONINAE to CICADELLINAE). **Young** 1977a (= *Paromenia*, n. syn.).

Scaroidana
New species: **DeLong** 1982c. **Kramer** 1963a.
Records: **Blocker** 1979a (in key to genera of New World IASSINAE). **DeLong & Freytag** 1972a (to IASSINIAE [sic], = IASSINAE). **Kramer** 1963a (key to species).

Scaropsia
New species: **Blocker** 1979a.

Scenergates
Records: **Dubovskij** 1978a (Uzbekistan, Zarafshan Valley). **Hamilton, K. G. A.** 1975b (APHRODINAE, HECALINI). **Korolevskaya** 1979b (Tadzhikistan). **Nast** 1982a (catalogued; DELTOCEPHALINAE).

Schildola
Record: **Young** 1977a (n. comb. from *Microgoniella*; key to species).

Schistogonalia
New species: **Young** 1977a.
Record: **Young** 1977a (key to species).

Schizandrasca
Records: **Anufriev** 1972a (n. comb. from *Alebroides*); 1978a (USSR: Maritime Territory). **Dworakowska** 1982b (n. comb. from *Empoasca*). **Lee, C. E.** 1979a (Korea, redescribed).

Scleroracus
New species: **Dlabola** 1965c, 1967a. **Emeljanov** 1966a, 1972b. **Kuoh, C. L.** 1981a. **Medler** 1958a. **Mitjaev** 1968a. **Vilbaste** 1965b.
Records: **Anufriev** 1977b (Kurile Islands); 1978a (USSR: Maritime Territory). **Anufriev & Emeljanov** 1968a (Soviet Far East). **Beirne** 1956a (Canada; key to species). **DeLong** 1966a (Alaska; Muir Inlet). **Dlabola** 1958b (Caucasus); 1961b (central Asia); 1964c (Albania); 1965c, 1967b, 1967c, 1967d, 1968a, 1968b, 1970b (Mongolia). **Dubovskij** 1966a (Uzbekistan). **D'Urso** 1980b (Italy). **Elias** 1982a (Canada: Labrador, Umiakoviaruske site (fossil)). **Emeljanov** 1964c (European USSR; key to species); 1977a (Mongolia). **Hamilton, K. G. A.** 1972d (Canada: Manitoba); 1975b (APHRODINAE, DELTOCEPHALINI, ATHYSANINA). **Jankovic** 1966a (Serbia). **Jasinska** 1980a (Poland: Bledowska Wilderness). **Kartal** 1982a (Turkey). **Lauterer** 1957a (Czechoslovakia). **Lee, C. E.** 1979a (Korea, redescribed). **Lee, C. E. & Kwon** 1977b (Korea). **LeQuesne** 1962a, 1969c (Britain). **Logvinenko** 1984a (USSR: Ukraine). **Medler** 1958b (key to Nearctic species). **Mitjaev** 1968b (eastern Kazakhstan); 1971a (Kazakhstan; key to 5 species). **Nast** 1972a (catalogued); 1973a, 1976b (Poland; catalogued). **Nielson** 1968b (characterization of vector species). **Ossiannilsson** 1974a (Norway); 1983a (= *Ophiola*; Fennoscandia & Denmark). **Schulz** 1976a (Norway). **Vilbaste** 1958a, 1959b (Estonia); 1965b (Altai Mts.); 1968a (USSR: coastal regions); 1973a (notes on Flor's material); 1980a (Kamchatka); 1980b (Tuva).

Scopogonalia
New species: **Young** 1977a.
Record: **Young** 1977a (n. comb. from *Cicadella* and *Tettigonia*; key to species).

Scoposcartula
New species: **Young** 1977a.
Record: **Young** 1977a (n. comb. from *Cardioscarta* and *Tettigonia*; key to species).

Secopennis
Record: **Hamilton, K. G. A.** 1975b (APHRODINAE, DELTOCEPHALINI, DELTOCEPHALINA).

Segonalia
New species: **Young** 1977a.

Selachina
Records: **Emeljanov** 1962c (n. comb. from *Carchariacephalus*). **Hamilton, K. G. A.** 1975b (APHRODINAE; unassigned).

Selenocephalus
New species: **Dlabola** 1957a, 1957b, 1965c, 1974a, 1974c, 1974d, 1977a, 1981a. **Emeljanov** 1972d. **Lindberg** 1960b. **Linnavuori** 1962a, 1969a. **Rodrigues** 1968a.
Records: **Bonfils & Della Giustina** 1978a (Corsica). **Datta** 1973j (descriptive notes). **Cantoreanu** 1968b (Danube river delta); 1971c (Romania). **Dlabola** 1958b (Caucasus); 1964c (Albania); 1977c (Mediterranean). **Dubovskij** 1980a (USSR: western Turkmenia). **Emeljanov** 1964c (European USSR; key to species). **Hamilton, K. G. A.** 1975b (APHRODINAE, SELENOCEPHALINI). **Kalkandelen** 1974a (Turkey; key to species). **Kholmuminov & Dlabola** 1979a (USSR: Golodnostep Plain). **Lindberg** 1960a (Portugal); 1960b (Soviet Armenia; list of regional taxa). **Linnavuori** 1956d (Spanish Morocco); 1959a (Mt. Sibillini); 1959c (Mt. Picentini); 1962a (Israel; = *Levantotettix*, n. syn.); 1965a (Mediterranean). **Linnavuori & Al-Ne'amy** 1983a (in key to genera of African SELENOCEPHALINI). **Logvinenko** 1957b (USSR: Ukraine). **Quartau & Rodrigues** 1969a (Portugal). **Servadei** 1957a, 1960a, 1968a, 1971a (Italy). **Sharma, B.** 1977b (India: Jammu). **Theron** 1974b (n. comb. to *Distantia*).

Selenomorphus
New species: **Evans, J. W.** 1974a.

Selenopsis
Record: **Nielson** 1975a (transferred from COELIDIINAE to DELTOCEPHALINAE).

Selvitsa
New species: **Young** 1977a.
Record: **Young** 1977a (n. comb. from *Cicadella*, *Tettigonia* and *Tettigoniella*; key to species).

Semenovium
Record: **Dlabola** 1961b (Tadzhikistan). **Emeljanov** 1964c (European USSR). **Hamilton, K. G. A.** 1975b (APHRODINAE, PARADORYDIINI).

Sempia
Record: **Dworakowska** 1970m (n. comb. from *Erythroneura*).

Sequoiatettix
Records: **Bliven** 1958a (descriptive notes); 1963a (comments on sp. syn.). **Hamilton, K. G. A.** 1975b (APHRODINAE, DELTOCEPHALINI, PLATYMETOPIINA). **Nielson** 1966a (= *Colladonus*, n. syn).

Seriana
New species: **Dworakowska** 1971e, 1978a, 1979f, 1981e, 1981j. **Dworakowska, Nagaich & Singh** 1978a.
Records: **Ahmed, Manzoor** 1985c (n. comb. from *Zygina*). **Dworakowska** 1972g (n. comb. from *Empoasca*); 1976b (Taiwan); 1978a (n. comb. from *Typhlocyba*); 1979a (transfer of *S. atrosignata* to *Typhlocyba* rejected); 1979b (Vietnam); 1981j (n. comb. from *Typhlocyba*); 1984a (Malaysia). **Dworakowska, Nagaich & Singh** 1978a (n. comb. from *Typhlocyba*, *Zygina*, and *Zyginopsis*; n. sp. syn.). **Sohi & Dworakowska** 1984a (list of Indian species).

Serpa
Record: **Young** 1977a (redescribed).

Serratulus
New species: **Mahmood** 1967a.

Serratus
New species: **Linnavuori** 1959b.
Record: **Linnavuori & DeLong** 1978d (in key to genera of the *Copidodonus* group).

Serridonus
New species: **Linnavuori** 1959b.
Record: **Linnavuori & DeLong** 1978b (in key to genera of the *Bahita* group).

Sestrelicola
New species: **Remane & Asche** 1980a.
Record: **Nast** 1982a (catalogued; DELTOCEPHALINAE).

Shaddai
New species: **Dworakowska** 1977i. **Ma**, in **Chou, I. & Ma**, 1981a.
Records: **Dworakowska** 1971c (generic rank); 1977i (= *Sinalebra*, n. syn.). **Mahmood** 1967a (= *Alebra*). **Sharma, B.** 1979a (India: Nagaland, Kohima). **Sohi & Dworakowska** 1984a (list of Indian species).

Shamala
New species: **Dworakowska** 1980b, 1981g.
Records: **Dworakowska** 1982a (new subgenus proposed). **Sohi & Dworakowska** 1984a (list of Indian species). **Thapa** 1983a (Nepal).

Shirazia
New species: **Dlabola** 1977b.
Record: **Nast** 1982a (catalogued; DELTOCEPHALINAE).

Shonenus
Records: **Anufriev** 1971d (unnecessary replacement name). **Ishihara** 1958a (replacement for *Matsumurella*); 1966a (Kurile Islands). **Vilbaste** 1968a (USSR: coastal regions).

Sibovia
New species: **Young** 1977a.
Records: **Schroder** 1959a (diagnosis). **Young** 1977a (= *Ceratogoniella*, n. syn.; n. comb. from *Ceratogonia*, *Entogonia*, *Neokolla* and *Tettigonia*; n. sp. syn.; key to species).

Sichaea
   Records: **Davis, R. B.** 1975a (morphology; phyletic inferences). **Linnavuori** 1972b (transferred from ULOPINAE to LEDRINAE; key to species). **Theron** 1980a (notes on type specimens).
Sicistella
   Records: **Emeljanov** 1972d (n. comb. from *Orocastus*). **Hamilton, K. G. A.** 1975b (APHRODINAE, DELTOCEPHALINI, DELTOCEPHALINA). **Logvinenko** 1984a (as *Socistella*; USSR: Ukraine). **Nast** 1982a (catalogued; DELTOCEPHALINAE).
Sidelloides
   New species: **Evans, J. W.** 1972a.
Siderojassus
   New species: **Evans, J. W.** 1972b.
Sigista
   Records: **Emeljanov** 1966a (n. comb. from *Deltocephalus*). **Hamilton, K. G. A.** 1975b (APHRODINAE, DELTOCEPHALINI, DELTOCEPHALINA).
Signoretia
   New species: **Anufriev** 1971f. **Linnavuori** 1978a.
   Records: **Anufriev** 1971f (key to species). **Linnavuori** 1978a (key to species).
Similitopia
   Records: **Schroder** 1959a (n. comb. from *Oncometopia*); 1960b (comments on synonymy of type species). **Young** 1968a (subgenus of *Oncometopia*, n. status).
Sinalebra
   Record: **Dworakowska** 1977i (= *Shaddai*, n. syn.).
Sincholata
   New species: **DeLong** 1982f.
   Record: **DeLong** 1982f (n. comb. from *Osbornellus*).
Sinchonoa
   New species: **DeLong** 1980h. **Linnavuori & DeLong** 1978b.
   Record: **Cwikla & Blocker** 1981a (comparative notes).
Singapora
   New species: **Dworakowska** 1981e, 1983a. **Ghauri** 1974c, 1985b. **Mahmood** 1967a. **Chou, I. & Ma** 1981a. **Viraktamath & Dworakowska** 1979a.
   Records: **Dworakowska** 1970n (n. comb. from *Erythroneura* and *Zygina*); 1976b (China); 1980b, 1981e (India); 1983a (n. sp. syn.). **Dworakowska & Sohi** 1978a (n. sp. syn.). **Ghauri** 1985b (key to species). **Lee, C. E.** 1979a (Korea, redescribed). **Lee, C. E. & Kwon** 1977b (Korea). **Sohi & Dworakowska** 1984a (list of Indian species). **Viraktamath, C. A. & Dworakowska** 1979a (key to species).
Singhardina
   New species: **Mahmood** 1967a.
   Record: **Dworakowska** 1969g (subgenus of *Eurhadina*, n. status).
Sirosoma
   New species: **Ramakrishnan & Menon** 1974a.
   Record: **Ramakrishnan & Menon** 1974a (in key to genera of Oriental ERYTHRONEURINI).
Sisimitalia
   Record: **Young** 1977a (n. comb. from *Tettigonia*; key to species).
Sispocnis
   Records: **Anufriev** 1977b (generotype misidentified). **Hamilton, K. G. A.** 1980b (subgenus of *Pediopsoides*, n. status; identity of type species discussed). **Viraktamath, C. A.** 1981b (as subgenus of *Pediopsoides*).
Sitades
   Record: **Vilbaste** 1975a (generic synonymy discussed).
Ska
   New species: **Dworakowska** 1976b.
Smicrocotus
   Record: **Evans, J. W.** 1966a (n. comb. from *Kyphocotis*; n. sp. syn.; key to species).
Sobara
   Record: **Hamilton, K. G. A.** 1975b (APHRODINAE, DELTOCEPHALINI, PLATYMETOPIINA).
Sobrala
   New species: **Dworakowska** 1977i.
Sohinara
   Records: **Dworakowska** 1980d (= *Empoascanara*, n. syn.). **Ramakrishnan & Ghauri** 1979b (n. comb. from *Empoascanara*; in key to genera of *Empoascanara* complex).

Soibanga
   Record: **Datta & Dhar** 1984a (India: U. P.; descriptive notes).
Solanasca
   New species: **Ghauri** 1974a.
   Record: **Ghauri** 1974a (n. comb. from *Empoasca*; key to species).
Soleatus
   Record: **Cwikla & Blocker** 1981a (comparative notes).
Sonesimia
   New species: **Young** 1977a. **Cavichioli & Sakakibara** 1984a.
   Records: **Cavichioli & Sakakibara** 1984a (synopsis). **Young** 1977a (n. comb. from *Apogonalia* and *Tettigonia*; n. sp. syn.; key to species).
Sonronius
   Records: **Dlabola** 1965c, 1967b, 1967d, 1970b (Mongolia). **Emeljanov** 1964c (European USSR; key to species); 1977a (Mongolia). **Hamilton, K. G. A.** 1975b (APHRODINAE, DELTOCEPHALINI, MACROSTELINA). **Jankovic** 1971a (Serbia). **LeQuesne** 1969c (Britain). **Linnavuori** 1959b (in key to genera of MACROSTELINI). **Mitjaev** 1968b (eastern Kazakhstan); 1971a (Kazakhstan). **Moore, T. E. & Ross** 1957a (as subgenus of *Macrosteles*). **Nast** 1972a (catalogued); 1976a (Poland: Pieniny Mts.); 1976b (Poland; catalogued). **Okali** 1968a (Slovakia). **Ossiannilsson** 1983a (Fennoscandia & Denmark; key to species). **Vilbaste** 1965b (Altai Mts.); 1973a (notes on Flor's material); 1980a (Kamchatka); 1980b (Tuva).
Soortana
   Record: **Nielson** 1977a (n. sp. syn.).
Soosiulus
   New species: **Young** 1977a.
   Record: **Young** 1977a (n. comb. from *Cardioscarta*, *Poeciloscarta*, *Tettigonia* and *Tettigoniella*; key to species).
Soracte
   Record: **Evans, J. W.** 1966a (synopsis).
Sordana
   New species: **DeLong** 1976c.
   Record: **DeLong** 1976c (n. comb. from *Gypona*; n. sp. syn.).
Sorhoanus
   New species: **Dlabola** 1965c. **Emeljanov** 1964f, 1966a. **Kuoh, C. L.** 1981a, 1981b. **Metcalf** 1967b (new name). **Mitjaev** 1969c. **Vilbaste** 1965b.
   Records: **Anufriev** 1977a (Kurile Islands); 1978a (USSR: Maritime Territory). **Dlabola** 1958b (Caucasus); 1961b (central Asia); 1963c (notes on Haupt's material); 1964c (Albania); 1965c, 1966a, 1967b, 1967c, 1967d, 1968a, 1968b, 1970b (Mongolia). **Dubovskij** 1966a (Uzbekistan). **Dworakowska** 1969a, 1973d (Mongolia). **Emeljanov** 1964c (European USSR; key to species); 1966a (n. comb. to *Boreotettix*, *Lebradea* and *Rosenus*); 1977a (Mongolia). **Hamilton, K. G. A.** 1972d (Canada: Manitoba); 1975b (APHRODINAE, DELTOCEPHALINI, DELTOCEPHALINA); 1983b (transboreal taxon; *Arthaldeus*, *Cazenus* and *Lebradea* as subgenera). **Jasinska** 1980a (Poland: Bledowska Wilderness). **Lee, C. E. & Kwon** 1977b (Korea). **LeQuesne** 1969c (Britain). **Linnavuori** 1959a (Mt. Sibillini). **Logvinenko** 1957b, 1984a (USSR: Ukraine). **Mitjaev** 1968b (eastern Kazakhstan); 1971a (Kazakhstan; key to 8 species). **Nast** 1972a (catalogued); 1976a (Poland: Pieniny Mts.); 1976b (Poland; catalogued). **Ossiannilsson** 1983a (Fennoscandia & Denmark). **Servadei** 1968a (Italy). **Trolle** 1968a (Denmark). **Vilbaste** 1958a (Estonia); 1964a (Auwiesen Estlands); 1965b (Altai Mts.); 1968a (USSR: coastal regions); 1969a (notes on Matsumura's material); 1980a (Kamchatka); 1980b (Tuva; n. sp. syn.).
Sotanus
   New species: **Linnavuori** 1961a.
   Record: **Hamilton, K. G. A.** 1975b (APHRODINAE, DELTOCEPHALINI, ATHYSANINA).
Spangbergiella
   Records: **Hamilton, K. G. A.** 1975b, 1975d (APHRODINAE, HECALINI); 1976a (identity of Provancher material). **Linnavuori** 1956a (Brazil, Colombia, Paraguay); 1957a (in key to genera of Neotropical HECALINAE). **Linnavuori & DeLong** 1977c (DELTOCEPHALINAE, HECALINI); 1978c (Mexico); 1978e (list of Neotropical taxa).

Spaniglorlus
New species: **Nielson** 1979b.
Spanotartessus
New species: **Evans, F.** 1981a.
Record: **Evans, F.** 1981a (n. comb. from *Tartessus*).
Spartopyge
Records: **Cwikla & Blocker** 1981a (comparative notes). **Hamilton, K. G. A.** 1975b (APHRODINAE, DELTOCEPHALINI, DELTOCEPHALINA). **Young & Beirne** 1959a (n. comb. from *Flexamia*; key to species)
Spathanus
Records: **Cwikla & Blocker** 1981a (comparative notes). **Ellsbury & Nielson** 1978a (n. sp. syn.; intraspecific variation). **Hamilton, K. G. A.** 1975b (APHRODINAE, DELTOCEPHALINI, PLATYMETOPIINA).
Spathifer
Record: **Linnavuori** 1959b (redescribed).
Spatulostylus
New species: **Ramakrishnan & Menon** 1973a.
Records: **Ramakrishnan & Menon** 1973a (key to species); 1974a (in key to genera of Oriental ERYTHROENURINI).
Speudotettix
New species: **Emeljanov** 1962a.
Records: **Anufriev** 1977b (Kurile Islands); 1978a (USSR: Maritime Territory). **Dlabola** 1965c, 1967d, 1968a, 1968b, 1970b (Mongolia); 1981a (Iran). **D'Urso** 1980b (Italy). **Emeljanov** 1964c (European USSR). **Hamilton, K. G. A.** 1975b (APHRODINAE, DELTOCEPHALINI, ATHYSANINA). **Jasinska** 1980a (Poland: Bledowska Wilderness). **Kartal** 1983a (Turkey). **Kwon & Lee** 1979b (Korea). **Lee, C. E.** 1979a (Korea, redescribed). **Lee, C. E. & Kwon** 1977b (Korea). **LeQuesne** 1969c (Britain). **Linnavuori** 1959c (Mt. Picentini). **Logvinenko** 1957b (USSR: Ukraine). **Mitjaev** 1968b (eastern Kazakhstan); 1971a (Kazakhstan). **Nast** 1972a (catalogued); 1976a (Poland: Pieniny Mts.); 1976b (Poland; catalogued). **Nielson** 1968b (characterization of vector species). **Ossiannilsson** 1974a (Norway); 1983a (Fennoscandia & Denmark). **Schulz** 1976a (Norway). **Servadei** 1960a (as *Pseudotettix*); 1968a, 1972a (Italy). **Vilbaste** 1959b (Estonia); 1962b (eastern part of the Caspian lowlands); 1965b (Altai Mts.); 1980a (Kamchatka); 1980b (Tuva).
Sphaeropogonia
Records: **Young** 1965b (lectotype designations). **Young & Lauterer** 1966a (lectotype designation). **Young & Nast** 1963a (notes on type material).
Sphinctogonia
Record: **Young** 1965b (lectotype designations).
Spinolidia
New species: **Nielson** 1982e.
Record: **Nielson** 1982e (n. comb. from *Coelidia*; key to species).
Spinulana
New species: **DeLong** 1967a.
Record: **Cwikla & Blocker** 1981a (comparative notes).
Splonia
New species: **Young** 1968a.
Records: **Young** 1968a (n. gen. syn.; n. sp. syn.; n. comb. from *Rhaphirrhinus* (sic); key to species). **Young & Beier** 1963a (notes of extant specimen).
Sprundigia
New species: **Nielson** 1979b.
Stalinabada
New species: **Dlabola** 1961b.
Records: **Hamilton, K. G. A.** 1975b (APHRODINAE, APHRODINI, DORATURINA). **Nast** 1972a (=*Aconura*, n. syn.).
Stalolidia
New species: **Nielson** 1979b.
Record: **Nielson** 1979b (n. comb. from *Coelidia*; n. sp. syn.; key to species).
Stegelytra
New species: **Dlabola** 1974a. **Linnavuori** 1962a.
Records: **Dlabola** 1963c (notes on Haupt's material); 1974a (COELIDIINAE); 1981a (Iran). **Linnavuori** 1965a (Spanish Morocco). **Lodos & Kalkandelen** 1983a (Turkey). **Nielson** 1975a (referred to STEGELYTRINAE). **Quartau & Rodrigues** 1969a (Portugal). **Servadei** 1958a (Italy).

Stehlikiana
New species: **Young** 1977a.
Record: **Young** 1977a (n. comb. from *Cardioscarta*, *Tettigonia* and *Tettigoniella*; n. sp. syn.; key to species).
Stehliksia
Records: **Dworakowska** 1972f (n. comb. from *Zygina*). **Dworakowska & Trolle** 1976a (Kenya).
Stellena
Record: **Theron** 1973a (n. comb. from *Macrosteles*).
Stenagallia
Records: **Kramer** 1964a (in key to genera of New World AGALLIINAE). **Linnavuori & DeLong** 1977c (Chile).
Stenalsella
New species: **Evans, J. W.** 1966a.
Stenipo
Records: **Evans, J. W.** 1966a (synopsis).
Stenocoelidia
Records: **Beirne** 1956a (Canada: southern British Columbia). **Kramer** 1964d (=*Neocoelidia*, n. syn.).
Stenocotis
Record: **Evans, J. W.** 1966a (in key to genera of of STENOCOTINI; n. sp. syn.).
Stenogiffardia
New species: **Evans, J. W.** 1977a.
Stenometohardya
New species: **Dlabola** 1981a.
Stenometopiellus
New species: **Dlabola** 1957b, 1961b, 1967a, 1967d. **Emeljanov** 1964c, 1964f, 1972b. **Logvinenko** 1962a. **Mitjaev** 1971a, 1980a. **Vilbaste** 1961a, 1980b.
Records: **Beirne** 1956a (Canada: southern Alberta). **Bonfils & Della Giustina** 1978a (Corsica). **Dlabola** 1957a (Turkey; *Diplocolenoidea* as subgenus); 1961b (central Asia); 1963c (notes on Haupt's material); 1965c, 1967b, 1970b (Mongolia); 1971a (Turkey); 1972a (Afghanistan); 1981a (Iran). **Dubovskij** 1970b (Uzbekistan); 1978a (Uzbekistan, Zarafshan Valley); 1980a (USSR: western Turkmenia). **Emeljanov** 1964c (European USSR; key to species); 1977a (Mongolia). **Hamilton, K. G. A.** 1975b (APHRODINAE, DELTOCEPHALINI, ATHYSANINA). **Kalkandelen** 1974a (Turkey). **Korolevskaya** 1973a (Tadzhikistan). **Lindberg** 1960b (as *Stenopiellus*; Soviet Armenia). **Mitjaev** 1968b (eastern Kazakhstan); 1971a (Kazakhstan; key to 8 species); 1971b (southern Kazakhstan). **Vilbaste** 1962b (eastern part of the Caspian lowlands); 1965b (Altai Mts.); 1980b (Tuva).
Stenometopius
New species: **Evans, J. W.** 1966a.
Records: **Evans, J. W.** 1973a (NE New Guinea). **Linnavuori & DeLong** 1978c (as *Hododoecus*; =*Hodoedocus*, n. syn.).
Stenomiella
New species: **Linnavuori** 1969a.
Records: **Linnavuori** 1978d (comparative notes). **Webb, M. D.** 1981a (in key to genera of PARABOLOPONINAE).
Stenopsoides
Records: **Evans, J. W.** 1966a (synopsis). **Hamilton, K. G. A.** 1980b (comparative notes).
Stenoscopus
Record: **Evans, J. W.** 1966a (synopsis).
Stenotartessus
New species: **Evans, F.** 1966a.
Records: **Evans, J. W.** 1981a (=*Newmaniana*). **Evans, J. W.** 1966a (=*Newmaniana*, n. syn.).
Stenotortor
New species: **Evans, J. W.** 1973a.
Stephanolla
New species: **Young** 1977a.
Record: **Young** 1977a (n. comb. from *Tettigonia*; key to species).
Stictocoris
Records: **Cantoreanu** 1971c (Romania). **Dlabola** 1957a (Turkey); 1958b (Caucasus); 1964c (Albania). **Dubovskij** 1966a (Uzbekistan). **Emeljanov** 1964c (European USSR). **Hamilton, K. G. A.** 1975b (APHRODINAE, DELTOCEPHALINI, PLATYMETOPIINA). **Kalkandelen** 1974a (Turkey). **Lauterer** 1957a (Czechoslovakia). **Logvinenko** 1957b (USSR: Ukraine). **Mitjaev** 1968b (eastern Kazakhstan); 1971a (Kazakhstan).

Nast 1972a (catalogued); 1976a (Poland: Pieniny Mts.); 1976b (Poland; catalogued). Ossiannilsson 1983a (Fennoscandia & Denmark). Servadei 1968a, 1971a (Italy). Vilbaste 1965b (Altai Mts.); 1980b (Tuva). Zachvatkin 1953b (central Russia; Oka district).

Stictoscarta
New species: Young 1968a.
Records: Young 1965c (lectotype designation); 1968a (n. comb. from *Aulacizes*; n. sp. syn.; key to species) Young & Lauterer 1966a (lectotype designation). Young & Nast 1963a (notes on type material).

Stirellus
New species: DeLong & Linnavuori 1978b, 1979a. Emeljanov 1968b. Linnavuori 1959b, 1965a. Mahmood, Sultana & Waheed 1972a.
Records: Cheng, Y. J. 1980a (Paraguay). Dlabola 1965d (Caucasus); 1971a, 1972a (Afghanistan; n. comb. from *Aconura*). Dubovskij 1966a (Uzbekistan; n. comb. from *Trebellius*). Emeljanov 1962b (misinterpretation of generic synonymy); 1966b (DELTOCEPHALINAE, STIRELLINI); 1968b (n. comb. from *Thamnotettix*; n. sp. syn.). Hamilton, K. G. A. 1972d (Canada: Manitoba); 1975b (APHRODINAE, STIRELLINI). Linnavuori 1959b (=*Penestirellus*, n. syn.; n. comb. from *Athysanus* and *Deltocephalus*; key to species); 1964e (Egypt); 1965a (Libya); 1973c (Cuba); 1975c (Micronesia: Bonin Islands). Theron 1975a (n. comb. from *Aconura*). Vilbaste 1965a (revised generic synonymy). Whitcomb, Kramer & Coan 1972a (n. sp. syn.).

Stonasla
New species: Webb, M. D. 1980a.
Record: Davis, R. B. 1975a (morphology; phyletic inferences). Synave & Dlabola 1976a (St. Helene).

Stoneana
Record: Cwikla & Blocker 1981a (comparative notes).

Stragania
New species: Bliven 1957a. Linnavuori 1956a, 1957b.
Records: Beirne 1956a (Canada; key to species). Blocker 1979a (n. sp. syn.; key to species); 1979a (in key to genera of New World IASSINAE; n. sp. syn.). Hamilton, K. G. A. 1972d (Canada: Manitoba). Ishihara 1958a (Japan: north Honshu). Knight, W. J. 1983b (n. comb. from *Batracomorphus*). Linnavuori 1956a (synopsis of Neotropical taxa); 1957b (as subgenus of *Batracomorphus*). Linnavuori & Quartau 1975a (notes on affinities with *Batracomorphus*). Ramos Elorduy 1972c (Mexico). Suehiro 1961a (Hawaii; Kauai). Valley & Wheeler 1985a (immature stages illustrated).

Straganiassus
Records: Anufriev 1971g (n. comb. from *Stragania*); 1978a (n. comb. from *Macropsis*; n. sp. syn.). Nast 1981a (*Straganiassus* Anufriev 1978a a different generic taxon, *Anufrieviella*). Viraktamath, C. A. 1979a (= *Iassus*, n. syn.).

Streptanulus
Record: Hamilton, K. G. A. 1975b (APHRODINAE, DELTOCEPHALINI, ATHYSANINA).

Streptanus
New species: Dlabola 1957c, 1967a, 1984a. Emeljanov 1964c. Mitjaev 1967c. Vilbaste 1965b. Zachvatkin 1948b.
Records: Anufriev 1977b (Kurile Islands); 1979a (USSR: Maritime Territory). Beirne 1956a (Canada; key to species). Dlabola 1960a (notes on Horvath's material); 1964c (Albania); 1965c, 1966a, 1967b, 1967c, 1967d, 1968a (Mongolia); 1977c (Mediterranean); 1981a (Iran). Dubovskij 1966a (Uzbekistan). Emeljanov 1962a (new subgenus proposed); 1964c (European USSR; key to species); 1966a (comparative notes); 1977a (Mongolia). Gravestein 1965a (The Netherlands: Terschelling I.). Hamilton, K. G. A. 1972d (Canada: Manitoba); 1975b (APHRODINAE, DELTOCEPHALINI, ATHYSANINA). Jankovic 1966a (Serbia). Jasinska 1980a (Poland: Bledowska Wilderness). Kalkandelen 1974a (Turkey). Logvinenko 1957b, 1962e, 1984a (USSR: Ukraine). Mitjaev 1968b (eastern Kazakhstan); 1971a (Kazakhstan; key to 2 species). Nast 1972a (catalogued); 1976a (Poland: Pieniny Mts.); 1976b (Poland; catalogued). Ossiannilsson 1974a (Norway); 1983a (Fennoscandia & Denmark; key to species). Schulz 1976a (Norway). Servadei 1960a, 1968a (Italy). Vilbaste 1964a (Auwiesen Estlands); 1965b (Altai Mts.); 1969b (Taimyr); 1973a (notes on Flor's material); 1980a (Kamchatka); 1980b (Tuva).

Streptopyx
New species: Dlabola 1984a. Linnavuori 1958a.
Record: Della Giustina 1983a (France). Servadei 1969a (Italy).

Strictogonia
Record: Young & Lauterer 1966a (lectotype designation).

Stroggylocephalus
Records: Anufriev 1977b (Kurile Islands); 1978a (USSR: Maritime Territory). Cantoreanu 1959a (Romania); 1965d (as *Strorgylocephalus*; Romania). Dubovskij 1966a (Uzbekistan). Emeljanov 1964c (European USSR; key to species); 1977a (Mongolia). Hamilton, K. G. A. 1975b (APHRODINAE, APHRODINI); 1975d (Canada: Manitoba); 1976a (identity of Provancher material). Lee, C. E. 1979a (Korea, redescribed). Lee, C. E. & Kwon 1977b (Korea). LeQuesne 1965c (Britain). Logvinenko 1957b (as *Strongilocephalus*; USSR: Ukraine). Nast 1972a (catalogued); 1976a (Poland: Pieniny Mts.); 1976b (Poland; catalogued). Ossiannilsson 1981a (Fennoscandia & Denmark; key to species). Servadei 1968a (as *Stronggylocephalus*; Italy). Vilbaste 1958a (as *Strongylocephalus*; Estonia); 1964a (Auwiesen Estlands); 1965b (Altai Mts.); 1968a (USSR: coastal regions); 1973a (notes on Flor's material).

Strongylomma
Records: Linnavuori 1970a (photograph of type specimen; IDIOCERINAE). Webb, M. D. 1983b (erroneously recorded from South Africa, actually from South America).

Stymphalus
New species: Linnavuori 1961a.
Records: Anufriev 1978a (USSR: Maritime Territory). Dlabola 1958a (southern Europe; n. sp. syn.); 1977c (Mediterranean); 1981a (Iran). Dworakowska 1973d (Korea). Hamilton, K. G. A. 1975b (APHRODINAE, PARABOLOPINI). Heller & Linnavuori 1968a (Ethiopia). Lee, C. E. 1979a (Korea, redescribed). Linnavuori 1962a (Israel); 1978d (= *Varta*, n. syn.; key to African and Palaearctic species). Lodos & Kalkandelen 1985c (Turkey). Theron 1980a (redescribed); 1982b (South Africa). Vilbaste 1968a (USSR: coastal regions); 1975a (placement for *Deltocephalus rubrolineatus*).

Suarezia
Records: Cwikla & Blocker 1981a (comparative notes). Linnavuori & DeLong 1978b (n. comb. from *Menosoma*).

Subbanara
Records: Dworakowska 1980d (= *Empoascanara*, n. syn.). Ramakrishnan & Ghauri 1979a (in key to genera of Oriental ERYTHRONEURINI).

Subhimalus
New species: Ghauri 1971c. Ramakrishnan 1983a.

Subrasaca
New species: Young 1977a.
Record: Young 1977a (n. comb. from *Cardioscarta*, *Microgoniella* and *Tettigonia*; n. sp. syn.; key to species).

Sudra
New species: Kramer 1964c.
Record: Kramer 1964c (key to species).

Sujitettix
New species: Ramakrishnan & Menon 1972a.
Records: Anufriev 1972a (as subgenus of *Apheliona*). Dworakowska 1970a (= *Apheliona*, n. syn.); 1977d (= *Apheliona*).

Sulamicerus
Records: Dlabola 1974a (n. comb. from *Idiocerus*); 1974b (n. comb. from *Idiocerinus* and *Idiocerus*); 1981a (Iran). Lodos & Kalkandelen 1982a (Turkey).

Sulcana
New species: DeLong & Freytag 1966b.
Record: DeLong & Freytag 1972a (in key to genera of GYPONINAE).

Sunda
Record: Vilbaste 1965a (= *Doratulina*, n. syn.).

Sundapteryx
Records: **Dlabola** 1971a (Pakistan); 1972a (Afghanistan). **Dworakowska** 1970h (n. comb. from *Chlorita* and *Empoasca*; n. sp. syn.); 1972g (China). **Dworakowska & Viraktamath** 1975a (=*Amrasca*, n. syn.). **Linnavuori** 1975c (n. comb. from *Empoasca*; n. sp. syn.; Micronesia).

Sundara
New species: **Dworakowska** 1977i. **Ramakrishnan & Menon** 1972b.
Records: **Ahmed, Manzoor** 1985c (Pakistan). **Dworakowska** 1977i (n. comb. from *Typhlocyba*; n. sp. syn.); 1979c (n. comb. from *Zyginella*). **Dworakowska & Sohi** 1978a (n. comb. from *Typhlocyba*; n. sp. syn.). **Dworakowska, Sohi & Viraktamath** 1980a (in key to Indian species of ZYGINELLINI). **Ramakrishnan & Menon** 1972b (in key to genera of Indian TYPHLOCYBINI). **Sohi & Dworakowska** 1984a (list of Indian species).

Svanetia
New species: **Schengeliya & Dlabola** 1964a.
Record: **Hamilton, K. G. A.** 1975b (APHRODINAE; unassigned).

Swarajnara
Records: **Dworakowska** 1980d (=*Empoascanara*, n. syn.). **Ramakrishnan & Ghauri** 1979b (in key to genera of the Oriental *Empoascanara* complex).

Sylhetia
New species: **Ahmed, Manzoor** 1972a.
Record: **Dworakowska** 1979c (ZYGINELLINI).

Symphypyga
New species: **Dlabola** 1966a. **Dubovskij** 1966a, 1968b. **Emeljanov** 1964c, 1972b. **Kameswara Rao & Ramakrishnan** 1983b.
Records: **Datta** 1973c (descriptive notes). **Davis, R. B.** 1975a (morphology; phyletic inferences). **Dlabola** 1961b (central Asia); 1967b (Mongolia); 1981a (Iran). **Dubovskij** 1978a (Uzbekistan, Zarafshan Valley); 1980a (USSR: western Turkmenia). **Emeljanov** 1964c (European USSR; key to species); 1975a (in key to genera of ADELUNGIINI); 1977a (Mongolia). **Korolevskaya** 1975a (Tadzhikistan). **Lindberg** 1960b (Soviet Armenia). **Mitjaev** 1971a (Kazakhstan).

Syncharina
Record: **Young** 1977a (n. comb. from *Draeculacephala* and *Tettigonia*; n. sp. syn.; key to species).

Synogonia
Records: **Kramer** 1976c (to NIRVANINAE). **Young** 1977a (excluded from CICADELLINAE).

Synophropsis
Records: **Bonfils & Lauriant** 1975a (France). **Della Giustina** 1983a (France). **Dlabola** 1958a (southern Europe); 1965d (Jordan); 1971b (Turkey); 1977c (Mediterranean); 1981a (Iran). **Emeljanov** 1964c (European USSR). **Hamilton, K. G. A.** 1975b (APHRODINAE, DELTOCEPHALINI, PLATYMETOPIINA). **Linnavuori** 1962a (Israel); 1965a (Turkey). **Logvinenko** 1963a (USSR: Ukraine. **Servadei** 1968a (Italy).

Syringius
Record: **Hamilton, K. G. A.** 1975b (APHRODINAE, DELTOCEPHALINI, PLATYMETOPIINA).

Syringophora
Record: **Young** 1968a (=*Splonia*, n. syn.).

Szymczakowskia
New species: **Dworakowska** 1974a, 1976b.

Tacora
New species: **Young** 1977a.
Records: **Young** 1977a (redescribed; key to species). **Young & Lauterer** 1966a (lectotype designation).

Taeniocerus
Records: **Dlabola** 1974b (n. comb. from *Idiocerus*); 1981a (Iran). **Lodos & Kalkandelen** 1982a (Turkey).

Tafalka
New species: **Dworakowska** 1979c.

Tahara
New species: **Nielson** 1977a.
Record: **Nielson** 1977a (key to species).

Taharana
New species: **Nielson** 1982e.
Record: **Nielson** 1982e (n. comb. from *Coelidia* and *Jassus*; key to species).

Tahuampa
New species: **Young** 1977a.
Record: **Young** 1977a (n. comb. from *Tettigonia*; key to species).

Tahura
New species: **Kramer** 1976c.
Records: **Kramer** 1976c (NIRVANINAE). **Young** 1965a (MILEEWANINI); 1977a (excluded from CICADELLINAE).

Takagiana
New species: **Dworakowska** 1974a.

Takagiella
New species: **Kwon & Lee** 1979c.
Record: **Hamilton, K. G. A.** 1975b (APHRODINAE, DELTOCEPHALINI, DELTOCEPHALINA). **Lee, C. E.** 1979a (Korea, redescribed). **Vilbaste** 1969a (n. comb. from *Deltocephalus*)

Takama
New species: **Dworakowska** 1979a, 1980b. **Dworakowska & Viraktamath** 1975a.
Record: **Sohi & Dworakowska** 1984a (list of Indian species).

Tamaricades
Records: **Emeljanov** 1977a (Mongolia; n. comb. from *Goniagnathus*; generic rank). **Hamilton, K. G. A.** 1975b (APHRODINAE, SELENOCEPHALINI).

Tamaricella
New species: **Dlabola** 1981a. **Dworakowska** 1971b. **Logvinenko** 1981c. **Mitjaev** 1975a.
Records: **Ahmed, Manzoor** 1985c (Pakistan). **Della Giustina** 1983a (France). **Dlabola** 1958c (in key to genera of Palaearctic TYPHLOCYBINAE). **Dubovskij** 1978a (Uzbekistan, Zarafshan Valley); 1980a (USSR: western Turkmenia). **Dworakowska** 1971b (generic status; n. comb. from *Chlorita, Erythroneura, Heliona, Helionidia* and *Zygina*). **Emeljanov** 1977a (Mongolia). **Kholmuminov & Dlabola** 1979a (USSR: Golodnostep Plain). **Lodos & Kalkandelen** 1984c (Turkey). **Logvinenko** 1984a (USSR: Ukraine). **Sohi & Dworakowska** 1984a (list of Indian species).

Tambila
Record: **Evans, J. W.** 1972a (comparative notes).

Tantogonalia
New species: **Young** 1977a.
Record: **Young** 1977a (key to species).

Tantulidia
Record: **Nielson** 1979a (n. comb. from *Coelidia* and *Tinobregmus*; key to species).

Tapajosa
New species: **Young** 1968a.
Records: **Young** 1968a (n. comb. from *Cuerna, Oncometopia* and *Tettigonia*; n. sp. syn.; key to species). **Young & Lauterer** 1966a (lectotype designation).

Taperhina
New species: **Linnavuori** 1959b. **Linnavuori & DeLong** 1978b.
Records: **Cwikla & Blocker** 1981a (as *Taberinha*; comparative notes). **Linnavuori & Heller** 1961a (as *Taberinha*; Peru).

Tapetia
New species: **Emeljanov** 1964b.
Record: **Hamilton, K. G. A.** 1975b (APHRODINAE; unassigned). **Korolevskaya** 1979b (Tadzhikistan).

Tartessella
Records: **Evans, F.** 1981a (n. comb. from *Tartessus*). **Evans, J. W.** 1966a (synopsis).

Tartessoides
New species: **Evans, J. W.** 1966a.
Record: **Evans, F.** 1981a (synopsis).

Tartessops
New species: **Evans, F.** 1981a.
Record: **Evans, F.** 1981a (n. comb. from *Bythoscopus*).

Tartessus
New species: **Evans, J. W.** 1966a. **Linnavuori** 1960a.
Records: **Datta & Dhar** 1984a (descriptive notes). **Evans, F.** 1981a ((n. comb. to *Alotartessus, Austrotartessus, Borditartessus, Brunotartessus, Calotartessus, Distantessus, Microtartessus, Milotartessus, Neotartessus, Philotartessus, Plexitartessus, Protartessus, Spanotartessus, Tartessella,*

Tenuitartessus, Triviotartessus). **Evans, J. W.** 1974a (n. sp. syn.). **Ishihara** 1965c (Formosa). **Linnavuori** 1956e (identities of Spangberg and Stal species); 1960a (Micronesia; key to species); 1975c (Micronesia: Guam, Palau, Yap).

Taslopa
Records: **Davis, R. B.** 1975a (morphology; phyletic inferences). **Evans, J. W.** 1966a (synopsis).

Tasmanotettix
Record: **Evans, J. W.** 1966a (= *Thamnophryne*, n. syn.).

Tasnimocerus
New species: **Ghauri** 1975c.

Tataka
New species: **Dworakowska** 1974a, 1979c.
Record: **Dworakowska** 1979c (= *Orucyba*, n. syn.).

Taurotettix
Records: **Dlabola** 1960a (descriptive notes). **Emeljanov** 1962a (comparative notes); 1964c (European USSR); 1977a (Mongolia). **Hamilton, K. G. A.** 1975b (APHRODINAE; unassigned). **Mitjaev** 1968b (eastern Kazakhstan); 1971a (Kazakhstan; in key to genera of DELTOCEPHALINAE). **Vilbaste** 1962b (eastern part of the Caspian lowlands); 1980b (Tuva; comments *vis a vis Callistophia*).

Tautocerus
New species: **Anufriev** 1971a. **Kwon** 1985a.
Records: **Dlabola** 1974b (in key to genera of Palaearctic TYPHLOCYBINAE). **Kwon** 1985a (Korea; key to species). **Lee, C. E.** 1979a (Korea, redescribed).

Tautoneura
New species: **Anufriev** 1969b. **Dworakowska** 1977f, 1979a, 1979b, 1980b, 1981e, 1981j, 1984a. **Ma** 1983a. **Thapa** 1984b.
Records: **Dworakowska** 1979a (n. comb. from *Typhlocyba*); 1981j (revision of sp. syn.). **Dworakowska & Sohi** 1978a (n. comb. from *Zygina* and *Zyginopsis*; n. sp. syn.). **Sohi & Dworakowska** 1984a (list of Indian species).

Tblisica
New species: **Dlabola** 1958b, 1974a, 1984a. **Kartal** 1983a.
Records: **Dlabola** 1981a (Iran). **Hamilton, K. G. A.** 1975b (APHRODINAE, DELTOCEPHALINI, PLATYMETOPIINA).

Teleogonia
New species: **Young** 1977a.
Records: **Young** 1977a (= *Pygometopia*, n. syn.; n. comb. from *Oncometopia* and *Tettigonia*; key to species). **Young & Nast** 1963a (notes on type specimens).

Teletusa
Records: **Young** 1965a (lectotype designation); 1968a (= *Myogonia*, n. syn; n. comb. from *Myogonia* and *Tettigonia*; key to species). **Young & Lauterer** 1966a (lectotype designations). **Young & Nast** 1963a (notes on type specimens).

Telopetulcus
New species: **Evans, J. W.** 1972b.

Telusus
Record: **Hamilton, K. G. A.** 1975b (APHRODINAE, DELTOCEPHALINI, DELTOCEPHALINA).

Tenuacia
New species: **DeLong** 1977c.

Tenucephalus
New species: **DeLong** 1982c. **Linnavuori & DeLong** 1976a, 1977b, 1978c.
Records: **Cwikla & Blocker** 1981a (comparative notes). **Linnavuori** 1957a (in key to genera of Neotropical HECALINAE).

Tenuisanus
Record: **Cwikla & Blocker** 1981a (comparative notes).

Tenuitartessus
New species: **Evans, F.** 1981a.
Record: **Evans, F.** 1981a (n. comb. from *Tartessus*).

Terulia
New species: **Nielson** 1979b, 1982d.
Records: **Nielson** 1979b (n. comb. from *Coelidia*; n. sp. syn.; key to species); 1982d (revised key to species).

Tetartostylus
New species: **Heller & Linnavuori** 1968a. **Linnavuori** 1961a. **Theron** 1973b, 1982b. **Nast** 1965a.
Records: **Cantoreanu** 1961a (Romania). **Dlabola** 1958a (southern Europe); 1964c (Albania). **Emeljanov** 1964c (European USSR). **Hamilton, K. G. A.** 1975b (APHRODINAE, DELTOCEPHALINI, PLATYMETOPIINA). **Jankovic** 1971a (Serbia). **Kalkandelen** 1974a (Turkey). **Linnavuori** 1961a (South Africa; n. comb. from *Chlorotettix*). **Logvinenko** 1984a (USSR: Ukraine). **Nast** 1985a (n. comb. from *Limotettix*). **Servadei** 1968a (Italy). **Theron** 1973b (redescribed; key to species).

Tetelloides
Record: **Linnavuori** 1979a (= *Baramapulana*, n. syn).

Tetigela and **Tetigella**
New species: **Schroder** 1959a.
Records: **Cantoreanu** 1965d (Romania); 1968b (Danube river delta); 1971c (Romania). **Datta** 1980a (descriptive notes). **Datta & Pramanik** 1965d (as *Tettigela*); 1977a (India: A. P., Subansiri division). **Ishihara** 1958a (Japan: north Honshu); 1961a (Thailand); 1965c (Formosa). **Lee, C. E. & Kwon** 1977b (Korea). **Lindberg** 1960b (Soviet Armenia). **Servadei** 1968a, 1969a, 1971a, 1972a (Italy). **Wagner** 1959c (Greece). **Young** 1977a (species of uncertain placement).

Tettigella
New species: **Metcalf** 1965a (new name).

Tettigonia
New species: **Metcalf** 1965a (new name).
Records: **ICZN** 1954a (*Tettigonia* validated under Plenary Powers as a generic name in Orthoptera, Opinion 299). **Ishihara** 1967b (identity of *T. albomarginata*). **Ramos Elorduy** 1972c (Mexico). **Young** 1963a, 1964a, 1965b, 1965c, 1974a (lectotype designations); 1977a (species of uncertain position). **Young & Beier** 1963a (records of extant specimens; lectotype designations). **Young & Lauterer** 1966a (lectotype designations).

Tettigoniella
Records: **Datta** 1973k, 1973m (descriptive notes). **Linnavuori** 1979a (*T. bimaculata*, uncertain placement). **Ramachandra Rao** 1973a (India: Poona). **Sharma, B.** 1977b (India: Jammu). **Young** 1965c (lectotype designation); 1977a (species of uncertain placement). **Young & Lauterer** 1964a, 1966a (lectotype designations).

Tettisamia
Record: **Young** 1977a (n. comb. from *Tettigonia*; n. sp. syn.).

Tettiselva
New species: **Young** 1977a.

Texananus
New species: **DeLong & Martinson** 1973e. **Linnavuori** 1959b.
Records: **Beirne** 1956a (Canada; key to species). **Datta & Dhar** 1984a (descriptive notes). **Hamilton, K. G. A.** 1972d (Canada: Manitoba); 1975b (APHRODINAE, DELTOCEPHALINI, PLATYMETOPIINA). **Linnavuori** 1959b (key to subgenera and species). **Nielson** 1968b (characterization of vector species). **Scudder** 1961a (Canada: British Columbia).

Teyasteles
New species: **Heller & Linnavuori** 1968a.
Records: **Linnavuori** 1969a (n. comb. from *Cicadula*). **Theron** 1973a (redescription; placement in MACROSTELINI questioned).

Thagria
New species: **Kwon & Lee** 1979b. **Nielson** 1980a, 1980b, 1980c, 1980d, 1982c.
Records: **Lee, C. E.** 1979a (Korea, redescribed). **Nielson** 1977a (n. gen. syn.; key to species); 1980d (n. sp. syn.).

Thaia
New species: **Dworakowska** 1970c, 1972k, 1976b, 1979b, 1980b, 1984a. **Dworakowska & Viraktamath** 1979a. **Ghauri** 1962a. **Kuoh, C. L.** 1982a. **Thapa & Sohi** 1982a.
Records: **Ahmed, Manzoor** 1970b (East Pakistan); 1983b (remarks on shape of male styles). **Dworakowska** 1970c (= *Hardiana*, n. syn.; n. comb. from *Hardiana*); 1972g (Cochinchina); 1976b (southeast Aisa; *Nlunga* as subgenus, n. status); 1977f (north India); 1979b (new subgenus proposed); 1984a (Malaysia). **Dworakowska & Viraktamath** 1979a (n. comb. from *Sirosoma*). **Ghauri** 1972e (Pakistan). **Mahmood** 1967a (in key to genera of Oriental ERYTHRONEURINI). **Sohi** 1976a (n. comb. from

Hardiana). **Sohi & Dworakowska** 1984a (list of Indian species). **Thapa & Sohi** 1982a (summary of records for Nepal).

Thailocyba
New species: **Mahmood** 1967a.
Record: **Dworakowska** 1979c (ZYGINELLINI).

Thailus
New species: **Mahmood** 1967a.

Thalattoscopus
Records: **Knight, W. J.** 1983b (removed from synonymy with *Batracomorphus*). **Linnavuori & Quartau** 1975a (= *Batracomorphus*, n. syn.).

Thamnophryne
Record: **Evans, J. W.** 1966a (= *Tasmanotettix*, n. syn.; n. sp. syn.).

Thamnotettix
New species: **Dlabola** 1965d, 1965e, 1971b, 1974c, 1974d. **Emeljanov** 1962a. **Linnavuori** 1958a.
Records: **Anufriev** 1978a (USSR: Maritime Territory). **Beirne** 1956a (Canada). **Bonfils & Della Giustina** 1978a (Corsica). **Cantoreanu** 1961a, 1971c (Romania). **Datta** 1972f (descriptive notes). **Della Giustina** 1983a (France). **Dlabola** 1958a (southern Europe); 1958b (Caucasus); 1960a (notes on Horvath's material); 1961b (central Asia); 1963c (notes on Haupt's material); 1964a (Albania); 1965c (Mongolia); 1965d (Jordan; key to species of the *confinis* group); 1967b, 1967c, 1967d, 1968a, 1968b, 1970b (Mongolia); 1971b (Iran and Turkey); 1974c (synopsis); 1977c (Mediterranean); 1981a (Iran). **D'Urso** 1980b (Italy). **Dworakowska** 1973d (Mongolia). **Emeljanov** 1964c (European USSR); 1977a (Mongolia). **Hamilton, K. G. A.** 1972d (Canada: Manitoba); 1975b (APHRODINAE, DELTOCEPHALINI, CICADULINA); 1976a (notes on Provancher's material); 1983b (transboreal taxon). **Ishihara** 1966a (Kurile Islands). **Jankovic** 1966a, 1971a (Serbia). **Jasinska** 1980a (Poland: Bledowska Wilderness). **Kartal** 1982a (Turkey); 1983a (Turkey; *T. creticus* redescribed). **Lee, C. E. & Kwon** 1977b (Korea). **LeQuesne** 1969c (Britain). **Lindberg** 1960a (Portugal); 1960b (Soviet Armenia); 1963a (Morocco). **Linnavuori** 1959b (Neotropical species of *incertae sedis*); 1959c (Mt. Picentini); 1961a (n. comb. to *Circulifer*); 1962a (Israel). **Linnavuori & DeLong** 1977b (species of *incertae sedis*). **Logvinenko** 1957b, 1957c (USSR: Ukraine). **Mitjaev** 1968b (eastern Kazakhstan); 1971a (Kazakhstan). **Nast** 1972a (catalogued); 1976a (Poland: Pieniny Mts.); 1976b (Poland; catalogued). **Ossiannilsson** 1974a (Norway); 1983a (Fennoscandia & Denmark; key to species). **Quartau & Rodrigues** 1969a (Portugal). **Schulz** 1976a (Norway). **Servadei** 1957a, 1958a, 1960a, 1968a, 1971a, 1972a (Italy). **Theron** 1970b (redescription of Cogan's species); 1974b (n. comb. to *Neoaliturus* and *Recilia*). **Vilbaste** 1969a (notes on Matsumura's material); 1973a (notes on Flor's material); 1975a (notes on Motschulsky's material); 1976a (Matsumura's species referred to *Aconurella, Cechenotettix, Cicadula, Ericotettix, Heliotettix, Mocydiopsis* and *Neoaliturus*; some n. sp. syn. established); 1980a (Kamchatka); 1980b (Tuva). **Wagner** 1959c (Greece). **Zachvatkin** 1953b (central Russia; Oka district).

Thampoa
New species: **Mahmood** 1967a.

Tharra
New species: **Evans, J. W.** 1966a. **Linnavuori** 1960a, 1960b. **Nielson** 1982b.
Records: **Evans, J. W.** 1974a (New Caledonia; n. comb. from *Jassus*). **Eyles & Linnavuori** 1974a (Cook Islands; Niue Island). **Knight, W. J.** 1982a (Rennell Island). **Linnavuori** 1960a (Micronesia; key to species); 1960b (Fiji; key to species); 1975c (Micronesia: Palau). **Nast** 1982a (catalogued; COELIDIINAE). **Nielson** 1975a (n. gen. syn.; n. comb. from *Coelidia, Jassoidula, Muirella* and *Nisistrana*; n. sp. syn.; key to species).

Thatuna
Records: **Oman** 1971a (female described). **Oman & Musgrave** 1975a (redefined).

Thaumatopoides
Records: **Linnavuori** 1977a (in key to genera of African PENTHIMIINAE). **Theron** 1976a (n. comb. from *Penthimia*; n. sp. syn.).

Thaumatoscopus
Records: **Evans, J. W.** 1966a (synopsis); 1972a (n. comb. to *Vulturnus*; n. sp. syn.; PENTHIMIINAE).

Theasca
New species: **Dworakowska** 1972c.

Theronopus
New species: **Webb, M. D.** 1983b.
Record: **Webb, M. D.** 1983b (key to species).

Thlasia
New species: **Linnavuori** 1972b.

Thomsoniella
Records: **Dlabola** 1957b, 1972a (Afghanistan).

Thymbrella
New species: **Evans, J. W.** 1969a.

Thymbris
New species: **Evans, J. W.** 1966a, 1969a.
Record: **Evans, J. W.** 1966a (n. comb. from *Rhothidus*).

Tiaja
New species: **Gill & Oman** 1982a. **Oman** 1972a. **Sawbridge** 1975b.
Record: **Davis, R. B.** 1975a (morphology; phyletic inferences).

Tialidia
New species: **Nielson** 1982e.
Record: **Nielson** 1982e (key to species).

Tiaratus
New species: **Emeljanov** 1961a
Records: **Dlabola** 1965c, 1967b, 1967d, 1968b, 1970b (Mongolia). **Dworakowska** 1969a (Mongolia). **Emeljanov** 1977a (Mongolia). **Hamilton, K. G. A.** 1975b (APHRODINAE, DELTOCEPHALINI, PLATYMETOPIINA). **Mitjaev** 1968b (eastern Kazakhstan); 1971a (Kazakhstan). **Vilbaste** 1965b (Altai Mts.); 1980b (Tuva).

Tichocoelidia
New species: **Kramer** 1962a.

Tideltellus
Records: **Hamilton, K. G. A.** 1975b (APHRODINAE, DELTOCEPHALINI, DELTOCEPHALINA). **Kramer** 1971d (n. comb. from *Deltocephalus*).

Tigriculus
New species: **Dlabola** 1961b.
Records: **Dubovskij** 1966a (Uzbekistan). **Hamilton, K. G. A.** 1975b (APHRODINAE, DELTOCEPHALINI, DELTOCEPHALINA). **Nast** 1972a (catalogued).

Tinderella
New species: **Webb, M. D.** 1983a.

Tingolix
New species: **Linnavuori & DeLong** 1978d.
Record: **Cwikla & Blocker** 1981a (comparative notes).

Tingopyx
New species: **Linnavuori & DeLong** 1978d.
Record: **Cwikla & Blocker** 1981a (comparative notes).

Tinobregmus
Records: **Nielson** 1975a (n. sp. syn.). **Ramos Elorduy** 1972c (Mexico).

Tinocripus
New species: **Nielson** 1982e.
Record: **Nielson** 1982e (key to species).

Tipuana
New species: **Young** 1977a.
Record: **Young** 1977a (n. comb. from *Diedrocephala*; key to species).

Titiella
Records: **Linnavuori** 1972b (in key to genera of African LEDRINAE). **Theron** 1976a (n. comb. from *Hangklipia*).

Tlagonalia
Record: **Young** 1977a (n. comb. from *Tettigonia*).

Toba
Record: **Nielson** 1975a (removed from COELIDIINAE; unassigned).

Togacephalus
  Records: **Hamilton, K. G. A.** 1975b (APHRODINAE, DELTOCEPHALINI, DELTOCEPHALINA). **Kwon & Lee** 1979b (as subgenus of *Recilia*). **Lee, C. E. & Kwon** 1977b (Korea).
Togaricrania
  New species: **Ma, in Chou, I. & Ma** 1981a.
Togaritettix
  Records: **Ahmed, Manzoor** 1983b (remarks on wing venation); 1985c (n. comb. from *Mahmoodiana*). **Dworakowska** 1971d (=*Motschulskyia*, in part). **Lee, C. E.** 1979a (Korea, redescribed). **Lee, C. E. & Kwon** 1977b (Korea). **Ramakrishnan & Menon** 1971a (India; redescribed).
Tolasella
  New species: **Evans, J. W.** 1972a.
  Record: **Evans, J. W.** 1972a (n. comb. from *Neodartus*).
Toldoanus
  Records: **Linnavuori** 1956a (Brazil); 1959b (Argentina, Bolivia, Paraguay).
Tolua
  Record: **Young** 1968a (=*Amblydisca*, n. syn.).
Tomaloides
  New species: **Evans, J. W.** 1972a.
Tomopennis
  New species: **Maldonado-Capriles** 1984a.
  Record: **Maldonado-Capriles** 1984a (in key to Neotropical genera of IDIOCERINAE).
Tongdotettix
  New species: **Kwon** 1980b.
  Record: **Nast** 1982a (catalogued; DELTOCEPHALINAE).
Torenadoga
  New species: **Blocker** 1979a.
Toroa
  New species: **Ahmed, Manzoor** 1979a.
Toropsis
  Record: **Hamilton, K. G. A.** 1980b (n. comb. from *Macropsis*).
Torresabela
  Record: **Young** 1977a (n. comb. from *Cicadella* and *Tettigonia*; n. sp. syn.).
Tortigonalia
  New species: **Young** 1977a.
  Record: **Young** 1977a (key to species).
Tortor
  New species: **Evans, J. W.** 1966a.
  Record: **Evans, J. W.** 1966a (n. comb. from *Kana*; n. sp. syn.).
Tortotettix
  New species: **Theron** 1982b.
Tozzita
  New species: **Kramer** 1964d, 1967b.
Trachygonalia
  Record: **Young** 1977a (n. comb. from *Tettigonia*).
Trebellius
  Record: **Hamilton, K. G. A.** 1975b (APHRODINAE, STIRELLINI).
Tremulicerus
  Records: **Anufriev** 1977b (Kurile Islands). **Dlabola** 1974b (in key to genera of Palaearctic IDIOCERINAE; n. comb. from *Idiocerus*); 1977b (Iran); 1977c (Mediterranean); 1981a (Iran). **Hamilton, K. G. A.** 1980a (=*Idiocerus*, n. syn). **Lodos & Kalkandelen** 1981a (Turkey). **Ossiannilsson** 1981a (Fennoscandia & Denmark; key to species). **Vilbaste** 1980b (Tuva).
Tretogonia
  New species: **Metcalf** 1965a (new name). **Young** 1968a.
  Records: **Lauterer & Schroder** 1970a (lectotype designations). **Young** 1968a (n. comb. from *Amblydisca* and *Tettigonia*; n. sp. syn.; key to species). **Young & Lauterer** 1966a (lectotype designations). **Young & Nast** 1963a (notes on type specimens).
Trichogonia
  New species: **Young** 1977a.
  Records: **Young** 1965b, 1965c (lectotype designations); 1977a (n. comb. from *Tettigonia*; key to species). **Young & Nast** 1963a (notes on type specimens).
Triquetolidia
  New species: **Nielson** 1982e.

Triviotartessus
  New species: **Evans, F.** 1981a.
  Record: **Evans, F.** 1981a (n. comb. from *Tartessus*).
Trocnada
  Records: **Evans, J. W.** 1966a (synopsis; n. sp. syn.). **Linnavuori & Quartau** 1975a (generotype characterized).
Trocnadella
  Records: **Datta** 1973l (descriptive notes). **Viraktamath, C. A.** (n. comb. from *Iassus*).
Tropicanus
  New species: **Linnavuori** 1959b.
  Records: **Cheng, Y. J.** 1980a (Paraguay). **Hamilton, K. G. A.** 1975b (APHRODINAE, DELTOCEPHALINI, PLATYMETOPIINA). **Linnavuori** 1959b (key to subgenera and species). **Linnavuori & DeLong** 1977b (n. comb. from *Menosoma*).
Trypanalebra
  New species: **Young** 1957c.
  Records: **Linnavuori** 1956a (Colombia). **Young** 1957c (n. comb. from *Protalebra*; key to species).
Tsavopsis
  New species: **Linnavuori** 1978b.
Tuakamara
  New species: **Webb, M. D.** 1980a.
Tuberana
  New species: **DeLong & Freytag** 1971a.
  Record: **DeLong & Freytag** 1972a (in key to genera of GYPONINAE).
Tubiga
  New species: **Young** 1977a.
Tubulanus
  New species: **Cheng, Y. J.** 1980a. **Linnavuori** 1959b, 1973b.
  Records: **Linnavuori** 1956a (Brazil); 1959b (key to species). **Linnavuori & DeLong** 1976a (=*Atanus*, n. syn.). **Menezes** 1973b (name preoccupied; *Linnatanus*, replacement).
Tumeus
  Record: **Hamilton, K. G. A.** 1975b (APHRODINAE, DELTOCEPHALINI, PLATYMETOPIINA).
Tumocerus
  New species: **Webb, M. D.** 1983a.
  Records: **Evans, J. W.** 1966a (synopsis). **Webb, M. D.** (n. comb. from *Austrocerus*; n. sp. syn.; key to species).
Tumupasa
  New species: **Linnavuori** 1959b.
Tungurahuala
  New species: **Kramer** 1965a.
Turitia
  Record: **Linnavuori** 1972b (unknown; Schumacher's description repeated).
Turrutus
  Records: **Anufriev** 1978a (USSR: Maritime Territory). **Cantoreanu** 1959a, 1965d, 1971c (Romania). **Dlabola** 1965c, 1967b, 1967c, 1968b, 1970b (Mongolia). **Dworakowska** 1969a (Mongolia). **Emeljanov** 1964c (European USSR); 1977a (Mongolia). **Gyllensvard** 1969a (Sweden). **Hamilton, K. G. A.** 1975b (APHRODINAE, DELTOCEPHALINI, DELTOCEPHALINA). **LeQuesne** 1969c (Britain). **Mitjaev** 1968b (eastern Kazakhstan); 1971a (Kazakhstan). **Nast** 1972a (catalogued); 1976b (Poland; catalogued). **Ossiannilsson** 1983a (Fennoscandia & Denmark). **Vilbaste** 1964a (Auwiesen Estlands); 1965b (Altai Mts.); 1973a (notes on Flor's material); 1980b (Tuva).
Tuzinka
  New species: **Dworakowska & Viraktamath** 1979a.
  Record: **Sohi & Dworakowska** 1984a (list of Indian species)
Twiningia
  Records: **Beirne** 1956a (Canada; key to species). **Hamilton, K. G. A.** 1975b (APHRODINAE, DELTOCEPHALINI, PLATYMETOPIINA).
Tylozygus
  Records: **Beirne** 1956a (Canada). **Ramos Elorduy** 1972c (Mexico). **Schroder** 1959a (diagnosis). **Young** 1977a (n. sp. syn.; key to species).
Typhlocyba
  New species: **Ahmed, Manzoor** 1971c, 1972a. **Ahmed, Manzoor, Naheed & Samad** 1981a. **Christian** 1960a. **Datta & Ghosh** 1973b. **Dlabola** 1971a. **Dworakowska** 1967b, 1977i,

1978b, 1979d, 1981g. **Gyllensvard** 1964a. **Hamilton, K. G. A.** 1983a, 1983b. **Ishihara** 1958a. **Logvinenko** 1967a. **Sharma, B.** 1984a. **Vidano** 1961a, 1961d. **Zachvatkin** 1948a.
Records: **Anufriev** 1973a (redefined); 1978a (USSR: Maritime Territory). **Beirne** 1956 (Canada; key to species). **Cantoreanu** 1959a, 1965b (Romania). **Datta** 1973a (descriptive notes). **Dlabola** 1957a (Turkey); 1958c (in key to genera of Palaearctic TYPHLOCYBINAE; 1967b (Mongolia); 1972a (Afghanistan). **Dworakowska** 1970k (n. comb. to *Hauptidia*); 1972e (*Zygina* as subgenus); 1977b (notes on Dlabola's illustrations; China). **Emeljanov** 1964c (European USSR; key to species); 1977a (Mongolia). **Evans, J. W.** 1966a (Australia; adventive). **Gunthart, H.** 1971a (Switzerland). **Gyllensvard** 1963a, 1965a (Sweden). **Hamilton, K. G. A.** 1983a (*Empoa* as subgenus; n. comb. from *Empoa*).; 1983b (transboreal taxon). **Jankovic** 1966a (Serbia). **Kirejtshuk** 1977a (USSR: Kharkov region). **Knight, W. J.** 1976a (New Zealand; adventive). **Lauterer** 1957a (Czechoslovakia). **LeQuesne & Payne** 1981a (Britain). **Lindberg** 1960a (Portugal). **Linnavuori** 1959c (Mt. Picentini); 1962a (Israel). **Linnavuori & DeLong** 1977c (Chile; adventive). **Lodos & Kalkandelen** 1984a (Turkey). **Logvinenko** 1957b (USSR: Ukraine); 1959a (Transcarpathian region); 1963a (USSR: Ukraine); 1967a (n. comb. to *Zygina*). **Mahmood** 1967a (in key to genera of Oriental TYPHLOCYBINAE). **Mitjaev** 1963b (Kazakhstan); 1967b (sp. transferred to *Linnavuoriana*); 1968b (eastern Kazakhstan); 1971a (Kazakhstan; in key to genera of TYPHLOCYBINAE). **Nast** 1972a (catalogued); 1976a (Poland: Pieniny Mts.); 1976b (Poland; catalogued). **Ossiannilsson** 1981a (Fennoscandia & Denmark; key to species). **Quartau** 1979a (Azores). **Servadei** 1957a, 1960a, 1968a, 1969a (Italy). **Sharma, B.** 1984a (India: J. & K., Kishtwar). **Sohi** 1977a (northwest India). **Sohi & Dworakowska** 1984a (list of Indian species). **Thapa** 1983a (Nepal). **Trolle** 1968a (Denmark). **Varty** 1967a (Canada). **Vidano** 1960a (new subgenus proposed); 1961a, 1961b (*Edwardsiana* as subgenus). **Viggiana** 1971a (key to species). **Vilbaste** 1973a (identity of Flor's material); 1975a (identity of Motschulsky's material); 1976a (Matsumura's species referred to *Eupteryx, Flammigeroidia, Youngiada* and *Zyginidia*; some sp. syn. noted); 1980b (Tuva). **Wilson, M. R.** 1978a (key characters of 5th instars).

Typhlocybella
Record: **Ruppel** 1966a (Colombia).

Tzitzikamaia
New species: **Linnavuori** 1961a.
Record: **Theron** 1974b (n. comb. from *Euscelis*).

Ujna
Record: **Young** 1965c (lectotype designation).

Ulopa
New species: **Dlabola** 1957b, 1969a. **Linnavuori** 1972a.
Records: **Cantoreanu** 1971c (Romania). **Datta** 1973l (descriptive notes). **Davis, R. B.** 1975a (morphology; phyletic inferences). **Dlabola** 1958b (Caucasus); 1961b (central Asia); 1964c (Albania); 1981a (Iran). **D'Urso** 1980b (Italy). **Emeljanov** 1964c (European USSR; key to species). **LeQuesne** 1964d (macropterous morph); 1965c (Britain). **Lindberg** 1960a (Portugal); 1960b (Soviet Armenia). **Linnavuori** 1956d (Spain); 1962a (Israel); 1972a (in key to genera of African ULOPINAE). **Lodos & Kalkandelen** 1981a (Turkey). **Logvinenko** 1957b (USSR: Ukraine). **Morris** 1972a (Britain). **Nast** 1955a (Poland); 1972a (catalogued); 1976a (Poland: Pieniny Mts.); 1976b (Poland; catalogued). **Orosz** 1977a (revised sp. syn.). **Ossiannilsson** 1981a (Fennoscandia & Denmark). **Servadei** 1968a, 1971a (Italy). **Vilbaste** 1973a (notes on Flor's material).

Ulopella
New species: **Van Stalle** 1983a.
Record: **Linnavuori** 1972a (n. comb. from *Daimachus*; in key to genera of African ULOPINAE). **Van Stalle** 1983c (Zaire).

Uloprora
Record: **Evans, J. W.** 1966a (synopsis).

Umesaona
New species: **Ishihara** 1961a.
Record: **Vilbaste** 1965a (=*Doratulina*, n. syn.).

Unerus
New species: **Freytag** 1983a. **Linnavuori** 1959b.
Record: **Linnavuori** 1959b (generic rank, n. status; new subgenus proposed; n. sp. syn.; key to subgenera and species); 1975d (Peru).

Unguitartessus
New species: **Evans, F.** 1981a.

Unitra
New species: **Dworakowska** 1974a.

Unoka
Records: **Beirne** 1956a (Canada: southern British Columbia). **Hamilton, K. G. A.** 1975b (APHRODINAE, DELTOCEPHALINI, DELTOCEPHALINA).

Uperogonalia
New species: **Young** 1977a.

Urganus
New species: **Dlabola** 1965c.
Records: **Anufriev** 1978a (USSR: Maritime Territory). **Dlabola** 1967c, 1968a, 1970b (Mongolia). **Dworakowska** 1969a (Mongolia). **Emeljanov** 1966a (USSR: Amur and Chita districts); 1977a (Mongolia). **Hamilton, K. G. A.** 1975b (APHRODINAE, DELTOCEPHALINI, DELTOCEPHALINA). **Lee, C. E.** 1979a (Korea, redescribed). **Lee, C. E. & Kwon** 1977b (Korea). **Vilbaste** 1969a (notes on Matsumura's material; n. sp. syn.); 1980b (Tuva).

Urmila
New species: **Dworakowska** 1981j.
Record: **Sohi & Dworakowska** 1984a (list of Indian species).

Usanus
Record: **Cwikla & Blocker** 1981a (comparative notes).

Usharia
New species: **Dworakowska** 1977f, 1977i, 1980b, 1984a. **Thapa** 1985a.
Record: **Sohi & Dworakowska** 1984a (list of Indian species).

Ussuriasca
Record: **Anufriev** 1978a (USSR: Maritime Territory; in key to genera of EMPOASCINI).

Usuironus
Record: **Hamilton, K. G. A.** 1975b (APHRODINAE, DELTOCEPHALINI, ATHYSANINA).

Uzeldikra
Records: **Dworakowska** 1971d (n. comb. from *Empoasca*); 1980b (southern India); 1984a (Malaysia: Penang I.). **Sharma, B.** 1978a (=*Hameedia*, n. syn.; =*Pusaneura*, n. syn.). **Sohi & Dworakowska** 1984a (list of Indian species).

Uzelina
New species: **Linnavuori** 1977a.
Records: **Evans, J. W.** 1972a (comparative notes). **Linnavuori** 1977a (key to subgenera and species).

Vangama
Record: **Datta** 1973k (descriptive notes).

Varicopsella
New species: **Hamilton, K. G. A.** 1980b. **Viraktamath, C. A.** 1981b.
Record: **Hamilton, K. G. A.** 1980b (n. comb. from *Macropsis*; in key to genera and subgenera of MACROPSINI).

Varta
New species: **Ramachandra Rao** 1973a.
Record: **Linnavuori** 1978d (=*Stymphalus*, n. syn.).

Vecaulis
Record: **Theron** 1975a (n. comb. from *Deltocephalus*).

Velu
New species: **Dworakowska** 1980b. **Ghauri** 1963e.
Records: **Ahmed, Manzoor** 1969b, 1985c (Pakistan). **Ahmed, Manzoor, Samad & Naheed** 1981a (West Pakistan). **Dworakowska & Viraktamath** 1978a (southern India). **Sohi & Dworakowska** 1984a (list of Indian species).

Verdanulus
Record: **Hamilton, K. G. A.** 1975b (APHRODINAE, DELTOCEPHALINI, DELTOCEPHALINA).

Verdanus
Records: **Beirne** 1956a (Canada). **Emeljanov** 1966a (as subgenus of *Diplocolenus*). **Hamilton, K. G. A.** 1972d (Canada: Manitoba); 1975b (APHRODINAE, DELTOCEPHALINI, DELTOCEPHALINA). **Knight, W. J.** 1974c (as subgenus of *Diplocolenus*). **Nast** 1972a

(catalogued); 1976b (Poland; catalogued). **Ossiannilsson** 1983a (Fennoscandia & Denmark; key to species). **Vilbaste** 1959b (Estonia); 1964a (Auwiesen Estlands); 1965b (Altai Mts.); 1969b (Taimyr); 1973a (notes on Flor's material); 1980a (Kamchatka); 1980b (Tuva).

Vermara
New species: **Dworakowska** 1980b.
Record: **Sohi & Dworakowska** 1984a (list of Indian species).

Vernobia
New species: **Nielson** 1979b.

Versigonalia
New species: **Young** 1977a.
Record: **Young** 1977a (n. comb. from *Tettigonia*; n. sp. syn.; key to species).

Vertanus
Record: **Hamilton, K. G. A.** 1975b (APHRODINAE, DELTOCEPHALINI, PLATYMETOPIINA).

Vertigella
New species: **Evans, J. W.** 1972a, 1973a.

Vicosa
New species: **Linnavuori & DeLong** 1978c.
Record: **Cwikla & Blocker** 1981a (comparative notes).

Vidanoana
Record: **Young** 1977a (n. comb. from *Microgoniella*).

Vietnara
Records: **Dworakowska** 1980d (=*Empoascanara*, n. syn.). **Ramakrishnan & Ghauri** 1979b (n. comb. from *Empoascanara*).

Vilargus
Record: **Theron** 1975a (n. comb. from *Deltocephalus*).

Vilbasteana
Record: **Anufriev** 1970d (n. comb. from *Dikraneura*).

Virganana
New species: **DeLong & Thambimuttu** 1973b.
Records: **Cwikla & Blocker** 1981a (comparative notes). **Linnavuori & DeLong** 1977c (Chile; in key to genera of the *Faltala* group).

Viridasca
Records: **Dlabola** 1974b (n. comb. from *Bythoscopus*); 1981a (Iran). **Lodos & Kalkandelen** 1981a (Turkey). **Sharma, B.** 1977b (India: Jammu).

Viriosana
Record: **Hamilton, K. G. A.** 1975b (APHRODINAE, DELTOCEPHALINI, ATHYSANINA).

Volusenus
Record: **Vilbaste** 1965a (=*Doratulina*, n. syn.).

Vulturnellus
New species: **Evans, J. W.** 1966a.
Record: **Evans, J. W.** 1972a (=*Neodartus*, n. syn.).

Vulturnus
New species: **Evans, J. W.** 1972a, 1973a. **Linnavuori** 1960b.
Records: **Evans, J. W.** 1966a (synopsis); 1972a (n. comb. from *Penthimia* and *Thaumatoscopus*); 1977a (n. comb. from *Neodartus*; n. comb. to *Neovulturnus*).

Wadkufia
New species: **Linnavuori** 1965a.
Record: **Nielson** 1975a (referred to DELTOCEPHALINAE).

Wagneriala
New species: **Dworakowska** 1979e.
Records: **Anufriev** 1970g (replacement for *Wagneriana*, preoccupied). **Emeljanov** 1977a (Mongolia). **Jansky** 1985a (Slovakia). **Logvinenko** 1984a (USSR: Ukraine). **Ossiannilsson** 1981a (Fennoscandia & Denmark; key to species). **Vilbaste** 1980b (Tuva).

Wagneriana
Record: **Anufriev** 1970d (n. comb. from *Notus*).

Wagneripteryx
New species: **Anufriev** 1971b, 1971d. **Vilbaste** 1965b.
Records: **Anufriev** 1971b (key to species). **Dlabola** 1958b (Caucasus). **Dworakowska** 1967b (Mongolia); 1972a (=*Aguriahana*, n. syn.). **Emeljanov** 1964c (European USSR). **Servadei** 1960a (Italy). **Vilbaste** 1965b (Altai Mts.); 1968b (USSR coastal regions).

Wagneriunia
New species: **Dworakowska** 1969c.

Wakaya
New species: **Linnavuori** 1960b.

Wanritettix
Records: **Hamilton, K. G. A.** 1975b (APHRODINAE, DELTOCEPHALINI, PLATYMETOPIINA). **Vilbaste** 1969a (n. comb. from *Thamnotettix*).

Warodia
New species: **Dworakowska** 1982a.
Records: **Dworakowska** 1970f (n. comb. from *Typhlocyba*); 1982a (n. sp. syn.).

Watanabella
New species: **Choe** 1981a.
Records: **Anufriev** 1971a (Sakhalin); 1977b (Kurile Islands); 1978a (USSR: Maritime Territory). **Hamilton, K. G. A.** 1975b (APHRODINAE, DELTOCEPHALINI, PLATYMETOPIINA). **Vilbaste** 1969a (n. comb. from *Thamnotettix*).

Watara
Records: **Ahmed, Manzoor** 1985c (Pakistan). **Dworakowska** 1977f (n. comb. from *Platytettix*). **Sohi & Dworakowska** 1984a (list of Indian species).

Webaskola
New species: **Blocker** 1979a.
Record: **Blocker** 1979a (n. comb. from *Batracomorphus*).

Webbanara
Records: **Dworakowska** 1980d (=*Empoascanara*, n. syn.). **Ramakrishnan & Ghauri** 1979b (n. comb. from *Empoascanara*).

Wemba
New species: **Dworakowska** 1974a. **Dworakowska & Trolle** 1976a.

Westindica
Record: **Dworakowska** 1980d (=*Empoascanara*, n. syn.).

Wiata
New species: **Dworakowska** 1972f, 1977i, 1981i.
Records: **Dworakowska** 1979c (ZYGINELLINI); 1981i (new subgenus proposed).

Willeiana
New species: **Young** 1977a.
Record: **Young** 1977a (n. comb. from *Cardioscarta*, *Cicadella* and *Tettigonia*; n. sp. syn.; key to species).

Wiloatma
New species: **Webb, M. D.** 1983a.
Record: **Webb, M. D.** 1983a (key to species).

Witera
New species: **Dworakowska** 1981j.
Record: **Sohi & Dworakowska** 1984 (list of Indian species).

Woldana
New species: **DeLong** 1980d.

Wolfella
New species: **Boulard** 1969a, 1975a. **Kramer** 1965c. **Linnavuori** 1969a.
Records: **Knight, W. J.** 1966a (description of male). **Kramer** 1965c (key to species). **Linnavuori** 1969a (Congo); 1972b (in key to genera of African HYLICIDAE).

Woodella
New species: **Evans, J. W.** 1966a.

Wutingia
Record: **Young & Lauterer** 1966a (lectotype designation).

Xedreota
Record: **Kramer** 1966a (n. comb. from *Xerophloea*; in key to genera of New World LEDRINAE).

Xenocoelidia
New species: **DeLong & Kolbe** 1975b. **Kramer** 1959a, 1967b.
Records: **Kramer** 1964a (n. comb. from *Neocoelidia*); 1967b (key to species).

Xenogonalia
Record: **Young** 1977a (n. comb. from *Cicadella*).

Xerophloea
Records: **Beirne** 1956a (Canada). **Hamilton, K. G. A.** 1972d (Canada: Manitoba); 1975c (key to species). **Kramer** 1966a (in key to genera of New World LEDRINAE). **Linnavuori** 1959b (XAROPHLOEINAE). **Linnavuori & DeLong** 1977c (Chile). **Nielson** 1962b (key to species).

Xestocephalus
New species: **Choe** 1981a. **Cwikla** 1985a. **DeLong** 1980b, 1982b, 1982g. **DeLong & Linnavuori** 1978a. **DeLong, Wolda & Estribi** 1980a, 1983b. **Evans, J. W.** 1966a. **Heller & Linnavuori** 1968a. **Ishihara** 1961b, 1964b. **Kwon** 1981a. **Lee,**

C. E. & Kwon 1977a. **Linnavuori** 1957c, 1959b, 1960a, 1960b, 1969a, 1973c. **Logvinenko** 1981a.
Records: **Anufriev** 1978a (USSR: Maritime Territory). **Beirne** 1956a (Canada; key to species). **Capco** 1960a (Philippines; key to species). **Cwikla** 1985a (n. sp. syn.; key to species of North America, Central America and the West Indies). **Davis, R. B.** 1975a (morphology; phyletic inferences). **Evans, J. W.** 1966a (Australia; synopsis); 1974a (New Caledonia ?). **Eyles & Linnavuori** 1974a (Niue Island). **Hamilton, K. G. A.** 1972d (Canada: Manitoba); 1975b and 1983c (APHRODINAE, APHRODINI, XESTOCEPHALINA); 1976a (identity of Provancher material). **Ishihara** 1961a (Thailand); 1961b (Japan; key to species). **Knight, W. J.** 1974b (New Zealand); 1976b (Chatham Island) 1982a (Rennell Island). **Kwon & Lee** 1980c (Korea). **Lee, C. E.** 1979a (Korea, redescribed). **Lee, C. E. & Kwon** 1977b (Korea). **Linnavuori** 1959b (key to Neotropical species); 1960a (Micronesia; key to species); 1960b (Fiji; key to species); 1973c (Cuba); 1975c (Micronesia: Palau); 1979b (Africa; key to species). **Linnavuori & DeLong** 1977c (Chile). **Linnavuori & Heller** 1961a (Peru). **Lodos & Kalkandelen** 1983a (Turkey). **Vilbaste** 1968a (USSR: coastal regions); 1975a (identity of Motschulsky's material).

Xiqilliba
New species: **Kramer** 1964d. **Linnavuori** 1965b.
Record: **Kramer** 1967b (n. sp. syn.).

Xyphon
Record: **Hamilton, K. G. A.** 1985a (n. comb. from *Carneocephala*).

Yachandra
New species: **Webb, M. D.** 1983b.
Record: **Webb, M. D.** 1983b (n. comb. from *Idiocerus*).

Yakunopona
Record: **Hamilton, K. G. A.** 1975b (APHRODINAE, PARABOLOPONINI).

Yamatotettix
Records: **Hamilton, K. G. A.** 1975b (APHRODINAE, DELTOCEPHALINI, MACROSTELINA). **Ishihara** 1961a (Thailand). **Lee, C. E.** 1979a (Korea, redescribed). **Lee, C. E. & Kwon** 1977b (Korea, Taiwan; MACROSTELINA).

Yanocephalus
New species: **Dworakowska** 1973d.
Records: **Anufriev** 1978a (USSR: Maritime Territory). **Choe** 1985a (Korea). **Hamilton, K. G. A.** 1975b (APHRODINAE, DELTOCEPHALINI, DELTOCEPHALINA). **Lee, C. E.** 1979a (Korea, redescribed). **Lee, C. E. & Kwon** 1977b (Korea). **Vilbaste** 1967a (n. sp. syn.); 1968a (USSR: coastal regions).

Yaouandea
New species: **Linnavuori** 1979b.

Yasumatsuus
Records: **Ishihara** 1971a (n. comb. from *Kolla*). **Young** 1979a (= *Cofana*, n. syn.).

Yisiona
New species: **Kuoh, C. L.** 1981b.

Yochlia
New species: **Nielson** 1979b.
Record: **Nielson** 1979b (key to species).

Yotala
Records: **Young** 1968a (redescribed). **Young & Soos** 1964a (lectotype designation).

Youngama
New species: **Ahmed, Manzoor** 1969d.
Record: **Ahmed, Manzoor** 1985c (Pakistan).

Youngia
Record: **Dlabola** 1958c (in key to genera of Palaearctic TYPHLOCYBINAE; preoccupied name, see *Youngiada*).

Youngiada
New species: **Linnavuori** 1962a.
Records: **Dlabola** 1981a (Turkey); **Linnavuori** 1965a (Tunisia, Turkey). **Lodos & Kalkandelen** 1984a (Turkey). **Vilbaste** 1976a (identity of Matsumura's material). **Wilson, M. R.** 1978a (key characters for 5th instars).

Youngolidia
New species: **Nielson** 1983e.
Record: **Nielson** 1983e (n. comb. from *Coelidia*; key to species).

Yunga
New species: **Young** 1968a.
Records: **Young** 1968a (key to species). **Young & Lauterer** 1966a (lectotype designation). **Young & Nast** 1963a (notes on type material).

Yungasia
New species: **Linnavuori** 1959b. **Linnavuori & DeLong** 1976a. **Linnavuori & Heller** 1961a.

Yuraca
New species: **Linnavuori & DeLong** 1978d.
Record: **Cwikla & Blocker** 1981a (comparative notes).

Zabrosa
New species: **Linnavuori & DeLong** 1978c. **Menezes** 1973b.
Records: **Hamilton, K. G. A.** 1975b (APHRODINAE, DELTOCEPHALINI, PLATYMETOPIINA). **Linnavuori** 1959b (redescribed).

Zaletta
New species: **Webb, M. D.** 1983a.
Records: **Evans, J. W.** 1966a (synopsis). **Webb, M. D.** 1983a (n. comb. from *Idiocerus*; n. sp. syn.; key to species).

Zanjoneura
New species: **Ghauri** 1974d.
Record: **Ghauri** 1974d (in key to genera of African ERYTHRONEURINI; n. comb. from *Frutioidia*).

Zapycna
New species: **Emeljanov** 1968a.
Records: **Emeljanov** 1977a (Mongolia; n. comb. from *Diacra*; n. sp. syn.). **Hamilton, K. G. A.** 1975b (APHRODINAE, APHRODINI, ACHAETICINA).

Zaruma
New species: **Young** 1977a.
Records: **Young** 1977a (key to species). **Young & Lauterer** 1966a (lectotype designation).

Zelenius
Records: **Emeljanov** 1966a (n. comb. from *Boreotettix*, *Rosenus* and *Sorhoanus*). **Hamilton, K. G. A.** 1975b (APHRODINAE, DELTOCEPHALINI, DELTOCEPHALINA).

Zelopsis
New species: **Evans, J. W.** 1966a.
Record: **Knight, W. J.** 1974b (New Zealand; redescribed).

Zercanus
New species: **Dlabola** 1965d.
Record: **Hamilton, K. G. A.** 1975b (APHRODINAE, DELTOCEPHALINI, ATHYSANINA).

Ziczacella
New species: **Dworakowska** 1979b.
Records: **Dworakowska** 1970n (as *Zigzacella*; n. comb. from *Erythroneura* and *Zygina*; generic rank); 1979b (n. comb. from *Erythroneura*); 1981d (synopsis).. **Lee, C. E.** 1979a (Korea, redescribed) **Lee, C. E. & Kwon** 1977b (Korea). **Logvinenko** 1984a (USSR: Ukraine). **Vilbaste** 1980b (Tuva).

Zilkaria
New species: **Menezes** 1974a.
Record: **Cwikla & Blocker** 1981a (comparative notes).

Zinga
New species: **Dworakowska** 1972b.

Zinislopa
New species: **Webb, M. D.** 1983a.

Zinneca
Records: **Hamilton, K. G. A.** 1980b (=*Oncopsis*, n. gen. syn.; n. sp. syn.); 1983d (=*Oncopsis*).

Zioninus
Record: **Hamilton, K. G. A.** 1975b (APHRODINAE, DELTOCEPHALINI, PLATYMETOPIINA).

Zizyphoides
New species: **Bhattacharya** 1973d. **Linnavuori** 1969a. **Ramachandra Rao** 1973a.
Records: **Dlabola** 1981a (Iran). **Ghauri** 1963f (n. comb. from *Rhombopsis*). **Hamilton, K. G. A.** 1975b (APHRODINAE, PARABOLOPONINI). **Linnavuori** 1978d (=*Paganalia*, n. syn.). **Sharma, B.** 1977b (as *Zyzyphoides*; India: Jammu). **Webb, M. D.** 1981a (=*Dryadomorpha*; n. comb. to *Mahalana* and *Stirellus*; n. sp. syn.).

Zonana
New species: **DeLong & Freytag** 1963b.

Record: **DeLong & Freytag** 1972a (in key to genera of GYPONINAE).

Zorka
New species: **Dworakowska** 1970f, 1977i.

Zubara
New species: **Al-Ne'amy & Linnavuori** 1982a.

Zygina
New species: **Ahmed, Manzoor** 1969a, 1970b, 1971b. **Ahmed, Manzoor & Samad** 1980a. **Datta** 1969a. **Datta & Ghosh** 1973a. **Dlabola** 1961b, 1967e, 1977b. **Dworakowska** 1968d, 1981e. **Gerard** 1972a. **Ghauri** 1963a, 1963b, 1963c, 1964c, 1980a. **Kirejtshuk** 1975a. **Knight, W. J.** 1976a, 1976b. **Mitjaev** 1969b, 1971a, 1975a. **Ramakrishnan & Menon** 1974a. **Samad & Ahmed** 1979b. **Sawai Singh** 1969a. **Vidano** 1964b.

Records: **Ahmed, Manzoor** 1983b (remarks on shape of male style); 1985c (Pakistan; n. comb. to *Seriana*). **Anufriev** 1978a (USSR: Maritime Territory). **Beardsley & Funasaki** 1976a (Hawaii). **Bonfils & Della Giustina** 1978a (Corsica). **Cantoreanu** 1965a (Romania). **Cardosa** 1974a (Portugal). **Dlabola** 1958a (southern Europe); 1958b (Caucasus); 1958c (in key to genera of Palaearctic TYPHLOCYBINAE); 1959b (central and southern Europe); 1961b (central Asia); 1964a (Afghanistan); 1964b (Sudan); 1964c (Albania); 1965d (Jordan); 1970b (Mongolia); 1977a (Griechenland; n. sp. syn.); 1981a (Iran). **Dubovskij** 1966a (Uzbekistan; key to species); 1978a (Uzbekistan, Zarafshan Valley). **Dworakowska** 1970j (new subgenus proposed; n. comb. from *Erythroneura, Flammigeroidia* and *Typhlocyba*); 1977b (n. sp. syn.). **Dworakowska & Sohi** 1978b (India: H. P., Kulu). **Emeljanov** 1964c (European USSR; key to species); 1977a (Mongolia). **Evans, J. W.** 1966a (New Zealand; key to species). **Gerard** 1972a (n. comb. from *Typhlocyba;* lectotype designation). **Ghauri** 1963b (New Zealand; key to species). **Gunthart, H.** 1971a (Switzerland). **Hamilton, K. G. A.** 1983b (Canada: British Columbia, and USA: Washington, adventive). **Jankovic** 1966a (Serbia). **Jansky** 1984a (Slovakia). **Jasinska** 1980a (Poland: Bledowska Wilderness). **Kirejtshuk** 1977a (USSR: Kharkov region). **Korolevskaya** 1978a (Tadzhikistan). **Lauterer** 1980a (Moravia, Slovakia); 1984a (Moravia).. **Lee, C. E.** 1979a (Korea, redescribed) **Lee, C. E. & Kwon** 1977b (Korea). **LeQuesne & Payne** 1981a (Britain). **Lindberg** 1960b (Soviet Armenia). **Lodos** 1982a (Turkey). **Lodos & Kalkandelen** 1984d (Turkey). **Logvinenko** 1984a (USSR: Ukraine). **Mitjaev** 1963b, 1963c (Kazakhstan); 1968b (eastern Kazakhstan); 1971a (Kazakhstan; key to species). **Nast** 1972a (catalogued); 1976a (Poland: Pieniny Mts.); 1976b (Poland; catalogued). **Ossiannilsson** 1981a (Fennoscandia & Denmark; key to species). **Quartau & Rodrigues** 1969a (Portugal). **Ramakrishnan & Ghauri** 1979b (n. comb. to *Indoformosa*). **Remane** 1961d (Germany). **Servadei** 1960a, 1968a, 1969a (Italy). **Sharma, B.** 1977b (India: Jammu); 1984a (India: J. & K., Kishtwar). **Sohi** 1976a (n. comb. to *Empoascanara*; n. comb. to *Zyginidia*). **Sohi** 1976a (species transferred to *Empoascanara* and *Zyginidia*). **Sohi & Dworakowska** 1984a (list of Indian species). **Vidano** 1959a (identity of *Z. rhammi*). **Vilbaste** 1962b (eastern part of the Caspian lowlands); 1979a (Vooremaa hardwood/spruce forest); 1980b (Tuva). **Wilson, M. R.** 1978a (key characters of 5th instars).

Zyginella
New species: **Dlabola** 1974a. **Dworakowska** 1977i. **Dworakowska, Sohi & Viraktamath** 1980a. **Heller & Linnavuori** 1968a. **Linnavuori** 1962a.

Records: **Anufriev** 1978a (USSR: Maritime Territory). **Dlabola** 1958a (southern Europe). **Dworakowska** 1969d (n. sp. syn.); 1970h (n. gen. syn.; n. comb. from *Conometopius, Eupteryx, Pyramidotettix* and *Remmia*; lectotype designation; n. sp. syn.); 1979c (summary of records). **Emeljanov** 1964c (European USSR). **Ishihara** 1958a (Japan: north Honshu). **Knight, W. J.** 1982a (Rennell Island). **Linnavuori** 1965a (Turkey). **Lodos & Kalkandelen** 1984a (Turkey). **Nast** 1972a (catalogued); 1976b (Poland; catalogued). **Servadei** 1968a (Italy). **Chou, I. & Zhang** 1985a (in key to genera of ZYGINELLINI; key to Chinese species).

Zyginidia
New species: **Dworakowska** 1970l. **Kalkandelen** 1985a. **Meusnier** 1982a. **Moravskaja** 1948a. **Vidano** 1981a. **Zachvatkin** 1948b, 1953c.

Records: **Ahmed, Manzoor** 1983b (remarks on shape of male styles); 1985c (Pakistan). **Bonfils & Della Giustina** 1978a (Corsica). **Cardosa** 1974a (Portugal). **Della Giustina** 1983a (France). **Dlabola** 1958a (southern Europe); 1971a (Afghanistan and Turkey); 1972a (Afghanistan); 1977c (Mediterranean); 1981a (Iran, Turkey). **Dubovskij** 1980a (USSR: western Turkmenia). **D'Urso** 1980b (Italy). **Dworakowska** 1970k (n. comb. to *Hauptidia*); 1970l (redescribed; n. comb. from *Erythroneura*; n. sp. syn.). **Jasinska** 1980a (Poland: Bledowska Wilderness). **Kalkandelen** 1985a (synopsis). **Korolevskaya** 1978a (Tadzhikistan). **Lauterer** 1984a (Moravia). **LeQuesne & Payne** 1981a (Britain). **Lindberg** 1960a (Portugal); 1960b (Soviet Armenia). **Lodos** 1982a (Turkey). **Logvinenko** 1957b, 1984a (USSR: Ukraine). **Moravskaja** 1948a (notes; systematics). **Nast** 1972a (catalogued); 1976a (Poland: Pieniny Mts.); 1976b (Poland; catalogued). **Ossiannilsson** 1981a (Fennoscandia & Denmark; key to species). **Quartau & Rodrigues** 1969a (Portugal). **Sohi** 1976a (n. comb. from *Zygina*); 1977a (northwestern India). **Sohi & Dworakowska** 1984a (list of Indian species). **Vidano & Arzone** 1985a (*Z. pullula*; biology, morphology, distribution). **Vilbaste** 1976a (identity of Matsumura's material).

Zyginoides
New species: **Dworakowska** 1972g. **Dworakowska & Lauterer** 1975a. **Dworakowska & Sohi** 1978a.

Records: **Dworakowska** 1972g (new subgenus proposed; n. comb. from *Motschulskia* and *Typhlocyba*; n. sp. syn.); 1976b (W. Flores); 1977f (=*Pakeasta*, n. syn.; n. sp. syn.); 1981c (=*Diomma*). **Vilbaste** 1975a (=*Diomma*, n. syn.).

Zyginopsis
New species: **Dworakowska** 1981e. **Ramakrishnan & Menon** 1973a.

Records: **Ahmed, Manzoor** 1985c (Pakistan). **Dworakowska** 1981e (India: M. P., Jabalpur). **Dworakowska & Sohi** 1978a (n. comb. from *Erythroneura*; n. sp. syn.). **Ramakrishnan & Menon** 1973a (key to species); 1974a (in key to genera of Indian ERYTHRONEURINI). **Sohi & Dworakowska** 1984a (list of Indian species).

Zyzzogeton
Records: **Young** 1965b, 1965c (lectotype designations); 1968a (n. comb. from *Ledra*; n. sp. syn.).